Student Solutions Manual

to accompany

Chemistry

The Molecular Nature of Matter and Change with Advanced Topics

Eighth Edition

Martin S. Silberberg

and

Patricia G. Amateis
Virginia Tech

Prepared by
Mara Vorachek-Warren
St. Charles Community College

Mc Graw Hill Education

STUDENT SOLUTIONS MANUAL TO ACCOMPANY

CHEMISTRY: THE MOLECULAR NATURE OF MATTER AND CHANGE WITH ADVANCED TOPICS, EIGHTH EDITION

Published by McGraw-Hill Education, 2 Penn Plaza, New York, NY 10121. Copyright © 2018 by McGraw-Hill Education. All rights reserved. Printed in the United States of America. No part of this publication may be reproduced or distributed in any form or by any means, or stored in a database or retrieval system, without the prior written consent of McGraw-Hill Education, including, but not limited to, in any network or other electronic storage or transmission, or broadcast for distance learning.

Some ancillaries, including electronic and print components, may not be available to customers outside the United States.

This book is printed on acid-free paper.

1 2 3 4 5 6 QVS 21 20 19 18 17

ISBN 978-1-259-98292-7
MHID 1-259-98292-0

All credits appearing on page or at the end of the book are considered to be an extension of the copyright page.

The Internet addresses listed in the text were accurate at the time of publication. The inclusion of a website does not indicate an endorsement by the authors or McGraw-Hill Education, and McGraw-Hill Education does not guarantee the accuracy of the information presented at these sites.

mheducation.com/highered

CONTENTS

PREFACE

WELCOME TO YOUR STUDENT SOLUTIONS MANUAL

Your Student Solutions Manual (SSM) includes detailed solutions for the Follow-up Problems, selected Boxed Reading Problems, and highlighted End-of-Chapter Problems in the eighth edition of *Chemistry: The Molecular Nature of Matter and Change* by Martin Silberberg and Patricia Amateis.

You should use the SSM in your study of chemistry as a study tool:

- to better understand the reasoning behind problem solutions. The plan-solution format illustrates the problem-solving thought process for the Follow-up Problems and for selected End-of-Chapter Problems.
- to better understand how concepts are applied to different types of problems. You will notice that concepts learned in previous chapters will reappear as you progress through the textbook and SSM.
- to check your problem solutions. Solutions often provide comments on the solution process as well as the answer.

To succeed in general chemistry, you must develop skills in problem solving. Not only does this mean being able to follow a solution path and reproduce it on your own, but also to analyze problems you have never seen before and develop your own solution strategy. Chemistry problems are story problems that bring together both conceptual understanding and mathematical reasoning. The analysis of new problems is the most challenging step for general chemistry students. When you find that you are having difficulty starting problems, do not be discouraged. This is an opportunity to learn new skills that will benefit you in future courses and your future career, and perseverance is one of these skills. The following two strategies, tested and found successful by many students, may help you develop the skills you need.

The first strategy is to become aware of your own thought processes as you solve problems. As you solve a problem, make notes in the margin concerning your thoughts. Why are you doing each step and what questions do you ask yourself during the solution? After you complete the problem, review your notes and make an outline of the process you used in the solution while reviewing the reasoning behind the solution. Write a paragraph describing the solution process you used.

The second strategy helps you develop the ability to transfer a solution process to a new problem. After solving a problem, rewrite the problem to ask a different question. One way to rewrite the question is to ask the question backwards—find what has been given in the problem from the answer to the problem. Another approach is to change the conditions—for instance, ask yourself what if the temperature is higher or there is twice as much carbon dioxide present? Alternatively, try changing the reaction or process taking place—what if the substance is melting instead of boiling?

At times, you may find slight differences between your calculated answer and the one in the SSM. Two reasons may account for the differences. First, SSM calculations do not round answers until the final step, impacting the numerical answer. Note that in preliminary calculations extra digits are retained and shown in the intermediate answers. The final answers are rounded using the rules for arithmetic operations covered in Section 1.5 of the textbook. Another reason for discrepancies may be that your solution route was different from the one given in the SSM. Valid alternate paths exist for many problems, but the SSM does not have space to show all alternate solutions. So, trust your solution as long as the discrepancy with the SSM answer is small, and use the different solution route to understand the concepts used in the problem.

I would like to thank Robin Reed and the staff at McGraw-Hill for their assistance in completing this project.

Mara Vorachek-Warren

St. Charles Community College, Cottleville, MO

CHAPTER 1 KEYS TO THE STUDY OF CHEMISTRY

FOLLOW–UP PROBLEMS

1.1A   Plan: The real question is "Does the substance change composition or just change form?" A change in composition is a chemical change while a change in form is a physical change.
Solution:
The figure on the left shows red atoms and molecules composed of one red atom and one blue atom. The figure on the right shows a change to blue atoms and molecules containing two red atoms. The change is **chemical** since the substances themselves have changed in composition.

1.1B   Plan: The real question is "Does the substance change composition or just change form?" A change in composition is a chemical change while a change in form is a physical change.
Solution:
The figure on the left shows red atoms that are close together, in the solid state. The figure on the right shows red atoms that are far apart from each other, in the gaseous state. The change is **physical** since the substances themselves have not changed in composition.

1.2A   Plan: The real question is "Does the substance change composition or just change form?" A change in composition is a chemical change while a change in form is a physical change.
Solution:
a) Both the solid and the vapor are iodine, so this must be a **physical** change.
b) The burning of the gasoline fumes produces energy and products that are different gases. This is a **chemical** change.
c) The scab forms due to a **chemical** change.

1.2B   Plan: The real question is "Does the substance change composition or just change form?" A change in composition is a chemical change while a change in form is a physical change.
Solution:
a) Clouds form when gaseous water (water vapor) changes to droplets of liquid water. This is a **physical** change.
b) When old milk sours, the compounds in milk undergo a reaction to become different compounds (as indicated by a change in the smell, the taste, the texture, and the consistency of the milk). This is a **chemical** change.
c) Both the solid and the liquid are butter, so this must be a **physical** change.

1.3A   Plan: We need to find the amount of time it takes for the professor to walk 10,500 m. We know how many miles she can walk in 15 min (her speed), so we can convert the distance the professor walks to miles and use her speed to calculate the amount of time it will take to walk 10,500 m.
Solution:
$$\text{Time (min)} = 10{,}500 \text{ m} \left(\frac{1 \text{ km}}{1000 \text{ m}}\right)\left(\frac{1 \text{ mi}}{1.609 \text{ km}}\right)\left(\frac{15 \text{ min}}{1 \text{ mi}}\right) = 97.8869 = \textbf{98 min}$$

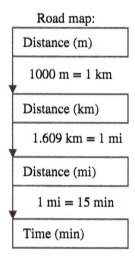

Road map:

Distance (m)

1000 m = 1 km

Distance (km)

1.609 km = 1 mi

Distance (mi)

1 mi = 15 min

Time (min)

1.3B   Plan: We need to find the number of virus particles that can line up side by side in a 1 inch distance. We know the diameter of a virus in nm units. If we convert the 1 inch distance to nm, we can use the diameter of the virus to calculate the number of virus particles we can line up over a 1 inch distance.
Solution:

$$\text{No. of virus particles} = 1.0 \text{ in} \left(\frac{2.54 \text{ cm}}{1 \text{ in}}\right)\left(\frac{1 \times 10^{7} \text{ nm}}{1 \text{ cm}}\right)\left(\frac{1 \text{ virus particle}}{30 \text{ nm}}\right) = 8.4667 \times 10^{5} = \mathbf{8.5 \times 10^{5} \text{ virus particles}}$$

Road map:

Length (in)

1 in = 2.54 cm

Length (cm)

1 cm = 1x10⁷ nm

Length (nm)

30 nm = 1 particle

No. of particles

1.4A   Plan: The diameter in nm is used to obtain the radius in nm, which is converted to the radius in dm. The volume of the ribosome in dm³ is then determined using the equation for the volume of a sphere given in the problem. This volume may then be converted to volume in μL.
Solution:

$$\text{Radius (dm)} = \frac{\text{diameter}}{2} = \left(\frac{21.4 \text{ nm}}{2}\right)\left(\frac{1 \text{ m}}{1 \times 10^{9} \text{ nm}}\right)\left(\frac{1 \text{ dm}}{0.1 \text{ m}}\right) = 1.07 \times 10^{-7} \text{ dm}$$

$$\text{Volume (dm}^3) = \frac{4}{3}\pi r^3 = \frac{4}{3}(3.14159)(1.07 \times 10^{-7} \text{ dm})^3 = 5.13145 \times 10^{-21} = \mathbf{5.13 \times 10^{-21} \text{ dm}^3}$$

$$\text{Volume (μL)} = (5.13145 \times 10^{-21} \text{ dm}^3)\left(\frac{1 \text{ L}}{(1 \text{ dm})^3}\right)\left(\frac{1 \text{ μL}}{10^{-6} \text{ L}}\right) = 5.13145 \times 10^{-15} = \mathbf{5.13 \times 10^{-15} \text{ μL}}$$

Road map:

```
┌─────────────────────┐
│ Diameter (dm)       │
└─────────────────────┘
      diameter = 2r
┌─────────────────────┐
│ Radius (dm)         │
└─────────────────────┘
      V = 4/3πr³
┌─────────────────────┐
│ Volume (dm³)        │
└─────────────────────┘
      1 dm³ = 1 L
      1 L = 10⁶ μL
┌─────────────────────┐
│ Volume (μL)         │
└─────────────────────┘
```

1.4B   <u>Plan:</u> We need to convert gallon units to liter units. If we first convert gallons to dm^3, we can then convert to L.
<u>Solution:</u>

$$\text{Volume (L)} = 8400 \text{ gal}\left(\frac{3.785 \text{ dm}^3}{1 \text{ gal}}\right)\left(\frac{1 \text{ L}}{1 \text{ dm}^3}\right) = 31,794 = \mathbf{32,000 \text{ L}}$$

Road map:

```
┌─────────────────────┐
│ Volume (gal)        │
└─────────────────────┘
    1 gal = 3.785 dm³
┌─────────────────────┐
│ Volume (dm³)        │
└─────────────────────┘
      1 dm³ = 1 L
┌─────────────────────┐
│ Volume (L)          │
└─────────────────────┘
```

1.5A   <u>Plan:</u> The time is given in hours and the rate of delivery is in drops per second. Conversions relating hours to seconds are needed. This will give the total number of drops, which may be combined with their mass to get the total mass. The mg of drops will then be changed to kilograms.
<u>Solution:</u>

$$\text{Mass (kg)} = 8.0 \text{ h}\left(\frac{60 \text{ min}}{1 \text{ h}}\right)\left(\frac{60 \text{ s}}{1 \text{ min}}\right)\left(\frac{1.5 \text{ drops}}{1 \text{ s}}\right)\left(\frac{65 \text{ mg}}{1 \text{ drop}}\right)\left(\frac{10^{-3} \text{ g}}{1 \text{ mg}}\right)\left(\frac{1 \text{ kg}}{10^3 \text{ g}}\right) = 2.808 = \mathbf{2.8 \text{ kg}}$$

Road map:

```
┌─────────────────────────┐
│ Time (hr)               │
└─────────────────────────┘
    1 hr = 60 min
┌─────────────────────────┐
│ Time (min)              │
└─────────────────────────┘
     1 min = 60 s
┌─────────────────────────┐
│ Time (s)                │
└─────────────────────────┘
     1 s = 1.5 drops
```

```
┌─────────────────────┐
│  No. of drops       │
└─────────────────────┘
  │  1 drop = 65 mg
  ▼
┌─────────────────────┐
│  Mass (mg) of solution │
└─────────────────────┘
  │  1 mg = 10³ g
  ▼
┌─────────────────────┐
│  Mass (g) of solution │
└─────────────────────┘
  │  10³ g = 1 kg
  ▼
┌─────────────────────┐
│  Mass (kg) of solution │
└─────────────────────┘
```

1.5B   Plan: We have the mass of apples in kg and need to find the mass of potassium in those apples in g. The number of apples per pound and the mass of potassium per apple are given. Convert the mass of apples in kg to pounds. Then use the number of apples per pound to calculate the number of apples. Use the mass of potassium in one apple to calculate the mass (mg) of potassium in the group of apples. Finally, convert the mass in mg to g.

Solution:

$$\text{Mass (g)} = 3.25 \text{ kg}\left(\frac{1 \text{ lb}}{0.4536 \text{ kg}}\right)\left(\frac{3 \text{ apples}}{1 \text{ lb}}\right)\left(\frac{159 \text{ mg potassium}}{1 \text{ apple}}\right)\left(\frac{1 \text{ g}}{10^3 \text{ mg}}\right) = 3.4177 = \mathbf{3.42 \text{ g}}$$

Road map:

```
┌─────────────────────┐
│  Mass (kg) of apples │
└─────────────────────┘
  │  0.4536 kg = 1 lb
  ▼
┌─────────────────────┐
│  Mass (lb) of apples │
└─────────────────────┘
  │  1 lb = 3 apples
  ▼
┌─────────────────────┐
│  No. of apples      │
└─────────────────────┘
  │  1 apple = 159 mg potassium
  ▼
┌─────────────────────┐
│  Mass (mg) potassium │
└─────────────────────┘
  │  10³ mg = 1 g
  ▼
┌─────────────────────┐
│  Mass (g) potassium │
└─────────────────────┘
```

1.6A   Plan: We know the area of a field in m^2. We need to know how many bottles of herbicide will be needed to treat that field. The volume of each bottle (in fl oz) and the volume of herbicide needed to treat 300 ft^2 of field are given. Convert the area of the field from m^2 to ft^2 (don't forget to square the conversion factor when converting from squared units to squared units!). Then use the given conversion factors to calculate the number of bottles of herbicide needed. Convert first from ft^2 of field to fl oz of herbicide (because this conversion is from a squared unit to a non-squared unit, we do not need to square the conversion factor). Then use the number of fl oz per bottle to calculate the number of bottles needed.

Solution:

$$\text{No. of bottles} = 2050 \ \text{m}^2 \left(\frac{1 \ \text{ft}^2}{(0.3048)^2 \ \text{m}^2} \right) \left(\frac{1.5 \ \text{fl oz}}{300 \ \text{ft}^2} \right) \left(\frac{1 \ \text{bottle}}{16 \ \text{fl oz}} \right) = 6.8956 = \textbf{7 bottles}$$

Road map:

Area (m²)

$(0.3048)^2 \ \text{m}^2 = 1 \ \text{ft}^2$

Area (ft²)

$300 \ \text{ft}^2 = 1.5 \ \text{fl oz}$

Volume (fl oz)

$16 \ \text{oz} = 1 \ \text{bottle}$

No. of bottles

1.6B   Plan: Calculate the mass of mercury in g. Convert the surface area of the lake from mi² to ft². Find the volume of the lake in ft³ by multiplying the surface area (in ft²) by the depth (in ft). Then convert the volume of the lake to mL by converting first from ft³ to m³, then from m³ to cm³, and from cm³ to mL. Finally, divide the mass in g by the volume in mL to find the mass of mercury in each mL of the lake.

Solution:

$$\text{Mass (g)} = 75,000 \ \text{kg} \left(\frac{1000 \ \text{g}}{1 \ \text{kg}} \right) = 7.5 \ \text{x} \ 10^7 \ \text{g}$$

$$\text{Volume (mL)} = 4.5 \ \text{mi}^2 \left(\frac{(5280)^2 \ \text{ft}^2}{1 \ \text{mi}^2} \right) (35 \ \text{ft}) \left(\frac{0.02832 \ \text{m}^3}{1 \ \text{ft}^3} \right) \left(\frac{1 \ \text{x} \ 10^6 \ \text{cm}^3}{1 \ \text{m}^3} \right) \left(\frac{1 \ \text{mL}}{1 \ \text{cm}^3} \right) = 1.24349 \text{x} 10^{14} \ \text{mL}$$

$$\text{Mass (g) of mercury per mL} = \frac{7.5 \ \text{x} \ 10^7 \ \text{g}}{1.24349 \ \text{x} \ 10^{14} \ \text{mL}} = 6.0314 \text{x} 10^{-7} = \textbf{6.0x10}^{-7} \ \textbf{g/mL}$$

Road map:

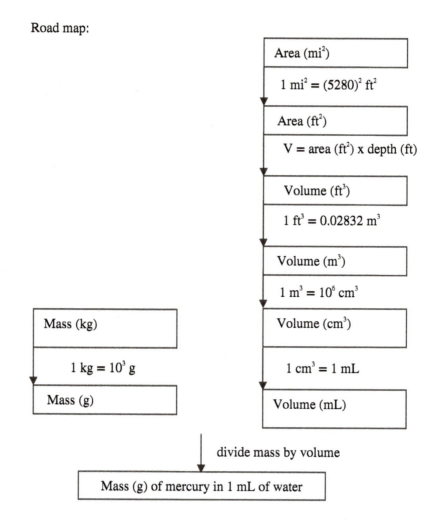

1.7A   <u>Plan:</u> Find the mass of Venus in g. Calculate the radius of Venus by dividing its diameter by 2. Convert the radius from km to cm. Use the radius to calculate the volume of Venus. Finally, find the density of Venus by dividing the mass of Venus (in g) by the volume of Venus (in cm^3).
<u>Solution:</u>

$$\text{Mass (g)} = 4.9 \times 10^{24} \text{ kg} \left(\frac{10^3 \text{ g}}{1 \text{ kg}}\right) = 4.9 \times 10^{27} \text{ g}$$

$$\text{Radius (cm)} = \left(\frac{12,100 \text{ km}}{2}\right)\left(\frac{10^3 \text{ m}}{1 \text{ km}}\right)\left(\frac{10^2 \text{ cm}}{1 \text{ m}}\right) = 6.05 \; 10^8 \text{ cm}$$

$$\text{Volume (cm}^3) = \frac{4}{3}\pi r^3 = \frac{4}{3}(3.14159)(6.05 \times 10^8 \text{ cm})^3 = 9.27587 \times 10^{26} \text{ cm}^3$$

$$\text{Density (g/cm}^3) = \frac{4.9 \times 10^{27} \text{g}}{9.27587 \times 10^{26} \text{cm}^3} = 5.28252 = \textbf{5.3 g/cm}^3$$

Road map:

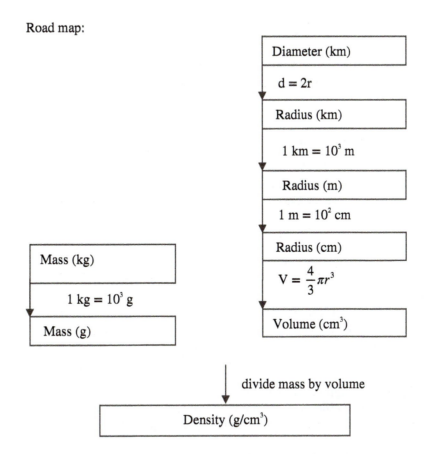

Diameter (km)

$d = 2r$

Radius (km)

$1 \text{ km} = 10^3 \text{ m}$

Radius (m)

$1 \text{ m} = 10^2 \text{ cm}$

Radius (cm)

$V = \dfrac{4}{3}\pi r^3$

Volume (cm³)

Mass (kg)

$1 \text{ kg} = 10^3 \text{ g}$

Mass (g)

divide mass by volume

Density (g/cm³)

1.7B   Plan: The volume unit may be factored away by multiplying by the density. Then it is simply a matter of changing grams to kilograms.
Solution:

$$\text{Mass (kg)} = \left(4.6 \text{ cm}^3\right)\left(\frac{7.5 \text{ g}}{\text{cm}^3}\right)\left(\frac{1 \text{ kg}}{1000 \text{ g}}\right) = 0.0345 = \textbf{0.034 kg}$$

Road map:

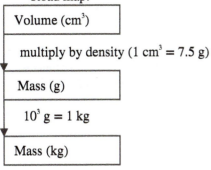

Volume (cm³)

multiply by density (1 cm³ = 7.5 g)

Mass (g)

$10^3 \text{ g} = 1 \text{ kg}$

Mass (kg)

1.8A   Plan: Using the relationship between the Kelvin and Celsius scales, change the Kelvin temperature to the Celsius temperature. Then convert the Celsius temperature to the Fahrenheit value using the relationship between these two scales.
Solution:

$T \text{ (in °C)} = T \text{ (in K)} - 273.15 = 234 \text{ K} - 273.15 = -39.15 = \textbf{-39°C}$

$T \text{ (in °F)} = \dfrac{9}{5} T \text{ (in °C)} + 32 = \dfrac{9}{5} \left(-39.15°C\right) + 32 = -38.47 = \textbf{-38°F}$

<u>Check:</u> Since the Kelvin temperature is below 273, the Celsius temperature must be negative. The low Celsius value gives a negative Fahrenheit value.

1.8B <u>Plan:</u> Convert the Fahrenheit temperature to the Celsius value using the relationship between these two scales. Then use the relationship between the Kelvin and Celsius scales to change the Celsius temperature to the Kelvin temperature.
<u>Solution:</u>

$$T \text{ (in °C)} = \frac{5}{9}(T \text{ (in °F)} - 32) = \frac{5}{9}(2325 \text{ °F} - 32) = 1273.8889 = 1274 \text{ °C}$$

$$T \text{ (in K)} = T \text{ (in °C)} + 273.15 = 1274 \text{ °C} + 273.15 = 1547.15 = 1547 \text{ K}$$

<u>Check:</u> Since the Fahrenheit temperature is large and positive, both the Celsius and Kelvin temperatures should also be positive. Because the Celsius temperature is greater than 273, the Kelvin temperature should be greater than 273, which it is.

1.9A <u>Plan:</u> Determine the significant figures by counting the digits present and accounting for the zeros. Zeros between non-zero digits are significant, as are trailing zeros to the right of a decimal point. Trailing zeros to the left of a decimal point are only significant if the decimal point is present.
<u>Solution:</u>
a) 31.070 mg; **five** significant figures
b) 0.06060 g; **four** significant figures
c) 850.°C; **three** significant figures — note the decimal point that makes the zero significant.
<u>Check:</u> All significant zeros must come after a significant digit.

1.9B <u>Plan:</u> Determine the significant figures by counting the digits present and accounting for the zeros. Zeros between non-zero digits are significant, as are trailing zeros to the right of a decimal point. Trailing zeros to the left of a decimal point are only significant if the decimal point is present.
<u>Solution:</u>
a) 2.000×10^2 mL; **four** significant figures
b) 3.9×10^{-6} m; **two** significant figures — note that none of the zeros are significant.
c) 4.01×10^{-4} L; **three** significant figures
<u>Check:</u> All significant zeros must come after a significant digit.

1.10A <u>Plan:</u> Use the rules presented in the text. Add the two values in the numerator before dividing. The time conversion is an exact conversion and, therefore, does not affect the significant figures in the answer.
<u>Solution:</u>
The addition of 25.65 mL and 37.4 mL gives an answer where the last significant figure is the one after the decimal point (giving three significant figures total):
$$25.65 \text{ mL} + 37.4 \text{ mL} = 63.05 \text{ (would round to 63.0 if not an intermediate step)}$$
When a four significant figure number divides a three significant figure number, the answer must round to three significant figures. An exact number (1 min / 60 s) will have no bearing on the number of significant figures.

$$\frac{63.05 \text{ mL}}{73.55 \text{ s}\left(\dfrac{1 \text{ min}}{60 \text{ s}}\right)} = 51.4344 = \textbf{51.4 mL/min}$$

1.10B <u>Plan:</u> Use the rules presented in the text. Subtract the two values in the numerator and multiply the numbers in the denominator before dividing.
<u>Solution:</u>
The subtraction of 35.26 from 154.64 gives an answer in which the last significant figure is two places after the decimal point (giving five significant figures total):
$$154.64 \text{ g} - 35.26 \text{ g} = 119.38 \text{ g}$$

The multiplication of 4.20 cm (three significant figures) by 5.12 cm (three significant figures) by 6.752 cm (four significant figures) gives a number with three significant figures.

4.20 cm x 5.12 cm x 6.752 cm = 145.1950 (would round to 145 cm^3 if not an intermediate step)

When a three significant figure number divides a five significant figure number, the answer must round to three significant figures.

$$\frac{119.38 \text{ g}}{145.1950 \text{ cm}^3} = 0.82220 = \textbf{0.822 g/cm}^3$$

END–OF–CHAPTER PROBLEMS

1.2   Plan: Apply the definitions of the states of matter to a container. Next, apply these definitions to the examples. Gas molecules fill the entire container; the volume of a gas is the volume of the container. Solids and liquids have a definite volume. The volume of the container does not affect the volume of a solid or liquid.
        Solution:
        a) The helium fills the volume of the entire balloon. The addition or removal of helium will change the volume of a balloon. Helium is a **gas**.
        b) At room temperature, the mercury does not completely fill the thermometer. The surface of the **liquid** mercury indicates the temperature.
        c) The soup completely fills the bottom of the bowl, and it has a definite surface. The soup is a **liquid**, though it is possible that solid particles of food will be present.

1.4   Plan: Define the terms and apply these definitions to the examples.
        Solution:
        Physical property – A characteristic shown by a substance itself, without interacting with or changing into other substances.
        Chemical property – A characteristic of a substance that appears as it interacts with, or transforms into, other substances.
        a) The change in color (yellow–green and silvery to white), and the change in physical state (gas and metal to crystals) are examples of **physical properties**. The change in the physical properties indicates that a chemical change occurred. Thus, the interaction between chlorine gas and sodium metal producing sodium chloride is an example of a **chemical property**.
        b) The sand and the iron are still present. Neither sand nor iron became something else. Colors along with magnetism are **physical properties**. No chemical changes took place, so there are no chemical properties to observe.

1.6   Plan: Apply the definitions of chemical and physical changes to the examples.
        Solution:
        a) Not a chemical change, but a **physical change** — simply cooling returns the soup to its original form.
        b) There is a **chemical change** — cooling the toast will not "un–toast" the bread.
        c) Even though the wood is now in smaller pieces, it is still wood. There has been no change in composition, thus this is a **physical change**, and not a chemical change.
        d) This is a **chemical change** converting the wood (and air) into different substances with different compositions. The wood cannot be "unburned."

1.8   Plan: A system has a higher potential energy before the energy is released (used).
        Solution:
        a) The exhaust is lower in energy than the fuel by an amount of energy equal to that released as the fuel burns. The **fuel** has a higher potential energy.
        b) **Wood**, like the fuel, is higher in energy by the amount released as the wood burns.

1.13   Lavoisier measured the total mass of the reactants and products, not just the mass of the solids. The total mass of the reactants and products remained constant. His measurements showed that a gas was involved in the reaction. He called this gas oxygen (one of his key discoveries).

1.16   A well-designed experiment must have the following essential features:
        1) There must be two variables that are expected to be related.
        2) There must be a way to control all the variables, so that only one at a time may be changed.
        3) The results must be reproducible.

1.20   Plan: Density = $\dfrac{mass}{volume}$. An increase in mass or a decrease in volume will increase the density. A decrease
        in density will result if the mass is decreased or the volume increased.
        Solution:
        a) Density **increases**. The mass of the chlorine gas is not changed, but its volume is smaller.
        b) Density **remains the same**. Neither the mass nor the volume of the solid has changed.
        c) Density **decreases**. Water is one of the few substances that expands on freezing. The mass is constant, but the volume increases.
        d) Density **increases**. Iron, like most materials, contracts on cooling; thus the volume decreases while the mass does not change.
        e) Density **remains the same**. The water does not alter either the mass or the volume of the diamond.

1.23   Plan: Review the definitions of extensive and intensive properties.
        Solution:
        An extensive property depends on the amount of material present. An intensive property is the same regardless of how much material is present.
        a) Mass is an **extensive property**. Changing the amount of material will change the mass.
        b) Density is an **intensive property**. Changing the amount of material changes both the mass and the volume, but the ratio (density) remains fixed.
        c) Volume is an **extensive property**. Changing the amount of material will change the size (volume).
        d) The melting point is an **intensive property**. The melting point depends on the substance, not on the amount of substance.

1.24   Plan: Review the table of conversions in the chapter or inside the back cover of the book. Write the conversion factor so that the unit initially given will cancel, leaving the desired unit.
        Solution:
        a) To convert from in² to cm², use $\dfrac{(2.54\ \text{cm})^2}{(1\ \text{in})^2}$; to convert from cm² to m², use $\dfrac{(1\ \text{m})^2}{(100\ \text{cm})^2}$

        b) To convert from km² to m², use $\dfrac{(1000\ \text{m})^2}{(1\ \text{km})^2}$; to convert from m² to cm², use $\dfrac{(100\ \text{cm})^2}{(1\ \text{m})^2}$

        c) This problem requires two conversion factors: one for distance and one for time. It does not matter which conversion is done first. Alternate methods may be used.
        To convert distance, mi to m, use:

$$\left(\frac{1.609\ \text{km}}{1\ \text{mi}}\right)\left(\frac{1000\ \text{m}}{1\ \text{km}}\right) = 1.609 \times 10^3\ \text{m/mi}$$

        To convert time, h to s, use:

$$\left(\frac{1\ \text{h}}{60\ \text{min}}\right)\left(\frac{1\ \text{min}}{60\ \text{s}}\right) = 1\ \text{h/3600 s}$$

        Therefore, the complete conversion factor is $\left(\dfrac{1.609 \times 10^3\ \text{m}}{1\ \text{mi}}\right)\left(\dfrac{1\ \text{h}}{3600\ \text{s}}\right) = \dfrac{0.4469\ \text{m} \cdot \text{h}}{\text{mi} \cdot \text{s}}$.

        Do the units cancel when you start with a measurement of mi/h?

d) To convert from pounds (lb) to grams (g), use $\dfrac{1000\ \text{g}}{2.205\ \text{lb}}$.

To convert volume from ft^3 to cm^3 use, $\left(\dfrac{(1\ \text{ft})^3}{(12\ \text{in})^3}\right)\left(\dfrac{(1\ \text{in})^3}{(2.54\ \text{cm})^3}\right) = 3.531 \times 10^{-5}\ \text{ft}^3/\text{cm}^3$.

1.26 **Plan:** Use conversion factors from the inside back cover: 1 pm = 10^{-12} m; 10^{-9} m = 1 nm.
Solution:

$$\text{Radius (nm)} = (1430\ \text{pm})\left(\dfrac{10^{-12}\ \text{m}}{1\ \text{pm}}\right)\left(\dfrac{1\ \text{nm}}{10^{-9}\ \text{m}}\right) = \textbf{1.43 nm}$$

1.28 **Plan:** Use conversion factors: 0.01 m = 1 cm; 2.54 cm = 1 in.
Solution:

$$\text{Length (in)} = (100.\ \text{m})\left(\dfrac{1\ \text{cm}}{0.01\ \text{m}}\right)\left(\dfrac{1\ \text{in}}{2.54\ \text{cm}}\right) = 3.9370 \times 10^3 = \textbf{3.94} \times \textbf{10}^3\ \textbf{in}$$

1.30 **Plan:** Use conversion factors $(1\ \text{cm})^2 = (0.01\ \text{m})^2$; $(1000\ \text{m})^2 = (1\ \text{km})^2$ to express the area in km^2. To calculate the cost of the patch, use the conversion factor: $(2.54\ \text{cm})^2 = (1\ \text{in})^2$.
Solution:

a) $\text{Area (km}^2) = (20.7\ \text{cm}^2)\left(\dfrac{(0.01\ \text{m})^2}{(1\ \text{cm})^2}\right)\left(\dfrac{(1\ \text{km})^2}{(1000\ \text{m})^2}\right) = \textbf{2.07} \times \textbf{10}^{-9}\ \textbf{km}^2$

b) $\text{Cost} = (20.7\ \text{cm}^2)\left(\dfrac{(1\ \text{in})^2}{(2.54\ \text{cm})^2}\right)\left(\dfrac{\$3.25}{1\ \text{in}^2}\right) = 10.4276 = \textbf{\$10.43}$

1.32 **Plan:** Use conversion factor 1 kg = 2.205 lb. The following assumes a body weight of 155 lbs. Use your own body weight in place of the 155 lbs.
Solution:

$$\text{Body weight (kg)} = (155\ \text{lb})\left(\dfrac{1\ \text{kg}}{2.205\ \text{lb}}\right) = \textbf{70.3 kg}$$

Answers will vary, depending on the person's mass.

1.34 **Plan:** Mass in g is converted to kg in part a) with the conversion factor 1000 g = 1 kg; mass in g is converted to lb in part b) with the conversion factors 1000 g = 1 kg; 1 kg = 2.205 lb. Volume in cm^3 is converted to m^3 with the conversion factor $(1\ \text{cm})^3 = (0.01\ \text{m})^3$ and to ft^3 with the conversion factors $(2.54\ \text{cm})^3 = (1\ \text{in})^3$; $(12\ \text{in})^3 = (1\ \text{ft})^3$. The conversions may be performed in any order.
Solution:

a) $\text{Density (kg/m}^3) = \left(\dfrac{5.52\ \text{g}}{\text{cm}^3}\right)\left(\dfrac{(1\ \text{cm})^3}{(0.01\ \text{m})^3}\right)\left(\dfrac{1\ \text{kg}}{1000\ \text{g}}\right) = \textbf{5.52} \times \textbf{10}^3\ \textbf{kg/m}^3$

b) $\text{Density (lb/ft}^3) = \left(\dfrac{5.52\ \text{g}}{\text{cm}^3}\right)\left(\dfrac{(2.54\ \text{cm})^3}{(1\ \text{in})^3}\right)\left(\dfrac{(12\ \text{in})^3}{(1\ \text{ft})^3}\right)\left(\dfrac{1\ \text{kg}}{1000\ \text{g}}\right)\left(\dfrac{2.205\ \text{lb}}{1\ \text{kg}}\right) = 344.661 = \textbf{345 lb/ft}^3$

1.36 **Plan:** Use the conversion factors $(1\ \mu\text{m})^3 = (1 \times 10^{-6}\ \text{m})^3$; $(1 \times 10^{-3}\ \text{m})^3 = (1\ \text{mm})^3$ to convert to mm^3. To convert to L, use the conversion factors $(1\ \mu\text{m})^3 = (1 \times 10^{-6}\ \text{m})^3$; $(1 \times 10^{-2}\ \text{m})^3 = (1\ \text{cm})^3$; 1 cm^3 = 1 mL; 1 mL = 1×10^{-3} L.

Solution:

a) Volume $(mm^3) = \left(\dfrac{2.56\ \mu m^3}{cell}\right)\left(\dfrac{(1\times10^{-6}\ m)^3}{(1\ \mu m)^3}\right)\left(\dfrac{(1\ mm)^3}{(1\times10^{-3}\ m)^3}\right) = \mathbf{2.56\times10^{-9}\ mm^3/cell}$

b) Volume $(L) = (10^5\ cells)\left(\dfrac{2.56\ \mu m^3}{cell}\right)\left(\dfrac{(1\times10^{-6}\ m)^3}{(1\ \mu m)^3}\right)\left(\dfrac{(1\ cm)^3}{(1\times10^{-2}\ m)^3}\right)\left(\dfrac{1\ mL}{1\ cm^3}\right)\left(\dfrac{1\times10^{-3}\ L}{1\ mL}\right)$

$= 2.56\times10^{-10} = \mathbf{10^{-10}\ L}$

1.38   Plan: The mass of the mercury in the vial is the mass of the vial filled with mercury minus the mass of the empty vial. Use the density of mercury and the mass of the mercury in the vial to find the volume of mercury and thus the volume of the vial. Once the volume of the vial is known, that volume is used in part b. The density of water is used to find the mass of the given volume of water. Add the mass of water to the mass of the empty vial.
Solution:
a) Mass (g) of mercury = mass of vial and mercury – mass of vial = 185.56 g – 55.32 g = 130.24 g

Volume (cm^3) of mercury = volume of vial = $(130.24\ g)\left(\dfrac{1\ cm^3}{13.53\ g}\right) = 9.626016 = \mathbf{9.626\ cm^3}$

b) Volume (cm^3) of water = volume of vial = $9.626016\ cm^3$

Mass (g) of water = $(9.626016\ cm^3)\left(\dfrac{0.997\ g}{1\ cm^3}\right) = 9.59714$ g water

Mass (g) of vial filled with water = mass of vial + mass of water = 55.32 g + 9.59714 g = 64.91714 = **64.92 g**

1.40   Plan: Calculate the volume of the cube using the relationship Volume = (length of side)^3. The length of side in mm must be converted to cm so that volume will have units of cm^3. Divide the mass of the cube by the volume to find density.
Solution:

Side length $(cm) = (15.6\ mm)\left(\dfrac{10^{-3}\ m}{1\ mm}\right)\left(\dfrac{1\ cm}{10^{-2}\ m}\right) = 1.56\ cm$ (convert to cm to match density unit)

Al cube volume $(cm^3) = (\text{length of side})^3 = (1.56\ cm)^3 = 3.7964\ cm^3$

Density $(g/cm^3) = \dfrac{mass}{volume} = \dfrac{10.25\ g}{3.7964\ cm^3} = 2.69993 = \mathbf{2.70\ g/cm^3}$

1.42   Plan: Use the equations given in the text for converting between the three temperature scales.
Solution:

a) T (in °C) = $[T$ (in °F) – 32$]\dfrac{5}{9} = [68°F - 32]\dfrac{5}{9} = \mathbf{20.°C}$

T (in K) = T (in °C) + 273.15 = 20.°C + 273.15 = 293.15 = **293 K**

b) T (in K) = T (in °C) + 273.15 = –164°C + 273.15 = 109.15 = **109 K**

T (in °F) = $\dfrac{9}{5}T$ (in °C) + 32 = $\dfrac{9}{5}$(–164°C) + 32 = –263.2 = **–263°F**

c) T (in °C) = T (in K) – 273.15 = 0 K – 273.15 = –273.15 = **–273°C**

T (in °F) = $\dfrac{9}{5}T$ (in °C) + 32 = $\dfrac{9}{5}$(–273.15°C) + 32 = –459.67 = **–460.°F**

1.45   Plan: Use 1 nm = 10^{-9} m to convert wavelength in nm to m. To convert wavelength in pm to Å, use 1 pm = 0.01 Å.

Solution:

a) Wavelength (m) $= \left(247 \text{ nm}\right)\left(\dfrac{10^{-9} \text{ m}}{1 \text{ nm}}\right) = \mathbf{2.47 \times 10^{-7}}$ **m**

b) Wavelength (Å) $= \left(6760 \text{ pm}\right)\left(\dfrac{0.01 \text{ Å}}{1 \text{ pm}}\right) = \mathbf{67.6 \text{ Å}}$

1.52   Plan: Review the rules for significant zeros.
Solution:
a) No significant zeros (leading zeros are not significant)
b) No significant zeros (leading zeros are not significant)
c) 0.0041<u>0</u> (terminal zeros to the right of the decimal point are significant)
d) 4.<u>0100</u> x 10^4 (zeros between nonzero digits and terminal zeros to the right of the decimal point are significant)

1.54   Plan: Review the rules for rounding.
Solution: (significant figures are underlined)
a) 0.000<u>3</u>554: the extra digits are 54 at the end of the number. When the digit to be removed is 5 and that 5 is followed by nonzero numbers, the last digit kept is increased by 1: **0.00036**
b) <u>35.8</u>348: the extra digits are 48. Since the digit to be removed (4) is less than 5, the last digit kept is unchanged: **35.83**
c) <u>22.4</u>555: the extra digits are 555. When the digit to be removed is 5 and that 5 is followed by nonzero numbers, the last digit kept is increased by 1: **22.5**

1.56   Plan: Review the rules for rounding.
Solution:
19 rounds to 20: the digit to be removed (9) is greater than 5 so the digit kept is increased by 1.
155 rounds to 160: the digit to be removed is 5 and the digit to be kept is an odd number, so that digit kept is increased by 1.
8.3 rounds to 8: the digit to be removed (3) is less than 5 so the digit kept remains unchanged.
3.2 rounds to 3: the digit to be removed (2) is less than 5 so the digit kept remains unchanged.
2.9 rounds to 3: the digit to be removed (9) is greater than 5 so the digit kept is increased by 1.
4.7 rounds to 5: the digit to be removed (7) is greater than 5 so the digit kept is increased by 1.

$$\left(\dfrac{20 \times 160 \times 8}{3 \times 3 \times 5}\right) = 568.89 = \mathbf{6 \times 10^2}$$

Since there are numbers in the calculation with only one significant figure, the answer can be reported only to one significant figure. (Note that the answer is 560 with the original number of significant digits.)

1.58   Plan: Use a calculator to obtain an initial value. Use the rules for significant figures and rounding to get the final answer.
Solution:

a) $\dfrac{\left(2.795 \text{ m}\right)\left(3.10 \text{ m}\right)}{6.48 \text{ m}} = 1.3371 = \mathbf{1.34 \text{ m}}$ (maximum of 3 significant figures allowed since two of the original

numbers in the calculation have only 3 significant figures)

b) $V = \left(\dfrac{4}{3}\right)\pi\left(17.282 \text{ mm}\right)^3 = 21,620.74 = \mathbf{21,621 \text{ mm}^3}$ (maximum of 5 significant figures allowed)

c) 1.110 cm + 17.3 cm + 108.2 cm + 316 cm = 442.61 = **443 cm** (no digits allowed to the right of the decimal since 316 has no digits to the right of the decimal point)

1.60 Plan: Review the procedure for changing a number to scientific notation. There can be only 1 nonzero digit to the left of the decimal point in correct scientific notation. Moving the decimal point to the left results in a positive exponent while moving the decimal point to the right results in a negative exponent.
Solution:
a) **1.310000×10^5** (Note that all zeros are significant.)
b) **4.7×10^{-4}** (No zeros are significant.)
c) **2.10006×10^5**
d) **2.1605×10^3**

1.62 Plan: Review the examples for changing a number from scientific notation to standard notation. If the exponent is positive, move the decimal back to the right; if the exponent is negative, move the decimal point back to the left.
Solution:
a) **5550** (Do not use terminal decimal point since the zero is not significant.)
b) **10070.** (Use terminal decimal point since final zero is significant.)
c) **0.000000885**
d) **0.003004**

1.64 Plan: In most cases, this involves a simple addition or subtraction of values from the exponents. There can be only 1 nonzero digit to the left of the decimal point in correct scientific notation.
Solution:
a) **8.025×10^4** (The decimal point must be moved an additional 2 places to the left: $10^2 + 10^2 = 10^4$)
b) **1.0098×10^{-3}** (The decimal point must be moved an additional 3 places to the left: $10^3 + 10^{-6} = 10^{-3}$)
c) **7.7×10^{-11}** (The decimal point must be moved an additional 2 places to the right: $10^{-2} + 10^{-9} = 10^{-11}$)

1.66 Plan: Calculate a temporary answer by simply entering the numbers into a calculator. Then you will need to round the value to the appropriate number of significant figures. Cancel units as you would cancel numbers, and place the remaining units after your numerical answer.
Solution:

a) $\dfrac{\left(6.626 \times 10^{-34} \text{ J/s}\right)\left(2.9979 \times 10^8 \text{ m/s}\right)}{489 \times 10^{-9} \text{ m}} = 4.062185 \times 10^{-19}$

= **4.06×10^{-19} J** (489×10^{-9} m limits the answer to 3 significant figures; units of m and s cancel)

b) $\dfrac{\left(6.022 \times 10^{23} \text{ molecules/mol}\right)\left(1.23 \times 10^2 \text{ g}\right)}{46.07 \text{ g/mol}} = 1.6078 \times 10^{24}$

= **1.61×10^{24} molecules** (1.23×10^2 g limits answer to 3 significant figures; units of mol and g cancel)

c) $\left(6.022 \times 10^{23} \text{ atoms/mol}\right)\left(2.18 \times 10^{-18} \text{ J/atom}\right)\left(\dfrac{1}{2^2} - \dfrac{1}{3^2}\right) = 1.82333 \times 10^5$

= **1.82×10^5 J/mol** (2.18×10^{-18} J/atom limits answer to 3 significant figures; unit of atoms cancels)

1.68 Plan: Exact numbers are those which have no uncertainty. Unit definitions and number counts of items in a group are examples of exact numbers.
Solution:
a) The height of Angel Falls is a measured quantity. This is **not** an exact number.
b) The number of planets in the solar system is a number count. This **is** an exact number.
c) The number of grams in a pound is not a unit definition. This is **not** an exact number.
d) The number of millimeters in a meter is a definition of the prefix "milli–." This **is** an exact number.

1.70   <u>Plan:</u> Observe the figure, and estimate a reading the best you can.
        <u>Solution:</u>
        The scale markings are 0.2 cm apart. The end of the metal strip falls between the mark for 7.4 cm and 7.6 cm. If we assume that one can divide the space between markings into fourths, the uncertainty is one-fourth the separation between the marks. Thus, since the end of the metal strip falls between 7.45 and 7.55 we can report its length as **7.50 ± 0.05 cm**. (Note: If the assumption is that one can divide the space between markings into halves only, then the result is 7.5 ± 0.1 cm.)

1.72   <u>Plan:</u> Calculate the average of each data set. Remember that accuracy refers to how close a measurement is to the actual or true value while precision refers to how close multiple measurements are to each other.
        <u>Solution:</u>
        a) $I_{avg} = \dfrac{8.72 \text{ g} + 8.74 \text{ g} + 8.70 \text{ g}}{3} = 8.7200 = \textbf{8.72 g}$

        $II_{avg} = \dfrac{8.56 \text{ g} + 8.77 \text{ g} + 8.83 \text{ g}}{3} = 8.7200 = \textbf{8.72 g}$

        $III_{avg} = \dfrac{8.50 \text{ g} + 8.48 \text{ g} + 8.51 \text{ g}}{3} = 8.4967 = \textbf{8.50 g}$

        $IV_{avg} = \dfrac{8.41 \text{ g} + 8.72 \text{ g} + 8.55 \text{ g}}{3} = 8.5600 = \textbf{8.56 g}$

        Sets **I** and **II** are most accurate since their average value, 8.72 g, is closest to the true value, 8.72 g.
        b) To get an idea of precision, calculate the range of each set of values: largest value – smallest value. A small range is an indication of good precision since the values are close to each other.
        I_{range} = 8.74 g – 8.70 g = 0.04 g
        II_{range} = 8.83 g – 8.56 g = 0.27 g
        III_{range} = 8.51 g – 8.48 g = 0.03 g
        IV_{range} = 8.72 g – 8.41 g = 0.31 g
        Set III is the most precise (smallest range), but is the least accurate (the average is the farthest from the actual value).
        c) **Set I** has the best combination of high accuracy (average value = actual value) and high precision (relatively small range).
        d) **Set IV** has both low accuracy (average value differs from actual value) and low precision (has the largest range).

1.74   <u>Plan:</u> If it is necessary to force something to happen, the potential energy will be higher.
        <u>Solution:</u>

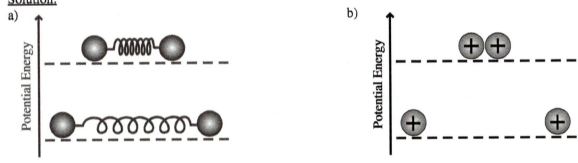

        a) The balls on the relaxed spring have a lower potential energy and are more stable. The balls on the compressed spring have a higher potential energy, because the balls will move once the spring is released. This configuration is less stable.
        b) The two + charges apart from each other have a lower potential energy and are more stable. The two + charges near each other have a higher potential energy, because they repel one another. This arrangement is less stable.

1.76   <u>Plan:</u> Use the concentrations of bromine given.
       <u>Solution:</u>

$$\frac{\text{Mass bromine in Dead Sea}}{\text{Mass bromine in seawater}} = \frac{0.50 \text{ g/L}}{0.065 \text{ g/L}} = \textbf{7.7/1}$$

1.78   <u>Plan:</u> In each case, calculate the overall density of the ball and contents and compare to the density of air. The volume of the ball in cm^3 is converted to units of L to find the density of the ball itself in g/L. The densities of the ball and the gas in the ball are additive because the volume of the ball and the volume of the gas are the same.
       <u>Solution:</u>
a) Density of evacuated ball: the mass is only that of the sphere itself:

$$\text{Volume of ball (L)} = \left(560 \text{ cm}^3\right)\left(\frac{1 \text{ mL}}{1 \text{ cm}^3}\right)\left(\frac{10^{-3} \text{ L}}{1 \text{ mL}}\right) = 0.560 = 0.56 \text{ L}$$

$$\text{Density of evacuated ball} = \frac{\text{mass}}{\text{volume}} = \frac{0.12 \text{ g}}{0.560} = 0.21 \text{ g/L}$$

The evacuated ball will **float** because its density is less than that of air.
b) Because the density of CO_2 is greater than that of air, a ball filled with CO_2 will **sink**.
c) Density of ball + density of hydrogen = 0.0899 + 0.21 g/L = 0.30 g/L
The ball will **float** because the density of the ball filled with hydrogen is less than the density of air.
d) Because the density of O_2 is greater than that of air, a ball filled with O_2 will **sink**.
e) Density of ball + density of nitrogen = 0.21 g/L + 1.165 g/L = 1.38 g/L
The ball will **sink** because the density of the ball filled with nitrogen is greater than the density of air.

f) To sink, the total mass of the ball and gas must weigh $(0.560 \text{ L})\left(\frac{1.189 \text{ g}}{1 \text{ L}}\right) = 0.66584 \text{ g}$

For the evacuated ball:
0.66584 – 0.12 g = 0.54585 = **0.55 g**. More than 0.55 g would have to be added to make the ball sink.
For ball filled with hydrogen:

$$\text{Mass of hydrogen in the ball} = (0.56 \text{ L})\left(\frac{0.0899 \text{ g}}{1 \text{ L}}\right) = 0.0503 \text{ g}$$

Mass of hydrogen and ball = 0.0503 g + 0.12 g = 0.17 g
0.66584 – 0.17 g = 0.4958 = **0.50 g**. More than 0.50 g would have to be added to make the ball sink.

1.80   <u>Plan:</u> Convert the surface area to m^2 and then use the surface area and the depth to determine the volume of the oceans (area x depth = volume) in m^3. The volume is then converted to liters, and finally to the mass of gold using the density of gold in g/L. Once the mass of the gold is known, its density is used to find the volume of that amount of gold. The mass of gold is converted to troy oz and the price of gold per troy oz gives the total price.
       <u>Solution:</u>

a) $\text{Area of ocean (m}^2) = \left(3.63 \times 10^8 \text{ km}^2\right)\left(\frac{(1000 \text{ m})^2}{(1 \text{ km})^2}\right) = 3.63 \times 10^{14} \text{ m}^2$

$\text{Volume of ocean (m}^3) = (\text{area})(\text{depth}) = (3.63 \times 10^{14} \text{ m}^2)(3800 \text{ m}) = 1.3794 \times 10^{18} \text{ m}^3$

$\text{Mass of gold (g)} = \left(1.3794 \times 10^{18} \text{ m}^3\right)\left(\frac{1 \text{ L}}{10^{-3} \text{ m}^3}\right)\left(\frac{5.8 \times 10^{-9} \text{ g}}{\text{L}}\right) = 8.00052 \times 10^{12} = \textbf{8.0} \times \textbf{10}^{12} \textbf{ g}$

b) Use the density of gold to convert mass of gold to volume of gold:

$\text{Volume of gold (m}^3) = \left(8.00052 \times 10^{12} \text{ g}\right)\left(\frac{1 \text{ cm}^3}{19.3 \text{ g}}\right)\left(\frac{(0.01 \text{ m})^3}{(1 \text{ cm})^3}\right) = 4.14535 \times 10^5 = \textbf{4.1} \times \textbf{10}^5 \textbf{ m}^3$

c) $\text{Value of gold} = \left(8.00052 \times 10^{12} \text{ g}\right)\left(\frac{1 \text{ tr. oz}}{31.1 \text{ g}}\right)\left(\frac{\$1595}{1 \text{ tr. oz.}}\right) = 4.1032 \times 10^{14} = \textbf{\$4.1} \times \textbf{10}^{14}$

1.82   Plan: Use the equations for temperature conversion given in the chapter. The mass of nitrogen is conserved when the gas is liquefied; the mass of the nitrogen gas equals the mass of the liquid nitrogen. Use the density of nitrogen gas to find the mass of the nitrogen; then use the density of liquid nitrogen to find the volume of that mass of liquid nitrogen.
Solution:
a) T (in °C) = T (in K) – 273.15 = 77.36 K – 273.15 = **–195.79°C**

b) T (in °F) = $\frac{9}{5}T$(in °C) + 32 = $\frac{9}{5}$(–195.79°C) + 32 = –320.422 = **–320.42°F**

c) Mass of liquid nitrogen = mass of gaseous nitrogen = $(895.0 \text{ L})\left(\frac{4.566 \text{ g}}{1 \text{ L}}\right)$ = 4086.57 g N_2

Volume of liquid N_2 = $(4086.57 \text{ g})\left(\frac{1 \text{ L}}{809 \text{ g}}\right)$ = 5.0514 = **5.05 L**

1.83   Plan: For part a), convert mi to m and h to s. For part b), time is converted from h to min and length from mi to km. For part c), convert the distance in ft to mi and use the average speed in mi/h to find the time necessary to cover the given distance.
Solution:

a) Speed (m/s) = $\left(\frac{5.9 \text{ mi}}{h}\right)\left(\frac{1000 \text{ m}}{0.62 \text{ mi}}\right)\left(\frac{1 \text{ h}}{3600 \text{ s}}\right)$ = 2.643 = **2.6 m/s**

b) Distance (km) = $(98 \text{ min})\left(\frac{1 \text{ h}}{60 \text{ min}}\right)\left(\frac{5.9 \text{ mi}}{h}\right)\left(\frac{1 \text{ km}}{0.62 \text{ mi}}\right)$ = 15.543 = **16 km**

c) Time (h) = $(4.75 \times 10^4 \text{ ft})\left(\frac{1 \text{ mi}}{5280 \text{ ft}}\right)\left(\frac{1 \text{ h}}{5.9 \text{ mi}}\right)$ = 1.5248 = 1.5 h

If she starts running at 11:15 am, 1.5 hours later the time is **12:45 pm**.

1.85   Plan: In visualizing the problem, the two scales can be set next to each other.
Solution:
There are 50 divisions between the freezing point and boiling point of benzene on the °X scale and 74.6 divisions

(80.1°C – 5.5°C) on the °C scale. So °X = $\left(\frac{50°X}{74.6°C}\right)°C$

This does not account for the offset of 5.5 divisions in the °C scale from the zero point on the °X scale.

So °X = $\left(\frac{50°X}{74.6°C}\right)(°C – 5.5°C)$

Check: Plug in 80.1°C and see if result agrees with expected value of 50°X.

So °X = $\left(\frac{50°X}{74.6°C}\right)(80.1°C – 5.5°C)$ = 50°X

Use this formula to find the freezing and boiling points of water on the °X scale.

fp_{water} °X = $\left(\frac{50°X}{74.6°C}\right)(0.00°C – 5.5°C)$ = 3.68°X = **–3.7°X**

bp_{water} °X = $\left(\frac{50°X}{74.6°C}\right)(100.0°C – 5.5°C)$ = **63.3°X**

1.86   Plan: Determine the total mass of Earth's crust in metric tons (t) by finding the volume of crust (surface area x depth) in km^3 and then in cm^3 and then using the density to find the mass of this volume, using conversions from the inside back cover. The mass of each individual element comes from the concentration of that element multiplied by the mass of the crust.

<u>Solution:</u>

Volume of crust (km^3) = area x depth = $(35 \text{ km})(5.10 \times 10^8 \text{ km}^2)$ = $1.785 \times 10^{10} \text{ km}^3$

Volume of crust (cm^3) = $(1.785 \times 10^{10} \text{ km}^3)\left(\frac{(1000 \text{ m})^3}{(1 \text{ km})^3}\right)\left(\frac{(1 \text{ cm})^3}{(0.01 \text{ m})^3}\right)$ = $1.785 \times 10^{25} \text{ cm}^3$

Mass of crust (t) = $(1.785 \times 10^{25} \text{ cm}^3)\left(\frac{2.8 \text{ g}}{1 \text{ cm}^3}\right)\left(\frac{1 \text{ kg}}{1000 \text{ g}}\right)\left(\frac{1 \text{ t}}{1000 \text{ kg}}\right)$ = $4.998 \times 10^{19} \text{ t}$

Mass of oxygen (g) = $(4.998 \times 10^{19} \text{ t})\left(\frac{4.55 \times 10^5 \text{ g oxygen}}{1 \text{ t}}\right)$ = 2.2741×10^{25} = $\mathbf{2.3 \times 10^{25} \text{ g oxygen}}$

Mass of silicon (g) = $(4.998 \times 10^{19} \text{ t})\left(\frac{2.72 \times 10^5 \text{ g silicon}}{1 \text{ t}}\right)$ = 1.3595×10^{25} = $\mathbf{1.4 \times 10^{25} \text{ g silicon}}$

Mass of ruthenium = mass of rhodium = $(4.998 \times 10^{19} \text{ t})\left(\frac{1 \times 10^{-4} \text{ g element}}{1 \text{ t}}\right)$

$= 4.998 \times 10^{15}$ = $\mathbf{5 \times 10^{15} \text{ g each of ruthenium and rhodium}}$

CHAPTER 2 THE COMPONENTS OF MATTER

FOLLOW–UP PROBLEMS

2.1A   <u>Plan:</u> An element has only one kind of atom; a compound is composed of at least two kinds of atoms. A mixture consists of two or more substances mixed together in the same container.
<u>Solution:</u>
(a) There is only one type of atom (blue) present, so this is an **element**.
(b) Two different atoms (brown and green) appear in a fixed ratio of 1/1, so this is a **compound**.
(c) These molecules consist of one type of atom (orange), so this is an **element**.

2.1B   <u>Plan:</u> An element has only one kind of atom; a compound is composed of at least two kinds of atoms.
<u>Solution:</u>
The circle on the left contains molecules with either only orange atoms or only blue atoms. This is a **mixture of two different elements**. In the circle on the right, the molecules are composed of one orange atom and one blue atom so this is a **compound**.

2.2A   <u>Plan:</u> Use the mass fraction of iron in fool's gold to find the mass of fool's gold that contains 86.2 g of iron. Subtract the amount of iron in the mass fraction from the amount of fool's gold in the mass fraction to obtain the mass of sulfur in that amount of fool's gold. Find the mass fraction of sulfur in fool's gold and multiply the amount of fool's gold by the mass fraction of sulfur to determine the mass of sulfur in the sample.
<u>Solution:</u>

$$\text{Mass (g) of fool's gold} = 86.2 \text{ g iron} \times \left(\frac{110.0 \text{ g fool's gold}}{51.2 \text{ g iron}} \right) = 185.195 \text{ g fool's gold}$$

$$\text{Mass (g) of sulfur in 110.0 g of fool's gold} = 110.0 \text{ g fool's gold} - 51.2 \text{ g iron} = 58.8 \text{ g sulfur}$$

$$\text{Mass (g) of sulfur} = 185.195 \text{ g fool's gold} \times \left(\frac{58.8 \text{ g sulfur}}{110.0 \text{ g fool's gold}} \right) = 98.995 = \textbf{99.0 g sulfur}$$

2.2B   <u>Plan:</u> Subtract the amount of silver from the amount of silver bromide to find the mass of bromine in 26.8 g of silver bromide. Use the mass fraction of silver in silver bromide to find the mass of silver in 3.57 g of silver bromide. Use the mass fraction of bromine in silver bromide to find the mass of bromine in 3.57 g of silver bromide.
<u>Solution:</u>

$$\text{Mass (g) of bromine in 26.8 g silver bromide} = 26.8 \text{ g silver bromide} - 15.4 \text{ g silver} = 11.4 \text{ g bromine}$$

$$\text{Mass (g) of silver in 3.57 g silver bromide} = 3.57 \text{ g silver bromide} \left(\frac{15.4 \text{ g silver}}{26.8 \text{ g silver bromide}} \right) = \textbf{2.05 g silver}$$

$$\text{Mass (g) of bromine in 3.57 g silver bromide} = 3.57 \text{ g silver bromide} \left(\frac{11.4 \text{ g bromine}}{26.8 \text{ g silver bromide}} \right) = \textbf{1.52 g bromine}$$

2.3A   <u>Plan:</u> The law of multiple proportions states that when two elements react to form two compounds, the different masses of element B that react with a fixed mass of element A is a ratio of small whole numbers. The law of definite composition states that the elements in a compound are present in fixed parts by mass. The law of mass conversation states that the total mass before and after a reaction is the same.
<u>Solution:</u>
The law of **mass conservation** is illustrated because the number of atoms does not change as the reaction proceeds (there are 14 red spheres and 12 black spheres before and after the reaction occurs). The law of **multiple proportions** is illustrated because two compounds are formed as a result of the reaction. One of the compounds

has a ratio of 2 red spheres to 1 black sphere. The other has a ratio of 1 red sphere to 1 black sphere. The law of **definite proportions** is illustrated because each compound has a fixed ratio of red-to-black atoms.

2.3B   Plan: The law of multiple proportions states that when two elements react to form two compounds, the different masses of element B that react with a fixed mass of element A is a ratio of small whole numbers.
Solution:
Only **Sample B** shows two different bromine-fluorine compounds. In one compound there are three fluorine atoms for every one bromine atom; in the other compound, there is one fluorine atom for every bromine atom.

2.4A   Plan: The subscript (atomic number = Z) gives the number of protons, and for an atom, the number of electrons. The atomic number identifies the element. The superscript gives the mass number (A) which is the total of the protons plus neutrons. The number of neutrons is simply the mass number minus the atomic number ($A - Z$).
Solution:
^{46}Ti $Z = 22$ and $A = 46$, there are **22 p$^+$** and **22 e$^-$** and $46 - 22 =$ **24 n^0**
^{47}Ti $Z = 22$ and $A = 47$, there are **22 p$^+$** and **22 e$^-$** and $47 - 22 =$ **25 n^0**
^{48}Ti $Z = 22$ and $A = 48$, there are **22 p$^+$** and **22 e$^-$** and $48 - 22 =$ **26 n^0**
^{49}Ti $Z = 22$ and $A = 49$, there are **22 p$^+$** and **22 e$^-$** and $49 - 22 =$ **27 n^0**
^{50}Ti $Z = 22$ and $A = 50$, there are **22 p$^+$** and **22 e$^-$** and $50 - 22 =$ **28 n^0**

2.4B   Plan: The subscript (atomic number = Z) gives the number of protons, and for an atom, the number of electrons. The atomic number identifies the element. The superscript gives the mass number (A) which is the total of the protons plus neutrons. The number of neutrons is simply the mass number minus the atomic number ($A - Z$).
Solution:
a) $Z = 5$ and $A = 11$, there are 5 p$^+$ and 5 e$^-$ and $11 - 5 = 6$ n^0; Atomic number = 5 = **B**.
b) $Z = 20$ and $A = 41$, there are 20 p$^+$ and 20 e$^-$ and $41 - 20 = 21$ n^0; Atomic number = 20 = **Ca**.
c) $Z = 53$ and $A = 131$, there are 53 p$^+$ and 53 e$^-$ and $131 - 53 = 78$ n^0; Atomic number = 53 = **I**.

2.5A   Plan: First, divide the percent abundance value (found in Figure B2.2C, Tools of the Laboratory, p. 57) by 100 to obtain the fractional value for each isotope. Multiply each isotopic mass by the fractional value, and add the resulting masses to obtain neon's atomic mass.
Solution:
Atomic Mass = (^{20}Ne mass) (fractional abundance of ^{20}Ne) + (^{21}Ne mass) (fractional abundance of ^{21}Ne) +
              (^{22}Ne mass) (fractional abundance of ^{22}Ne)
      ^{20}Ne = (19.99244 amu)(0.9048) = 18.09 amu
      ^{21}Ne = (20.99385 amu)(0.0027) = 0.057 amu
      ^{22}Ne = (21.99139 amu)(0.0925) = 2.03 amu
                            20.177 amu = **20.18 amu**

2.5B   Plan: To find the percent abundance of each B isotope, let x equal the fractional abundance of ^{10}B and $(1 - x)$ equal the fractional abundance of ^{11}B. Remember that atomic mass = isotopic mass of ^{10}B x fractional abundance) + (isotopic mass of ^{11}B x fractional abundance).
Solution:
Atomic Mass = (^{10}B mass) (fractional abundance of ^{10}B) + (^{11}B mass) (fractional abundance of ^{11}B)
Amount of ^{10}B + Amount ^{11}B = 1   (setting ^{10}B = x gives ^{11}B = 1 – x)
        10.81 amu = (10.0129 amu)(x) + (11.0093 amu) (1 – x)
        10.81 amu = 11.0093 – 11.0093x + 10.0129 x
        10.81 amu = 11.0093 – 0.9964 x
          –0.1993 = – 0.9964x
          x = 0.20;   1 – x = 0.80
    (10.81 – 11.0093 limits the answer to 2 significant figures)
        Fraction x 100% = percent abundance.
        % abundance of ^{10}B = **20.%**; % abundance of ^{11}B = **80.%**

2.6A   Plan: Use the provided atomic numbers (the Z numbers) to locate these elements on the periodic table. The name of the element is on the periodic table or on the list of elements inside the front cover of the textbook. Use the periodic table to find the group/column number (listed at the top of each column) and the period/row number (listed at the left of each row) in which the element is located. Classify the element from the color coding in the periodic table.
        Solution:
        (a) Z = 14: Silicon, Si; Group 4A(14) and Period 3; metalloid
        (b) Z = 55: Cesium, Cs; Group 1A(1) and Period 6; main-group metal
        (c) Z = 54: Xenon, Xe; Group 8A(18) and Period 5; nonmetal

2.6B   Plan: Use the provided atomic numbers (the Z numbers) to locate these elements on the periodic table. The name of the element is on the periodic table or on the list of elements inside the front cover of the textbook. Use the periodic table to find the group/column number (listed at the top of each column) and the period/row number (listed at the left of each row) in which the element is located. Classify the element from the color coding in the periodic table.
        Solution:
        (a) Z = 12: Magnesium, Mg; Group 2A(2) and Period 3; main-group metal
        (b) Z = 7: Nitrogen, N; Group 5A(15) and Period 2; nonmetal
        (c) Z = 30: Zinc, Zn; Group 2B(12) and Period 4; transition metal

2.7A   Plan: Locate these elements on the periodic table and predict what ions they will form. For A-group cations (metals), ion charge = group number; for anions (nonmetals), ion charge = group number − 8. Or, relate the element's position to the nearest noble gas. Elements after a noble gas lose electrons to become positive ions, while those before a noble gas gain electrons to become negative ions.
        Solution:
        a) $_{16}S^{2-}$ [Group 6A(16); 6 − 8 = −2]; sulfur needs to gain 2 electrons to match the number of electrons in $_{18}Ar$.
        b) $_{37}Rb^{+}$ [Group 1A(1)]; rubidium needs to lose 1 electron to match the number of electrons in $_{36}Kr$.
        c) $_{56}Ba^{2+}$ [Group 2A(2)]; barium needs to lose 2 electrons to match the number of electrons in $_{54}Xe$.

2.7B   Plan: Locate these elements on the periodic table and predict what ions they will form. For A-group cations (metals), ion charge = group number; for anions (nonmetals), ion charge = group number − 8. Or, relate the element's position to the nearest noble gas. Elements after a noble gas lose electrons to become positive ions, while those before a noble gas gain electrons to become negative ions.
        Solution:
        a) $_{38}Sr^{2+}$ [Group 2A(2)]; strontium needs to lose 2 electrons to match the number of electrons in $_{36}Kr$.
        b) $_{8}O^{2-}$ [Group 6A(16); 6 − 8 = −2]; oxygen needs to gain 2 electrons to match the number of electrons in $_{10}Ne$.
        c) $_{55}Cs^{+}$ [Group 1A(1)]; cesium needs to lose 1 electron to match the number of electrons in $_{54}Xe$.

2.8A   Plan: When dealing with ionic binary compounds, the first name is that of the metal and the second name is that of the nonmetal. If there is any doubt, refer to the periodic table. The metal name is unchanged, while the nonmetal has an -ide suffix added to the nonmetal root.
        Solution:
        a) Zinc is the metal and oxygen is the nonmetal: **zinc oxide.**
        b) Silver is the metal and bromine is the nonmetal: **silver bromide.**
        c) Lithium is the metal and chlorine is the nonmetal: **lithium chloride**.
        d) Aluminum is the metal and sulfur is the nonmetal: **aluminum sulfide.**

2.8B   Plan: When dealing with ionic binary compounds, the first name is that of the metal and the second name is that of the nonmetal. If there is any doubt, refer to the periodic table. The metal name is unchanged, while the nonmetal has an -ide suffix added to the nonmetal root.
        Solution:
        a) Potassium is the metal and sulfur is the nonmetal: **potassium sulfide.**
        b) Barium is the metal and iodine is the nonmetal: **barium iodide.**
        c) Cesium is the metal and nitrogen is the nonmetal: **cesium nitride.**
        d) Sodium is the metal and hydrogen is the nonmetal: **sodium hydride.**

2.9A   Plan: Use the charges of the ions to predict the lowest ratio leading to a neutral compound. The sum of the total charges must be 0.
Solution:
a)   Zinc should form Zn^{2+} and oxygen should form O^{2-}; these will combine to give **ZnO**. The charges cancel $[+2 + -2 = 0]$, so this is an acceptable formula.
b)   Silver should form Ag^+ and bromine should form Br^-; these will combine to give **AgBr**. The charges cancel $[+1 + -1 = 0]$, so this is an acceptable formula.
c)   Lithium should form Li^+ and chlorine should form Cl^-; these will combine to give **LiCl**. The charges cancel $[+1 + -1 = 0]$, so this is an acceptable formula.
d)   Aluminum should form Al^{3+} and sulfur should form S^{2-}; to produce a neutral combination the formula is **Al_2S_3**. This way the charges will cancel $[2(+3) + 3(-2) = 0]$, so this is an acceptable formula.

2.9B   Plan: Use the charges of the ions to predict the lowest ratio leading to a neutral compound. The sum of the total charges must be 0.
Solution:
a) Potassium should form K^+ and sulfur should form S^{2-}; these will combine to give **K_2S**. The charges cancel $[2(+1) + -2 = 0]$, so this is an acceptable formula.
b) Barium should form Ba^{2+} and iodine should form I^-, these will combine to give **BaI_2**. The charges cancel $[+2 + 2(-1) = 0]$, so this is an acceptable formula.
c) Cesium should form Cs^+ and nitrogen should form N^{3-}; these will combine to give **Cs_3N**. The charges cancel $[3(+1) + -3 = 0]$, so this is an acceptable formula.
d) Sodium should form Na^+ and hydrogen should form H^-; to produce a neutral combination the formula is **NaH**. This way the charges will cancel $(+1 + -1 = 0)$, so this is an acceptable formula.

2.10A   Plan: Determine the names or symbols of each of the species present. Then combine the species to produce a name or formula. The metal or positive ions are written first. Review the rules for nomenclature covered in the chapter. For metals, like many transition metals, that can form more than one ion each with a different charge, the ionic charge of the metal ion is indicated by a Roman numeral within parentheses immediately following the metal's name.
Solution:
a) The Roman numeral means that the lead is Pb^{4+}; oxygen produces the usual O^{2-}. The neutral combination is $[+4 + 2(-2) = 0]$, so the formula is **PbO_2**.
b) Sulfide (Group 6A(16)), like oxide, is -2 ($6 - 8 = -2$). This is split between two copper ions, each of which must be $+1$. This is one of the two common charges for copper ions. The $+1$ charge on the copper is indicated with a Roman numeral. This gives the name **copper(I) sulfide** (common name = cuprous sulfide).
c) Bromine (Group 7A(17)), like other elements in the same column of the periodic table, forms a -1 ion. Two of these ions require a total of $+2$ to cancel them out. Thus, the iron must be $+2$ (indicated with a Roman numeral). This is one of the two common charges on iron ions. This gives the name **iron(II) bromide** (or ferrous bromide).
d) The mercuric ion is Hg^{2+}, and two -1 ions (Cl^-) are needed to cancel the charge. This gives the formula **$HgCl_2$**.

2.10B   Plan: Determine the names or symbols of each of the species present. Then combine the species to produce a name or formula. The metal or positive ions are written first. Review the rules for nomenclature covered in the chapter. For metals, like many transition metals, that can form more than one ion each with a different charge, the ionic charge of the metal ion is indicated by a Roman numeral within parentheses immediately following the metal's name.
Solution:
a) The Roman numeral means that the copper ion is Cu^{2+}; nitride produces the usual N^{3-}. The neutral combination is $[3(+2) + 2(-3) = 0]$, so the formula is **Cu_3N_2**. Three Cu^{2+} ions balance two N^{3-} ions.
b) The anion is iodide, I^-, and the formula shows two I^-. Therefore the cation must be Pb^{2+}, lead(II) ion: PbI_2 is **lead(II) iodide**. (The common name is plumbous iodide.)
c) Chromic is the common name for chromium(III) ion, Cr^{3+}; sulfide ion is S^{2-}. To balance the charges, the formula is **Cr_2S_3**. [The systematic name is chromium(III) sulfide.]
d) The anion is oxide, O^{2-}, which requires that the cation be Fe^{2+}. The name is **iron(II) oxide**. (The common name is ferrous oxide.)

2.11A   Plan: Determine the names or symbols of each of the species present. Then combine the species to produce a name or formula. The metal or positive ions always go first.

Solution:

a) The cupric ion, Cu^{2+}, requires two nitrate ions, NO_3^-, to cancel the charges. Trihydrate means three water molecules. These combine to give **$Cu(NO_3)_2 \cdot 3H_2O$**.

b) The zinc ion, Zn^{2+}, requires two hydroxide ions, OH^-, to cancel the charges. These combine to give **$Zn(OH)_2$**.

c) Lithium only forms the Li^+ ion, so Roman numerals are unnecessary. The cyanide ion, CN^-, has the appropriate charge. These combine to give **lithium cyanide**.

2.11B   Plan: Determine the names or symbols of each of the species present. Then combine the species to produce a name or formula. The metal or positive ions always go first.

Solution:

a) Two ammonium ions, NH_4^+, are needed to balance the charge on one sulfate ion, SO_4^{2-}. These combine to give **$(NH_4)_2SO_4$**.

b) The nickel ion is combined with two nitrate ions, NO_3^-, so the charge on the nickel ion is 2+, Ni^{2+}. There are 6 water molecules (hexahydrate). Therefore, the name is **nickel(II) nitrate hexahydrate**.

c) Potassium forms the K^+ ion. The bicarbonate ion, HCO_3^-, has the appropriate charge to balance out one potassium ion. Therefore, the formula of this compound is **$KHCO_3$**.

2.12A   Plan: Determine the names or symbols of each of the species present. Then combine the species to produce a name or formula. The metal or positive ions always go first. Make corrections accordingly.

Solution:

a) The ammonium ion is NH_4^+ and the phosphate ion is PO_4^{3-}. To give a neutral compound they should combine $[3(+1) + (-3) = 0]$ to give the correct formula **$(NH_4)_3PO_4$**.

b) Aluminum gives Al^{3+} and the hydroxide ion is OH^-. To give a neutral compound they should combine $[+3 + 3(-1) = 0]$ to give the correct formula **$Al(OH)_3$**. Parentheses are required around the polyatomic ion.

c) Manganese is Mn, and Mg, in the formula, is magnesium. Magnesium only forms the Mg^{2+} ion, so Roman numerals are unnecessary. The other ion is HCO_3^-, which is called the hydrogen carbonate (or bicarbonate) ion. The correct name is **magnesium hydrogen carbonate** or **magnesium bicarbonate**.

2.12B   Plan: Determine the names or symbols of each of the species present. Then combine the species to produce a name or formula. The metal or positive ions always go first. Make corrections accordingly.

Solution:

a) Either use the "-ic" suffix or the "(III)" but not both. Nitr*ide* is N^{3-}, and nitr*ate* is NO_3^-. This gives the correct name: **chromium(III) nitrate** (the common name is chromic nitrate).

b) Cadmium is Cd, and Ca, in the formula, is calcium. Nitr*ate* is NO_3^-, and nitr*ite* is NO_2^-. The correct name is **calcium nitrite**.

c) Potassium is K, and P, in the formula, is phosphorus. *Per*chlorate is ClO_4^-, and chlorate is ClO_3^-. Additionally, parentheses are not needed when there is only one of a given polyatomic ion. The correct formula is **$KClO_3$**.

2.13A   Plan: Use the name of the acid to determine the name of the anion of the acid. The name *hydro_____ic acid* indicates that the anion is a monatomic nonmetal. The name *_____ic acid* indicates that the anion is an oxoanion with an – ate ending. The name *_____ous acid* indicates that the anion is an oxoanion with an –ite ending.

Solution:

a) Chloric acid is derived from the **chlorate ion, ClO_3^-**. The –1 charge on the ion requires one hydrogen. These combine to give the formula $HClO_3$.

b) Hydrofluoric acid is derived from the **fluoride** ion, **F^-**. The –1 charge on the ion requires one hydrogen. These combine to give the formula HF.

c) Acetic acid is derived from the **acetate ion**, which may be written as **CH_3COO^-** or as **$C_2H_3O_2^-$**. The –1 charge means that one H is needed. These combine to give the formula CH_3COOH or $HC_2H_3O_2$.

d) Nitrous acid is derived from the **nitrite ion, NO_2^-**. The –1 charge on the ion requires one hydrogen. These combine to give the formula HNO_2.

2.13B   Plan: Remove a hydrogen ion to determine the formula of the anion. Identify the corresponding name of the anion and use the name of the anion to name the acid. For the oxoanions, the -ate suffix changes to -ic acid and the -ite suffix changes to -ous acid. For the monatomic nonmetal anions, the name of the acid includes a hydro- prefix and the –ide suffix changes to –ic acid.
Solution:
a) Removing a hydrogen ion from the formula H_2SO_3 gives the oxoanion HSO_3^-, **hydrogen sulfite**; removing two hydrogen ions gives the oxoanion SO_3^{2-}, **sulfite**. To name the acid, the "-ite of "sulfite" must be replaced with "-ous". The corresponding name is **sulfurous acid**.
b) HBrO is an oxoacid containing the BrO^- ion (**hypobromite** ion). To name the acid, the "-ite" must be replaced with "-ous". This gives the name: **hypobromous acid**.
c) $HClO_2$ is an oxoacid containing the ClO_2^- ion (**chlorite** ion). To name the acid, the "-ite" must be replaced with "-ous". This gives the name: **chlorous acid**.
d) HI is a binary acid containing the I^- ion (**iodide** ion). To name the acid, a "hydro-" prefix is used, and the "-ide" must be replaced with "-ic". This gives the name: **hydroiodic acid**.

2.14A   Plan: Determine the names or symbols of each of the species present. Since these are binary compounds consisting of two nonmetals, the number of each type of atom is indicated with a Greek prefix.
Solution:
a) **Sulfur trioxide** — one sulfur and three (tri) oxygens, as oxide, are present.
b) **Silicon dioxide** — one silicon and two (di) oxygens, as oxide, are present.
c) N_2O Nitrogen has the prefix "di" = 2, and oxygen has the prefix "mono" = 1 (understood in the formula).
d) SeF_6 Selenium has no prefix (understood as = 1), and the fluoride has the prefix "hexa" = 6.

2.14B   Plan: Determine the names or symbols of each of the species present. Since these are binary compounds consisting of two nonmetals, the number of each type of atom is indicated with a Greek prefix.
Solution:
a) **Sulfur dichloride** — one sulfur and two (di) chlorines, as chloride, are present.
b) **Dinitrogen pentoxide** — two (di) nitrogen and five (penta) oxygens, as oxide, are present. Note that the "a" in "penta" is dropped when this prefix is combined with "oxide".
c) BF_3 Boron doesn't have a prefix, so there is one boron atom present. Fluoride has the prefix "tri" = 3.
d) IBr_3 Iodine has no prefix (understood as = 1), and the bromide has the prefix "tri" = 3.

2.15A   Plan: Determine the names or symbols of each of the species present. For compounds between nonmetals, the number of atoms of each type is indicated by a Greek prefix. If both elements in the compound are in the same group, the one with the higher period number is named first.
Solution:
a) Suffixes are not used in the common names of the nonmetal listed first in the formula. Sulfur does not qualify for the use of a suffix. Chlorine correctly has an "ide" suffix. There are two of each nonmetal atom, so both names require a "di" prefix. This gives the name **disulfur dichloride**.
b) Both elements are nonmetals, and there is just one nitrogen and one oxygen. These combine to give the formula **NO**.
c) Br has a higher period number than Cl and should be named first. The three chlorides are correctly named. The correct name is **bromine trichloride**.

2.15B   Plan: Determine the names or symbols of each of the species present. For compounds between nonmetals, the number of atoms of each type is indicated by a Greek prefix. If both elements in the compound are in the same group, the one with the higher period number is named first.
Solution:
a) The name of the element phosphorus ends in –us, not –ous. Additionally, the prefix *hexa-* is shortened to *hex-* before oxide. The correct name is **tetraphosphorus hexoxide**.
b) Because sulfur is listed first in the formula (and has a lower group number), it should be named first. The fluorine should come second in the name, modified with an –*ide* ending. The correct name is **sulfur hexafluoride**.
c) Nitrogen's symbol is N, not Ni. Additionally, the second letter of an element symbol should be lowercase (Br, not BR). The correct formula is **NBr_3**.

2.16A   Plan: First, write a formula to match the name. Next, multiply the number of each type of atom by the atomic mass of that atom. Sum all the masses to get an overall mass.
Solution:
a) The peroxide ion is O_2^{2-}, which requires two hydrogen atoms to cancel the charge: H_2O_2.
Molecular mass = (2 x 1.008 amu) + (2 x 16.00 amu) = 34.016 = **34.02 amu**.
b) Two Cs^{+1} ions are required to balance the charge on one CO_3^{2-} ion: Cs_2CO_3;
formula mass = (2 x 132.9 amu) + (1 x 12.01 amu) + (3 x 16.00 amu) = 325.81 = **325.8 amu**.

2.16B   Plan: First, write a formula to match the name. Next, multiply the number of each type of atom by the atomic mass of that atom. Sum all the masses to get an overall mass.
Solution:
a) Sulfuric acid contains the sulfate ion, SO_4^{2-}, which requires two hydrogen atoms to cancel the charge: H_2SO_4;
molecular mass = (2 x 1.008 amu) + 32.06 amu + (4 x 16.00 amu) = 98.076 = **98.08 amu**.
b) The sulfate ion, SO_4^{2-}, requires two +1 potassium ions, K^+, to give K_2SO_4;
formula mass = (2 x 39.10 amu) + 32.06 amu + (4 x 16.00 amu) = **174.26 amu**.

2.17A   Plan: Since the compounds only contain two elements, finding the formulas by counting each type of atom and developing a ratio. Name the compounds. Multiply the number of each type of atom by the atomic mass of that atom. Sum all the masses to get an overall mass.
Solution:
a) There are two brown atoms (sodium) for every red (oxygen). The compound contains a metal with a nonmetal. Thus, the compound is **sodium oxide**, with the formula Na_2O. The formula mass is twice the mass of sodium plus the mass of oxygen:
2 (22.99 amu) + (16.00 amu) = **61.98 amu**
b) There is one blue (nitrogen) and two reds (oxygen) in each molecule. The compound only contains nonmetals. Thus, the compound is **nitrogen dioxide**, with the formula NO_2. The molecular mass is the mass of nitrogen plus twice the mass of oxygen: (14.01 amu) + 2 (16.00 amu) = **46.01 amu**.

2.17B   Plan: Since the compounds only contain two elements, finding the formulas by counting each type of atom and developing a ratio. Name the compounds. Multiply the number of each type of atom by the atomic mass of that atom. Sum all the masses to get an overall mass.
Solution:
a) There is one gray (magnesium) for every two green (chlorine). The compound contains a metal with a nonmetal. Thus, the compound is **magnesium chloride**, with the formula $MgCl_2$. The formula mass is the mass of magnesium plus twice the mass of chlorine: (24.31 amu) + 2 (35.45 amu) = **95.21 amu**
b) There is one green (chlorine) and three golds (fluorine) in each molecule. The compound only contains nonmetals. Thus, the compound is **chlorine trifluoride**, with the formula ClF_3. The molecular mass is the mass of chlorine plus three times the mass of fluorine: (35.45 amu) + 3 (19.00 amu) = **92.45 amu**.

TOOLS OF THE LABORATORY BOXED READING PROBLEMS

B2.1   Plan: There is one peak for each type of Cl atom and peaks for the Cl_2 molecule. The m/e ratio equals the mass divided by 1+.
Solution:
a) There is one peak for the ^{35}Cl atom and another peak for the ^{37}Cl atom. There are three peaks for the three possible Cl_2 molecules: $^{35}Cl^{35}Cl$ (both atoms are mass 35), $^{37}Cl^{37}Cl$ (both atoms are mass 37), and $^{35}Cl^{37}Cl$ (one atom is mass 35 and one is mass 37). So the mass of chlorine will have **5 peaks**.

b) Peak	m/e ratio	
^{35}Cl	**35**	lightest particle
^{37}Cl	37	
$^{35}Cl^{35}Cl$	70 (35 + 35)	
$^{35}Cl^{37}Cl$	72 (35 + 37)	
$^{37}Cl^{37}Cl$	**74** (35 + 37)	heaviest particle

B2.2 **Plan:** Each peak in the mass spectrum of carbon represents a different isotope of carbon. The heights of the peaks correspond to the natural abundances of the isotopes.

Solution:

Carbon has three naturally occurring isotopes: ^{12}C, ^{13}C, and ^{14}C. ^{12}C has an abundance of 98.89% and would have the tallest peak in the mass spectrum as the most abundant isotope. ^{13}C has an abundance of 1.11% and thus would have a significantly shorter peak; the shortest peak in the mass spectrum would correspond to the least abundant isotope, ^{14}C, the abundance of which is less than 0.01%. Peak Y, as the tallest peak, has a m/e ratio of 12 (^{12}C); **X**, the shortest peak, has a m/e ratio of 14(^{14}C). Peak Z corresponds to ^{13}C with a m/e ratio of **13**.

B2.3 **Plan:** Review the discussion on separations.

Solution:

a) Salt dissolves in water and pepper does not. Procedure: add water to mixture and filter to remove solid pepper. Evaporate water to recover solid salt.

b) The water/soot mixture can be filtered; the water will flow through the filter paper, leaving the soot collected on the filter paper.

c) Allow the mixture to warm up, and then pour off the melted ice (water); or, add water, and the glass will sink and the ice will float.

d) Heat the mixture; the alcohol will boil off (distill), while the sugar will remain behind.

e) The spinach leaves can be extracted with a solvent that dissolves the pigments. Chromatography can be used to separate one pigment from the other.

END–OF–CHAPTER PROBLEMS

2.1 **Plan:** Refer to the definitions of an element and a compound.

Solution:

Unlike compounds, elements cannot be broken down by chemical changes into simpler materials. Compounds contain different types of atoms; there is only one type of atom in an element.

2.4 **Plan:** Remember that an element contains only one kind of atom while a compound contains at least two different elements (two kinds of atoms) in a fixed ratio. A mixture contains at least two different substances in a composition that can vary.

Solution:

a) The presence of more than one element (calcium and chlorine) makes this pure substance a **compound**.

b) There are only atoms from one element, sulfur, so this pure substance is an **element**.

c) This is a combination of two compounds and has a varying composition, so this is a **mixture**.

d) The presence of more than one type of atom means it cannot be an element. The specific, not variable, arrangement means it is a **compound**.

2.12 **Plan:** Restate the three laws in your own words.

Solution:

a) The law of mass conservation applies to all substances — **elements, compounds, and mixtures**. Matter can neither be created nor destroyed, whether it is an element, compound, or mixture.

b) The law of definite composition applies to **compounds** only, because it refers to a constant, or definite, composition of elements within a compound.

c) The law of multiple proportions applies to **compounds** only, because it refers to the combination of elements to form compounds.

2.14 **Plan:** Review the three laws: law of mass conservation, law of definite composition, and law of multiple proportions.

Solution:

a) **Law of Definite Composition** — The compound potassium chloride, KCl, is composed of the same elements and same fraction by mass, regardless of its source (Chile or Poland).

b) **Law of Mass Conservation** — The mass of the substances inside the flashbulb did not change during the chemical reaction (formation of magnesium oxide from magnesium and oxygen).

c) **Law of Multiple Proportions** — Two elements, O and As, can combine to form two different compounds that have different proportions of As present.

2.16   <u>Plan:</u> Review the definition of percent by mass.
<u>Solution:</u>
a) **No**, the <u>mass</u> <u>percent</u> of each element in a compound is fixed. The percentage of Na in the compound NaCl is 39.34% (22.99 amu/58.44 amu), whether the sample is 0.5000 g or 50.00 g.
b) **Yes**, the <u>mass</u> of each element in a compound depends on the mass of the compound. A 0.5000 g sample of NaCl contains 0.1967 g of Na (39.34% of 0.5000 g), whereas a 50.00 g sample of NaCl contains 19.67 g of Na (39.34% of 50.00 g).

2.18   <u>Plan:</u> Review the mass laws: law of mass conservation, law of definite composition, and law of multiple proportions. For each experiment, compare the mass values before and after each reaction and examine the ratios of the mass of white compound to the mass of colorless gas.
<u>Solution:</u>
Experiment 1: mass before reaction = 1.00 g;   mass after reaction = 0.64 g + 0.36 g = 1.00 g
Experiment 2: mass before reaction = 3.25 g;   mass after reaction = 2.08 g + 1.17 g = 3.25 g
Both experiments demonstrate the **law of mass conservation** since the total mass before reaction equals the total mass after reaction.
Experiment 1: mass white compound/mass colorless gas = 0.64 g/0.36 g = 1.78
Experiment 2: mass white compound/mass colorless gas = 2.08 g/1.17 g = 1.78
Both Experiments 1 and 2 demonstrate the **law of definite composition** since the compound has the same composition by mass in each experiment.

2.20   <u>Plan:</u> Fluorite is a mineral containing only calcium and fluorine. The difference between the mass of fluorite and the mass of calcium gives the mass of fluorine. Mass fraction is calculated by dividing the mass of element by the mass of compound (fluorite) and mass percent is obtained by multiplying the mass fraction by 100.
<u>Solution:</u>
a) Mass (g) of fluorine = mass of fluorite – mass of calcium = 2.76 g – 1.42 g = **1.34 g fluorine**

b) Mass fraction of Ca = $\dfrac{\text{mass Ca}}{\text{mass fluorite}}$ = $\dfrac{1.42 \text{ g Ca}}{2.76 \text{ g fluorite}}$ = 0.51449 = **0.514**

Mass fraction of F = $\dfrac{\text{mass F}}{\text{mass fluorite}}$ = $\dfrac{1.34 \text{ g F}}{2.76 \text{ g fluorite}}$ = 0.48551 = **0.486**

c) Mass percent of Ca = 0.51449 x 100 = 51.449 = **51.4%**
Mass percent of F = 0.48551 x 100 = 48.551 = **48.6%**

2.22   <u>Plan:</u> Dividing the mass of magnesium by the mass of the oxide gives the ratio. Multiply the mass of the second sample of magnesium oxide by this ratio to determine the mass of magnesium.
<u>Solution:</u>
a) If 1.25 g of MgO contains 0.754 g of Mg, then the mass ratio (or fraction) of magnesium in the oxide

compound is $\dfrac{\text{mass Mg}}{\text{mass MgO}}$ = $\dfrac{0.754 \text{ g Mg}}{1.25 \text{ g MgO}}$ = 0.6032 = **0.603**.

b) Mass (g) of magnesium = $\left(534 \text{ g MgO}\right)\left(\dfrac{0.6032 \text{ g Mg}}{1 \text{ g MgO}}\right)$ = 322.109 = **322 g magnesium**

2.24   <u>Plan:</u> Since copper is a metal and sulfur is a nonmetal, the sample contains 88.39 g Cu and 44.61 g S. Calculate the mass fraction of each element in the sample by dividing the mass of element by the total mass of compound. Multiply the mass of the second sample of compound in grams by the mass fraction of each element to find the mass of each element in that sample.

Solution:

$$\text{Mass (g) of compound} = 88.39 \text{ g copper} + 44.61 \text{ g sulfur} = 133.00 \text{ g compound}$$

$$\text{Mass fraction of copper} = \left(\frac{88.39 \text{ g copper}}{133.00 \text{ g compound}}\right) = 0.664586$$

$$\text{Mass (g) of copper} = (5264 \text{ kg compound})\left(\frac{10^3 \text{ g compound}}{1 \text{ kg compound}}\right)\left(\frac{0.664586 \text{ g copper}}{1 \text{ g compound}}\right)$$

$$= 3.49838 \times 10^6 = \mathbf{3.498 \times 10^6 \text{ g copper}}$$

$$\text{Mass fraction of sulfur} = \left(\frac{44.61 \text{ g sulfur}}{133.00 \text{ g compound}}\right) = 0.335414$$

$$\text{Mass (g) of sulfur} = (5264 \text{ kg compound})\left(\frac{10^3 \text{ g compound}}{1 \text{ kg compound}}\right)\left(\frac{0.335414 \text{ g sulfur}}{1 \text{ g compound}}\right)$$

$$= 1.76562 \times 10^6 = \mathbf{1.766 \times 10^6 \text{ g sulfur}}$$

2.26   Plan: The law of multiple proportions states that if two elements form two different compounds, the relative amounts of the elements in the two compounds form a whole-number ratio. To illustrate the law we must calculate the mass of one element to one gram of the other element for each compound and then compare this mass for the two compounds. The law states that the ratio of the two masses should be a small whole-number ratio such as 1:2, 3:2, 4:3, etc.
Solution:

Compound 1: $\dfrac{47.5 \text{ mass \% S}}{52.5 \text{ mass \% Cl}} = 0.90476 = 0.905$

Compound 2: $\dfrac{31.1 \text{ mass \% S}}{68.9 \text{ mass \% Cl}} = 0.451379 = 0.451$

Ratio: $\dfrac{0.90476}{0.451379} = 2.0044 = 2.00:1.00$

Thus, the ratio of the mass of sulfur per gram of chlorine in the two compounds is a small whole-number ratio of 2:1, which agrees with the law of multiple proportions.

2.29   Plan: Determine the mass percent of sulfur in each sample by dividing the grams of sulfur in the sample by the total mass of the sample and multiplying by 100. The coal type with the smallest mass percent of sulfur has the smallest environmental impact.
Solution:

$$\text{Mass \% in Coal A} = \left(\frac{11.3 \text{ g sulfur}}{378 \text{ g sample}}\right)(100\%) = 2.9894 = 2.99\% \text{ S (by mass)}$$

$$\text{Mass \% in Coal B} = \left(\frac{19.0 \text{ g sulfur}}{495 \text{ g sample}}\right)(100\%) = 3.8384 = 3.84\% \text{ S (by mass)}$$

$$\text{Mass \% in Coal C} = \left(\frac{20.6 \text{ g sulfur}}{675 \text{ g sample}}\right)(100\%) = 3.0519 = 3.05\% \text{ S (by mass)}$$

Coal A has the smallest environmental impact.

2.31   Plan: This question is based on the law of definite composition. If the compound contains the same types of atoms, they should combine in the same way to give the same mass percentages of each of the elements.
Solution:
Potassium nitrate is a compound composed of three elements — potassium, nitrogen, and oxygen — in a specific ratio. If the ratio of these elements changed, then the compound would be changed to a different compound, for example, to potassium nitrite, with different physical and chemical properties. Dalton postulated that atoms

of an element are identical, regardless of whether that element is found in India or Italy. Dalton also postulated that compounds result from the chemical combination of specific ratios of different elements. Thus, Dalton's theory explains why potassium nitrate, a compound comprised of three different elements in a specific ratio, has the same chemical composition regardless of where it is mined or how it is synthesized.

2.32   Plan: Review the discussion of the experiments in this chapter.
Solution:
Millikan determined the minimum *charge* on an oil drop and that the minimum charge was equal to the charge on one electron. Using Thomson's value for the *mass/charge ratio* of the electron and the determined value for the charge on one electron, Millikan calculated the mass of an electron (charge/(charge/mass)) to be 9.109×10^{-28} g.

2.36   Plan: Re-examine the definitions of atomic number and the mass number.
Solution:
The atomic number is the number of protons in the nucleus of an atom. When the atomic number changes, the identity of the element also changes. The mass number is the total number of protons and neutrons in the nucleus of an atom. Since the identity of an element is based on the number of protons and not the number of neutrons, the mass number can vary (by a change in number of neutrons) without changing the identity of the element.

2.39   Plan: The superscript is the mass number, the sum of the number of protons and neutrons. Consult the periodic table to get the atomic number (the number of protons). The mass number – the number of protons = the number of neutrons. For atoms, the number of protons and electrons are equal.
Solution:

Isotope	Mass Number	# of Protons	# of Neutrons	# of Electrons
^{36}Ar	36	18	18	18
^{38}Ar	38	18	20	18
^{40}Ar	40	18	22	18

2.41   Plan: The superscript is the mass number (A), the sum of the number of protons and neutrons; the subscript is the atomic number (Z, number of protons). The mass number – the number of protons = the number of neutrons. For atoms, the number of protons = the number of electrons.
Solution:
a) $^{16}_{8}$O and $^{17}_{8}$O have the **same number of protons and electrons** (8), but different numbers of neutrons.
$^{16}_{8}$O and $^{17}_{8}$O are isotopes of oxygen, and $^{16}_{8}$O has $16 - 8 = 8$ neutrons whereas $^{17}_{8}$O has $17 - 8 = 9$ neutrons. **Same Z value**

b) $^{40}_{18}$Ar and $^{41}_{19}$K have the **same number of neutrons** (Ar: $40 - 18 = 22$; K: $41 - 19 = 22$) but different numbers of protons and electrons (Ar = 18 protons and 18 electrons; K = 19 protons and 19 electrons). **Same N value**

c) $^{60}_{27}$Co and $^{60}_{28}$Ni have different numbers of protons, neutrons, and electrons. Co: 27 protons, 27 electrons, and $60 - 27 = 33$ neutrons; Ni: 28 protons, 28 electrons and $60 - 28 = 32$ neutrons. However, both have a mass number of 60. **Same A value**

2.43   Plan: Combine the particles in the nucleus (protons + neutrons) to give the mass number (superscript, A). The number of protons gives the atomic number (subscript, Z) and identifies the element.
Solution:
a) $A = 18 + 20 = 38$; $Z = 18$; $^{38}_{18}$**Ar**
b) $A = 25 + 30 = 55$; $Z = 25$; $^{55}_{25}$**Mn**
c) $A = 47 + 62 = 109$; $Z = 47$; $^{109}_{47}$**Ag**

2.45   Plan: Determine the number of each type of particle. The superscript is the mass number (A) and the subscript is the atomic number (Z, number of protons). The mass number – the number of protons = the number of neutrons. For atoms, the number of protons = the number of electrons. The protons and neutrons are in the nucleus of the atom.

Solution:

a) $^{48}_{22}\text{Ti}$

22 protons

22 electrons

48 – 22 = 26 neutrons

b) $^{79}_{34}\text{Se}$

34 protons

34 electrons

79 – 34 = 45 neutrons

c) $^{11}_{5}\text{B}$

5 protons

5 electrons

11 – 5 = 6 neutrons

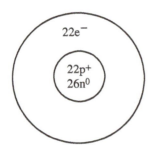
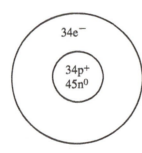
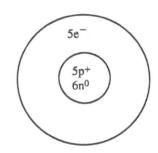

2.47 Plan: To calculate the atomic mass of an element, take a weighted average based on the natural abundance of the isotopes: (isotopic mass of isotope 1 x fractional abundance) + (isotopic mass of isotope 2 x fractional abundance).

Solution:

Atomic mass of gallium $= (68.9256 \text{ amu})\left(\dfrac{60.11\%}{100\%}\right) + (70.9247 \text{ amu})\left(\dfrac{39.89\%}{100\%}\right) = 69.7230 = \textbf{69.72 amu}$

2.49 Plan: To find the percent abundance of each Cl isotope, let x equal the fractional abundance of ^{35}Cl and (1 – x) equal the fractional abundance of ^{37}Cl since the sum of the fractional abundances must equal 1. Remember that atomic mass = (isotopic mass of ^{35}Cl x fractional abundance) + (isotopic mass of ^{37}Cl x fractional abundance).

Solution:

Atomic mass = (isotopic mass of ^{35}Cl x fractional abundance) + (isotopic mass of ^{37}Cl x fractional abundance)

35.4527 amu = 34.9689 amu(x) + 36.9659 amu(1 – x)

35.4527 amu = 34.9689 amu(x) + 36.9659 amu – 36.9659 amu(x)

35.4527 amu = 36.9659 amu – 1.9970 amu(x)

1.9970 amu(x) = 1.5132 amu

x = 0.75774 and 1 – x = 1 – 0.75774 = 0.24226

% abundance ^{35}Cl = **75.774%**   % abundance ^{37}Cl = **24.226%**

2.52 Plan: Review the section in the chapter on the periodic table.

Solution:

a) In the modern periodic table, the elements are arranged in order of increasing atomic **number**.

b) Elements in a **column or group** (or family) have similar chemical properties, not those in the same period or row.

c) Elements can be classified as **metals**, metalloids, or nonmetals.

2.55 Plan: Review the properties of these two columns in the periodic table.

Solution:

The alkali metals (Group 1A(1)) are metals and readily lose one electron to form cations whereas the halogens (Group 7A(17)) are nonmetals and readily gain one electron to form anions.

2.56 Plan: Locate each element on the periodic table. The Z value is the atomic number of the element. Metals are to the left of the "staircase," nonmetals are to the right of the "staircase," and the metalloids are the elements that lie along the "staircase" line.

Solution:

a) Germanium   Ge   4A(14) metalloid

b) Phosphorus	P	5A(15)	nonmetal
c) Helium	He	8A(18)	nonmetal
d) Lithium	Li	1A(1)	metal
e) Molybdenum	Mo	6B(6)	metal

2.58 Plan: Review the section in the chapter on the periodic table. Remember that alkaline earth metals are in Group 2A(2), the halogens are in Group 7A(17), and the metalloids are the elements that lie along the "staircase" line; periods are horizontal rows.
Solution:
a) The symbol and atomic number of the heaviest alkaline earth metal are **Ra** and **88**.
b) The symbol and atomic number of the lightest metalloid in Group 4A(14) are **Si** and **14**.
c) The symbol and atomic mass of the coinage metal whose atoms have the fewest electrons are **Cu** and **63.55 amu**.
d) The symbol and atomic mass of the halogen in Period 4 are **Br** and **79.90 amu**.

2.60 Plan: Review the section of the chapter on the formation of ionic compounds.
Solution:
Reactive metals and nometals will form **ionic** bonds, in which one or more electrons are transferred from the metal atom to the nonmetal atom to form a cation and an anion, respectively. The oppositely charged ions attract, forming the ionic bond.

2.63 Plan: Assign charges to each of the ions. Since the sizes are similar, there are no differences due to the sizes.
Solution:
Coulomb's law states the energy of attraction in an ionic bond is directly proportional to the *product of charges* and inversely proportional to the *distance between charges*. The *product of charges* in MgO (+2 x –2 = –4) is greater than the *product of charges* in LiF (+1 x –1 = –1). Thus, **MgO** has stronger ionic bonding.

2.66 Plan: Locate these groups on the periodic table and assign charges to the ions that would form.
Solution:
The monatomic ions of Group 1A(1) have a +1 charge (e.g., Li^+, Na^+, and K^+) whereas the monatomic ions of Group 7A(17) have a –1 charge (e.g., F^-, Cl^-, and Br^-). Elements gain or lose electrons to form ions with the same number of electrons as the nearest noble gas. For example, Na loses one electron to form a cation with the same number of electrons as Ne. The halogen F gains one electron to form an anion with the same number of electrons as Ne.

2.68 Plan: A metal and a nonmetal will form an ionic compound. Locate these elements on the periodic table and predict their charges.
Solution:
Potassium sulfide (K_2S) is an ionic compound formed from a metal (potassium) and a nonmetal (sulfur). Potassium atoms transfer electrons to sulfur atoms. Each potassium atom loses one electron to form an ion with +1 charge and the same number of electrons (18) as the noble gas argon. Each sulfur atom gains two electrons to form an ion with a –2 charge and the same number of electrons (18) as the noble gas argon. The oppositely charged ions, K^+ and S^{2-}, attract each other to form an ionic compound with the ratio of two K^+ ions to one S^{2-} ion. The total number of electrons lost by the potassium atoms equals the total number of electrons gained by the sulfur atoms.

2.70 Plan: Locate these elements on the periodic table and predict what ions they will form. For A group cations (metals), ion charge = group number; for anions (nonmetals), ion charge = group number minus 8.
Solution:
Potassium (K) is in Group 1A(1) and forms the K^+ ion. Bromine (Br) is in Group 7A(17) and forms the Br^- ion (7 – 8 = –1).

2.72 Plan: Use the number of protons (atomic number) to identify the element. Add the number of protons and neutrons together to get the mass number. Locate the element on the periodic table and assign its group and period number.

Solution:
a) Oxygen (atomic number = 8)   mass number = 8p + 9n = 17   Group 6A(16)   Period 2
b) Fluorine (atomic number = 9)   mass number = 9p + 10n = 19   Group 7A(17)   Period 2
c) Calcium (atomic number = 20) mass number = 20p + 20n = 40 Group 2A(2)   Period 4

2.74   Plan: Determine the charges of the ions based on their position on the periodic table. For A group
cations (metals), ion charge = group number; for anions (nonmetals), ion charge = group number minus 8.
Next, determine the ratio of the charges to get the ratio of the ions.
Solution:
Lithium [Group 1A(1)] forms the Li^+ ion; oxygen [Group 6A(16)] forms the O^{2-} ion (6 – 8 = –2). The ionic
compound that forms from the combination of these two ions must be electrically neutral, so two Li^+ ions
combine with one O^{2-} ion to form the compound Li_2O. There are twice as many Li^+ ions as O^{2-} ions in a sample of
Li_2O.

Number of O^{2-} ions = $(8.4 \times 10^{21}\ Li^+\ \text{ions})\left(\dfrac{1\ O^{2-}\ \text{ion}}{2\ Li^+\ \text{ions}}\right)$ = **4.2x10²¹ O^{2-} ions**

2.76   Plan: The key is the size of the two alkali metal ions. The charges on the sodium and potassium ions are the same
as both are in Group 1A(1), so there will be no difference due to the charge. The chloride ions are the same in size
and charge, so there will be no difference due to the chloride ion.
Solution:
Coulomb's law states that the energy of attraction in an ionic bond is directly proportional to the *product of
charges* and inversely proportional to the *distance between charges*. The *product of the charges* is the same in both
compounds because both sodium and potassium ions have a +1 charge. Attraction increases as distance decreases,
so the ion with the smaller radius, Na^+, will form a stronger ionic interaction (**NaCl**).

2.78   Plan: Review the definition of molecular formula.
Solution:
The subscripts in the formula, MgF_2, give the number of ions in a formula unit of the ionic compound. The
subscripts indicate that there are two F^- ions for every one Mg^{2+} ion. Using this information and the mass of each
element, we could calculate the percent mass of each element.

2.80   Plan: Review the concepts of atoms and molecules.
Solution:
The mixture is similar to the sample of hydrogen peroxide in that both contain 20 billion oxygen atoms and 20
billion hydrogen atoms since both O_2 and H_2O_2 contain 2 oxygen atoms per molecule and both H_2 and H_2O_2
contain 2 hydrogen atoms per molecule. They differ in that they contain different types of molecules: H_2O_2
molecules in the hydrogen peroxide sample and H_2 and O_2 molecules in the mixture. In addition, the mixture
contains 20 billion molecules (10 billion H_2 molecules + 10 billion O_2 molecules) while the hydrogen peroxide
sample contains 10 billion molecules.

2.84   Plan: Locate each of the individual elements on the periodic table, and assign charges to each of the ions.
For A group cations (metals), ion charge = group number; for anions (nonmetals), ion charge = group number
minus 8. Find the smallest number of each ion that gives a neutral compound. To name ionic compounds with
metals that form only one ion, name the metal, followed by the nonmetal name with an -ide suffix.
Solution:
a) Sodium is a metal that forms a +1 (Group 1A) ion and nitrogen is a nonmetal that forms a –3 ion
(Group **5A**, 5 – 8 = –3).

$\qquad\qquad\qquad$ +3 –3
$\qquad$ +1 –3 $\qquad$ +1
$\qquad$ Na N $\qquad$ Na_3N $\qquad\qquad$ The compound is **Na_3N, sodium nitride**.
b) Oxygen is a nonmetal that forms a –2 ion (Group **6A**, 6 – 8 = –2) and strontium is a metal that forms a +2 ion
(Group **2A**). $\qquad\qquad$ +2 –2
$\qquad\qquad$ Sr O $\qquad$ The compound is **SrO, strontium oxide**.

c) Aluminum is a metal that forms a +3 ion (Group 3A) and chlorine is a nonmetal that forms a –1 ion (Group 7A, 7– 8 = –1).

$$+3\ –3$$
$$+3\ –1 \qquad\qquad +3\ –1$$
$$\text{Al Cl} \qquad\qquad \text{AlCl}_3 \qquad \text{The compound is } \mathbf{AlCl_3, \ aluminum \ chloride.}$$

2.86   Plan: Based on the atomic numbers (the subscripts) locate the elements on the periodic table. Once the atomic numbers are located, identify the element and based on its position, assign a charge. For A group cations (metals), ion charge = group number; for anions (nonmetals), ion charge = group number minus 8. Find the smallest number of each ion that gives a neutral compound. To name ionic compounds with metals that form only one ion, name the metal, followed by the nonmetal name with an -ide suffix.
Solution:
a) $_{12}$L is the element Mg (Z = 12). Magnesium [Group 2A(2)] forms the Mg^{2+} ion. $_9$M is the element F (Z = 9). Fluorine [Group 7A(17)] forms the F^- ion (7 – 8 = –1). The compound formed by the combination of these two elements is $\mathbf{MgF_2}$, **magnesium fluoride**.
b) $_{30}$L is the element Zn (Z = 30). Zinc forms the Zn^{2+} ion (see Table 2.3). $_{16}$M is the element S (Z = 16). Sulfur [Group 6A(16)] will form the S^{2-} ion (6 – 8 = –2). The compound formed by the combination of these two elements is **ZnS, zinc sulfide**.
c) $_{17}$L is the element Cl (Z = 17). Chlorine [Group 7A(17)] forms the Cl^- ion (7 – 8 = –1). $_{38}$M is the element Sr (Z = 38). Strontium [Group 2A(2)] forms the Sr^{2+} ion. The compound formed by the combination of these two elements is $\mathbf{SrCl_2}$, **strontium chloride**.

2.88   Plan: Review the rules for nomenclature covered in the chapter. For ionic compounds, name the metal, followed by the nonmetal name with an -ide suffix. For metals, like many transition metals, that can form more than one ion each with a different charge, the ionic charge of the metal ion is indicated by a Roman numeral within parentheses immediately following the metal's name.
Solution:
a) tin(IV) chloride = $\mathbf{SnCl_4}$ The (IV) indicates that the metal ion is Sn^{4+} which requires 4 Cl^- ions for a neutral compound.
b) $FeBr_3$ = **iron(III) bromide** (common name is ferric bromide); the charge on the iron ion is +3 to match the –3 charge of 3 Br^- ions. The +3 charge of the Fe is indicated by (III). $\qquad\qquad$ +6 –6
c) cuprous bromide = **CuBr** (cuprous is +1 copper ion, cupric is +2 copper ion). $\qquad$ +3 –2
d) Mn_2O_3 = **manganese(III) oxide** Use (III) to indicate the +3 ionic charge of Mn: $\quad Mn_2O_3$

2.90   Plan: Review the rules for nomenclature covered in the chapter. For ionic compounds containing polyatomic ions, name the metal, followed by the name of the polyatomic ion. Hydrates, compounds with a specific number of water molecules associated with them, are named with a prefix before the word hydrate to indicate the number of water molecules.
Solution:
a) Na_2HPO_4 = **sodium hydrogen phosphate** Sodium [Group 1A(1)] forms the Na^+ ion; HPO_4^{2-} is the hydrogen phosphate ion.
b) ammonium perchlorate = $\mathbf{NH_4ClO_4}$ Ammonium is the polyatomic ion NH_4^+ and perchlorate is the polyatomic ion ClO_4^-. One NH_4^+ is required for every one ClO_4^- ion.
c) $Pb(C_2H_3O_2)_2 \cdot 3H_2O$ = **lead(II) acetate trihydrate** The $C_2H_3O_2^-$ ion has a –1 charge (see Table 2.5); since there are two of these ions, the lead ion has a +2 charge which must be indicated with the Roman numeral II. The $\cdot3H_2O$ indicates a hydrate in which the number of H_2O molecules is indicated by the prefix tri-.
d) $NaNO_2$ = **sodium nitrite** NO_2^- is the nitrite polyatomic ion.

2.92   Plan: Review the rules for nomenclature covered in the chapter. For metals, like many transition metals, that can form more than one ion each with a different charge, the ionic charge of the metal ion is indicated by a Roman numeral within parentheses immediately following the metal's name. Compounds must be neutral.
Solution:
a) Barium [Group 2A(2)] forms Ba^{2+} and oxygen [Group 6A(16)] forms O^{2-} (6 – 8 = –2) so the neutral compound forms from one Ba^{2+} ion and one O^{2-} ion. Correct formula is **BaO**.

2-15

b) Iron(II) indicates Fe^{2+} and nitrate is NO_3^- so the neutral compound forms from one iron(II) ion and two nitrate ions. Correct formula is **$Fe(NO_3)_2$**.

c) Mn is the symbol for manganese. Mg is the correct symbol for magnesium. Correct formula is **MgS**. Sulfide is the S^{2-} ion and sulfite is the SO_3^{2-} ion.

2.94   <u>Plan:</u> Acids donate H^+ ion to the solution, so the acid is a combination of H^+ and a negatively charged ion. Binary acids (H plus one other nonmetal) are named hydro- + nonmetal root + -ic acid. Oxoacids (H + an oxoanion) are named by changing the suffix of the oxoanion: -ate becomes -ic acid and -ite becomes -ous acid.

   <u>Solution:</u>
   a) Hydrogen carbonate is HCO_3^-, so its source acid is **H_2CO_3**. The name of the acid is **carbonic acid** (-ate becomes –ic acid).
   b) **HIO_4, periodic acid**. IO_4^- is the periodate ion: -ate becomes –ic acid.
   c) Cyanide is CN^-; its source acid is **HCN hydrocyanic acid** (binary acid).
   d) **H_2S, hydrosulfuric acid** (binary acid).

2.96   <u>Plan:</u> Use the formulas of the polyatomic ions. Recall that oxoacids are named by changing the suffix of the oxoanion: -ate becomes -ic acid and -ite becomes -ous acid. Compounds must be neutral.
   <u>Solution:</u>
   a) ammonium ion = **NH_4^+**   ammonia = **NH_3**
   b) magnesium sulfide = **MgS**   magnesium sulfite = **$MgSO_3$**   magnesium sulfate = **$MgSO_4$**
   Sulfide = S^{2-}; sulfite = SO_3^{2-}; sulfate = SO_4^{2-}.
   c) hydrochloric acid = **HCl**   chloric acid = **$HClO_3$**   chlorous acid = **$HClO_2$**
   Binary acids (H plus one other nonmetal) are named hydro- + nonmetal root + -ic acid. Chloric indicates the polyatomic ion ClO_3^- while chlorous indicates the polyatomic ion ClO_2^-.
   d) cuprous bromide = **CuBr**   cupric bromide = **$CuBr_2$**
   The suffix -ous indicates the lower charge, +1, while the suffix -ic indicates the higher charge, +2.

2.98   <u>Plan:</u> This compound is composed of two nonmetals. The element with the lower group number is named first. Greek numerical prefixes are used to indicate the number of atoms of each element in the compound.
   <u>Solution:</u>
   disulfur tetrafluoride   **S_2F_4**   Di- indicates two S atoms and tetra- indicates four F atoms.

2.100   <u>Plan:</u> Review the nomenclature rules in the chapter. For ionic compounds, name the metal, followed by the nonmetal name with an -ide suffix. For metals, like many transition metals, that can form more than one ion each with a different charge, the ionic charge of the metal ion is indicated by a Roman numeral within parentheses immediately following the metal's name. Binary acids (H plus one other nonmetal) are named hydro- + nonmetal root + -ic acid.
   <u>Solution:</u>
   a) Calcium(II) dichloride, $CaCl_2$: The name becomes **calcium chloride** because calcium does not require "(II)" since it only forms +2 ions. Prefixes like di- are only used in naming covalent compounds between nonmetal elements.
   b) Copper(II) oxide, Cu_2O: The charge on the oxide ion is O^{2-}, which makes each copper a Cu^+. The name becomes **copper(I) oxide** to match the charge on the copper.
   c) Stannous fluoride, SnF_4: Stannous refers to Sn^{2+}, but the tin in this compound is Sn^{4+} due to the charge on the fluoride ion. The tin(IV) ion is the stannic ion; this gives the name **stannic fluoride or tin(IV) fluoride**.
   d) Hydrogen chloride acid, HCl: Binary acids consist of the root name of the nonmetal (chlor in this case) with a hydro- prefix and an -ic suffix. The word acid is also needed. This gives the name **hydrochloric acid**.

2.102   <u>Plan:</u> Review the rules for nomenclature covered in the chapter. For ionic compounds containing polyatomic ions, name the metal, followed by the name of the polyatomic ion. The molecular (formula) mass is the sum of the atomic masses of all of the atoms.
   <u>Solution:</u>
   a) **$(NH_4)_2SO_4$**   ammonium is NH_4^+ and sulfate is SO_4^{2-}
                N   =   2(14.01 amu)   =   28.02 amu
                H   =   8(1.008 amu)   =   8.064 amu

S	=	1(32.06 amu)	=	32.06 amu
O	=	4(16.00 amu)	=	64.00 amu
				132.14 amu

b) **NaH₂PO₄**   sodium is Na⁺ and dihydrogen phosphate is $H_2PO_4^-$

Na	=	1(22.99 amu)	=	22.99 amu
H	=	2(1.008 amu)	=	2.016 amu
P	=	1(30.97 amu)	=	30.97 amu
O	=	4(16.00 amu)	=	64.00 amu
				119.98 amu

c) **KHCO₃**   potassium is K⁺ and bicarbonate is HCO_3^-

K	=	1(39.10 amu)	=	39.10 amu
H	=	1(1.008 amu)	=	1.008 amu
C	=	1(12.01 amu)	=	12.01 amu
O	=	3(16.00 amu)	=	48.00 amu
				100.12 amu

2.104   Plan: Convert the names to the appropriate chemical formulas. The molecular (formula) mass is the sum of the masses of each atom times its atomic mass.
Solution:
a) dinitrogen pentoxide N_2O_5 (di- = 2 and penta- = 5)

N	=	2(14.01 amu)	=	28.02 amu
O	=	5(16.00 amu)	=	80.00 amu
				108.02 amu

b) lead(II) nitrate $Pb(NO_3)_2$ (lead(II) is Pb^{2+} and nitrate is NO_3^-)

Pb	=	1(207.2 amu)	=	207.2 amu
N	=	2(14.01 amu)	=	28.02 amu
O	=	6(16.00 amu)	=	96.00 amu
				331.2 amu

c) calcium peroxide CaO_2 (calcium is Ca^{2+} and peroxide is O_2^{2-})

Ca	=	1(40.08 amu)	=	40.08 amu
O	=	2(16.00 amu)	=	32.00 amu
				72.08 amu

2.106   Plan: Break down each formula to the individual elements and count the number of atoms of each element by observing the subscripts. The molecular (formula) mass is the sum of the atomic masses of all of the atoms.
Solution:
a) There are **12 atoms of oxygen** in $Al_2(SO_4)_3$. The molecular mass is:

Al	=	2(26.98 amu)	=	53.96 amu
S	=	3(32.06 amu)	=	96.18 amu
O	=	12(16.00 amu)	=	192.00 amu
				342.14 amu

b) There are **9 atoms of hydrogen** in $(NH_4)_2HPO_4$. The molecular mass is:

N	=	2(14.01 amu)	=	28.02 amu
H	=	9(1.008 amu)	=	9.072 amu
P	=	1(30.97 amu)	=	30.97 amu
O	=	4(16.00 amu)	=	64.00 amu
				132.06 amu

c) There are **8 atoms of oxygen** in $Cu_3(OH)_2(CO_3)_2$. The molecular mass is:

Cu	=	3(63.55 amu)	=	190.65 amu
O	=	8(16.00 amu)	=	128.00 amu
H	=	2(1.008 amu)	=	2.016 amu
C	=	2(12.01 amu)	=	24.02 amu
				344.69 amu

2.108   Plan: Use the chemical symbols and count the atoms of each type to give a molecular formula. Use the nomenclature rules in the chapter to derive the name. The molecular (formula) mass is the sum of the masses of each atom times its atomic mass.
Solution:
a) Formula is SO_3. Name is **sulfur trioxide** (the prefix tri- indicates 3 oxygen atoms).

S	=	1(32.06 amu)	=	32.06 amu
O	=	3(16.00 amu)	=	48.00 amu
				80.06 amu

b) Formula is C_3H_8. Since it contains only carbon and hydrogen it is a hydrocarbon and with three carbons its name is **propane**.

C	=	3(12.01 amu)	=	36.03 amu
H	=	8(1.008 amu)	=	8.064 amu
				44.09 amu

2.112   Plan: Review the discussion on separations.
Solution:
Separating the components of a mixture requires physical methods only; that is, no chemical changes (no changes in composition) take place and the components maintain their chemical identities and properties throughout. Separating the components of a compound requires a chemical change (change in composition).

2.115   Plan: Review the definitions of homogeneous and heterogeneous. The key is that a homogeneous mixture has a uniform composition while a heterogeneous mixture does not. A mixture consists of two or more substances physically mixed together while a compound is a pure substance.
Solution:
a) Distilled water is a **compound** that consists of H_2O molecules only.
b) Gasoline is a **homogeneous mixture** of hydrocarbon compounds of uniform composition that can be separated by physical means (distillation).
c) Beach sand is a **heterogeneous mixture** of different size particles of minerals and broken bits of shells.
d) Wine is a **homogeneous mixture** of water, alcohol, and other compounds that can be separated by physical means (distillation).
e) Air is a **homogeneous mixture** of different gases, mainly N_2, O_2, and Ar.

2.117   Plan: Review the discussion on separations.
Solution:
a) Filtration — separating the mixture on the basis of differences in particle size. The water moves through the holes in the colander but the larger pasta cannot.
b) Extraction — The colored impurities are extracted into a solvent that is rinsed away from the raw sugar (or **chromatography**). A sugar solution is passed through a column in which the impurities stick to the stationary phase and the sugar moves through the column in the mobile phase.

2.119   Plan: Use the equation for the volume of a sphere in part a) to find the volume of the nucleus and the volume of the atom. Calculate the fraction of the atom volume that is occupied by the nucleus. For part b), calculate the total mass of the two electrons; subtract the electron mass from the mass of the atom to find the mass of the nucleus. Then calculate the fraction of the atom's mass contributed by the mass of the nucleus.
Solution:
a) Volume (m^3) of nucleus $= \frac{4}{3}\pi r^3 = \frac{4}{3}\pi \left(2.5 \times 10^{-15}\ m\right)^3 = 6.54498 \times 10^{-44}\ m^3$

Volume (m^3) of atom $= \frac{4}{3}\pi r^3 = \frac{4}{3}\pi \left(3.1 \times 10^{-11}\ m\right)^3 = 1.24788 \times 10^{-31}\ m^3$

Fraction of volume $= \dfrac{\text{volume of Nucleus}}{\text{volume of Atom}} = \dfrac{6.54498 \times 10^{-44}\ m^3}{1.24788 \times 10^{-31}\ m^3} = 5.2449 \times 10^{-13} = \mathbf{5.2 \times 10^{-13}}$

b) Mass of nucleus = mass of atom – mass of electrons

$$= 6.64648 \times 10^{-24} \text{ g} - 2(9.10939 \times 10^{-28} \text{ g}) = 6.64466 \times 10^{-24} \text{ g}$$

$$\text{Fraction of mass} = \frac{\text{mass of Nucleus}}{\text{mass of Atom}} = \frac{\left(6.64466 \times 10^{-24} \text{ g}\right)}{\left(6.64648 \times 10^{-24} \text{ g}\right)} = 0.99972617 = \textbf{0.999726}$$

As expected, the volume of the nucleus relative to the volume of the atom is small while its relative mass is large.

2.120   <u>Plan:</u> Use Coulomb's law which states that the energy of attraction in an ionic bond is directly proportional to the *product of charges* and inversely proportional to the *distance between charges*. Choose the largest ionic charges and smallest radii for the strongest ionic bonding and the smallest ionic charges and largest radii for the weakest ionic bonding.
<u>Solution:</u>
Strongest ionic bonding: **MgO**. Mg^{2+}, Ba^{2+}, and O^{2-} have the largest charges. Attraction increases as distance decreases, so the positive ion with the smaller radius, Mg^{2+}, will form a stronger ionic bond than the larger ion Ba^{2+}. Weakest ionic bonding: **RbI**. K^+, Rb^+, Cl^-, and I^- have the smallest charges. Attraction decreases as distance increases, so the ions with the larger radii, Rb^+ and I^-, will form the weakest ionic bond.

2.124   <u>Plan:</u> Determine the percent oxygen in each oxide by subtracting the percent nitrogen from 100%. Express the percentage in amu and divide by the atomic mass of the appropriate elements. Then divide each amount by the smaller number and convert to the simplest whole-number ratio. To find the mass of oxygen per 1.00 g of nitrogen, divide the mass percentage of oxygen by the mass percentage of nitrogen.
<u>Solution:</u>
a)   I $(100.00 - 46.69 \text{ N})\% = 53.31\% \text{ O}$

$$\left(\frac{46.69 \text{ amu N}}{14.01 \text{ amu N}}\right) = 3.3326 \text{ N} \qquad\qquad \left(\frac{53.31 \text{ amu O}}{16.00 \text{ amu O}}\right) = 3.3319 \text{ O}$$

$$\frac{3.3326 \text{ N}}{3.3319} = 1.0002 \text{ N} \qquad\qquad \frac{3.3319 \text{ O}}{3.3319} = 1.0000 \text{ O}$$

The simplest whole-number ratio is **1:1 N:O**.

II $(100.00 - 36.85 \text{ N})\% = 63.15\% \text{ O}$

$$\left(\frac{36.85 \text{ amu N}}{14.01 \text{ amu N}}\right) = 2.6303 \text{ N} \qquad\qquad \left(\frac{63.15 \text{ amu O}}{16.00 \text{ amu O}}\right) = 3.9469 \text{ O}$$

$$\frac{2.6303 \text{ N}}{2.6303} = 1.0000 \text{ mol N} \qquad\qquad \frac{3.9469 \text{ O}}{2.6303} = 1.5001 \text{ O}$$

The simplest whole-number ratio is 1:1.5 N:O = **2:3 N:O**.

III $(100.00 - 25.94 \text{ N})\% = 74.06\% \text{ O}$

$$\left(\frac{25.94 \text{ amu N}}{14.01 \text{ amu N}}\right) = 1.8515 \text{ N} \qquad\qquad \left(\frac{74.06 \text{ amu O}}{16.00 \text{ amu O}}\right) = 4.6288 \text{ O}$$

$$\frac{1.8515 \text{ N}}{1.8515} = 1.0000 \text{ N} \qquad\qquad \frac{4.6288 \text{ O}}{1.8515} = 2.5000 \text{ O}$$

The simplest whole-number ratio is 1:2.5 N:O = **2:5 N:O**.

b)   I $$\left(\frac{53.31 \text{ amu O}}{46.69 \text{ amu N}}\right) = 1.1418 = \textbf{1.14 g O}$$

II $\left(\dfrac{63.15 \text{ amu O}}{36.85 \text{ amu N}}\right) = 1.7137 = \mathbf{1.71 \text{ g O}}$

III $\left(\dfrac{74.06 \text{ amu O}}{25.94 \text{ amu N}}\right) = 2.8550 = \mathbf{2.86 \text{ g O}}$

2.128   <u>Plan:</u> To find the mass percent divide the mass of each substance in mg by the amount of seawater in mg and multiply by 100. The percent of an ion is the mass of that ion divided by the total mass of ions.
<u>Solution:</u>

a) Mass (mg) of seawater = $(1 \text{ kg})\left(\dfrac{1000 \text{ g}}{1 \text{ kg}}\right)\left(\dfrac{1000 \text{ mg}}{1 \text{ g}}\right) = 1 \times 10^6 \text{ mg}$

Mass % = $\left(\dfrac{\text{mass of substance}}{\text{mass of seawater}}\right)(100\%)$

Mass % Cl$^-$ = $\left(\dfrac{18{,}980 \text{ mg Cl}^-}{1 \times 10^6 \text{ mg seawater}}\right)(100\%) = \mathbf{1.898\% \text{ Cl}^-}$

Mass % Na$^+$ = $\left(\dfrac{10.560 \text{ mg Na}^+}{1 \times 10^6 \text{ mg seawater}}\right)(100\%) = \mathbf{1.056\% \text{ Na}^+}$

Mass % SO$_4^{2-}$ = $\left(\dfrac{2650 \text{ mg SO}_4^{2-}}{1 \times 10^6 \text{ mg seawater}}\right)(100\%) = \mathbf{0.265\% \text{ SO}_4^{2-}}$

Mass % Mg^{2+} = $\left(\dfrac{1270 \text{ mg Mg}^{2+}}{1 \times 10^6 \text{ mg seawater}}\right)(100\%) = \mathbf{0.127\% \text{ Mg}^{2+}}$

Mass % Ca^{2+} = $\left(\dfrac{400 \text{ mg Ca}^{2+}}{1 \times 10^6 \text{ mg seawater}}\right)(100\%) = \mathbf{0.04\% \text{ Ca}^{2+}}$

Mass % K$^+$ = $\left(\dfrac{380 \text{ mg K}^+}{1 \times 10^6 \text{ mg seawater}}\right)(100\%) = \mathbf{0.038\% \text{ K}^+}$

Mass % HCO$_3^-$ = $\left(\dfrac{140 \text{ mg HCO}_3^-}{1 \times 10^6 \text{ mg seawater}}\right)(100\%) = \mathbf{0.014\% \text{ HCO}_3^-}$

The mass percents do not add to 100% since the majority of seawater is H_2O.
b) Total mass of ions in 1 kg of seawater
= 18,980 mg + 10,560 mg + 2650 mg + 1270 mg + 400 mg + 380 mg + 140 mg = 34,380 mg

% Na$^+$ = $\left(\dfrac{10{,}560 \text{ mg Na}^+}{34{,}380 \text{ mg total ions}}\right)(100) = 30.71553 = \mathbf{30.72\%}$

c) Alkaline earth metal ions are Mg^{2+} and Ca^{2+} (Group 2 ions).
Total mass % = 0.127% Mg^{2+} + 0.04% Ca^{2+} = 0.167%
Alkali metal ions are Na$^+$ and K$^+$ (Group 1 ions). Total mass % = 1.056% Na$^+$ + 0.038% K$^+$ = 1.094%

$\dfrac{\text{Mass \% of alkali metal ions}}{\text{Mass \% of alkaline earth metal ions}} = \dfrac{1.094\%}{0.167\%} = 6.6$

Total mass percent for alkali metal ions is **6.6 times greater** than the total mass percent for alkaline earth metal ions. Sodium ions (alkali metal ions) are dominant in seawater.
d) Anions are Cl$^-$, SO$_4^{2-}$, and HCO$_3^-$.
Total mass % = 1.898% Cl$^-$ + 0.265% SO$_4^{2-}$ + 0.014% HCO$_3^-$ = 2.177% anions
Cations are Na$^+$, Mg^{2+}, Ca^{2+}, and K$^+$.

Total mass % = 1.056% Na^+ + 0.127% Mg^{2+} + 0.04% Ca^{2+} + 0.038% K^+ = 1.2610 = **1.26% cations**
The mass fraction of **anions** is larger than the mass fraction of cations. Is the solution neutral since the mass of anions exceeds the mass of cations? Yes, although the mass is larger, the number of positive charges equals the number of negative charges.

2.131   <u>Plan:</u> First, count each type of atom present to produce a molecular formula. The molecular (formula) mass is the sum of the atomic masses of all of the atoms. Divide the mass of each element in the compound by the molecular mass and multiply by 100 to obtain the mass percent of each element.
<u>Solution:</u>
The molecular formula of succinic acid is $C_4H_6O_4$.

C	=	4(12.01 amu)	=	48.04 amu
H	=	6(1.008 amu)	=	6.048 amu
O	=	4 (16.00 amu)	=	<u>64.00 amu</u>
				118.09 amu

$$\% \, C = \left(\frac{48.04 \text{ amu C}}{118.088 \text{ amu}}\right)100\% = 40.6815 = \mathbf{40.68\% \ C}$$

$$\% \, H = \left(\frac{6.048 \text{ amu H}}{118.088 \text{ amu}}\right)100\% = 5.1216 = \mathbf{5.122\% \ H}$$

$$\% \, O = \left(\frac{64.00 \text{ amu O}}{118.088 \text{ amu}}\right)100\% = 54.1969 = \mathbf{54.20\% \ O}$$

<u>Check:</u> Total = (40.68 + 5.122 + 54.20)% = 100.00% The answer checks.

2.134   <u>Plan:</u> List all possible combinations of the isotopes. Determine the masses of each isotopic composition. The molecule consisting of the lower abundance isotopes (N-15 and O-18) is the least common, and the one containing only the more abundant isotopes (N-14 and O-16) will be the most common.
<u>Solution:</u>
a)   b)

Formula	Mass (amu)	
$^{15}N_2{}^{18}O$	2(15 amu N) + 18 amu O = 48	**least common**
$^{15}N_2{}^{16}O$	2(15 amu N) + 16 amu O = 46	
$^{14}N_2{}^{18}O$	2(14 amu N) + 18 amu O = 46	
$^{14}N_2{}^{16}O$	2(14 amu N) + 16 amu O = 44	**most common**
$^{15}N^{14}N^{18}O$	1(15 amu N) + 1(14 amu N) + 18 amu O = 47	
$^{15}N^{14}N^{16}O$	1(15 amu N) + 1(14 amu N) + 16 amu O = 45	

2.136   <u>Plan:</u> To find the formula mass of potassium fluoride, add the atomic masses of potassium and fluorine. Fluorine has only one naturally occurring isotope, so the mass of this isotope equals the atomic mass of fluorine. The atomic mass of potassium is the weighted average of the two isotopic masses: (isotopic mass of isotope 1 x fractional abundance) + (isotopic mass of isotope 2 x fractional abundance).
<u>Solution:</u>
Average atomic mass of K =
         (isotopic mass of ^{39}K x fractional abundance) + (isotopic mass of ^{41}K x fractional abundance)
Average atomic mass of K = (38.9637 amu)$\left(\dfrac{93.258\%}{100\%}\right)$ + (40.9618 amu)$\left(\dfrac{6.730\%}{100\%}\right)$ = 39.093 amu

The formula for potassium fluoride is KF, so its molecular mass is (39.093 + 18.9984) = **58.091 amu**

2.138   <u>Plan:</u> One molecule of NO is released per atom of N in the medicine. Divide the total mass of NO released by the molecular mass of the medicine and multiply by 100 for mass percent.

Solution:

NO = (14.01 + 16.00) amu = 30.01 amu

Nitroglycerin:

$C_3H_5N_3O_9$ = 3(12.01 amu C) + 5(1.008 amu H) + 3(14.01 amu N) + 9(16.00 amu O) = 227.10 amu

In $C_3H_5N_3O_9$ (molecular mass = 227.10 amu), there are 3 atoms of N; since 1 molecule of NO is released per atom of N, this medicine would release 3 molecules of NO. The molecular mass of NO = 30.01 amu.

$$\text{Mass percent of NO} = \frac{\text{total mass of NO}}{\text{mass of compound}}(100) = \frac{3(30.01 \text{ amu})}{227.10 \text{ amu}}(100) = 39.6433 = \mathbf{39.64\%}$$

Isoamyl nitrate:

$C_5H_{11}NO_3$ = 5(12.01 amu C) + 11(1.008 amu H) + 1(14.01 amu N) + 3(16.00 amu O) = 133.15 amu

In $(CH_3)_2CHCH_2CH_2ONO_2$ (molecular mass = 133.15 amu), there is one atom of N; since 1 molecule of NO is released per atom of N, this medicine would release 1 molecule of NO.

$$\text{Mass percent of NO} = \frac{\text{total mass of NO}}{\text{mass of compound}}(100) = \frac{1(30.01 \text{ amu})}{133.15 \text{ amu}}(100) = 22.5385 = \mathbf{22.54\%}$$

2.139   Plan: First, count each type of atom present to produce a molecular formula. Determine the mass fraction of each element. Mass fraction = $\dfrac{\text{total mass of the element}}{\text{molecular mass of TNT}}$. The mass of TNT multiplied by the mass fraction of each element gives the mass of that element.

Solution:

The molecular formula for TNT is $C_7H_5O_6N_3$. The molecular mass of TNT is:

C	=	7(12.01 amu)	=	84.07 amu
H	=	5(1.008 amu)	=	5.040 amu
O	=	6(16.00 amu)	=	96.00 amu
N	=	3(14.01 amu)	=	42.03 amu
				227.14 amu

The mass fraction of each element is:

$$C = \frac{84.07 \text{ amu}}{227.14 \text{ amu}} = 0.3701 \text{ C} \qquad H = \frac{5.040 \text{ amu}}{227.14 \text{ amu}} = 0.02219 \text{ H}$$

$$O = \frac{96.00 \text{ amu}}{227.14 \text{ amu}} = 0.4226 \text{ O} \qquad N = \frac{42.03 \text{ amu}}{227.14 \text{ amu}} = 0.1850 \text{ N}$$

Masses of each element in 1.00 lb of TNT = mass fraction of element x 1.00 lb.

Mass (lb) C = 0.3701 x 1.00 lb = **0.370 lb C**

Mass (lb) H = 0.02219 x 1.00 lb = **0.0222 lb H**

Mass (lb) O = 0.4226 x 1.00 lb = **0.423 lb O**

Mass (lb) N = 0.1850 x 1.00 lb = **0.185 lb N**

2.144   Plan: A change is physical when there has been a change in physical form but not a change in composition. In a chemical change, a substance is converted into a different substance.

Solution:

1) Initially, all the molecules are present in blue-blue or red-red pairs. After the change, there are no red-red pairs, and there are now red-blue pairs. Changing some of the pairs means there has been a **chemical change**.

2) There are two blue-blue pairs and four red-blue pairs both before and after the change, thus no chemical change occurred. The different types of molecules are separated into different boxes. This is a **physical change**.

3) The identity of the box contents has changed from pairs to individuals. This requires a **chemical change**.

4) The contents have changed from all pairs to all triplets. This is a change in the identity of the particles, thus, this is a **chemical change**.

5) There are four red-blue pairs both before and after, thus there has been no change in the identity of the individual units. There has been a **physical change**.

CHAPTER 3 STOICHIOMETRY OF FORMULAS AND EQUATIONS

FOLLOW–UP PROBLEMS

3.1A   <u>Plan:</u> The mass of carbon must be changed from mg to g. The molar mass of carbon can then be used to determine the number of moles.
<u>Solution:</u>

$$\text{Moles of carbon} = 315 \text{ mg C} \left(\frac{10^{-3} \text{ g}}{1 \text{ mg}}\right)\left(\frac{1 \text{ mol C}}{12.01 \text{ g C}}\right) = 2.6228 \times 10^{-2} = \mathbf{2.62 \times 10^{-2} \text{ mol C}}$$

Road map:

Mass (mg) of C

$10^3 \text{ mg} = 1 \text{ g}$

Mass (g) of C

Divide by $\mathcal{M}$ (g/mol)

Amount (moles) of C

3.1B   <u>Plan:</u> The number of moles of aluminum must be changed to g. Then the mass of aluminum per can can be used to calculate the number of soda cans that can be made from 52 mol of Al.
<u>Solution:</u>

$$\text{Number of soda cans} = 52 \text{ mol Al} \left(\frac{26.98 \text{ g Al}}{1 \text{ mol Al}}\right)\left(\frac{1 \text{ soda can}}{14 \text{ g Al}}\right) = 100.21 = \mathbf{100 \text{ soda cans}}$$

Road map:

Amount (mol) of Al

Multiply by $\mathcal{M}$ (g/mol)
(1 mol Al = 26.98 g Al)

Mass (g) of Al

14 g Al = 1 soda can

Number of cans

3.2A   <u>Plan:</u> Avogadro's number is needed to convert the number of nitrogen molecules to moles. Since nitrogen molecules are diatomic (composed of two N atoms), the moles of molecules must be multiplied by 2 to obtain moles of atoms.
<u>Solution:</u>

$$\text{Moles of N atoms} = \left(9.72 \times 10^{21} \text{ N}_2 \text{ molecules}\right)\left(\frac{1 \text{ mol N}_2}{6.022 \times 10^{23} \text{ N}_2 \text{ molecules}}\right)\left(\frac{2 \text{ N atoms}}{1 \text{ mol N}_2}\right)$$

$$= 3.2281634 \times 10^{-2} = \mathbf{3.23 \times 10^{-2} \text{ mol N}}$$

Road map:

```
┌─────────────────────────┐
│ No. of N₂ molecules     │
└─────────────────────────┘
   Divide by Avogadro's
   number (molecules/mol)
┌─────────────────────────┐
│ Amount (moles) of N₂    │
└─────────────────────────┘
   Use chemical formula
   (1 mol N₂ = 2 mol N)
┌─────────────────────────┐
│ Amount (moles) of N     │
└─────────────────────────┘
```

3.2B   Plan: Avogadro's number is needed to convert the number of moles of He to atoms.
        Solution:

$$\text{Number of He atoms} = 325 \text{ mol He} \left(\frac{6.022 \times 10^{23} \text{ He atoms}}{1 \text{ mol He}} \right) = 1.9572 \times 10^{26} = \mathbf{1.96 \times 10^{26} \text{ He atoms}}$$

Road map:

```
┌─────────────────────────┐
│ Amount (mol) of He      │
└─────────────────────────┘
   Multiply by Avogadro's
   number
   (1 mol He = 6.022×10²³ He atoms)
┌─────────────────────────┐
│ Number of He atoms      │
└─────────────────────────┘
```

3.3A   Plan: Avogadro's number is needed to convert the number of atoms to moles. The molar mass of manganese can
        then be used to determine the number of grams.
        Solution:

$$\text{Mass (g) of Mn} = \left(3.22 \times 10^{20} \text{ Mn atoms} \right) \left(\frac{1 \text{ mol Mn}}{6.022 \times 10^{23} \text{ Mn atoms}} \right) \left(\frac{54.94 \text{ g Mn}}{1 \text{ mol Mn}} \right)$$

$$= 2.9377 \times 10^{-2} = \mathbf{2.94 \times 10^{-2} \text{ g Mn}}$$

Road map:

```
┌─────────────────────────┐
│ No. of Mn atoms         │
└─────────────────────────┘
   Divide by Avogadro's
   number (molecules/mol)
┌─────────────────────────┐
│ Amount (moles) of Mn    │
└─────────────────────────┘
   Multiply by ℳ (g/mol)
┌─────────────────────────┐
│ Mass (g) of Mn          │
└─────────────────────────┘
```

3.3B   Plan: Use the molar mass of copper to calculate the number of moles of copper present in a penny. Avogadro's
        number is then needed to convert the number of moles of Cu to Cu atoms.

Solution:

$$\text{Number of Cu atoms} = 0.0625 \text{ g Cu} \left(\frac{1 \text{ mol Cu}}{63.55 \text{ g Cu}}\right)\left(\frac{6.022 \times 10^{23} \text{ Cu atoms}}{1 \text{ mol Cu}}\right)$$

$$= 5.9225 \times 10^{20} = \mathbf{5.92 \times 10^{20} \text{ Cu atoms}}$$

Road map:

Mass (g) of Cu

Divide by $\mathcal{M}$ (g/mol)
(1 mol Cu = 63.55 g Cu)

Amount (moles) of Cu

Multiply by Avogadro's number
(1 mol Cu = 6.022×10^{23} Cu atoms)

No. of Cu atoms

3.4A   Plan: Avogadro's number is used to change the number of formula units to moles. Moles may be changed to mass using the molar mass of sodium fluoride, which is calculated from its formula.
Solution:

The formula of sodium fluoride is NaF.

$\mathcal{M}$ of NaF = (1 x $\mathcal{M}$ of Na) + (1 x $\mathcal{M}$ of F) = 22.99 g/mol + 19.00 g/mol = 41.99 g/mol

$$\text{Mass (g) of NaF} = \left(1.19 \times 10^{19} \text{ NaF formula units}\right)\left(\frac{1 \text{ mol NaF}}{6.022 \times 10^{23} \text{ NaF formula units}}\right)\left(\frac{41.99 \text{ g NaF}}{1 \text{ mol NaF}}\right)$$

$$= 8.29759 \times 10^{-4} = \mathbf{8.30 \times 10^{-4} \text{ g NaF}}$$

Road map:

No. of NaF formula units

Divide by Avogadro's
number (molecules/mol)

Amount (moles) of NaF

Multiply by $\mathcal{M}$ (g/mol)

Mass (g) of NaF

3.4B   Plan: Convert the mass of calcium chloride from pounds to g. Use the molar mass to calculate the number of moles of calcium chloride in the sample. Finally, use Avogadro's number to change the number of moles to formula units.
Solution:

$\mathcal{M}$ of $CaCl_2$ = (1 x $\mathcal{M}$ of Ca) + (2 x $\mathcal{M}$ of Cl) = 40.08 g/mol + 2(35.45 g/mol) = 110.98 g/mol

$$\text{Number of formula units of } CaCl_2 = 400 \text{ lb} \left(\frac{453.6 \text{ g}}{1 \text{ lb}}\right)\left(\frac{1 \text{ mol } CaCl_2}{110.98 \text{ g } CaCl_2}\right)\left(\frac{6.022 \times 10^{23} \text{ formula units } CaCl_2}{1 \text{ mol } CaCl_2}\right)$$

$$= 9.8453 \times 10^{26} = \mathbf{1 \times 10^{27} \text{ formula units } CaCl_2}$$

Road map:

Mass (lb) of $CaCl_2$

1 lb = 453.6 g

Mass (g) of $CaCl_2$

Divide by $\mathcal{M}$ (g/mol)
(1 mol $CaCl_2$ = 110.98 g $CaCl_2$)

Amount (mol) of $CaCl_2$

Multiply by Avogadro's number
(1 mol $CaCl_2$ = 6.022x10²³ $CaCl_2$ formula units)

No. of formula units of $CaCl_2$

3.5A   Plan: Avogadro's number is used to change the number of molecules to moles. Moles may be changed to mass by multiplying by the molar mass. The molar mass of tetraphosphorus decoxide is obtained from its chemical formula. Each molecule has four phosphorus atoms, so the total number of atoms is four times the number of molecules.
Solution:
a) Tetra = 4, and deca = 10 to give P_4O_{10}.
The molar mass, $\mathcal{M}$, is the sum of the atomic weights, expressed in g/mol:

$$P = 4(30.97) = 123.88 \text{ g/mol}$$
$$O = 10(16.00) = \underline{160.00 \text{ g/mol}}$$
$$= 283.88 \text{ g/mol of } P_4O_{10}$$

$$\text{Mass (g) of } P_4O_{10} = \left(4.65 \times 10^{22} \text{ molecules } P_4O_{10}\right)\left(\frac{1 \text{ mol}}{6.022 \times 10^{23} \text{ molecules}}\right)\left(\frac{283.88 \text{ g}}{1 \text{ mol}}\right)$$

$$= 21.9203 = \textbf{21.9 g } \mathbf{P_4O_{10}}$$

b) Number of P atoms = $4.65 \times 10^{22} \text{ molecules } P_4O_{10}\left(\frac{4 \text{ atoms P}}{1 \text{ } P_4O_{10} \text{ molecule}}\right) = \textbf{1.86x10}^{\textbf{23}} \textbf{ P atoms}$

3.5B   Plan: The mass of calcium phosphate is converted to moles of calcium phosphate by dividing by the molar mass. Avogadro's number is used to change the number of moles to formula units. Each formula unit has two phosphate ions, so the total number of phosphate ions is two times the number of formula units.
Solution:
a) The formula of calcium phosphate is $Ca_3(PO_4)_2$.
The molar mass, $\mathcal{M}$, is the sum of the atomic weights, expressed in g/mol:
$\mathcal{M}$ = (3 x $\mathcal{M}$ of Ca) + (2 x $\mathcal{M}$ of P) + (8 x $\mathcal{M}$ of O)
   = (3 x 40.08 g/mol Ca) + (2 x 30.97 g/mol P) + (8 x 16.00 g/mol O)
   = 310.18 g/mol $Ca_3(PO_4)_2$

$$\text{No. of formula units } Ca_3(PO_4)_2 = 75.5 \text{ g } Ca_3(PO_4)_2\left(\frac{1 \text{ mol } Ca_3(PO_4)_2}{310.18 \text{ g } Ca_3(PO_4)_2}\right)\left(\frac{6.022 \times 10^{23} \text{ formula units } Ca_3(PO_4)_2}{1 \text{ mol } Ca_3(PO_4)_2}\right)$$
$$= 1.4658 \times 10^{23} = \textbf{1.47x10}^{\textbf{23}} \textbf{ formula units } \mathbf{Ca_3(PO_4)_2}$$

b) No. of phosphate (PO_4^{3-}) ions = $1.47 \times 10^{23} \text{ formula units } Ca_3(PO_4)_2\left(\frac{2 \text{ } PO_4^{3-} \text{ ions}}{1 \text{ formula unit } Ca_3(PO_4)_2}\right)$

$$= \textbf{2.94x10}^{\textbf{23}} \textbf{ phosphate ions}$$

3.6A   <u>Plan:</u> Calculate the molar mass of glucose. The total mass of carbon in the compound divided by the molar mass of the compound, multiplied by 100% gives the mass percent of C.
<u>Solution:</u>
The formula for glucose is $C_6H_{12}O_6$. There are 6 atoms of C per each formula.
Molar mass of $C_6H_{12}O_6$ = (6 x $\mathscr{M}$ of C) + (12 x $\mathscr{M}$ of H) + (6 x $\mathscr{M}$ of O)
                = (6 x 12.01 g/mol) + (12 x 1.008 g/mol) + (6 x 16.00 g/mol)
                = 180.16 g/mol

$$\text{Mass \% of C} = \frac{\text{total mass of C}}{\text{molar mass of } C_6H_{12}O_6}(100) = \frac{6 \times 12.01 \text{ g/mol}}{180.16 \text{ g/mol}}(100) = 39.9978 = \textbf{40.00\% C}$$

3.6B   <u>Plan:</u> Calculate the molar mass of CCl_3F. The total mass of chlorine in the compound divided by the molar mass of the compound, multiplied by 100% gives the mass percent of Cl.
<u>Solution:</u>
The formula is CCl_3F. There are 3 atoms of Cl per each formula.
Molar mass of CCl_3F = (1 x $\mathscr{M}$ of C) + (3 x $\mathscr{M}$ of Cl) + (1 x $\mathscr{M}$ of F)
             = (1 x 12.01 g/mol) + (3 x 35.45 g/mol) + (1 x 19.00 g/mol)
             = 137.36 g/mol

$$\text{Mass \% of Cl} = \frac{\text{total mass of Cl}}{\text{molar mass of } CCl_3F}(100) = \frac{3 \times 35.45 \text{ g/mol}}{137.36 \text{ g/mol}}(100) = 77.4243 = \textbf{77.42\% Cl}$$

3.7A   <u>Plan:</u> Multiply the mass of the sample by the mass fraction of C found in the preceding problem.
<u>Solution:</u>

$$\text{Mass (g) of C} = 16.55 \text{ g } C_6H_{12}O_6 \left(\frac{72.06 \text{ g C}}{180.16 \text{ g } C_6H_{12}O_6} \right) = 6.6196 = \textbf{6.620 g C}$$

3.7B   <u>Plan:</u> Multiply the mass of the sample by the mass fraction of Cl found in the preceding problem.
<u>Solution:</u>

$$\text{Mass (g) of Cl} = 112 \text{ g } CCl_3F \left(\frac{106.35 \text{ g Cl}}{137.36 \text{ g } CCl_3F} \right) = 86.6900 = \textbf{86.7 g Cl}$$

3.8A   <u>Plan:</u> We are given fractional amounts of the elements as subscripts. Convert the fractional amounts to whole numbers by dividing each number by the smaller number and then multiplying by the smallest integer that will turn both subscripts into integers.
<u>Solution:</u>
Divide each subscript by the smaller value, 0.170: $B_{\frac{0.170}{0.170}}O_{\frac{0.255}{0.170}} = B_1O_{1.5}$
Multiply the subscripts by 2 to obtain integers: $B_{1 \times 2}O_{1.5 \times 2} = \textbf{B}_2\textbf{O}_3$

3.8B   <u>Plan:</u> We are given fractional amounts of the elements as subscripts. Convert the fractional amounts to whole numbers by dividing each number by the smaller number and then multiplying by the smallest integer that will turn both subscripts into integers.
<u>Solution:</u>
Divide each subscript by the smaller value, 6.80: $C_{\frac{6.80}{6.80}}H_{\frac{18.1}{6.80}} = C_1H_{2.67}$

Multiply the subscripts by 3 to obtain integers: $C_{1 \times 3}H_{2.66 \times 3} = \textbf{C}_3\textbf{H}_8$

3.9A **Plan:** Calculate the number of moles of each element in the sample by dividing by the molar mass of the corresponding element. The calculated numbers of moles are the fractional amounts of the elements and can be used as subscripts in a chemical formula. Convert the fractional amounts to whole numbers by dividing each number by the smallest subscripted number.

Solution:

$$\text{Moles of H} = 1.23 \text{ g H} \left(\frac{1 \text{ mol H}}{1.008 \text{ g H}} \right) = 1.2202 \text{ mol H}$$

$$\text{Moles of P} = 12.64 \text{ g P} \left(\frac{1 \text{ mol P}}{30.97 \text{ g P}} \right) = 0.40814 \text{ mol P}$$

$$\text{Moles of O} = 26.12 \text{ g O} \left(\frac{1 \text{ mol O}}{16.00 \text{ g O}} \right) = 1.6325 \text{ mol O}$$

Divide each subscript by the smaller value, 0.408: $H_{\frac{1.2202}{0.40814}} P_{\frac{0.40814}{0.40814}} O_{\frac{1.6325}{0.40814}} = \textbf{H}_3\textbf{PO}_4$, this is **phosphoric acid**.

3.9B **Plan:** The moles of sulfur may be calculated by dividing the mass of sulfur by the molar mass of sulfur. The moles of sulfur and the chemical formula will give the moles of M. The mass of M divided by the moles of M will give the molar mass of M. The molar mass of M can identify the element.

Solution:

$$\text{Moles of S} = \left(2.88 \text{ g S} \right) \left(\frac{1 \text{ mol S}}{32.06 \text{ g S}} \right) = 0.089832 \text{ mol S}$$

$$\text{Moles of M} = \left(0.089832 \text{ mol S} \right) \left(\frac{2 \text{ mol M}}{3 \text{ mol S}} \right) = 0.059888 \text{ mol M}$$

$$\text{Molar mass of M} = \frac{3.12 \text{ g M}}{0.059888 \text{ mol M}} = 52.0972 = 52.1 \text{ g/mol}$$

The element is Cr (52.00 g/mol); M is **Chromium** and M_2S_3 is **chromium(III) sulfide**.

3.10A **Plan:** If we assume there are 100 grams of this compound, then the masses of carbon and hydrogen, in grams, are numerically equivalent to the percentages. Divide the atomic mass of each element by its molar mass to obtain the moles of each element. Dividing each of the moles by the smaller value gives the simplest ratio of C and H. The smallest multiplier to convert the ratios to whole numbers gives the empirical formula. To obtain the molecular formula, divide the given molar mass of the compound by the molar mass of the empirical formula to find the whole-number by which the empirical formula is multiplied.

Solution:

Assuming 100 g of compound gives 95.21 g C and 4.79 g H:

$$\text{Moles of C} = 95.21 \text{ g C} \left(\frac{1 \text{ mol C}}{12.01 \text{ g C}} \right) = 7.92756 \text{ mol C}$$

$$\text{Mole of H} = 4.79 \text{ g H} \left(\frac{1 \text{ mol H}}{1.008 \text{ g H}} \right) = 4.75198 \text{ mol H}$$

Divide each of the moles by 4.75198, the smaller value:

$$C_{\frac{7.92756}{4.75198}} H_{\frac{4.75198}{4.75198}} = C_{1.6683} H_1$$

The value 1.668 is 5/3, so the moles of C and H must each be multiplied by 3. If it is not obvious that the value is near 5/3, use a trial and error procedure whereby the value is multiplied by the successively larger integer until a value near an integer results. This gives C_5H_3 as the empirical formula. The molar mass of this formula is:

$$(5 \times 12.01 \text{ g/mol}) + (3 \times 1.008 \text{ g/mol}) = 63.074 \text{ g/mol}$$

$$\text{Whole-number multiple} = \frac{\text{molar mass of compound}}{\text{molar mass of empirical formula}} = \frac{252.30 \text{ g/mol}}{63.074 \text{ g/mol}} = 4$$

Thus, the empirical formula must be multiplied by 4 to give $4(C_5H_3) = C_{20}H_{12}$ as the molecular formula of benzo[a]pyrene.

3.10B   Plan: If we assume there are 100 grams of this compound, then the masses of carbon, hydrogen, nitrogen, and oxygen, in grams, are numerically equivalent to the percentages. Divide the atomic mass of each element by its molar mass to obtain the moles of each element. Dividing each of the moles by the smaller value gives the simplest ratio of C, H, N, and O. To obtain the molecular formula, divide the given molar mass of the compound by the molar mass of the empirical formula to find the whole-number by which the empirical formula is multiplied.
Solution:
Assuming 100 g of compound gives 49.47 g C, 5.19 g H, 28.86 g N, and 16.48 g O:

$$\text{Moles of C} = 49.47 \text{ g C}\left(\frac{1 \text{ mol C}}{12.01 \text{ g C}}\right) = 4.1191 \text{ mol C}$$

$$\text{Moles of H} = 5.19 \text{ g H}\left(\frac{1 \text{ mol H}}{1.008 \text{ g H}}\right) = 5.1488 \text{ mol H}$$

$$\text{Moles of N} = 28.86 \text{ g N}\left(\frac{1 \text{ mol N}}{14.01 \text{ g N}}\right) = 2.0600 \text{ mol N}$$

$$\text{Moles of O} = 16.48 \text{ g O}\left(\frac{1 \text{ mol O}}{16.00 \text{ g O}}\right) = 1.0300 \text{ mol O}$$

Divide each subscript by the smaller value, 1.030: $C_{\frac{4.1191}{1.0300}}H_{\frac{5.1488}{1.0300}}N_{\frac{2.0600}{1.0300}}O_{\frac{1.0300}{1.0300}} = C_4H_5N_2O$

This gives $C_4H_5N_2O$ as the empirical formula. The molar mass of this formula is:

$$(4 \times 12.01 \text{ g/mol}) + (5 \times 1.008 \text{ g/mol}) + (2 \times 14.01 \text{ g/mol}) + (1 \times 16.00 \text{ g/mol}) = 97.10 \text{ g/mol}$$

The molar mass of caffeine is 194.2 g/mol, which is larger than the empirical formula mass of 97.10 g/mol, so the molecular formula must be a whole-number multiple of the empirical formula.

$$\text{Whole-number multiple} = \frac{\text{molar mass of compound}}{\text{molar mass of empirical formula}} = \frac{194.2 \text{ g/mol}}{97.10 \text{ g/mol}} = 2$$

Thus, the empirical formula must be multiplied by 2 to give $2(C_4H_5N_2O) = C_8H_{10}N_4O_2$ as the molecular formula of caffeine.

3.11A   Plan: The carbon in the sample is converted to carbon dioxide, the hydrogen is converted to water, and the remaining material is chlorine. The grams of carbon dioxide and the grams of water are both converted to moles. One mole of carbon dioxide gives one mole of carbon, while one mole of water gives two moles of hydrogen. Using the molar masses of carbon and hydrogen, the grams of each of these elements in the original sample may be determined. The original mass of sample minus the masses of carbon and hydrogen gives the mass of chlorine. The mass of chlorine and the molar mass of chlorine will give the moles of chlorine. Once the moles of each of the elements have been calculated, divide by the smallest value, and, if necessary, multiply by the smallest number required to give a set of whole numbers for the empirical formula. Compare the molar mass of the empirical formula to the molar mass given in the problem to find the molecular formula.

Solution:

Determine the moles and the masses of carbon and hydrogen produced by combustion of the sample.

$$0.451 \text{ g CO}_2 \left(\frac{1 \text{ mol CO}_2}{44.01 \text{ g CO}_2} \right) \left(\frac{1 \text{ mol C}}{1 \text{ mol CO}_2} \right) = 0.010248 \text{ mol C} \left(\frac{12.01 \text{ g C}}{1 \text{ mol C}} \right) = 0.12307 \text{ g C}$$

$$0.0617 \text{ g H}_2\text{O} \left(\frac{1 \text{ mol H}_2\text{O}}{18.016 \text{ g H}_2\text{O}} \right) \left(\frac{2 \text{ mol H}}{1 \text{ mol H}_2\text{O}} \right) = 0.0068495 \text{ mol H} \left(\frac{1.008 \text{ g H}}{1 \text{ mol H}} \right) = 0.006904 \text{ g H}$$

The mass of chlorine is given by: 0.250 g sample – (0.12307 g C + 0.006904 g H) = 0.12003 g Cl

The moles of chlorine are:

$$0.12003 \text{ g Cl} \left(\frac{1 \text{ mol Cl}}{35.45 \text{ g Cl}} \right) = 0.0033859 \text{ mol Cl. This is the smallest number of moles.}$$

Divide each mole value by the lowest value, 0.0033850: $C_{\frac{0.010248}{0.0033859}} H_{\frac{0.0068495}{0.0033859}} Cl_{\frac{0.0033859}{0.0033859}} = C_3H_2Cl$

The empirical formula has the following molar mass:

(3 x 12.01 g/mol) + (2 x 1.008 g/mol) + (35.45 g/mol) = 73.496 g/mol C_3H_2Cl

Whole-number multiple = $\dfrac{\text{molar mass of compound}}{\text{molar mass of empirical formula}}$ = $\dfrac{146.99 \text{ g/mol}}{73.496 \text{ g/mol}}$ = 2

Thus, the molecular formula is two times the empirical formula, $2(C_3H_2Cl) = \mathbf{C_6H_4Cl_2}$.

3.11B   <u>Plan:</u> The carbon in the sample is converted to carbon dioxide, the hydrogen is converted to water, and the remaining material is oxygen. The grams of carbon dioxide and the grams of water are both converted to moles. One mole of carbon dioxide gives one mole of carbon, while one mole of water gives two moles of hydrogen. Using the molar masses of carbon and hydrogen, the grams of each of these elements in the original sample may be determined. The original mass of sample minus the masses of carbon and hydrogen gives the mass of oxygen. The mass of oxygen and the molar mass of oxygen will give the moles of oxygen. Once the moles of each of the elements have been calculated, divide by the smallest value, and, if necessary, multiply by the smallest number required to give a set of whole numbers for the empirical formula. Compare the molar mass of the empirical formula to the molar mass given in the problem to find the molecular formula.
<u>Solution:</u>

Determine the moles and the masses of carbon and hydrogen produced by combustion of the sample.

$$3.516 \text{ g CO}_2 \left(\frac{1 \text{ mol CO}_2}{44.01 \text{ g CO}_2} \right) \left(\frac{1 \text{ mol C}}{1 \text{ mol CO}_2} \right) = 0.079891 \text{ mol C} \left(\frac{12.01 \text{ g C}}{1 \text{ mol C}} \right) = 0.95949 \text{ g C}$$

$$1.007 \text{ g H}_2\text{O} \left(\frac{1 \text{ mol H}_2\text{O}}{18.02 \text{ g H}_2\text{O}} \right) \left(\frac{2 \text{ mol H}}{1 \text{ mol H}_2\text{O}} \right) = 0.11176 \text{ mol H} \left(\frac{1.008 \text{ g H}}{1 \text{ mol H}} \right) = 0.11266 \text{ g H}$$

The mass of oxygen is given by: 1.200 g sample – (0.95949 g C + 0.11266 g H) = 0.12785 g O
The moles of C and H are calculated above. The moles of oxygen are:

$$0.12785 \text{ g O} \left(\frac{1 \text{ mol O}}{16.00 \text{ g O}} \right) = 0.0079906 \text{mol O. This is the smallest number of moles.}$$

Divide each subscript by the smallest value, 0.00800: $C_{\frac{0.079891}{0.0079906}} H_{\frac{0.11176}{0.0079906}} O_{\frac{0.0079906}{0.0079906}} = C_{10}H_{14}O$

This gives $C_{10}H_{14}O$ as the empirical formula. The molar mass of this formula is:

(10 x 12.01 g/mol) + (14 x 1.008 g/mol) + (1 x 16.00 g/mol) = 150.21 g/mol

The molar mass of the steroid is 300.42 g/mol, which is larger than the empirical formula mass of 150.21 g/mol, so the molecular formula must be a whole-number multiple of the empirical formula.

$$\text{Whole-number multiple} = \frac{\text{molar mass of compound}}{\text{molar mass of empirical formula}} = \frac{300.42 \text{ g/mol}}{150.21 \text{ g/mol}} = 2$$

Thus, the empirical formula must be multiplied by 2 to give $2(C_{10}H_{14}O) = \mathbf{C_{20}H_{28}O_2}$ as the molecular formula of the steroid.

3.12A   Plan: In each part it is necessary to determine the chemical formulas, including the physical states, for both the reactants and products. The formulas are then placed on the appropriate sides of the reaction arrow. The equation is then balanced.
Solution:

a) Sodium is a metal (solid) that reacts with water (liquid) to produce hydrogen (gas) and a solution of sodium hydroxide (aqueous). Sodium is Na; water is H_2O; hydrogen is H_2; and sodium hydroxide is NaOH.

$$Na(s) + H_2O(l) \rightarrow H_2(g) + NaOH(aq) \text{ is the equation.}$$

Balancing will precede one element at a time. One way to balance hydrogen gives:

$$Na(s) + 2H_2O(l) \rightarrow H_2(g) + 2NaOH(aq)$$

Next, the sodium will be balanced:

$$\mathbf{2Na(s) + 2H_2O(l) \rightarrow H_2(g) + 2NaOH(aq)}$$

On inspection, we see that the oxygen is already balanced.

b) Aqueous nitric acid reacts with calcium carbonate (solid) to produce carbon dioxide (gas), water (liquid), and aqueous calcium nitrate. Nitric acid is HNO_3; calcium carbonate is $CaCO_3$; carbon dioxide is CO_2; water is H_2O; and calcium nitrate is $Ca(NO_3)_2$. The starting equation is

$$HNO_3(aq) + CaCO_3(s) \rightarrow CO_2(g) + H_2O(l) + Ca(NO_3)_2(aq)$$

Initially, Ca and C are balanced. Proceeding to another element, such as N, or better yet the group of elements in NO_3^- gives the following partially balanced equation:

$$\mathbf{2HNO_3(aq) + CaCO_3(s) \rightarrow CO_2(g) + H_2O(l) + Ca(NO_3)_2(aq)}$$

Now, all the elements are balanced.

c) We are told all the substances involved are gases. The reactants are phosphorus trichloride and hydrogen fluoride, while the products are phosphorus trifluoride and hydrogen chloride. Phosphorus trifluoride is PF_3; phosphorus trichloride is PCl_3; hydrogen fluoride is HF; and hydrogen chloride is HCl. The initial equation is:

$$PCl_3(g) + HF(g) \rightarrow PF_3(g) + HCl(g)$$

Initially, P and H are balanced. Proceed to another element (either F or Cl); if we will choose Cl, it balances as:

$$PCl_3(g) + HF(g) \rightarrow PF_3(g) + 3HCl(g)$$

The balancing of the Cl unbalances the H, this should be corrected by balancing the H as:

$$\mathbf{PCl_3(g) + 3HF(g) \rightarrow PF_3(g) + 3HCl(g)}$$

Now, all the elements are balanced.

3.12B   Plan: In each part it is necessary to determine the chemical formulas, including the physical states, for both the reactants and products. The formulas are then placed on the appropriate sides of the reaction arrow. The equation is then balanced.
Solution:

a) We are told that nitroglycerine is a liquid reactant, and that all the products are gases. The formula for nitroglycerine is given. Carbon dioxide is CO_2; water is H_2O; nitrogen is N_2; and oxygen is O_2. The initial equation is:

$$C_3H_5N_3O_9(l) \rightarrow CO_2(g) + H_2O(g) + N_2(g) + O_2(g)$$

Counting the atoms shows no atoms are balanced.
One element should be picked and balanced. Any element except oxygen will work. Oxygen will not work in this case because it appears more than once on one side of the reaction arrow. We will start with carbon. Balancing C gives:

$$C_3H_5N_3O_9(l) \rightarrow 3CO_2(g) + H_2O(g) + N_2(g) + O_2(g)$$

Now balancing the hydrogen gives:

$$C_3H_5N_3O_9(l) \rightarrow 3CO_2(g) + 5/2H_2O(g) + N_2(g) + O_2(g)$$

Similarly, if we balance N we get:

$$C_3H_5N_3O_9(l) \rightarrow 3CO_2(g) + 5/2H_2O(g) + 3/2N_2(g) + O_2(g)$$

Clear the fractions by multiplying everything except the unbalanced oxygen by 2:

$$2C_3H_5N_3O_9(l) \rightarrow 6CO_2(g) + 5H_2O(g) + 3N_2(g) + O_2(g)$$

This leaves oxygen to balance. Balancing oxygen gives:

$$2C_3H_5N_3O_9(l) \rightarrow 6CO_2(g) + 5H_2O(g) + 3N_2(g) + 1/2O_2(g)$$

Again clearing fractions by multiplying everything by 2 gives:

$$\mathbf{4C_3H_5N_3O_9(l) \rightarrow 12CO_2(g) + 10H_2O(g) + 6N_2(g) + O_2(g)}$$

Now all the elements are balanced.

b) Potassium superoxide (KO_2) is a solid. Carbon dioxide (CO_2) and oxygen (O_2) are gases. Potassium carbonate (K_2CO_3) is a solid. The initial equation is:

$$KO_2(s) + CO_2(g) \rightarrow O_2(g) + K_2CO_3(s)$$

Counting the atoms indicates that the carbons are balanced, but none of the other atoms are balanced. One element should be picked and balanced. Any element except oxygen will work (oxygen will be more challenging to balance because it appears more than once on each side of the reaction arrow). Because the carbons are balanced, we will start with potassium. Balancing potassium gives:

$$2KO_2(s) + CO_2(g) \rightarrow O_2(g) + K_2CO_3(s)$$

Now all elements except for oxygen are balanced. Balancing oxygen by adding a coefficient in front of the O_2 gives:

$$2KO_2(s) + CO_2(g) \rightarrow 3/2O_2(g) + K_2CO_3(s)$$

Clearing the fractions by multiplying everything by 2 gives:

$$\mathbf{4KO_2(s) + 2CO_2(g) \rightarrow 3O_2(g) + 2K_2CO_3(s)}$$

Now all the elements are balanced.

c) Iron(III) oxide (Fe_2O_3) is a solid, as is iron metal (Fe). Carbon monoxide (CO) and carbon dioxide (CO_2) are gases. The initial equation is:

$$Fe_2O_3(s) + CO(g) \rightarrow Fe(s) + CO_2(g)$$

Counting the atoms indicates that the carbons are balanced, but none of the other atoms are balanced. One element should be picked and balanced. Because oxygen appears in more than one compound on one side of the reaction arrow, it is best not to start with that element. Because the carbons are balanced, we will start with iron. Balancing iron gives:

$$Fe_2O_3(s) + CO(g) \rightarrow 2Fe(s) + CO_2(g)$$

Now all the atoms but oxygen are balanced. There are 4 oxygen atoms on the left hand side of the reaction arrow and 2 oxygen atoms on the right hand side of the reaction arrow. In order to balance the oxygen, we want to change the coefficients in front of the carbon-containing compounds (if we changed the coefficient in front of the iron(III) oxide, the iron atoms would no longer be balanced). To maintain the balance of carbons, the coefficients in front of the carbon monoxide and the carbon dioxide must be the same. On the left hand side of the equation, there are 3 oxygens in Fe_2O_3 plus 1X oxygen atoms from the CO (where X is the coefficient in the balanced equation). On the right hand side of the equation, there are 2X oxygen atoms. The number of oxygen atoms on both sides of the equation should be the same:

$$3 + 1X = 2X$$
$$3 = X$$

Balancing oxygen by adding a coefficient of 3 in front of the CO and CO_2 gives:

$$\mathbf{Fe_2O_3(s) + 3CO(g) \rightarrow 2Fe(s) + 3CO_2(g)}$$

Now all the elements are balanced.

3.13A Plan: Count the number of each type of atom in each molecule to write the formulas of the reactants and-products.
Solution:

$$6CO(g) + 3O_2(g) \rightarrow 6CO_2(g)$$

or, $$\mathbf{2CO(g) + O_2(g) \rightarrow 2CO_2(g)}$$

3.13B Plan: Count the number of each type of atom in each molecule to write the formulas of the reactants and-products.
Solution:

$$6H_2(g) + 2N_2(g) \rightarrow 4NH_3(g)$$

or, $$\mathbf{3H_2(g) + N_2(g) \rightarrow 2NH_3(g)}$$

3.14A Plan: The reaction, like all reactions, needs a balanced chemical equation. The balanced equation gives the molar ratio between the moles of iron and moles of iron(III) oxide.
Solution:
The names and formulas of the substances involved are: iron(III) oxide, Fe_2O_3, and aluminum, Al, as reactants, and aluminum oxide, Al_2O_3, and iron, Fe, as products. The iron is formed as a liquid; all other substances are solids. The equation begins as:

$$Fe_2O_3(s) + Al(s) \rightarrow Al_2O_3(s) + Fe(l)$$

There are 2 Fe, 3 O, and 1 Al on the reactant side and 1 Fe, 3 O, and 2 Al on the product side.

Balancing aluminum: $$Fe_2O_3(s) + 2Al(s) \rightarrow Al_2O_3(s) + Fe(l)$$

Balancing iron: $$Fe_2O_3(s) + 2Al(s) \rightarrow Al_2O_3(s) + 2Fe(l)$$

$$\text{Moles of } Fe_2O_3 = (3.60\text{x}10^3 \text{ mol Fe})\left(\frac{1 \text{ mol } Fe_2O_3}{2 \text{ mol Fe}}\right) = \mathbf{1.80\text{x}10^3 \text{ mol } Fe_2O_3}$$

Road map:

Amount (moles) of Fe

Molar ratio
(2 mol Fe = 1 mol Fe_2O_3)

Amount (moles) of Fe_2O_3

3.14B Plan: The reaction, like all reactions, needs a balanced chemical equation. The balanced equation gives the molar ratio between the moles of aluminum and moles of silver sulfide.
Solution:
The names and formulas of the substances involved are: silver sulfide, Ag_2S, and aluminum, Al, as reactants; and aluminum sulfide, Al_2S_3, and silver, Ag, as products. All reactants and compounds are solids. The equation begins as:

$$Ag_2S(s) + Al(s) \rightarrow Al_2S_3(s) + Ag(s)$$

There are 2 Ag, 1S, and 1 Al on the reactant side and 2Al, 3 S, and 1 Ag on the product side.

Balancing sulfur: $$3Ag_2S(s) + Al(s) \rightarrow Al_2S_3(s) + Ag(s)$$

Balancing silver: $$3Ag_2S(s) + Al(s) \rightarrow Al_2S_3(s) + 6Ag(s)$$

Balancing aluminum: $$3Ag_2S(s) + 2Al(s) \rightarrow Al_2S_3(s) + 6Ag(s)$$

$$\text{Moles of Al} = 0.253 \text{ mol } Ag_2S \left(\frac{2 \text{ mol Al}}{3 \text{ mol } Ag_2S}\right) = 0.1687 = \mathbf{0.169 \text{ mol Al}}$$

Road map:

```
┌─────────────────────────────┐
│ Amount (moles) of Ag₂S      │
└─────────────────────────────┘
   │
   │  Molar ratio
   │  (3 mol Ag₂S = 2 mol Al)
   ▼
┌─────────────────────────────┐
│ Amount (moles) of Al        │
└─────────────────────────────┘
```

3.15A Plan: Divide the formula units of aluminum oxide by Avogadro's number to obtain moles of compound. The balanced equation gives the molar ratio between moles of iron(III) oxide and moles of iron.

<u>Solution:</u>

$$\text{Moles of Fe} = \left(1.85 \times 10^{25} \text{ Fe}_2\text{O}_3 \text{ formula units}\right)\left(\frac{1 \text{ mol Fe}_2\text{O}_3}{6.022 \times 10^{23} \text{ Fe}_2\text{O}_3 \text{ formula units}}\right)\left(\frac{2 \text{ mol Fe}}{1 \text{ mol Fe}_2\text{O}_3}\right)$$

$$61.4414 = \textbf{61.4 mol Fe}$$

Road map:

```
┌─────────────────────────────────┐
│ No. of Fe₂O₃ formula units      │
└─────────────────────────────────┘
   │
   │  Divide by Avogadro's number
   │  (6.022 x 10²³ Fe₂O₃ formula units = 1 mol Fe₂O₃)
   ▼
┌─────────────────────────────────┐
│ Amount (moles) of Fe₂O₃         │
└─────────────────────────────────┘
   │
   │  Molar ratio
   │  (1 mol Fe₂O₃ = 2 mol Fe)
   ▼
┌─────────────────────────────────┐
│ Amount (moles) of Fe            │
└─────────────────────────────────┘
```

3.15B Plan: Divide the mass of silver sulfide by its molar mass to obtain moles of the compound. The balanced equation gives the molar ratio between moles of silver sulfide and moles of silver.

<u>Solution:</u>

$$\text{Moles of Ag} = 32.6 \text{ g Ag}_2\text{S}\left(\frac{1 \text{ mol Ag}_2\text{S}}{247.9 \text{ g Ag}_2\text{S}}\right)\left(\frac{6 \text{ mol Ag}}{3 \text{ mol Ag}_2\text{S}}\right) = 0.2630 = \textbf{0.263 mol Ag}$$

Road map:

```
┌─────────────────────────────┐
│ Mass (g) of Ag₂S            │
└─────────────────────────────┘
   │
   │  Divide by ℳ (g/mol)
   │  (247.9 g Ag₂S = 1 mol Ag₂S)
   ▼
┌─────────────────────────────┐
│ Amount (moles) of Ag₂S      │
└─────────────────────────────┘
   │
   │  Molar ratio
   │  (3 mol Ag₂S = 6 mol Ag)
   ▼
┌─────────────────────────────┐
│ Amount (moles) of Ag        │
└─────────────────────────────┘
```

3.16A   Plan: The mass of aluminum oxide must be converted to moles by dividing by its molar mass. The balanced chemical equation (follow-up problem 3.14A) shows there are two moles of aluminum for every mole of aluminum oxide. Multiply the moles of aluminum by Avogadro's number to obtain atoms of Al.
Solution:

$$\text{Atoms of Al} = \left(1.00 \text{ g Al}_2\text{O}_3\right)\left(\frac{1 \text{ mol Al}_2\text{O}_3}{101.96 \text{ g Al}_2\text{O}_3}\right)\left(\frac{2 \text{ mol Al}}{1 \text{ mol Al}_2\text{O}_3}\right)\left(\frac{6.022 \times 10^{23} \text{ atoms Al}}{1 \text{ mol Al}}\right)$$

$$= 1.18125 \times 10^{22} = \textbf{1.18} \times \textbf{10}^{\textbf{22}} \textbf{ atoms Al}$$

Road map:

Mass (g) of Al_2O_3

Divide by $\mathcal{M}$ (g/mol)

Amount (moles) of Al_2O_3

Molar ratio

Amount (moles) of Al

Multiply by Avogadro's number

Number of Al atoms

3.16B   Plan: The mass of aluminum sulfide must be converted to moles by dividing by its molar mass. The balanced chemical equation (follow-up problem 3.14B) shows there are two moles of aluminum for every mole of aluminum sulfide. Multiply the moles of aluminum by its molar mass to obtain the mass (g) of aluminum.
Solution:

$$\text{Mass (g) of Al} = 12.1 \text{ g Al}_2\text{S}_3 \left(\frac{1 \text{ mol Al}_2\text{S}_3}{150.14 \text{ g Al}_2\text{S}_3}\right)\left(\frac{2 \text{ mol Al}}{1 \text{ mol Al}_2\text{S}_3}\right)\left(\frac{26.98 \text{ g Al}}{1 \text{ mol Al}}\right) = 4.3487 = \textbf{4.35 g Al}$$

Road map:

Mass (g) of Al_2S_3

Divide by $\mathcal{M}$ (g/mol)
(150.14 g Al_2S_3 = 1 mol Al_2S_3)

Amount (moles) of Al_2S_3

Molar ratio
(1 mol Al_2S_3 = 2 mol Al)

Amount (moles) of Al

Multiply by $\mathcal{M}$ (g/mol)
(1 mol Al = 26.98 g Al)

Mass (g) of Al

$$\text{Moles of N atoms} = \left(1.44575\text{x}10^{-3} \text{ mol NH}_4\text{NO}_2\right)\left(\frac{2 \text{ mol N atoms}}{1 \text{ mol NH}_4\text{NO}_2}\right) = 2.8915 \times 10^{-3} \text{ mol N atoms}$$

$$\text{Number of N atoms} = \left(2.8915\text{x}10^{-3} \text{ mol N atoms}\right)\frac{\left(6.022 \times 10^{23} \text{ N atoms}\right)}{1 \text{ mol N atoms}}$$

$$= 1.74126\text{x}10^{21} = \mathbf{1.74x10^{21}\,N\ atoms}$$

3.18   Plan: The formula of each compound must be determined from its name. The molar mass for each formula comes from the formula and atomic masses from the periodic table. Determine the molar mass of each substance, then perform the appropriate molar conversions. In part a), multiply the moles by the molar mass of the compound to find the mass of the sample. In part b), divide the number of molecules by Avogadro's number to find moles; multiply the number of moles by the molar mass to obtain the mass. In part c), divide the mass by the molar mass to find moles of compound and multiply moles by Avogadro's number to find the number of formula units. In part d), use the fact that each formula unit contains 1 Na ion, 1 perchlorate ion, 1 Cl atom, and 4 O atoms.
Solution:

a) Carbonate is a polyatomic anion with the formula, CO_3^{2-}. Copper(I) indicates Cu^+. The correct formula for this ionic compound is Cu_2CO_3.

$\mathcal{M}$ of Cu_2CO_3 = (2 x $\mathcal{M}$ of Cu) + (1 x $\mathcal{M}$ of C) + (3 x $\mathcal{M}$ of O)

$\quad\quad$ = (2 x 63.55 g/mol Cu) + (1 x 12.01 g/mol C) + (3 x 16.00 g/mol O) = 187.11 g/mol of Cu_2CO_3

$$\text{Mass (g) of Cu}_2\text{CO}_3 = \left(8.35 \text{ mol Cu}_2\text{CO}_3\right)\left(\frac{187.11 \text{ g Cu}_2\text{CO}_3}{1 \text{ mol Cu}_2\text{CO}_3}\right) = 1562.4 = \mathbf{1.56\ x\ 10^3\ g\ Cu_2CO_3}$$

b) Dinitrogen pentaoxide has the formula N_2O_5. Di- indicates 2 N atoms and penta- indicates 5 O atoms.

$\mathcal{M}$ of N_2O_5 = (2 x $\mathcal{M}$ of N) + (5 x $\mathcal{M}$ of O)

$\quad\quad$ = (2 x 14.01 g/mol N) + (5 x 16.00 g/mol O) = 108.02 g/mol of N_2O_5

$$\text{Moles of N}_2\text{O}_5 = \left(4.04\text{x}10^{20} \text{ N}_2\text{O}_5 \text{ molecules}\right)\left(\frac{1 \text{ mol N}_2\text{O}_5}{6.022\text{x}10^{23} \text{ N}_2\text{O}_5 \text{ molecules}}\right) = 6.7087 \times 10^{-4} \text{ mol N}_2\text{O}_5$$

$$\text{Mass (g) of N}_2\text{O}_5 = \left(6.7087\text{x}10^{-4} \text{ mol N}_2\text{O}_5\right)\left(\frac{108.02 \text{ g N}_2\text{O}_5}{1 \text{ mol N}_2\text{O}_5}\right) = 0.072467 = \mathbf{0.0725\ g\ N_2O_5}$$

c) The correct formula for this ionic compound is $NaClO_4$; Na has a charge of +1 (Group 1 ion) and the perchlorate ion is ClO_4^-.

$\mathcal{M}$ of $NaClO_4$ = (1x $\mathcal{M}$ of Na) + (1 x $\mathcal{M}$ of Cl) + (4 x $\mathcal{M}$ of O)

$\quad\quad$ = (1 x 22.99 g/mol Na) + (1 x 35.45 g/mol Cl) + (4 x 16.00 g/mol O) = 122.44 g/mol of $NaClO_4$

$$\text{Moles of NaClO}_4 = \left(78.9 \text{ g NaClO}_4\right)\left(\frac{1 \text{ mol NaClO}_4}{122.44 \text{ g NaClO}_4}\right) = 0.644397 = \mathbf{0.644\ mol\ NaClO_4}$$

FU = formula units

$$\text{FU of NaClO}_4 = \left(0.644397 \text{ mol NaClO}_4\right)\left(\frac{6.022 \times 10^{23} \text{ FU NaClO}_4}{1 \text{ mol NaClO}_4}\right)$$

$$= 3.88056\text{x}10^{23} = \mathbf{3.88x10^{23}\ FU\ NaClO_4}$$

d) Number of Na^+ ions = $\left(3.88056\text{x}10^{23} \text{ FU NaClO}_4\right)\left(\dfrac{1 \text{ Na}^+ \text{ ion}}{1 \text{ FU NaClO}_4}\right) = \mathbf{3.88x10^{23}\ Na^+\ ions}$

Number of ClO_4^- ions = $\left(3.88056\text{x}10^{23} \text{ FU NaClO}_4\right)\left(\dfrac{1 \text{ ClO}_4^- \text{ ion}}{1 \text{ FU NaClO}_4}\right) = \mathbf{3.88x10^{23}\ ClO_4^-\ ions}$

$$\text{Number of Cl atoms} = \left(3.88056\text{x}10^{23} \text{ FU NaClO}_4\right)\left(\frac{1 \text{ Cl atom}}{1 \text{ FU NaClO}_4}\right) = \mathbf{3.88\text{x}10^{23} \text{ Cl atoms}}$$

$$\text{Number of O atoms} = \left(3.88056\text{x}10^{23} \text{ FU NaClO}_4\right)\frac{4 \text{ O atoms}}{1 \text{ FU NaClO}_4} = \mathbf{1.55\text{x}10^{24} \text{ O atoms}}$$

3.20   Plan: Determine the formula and the molar mass of each compound. The formula gives the relative number of moles of each element present. Multiply the number of moles of each element by its molar mass to find the total mass of element in 1 mole of compound. Mass percent = $\dfrac{\text{total mass of element}}{\text{molar mass of compound}}(100)$.

Solution:

a) Ammonium bicarbonate is an ionic compound consisting of ammonium ions, NH_4^+ and bicarbonate ions, HCO_3^-. The formula of the compound is NH_4HCO_3.

$\mathcal{M}$ of NH_4HCO_3 = (1 x $\mathcal{M}$ of N) + (5 x $\mathcal{M}$ of H) + (1 x $\mathcal{M}$ of C) + (3 x $\mathcal{M}$ of O)
        = (1 x 14.01 g/mol N) + (5 x 1.008 g/mol H) + (1 x 12.01 g/mol C) + (3 x 16.00 g/mol O)
        = 79.06 g/mol of NH_4HCO_3
There are 5 moles of H in 1 mole of NH_4HCO_3.

$$\text{Mass (g) of H} = (5 \text{ mol H})\left(\frac{1.008 \text{ g H}}{1 \text{ mol H}}\right) = 5.040 \text{ g H}$$

$$\text{Mass percent} = \frac{\text{total mass H}}{\text{molar mass of compound}}(100) = \frac{5.040 \text{ g H}}{79.06 \text{ g } NH_4HCO_3}(100) = 6.374905 = \mathbf{6.375\% \text{ H}}$$

b) Sodium dihydrogen phosphate heptahydrate is a salt that consists of sodium ions, Na^+, dihydrogen phosphate ions, $H_2PO_4^-$, and seven waters of hydration. The formula is $NaH_2PO_4 \bullet 7H_2O$. Note that the waters of hydration are included in the molar mass.

$\mathcal{M}$ of $NaH_2PO_4 \bullet 7H_2O$ = (1 x $\mathcal{M}$ of Na) + (16 x $\mathcal{M}$ of H) + (1 x $\mathcal{M}$ of P) + (11 x $\mathcal{M}$ of O)
        = (1 x 22.99 g/mol Na) + (16 x 1.008 g/mol H) + (1 x 30.97 g/mol P) + (11 x 16.00 g/mol O)
        = 246.09 g/mol $NaH_2PO_4 \bullet 7H_2O$
There are 11 moles of O in 1 mole of $NaH_2PO_4 \bullet 7H_2O$.

$$\text{Mass (g) of O} = (11 \text{ mol O})\left(\frac{16.00 \text{ g O}}{1 \text{ mol O}}\right) = 176.00 \text{ g O}$$

$$\text{Mass percent} = \frac{\text{total mass O}}{\text{molar mass of compound}}(100) = \frac{176.00 \text{ g O}}{246.09 \text{ g } NaH_2PO_4 \bullet 7H_2O}(100)$$
$$= 71.51855 = \mathbf{71.52\% \text{ O}}$$

3.22   Plan: Determine the formula and the molar mass of each compound. The formula gives the relative number of moles of each element present. Multiply the number of moles of each element by its molar mass to find the total mass of element in 1 mole of compound. Mass fraction = $\dfrac{\text{total mass of element}}{\text{molar mass of compound}}$.

Solution:

a) Cesium acetate is an ionic compound consisting of Cs^+ cations and $C_2H_3O_2^-$ anions. (Note that the formula for acetate ions can be written as either $C_2H_3O_2^-$ or CH_3COO^-.) The formula of the compound is $CsC_2H_3O_2$.

$\mathcal{M}$ of $CsC_2H_3O_2$ = (1 x $\mathcal{M}$ of Cs) + (2 x $\mathcal{M}$ of C) + (3 x $\mathcal{M}$ of H) + (2 x $\mathcal{M}$ of O)
        = (1 x 132.9 g/mol Cs) + (2 x 12.01 g/mol C) + (3 x 1.008 g/mol H) + (2 x 16.00 g/mol O)
        = 191.9 g/mol of $CsC_2H_3O_2$
There are 2 moles of C in 1 mole of $CsC_2H_3O_2$.

$$\text{Mass (g) of C} = (2 \text{ mol C})\left(\frac{12.01 \text{ g C}}{1 \text{ mol C}}\right) = 24.02 \text{ g C}$$

$$\text{Mass fraction} = \frac{\text{total mass C}}{\text{molar mass of compound}} = \frac{24.02 \text{ g C}}{191.9 \text{ g CsC}_2\text{H}_3\text{O}_2} = 0.125169 = \textbf{0.1252 mass fraction C}$$

b) Uranyl sulfate trihydrate is a salt that consists of uranyl ions, UO_2^{2+}, sulfate ions, SO_4^{2-}, and three waters of hydration. The formula is $UO_2SO_4 \cdot 3H_2O$. Note that the waters of hydration are included in the molar mass.

$\mathcal{M}$ of $UO_2SO_4 \cdot 3H_2O$ = (1 x $\mathcal{M}$ of U) + (9 x $\mathcal{M}$ of O) + (1 x $\mathcal{M}$ of S) + (6 x $\mathcal{M}$ of H)
= (1 x 238.0 g/mol U) + (9 x 16.00 g/mol O) + (1 x 32.06 g/mol S) + (6 x 1.008 g/mol H)
= 420.1 g/mol of $UO_2SO_4 \cdot 3H_2O$

There are 9 moles of O in 1 mole of $UO_2SO_4 \cdot 3H_2O$.

$$\text{Mass (g) of O} = (9 \text{ mol O})\left(\frac{16.00 \text{ g O}}{1 \text{ mol O}}\right) = 144.0 \text{ g O}$$

$$\text{Mass fraction} = \frac{\text{total mass O}}{\text{molar mass of compound}} = \frac{144.0 \text{ g O}}{420.1 \text{ g } UO_2SO_4 \cdot 3H_2O} = 0.3427755 = \textbf{0.3428 mass fraction O}$$

3.25 **Plan:** Determine the formula of cisplatin from the figure, and then calculate the molar mass from the formula. Divide the mass given by the molar mass to find moles of cisplatin. Since 1 mole of cisplatin contains 6 moles of hydrogen atoms, multiply the moles given by 6 to obtain moles of hydrogen and then multiply by Avogadro's number to obtain the number of atoms.
Solution:
The formula for cisplatin is $Pt(Cl)_2(NH_3)_2$.

$\mathcal{M}$ of $Pt(Cl)_2(NH_3)_2$ = (1 x $\mathcal{M}$ of Pt) + (2 x $\mathcal{M}$ of Cl) + (2 x $\mathcal{M}$ of N) + (6 x $\mathcal{M}$ of H)
= (1 x 195.1 g/mol Pt) + (2 x 35.45 g/mol Cl) + (2 x 14.01 g/mol N) + (6 x 1.008 g/mol H)
= 300.1 g/mol of $Pt(Cl)_2(NH_3)_2$

$$\text{a) Moles of cisplatin} = (285.3 \text{ g cisplatin})\left(\frac{1 \text{ mol cisplatin}}{300.1 \text{ g cisplatin}}\right) = 0.9506831 = \textbf{0.9507 mol cisplatin}$$

$$\text{b) Moles of H atoms} = (0.98 \text{ mol cisplatin})\left(\frac{6 \text{ mol H}}{1 \text{ mol cisplatin}}\right) = 5.88 \text{ mol H atoms}$$

$$\text{Number of H atoms} = (5.88 \text{ mol H atoms})\left(\frac{6.022 \times 10^{23} \text{ H atoms}}{1 \text{ mol H atoms}}\right) = 3.540936 \times 10^{24} = \textbf{3.5} \times \textbf{10}^{24} \textbf{ H atoms}$$

3.27 **Plan:** Determine the molar mass of rust. Convert mass in kg to mass in g and divide by the molar mass to find the moles of rust. Since each mole of rust contains 1 mole of Fe_2O_3, multiply the moles of rust by 1 to obtain moles of Fe_2O_3. Multiply the moles of Fe_2O_3 by 2 to obtain moles of Fe (1:2 Fe_2O_3:Fe mole ratio) and multiply by the molar mass of Fe to convert to mass.
Solution:
a) $\mathcal{M}$ of $Fe_2O_3 \cdot 4H_2O$ = (2 x $\mathcal{M}$ of Fe) + (7 x $\mathcal{M}$ of O) + (8 x $\mathcal{M}$ of H)
= (2 x 55.85 g/mol Fe) + (7 x 16.00 g/mol O) + (8 x 1.008 g/mol H) = 231.76 g/mol

$$\text{Mass (g) of rust} = (45.2 \text{ kg rust})\left(\frac{10^3 \text{ g}}{1 \text{ kg}}\right) = 4.52 \times 10^4 \text{ g}$$

$$\text{Moles of rust} = (4.52 \times 10^4 \text{ g rust})\left(\frac{1 \text{ mol rust}}{231.76 \text{ g rust}}\right) = 195.029 = \textbf{195 mol rust}$$

b) The formula shows that there is 1 mole of Fe_2O_3 for every mole of rust, so there are also **195 mol of Fe_2O_3**.

$$\text{c) Moles of iron} = (195.029 \text{ mol } Fe_2O_3)\left(\frac{2 \text{ mol Fe}}{1 \text{ mol } Fe_2O_3}\right) = 390.058 \text{ mol Fe}$$

$$\text{Mass (g) of iron} = \left(390.058 \text{ mol Fe}\right)\left(\frac{55.85 \text{ g Fe}}{1 \text{ mol Fe}}\right) = 21784.74 = \textbf{2.18x10}^4 \text{ g Fe}$$

3.29   Plan: Determine the formula and the molar mass of each compound. The formula gives the relative number of moles of nitrogen present. Multiply the number of moles of nitrogen by its molar mass to find the total mass of nitrogen in 1 mole of compound. Divide the total mass of nitrogen by the molar mass of compound and multiply by 100 to determine mass percent. Mass percent = $\dfrac{(\text{mol N}) \times (\text{molar mass N})}{\text{molar mass of compound}}(100)$. Then rank the values in order of decreasing mass percent N.

Solution:

Name	Formula	Molar Mass (g/mol)
Potassium nitrate	KNO_3	101.11
Ammonium nitrate	NH_4NO_3	80.05
Ammonium sulfate	$(NH_4)_2SO_4$	132.14
Urea	$CO(NH_2)_2$	60.06

$$\text{Mass \% N in potassium nitrate} = \frac{\left(1 \text{ mol N}\right)\left(14.01 \text{ g/mol N}\right)}{101.11 \text{ g/mol}} \times 100 = 13.856196 = \textbf{13.86\% N}$$

$$\text{Mass \% N in ammonium nitrate} = \frac{\left(2 \text{ mol N}\right)\left(14.01 \text{ g/mol N}\right)}{80.05 \text{ g/mol}} \times 100 = 35.003123 = \textbf{35.00\% N}$$

$$\text{Mass \% N in ammonium sulfate} = \frac{\left(2 \text{ mol N}\right)\left(14.01 \text{ g/mol N}\right)}{132.14 \text{ g/mol}} \times 100 = 21.20478 = \textbf{21.20\% N}$$

$$\text{Mass \% N in urea} = \frac{\left(2 \text{ mol N}\right)\left(14.01 \text{ g/mol N}\right)}{60.06 \text{ g/mol}} \times 100 = 46.6533 = \textbf{46.65\% N}$$

Rank is $\textbf{CO(NH}_2\textbf{)}_2 > \textbf{NH}_4\textbf{NO}_3 > \textbf{(NH}_4\textbf{)}_2\textbf{SO}_4 > \textbf{KNO}_3$

3.30   Plan: The volume must be converted from cubic feet to cubic centimeters. The volume and the density will give the mass of galena which is then divided by molar mass to obtain moles. Part b) requires a conversion from cubic decimeters to cubic centimeters. The density allows a change from volume in cubic centimeters to mass which is then divided by the molar mass to obtain moles; the amount in moles is multiplied by Avogadro's number to obtain formula units of PbS which is also the number of Pb atoms due to the 1:1 PbS:Pb mole ratio.

Solution:

Lead(II) sulfide is composed of Pb^{2+} and S^{2-} ions and has a formula of PbS.

$\mathcal{M}$ of PbS = (1 x $\mathcal{M}$ of Pb) + (1 x $\mathcal{M}$ of S) = (1 x 207.2 g/mol Pb) + (1 x 32.06 g/mol S) = 239.3 g/mol

$$\text{a) Volume (cm}^3) = \left(1.00 \text{ ft}^3 \text{ PbS}\right)\left(\frac{(12 \text{ in})^3}{(1 \text{ ft})^3}\right)\left(\frac{(2.54 \text{ cm})^3}{(1 \text{ in})^3}\right) = 28316.85 \text{ cm}^3$$

$$\text{Mass (g) of PbS} = \left(28316.85 \text{ cm}^3 \text{ PbS}\right)\left(\frac{7.46 \text{ g PbS}}{1 \text{ cm}^3}\right) = 211243.7 \text{ g PbS}$$

$$\text{Moles of PbS} = \left(211243.7 \text{ g PbS}\right)\left(\frac{1 \text{ mol PbS}}{239.3 \text{ g PbS}}\right) = 882.7568 = \textbf{883 mol PbS}$$

$$\text{b) Volume (cm}^3) = \left(1.00 \text{ dm}^3 \text{ PbS}\right)\left(\frac{(0.1 \text{ m})^3}{(1 \text{ dm})^3}\right)\left(\frac{(1 \text{ cm})^3}{(10^{-2} \text{ m})^3}\right) = 1.00\text{x}10^3 \text{ cm}^3$$

$$\text{Mass (g) of PbS} = \left(1.00 \text{x} 10^3 \text{ cm}^3 \text{ PbS}\right)\left(\frac{7.46 \text{ g PbS}}{1 \text{ cm}^3}\right) = 7460 \text{ g PbS}$$

$$\text{Moles of PbS} = \left(7460 \text{ g PbS}\right)\left(\frac{1 \text{ mol PbS}}{239.3 \text{ g PbS}}\right) = 31.17426 \text{ mol PbS}$$

$$\text{Moles of Pb} = \left(31.17426 \text{ mol PbS}\right)\left(\frac{1 \text{ mol Pb}}{1 \text{ mol PbS}}\right) = 31.17426 \text{ mol Pb}$$

$$\text{Number of lead atoms} = \left(31.17426 \text{ mol Pb}\right)\left(\frac{6.022 \text{x} 10^{23} \text{ Pb atoms}}{1 \text{ mol Pb}}\right) = 1.87731 \text{x} 10^{25} = \mathbf{1.88 \text{x} 10^{25} \text{ Pb atoms}}$$

3.34   Plan: Remember that the molecular formula tells the *actual* number of moles of each element in one mole of compound.
Solution:
a) No, this information does not allow you to obtain the molecular formula. You can obtain the empirical formula from the number of moles of each type of atom in a compound, but not the molecular formula.
b) Yes, you can obtain the molecular formula from the mass percentages and the total number of atoms.
Plan:

1) Assume a 100.0 g sample and convert masses (from the mass % of each element) to moles using molar mass.
2) Identify the element with the lowest number of moles and use this number to divide into the number of moles for each element. You now have at least one elemental mole ratio (the one with the smallest number of moles) equal to 1.00 and the remaining mole ratios that are larger than one.
3) Examine the numbers to determine if they are whole numbers. If not, multiply each number by a whole-number factor to get whole numbers for each element. You will have to use some judgment to decide when to round. Write the empirical formula using these whole numbers.
4) Check the total number of atoms in the empirical formula. If it equals the total number of atoms given then the empirical formula is also the molecular formula. If not, then divide the total number of atoms given by the total number of atoms in the empirical formula. This should give a whole number. Multiply the number of atoms of each element in the empirical formula by this whole number to get the molecular formula. If you do not get a whole number when you divide, return to step 3 and revise how you multiplied and rounded to get whole numbers for each element.

Roadmap:

Mass (g) of each element (express mass percent directly as grams)

Divide by $\mathcal{M}$ (g/mol)

Amount (mol) of each element

Use numbers of moles as subscripts

Preliminary empirical formula

Change to integer subscripts

Empirical formula

Divide total number of atoms in molecule by the number of atoms in the empirical formula and multiply the empirical formula by that factor

Molecular formula

c) Yes, you can determine the molecular formula from the mass percent and the number of atoms of one element in a compound. Plan:
1) Follow steps 1–3 in part b).
2) Compare the number of atoms given for the one element to the number in the empirical formula. Determine the factor the number in the empirical formula must be multiplied by to obtain the given number of atoms for that element. Multiply the empirical formula by this number to get the molecular formula.

Roadmap:
(Same first three steps as in b).

Empirical formula

Divide the number of atoms of the one element in the molecule by the number of atoms of that element in the empirical formula and multiply the empirical formula by that factor

Molecular formula

d) No, the mass % will only lead to the empirical formula.
e) Yes, a structural formula shows all the atoms in the compound. Plan: Count the number of atoms of each type of element and record as the number for the molecular formula.
Roadmap:

Structural formula

Count the number of atoms of each element and use these numbers as subscripts

Molecular formula

3.36   Plan: Examine the number of atoms of each type in the compound. Divide all atom numbers by the common factor that results in the lowest whole-number values. Add the molar masses of the atoms to obtain the empirical formula mass.
Solution:
a) C_2H_4 has a ratio of 2 carbon atoms to 4 hydrogen atoms, or 2:4. This ratio can be reduced to 1:2, so that the empirical formula is CH_2. The empirical formula mass is 12.01 g/mol C + 2(1.008 g/mol H) = **14.03 g/mol**.
b) The ratio of atoms is 2:6:2, or 1:3:1. The empirical formula is CH_3O and its empirical formula mass is 12.01 g/mol C + 3(1.008 g/mol H) + 16.00 g/mol O = **31.03 g/mol**.
c) Since, the ratio of elements cannot be further reduced, the molecular formula and empirical formula are the same, N_2O_5. The formula mass is 2(14.01 g/mol N) + 5(16.00 g/mol O) = **108.02 g/mol**.
d) The ratio of elements is 3 atoms of barium to 2 atoms of phosphorus to 8 atoms of oxygen, or 3:2:8. This ratio cannot be further reduced, so the empirical formula is also $Ba_3(PO_4)_2$, with a formula mass of 3(137.3 g/mol Ba) + 2(30.97 g/mol P) + 8(16.00 g/mol O) = **601.8 g/mol**.
e) The ratio of atoms is 4:16, or 1:4. The empirical formula is TeI_4, and the formula mass is 127.6 g/mol Te + 4(126.9 g/mol I) = **635.2 g/mol**.

3.38   Plan: Use the chemical symbols and count the atoms of each type to obtain the molecular formula. Divide the molecular formula by the largest common factor to give the empirical formula. Use nomenclature rules to derive the name. This compound is composed of two nonmetals. The naming rules for binary covalent compounds indicate that the element with the lower group number is named first. Greek numerical prefixes are used to indicate the number of atoms of each element in the compound. The molecular (formula) mass is the sum of the atomic masses of all of the atoms.

Solution:

The compound has 2 sulfur atoms and 2 chlorine atoms and a molecular formula of S_2Cl_2. The compound's name is **disulfur dichloride**. Sulfur is named first since it has the lower group number. The prefix di- is used for both elements since there are 2 atoms of each element. The empirical formula is $S_{\frac{2}{2}}Cl_{\frac{2}{2}}$ or **SCl**.

$\mathcal{M}$ of S_2Cl_2 = (2 x $\mathcal{M}$ of S) + (2 x $\mathcal{M}$ of Cl) = (2 x 32.06 g/mol S) + (2 x 35.45 g/mol Cl) = **135.02 g/mol**

3.40   Plan: Determine the molar mass of each empirical formula. The subscripts in the molecular formula are whole-number multiples of the subscripts in the empirical formula. To find this whole number, divide the molar mass of the compound by its empirical formula mass. Multiply each subscript in the empirical formula by the whole number.
Solution:

Only approximate whole-number values are needed.

a) CH_2 has empirical mass equal to 12.01 g/mol C + 2(1.008 g/mol C) = 14.03 g/mol

$$\text{Whole-number multiple} = \frac{\text{molar mass of compound}}{\text{empirical formula mass}} = \left(\frac{42.08 \text{ g/mol}}{14.03 \text{ g/mol}}\right) = 3$$

Multiplying the subscripts in CH_2 by 3 gives $\mathbf{C_3H_6}$.

b) NH_2 has empirical mass equal to 14.01 g/mol N + 2(1.008 g/mol H) = 16.03 g/mol

$$\text{Whole-number multiple} = \frac{\text{molar mass of compound}}{\text{empirical formula mass}} = \left(\frac{32.05 \text{ g/mol}}{16.03 \text{ g/mol}}\right) = 2$$

Multiplying the subscripts in NH_2 by 2 gives $\mathbf{N_2H_4}$.

c) NO_2 has empirical mass equal to 14.01 g/mol N + 2(16.00 g/mol O) = 46.01 g/mol

$$\text{Whole-number multiple} = \frac{\text{molar mass of compound}}{\text{empirical formula mass}} = \left(\frac{92.02 \text{ g/mol}}{46.01 \text{ g/mol}}\right) = 2$$

Multiplying the subscripts in NO_2 by 2 gives $\mathbf{N_2O_4}$.

d) CHN has empirical mass equal to 12.01 g/mol C + 1.008 g/mol H + 14.01 g/mol N = 27.03 g/mol

$$\text{Whole-number multiple} = \frac{\text{molar mass of compound}}{\text{empirical formula mass}} = \left(\frac{135.14 \text{ g/mol}}{27.03 \text{ g/mol}}\right) = 5$$

Multiplying the subscripts in CHN by 5 gives $\mathbf{C_5H_5N_5}$.

3.42   Plan: The empirical formula is the smallest whole-number ratio of the atoms or moles in a formula. All data must be converted to moles of an element by dividing mass by the molar mass. Divide each mole number by the smallest mole number to convert the mole ratios to whole numbers.
Solution:

a) 0.063 mol Cl and 0.22 mol O: preliminary formula is $Cl_{0.063}O_{0.22}$

Converting to integer subscripts (dividing all by the smallest subscript):

$$Cl_{\frac{0.063}{0.063}}O_{\frac{0.22}{0.063}} \rightarrow Cl_1O_{3.5}$$

The formula is $Cl_1O_{3.5}$, which in whole numbers (x 2) is $\mathbf{Cl_2O_7}$.

b) Find moles of elements by dividing by molar mass:

$$\text{Moles of Si} = \left(2.45 \text{ g Si}\right)\left(\frac{1 \text{ mol Si}}{28.09 \text{ g Si}}\right) = 0.08722 \text{ mol Si}$$

$$\text{Moles of Cl} = \left(12.4 \text{ g Cl}\right)\left(\frac{1 \text{ mol Cl}}{35.45 \text{ g Cl}}\right) = 0.349788 \text{ mol Cl}$$

Preliminary formula is $Si_{0.08722}Cl_{0.349788}$

Converting to integer subscripts (dividing all by the smallest subscript):

$$Si_{\frac{0.08722}{0.08722}}Cl_{\frac{0.349788}{0.08722}} \rightarrow Si_1Cl_4$$

The empirical formula is **$SiCl_4$**.

c) Assume a 100 g sample and convert the masses to moles by dividing by the molar mass:

$$\text{Moles of C} = \left(100 \text{ g}\right)\left(\frac{27.3 \text{ parts C by mass}}{100 \text{ parts by mass}}\right)\left(\frac{1 \text{ mol C}}{12.01 \text{ g C}}\right) = 2.2731 \text{ mol C}$$

$$\text{Moles of O} = \left(100 \text{ g}\right)\left(\frac{72.7 \text{ parts O by mass}}{100 \text{ parts by mass}}\right)\left(\frac{1 \text{ mol O}}{16.00 \text{ g O}}\right) = 4.5438 \text{ mol O}$$

Preliminary formula is $C_{2.2731}O_{4.5438}$

Converting to integer subscripts (dividing all by the smallest subscript):

$$C_{\frac{2.2731}{2.2731}}O_{\frac{4.5438}{2.2731}} \rightarrow C_1O_2$$

The empirical formula is **CO_2**.

3.44 Plan: The percent oxygen is 100% minus the percent nitrogen. Assume 100 grams of sample, and then the moles of each element may be found by dividing the mass of each element by its molar mass. Divide each of the moles by the smaller value, and convert to whole numbers to get the empirical formula. The subscripts in the molecular formula are whole-number multiples of the subscripts in the empirical formula. To find this whole number, divide the molar mass of the compound by its empirical formula mass. Multiply each subscript in the empirical formula by the whole number.
Solution:
a) % O = 100% − % N = 100% − 30.45% N = 69.55% O

Assume a 100 g sample and convert the masses to moles by dividing by the molar mass:

$$\text{Moles of N} = \left(100 \text{ g}\right)\left(\frac{30.45 \text{ parts N by mass}}{100 \text{ parts by mass}}\right)\left(\frac{1 \text{ mol N}}{14.01 \text{ g N}}\right) = 2.1734 \text{ mol N}$$

$$\text{Moles of O} = \left(100 \text{ g}\right)\left(\frac{69.55 \text{ parts O by mass}}{100 \text{ parts by mass}}\right)\left(\frac{1 \text{ mol O}}{16.00 \text{ g O}}\right) = 4.3469 \text{ mol O}$$

Preliminary formula is $N_{2.1734}O_{4.3469}$

Converting to integer subscripts (dividing all by the smallest subscript):

$$N_{\frac{2.1734}{2.1734}}O_{\frac{4.3469}{2.1734}} \rightarrow N_1O_2$$

The empirical formula is **NO_2**.

b) Formula mass of empirical formula = 14.01 g/mol N + 2(16.00 g/mol O) = 46.01 g/mol

$$\text{Whole-number multiple} = \frac{\text{molar mass of compound}}{\text{empirical formula mass}} = \left(\frac{90 \text{ g/mol}}{46.01 \text{ g/mol}}\right) = 2$$

Multiplying the subscripts in NO_2 by 2 gives **N_2O_4** as the molecular formula.

Note: Only an approximate value of the molar mass is needed.

3.46   Plan: The moles of the metal are known, and the moles of fluorine atoms may be found in part a) from the M:F mole ratio in the compound formula. In part b), convert moles of F atoms to mass and subtract the mass of F from the mass of MF_2 to find the mass of M. In part c), divide the mass of M by moles of M to determine the molar mass of M which can be used to identify the element.
Solution:
a) Determine the moles of fluorine.

$$\text{Moles of F} = \left(0.600 \text{ mol M}\right)\left(\frac{2 \text{ mol F}}{1 \text{ mol M}}\right) = \textbf{1.20 mol F}$$

b) Determine the mass of M.

$$\text{Mass of F} = \left(1.20 \text{ mol F}\right)\left(\frac{19.00 \text{ g F}}{1 \text{ mol F}}\right) = 22.8 \text{ g F}$$

$$\text{Mass (g) of M} = MF_2(g) - F(g) = 46.8 \text{ g} - 22.8 \text{ g} = \textbf{24.0 g M}$$

c) The molar mass is needed to identify the element.

$$\text{Molar mass of M} = \frac{24.0 \text{ g M}}{0.600 \text{ mol M}} = 40.0 \text{ g/mol}$$

The metal with the closest molar mass to 40.0 g/mol is **calcium**.

3.48   Plan: In combustion analysis, finding the moles of carbon and hydrogen is relatively simple because all of the carbon present in the sample is found in the carbon of CO_2, and all of the hydrogen present in the sample is found in the hydrogen of H_2O. Convert the mass of CO_2 to moles and use the ratio between CO_2 and C to find the moles and mass of C present. Do the same to find the moles and mass of H from H_2O. The moles of oxygen are more difficult to find, because additional O_2 was added to cause the combustion reaction. Subtracting the masses of C and H from the mass of the sample gives the mass of O. Convert the mass of O to moles of O. Take the moles of C, H, and O and divide by the smallest value to convert to whole numbers to get the empirical formula. Determine the empirical formula mass and compare it to the molar mass given in the problem to see how the empirical and molecular formulas are related. Finally, determine the molecular formula.
Solution:

$$\text{Moles of C} = \left(0.449 \text{ g CO}_2\right)\left(\frac{1 \text{ mol CO}_2}{44.01 \text{ g CO}_2}\right)\left(\frac{1 \text{ mol C}}{1 \text{ mol CO}_2}\right) = 0.010202 \text{ mol C}$$

$$\text{Mass (g) of C} = \left(0.010202 \text{ mol C}\right)\left(\frac{12.01 \text{ g C}}{1 \text{ mol C}}\right) = 0.122526 \text{ g C}$$

$$\text{Moles of H} = \left(0.184 \text{ g H}_2O\right)\left(\frac{1 \text{ mol H}_2O}{18.02 \text{ g H}_2O}\right)\left(\frac{2 \text{ mol H}}{1 \text{ mol H}_2O}\right) = 0.020422 \text{ mol H}$$

$$\text{Mass (g) of H} = \left(0.020422 \text{ mol H}\right)\left(\frac{1.008 \text{ g H}}{1 \text{ mol H}}\right) = 0.020585 \text{ g H}$$

Mass (g) of O = Sample mass – (mass of C + mass of H)
$$= 0.1595 \text{ g} - (0.122526 \text{ g C} + 0.020585 \text{ g H}) = 0.016389 \text{ g O}$$

$$\text{Moles of O} = \left(0.016389 \text{ g O}\right)\left(\frac{1 \text{ mol O}}{16.00 \text{ g O}}\right) = 0.0010243 \text{ mol O}$$

Preliminary formula = $C_{0.010202}H_{0.020422}O_{0.0010243}$

Converting to integer subscripts (dividing all by the smallest subscript):

$$C_{\frac{0.010202}{0.0010243}} H_{\frac{0.020422}{0.0010243}} O_{\frac{0.0010243}{0.0010243}} \rightarrow C_{10}H_{20}O_1$$

Empirical formula = $C_{10}H_{20}O$

Empirical formula mass = $10(12.01 \text{ g/mol C}) + 20(1.008 \text{ g/mol H}) + 1(16.00 \text{ g/mol O}) = 156.26$ g/mol

The empirical formula mass is the same as the given molar mass so the empirical and molecular formulas are the same. The molecular formula is $\mathbf{C_{10}H_{20}O}$.

3.51   Plan: The empirical formula is the smallest whole-number ratio of the atoms or moles in a formula. Assume 100 grams of cortisol so the percentages are numerically equivalent to the masses of each element. Convert each of the masses to moles by dividing by the molar mass of each element involved. Divide each mole number by the smallest mole number to convert the mole ratios to whole numbers. The subscripts in the molecular formula are whole-number multiples of the subscripts in the empirical formula. To find this whole number, divide the molar mass of the compound by its empirical formula mass. Multiply each subscript in the empirical formula by the whole number.
Solution:

$$\text{Moles of C} = \left(69.6 \text{ g C}\right)\left(\frac{1 \text{ mol C}}{12.01 \text{ g C}}\right) = 5.7952 \text{ mol C}$$

$$\text{Moles of H} = \left(8.34 \text{ g H}\right)\left(\frac{1 \text{ mol H}}{1.008 \text{ g H}}\right) = 8.2738 \text{ mol H}$$

$$\text{Moles of O} = \left(22.1 \text{ g O}\right)\left(\frac{1 \text{ mol O}}{16.00 \text{ g O}}\right) = 1.38125 \text{ mol O}$$

Preliminary formula is $C_{5.7952}H_{8.2738}O_{1.38125}$

Converting to integer subscripts (dividing all by the smallest subscript):

$$C_{\frac{5.7952}{1.38125}} H_{\frac{8.2738}{1.38125}} O_{\frac{1.38125}{1.38125}} \rightarrow C_{4.2}H_6O_1$$

The carbon value is not close enough to a whole number to round the value. The smallest number that 4.20 may be multiplied by to get close to a whole number is 5. (You may wish to prove this to yourself.) All three ratios need to be multiplied by five: $5(C_{4.2}H_6O_1) = C_{21}H_{30}O_5$.
The empirical formula mass is = $21(12.01 \text{ g/mol C}) + 30(1.008 \text{ g/mol H}) + 5(16.00 \text{ g/mol O}) = 362.45$ g/mol

$$\text{Whole-number multiple} = \frac{\text{molar mass of compound}}{\text{empirical formula mass}} = \left(\frac{362.47 \text{ g/mol}}{362.45 \text{ g/mol}}\right) = 1$$

The empirical formula mass and the molar mass given are the same, so the empirical and the molecular formulas are the same. The molecular formula is $\mathbf{C_{21}H_{30}O_5}$.

3.53   A balanced chemical equation describes:
1) The identities of the reactants and products.
2) The molar (and molecular) ratios by which reactants form products.
3) The physical states of all substances in the reaction.

3.56   Plan: Examine the diagram and label each formula. We will use A for red atoms and B for green atoms.
Solution:
The reaction shows A_2 and B_2 diatomic molecules forming AB molecules. Equal numbers of A_2 and B_2 combine to give twice as many molecules of AB. Thus, the reaction is $A_2 + B_2 \rightarrow 2 \text{ AB}$. This is the balanced equation in **b**.

3.58   Plan: Balancing is a trial-and-error procedure. Balance one element at a time, placing coefficients where needed to have the same number of atoms of a particular element on each side of the equation. The smallest whole-number coefficients should be used.

Solution:

a) __Cu(s) + __ S$_8$(s) → __Cu$_2$S(s)

Balance the S first, because there is an obvious deficiency of S on the right side of the equation. The 8 S atoms in S$_8$ require the coefficient 8 in front of Cu$_2$S:

__Cu(s) + __S$_8$(s) → 8Cu$_2$S(s)

Then balance the Cu. The 16 Cu atoms in Cu$_2$S require the coefficient 16 in front of Cu:

16Cu(s) + S$_8$(s) → 8Cu$_2$S(s)

b) __P$_4$O$_{10}$(s) + __H$_2$O(l) → __H$_3$PO$_4$(l)

Balance the P first, because there is an obvious deficiency of P on the right side of the equation. The 4 P atoms in P$_4$O$_{10}$ require a coefficient of 4 in front of H$_3$PO$_4$:

___P$_4$O$_{10}$(s) + __H$_2$O(l) → 4H$_3$PO$_4$(l)

Balance the H next, because H is present in only one reactant and only one product. The 12 H atoms in 4H$_3$PO$_4$ on the right require a coefficient of 6 in front of H$_2$O:

___ P$_4$O$_{10}$(s) + 6H$_2$O(l) → 4H$_3$PO$_4$(l)

Balance the O last, because it appears in both reactants and is harder to balance. There are 16 O atoms on each side:

P$_4$O$_{10}$(s) + 6H$_2$O(l) → 4H$_3$PO$_4$(l)

c) __B$_2$O$_3$(s) + __ NaOH(aq) → __Na$_3$BO$_3$(aq) + __H$_2$O(l)

Balance oxygen last because it is present in more than one place on each side of the reaction. The 2 B atoms in B$_2$O$_3$ on the left require a coefficient of 2 in front of Na$_3$BO$_3$ on the right:

__B$_2$O$_3$(s) + __NaOH(aq) → 2Na$_3$BO$_3$(aq) + __H$_2$O(l)

The 6 Na atoms in 2Na$_3$BO$_3$ on the right require a coefficient of 6 in front of NaOH on the left:

__B$_2$O$_3$(s) + 6NaOH(aq) → 2Na$_3$BO$_3$(aq) + __H$_2$O(l)

The 6 H atoms in 6NaOH on the left require a coefficent of 3 in front of H$_2$O on the right:

__B$_2$O$_3$(s) + 6NaOH(aq) → 2Na$_3$BO$_3$(aq) + 3H$_2$O(l)

The oxygen is now balanced with 9 O atoms on each side:

B$_2$O$_3$(s) + 6NaOH(aq) → 2Na$_3$BO$_3$(aq) + 3H$_2$O(l)

d) __CH$_3$NH$_2$(g) + __O$_2$(g) → __CO$_2$(g) + __H$_2$O(g) + __N$_2$(g)

There are 2 N atoms on the right in N$_2$, so a coefficient of 2 is required in front of CH$_3$NH$_2$ on the left:

2CH$_3$NH$_2$(g) + __O$_2$(g) → __CO$_2$(g) + __H$_2$O(g) + __N$_2$(g)

There are now 10 H atoms in 2CH$_3$NH$_2$ on the left so a coefficient of 5 is required in front of H$_2$O on the right:

2CH$_3$NH$_2$(g) + __O$_2$(g) → __CO$_2$(g) + 5H$_2$O(g) + __N$_2$(g)

The 2 C atoms on the left require a coefficient of 2 in front of CO$_2$ on the right:

2CH$_3$NH$_2$(g) + __O$_2$(g) → 2CO$_2$(g) + 5H$_2$O(g) + __N$_2$(g)

The 9 O atoms on the right (4 O atoms in 2CO$_2$ plus 5 in 5H$_2$O) require a coefficient of 9/2 in front of O$_2$ on the left:

2CH$_3$NH$_2$(g) + 9/2O$_2$(g) → 2CO$_2$(g) + 5H$_2$O(g) + __N$_2$(g)

Multiply all coefficients by 2 to obtain whole numbers:

4CH$_3$NH$_2$(g) + 9O$_2$(g) → 4CO$_2$(g) + 10H$_2$O(g) + 2N$_2$(g)

3.60   Plan: Balance one element at a time, placing coefficients where needed to have the same number of atoms of a particular element on each side of the equation. The smallest whole-number coefficients should be used.

Solution:

a) __SO$_2$(g) + __O$_2$(g) → __SO$_3$(g)

There are 4 O atoms on the left and 3 O atoms on the right. Since there is an odd number of O atoms on the right, place a coefficient of 2 in front of SO$_3$ for an even number of 6 O atoms on the right:

__SO$_2$(g) + __O$_2$(g) → 2SO$_3$(g)

Since there are now 2 S atoms on the right, place a coefficient of 2 in front of SO$_2$ on the left. There are now 6 O atoms on each side:

2SO$_2$(g) + O$_2$(g) → 2SO$_3$(g)

b) __Sc$_2$O$_3$(s) + __H$_2$O(l) → __ Sc(OH)$_3$(s)

The 2 Sc atoms on the left require a coefficient of 2 in front of Sc(OH)$_3$ on the right:

__Sc$_2$O$_3$(s) + __H$_2$O(l) → 2Sc(OH)$_3$(s)

The 6 H atoms in 2Sc(OH)$_3$ on the right require a coefficient of 3 in front of H$_2$O on the left. There are now 6 O atoms on each side:

Sc$_2$O$_3$(s) + 3H$_2$O(l) → 2Sc(OH)$_3$(s)

c) __H$_3$PO$_4$(aq) + __NaOH(aq) → __Na$_2$HPO$_4$(aq) + __H$_2$O(l)

The 2 Na atoms in Na$_2$HPO$_4$ on the right require a coefficient of 2 in front of NaOH on the left:

__H$_3$PO$_4$(aq) + 2NaOH(aq) → __Na$_2$HPO$_4$(aq) + __H$_2$O(l)

There are 6 O atoms on the right (4 in H$_3$PO$_4$ and 2 in 2NaOH); there are 4 O atoms in Na$_2$HPO$_4$ on the right so a coefficient of 2 in front of H$_2$O will result in 6 O atoms on the right:

__H$_3$PO$_4$(aq) + 2NaOH(aq) → __Na$_2$HPO$_4$(aq) + 2H$_2$O(l)

Now there are 4 H atoms on each side:

H$_3$PO$_4$(aq) + 2NaOH(aq) → Na$_2$HPO$_4$(aq) + 2H$_2$O(l)

d) __C$_6$H$_{10}$O$_5$(s) + __O$_2$(g) → __CO$_2$(g) + __H$_2$O(g)

The 6 C atoms in C$_6$H$_{10}$O$_5$ on the left require a coefficient of 6 in front of CO$_2$ on the right:

__C$_6$H$_{10}$O$_5$(s) + __O$_2$(g) → 6CO$_2$(g) + __H$_2$O(g)

The 10 H atoms in C$_6$H$_{10}$O$_5$ on the left require a coefficient of 5 in front of H$_2$O on the right:

__C$_6$H$_{10}$O$_5$(s) + __O$_2$(g) → 6CO$_2$(g) + 5H$_2$O(g)

There are 17 O atoms on the right (12 in 6CO$_2$ and 5 in 5H$_2$O); there are 5 O atoms in C$_6$H$_{10}$O$_5$ so a coefficient of 6 in front of O$_2$ will bring the total of O atoms on the left to 17:

C$_6$H$_{10}$O$_5$(s) + 6O$_2$(g) → 6CO$_2$(g) + 5H$_2$O(g)

3.62　Plan: The names must first be converted to chemical formulas. Balancing is a trial-and-error procedure. Balance one element at a time, placing coefficients where needed to have the same number of atoms of a particular element on each side of the equation. The smallest whole-number coefficients should be used. Remember that oxygen is diatomic.

Solution:

a) Gallium (a solid) and oxygen (a gas) are reactants and solid gallium(III) oxide is the only product:

__Ga(s) + __O$_2$(g) → __Ga$_2$O$_3$(s)

A coefficient of 2 in front of Ga on the left is needed to balance the 2 Ga atoms in Ga$_2$O$_3$:

2Ga(s) + __O$_2$(g) → __Ga$_2$O$_3$(s)

The 3 O atoms in Ga$_2$O$_3$ on the right require a coefficient of 3/2 in front of O$_2$ on the left:

2Ga(s) + 3/2O$_2$(g) → __Ga$_2$O$_3$(s)

Multiply all coefficients by 2 to obtain whole numbers:

4Ga(s) + 3O$_2$(g) → 2Ga$_2$O$_3$(s)

b) Liquid hexane and oxygen gas are the reactants while carbon dioxide gas and gaseous water are the products:

__C$_6$H$_{14}$(l) + __O$_2$(g) → __CO$_2$(g) + __H$_2$O(g)

The 6 C atoms in C_6H_{14} on the left require a coefficient of 6 in front of CO_2 on the right:

$\underline{}C_6H_{14}(l) + \underline{}O_2(g) \rightarrow \underline{6}CO_2(g) + \underline{}H_2O(g)$

The 14 H atoms in C_6H_{14} on the left require a coefficient of 7 in front of H_2O on the right:

$\underline{}C_6H_{14}(l) + \underline{}O_2(g) \rightarrow \underline{6}CO_2(g) + \underline{7}H_2O(g)$

The 19 O atoms on the right (12 in $6CO_2$ and 7 in $7H_2O$) require a coefficient of 19/2 in front of O_2 on the left:
Multiply all coefficients by 2 to obtain whole numbers:

$\mathbf{2C_6H_{14}(l) + 19O_2(g) \rightarrow 12CO_2(g) + 14H_2O(g)}$

c) Aqueous solutions of calcium chloride and sodium phosphate are the reactants; solid calcium phosphate and an aqueous solution of sodium chloride are the products:

$\underline{}CaCl_2(aq) + \underline{}Na_3PO_4(aq) \rightarrow \underline{}Ca_3(PO_4)_2(s) + \underline{}NaCl(aq)$

The 3 Ca atoms in $Ca_3(PO_4)_2$ on the right require a coefficient of 3 in front of $CaCl_2$ on the left:

$\underline{3}CaCl_2(aq) + \underline{}Na_3PO_4(aq) \rightarrow \underline{}Ca_3(PO_4)_2(s) + \underline{}NaCl(aq)$

The 6 Cl atoms in $3CaCl_2$ on the left require a coefficient of 6 in front of NaCl on the right:

$\underline{3}CaCl_2(aq) + \underline{}Na_3PO_4(aq) \rightarrow \underline{}Ca_3(PO_4)_2(s) + \underline{6}NaCl(aq)$

The 6 Na atoms in 6NaCl on the right require a coefficient of 2 in front of Na_3PO_4 on the left:

$\underline{3}CaCl_2(aq) + \underline{2}Na_3PO_4(aq) \rightarrow \underline{}Ca_3(PO_4)_2(s) + \underline{6}NaCl(aq)$

There are now 2 P atoms on each side:

$\mathbf{3CaCl_2(aq) + 2Na_3PO_4(aq) \rightarrow Ca_3(PO_4)_2(s) + 6NaCl(aq)}$

3.67   Plan: First, write a balanced chemical equation. Since A is the limiting reagent (B is in excess), A is used to determine the amount of C formed, using the mole ratio between reactant A and product C.
Solution:
Plan: The balanced equation is aA + bB → cC. Divide the mass of A by its molar mass to obtain moles of A. Use the molar ratio from the balanced equation to find the moles of C. Multiply moles of C by its molar mass to obtain mass of C.
Roadmap:

Mass (g) of A

Divide by $\mathscr{M}$ (g/mol)

Amount (mol) of A

Molar ratio between A and C

Amount (moles) of C

Multiply by $\mathscr{M}$ (g/mol)

Mass (g) of C

3.69   Plan: Always check to see if the initial equation is balanced. If the equation is not balanced, it should be balanced before proceeding. Use the mole ratio from the balanced chemical equation to determine the moles of Cl_2 produced. The equation shows that 1 mole of Cl_2 is produced for every 4 moles of HCl that react. Multiply the moles of Cl_2 produced by the molar mass to convert to mass in grams.
Solution:

$4HCl(aq) + MnO_2(s) \rightarrow MnCl_2(aq) + 2H_2O(g) + Cl_2(g)$

a) Moles of $Cl_2 = \left(1.82 \text{ mol HCl}\right)\left(\dfrac{1 \text{ mol Cl}_2}{4 \text{ mol HCl}}\right) = \mathbf{0.455 \text{ mol Cl}_2}$

b) Mass (g) of $Cl_2 = (0.455 \text{ mol } Cl_2)\left(\dfrac{70.90 \text{ g } Cl_2}{1 \text{ mol } Cl_2}\right) = 32.2595 = \textbf{32.3 g } Cl_2$

3.71 **Plan:** Convert the kilograms of oxygen to grams of oxygen and then moles of oxygen by dividing by its molar mass. Use the moles of oxygen and the mole ratio from the balanced chemical equation to determine the moles of KNO_3 required. Multiply the moles of KNO_3 by its molar mass to obtain the mass in grams.
Solution:

a) Mass (g) of $O_2 = (56.6 \text{ kg } O_2)\left(\dfrac{10^3 \text{ g}}{1 \text{ kg}}\right) = 5.66\text{x}10^4 \text{ g } O_2$

Moles of $O_2 = (5.66\text{x}10^4 \text{ g } O_2)\left(\dfrac{1 \text{ mol } O_2}{32.00 \text{ g } O_2}\right) = 1.76875\text{x}10^3 \text{ mol } O_2$

Moles of $KNO_3 = (1.76875 \text{ mol } O_2)\left(\dfrac{4 \text{ mol } KNO_3}{5 \text{ mol } O_2}\right) = 1415 = \textbf{1.42x}10^3 \textbf{ mol } KNO_3$

b) Mass (g) of $KNO_3 = (1415 \text{ mol } KNO_3)\left(\dfrac{101.11 \text{ g } KNO_3}{1 \text{ mol } KNO_3}\right) = 143070.65 = \textbf{1.43x}10^5 \textbf{ g } KNO_3$

Combining all steps gives:

Mass (g) of $KNO_3 = (56.6 \text{ kg } O_2)\left(\dfrac{10^3 \text{ g}}{1 \text{ kg}}\right)\left(\dfrac{1 \text{ mol } O_2}{32.00 \text{ g } O_2}\right)\left(\dfrac{4 \text{ mol } KNO_3}{5 \text{ mol } O_2}\right)\left(\dfrac{101.11 \text{ g } KNO_3}{1 \text{ mol } KNO_3}\right)$

$= 143070.65 = \textbf{1.43x}10^5 \textbf{ g } KNO_3$

3.73 **Plan:** First, balance the equation. Convert the grams of diborane to moles of diborane by dividing by its molar mass. Use mole ratios from the balanced chemical equation to determine the moles of the products. Multiply the mole amount of each product by its molar mass to obtain mass in grams.
Solution:

The balanced equation is: $B_2H_6(g) + 6H_2O(l) \rightarrow 2H_3BO_3(s) + 6H_2(g)$.

Moles of $B_2H_6 = (43.82 \text{ g } B_2H_6)\left(\dfrac{1 \text{ mol } B_2H_6}{27.67 \text{ g } B_2H_6}\right) = 1.583665 \text{ mol } B_2H_6$

Moles of $H_3BO_3 = (1.583665 \text{ mol } B_2H_6)\left(\dfrac{2 \text{ mol } H_3BO_3}{1 \text{ mol } B_2H_6}\right) = 3.16733 \text{ mol } H_3BO_3$

Mass (g) of $H_3BO_3 = (3.16733 \text{ mol } H_3BO_3)\left(\dfrac{61.83 \text{ g } H_3BO_3}{1 \text{ mol } H_3BO_3}\right) = 195.83597 = \textbf{195.8 g } H_3BO_3$

Combining all steps gives:

Mass (g) of $H_3BO_3 = (43.82 \text{ g } B_2H_6)\left(\dfrac{1 \text{ mol } B_2H_6}{27.67 \text{ g } B_2H_6}\right)\left(\dfrac{2 \text{ mol } H_3BO_3}{1 \text{ mol } B_2H_6}\right)\left(\dfrac{61.83 \text{ g } H_3BO_3}{1 \text{ mol } H_3BO_3}\right)$

$= 195.83597 = \textbf{195.8 g } H_3BO_3$

Moles of $H_2 = (1.583665 \text{ mol } B_2H_6)\left(\dfrac{6 \text{ mol } H_2}{1 \text{ mol } B_2H_6}\right) = 9.50199 \text{ mol } H_2$

Mass (g) of $H_2 = (9.50199 \text{ mol } H_2)\left(\dfrac{2.016 \text{ g } H_2}{1 \text{ mol } H_2}\right) = 19.15901 \text{ g } H_2 = \textbf{19.16 g } H_2$

Combining all steps gives:

$$\text{Mass (g) of } H_2 = \left(43.82 \text{ g } B_2H_6\right)\left(\frac{1 \text{ mol } B_2H_6}{27.67 \text{ g } B_2H_6}\right)\left(\frac{6 \text{ mol } H_2}{1 \text{ mol } B_2H_6}\right)\left(\frac{2.016 \text{ g } H_2}{1 \text{ mol } H_2}\right) = 19.15601 = \textbf{19.16 g } H_2$$

3.75   Plan: Write the balanced equation by first writing the formulas for the reactants and products. Convert the mass of phosphorus to moles by dividing by the molar mass, use the mole ratio between phosphorus and chlorine from the balanced chemical equation to obtain moles of chlorine, and finally divide the moles of chlorine by its molar mass to obtain amount in grams.
Solution:
Reactants: formula for phosphorus is given as P_4 and formula for chlorine gas is Cl_2 (chlorine occurs as a diatomic molecule). Product: formula for phosphorus pentachloride (the name indicates one phosphorus atom and five chlorine atoms) is PCl_5.

Equation: $P_4 + Cl_2 \rightarrow PCl_5$

Balancing the equation: $P_4 + 10Cl_2 \rightarrow 4PCl_5$

$$\text{Moles of } P_4 = \left(455 \text{ g } P_4\right)\left(\frac{1 \text{ mol } P_4}{123.88 \text{ g } P_4}\right) = 3.67291 \text{ mol } P_4$$

$$\text{Moles of } Cl_2 = \left(3.67291 \text{ mol } P_4\right)\left(\frac{10 \text{ mol } Cl_2}{1 \text{ mol } P_4}\right) = 36.7291 \text{ mol } Cl_2$$

$$\text{Mass (g) of } Cl_2 = \left(36.7291 \text{ mol } Cl_2\right)\left(\frac{70.90 \text{ g } Cl_2}{1 \text{ mol } Cl_2}\right) = 2604.09 = \textbf{2.60x10}^3 \textbf{ g } Cl_2$$

Combining all steps gives:

$$\text{Mass (g) of } Cl_2 = \left(455 \text{ g } P_4\right)\left(\frac{1 \text{ mol } P_4}{123.88 \text{ g } P_4}\right)\left(\frac{10 \text{ mol } Cl_2}{1 \text{ mol } P_4}\right)\left(\frac{70.90 \text{ g } Cl_2}{1 \text{ mol } Cl_2}\right) = 2604.09267 = \textbf{2.60x10}^3 \textbf{ g } Cl_2$$

3.77   Plan: Begin by writing the chemical formulas of the reactants and products in each step. Next, balance each of the equations. Combine the equations for the separate steps by adjusting the equations so the intermediate (iodine monochloride) cancels. Finally, change the mass of product from kg to grams to moles by dividing by the molar mass and use the mole ratio between iodine and product to find the moles of iodine. Multiply moles by the molar mass of iodine to obtain mass of iodine.
Solution:

a) *Step 1* $I_2(s) + Cl_2(g) \rightarrow 2ICl(s)$

    Step 2 $ICl(s) + Cl_2(g) \rightarrow ICl_3(s)$

b) Multiply the coefficients of the second equation by 2, so that $ICl(s)$, an intermediate product, can be eliminated from the overall equation.

    $I_2(s) + Cl_2(g) \longrightarrow 2ICl(s)$

    $\underline{2ICl(s) + 2Cl_2(g) \longrightarrow 2ICl_3(s)}$

    $I_2(s) + Cl_2(g) + \cancel{2ICl(s)} + 2Cl_2(g) \rightarrow \cancel{2ICl(s)} + 2ICl_3(s)$

    Overall equation: $\textbf{I}_2\textbf{(s) + 3Cl}_2\textbf{(g)} \rightarrow \textbf{2ICl}_3\textbf{(s)}$

c) $$\text{Mass (g) of } ICl_3 = \left(2.45 \text{ kg } ICl_3\right)\left(\frac{10^3 \text{ g}}{1 \text{ kg}}\right) = 2450 \text{ g } ICl_3$$

$$\text{Moles of } ICl_3 = \left(2450 \text{ g } ICl_3\right)\left(\frac{1 \text{ mol } ICl_3}{233.2 \text{ g } ICl_3}\right) = 10.506 \text{ mol } ICl_3$$

$$\text{Moles of } I_2 = (10.506 \text{ mol } ICl_3)\left(\frac{1 \text{ mol } I_2}{2 \text{ mol } ICl_3}\right) = 5.253 \text{ mol } I_2$$

$$\text{Mass (g) of } I_2 = (5.253 \text{ mol } I_2)\left(\frac{253.8 \text{ g } I_2}{1 \text{ mol } I_2}\right) = 1333.211 = \mathbf{1.33x10^3 \text{ g } I_2}$$

Combining all steps gives:

$$\text{Mass (g) of } I_2 = (2.45 \text{ kg } ICl_3)\left(\frac{10^3 \text{ g}}{1 \text{ kg}}\right)\left(\frac{1 \text{ mol } ICl_3}{233.2 \text{ g } ICl_3}\right)\left(\frac{1 \text{ mol } I_2}{2 \text{ mol } ICl_3}\right)\left(\frac{253.8 \text{ g } I_2}{1 \text{ mol } I_2}\right) = 1333.211 = \mathbf{1.33x10^3 \text{ g } I_2}$$

3.79   Plan: Write the balanced reaction between A_2 and B_2 to form AB_3. To find the limiting reactant, find the number of molecules of product that would form from the numbers of molecules of each reactant.
Solution:
a) The reaction is $A_2 + 3B_2 \rightarrow 2AB_3$

$$\text{Molecules of } AB_3 \text{ from } A_2 = (3 \text{ } A_2 \text{ molecules})\left(\frac{2 \text{ } AB_3 \text{ molecules}}{1 \text{ } A_2 \text{ molecule}}\right) = 6 \text{ } AB_3 \text{ molecules}$$

$$\text{Molecules of } AB_3 \text{ from } B_2 = (6 \text{ } B_2 \text{ molecules})\left(\frac{2 \text{ } AB_3 \text{ molecules}}{3 \text{ } B_2 \text{ molecules}}\right) = 4 \text{ } AB_3 \text{ molecules}$$

B_2 **is the limiting reactant** since fewer molecules of product form from B_2.

b) Using the molar ratio between B_2 and AB_3: $(6 \text{ } B_2 \text{ molecules})\left(\dfrac{2 \text{ } AB_3 \text{ molecules}}{3 \text{ } B_2 \text{ molecules}}\right) = \mathbf{4 \text{ } AB_3 \text{ molecules}}$

3.81   Plan: Write the balanced reaction. Calculate the amount (mol) of CH_3OH produced from the amount   (mol) of each reactant. The smaller amount of product is the correct answer.
Solution:
The reaction is $CO(g) + 2H_2(g) \rightarrow CH_3OH(l)$.

$$\text{Amount (mol) of } CH_3OH \text{ from } CO = (4.5 \text{ mol } CO)\left(\frac{1 \text{ mol } CH_3OH}{1 \text{ mol } CO}\right) = 4.5 \text{ mol } CH_3OH$$

$$\text{Amount (mol) of } CH_3OH \text{ from } H_2 = (7.2 \text{ mol } H_2)\left(\frac{1 \text{ mol } CH_3OH}{2 \text{ mol } H_2}\right) = 3.6 \text{ mol } CH_3OH$$

H_2 is the limiting reactant and **3.6 mol of CH_3OH** are produced.

3.83   Plan: Convert the given mass of each reactant to moles by dividing by the molar mass of that reactant. Use the mole ratio from the balanced chemical equation to find the moles of CaO formed from each reactant, assuming an excess of the other reactant. The reactant that produces fewer moles of CaO is the limiting reactant. Convert the moles of CaO obtained from the limiting reactant to grams using the molar mass.
Solution:
$2Ca(s) + O_2(g) \rightarrow 2CaO(s)$

a) $\text{Moles of } Ca = (4.20 \text{ g } Ca)\left(\dfrac{1 \text{ mol } Ca}{40.08 \text{ g } Ca}\right) = 0.104790 \text{ mol } Ca$

$$\text{Moles of CaO from Ca} = (0.104790 \text{ mol } Ca)\left(\frac{2 \text{ mol CaO}}{2 \text{ mol } Ca}\right) = 0.104790 = \mathbf{0.105 \text{ mol CaO}}$$

b) $\text{Moles of } O_2 = (2.80 \text{ g } O_2)\left(\dfrac{1 \text{ mol } O_2}{32.00 \text{ g } O_2}\right) = 0.0875 \text{ mol } O_2$

$$\text{Moles of CaO from } O_2 = \left(0.0875 \text{ mol } O_2\right)\left(\frac{2 \text{ mol CaO}}{1 \text{ mol } O_2}\right) = 0.17500 = \textbf{0.175 mol CaO}$$

c) **Calcium** is the limiting reactant since it will form less calcium oxide.

d) The mass of CaO formed is determined by the limiting reactant, Ca.

$$\text{Mass (g) of CaO} = \left(0.104790 \text{ mol CaO}\right)\left(\frac{56.08 \text{ g CaO}}{1 \text{ mol CaO}}\right) = 5.8766 = \textbf{5.88 g CaO}$$

Combining all steps gives:

$$\text{Mass (g) of CaO} = \left(4.20 \text{ g Ca}\right)\left(\frac{1 \text{ mol Ca}}{40.08 \text{ g Ca}}\right)\left(\frac{2 \text{ mol CaO}}{2 \text{ mol Ca}}\right)\left(\frac{56.08 \text{ g CaO}}{1 \text{ mol CaO}}\right) = 5.8766 = \textbf{5.88 g CaO}$$

3.85   **Plan:** First, balance the chemical equation. To determine which reactant is limiting, calculate the amount of HIO_3 formed from each reactant, assuming an excess of the other reactant. The reactant that produces less product is the limiting reagent. Use the limiting reagent and the mole ratio from the balanced chemical equation to determine the amount of HIO_3 formed and the amount of the excess reactant that reacts. The difference between the amount of excess reactant that reacts and the initial amount of reactant supplied gives the amount of excess reactant remaining.
Solution:
The balanced chemical equation for this reaction is:

$$2ICl_3 + 3H_2O \rightarrow ICl + HIO_3 + 5HCl$$

Hint: Balance the equation by starting with oxygen. The other elements are in multiple reactants and/or products and are harder to balance initially.
Finding the moles of HIO_3 from the moles of ICl_3 (if H_2O is limiting):

$$\text{Moles of } ICl_3 = \left(635 \text{ g } ICl_3\right)\left(\frac{1 \text{ mol } ICl_3}{233.2 \text{ g } ICl_3}\right) = 2.722985 \text{ mol } ICl_3$$

$$\text{Moles of } HIO_3 \text{ from } ICl_3 = \left(2.722985 \text{ mol } ICl_3\right)\left(\frac{1 \text{ mol } HIO_3}{2 \text{ mol } ICl_3}\right) = 1.361492 = 1.36 \text{ mol } HIO_3$$

Finding the moles of HIO_3 from the moles of H_2O (if ICl_3 is limiting):

$$\text{Moles of } H_2O = \left(118.5 \text{ g } H_2O\right)\left(\frac{1 \text{ mol } H_2O}{18.02 \text{ g } H_2O}\right) = 6.57603 \text{ mol } H_2O$$

$$\text{Moles } HIO_3 \text{ from } H_2O = \left(6.57603 \text{ mol } H_2O\right)\left(\frac{1 \text{ mol } HIO_3}{3 \text{ mol } H_2O}\right) = 2.19201 = 2.19 \text{ mol } HIO_3$$

ICl_3 is the limiting reagent and will produce **1.36 mol HIO_3**.

$$\text{Mass (g) of } HIO_3 = \left(1.361492 \text{ mol } HIO_3\right)\left(\frac{175.9 \text{ g } HIO_3}{1 \text{ mol } HIO_3}\right) = 239.486 = \textbf{239 gHIO}_3$$

Combining all steps gives:

$$\text{Mass (g) of } HIO_3 = \left(635 \text{ g } ICl_3\right)\left(\frac{1 \text{ mol } ICl_3}{233.2 \text{ g } ICl_3}\right)\left(\frac{1 \text{ mol } HIO_3}{2 \text{ mol } ICl_3}\right)\left(\frac{175.9 \text{ g } HIO_3}{1 \text{ mol } HIO_3}\right) = 239.486 = \textbf{239 g HIO}_3$$

The remaining mass of the excess reagent can be calculated from the amount of H_2O combining with the limiting reagent.

$$\text{Moles of } H_2O \text{ required to react with 635 g } ICl_3 = \left(2.722985 \text{ mol } ICl_3\right)\left(\frac{3 \text{ mol } H_2O}{2 \text{ mol } ICl_3}\right) = 4.0844775 \text{ mol } H_2O$$

Mass (g) of H_2O required to react with 635 g ICl_3 = $\left(4.0844775 \text{ mol } H_2O\right)\left(\dfrac{18.02 \text{ g } H_2O}{1 \text{ mol } H_2O}\right)$

$$= 73.6023 = 73.6g \text{ } H_2O \text{ reacted}$$

Remaining H_2O = 118.5 g – 73.6 g = **44.9 g H_2O**

3.87   Plan: Write the balanced equation; the formula for carbon is C, the formula for oxygen is O_2 and the formula for carbon dioxide is CO_2. To determine which reactant is limiting, calculate the amount of CO_2 formed from each reactant, assuming an excess of the other reactant. The reactant that produces less product is the limiting reagent. Use the limiting reagent and the mole ratio from the balanced chemical equation to determine the amount of CO_2 formed and the amount of the excess reactant that reacts. The difference between the amount of excess reactant that reacts and the initial amount of reactant supplied gives the amount of excess reactant remaining.
Solution:

The balanced equation is: $C(s) + O_2(g) \rightarrow CO_2(g)$

Finding the moles of CO_2 from the moles of carbon (if O_2 is limiting):

Moles of CO_2 from C = $\left(0.100 \text{ mol C}\right)\left(\dfrac{1 \text{ mol } CO_2}{1 \text{ mol C}}\right)$ = 0.100 mol CO_2

Finding the moles of CO_2 from the moles of oxygen (if C is limiting):

Moles of O_2 = $\left(8.00 \text{ g } O_2\right)\left(\dfrac{1 \text{ mol } O_2}{32.00 \text{ g } O_2}\right)$ = 0.250 mol O_2

Moles of CO_2 from O_2 = $\left(0.250 \text{ mol } O_2\right)\left(\dfrac{1 \text{ mol } CO_2}{1 \text{ mol } O_2}\right)$ = 0.25000 = 0.250 mol CO_2

Carbon is the limiting reactant and will be used to determine the amount of CO_2 that will form.

Mass (g) of CO_2 = $\left(0.100 \text{ mol } CO_2\right)\left(\dfrac{44.01 \text{ g } CO_2}{1 \text{ mol } CO_2}\right)$ = 4.401 = **4.40 g CO_2**

Since carbon is limiting, the **O_2 is in excess**. The amount remaining depends on how much combines with the limiting reagent.

Moles of O_2 required to react with 0.100 mol of C = $\left(0.100 \text{ mol C}\right)\left(\dfrac{1 \text{ mol } O_2}{1 \text{ mol C}}\right)$ = 0.100 mol O_2

Mass (g) of O_2 required to react with 0.100 mol of C = $\left(0.100 \text{ mol } O_2\right)\left(\dfrac{32.00 \text{ mol } O_2}{1 \text{ mol } O_2}\right)$ = 3.20 g O_2

Remaining O_2 = 8.00 g – 3.20 g = **4.80 g O_2**

3.89   Plan: The question asks for the mass of each substance present at the end of the reaction. "Substance" refers to both reactants and products. Solve this problem using multiple steps. Recognizing that this is a limiting reactant problem, first write a balanced chemical equation. To determine which reactant is limiting, calculate the amount of any product formed from each reactant, assuming an excess of the other reactant. The reactant that produces less product is the limiting reagent. Any product can be used to predict the limiting reactant; in this case, $AlCl_3$ is used. Use the limiting reagent and the mole ratio from the balanced chemical equation to determine the amount of both products formed and the amount of the excess reactant that reacts. The difference between the amount of excess reactant that reacts and the initial amount of reactant supplied gives the amount of excess reactant remaining.
Solution:

The balanced chemical equation is:

$$Al(NO_2)_3(aq) + 3NH_4Cl(aq) \rightarrow AlCl_3(aq) + 3N_2(g) + 6H_2O(l)$$

Now determine the limiting reagent. We will use the moles of $AlCl_3$ produced to determine which is limiting. Finding the moles of $AlCl_3$ from the moles of $Al(NO_2)_3$ (if NH_4Cl is limiting):

$$\text{Moles of } Al(NO_2)_3 = \left(72.5 \text{ g } Al(NO_2)_3\right)\left(\frac{1 \text{ mol } Al(NO_2)_3}{165.01 \text{ g } Al(NO_2)_3}\right) = 0.439367 \text{ mol } Al(NO_2)_3$$

$$\text{Moles of } AlCl_3 \text{ from } Al(NO_2)_3 = \left(0.439367 \text{ mol } Al(NO_2)_3\right)\left(\frac{1 \text{ mol } AlCl_3}{1 \text{ mol } Al(NO_2)_3}\right) = 0.439367 = 0.439 \text{ mol } AlCl_3$$

Finding the moles of $AlCl_3$ from the moles of NH_4Cl (if $Al(NO_2)_3$ is limiting):

$$\text{Moles of } NH_4Cl = \left(58.6 \text{ g } NH_4Cl\right)\left(\frac{1 \text{ mol } NH_4Cl}{53.49 \text{ g } NH_4Cl}\right) = 1.09553 \text{ mol } NH_4Cl$$

$$\text{Moles of } AlCl_3 \text{ from } NH_4Cl = \left(1.09553 \text{ mol } NH_4Cl\right)\left(\frac{1 \text{ mol } AlCl_3}{3 \text{ mol } NH_4Cl}\right) = 0.365177 = 0.365 \text{ mol } AlCl_3$$

Ammonium chloride is the limiting reactant, and it is used for all subsequent calculations.
Mass of substances after the reaction:
$Al(NO_2)_3$:
Mass (g) of $Al(NO_2)_3$ (the excess reactant) required to react with 58.6 g of NH_4Cl =

$$\left(1.09553 \text{ mol } NH_4Cl\right)\left(\frac{1 \text{ mol } Al(NO_2)_3}{3 \text{ mol } NH_4Cl}\right)\left(\frac{165.01 \text{ g } Al(NO_2)_3}{1 \text{ mol } Al(NO_2)_3}\right) = 60.2579 = 60.3 \text{ g } Al(NO_2)_3$$

$Al(NO_2)_3$ remaining: 72.5 g – 60.3 g = **12.2 g $Al(NO_2)_3$**
NH_4Cl: **None left** since it is the limiting reagent.
$AlCl_3$:

$$\text{Mass (g) of } AlCl_3 = \left(0.365177 \text{ mol } AlCl_3\right)\left(\frac{133.33 \text{ g } AlCl_3}{1 \text{ mol } AlCl_3}\right) = 48.689 = \textbf{48.7 g } AlCl_3$$

N_2:

$$\text{Mass (g) of } N_2 = \left(1.09553 \text{ mol } NH_4Cl\right)\left(\frac{3 \text{ mol } N_2}{3 \text{ mol } NH_4Cl}\right)\left(\frac{28.02 \text{ g } N_2}{1 \text{ mol } N_2}\right) = 30.697 = \textbf{30.7 g } N_2$$

H_2O:

$$\text{Mass (g) of } H_2O = \left(1.09553 \text{ mol } NH_4Cl\right)\left(\frac{6 \text{ mol } H_2O}{3 \text{ mol } NH_4Cl}\right)\left(\frac{18.02 \text{ g } H_2O}{1 \text{ mol } H_2O}\right) = 39.483 = \textbf{39.5 g } H_2O$$

3.91   Plan: Express the yield of each step as a fraction of 1.00; multiply the fraction of the first step by that of the second step and then multiply by 100 to get the overall percent yield.
Solution:
73% = 0.73; 68% = 0.68
(0.73 x 0.68) x 100 = 49.64 = **50.%**

3.93   Plan: Write and balance the chemical equation using the formulas of the substances. Determine the theoretical yield of the reaction from the mass of tungsten(VI) oxide. To do that, convert the mass of tungsten(VI) oxide to moles by dividing by its molar mass and then use the mole ratio between tungsten(VI) oxide and water to determine the moles and then mass of water that should be produced. Use the density of water to determine the actual yield of water in grams. The actual yield divided by the theoretical yield just calculated (with the result multiplied by 100%) gives the percent yield.

Solution:

The balanced chemical equation is:

$WO_3(s) + 3H_2(g) \rightarrow W(s) + 3H_2O(l)$

Determining the theoretical yield of H_2O:

Moles of $WO_3 = \left(45.5 \text{ g } WO_3\right)\left(\dfrac{1 \text{ mol } WO_3}{231.9 \text{ g } WO_3}\right) = 0.1962053 \text{ mol } WO_3$

Mass (g) of H_2O (theoretical yield) $= \left(0.1962053 \text{ mol } WO_3\right)\left(\dfrac{3 \text{ mol } H_2O}{1 \text{ mol } WO_3}\right)\left(\dfrac{18.02 \text{ g } H_2O}{1 \text{ mol } H_2O}\right) = 10.60686 \text{ g } H_2O$

Determining the actual yield of H_2O:

Mass (g) of H_2O (actual yield) $= \left(9.60 \text{ mL } H_2O\right)\left(\dfrac{1.00 \text{ g } H_2O}{1 \text{ mL } H_2O}\right) = 9.60 \text{ g } H_2O$

% yield $= \left(\dfrac{\text{actual Yield}}{\text{theoretical Yield}}\right) \times 100\% = \left(\dfrac{9.60 \text{ g } H_2O}{10.60686 \text{ g } H_2O}\right) \times 100\% = 90.5075 = \mathbf{90.5\%}$

3.95 **Plan:** Write the balanced chemical equation. Since quantities of two reactants are given, we must determine which is the limiting reactant. To determine which reactant is limiting, calculate the amount of any product formed from each reactant, assuming an excess of the other reactant. The reactant that produces less product is the limiting reagent. Any product can be used to predict the limiting reactant; in this case, CH_3Cl is used. Only 75.0% of the calculated amounts of products actually form, so the actual yield is 75% of the theoretical yield.
Solution:

The balanced equation is: $CH_4(g) + Cl_2(g) \rightarrow CH_3Cl(g) + HCl(g)$

Determining the limiting reactant:

Finding the moles of CH_3Cl from the moles of CH_4 (if Cl_2 is limiting):

Moles of $CH_4 = \left(20.5 \text{ g } CH_4\right)\left(\dfrac{1 \text{ mol } CH_4}{16.04 \text{ g } CH_4}\right) = 1.278055 \text{ mol } CH_4$

Moles of CH_3Cl from $CH_4 = \left(1.278055 \text{ mol } CH_4\right)\left(\dfrac{1 \text{ mol } CH_3Cl}{1 \text{ mol } CH_4}\right) = 1.278055 \text{ mol } CH_3Cl$

Finding the moles of CH_3Cl from the moles of Cl_2 (if CH_4 is limiting):

Moles of $Cl_2 = \left(45.0 \text{ g } Cl_2\right)\left(\dfrac{1 \text{ mol } Cl_2}{70.90 \text{ g } Cl_2}\right) = 0.634697 \text{ mol } Cl_2$

Moles of CH_3Cl from $Cl_2 = \left(0.634697 \text{ mol } Cl_2\right)\left(\dfrac{1 \text{ mol } CH_3Cl}{1 \text{ mol } Cl_2}\right) = 0.634697 \text{ mol } CH_3Cl$

Chlorine is the limiting reactant and is used to determine the theoretical yield of CH_3Cl:

Mass (g) of CH_3Cl (theoretical yield) $= \left(0.634697 \text{ mol } CH_3Cl\right)\left(\dfrac{50.48 \text{ g } CH_3Cl}{1 \text{ mol } CH_3Cl}\right) = 32.0395 \text{ g } CH_3Cl$

% yield $= \left(\dfrac{\text{actual Yield}}{\text{theoretical Yield}}\right) \times 100\%$

Actual yield (g) of $CH_3Cl = \dfrac{\% \text{ yield}}{100\%}(\text{theoretical yield}) = \dfrac{75\%}{100\%}(32.0395 \text{ g } CH_3Cl) = 24.02962 = \mathbf{24.0 \text{ g } CH_3Cl}$

3.97   Plan: Write the balanced equation; the formula for fluorine is F_2, the formula for carbon tetrafluoride is CF_4, and the formula for nitrogen trifluoride is NF_3. To determine which reactant is limiting, calculate the amount of CF_4 formed from each reactant, assuming an excess of the other reactant. The reactant that produces less product is the limiting reagent. Use the limiting reagent and the mole ratio from the balanced chemical equation to determine the mass of CF_4 formed.
Solution:
The balanced chemical equation is:
$(CN)_2(g) + 7F_2(g) \rightarrow 2CF_4(g) + 2NF_3(g)$
Determining the limiting reactant:
Finding the moles of CF_4 from the moles of $(CN)_2$ (if F_2 is limiting):

$$\text{Moles of } CF_4 \text{ from } (CN)_2 = \left(60.0 \text{ g } (CN)_2\right)\left(\frac{1 \text{ mol } (CN)_2}{52.04 \text{ g } (CN)_2}\right)\left(\frac{2 \text{ mol } CF_4}{1 \text{ mol } (CN)_2}\right) = 2.30592 \text{ mol } CF_4$$

Finding the moles of CF_4 from the moles of F_2 (if $(CN)_2$ is limiting):

$$\text{Moles of } CF_4 \text{ from } F_2 = \left(60.0 \text{ g } F_2\right)\left(\frac{1 \text{ mol } F_2}{38.00 \text{ g } F_2}\right)\left(\frac{2 \text{ mol } CF_4}{7 \text{ mol } F_2}\right) = 0.4511278 \text{ mol } CF_4$$

F_2 is the limiting reactant, and will be used to calculate the amount of CF_4 produced.

$$\text{Mass (g) of } CF_4 = \left(60.0 \text{ g } F_2\right)\left(\frac{1 \text{ mol } F_2}{38.00 \text{ g } F_2}\right)\left(\frac{2 \text{ mol } CF_4}{7 \text{ mol } F_2}\right)\left(\frac{88.01 \text{ g } CF_4}{1 \text{ mol } CF_4}\right) = 39.70376 = \mathbf{39.7 \text{ g } CF_4}$$

3.98   Plan: Write and balance the chemical reaction. Remember that both chlorine and oxygen exist as diatomic molecules. Use the mole ratio between oxygen and dichlorine monoxide to find the moles of dichlorine monoxide that reacted. Multiply the amount in moles by Avogadro's number to convert to number of molecules.
Solution:
a) Both oxygen and chlorine are diatomic. **Scene A** best represents the product mixture as there are O_2 and Cl_2 molecules in Scene A. Scene B shows oxygen and chlorine atoms and Scene C shows atoms and molecules. Oxygen and chlorine atoms are NOT products of this reaction.
b) The balanced reaction is $\mathbf{2Cl_2O(g) \rightarrow 2Cl_2(g) + O_2(g)}$.
c) There is a 2:1 mole ratio between Cl_2 and O_2. In Scene A, there are 6 green molecules and 3 red molecules. Since twice as many Cl_2 molecules are produced as there are O_2 molecules produced, the red molecules are the O_2 molecules.

$$\text{Moles of } Cl_2O = \left(3 \text{ } O_2 \text{ molecules}\right)\left(\frac{2 \text{ O atoms}}{1 \text{ } O_2 \text{ molecule}}\right)\left(\frac{0.050 \text{ mol O atoms}}{1 \text{ O atom}}\right)\left(\frac{1 \text{ mol } O_2 \text{ molecules}}{2 \text{ mol O atoms}}\right)\left(\frac{2 \text{ mol } Cl_2O}{1 \text{ mol } O_2}\right)$$

$$= 0.30 \text{ mol } Cl_2O$$

$$\text{Molecules of } Cl_2O = \left(0.30 \text{ mol } Cl_2O\right)\left(\frac{6.022 \times 10^{23} \text{ } Cl_2O \text{ molecules}}{1 \text{ mol } Cl_2O}\right)$$

$$= 1.8066 \times 10^{23} = \mathbf{1.8 \times 10^{23} \text{ } Cl_2O \text{ molecules}}$$

3.105   Plan: The moles of narceine and the moles of water are required. We can assume any mass of narceine hydrate (we will use 100 g), and use this mass to determine the moles of hydrate. The moles of water in the hydrate is obtained by taking 10.8% of the 100 g mass of hydrate and converting the mass to moles of water. Divide the moles of water by the moles of hydrate to find the value of x.
Solution:
Assuming a 100 g sample of narceine hydrate:

$$\text{Moles of narceine hydrate} = \left(100 \text{ g narceine hydrate}\right)\left(\frac{1 \text{ mol narceine hydrate}}{499.52 \text{ g narceine hydrate}}\right)$$

$$= 0.20019 \text{ mol narceine hydrate}$$

$$\text{Mass (g) of } H_2O = (100 \text{ g narceine hydrate})\left(\frac{10.8\% \ H_2O}{100\% \text{ narceine hydrate}}\right) = 10.8 \text{ g } H_2O$$

$$\text{Moles of } H_2O = (10.8 \text{ g } H_2O)\left(\frac{1 \text{ mol } H_2O}{18.02 \text{ g } H_2O}\right) = 0.59933 \text{ mol } H_2O$$

$$x = \frac{\text{moles of } H_2O}{\text{moles of hydrate}} = \frac{0.59933 \text{ mol}}{0.20019 \text{ mol}} = 3$$

Thus, there are three water molecules per mole of hydrate. The formula for narceine hydrate is **narceine • 3H₂O**.

3.106   Plan: Determine the formula and the molar mass of each compound. The formula gives the relative numbers of moles of each element present. Multiply the number of moles of each element by its molar mass to find the total mass of element in 1 mole of compound. Mass percent = $\dfrac{\text{total mass of element}}{\text{molar mass of compound}}(100)$. List the compounds from the highest %H to the lowest.

Solution:

Name	Chemical formula	Molar mass (g/mol)	Mass percent $H = \dfrac{\text{moles of } H \ \text{x molar mass}}{\text{molar mass of compound}}(100)$
Ethane	C_2H_6	30.07	$\dfrac{6 \text{ mol}(1.008 \text{ g/mol})}{30.07 \text{ g}}(100) = 20.11\% \ H$
Propane	C_3H_8	44.09	$\dfrac{8 \text{ mol}(1.008 \text{ g/mol})}{44.09 \text{ g}}(100) = 18.29\% \ H$
Benzene	C_6H_6	78.11	$\dfrac{6 \text{ mol}(1.008 \text{ g/mol})}{78.11 \text{ g}}(100) = 7.743\% \ H$
Ethanol	C_2H_5OH	46.07	$\dfrac{6 \text{ mol}(1.008 \text{ g/mol})}{46.07 \text{ g}}(100) = 13.13\% \ H$
Cetyl palmitate	$C_{32}H_{64}O_2$	480.83	$\dfrac{64 \text{ mol}(1.008 \text{ g/mol})}{480.83 \text{ g}}(100) = 13.42\% \ H$

The hydrogen percentage decreases in the following order:

Ethane > Propane > Cetyl palmitate > Ethanol > Benzene

3.111   Plan: The key to solving this problem is determining the overall balanced equation. Each individual step must be set up and balanced first. The separate equations can then be combined to get the overall equation. The mass of iron is converted to moles of iron by dividing by the molar mass, and the mole ratio from the balanced equation is used to find the moles and then the mass of CO required to produce that number of moles of iron.

Solution:

a) In the first step, ferric oxide (ferric denotes Fe^{3+}) reacts with carbon monoxide to form Fe_3O_4 and carbon dioxide:

$$3Fe_2O_3(s) + CO(g) \rightarrow 2Fe_3O_4(s) + CO_2(g) \quad (1)$$

In the second step, Fe_3O_4 reacts with more carbon monoxide to form ferrous oxide:

$$Fe_3O_4(s) + CO(g) \rightarrow 3FeO(s) + CO_2(g) \quad (2)$$

In the third step, ferrous oxide reacts with more carbon monoxide to form molten iron:

$$FeO(s) + CO(g) \rightarrow Fe(l) + CO_2(g) \quad (3)$$

Common factors are needed to allow these equations to be combined. The intermediate products are Fe_3O_4 and FeO, so multiply equation (2) by 2 to cancel Fe_3O_4 and equation (3) by 6 to cancel FeO:

$$3Fe_2O_3(s) + CO(g) \rightarrow \cancel{2Fe_3O_4(s)} + CO_2(g)$$
$$\cancel{2Fe_3O_4(s)} + 2CO(g) \rightarrow \cancel{6FeO(s)} + 2CO_2(g)$$
$$\cancel{6FeO(s)} + 6CO(g) \rightarrow 6Fe(l) + 6CO_2(g)$$
$$\overline{3Fe_2O_3(s) + 9CO(g) \rightarrow 6Fe(l) + 9CO_2(g)}$$

Then divide by 3 to obtain the smallest integer coefficients:

$$Fe_2O_3(s) + 3CO(g) \rightarrow 2Fe(s) + 3CO_2(g)$$

b) A metric ton is equal to 1000 kg.

Converting 45.0 metric tons of Fe to mass in grams:

$$\text{Mass (g) of Fe} = \left(45.0 \text{ ton Fe}\right)\left(\frac{10^3 \text{ kg}}{1 \text{ ton}}\right)\left(\frac{10^3 \text{ g}}{1 \text{ kg}}\right) = 4.50 \times 10^7 \text{ g Fe}$$

$$\text{Moles of Fe} = \left(4.50 \times 10^7 \text{ g Fe}\right)\left(\frac{1 \text{ mol Fe}}{55.85 \text{ g Fe}}\right) = 8.05730 \times 10^5 \text{ mol Fe}$$

$$\text{Mass (g) of CO} = \left(8.05730 \times 10^5 \text{ mol Fe}\right)\left(\frac{3 \text{ mol CO}}{2 \text{ mol Fe}}\right)\left(\frac{28.01 \text{ g CO}}{1 \text{ mol CO}}\right) = 3.38527 \times 10^7 = \textbf{3.39} \times \textbf{10}^\textbf{7} \textbf{ g CO}$$

3.113   Plan: If 100.0 g of dinitrogen tetroxide reacts with 100.0 g of hydrazine (N_2H_4), what is the theoretical yield of nitrogen if no side reaction takes place? First, we need to identify the limiting reactant. To determine which reactant is limiting, calculate the amount of nitrogen formed from each reactant, assuming an excess of the other reactant. The reactant that produces less product is the limiting reagent. Use the limiting reagent and the mole ratio from the balanced chemical equation to determine the theoretical yield of nitrogen. Then determine the amount of limiting reactant required to produce 10.0 grams of NO. Reduce the amount of limiting reactant by the amount used to produce NO. The reduced amount of limiting reactant is then used to calculate an "actual yield." The "actual" and theoretical yields will give the maximum percent yield.
Solution:

The balanced reaction is $2N_2H_4(l) + N_2O_4(l) \rightarrow 3N_2(g) + 4H_2O(g)$

Determining the limiting reactant:

Finding the moles of N_2 from the amount of N_2O_4 (if N_2H_4 is limiting):

$$\text{Moles of } N_2 \text{ from } N_2O_4 = \left(100.0 \text{ g } N_2O_4\right)\left(\frac{1 \text{ mol } N_2O_4}{92.02 \text{ g } N_2O_4}\right)\left(\frac{3 \text{ mol } N_2}{1 \text{ mol } N_2O_4}\right) = 3.26016 \text{ mol } N_2$$

Finding the moles of N_2 from the amount of N_2H_4 (if N_2O_4 is limiting):

$$N_2 \text{ from } N_2H_4 = \left(100.0 \text{ g } N_2H_4\right)\left(\frac{1 \text{ mol } N_2H_4}{32.05 \text{ g } N_2H_4}\right)\left(\frac{3 \text{ mol } N_2}{2 \text{ mol } N_2H_4}\right) = 4.68019 \text{ mol } N_2$$

N_2O_4 is the limiting reactant.

$$\text{Theoretical yield of } N_2 = \left(100.0 \text{ g } N_2O_4\right)\left(\frac{1 \text{ mol } N_2O_4}{92.02 \text{ g } N_2O_4}\right)\left(\frac{3 \text{ mol } N_2}{1 \text{ mol } N_2O_4}\right)\left(\frac{28.02 \text{ g } N_2}{1 \text{ mol } N_2}\right) = 91.3497 \text{ g } N_2$$

How much of the limiting reactant is used to produce 10.0 g NO?

$$N_2H_4(l) + 2N_2O_4(l) \rightarrow 6NO(g) + 2H_2O(g)$$

$$\text{Mass (g) of } N_2O_4 \text{ used} = \left(10.0 \text{ g NO}\right)\left(\frac{1 \text{ mol NO}}{30.01 \text{ g NO}}\right)\left(\frac{2 \text{ mol } N_2O_4}{6 \text{ mol NO}}\right)\left(\frac{92.02 \text{ g } N_2O_4}{1 \text{ mol } N_2O_4}\right)$$

$$= 10.221 \text{ g } N_2O_4$$

Amount of N_2O_4 available to produce N_2 = 100.0 g N_2O_4 – mass of N_2O_4 required to produce 10.0 g NO

$$= 100.0 \text{ g} - 10.221 \text{ g} = 89.779 \text{ g } N_2O_4$$

Determine the "actual yield" of N_2 from 89.779 g N_2O_4:

"Actual yield" of N_2 = $(89.779 \text{ g } N_2O_4)\left(\dfrac{1 \text{ mol } N_2O_4}{92.02 \text{ g } N_2O_4}\right)\left(\dfrac{3 \text{ mol } N_2}{1 \text{ mol } N_2O_4}\right)\left(\dfrac{28.02 \text{ g } N_2}{1 \text{ mol } N_2}\right)$

$$= 82.01285 \text{ g } N_2$$

Theoretical yield = $\left(\dfrac{\text{actual yield}}{\text{theoretical yield}}\right)(100) = \left(\dfrac{82.01285}{91.3497}\right)(100) = 89.7790 = \mathbf{89.8\%}$

3.115   <u>Plan:</u> Identify the product molecules and write the balanced equation. To determine the limiting reactant for part b), examine the product circle to see which reactant remains in excess and which reactant was totally consumed. For part c), use the mole ratios in the balanced equation to determine the number of moles of product formed by each reactant, assuming the other reactant is in excess. The reactant that produces fewer moles of product is the limiting reactant. Use the mole ratio between the two reactants to determine the moles of excess reactant required to react with the limiting reactant. The difference between the initial moles of excess reactant and the moles required for reaction is the moles of excess reactant that remain.
<u>Solution:</u>

a) The contents of the boxes give:

       $AB_2 + B_2 \rightarrow AB_3$

Balancing the reaction gives:

       $\mathbf{2AB_2 + B_2 \rightarrow 2AB_3}$

b) Two B_2 molecules remain after reaction so B_2 is in excess. All of the AB_2 molecules have reacted so $\mathbf{AB_2}$ is the limiting reactant.

c) Finding the moles of AB_3 from the moles of AB_2 (if B_2 is limiting):

Moles of AB_3 from AB_2 = $(5.0 \text{ mol } AB_2)\left(\dfrac{2 \text{ mol } AB_3}{2 \text{ mol } AB_2}\right) = 5.0 \text{ mol } AB_3$

Finding the moles of AB_3 from the moles of B_2 (if AB_2 is limiting):

Moles of AB_3 from B_2 = $(3.0 \text{ mol } B_2)\left(\dfrac{2 \text{ mol } AB_3}{1 \text{ mol } B_2}\right) = 6.0 \text{ mol } AB_3$

AB_2 is the limiting reagent and **5.0 mol of AB_3** is formed.

d) Moles of B_2 that react with 5.0 mol AB_2 = $(5.0 \text{ mol } AB_2)\left(\dfrac{1 \text{ mol } B_2}{2 \text{ mol } AB_2}\right) = 2.5 \text{ mol } B_2$

The unreacted B_2 is 3.0 mol – 2.5 mol = **0.5 mol B_2**.

3.116   <u>Plan:</u> Write the formulas in the form $C_xH_yO_z$. Reduce the formulas to obtain the empirical formulas. Add the atomic masses in that empirical formula to obtain the molecular mass.
<u>Solution:</u>

Compound A: $C_4H_{10}O_2 = C_2H_5O$ Compound B: C_2H_4O

Compound C: $C_4H_8O_2 = C_2H_4O$ Compound D: $C_6H_{12}O_3 = C_2H_4O$

Compound E: $C_5H_8O_2$

Compounds **B, C and D** all have the same empirical formula, C_2H_4O. The molecular mass of this formula is (2 x 12.01 g/mol C) + (4 x 1.008 g/mol H) + (1 x 16.00 g/mol O) = **44.05 g/mol**.

3.117   <u>Plan:</u> Write the balanced chemical equation. Since quantities of two reactants are given, we must determine which is the limiting reactant. To determine which reactant is limiting, calculate the amount of product formed from each reactant, assuming an excess of the other reactant. The reactant that produces less product is the limiting reagent. Use the limiting reactant to determine the theoretical yield of product. The actual yield is given. The actual yield divided by the theoretical yield just calculated (with the result multiplied by 100%) gives the percent yield.

Solution:

Determine the balanced chemical equation:

$ZrOCl_2 \cdot 8H_2O(s) + 4H_2C_2O_4 \cdot 2H_2O(s) + 4KOH(aq) \rightarrow K_2Zr(C_2O_4)_3(H_2C_2O_4) \cdot H_2O(s) + 2KCl(aq) + 20H_2O(l)$

Determining the limiting reactant:

Finding the moles of product from the amount of $ZrOCl_2 \cdot 8H_2O$ (if $H_2C_2O_4 \cdot 2H_2O$ is limiting):

Moles of product from $ZrOCl_2 \cdot 8H_2O$ =

$$\left(1.68 \text{ g } ZrOCl_2 \cdot 8H_2O\right)\left(\frac{1 \text{ mol } ZrOCl_2 \cdot 8H_2O}{322.25 \text{ g } ZrOCl_2 \cdot 8H_2O}\right)\left(\frac{1 \text{ mol product}}{1 \text{ mol } ZrOCl_2 \cdot 8H_2O}\right) = 0.00521334 \text{ mol product}$$

Finding the moles of product from the amount of $H_2C_2O_4 \cdot 2H_2O$ (if $ZrOCl_2 \cdot 8H_2O$ is limiting):

Moles of product from $H_2C_2O_4 \cdot 2H_2O$ =

$$\left(5.20 \text{ g } H_2C_2O_4 \cdot 2H_2O\right)\left(\frac{1 \text{ mol } H_2C_2O_4 \cdot H_2O}{126.07 \text{ g } H_2C_2O_4 \cdot 2H_2O}\right)\left(\frac{1 \text{ mol product}}{4 \text{ mol } H_2C_2O_4 \cdot 2H_2O}\right) = 0.0103117 \text{ mol product}$$

It is not necessary to find the moles of product from KOH because KOH is stated to be in excess.

The $ZrOCl_2 \cdot 8H_2O$ is the limiting reactant, and will be used to calculate the theoretical yield:

$$\text{Mass (g) of product} = \left(0.00521334 \text{ mol product}\right)\left(\frac{541.53 \text{ g product}}{1 \text{ mol product}}\right) = 2.82318 \text{ g product}$$

Calculating the percent yield:

$$\text{Percent yield} = \left(\frac{\text{actual Yield}}{\text{theoretical Yield}}\right) \times 100\% = \left(\frac{1.25 \text{ g}}{2.82318 \text{ g}}\right) \times 100\% = 44.276 = \textbf{44.3\% yield}$$

3.126   Plan: Deal with the methane and propane separately, and combine the results. Balanced equations are needed for each hydrocarbon. The total mass and the percentages will give the mass of each hydrocarbon. The mass of each hydrocarbon is changed to moles, and through the balanced chemical equation the amount of CO_2 produced by each gas may be found. Summing the amounts of CO_2 gives the total from the mixture. For part b), let x and 252 – x represent the masses of CH_4 and C_3H_8, respectively.

Solution:

a) The balanced chemical equations are:

    Methane: $CH_4(g) + 2O_2(g) \rightarrow CO_2(g) + 2H_2O(l)$

    Propane: $C_3H_8(g) + 5O_2(g) \rightarrow 3CO_2(g) + 4H_2O(l)$

Mass (g) of CO_2 from each:

$$\text{Methane: } \left(200. \text{ g Mixture}\right)\left(\frac{25.0\%}{100\%}\right)\left(\frac{1 \text{ mol } CH_4}{16.04 \text{ g } CH_4}\right)\left(\frac{1 \text{ mol } CO_2}{1 \text{ mol } CH_4}\right)\left(\frac{44.01 \text{ g } CO_2}{1 \text{ mol } CO_2}\right) = 137.188 \text{ g } CO_2$$

$$\text{Propane: } \left(200. \text{ g Mixture}\right)\left(\frac{75.0\%}{100\%}\right)\left(\frac{1 \text{ mol } C_3H_8}{44.09 \text{ g } C_3H_8}\right)\left(\frac{3 \text{ mol } CO_2}{1 \text{ mol } C_3H_8}\right)\left(\frac{44.01 \text{ g } CO_2}{1 \text{ mol } CO_2}\right) = 449.183 \text{ g } CO_2$$

Total CO_2 = 137.188 g + 449.183 g = 586.371 = **586 g CO_2**

b) Since the mass of CH_4 + the mass of C_3H_8 = 252 g, let x = mass of CH_4 in the mixture and 252 – x = mass of C_3H_8 in the mixture. Use mole ratios to calculate the amount of CO_2 formed from x amount of CH_4 and the amount of CO_2 formed from 252 – x amount of C_3H_8.

The total mass of CO_2 produced = 748 g.

The total moles of CO_2 produced $= \left(748 \text{ g } CO_2\right)\left(\dfrac{1 \text{ mol } CO_2}{44.01 \text{ g } CO_2}\right) = 16.996 \text{ mol } CO_2$

$16.996 \text{ mol } CO_2 = \left(x \text{ g } CH_4\right)\left(\dfrac{1 \text{ mol } CH_4}{16.04 \text{ g } CH_4}\right)\left(\dfrac{1 \text{ mol } CO_2}{1 \text{ mol } CH_4}\right) + \left(252 - x \text{ g } C_3H_8\right)\left(\dfrac{1 \text{ mol } C_3H_8}{44.09 \text{ g } C_3H_8}\right)\left(\dfrac{3 \text{ mol } CO_2}{1 \text{ mol } C_3H_8}\right)$

$16.996 \text{ mol } CO_2 = \dfrac{x}{16.04} \text{ mol } CO_2 + \dfrac{3(252 - x)}{44.09} \text{ mol } CO_2$

$16.996 \text{ mol } CO_2 = \dfrac{x}{16.04} \text{ mol } CO_2 + \dfrac{756 - 3x}{44.09} \text{ mol } CO_2$

$16.996 \text{ mol } CO_2 = 0.06234x \text{ mol } CO_2 + (17.147 - 0.06804x \text{ mol } CO_2)$

$\qquad\qquad\qquad = 17.147 - 0.0057x$

$x = 26.49 \text{ g } CH_4 \qquad\qquad 252 - x = 252 \text{ g} - 26.49 \text{ g} = 225.51 \text{ g } C_3H_8$

Mass % $CH_4 = \dfrac{\text{mass of } CH_4}{\text{mass of mixture}}(100) = \dfrac{26.49 \text{ g } CH_4}{252 \text{ g mixture}}(100) = \textbf{10.5\% } \mathbf{CH_4}$

Mass % $C_3H_8 = \dfrac{\text{mass of } C_3H_8}{\text{mass of mixture}}(100) = \dfrac{225.51 \text{ g } C_3H_8}{252 \text{ g mixture}}(100) = \textbf{89.5\% } \mathbf{C_3H_8}$

3.127   Plan: If we assume a 100-gram sample of fertilizer, then the 30:10:10 percentages become the masses, in grams, of N, P_2O_5, and K_2O. These masses may be changed to moles of substance, and then to moles of each element. To get the desired x:y:1.0 ratio, divide the moles of each element by the moles of potassium.
Solution:
A 100-gram sample of 30:10:10 fertilizer contains 30 g N, 10 g P_2O_5, and 10 g K_2O.

Moles of N $= \left(30 \text{ g N}\right)\left(\dfrac{1 \text{ mol N}}{14.01 \text{ g N}}\right) = 2.1413 \text{ mol N}$

Moles of P $= \left(10 \text{ g } P_2O_5\right)\left(\dfrac{1 \text{ mol } P_2O_5}{141.94 \text{ g } P_2O_5}\right)\left(\dfrac{2 \text{ mol P}}{1 \text{ mol } P_2O_5}\right) = 0.14090 \text{ mol P}$

Moles of K $= \left(10 \text{ g } K_2O\right)\left(\dfrac{1 \text{ mol } K_2O}{94.20 \text{ g } K_2O}\right)\left(\dfrac{2 \text{ mol K}}{1 \text{ mol } K_2O}\right) = 0.21231 \text{ mol K}$

This gives a N:P:K ratio of 2.1413:0.14090:0.21231
The ratio must be divided by the moles of K and rounded.

$\dfrac{2.1413 \text{ mol N}}{0.21231} = 10.086 \qquad\qquad \dfrac{0.14090 \text{ mol P}}{0.21231} = 0.66365 \qquad\qquad \dfrac{0.21231 \text{ mol K}}{0.21231} = 1$

$\qquad\qquad$ 10.086:0.66365:1.000 $\qquad$ or $\qquad$ **10:0.66:1.0**

3.132   Plan: Determine the molecular formula from the figure. Once the molecular formula is known, use the periodic table to determine the molar mass. Convert the volume of lemon juice in part b) from qt to mL and use the density to convert from mL to mass in g. Take 6.82% of that mass to find the mass of citric acid and use the molar mass to convert to moles.
Solution:

a) The formula of citric acid obtained by counting the number of carbon atoms, oxygen atoms, and hydrogen atoms is $\mathbf{C_6H_8O_7}$.
$\qquad$ Molar mass $= (6 \times 12.01 \text{ g/mol C}) + (8 \times 1.008 \text{ g/mol H}) + (7 \times 16.00 \text{ g/mol O}) = \textbf{192.12 g/mol}$

b) Converting volume of lemon juice in qt to mL:

$$\text{Volume (mL) of lemon juice} = (1.50 \text{ qt})\left(\frac{1 \text{ L}}{1.057 \text{ qt}}\right)\left(\frac{1 \text{ mL}}{10^{-3} \text{ L}}\right) = 1419.111 \text{ mL}$$

Converting volume to mass in grams:

$$\text{Mass (g) of lemon juice} = (1419.111 \text{ mL})\left(\frac{1.09 \text{ g}}{\text{mL}}\right) = 1546.831 \text{ g lemon juice}$$

$$\text{Mass (g) of } C_6H_8O_7 = (1546.831 \text{ g lemon juice})\left(\frac{6.82\% \ C_6H_8O_7}{100\% \text{ lemon juice}}\right) = 105.494 \text{ g } C_6H_8O_7$$

$$\text{Moles of } C_6H_8O_7 = (105.494 \text{ g } C_6H_8O_7)\left(\frac{1 \text{ mol } C_6H_8O_7}{192.12 \text{ g } C_6H_8O_7}\right) = 0.549104 = \textbf{0.549 mol } C_6H_8O_7$$

3.133   Plan: Determine the formulas of each reactant and product, then balance the individual equations. Remember that nitrogen and oxygen are diatomic. Combine the three smaller equations to give the overall equation, where some substances serve as intermediates and will cancel. Use the mole ratio between nitrogen and nitric acid in the overall equation to find the moles and then mass of nitric acid produced. The amount of nitrogen in metric tons must be converted to mass in grams to convert the mass of nitrogen to moles.
Solution:
a) Nitrogen and oxygen combine to form nitrogen monoxide:

$$N_2(g) + O_2(g) \rightarrow 2NO(g)$$

Nitrogen monoxide reacts with oxygen to form nitrogen dioxide:

$$2NO(g) + O_2(g) \rightarrow 2NO_2(g)$$

Nitrogen dioxide combines with water to form nitric acid and nitrogen monoxide:

$$3NO_2(g) + H_2O(g) \rightarrow 2HNO_3(aq) + NO(g)$$

b) Combining the reactions may involve adjusting the equations in various ways to cancel out as many materials as possible other than the reactants added and the desired products.

$2 \times (N_2(g) + O_2(g) \rightarrow 2NO(g)) =$ $\qquad$ $2N_2(g) + 2O_2(g) \rightarrow \cancel{4NO(g)}$

$3 \times (2NO(g) + O_2(g) \rightarrow 2NO_2(g)) =$ $\qquad$ $\cancel{6NO(g)} + 3O_2(g) \rightarrow \cancel{6NO_2(g)}$

$2 \times (3NO_2(g) + H_2O(g) \rightarrow 2HNO_3(aq) + NO(g)) =$ $\cancel{6NO_2(g)} + 2H_2O(g) \rightarrow 4HNO_3(aq) + \cancel{2NO(g)}$

Multiplying the above equations as shown results in the 6 moles of NO on each side and the 6 moles of NO$_2$ on each side canceling. Adding the equations gives:

$$2N_2(g) + 5O_2(g) + 2H_2O(g) \rightarrow 4HNO_3(aq)$$

c) $\text{Mass (g) of } N_2 = (1350 \text{ t } N_2)\left(\frac{10^3 \text{ kg}}{1\text{t}}\right)\left(\frac{10^3 \text{ g}}{1 \text{ kg}}\right) = 1.35 \times 10^9 \text{ g } N_2$

$\text{Moles of } N_2 = (1.35 \times 10^9 \text{ g } N_2)\left(\frac{1 \text{ mol } N_2}{28.02 \text{ g } N_2}\right) = 4.817987 \times 10^7 \text{ mol } N_2$

$\text{Mass (g) of } HNO_3 = (4.817987 \times 10^7 \text{ mol } N_2)\left(\frac{4 \text{ mol } HNO_3}{2 \text{ mol } N_2}\right)\left(\frac{63.02 \text{ g } HNO_3}{1 \text{ mol } HNO_3}\right) = 6.072591 \times 10^9 \text{ g } HNO_3$

Metric tons $HNO_3 =$

$$(6.072591 \times 10^9 \text{ g } HNO_3)\left(\frac{1 \text{ kg}}{10^3 \text{ g}}\right)\left(\frac{1\text{t}}{10^3 \text{ kg}}\right) = 6.072591 \times 10^3 = \textbf{6.07} \times \textbf{10}^3 \textbf{ metric tons } HNO_3$$

3.134   Plan: Write and balance the chemical reaction. Use the mole ratio to find the amount of product that should be produced and take 66% of that amount to obtain the actual yield.
Solution:
$2NO(g) + O_2(g) \rightarrow 2NO_2(g)$
With 6 molecules of NO and 3 molecules of O_2 reacting, 6 molecules of NO_2 can be produced. If the reaction only has a 66% yield, then $(0.66)(6) = 4$ molecules of NO_2 will be produced. **Circle A** shows the formation of 4 molecules of NO_2. Circle B also shows the formation of 4 molecules of NO_2 but also has 2 unreacted molecules of NO and 1 unreacted molecule of O_2. Since neither reactant is limiting, there will be no unreacted reactant remaining after the reaction is over.

3.136   Plan: Use the mass percent to find the mass of heme in the sample; use the molar mass to convert the mass of heme to moles. Then find the mass of Fe in the sample by using the mole ratio between heme and iron. The mass of hemin is found by using the mole ratio between heme and hemoglobin.
Solution:

a) Mass (g) of heme $= \left(0.65 \text{ g hemoglobin}\right)\left(\dfrac{6.0\% \text{ heme}}{100\% \text{ hemoglobin}}\right) = \textbf{0.039g heme}$

b) Moles of heme $= \left(0.039 \text{ g heme}\right)\left(\dfrac{1 \text{ mol heme}}{616.49 \text{ g heme}}\right) = 6.32614 \times 10^{-5} = \textbf{6.3} \times \textbf{10}^{-5} \textbf{ mol heme}$

c) Mass (g) of Fe $= \left(6.32614 \times 10^{-5} \text{ mol heme}\right)\left(\dfrac{1 \text{ mol Fe}}{1 \text{ mol heme}}\right)\left(\dfrac{55.85 \text{ g Fe}}{1 \text{ mol Fe}}\right)$

$= 3.5331 \times 10^{-3} = \textbf{3.5} \times \textbf{10}^{-3} \textbf{ g Fe}$

d) Mass (g) of hemin $= \left(6.32614 \times 10^{-5} \text{ mol heme}\right)\left(\dfrac{1 \text{ mol hemin}}{1 \text{ mol heme}}\right)\left(\dfrac{651.94 \text{ g hemin}}{1 \text{ mol hemin}}\right)$

$= 4.1243 \times 10^{-2} = \textbf{4.1} \times \textbf{10}^{-2} \textbf{ g hemin}$

3.138   Plan: Determine the formula and the molar mass of each compound. The formula gives the relative number of moles of nitrogen present. Multiply the number of moles of nitrogen by its molar mass to find the total mass of nitrogen in 1 mole of compound. Mass percent $= \dfrac{\text{total mass of element}}{\text{molar mass of compound}}(100)$. For part b), convert mass of ornithine to moles, use the mole ratio between ornithine and urea to find the moles of urea, and then use the ratio between moles of urea and nitrogen to find the moles and mass of nitrogen produced.
Solution:
a) Urea: CH_4N_2O, $\mathcal{M} = 60.06$ g/mol
There are 2 moles of N in 1 mole of CH_4N_2O.

Mass (g) of N $= (2 \text{ mol N})\left(\dfrac{14.01 \text{ g N}}{1 \text{ mol N}}\right) = 28.02 \text{ g N}$

Mass percent $= \dfrac{\text{total mass N}}{\text{molar mass of compound}}(100) = \dfrac{28.02 \text{ g N}}{60.06 \text{ g } CH_4N_2O}(100) = 46.6533 = \textbf{46.65\% N in urea}$

Arginine: $C_6H_{15}N_4O_2$, $\mathcal{M} = 175.22$ g/mol
There are 4 moles of N in 1 mole of $C_6H_{15}N_4O_2$.

Mass (g) of N $= (4 \text{ mol N})\left(\dfrac{14.01 \text{ g N}}{1 \text{ mol N}}\right) = 56.04 \text{ g N}$

Mass percent $= \dfrac{\text{total mass N}}{\text{molar mass of compound}}(100) = \dfrac{56.04 \text{ g N}}{175.22 \text{ g } C_6H_{15}N_4O_2}(100)$

$= 31.98265 = \textbf{31.98\% N in arginine}$

Ornithine: $C_5H_{13}N_2O_2$, $\mathcal{M} = 133.17$ g/mol

There are 2 moles of N in 1 mole of $C_5H_{13}N_2O_2$.

$$\text{Mass (g) of N} = (2 \text{ mol N})\left(\frac{14.01 \text{ g N}}{1 \text{ mol N}}\right) = 28.02 \text{ g N}$$

$$\text{Mass percent} = \frac{\text{total mass N}}{\text{molar mass of compound}}(100) = \frac{28.02 \text{ g N}}{133.17 \text{ g } C_5H_{13}N_2O_2}(100)$$

$$= 21.04077 = \textbf{21.04\% N in ornithine}$$

b) Moles of urea $= \left(135.2 \text{ g } C_5H_{13}N_2O_2\right)\left(\frac{1 \text{ mol } C_5H_{13}N_2O_2}{133.17 \text{ g } C_5H_{13}N_2O_2}\right)\left(\frac{1 \text{ mol } CH_4N_2O}{1 \text{ mol } C_5H_{13}N_2O_2}\right) = 1.015244 \text{ mol urea}$

$$\text{Mass (g) of nitrogen} = \left(1.015244 \text{ mol } CH_4N_2O\right)\left(\frac{2 \text{ mol N}}{1 \text{ mol } CH_4N_2O}\right)\left(\frac{14.01 \text{ g N}}{1 \text{ mol N}}\right) = 28.447 = \textbf{28.45 g N}$$

3.140 Plan: Determine the molar mass of each product and use the equation for percent atom economy.

Solution:

Molar masses of product: N_2H_4: 32.05 g/mol   NaCl: 58.44 g/mol H_2O: 18.02 g/mol

$$\% \text{ atom economy} = \frac{\text{no. of moles x molar mass of desired products}}{\text{sum of }\left(\text{no. of moles x molar mass}\right)\text{ for all products}} \text{ x } 100\%$$

% atom economy =

$$\frac{(1 \text{ mol})\left(\text{molar mass of } N_2H_4\right)}{(1 \text{ mol})\left(\text{molar mass of } N_2H_4\right) + (1 \text{ mol})\left(\text{molar mass of NaCl}\right) + (1 \text{ mol})\left(\text{molar mass of } H_2O\right)} \text{ x } 100\%$$

$$= \frac{\left(1 \text{ mol}\right)\left(32.05 \text{ g/mol}\right)}{(1 \text{ mol})(32.05 \text{ g/mol}) + (1 \text{ mol})(58.44 \text{ g/mol}) + (1 \text{ mol})(18.02 \text{ g/mol})} \text{ x } 100\%$$

$$= 29.5364 = \textbf{29.54\% atom economy}$$

3.142 Plan: Convert the mass of ethanol to moles, and use the mole ratio between ethanol and diethyl ether to determine the theoretical yield of diethyl ether. The actual yield divided by the theoretical yield just calculated (with the result multiplied by 100%) gives the percent yield. The difference between the actual and theoretical yields is related to the quantity of ethanol that did not produce diethyl ether, forty-five percent of which produces ethylene instead. Use the mole ratio between ethanol and ethylene to find the mass of ethylene produced by the forty-five percent of ethanol that did not produce diethyl ether.

Solution:

a) The determination of the theoretical yield:

Mass (g) of diethyl ether =

$$\left(50.0 \text{ g } CH_3CH_2OH\right)\left(\frac{1 \text{ mol } CH_3CH_2OH}{46.07 \text{ g } CH_3CH_2OH}\right)\left(\frac{1 \text{ mol } CH_3CH_2OCH_2CH_3}{2 \text{ mol } CH_3CH_2OH}\right)\left(\frac{74.12 \text{ g } CH_3CH_2OCH_2CH_3}{1 \text{ mol } CH_3CH_2OCH_2CH_3}\right)$$

$$= 40.2214 \text{ g diethyl ether}$$

Determining the percent yield:

$$\text{Percent yield} = \left(\frac{\text{actual Yield}}{\text{theoretical Yield}}\right) \text{ x } 100\% = \left(\frac{35.9 \text{ g}}{40.2214 \text{ g}}\right) \text{ x } 100\% = 89.2560 = \textbf{89.3\% yield}$$

b) To determine the amount of ethanol not producing diethyl ether, we will use the difference between the theoretical yield and actual yield to determine the amount of diethyl ether that did not form and hence, the amount of ethanol that did not produce the desired product. Forty-five percent of this amount will be used to determine the amount of ethylene formed.

Mass difference = theoretical yield − actual yield = 40.2214 g − 35.9 g = 4.3214 g diethyl ether that did not form

Mass (g) of ethanol not producing diethyl ether =

$$\left(4.3214 \text{ g CH}_3\text{CH}_2\text{OCH}_2\text{CH}_3\right)\left(\frac{1 \text{ mol CH}_3\text{CH}_2\text{OCH}_2\text{CH}_3}{74.12 \text{ g CH}_3\text{CH}_2\text{OCH}_2\text{CH}_3}\right)\left(\frac{2 \text{ mol CH}_3\text{CH}_2\text{OH}}{1 \text{ mol CH}_3\text{CH}_2\text{OCH}_2\text{CH}_3}\right)\left(\frac{46.07 \text{ g CH}_3\text{CH}_2\text{OH}}{1 \text{ mol CH}_3\text{CH}_2\text{OH}}\right)$$

$$= 5.37202 \text{ g ethanol}$$

Mass of ethanol producing ethylene = $\left(5.37202 \text{ g CH}_3\text{CH}_2\text{OH}\right)\left(\frac{45.0\%}{100\%}\right) = 2.417409$ g ethanol

Mass (g) of ethylene =

$$\left(2.417409 \text{ g CH}_3\text{CH}_2\text{OH}\right)\left(\frac{1 \text{ mol CH}_3\text{CH}_2\text{OH}}{46.07 \text{ g CH}_3\text{CH}_2\text{OH}}\right)\left(\frac{1 \text{ mol C}_2\text{H}_4}{1 \text{ mol CH}_3\text{CH}_2\text{OH}}\right)\left(\frac{28.05 \text{ g C}_2\text{H}_4}{1 \text{ mol C}_2\text{H}_4}\right)$$

$$= 1.47185 = \textbf{1.47 g ethylene}$$

3.144   Plan: For part a), use the given solubility of the salt to find the mass that is soluble in the given volume of water. For part b), convert the mass of dissolved salt in part a) to moles of salt and then to moles of cocaine and then to mass of cocaine. Use the solubility of cocaine to find the volume of water needed to dissolve this mass of cocaine.
Solution:

a) Mass (g) of dissolved salt = $\left(50.0 \text{ mL H}_2\text{O}\right)\left(\frac{10^{-3} \text{ L H}_2\text{O}}{1 \text{ mL H}_2\text{O}}\right)\left(\frac{2.50 \text{ kg salt}}{1 \text{ L H}_2\text{O}}\right)\left(\frac{10^3 \text{ g}}{1 \text{ kg}}\right) = \textbf{125 g salt}$

b) Moles of dissolved salt = $\left(125 \text{ g salt}\right)\left(\frac{1 \text{ mol salt}}{339.81 \text{ g salt}}\right) = 0.367853$ mol salt

Mass (g) cocaine = $\left(0.367853 \text{ mol salt}\right)\left(\frac{1 \text{ mol cocaine}}{1 \text{ mol salt}}\right)\left(\frac{303.35 \text{ g cocaine}}{1 \text{ mol cocaine}}\right) = 111.588$ g cocaine

Volume (L) of water needed to dissolve the cocaine = $\left(111.588 \text{ g cocaine}\right)\left(\frac{1 \text{ L}}{1.70 \text{ g cocaine}}\right) = 65.64$ L

Additional water needed = total volume needed − original volume of water

$$= 65.64 \text{ L} − 0.0500 \text{ L} = 65.59 = \textbf{65.6 L H}_2\textbf{O}$$

CHAPTER 4 THREE MAJOR CLASSES OF CHEMICAL REACTIONS

FOLLOW–UP PROBLEMS

4.1A <u>Plan:</u> Examine each compound to see what ions, and how many of each, result when the compound is dissolved in water and match one compound's ions to those in the beaker. Use the total moles of particles and the molar ratio in the compound's formula to find moles and then mass of compound.
<u>Solution:</u>

a) $LiBr(s) \xrightarrow{H_2O} Li^+(aq) + Br^-(aq)$

$Cs_2CO_3(s) \xrightarrow{H_2O} 2Cs^+(aq) + CO_3^{2-}(aq)$

$BaCl_2(s) \xrightarrow{H_2O} Ba^{2+}(aq) + 2Cl^-(aq)$

Since the beaker contains +2 ions and twice as many –1 ions, the electrolyte is **$BaCl_2$**.

b) Mass (g) of $BaCl_2 = \left(3\ Ba^{2+}\ \text{particles}\right)\left(\dfrac{0.05\ \text{mol}\ Ba^{2+}\ \text{ions}}{1\ Ba^{2+}\ \text{particle}}\right)\left(\dfrac{1\ \text{mol}\ BaCl_2}{1\ \text{mol}\ Ba^{2+}\ \text{ions}}\right)\left(\dfrac{208.2\ \text{g}\ BaCl_2}{1\ \text{mol}\ BaCl_2}\right)$

$= 31.23 = $ **31.2 g $BaCl_2$**

4.1B <u>Plan:</u> Write the formula for sodium phosphate and then write a balanced equation showing the ions that result when sodium phosphate is placed in water. Use the balanced equation to determine the number of ions that result when 2 formula units of sodium phosphate are placed in water. The molar ratio from the balanced equation gives the relationship between the moles of sodium phosphate and the moles of ions produced. Use this molar ratio to calculate the moles of ions produced when 0.40 mol of sodium phosphate is placed in water.
<u>Solution:</u>

a) The formula for sodium phosphate is Na_3PO_4. When the compound is placed in water, 4 ions are produced for each formula unit of sodium phosphate: three sodium ions, Na^+, and 1 phosphate ion, PO_4^{3-}.

$Na_3PO_4(s) \xrightarrow{H_2O} 3Na^+(aq) + PO_4^{3-}(aq)$

If two formula units of sodium phosphate are placed in water, twice as many ions should be produced:

$2Na_3PO_4(s) \xrightarrow{H_2O} 6Na^+(aq) + 2PO_4^{3-}(aq)$

Any drawing should include 2 phosphate ions (each with a 3– charge) and 6 sodium ions (each with a 1+ charge).

b) Moles of ions $= 0.40\ \text{mol}\ Na_3PO_4\left(\dfrac{4\ \text{mol ions}}{1\ \text{mol}\ Na_3PO_4}\right) = $ **1.6 moles of ions**

4.2A <u>Plan:</u> Write an equation showing the dissociation of one mole of compound into its ions. Use the given information to find the moles of compound; use the molar ratio between moles of compound and moles of ions in the dissociation equation to find moles of ions.
<u>Solution:</u>

a) One mole of $KClO_4$ dissociates to form one mole of potassium ions and one mole of perchlorate ions.

$KClO_4(s) \xrightarrow{H_2O} K^+(aq) + ClO_4^-(aq)$

Therefore, 2 moles of solid $KClO_4$ produce **2 mol of K^+** ions and **2 mol of ClO_4^-** ions.

b) $Mg(C_2H_3O_2)_2(s) \xrightarrow{H_2O} Mg^{2+}(aq) + 2C_2H_3O_2^-(aq)$

First convert grams of $Mg(C_2H_3O_2)_2$ to moles of $Mg(C_2H_3O_2)_2$ and then use molar ratios to determine the moles of each ion produced.

Moles of $Mg(C_2H_3O_2)_2 = \left(354\ \text{g}\ Mg(C_2H_3O_2)_2\right)\left(\dfrac{1\ \text{mol}\ Mg(C_2H_3O_2)_2}{142.40\ \text{g}\ Mg(C_2H_3O_2)_2}\right) = 2.48596\ \text{mol}$

The dissolution of 2.48596 mol $Mg(C_2H_3O_2)_2(s)$ produces **2.49 mol Mg^{2+}** and (2 x 2.48596) = **4.97 mol $C_2H_3O_2^-$**

c) $(NH_4)_2CrO_4(s) \xrightarrow{H_2O} 2NH_4^+(aq) + CrO_4^{2-}(aq)$

First convert formula units to moles.

$$\text{Moles of } (NH_4)_2CrO_4 = \left(1.88 \times 10^{24}\,FU\right)\left(\frac{1\,\text{mol}\,(NH_4)_2CrO_4}{6.022 \times 10^{23}\,FU}\right) = 3.121886\,\text{mol}$$

The dissolution of 3.121886 mol $(NH_4)_2CrO_4(s)$ produces $(2 \times 3.121886) =$ **6.24 mol NH_4^+** and **3.12 mol CrO_4^{2-}**.

4.2B   Plan: Write an equation showing the dissociation of one mole of compound into its ions. Use the given information to find the moles of compound; use the molar ratio between moles of compound and moles of ions in the dissociation equation to find moles of ions.
Solution:

a) One mole of Li_2CO_3 dissociates to form two moles of lithium ions and one mole of carbonate ions.

$$Li_2CO_3(s) \xrightarrow{H_2O} 2Li^+(aq) + CO_3^{2-}(aq)$$

Therefore, 4 moles of solid Li_2CO_3 produce **8 mol of Li^+** ions and **4 mol of CO_3^{2-}** ions.

b) $Fe_2(SO_4)_3(s) \xrightarrow{H_2O} 2Fe^{3+}(aq) + 3SO_4^{2-}(aq)$

First convert grams of $Fe_2(SO_4)_3$ to moles of $Fe_2(SO_4)_3$ and then use molar ratios to determine the moles of each ion produced.

$$\text{Moles of } Fe_2(SO_4)_3 = (112\,g\,Fe_2(SO_4)_3)\left(\frac{1\,\text{mol}\,Fe_2(SO_4)_3}{399.88\,g\,Fe_2(SO_4)_3}\right) = 0.2801 = 0.280\,\text{mol}\,Fe_2(SO_4)_3$$

The dissolution of 0.280 mol $Fe_2(SO_4)_3(s)$ produces **0.560 mol Fe^{3+}** (2×0.2801) and **0.840 mol SO_4^{2-}** (3×0.2801)

c) $Al(NO_3)_3(s) \xrightarrow{H_2O} Al^{3+}(aq) + 3NO_3^-(aq)$

First convert formula units of $Al(NO_3)_3$ to moles of $Al(NO_3)_3$ and then use molar ratios to determine the moles of each ion produced.

$$\text{Moles of } Al(NO_3)_3 = (8.09 \times 10^{22}\,\text{formula units}\,Al(NO_3)_3)\left(\frac{1\,\text{mol}\,Al(NO_3)_3}{6.022 \times 10^{23}\,\text{formula units}\,Al(NO_3)_3}\right)$$

$$= 0.1343\,\text{mol}\,Al(NO_3)_3$$

The dissolution of 0.134 mol $Al(NO_3)_3$ (s) produces **0.134 mol Al^{3+}** and **0.403 mol NO_3^-** (3×0.1343)

4.3A   Plan: Convert the volume from mL to liters. Convert the mass to moles by dividing by the molar mass of KI. Divide the moles by the volume in liters to calculate molarity.
Solution:

$$\text{Amount (moles) of KI} = 6.97\,g\,KI\left(\frac{1\,\text{mol}\,KI}{166.0\,g\,KI}\right) = 0.041988\,\text{mol}\,KI$$

$$M = \left(\frac{0.041988\,\text{mol}\,KI}{100.\,mL}\right)\left(\frac{1000\,mL}{1\,L}\right) = 0.41988 = \textbf{0.420}\,\textbf{\textit{M}}$$

4.3B   Plan: Convert the volume from mL to liters. Convert the mass to moles by first converting mg to g and then dividing by the molar mass of $NaNO_3$. Divide the moles by the volume in liters to calculate molarity.
Solution:

$$\text{Amount (moles) of } NaNO_3 = 175\,mg\,NaNO_3\left(\frac{1\,g}{1000\,mg}\right)\left(\frac{1\,\text{mol}\,NaNO_3}{85.00\,g\,NaNO_3}\right) = 0.0020588\,\text{mol}\,NaNO_3$$

$$M = \left(\frac{0.0020588\,\text{mol}\,NaNO_3}{15.0\,mL}\right)\left(\frac{1000\,mL}{1\,L}\right) = 0.13725 = \textbf{0.137}\,\textbf{\textit{M}}$$

4.4A Plan: Divide the mass of sucrose by its molar mass to change the grams to moles. Divide the moles of sucrose by the molarity to obtain the volume of solution.
Solution:
Volume (L) of solution =

$$\left(135 \text{ g } C_{12}H_{22}O_{11}\right)\left(\frac{1 \text{ mol } C_{12}H_{22}O_{11}}{342.30 \text{ g } C_{12}H_{22}O_{11}}\right)\left(\frac{1 \text{ L}}{3.30 \text{ mol } C_{12}H_{22}O_{11}}\right) = 0.11951 = \textbf{0.120 L}$$

Road map:

Mass (g) of sucrose

Divide by $\mathcal{M}$ (g/mol)

Amount (moles) of sucrose

Divide by M (mol/L)

Volume (L) of solution

4.4B Plan: Convert the volume from mL to L; then multiply by the molarity of the solution to obtain the moles of H_2SO_4.
Solution:

$$\text{Amount (mol) of } H_2SO_4 = 40.5 \text{ mL } H_2SO_4 \left(\frac{1 \text{ L}}{1000 \text{ mL}}\right)\left(\frac{0.128 \text{ mol } H_2SO_4}{1 \text{ L}}\right) = 0.005184 = \textbf{0.00518 mol } H_2SO_4$$

Road map:

Volume (mL) of soln

1000 mL = 1 L

Volume (L) of soln

Multiply by M
(1 L soln = 0.128 mol H_2SO_4)

Amount (mol) H_2SO_4

4.5A Plan: Multiply the volume and molarity to calculate the number of moles of sodium phosphate in the solution. Write the formula of sodium phosphate. Determine the number of each type of ion that is included in each formula unit. Use this information to determine the amount of each type of ion in the described solution.
Solution:

$$\text{Amount (mol) of } Na_3PO_4 = 1.32 \text{ L} \left(\frac{0.55 \text{ mol } Na_3PO_4}{1 \text{ L}}\right) = 0.7260 = 0.73 \text{ mol } Na_3PO_4$$

In each formula unit of Na_3PO_4, there are 3 Na^+ ions and 1 PO_4^{3-} ion.

$$\text{Amount (mol) of } Na^+ = 0.7260 \text{ mol } Na_3PO_4 \left(\frac{3 \text{ mol } Na^+}{1 \text{ mol } Na_3PO_4}\right) = 2.178 = \textbf{2.2 mol } Na^+$$

$$\text{Amount (mol) of } PO_4^{3-} = 0.73 \text{ mol } Na_3PO_4 \left(\frac{1 \text{ mol } PO_4^{3-}}{1 \text{ mol } Na_3PO_4}\right) = \textbf{0.73 mol } PO_4^{3-}$$

4.5B Plan: Write the formula of aluminum sulfate. Determine the number of aluminum ions in each formula unit. Calculate the number of aluminum ions in the sample of aluminum sulfate. Convert the volume from mL to L. Divide the number of moles of aluminum ion by the volume in L to calculate the molarity of the solution.
Solution:
In each formula unit of $Al_2(SO_4)_3$, there are 2 Al^{3+} ions.

$$\text{Amount (mol) of } Al^{3+} = 1.25 \text{ mol } Al_2(SO_4)_3 \left(\frac{2 \text{ mol } Al^{3+}}{1 \text{ mol } Al_2(SO_4)_3} \right) = 2.50 \text{ mol } Al^{3+}$$

$$M = \left(\frac{2.50 \text{ mol } Al^{3+}}{875 \text{ mL}} \right) \left(\frac{1000 \text{ mL}}{1 \text{ L}} \right) = \textbf{2.86 } \textbf{\textit{M}}$$

4.6A Plan: Determine the new volume from the dilution equation $(M_{conc})(V_{conc}) = (M_{dil})(V_{dil})$.
Solution:

$$V_{dil} = \frac{M_{conc}V_{conc}}{M_{dil}} = \frac{(4.50 \text{ M})(60.0 \text{ mL})}{(1.25 \text{ } M)} = \textbf{216 mL}$$

4.6B Plan: Determine the new concentration from the dilution equation $(M_{conc})(V_{conc}) = (M_{dil})(V_{dil})$. Convert the molarity (mol/L) to g/mL in two steps (one step is moles to grams, and the other step is L to mL).
Solution:

$$M_{dil} = \frac{M_{conc}V_{conc}}{V_{dil}} = \frac{(7.50 \text{ M})(25.0 \text{ m}^3)}{500. \text{ m}^3} = 0.375 \text{ } M$$

$$\text{Concentration (g/mL)} = \left(\frac{0.375 \text{ mol } H_2SO_4}{1 \text{ L}} \right) \left(\frac{98.08 \text{ g } H_2SO_4}{1 \text{ mol } H_2SO_4} \right) \left(\frac{10^{-3} \text{ L}}{1 \text{ mL}} \right)$$

$$= 0.036780 = \textbf{3.68 x } \textbf{10}^{-2} \textbf{ g/mL} \text{ solution}$$

4.7A Plan: Count the number of particles in each solution per unit volume.
Solution:
Solution A has 6 particles per unit volume while Solution B has 12 particles per unit volume. Solution B is more concentrated than Solution A. To obtain Solution B, the total volume of Solution A was reduced by half:

$$V_{conc} = \frac{N_{dil}V_{dil}}{N_{conc}} = \frac{(6 \text{ particles})(1.0 \text{ mL})}{(12 \text{ particles})} = 0.50 \text{ mL}$$

Solution C has 4 particles and is thus more dilute than Solution A. To obtain Solution C, ½ the volume of solvent must be added for every volume of Solution A:

$$V_{dil} = \frac{N_{conc}V_{conc}}{N_{dil}} = \frac{(6 \text{ particles})(1.0 \text{ mL})}{(4 \text{ particles})} = 1.5 \text{ mL}$$

4.7B Plan: Count the number of particles in each solution per unit volume. Determine the final volume of the solution. Use the dilution equation, $(N_{conc})(V_{conc}) = (N_{dil})(V_{dil})$, to determine the number of particles that will be present when 300. mL of solvent is added to the 100. mL of solution represented in circle A. (Because M is directly proportional to the number of particles in a given solution, we can replace the molarity terms in the dilution equation with terms representing the number of particles.)
Solution:
There are 12 particles in circle A, 3 particles in circle B, 4 particles in circle C, and 6 particles in circle D. The concentrated solution (circle A) has a volume of 100. mL. 300. mL of solvent are added, so the volume of the diluted solution is 400. mL.

$$N_{dil} = \frac{N_{conc}V_{conc}}{V_{dil}} = \frac{(12 \text{ particles})(100.0 \text{ mL})}{(400. \text{ mL})} = 3 \text{ particles}$$

There are 3 particles in circle B, so **circle B** represents the diluted solution.

4.8A   Plan: Determine the ions present in each substance on the reactant side and write new cation-anion combinations. Use Table 4.1 to determine if either of the combinations of ions is not soluble. If a precipitate forms there will be a reaction and chemical equations may be written. The molecular equation simply includes the formulas of the substances and balancing. In the total ionic equation, all soluble substances are written as separate ions. The net ionic equation comes from the total ionic equation by eliminating all substances appearing in identical form (spectator ions) on each side of the reaction arrow.
Solution:
a) The resulting ion combinations that are possible are iron(III) phosphate and cesium chloride. According to Table 4.1, iron(III) phosphate is insoluble, so a reaction occurs. We see that cesium chloride is soluble.
Total ionic equation:

$$Fe^{3+}(aq) + 3Cl^-(aq) + 3Cs^+(aq) + PO_4^{3-}(aq) \rightarrow FePO_4(s) + 3Cl^-(aq) + 3Cs^+(aq)$$

Net ionic equation:

$$Fe^{3+}(aq) + PO_4^{3-}(aq) \rightarrow FePO_4(s)$$

b) The resulting ion combinations that are possible are sodium nitrate (soluble) and cadmium hydroxide (insoluble). A reaction occurs.
Total ionic equation:

$$2Na^+(aq) + 2OH^-(aq) + Cd^{2+}(aq) + 2NO_3^-(aq) \rightarrow Cd(OH)_2(s) + 2Na^+(aq) + 2NO_3^-(aq)$$

Note: The coefficients for Na^+ and OH^- are necessary to balance the reaction and must be included.
Net ionic equation:

$$Cd^{2+}(aq) + 2OH^-(aq) \rightarrow Cd(OH)_2(s)$$

c) The resulting ion combinations that are possible are magnesium acetate (soluble) and potassium bromide (soluble). No reaction occurs.

4.8B   Plan: Determine the ions present in each substance on the reactant side and write new cation-anion combinations. Use Table 4.1 to determine if either of the combinations of ions is not soluble. If a precipitate forms there will be a reaction and chemical equations may be written. The molecular equation simply includes the formulas of the substances and balancing. In the total ionic equation, all soluble substances are written as separate ions. The net ionic equation comes from the total ionic equation by eliminating all substances appearing in identical form (spectator ions) on each side of the reaction arrow.
Solution:

a) The resulting ion combinations that are possible are silver chloride (insoluble, an exception) and barium nitrate (soluble). A reaction occurs.
Total ionic equation:

$$2Ag^+(aq) + 2NO_3^-(aq) + Ba^{2+}(aq) + 2Cl^-(aq) \rightarrow 2AgCl(s) + Ba^{2+}(aq) + 2NO_3^-(aq)$$

Net ionic equation:

$$Ag^+(aq) + Cl^-(aq) \rightarrow AgCl(s)$$

b) The resulting ion combinations that are possible are ammonium sulfide (soluble) and potassium carbonate (soluble). No reaction occurs.
c) The resulting ion combinations that are possible are lead(II) sulfate (insoluble, an exception) and nickel(II) nitrate (soluble). A reaction occurs.
Total ionic equation:

$$Ni^{2+}(aq) + SO_4^{2-}(aq) + Pb^{2+}(aq) + 2NO_3^-(aq) \rightarrow PbSO_4(s) + Ni^{2+}(aq) + 2NO_3^-(aq)$$

Net ionic equation:

$$Pb^{2+}(aq) + SO_4^{2-}(aq) \rightarrow PbSO_4 (s)$$

4.9A   Plan: Look at the ions (number and charge) produced when each of the given compounds dissolves in water and find the match to the ions shown in the beaker. Once the ions in each beaker are known, write new cation-anion combinations and use Table 4.1 to determine if any of the combination of ions is not soluble. If a precipitate forms there will be a reaction and chemical equations may be written. The molecular equation simply includes the formulas of the substances and must be balanced. In the total ionic equation, all soluble substances are written as separate ions. The net ionic equation comes from the total ionic equation by eliminating all substances appearing in identical form (spectator ions) on each side of the reaction arrow.

Solution:

a) Beaker A has four ions with a +2 charge and eight ions with a –1 charge. The beaker contains dissolved **Zn(NO₃)₂** which dissolves to produce Zn^{2+} and NO_3^- ions in a 1:2 ratio. The compound $PbCl_2$ also has a +2 ion and –1 ion in a 1:2 ratio but $PbCl_2$ is insoluble so ions would not result from this compound.

b) Beaker B has three ions with a +2 charge and six ions with a –1 charge. The beaker contains dissolved **Ba(OH)₂** which dissolves to produce Ba^{2+} and OH^- ions in a 1:2 ratio. $Cd(OH)_2$ also has a +2 ion and a –1 ion in a 1:2 ratio but $Cd(OH)_2$ is insoluble so ions would not result from this compound.

c) The resulting ion combinations that are possible are zinc hydroxide (insoluble) and barium nitrate (soluble). The precipitate formed is **Zn(OH)₂**. The spectator ions are **Ba^{2+} and NO_3^-**.

Balanced molecular equation: $Zn(NO_3)_2(aq) + Ba(OH)_2(aq) \rightarrow Zn(OH)_2(s) + Ba(NO_3)_2(aq)$

Total ionic equation:

$$Zn^{2+}(aq) + 2NO_3^-(aq) + Ba^{2+}(aq) + 2OH^-(aq) \rightarrow Zn(OH)_2(s) + Ba^{2+}(aq) + 2NO_3^-(aq)$$

Net ionic equation:

$$Zn^{2+}(aq) + 2OH^-(aq) \rightarrow Zn(OH)_2(s)$$

d) Since there are only six OH^- ions and four Zn^{2+} ions, the OH^- is the limiting reactant.

$$\text{Mass of } Zn(OH)_2 = (6 \text{ OH}^- \text{ ions}) \left(\frac{0.050 \text{ mol OH}^-}{1 \text{OH}^- \text{ particle}} \right) \left(\frac{1 \text{ mol } Zn(OH)_2}{2 \text{ mol OH}^-} \right) \left(\frac{99.40 \text{ g } Zn(OH)_2}{1 \text{ mol } Zn(OH)_2} \right)$$

$$= 14.9100 = \textbf{15 g Zn(OH)}_2$$

4.9B   Plan: Look at the ions (number and charge) produced when each of the given compounds dissolves in water and find the match to the ions shown in the beaker. Once the ions in each beaker are known, write new cation-anion combinations and use Table 4.1 to determine if any of the combination of ions is not soluble. If a precipitate forms there will be a reaction and chemical equations may be written. The molecular equation simply includes the formulas of the substances and must be balanced. In the total ionic equation, all soluble substances are written as separate ions. The net ionic equation comes from the total ionic equation by eliminating all substances appearing in identical form (spectator ions) on each side of the reaction arrow.

Solution:

a) Beaker A has eight ions with a +1 charge and four ions with a –2 charge. The beaker contains dissolved **Li₂CO₃** which dissolves to produce Li^+ and CO_3^{2-} ions in a 2:1 ratio. The compound Ag_2SO_4 also has a +1 ion and –2 ion in a 2:1 ratio but Ag_2SO_4 is insoluble so ions would not result from this compound.

b) Beaker B has three ions with a +2 charge and six ions with a –1 charge. The beaker contains dissolved **CaCl₂** which dissolves to produce Ca^{2+} and Cl^- ions in a 1:2 ratio. $Ni(OH)_2$ also has a +2 ion and a –1 ion in a 1:2 ratio but $Cd(OH)_2$ is insoluble so ions would not result from this compound.

c) The resulting ion combinations that are possible are calcium carbonate (insoluble) and lithium chloride (soluble). The precipitate formed is **CaCO₃**. The spectator ions are **Li^+ and Cl^-**.

Balanced molecular equation: $Li_2CO_3(aq) + CaCl_2(aq) \rightarrow CaCO_3(s) + 2LiCl(aq)$

Total ionic equation:

$$2Li^+(aq) + CO_3^{2-}(aq) + Ca^{2+}(aq) + 2Cl^-(aq) \rightarrow CaCO_3(s) + 2Li^+(aq) + 2Cl^-(aq)$$

Net ionic equation:

$$Ca^{2+}(aq) + CO_3^{2-}(aq) \rightarrow CaCO_3(s)$$

d) Since there are only four CO_3^{2-} ions and three Ca^{2+} ions, the Ca^{2+} is the limiting reactant.

$$\text{Mass of } CaCO_3 = (3 \text{ Ca}^{2+} \text{ ions}) \left(\frac{0.20 \text{ mol Ca}^{2+}}{1 \text{Ca}^{2+} \text{ particle}} \right) \left(\frac{1 \text{ mol } CaCO_3}{1 \text{ mol Ca}^{2+}} \right) \left(\frac{100.09 \text{ g } CaCO_3}{1 \text{ mol } CaCO_3} \right) = 60.0540 = \textbf{60. g CaCO}_3$$

4.10A   Plan: We are given the molarity and volume of calcium chloride solution, and we must find the volume of sodium phosphate solution that will react with this amount of calcium chloride. After writing the balanced equation, we find the amount (mol) of calcium chloride from its molarity and volume and use the molar ratio to find the amount (mol) of sodium phosphate required to react with the calcium chloride. Finally, we use the molarity of the sodium phosphate solution to convert the amount (mol) of sodium phosphate to volume (L).

Solution:

The balanced equation is: $3CaCl_2(aq) + 2Na_3PO_4(aq) \rightarrow Ca_3(PO_4)_2(s) + 6NaCl(aq)$

Finding the volume (L) of Na_3PO_4 needed to react with the $CaCl_2$:

$$\text{Volume (L) of } Na_3PO_4 = 0.300 \text{ L } CaCl_2 \left(\frac{0.175 \text{ moles } CaCl_2}{1 \text{ L}}\right)\left(\frac{2 \text{ mol } Na_3PO_4}{3 \text{ mol } CaCl_2}\right)\left(\frac{1 \text{ L}}{0.260 \text{ mol } Na_3PO_4}\right)$$

$$= 0.1346 = \textbf{0.135 L } Na_3PO_4$$

4.10B   Plan: We are given the mass of silver chloride produced in the reaction of silver nitrate and sodium chloride, and we must find the molarity of the silver nitrate solution. After writing the balanced equation, we find the amount (mol) of silver nitrate that produces the precipitate by dividing the mass of silver chloride produced by its molar mass and then using the molar ratios from the balanced equation. Then we calculate the molarity by dividing the moles of silver nitrate by the volume of the solution (in L).

The balanced equation is: $AgNO_3(aq) + NaCl(aq) \rightarrow AgCl(s) + NaNO_3(aq)$

$$\text{Amount (mol) of } AgNO_3 = 0.148 \text{ g } AgCl \left(\frac{1 \text{ mol } AgCl}{143.4 \text{ g } AgCl}\right)\left(\frac{1 \text{ mol } AgNO_3}{1 \text{ mol } AgCl}\right) = 0.0010321 = 1.03 \times 10^{-3} \text{ mol } AgNO_3$$

$$\text{Molarity } (M) \text{ of } AgNO_3 = \left(\frac{1.0321 \times 10^{-3} \text{ mol } AgNO_3}{45.0 \text{ mL}}\right)\left(\frac{1000 \text{ mL}}{1 \text{ L}}\right) = 2.2935 \times 10^{-2} = \textbf{2.29} \times \textbf{10}^{-2} \textbf{ M}$$

4.11A   Plan: Multiply the volume in liters of each solution by its molarity to obtain the moles of each reactant. Write a balanced equation. Use molar ratios from the balanced equation to determine the moles of lead(II) chloride that may be produced from each reactant. The reactant that generates the smaller number of moles is limiting. Change the moles of lead(II) chloride from the limiting reactant to the grams of product using the molar mass of lead(II) chloride.

Solution:

(a) The balanced equation is:

$$Pb(C_2H_3O_2)_2(aq) + 2NaCl(aq) \rightarrow PbCl_2(s) + 2NaC_2H_3O_2(aq)$$

$$\text{Moles of } Pb(C_2H_3O_2)_2 = (268 \text{ mL})\left(\frac{10^{-3} \text{ L}}{1 \text{ mL}}\right)\left(\frac{1.50 \text{ mol } Pb(C_2H_3O_2)_2}{1 \text{ L}}\right) = 0.402 \text{ mol } Pb(C_2H_3O_2)_2$$

$$\text{Moles of } NaCl = (130 \text{ mL})\left(\frac{10^{-3} \text{ L}}{1 \text{ mL}}\right)\left(\frac{3.40 \text{ mol } NaCl}{1 \text{ L}}\right) = 0.442 \text{ mol } NaCl$$

$$\text{Moles of } PbCl_2 \text{ from } Pb(C_2H_3O_2)_2 = (0.402 \text{ mol } Pb(C_2H_3O_2)_2)\left(\frac{1 \text{ mol } PbCl_2}{1 \text{ mol } Pb(C_2H_3O_2)_2}\right)$$

$$= 0.402 \text{ mol } PbCl_2$$

$$\text{Moles of } PbCl_2 \text{ from } NaCl = (0.442 \text{ mol } NaCl)\left(\frac{1 \text{ mol } PbCl_2}{2 \text{ mol } NaCl}\right) = 0.221 \text{ mol } PbCl_2$$

The NaCl is limiting. The mass of $PbCl_2$ may now be determined using the molar mass.

$$\text{Mass (g) of } PbCl_2 = (0.221 \text{ mol } PbCl_2)\left(\frac{278.1 \text{ g } PbCl_2}{1 \text{ mol } PbCl_2}\right) = 61.4601 = \textbf{61.5 g } PbCl_2$$

(b) Ac is used to represent $C_2H_3O_2$:

Amount (mol)	$Pb(Ac)_2$	+	$2NaCl$	$\rightarrow$	$PbCl_2$	+	$2NaAc$
Initial	0.402		0.442		0		0
Change	−0.221		−0.442		+0.221		+0.442
Final	0.181		0		0.221		0.442

4.11B **Plan:** Write a balanced equation. Multiply the volume in liters of each solution by its molarity to obtain the moles of each reactant. Use molar ratios from the balanced equation to determine the moles of mercury(II) sulfide that may be produced from each reactant. The reactant that generates the smaller number of moles is limiting. Change the moles of mercury(II) sulfide from the limiting reactant to the grams of product using the molar mass of mercury(II) sulfide.
Solution:

(a)The balanced equation is:

$$Hg(NO_3)_2(aq) + 2Na_2S(aq) \rightarrow HgS(s) + 2NaNO_3(aq)$$

$$\text{Moles of } Hg(NO_3)_2 = (0.050\ L)\left(\frac{0.010\ \text{mol } Hg(NO_3)_2}{1\ L}\right) = 5.0 \times 10^{-4}\ \text{mol } Hg(NO_3)_2$$

$$\text{Moles of } Na_2S = (0.020\ L)\left(\frac{0.10\ \text{mol } Na_2S}{1\ L}\right) = 2.0 \times 10^{-3}\ \text{mol } Na_2S$$

$$\text{Moles of HgS from } Hg(NO_3)_2 = (5.0 \times 10^{-4}\ \text{mol } Hg(NO_3)_2)\left(\frac{1\ \text{mol HgS}}{1\ \text{mol } Hg(NO_3)_2}\right) = 5.0 \times 10^{-4}\ \text{mol HgS}$$

$$\text{Moles of HgS from } Na_2S = (2.0 \times 10^{-3}\ \text{mol } Na_2S)\left(\frac{1\ \text{mol HgS}}{1\ \text{mol } Na_2S}\right) = 2.0 \times 10^{-3}\ \text{mol HgS}$$

The $Hg(NO_3)_2$ is limiting. The mass of HgS may now be determined using the molar mass.

$$\text{Mass (g) of HgS} = (2.0 \times 10^{-3}\ \text{mol HgS})\left(\frac{232.7\ \text{g HgS}}{1\ \text{mol HgS}}\right) = 0.11635 = \textbf{0.12 g HgS}$$

(b)

Amount (mol)	$Hg(NO_3)_2$	+ Na_2S	$\rightarrow$	HgS	+ $2NaNO_3$
Initial	5.0×10^{-4}	2.0×10^{-3}		0	0
Change	-5.0×10^{-4}	-5.0×10^{-4}		$+5.0 \times 10^{-4}$	$+1.0 \times 10^{-3}$
Final	0	1.5×10^{-3}		5.0×10^{-4}	1.0×10^{-3}

4.12A **Plan:** Convert the given volume from mL to L and multiply by the molarity (mol/L) to find moles of $Ca(OH)_2$. Each mole of the strong base $Ca(OH)_2$ will produce two moles of hydroxide ions. Finally, multiply the amount (mol) of hydroxide ions by Avogadro's number to calculate the number of hydroxide ions produced.
Solution:

$$Ca(OH)_2(s) \xrightarrow{H_2O} Ca^{2+}(aq) + 2OH^-(aq)$$

No. of OH^- ions produced =

$$451\ mL\ Ca(OH)_2 \left(\frac{1\ L}{1000\ mL}\right)\left(\frac{0.0120\ \text{mole } Ca(OH)_2}{1\ L}\right)\left(\frac{2\ \text{mol } OH^-}{1\ \text{mole } Ca(OH)_2}\right)\left(\frac{6.022 \times 10^{23}\ OH^-\ \text{ions}}{1\ \text{mole } OH^-}\right)$$

$$= 6.5182 \times 10^{21} = \textbf{6.52} \times \textbf{10}^{\textbf{21}}\ \textbf{OH}^-\ \textbf{ions}$$

4.12B **Plan:** Convert the given volume from mL to L and multiply by the molarity (mol/L) to find moles of HCl. Each mole of the strong acid HCl will produce one mole of hydrogen ions. Finally, multiply the amount (mol) of hydrogen ions by Avogadro's number to calculate the number of hydrogen ions produced.
Solution:

$$HCl(g) \xrightarrow{H_2O} H^+(aq) + Cl^-(aq)$$

No. of H^+ ions produced = $65.5\ mL\ HCl \left(\frac{1\ L}{1000\ mL}\right)\left(\frac{0.722\ \text{mole HCl}}{1\ L}\right)\left(\frac{1\ \text{mol } H^+}{1\ \text{mole HCl}}\right)\left(\frac{6.022 \times 10^{23}\ H^+\ \text{ions}}{1\ \text{mole } H^+}\right)$

$$= 2.8479 \times 10^{22} = \textbf{2.85} \times \textbf{10}^{\textbf{22}}\ \textbf{H}^+\ \textbf{ions}$$

4.13A **Plan:** According to Table 4.2, both reactants are strong and therefore completely dissociate in water. Thus, the key reaction is the formation of water. The other product of the reaction is soluble.
Solution:
Molecular equation: $2HNO_3(aq) + Ca(OH)_2(aq) \rightarrow Ca(NO_3)_2(aq) + 2H_2O(l)$
Total ionic equation:

$$2H^+(aq) + 2NO_3^-(aq) + Ca^{2+}(aq) + 2OH^-(aq) \rightarrow Ca^{2+}(aq) + 2NO_3^-(aq) + 2H_2O(l)$$

Net ionic equation: $2H^+(aq) + 2OH^-(aq) \rightarrow 2H_2O(l)$ which simplifies to $\mathbf{H^+(aq) + OH^-(aq) \rightarrow H_2O(l)}$

4.13B **Plan:** According to Table 4.2, both reactants are strong and therefore completely dissociate in water. Thus, the key reaction is the formation of water. The other product of the reaction is soluble.
Solution:
Molecular equation: $HI(aq) + LiOH(aq) \rightarrow LiI(aq) + H_2O(l)$
Total ionic equation: $H^+(aq) + I^-(aq) + Li^+(aq) + OH^-(aq) \rightarrow Li^+(aq) + I^-(aq) + H_2O(l)$
Net ionic equation: $\mathbf{H^+(aq) + OH^-(aq) \rightarrow H_2O(l)}$

4.14A **Plan:** The reactants are a weak acid and a strong base. The acidic species is H^+ and a proton is transferred to OH^- from the acid. The only spectator ion is the cation of the base.
Solution:
Molecular equation:
$2HNO_2(aq) + Sr(OH)_2(aq) \rightarrow Sr(NO_2)_2(aq) + 2H_2O(l)$
Ionic equation:
$2HNO_2(aq) + Sr^{2+}(aq) + 2OH^-(aq) \rightarrow Sr^{2+}(aq) + 2NO_2^-(aq) + 2H_2O(l)$
Net ionic equation:
$2HNO_2(aq) + 2OH^-(aq) \rightarrow 2NO_2^-(aq) + 2H_2O(l)$ or
$\qquad \mathbf{HNO_2(aq) + OH^-(aq) \rightarrow NO_2^-(aq) + H_2O(l)}$

$\mathbf{HNO_2(aq) + OH^-(aq) \rightarrow NO_2^-(aq) + H_2O(l)}$

The salt is $\mathbf{Sr(NO_2)_2}$, **strontium nitrite**, and the spectator ion is $\mathbf{Sr^{2+}}$.

4.14B **Plan:** The reactants are a strong acid and the salt of a weak base. The acidic species is H_3O^+ and a proton is transferred to the weak base HCO_3^- to form H_2CO_3, which then decomposes to form CO_2 and water.
Solution:

Molecular equation:
$2HBr(aq) + Ca(HCO_3)_2(aq) \rightarrow CaBr_2(aq) + 2H_2CO_3(aq)$
Ionic equation:
$2H_3O^+(aq) + 2Br^-(aq) + Ca^{2+}(aq) + 2HCO_3^-(aq) \rightarrow 2CO_2(g) + 4H_2O(l) + 2Br^-(aq) + Ca^{2+}(aq)$
Net ionic equation:
$2H_3O^+(aq) + 2HCO_3^-(aq) \rightarrow 2CO_2(g) + 4H_2O(l)$ or $\qquad \mathbf{H_3O^+(aq) + HCO_3^-(aq) \rightarrow CO_2(g) + 2H_2O(l)}$

$\mathbf{H_3O^+(aq) + HCO_3^-(aq) \rightarrow CO_2(g) + 2H_2O(l)}$

The salt is $\mathbf{CaBr_2}$, **calcium bromide**.

4.15A **Plan:** Write a balanced equation. Determine the moles of HCl by multiplying its molarity by its volume, and, through the balanced chemical equation and the molar mass of aluminum hydroxide, determine the mass of aluminum hydroxide required for the reaction.
Solution:
The balanced equation is: $Al(OH)_3(s) + 3HCl(aq) \rightarrow AlCl_3(aq) + 3H_2O(l)$

$$\text{Mass(g) of } Al(OH)_3 = 3.4 \times 10^{-2} \text{ L HCl} \left(\frac{0.10 \text{ mol HCl}}{1 \text{ L}} \right)\left(\frac{1 \text{ mol } Al(OH)_3}{3 \text{ mol HCl}} \right)\left(\frac{78.00 \text{ g } Al(OH)_3}{1 \text{ mol } Al(OH)_3} \right)$$

$$= 0.08840 = \mathbf{0.088 \text{ g } Al(OH)_3}$$

4.15B   Plan: Write a balanced equation. Determine the moles of NaOH by multiplying its molarity by its volume, and, through the balanced chemical equation and the molar mass of acetylsalicylic acid, determine the mass of acetylsalicylic acid in the tablet.
Solution:
The balanced equation is: $HC_9H_7O_4(aq) + NaOH(aq) \rightarrow NaC_9H_7O_4(aq) + H_2O(l)$

$$Mass(g)\ of\ HC_9H_7O_4 = 14.10\ mL\ NaOH \left(\frac{1\ L}{1000\ mL}\right)\left(\frac{0.128\ mol\ NaOH}{1\ L}\right)\left(\frac{1\ mol\ HC_9H_7O_4}{1\ mol\ NaOH}\right)\left(\frac{180.15\ g\ HC_9H_7O_4}{1\ mol\ HC_9H_7O_4}\right)$$

$$= 0.3251 = \textbf{0.325 g } HC_9H_7O_4$$

4.16A   Plan: Write a balanced chemical equation. Determine the moles of HCl by multiplying its molarity by its volume, and, through the balanced chemical equation, determine the moles of $Ba(OH)_2$ required for the reaction. The amount of base in moles is divided by its molarity to find the volume.
Solution:
The molarity of the HCl solution is 0.1000 M. However, the molar ratio is not 1:1 as in the example problem. According to the balanced equation, the ratio is 2 moles of acid per 1 mole of base:
$2HCl(aq) + Ba(OH)_2(aq) \rightarrow BaCl_2(aq) + 2H_2O(l)$
Volume (L) of $Ba(OH)_2$ solution =

$$\left(50.00\ mL\right)\left(\frac{1\ L}{1000\ mL}\right)\left(\frac{0.1000\ mol\ HCl}{1\ L}\right)\left(\frac{1\ mol\ Ba(OH)_2}{2\ mol\ HCl}\right)\left(\frac{1\ L}{0.1292\ mol\ Ba(OH)_2}\right)$$

$$= 0.0193498 = \textbf{0.01935 L } Ba(OH)_2 \textbf{ solution}$$

4.16B   Plan: Write a balanced chemical equation. Determine the moles of H_2SO_4 by multiplying its molarity by its volume, and, through the balanced chemical equation, determine the moles of KOH required for the reaction. The amount of base in moles is divided by its volume (in L) to find the molarity.
Solution:
The balanced equation is: $KOH(aq) + HNO_3(aq) \rightarrow KNO_3(aq) + H_2O(l)$

$$Amount\ (mol)\ of\ KOH = 20.00\ mL\ HNO_3\left(\frac{1\ L}{1000\ mL}\right)\left(\frac{0.2452\ mol\ HNO_3}{1\ L}\right)\left(\frac{1\ mol\ KOH}{1\ mol\ HNO_3}\right) = 0.004904\ mol\ KOH$$

$$Molarity = \left(\frac{0.004904\ mol\ KOH}{18.15\ mL}\right)\left(\frac{1000\ mL}{1\ L}\right) = 0.270193 = \textbf{0.2702 } M$$

4.17A   Plan: Apply Table 4.4 to the compounds. Do not forget that the sum of the O.N.'s (oxidation numbers) for a compound must sum to zero, and for a polyatomic ion, the sum must equal the charge on the ion.
Solution:
a) **Sc = +3 O = –2** In most compounds, oxygen has a –2 O.N., so oxygen is often a good starting point. If each oxygen atom has a –2 O.N., then each scandium must have a +3 oxidation state so that the sum of O.N.'s equals zero: 2(+3) + 3(–2) = 0.
b) **Ga = +3 Cl = –1** In most compounds, chlorine has a –1 O.N., so chlorine is a good starting point. If each chlorine atom has a –1 O.N., then the gallium must have a +3 oxidation state so that the sum of O.N.'s equals zero: 1(+3) + 3(–1) = 0.
c) **H = +1 P = +5 O = –2** The hydrogen phosphate ion is HPO_4^{2-}. Again, oxygen has a –2 O.N. Hydrogen has a +1 O.N. because it is combined with nonmetals. The sum of the O.N.'s must equal the ionic charge, so the following algebraic equation can be solved for P: 1(+1) + 1(P) + 4(–2) = –2; O.N. for P = +5.
d) **I = +3 F = –1** The formula of iodine trifluoride is IF_3. In all compounds, fluorine has a –1 O.N., so fluorine is often a good starting point. If each fluorine atom has a –1 O.N., then the iodine must have a +3 oxidation state so that the sum of O.N.'s equal zero: 1(+3) + 3(–1) = 0.

4.17B   Plan: Apply Table 4.4 to the compounds. Do not forget that the sum of the O.N.'s (oxidation numbers) for a compound must sum to zero, and for a polyatomic ion, the sum must equal the charge on the ion.

a) **K = +1 C = +4 O = –2** In all compounds, potassium has a +1 O.N, and in most compounds, oxygen has a –2 O.N. If each potassium has a +1 O.N. and each oxygen has a –2 O.N., carbon must have a +4 oxidation state so that the sum of the O.N.'s equals zero: $2(+1) + 1(+4) + 3(-2) = 0$.

b) **N = –3 H = +1** When it is combined with a nonmetal, like N, hydrogen has a +1 O.N. If hydrogen has a +1 O.N., nitrogen must have a –3 O.N. so the sum of the O.N.'s equals +1, the charge on the polyatomic ion: $1(-3) + 4(+1) = +1$.

c) **Ca = +2 P = –3** Calcium, a group 2A metal, has a +2 O.N. in all compounds. If calcium has a +2 O.N., the phosphorus must have a –3 O.N. so the sum of the O.N.'s equals zero: $3(+2) + 2(-3) = 0$.

d) **S = +4 Cl = –1** Chlorine, a group 7A nonmetal, has a –1 O.N. when it is in combination with any nonmetal except O or other halogens lower in the group. If chlorine has an O.N. of –1, the sulfur must have an O.N. of +4 so that the sum of the O.N.'s equals zero: $1(+4) + 4(-1) = 0$.

4.18A   Plan: Apply Table 4.4 to determine the oxidation numbers for all the compounds in the reaction. Do not forget that the sum of the O.N.'s (oxidation numbers) for a compound must sum to zero, and for a polyatomic ion, the sum must equal the charge on the ion. After determining oxidation numbers for all atoms in the reaction, identify the atoms for which the oxidation numbers have changed from the left hand side of the equation to the right hand side of the equation. If the oxidation number of a particular atom increases, that atom has been oxidized, and the compound, element, or ion containing that atom on the reactant side of the equation is the reducing agent. If the oxidation number of a particular atom decreases, that atom has been reduced, and the compound, element, or ion containing that atom on the reactant side of the equation is the oxidizing agent.
Solution:
a) Oxidation numbers in NCl_3: N = +3, Cl = –1
Oxidation numbers in H_2O: H = +1, O = –2
Oxidation numbers in NH_3: N = –3, H = +1
Oxidation numbers in HOCl: H = +1, O = –2, Cl = +1
The oxidation number of nitrogen decreases from +3 to –3, so N is reduced and **NCl_3 is the oxidizing agent**.
The oxidation number of chlorine increases from –1 to +1, so Cl is oxidized and **NCl_3 is also the reducing agent**.
b) Oxidation numbers in $AgNO_3$: N = +1, N = +5, O = –2
Oxidation numbers in NH_4I: N = –3, H = +1, I = –1
Oxidation numbers in AgI: Ag = +1, I = –1
Oxidation numbers in NH_4NO_3: N (in NH_4^+) = –3, H = +1, N (in NO_3^-) = +5, O = –2
None of the oxidation numbers change, so **this is not an oxidation-reduction reaction**.
c) Oxidation numbers in H_2S: H = +1, S = –2
Oxidation numbers in O_2: O = 0
Oxidation numbers in SO_2: S = +4, O = –2
Oxidation numbers in H_2O: H = +1, O = –2
The oxidation number of oxygen decreases from 0 to –2, so O is reduced and **O_2 is the oxidizing agent**.
The oxidation number of sulfur increases from –2 to +4, so S is oxidized and **H_2S is the reducing agent**.

4.18B   Plan: Apply Table 4.4 to determine the oxidation numbers for all the compounds in the reaction. Do not forget that the sum of the O.N.'s (oxidation numbers) for a compound must sum to zero, and for a polyatomic ion, the sum must equal the charge on the ion. After determining oxidation numbers for all atoms in the reaction, identify the atoms for which the oxidation numbers have changed from the left hand side of the equation to the right hand side of the equation. If the oxidation number of a particular atom increases, that atom has been oxidized, and the compound, element, or ion containing that atom on the reactant side of the equation is the reducing agent. If the oxidation number of a particular atom decreases, that atom has been reduced, and the compound, element, or ion containing that atom on the reactant side of the equation is the oxidizing agent.
Solution:
a) Oxidation numbers in SiO_2: Si = +4, O = –2
Oxidation numbers in HF: H = +1, F = –1
Oxidation numbers in SiF_4: Si = +4, F = –1
Oxidation numbers in H_2O: H = +1, O = –2
None of the oxidation numbers change, so **this is not an oxidation-reduction reaction**.

4.3 Plan: Solutions that conduct an electric current contain electrolytes.
Solution:
Ions must be present in an aqueous solution for it to conduct an electric current. Ions come from ionic compounds or from other electrolytes such as acids and bases.

4.6 Plan: Write the formula for magnesium nitrate and note the ratio of magnesium ions to nitrate ions.
Solution:
Upon dissolving the salt in water, magnesium nitrate, $Mg(NO_3)_2$, would dissociate to form one Mg^{2+} ion for every two NO_3^- ions, thus forming twice as many nitrate ions. **Scene B** best represents a volume of magnesium nitrate solution. Only Scene B has twice as many nitrate ions (red circles) as magnesium ions (blue circles).

4.13 Plan: Remember that molarity is moles of solute/volume of solution.
Solution:
Volumes may **not** be additive when two different solutions are mixed, so the final volume may be slightly different from 1000.0 mL. The correct method would state, "Take 100.0 mL of the 10.0 M solution and add water until the total volume is 1000 mL."

4.14 Plan: Compounds that are soluble in water tend to be ionic compounds or covalent compounds that have polar bonds. Many ionic compounds are soluble in water because the attractive force between the oppositely charged ions in an ionic compound are replaced with an attractive force between the polar water molecule and the ions when the compound is dissolved in water. Covalent compounds with polar bonds are often soluble in water since the polar bonds of the covalent compound interact with those in water.
Solution:
a) Benzene, a covalent compound, is likely to be **insoluble** in water because it is nonpolar and water is polar.
b) Sodium hydroxide (NaOH) is an ionic compound and is therefore likely to be **soluble** in water.
c) Ethanol (CH_3CH_2OH) will likely be **soluble** in water because it contains a polar –OH bond like water.
d) Potassium acetate ($KC_2H_3O_2$) is an ionic compound and will likely be **soluble** in water.

4.16 Plan: Substances whose aqueous solutions conduct an electric current are electrolytes such as ionic compounds, acids, and bases.
Solution:
a) Cesium bromide, CsBr, is a soluble ionic compound, and a solution of this salt in water contains Cs^+ and Br^- ions. Its solution **conducts** an electric current.
b) HI is a strong acid that dissociates completely in water. Its aqueous solution contains H^+ and I^- ions, so it **conducts** an electric current.

4.18 Plan: To determine the total moles of ions released, write an equation that shows the compound dissociating into ions with the correct molar ratios. Convert mass and formula units to moles of compound and use the molar ratio to convert moles of compound to moles of ions.
Solution:
a) Each mole of NH_4Cl dissolves in water to form 1 mole of NH_4^+ ions and 1 mole of Cl^- ions, or a total of 2 moles of ions: $NH_4Cl(s) \rightarrow NH_4^+(aq) + Cl^-(aq)$.

$$\text{Moles of ions} = \left(0.32 \text{ mol } NH_4Cl\right)\left(\frac{2 \text{ mol ions}}{1 \text{ mol } NH_4Cl}\right) = \textbf{0.64 mol of ions}$$

b) Each mole of $Ba(OH)_2 \cdot 8H_2O$ forms 1 mole of Ba^{2+} ions and 2 moles of OH^- ions, or a total of 3 moles of ions: $Ba(OH)_2 \cdot 8H_2O(s) \rightarrow Ba^{2+}(aq) + 2OH^-(aq)$. The waters of hydration become part of the larger bulk of water. Convert mass to moles using the molar mass.

$$\text{Moles of ions} = \left(25.4 \text{ g } Ba(OH)_2 \cdot 8H_2O\right)\left(\frac{1 \text{ mol } Ba(OH)_2 \cdot 8H_2O}{315.4 \text{ g } Ba(OH)_2 \cdot 8H_2O}\right)\left(\frac{3 \text{ mol ions}}{1 \text{ mol } Ba(OH)_2 \cdot 8H_2O}\right)$$

$$= 0.2415980 = \textbf{0.242 mol of ions}$$

c) Each mole of LiCl produces 2 moles of ions (1 mole of Li^+ ions and 1 mole of Cl^- ions): $LiCl(s) \rightarrow Li^+(aq) + Cl^-(aq)$. Recall that a mole contains 6.022 x 10^{23} entities, so a mole of LiCl contains 6.022 x 10^{23} units of LiCl, more easily expressed as formula units.

$$\text{Moles of ions} = \left(3.55 \times 10^{19}\,\text{FU LiCl}\right)\left(\frac{1\,\text{mol LiCl}}{6.022 \times 10^{23}\,\text{FU LiCl}}\right)\left(\frac{2\,\text{mol ions}}{1\,\text{mol LiCl}}\right)$$

$$= 1.17901 \times 10^{-4} = \mathbf{1.18 \times 10^{-4}\ mol\ of\ ions}$$

4.20 **Plan:** To determine the total moles of ions released, write an equation that shows the compound dissociating into ions with the correct molar ratios. Convert mass and formula units to moles of compound and use the molar ratio to convert moles of compound to moles of ions.
Solution:
a) Each mole of K_3PO_4 forms 3 moles of K^+ ions and 1 mole of PO_4^{3-} ions, or a total of 4 moles of ions:
$K_3PO_4(s) \rightarrow 3K^+(aq) + PO_4^{3-}(aq)$

$$\text{Moles of ions} = \left(0.75\,\text{mol }K_3PO_4\right)\left(\frac{4\,\text{mol ions}}{1\,\text{mol }K_3PO_4}\right) = \mathbf{3.0\ mol\ of\ ions.}$$

b) Each mole of $NiBr_2 \cdot 3H_2O$ forms 1 mole of Ni^{2+} ions and 2 moles of Br^- ions, or a total of 3 moles of ions: $NiBr_2 \cdot 3H_2O(s) \rightarrow Ni^{2+}(aq) + 2Br^-(aq)$. The waters of hydration become part of the larger bulk of water. Convert mass to moles using the molar mass.

$$\text{Moles of ions} = \left(6.88 \times 10^{-3}\,\text{g }NiBr_2 \cdot 3H_2O\right)\left(\frac{1\,\text{mol }NiBr_2 \cdot 3H_2O}{272.54\,\text{g }NiBr_2 \cdot 3H_2O}\right)\left(\frac{3\,\text{mol ions}}{1\,\text{mol }NiBr_2 \cdot 3H_2O}\right)$$

$$= 7.5732 \times 10^{-5} = \mathbf{7.57 \times 10^{-5}\ mol\ of\ ions}$$

c) Each mole of $FeCl_3$ forms 1 mole of Fe^{3+} ions and 3 moles of Cl^- ions, or a total of 4 moles of ions: $FeCl_3(s) \rightarrow Fe^{3+}(aq) + 3Cl^-(aq)$. Recall that a mole contains 6.022×10^{23} entities, so a mole of $FeCl_3$ contains 6.022×10^{23} units of $FeCl_3$, more easily expressed as formula units.

$$\text{Moles of ions} = \left(2.23 \times 10^{22}\,\text{FU }FeCl_3\right)\left(\frac{1\,\text{mol }FeCl_3}{6.022 \times 10^{23}\,\text{FU }FeCl_3}\right)\left(\frac{4\,\text{mol ions}}{1\,\text{mol }FeCl_3}\right)$$

$$= 0.148124 = \mathbf{0.148\ mol\ of\ ions}$$

4.22 **Plan:** In all cases, use the known quantities and the definition of molarity $\left(M = \dfrac{\text{moles solute}}{\text{L of solution}}\right)$ to find the unknown quantity. Volume must be expressed in liters. The molar mass is used to convert moles to grams. The chemical formulas must be written to determine the molar mass. (a) You will need to convert milliliters to liters, multiply by the molarity to find moles, and convert moles to mass in grams. (b) Convert mass of solute to moles and volume from mL to liters. Divide the moles by the volume. (c) Multiply the molarity by the volume.
Solution:
a) Calculating moles of solute in solution:

$$\text{Moles of }Ca(C_2H_3O_2)_2 = \left(185.8\,\text{mL}\right)\left(\frac{10^{-3}\,\text{L}}{1\,\text{mL}}\right)\left(\frac{0.267\,\text{mol }Ca(C_2H_3O_2)_2}{1\,\text{L}}\right) = 0.0496086\,\text{mol }Ca(C_2H_3O_2)_2$$

Converting from moles of solute to grams:

$$\text{Mass (g) of }Ca(C_2H_3O_2)_2 = \left(0.0496086\,\text{mol }Ca(C_2H_3O_2)_2\right)\left(\frac{158.17\,\text{g }Ca(C_2H_3O_2)_2}{1\,\text{mol }Ca(C_2H_3O_2)_2}\right)$$

$$= 7.84659 = \mathbf{7.85\ g\ }Ca(C_2H_3O_2)_2$$

b) Converting grams of solute to moles:

$$\text{Moles of KI} = \left(21.1\,\text{g KI}\right)\left(\frac{1\,\text{mol KI}}{166.0\,\text{g KI}}\right) = 0.127108\,\text{moles KI}$$

$$\text{Volume (L)} = \left(500.\,\text{mL}\right)\left(\frac{10^{-3}\,\text{L}}{1\,\text{mL}}\right) = 0.500\,\text{L}$$

$$\text{Molarity of KI} = \frac{0.127108 \text{ mol KI}}{0.500 \text{ L}} = 0.254216 = \textbf{0.254 } \textit{M} \textbf{ KI}$$

c) Moles of NaCN = $\left(145.6 \text{ L}\right)\left(\dfrac{0.850 \text{ mol NaCN}}{1 \text{ L}}\right) = 123.76 = \textbf{124 mol NaCN}$

4.24   <u>Plan:</u> In all cases, use the known quantities and the definition of molarity $\left(M = \dfrac{\text{moles solute}}{\text{L of solution}}\right)$ to find the unknown quantity. Volume must be expressed in liters. The molar mass is used to convert moles to grams. The chemical formulas must be written to determine the molar mass. (a) Convert volume in milliliters to liters, multiply the volume by the molarity to obtain moles of solute, and convert moles to mass in grams. (b) The simplest way will be to convert the milligrams to millimoles. Molarity may not only be expressed as moles/L, but also as mmoles/mL. (c) Convert the milliliters to liters and find the moles of solute and moles of ions by multiplying the volume and molarity. Use Avogadro's number to determine the number of ions present.
<u>Solution:</u>
a) Calculating moles of solute in solution:

$$\text{Moles of } K_2SO_4 = \left(475 \text{ mL}\right)\left(\frac{10^{-3} \text{ L}}{1 \text{ mL}}\right)\left(\frac{5.62 \times 10^{-2} \text{ mol } K_2SO_4}{L}\right) = 0.026695 \text{ mol } K_2SO_4$$

Converting moles of solute to mass:

$$\text{Mass (g) of } K_2SO_4 = \left(0.026695 \text{ mol } K_2SO_4\right)\left(\frac{174.26 \text{ g } K_2SO_4}{1 \text{ mol } K_2SO_4}\right) = 4.6519 = \textbf{4.65 g } K_2SO_4$$

b) Calculating mmoles of solute:

$$\text{mmoles of } CaCl_2 = \left(\frac{7.25 \text{ mg } CaCl_2}{1 \text{ mL}}\right)\left(\frac{1 \text{ mmol } CaCl_2}{110.98 \text{ mg } CaCl_2}\right) = 0.065327 \text{ mmoles } CaCl_2$$

Calculating molarity:

$$\text{Molarity of } CaCl_2 = \left(\frac{0.065327 \text{ mmol } CaCl_2}{1 \text{ mL}}\right) = 0.065327 = \textbf{0.0653 } \textit{M} \textbf{ CaCl}_2$$

If you feel that molarity must be moles/liters then the calculation becomes:

$$\text{Molarity of } CaCl_2 = \left(\frac{7.25 \text{ mg } CaCl_2}{1 \text{ mL}}\right)\left(\frac{10^{-3} \text{ g}}{1 \text{ mg}}\right)\left(\frac{1 \text{ mL}}{10^{-3} \text{ L}}\right)\left(\frac{1 \text{ mol } CaCl_2}{110.98 \text{ g } CaCl_2}\right) = 0.065327 = \textbf{0.0653 } \textit{M} \textbf{ CaCl}_2$$

Notice that the two central terms cancel each other.

c) Converting volume in L to mL:

$$\text{Volume (L)} = \left(1 \text{ mL}\right)\left(\frac{10^{-3} \text{ L}}{1 \text{ mL}}\right) = 0.001 \text{ L}$$

Calculating moles of solute and moles of ions:

$$\text{Moles of } MgBr_2 = \left(0.001 \text{ L}\right)\left(\frac{0.184 \text{ mol } MgBr_2}{1 \text{ L}}\right) = 1.84 \times 10^{-4} \text{ mol } MgBr_2$$

$$\text{Moles of } Mg^{2+} \text{ ions} = \left(1.84 \times 10^{-4} \text{ mol } MgBr_2\right)\left(\frac{1 \text{ mol } Mg^{2+}}{1 \text{ mol } MgBr_2}\right) = 1.84 \times 10^{-4} \text{ mol } Mg^{2+} \text{ ions}$$

$$\text{Number of } Mg^{2+} \text{ ions} = \left(1.84 \times 10^{-4} \, Mg^{2+} \text{ ions}\right)\left(\frac{6.022 \times 10^{23} \, Mg^{2+} \text{ ions}}{1 \text{ mol } Mg^{2+} \text{ ions}}\right)$$

$$= 1.1080 \times 10^{20} = \textbf{1.11} \times \textbf{10}^{\textbf{20}} \textbf{ Mg}^{2+} \textbf{ ions}$$

4.26 **Plan:** To determine the total moles of ions released, write an equation that shows the compound dissociating into ions with the correct molar ratios. Convert the information given to moles of compound and use the molar ratio to convert moles of compound to moles of ions. Avogadro's number is used to convert moles of ions to numbers of ions.
Solution:
a) Each mole of $AlCl_3$ forms 1 mole of Al^{3+} ions and 3 moles of Cl^- ions: $AlCl_3(s) \rightarrow Al^{3+}(aq) + 3Cl^-(aq)$. Molarity and volume must be converted to moles of $AlCl_3$.

$$\text{Moles of } AlCl_3 = (130.\ mL)\left(\frac{10^{-3}\ L}{1\ mL}\right)\left(\frac{0.45\ mol\ AlCl_3}{L}\right) = 0.0585\ mol\ AlCl_3$$

$$\text{Moles of } Al^{3+} = (0.0585\ mol\ AlCl_3)\left(\frac{1\ mol\ Al^{3+}}{1\ mol\ AlCl_3}\right) = 0.0585 = \mathbf{0.058\ mol\ Al^{3+}}$$

$$\text{Number of } Al^{3+}\ ions = (0.0585\ mol\ Al^{3+})\left(\frac{6.022 \times 10^{23}\ Al^{3+}}{1\ mol\ Al^{3+}}\right) = 3.52287 \times 10^{22} = \mathbf{3.5 \times 10^{22}\ Al^{3+}\ ions}$$

$$\text{Moles of } Cl^- = (0.0585\ mol\ AlCl_3)\left(\frac{3\ mol\ Cl^-}{1\ mol\ AlCl_3}\right) = 0.1755 = \mathbf{0.18\ mol\ Cl^-}$$

$$\text{Number of } Cl^-\ ions = (0.1755\ mol\ Cl^-)\left(\frac{6.022 \times 10^{23}\ Cl^-}{1\ mol\ Cl^-}\right) = 1.05686 \times 10^{23} = \mathbf{1.1 \times 10^{23}\ Cl^-\ ions}$$

b) Each mole of Li_2SO_4 forms 2 moles of Li^+ ions and 1 mole of SO_4^{2-} ions: $Li_2SO_4(s) \rightarrow 2Li^+(aq) + SO_4^{2-}(aq)$.

$$\text{Moles of } Li_2SO_4 = (9.80\ mL)\left(\frac{10^{-3}\ L}{1\ mL}\right)\left(\frac{2.59\ g\ Li_2SO_4}{1\ L}\right)\left(\frac{1\ mol\ Li_2SO_4}{109.94\ g\ Li_2SO_4}\right) = 2.3087 \times 10^{-4}\ mol\ Li_2SO_4$$

$$\text{Moles of } Li^+ = (2.3087 \times 10^{-4}\ mol\ Li_2SO_4)\left(\frac{2\ mol\ Li^+}{1\ mol\ Li_2SO_4}\right) = 4.6174 \times 10^{-4} = \mathbf{4.62 \times 10^{-4}\ mol\ Li^+}$$

$$\text{Number of } Li^+\ ions = (4.6174 \times 10^{-4}\ mol\ Li^+)\left(\frac{6.022 \times 10^{23}\ Li^+}{1\ mol\ Li^+}\right) = 2.7806 \times 10^{20} = \mathbf{2.78 \times 10^{20}\ Li^+\ ions}$$

$$\text{Moles of } SO_4^{2-} = (2.3087 \times 10^{-4}\ mol\ Li_2SO_4)\left(\frac{1\ mol\ SO_4^{2-}}{1\ mol\ Li_2SO_4}\right) = 2.3087 \times 10^{-4} = \mathbf{2.31 \times 10^{-4}\ mol\ SO_4^{2-}}$$

$$\text{Number of } SO_4^{2-}\ ions = (2.3087 \times 10^{-4}\ mol\ SO_4^{2-})\left(\frac{6.022 \times 10^{23}\ SO_4^{2-}}{1\ mol\ SO_4^{2-}}\right)$$
$$= 1.39030 \times 10^{20} = \mathbf{1.39 \times 10^{20}\ SO_4^{2-}\ ions}$$

c) Each mole of KBr forms 1 mole of K^+ ions and 1 mole of Br^- ions: $KBr(s) \rightarrow K^+(aq) + Br^-(aq)$.

$$\text{Moles of } KBr = (245\ mL)\left(\frac{10^{-3}\ L}{1\ mL}\right)\left(\frac{3.68 \times 10^{22}\ FU\ KBr}{L}\right)\left(\frac{1\ mol\ KBr}{6.022 \times 10^{23}\ FU\ KBr}\right) = 0.01497\ mol\ KBr$$

$$\text{Moles of } K^+ = (0.01497\ mol\ KBr)\left(\frac{1\ mol\ K^+}{1\ mol\ KBr}\right) = 0.01497 = \mathbf{1.50 \times 10^{-2}\ mol\ K^+}$$

$$\text{Number of } K^+\ ions = (0.01497\ mol\ K^+)\left(\frac{6.022 \times 10^{23}\ K^+}{1\ mol\ K^+}\right) = 9.016 \times 10^{21} = \mathbf{9.02 \times 10^{21}\ K^+\ ions}$$

$$\text{Moles of } Br^- = (0.01497\ mol\ KBr)\left(\frac{1\ mol\ Br^-}{1\ mol\ KBr}\right) = 0.01497 = \mathbf{1.50 \times 10^{-2}\ mol\ Br^-}$$

$$\text{Number of Br}^- \text{ ions} = \left(0.01497 \text{ mol Br}^-\right)\left(\frac{6.022 \times 10^{23} \text{ Br}^-}{1 \text{ mol Br}^-}\right) = 9.016 \times 10^{21} = \textbf{9.02} \times \textbf{10}^{\textbf{21}} \text{ \textbf{Br}}^- \text{ \textbf{ions}}$$

4.28   <u>Plan:</u> These are dilution problems. Dilution problems can be solved by converting to moles and using the new volume; however, it is much easier to use $M_1V_1 = M_2V_2$. The dilution equation does not require a volume in liters; it only requires that the volume units match. In part c), it is necessary to find the moles of sodium ions in each separate solution, add these two mole amounts, and divide by the total volume of the two solutions.
<u>Solution:</u>
a) $M_1 = 0.250 \ M$ KCl $V_1 = 37.00$ mL $M_2 = ?$ $V_2 = 150.00$ mL
$M_1V_1 = M_2V_2$

$$M_2 = \frac{M_1 \times V_1}{V_2} = \frac{(0.250 \ M)(37.00 \text{ mL})}{150.00 \text{ mL}} = 0.061667 = \textbf{0.0617} \ \textbf{\textit{M}} \text{ \textbf{KCl}}$$

b) $M_1 = 0.0706 \ M$ $(NH_4)_2SO_4$ $V_1 = 25.71$ mL $M_2 = ?$ $V_2 = 500.00$ mL
$M_1V_1 = M_2V_2$

$$M_2 = \frac{M_1 \times V_1}{V_2} = \frac{(0.0706 \ M)(25.71 \text{ mL})}{500.00 \text{ mL}} = 0.003630 = \textbf{0.00363} \ \textbf{\textit{M}} \ \textbf{(NH}_4\textbf{)}_2\textbf{SO}_4$$

c) Moles of Na^+ from NaCl solution $= \left(3.58 \text{ mL}\right)\left(\frac{10^{-3} \text{ L}}{1 \text{ mL}}\right)\left(\frac{0.348 \text{ mol NaCl}}{1 \text{ L}}\right)\left(\frac{1 \text{ mol Na}^+}{1 \text{ mol NaCl}}\right)$

$$= 0.00124584 \text{ mol Na}^+$$

Moles of Na^+ from Na_2SO_4 solution $= \left(500. \text{ mL}\right)\left(\frac{10^{-3} \text{ L}}{1 \text{ mL}}\right)\left(\frac{6.81 \times 10^{-2} \text{ mol Na}_2SO_4}{1 \text{ L}}\right)\left(\frac{2 \text{ mol Na}^+}{1 \text{ mol Na}_2SO_4}\right)$

$$= 0.0681 \text{ mol Na}^+$$

Total moles of Na^+ ions $= 0.00124584$ mol Na^+ ions $+ 0.0681$ mol Na^+ ions $= 0.06934584$ mol Na^+ ions
Total volume $= 3.58$ mL $+ 500.$ mL $= 503.58$ mL $= 0.50358$ L

$$\text{Molarity of Na}^+ = \frac{\text{total moles Na}^+ \text{ ions}}{\text{total volume}} = \frac{0.06934584 \text{ mol Na}^+ \text{ ions}}{0.50358 \text{ L}} = 0.1377057 = \textbf{0.138} \ \textbf{\textit{M}} \text{ \textbf{Na}}^+ \text{ \textbf{ions}}$$

4.30   <u>Plan:</u> Use the density of the solution to find the mass of 1 L of solution. Volume in liters must be converted to volume in mL. The 70.0% by mass translates to 70.0 g solute/100 g solution and is used to find the mass of HNO_3 in 1 L of solution. Convert mass of HNO_3 to moles to obtain moles/L, molarity.
<u>Solution:</u>

a) Mass (g) of 1 L of solution $= \left(1 \text{ L solution}\right)\left(\frac{1 \text{ mL}}{10^{-3} \text{ L}}\right)\left(\frac{1.41 \text{ g solution}}{1 \text{ mL}}\right) = 1410 \text{ g solution}$

Mass (g) of HNO_3 in 1 L of solution $= \left(1410 \text{ g solution}\right)\left(\frac{70.0 \text{ g HNO}_3}{100 \text{ g solution}}\right) = \textbf{987 g HNO}_3\textbf{/L}$

b) Moles of $HNO_3 = \left(987 \text{ g HNO}_3\right)\left(\frac{1 \text{ mol HNO}_3}{63.02 \text{ g HNO}_3}\right) = 15.6617 \text{ mol HNO}_3$

Molarity of $HNO_3 = \left(\frac{15.6617 \text{ mol HNO}_3}{1 \text{ L solution}}\right) = 15.6617 = \textbf{15.7} \ \textbf{\textit{M}} \text{ \textbf{HNO}}_3$

4.33   <u>Plan:</u> The first part of the problem is a simple dilution problem $(M_1V_1 = M_2V_2)$. The volume in units of gallons can be used. In part b), convert mass of HCl to moles and use the molarity to find the volume that contains that number of moles.

Solution:

a) $M_1 = 11.7\ M$ $\qquad$ $V_1 = ?$ $\qquad$ $M_2 = 3.5\ M$ $\qquad$ $V_2 = 3.0$ gal

$$V_1 = \frac{M_2 \times V_2}{M_1} = \frac{(3.5\ M)(3.0\ \text{gal})}{11.7\ M} = 0.897436\ \text{gal} = \textbf{0.90 gal}$$

Instructions: Be sure to wear goggles to protect your eyes! Pour approximately 2.0 gal of water into the container. Add slowly and with mixing 0.90 gal of 11.7 M HCl into the water. Dilute to 3.0 gal with water.

b) Converting from mass of HCl to moles of HCl:

$$\text{Moles of HCl} = \left(9.66\ \text{g HCl}\right)\left(\frac{1\ \text{mol HCl}}{36.46\ \text{g HCl}}\right) = 0.264948\ \text{mol HCl}$$

Converting from moles of HCl to volume:

$$\text{Volume (mL) of solution} = \left(0.264948\ \text{mol HCl}\right)\left(\frac{1\ \text{L}}{11.7\ \text{mol HCl}}\right)\left(\frac{1\ \text{mL}}{10^{-3}\ \text{L}}\right)$$

$$= 22.64513 = \textbf{22.6 mL muriatic acid solution}$$

4.36 Plan: Review the definition of spectator ions.
Solution:
Ions in solution that do not participate in the reaction do not appear in a net ionic equation. These spectator ions remain as dissolved ions throughout the reaction. These ions are only present to balance charge.

4.40 Plan: Use the solubility rules to predict the products of this reaction. Ions not involved in the precipitate are spectator ions and are not included in the net ionic equation.
Solution:
Assuming that the left beaker is $AgNO_3$ (because it has gray Ag^+ ions) and the right must be NaCl, then the NO_3^- is blue, the Na^+ is brown, and the Cl^- is green. (Cl^- must be green since it is present with Ag^+ in the precipitate in the beaker on the right.)
Molecular equation: $AgNO_3(aq) + NaCl(aq) \rightarrow AgCl(s) + NaNO_3(aq)$
Total ionic equation: $Ag^+(aq) + NO_3^-(aq) + Na^+(aq) + Cl^-(aq) \rightarrow AgCl(s) + Na^+(aq) + NO_3^-(aq)$
Net ionic equation: $Ag^+(aq) + Cl^-(aq) \rightarrow AgCl(s)$

4.41 Plan: Write the new cation-anion combinations as the products of the reaction and use the solubility rules to determine if any of the new combinations are insoluble. The total ionic equation shows all soluble ionic substances dissociated into ions. The spectator ions are the ions that are present in the soluble ionic compound. The spectator ions are omitted from the net ionic equation.
Solution:
a) Molecular: $Hg_2(NO_3)_2(aq) + 2KI(aq) \rightarrow Hg_2I_2(s) + 2KNO_3(aq)$
$\qquad$ Total ionic: $Hg_2^{2+}(aq) + 2NO_3^-(aq) + 2K^+(aq) + 2I^-(aq) \rightarrow Hg_2I_2(s) + 2K^+(aq) + 2NO_3^-(aq)$
$\qquad$ Net ionic: $Hg_2^{2+}(aq) + 2I^-(aq) \rightarrow Hg_2I_2(s)$
$\qquad$ Spectator ions are K^+ and NO_3^-.
b) Molecular: $FeSO_4(aq) + Sr(OH)_2(aq) \rightarrow Fe(OH)_2(s) + SrSO_4(s)$
$\qquad$ Total ionic: $Fe^{2+}(aq) + SO_4^{2-}(aq) + Sr^{2+}(aq) + 2OH^-(aq) \rightarrow Fe(OH)_2(s) + SrSO_4(s)$
$\qquad$ Net ionic: This is the same as the total ionic equation because there are no spectator ions.

4.43 Plan: A precipitate forms if reactant ions can form combinations that are insoluble, as determined by the solubility rules in Table 4.1. Create cation-anion combinations other than the original reactants and determine if they are insoluble. Any ions not involved in a precipitate are spectator ions and are omitted from the net ionic equation.
Solution:
a) $NaNO_3(aq) + CuSO_4(aq) \rightarrow Na_2SO_4(aq) + Cu(NO_3)_2(aq)$
No precipitate will form. The ions Na^+ and SO_4^{2-} will not form an insoluble salt according to the first solubility rule which states that all common compounds of Group 1A ions are soluble. The ions Cu^{2+} and NO_3^- will not form an insoluble salt according to the solubility rule #2: All common nitrates are soluble. There is no reaction.

b) A precipitate will form because silver ions, Ag^+, and bromide ions, Br^-, will combine to form a solid salt, silver bromide, AgBr. The ammonium and nitrate ions do not form a precipitate.

Molecular: $NH_4Br(aq) + AgNO_3(aq) \rightarrow AgBr(s) + NH_4NO_3(aq)$

Total ionic: $NH_4^+(aq) + Br^-(aq) + Ag^+(aq) + NO_3^-(aq) \rightarrow AgBr(s) + NH_4^+(aq) + NO_3^-(aq)$

Net ionic: $Ag^+(aq) + Br^-(aq) \rightarrow AgBr(s)$

4.45 **Plan:** A precipitate forms if reactant ions can form combinations that are insoluble, as determined by the solubility rules in Table 4.1. Create cation-anion combinations other than the original reactants and determine if they are insoluble. Any ions not involved in a precipitate are spectator ions and are omitted from the net ionic equation.
Solution:
a) New cation-anion combinations are potassium nitrate (KNO_3) and iron(III) chloride ($FeCl_3$). The solubility rules state that all common nitrates and chlorides (with some exceptions) are soluble, so no precipitate forms.

$$3KCl(aq) + Fe(NO_3)_3(aq) \rightarrow 3KNO_3(aq) + FeCl_3(aq)$$

b) New cation-anion combinations are ammonium chloride and barium sulfate. The solubility rules state that most chlorides are soluble; however, another rule states that sulfate compounds containing barium are insoluble. Barium sulfate is a precipitate and its formula is $BaSO_4$.

Molecular: $(NH_4)_2SO_4(aq) + BaCl_2(aq) \rightarrow BaSO_4(s) + 2NH_4Cl(aq)$

Total ionic: $2NH_4^+(aq) + SO_4^{2-}(aq) + Ba^{2+}(aq) + 2Cl^-(aq) \rightarrow BaSO_4(s) + 2NH_4^+(aq) + 2Cl^-(aq)$

Net ionic: $Ba^{2+}(aq) + SO_4^{2-}(aq) \rightarrow BaSO_4(s)$

4.47 **Plan:** Write a balanced equation for the chemical reaction described in the problem. By applying the solubility rules to the two possible products ($NaNO_3$ and PbI_2), determine that PbI_2 is the precipitate. By using molar relationships, determine how many moles of $Pb(NO_3)_2$ are required to produce 0.628 g of PbI_2. The molarity is calculated by dividing moles of $Pb(NO_3)_2$ by its volume in liters.
Solution:

The reaction is: $Pb(NO_3)_2(aq) + 2NaI(aq) \rightarrow PbI_2(s) + 2NaNO_3(aq)$.

$$\text{Moles of } Pb(NO_3)_2 = \left(0.628 \text{ g } PbI_2\right)\left(\frac{1 \text{ mol } PbI_2}{461.0 \text{ g } PbI_2}\right)\left(\frac{1 \text{ mol } Pb(NO_3)_2}{1 \text{ mol } PbI_2}\right) = 0.001362256 \text{ mol } Pb(NO_3)_2$$

Moles of Pb^{2+} = moles of $Pb(NO_3)_2$ = 0.001362256 mol Pb^{2+}

$$\text{Molarity of } Pb^{2+} = \frac{\text{moles } Pb^{2+}}{\text{volume of } Pb^{2+}} = \frac{0.001362256 \text{ mol}}{38.5 \text{ mL}}\left(\frac{1 \text{ mL}}{10^{-3} \text{ L}}\right) = 0.035383 = \mathbf{0.0354 \textit{ M } Pb^{2+}}$$

4.49 **Plan:** The first step is to write and balance the chemical equation for the reaction. Multiply the molarity and volume of each of the reactants to determine the moles of each. To determine which reactant is limiting, calculate the amount of barium sulfate formed from each reactant, assuming an excess of the other reactant. The reactant that produces less product is the limiting reagent. Use the limiting reagent and the mole ratio from the balanced chemical equation to determine the mass of barium sulfate formed.
Solution:
The balanced chemical equation is:

$BaCl_2(aq) + Na_2SO_4(aq) \rightarrow BaSO_4(s) + 2NaCl(aq)$

$$\text{Moles of } BaCl_2 = \left(35.0 \text{ mL}\right)\left(\frac{10^{-3} \text{ L}}{1 \text{ mL}}\right)\left(\frac{0.160 \text{ mol } BaCl_2}{1 \text{ L}}\right) = 0.00560 \text{ mol } BaCl_2$$

Finding the moles of $BaSO_4$ from the moles of $BaCl_2$ (if Na_2SO_4 is in excess):

$$\text{Moles of } BaSO_4 \text{ from } BaCl_2 = \left(0.00560 \text{ mol } BaCl_2\right)\left(\frac{1 \text{ mol } BaSO_4}{1 \text{ mol } BaCl_2}\right) = 0.00560 \text{ mol } BaSO_4$$

$$\text{Moles of } Na_2SO_4 = \left(58.0 \text{ mL}\right)\left(\frac{10^{-3} \text{ L}}{1 \text{ mL}}\right)\left(\frac{0.065 \text{ mol } Na_2SO_4}{1 \text{ L}}\right) = 0.00377 \text{ mol } Na_2SO_4$$

Finding the moles of $BaSO_4$ from the moles of Na_2SO_4 (if $BaCl_2$ is in excess):

$$\text{Moles } BaSO_4 \text{ from } Na_2SO_4 = \left(0.00377 \text{ moL } Na_2SO_4\right)\left(\frac{1 \text{ mol } BaSO_4}{1 \text{ mol } Na_2SO_4}\right) = 0.00377 \text{ mol } BaSO_4$$

Sodium sulfate is the limiting reactant.
Converting from moles of $BaSO_4$ to mass:

$$\text{Mass (g) of } BaSO_4 = \left(0.0377 \text{ moL } BaSO_4\right)\left(\frac{233.4 \text{ g } BaSO_4}{1 \text{ mol } BaSO_4}\right) = 0.879918 = \textbf{0.88 g } BaSO_4$$

4.51   Plan: A precipitate forms if reactant ions can form combinations that are insoluble, as determined by the solubility rules in Table 4.1. Create cation-anion combinations other than the original reactants and determine if they are insoluble. Any ions not involved in a precipitate are spectator ions and are omitted from the net ionic equation. Use the molar ratio in the balanced net ionic equation to calculate the mass of product.
Solution:
a) The yellow spheres cannot be ClO_4^- or NO_3^- as these ions form only soluble compounds. So the yellow sphere must be SO_4^{2-}. The only sulfate compounds possible that would be insoluble are Ag_2SO_4 and $PbSO_4$. The precipitate has a 1:1 ratio between its ions. Ag_2SO_4 has a 2:1 ratio between its ions. Therefore the blue spheres are Pb^{2+} and the yellow spheres are SO_4^{2-}. The precipitate is thus **$PbSO_4$**.
b) The net ionic equation is $Pb^{2+}(aq) + SO_4^{2-}(aq) \rightarrow PbSO_4(s)$.

$$\text{c) Mass (g) of } PbSO_4 = \left(10 \text{ } Pb^{2+} \text{ spheres}\right)\left(\frac{5.0 \times 10^{-4} \text{ mol } Pb^{2+}}{1 \text{ } Pb^{2+} \text{ sphere}}\right)\left(\frac{1 \text{ mol } PbSO_4}{1 \text{ mol } Pb^{2+}}\right)\left(\frac{303.3 \text{ g } PbSO_4}{1 \text{ mol } PbSO_4}\right)$$

$$= 1.5165 = \textbf{1.5 g } PbSO_4$$

4.53   Plan: Multiply the molarity of the $CaCl_2$ solution by its volume in liters to determine the number of moles of $CaCl_2$ reacted. Write the precipitation reaction using M to represent the alkali metal. Use the moles of $CaCl_2$ reacted and the molar ratio in the balanced equation to find the moles of alkali metal carbonate. Divide the mass of that carbonate by the moles to obtain the molar mass of the carbonate. Subtract the mass of CO_3 from the molar mass to obtain the molar mass of the alkali metal, which can be used to identify the alkali metal and the formula and name of the compound.
Solution:
The reaction is: $M_2CO_3(aq) + CaCl_2(aq) \rightarrow CaCO_3(s) + 2MCl(aq)$.
Since the alkali metal M forms a +1 ion and carbonate is a –2 ion, the formula is M_2CO_3.

$$\text{Moles of } CaCl_2 = (31.10 \text{ mL})\left(\frac{10^{-3} \text{ L}}{1 \text{ mL}}\right)\left(\frac{0.350 \text{ mol } CaCl_2}{1 \text{ L}}\right) = 0.010885 \text{ mol } CaCl_2$$

$$\text{Moles of } M_2CO_3 = \left(0.010885 \text{ mol } CaCl_2\right)\left(\frac{1 \text{ mol } M_2CO_3}{1 \text{ mol } CaCl_2}\right) = 0.010885 \text{ mol } M_2CO_3$$

$$\text{Molar mass (g/mol) of } M_2CO_3 = \frac{\text{mass of } M_2CO_3}{\text{moles of } M_2CO_3} = \frac{1.50 \text{ g}}{0.010885 \text{ mol}} = 137.804 \text{ g/mol}$$

Molar mass (g/mol) of M_2 = molar mass of M_2CO_3 – molar mass of CO_3
= 137.804 g/mol – 60.01 g/mol = 77.79 g/mol
Molar mass (g/mol) of M = 77.79 g/mol/2 = 38.895 g/mol = 38.9 g/mol
This molar mass is closest to that of potassium; the formula of the compound is **K_2CO_3, potassium carbonate**.

4.55   Plan: Write a balanced equation for the reaction. Find the moles of $AgNO_3$ by multiplying the molarity and volume of the $AgNO_3$ solution; use the molar ratio in the balanced equation to find the moles of Cl^- present in the 25.00 mL sample. Then, convert moles of Cl^- into grams, and convert the sample volume into grams using the given density. The mass percent of Cl^- is found by dividing the mass of Cl^- by the mass of the sample volume and multiplying by 100.
Solution:
The balanced equation is $AgNO_3(aq) + Cl^-(aq) \rightarrow AgCl(s) + NO_3^-(aq)$.

$$\text{Moles of AgNO}_3 = \left(53.63 \text{ mL}\right)\left(\frac{10^{-3} \text{ L}}{1 \text{ mL}}\right)\left(\frac{0.2970 \text{ mol AgNO}_3}{\text{L}}\right) = 0.01592811 \text{ mol AgNO}_3$$

$$\text{Mass (g) of Cl}^- = \left(0.01592811 \text{ mol AgNO}_3\right)\left(\frac{1 \text{ mol Cl}^-}{1 \text{ mol AgNO}_3}\right)\left(\frac{35.45 \text{ g Cl}}{1 \text{ mol Cl}^-}\right) = 0.56465 \text{ g Cl}^-$$

$$\text{Mass (g) of seawater sample} = \left(25.00 \text{ mL}\right)\left(\frac{1.024 \text{ g}}{\text{mL}}\right) = 25.60 \text{ g sample}$$

$$\text{Mass \% Cl}^- = \frac{\text{mass Cl}^-}{\text{mass sample}} \times 100\% = \frac{0.56465 \text{ g Cl}^-}{25.60 \text{ g sample}} \times 100\% = 2.20566 = \mathbf{2.206\% \text{ Cl}^-}$$

4.61   Plan: Formation of a gaseous product or a precipitate drives a reaction to completion.
Solution:
a) The formation of a gas, $SO_2(g)$, and formation of water drive this reaction to completion, because both products remove reactants from solution.
b) The formation of a precipitate, $Ba_3(PO_4)_2(s)$, will cause this reaction to go to completion. This reaction is one between an acid and a base, so the formation of water molecules through the combination of H^+ and OH^- ions also drives the reaction.

4.63   Plan: The acids in this problem are all strong acids, so you can assume that all acid molecules dissociate completely to yield H^+ ions and associated anions. One mole of $HClO_4$, HNO_3 and HCl each produce one mole of H^+ upon dissociation, so moles H^+ = moles acid. Calculate the moles of acid by multiplying the molarity (moles/L) by the volume in liters.
Solution:
a) $HClO_4(aq) \rightarrow H^+(aq) + ClO_4^-(aq)$

$$\text{Moles H}^+ = \text{mol HClO}_4 = \left(1.40 \text{ L}\right)\left(\frac{0.25 \text{ mol}}{1 \text{ L}}\right) = \mathbf{0.35 \text{ mol H}^+}$$

b) $HNO_3(aq) \rightarrow H^+(aq) + NO_3^-(aq)$

$$\text{Moles H}^+ = \text{mol HNO}_3 = \left(6.8 \text{ mL}\right)\left(\frac{10^{-3} \text{ L}}{1 \text{ mL}}\right)\left(\frac{0.92 \text{ mol}}{1 \text{ L}}\right) = 6.256 \times 10^{-3} = \mathbf{6.3 \times 10^{-3} \text{ mol H}^+}$$

c) $HCl(aq) \rightarrow H^+(aq) + Cl^-(aq)$

$$\text{Moles H}^+ = \text{mol HCl} = \left(2.6 \text{ L}\right)\left(\frac{0.085 \text{ mol}}{1 \text{ L}}\right) = 0.221 = \mathbf{0.22 \text{ mol H}^+}$$

4.65   Plan: Remember that strong acids and bases can be written as ions in the total ionic equation but weak acids and bases cannot be written as ions. Omit spectator ions from the net ionic equation.
Solution:
a) KOH is a strong base and HBr is a strong acid; both may be written in dissociated form. KBr is a soluble compound since all Group 1A(1) compounds are soluble.
Molecular equation: $KOH(aq) + HBr(aq) \rightarrow KBr(aq) + H_2O(l)$
Total ionic equation: $K^+(aq) + OH^-(aq) + H^+(aq) + Br^-(aq) \rightarrow K^+(aq) + Br^-(aq) + H_2O(l)$
Net ionic equation: $OH^-(aq) + H^+(aq) \rightarrow H_2O(l)$
The spectator ions are $K^+(aq)$ and $Br^-(aq)$.
b) NH_3 is a weak base and is written in the molecular form. HCl is a strong acid and is written in the dissociated form (as ions). NH_4Cl is a soluble compound, because all ammonium compounds are soluble.
Molecular equation: $NH_3(aq) + HCl(aq) \rightarrow NH_4Cl(aq)$
Total ionic equation: $NH_3(aq) + H^+(aq) + Cl^-(aq) \rightarrow NH_4^+(aq) + Cl^-(aq)$
Net ionic equation: $NH_3(aq) + H^+(aq) \rightarrow NH_4^+(aq)$
Cl^- is the only spectator ion.

4.67   Plan: Write an acid-base reaction between $CaCO_3$ and HCl. Remember that HCl is a strong acid.
        Solution:
        Calcium carbonate dissolves in HCl(aq) because the carbonate ion, a base, reacts with the acid to form H_2CO_3
        which decomposes into $CO_2(g)$ and $H_2O(l)$.
        $$CaCO_3(s) + 2HCl(aq) \rightarrow CaCl_2(aq) + H_2CO_3(aq)$$
        Total ionic equation:
        $$CaCO_3(s) + 2H^+(aq) + 2Cl^-(aq) \rightarrow Ca^{2+}(aq) + 2Cl^-(aq) + H_2O(l) + CO_2(g)$$
        Net ionic equation:
        $$CaCO_3(s) + 2H^+(aq) \rightarrow Ca^{2+}(aq) + H_2O(l) + CO_2(g)$$

4.69   Plan: Convert the mass of calcium carbonate to moles, and use the mole ratio in the balanced chemical equation
        to find the moles of hydrochloric acid required to react with these moles of calcium carbonate. Use the molarity of
        HCl to find the volume that contains this number of moles.
        Solution:
        $$2HCl(aq) + CaCO_3(s) \rightarrow CaCl_2(aq) + CO_2(g) + H_2O(l)$$
        Converting from grams of $CaCO_3$ to moles:

        $$\text{Moles of } CaCO_3 = \left(16.2 \text{ g } CaCO_3\right)\left(\frac{1 \text{ mol } CaCO_3}{100.09 \text{ g } CaCO_3}\right) = 0.161854 \text{ mol } CaCO_3$$

        Converting from moles of $CaCO_3$ to moles of HCl:

        $$\text{Moles of HCl} = \left(0.161854 \text{ mol } CaCO_3\right)\left(\frac{2 \text{ mol HCl}}{1 \text{ mol } CaCO_3}\right) = 0.323708 \text{ mol HCl}$$

        Converting from moles of HCl to volume:

        $$\text{Volume (mL) of HCl} = \left(0.323708 \text{ mol HCl}\right)\left(\frac{1 \text{ L}}{0.383 \text{ mol HCl}}\right)\left(\frac{1 \text{ mL}}{10^{-3} \text{ L}}\right) = 845.1906 = \textbf{845 mL HCl solution}$$

4.71   Plan: Write a balanced equation. Find the moles of KOH from the molarity and volume information and use the
        molar ratio in the balanced equation to find the moles of acid present. Divide the moles of acid by its volume to
        determine the molarity.
        Solution:
        The reaction is: $KOH(aq) + CH_3COOH(aq) \rightarrow CH_3COOK(aq) + H_2O(l)$

        $$\text{Moles of KOH} = \left(25.98 \text{ mL}\right)\left(\frac{10^{-3} \text{ L}}{1 \text{ mL}}\right)\left(\frac{0.1180 \text{ mol KOH}}{L}\right) = 0.00306564 \text{ mol KOH}$$

        $$\text{Moles of } CH_3COOH = \left(0.00306564 \text{ mol KOH}\right)\left(\frac{1 \text{mol } CH_3COOH}{1 \text{ mol KOH}}\right) = 0.00306564 \text{ mol } CH_3COOH$$

        $$\text{Molarity of } CH_3COOH = \left(\frac{0.00306564 \text{ mol } CH_3COOH}{52.50 \text{ mL}}\right)\left(\frac{1 \text{ mL}}{10^{-3} \text{ L}}\right) = 0.05839314 = \textbf{0.05839 } \textit{M} \textbf{ CH}_3\textbf{COOH}$$

4.82   Plan: An oxidizing agent gains electrons and therefore has an atom whose oxidation number decreases during the
        reaction. Use the Rules for Assigning an Oxidation Number to assign S in H_2SO_4 an O.N. and see if this oxidation
        number changes during the reaction. An acid transfers a proton during reaction.
        Solution:

        a) In H_2SO_4, hydrogen has an O.N. of +1, for a total of +2; oxygen has an O.N. of –2 for a total of –8. The S has an
        O.N. of +6. In SO_2, the O.N. of oxygen is –2 for a total of –4 and S has an O.N. of +4. So the S has been reduced
        from +6 to +4 and is an **oxidizing agent**. Iodine is oxidized during the reaction.
        b) The oxidation number of S is +6 in H_2SO_4; in $BaSO_4$, Ba has an O.N. of +2, the four oxygen atoms have a total
        O.N. of –8, and S is again +6. Since the oxidation number of S (or any of the other atoms) did not change, this is not
        a redox reaction. H_2SO_4 transfers a proton to F^- to produce HF, so it acts as an **acid**.

4.84   Plan: Consult the Rules for Assigning an Oxidation Number. The sum of the O.N. values for the atoms in a molecule equals zero, while the sum of the O.N. values for the atoms in an ion equals the ion's charge.
Solution:
a) CF_2Cl_2. The rules dictate that F and Cl each have an O.N. of –1; two F and two Cl yield a sum of –4, so the O.N. of C must be +4. **C = +4**
b) $Na_2C_2O_4$. The rule dictates that Na [Group 1A(1)] has a +1 O.N.; rule 5 dictates that O has a –2 O.N.; two Na and four O yield a sum of –6 [2(+1) + 4(–2)]. Therefore, the total of the O.N.s on the two C atoms is +6 and each C is +3. **C = +3**
c) HCO_3^-. H is combined with nonmetals and has an O.N. of +1; O has an O.N. of –2 and three O have a sum of –6. To have an overall oxidation state equal to –1, C must be +4 because (+1) + (+4) + (–6) = –1. **C = +4**
d) C_2H_6. Each H has an O.N. of +1; six H gives +6. The sum of O.N.s for the two C atoms must be –6, so each C is –3. **C = –3**

4.86   Plan: Consult the Rules for Assigning an Oxidation Number. The sum of the O.N. values for the atoms in a molecule equals zero, while the sum of the O.N. values for the atoms in an ion equals the ion's charge.
Solution:
a) NH_2OH. Hydrogen has an O.N. of +1, for a total of +3 for the three hydrogen atoms. Oxygen has an O.N. of –2. The O.N. of N must be –1 since [(–1) + (+3) + (–2)] = 0. **N = –1**
b) N_2F_4. The O.N. of each fluorine is –1 for a total of –4; the sum of the O.N.s for the two N atoms must be +4, so each N has an O.N. of +2. **N = +2**
c) NH_4^+. The O.N. of each hydrogen is +1 for a total of +4; the O.N. of nitrogen must be –3 since the overall sum of the O.N.s must be +1: [(–3) + (+4)] = +1 **N = –3**
d) HNO_2. The O.N. of hydrogen is +1 and that of each oxygen is –2 for a total of –4 from the oxygens. The O.N. of nitrogen must be +3 since [(+1) + (+3) + (–4)] = 0. **N = +3**

4.88   Plan: Consult the Rules for Assigning an Oxidation Number. The sum of the O.N. values for the atoms in a molecule equals zero, while the sum of the O.N. values for the atoms in an ion equals the ion's charge.
Solution:
a) AsH_3. H is combined with a nonmetal, so its O.N. is +1 (Rule 3). Three H atoms have a sum of +3. To have a sum of 0 for the molecule, As has an O.N. of –3. **As = –3**
b) $H_2AsO_4^-$. The O.N. of H in this compound is +1, for a total of +2. The O.N. of each oxygen is –2, for a total of –8. As has an O.N. of **+5** since [(+2) + (+5) + (–8)] = –1, the charge of the ion. **As = +5**
c) $AsCl_3$. Each chlorine has an O.N. of –1, for a total of –3. The O.N. of As is +3. **As = +3**

4.90   Plan: Consult the Rules for Assigning an Oxidation Number. The sum of the O.N. values for the atoms in a molecule equals zero, while the sum of the O.N. values for the atoms in an ion equals the ion's charge.
Solution:
a) MnO_4^{2-}. The O.N. of each oxygen is –2, for a total of –8; the O.N. of Mn must be +6 since [(+6) + (–8)] = –2, the charge of the ion. **Mn = +6**
b) Mn_2O_3. The O.N. of each oxygen is –2, for a total of –6; the sum of the O.N.s of the two Mn atoms must be +6. The O.N. of each manganese is +3. **Mn = +3**
c) $KMnO_4$. The O.N. of potassium is +1 and the O.N. of each oxygen is –2, for a total of –8. The O.N. of Mn is +7 since [(+1) + (+7) + (–8)] = 0. **Mn = +7**

4.92   Plan: First, assign oxidation numbers to all atoms following the rules. The reactant that is the reducing agent contains an atom that is oxidized (O.N. increases from the left side to the right side of the equation). The reactant that is the oxidizing agent contains an atom that is reduced (O.N. decreases from the left side to the right side of the equation). Recognize that the agent is the compound that contains the atom that is oxidized or reduced, not just the atom itself.

<u>Solution:</u>

a)   +2 +6 −8   −8   −4   +2

    +1 +3 −2   +7 −2   +1   +2   +4 −2   +1 −2

$5H_2C_2O_4(aq) + 2MnO_4^-(aq) + 6H^+(aq) \rightarrow 2Mn^{2+}(aq) + 10CO_2(g) + 8H_2O(l)$

Mn in MnO_4^- changes from +7 to +2 (reduction). Therefore, **MnO_4^- is the oxidizing agent.** C in $H_2C_2O_4$ changes from +3 to +4 (oxidation), so **$H_2C_2O_4$ is the reducing agent.**

b)   −6   +2

   0   +1   +5 −2   +2   +2 −2   +1 −2

$3Cu(s) + 8H^+(aq) + 2NO_3^-(aq) \rightarrow 3Cu^{2+}(aq) + 2NO(g) + 4H_2O(l)$

Cu changes from 0 to +2 (is oxidized) and **Cu is the reducing agent.** N changes from +5 (in NO_3^-) to +2 (in NO) and is reduced, so **NO_3^- is the oxidizing agent.**

4.94   <u>Plan:</u> First, assign oxidation numbers to all atoms following the rules. The reactant that is the reducing agent contains an atom that is oxidized (O.N. increases from the left side to the right side of the equation). The reactant that is the oxidizing agent contains an atom that is reduced (O.N. decreases from the left side to the right side of the equation). Recognize that the agent is the compound that contains the atom that is oxidized or reduced, not just the atom itself.

       <u>Solution:</u>

a)   −6   −6   −4   +2

  +1   −1   0   +5 −2   +4 −1   +4 −2   +1 −2

$8H^+(aq) + 6Cl^-(aq) + Sn(s) + 4NO_3^-(aq) \rightarrow SnCl_6^{2-}(aq) + 4NO_2(g) + 4H_2O(l)$

Nitrogen changes from an O.N. of +5 in NO_3^- to +4 in NO_2 (is reduced) so **NO_3^- is the oxidizing agent.** Sn changes from an O.N. of 0 to an O.N. of +4 in $SnCl_6^{2-}$ (is oxidized) so **Sn is the reducing agent.**

b) −8   +2

  +7 −2   −1   +1   0   +2   +1 −2

$2MnO_4^-(aq) + 10Cl^-(aq) + 16H^+(aq) \rightarrow 5Cl_2(g) + 2Mn^{2+}(aq) + 8H_2O(l)$

Manganese changes from an O.N. of +7 in MnO_4^- to an O.N. of +2 in Mn^{2+} (is reduced) so **MnO_4^- is the oxidizing agent.** Chlorine changes its O.N. from −1 in Cl^- to 0 as the element Cl_2 (is oxidized) so **Cl^- is the reducing agent.**

4.96   <u>Plan:</u> Find the moles of MnO_4^- from the molarity and volume information. Use the molar ratio in the balanced equation to find the moles of H_2O_2. Multiply the moles of H_2O_2 by its molar mass to determine the mass of H_2O_2 present. Mass percent is calculated by dividing the mass of H_2O_2 by the mass of the sample and multiplying by 100. Assign oxidation numbers to all atoms following the rules. The reactant that is the reducing agent contains an atom that is oxidized (O.N. increases from the left side to the right side of the equation).

       <u>Solution:</u>

a) Moles of $MnO_4^- = \left(43.2\ \text{mL}\right)\left(\dfrac{10^{-3}\ \text{L}}{1\ \text{mL}}\right)\left(\dfrac{0.105\ \text{mol}\ MnO_4^-}{\text{L}}\right) = 4.536 \times 10^{-3} = \textbf{4.54} \times \textbf{10}^{-3}\ \textbf{mol}\ \textbf{MnO}_4^-$

b) Moles of $H_2O_2 = \left(4.536 \times 10^{-3}\ MnO_4^-\right)\left(\dfrac{5\ \text{mol}\ H_2O_2}{2\ \text{mol}\ MnO_4^-}\right) = 0.01134 = \textbf{0.0113 mol}\ \textbf{H}_2\textbf{O}_2$

c) Mass (g) of $H_2O_2 = \left(0.01134\ \text{mol}\ H_2O_2\right)\left(\dfrac{34.02\ \text{g}\ H_2O_2}{1\ \text{mol}\ H_2O_2}\right) = 0.3857868 = \textbf{0.386 g}\ \textbf{H}_2\textbf{O}_2$

d) Mass percent of $H_2O_2 = \dfrac{\text{mass of}\ H_2O_2}{\text{mass of sample}}(100) = \dfrac{0.3857868\ \text{g}\ H_2O_2}{14.8\ \text{g sample}}(100) = 2.606668 = \textbf{2.61\%}\ \textbf{H}_2\textbf{O}_2$

e)   −8   +2 −2   +2
    +7 −2   +1 −1   +1   0   +2   +1 −2
$2MnO_4^-(aq) + 5H_2O_2(aq) + 6H^+(aq) \rightarrow 5O_2(g) + 2Mn^{2+}(aq) + 8H_2O(l)$
The O.N. of oxygen increases from −1 in H_2O_2 to 0 in O_2 and is therefore oxidized while the O.N. of
Mn decreases from +7 in MnO_4^- to +2 in Mn^{2+} and is reduced. **H_2O_2 is the reducing agent.**

4.102   <u>Plan:</u> Recall that two reactants are combined to form one product in a combination reaction. A reaction is a redox
reaction only if the oxidation numbers of some atoms change during the reaction.
<u>Solution:</u>
A common example of a combination/redox reaction is the combination of a metal and nonmetal to form an ionic
salt, such as $2Mg(s) + O_2(g) \rightarrow 2MgO(s)$ in which magnesium is oxidized and oxygen is reduced. A common
example of a combination/non-redox reaction is the combination of a metal oxide and water to form an acid, such
as: $CaO(s) + H_2O(l) \rightarrow Ca(OH)_2(aq)$.

4.103   <u>Plan:</u> In a combination reaction, two or more reactants form one product. In a decomposition reaction, one
reactant forms two or more products. In a displacement reaction, atoms or ions exchange places. Balance the
reactions by inspection.
<u>Solution:</u>
a) $Ca(s) + 2H_2O(l) \rightarrow Ca(OH)_2(aq) + H_2(g)$
   Displacement: one Ca atom displaces 2 H atoms.
b) $2NaNO_3(s) \rightarrow 2NaNO_2(s) + O_2(g)$
   Decomposition: one reactant breaks into two products.
c) $C_2H_2(g) + 2H_2(g) \rightarrow C_2H_6(g)$
   Combination: two reactants combine to form one product.

4.105   <u>Plan:</u> In a combination reaction, two or more reactants form one product. In a decomposition reaction, one
reactant forms two or more products. In a displacement reaction, atoms or ions exchange places. Balance the
reactions by inspection.
<u>Solution:</u>
a) $2Sb(s) + 3Cl_2(g) \rightarrow 2SbCl_3(s)$
   Combination: two reactants combine to form one product.
b) $2AsH_3(g) \rightarrow 2As(s) + 3H_2(g)$
   Decomposition: one reactant breaks into two products.
c) $Zn(s) + Fe(NO_3)_2(aq) \rightarrow Zn(NO_3)_2(aq) + Fe(s)$
   Displacement: one Zn displaces one Fe atom.

4.107   <u>Plan:</u> In a combination reaction, two or more reactants form one product. Two elements as reactants often results
in a combination reaction. In a decomposition reaction, one reactant forms two or more products; one reactant only
often indicates a decomposition reaction. In a displacement reaction, atoms or ions exchange places. An element
and a compound as reactants often indicate a displacement reaction. Balance the reactions by inspection.
<u>Solution:</u>
a) The combination between a metal and a nonmetal gives a binary ionic compound.

   $Sr(s) + Br_2(l) \rightarrow SrBr_2(s)$

b) Many metal oxides release oxygen gas upon thermal decomposition.

   $2Ag_2O(s) \xrightarrow{\Delta} 4Ag(s) + O_2(g)$

c) This is a displacement reaction. Mn is a more reactive metal and displaces Cu^{2+} from solution.

   $Mn(s) + Cu(NO_3)_2(aq) \rightarrow Mn(NO_3)_2(aq) + Cu(s)$

4.109   <u>Plan:</u> In a combination reaction, two or more reactants form one product. Two elements as reactants often results in a combination reaction. In a decomposition reaction, one reactant forms two or more products; one reactant only often indicates a decomposition reaction. In a displacement reaction, atoms or ions exchange places. An element and a compound as reactants often indicate a displacement reaction. Balance the reactions by inspection.
<u>Solution:</u>
a) The combination of two nonmetals gives a covalent compound.

$N_2(g) + 3H_2(g) \rightarrow 2NH_3(g)$

b) Some compounds undergo thermal decomposition to simpler substances.

$2NaClO_3(s) \xrightarrow{\Delta} 2NaCl(s) + 3O_2(g)$

c) This is a displacement reaction. Active metals like Ba can displace hydrogen from water.

$Ba(s) + 2H_2O(l) \rightarrow Ba(OH)_2(aq) + H_2(g)$

4.111   <u>Plan:</u> In a combination reaction, two or more reactants form one product. Two elements as reactants often results in a combination reaction. In a decomposition reaction, one reactant forms two or more products; one reactant only often indicates a decomposition reaction. In a displacement reaction, atoms or ions exchange places. An element and a compound as reactants often indicate a displacement reaction. Balance the reactions by inspection.
<u>Solution:</u>
a) Cs, a metal, and I_2, a nonmetal, combine to form the binary ionic compound, CsI.

$2Cs(s) + I_2(s) \rightarrow 2CsI(s)$

b) Al is a stronger reducing agent than Mn and is able to displace Mn from solution, i.e., cause the reduction from $Mn^{2+}(aq)$ to $Mn^0(s)$.

$2Al(s) + 3MnSO_4(aq) \rightarrow Al_2(SO_4)_3(aq) + 3Mn(s)$

c) This is a combination reaction in which sulfur dioxide, SO_2, a nonmetal oxide, combines with oxygen, O_2, to form the higher oxide, SO_3.

$2SO_2(g) + O_2(g) \xrightarrow{\Delta} 2SO_3(g)$

It is not clear from the problem, but energy must be added to force this reaction to proceed.
d) Butane is a four carbon hydrocarbon with the formula C_4H_{10}. It burns in the presence of oxygen, O_2, to form carbon dioxide gas and water vapor. Although this is a redox reaction that could be balanced using the oxidation number method, it is easier to balance by considering only atoms on either side of the equation. First, balance carbon and hydrogen (because they only appear in one species on each side of the equation), and then balance oxygen.

$2C_4H_{10}(g) + 13O_2(g) \rightarrow 8CO_2(g) + 10H_2O(g)$

e) Total ionic equation in which soluble species are shown dissociated into ions:

$2Al(s) + 3Mn^{2+}(aq) + 3SO_4^{2-}(aq) \rightarrow 2Al^{3+}(aq) + 3SO_4^{2-}(aq) + 3Mn(s)$

Net ionic equation in which the spectator ions are omitted:

$2Al(s) + 3Mn^{2+}(aq) \rightarrow 2Al^{3+}(aq) + 3Mn(s)$

Note that the molar coefficients are not simplified because the number of electrons lost (6 e⁻) must equal the electrons gained (6 e⁻).

4.113   <u>Plan:</u> Write a balanced equation that shows the decomposition of HgO to its elements. Convert the mass of HgO to moles and use the molar ratio from the balanced equation to find the moles and then the mass of O_2. Perform the same calculation to find the mass of the other product.
<u>Solution:</u>

a) The balanced chemical equation is $2HgO(s) \xrightarrow{\Delta} 2Hg(l) + O_2(g)$.

$$\text{Moles of HgO} = \left(4.27 \text{ kg HgO}\right)\left(\frac{10^3 \text{ g}}{1 \text{ kg}}\right)\left(\frac{1 \text{ mol HgO}}{216.6 \text{ g HgO}}\right) = 19.71376 \text{ mol HgO}$$

$$\text{Moles of } O_2 = \left(19.71376 \text{ mol HgO}\right)\left(\frac{1 \text{ mol } O_2}{2 \text{ mol HgO}}\right) = 9.85688 \text{ mol } O_2$$

$$\text{Mass (g) of O}_2 = \left(9.85688 \text{ mol O}_2\right)\left(\frac{32.00 \text{ g O}_2}{1 \text{ mol O}_2}\right) = 315.420 = \textbf{315 g O}_2$$

b) The other product is **mercury**.

$$\text{Moles of Hg} = \left(19.71376 \text{ mol HgO}\right)\left(\frac{2 \text{ mol Hg}}{2 \text{ mol HgO}}\right) = 19.71376 \text{ mol Hg}$$

$$\text{Mass (kg) Hg} = \left(19.71376 \text{ mol Hg}\right)\left(\frac{200.6 \text{ g Hg}}{1 \text{ mol Hg}}\right)\left(\frac{1 \text{ kg}}{10^3 \text{ g}}\right) = 3.95458 = \textbf{3.95 kg Hg}$$

4.115 Plan: To determine the reactant in excess, write the balanced equation (metal + $O_2 \rightarrow$ metal oxide), convert reactant masses to moles, and use molar ratios to see which reactant makes the smaller ("limiting") amount of product. Use the limiting reactant to calculate the amount of product formed. Use the molar ratio to find the amount of excess reactant required to react with the limiting reactant; the amount of excess reactant that remains is the initial amount of excess reactant minus the amount required for the reaction.
Solution:
The balanced equation is $4Li(s) + O_2(g) \rightarrow 2Li_2O(s)$.

a) Moles of Li_2O if Li limiting $= \left(1.62 \text{ g Li}\right)\left(\frac{1 \text{ mol Li}}{6.941 \text{ g Li}}\right)\left(\frac{2 \text{ mol Li}_2O}{4 \text{ mol Li}}\right) = 0.1166979 \text{ mol Li}_2O$

Moles of Li_2O if O_2 limiting $= \left(6.50 \text{ g O}_2\right)\left(\frac{1 \text{ mol O}_2}{32.00 \text{ g O}_2}\right)\left(\frac{2 \text{ mol Li}_2O}{1 \text{ mol O}_2}\right) = 0.40625 \text{ mol Li}_2O$

Li is the limiting reactant since it produces the smaller amount of product; **O_2 is in excess**.
b) Using Li as the limiting reagent, $0.1166979 = \textbf{0.117 mol Li}_2\textbf{O}$ is formed.
c) Li is limiting, thus there will be none remaining (**0 g Li**).

$$\text{Mass (g) of Li}_2O = \left(0.1166979 \text{ mol Li}_2O\right)\left(\frac{29.88 \text{ g Li}_2O}{1 \text{ mol Li}_2O}\right) = 3.4869 = \textbf{3.49 g Li}_2\textbf{O}$$

$$\text{Mass (g) of O}_2 \text{ reacted} = \left(1.62 \text{ g Li}\right)\left(\frac{1 \text{ mol Li}}{6.941 \text{ g Li}}\right)\left(\frac{1 \text{ mol O}_2}{4 \text{ mol Li}}\right)\left(\frac{32.00 \text{ g O}_2}{1 \text{ mol O}_2}\right) = 1.867166 \text{ g O}_2$$

Remaining O_2 = initial amount – amount reacted = 6.50 g O_2 – 1.867166 g O_2 = 4.632834 = **4.63 g O$_2$**

4.117 Plan: Since mass must be conserved, the original amount of mixture – amount of residue = mass of oxygen produced. Write a balanced equation and use molar ratios to convert from the mass of oxygen produced to the amount of $KClO_3$ reacted. Mass percent is calculated by dividing the mass of $KClO_3$ by the mass of the sample and multiplying by 100.
Solution:

$$2KClO_3(s) \xrightarrow{\Delta} 2KCl(s) + 3O_2(g)$$

Mass (g) of O_2 produced = mass of mixture – mass of residue = 0.950 g – 0.700 g = 0.250 g O_2

$$\text{Mass (g) of KClO}_3 = \left(0.250 \text{ g O}_2\right)\left(\frac{1 \text{ mol O}_2}{32.00 \text{ g O}_2}\right)\left(\frac{2 \text{ mol KClO}_3}{3 \text{ mol O}_2}\right)\left(\frac{122.55 \text{ g KClO}_3}{1 \text{ mol KClO}_3}\right) = 0.63828125 \text{ g KClO}_3$$

$$\text{Mass \% KClO}_3 = \frac{\text{mass of KClO}_3}{\text{mass of sample}}(100\%) = \frac{0.63828125 \text{ g KClO}_3}{0.950 \text{ g sample}}(100\%) = 67.1875 = \textbf{67.2\% KClO}_3$$

4.119 Plan: Write the balanced equation for the displacement reaction, convert reactant masses to moles, and use molar ratios to see which reactant makes the smaller ("limiting") amount of product. Use the limiting reactant to calculate the amount of product formed.

Solution:
The balanced reaction is $2Al(s) + Fe_2O_3(s) \rightarrow 2Fe(l) + Al_2O_3(s)$.

$$\text{Moles of Fe if Al is limiting} = (1.50 \text{ kg Al})\left(\frac{10^3 \text{ g}}{1 \text{ kg}}\right)\left(\frac{1 \text{ mol Al}}{26.98 \text{ g Al}}\right)\left(\frac{2 \text{ mol Fe}}{2 \text{ mol Al}}\right) = 55.59674 \text{ mol Fe}$$

$$\text{Moles of Fe if } Fe_2O_3 \text{ is limiting} = (25.0 \text{ mol } Fe_2O_3)\left(\frac{2 \text{ mol Fe}}{1 \text{ mol } Fe_2O_3}\right) = 50.0 \text{ mol Fe}$$

Fe_2O_3 is the limiting reactant since it produces the smaller amount of Fe; 50.0 moles of Fe forms.

$$\text{Mass (g) of Fe} = (50.0 \text{ mol Fe})\left(\frac{55.85 \text{ g Fe}}{1 \text{ mol Fe}}\right) = 2792.5 = \mathbf{2790 \text{ g Fe} = 2.79 \text{ kg Fe}}$$

4.120 Plan: In ionic compounds, iron has two common oxidation states, +2 and +3. First, write the balanced equations for the formation and decomposition of compound A. Then, determine which reactant is limiting by converting reactant masses to moles, and using molar ratios to see which reactant makes the smaller ("limiting") amount of product. From the amount of the limiting reactant calculate how much compound B will form.
Solution:
Compound A is iron chloride with iron in the higher, +3, oxidation state. Thus, the formula for compound A is $FeCl_3$ and the correct name is iron(III) chloride. The balanced equation for formation of $FeCl_3$ is:

$$2Fe(s) + 3Cl_2(g) \rightarrow 2FeCl_3(s)$$

To find out whether 50.6 g Fe or 83.8 g Cl_2 limits the amount of product, calculate the number of moles of iron(III) chloride that could form based on each reactant.

$$\text{Moles of } FeCl_3 \text{ if Fe is limiting} = (50.6 \text{ g Fe})\left(\frac{1 \text{ mol Fe}}{55.85 \text{ g Fe}}\right)\left(\frac{2 \text{ mol } FeCl_3}{2 \text{ mol Fe}}\right) = 0.905998 \text{ mol } FeCl_3$$

$$\text{Moles of } FeCl_3 \text{ if } Cl_2 \text{ is limiting} = (83.8 \text{ g } Cl_2)\left(\frac{1 \text{ mol } Cl_2}{70.90 \text{ g } Cl_2}\right)\left(\frac{2 \text{ mol } FeCl_3}{3 \text{ mol } Cl_2}\right) = 0.787964 \text{ mol } FeCl_3$$

Since fewer moles of $FeCl_3$ are produced from the available amount of chlorine, the chlorine is the limiting reactant, producing 0.787964 mol of $FeCl_3$. The $FeCl_3$ decomposes to $FeCl_2$ with iron in the +2 oxidation state. $FeCl_2$ is compound B. The balanced equation for the decomposition of $FeCl_3$ is:

$$2FeCl_3(s) \rightarrow 2FeCl_2(s) + Cl_2(g)$$

$$\text{Mass (g) of } FeCl_2 = (0.787964 \text{ mol } FeCl_3)\left(\frac{2 \text{ mol } FeCl_2}{2 \text{ mol } FeCl_3}\right)\left(\frac{126.75 \text{ g } FeCl_2}{1 \text{ mol } FeCl_2}\right) = 99.874 = \mathbf{99.9 \text{ g } FeCl_2}$$

4.125 Plan: Review the concept of dynamic equilibrium.
Solution:

$$2NO(g) + Br_2(g) \leftrightarrow 2NOBr(g)$$

On a molecular scale, chemical reactions are dynamic. If NO and Br_2 are placed in a container, molecules of NO and molecules of Br_2 will react to form NOBr. Some of the NOBr molecules will decompose and the resulting NO and Br_2 molecules will recombine with different NO and Br_2 molecules to form more NOBr. In this sense, the reaction is dynamic because the original NO and Br_2 pairings do not remain permanently attached to each other. Eventually, the rate of the forward reaction (combination of NO and Br_2) will equal the rate of the reverse reaction (decomposition of NOBr) at which point the reaction is said to have reached equilibrium. If you could take "snapshot" pictures of the molecules at equilibrium, you would see a constant number of reactant (NO, Br_2) and product molecules (NOBr), but the pairings would not stay the same.

4.127 Plan: Ferrous ion is Fe^{2+}. Write a reaction to show the conversion of Fe to Fe^{2+}. Convert the mass of Fe in a 125-g serving to the mass of Fe in a 737-g sample. Use molar mass to convert mass of Fe to moles of Fe and use Avogadro's number to convert moles of Fe to moles of ions.

Solution:

a) Fe oxidizes to Fe^{2+} with a loss of 2 electrons. The H^+ in the acidic food is reduced to H_2 with a gain of 2 electrons. The balanced reaction is:

$$Fe(s) + 2H^+(aq) \rightarrow Fe^{2+}(aq) + H_2(g)$$

O.N.:   0   +1   +2   0

b) Mass (g) of Fe in the jar of tomato sauce $= \left(737 \text{ g sauce}\right)\left(\dfrac{49 \text{ mg Fe}}{125 \text{ g sauce}}\right)\left(\dfrac{10^{-3} \text{ g}}{1 \text{ mg}}\right) = 0.288904 \text{ g Fe}$

Number of Fe^{2+} ions $= \left(0.288904 \text{ g Fe}\right)\left(\dfrac{1 \text{ mol Fe}}{55.85 \text{ g Fe}}\right)\left(\dfrac{1 \text{ mol Fe}^{2+}}{1 \text{ mol Fe}}\right)\left(\dfrac{6.022 \times 10^{23} \text{ Fe}^{2+}\text{ions}}{1 \text{ mol Fe}^{2+}}\right)$

$= 3.11509 \times 10^{21} = \mathbf{3.1 \times 10^{21} \ Fe^{2+}}$ **ions** per jar of sauce

4.129   **Plan:** Convert the mass of glucose to moles and use the molar ratios from the balanced equation to find the moles of ethanol and CO_2. The amount of ethanol is converted from moles to grams using its molar mass. The amount of CO_2 is converted from moles to volume in liters using the conversion factor given.

Solution:

Moles of $C_2H_5OH = \left(100. \text{ g } C_6H_{12}O_6\right)\left(\dfrac{1 \text{ mol } C_6H_{12}O_6}{180.16 \text{ g } C_6H_{12}O_6}\right)\left(\dfrac{2 \text{ mol } C_2H_5OH}{1 \text{ mol } C_6H_{12}O_6}\right) = 1.11012 \text{ mol } C_2H_5OH$

Mass (g) of $C_2H_5OH = \left(1.11012 \text{ mol } C_2H_5OH\right)\left(\dfrac{46.07 \text{ g } C_2H_5OH}{1 \text{ mol } C_2H_5OH}\right) = 51.143 = \mathbf{51.1 \text{ g } C_2H_5OH}$

Moles of $CO_2 = \left(100. \text{ g } C_6H_{12}O_6\right)\left(\dfrac{1 \text{ mol } C_6H_{12}O_6}{180.16 \text{ g } C_6H_{12}O_6}\right)\left(\dfrac{2 \text{ mol } CO_2}{1 \text{ mol } C_6H_{12}O_6}\right) = 1.11012 \text{ mol } CO_2$

Volume (L) of $CO_2 = \left(1.11012 \text{ mol } CO_2\right)\left(\dfrac{22.4 \text{ L } CO_2}{1 \text{ mol } CO_2}\right) = 24.8667 = \mathbf{24.9 \text{ L } CO_2}$

4.131   **Plan:** Remember that spectator ions are omitted from net ionic equations. Assign oxidation numbers to all atoms in the titration reaction, following the rules. The reactant that is the reducing agent contains an atom that is oxidized (O.N. increases from the left side to the right side of the equation). The reactant that is the oxidizing agent contains an atom that is reduced (O.N. decreases from the left side to the right side of the equation). Find the number of moles of $KMnO_4$ from the molarity and volume and use molar ratios in the balanced equations to find the moles and then mass of $CaCl_2$. Mass percent is calculated by dividing the mass of $CaCl_2$ by the mass of the sample and multiplying by 100.

Solution:

a) The reaction is: $Na_2C_2O_4(aq) + CaCl_2(aq) \rightarrow CaC_2O_4(s) + 2NaCl(aq)$

The total ionic equation in which soluble substances are dissociated into ions is:

$$2Na^+(aq) + C_2O_4^{2-}(aq) + Ca^{2+}(aq) + 2Cl^-(aq) \rightarrow CaC_2O_4(s) + 2Na^+(aq) + 2Cl^-(aq)$$

Omitting Na^+ and Cl^- as spectator ions gives the net ionic equation:

$$\mathbf{Ca^{2+}(aq) + C_2O_4^{2-}(aq) \rightarrow CaC_2O_4(s)}$$

b) You may recognize this reaction as a redox titration, because the permanganate ion, MnO_4^-, is a common oxidizing agent. The MnO_4^- oxidizes the oxalate ion, $C_2O_4^{2-}$ to CO_2. Mn changes from +7 to +2 (reduction) and C changes from +3 to +4 (oxidation). The equation that describes this process is:

$$H_2C_2O_4(aq) + K^+(aq) + MnO_4^-(aq) \rightarrow Mn^{2+}(aq) + CO_2(g) + K^+(aq)$$

$H_2C_2O_4$ is a weak acid, so it cannot be written in a fully dissociated form. $KMnO_4$ is a soluble salt, so it can be written in its dissociated form. $K^+(aq)$ is omitted in the net ionic equation because it is a spectator ion. We will balance the equation using the oxidation number method. First assign oxidation numbers to all elements in the reaction:

$$\begin{array}{ccc} +2\ +6\ -8 & -8 & -4 \\ +1\ +3\ -2 & +7\ -2 & +2 \qquad +4\ -2 \\ H_2C_2O_4(aq) + & MnO_4^-(aq) \rightarrow & Mn^{2+}(aq) + 2CO_2(g) \end{array}$$

Identify the oxidized and reduced species and multiply one or both species by the appropriate factors to make the electrons lost equal the electrons gained. Each C in $H_2C_2O_4$ increases in O.N. from +3 to +4 for a total gain of 2 electrons. Mn in MnO_4^- decreases in O.N. from +7 to +2 for a loss of 5 electrons. Multiply C by 5 and Mn by 2 to get an electron loss = electron gain of 10 electrons.

$$5H_2C_2O_4(aq) + 2MnO_4^-(aq) \rightarrow 2Mn^{2+}(aq) + 10CO_2(g)$$

Adding water and $H^+(aq)$ to finish balancing the equation is appropriate since the reaction takes place in acidic medium. Add $8H_2O(l)$ to right side of equation to balance the oxygen and then add $6H^+(aq)$ to the left to balance hydrogen.

$$\mathbf{5H_2C_2O_4(aq) + 2MnO_4^-(aq) + 6H^+(aq) \rightarrow 10CO_2(g) + 2Mn^{2+}(aq) + 8H_2O(l)}$$

c) Mn in MnO_4^- decreases in O.N. from +7 to +2 and is reduced. **$KMnO_4$ is the oxidizing agent.**
d) C in $H_2C_2O_4$ increases in O.N. from +3 to +4 and is oxidized. **$H_2C_2O_4$ is the reducing agent.**
e) The balanced equations provide the accurate molar ratios between species.

$$\text{Moles of } MnO_4^- = \left(37.68 \text{ mL}\right)\left(\frac{10^{-3} \text{ L}}{1 \text{ mL}}\right)\left(\frac{0.1019 \text{ mol } KMnO_4}{1 \text{ L}}\right) = 0.0038396 \text{ mol } MnO_4^-$$

$$\text{Moles of } H_2C_2O_4 = \left(0.0038396 \text{ mol } MnO_4^-\right)\left(\frac{5 \text{ mol } H_2C_2O_4}{2 \text{ mol } MnO_4^-}\right) = 0.009599 \text{ mol } H_2C_2O_4$$

$$\text{Mass (g) } CaCl_2 = \left(0.009599 \text{ mol } H_2C_2O_4\right)\left(\frac{1 \text{ mol } CaCl_2}{1 \text{ mol } H_2C_2O_4}\right)\left(\frac{110.98 \text{ g } CaCl_2}{1 \text{ mol } CaCl_2}\right) = 1.06530 \text{ g } CaCl_2$$

$$\text{Mass percent } CaCl_2 = \frac{\text{mass of } CaCl_2}{\text{mass of sample}}(100) = \frac{1.06530 \text{ g } CaCl_2}{1.9348 \text{ g mixture}}(100) = 55.05995 = \mathbf{55.06\% \text{ } CaCl_2}$$

4.134   Plan: Write a balanced equation for the reaction. This is an acid-base reaction between HCl and the base CO_3^{2-} to form H_2CO_3 which then decomposes into CO_2 and H_2O. Find the moles of HCl from the molarity and volume information and use the molar ratio in the balanced equation to find the moles and mass of dolomite. To find mass %, divide the mass of dolomite by the mass of soil and multiply by 100.
Solution:
The balanced equation for this reaction is:

$$CaMg(CO_3)_2(s) + 4HCl(aq) \rightarrow Ca^{2+}(aq) + Mg^{2+}(aq) + 2H_2O(l) + 2CO_2(g) + 4Cl^-(aq)$$

$$\text{Moles of HCl} = \left(33.56 \text{ mL}\right)\left(\frac{10^{-3} \text{ L}}{1 \text{ mL}}\right)\left(\frac{0.2516 \text{ mol HCl}}{1 \text{ L}}\right) = 0.0084437 \text{ mol HCl}$$

$$\text{Mass } CaMg(CO_3)_2 = \left(0.0084437 \text{ mol HCl}\right)\left(\frac{1 \text{ mol } CaMg(CO_3)_2}{4 \text{ mol HCl}}\right)\left(\frac{184.41 \text{ g } CaMg(CO_3)_2}{1 \text{ mol } CaMg(CO_3)_2}\right)$$

$$= 0.389276 \text{ g } CaMg(CO_3)_2$$

$$\text{Mass percent } CaMg(CO_3)_2 = \frac{\text{mass } CaMg(CO_3)_2}{\text{mass soil}}(100) = \frac{0.389276 \text{ g } CaMg(CO_3)_2}{13.86 \text{ g soil}}(100)$$

$$= 2.80863 = \mathbf{2.809\% \text{ } CaMg(CO_3)_2}$$

4.136   Plan: For part a), assign oxidation numbers to each element; the oxidizing agent has an atom whose oxidation number decreases while the reducing agent has an atom whose oxidation number increases. For part b), use the molar ratios, beginning with step 3, to find the moles of NO_2, then moles of NO, then moles of NH_3 required to produce the given mass of HNO_3.

Solution:

a) *Step 1*

$$\overset{+3}{}\qquad\qquad\qquad\overset{+2}{}$$

$$\overset{-3\ +1}{}\qquad\overset{0}{}\qquad\overset{+2\ -2}{}\qquad\overset{+1\ -2}{}$$

$$4NH_3(g) + 5O_2(g) \rightarrow 4NO(g) + 6H_2O(l)$$

N is oxidized from –3 in NH_3 to +2 in NO; O is reduced from 0 in O_2 to to –2 in NO.

Oxidizing agent = O_2   **Reducing agent = NH_3**

Step 2   –4

$$\overset{+2\ -2}{}\qquad\overset{0}{}\qquad\overset{+4\ -2}{}$$

$$2NO(g) + O_2(g) \rightarrow 2NO_2(g)$$

N is oxidized from +2 in NO to +4 in NO_2; O is reduced from 0 in O_2 to –2 in NO_2.

Oxidizing agent = O_2   **Reducing agent = NO**

Step 3

$$\overset{-4}{}\qquad\overset{+2}{}\qquad\qquad\overset{-6}{}$$

$$\overset{+4\ -2}{}\qquad\overset{+1\ -2}{}\qquad\overset{+1+5\ -2}{}\qquad\overset{+2\ -2}{}$$

$$3NO_2(g) + H_2O(l) \rightarrow 2HNO_3(l) + NO(g)$$

N is oxidized from +4 in NO_2 to +5 in HNO_3; N is reduced from +4 in NO_2 to +2 in NO.

Oxidizing agent = NO_2   **Reducing agent = NO_2**

b) Moles of NO_2 = $\left(3.0\times10^4 \text{ kg HNO}_3\right)\left(\dfrac{10^3 \text{ g}}{1 \text{ kg}}\right)\left(\dfrac{1 \text{ mol HNO}_3}{63.02 \text{ g HNO}_3}\right)\left(\dfrac{3 \text{ mol NO}_2}{2 \text{ mol HNO}_3}\right)$ = 7.14059×10^5 mol NO_2

Moles of NO = $\left(7.14059\times10^5 \text{ mol NO}_2\right)\left(\dfrac{2 \text{ mol NO}}{2 \text{ mol NO}_2}\right)$ = 7.14059×10^5 mol NO

Moles of NH_3 = $\left(7.14059\times10^5 \text{ mol NO}\right)\left(\dfrac{4 \text{ mol NH}_3}{4 \text{ mol NO}}\right)$ = 7.14059×10^5 mol NH_3

Mass (kg) of NH_3 = $\left(7.14059\times10^5 \text{ mol NH}_3\right)\left(\dfrac{17.03 \text{ g NH}_3}{1 \text{ mol NH}_3}\right)\left(\dfrac{1 \text{ kg}}{10^3 \text{ g}}\right)$ = 1.21604×10^4 = **1.2×10^4 kg NH_3**

4.139   Plan: Write a balanced equation and use the molar ratio between Na_2O_2 and CO_2 to convert the amount of Na_2O_2 given to the amount of CO_2 that reacts with that amount. Convert that amount of CO_2 to liters of air.
Solution:
The reaction is: $2Na_2O_2(s) + 2CO_2(g) \rightarrow 2Na_2CO_3(s) + O_2(g)$.

Mass (g) of CO_2 = $\left(80.0 \text{ g Na}_2O_2\right)\left(\dfrac{1 \text{ mol Na}_2O_2}{77.98 \text{ g Na}_2O_2}\right)\left(\dfrac{2 \text{ mol CO}_2}{2 \text{ mol Na}_2O_2}\right)\left(\dfrac{44.01 \text{ g CO}_2}{1 \text{ mol CO}_2}\right)$ = 45.150 g CO_2

Volume (L) of air = $\left(45.150 \text{ g CO}_2\right)\left(\dfrac{\text{L air}}{0.0720 \text{ g CO}_2}\right)$ = 627.08 = **627 L air**

4.141   Plan: Use the mass percents to find the mass of each component needed for 1.00 kg of glass. Write balanced reactions for the decomposition of sodium carbonate and calcium carbonate to produce an oxide and CO_2. Use the molar ratios in these reactions to find the moles and mass of each carbonate required to produce the amount of oxide in the glass sample.

A 1.00-kg piece of glass of composition 75% SiO_2, 15% Na_2O, and 10% CaO would contain:

Mass (kg) of SiO_2 = $(1.00 \text{ kg glass})\left(\dfrac{75\% \ SiO_2}{100\% \ \text{glass}}\right)$ = **0.75 kg SiO_2**

Mass (kg) of Na_2O = $(1.00 \text{ kg glass})\left(\dfrac{15\% \ Na_2O}{100\% \ \text{glass}}\right)$ = 0.15 kg Na_2O

Mass (kg) of CaO = $(1.00 \text{ kg glass})\left(\dfrac{10.\% \ CaO}{100\% \ \text{glass}}\right)$ = 0.10 kg CaO

In this example, the SiO_2 is added directly while the sodium oxide comes from decomposition of sodium carbonate and the calcium oxide from decomposition of calcium carbonate:

$$Na_2CO_3(s) \rightarrow Na_2O(s) + CO_2(g)$$
$$CaCO_3(s) \rightarrow CaO(s) + CO_2(g)$$

Mass (kg) of Na_2CO_3 =

$$\left(0.15 \text{ kg } Na_2O\right)\left(\dfrac{10^3 \text{ g}}{1 \text{ kg}}\right)\left(\dfrac{1 \text{ mol } Na_2O}{61.98 \text{ g } Na_2O}\right)\left(\dfrac{1 \text{ mol } Na_2CO_3}{1 \text{ mol } Na_2O}\right)\left(\dfrac{105.99 \text{ g } Na_2CO_3}{1 \text{ mol } Na_2CO_3}\right)\left(\dfrac{1 \text{ kg}}{10^3 \text{ g}}\right)$$

$$= 0.25651 = \textbf{0.26 kg } Na_2CO_3$$

Mass (kg) of $CaCO_3$ = $\left(0.10 \text{ kg CaO}\right)\left(\dfrac{10^3 \text{ g}}{1 \text{ kg}}\right)\left(\dfrac{1 \text{ mol CaO}}{56.08 \text{ g CaO}}\right)\left(\dfrac{1 \text{ mol } CaCO_3}{1 \text{ mol CaO}}\right)\left(\dfrac{100.09 \text{ g } CaCO_3}{1 \text{ mol } CaCO_3}\right)\left(\dfrac{1 \text{ kg}}{10^3 \text{ g}}\right)$

$$= 0.178477 = \textbf{0.18 kg } CaCO_3$$

4.144 **Plan:** Write balanced reactions and use molar ratios to find the moles of product. To find mass %, divide the mass of thyroxine by the mass of extract and multiply by 100.
Solution:
a) There is not enough information to write complete chemical equations, but the following equations can be written:

$$C_{15}H_{11}I_4NO_4(s) + Na_2CO_3(s) \rightarrow 4I^-(aq) + \text{other products}$$
$$I^-(aq) + Br_2(l) + HCl(aq) \rightarrow IO_3^-(aq) + \text{other products}$$

Moles of IO_3^- = $\left(1 \text{ mol } C_{15}H_{11}I_4NO_4\right)\left(\dfrac{4 \text{ mol } I^-}{1 \text{ mol } C_{15}H_{11}I_4NO_4}\right)\left(\dfrac{1 \text{ mol } IO_3^-}{1 \text{ mol } I^-}\right)$ = **4 moles IO_3^- are produced**

b)   +5 –2   +1   –1   0   +1 –2
$$IO_3^-(aq) + H^+(aq) + I^-(aq) \rightarrow I_2(aq) + H_2O(l)$$

This is a difficult equation to balance because the iodine species are both reducing and oxidizing. The O.N. of iodine decreases from +5 in IO_3^- to 0 in I_2 and so gains 5 electrons; the O.N. of iodine increases from –1 in I^- to 0 in I_2 and so loses 1 electron. Start balancing the equation by placing a coefficient of 5 in front of $I^-(aq)$, so the electrons lost equal the electrons gained. Do <u>not</u> place a 5 in front of $I_2(aq)$, because not all of the $I_2(aq)$ comes from oxidation of $I^-(aq)$. Some of the $I_2(aq)$ comes from the reduction of $IO_3^-(aq)$. Place a coefficient of 3 in front of $I_2(aq)$ to correctly balance iodine:

$$IO_3^-(aq) + H^+(aq) + 5I^-(aq) \rightarrow 3I_2(aq) + H_2O(l)$$

The reaction is now balanced from a redox standpoint, so finish balancing the reaction by balancing the oxygen and hydrogen.

$$IO_3^-(aq) + 6H^+(aq) + 5I^-(aq) \rightarrow 3I_2(aq) + 3H_2O(l)$$

IO_3^- is the oxidizing agent, and I^- is the reducing agent.

Moles of I_2 produced per mole of thyroxine $= \left(1 \text{ mol } C_{15}H_{11}I_4NO_4\right)\left(\dfrac{4 \text{ mol } IO_3^-}{1 \text{ mol } C_{15}H_{11}I_4NO_4}\right)\left(\dfrac{3 \text{ mol } I_2}{1 \text{ mol } IO_3^-}\right)$

$$= \textbf{12 moles of } I_2 \textbf{ are produced per mole of thyroxine}$$

c) Using Thy to represent thyroxine:

The balanced equation for this reaction is $I_2(aq) + 2S_2O_3^{2-}(aq) \rightarrow 2I^-(aq) + S_4O_6^{2-}(aq)$.

Moles of $S_2O_3^{2-} = \left(17.23 \text{ mL}\right)\left(\dfrac{10^{-3} \text{ L}}{1 \text{ mL}}\right)\left(\dfrac{0.1000 \text{ mol } S_2O_3^{2-}}{L}\right) = 0.001723 \text{ mol } S_2O_3^{2-}$

Moles of Thy $= \left(0.001723 \text{ mol } S_2O_3^{2-}\right)\left(\dfrac{1 \text{ mol } I_2}{2 \text{ mol } S_2O_3^{2-}}\right)\left(\dfrac{1 \text{ mol Thy}}{12 \text{ mol } I_2}\right) = 7.179167 \times 10^{-5} \text{ mol Thy}$

Mass (g) of Thy $= \left(7.179167 \times 10^{-5} \text{ mol Thy}\right)\left(\dfrac{776.8 \text{ g Thy}}{1 \text{ mol Thy}}\right) = 0.055767769 \text{ g Thy}$

Mass % Thy $= \dfrac{\text{mass of Thy}}{\text{mass of extract}}(100) = \dfrac{0.055767769 \text{ g Thy}}{0.4332 \text{ g}}(100) = 12.8734 = \textbf{12.87\% thyroxine}$

4.147 Plan: Balance the equation to obtain the correct molar ratios. Use the mass percents to find the mass of each reactant in a 1.00 g sample, convert the mass of each reactant to moles, and use the molar ratios to find the limiting reactant and the amount of CO_2 produced. Convert moles of CO_2 produced to volume using the given conversion factor.
Solution:
a) Here is a suggested method for balancing the equation.
— Since PO_4^{2-} remains as a unit on both sides of the equation, treat it as a unit when balancing.
— On first inspection, one can see that Na needs to be balanced by adding a "2" in front of $NaHCO_3$. This then affects the balance of C, so add a "2" in front of CO_2.
— Hydrogen is not balanced, so change the coefficient of water to "2," as this will have the least impact on the other species.
— Verify that the other species are balanced.

$$Ca(H_2PO_4)_2(s) + 2NaHCO_3(s) \xrightarrow{\Delta} 2CO_2(g) + 2H_2O(g) + CaHPO_4(s) + Na_2HPO_4(s)$$

Determine whether $Ca(H_2PO_4)_2$ or $NaHCO_3$ limits the production of CO_2. In each case calculate the moles of CO_2 that might form.

Mass (g) of $NaHCO_3 = \left(1.00 \text{ g}\right)\left(\dfrac{31.0\%}{100\%}\right) = 0.31 \text{ g } NaHCO_3$

Mass (g) of $Ca(H_2PO_4)_2 = \left(1.00 \text{ g}\right)\left(\dfrac{35.0\%}{100\%}\right) = 0.35 \text{ g } Ca(H_2PO_4)_2$

Moles of CO_2 if $NaHCO_3$ is limiting $= \left(0.31 \text{ g } NaHCO_3\right)\left(\dfrac{1 \text{ mol } NaHCO_3}{84.01 \text{ g } NaHCO_3}\right)\left(\dfrac{2 \text{ mol } CO_2}{2 \text{ mol } NaHCO_3}\right)$

$$= 3.690 \times 10^{-3} \text{ mol } CO_2$$

Moles of CO_2 if $Ca(H_2PO_4)_2$ is limiting $= \left(0.35 \text{ g } Ca(H_2PO_4)_2\right)\left(\dfrac{1 \text{ mol } Ca(H_2PO_4)_2}{234.05 \text{ g } Ca(H_2PO_4)_2}\right)\left(\dfrac{2 \text{ mol } CO_2}{1 \text{ mol } Ca(H_2PO_4)_2}\right)$

$$= 2.9908 \times 10^{-3} \text{ mol } CO_2$$

Since $Ca(H_2PO_4)_2$ produces the smaller amount of product, it is the limiting reactant and $\textbf{3.0} \times \textbf{10}^{-3}$ **mol CO_2** will be produced.

b) Volume (L) of $CO_2 = \left(2.9908 \times 10^{-3} \text{ mol } CO_2\right)\left(\dfrac{37.0 \text{ L}}{1 \text{ mol } CO_2}\right) = 0.1106596 = \textbf{0.11 L } CO_2$

4.149 Plan: To determine the empirical formula, find the moles of each element present and divide by the smallest number of moles to get the smallest ratio of atoms. To find the molecular formula, divide the molar mass by the mass of the empirical formula to find the factor by which to multiple the empirical formula. Write the balanced acid-base reaction for part c) and use the molar ratio in that reaction to find the mass of bismuth(III) hydroxide.
Solution:

a) Determine the moles of each element present. The sample was burned in an unknown amount of O_2, therefore, the moles of oxygen must be found by a different method.

$$\text{Moles of C} = \left(0.1880 \text{ g CO}_2\right)\left(\frac{1 \text{ mol CO}_2}{44.01 \text{ g CO}_2}\right)\left(\frac{1 \text{ mol C}}{1 \text{ mol CO}_2}\right) = 4.271756 \times 10^{-3} \text{ mol C}$$

$$\text{Moles of H} = \left(0.02750 \text{ g H}_2\text{O}\right)\left(\frac{1 \text{ mol H}_2\text{O}}{18.02 \text{ g H}_2\text{O}}\right)\left(\frac{2 \text{ mol H}}{1 \text{ mol H}_2\text{O}}\right) = 3.052164 \times 10^{-3} \text{ mol H}$$

$$\text{Moles of Bi} = \left(0.1422 \text{ g Bi}_2\text{O}_3\right)\left(\frac{1 \text{ mol Bi}_2\text{O}_3}{466.0 \text{ g Bi}_2\text{O}_3}\right)\left(\frac{2 \text{ mol Bi}}{1 \text{ mol Bi}_2\text{O}_3}\right) = 6.103004 \times 10^{-4} \text{ mol Bi}$$

Subtracting the mass of each element present from the mass of the sample will give the mass of oxygen originally present in the sample. This mass is used to find the moles of oxygen.

$$\text{Mass (g) of C} = \left(4.271756 \times 10^{-3} \text{ mol C}\right)\left(\frac{12.01 \text{ g C}}{1 \text{ mol C}}\right) = 0.0513038 \text{ g C}$$

$$\text{Mass (g) of H} = \left(3.052164 \times 10^{-3} \text{ mol H}\right)\left(\frac{1.008 \text{ g H}}{1 \text{ mol H}}\right) = 0.0030766 \text{ g H}$$

$$\text{Mass (g) of Bi} = \left(6.103004 \times 10^{-4} \text{ mol Bi}\right)\left(\frac{209.0 \text{ g Bi}}{1 \text{ mol Bi}}\right) = 0.127553 \text{ g Bi}$$

Mass (g) of O = mass of sample – (mass C + mass H + mass Bi)

= 0.22105 g sample – (0.0513038 g C + 0.0030766 g H + 0.127553 g Bi) = 0.0391166 g O

$$\text{Moles of O} = \left(0.0391166 \text{ g O}\right)\left(\frac{1 \text{ mol O}}{16.00 \text{ g O}}\right) = 2.44482 \times 10^{-4} \text{ mol O}$$

Divide each of the moles by the smallest value (moles Bi).

$$C = \frac{4.271756 \times 10^{-3}}{6.103004 \times 10^{-4}} = 7 \qquad\qquad H = \frac{3.052164 \times 10^{-3}}{6.103004 \times 10^{-4}} = 5$$

$$O = \frac{2.4448 \times 10^{-3}}{6.103004 \times 10^{-4}} = 4 \qquad\qquad Bi = \frac{6.103004 \times 10^{-4}}{6.103004 \times 10^{-4}} = 1$$

Empirical formula = **C$_7$H$_5$O$_4$Bi**

b) The empirical formula mass is 362 g/mol. Therefore, there are 1086/362 = 3 empirical formula units per molecular formula making the molecular formula = 3 x C$_7$H$_5$O$_4$Bi = **C$_{21}$H$_{15}$O$_{12}$Bi$_3$**.

c) $\text{Bi(OH)}_3(s) + 3\text{HC}_7\text{H}_5\text{O}_3(aq) \rightarrow \text{Bi(C}_7\text{H}_5\text{O}_3)_3(s) + 3\text{H}_2\text{O}(l)$

d) $\text{Moles of C}_{21}\text{H}_{15}\text{O}_{12}\text{Bi}_3 = \left(0.600 \text{ mg C}_{21}\text{H}_{15}\text{O}_{12}\text{Bi}_3\right)\left(\frac{10^{-3} \text{ g}}{1 \text{ mg}}\right)\left(\frac{1 \text{ mol C}_{21}\text{H}_{15}\text{O}_{12}\text{Bi}_3}{1086 \text{ g C}_{21}\text{H}_{15}\text{O}_{12}\text{Bi}_3}\right)$

$= 5.52486 \times 10^{-4} \text{ mol C}_{21}\text{H}_{15}\text{O}_{12}\text{Bi}_3$

Mass (mg) of Bi(OH)$_3$ =

$\left(5.52486 \times 10^{-7} \text{ mol C}_{21}\text{H}_{15}\text{O}_{12}\text{Bi}_3\right)\left(\frac{3 \text{ mol Bi}}{1 \text{ mol C}_{21}\text{H}_{15}\text{O}_{12}\text{Bi}_3}\right)\left(\frac{1 \text{ mol Bi(OH)}_3}{1 \text{ mol Bi}}\right)\left(\frac{260.0 \text{ g Bi(OH)}_3}{1 \text{ mol Bi(OH)}_3}\right)\left(\frac{1 \text{ mg}}{10^{-3} \text{ g}}\right)\left(\frac{100\%}{88.0\%}\right)$

= 0.48970 = **0.490 mg Bi(OH)$_3$**

4.151   Plan: Write balanced equations. Use the density to convert volume of fuel to mass of fuel and then use the molar ratios to convert mass of each fuel to the mass of oxygen required for the reaction. Use the conversion factor given to convert mass of oxygen to volume of oxygen.
Solution:
a) Complete combustion of hydrocarbons involves heating the hydrocarbon in the presence of oxygen to produce carbon dioxide and water.

Ethanol: $C_2H_5OH(l) + 3O_2(g) \rightarrow 2CO_2(g) + 3H_2O(l)$
Gasoline: $2C_8H_{18}(l) + 25O_2(g) \rightarrow 16CO_2(g) + 18H_2O(g)$

b) The mass of each fuel must be found:

$$\text{Mass (g) of gasoline} = (1.00 \text{ L})\left(\frac{90\%}{100\%}\right)\left(\frac{1 \text{ mL}}{10^{-3} \text{ L}}\right)\left(\frac{0.742 \text{ g}}{1 \text{ mL}}\right) = 667.8 \text{ g gasoline}$$

$$\text{Mass (g) of ethanol} = (1.00 \text{ L})\left(\frac{10\%}{100\%}\right)\left(\frac{1 \text{ mL}}{10^{-3} \text{ L}}\right)\left(\frac{0.789 \text{ g}}{1 \text{ mL}}\right) = 78.9 \text{ g ethanol}$$

$$\text{Mass (g) of } O_2 \text{ to react with gasoline} = (667.8 \text{ g } C_8H_{18})\left(\frac{1 \text{ mol } C_8H_{18}}{114.22 \text{ g } C_8H_{18}}\right)\left(\frac{25 \text{ mol } O_2}{2 \text{ mol } C_8H_{18}}\right)\left(\frac{32.00 \text{ g } O_2}{1 \text{ mol } O_2}\right)$$

$$= 2338.64 \text{ g } O_2$$

$$\text{Mass (g) of } O_2 \text{ to react with ethanol} = (78.9 \text{ g } C_2H_5OH)\left(\frac{1 \text{ mol } C_2H_5OH}{46.07 \text{ g } C_2H_5OH}\right)\left(\frac{3 \text{ mol } O_2}{1 \text{ mol } C_2H_5OH}\right)\left(\frac{32.00 \text{ g } O_2}{1 \text{ mol } O_2}\right)$$

$$= 164.41 \text{ g } O_2$$

Total mass (g) of O_2 = 2338.64 g O_2 + 164.41 g O_2 = 2503.05 = **2.50x10³ g O_2**

$$\text{c) Volume (L) of } O_2 = (2503.05 \text{ g } O_2)\left(\frac{1 \text{ mol } O_2}{32.00 \text{ g } O_2}\right)\left(\frac{22.4 \text{ L}}{1 \text{ mol } O_2}\right) = 1752.135 = \textbf{1.75x10}^3 \textbf{ L } O_2$$

$$\text{d) Volume (L) of air} = (1752.135 \text{ L } O_2)\left(\frac{100\%}{20.9\%}\right) = 8383.42 = \textbf{8.38x10}^3 \textbf{ L air}$$

4.153   Plan: From the molarity and volume of the base NaOH, find the moles of NaOH and use the molar ratios from the two balanced equations to convert the moles of NaOH to moles of HBr to moles of vitamin C. Use the molar mass of vitamin C to convert moles to grams.
Solution:

$$\text{Moles of NaOH} = (43.20 \text{ mL NaOH})\left(\frac{10^{-3} \text{ L}}{1 \text{ mL}}\right)\left(\frac{0.1350 \text{ mol NaOH}}{1 \text{ L}}\right) = 0.005832 \text{ mol NaOH}$$

$$\text{Mass (g) of vitamin C} = (0.005832 \text{ mol NaOH})\left(\frac{1 \text{ mol HBr}}{1 \text{ mol NaOH}}\right)\left(\frac{1 \text{ mol } C_6H_8O_6}{2 \text{ mol HBr}}\right)\left(\frac{176.12 \text{ g } C_6H_8O_6}{1 \text{ mol } C_6H_8O_6}\right)\left(\frac{1 \text{ mg}}{10^{-3} \text{ g}}\right)$$

$$= 513.5659 = 513.6 \text{ mg } C_6H_8O_6$$

Yes, the tablets have the quantity advertised.

CHAPTER 5 GASES AND THE KINETIC-MOLECULAR THEORY

FOLLOW–UP PROBLEMS

5.1A   Plan: Use the equation for gas pressure in an open-end manometer to calculate the pressure of the gas. Use conversion factors to convert pressure in mmHg to units of torr, pascals and lb/in^2.
Solution:
Because $P_{gas} < P_{atm}$, $P_{gas} = P_{atm} - \Delta h$
P_{gas} = 753.6 mmHg – 174.0 mmHg = 579.6 mmHg

$$\text{Pressure (torr)} = \left(579.6 \text{ mmHg}\right)\left(\frac{1 \text{ torr}}{1 \text{ mmHg}}\right) = \textbf{579.6 torr}$$

$$\text{Pressure (Pa)} = \left(579.6 \text{ mmHg}\right)\left(\frac{1 \text{ atm}}{760 \text{ mmHg}}\right)\left(\frac{1.01325 \times 10^5 \text{Pa}}{1 \text{atm}}\right) = 7.727364 \times 10^4 = \textbf{7.727} \times \textbf{10}^4 \textbf{ Pa}$$

$$\text{Pressure (lb/in}^2\text{)} = \left(579.6 \text{ mmHg}\right)\left(\frac{1 \text{ atm}}{760 \text{ mmHg}}\right)\left(\frac{14.7 \text{lb/in}^2}{1 \text{ atm}}\right) = 11.21068 = \textbf{11.2 lb/in}^2$$

5.1B   Plan: Convert the atmospheric pressure to torr. Use the equation for gas pressure in an open-end manometer to calculate the pressure of the gas. Use conversion factors to convert pressure in torr to units of mmHg, pascals and lb/in^2.
Solution:
Because $P_{gas} > P_{atm}$, $P_{gas} = P_{atm} + \Delta h$
$$P_{gas} = (0.9475 \text{ atm})\left(\frac{760 \text{ torr}}{1 \text{ atm}}\right) + 25.8 \text{ torr} = 745.9 \text{ torr}$$

$$\text{Pressure (mmHg)} = (745.9 \text{ torr})\left(\frac{1 \text{ mmHg}}{1 \text{ torr}}\right) = \textbf{745.9 mmHg}$$

$$\text{Pressure (Pa)} = (745.9 \text{ mmHg})\left(\frac{1 \text{ atm}}{760 \text{ mmHg}}\right)\left(\frac{1.01325 \times 10^5 \text{ Pa}}{1 \text{ atm}}\right) = 9.94452 \times 10^4 = \textbf{9.945} \times \textbf{10}^4 \textbf{ Pa}$$

$$\text{Pressure (lb/in}^2\text{)} = (745.9 \text{ mmHg})\left(\frac{1 \text{ atm}}{760 \text{ mmHg}}\right)\left(\frac{14.7 \text{ lb/in}^2}{1 \text{ atm}}\right) = 14.427 = \textbf{14.4 lb/in}^2$$

5.2A   Plan: Given in the problem is an initial volume, initial pressure, and final volume for the argon gas. The final pressure is to be calculated. The temperature and amount of gas are fixed. Rearrange the ideal gas law to the appropriate form and solve for P_2. Once solved for, P_2 must be converted from atm units to kPa units.
Solution:
P_1 = 0.871 atm; V_1 = 105 mL
P_2 = unknown V_2 = 352 mL
$$\frac{P_1 V_1}{n_1 T_1} = \frac{P_2 V_2}{n_2 T_2} \qquad \text{At fixed } n \text{ and } T:$$
$$P_1 V_1 = P_2 V_2$$
$$P_2 \text{ (atm)} = \frac{P_1 V_1}{V_2} = \frac{(0.871 \text{ atm})(105 \text{ mL})}{(352 \text{ mL})} = 0.259815 = \textbf{0.260 atm}$$

$$P_2 \text{ (kPa)} = (0.260 \text{ atm})\left(\frac{101.325 \text{ kPa}}{1 \text{ atm}}\right) = \textbf{26.3 kPa}$$

5.2B Plan: Given in the problem is an initial volume, initial pressure, and final pressure for the oxygen gas. The final volume is to be calculated. The temperature and amount of gas are fixed. Convert the final pressure to atm units. Rearrange the ideal gas law to the appropriate form and solve for V_2.

Solution:

$P_1 = 122$ atm; $\qquad V_1 = 651$ L

$P_2 = 745$ mmHg $\qquad V_2 =$ unknown

$\dfrac{P_1 V_1}{n_1 T_1} = \dfrac{P_2 V_2}{n_2 T_2}$ $\qquad$ At fixed n and T:

$P_1 V_1 = P_2 V_2$

$P_2 \text{ (atm)} = (745 \text{ mmHg})\left(\dfrac{1 \text{ atm}}{760 \text{ mmHg}}\right) = 0.980263 \text{ atm}$

$V_2 \text{ (atm)} = \dfrac{P_1 V_1}{P_2} = \dfrac{(122 \text{ atm})(651 \text{ L})}{(0.980263 \text{ atm})} = 8.1021 \times 10^4 = \mathbf{8.10 \times 10^4 \text{ L}}$

5.3A Plan: Convert the temperatures to kelvin units and the initial pressure to units of torr. Examine the ideal gas law, noting the fixed variables and those variables that change. R is always constant so $\dfrac{P_1 V_1}{n_1 T_1} = \dfrac{P_2 V_2}{n_2 T_2}$. In this problem, P and T are changing, while n and V remain fixed.

Solution:

$T_1 = 23°C$ $\qquad\qquad\qquad T_2 = 100°C$

$P_1 = 0.991$ atm $\qquad\qquad\quad P_2 =$ unknown

n and V remain constant

Converting T_1 from °C to K: $23°C + 273.15 = 296.15$ K

Converting T_2 from °C to K: $100°C + 273.15 = 373.15$ K

$P_1 \text{ (torr)} = (0.991 \text{ atm})\left(\dfrac{760 \text{ torr}}{1 \text{ atm}}\right) = 753.16 \text{ torr}$

Arranging the ideal gas law and solving for P_2:

$\dfrac{P_1 \cancel{V_1}}{\cancel{n_1} T_1} = \dfrac{P_2 \cancel{V_2}}{\cancel{n_2} T_2}$ $\qquad$ or $\qquad$ $\dfrac{P_1}{T_1} = \dfrac{P_2}{T_2}$

$P_2 \text{ (torr)} = P_1 \dfrac{T_2}{T_1} = (753.16 \text{ torr})\left(\dfrac{373.15 \text{ K}}{296.15 \text{ K}}\right) = 948.98 = \mathbf{949 \text{ torr}}$

Because the pressure in the tank (949 torr) is less than the pressure at which the safety valve will open (1.00×10^3 torr), **the safety valve will not open.**

5.3B Plan: This is Charles's law: at constant pressure and with a fixed amount of gas, the volume of a gas is directly proportional to the absolute temperature of the gas. The temperature must be lowered to reduce the volume of a gas. Arrange the ideal gas law, solving for T_2 at fixed n and P. Temperature must be converted to kelvin units.

Solution:

$V_1 = 32.5$ L $\qquad\qquad\qquad V_2 = 28.6$ L

$T_1 = 40°C$ (convert to K) $\qquad T_2 =$ unknown

n and P remain constant

Converting T from °C to K: $T_1 = 40 \text{ °C} + 273.15 = 313.15$K

Arranging the ideal gas law and solving for T_2:

$\dfrac{\cancel{P_1} V_1}{\cancel{n_1} T_1} = \dfrac{\cancel{P_2} V_2}{\cancel{n_2} T_2}$ $\qquad$ or $\qquad$ $\dfrac{V_1}{T_1} = \dfrac{V_2}{T_2}$

$T_2 = T_1 \dfrac{V_2}{V_1} = (313.15 \text{ K})\left(\dfrac{28.6 \text{ L}}{32.5 \text{ L}}\right) = 275.572 \text{ K} - 273.15 = 2.422 = \mathbf{2°C}$

5.4A   Plan: In this problem, the amount of gas is decreasing. Since the container is rigid, the volume of the gas will not change with the decrease in moles of gas. The temperature is also constant. So, the only change will be that the pressure of the gas will decrease since fewer moles of gas will be present after removal of the 5.0 g of ethylene. Rearrange the ideal gas law to the appropriate form and solve for P_2. Since the ratio of moles of ethylene is equal to the ratio of grams of ethylene, there is no need to convert the grams to moles. (This is illustrated in the solution by listing the molar mass conversion twice.)
Solution:
$P_1 = 793$ torr; $P_2 = ?$ mass$_1$ = 35.0 g;   mass$_2$ = 35.0 – 5.0 = 30.0 g

$$\frac{P_1 V_1}{n_1 T_1} = \frac{P_2 V_2}{n_2 T_2} \qquad \text{At fixed } V \text{ and } T:$$

$$\frac{P_1}{n_1} = \frac{P_2}{n_2}$$

$$P_2 = \frac{P_1 n_2}{n_1} = (793 \text{ torr}) \frac{\left(30.0 \text{ g } C_2H_4\right)\left(\dfrac{1 \text{ mol } C_2H_4}{28.05 \text{ g } C_2H_4}\right)}{\left(35.0 \text{ g } C_2H_4\right)\left(\dfrac{1 \text{ mol } C_2H_4}{28.05 \text{ g } C_2H_4}\right)} = 679.714 = \textbf{680. torr}$$

5.4B   Plan: Examine the ideal gas law, noting the fixed variables and those variables that change. R is always constant so $\dfrac{P_1 V_1}{n_1 T_1} = \dfrac{P_2 V_2}{n_2 T_2}$. In this problem, n and V are changing, while P and T remain fixed.

Solution:
$m_1 = 1.26$ g N_2 $m_2 = 1.26$ g N_2 + 1.26 g He
$V_1 = 1.12$ L $V_2 =$ unknown
P and T remain constant

Converting m_1 (mass) to n_1 (moles): $(1.26 \text{ g } N_2)\left(\dfrac{1 \text{ mol } N_2}{28.02 \text{ g } N_2}\right) = 0.044968 \text{ mol } N_2 = n_1$

Converting m_2 (mass) to n_2 (moles): $0.044968 \text{ mol } N_2 + (1.26 \text{ g He})\left(\dfrac{1 \text{ mol He}}{4.003 \text{ g He}}\right)$

$$= 0.044968 \text{ mol } N_2 + 0.31476 \text{ mol He} = 0.35973 \text{ mol gas} = n_2$$

Arranging the ideal gas law and solving for V_2:

$$\frac{\cancel{P_1} V_1}{n_1 \cancel{T_1}} = \frac{\cancel{P_2} V_2}{n_2 \cancel{T_2}} \quad \text{or} \quad \frac{V_1}{n_1} = \frac{V_2}{n_2}$$

$$V_2 = V_1 \frac{P_2}{P_1} = (1.12 \text{ L})\left(\frac{0.35973 \text{ mol}}{0.044968 \text{ mol}}\right) = 8.9597 = \textbf{8.96 L}$$

5.5A   Plan: Convert the temperatures to kelvin. Examine the ideal gas law, noting the fixed variables and those variables that change. R is always constant so $\dfrac{P_1 V_1}{n_1 T_1} = \dfrac{P_2 V_2}{n_2 T_2}$. In this problem, P, V, and T are changing, while n remains fixed.

Solution:
$T_1 = 23°C$ $T_2 = 18°C$
$P_1 = 755$ mmHg $P_2 =$ unknown
$V_1 = 2.55$ L $V_2 = 4.10$ L
n remains constant
Converting T_1 from °C to K: 23°C + 273.15 = 296.15 K

Converting T_2 from °C to K: 18°C + 273.15 = 291.15 K

Arranging the ideal gas law and solving for P_2:

$$\frac{P_1V_1}{\cancel{n_1}T_1} = \frac{P_2V_2}{\cancel{n_2}T_2} \quad \text{or} \quad \frac{P_1V_1}{T_1} = \frac{P_2V_2}{T_2}$$

$$P_2 \text{ (mmHg)} = P_1\frac{V_1T_2}{V_2T_1} = (755 \text{ mmHg})\left(\frac{(2.55 \text{ L})(291.15 \text{ K})}{(4.10 \text{ L})(296.15 \text{ K})}\right) = 461.645 = \textbf{462 mmHg}$$

5.5B   Plan: Convert the temperatures to kelvin. Examine the ideal gas law, noting the fixed variables and those variables that change. R is always constant so $\frac{P_1V_1}{n_1T_1} = \frac{P_2V_2}{n_2T_2}$. In this problem, P, V, and T are changing, while n remains fixed.

Solution:

$T_1 = 28°C$ $T_2 = 21°C$

$P_1 = 0.980 \text{ atm}$ $P_2 = 1.40 \text{ atm}$

$V_1 = 2.2 \text{ L}$ $V_2 = \text{unknown}$

n remains constant

Converting T_1 from °C to K: 28°C + 273.15 = 301.15 K

Converting T_2 from °C to K: 21°C + 273.15 = 294.15 K

Arranging the ideal gas law and solving for V_2:

$$\frac{P_1V_1}{\cancel{n_1}T_1} = \frac{P_2V_2}{\cancel{n_2}T_2} \quad \text{or} \quad \frac{P_1V_1}{T_1} = \frac{P_2V_2}{T_2}$$

$$V_2 \text{ (L)} = V_1\frac{P_1T_2}{P_2T_1} = (2.2 \text{ L})\left(\frac{(0.980 \text{ atm })(294.15 \text{ K})}{(1.40 \text{ atm})(301.15 \text{ K})}\right) = 1.5042 = \textbf{1.5 L}$$

5.6A   Plan: From Sample Problem 5.6 the temperature of 21°C and volume of 438 L are given. The pressure is 1.37 atm and the unknown is the moles of oxygen gas. Use the ideal gas equation $PV = nRT$ to calculate the number of moles of gas. Multiply moles by molar mass to obtain mass.

Solution:

$PV = nRT$

$$n = \frac{PV}{RT} = \frac{(1.37 \text{ atm})(438 \text{ L})}{\left(\dfrac{0.0821 \text{ atm} \bullet \text{L}}{\text{mol} \bullet \text{K}}\right)((273.15+21)\text{K})} = 24.8475 \text{ mol } O_2$$

$$\text{Mass (g) of } O_2 = (24.8475 \text{ mol } O_2)\left(\frac{32.00 \text{ g } O_2}{1 \text{ mol } O_2}\right) = 795.12 = \textbf{795 g } \textbf{O}_2$$

5.6B   Plan: Convert the mass of helium to moles, the temperature to kelvin units, and the pressure to atm units. Use the ideal gas equation $PV = nRT$ to calculate the volume of the gas.

Solution:

$P = 731 \text{ mmHg}$ $V = \text{unknown}$

$m = 3950 \text{ kg He}$ $T = 20°C$

Converting m (mass) to n (moles): $(3950 \text{ kg He})\left(\dfrac{1000 \text{ g}}{1 \text{ kg}}\right)\left(\dfrac{1 \text{ mol He}}{4.003 \text{ g He}}\right) = 9.8676\text{x}10^5 \text{ mol} = n$

Converting T from °C to K: 20°C + 273.15 = 293.15 K

Converting P from mmHg to atm: $(731 \text{ mmHg})\left(\dfrac{1 \text{ atm}}{760 \text{ mmHg}}\right) = 0.962 \text{ atm}$

$$PV = nRT$$

$$V = \frac{nRT}{P} = \frac{(9.8676\times10^5 \text{ mol})\left(0.0821 \dfrac{\text{atm}\bullet\text{L}}{\text{mol}\bullet\text{K}}\right)(293.15 \text{ K})}{(0.962 \text{ atm})} = 2.4687\times10^7 = \mathbf{2.47\times10^7 \text{ L}}$$

5.7A   Plan: Balance the chemical equation. The pressure is constant and, according to the picture, the volume approximately doubles. The volume change may be due to the temperature and/or a change in moles. Examine the balanced reaction for a possible change in number of moles. Rearrange the ideal gas law to the appropriate form and solve for the variable that changes.
Solution:
The balanced chemical equation must be $2CD \rightarrow C_2 + D_2$
Thus, the number of mole of gas does not change (2 moles both before and after the reaction). Only the temperature remains as a variable to cause the volume change. Let V_1 = the initial volume and $2V_1$ = the final volume V_2.
$T_1 = (-73 + 273.15) \text{ K} = 200.15 \text{ K}$

$$\frac{P_1 V_1}{n_1 T_1} = \frac{P_2 V_2}{n_2 T_2} \qquad \text{At fixed } n \text{ and } P:$$

$$\frac{V_1}{T_1} = \frac{V_2}{T_2}$$

$$T_2 = \frac{V_2 T_1}{V_1} = \frac{(2V_1)(200.15 \text{ K})}{(V_1)} = 400.30 \text{ K} - 273.15 = 127.15 = \mathbf{127°C}$$

5.7B   Plan: The pressure is constant and, according to the picture, the volume approximately decreases by a factor of 2 (the final volume is approximately one half the original volume). The volume change may be due to the temperature change and/or a change in moles. Consider the change in temperature. Examine the balanced reactions for a possible change in number of moles. Think about the relationships between the variables in the ideal gas law in order to determine the effect of temperature and moles on gas volume.
Solution:
Converting T_1 from °C to K: 199°C + 273.15 = 472.15 K
Converting T_2 from °C to K: –155°C + 273.15 = 118.15 K
According to the ideal gas law, temperature and volume are directly proportional. The temperature decreases by a factor of 4, which should cause the volume to also decrease by a factor of 4. Because the volume only decreases by a factor of 2, the number of moles of gas must have *increased* by a factor of 2 (moles of gas and volume are also directly proportional).
1/4 (decrease in V from the decrease in T) x 2 (increase in V from the increase in n)
= 1/2 (a decrease in V by a factor of 2)
Thus, we need to find a reaction in which the number of moles of gas increases by a factor of 2.
In equation (1), 3 moles of gas yield 2 moles of gas.
In equation (2), 2 moles of gas yield 4 moles of gas.
In equation (3), 1 mole of gas yields 3 moles of gas.
In equation (4), 2 moles of gas yield 2 moles of gas.
Because the number of moles of gas doubles in **equation (2)**, that equation best describes the reaction in the figure in this problem.

5.8A   Plan: Density of a gas can be calculated using a version of the ideal gas equation, $d = (P\mathcal{M})/(RT)$. Two calculations are required, one with $T = 0°C = 273.15$ K and $P = 380$ torr and the other at STP which is defined as $T = 273$ K and $P = 1$ atm.

Solution:
Density at $T = 273$ K and $P = 380$ torr:

$$d = \frac{(380 \text{ torr})(44.01 \text{ g/mol})}{\left(\dfrac{0.0821 \text{ atm} \bullet \text{L}}{\text{mol} \bullet \text{K}}\right)(273.15 \text{ K})}\left(\frac{1 \text{ atm}}{760 \text{ torr}}\right) = 0.98124 = \mathbf{0.981 \text{ g/L}}$$

Density at $T = 273$ K and $P = 1$ atm. (Note: The 1 atm is an exact number and does not affect the significant figures in the answer.)

$$d = \frac{(44.01 \text{ g/mol})(1 \text{ atm})}{\left(\dfrac{0.0821 \text{ atm} \bullet \text{L}}{\text{mol} \bullet \text{K}}\right)(273.15 \text{ K})} = 1.96249 = \mathbf{1.96 \text{ g/L}}$$

The density of a gas increases proportionally to the increase in pressure.

5.8B   Plan: Density of a gas can be calculated using a version of the ideal gas equation, $d = P\mathcal{M}/RT$
Solution:
Density of NO_2 at $T = 297.15$ K (24°C + 273.15) and $P = 0.950$ atm:

$$d = \frac{(0.950 \text{ atm})(46.01 \text{ g/mol})}{\left(0.0821 \dfrac{\text{atm} \bullet \text{L}}{\text{mol} \bullet \text{K}}\right)(297.15 \text{ K})} = 1.7917 = \mathbf{1.79 \text{ g/L}}$$

Nitrogen dioxide is **more dense** than dry air at the same conditions (density of dry air = 1.13 g/L).

5.9A   Plan: Calculate the mass of the gas by subtracting the mass of the empty flask from the mass of the flask containing the condensed gas. The volume, pressure, and temperature of the gas are known.

The relationship $d = P\mathcal{M}/RT$ is rearranged to give $\mathcal{M} = \dfrac{dRT}{P}$ or $\mathcal{M} = \dfrac{mRT}{PV}$

Solution:
Mass (g) of gas = mass of flask + vapor – mass of flask = 68.697 – 68.322 = 0.375 g

$T = 95.0°C + 273.15 = 368.15$ K

$P = (740. \text{ torr})\left(\dfrac{1 \text{ atm}}{760 \text{ torr}}\right) = 0.973684$ atm

$V = 149$ mL = 0.149 L

$$\mathcal{M} = \frac{mRT}{PV} = \frac{(0.375 \text{ g})\left(\dfrac{0.0821 \text{ atm} \bullet \text{L}}{\text{mol} \bullet \text{K}}\right)(368.15 \text{ K})}{(0.973684 \text{ atm})(0.149 \text{ L})} = 78.125 = \mathbf{78.1 \text{ g}}$$

5.9B   Plan: Calculate the mass of the gas by subtracting the mass of the empty glass bulb from the mass of the bulb containing the gas. The volume, pressure, and temperature of the gas are known. The relationship $d = P\mathcal{M}/RT$ is

rearranged to give $\mathcal{M} = \dfrac{dRT}{P}$ or $\mathcal{M} = \dfrac{mRT}{PV}$. Use the molar mass of the gas to determine its identity.

Solution:
Mass (g) of gas = mass of bulb + gas – mass of bulb = 82.786 – 82.561 = 0.225 g

$T = 22°C + 273.15 = 295.15$ K

$P = (733 \text{ mmHg})\left(\dfrac{1 \text{ atm}}{760 \text{ mmHg}}\right) = 0.965$ atm

$V = 350. \text{ mL} = 0.350$ L

$$\mathcal{M} = \frac{mRT}{PV} = \frac{(0.225 \text{ g})\left(0.0821 \frac{\text{atm} \cdot \text{L}}{\text{mol} \cdot \text{K}}\right)(295.15 \text{ K})}{(0.965 \text{ atm })(0.350 \text{ L})} = 16.1426 = 16.1 \text{ g/mol}$$

Methane has a molar mass of 16.04 g/mol. Nitrogen monoxide has a molar mass of 30.01 g/mol. The gas that has a molar mass that matches the calculated value is **methane**.

5.10A   Plan: Calculate the number of moles of each gas present and then the mole fraction of each gas. The partial pressure of each gas equals the mole fraction times the total pressure. Total pressure equals 1 atm since the problem specifies STP. This pressure is an exact number, and will not affect the significant figures in the answer
Solution:

$$\text{Moles of He} = (5.50 \text{ g He})\left(\frac{1 \text{ mol He}}{4.003 \text{ g He}}\right) = 1.373970 \text{ mol He}$$

$$\text{Moles of Ne} = (15.0 \text{ g Ne})\left(\frac{1 \text{ mol Ne}}{20.18 \text{ g Ne}}\right) = 0.743310 \text{ mol Ne}$$

$$\text{Moles of Kr} = (35.0 \text{ g Kr})\left(\frac{1 \text{ mol Kr}}{83.80 \text{ g Kr}}\right) = 0.417661 \text{ mol Kr}$$

Total number of moles of gas = 1.373970 + 0.743310 + 0.417661 = 2.534941 mol
$$P_A = X_A \times P_{total}$$

$$P_{He} = \left(\frac{1.37397 \text{ mol He}}{2.534941 \text{ mol}}\right)(1 \text{ atm}) = 0.54201 = \textbf{0.542 atm He}$$

$$P_{Ne} = \left(\frac{0.74331 \text{ mol Ne}}{2.534941 \text{ mol}}\right)(1 \text{ atm}) = 0.29323 = \textbf{0.293 atm Ne}$$

$$P_{Kr} = \left(\frac{0.41766 \text{ mol Kr}}{2.534941 \text{ mol}}\right)(1 \text{ atm}) = 0.16476 = \textbf{0.165 atm Kr}$$

5.10B   Plan: Use the formula $P_A = X_A \times P_{total}$ to calculate the mole fraction of He. Multiply the mole fraction by 100% to calculate the mole percent of He.
Solution:
$$P_{He} = X_{He} \times P_{total}$$

$$\text{Mole percent He} = X_{He}(100\%) = \left(\frac{P_{He}}{P_{total}}\right)(100\%) = \left(\frac{143 \text{ atm}}{204 \text{ atm}}\right)(100\%) = \textbf{70.1\%}$$

5.11A   Plan: The gas collected over the water will consist of H_2 and H_2O gas molecules. The partial pressure of the water can be found from the vapor pressure of water at the given temperature given in the text. Subtracting this partial pressure of water from total pressure gives the partial pressure of hydrogen gas collected over the water. Calculate the moles of hydrogen gas using the ideal gas equation. The mass of hydrogen can then be calculated by converting the moles of hydrogen from the ideal gas equation to grams.
Solution:
From the table in the text, the partial pressure of water is 13.6 torr at 16°C.
$$P = 752 \text{ torr} - 13.6 \text{ torr} = 738.4 = \textbf{738 torr } \mathbf{H_2}$$
The unrounded partial pressure (738.4 torr) will be used to avoid rounding error.

$$\text{Moles of hydrogen} = n = \frac{PV}{RT} = \frac{(738.4 \text{ torr})(1495 \text{ mL})}{\left(\frac{0.0821 \text{ atm} \cdot \text{L}}{\text{mol} \cdot \text{K}}\right)((273.15 + 16)\text{K})}\left(\frac{1 \text{ atm}}{760 \text{ torr}}\right)\left(\frac{10^{-3}\text{L}}{1 \text{ mL}}\right)$$

$$= 0.061186 \text{ mol } H_2$$

$$\text{Mass (g) of hydrogen} = (0.061186 \text{ mol } H_2)\left(\frac{2.016 \text{ g } H_2}{1 \text{ mol } H_2}\right) = 0.123351 = \textbf{0.123 g } \mathbf{H_2}$$

5.11B Plan: The gas collected over the water will consist of O_2 and H_2O gas molecules. The partial pressure of the water can be found from the vapor pressure of water at the given temperature given in the text. Subtracting this partial pressure of water from total pressure gives the partial pressure of oxygen gas collected over the water. Calculate the moles of oxygen gas using the ideal gas equation. The mass of oxygen can then be calculated by converting the moles of oxygen from the ideal gas equation to grams.

Solution:

From the table in the text, the partial pressure of water is 17.5 torr at 20°C.

$$P = 748 \text{ torr} - 17.5 \text{ torr} = 730.5 = \mathbf{730. \text{ torr } O_2}$$

$$\text{Moles of oxygen} = n = \frac{PV}{RT} = \frac{(730. \text{ torr})(307 \text{ mL})}{\left(0.0821 \frac{\text{atm} \bullet \text{L}}{\text{mol} \bullet \text{K}}\right)(293.15 \text{ K})} \left(\frac{1 \text{ atm}}{760 \text{ torr}}\right)\left(\frac{1 \text{ L}}{1000 \text{ mL}}\right)$$

$$= 0.012252 \text{ mol } O_2$$

$$\text{Mass (g) of oxygen} = (0.012252 \text{ mol } O_2)\left(\frac{32.00 \text{ g } O_2}{1 \text{ mol } O_2}\right) = 0.39207 = \mathbf{0.392 \text{ g } O_2}$$

5.12A Plan: Write a balanced equation for the reaction. Calculate the moles of $HCl(g)$ from the starting amount of sodium chloride using the stoichiometric ratio from the balanced equation. Find the volume of the $HCl(g)$ from the molar volume at STP.

Solution:

The balanced equation is $H_2SO_4(aq) + 2NaCl(aq) \rightarrow Na_2SO_4(aq) + 2HCl(g)$.

$$\text{Moles of HCl} = (0.117 \text{ kg NaCl})\left(\frac{10^3 \text{ g}}{1 \text{ kg}}\right)\left(\frac{1 \text{ mol NaCl}}{58.44 \text{ g NaCl}}\right)\left(\frac{2 \text{ mol HCl}}{2 \text{ mol NaCl}}\right) = 2.00205 \text{ mol HCl}$$

$$\text{Volume (mL) of HCl} = (2.00205 \text{ mol HCl})\left(\frac{22.4 \text{ L}}{1 \text{ mol HCl}}\right)\left(\frac{1 \text{ mL}}{10^{-3} \text{ L}}\right)$$

$$= 4.4846 \times 10^4 = \mathbf{4.48 \times 10^4 \text{ mL HCl}}$$

5.12B Plan: Write a balanced equation for the reaction. Use the ideal gas law to calculate the moles of $CO_2(g)$ scrubbed. Use the molar ratios from the balanced equation to calculate the moles of lithium hydroxide needed to scrub that amount of CO_2. Finally, use the molar mass of lithium hydroxide to calculate the mass of lithium hydroxide required.

Solution:

The balanced equation is $2LiOH(s) + CO_2(g) \rightarrow Li_2CO_3(s) + H_2O(l)$.

$$\text{Amount (mol) of } CO_2 \text{ scrubbed} = n = \frac{PV}{RT} = \frac{(0.942 \text{ atm})(215 \text{ L})}{\left(0.0821 \frac{\text{atm} \bullet \text{L}}{\text{mol} \bullet \text{K}}\right)(296.15 \text{ K})} = 8.3298 = 8.3298 \text{ mol } CO_2$$

$$\text{Mass (g) of LiOH} = 8.3298 \text{ mol } CO_2 \left(\frac{2 \text{ mol LiOH}}{1 \text{ mol } CO_2}\right)\left(\frac{23.95 \text{ g LiOH}}{1 \text{ mol LiOH}}\right) = 398.9973 = \mathbf{399 \text{ g LiOH}}$$

5.13A Plan: Balance the equation for the reaction. Determine the limiting reactant by finding the moles of each reactant from the ideal gas equation, and comparing the values. Calculate the moles of remaining excess reactant. This is the only gas left in the flask, so it is used to calculate the pressure inside the flask.

Solution:

The balanced equation is $NH_3(g) + HCl(g) \rightarrow NH_4Cl(s)$.

The stoichiometric ratio of NH_3 to HCl is 1:1, so the reactant present in the lower quantity of moles is the limiting reactant.

$$\text{Moles of ammonia} = \frac{PV}{RT} = \frac{(0.452 \text{ atm})(10.0 \text{ L})}{\left(\frac{0.0821 \text{ atm} \bullet \text{L}}{\text{mol} \bullet \text{K}}\right)((273.15 + 22)\text{K})} = 0.18653 \text{ mol } NH_3$$

$$\text{Moles of hydrogen chloride} = \frac{PV}{RT} = \frac{(7.50\ \text{atm})(155\ \text{mL})}{\left(\dfrac{0.0821\ \text{atm} \bullet \text{L}}{\text{mol} \bullet \text{K}}\right)(271.15\ \text{K})}\left(\frac{10^{-3}\ \text{L}}{1\ \text{mL}}\right) = 0.052220\ \text{mol HCl}$$

The HCl is limiting so the moles of ammonia gas left after the reaction would be
$0.18653 - 0.052220 = 0.134310$ mol NH_3.

$$\text{Pressure (atm) of ammonia} = \frac{nRT}{V} = \frac{(0.134310\ \text{mol})\left(\dfrac{0.0821\ \text{atm} \bullet \text{L}}{\text{mol} \bullet \text{K}}\right)((273.15 + 22)\text{K})}{(10.0\ \text{L})}$$

$$= 0.325458 = \textbf{0.325 atm NH}_3$$

5.13B <u>Plan:</u> Balance the equation for the reaction. Use the ideal gas law to calculate the moles of fluorine that react. Determine the limiting reactant by determining the moles of product that can be produced from each of the reactants and comparing the values. Use the moles of IF_5 produced and the ideal gas law to calculate the volume of gas produced.

<u>Solution:</u>
The balanced equation is $I_2(s) + 5F_2\ (g) \rightarrow 2IF_5(g)$.

$$\text{Amount (mol) of } F_2 \text{ that reacts} = n = \frac{PV}{RT} = \frac{(0.974\ \text{atm})(2.48\ \text{L})}{\left(0.0821\ \dfrac{\text{atm} \bullet \text{L}}{\text{mol} \bullet \text{K}}\right)(291.15\ \text{K})} = 0.10105\ \text{mol } F_2$$

$$\text{Amount (mol) of } IF_5 \text{ produced from } F_2 = 0.10105\ \text{mol } F_2 \left(\frac{2\ \text{mol } IF_5}{5\ \text{mol } F_2}\right) = 0.040421\ \text{mol } IF_5$$

$$\text{Amount (mol) of } IF_5 \text{ produced from } I_2 = 4.16\ \text{g } I_2 \left(\frac{1\ \text{mol } I_2}{253.8\ \text{g } I_2}\right)\left(\frac{2\ \text{mol } IF_5}{1\ \text{mol } I_2}\right) = 0.032782\ \text{mol } IF_5$$

Because a smaller number of moles is produced from the I_2, I_2 is limiting and 0.032782 mol of IF_5 are produced.

$$\text{Volume (L) of } IF_5 = \frac{nRT}{P} = \frac{(0.032782\ \text{mol})\left(0.0821\ \dfrac{\text{atm} \bullet \text{L}}{\text{mol} \bullet \text{K}}\right)(378.15\ \text{K})}{(0.935\ \text{atm})} = 1.08850 = \textbf{1.09 L}$$

5.14A <u>Plan:</u> Graham's law can be used to solve for the effusion rate of the ethane since the rate and molar mass of helium are known, along with the molar mass of ethane. In the same way that running slower increases the time to go from one point to another, so the rate of effusion decreases as the time increases. The rate can be expressed as $1/\text{time}$.

<u>Solution:</u>

$$\frac{\text{Rate He}}{\text{Rate } C_2H_6} = \sqrt{\frac{\text{Molar mass}_{C_2H_6}}{\text{Molar mass}_{He}}}$$

$$\frac{\left(\dfrac{0.010\ \text{mol He}}{1.25\ \text{min}}\right)}{\left(\dfrac{0.010\ \text{mol } C_2H_6}{t_{C_2H_6}}\right)} = \sqrt{\frac{(30.07\ \text{g/mol})}{(4.003\ \text{g/mol})}}$$

$$0.800\ t = 2.74078$$
$$t = 3.42597 = \textbf{3.43 min}$$

5.14B <u>Plan:</u> Graham's law can be used to solve for the molar mass of the unknown gas since the rates of both gases and the molar mass of argon are known. Rate can be expressed as the volume of gas that effuses per unit time.

Solution:

Rate of Ar = 13.8 mL/time

Rate of unknown gas = 7.23 mL/time

Mass of Ar = 39.95 g/mol

$$\frac{\text{Rate of Ar}}{\text{Rate of unknown gas}} = \left(\sqrt{\frac{\text{Molar mass}_{\text{unknown gas}}}{\text{Molar mass}_{\text{Ar}}}} \right)$$

$$\left(\frac{\text{Rate of Ar}}{\text{Rate of unknown gas}} \right)^2 = \left(\frac{\text{Molar mass}_{\text{unknown gas}}}{\text{Molar mass}_{\text{Ar}}} \right)$$

$$\mathcal{M}_{\text{unknown gas}} = (\mathcal{M}_{\text{Ar}}) \left(\frac{\text{Rate of Ar}}{\text{Rate of unknown gas}} \right)^2$$

$$\mathcal{M}_{\text{unknown gas}} = (39.95 \text{ g/mol}) \left(\frac{13.8 \text{ mL/time}}{7.23 \text{ mL/time}} \right)^2 = \textbf{146 g/mol}$$

CHEMICAL CONNECTIONS BOXED READING PROBLEMS

B5.1   Plan: Examine the change in density of the atmosphere as altitude changes.
Solution:
The density of the atmosphere decreases with increasing altitude. High density causes more drag on the aircraft.
At high altitudes, low density means that there are relatively few gas particles present to collide with the aircraft.

B5.2   Plan: The conditions that result in deviations from ideal behavior are high pressure and low temperature. At high
pressure, the volume of the gas decreases, the distance between particles decreases, and attractive forces between gas
particles have a greater effect. A low temperature slows the gas particles, also increasing the affect of attractive
forces between particles.
Solution:
Since the pressure on Saturn is significantly higher and the temperature significantly lower than that on Venus,
atmospheric gases would deviate more from ideal gas behavior on **Saturn**.

B5.3   Plan: To find the volume percent of argon, multiply its mole fraction by 100. The partial pressure of argon gas can
be found by using the relationship $P_{\text{Ar}} = X_{\text{Ar}} \times P_{\text{total}}$. The mole fraction of argon is given in Table B5.1.
Solution:
Volume percent = mole fraction x 100 = 0.00934 x 100 = **0.934 %**
The total pressure at sea level is 1.00 atm = 760 torr.
$P_{\text{Ar}} = X_{\text{Ar}} \times P_{\text{total}}$ = 0.00934 x 760 torr = 7.0984 = **7.10 torr**

B5.4   Plan: To find the moles of gas, convert the mass of the atmosphere from t to g and divide by the molar mass of air.
Knowing the moles of air, the volume can be calculated at the specified pressure and temperature by using the
ideal gas law.
Solution:

a) Moles of gas = $(5.14 \times 10^{15} \text{ t}) \left(\frac{1000 \text{ kg}}{1 \text{ t}} \right) \left(\frac{1000 \text{ g}}{1 \text{ kg}} \right) \left(\frac{1 \text{ mol}}{28.8 \text{ g}} \right)$

$$= 1.78472 \times 10^{20} = \textbf{1.78} \times \textbf{10}^{20} \textbf{ mol}$$

b) $PV = nRT$

$$V = \frac{nRT}{P} = \frac{(1.78472 \times 10^{20} \text{ mol}) \left(0.0821 \frac{\text{L} \cdot \text{atm}}{\text{mol} \cdot \text{K}} \right) \left((273.15 + 25) \text{K} \right)}{(1 \text{ atm})} = 4.36866 \times 10^{21} = \textbf{4} \times \textbf{10}^{21} \textbf{ L}$$

END–OF–CHAPTER PROBLEMS

5.1 Plan: Review the behavior of the gas phase vs. the liquid phase.
Solution:
a) The volume of the liquid remains constant, but the volume of the gas increases to the volume of the larger container.
b) The volume of the container holding the gas sample increases when heated, but the volume of the container holding the liquid sample remains essentially constant when heated.
c) The volume of the liquid remains essentially constant, but the volume of the gas is reduced.

5.6 Plan: The ratio of the heights of columns of mercury and water are inversely proportional to the ratio of the densities of the two liquids. Convert the height in mm to height in cm.
Solution:

$$\frac{h_{H_2O}}{h_{Hg}} = \frac{d_{Hg}}{d_{H_2O}}$$

$$h_{H_2O} = \frac{d_{Hg}}{d_{H_2O}} \times h_{Hg} = \left(\frac{13.5 \text{ g/mL}}{1.00 \text{ g/mL}}\right)(730 \text{ mmHg})\left(\frac{10^{-3} \text{ m}}{1 \text{ mm}}\right)\left(\frac{1 \text{ cm}}{10^{-2} \text{ m}}\right) = 985.5 = \mathbf{990 \text{ cm } H_2O}$$

5.8 Plan: Use the conversion factors between pressure units:
1 atm = 760 mmHg = 760 torr = 101.325 kPa = 1.01325 bar
Solution:

a) Converting from atm to mmHg: $P\text{(mmHg)} = (0.745 \text{ atm})\left(\dfrac{760 \text{ mmHg}}{1 \text{ atm}}\right) = 566.2 = \mathbf{566 \text{ mmHg}}$

b) Converting from torr to bar: $P\text{(bar)} = (992 \text{ torr})\left(\dfrac{1.01325 \text{ bar}}{760 \text{ torr}}\right) = 1.32256 = \mathbf{1.32 \text{ bar}}$

c) Converting from kPa to atm: $P\text{(atm)} = (365 \text{ kPa})\left(\dfrac{1 \text{ atm}}{101.325 \text{ kPa}}\right) = 3.60227 = \mathbf{3.60 \text{ atm}}$

d) Converting from mmHg to kPa: $P\text{(kPa)} = (804 \text{ mmHg})\left(\dfrac{101.325 \text{ kPa}}{760 \text{ mmHg}}\right) = 107.191 = \mathbf{107 \text{ kPa}}$

5.10 Plan: This is an open-end manometer. Since the height of the mercury column in contact with the gas is higher than the column in contact with the air, the gas is exerting less pressure on the mercury than the air. Therefore the pressure corresponding to the height difference (Δh) between the two arms is subtracted from the atmospheric pressure. Since the height difference is in units of cm and the barometric pressure is given in units of torr, cm must be converted to mm and then torr before the subtraction is performed. The overall pressure is then given in units of atm.
Solution:

$$(2.35 \text{ cm})\left(\frac{10^{-2} \text{ m}}{1 \text{ cm}}\right)\left(\frac{1 \text{ mm}}{10^{-3} \text{ m}}\right)\left(\frac{1 \text{ torr}}{1 \text{ mmHg}}\right) = 23.5 \text{ torr}$$

738.5 torr – 23.5 torr = 715.0 torr

$$P\text{(atm)} = (715.0 \text{ torr})\left(\frac{1 \text{ atm}}{760 \text{ torr}}\right) = 0.940789 = \mathbf{0.9408 \text{ atm}}$$

5.12 Plan: This is a closed-end manometer. The difference in the height of the Hg (Δh) equals the gas pressure. The height difference is given in units of m and must be converted to mmHg and then to atm.
Solution:

$$P(\text{atm}) = \left(0.734 \text{ mHg}\right)\left(\frac{1 \text{ mmHg}}{10^{-3} \text{ mHg}}\right)\left(\frac{1 \text{ atm}}{760 \text{ mmHg}}\right) = 0.965789 = \mathbf{0.966 \text{ atm}}$$

5.18 Plan: Examine the ideal gas law; volume and temperature are constant and pressure and moles are variable.
Solution:

$$PV = nRT \qquad P = n\frac{RT}{V} \qquad R, T, \text{ and } V \text{ are constant}$$

$P = n \times \text{constant}$

At constant temperature and volume, the pressure of the gas is directly proportional to the amount of gas in moles.

5.20 Plan: Use the relationship $\dfrac{P_1V_1}{n_1T_1} = \dfrac{P_2V_2}{n_2T_2}$ or $V_2 = \dfrac{P_1V_1n_2T_2}{P_2n_1T_1}$.

Solution:
a) As the pressure on a fixed amount of gas (n is fixed) increases at constant temperature (T is fixed), the molecules move closer together, decreasing the volume. When the pressure is tripled, the **volume decreases to one-third of the original volume** at constant temperature (Boyle's law).

$$V_2 = \frac{P_1V_1n_2T_2}{P_2n_1T_1} = \frac{(P_1)(V_1)(1)(1)}{(3P_1)(1)(1)} \qquad V_2 = \tfrac{1}{3}V_1$$

b) As the temperature of a fixed amount of gas (n is fixed) increases at constant pressure (P is fixed), the gas molecules gain kinetic energy. With higher energy, the gas molecules collide with the walls of the container with greater force, which increases the size (volume) of the container. If the temperature is increased by a factor of 3.0 (at constant pressure) then the **volume will increase by a factor of 3.0** (Charles's law).

$$V_2 = \frac{P_1V_1n_2T_2}{P_2n_1T_1} = \frac{(1)(V_1)(1)(3T_1)}{(1)(1)(T_1)} \qquad V_2 = 3V_1$$

c) As the number of molecules of gas increases at constant pressure and temperature (P and T are fixed), the force they exert on the container increases. This results in an increase in the volume of the container. Adding 3 moles of gas to 1 mole increases the number of moles by a factor of 4, thus the **volume increases by a factor of 4** (Avogadro's law).

$$V_2 = \frac{P_1V_1n_2T_2}{P_2n_1T_1} = \frac{(1)(V_1)(4n_1)(1)}{(1)(n_1)(1)} \qquad V_2 = 4V_1$$

5.22 Plan: Use the relationship $\dfrac{P_1V_1}{T_1} = \dfrac{P_2V_2}{T_2}$ or $V_2 = \dfrac{P_1V_1T_2}{P_2T_1}$. R and n are fixed.

Solution:
a) The temperature is decreased by a factor of 2, so **the volume is decreased by a factor of 2** (Charles's law).

$$V_2 = \frac{P_1V_1T_2}{P_2T_1} = \frac{(1)(V_1)(400 \text{ K})}{(1)(800 \text{ K})} \qquad V_2 = \tfrac{1}{2} V_1$$

b) $T_1 = 250°C + 273.15 = 523.15 \text{ K}$ $\qquad\qquad\qquad T_2 = 500°C + 273.15 = 773.15 \text{ K}$
The temperature increases by a factor of $773.15/523.15 = 1.48$, so **the volume is increased by a factor of 1.48**

(Charles's law). $\qquad\qquad V_2 = \dfrac{P_1V_1T_2}{P_2T_1} = \dfrac{(1)(V_1)(773.15 \text{ K})}{(1)(523.15 \text{ K})} \qquad V_2 = 1.48V_1$

c) The pressure is increased by a factor of 3, so **the volume decreases by a factor of 3** (Boyle's law).

$$V_2 = \frac{P_1V_1T_2}{P_2T_1} = \frac{(2 \text{ atm})(V_1)(1)}{(6 \text{ atm})(1)} \qquad V_2 = \tfrac{1}{3}V_1$$

5.24   Plan: Examine the ideal gas law, noting the fixed variables and those variables that change. R is always constant so $\dfrac{P_1V_1}{n_1T_1} = \dfrac{P_2V_2}{n_2T_2}$. In this problem, P and V are changing, while n and T remain fixed.

Solution:

$V_1 = 1.61$ L $V_2 =$ unknown
$P_1 = 734$ torr $P_2 = 0.844$ atm
n and T remain constant

Converting P_1 from torr to atm: $(734 \text{ torr})\left(\dfrac{1 \text{ atm}}{760 \text{ torr}}\right) = 0.966$ atm

Arranging the ideal gas law and solving for V_2:

$$\dfrac{P_1V_1}{\cancel{n_1}\cancel{T_1}} = \dfrac{P_2V_2}{\cancel{n_2}\cancel{T_2}} \quad \text{or} \quad P_1V_1 = P_2V_2$$

$$V_2 = V_1\dfrac{P_1}{P_2} = (1.61 \text{ L})\left(\dfrac{0.966 \text{ atm}}{0.844 \text{ atm}}\right) = \textbf{1.84 L}$$

5.26   Plan: This is Charles's law: at constant pressure and with a fixed amount of gas, the volume of a gas is directly proportional to the absolute temperature of the gas. The temperature must be lowered to reduce the volume of a gas. Arrange the ideal gas law, solving for T_2 at fixed n and P. Temperature must be converted to kelvin.

Solution:

$V_1 = 9.10$ L $V_2 = 2.50$ L
$T_1 = 198°C$ (convert to K) $T_2 =$ unknown
n and P remain constant

Converting T from °C to K: $T_1 = 198°C + 273.15 = 471.15$K

Arranging the ideal gas law and solving for T_2:

$$\dfrac{\cancel{P_1}V_1}{\cancel{n_1}T_1} = \dfrac{\cancel{P_2}V_2}{\cancel{n_2}T_2} \quad \text{or} \quad \dfrac{V_1}{T_1} = \dfrac{V_2}{T_2}$$

$$T_2 = T_1\dfrac{V_2}{V_1} = 471.15 \text{ K}\left(\dfrac{2.50 \text{ L}}{9.10 \text{ L}}\right) = 129.437 \text{ K} - 273.15 = -143.713 = \textbf{-144°C}$$

5.28   Plan: Examine the ideal gas law, noting the fixed variables and those variables that change. R is always constant so $\dfrac{P_1V_1}{n_1T_1} = \dfrac{P_2V_2}{n_2T_2}$. In this problem, P and T are changing, while n and V remain fixed.

Solution:

$T_1 = 25$ °C $T_2 = 195$ °C
$P_1 = 177$ atm $P_2 =$ unknown
n and V remain constant

Converting T_1 from °C to K: 25 °C $+ 273.15 = 298.15$ K
Converting T_2 from °C to K: 195 °C $+ 273.15 = 468.15$ K
Arranging the ideal gas law and solving for P_2:

$$\dfrac{P_1\cancel{V_1}}{\cancel{n_1}T_1} = \dfrac{P_2\cancel{V_2}}{\cancel{n_2}T_2} \quad \text{or} \quad \dfrac{P_1}{T_1} = \dfrac{P_2}{T_2}$$

$$P_2 = P_1\dfrac{T_2}{T_1} = (177 \text{ atm})\left(\dfrac{468.15 \text{ K}}{298.15 \text{ K}}\right) = 277.92 = \textbf{278 atm}$$

5.30   Plan: Examine the ideal gas law, noting the fixed variables and those variables that change. R is always constant so $\dfrac{P_1V_1}{n_1T_1} = \dfrac{P_2V_2}{n_2T_2}$. In this problem, n and V are changing, while P and T remain fixed.

<u>Solution:</u>

$m_1 = 1.92$ g He $\qquad\qquad\qquad\qquad m_2 = 1.92$ g $- 0.850$ g $= 1.07$ g He

$V_1 = 12.5$ L $\qquad\qquad\qquad\qquad\qquad V_2 =$ unknown

P and T remain constant

Converting m_1 (mass) to n_1 (moles): $(1.92$ g He$)\left(\dfrac{1 \text{ mol He}}{4.003 \text{ g He}}\right) = 0.47964$ mol He $= n_1$

Converting m_2 (mass) to n_2 (moles): $(1.07$ g He$)\left(\dfrac{1 \text{ mol He}}{4.003 \text{ g He}}\right) = 0.26730$ mol He $= n_2$

Arranging the ideal gas law and solving for V_2:

$$\dfrac{\cancel{P_1}V_1}{n_1\cancel{T_1}} = \dfrac{\cancel{P_2}V_2}{n_2\cancel{T_2}} \quad \text{or} \quad \dfrac{V_1}{n_1} = \dfrac{V_2}{n_2}$$

$$V_2 = V_1\dfrac{n_2}{n_1} = (12.5 \text{ L})\left(\dfrac{0.26730 \text{ mol He}}{0.47964 \text{ mol He}}\right) = 6.9661 = \mathbf{6.97 \text{ L}}$$

5.32   <u>Plan:</u> Since the volume, temperature, and pressure of the gas are changing, use the combined gas law. Arrange the ideal gas law, solving for V_2 at fixed n. STP is 0°C (273 K) and 1 atm (101.325 kPa)

<u>Solution:</u>

$P_1 = 153.3$ kPa $\qquad\qquad\qquad\qquad P_2 = 101.325$ kPa

$V_1 = 25.5$ L $\qquad\qquad\qquad\qquad\qquad V_2 =$ unknown

$T_1 = 298$ K $\qquad\qquad\qquad\qquad\qquad T_2 = 273$ K

n remains constant

Arranging the ideal gas law and solving for V_2:

$$\dfrac{P_1V_1}{\cancel{n_1}T_1} = \dfrac{P_2V_2}{\cancel{n_2}T_2} \quad \text{or} \quad \dfrac{P_1V_1}{T_1} = \dfrac{P_2V_2}{T_2}$$

$$V_2 = V_1\left(\dfrac{T_2}{T_1}\right)\left(\dfrac{P_1}{P_2}\right) = (25.5 \text{ L})\left(\dfrac{273 \text{ K}}{298 \text{ K}}\right)\left(\dfrac{153.3 \text{ kPa}}{101.325 \text{ kPa}}\right) = 35.3437 = \mathbf{35.3 \text{ L}}$$

5.34   <u>Plan:</u> Given the volume, pressure, and temperature of a gas, the number of moles of the gas can be calculated using the ideal gas law, solving for n. The gas constant, $R = 0.0821$ L•atm/mol•K, gives pressure in atmospheres and temperature in Kelvin. The given pressure in torr must be converted to atmospheres and the temperature converted to kelvins.

<u>Solution:</u>

$P = 328$ torr (convert to atm) $\qquad\qquad V = 5.0$ L

$T = 37$°C $\qquad\qquad\qquad\qquad\qquad\qquad n =$ unknown

Converting P from torr to atm: $\qquad P = (328$ torr$)\left(\dfrac{1 \text{ atm}}{760 \text{ torr}}\right) = 0.43158$ atm

Converting T from °C to K: $\qquad\quad T = 37$°C $+ 273.15 = 310.15$ K

$PV = nRT$

Solving for n:

$$n = \dfrac{PV}{RT} = \dfrac{(0.43158 \text{ atm})(5.0 \text{ L})}{\left(0.0821\dfrac{\text{L} \bullet \text{atm}}{\text{mol} \bullet \text{K}}\right)(310.15 \text{ K})} = 0.084745 = \mathbf{0.085 \text{ mol chlorine}}$$

5.36   <u>Plan:</u> Solve the ideal gas law for moles and convert to mass using the molar mass of ClF_3. The gas constant, $R = 0.0821$ L•atm/mol•K, gives volume in liters, pressure in atmospheres, and temperature in Kelvin so volume must be converted to L, pressure to atm, and temperature to K.

<u>Solution:</u>

$V = 357$ mL $\qquad\qquad\qquad\qquad\qquad T = 45$°C

$P = 699$ mmHg $\qquad\qquad\qquad\qquad\quad n =$ unknown

Converting V from mL to L: $\quad V = (357\text{ mL})\left(\dfrac{10^{-3}\text{ L}}{1\text{ mL}}\right) = 0.357\text{ L}$

Converting T from °C to K: $\quad T = 45°C + 273.15 = 318.15\text{ K}$

Converting P from mmHg to atm: $\quad P = (699\text{ mmHg})\left(\dfrac{1\text{ atm}}{760\text{ mmHg}}\right) = 0.91974\text{ atm}$

$PV = nRT$
Solving for n:

$$n = \frac{PV}{RT} = \frac{(0.91974\text{ atm})(0.357\text{ L})}{\left(0.0821\dfrac{\text{L}\bullet\text{atm}}{\text{mol}\bullet\text{K}}\right)(318.15\text{ K})} = 0.012571\text{ mol ClF}_3$$

$$\text{Mass ClF}_3 = (0.012571\text{ mol ClF}_3)\left(\frac{92.45\text{ g ClF}_3}{1\text{ mol ClF}_3}\right) = 1.16216 = \textbf{1.16 g ClF}_3$$

5.39   Plan: Assuming that while rising in the atmosphere the balloon will neither gain nor lose gas molecules, the number of moles of gas calculated at sea level will be the same as the number of moles of gas at the higher altitude (n is fixed). Volume, temperature, and pressure of the gas are changing. Arrange the ideal gas law, solving for V_2 at fixed n. Given the sea-level conditions of volume, pressure, and temperature, and the temperature and pressure at the higher altitude for the gas in the balloon, we can set up an equation to solve for the volume at the higher altitude. Comparing the calculated volume to the given maximum volume of 835 L will tell us if the balloon has reached its maximum volume at this altitude. Temperature must be converted to kelvins and pressure in torr must be converted to atm for unit agreement.
Solution:
$P_1 = 745$ torr $\qquad\qquad\qquad\qquad P_2 = 0.066$ atm
$V_1 = 65$ L $\qquad\qquad\qquad\qquad\quad V_2 = $ unknown
$T_1 = 25°C + 273.15 = 298.15$ K $\qquad T_2 = -5°C + 273.15 = 268.15$ K
n remains constant

Converting P from torr to atm: $\quad P = (745\text{ torr})\left(\dfrac{1\text{ atm}}{760\text{ torr}}\right) = 0.98026\text{ atm}$

Arranging the ideal gas law and solving for V_2:

$$\frac{P_1V_1}{\cancel{n_1}T_1} = \frac{P_2V_2}{\cancel{n_2}T_2} \quad\text{or}\quad \frac{P_1V_1}{T_1} = \frac{P_2V_2}{T_2}$$

$$V_2 = V_1\left(\frac{T_2}{T_1}\right)\left(\frac{P_1}{P_2}\right) = (65\text{ L})\left(\frac{268.15\text{ K}}{298.15\text{ K}}\right)\left(\frac{0.98026\text{ atm}}{0.066\text{ atm}}\right) = 868.268 = \textbf{870 L}$$

The calculated volume of the gas at the higher altitude is more than the maximum volume of the balloon. **Yes,** the balloon will reach its maximum volume.
Check: Should we expect that the volume of the gas in the balloon should increase? At the higher altitude, the pressure decreases; this increases the volume of the gas. At the higher altitude, the temperature decreases, this decreases the volume of the gas. Which of these will dominate? The pressure decreases by a factor of $0.98/0.066 = 15$. If we label the initial volume V_1, then the resulting volume is $15V_1$. The temperature decreases by a factor of $298/268 = 1.1$, so the resulting volume is $V_1/1.1$ or $0.91V_1$. The increase in volume due to the change in pressure is greater than the decrease in volume due to change in temperature, so the volume of gas at the higher altitude should be greater than the volume at sea level.

5.41   The molar mass of H_2 is less than the average molar mass of air (mostly N_2, O_2, and Ar), so air is denser. To collect a beaker of $H_2(g)$, **invert** the beaker so that the air will be replaced by the lighter H_2. The molar mass of CO_2 is greater than the average molar mass of air, so $CO_2(g)$ is more dense. Collect the CO_2 holding the beaker **upright,** so the lighter air will be displaced out the top of the beaker.

5.45   <u>Plan:</u> Rearrange the ideal gas law to calculate the density of xenon from its molar mass at STP. Standard
temperature is 0°C (273 K) and standard pressure is 1 atm. Do not forget that the pressure at STP is exact and will
not affect the significant figures.
<u>Solution:</u>
P = 1 atm T = 273 K
$\mathcal{M}$ of Xe = 131.3 g/mol d = unknown
$PV = nRT$
Rearranging to solve for density:

$$d = P\mathcal{M}/RT = \frac{(1\ \text{atm})(131.3\ \text{g/mol})}{\left(0.0821\dfrac{\text{L}\bullet\text{atm}}{\text{mol}\bullet\text{K}}\right)(273\ \text{K})} = 5.8581 = \textbf{5.86 g/L}$$

5.47   <u>Plan:</u> Solve the ideal gas law for moles. Convert moles to mass using the molar mass of AsH_3 and divide this mass
by the volume to obtain density in g/L. Standard temperature is 0°C (273 K) and standard pressure is 1 atm. Do
not forget that the pressure at STP is exact and will not affect the significant figures.
<u>Solution:</u>
V = 0.0400 L T = 273 K
P = 1 atm n = unknown
$\mathcal{M}$ of AsH_3 = 77.94 g/mol
$PV = nRT$
Solving for n:

$$n = \frac{PV}{RT} = \frac{(1\ \text{atm})(0.0400\ \text{L})}{\left(0.0821\dfrac{\text{L}\bullet\text{atm}}{\text{mol}\bullet\text{K}}\right)(273\ \text{K})} = 1.78465\times10^{-3} = \textbf{1.78}\times\textbf{10}^{-3}\ \textbf{mol AsH}_3$$

Converting moles of AsH_3 to mass of AsH_3:

$$\text{Mass (g) of } AsH_3 = (1.78465\times10^{-3}\ \text{mol } AsH_3)\left(\frac{77.94\ \text{g } AsH_3}{1\ \text{mol } AsH_3}\right) = 0.1391\ \text{g } AsH_3$$

$$d = \frac{\text{mass}}{\text{volume}} = \frac{(0.1391\ \text{g})}{(0.0400\ \text{L})} = 3.4775 = \textbf{3.48 g/L}$$

5.49   <u>Plan:</u> Rearrange the formula $PV = (m/\mathcal{M})RT$ to solve for molar mass. Convert the mass in ng to g and volume
in μL to L. Temperature must be in Kelvin and pressure in atm.
<u>Solution:</u>
V = 0.206 μL T = 45°C + 273.15 = 318.15 K
P = 380 torr m = 206 ng
$\mathcal{M}$ = unknown

Converting P from torr to atm: $P = (380\ \text{torr})\left(\dfrac{1\ \text{atm}}{760\ \text{torr}}\right) = 0.510526\ \text{atm}$

Converting V from μL to L: $V = (0.206\ \text{μL})\left(\dfrac{10^{-6}\ \text{L}}{1\ \text{μL}}\right) = 2.06\times10^{-7}\ \text{L}$

Converting m from ng to g: $m = (206\ \text{ng})\left(\dfrac{10^{-9}\ \text{g}}{1\ \text{ng}}\right) = 2.06\times10^{-7}\ \text{g}$

$PV = (m/\mathcal{M})RT$
Solving for molar mass, $\mathcal{M}$:

$$\mathcal{M} = \frac{mRT}{PV} = \frac{(2.06\times10^{-7}\ \text{g})\left(0.0821\dfrac{\text{L}\bullet\text{atm}}{\text{mol}\bullet\text{K}}\right)(318.15\ \text{K})}{(0.510526\ \text{atm})(2.06\times10^{-7}\ \text{L})} = 51.163 = \textbf{51.2 g/mol}$$

5.51   Plan: Use the ideal gas law to determine the number of moles of Ar and of O_2. The gases are combined ($n_{total} = n_{Ar} + n_{O_2}$) into a 400 mL flask ($V$) at 27°C ($T$). Use the ideal gas law again to determine the total pressure from n_{total}, V, and T. Pressure must be in units of atm, volume in units of L and temperature in K.
Solution:
For Ar:
$V = 0.600$ L $T = 227°C + 273.15 = 500.15$ K
$P = 1.20$ atm $n =$ unknown
$PV = nRT$
Solving for n:
$$n = \frac{PV}{RT} = \frac{(1.20 \text{ atm})(0.600 \text{ L})}{\left(0.0821\dfrac{\text{L} \bullet \text{atm}}{\text{mol} \bullet \text{K}}\right)(500.15 \text{ K})} = 0.01753433 \text{ mol Ar}$$

For O_2:
$V = 0.200$ L $T = 127°C + 273.15 = 400.15$ K
$P = 501$ torr $n =$ unknown

Converting P from torr to atm: $P = (501 \text{ torr})\left(\dfrac{1 \text{ atm}}{760 \text{ torr}}\right) = 0.6592105$ atm

$PV = nRT$
Solving for n:
$$n = \frac{PV}{RT} = \frac{(0.6592105 \text{ atm})(0.200 \text{ L})}{\left(0.0821\dfrac{\text{L} \bullet \text{atm}}{\text{mol} \bullet \text{K}}\right)(400.15 \text{ K})} = 0.00401318 \text{ mol } O_2$$

$n_{total} = n_{Ar} + n_{O_2} = 0.01753433 \text{ mol} + 0.00401318 \text{ mol} = 0.02154751$ mol
For the mixture of Ar and O_2:
$V = 400$ mL $T = 27°C + 273.15 = 300.15$ K
$P =$ unknownn $n = 0.02154751$ mol

Converting V from mL to L: $V = (400 \text{ mL})\left(\dfrac{10^{-3} \text{ L}}{1 \text{ mL}}\right) = 0.400$ L

$PV = nRT$
Solving for P:
$$P_{mixture} = \frac{nRT}{V} = \frac{(0.02154751 \text{ mol})\left(0.0821\dfrac{\text{L} \bullet \text{atm}}{\text{mol} \bullet \text{K}}\right)(300.15 \text{ K})}{(0.400 \text{ L})} = 1.32745 = \mathbf{1.33 \text{ atm}}$$

5.53   Plan: Use the ideal gas law, solving for n to find the moles of O_2. Use the molar ratio from the balanced equation to determine the moles (and then mass) of phosphorus that will react with the oxygen. Standard temperature is 0°C (273 K) and standard pressure is 1 atm.
Solution:
$V = 35.5$ L $T = 273$ K
$P = 1$ atm $n =$ unknown
$PV = nRT$
Solving for n:
$$n = \frac{PV}{RT} = \frac{(1 \text{ atm})(35.5 \text{ L})}{\left(0.0821\dfrac{\text{L} \bullet \text{atm}}{\text{mol} \bullet \text{K}}\right)(273 \text{ K})} = 1.583881 \text{ mol } O_2$$

$$P_4(s) + 5O_2(g) \rightarrow P_4O_{10}(s)$$

$$\text{Mass } P_4 = (1.583881 \text{ mol } O_2)\left(\frac{1 \text{ mol } P_4}{5 \text{ mol } O_2}\right)\left(\frac{123.88 \text{ g } P_4}{1 \text{ mol } P_4}\right) = 39.24224 = \mathbf{39.2 \text{ g } P_4}$$

5.55   Plan: Since the amounts of two reactants are given, this is a limiting reactant problem. To find the mass of PH_3, write the balanced equation and use molar ratios to find the number of moles of PH_3 produced by each reactant. The smaller number of moles of product indicates the limiting reagent. Solve for moles of H_2 using the ideal gas law.
Solution:
Moles of hydrogen:

$V = 83.0$ L $\qquad\qquad$ $T = 273$ K

$P = 1$ atm $\qquad\qquad$ $n =$ unknown

$PV = nRT$

Solving for n:

$$n = \frac{PV}{RT} = \frac{(1\ \text{atm})(83.0\ \text{L})}{\left(0.0821\dfrac{\text{L}\bullet\text{atm}}{\text{mol}\bullet\text{K}}\right)(273\ \text{K})} = 3.7031584\ \text{mol H}_2$$

$$P_4(s) + 6H_2(g) \rightarrow 4PH_3(g)$$

$$PH_3 \text{ from } H_2 = (3.7031584\ \text{mol H}_2)\left(\frac{4\ \text{mol PH}_3}{6\ \text{mol H}_2}\right) = 2.4687723\ \text{mol PH}_3$$

$$PH_3 \text{ from } P_4 = (37.5\ \text{g P}_4)\left(\frac{1\ \text{mol P}_4}{123.88\ \text{g P}_4}\right)\left(\frac{4\ \text{mol PH}_3}{1\ \text{mol P}_4}\right) = 1.21085\ \text{mol PH}_3$$

P_4 is the limiting reactant because it forms less PH_3.

$$\text{Mass PH}_3 = (37.5\ \text{g P}_4)\left(\frac{1\ \text{mol P}_4}{123.88\ \text{g P}_4}\right)\left(\frac{4\ \text{mol PH}_3}{1\ \text{mol P}_4}\right)\left(\frac{33.99\ \text{g PH}_3}{1\ \text{mol PH}_3}\right) = 41.15676 = \textbf{41.2 g PH}_3$$

5.57   Plan: First, write the balanced equation. The moles of hydrogen produced can be calculated from the ideal gas law. The problem specifies that the hydrogen gas is collected over water, so the partial pressure of water vapor must be subtracted from the overall pressure given. Table 5.2 reports pressure at 26°C (25.2 torr) and 28°C (28.3 torr), so take the average of the two values to obtain the partial pressure of water at 27°C. Volume must be in units of liters, pressure in atm, and temperature in kelvins. Once the moles of hydrogen produced are known, the molar ratio from the balanced equation is used to determine the moles of aluminum that reacted.
Solution:

$V = 35.8$ mL $\qquad\qquad$ $T = 27°C + 273.15 = 300.15$ K

$P_{total} = 751$ mmHg $\qquad\qquad$ $n =$ unknown

$P_{water\ vapor} = (28.3 + 25.2)\ \text{torr}/2 = 26.75\ \text{torr} = 26.75\ \text{mmHg}$

$P_{hydrogen} = P_{total} - P_{water\ vapor} = 751\ \text{mmHg} - 26.75\ \text{mmHg} = 724.25\ \text{mmHg}$

Converting P from mmHg to atm: $P = (724.25\ \text{mmHg})\left(\dfrac{1\ \text{atm}}{760\ \text{mmHg}}\right) = 0.952960526\ \text{atm}$

Converting V from mL to L: $\qquad V = (35.8\ \text{mL})\left(\dfrac{10^{-3}\ \text{L}}{1\ \text{mL}}\right) = 0.0358\ \text{L}$

$PV = nRT$

Solving for n:

$$n = \frac{PV}{RT} = \frac{(0.952960526\ \text{atm})(0.0358\ \text{L})}{\left(0.0821\dfrac{\text{L}\bullet\text{atm}}{\text{mol}\bullet\text{K}}\right)(300.15\ \text{K})} = 0.001384447\ \text{mol H}_2$$

$$2Al(s) + 6HCl(aq) \rightarrow 2AlCl_3(aq) + 3H_2(g)$$

$$\text{Mass (g) of Al} = (0.001384447\ \text{mol H}_2)\left(\frac{2\ \text{mol Al}}{3\ \text{mol H}_2}\right)\left(\frac{26.98\ \text{g Al}}{1\ \text{mol Al}}\right) = 0.024902 = \textbf{0.0249 g Al}$$

5.61   Plan: The problem gives the mass, volume, temperature, and pressure of a gas; rearrange the formula $PV = (m/\mathcal{M})RT$ to solve for the molar mass of the gas. Temperature must be in Kelvin and pressure in atm. The problem also states that the gas is a hydrocarbon, which by, definition, contains only carbon and hydrogen atoms. We are also told that each molecule of the gas contains five carbon atoms so we can use this information and the calculated molar mass to find out how many hydrogen atoms are present and the formula of the compound.

Solution:

$V = 0.204$ L $T = 101°C + 273.15 = 374.15$ K

$P = 767$ torr $m = 0.482$ g

$\mathcal{M} = $ unknown

Converting P from torr to atm: $P = (767 \text{ torr})\left(\dfrac{1 \text{ atm}}{760 \text{ torr}}\right) = 1.009210526$ atm

$PV = (m/\mathcal{M})RT$

Solving for molar mass, $\mathcal{M}$:

$$\mathcal{M} = \frac{mRT}{PV} = \frac{(0.482 \text{ g})\left(0.0821 \dfrac{\text{L} \bullet \text{atm}}{\text{mol} \bullet \text{K}}\right)(374.15 \text{ K})}{(1.009210526 \text{ atm})(0.204 \text{ L})} = 71.9157 \text{ g/mol (unrounded)}$$

The mass of the five carbon atoms accounts for [5(12 g/mol)] = 60 g/mol; thus, the hydrogen atoms must make up the difference (72 – 60) = 12 g/mol. A value of 12 g/mol corresponds to 12 H atoms. (Since fractional atoms are not possible, rounding is acceptable.) Therefore, the molecular formula is $\mathbf{C_5H_{12}}$.

5.63   Plan: Since you have the pressure, volume, and temperature, use the ideal gas law to solve for n, the total moles of gas. Pressure must be in units of atmospheres and temperature in units of kelvins. The partial pressure of SO_2 can be found by multiplying the total pressure by the volume fraction of SO_2.

Solution:

a) $V = 21$ L $T = 45°C + 273.15 = 318.15$ K

$P = 850$ torr $n = $ unknown

Converting P from torr to atm: $P = (850 \text{ torr})\left(\dfrac{1 \text{ atm}}{760 \text{ torr}}\right) = 1.118421053$ atm

$PV = nRT$

Moles of gas $= n = \dfrac{PV}{RT} = \dfrac{(1.118421053 \text{ atm})(21 \text{ L})}{\left(0.0821 \dfrac{\text{L} \bullet \text{atm}}{\text{mol} \bullet \text{K}}\right)(318.15 \text{ K})} = 0.89919 = \mathbf{0.90 \text{ mol gas}}$

b) The equation $P_{SO_2} = X_{SO_2} \text{ x } P_{total}$ can be used to find partial pressure. The information given in ppm is a way of expressing the proportion, or fraction, of SO_2 present in the mixture. Since n is directly proportional to V, the *volume* fraction can be used in place of the *mole* fraction, X_{SO_2}. There are 7.95×10^3 parts SO_2 in a million parts of mixture, so volume fraction = $(7.95 \times 10^3 / 1 \times 10^6) = 7.95 \times 10^{-3}$.

$P_{D_2} = $ volume fraction x $P_{total} = (7.95 \times 10^{-3})$ (850. torr) $= 6.7575 = \mathbf{6.76 \text{ torr}}$

5.64   Plan: First, write the balanced equation. Convert mass of P_4S_3 to moles and use the molar ratio from the balanced equation to find the moles of SO_2 gas produced. Use the ideal gas law to find the volume of that amount of SO_2. Pressure must be in units of atm and temperature in kelvins.

Solution:

$P_4S_3(s) + 8O_2(g) \rightarrow P_4O_{10}(s) + 3SO_2(g)$

Moles $SO_2 = (0.800 \text{ g } P_4S_3)\left(\dfrac{1 \text{ mol } P_4S_3}{220.06 \text{ g } P_4S_3}\right)\left(\dfrac{3 \text{ mol } SO_2}{1 \text{ mol } P_4S_3}\right) = 0.010906 \text{ mol } SO_2$

Finding the volume of SO_2:

$V = $ unknown $T = 32°C + 273.15 = 305.15$ K

$P = 725$ torr $n = 0.010906$ mol

Converting P from torr to atm: $P = (725 \text{ torr})\left(\dfrac{1 \text{ atm}}{760 \text{ torr}}\right) = 0.953947368$ atm

$PV = nRT$

$-139.5 \text{ kJ} = -1280.14 - [\Delta H_f^\circ [SiCl_4(g)] + (-483.652 \text{ kJ})]$

$1140.64 \text{ kJ} = -\Delta H_f^\circ [SiCl_4(g)] + 483.652 \text{ kJ}$

$\Delta H_f^\circ [SiCl_4(g)] = -656.988 \text{ kJ/mol}$

The heats of reaction for the first two steps can now be calculated.

1) $Si(s) + 2Cl_2(g) \rightarrow SiCl_4(g)$

$\Delta H_{rxn}^\circ = \{1\ \Delta H_f^\circ [SiCl_4(g)]\} - \{1\ \Delta H_f^\circ [Si(s)] + 2\ \Delta H_f^\circ [Cl_2(g)]\}$

$= [(1 \text{ mol})(-656.988 \text{ kJ/mol})] - [(1 \text{ mol})(0 \text{ kJ/mol}) + (2 \text{ mol})(0 \text{ kJ/mol})] = -656.988 = \textbf{-657.0 kJ}$

2) $SiO_2(s) + 2C(graphite) + 2Cl_2(g) \rightarrow SiCl_4(g) + 2CO(g)$

$\Delta H_{rxn}^\circ = \{1\ \Delta H_f^\circ [SiCl_4(g)] + 2\ \Delta H_f^\circ [CO(g)]\}$

$\qquad - \{1\ \Delta H_f^\circ [SiO_2(g)] + 2\ \Delta H_f^\circ [C(graphite)] + 2\ \Delta H_f^\circ [Cl_2(g)]\}$

$\qquad = [(1 \text{ mol})(-656.988 \text{ kJ/mol}) + (2 \text{ mol})(-110.5 \text{ kJ/mol})]$

$\qquad\qquad - [(1 \text{ mol})(-910.9 \text{ kJ/mol}) + (2 \text{ mol})(0 \text{ kJ/mol}) + (2 \text{ mol})(0 \text{ kJ/mol})]$

$\qquad = 32.912 = \textbf{32.9 kJ}$

b) Adding reactions 2 and 3 yields:

(2) ~~SiO₂(s)~~ + 2C(graphite) + 2Cl$_2$(g) → ~~SiCl₄(g)~~ + 2CO(g) $\Delta H_{rxn}^\circ = \ $ 32.912 kJ

(3) ~~SiCl₄(g)~~ + 2H$_2$O(g) → ~~SiO₂(s)~~ + 4HCl(g) $\Delta H_{rxn}^\circ = -139.5$ kJ

$\overline{2C(graphite) + 2Cl_2(g) + 2H_2O(g) \rightarrow 2CO(g) + 4HCl(g) \qquad \Delta H_{rxn}^\circ = -106.588 \text{ kJ} = \textbf{-106.6 kJ}}$

Confirm this result by calculating ΔH_{rxn}° using Appendix B values.

$2C(graphite) + 2Cl_2(g) + 2H_2O(g) \rightarrow 2CO(g) + 4HCl(g)$

$\Delta H_{rxn}^\circ = \{2\ \Delta H_f^\circ [CO(g)] + 4\ \Delta H_f^\circ [HCl(g)]\} - \{2\ \Delta H_f^\circ [C(graphite)] + 2\ \Delta H_f^\circ [Cl_2(g)] + 2\ \Delta H_f^\circ [H_2O(g)]\}$

$\qquad = [(2 \text{ mol})(-110.5 \text{ kJ/mol}) + (4 \text{ mol})(-92.31 \text{ kJ})]$

$\qquad\qquad - [(2 \text{ mol})(0 \text{ kJ/mol}) + (2 \text{ mol})(0 \text{ kJ/mol}) + (2 \text{ mol})(-241.826 \text{ kJ/mol})]$

$\qquad = -106.588 = \textbf{-106.6 kJ}$

6.104   Plan: Use $PV = nRT$ to find the initial volume of nitrogen gas at 0°C and then the final volume at 819°C. Then the relationship $w = -P\Delta V$ can be used to calculate the work of expansion.

Solution:

a) $PV = nRT$

Initial volume at 0°C + 273 = 273 K = $V = \dfrac{nRT}{P} = \dfrac{(1 \text{ mol})\left(0.0821\dfrac{\text{L} \cdot \text{atm}}{\text{mol} \cdot \text{K}}\right)(273 \text{ K})}{(1.00 \text{ atm})} = 22.4133 \text{ L}$

Final volume at 819°C + 273 = 1092 K = $V = \dfrac{nRT}{P} = \dfrac{(1 \text{ mol})\left(0.0821\dfrac{\text{L} \cdot \text{atm}}{\text{mol} \cdot \text{K}}\right)(1092 \text{ K})}{(1.00 \text{ atm})} = 89.6532 \text{ L}$

$\Delta V = V_{final} - V_{initial} = 89.6532 \text{ L} - 22.4133 \text{ L} = 67.2399 \text{ L}$

$w = -P\Delta V = -(1 \text{ atm}) \times 67.2399 \text{ L} = -67.2399 \text{ atm} \cdot \text{L}$

$w \text{ (J)} = (-67.2399 \text{ atm} \cdot \text{L})\left(\dfrac{101.3 \text{ J}}{1 \text{ L} \cdot \text{atm}}\right) = -6811.40187 = \textbf{-6.81} \times \textbf{10}^3 \textbf{ J}$

b) $q = c \times \text{mass} \times \Delta T$

Mass (g) of $N_2 = (1 \text{ mol N}_2)\left(\dfrac{28.02 \text{ g}}{1 \text{ mol N}_2}\right) = 28.02 \text{ g}$

$\Delta T = \dfrac{q}{(c)(\text{mass})} = \dfrac{6.812553 \times 10^3 \text{ J}}{(28.02 \text{ g})(1.00 \text{ J/g} \cdot \text{K})} = 243.132 = 243 \text{ K} = \textbf{243°C}$

6.105   Plan: Note the numbers of moles of the reactants and products in the target equation and manipulate equations 1-5 and their ΔH_{rxn}° values so that these equations sum to give the target equation. Then the manipulated ΔH_{rxn}° values will add to give the ΔH_{rxn}° value of the target equation.

Solution:
Only reaction 3 contains $N_2O_4(g)$, and only reaction 1 contains $N_2O_3(g)$, so we can use those reactions as a starting point. N_2O_5 appears in both reactions 2 and 5, but note the physical states present: solid and gas. As a rough start, adding reactions 1, 3, and 5 yields the desired reactants and products, with some undesired intermediates:

Reverse (1)	$N_2O_3(g) \rightarrow NO(g) + NO_2(g)$	$\Delta H_{rxn}^{\circ} = -(-39.8\ kJ) = 39.8\ kJ$
Multiply (3) by 2	$4NO_2(g) \rightarrow 2N_2O_4(g)$	$\Delta H_{rxn}^{\circ} = 2(-57.2\ kJ) = -114.4\ kJ$
(5)	$N_2O_5(s) \rightarrow N_2O_5(g)$	$\Delta H_{rxn}^{\circ} = (54.1\ kJ) = 54.1\ kJ$

$$N_2O_3(g) + 4NO_2(g) + N_2O_5(s) \rightarrow NO(g) + NO_2(g) + 2N_2O_4(g) + N_2O_5(g)$$

To cancel out the $N_2O_5(g)$ intermediate, reverse equation 2. This also cancels out some of the undesired $NO_2(g)$ but adds $NO(g)$ and $O_2(g)$. Finally, add equation 4 to remove those intermediates:

Reverse (1)	$N_2O_3(g) \rightarrow \cancel{NO(g)} + \cancel{NO_2(g)}$	$\Delta H_{rxn}^{\circ} = -(-39.8\ kJ) = 39.8\ kJ$
Multiply (3) by 2	$\cancel{4NO_2(g)} \rightarrow 2N_2O_4(g)$	$\Delta H_{rxn}^{\circ} = 2(-57.2\ kJ) = -114.4\ kJ$
(5)	$N_2O_5(s) \rightarrow \cancel{N_2O_5(g)}$	$\Delta H_{rxn}^{\circ} = 54.1\ kJ$
Reverse (2)	$\cancel{N_2O_5(g)} \rightarrow \cancel{NO(g)} + \cancel{NO_2(g)} + \cancel{O_2(g)}$	$\Delta H_{rxn}^{\circ} = -(-112.5\ kJ) = 112.5$
(4)	$\cancel{2NO(g)} + \cancel{O_2(g)} \rightarrow \cancel{2NO_2(g)}$	$\Delta H_{rxn}^{\circ} = -114.2\ kJ$
Total:	$N_2O_3(g) + N_2O_5(s) \rightarrow 2N_2O_4(g)$	$\Delta H_{rxn}^{\circ} = \mathbf{-22.2\ kJ}$

6.106   Plan: The enthalpy change of a reaction is the sum of the heats of formation of the products minus the sum of the heats of formation of the reactants. Since the ΔH_f° values (Appendix B) are reported as energy per one mole, use the appropriate stoichiometric coefficient to reflect the higher number of moles. In this case, ΔH_{rxn}° of the second reaction is known and ΔH_f° of $N_2H_4(aq)$ must be calculated. For part b), calculate ΔH_{rxn}° for the reaction between $N_2H_4(aq)$ and O_2, using the value of ΔH_f° for $N_2H_4(aq)$ found in part a); then determine the moles of O_2 present by multiplying volume and molarity and multiply by the ΔH_{rxn}° for the reaction.

Solution:
a) $2NH_3(aq) + NaOCl(aq) \rightarrow N_2H_4(aq) + NaCl(aq) + H_2O(l)$

$\Delta H_{rxn}^{\circ} = \{1\ \Delta H_f^{\circ}\ [N_2H_4(aq)] + 1\ \Delta H_f^{\circ}\ [NaCl(aq)] + 1\ \Delta H_f^{\circ}\ [H_2O(l)]\}$
$$- \{2\ \Delta H_f^{\circ}\ [NH_3(aq)] + 1\ \Delta H_f^{\circ}\ [NaOCl(aq)]\}$$

Note that the Appendix B value for N_2H_4 is for $N_2H_4(l)$, not for $N_2H_4(aq)$, so this term must be calculated. In addition, Appendix B does not list a value for $NaCl(aq)$, so this term must be broken down into $\Delta H_f^{\circ}\ [Na^+(aq)]$ and $\Delta H_f^{\circ}\ [Cl^-(aq)]$.

$-151\ kJ = [\ \Delta H_f^{\circ}\ [N_2H_4(aq)] + (1\ mol)(-239.66\ kJ/mol) + (1\ mol)(-167.46\ kJ/mol) + (1\ mol)(-285.840\ kJ/mol)]$
$$- [(2\ mol)(-80.83\ kJ/mol) + (1\ mol)(-346\ kJ/mol)]$$

$-151\ kJ = [\ \Delta H_f^{\circ}\ [N_2H_4(aq)] + (-692.96\ kJ)] - [-507.66\ kJ]$

$-151\ kJ = [\ \Delta H_f^{\circ}\ [N_2H_4(aq)] + (-185.3\ kJ)$

$\Delta H_f^{\circ}\ [N_2H_4(aq)] = 34.3 = \mathbf{34\ kJ}$

b) Moles of $O_2 = \left(5.00 \times 10^3\ L\right)\left(\dfrac{2.50 \times 10^{-4}\ mol}{1\ L}\right) = 1.25\ mol\ O_2$

$N_2H_4(aq) + O_2(g) \rightarrow N_2(g) + 2H_2O(l)$

$\Delta H_{rxn}^{\circ} = \{1\ \Delta H_f^{\circ}\ [N_2(g)] + 2\ \Delta H_f^{\circ}\ [H_2O(l)]\} - \{1\ \Delta H_f^{\circ}\ [N_2H_4(aq)] + 1\ \Delta H_f^{\circ}\ [O_2(g)]\}$

$= [(1\ mol)(0\ kJ/mol) + (2\ mol)(-285.840\ kJ/mol)] - [(1\ mol)(34.3\ kJ/mol) + (1\ mol)(0\ kJ/mol)]$

$= -605.98\ kJ$

$$\text{Heat (kJ)} = (1.25 \text{ mol } O_2)\left(\frac{-605.98 \text{ kJ}}{1 \text{ mol } O_2}\right) = -757.475 = \textbf{-757 kJ}$$

6.108 Plan: First find the heat of reaction for the combustion of methane. The enthalpy change of a reaction is the sum of the heats of formation of the products minus the sum of the heats of formation of the reactants. Since the ΔH_f° values (Appendix B) are reported as energy per one mole, use the appropriate stoichiometric coefficient to reflect the higher number of moles. Convert the mass of methane to moles and multiply that mole number by the heat of combustion.

Solution:

a) The balanced chemical equation for this reaction is:

$CH_4(g) + 2O_2(g) \rightarrow CO_2(g) + 2H_2O(g)$

$\Delta H_{rxn}^\circ = \{1 \Delta H_f^\circ [CO_2(g)] + 2 \Delta H_f^\circ [H_2O(g)]\} - \{1 \Delta H_f^\circ [CH_4(g)] + 2 \Delta H_f^\circ [O_2(g)]\}$

$\quad = [(1 \text{ mol})(-393.5 \text{ kJ/mol}) + (2 \text{ mol})(-241.826 \text{ kJ/mol})] - [(1\text{mol})(-74.87 \text{ kJ/mol}) + (2 \text{ mol})(0.0 \text{ kJ/mol})]$

$\quad = -802.282 \text{ kJ}$

$$\text{Moles of } CH_4 = (25.0 \text{ g } CH_4)\left(\frac{1 \text{ mol}}{16.04 \text{ g } CH_4}\right) = 1.5586 \text{ mol } CH_4$$

$$\text{Heat (kJ)} = (1.5586 \text{ mol } CH_4)\left(\frac{-802.282 \text{ kJ}}{1 \text{ mol } CH_4}\right) = -1250.4 = \textbf{-1.25x10}^3 \textbf{ kJ}$$

b) The heat released by the reaction is "stored" in the gaseous molecules by virtue of their specific heat capacities, c, using the equation $q = c$ x mass x ΔT. The problem specifies heat capacities on a molar basis, so we modify the equation to use moles, instead of mass. The gases that remain at the end of the reaction are CO_2 and H_2O. All of the methane and oxygen molecules were consumed. However, the oxygen was added as a component of air, which is 78% N_2 and 21% O_2, and there is leftover N_2.

$$\text{Moles of } CO_2(g) = (1.5586 \text{ mol } CH_4)\left(\frac{1 \text{ mol } CO_2}{1 \text{ mol } CH_4}\right) = 1.5586 \text{ mol } CO_2(g)$$

$$\text{Moles of } H_2O(g) = (1.5586 \text{ mol } CH_4)\left(\frac{2 \text{ mol } H_2O}{1 \text{ mol } CH_4}\right) = 3.1172 \text{ mol } H_2O(g)$$

$$\text{Moles of } O_2(g) \text{ reacted} = (1.5586 \text{ mol } CH_4)\left(\frac{2 \text{ mol } O_2}{1 \text{ mol } CH_4}\right) = 3.1172 \text{ mol } O_2(g)$$

Mole fraction N_2 = (79%/100%) = 0.79
Mole fraction O_2 = (21%/100%) = 0.21

$$\text{Moles of } N_2(g) = (3.1172 \text{ mol } O_2 \text{ reacted})\left(\frac{0.79 \text{ mol } N_2}{0.21 \text{ mol } O_2}\right) = 11.72661 \text{ mol } N_2$$

$q = c$ x mass x ΔT

$$q = (1250.4 \text{ kJ})\left(\frac{10^3 \text{ J}}{1 \text{ kJ}}\right) = 1.2504\text{x}10^6 \text{ J}$$

$1.2504\text{x}10^6 \text{ J} = (1.5586 \text{ mol } CO_2)(57.2 \text{ J/mol°C})(T_{final} - 0.0)°C$

$\qquad\qquad\qquad + (3.1172 \text{ mol } H_2O)(36.0 \text{ J/mol°C})(T_{final} - 0.0)°C$

$\qquad\qquad\qquad\qquad + (11.72661 \text{ mol } N_2)(30.5 \text{ J/mol°C})(T_{final} - 0.0)°C$

$1.2504\text{x}10^6 \text{ J} = 89.15192 \text{ J/°C}(T_{final}) + 112.2192 \text{ J/°C}(T_{final}) + 357.6616 \text{ J/°C}(T_{final})$

$1.2504\text{x}10^6 \text{ J} = (559.03272 \text{ J/°C})T_{final}$

$T_{final} = (1.2504\text{x}10^6 \text{ J})/(559.0324 \text{ J/°C}) = 2236.72 = \textbf{2.24x10}^3\textbf{°C}$

CHAPTER 7 QUANTUM THEORY AND ATOMIC STRUCTURE

The value for the speed of light will be 3.00×10^8 m/s except when more significant figures are necessary, in which cases, 2.9979×10^8 m/s will be used.

FOLLOW–UP PROBLEMS

7.1A   Plan: Given the frequency of the light, use the equation $c = \lambda \nu$ to solve for wavelength.
Solution:

$$\lambda = \frac{c}{\nu} = \frac{3.00 \times 10^8 \, \text{m/s}}{7.23 \times 10^{14} \, \text{s}^{-1}} \left(\frac{1 \, \text{nm}}{10^{-9} \, \text{m}} \right) = 414.938 = \textbf{415 nm}$$

$$\lambda = \frac{c}{\nu} = \frac{3.00 \times 10^8 \, \text{m/s}}{7.23 \times 10^{14} \, \text{s}^{-1}} \left(\frac{1 \, \text{Å}}{10^{-10} \, \text{m}} \right) = 4149.38 = \textbf{4150 Å}$$

7.1B   Plan: Given the wavelength of the light, use the equation $c = \lambda \nu$ to solve for frequency. Remember that wavelength must be in units of m in this equation.
Solution:

$$\nu = \frac{c}{\lambda} = \left(\frac{3.00 \times 10^8 \, \text{m/s}}{940 \, \text{nm}} \right) \left(\frac{10^9 \, \text{nm}}{1 \, \text{m}} \right) = 3.1915 \times 10^{14} = \textbf{3.2} \times \textbf{10}^{14} \, \textbf{s}^{-1}$$

This is **infrared** radiation.

7.2A   Plan: Use the formula $E = h\nu$ to solve for the frequency. Then use the equation $c = \lambda \nu$ to solve for the wavelength.
Solution:

$$\nu = \frac{E}{h} = \left(\frac{8.2 \times 10^{-19} \, \text{J}}{6.626 \times 10^{-34} \, \text{J} \cdot \text{s}} \right) = 1.2375 \times 10^{15} = \textbf{1.2} \times \textbf{10}^{15} \, \textbf{s}^{-1}$$

(using the unrounded number in the next calculation to avoid rounding errors)

$$\lambda = \frac{c}{\nu} = \left(\frac{3.00 \times 10^8 \, \text{m/s}}{1.2375 \times 10^{15} \, \text{s}^{-1}} \right) \left(\frac{10^9 \, \text{nm}}{1 \, \text{m}} \right) = \textbf{240 nm}$$

7.2B   Plan: To calculate the energy for each wavelength we use the formula $E = hc/\lambda$.
Solution:

$$E = \frac{hc}{\lambda} = \frac{\left(6.626 \times 10^{-34} \, \text{J} \cdot \text{s} \right) \left(3.00 \times 10^8 \, \text{m/s} \right)}{1 \times 10^{-8} \, \text{m}} = 1.9878 \times 10^{-17} = \textbf{2} \times \textbf{10}^{-17} \, \textbf{J}$$

$$E = \frac{hc}{\lambda} = \frac{\left(6.626 \times 10^{-34} \, \text{J} \cdot \text{s} \right) \left(3.00 \times 10^8 \, \text{m/s} \right)}{5 \times 10^{-7} \, \text{m}} = 3.9756 \times 10^{-19} = \textbf{4} \times \textbf{10}^{-19} \, \textbf{J}$$

$$E = \frac{hc}{\lambda} = \frac{\left(6.626 \times 10^{-34} \, \text{J} \cdot \text{s} \right) \left(3.00 \times 10^8 \, \text{m/s} \right)}{1 \times 10^{-4} \, \text{m}} = 1.9878 \times 10^{-21} = \textbf{2} \times \textbf{10}^{-21} \, \textbf{J}$$

As the wavelength of light increases from ultraviolet to visible to infrared, the energy of the light decreases.

7.3A   Plan: Use the equation relating $\Delta E = -2.18 \times 10^{-18} \text{ J} \left(\dfrac{1}{n^2_{\text{final}}} - \dfrac{1}{n^2_{\text{initial}}} \right)$ to find the energy change; a photon in the IR (infrared) region is emitted when n has a final value of 3. Then use $E = hc/\lambda$ to find the wavelength of the photon.
Solution:

a) $\Delta E = -2.18 \times 10^{-18} \text{ J} \left(\dfrac{1}{n^2_{\text{final}}} - \dfrac{1}{n^2_{\text{initial}}} \right)$

$\Delta E = -2.18 \times 10^{-18} \text{ J} \left(\dfrac{1}{3^2} - \dfrac{1}{6^2} \right)$

$\Delta E = -1.8166667 \times 10^{-19} = \mathbf{-1.82 \times 10^{-19} \text{ J}}$

b) $E = \dfrac{hc}{\lambda}$

$\lambda = \dfrac{hc}{E} = \dfrac{(6.626 \times 10^{-34} \text{ J} \bullet \text{s})(3.00 \times 10^8 \text{ m/s})}{1.8166667 \times 10^{-19} \text{ J}} \left(\dfrac{1 \text{ Å}}{10^{-10} \text{ m}} \right) = 1.094202 \times 10^4 = \mathbf{1.09 \times 10^4 \text{ Å}}$

7.3B   Plan: Use the equation $E = hc/\lambda$ to find the energy change for this reaction. Then use the equation
$\Delta E = -2.18 \times 10^{-18} \text{ J} \left(\dfrac{1}{n^2_{\text{final}}} - \dfrac{1}{n^2_{\text{initial}}} \right)$ to find the final energy level to which the electron moved.
Solution:

a) $\Delta E = hc/\lambda = \left(\dfrac{(6.626 \times 10^{-34} \text{ J} \bullet \text{s})(3.00 \times 10^8 \text{ m/s})}{410. \text{ nm}} \right) \left(\dfrac{10^9 \text{ nm}}{1 \text{ m}} \right) = \Delta E = 4.8483 \times 10^{-19} = 4.85 \times 10^{-19} \text{ J}$

Because the photon is emitted, energy is being given off, so the sign of ΔE should be negative. Therefore, $\Delta E = -4.8483 \times 10^{-19}$ or -4.85×10^{-19} J rounded to three significant figures.

$\Delta E \text{ (kJ/mol)} = \left(\dfrac{-4.8483 \times 10^{-19} \text{ J}}{1 \text{ H atom}} \right) \left(\dfrac{6.022 \times 10^{23} \text{ H atoms}}{1 \text{ mol H}} \right) \left(\dfrac{1 \text{ kJ}}{1000 \text{ J}} \right) = -291.9646 = \mathbf{-292 \text{ kJ/mol}}$ (number of atoms is a positive number)

b) $\Delta E = -2.18 \times 10^{-18} \text{ J} \left(\dfrac{1}{n^2_{\text{final}}} - \dfrac{1}{n^2_{\text{initial}}} \right)$

$\dfrac{1}{n^2_{\text{final}}} = \dfrac{\Delta E}{-2.18 \times 10^{-18} \text{ J}} + \dfrac{1}{n^2_{\text{initial}}}$

$\dfrac{1}{n^2_{\text{final}}} = \dfrac{-4.8483 \times 10^{-19} \text{ J}}{-2.18 \times 10^{-18} \text{ J}} + \dfrac{1}{6^2}$

$\dfrac{1}{n^2_{\text{final}}} = 0.25018$

$n^2_{\text{final}} = 3.9972 = 4$ (The final energy level is an integer, so its square is also an integer.)

$\mathbf{n_{\text{final}} = 2}$

7.4A   Plan: With the equation for the de Broglie wavelength, $\lambda = h/mu$ and the given de Broglie wavelength, calculate the electron speed. The wavelength must be expressed in meters. Use the same formulas to calculate the speed of the golf ball. The mass of both the electron and the golf ball must be expressed in kg in their respective calculations.

Solution:

a) $\lambda = h/mu$

$$u = \frac{h}{m\lambda} = \frac{6.626 \times 10^{-34} \text{ J} \cdot \text{s}}{\left(9.11 \times 10^{-31} \text{ kg}\right)\left[\left(100. \text{ nm}\right)\left(\dfrac{10^{-9} \text{ m}}{1 \text{ nm}}\right)\right]}\left(\dfrac{\text{kg} \cdot \text{m}^2/\text{s}^2}{\text{J}}\right) = 7273.3 = \textbf{7.27} \times \textbf{10}^3 \text{ m/s}$$

b) Mass (kg) of the golf ball $= (45.9 \text{ g})\left(\dfrac{1 \text{ kg}}{1000 \text{ g}}\right) = 0.0459 \text{ kg}$

$$u = \frac{h}{m\lambda} = \frac{6.626 \times 10^{-34} \text{ J} \cdot \text{s}}{(0.0459 \text{ kg})\left[(100. \text{ nm})\left(\dfrac{10^{-9}}{1 \text{ nm}}\right)\right]}\left(\dfrac{\text{kg} \cdot \text{m}^2/\text{s}^2}{\text{J}}\right) = 1.4436 \times 10^{-25} = \textbf{1.44} \times \textbf{10}^{-25} \text{ m/s}$$

7.4B   Plan: Use the equation for the de Broglie wavelength, $\lambda = h/mu$ with the given mass and speed to calculate the de Broglie wavelength of the racquetball. The mass of the racquetball must be expressed in kg, and the speed must be expressed in m/s in the equation for the de Broglie wavelength.
Solution:

Mass (kg) of the racquetball $= (39.7 \text{ g})\left(\dfrac{1 \text{ kg}}{1000 \text{ g}}\right) = 0.0397 \text{ kg}$

Speed (m/s) of the racquetball $= \left(\dfrac{55 \text{ mi}}{\text{hr}}\right)\left(\dfrac{1 \text{ hr}}{3600 \text{ s}}\right)\left(\dfrac{1.609 \text{ km}}{\text{mi}}\right)\left(\dfrac{1000 \text{ m}}{1 \text{ km}}\right) = 24.5819 = 25 \text{ m/s}$

$\lambda = h/mu$

$$\lambda = \frac{h}{mu} = \frac{6.626 \times 10^{-34} \text{ J} \cdot \text{s}}{(0.0397 \text{ kg})(25 \text{ m/s})} = 6.67607 \times 10^{-34} = \textbf{6.7} \times \textbf{10}^{-34} \text{ m}$$

7.5A   Plan: Use the equation $\Delta x \cdot m\Delta u \geq \dfrac{h}{4\pi}$ to solve for the uncertainty (Δx) in position of the baseball.

Solution:

$$\Delta x \cdot m\Delta u \geq \frac{h}{4\pi}$$

$\Delta u = 1\%$ of $u = 0.0100(44.7 \text{ m/s}) = 0.447 \text{ m/s}$

$$\Delta x \geq \frac{h}{4\pi m\Delta u} = \frac{6.626 \times 10^{-34} \text{ J} \cdot \text{s}}{4\pi (0.142 \text{ kg})(0.447 \text{ m/s})} = 8.3070285 \times 10^{-34}$$

$\Delta x \geq \textbf{8.31} \times \textbf{10}^{-34} \text{ m}$

7.5B   Plan: Use the equation $\Delta x \cdot m\Delta u \geq \dfrac{h}{4\pi}$ to solve for the uncertainty (Δx) in position of the neutron.

Solution:

$$\Delta x \cdot m\Delta u \geq \frac{h}{4\pi}$$

$\Delta u = 1\%$ of $u = 0.0100(8 \times 10^7 \text{ m/s}) = 8 \times 10^5 \text{ m/s}$

$$\Delta x \geq \frac{h}{4\pi m\Delta u} = \frac{6.626 \times 10^{-34} \text{ J} \cdot \text{s}}{4\pi (1.67 \times 10^{-27} \text{ kg})(8 \times 10^5 \text{ m/s})} = 3.9467 \times 10^{-14} \text{ m}$$

$\Delta x \geq \textbf{4} \times \textbf{10}^{-14} \text{ m}$

7.6A   Plan: Following the rules for l (integer from 0 to $n-1$) and m_l (integer from $-l$ to $+l$), write quantum numbers for $n = 4$.
Solution:
For $n = 4$ $l = 0, 1, 2, 3$
For $l = 0$, $m_l = 0$
For $l = 1$, $m_l = -1, 0, 1$
For $l = 2$, $m_l = -2, -1, 0, 1, 2$
For $l = 3$, $m_l = -3, -2, -1, 0, 1, 2, 3$

7.6B   Plan: Following the rules for l (integer from 0 to $n-1$) and m_l (integer from $-l$ to $+l$), determine which value of the principal quantum number has five allowed levels of l.
Solution:
The number of possible l values is equal to n, so the $n = 5$ principal quantum number has five allowed values of l.
For $n = 5$ $l = 0, 1, 2, 3, 4$
For $l = 0$, $m_l = 0$
For $l = 1$, $m_l = -1, 0, 1$
For $l = 2$, $m_l = -2, -1, 0, 1, 2$
For $l = 3$, $m_l = -3, -2, -1, 0, 1, 2, 3$
For $l = 4$, $m_l = -4, -3, -2, -1, 0, 1, 2, 3, 4$

7.7A   Plan: Identify n and l from the sublevel designation. n is the integer in front of the sublevel letter. The sublevels are given a letter designation, in which s represents $l = 0$, p represents $l = 1$, d represents $l = 2$, f represents $l = 3$. Knowing the value for l, find the m_l values (integer from $-l$ to $+l$).
Solution:

Sublevel name	n value	l value	m_l values
2p	2	1	-1, 0, 1
5f	5	3	-3, -2, -1, 0, 1, 2, 3

7.7B   Plan: Identify n and l from the sublevel designation. n is the integer in front of the sublevel letter. The sublevels are given a letter designation, in which s represents $l = 0$, p represents $l = 1$, d represents $l = 2$, f represents $l = 3$. Knowing the value for l, find the m_l values (integer from $-l$ to $+l$).
Solution:

Sublevel name	n value	l value	m_l values
4d	4	2	-2, -1, 0, 1, 2
6s	6	0	0

The number of orbitals for each sublevel equals $2l + 1$. Sublevel 4d should have **5 orbitals** and sublevel 6s should have **1 orbital**. Both of these agree with the number of m_l values for the sublevel.

7.8A   Plan: Use the rules for designating quantum numbers to fill in the blanks.
For a given n, l can be any integer from 0 to $n-1$.
For a given l, m_l can be any integer from $-l$ to $+l$.
The sublevels are given a letter designation, in which s represents $l = 0$, p represents $l = 1$, d represents $l = 2$, f represents $l = 3$.
Solution:
The completed table is:

	n	l	m_l	Name
a)	4	1	0	4p
b)	2	1	0	2p
c)	3	2	-2	3d
d)	2	0	0	2s

7.8B   Plan: Use the rules for designating quantum numbers to determine what is wrong with the quantum number designations provided in the problem.
For a given n, l can be any integer from 0 to $n-1$.
For a given l, m_l can be any integer from $-l$ to $+l$.
The sublevels are given a letter designation, in which s represents $l = 0$, p represents $l = 1$, d represents $l = 2$, f represents $l = 3$.
Solution:
The provided table is:

	n	l	m_l	Name
a)	5	3	4	5f
b)	2	2	1	2d
c)	6	1	−1	6s

a) For $l = 3$, the allowed values for m_l are −3, −2, −1, 0, 1, 2, 3, not 4
b) For n = 2, $l = 0$ or 1 only, not 2; the sublevel is 2p, since $m_l = 1$.
c) The value $l = 1$ indicates the p sublevel, not the s; the sublevel name is 6p.

TOOLS OF THE LABORATORY BOXED READING PROBLEMS

B7.1   Plan: Plot absorbance on the y-axis and concentration on the x-axis. Since this is a linear plot, the graph is of the type $y = mx + b$, with m = slope and b = intercept. Any two points may be used to find the slope, and the slope is used to find the intercept. Once the equation for the line is known, the absorbance of the solution in part b) is used to find the concentration of the diluted solution, after which the dilution equation is used to find the molarity of the original solution.
Solution:
a) Absorbance vs. Concentration:

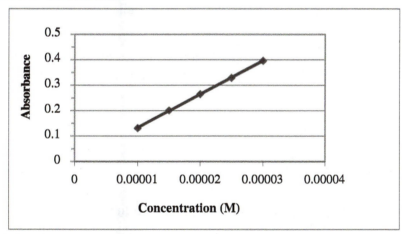

This is a linear plot, thus, using the first and last points given:

$$m = \frac{y_2 - y_1}{x_2 - x_1} = \frac{(0.396 - 0.131)}{(3.0\text{x}10^{-5} - 1.0\text{x}10^{-5})M} = 13{,}250 = \mathbf{1.3\text{x}10^4 M^{-1}}$$

Using the slope just calculated and any of the data points, the value of the intercept may be found.
$b = y - mx = 0.396 - (13{,}250\ M^{-1})(3.0\text{x}10^{-5}\ M) = -0.0015 = \mathbf{0.00}$ (absorbance has no units)
b) Use the equation just determined: $y = (13{,}250\ M^{-1})\ x + 0.00$.
$x = (y - 0.00)/(13{,}250\ M^{-1}) = (0.236/13{,}250\ M^{-1}) = 1.78113\text{x}10^{-5}\ M = \mathbf{1.8\text{x}10^{-5}\ M}$
This value is M_f in a dilution problem $(M_iV_i) = (M_fV_f)$ with $V_i = 20.0$ mL and $V_f = 150.$ mL.

$$M_i = \frac{(M_f)(V_f)}{(V_i)} = \frac{(1.78113\text{x}10^{-5}\ M)(150.\ mL)}{(20.0\ mL)} = 1.335849\text{x}10^{-4} = \mathbf{1.3\text{x}10^{-4}\ M}$$

B7.2   Plan: The color of light associated with each wavelength can be found from Figure 7.3. The frequency of each wavelength can be determined from the relationship $c = \lambda\nu$ or $\nu = \dfrac{c}{\lambda}$. The wavelength in nm must be converted to meters.

Solution:

a) red $\nu = \dfrac{3.00\text{x}10^8 \text{ m/s}}{671 \text{ nm}}\left(\dfrac{1 \text{ nm}}{10^{-9} \text{ m}}\right) = 4.4709\text{x}10^{14} = \mathbf{4.47\text{x}10^{14} \text{ s}^{-1}}$

b) blue $\nu = \dfrac{3.00\text{x}10^8 \text{ m/s}}{453 \text{ nm}}\left(\dfrac{1 \text{ nm}}{10^{-9} \text{ m}}\right) = 6.6225\text{x}10^{14} = \mathbf{6.62\text{x}10^{14} \text{ s}^{-1}}$

c) yellow-orange $\nu = \dfrac{3.00\text{x}10^8 \text{ m/s}}{589 \text{ nm}}\left(\dfrac{1 \text{ nm}}{10^{-9} \text{ m}}\right) = 5.0933786\text{x}10^{14} = \mathbf{5.09\text{x}10^{14} \text{ s}^{-1}}$

END–OF–CHAPTER PROBLEMS

7.2   Plan: Recall that the shorter the wavelength, the higher the frequency and the greater the energy. Figure 7.3 describes the electromagnetic spectrum by wavelength and frequency.

Solution:
a) Wavelength increases from left (10^{-2} nm) to right (10^{12} nm) in Figure 7.3. The trend in increasing wavelength is: **x-ray < ultraviolet < visible < infrared < microwave < radio wave**.
b) Frequency is inversely proportional to wavelength according to the equation $c = \lambda\nu$, so frequency has the opposite trend: **radio wave < microwave < infrared < visible < ultraviolet < x-ray**.
c) Energy is directly proportional to frequency according to the equation $E = h\nu$. Therefore, the trend in increasing energy matches the trend in increasing frequency: **radio wave < microwave < infrared < visible < ultraviolet < x-ray**.

7.7   Plan: Wavelength is related to frequency through the equation $c = \lambda\nu$. Recall that a Hz is a reciprocal second, or $1/s = s^{-1}$. Assume that the number "950" has three significant figures.

Solution:
$c = \lambda\nu$

$\lambda \text{ (m)} = \dfrac{c}{\nu} = \dfrac{3.00\text{x}10^8 \text{ m/s}}{\left(950. \text{ kHz}\right)\left(\dfrac{10^3 \text{ Hz}}{1 \text{ kHz}}\right)\left(\dfrac{\text{s}^{-1}}{\text{Hz}}\right)} = 315.789 = \mathbf{316 \text{ m}}$

$\lambda \text{ (nm)} = \dfrac{c}{\nu} = (315.789 \text{ m})\left(\dfrac{1 \text{ nm}}{10^{-9} \text{ m}}\right) = 3.15789\text{x}10^{11} = \mathbf{3.16\text{x}10^{11} \text{ nm}}$

$\lambda \text{ (Å)} = \dfrac{c}{\nu} = (315.789 \text{ m})\left(\dfrac{1 \text{ Å}}{10^{-10} \text{ m}}\right) = 3.158\text{x}10^{12} = \mathbf{3.16\text{x}10^{12} \text{ Å}}$

7.9   Plan: Frequency is related to energy through the equation $E = h\nu$. Note that $1 \text{ Hz} = 1 \text{ s}^{-1}$.

Solution:
$E = h\nu$
$E = (6.626\text{x}10^{-34} \text{ J}\bullet\text{s})(3.8\text{x}10^{10} \text{ s}^{-1}) = 2.51788\text{x}10^{-23} = \mathbf{2.5\text{x}10^{-23} \text{ J}}$

7.11   Plan: Energy is inversely proportional to wavelength ($E = \dfrac{hc}{\lambda}$). As wavelength decreases, energy increases.

Solution:
In terms of increasing energy the order is **red < yellow < blue**.

7.13   Plan: Wavelength is related to frequency through the equation $c = \lambda\nu$. Recall that a Hz is a reciprocal second, or $1/s = s^{-1}$.
Solution:

$$\nu = (s^{-1}) = \left(22.235 \text{ GHz}\right)\left(\frac{10^9 \text{ Hz}}{1 \text{ GHz}}\right)\left(\frac{s^{-1}}{\text{Hz}}\right) = 2.2235\text{x}10^{10} \text{ s}^{-1}$$

$$\lambda \text{ (nm)} = \frac{c}{\nu} = \frac{2.9979\text{x}10^8 \text{ m/s}}{2.2235\text{x}10^{10} \text{ s}^{-1}}\left(\frac{1 \text{ nm}}{10^{-9} \text{ m}}\right) = 1.3482797\text{x}10^7 = \mathbf{1.3483\text{x}10^7 \text{ nm}}$$

$$\lambda \text{ (Å)} = \frac{c}{\nu} = \frac{2.9979\text{x}10^8 \text{ m/s}}{2.2235\text{x}10^{10} \text{ s}^{-1}}\left(\frac{1 \text{ Å}}{10^{-10} \text{ m}}\right) = 1.3482797\text{x}10^8 = \mathbf{1.3483\text{x}10^8 \text{ Å}}$$

7.16   Plan: The least energetic photon in part a) has the longest wavelength (242 nm). The most energetic photon in part b) has the shortest wavelength (2200 Å). Use the relationship $c = \lambda\nu$ to find the frequency of the photons and relationship $E = \dfrac{hc}{\lambda}$ to find the energy.

Solution:
a) $c = \lambda\nu$

$$\nu = \frac{c}{\lambda} = \frac{3.00\text{x}10^8 \text{ m/s}}{242 \text{ nm}}\left(\frac{1 \text{ nm}}{10^{-9} \text{ m}}\right) = 1.239669\text{x}10^{15} = \mathbf{1.24\text{x}10^{15} \text{ s}^{-1}}$$

$$E = \frac{hc}{\lambda} = \frac{\left(6.626\text{x}10^{-34} \text{ J} \cdot \text{s}\right)\left(3.00\text{x}10^8 \text{ m/s}\right)}{242 \text{ nm}}\left(\frac{1 \text{ nm}}{10^{-9} \text{ m}}\right) = 8.2140\text{x}10^{-19} = \mathbf{8.21\text{x}10^{-19} \text{ J}}$$

b) $\nu = \dfrac{c}{\lambda} = \dfrac{3.00\text{x}10^8 \text{ m/s}}{2200 \text{ Å}}\left(\dfrac{1 \text{ Å}}{10^{-10} \text{ m}}\right) = 1.3636\text{x}10^{15} = \mathbf{1.4\text{x}10^{15} \text{ s}^{-1}}$

$$E = \frac{hc}{\lambda} = \frac{\left(6.626\text{x}10^{-34} \text{ J} \cdot \text{s}\right)\left(3.00\text{x}10^8 \text{ m/s}\right)}{2200 \text{ Å}}\left(\frac{1 \text{ Å}}{10^{-10} \text{ m}}\right) = 9.03545\text{x}10^{-19} = \mathbf{9.0\text{x}10^{-19} \text{ J}}$$

7.18   Bohr's key assumption was that the electron in an atom does not radiate energy while in a stationary state, and the electron can move to a different orbit by absorbing or emitting a photon whose energy is equal to the difference in energy between two states. These differences in energy correspond to the wavelengths in the known spectra for the hydrogen atoms. A Solar System model does not allow for the movement of electrons between levels.

7.20   Plan: The quantum number n is related to the energy level of the electron. An electron *absorbs* energy to change from lower energy (lower n) to higher energy (higher n), giving an absorption spectrum. An electron *emits* energy as it drops from a higher energy level (higher n) to a lower one (lower n), giving an emission spectrum.
Solution:
a) The electron is moving from a lower value of n (2) to a higher value of n (4): **absorption**
b) The electron is moving from a higher value of n (3) to a lower value of n (1): **emission**
c) The electron is moving from a higher value of n (5) to a lower value of n (2): **emission**
d) The electron is moving from a lower value of n (3) to a higher value of n (4): **absorption**

7.22 The Bohr model has successfully predicted the line spectra for the H atom and Be^{3+} ion since both are one-electron

species. The energies could be predicted from $E_n = \dfrac{-(Z^2)(2.18\times10^{-18}\ J)}{n^2}$ where Z is the atomic number for the

atom or ion. The line spectra for H would not match the line spectra for Be^{3+} since the H nucleus contains one proton while the Be^{3+} nucleus contains 4 protons (the Z values in the equation do not match); the force of attraction of the nucleus for the electron would be greater in the beryllium ion than in the hydrogen atom. This means that the pattern of lines would be similar, but at different wavelengths.

7.23 Plan: Calculate wavelength by substituting the given values into Equation 7.3, where $n_1 = 2$ and $n_2 = 5$ because $n_2 > n_1$. Although more significant figures could be used, five significant figures are adequate for this calculation.
Solution:

$$\frac{1}{\lambda} = R\left(\frac{1}{n_1^2} - \frac{1}{n_2^2}\right) \qquad R = 1.096776\times10^7\ m^{-1}$$

$n_1 = 2 \quad n_2 = 5$

$$\frac{1}{\lambda} = R\left(\frac{1}{n_1^2} - \frac{1}{n_2^2}\right) = \left(1.096776\times10^7\ m^{-1}\right)\left(\frac{1}{2^2} - \frac{1}{5^2}\right) = 2{,}303{,}229.6\ m^{-1}$$

$$\lambda\ (nm) = \left(\frac{1}{2{,}303{,}229.6\ m^{-1}}\right)\left(\frac{1\ nm}{10^{-9}\ m}\right) = 434.1729544 = \mathbf{434.17\ nm}$$

7.25 Plan: The Rydberg equation is needed. For the infrared series of the H atom, n_1 equals 3. The least energetic spectral line in this series would represent an electron moving from the next highest energy level, $n_2 = 4$. Although more significant figures could be used, five significant figures are adequate for this calculation.
Solution:

$$\frac{1}{\lambda} = R\left(\frac{1}{n_1^2} - \frac{1}{n_2^2}\right) = \left(1.096776\times10^7\ m^{-1}\right)\left(\frac{1}{3^2} - \frac{1}{4^2}\right) = 533{,}155\ m^{-1}$$

$$\lambda\ (nm) = \left(\frac{1}{533{,}155\ m^{-1}}\right)\left(\frac{1\ nm}{10^{-9}\ m}\right) = 1875.627 = \mathbf{1875.6\ nm}$$

7.27 Plan: To find the transition energy, use the equation for the energy of an electron transition and multiply by Avogadro's number to convert to energy per mole.
Solution:

$$\Delta E = \left(-2.18\times10^{-18}\ J\right)\left(\frac{1}{n_{final}^2} - \frac{1}{n_{initial}^2}\right)$$

$$\Delta E = \left(-2.18\times10^{-18}\ J\right)\left(\frac{1}{2^2} - \frac{1}{5^2}\right) = -4.578\times10^{-19}\ J/photon$$

$$\Delta E = \left(\frac{-4.578\times10^{-19}\ J}{photon}\right)\left(\frac{6.022\times10^{23}\ photons}{1\ mol}\right) = -2.75687\times10^5 = \mathbf{-2.76\times10^5\ J/mol}$$

The energy has a negative value since this electron transition to a lower n value is an emission of energy.

7.29 Plan: Determine the relative energy of the electron transitions. Remember that energy is directly proportional to frequency ($E = h\nu$).

Solution:
Looking at an energy chart will help answer this question.

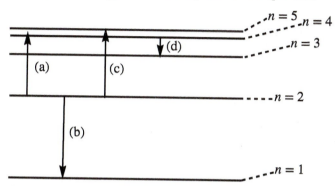

Frequency is proportional to energy so the smallest frequency will be d) $n = 4$ to $n = 3$; levels 3 and 4 have a smaller ΔE than the levels in the other transitions. The largest frequency is b) $n = 2$ to $n = 1$ since levels 1 and 2 have a larger ΔE than the levels in the other transitions. Transition a) $n = 2$ to $n = 4$ will be smaller than transition c) $n = 2$ to $n = 5$ since level 5 is a higher energy than level 4. In order of increasing frequency the transitions are **d < a < c < b.**

7.31 **Plan:** Use the Rydberg equation. Since the electron is in the ground state (lowest energy level), $n_1 = 1$. Convert the wavelength from nm to units of meters.
Solution:

$$\lambda = (97.20 \text{ nm})\left(\frac{10^{-9} \text{ m}}{1 \text{ nm}}\right) = 9.720 \times 10^{-8} \text{ m} \qquad \text{ground state: } n_1 = 1; \quad n_2 = ?$$

$$\frac{1}{\lambda} = \left(1.096776 \times 10^7 \text{ m}^{-1}\right)\left(\frac{1}{n_1^2} - \frac{1}{n_2^2}\right)$$

$$\frac{1}{9.720 \times 10^{-8} \text{ m}} = \left(1.096776 \times 10^7 \text{ m}^{-1}\right)\left(\frac{1}{1^2} - \frac{1}{n_2^2}\right)$$

$$0.93803 = \left(\frac{1}{1^2} - \frac{1}{n_2^2}\right)$$

$$\frac{1}{n_2^2} = 1 - 0.93803 = 0.06197$$

$$n_2^2 = 16.14$$

$$n_2 = 4$$

7.37 Macroscopic objects have significant mass. A large m in the denominator of $\lambda = h/mu$ will result in a very small wavelength. Macroscopic objects do exhibit a wavelike motion, but the wavelength is too small for humans to see it.

7.39 **Plan:** Use the de Broglie equation. Mass in lb must be converted to kg and velocity in mi/h must be converted to m/s because a joule is equivalent to kg•m²/s².
Solution:

a) Mass (kg) $= (232 \text{ lb})\left(\frac{1 \text{ kg}}{2.205 \text{ lb}}\right) = 105.2154 \text{ kg}$

Velocity (m/s) $= \left(\frac{19.8 \text{ mi}}{\text{h}}\right)\left(\frac{1 \text{ km}}{0.62 \text{ mi}}\right)\left(\frac{10^3 \text{ m}}{1 \text{ km}}\right)\left(\frac{1 \text{ h}}{3600 \text{ s}}\right) = 8.87097 \text{ m/s}$

$$\lambda = \frac{h}{mu} = \frac{(6.626\text{x}10^{-34}\,\text{J}\cdot\text{s})}{(105.2154\text{ kg})\left(8.87097\,\dfrac{\text{m}}{\text{s}}\right)}\left(\frac{\text{kg}\cdot\text{m}^2/\text{s}^2}{\text{J}}\right) = 7.099063\text{x}10^{-37} = \mathbf{7.10\text{x}10^{-37}\ m}$$

b) Uncertainty in velocity (m/s) = $\left(\dfrac{0.1\text{ mi}}{\text{h}}\right)\left(\dfrac{1\text{ km}}{0.62\text{ mi}}\right)\left(\dfrac{10^3\text{ m}}{1\text{ km}}\right)\left(\dfrac{1\text{ h}}{3600\text{ s}}\right) = 0.0448029$ m/s

$$\Delta x \bullet m\Delta v \geq \frac{h}{4\pi}$$

$$\Delta x \geq \frac{h}{4\pi m\Delta v} \geq \frac{(6.626\text{x}10^{-34}\,\text{J}\cdot\text{s})}{4\pi(105.2154\text{ kg})\left(\dfrac{0.0448029\text{ m}}{\text{s}}\right)}\left(\frac{\text{kg}\cdot\text{m}^2/\text{s}^2}{\text{J}}\right) \geq 1.11855\text{x}10^{-35} \geq \mathbf{1\text{x}10^{-35}\ m}$$

7.41   Plan: Use the de Broglie equation. Mass in g must be converted to kg and wavelength in Å must be converted to m because a joule is equivalent to kg•m²/s².
Solution:

Mass (kg) = $(56.5\text{ g})\left(\dfrac{1\text{ kg}}{10^3\text{ g}}\right) = 0.0565$ kg

Wavelength (m) = $\left(5400\text{ Å}\right)\left(\dfrac{10^{-10}\text{ m}}{1\text{ Å}}\right) = 5.4\text{x}10^{-7}$ m

$$\lambda = \frac{h}{mu}$$

$$u = \frac{h}{m\lambda} = \frac{(6.626\text{x}10^{-34}\,\text{J}\cdot\text{s})}{(0.0565\text{ kg})(5.4\text{x}10^{-7}\text{ m})}\left(\frac{\text{kg}\cdot\text{m}^2/\text{s}^2}{\text{J}}\right) = 2.1717\text{x}10^{-26} = \mathbf{2.2\text{x}10^{-26}\ m/s}$$

7.43   Plan: The de Broglie wavelength equation will give the mass equivalent of a photon with known wavelength and velocity. The term "mass equivalent" is used instead of "mass of photon" because photons are quanta of electromagnetic energy that have no mass. A light photon's velocity is the speed of light, 3.00x10⁸ m/s. Wavelength in nm must be converted to m.
Solution:

Wavelength (m) = $(589\text{ nm})\left(\dfrac{10^{-9}\text{ m}}{1\text{ nm}}\right) = 5.89\text{x}10^{-7}$ m

$$\lambda = \frac{h}{mu}$$

$$m = \frac{h}{\lambda u} = \frac{(6.626\text{x}10^{-34}\,\text{J}\cdot\text{s})}{(5.89\text{x}10^{-7}\text{ m})(3.00\text{x}10^8\text{ m/s})}\left(\frac{\text{kg}\cdot\text{m}^2/\text{s}^2}{\text{J}}\right) = 3.7499\text{x}10^{-36} = \mathbf{3.75\text{x}10^{-36}\ kg/photon}$$

7.47   A peak in the radial probability distribution at a certain distance means that the total probability of finding the electron is greatest within a thin spherical volume having a radius very close to that distance. Since principal quantum number (n) correlates with distance from the nucleus, the peak for $n = 2$ would occur at a greater distance from the nucleus than 0.529 Å. Thus, the probability of finding an electron at 0.529 Å is much greater for the 1s orbital than for the 2s.

7.48   a) Principal quantum number, n, relates to the size of the orbital. More specifically, it relates to the distance from the nucleus at which the probability of finding an electron is greatest. This distance is determined by the energy of the electron.

b) Angular momentum quantum number, l, relates to the shape of the orbital. It is also called the azimuthal quantum number.
c) Magnetic quantum number, m_l, relates to the orientation of the orbital in space in three-dimensional space.

7.49   Plan: The following letter designations correlate with the following l quantum numbers: $l = 0 = s$ orbital; $l = 1 = p$ orbital; $l = 2 = d$ orbital; $l = 3 = f$ orbital. Remember that allowed m_l values are $-l$ to $+l$. The number of orbitals of a particular type is given by the number of possible m_l values.
Solution:
a) There is only a single s orbital in any shell. $l = 1$ and $m_l = 0$: one value of $m_l = $ **one** s orbital.
b) There are five d orbitals in any shell. $l = 2$ and $m_l = -2, -1, 0, +1, +2$. Five values of $m_l = $ **five** d orbitals.
c) There are three p orbitals in any shell. $l = 1$ and $m_l = -1, 0, +1$. Three values of $m_l = $ **three** p orbitals.
d) If $n = 3$, $l = 0(s)$, $1(p)$, and $2(d)$. There is a $3s$ (1 orbital), a $3p$ set (3 orbitals), and a $3d$ set (5 orbitals) for a total of **nine** orbitals $(1 + 3 + 5 = 9)$.

7.51   Plan: Magnetic quantum numbers (m_l) can have integer values from $-l$ to $+l$. The l quantum number can have integer values from 0 to $n - 1$.
Solution:
a) $l = 2$ so $m_l = -2, -1, 0, +1, +2$
b) $n = 1$ so $l = 1 - 1 = 0$ and $m_l = 0$
c) $l = 3$ so $m_l = -3, -2, -1, 0, +1, +2, +3$

7.53   Plan: The s orbital is spherical; p orbitals have two lobes; the subscript x indicates that this orbital lies along the x-axis.
Solution:
a) s: spherical

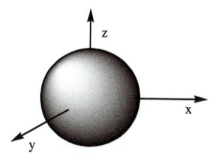

b) p_x: 2 lobes along the x-axis

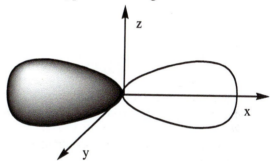

The variations in coloring of the p orbital are a consequence of the quantum mechanical derivation of atomic orbitals that are beyond the scope of this course.

7.55   Plan: The following letter designations for the various sublevels (orbitals) correlate with the following l quantum numbers: $l = 0 = s$ orbital; $l = 1 = p$ orbital; $l = 2 = d$ orbital; $l = 3 = f$ orbital. Remember that allowed m_l values are $-l$ to $+l$. The number of orbitals of a particular type is given by the number of possible m_l values.
Solution:

sublevel	allowable m_l	# of possible orbitals
a) d ($l = 2$)	$-2, -1, 0, +1, +2$	5
b) p ($l = 1$)	$-1, 0, +1$	3
c) f ($l = 3$)	$-3, -2, -1, 0, +1, +2, +3$	7

7.57   Plan: The integer in front of the letter represents the n value. The letter designates the l value: $l = 0 = s$ orbital; $l = 1 = p$ orbital; $l = 2 = d$ orbital; $l = 3 = f$ orbital. Remember that allowed m_l values are $-l$ to $+l$.
Solution:
a) For the $5s$ subshell, $n = 5$ and $l = 0$. Since $m_l = 0$, there is **one** orbital.
b) For the $3p$ subshell, $n = 3$ and $l = 1$. Since $m_l = -1, 0, +1$, there are **three** orbitals.
c) For the $4f$ subshell, $n = 4$ and $l = 3$. Since $m_l = -3, -2, -1, 0, +1, +2, +3$, there are **seven** orbitals.

7.59   Plan: Allowed values of quantum numbers: n = positive integers; l = integers from 0 to $n-1$; m_l = integers from $-l$ through 0 to $+l$.

Solution:

a) $n = 2$; $l = 0$; $m_l = -1$: With $n = 2$, l can be 0 or 1; with $l = 0$, the only allowable m_l value is 0. This combination is not allowed. To correct, either change the l or m_l value.

Correct: $n = 2$; $l = 1$; $m_l = -1$ or $n = 2$; $l = 0$; $m_l = 0$.

b) $n = 4$; $l = 3$; $m_l = -1$: With $n = 4$, l can be 0, 1, 2, or 3; with $l = 3$, the allowable m_l values are -3, -2, -1, 0, $+1$, $+2$, $+3$. Combination is allowed.

c) $n = 3$; $l = 1$; $m_l = 0$: With $n = 3$, l can be 0, 1, or 2; with $l = 1$, the allowable m_l values are -1, 0, $+1$. Combination is allowed.

d) $n = 5$; $l = 2$; $m_l = +3$: With $n = 5$, l can be 0, 1, 2, 3, or 4; with $l = 2$, the allowable m_l values are -2, -1, 0, $+1$, $+2$. $+3$ is not an allowable m_l value. To correct, either change l or m_l value.

Correct: $n = 5$; $l = 3$; $m_l = +3$ or $n = 5$; $l = 2$; $m_l = 0$.

7.62   Plan: For part a, use the values of the constants h, π, m_e, and a_0 to find the overall constant in the equation. Use the resulting equation to calculate ΔE in part b). Use the relationship $E = \dfrac{hc}{\lambda}$ to calculate the wavelength in part c). Remember that a joule is equivalent to $kg \bullet m^2/s^2$.

Solution:

a) $h = 6.626 \times 10^{-34}$ J•s; $m_e = 9.1094 \times 10^{-31}$ kg; $a_0 = 52.92 \times 10^{-12}$ m

$$E = -\frac{h^2}{8\pi^2 m_e a_0^2 n^2} = -\frac{h^2}{8\pi^2 m_e a_0^2}\left(\frac{1}{n^2}\right)$$

$$E = -\frac{\left(6.626 \times 10^{-34} \text{ J} \bullet \text{s}\right)^2}{8\pi^2 \left(9.1094 \times 10^{-31} \text{ kg}\right)\left(52.92 \times 10^{-12} \text{ m}\right)^2}\left(\frac{\text{kg} \bullet \text{m}^2/\text{s}^2}{\text{J}}\right)\left(\frac{1}{n^2}\right)$$

$$= -(2.17963 \times 10^{-18} \text{ J})\left(\frac{1}{n^2}\right) = -(2.180 \times 10^{-18} \text{ J})\left(\frac{1}{n^2}\right)$$

This is identical with the result from Bohr's theory. For the H atom, $Z = 1$ and Bohr's constant $= -2.18 \times 10^{-18}$ J. For the hydrogen atom, derivation using classical principles or quantum-mechanical principles yields the same constant.

b) The $n = 3$ energy level is higher in energy than the $n = 2$ level. Because the zero point of the atom's energy is defined as an electron's infinite distance from the nucleus, a larger negative number describes a lower energy level. Although this may be confusing, it makes sense that an energy *change* would be a positive number.

$$\Delta E = -(2.17963 \times 10^{-18} \text{ J})\left(\frac{1}{2^2} - \frac{1}{3^2}\right) = -3.02726388 \times 10^{-19} = \mathbf{3.027 \times 10^{-19} \text{ J}}$$

c) $E = \dfrac{hc}{\lambda}$

$$\lambda \text{ (m)} = \frac{hc}{E} = \frac{\left(6.626 \times 10^{-34} \text{ J} \bullet \text{s}\right)\left(2.9979 \times 10^8 \text{m/s}\right)}{\left(3.02726388 \times 10^{-19} \text{J}\right)} = 6.56173 \times 10^{-7} = \mathbf{6.562 \times 10^{-7} \text{ m}}$$

$$\lambda \text{ (nm)} = \left(6.56173 \times 10^{-7} \text{ m}\right)\left(\frac{1 \text{ nm}}{10^{-9} \text{ m}}\right) = 656.173 = \mathbf{656.2 \text{ nm}}$$

This is the wavelength for the observed red line in the hydrogen spectrum.

7.63   Plan: When light of sufficient frequency (energy) shines on metal, electrons in the metal break free and a current flows.

Solution:

a) The lines do not begin at the origin because an electron must absorb a minimum amount of energy before it has enough energy to overcome the attraction of the nucleus and leave the atom. This minimum energy is the energy of photons of light at the threshold frequency.

b) The lines for K and Ag do not begin at the same point. The amount of energy that an electron must absorb to leave the K atom is less than the amount of energy that an electron must absorb to leave the Ag atom, where the attraction between the nucleus and outer electron is stronger than in a K atom.

c) Wavelength is inversely proportional to energy. Thus, the metal that requires a larger amount of energy to be absorbed before electrons are emitted will require a shorter wavelength of light. Electrons in Ag atoms require more energy to leave, so Ag requires a shorter wavelength of light than K to eject an electron.

d) The slopes of the line show an increase in kinetic energy as the frequency (or energy) of light is increased. Since the slopes are the same, this means that for an increase of one unit of frequency (or energy) of light, the increase in kinetic energy of an electron ejected from K is the same as the increase in the kinetic energy of an electron ejected from Ag. After an electron is ejected, the energy that it absorbs above the threshold energy becomes the kinetic energy of the electron. For the same increase in energy above the threshold energy, for either K or Ag, the kinetic energy of the ejected electron will be the same.

7.66   Plan: The Bohr model has been successfully applied to predict the spectral lines for one-electron species other than H. Common one-electron species are small cations with all but one electron removed. Since the problem specifies a metal ion, assume that the possible choices are Li^{2+} or Be^{3+}. Use the relationship $E = h\nu$ to convert the frequency to energy and then solve Bohr's equation $E = (2.18 \times 10^{-18}\ J)\left(\dfrac{Z^2}{n^2}\right)$ to verify if a whole number for Z can be calculated. Recall that the negative sign is a convention based on the zero point of the atom's energy; it is deleted in this calculation to avoid taking the square root of a negative number.

Solution:

The highest energy line corresponds to the transition from $n = 1$ to $n = \infty$.

$E = h\nu = (6.626 \times 10^{-34}\ J\bullet s)\,(2.961 \times 10^{16}\ Hz)\,(s^{-1}/Hz) = 1.9619586 \times 10^{-17}\ J$

$E = (2.18 \times 10^{-18}\ J)\left(\dfrac{Z^2}{n^2}\right)$ $Z = $ charge of the nucleus

$Z^2 = \dfrac{En^2}{2.18 \times 10^{-18}\ J} = \dfrac{1.9619586 \times 10^{-17}(1^2)}{2.18 \times 10^{-18}\ J} = 8.99998$

Then $Z^2 = 9$ and $Z = 3$.

Therefore, the ion is Li^{2+} with an atomic number of 3.

7.68   Plan: The electromagnetic spectrum shows that the visible region goes from 400 to 750 nm (4000 Å to 7500 Å). Thus, wavelengths b, c, and d are for the three transitions in the visible series with $n_{final} = 2$. Wavelength a is in the ultraviolet region of the spectrum and the ultraviolet series has $n_{final} = 1$. Wavelength e is in the infrared region of the spectrum and the infrared series has $n_{final} = 3$. Use the Rydberg equation to find the $n_{initial}$ for each line. Convert the wavelengths from Å to units of m.

Solution:

$n = ? \rightarrow n = 1$; $\lambda = 1212.7$ Å   (shortest λ corresponds to the largest ΔE)

$\lambda\ (m) = (1212.7\ \text{Å})\left(\dfrac{10^{-10}\ m}{1\ \text{Å}}\right) = 1.2127 \times 10^{-7}\ m$

$\dfrac{1}{\lambda} = (1.096776 \times 10^7\ m^{-1})\left(\dfrac{1}{n_1^2} - \dfrac{1}{n_2^2}\right)$

$\left(\dfrac{1}{1.2127 \times 10^{-7}\ m}\right) = (1.096776 \times 10^7\ m^{-1})\left(\dfrac{1}{1^2} - \dfrac{1}{n_2^2}\right)$

$0.7518456 = \left(\dfrac{1}{1^2} - \dfrac{1}{n_2^2}\right)$

$\left(\dfrac{1}{n_2^2}\right) = 1 - 0.7518456$

$$\left(\frac{1}{n_2^2}\right) = 0.2481544$$

$$n_2^2 = 4.029749$$

$n_2 = 2$ for line (a) ($n = 2 \rightarrow n = 1$)

$n = ? \rightarrow n = 3$; $\lambda = 10{,}938$ Å (longest λ corresponds to the smallest ΔE)

$$\lambda \text{ (m)} = \left(10{,}938 \text{ Å}\right)\left(\frac{10^{-10} \text{ m}}{1 \text{ Å}}\right) = 1.0938 \times 10^{-6} \text{ m}$$

$$\frac{1}{\lambda} = \left(1.096776 \times 10^7 \text{ m}^{-1}\right)\left(\frac{1}{n_1^2} - \frac{1}{n_2^2}\right)$$

$$\left(\frac{1}{1.0938 \times 10^{-6} \text{ m}}\right) = \left(1.096776 \times 10^7 \text{ m}^{-1}\right)\left(\frac{1}{3^2} - \frac{1}{n_2^2}\right)$$

$$0.083357397 = \left(\frac{1}{3^2} - \frac{1}{n_2^2}\right)$$

$$0.083357397 = 0.111111111 - \frac{1}{n_2^2}$$

$$\left(\frac{1}{n_2^2}\right) = 0.11111111 - 0.083357396$$

$$\left(\frac{1}{n_2^2}\right) = 0.0277537151$$

$$n_2^2 = 36.03121$$

$n_2 = 6$ for line (e) ($n = 6 \rightarrow n = 3$)

For the other three lines, $n_1 = 2$.

For line (d), $n_2 = 3$ (largest $\lambda \rightarrow$ smallest ΔE).

For line (b), $n_2 = 5$ (smallest $\lambda \rightarrow$ largest ΔE).

For line (c), $n_2 = 4$.

7.72   Plan: Allowed values of quantum numbers: n = positive integers; l = integers from 0 to $n - 1$; m_l = integers from $-l$ through 0 to $+l$.
Solution:
a) The l value must be at least 1 for m_l to be -1, but cannot be greater than $n - 1 = 3 - 1 = 2$. Increase the l value to 1 or 2 to create an allowable combination.
b) The l value must be at least 1 for m_l to be $+1$, but cannot be greater than $n - 1 = 3 - 1 = 2$. Decrease the l value to 1 or 2 to create an allowable combination.
c) The l value must be at least 3 for m_l to be $+3$, but cannot be greater than $n - 1 = 7 - 1 = 6$. Increase the l value to 3, 4, 5, or 6 to create an allowable combination.
d) The l value must be at least 2 for m_l to be -2, but cannot be greater than $n - 1 = 4 - 1 = 3$. Increase the l value to 2 or 3 to create an allowable combination.

7.74   Plan: Ionization occurs when the electron is completely removed from the atom, or when $n_{final} = \infty$. We can use the equation for the energy of an electron transition to find the quantity of energy needed to remove completely the electron, called the ionization energy (IE). To obtain the ionization energy per mole of species, multiply by Avogadro's number. The charge on the nucleus must affect the IE because a larger nucleus would exert a greater pull on the escaping electron. The Bohr equation applies to H and other one-electron species. Use the expression to determine the ionization energy of B^{4+}. Then use the expression to find the energies of the transitions listed and use $E = \dfrac{hc}{\lambda}$ to convert energy to wavelength.

Solution:

a) $E = \left(-2.18\text{x}10^{-18}\text{ J}\right)\left(\dfrac{Z^2}{n^2}\right)$ $Z =$ atomic number

$\Delta E = \left(-2.18\text{x}10^{-18}\text{ J}\right)\left(\dfrac{1}{n_{\text{final}}^2} - \dfrac{1}{n_{\text{initial}}^2}\right)Z^2$

$\Delta E = \left(-2.18\text{x}10^{-18}\text{ J}\right)\left(\dfrac{1}{\infty^2} - \dfrac{1}{n_{\text{initial}}^2}\right)Z^2\left(\dfrac{6.022\text{x}10^{23}}{1\text{ mol}}\right)$

$\qquad = (1.312796\text{x}10^6)\ Z^2$ for $n = 1$

b) In the ground state $n = 1$, the initial energy level for the single electron in B^{4+}. Once ionized, $n = \infty$ is the final energy level.
$Z = 5$ for B^{4+}.

$\Delta E = \text{IE} = (1.312796\text{x}10^6)\ Z^2 = (1.312796\text{x}10^6\text{ J/mol})(5^2) = 3.28199\text{x}10^7 = \mathbf{3.28\text{x}10^7\text{ J/mol}}$

c) $n_{\text{final}} = \infty$, $n_{\text{initial}} = 3$, and $Z = 2$ for He^+.

$\Delta E = \left(-2.18\text{x}10^{-18}\text{J}\right)\left(\dfrac{1}{n_{\text{final}}^2} - \dfrac{1}{n_{\text{initial}}^2}\right)Z^2 = \left(-2.18\text{x}10^{-18}\text{J}\right)\left(\dfrac{1}{\infty^2} - \dfrac{1}{3^2}\right)2^2 = 9.68889\text{x}10^{-19}\text{ J}$

$E = \dfrac{hc}{\lambda}$

$\lambda \text{ (m)} = \dfrac{hc}{E} = \dfrac{\left(6.626\text{x}10^{-34}\text{ J}\bullet\text{s}\right)\left(3.00\text{x}10^8\text{ m/s}\right)}{9.68889\text{x}10^{-19}\text{ J}} = 2.051628\text{x}10^{-7}\text{ m}$

$\lambda \text{ (nm)} = \left(2.051628\text{x}10^{-7}\text{ m}\right)\left(\dfrac{1\text{ nm}}{10^{-9}\text{ m}}\right) = 205.1628 = \mathbf{205\text{ nm}}$

d) $n_{\text{final}} = \infty$, $n_{\text{initial}} = 2$, and $Z = 4$ for Be^{3+}.

$\Delta E = \left(-2.18\text{x}10^{-18}\text{ J}\right)\left(\dfrac{1}{\infty^2} - \dfrac{1}{n_{\text{initial}}^2}\right)Z^2 = \left(-2.18\text{x}10^{-18}\text{J}\right)\left(\dfrac{1}{\infty^2} - \dfrac{1}{2^2}\right)4^2 = 8.72\text{x}10^{-18}\text{ J}$

$\lambda \text{ (m)} = \dfrac{hc}{E} = \dfrac{\left(6.626\text{x}10^{-34}\text{ J}\bullet\text{s}\right)\left(3.00\text{x}10^8\text{ m/s}\right)}{8.72\text{x}10^{-18}\text{ J}} = 2.279587\text{x}10^{-8}\text{ m}$

$\lambda \text{ (nm)} = \left(2.279587\text{x}10^{-8}\text{ m}\right)\left(\dfrac{1\text{ nm}}{10^{-9}\text{ m}}\right) = 22.79587 = \mathbf{22.8\text{ nm}}$

7.76   Plan: Use the values and the equation given in the problem to calculate the appropriate values.
Solution:

a) $r_n = \dfrac{n^2 h^2 \varepsilon_0}{\pi m_e e^2}$

$r_1 = \dfrac{1^2\left(6.626\text{x}10^{-34}\text{ J}\bullet\text{s}\right)^2\left(8.854\text{x}10^{-12}\ \dfrac{C^2}{\text{J}\bullet\text{m}}\right)}{\pi\left(9.109\text{x}10^{-31}\text{ kg}\right)\left(1.602\text{x}10^{-19}\text{ C}\right)^2}\left(\dfrac{\text{kg}\bullet\text{m}^2/\text{s}^2}{\text{J}}\right) = 5.2929377\text{x}10^{-11} = \mathbf{5.293\text{x}10^{-11}\text{ m}}$

b) $r_{10} = \dfrac{10^2 \left(6.626\text{x}10^{-34}\ \text{J}\bullet\text{s}\right)^2 \left(8.854\text{x}10^{-12}\ \dfrac{\text{C}^2}{\text{J}\bullet\text{m}}\right)}{\pi \left(9.109\text{x}10^{-31}\ \text{kg}\right)\left(1.602\text{x}10^{-19}\ \text{C}\right)^2} \left(\dfrac{\text{kg}\bullet\text{m}^2/\text{s}^2}{\text{J}}\right) = 5.2929377\text{x}10^{-9} = \mathbf{5.293\text{x}10^{-9}\ m}$

7.78   <u>Plan:</u> Refer to Chapter 6 for the calculation of the amount of heat energy absorbed by a substance from its specific heat capacity and temperature change ($q = c$ x mass x ΔT). Using this equation, calculate the energy absorbed by the water. This energy equals the energy from the microwave photons. The energy of each photon can be calculated from its wavelength: $E = hc/\lambda$. Dividing the total energy by the energy of each photon gives the number of photons absorbed by the water.
<u>Solution:</u>
$q = c$ x mass x ΔT
$q = (4.184\ \text{J/g}°\text{C})(252\ \text{g})(98 - 20.)°\text{C} = 8.22407\text{x}10^4\ \text{J}$

$E = \dfrac{hc}{\lambda} = \dfrac{\left(6.626\text{x}10^{-34}\ \text{J}\bullet\text{s}\right)\left(3.00\text{x}10^8\ \text{m/s}\right)}{1.55\text{x}10^{-2}\ \text{m}} = 1.28245\text{x}10^{-23}\ \text{J/photon}$

Number of photons $= \left(8.22407\text{x}10^4\ \text{J}\right)\left(\dfrac{1\ \text{photon}}{1.28245\text{x}10^{-23}\ \text{J}}\right) = 6.41278\text{x}10^{27} = \mathbf{6.4\text{x}10^{27}\ photons}$

7.80   <u>Plan:</u> In general, to test for overlap of the two series, compare the longest wavelength in the "n" series with the shortest wavelength in the "$n+1$" series. The longest wavelength in any series corresponds to the transition between the n_1 level and the next level above it; the shortest wavelength corresponds to the transition between the n_1 level and the $n = \infty$ level. Use the relationship $\dfrac{1}{\lambda} = R\left(\dfrac{1}{n_1^2} - \dfrac{1}{n_2^2}\right)$ to calculate the wavelengths.

<u>Solution:</u>
$\dfrac{1}{\lambda} = R\left(\dfrac{1}{n_1^2} - \dfrac{1}{n_2^2}\right) = \left(1.096776\text{x}10^7\ \text{m}^{-1}\right)\left(\dfrac{1}{n_1^2} - \dfrac{1}{n_2^2}\right)$

a) The overlap between the $n_1 = 1$ series and the $n_1 = 2$ series would occur between the longest wavelengths for $n_1 = 1$ and the shortest wavelengths for $n_1 = 2$.
Longest wavelength in $n_1 = 1$ series has n_2 equal to 2.
$\dfrac{1}{\lambda} = \left(1.096776\text{x}10^7\ \text{m}^{-1}\right)\left(\dfrac{1}{1^2} - \dfrac{1}{2^2}\right) = 8{,}225{,}820\ \text{m}^{-1}$

$\lambda = \dfrac{1}{8{,}225{,}820\ \text{m}^{-1}} = 1.215684272\text{x}10^{-7} = \mathbf{1.215684\text{x}10^{-7}\ m}$
Shortest wavelength in the $n_1 = 2$ series:
$\dfrac{1}{\lambda} = \left(1.096776\text{x}10^7\ \text{m}^{-1}\right)\left(\dfrac{1}{2^2} - \dfrac{1}{\infty^2}\right) = 2{,}741{,}940\ \text{m}^{-1}$

$\lambda = \dfrac{1}{2{,}741{,}940\ \text{m}^{-1}} = 3.647052817\text{x}10^{-7} = \mathbf{3.647053\text{x}10^{-7}\ m}$
Since the longest wavelength for $n_1 = 1$ series is shorter than shortest wavelength for $n_1 = 2$ series, there is **no overlap** between the two series.
b) The overlap between the $n_1 = 3$ series and the $n_1 = 4$ series would occur between the longest wavelengths for $n_1 = 3$ and the shortest wavelengths for $n_1 = 4$.
Longest wavelength in $n_1 = 3$ series has n_2 equal to 4.
$\dfrac{1}{\lambda} = \left(1.096776\text{x}10^7\ \text{m}^{-1}\right)\left(\dfrac{1}{3^2} - \dfrac{1}{4^2}\right) = 533{,}155\ \text{m}^{-1}$

$$\lambda = \frac{1}{533,155 \text{ m}^{-1}} = 1.875627163\text{x}10^{-6} = \textbf{1.875627x10}^{-6} \textbf{ m}$$

Shortest wavelength in $n_1 = 4$ series has $n_2 = \infty$.

$$\frac{1}{\lambda} = \left(1.096776\text{x}10^{7} \text{ m}^{-1}\right)\left(\frac{1}{4^2} - \frac{1}{\infty^2}\right) = 685,485 \text{ m}^{-1}$$

$$\lambda = \frac{1}{685,485 \text{ m}^{-1}} = 1.458821127\text{x}10^{-6} = \textbf{1.458821x10}^{-6} \textbf{ m}$$

Since the $n_1 = 4$ series shortest wavelength is shorter than the $n_1 = 3$ series longest wavelength, the **series do overlap**.

c) Shortest wavelength in $n_1 = 5$ series has $n_2 = \infty$.

$$\frac{1}{\lambda} = \left(1.096776\text{x}10^{7} \text{ m}^{-1}\right)\left(\frac{1}{5^2} - \frac{1}{\infty^2}\right) = 438,710.4 \text{ m}^{-1}$$

$$\lambda = \frac{1}{438,710.4 \text{ m}^{-1}} = 2.27940801\text{x}10^{-6} = \textbf{2.279408x10}^{-6} \textbf{ m}$$

Calculate the first few longest lines in the $n_1 = 4$ series to determine if any overlap with the shortest wavelength in the $n_1 = 5$ series:

For $n_1 = 4$, $n_2 = 5$:

$$\frac{1}{\lambda} = \left(1.096776\text{x}10^{7} \text{ m}^{-1}\right)\left(\frac{1}{4^2} - \frac{1}{5^2}\right) = 246,774.6 \text{ m}^{-1}$$

$$\lambda = \frac{1}{246,774.6 \text{ m}^{-1}} = \textbf{4.052281x10}^{-6} \textbf{ m}$$

For $n_1 = 4$, $n_2 = 6$:

$$\frac{1}{\lambda} = \left(1.096776\text{x}10^{7} \text{ m}^{-1}\right)\left(\frac{1}{4^2} - \frac{1}{6^2}\right) = 380,825 \text{ m}^{-1}$$

$$\lambda = \frac{1}{380,825 \text{ m}^{-1}} = \textbf{2.625878x10}^{-6} \textbf{ m}$$

For $n_1 = 4$, $n_2 = 7$:

$$\frac{1}{\lambda} = \left(1.096776\text{x}10^{7} \text{ m}^{-1}\right)\left(\frac{1}{4^2} - \frac{1}{7^2}\right) = 461,653.2 \text{ m}^{-1}$$

$$\lambda = \frac{1}{461,653.2 \text{ m}^{-1}} = \textbf{2.166128x10}^{-6} \textbf{ m}$$

The wavelengths of the first **two lines** of the $n_1 = 4$ series are longer than the shortest wavelength in the $n_1 = 5$ series. Therefore, only the first **two lines** of the $n_1 = 4$ series overlap the $n_1 = 5$ series.

d) At longer wavelengths (i.e., lower energies), there is increasing overlap between the lines from different series (i.e., with different n_1 values). The hydrogen spectrum becomes more complex, since the lines begin to merge into a more-or-less continuous band, and much more care is needed to interpret the information.

7.82   <u>Plan:</u> The energy differences sought may be determined by looking at the energy changes in steps. The wavelength is calculated from the relationship $\lambda = \dfrac{hc}{E}$.

<u>Solution:</u>
a) The difference between levels 3 and 2 (E_{32}) may be found by taking the difference in the energies for the $3 \rightarrow 1$ transition (E_{31}) and the $2 \rightarrow 1$ transition (E_{21}).

$$E_{32} = E_{31} - E_{21} = (4.854\text{x}10^{-17} \text{ J}) - (4.098\text{x}10^{-17} \text{ J}) = \textbf{7.56x10}^{-18} \textbf{ J}$$

$$\lambda = \frac{hc}{E} = \frac{\left(6.626\text{x}10^{-34} \text{ J} \cdot \text{s}\right)\left(3.00\text{x}10^{8} \text{ m/s}\right)}{\left(7.56\text{x}10^{-18} \text{ J}\right)} = 2.629365\text{x}10^{-8} = \textbf{2.63x10}^{-8} \textbf{ m}$$

b) The difference between levels 4 and 1 (E_{41}) may be found by adding the energies for the $4 \rightarrow 2$ transition (E_{42}) and the $2 \rightarrow 1$ transition (E_{21}).

$E_{41} = E_{42} + E_{21} = (1.024\text{x}10^{-17}\text{ J}) + (4.098\text{x}10^{-17}\text{ J}) = \textbf{5.122x10}^{\textbf{-17}}\textbf{ J}$

$$\lambda = \frac{hc}{E} = \frac{\left(6.626\text{x}10^{-34}\text{ J}\bullet\text{s}\right)\left(3.00\text{x}10^{8}\text{ m/s}\right)}{\left(5.122\text{x}10^{-17}\text{ J}\right)} = 3.88091\text{x}10^{-9} = \textbf{3.881x10}^{\textbf{-9}}\textbf{ m}$$

c) The difference between levels 5 and 4 (E_{54}) may be found by taking the difference in the energies for the $5 \rightarrow 1$ transition (E_{51}) and the $4 \rightarrow 1$ transition (see part b)).

$E_{54} = E_{51} - E_{41} = (5.242\text{x}10^{-17}\text{ J}) - (5.122\text{x}10^{-17}\text{ J}) = \textbf{1.2x10}^{\textbf{-18}}\textbf{ J}$

$$\lambda = \frac{hc}{E} = \frac{\left(6.626\text{x}10^{-34}\text{ J}\bullet\text{s}\right)\left(3.00\text{x}10^{8}\text{ m/s}\right)}{\left(1.2\text{x}10^{-18}\text{ J}\right)} = 1.6565\text{x}10^{-7} = \textbf{1.66x10}^{\textbf{-7}}\textbf{ m}$$

7.84   Plan: For part a), use the equation for kinetic energy, $E_k = \frac{1}{2}mu^2$. For part b), use the relationship $E = hc/\lambda$ to find the energy of the photon absorbed. From that energy subtract the kinetic energy of the dislodged electron to obtain the work function.
Solution:
a) The energy of the electron is a function of its speed leaving the surface of the metal. The mass of the electron is $9.109\text{x}10^{-31}$ kg.

$$E_k = \frac{1}{2}mu^2 = \frac{1}{2}\left(9.109\text{x}10^{-31}\text{ kg}\right)\left(6.40\text{x}10^{5}\text{ m/s}\right)^2\left(\frac{\text{J}}{\text{kg}\bullet\text{m}^2/\text{s}^2}\right) = 1.86552\text{x}10^{-19} = \textbf{1.87x10}^{\textbf{-19}}\textbf{ J}$$

b) The minimum energy required to dislodge the electron (ϕ) is a function of the incident light. In this example, the incident light is higher than the threshold frequency, so the kinetic energy of the electron, E_k, must be subtracted from the total energy of the incident light, $h\nu$, to yield the work function, ϕ. (The number of significant figures given in the wavelength requires more significant figures in the speed of light.)

$$\lambda\text{ (m)} = (358.1\text{ nm})\left(\frac{10^{-9}\text{ m}}{1\,\text{nm}}\right) = 3.581\text{x}10^{-7}\text{ m}$$

$$E = hc/\lambda = \frac{\left(6.626\text{x}10^{-34}\text{ J}\bullet\text{s}\right)\left(2.9979\text{x}10^{8}\text{ m/s}\right)}{\left(3.581\text{x}10^{-7}\text{ m}\right)} = 5.447078\text{x}10^{-19}\text{ J}$$

$\Phi = h\nu - E_k = (5.447078\text{x}10^{-19}\text{ J}) - (1.86552\text{x}10^{-19}\text{ J}) = 3.581558\text{x}10^{-19} = \textbf{3.58x10}^{\textbf{-19}}\textbf{ J}$

7.86   Plan: Examine Figure 7.3 and match the given wavelengths to their colors. For each salt, convert the mass of salt to moles and multiply by Avogadro's number to find the number of photons emitted by that amount of salt (assuming that each atom undergoes one-electron transition). Use the relationship $E = \dfrac{hc}{\lambda}$ to find the energy of one photon and multiply by the total number of photons for the total energy of emission.
Solution:
a) Figure 7.3 indicates that the 641 nm wavelength of Sr falls in the **red** region and the 493 nm wavelength of Ba falls in the **green** region.
b) $SrCl_2$

$$\text{Number of photons} = \left(5.00\text{ g SrCl}_2\right)\left(\frac{1\text{ mol SrCl}_2}{158.52\text{ g SrCl}_2}\right)\left(\frac{6.022\text{x}10^{23}\text{ photons}}{1\text{ mol SrCl}_2}\right) = 1.8994449\text{x}10^{22}\text{ photons}$$

$$\lambda\text{ (m)} = (641\text{ nm})\left(\frac{10^{-9}\text{ m}}{1\,\text{nm}}\right) = 6.41\text{x}10^{-7}\text{ m}$$

$$E_{\text{photon}} = \frac{hc}{\lambda} = \frac{\left(6.626\text{x}10^{-34}\ \text{J}\bullet\text{s}\right)\left(3.00\text{x}10^{8}\ \text{m/s}\right)}{6.41\text{x}10^{-7}\ \text{m}}\left(\frac{1\ \text{kJ}}{10^{3}\ \text{J}}\right) = 3.10109\text{x}10^{-22}\ \text{kJ/photon}$$

$$E_{\text{total}} = \left(1.8994449\text{x}10^{22}\ \text{photons}\right)\left(\frac{3.10109\text{x}10^{-22}\ \text{kJ}}{1\ \text{photon}}\right) = 5.89035 = \textbf{5.89 kJ}$$

$BaCl_2$

$$\text{Number of photons} = \left(5.00\ \text{g BaCl}_2\right)\left(\frac{1\ \text{mol BaCl}_2}{208.2\ \text{g BaCl}_2}\right)\left(\frac{6.022\text{x}10^{23}\ \text{photons}}{1\ \text{mol BaCl}_2}\right) = 1.44620557\text{x}10^{22}\ \text{photons}$$

$$\lambda\ (\text{m}) = \left(493\ \text{nm}\right)\left(\frac{10^{-9}\ \text{m}}{1\,\text{nm}}\right) = 4.93\text{x}10^{-7}\ \text{m}$$

$$E_{\text{photon}} = \frac{hc}{\lambda} = \frac{\left(6.626\text{x}10^{-34}\ \text{J}\bullet\text{s}\right)\left(3.00\text{x}10^{8}\,\text{m/s}\right)}{4.93\text{x}10^{-7}\ \text{m}}\left(\frac{1\ \text{kJ}}{10^{3}\ \text{J}}\right) = 4.0320487\text{x}10^{-22}\ \text{kJ/photon}$$

$$E_{\text{total}} = \left(1.44620557\text{x}10^{22}\ \text{photons}\right)\left(\frac{4.0320487\text{x}10^{-22}\ \text{kJ}}{1\ \text{photon}}\right) = 5.83117 = \textbf{5.83 kJ}$$

7.88   Plan: Examine Figure 7.3 to find the region of the electromagnetic spectrum in which the wavelength lies. Compare the absorbance of the given concentration of Vitamin A to the absorbance of the given amount of fish-liver oil to find the concentration of Vitamin A in the oil.
Solution:
a) At this wavelength the sensitivity to absorbance of light by Vitamin A is maximized while minimizing interference due to the absorbance of light by other substances in the fish-liver oil.
b) The wavelength 329 nm lies in the **ultraviolet region** of the electromagnetic spectrum.
c) A known quantity of vitamin A ($1.67\text{x}10^{-3}$ g) is dissolved in a known volume of solvent (250. mL) to give a standard concentration with a known response (1.018 units). This can be used to find the unknown quantity of Vitamin A that gives a response of 0.724 units. An equality can be made between the two concentration-to-absorbance ratios.

$$\text{Concentration } (C_1,\ \text{g/mL}) \text{ of Vitamin A} = \left(\frac{1.67\text{x}10^{-3}\ \text{g}}{250.\ \text{mL}}\right) = 6.68\text{x}10^{-6}\ \text{g/mL Vitamin A}$$

Absorbance (A_1) of Vitamin A = 1.018 units.

Absorbance (A_2) of fish-liver oil = 0.724 units

Concentration (g/mL) of Vitamin A in fish-liver oil sample = C_2

$$\frac{A_1}{C_1} = \frac{A_2}{C_2}$$

$$C_2 = \frac{A_2 C_1}{A_1} = \frac{(0.724)\left(6.68\text{x}10^{-6}\ \text{g/mL}\right)}{(1.018)} = 4.7508\text{x}10^{-6}\ \text{g/mL Vitamin A}$$

$$\text{Mass (g) of Vitamin A in oil sample} = \left(500.\ \text{mL oil}\right)\left(\frac{4.7508\text{x}10^{-6}\ \text{g Vitamin A}}{1\ \text{mL oil}}\right) = 2.3754\text{x}10^{-3}\ \text{g Vitamin A}$$

$$\text{Concentration of Vitamin A in oil sample} = \frac{\left(2.3754\text{x}10^{-3}\ \text{g}\right)}{\left(0.1232\ \text{g Oil}\right)} = 1.92808\text{x}10^{-2} = \textbf{1.93x10}^{-2}\ \textbf{g Vitamin A/g oil}$$

7.92   Plan: First find the energy in joules from the light that shines on the text. Each watt is one joule/s for a total of 75 J; take 5% of that amount of joules and then 10% of that amount. Use $E = \dfrac{hc}{\lambda}$ to find the energy of one photon of light with a wavelength of 550 nm. Divide the energy that shines on the text by the energy of one photon to obtain the number of photons.

Solution:

The amount of energy is calculated from the wavelength of light:

$$\lambda \text{ (m)} = (550 \text{ nm})\left(\frac{10^{-9} \text{ m}}{1 \text{ nm}}\right) = 5.50 \times 10^{-7} \text{ m}$$

$$E = \frac{hc}{\lambda} = \frac{\left(6.626 \times 10^{-34} \text{ J} \cdot \text{s}\right)\left(3.00 \times 10^8 \text{ m/s}\right)}{5.50 \times 10^{-7} \text{ m}} = 3.614182 \times 10^{-19} \text{ J/photon}$$

$$\text{Amount of power from the bulb} = \left(75 \text{ W}\right)\left(\frac{1 \text{ J/s}}{1 \text{ W}}\right) = 75 \text{ J/s}$$

$$\text{Amount of power converted to light} = (75 \text{ J/s})\left(\frac{5\%}{100\%}\right) = 3.75 \text{ J/s}$$

$$\text{Amount of light shining on book} = (3.75 \text{ J/s})\left(\frac{10\%}{100\%}\right) = 0.375 \text{ J/s}$$

$$\text{Number of photons:} \left(\frac{0.375 \text{ J}}{\text{s}}\right)\left(\frac{1 \text{ photon}}{3.614182 \times 10^{-19} \text{ J}}\right) = 1.0376 \times 10^{18} = \mathbf{1.0 \times 10^{18} \text{ photons/s}}$$

7.95   Plan: In the visible series with $n_{final} = 2$, the transitions will end in either the 2s or 2p orbitals since those are the only two types of orbitals in the second main energy level. With the restriction that the angular momentum quantum number can change by only ± 1, the allowable transitions are from a p orbital to 2s ($l = 1$ to $l = 0$), from an s orbital to 2p ($l = 0$ to $l = 1$), and from a d orbital to 2p ($l = 2$ to $l = 1$). The problem specifies a change in *energy level*, so n_{init} must be 3, 4, 5, etc. (Although a change from 2p to 2s would result in a +1 change in l, this is not a change in energy level.)

Solution:

The first four transitions are as follows:

$3s \rightarrow 2p$
$3d \rightarrow 2p$
$4s \rightarrow 2p$
$3p \rightarrow 2s$

7.97   a) Lengh of box (m) = $\left(0.450 \text{ nm}\right)\left(\dfrac{10^{-9} \text{ m}}{1 \text{ nm}}\right) = 4.50 \times 10^{-10} \text{ m}$

$$\Delta E = (2n+1)\frac{h^2}{8mL^2} = (2(2)+1)\frac{\left(6.626 \times 10^{-34} \text{ J} \cdot \text{s}\right)^2}{8\left(9.109 \times 10^{-31} \text{ kg}\right)\left(4.50 \times 10^{-10} \text{ m}\right)^2} = 1.49 \times 10^{-18} \text{ J}$$

$$\lambda = hc/E = \frac{\left(6.626 \times 10^{-34} \text{ J} \cdot \text{s}\right)\left(3.00 \times 10^8 \text{ m}/\text{s}\right)}{\left(1.49 \times 10^{-18} \text{ J}\right)}\left(\frac{1 \text{ nm}}{10^{-9} \text{ m}}\right) = 133.4094 = \mathbf{133 \text{ nm}}$$

7.99   <u>Plan:</u> Equation 7.9 is used to find the difference in energy between two levels for a particle of mass m in a box of length L. The mass of the electron and an O_2 molecule must be converted to kg and the length of the box must be expressed in m.
        <u>Solution:</u>

a) Length of box $(m) = (0.10\ nm)\left(\dfrac{10^{-9}\ m}{1\ nm}\right) = 1.0x10^{-10}\ m$

$$\Delta E = (2n+1)\dfrac{h^2}{8mL^2} = (2(1)+1)\dfrac{\left(6.626x10^{-34}\ J \bullet s\right)^2}{8\left(9.109x10^{-31}\ kg\right)\left(1x10^{-10}\ m\right)^2} = \mathbf{1.8x10^{-17}\ J}$$

b) Length of box $(m) = (10.cm)\left(\dfrac{10^{-2}\ m}{1\ cm}\right) = 0.10\ m$

Mass of O_2 molecule $(kg) = \left(1\ O_2 molecule\right)\left(\dfrac{1\ mol\ O_2}{6.022X10^{23}\ O_2\ molecules}\right)\left(\dfrac{32.00\ g}{1\ mol\ O_2}\right)\left(\dfrac{10^{-3}\ kg}{1\ g}\right)$

$= 5.314x10^{-26}\ kg$

$$\Delta E = (2n+1)\dfrac{h^2}{8mL^2} = (2(1)+1)\dfrac{\left(6.626x10^{-34}\ J \bullet s\right)^2}{8\left(5.314x10^{-26}\ kg\right)\left(0.10 m\right)^2} = \mathbf{3.1x10^{-40}\ J}$$

c) ΔE for the electron is significantly larger than ΔE for the more massive O_2 molecule; as the mass m of the particle increases, the difference in energy between levels decreases. Thus, quantization is important only for particles of very small mass, such as electrons.

7.101   <u>Plan:</u> Examine the relationship between the energy of an energy level and the length of the one-dimensional box.
        <u>Solution:</u>

$n=1$ for the ground state energy level; m is a constant value for the mass of the electron. Notice that E and L are inversely proportional in the relationship $E_n = \dfrac{n^2 h^2}{8mL^2}$. The greater the value of L, the lower the energy of the ground state; conversely, the smaller the value of L, the higher the energy of the ground state. The electron has the lowest ground state energy in the box with the longest dimension, $\mathbf{10^{-4}\ m}$, and has the highest ground state energy in the smallest box, $\mathbf{10^{-10}\ m.}$

CHAPTER 8 ELECTRON CONFIGURATION AND CHEMICAL PERIODICITY

FOLLOW–UP PROBLEMS

8.1A   Plan: The superscripts can be added to indicate the number of electrons in the element, and hence its identity. A horizontal orbital diagram is written for simplicity, although it does not indicate the sublevels have different energies. Based on the orbital diagram, we identify the electron of interest and determine its four quantum numbers.
Solution:
The number of electrons = 2 + 2 + 4 = 8; element = **oxygen**, atomic number 8.
Orbital diagram:

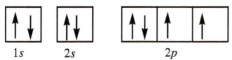

1s   2s   2p

The electron in the middle 2p orbital is the sixth electron (this electron would have entered in this position due to Hund's rule). This electron has the following quantum numbers: $n = 2, l = 1$ (for p orbital), $m_l = 0, m_s = +1/2$. By convention, +1/2 is assigned to the first electron in an orbital. Also, the first p orbital is assigned an m_l value of –1, the middle p orbital a m_l value of 0, and the last p orbital a m_l value of +1.

8.1B   Plan: Use the quantum numbers to determine the orbital into which the last electron is added and whether the electron is the first or second electron added to that orbital. Draw the orbital diagram that matches the quantum numbers. Then use the orbital diagram and the periodic table to identify the element. Finally, write the electron configuration for the element.
Solution:
The last electron added to the atom has the quantum numbers $n = 2, l = 1, m_l = 0$ and $m_s = +1/2$. This electron is in the second 2p orbital ($n = 2, l = 1$ tells us that the orbital is 2p; $m_l = 0$ tells us that the orbital is the 2nd of the three 2p orbitals). Additionally, because $m_s = +1/2$, we know the electron is the first electron in that orbital. The orbital diagram that matches this description is:
Orbital diagram:

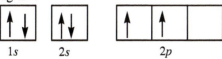

1s   2s   2p

There are a total of 6 electrons in the orbital diagram. **Carbon** is the element with 6 protons and, in its neutral state, 6 electrons. Its electron configuration is: $1s^2 2s^2 2p^2$.

8.2A   Plan: The atomic number gives the number of electrons. The order of filling may be inferred by the location of the element on the periodic table. The partial orbital diagrams shows only those electrons after the preceding noble gas except those used to fill inner d and f subshells. The number of inner electrons is simply the total number of electrons minus those electrons in the partial orbital diagram.
Solution:
a) For Ni ($Z = 28$), the full electron configuration is $1s^2 2s^2 2p^6 3s^2 3p^6 4s^2 3d^8$.
The condensed configuration is $[Ar]4s^2 3d^8$.
The partial orbital diagram for the valence electrons is

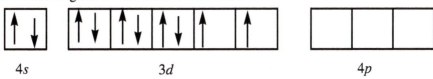

4s   3d   4p

There are 28 – 10(valence) = **18 inner electrons**.

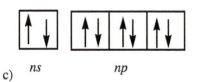 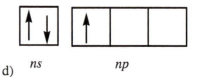

c)   *ns*   *np*   d)   *ns*   *np*

8.81   Plan: Substances are paramagnetic if they have unpaired electrons. To find the number of unpaired electrons look at the electron configuration expanded to include the different orientations of the orbitals, such as p_x and p_y and p_z. Remember that all orbitals in a *p*, *d*, or *f* set will each have one electron before electrons pair in an orbital. In the noble gas configurations, all electrons are paired because all orbitals are filled.
Solution:
a) Ga ($Z = 31$) = $[Ar]4s^2 3d^{10} 4p^1$. The *s* and *d* sublevels are filled, so all electrons are paired. The lone *p* electron is unpaired, so this element is **paramagnetic**.
b) Si ($Z = 14$) = $[Ne]3s^2 3p_x^1 3p_y^1 3p_z^0$. This element is **paramagnetic** with two unpaired electrons.

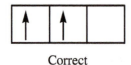   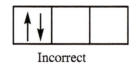

               Correct   Incorrect

c) Be ($Z = 4$) = $[He]2s^2$. The two *s* electrons are paired so Be is **not paramagnetic**.
d) Te ($Z = 52$) = $[Kr]5s^2 4d^{10} 5p_x^2 5p_y^1 5p_z^1$ is **paramagnetic** with two unpaired electrons in the *5p* set.

8.83   Plan: Substances are paramagnetic if they have unpaired electrons. Write the electron configuration of the atom and then remove the specified number of electrons. Remember that all orbitals in a *p*, *d*, or *f* set will each have one electron before electrons pair in an orbital. In the noble gas configurations, all electrons are paired because all orbitals are filled.
Solution:
a) V: $[Ar]4s^2 3d^3$; **V^{3+}: [Ar]3d^2** Transition metals first lose the *s* electrons in forming ions, so to form the +3 ion a vanadium atom loses two 4*s* electrons and one 3*d* electron. **Paramagnetic**

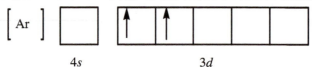

               4s   *3d*

b) Cd: $[Kr]5s^2 4d^{10}$; **Cd^{2+}: [Kr]4d^{10}** Cadmium atoms lose two electrons from the 5*s* orbital to form the +2 ion. **Diamagnetic**

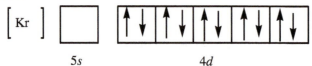

               5s   *4d*

c) Co: $[Ar]4s^2 3d^7$; **Co^{3+}: [Ar]3d^6** Cobalt atoms lose two 4*s* electrons and one 3*d* electron to form the +3 ion. **Paramagnetic**

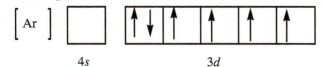

               4s   *3d*

d) Ag: $[Kr]5s^1 4d^{10}$; **Ag$^+$: [Kr]4d^{10}** Silver atoms lose the one electron in the 5*s* orbital to form the +1 ion. **Diamagnetic**

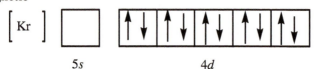

               5s   *4d*

8.85   Plan: Substances are diamagnetic if they have no unpaired electrons. Draw the partial orbital diagrams, remembering that all orbitals in d set will each have one electron before electrons pair in an orbital.
Solution:
You might first write the condensed electron configuration for Pd as $[Kr]5s^2 4d^8$. However, the partial orbital diagram is not consistent with diamagnetism.

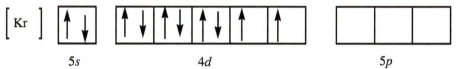

Promoting an s electron into the d sublevel (as in (c) $[Kr]5s^1 4d^9$) still leaves two electrons unpaired.

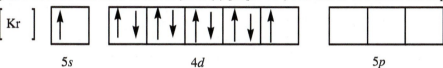

The only configuration that supports diamagnetism is **(b) $[Kr]4d^{10}$**.

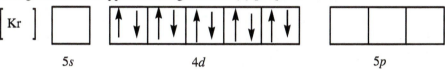

8.87   Plan: The size of ions increases down a group. For ions that are isoelectronic (have the same electron configuration) size decreases with increasing atomic number.
Solution:
a) Increasing size: **$Li^+ < Na^+ < K^+$**, size increases down Group 1A(1).
b) Increasing size: **$Rb^+ < Br^- < Se^{2-}$**, these three ions are isoelectronic with the same electron configuration as krypton. Size decreases with increasing atomic number in an isoelectronic series.
c) Increasing size: **$F^- < O^{2-} < N^{3-}$**, the three ions are isoelectronic with an electron configuration identical to neon. Size decreases with increasing atomic number in an isoelectronic series.

8.90   Plan: Write the electron configuration for each atom. Remove the specified number of electrons, given by the positive ionic charge, to write the configuration for the ions. Remember that electrons with the highest n value are removed first.
Solution:
Ce: $[Xe]6s^2 4f^1 5d^1$ Eu: $[Xe]6s^2 4f^7$
Ce^{4+}: $[Xe]$ Eu^{2+}: $[Xe]4f^7$
Ce^{4+} has a noble gas configuration and Eu^{2+} has a half-filled f subshell.

8.91   Plan: Write the formula of the oxoacid. Remember that in naming oxoacids (H + polyatomic ion), the suffix of the polyatomic changes: -ate becomes -ic acid and -ite becomes -ous acid. Determine the oxidation state of the nonmetal in the oxoacid; hydrogen has an O.N. of +1 and oxygen has an O.N. of –2. Based on the oxidation state of the nonmetal, and the oxidation state of the oxide ion (–2), the formula of the nonmetal oxide may be determined. The name of the nonmetal oxide comes from the formula; remember that nonmetal compounds use prefixes to indicate the number of each type of atom in the formula.
Solution:
a) hypochlorous acid = $HClO$ has Cl^+ so the oxide is **Cl_2O = dichlorine oxide** or **dichlorine monoxide**
b) chlorous acid = $HClO_2$ has Cl^{3+} so the oxide is **Cl_2O_3 = dichlorine trioxide**
c) chloric acid = $HClO_3$ has Cl^{5+} so the oxide is **Cl_2O_5 = dichlorine pentaoxide**
d) perchloric acid = $HClO_4$ has Cl^{7+} so the oxide is **Cl_2O_7 = dichlorine heptaoxide**
e) sulfuric acid = H_2SO_4 has S^{6+} so the oxide is **SO_3 = sulfur trioxide**
f) sulfurous acid = H_2SO_3 has S^{4+} so the oxide is **SO_2 = sulfur dioxide**
g) nitric acid = HNO_3 has N^{5+} so the oxide is **N_2O_5 = dinitrogen pentaoxide**

h) nitrous acid = HNO_2 has N^{3+} so the oxide is N_2O_3 = **dinitrogen trioxide**

i) carbonic acid = H_2CO_3 has C^{4+} so the oxide is CO_2 = **carbon dioxide**

j) phosphoric acid = H_3PO_4 has P^{5+} so the oxide is P_2O_5 = **diphosphorus pentaoxide**

8.94 <u>Plan:</u> Remember that isoelectronic species have the same electron configuration. Atomic radius decreases up a group and left to right across a period.
<u>Solution:</u>
a) A chemically unreactive Period 4 element would be Kr in Group 8A(18). Both the Sr^{2+} ion and Br^- ion are isoelectronic with Kr. Their combination results in **$SrBr_2$, strontium bromide.**
b) Ar is the Period 3 noble gas. Ca^{2+} and S^{2-} are isoelectronic with Ar. The resulting compound is **CaS, calcium sulfide.**
c) The smallest filled d subshell is the $3d$ shell, so the element must be in Period 4. Zn forms the Zn^{2+} ion by losing its two s subshell electrons to achieve a *pseudo–noble gas* configuration ($[Ar]3d^{10}$). The smallest halogen is fluorine, whose anion is F^-. The resulting compound is **ZnF_2, zinc fluoride.**
d) Ne is the smallest element in Period 2, but it is not ionizable. Li is the largest atom whereas F is the smallest atom in Period 2. The resulting compound is **LiF, lithium fluoride.**

8.95 <u>Plan:</u> Recall Hess's law: the enthalpy change of an overall process is the sum of the enthalpy changes of its individual steps. Both the ionization energies and the electron affinities of the elements are needed.
<u>Solution:</u>
a) F: ionization energy = 1681 kJ/mol   electron affinity = –328 kJ/mol

$F(g) \rightarrow F^+(g) + e^-$ $\Delta H = 1681$ kJ/mol
$F(g) + e^- \rightarrow F^-(g)$ $\Delta H = -328$ kJ/mol

Reverse the electron affinity reaction to give: $F^-(g) \rightarrow F(g) + e^-$ $\Delta H = +328$ kJ/mol
Summing the ionization energy reaction with the reversed electron affinity reaction (Hess's law):

$\cancel{F(g)} \rightarrow F^+(g) + e^-$ $\Delta H = 1681$ kJ/mol

$F^-(g) \rightarrow \cancel{F(g)} + e^-$ $\Delta H = +328$ kJ/mol
_____
$F^-(g) \rightarrow F^+(g) + 2\ e^-$ $\Delta H = \textbf{2009 kJ/mol}$

b) Na: ionization energy = 496 kJ/mol   electron affinity = –52.9 kJ/mol

$Na(g) \rightarrow Na^+(g) + e^-$ $\Delta H = 496$ kJ/mol

$Na(g) + e^- \rightarrow Na^-(g)$ $\Delta H = -52.9$ kJ/mol

Reverse the ionization reaction to give: $Na^+(g) + e^- \rightarrow Na(g)$ $\Delta H = -496$ kJ/mol
Summing the electron affinity reaction with the reversed ionization reaction (Hess's law):

$\cancel{Na(g)} + e^- \rightarrow Na^-(g)$ $\Delta H = -52.9$ kJ/mol

$Na^+(g) + e^- \rightarrow \cancel{Na(g)}$ $\Delta H = -496$ kJ/mol
_____
$Na^+(g) + 2\ e^- \rightarrow Na^-(g)$ $\Delta H = -548.9 = \textbf{–549 kJ/mol}$

8.97 <u>Plan:</u> Determine the electron configuration for iron, and then begin removing one electron at a time. Remember that all orbitals in a d set will each have one electron before electrons pair in an orbital, and electrons with the highest n value are removed first. Ions with all electrons paired are diamagnetic. Ions with at least one unpaired electron are paramagnetic. The more unpaired electrons, the greater the attraction to a magnetic field.
<u>Solution:</u>

Fe	$[Ar]4s^23d^6$	partially filled $3d$ = **paramagnetic**	number of unpaired electrons = 4
Fe^+	$[Ar]4s^13d^6$	partially filled $3d$ = **paramagnetic**	number of unpaired electrons = 5
Fe^{2+}	$[Ar]3\ d^6$	partially filled $3d$ = **paramagnetic**	number of unpaired electrons = 4
Fe^{3+}	$[Ar]3d^5$	partially filled $3d$ = **paramagnetic**	number of unpaired electrons = 5
Fe^{4+}	$[Ar]3d^4$	partially filled $3d$ = **paramagnetic**	number of unpaired electrons = 4
Fe^{5+}	$[Ar]3d^3$	partially filled $3d$ = **paramagnetic**	number of unpaired electrons = 3
Fe^{6+}	$[Ar]3d^2$	partially filled $3d$ = **paramagnetic**	number of unpaired electrons = 2

Fe^{7+} [Ar]$3d^1$ partially filled $3d$ = **paramagnetic** number of unpaired electrons = 1

Fe^{8+} [Ar]   filled orbitals   = **diamagnetic**   number of unpaired electrons = 0

Fe^{9+} [Ne]$3s^2 3p^5$ partially filled $3p$ = **paramagnetic** number of unpaired electrons = 1

Fe^{10+} [Ne]$3s^2 3p^4$ partially filled $3p$ = **paramagnetic** number of unpaired electrons = 2

Fe^{11+} [Ne]$3s^2 3p^3$ partially filled $3p$ = **paramagnetic** number of unpaired electrons = 3

Fe^{12+} [Ne]$3s^2 3p^2$ partially filled $3p$ = **paramagnetic** number of unpaired electrons = 2

Fe^{13+} [Ne]$3s^2 3p^1$ partially filled $3p$ = **paramagnetic** number of unpaired electrons = 1

Fe^{14+} [Ne]$3s^2$ filled orbitals   = **diamagnetic**   number of unpaired electrons = 0

Fe$^+$ and **Fe^{3+}** would both be most attracted to a magnetic field. They each have 5 unpaired electrons.

CHAPTER 9 MODELS OF CHEMICAL BONDING

FOLLOW–UP PROBLEMS

9.1A Plan: First, write out the condensed electron configuration, partial orbital diagram, and electron-dot structure for magnesium atoms and chlorine atoms. In the formation of the ions, each magnesium atom will lose two electrons to form the +2 ion and each chlorine atom will gain one electron to form the –1 ion. Write the condensed electron configuration, partial orbital diagram, and electron-dot structure for each of the ions. The formula of the compound is found by combining the ions in a ratio that gives a neutral compound.
Solution:
Condensed electron configurations:
$Mg ([Ne]3s^2) + Cl ([Ne]3s^23p^5) \rightarrow Mg^{2+} ([Ne]) + Cl^- ([Ne]3s^23p^6)$
In order to balance the charge (or the number of electrons lost and gained) two chlorine atoms are needed:
$Mg ([Ne]3s^2) + 2Cl ([Ne]3s^23p^5) \rightarrow Mg^{2+} ([Ne]) + 2Cl^- ([Ne]3s^23p^6)$
Partial orbital diagrams:

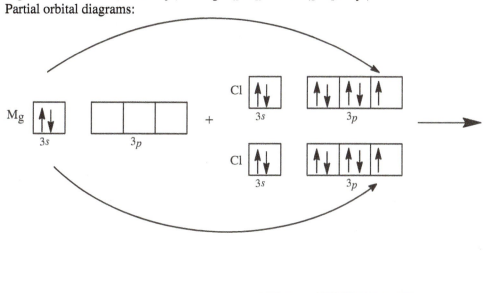

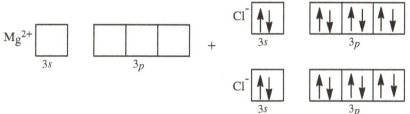

Lewis electron–dot symbols:

The formula of the compound would contain two chloride ions for each magnesium ion, **MgCl₂**.

9.1B Plan: First, write out the condensed electron configuration, partial orbital diagram, and electron-dot structure for calcium atoms and oxygen atoms. In the formation of the ions, each calcium atom will lose two electrons to form the +2 ion and each oxygen atom will gain two electrons to form the –2 ion. Write the condensed electron configuration, partial orbital diagram, and electron-dot structure for each of the ions. The formula of the compound is found by combining the ions in a ratio that gives a neutral compound.
Solution:
Condensed electron configurations:
Ca ([Ar]$4s^2$) + O ([He]$2s^2 2p^4$) → Ca^{2+} ([Ar]) + O^{2-} ([He]$2s^2 2p^6$)
The charge is balanced (the number of electrons lost equals the number of electrons gained).
Partial orbital diagrams:

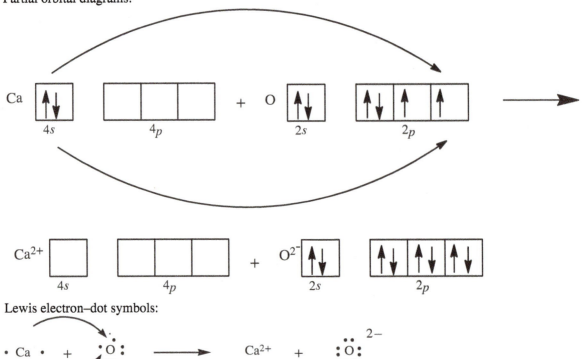

Lewis electron–dot symbols:

The formula of the compound would contain one oxide ion for each calcium ion, **CaO**.

9.2A Plan: Examine the charge of the ions involved in the compounds. Use periodic trends in ionic radii to determine the relative size of the ions in the compounds. Then apply Coulomb's law. According to Coulomb's law, for ions of similar size, lower charge leads to a smaller lattice energy. For ions with the same charge, larger size leads to smaller lattice energy.
Solution:
The compound with the smaller lattice energy is **BaF₂**. The only difference between these compounds is the size of the cation: the Ba^{2+} ion is larger than the Sr^{2+} ion (ionic size increases down the group). According to Coulomb's law, for ions with the same charge, larger size leads to smaller lattice energy.

9.2B Plan: Examine the charge of the ions involved in the compounds. Use periodic trends in ionic radii to determine the relative size of the ions in the compounds. Then apply Coulomb's law. According to Coulomb's law, for ions of similar size, higher charge leads to a larger lattice energy. For ions with the same charge, smaller size leads to larger lattice energy.
Solution:
Na₂O < MgF₂ < CaO   In Na₂O, the ions are Na^+ and O^{2-}; the ions in MgF₂ are Mg^{2+} and F^-. The products of the cation charge and anion charge are the same for these two compounds (1 x 2 and 2 x 1); however, since Na^+ is larger than Mg^{2+}, and O^{2-} is larger than F^-, Na₂O has the smaller lattice energy. In CaO, the ions are Ca^{2+} and O^{2-}.

The Ca^{2+} ion is larger than Na^+ or Mg^{2+} but the larger product of cation and anion charge (2 x 2 for CaO) results in a larger lattice energy for CaO. In nearly every case, charge is more important than size.

9.3A   Plan: a) All bonds are triple bonds from carbon to a second row element. The trend in bond length can be predicted from the fact that atomic radii decrease across a row, so oxygen will be smaller than nitrogen, which is smaller than carbon. Bond length will therefore decrease across the row while bond energy increases.
b) All the bonds are single bonds from phosphorus to a group 7A element. The trend in bond length can be predicted from the fact that atomic radii increase down a column, so fluorine will be smaller than bromine, which is smaller than iodine. Bond length will therefore *increase* down the column while bond energy increases.
Solution:
a) Bond length: $C\equiv C$ > $C\equiv N$ > $C\equiv O$ Bond strength: $C\equiv O$ > $C\equiv N$ > $C\equiv C$
b) Bond length: P–I > P–Br > P–F   Bond energy: P–F > P–Br > P–I
Check: (Use the tables in the chapter.)
Bond lengths from Table: $C\equiv C$, 121 pm > $C\equiv N$, 115 pm > $C\equiv O$, 113 pm
Bond energies from Table: $C\equiv O$, 1070 kJ/mol > $C\equiv N$, 891 kJ/mol > $C\equiv C$, 839 kJ/mol
Bond lengths from Table: P–I, 246 pm > P–Br, 222 pm > P–F, 156 pm
Bond energies from Table: P–F, 490 kJ/mol > P–Br, 272 kJ/mol > P–I, 184 kJ/mol
The values from the tables agree with the order predicted.

9.3B   Plan: a) All bonds are single bonds from silicon to a second row element. The trend in bond length can be predicted from the fact that atomic radii decrease across a row, so fluorine will be smaller than oxygen, which is smaller than carbon. Bond length will therefore decrease across the row while bond energy increases.
b) All the bonds are between two nitrogen atoms, but differ in the number of electrons shared between the atoms. The more electrons shared the shorter the bond length and the greater the bond energy.
Solution:
a) Bond length: Si–F < Si–O < Si–C   Bond strength: Si–C < Si–O < Si–F
b) Bond length: $N\equiv N$ < N=N < N–N   Bond energy: N–N < N=N < $N\equiv N$
Check: (Use the tables in the chapter.)
Bond lengths from Table: Si–F, 156 pm < Si–O, 161 pm < Si–C, 186 pm
Bond energies from Table: Si–C, 301 kJ < Si–O, 368 kJ < Si–F, 565 kJ
Bond lengths from Table: $N\equiv N$, 110 pm < N=N, 122 pm < N–N, 146 pm
Bond energies from Table: N–N, 160 kJ < N=N, 418 kJ < $N\equiv N$, 945 kJ
The values from the tables agree with the order predicted.

9.4A   Plan: Assume that all reactant bonds break and all product bonds form. Use the bond energy table to find the values for the different bonds. Sum the values for the reactants, and sum the values for the products (these are all negative values). Add the sum of the product values to the sum of the reactant values.
Solution:
Calculating $\Delta H°$ for the bonds broken:

$$1 \text{ x } N\equiv N = (1 \text{ mol})(945 \text{ kJ/mol}) = 945 \text{ kJ}$$

$$3 \text{ x } H–H = (3 \text{ mol})(432 \text{ kJ/mol}) = \underline{1296 \text{ kJ}}$$

$\Sigma \Delta H°_{\text{bonds broken}}$ = 2241 kJ

Calculating $\Delta H°$ for the bonds formed:

$$1 \text{ x } NH_3 = (3 \text{ mol})(–391 \text{ kJ/mol}) = –1173 \text{ kJ}$$

$$1 \text{ x } NH_3 = (3 \text{ mol})(–391 \text{ kJ/mol}) = \underline{–1173 \text{ kJ}}$$

$\Sigma \Delta H°_{\text{bonds formed}}$ = –2346 kJ

$$\Delta H°_{\text{rxn}} = \Sigma \Delta H°_{\text{bonds broken}} + \Sigma \Delta H°_{\text{bonds formed}}$$

$$\Delta H°_{\text{rxn}} = 2241 \text{ kJ} + (–2346 \text{ kJ}) = \textbf{–105 kJ}$$

9.4B Plan: Assume that all reactant bonds break and all product bonds form. Use the bond energy table to find the values for the different bonds. Sum the values for the reactants, and sum the values for the products (these are all negative values). Add the sum of the product values to the sum of the reactant values.
Solution:
Calculating $\Delta H°$ for the bonds broken:

$$2 \text{ x O–F} = (2 \text{ mol})(190 \text{ kJ/mol}) = 380 \text{ kJ}$$
$$2 \text{ x O–H} = (2 \text{ mol})(467 \text{ kJ/mol}) = \underline{934 \text{ kJ}}$$
$$\Sigma \Delta H°_{\text{bonds broken}} \qquad\qquad\qquad\qquad = 1314 \text{ kJ}$$

Calculating $\Delta H°$ for the bonds formed:

$$2 \text{ x H–F} = (2 \text{ mol})(-565 \text{ kJ/mol}) = -1130 \text{ kJ}$$
$$2 \text{ x O=O} = (1 \text{ mol})(-498 \text{ kJ/mol}) = \underline{-498 \text{ kJ}}$$
$$\Sigma \Delta H°_{\text{bonds formed}} \qquad\qquad\qquad\qquad = -1628 \text{ kJ}$$

$$\Delta H°_{\text{rxn}} = \Sigma \Delta H°_{\text{bonds broken}} + \Sigma \Delta H°_{\text{bonds formed}}$$
$$\Delta H°_{\text{rxn}} = 1314 \text{ kJ} + (-1628 \text{ kJ}) = \textbf{--314 kJ}$$

9.5A Plan: Bond polarity can be determined from the difference in electronegativity of the two atoms. The more electronegative atom holds the $\delta-$ charge and the other atom holds the $\delta+$ charge.
Solution:
a) From the electronegativity figure, EN of Cl = 3.0, EN of F = 4.0, EN of Br = 2.8. Cl–Cl will be the least polar since there is no difference between the electronegativity of the two chlorine atoms. Cl–F will be more polar than Cl–Br since the electronegativity difference between Cl and F
(4.0 – 3.0 = 1.0) is greater than the electronegativity difference between Cl and Br (3.0 – 2.8 = 0.2).

$$\overset{\delta+\ \ \delta-}{\text{ }}\quad\overset{\delta+\ \delta-}{\text{ }}$$
$$\text{Cl–Cl} < \text{Br–Cl} < \text{Cl-F}$$

b) EN of P = 2.1, EN of F = 4.0, EN of N = 3.0, EN of O = 3.5. The least polar is N–O with an EN difference of 0.5 (3.5 – 3.0 = 0.5), next is N–F with an EN difference of 1.0 (4.0 – 3.0 = 1.0), the highest is P–F with an EN difference of 1.9 (4.0 – 2.1 = 1.9) .

$$\overset{\delta+\ \ \delta-}{\text{ }}\quad\overset{\delta+\ \ \delta-}{\text{ }}\quad\overset{\delta+\ \ \delta-}{\text{ }}$$
$$\text{N–O} < \text{N–F} < \text{P–F}$$

9.5B Plan: Bond polarity can be determined from the difference in electronegativity of the two atoms. The more electronegative atom holds the $\delta-$ charge and the other atom holds the $\delta+$ charge.
Solution:
a) From the electronegativity figure, EN of Cl = 3.0, EN of F = 4.0, EN of Br = 2.8, and EN of Se = 2.4. Se–F will be more polar than Se–Cl, which will be more polar than Se–Br since the electronegativity difference between Se and F (4.0 – 2.4 = 1.6) is greater than the electronegativity difference between Se and Cl (3.0 – 2.4 = 0.6), which is greater than the electronegativity difference between Se and Br (2.8 – 2.4 = 0.4).

$$\text{Se–F} > \text{Se–Cl} > \text{Se–Br}$$

b) EN of Cl = 3.0, EN of B = 2.0, EN of F = 4.0, EN of S = 2.5. The most polar is F–B with an EN difference of 2.0, next is Cl–B with an EN difference of 1.0, and next is S–B with an EN of 0.5.

$$\text{F–B} > \text{Cl–B} > \text{S–B}$$

TOOLS OF THE LABORATORY BOXED READING PROBLEMS

B9.1   Plan: Bond strength increases as the number of electrons in the bond increases.
Solution:
The C=C bond would show absorption of IR at shorter wavelength (higher energy) due to it being a stronger bond than C–C. The C≡C bond would show absorption at a shorter wavelength (higher energy) than the C=C bond since the triple bond has a larger bond energy than the double bond.

B9.2   Plan: The frequencies of the vibrations are given and are converted to wavelength with the relationship $c = \lambda \nu$. The energy of each vibration is calculated by using $E = h\nu$.
Solution:

a) $c = \lambda \nu$ or $\lambda = \dfrac{c}{\nu}$

$\lambda \text{ (nm)} = \dfrac{c}{\nu} = \dfrac{3.00 \times 10^8 \text{ m/s}}{4.02 \times 10^{13}/\text{s}} \left(\dfrac{10^9 \text{ nm}}{1 \text{ m}} \right) = 7462.687 = \mathbf{7460 \text{ nm}}$ (symmetric stretch)

$\lambda \text{ (nm)} = \dfrac{c}{\nu} = \dfrac{3.00 \times 10^8 \text{ m/s}}{2.00 \times 10^{13}/\text{s}} \left(\dfrac{10^9 \text{ nm}}{1 \text{ m}} \right) = \mathbf{1.50 \times 10^4 \text{ nm}}$ (bending)

$\lambda \text{ (nm)} = \dfrac{c}{\nu} = \dfrac{3.00 \times 10^8 \text{ m/s}}{7.05 \times 10^{13}/\text{s}} \left(\dfrac{10^9 \text{ nm}}{1 \text{ m}} \right) = 4255.32 = \mathbf{4260 \text{ nm}}$ (asymmetrical stretch)

b) $E = h\nu$

$E = (6.626 \times 10^{-34} \text{ J} \cdot \text{s})(4.02 \times 10^{13} \text{ s}^{-1}) = 2.66365 \times 10^{-20} = \mathbf{2.66 \times 10^{-20} \text{ J}}$ (symmetric stretch)

$E = (6.626 \times 10^{-34} \text{ J} \cdot \text{s})(2.00 \times 10^{13} \text{ s}^{-1}) = 1.3252 \times 10^{-20} = \mathbf{1.33 \times 10^{-20} \text{ J}}$ (bending)

$E = (6.626 \times 10^{-34} \text{ J} \cdot \text{s})(7.05 \times 10^{13} \text{ s}^{-1}) = 4.6713 \times 10^{-20} = \mathbf{4.67 \times 10^{-20} \text{ J}}$ (asymmetrical stretch)

Bending requires the least amount of energy.

END–OF–CHAPTER PROBLEMS

9.1   a) Larger ionization energy decreases metallic character.
b) Larger atomic radius increases metallic character.
c) Larger number of outer electrons decreases metallic character.
d) Larger effective nuclear charge decreases metallic character.

9.4   Plan: Metallic behavior increases to the left and down a group in the periodic table.
Solution:
a) **Cs** is more metallic since it is further down the alkali metal group than Na.
b) **Rb** is more metallic since it is both to the left and down from Mg.
c) **As** is more metallic since it is further down Group 5A(15) than N.

9.6   Plan: Ionic bonding occurs between metals and nonmetals, covalent bonding between nonmetals, and metallic bonds between metals.
Solution:
a) Bond in CsF is **ionic** because Cs is a metal and F is a nonmetal.
b) Bonding in N_2 is **covalent** because N is a nonmetal.
c) Bonding in Na(s) is **metallic** because this is a monatomic, metal solid.

9.8   Plan: Ionic bonding occurs between metals and nonmetals, covalent bonding between nonmetals, and metallic bonds between metals.
Solution:
a) Bonding in O_3 would be **covalent** since O is a nonmetal.
b) Bonding in $MgCl_2$ would be **ionic** since Mg is a metal and Cl is a nonmetal.
c) Bonding in BrO_2 would be **covalent** since both Br and O are nonmetals.

9.10   Plan: Lewis electron-dot symbols show valence electrons as dots. Place one dot at a time on the four sides (this method explains the structure in b) and then pair up dots until all valence electrons are used. The group number of the main-group elements (Groups $1A(1)$-$8A(18)$) gives the number of valence electrons. Rb is Group $1A(1)$, Si is Group $4A(14)$, and I is Group $7A(17)$.
Solution:

a)   Rb$\cdot$ b) $\cdot\overset{\bullet}{\underset{\bullet}{Si}}\cdot$ c) $\overset{\bullet\bullet}{\underset{\bullet\bullet}{:I}}\cdot$

9.12   Plan: Lewis electron-dot symbols show valence electrons as dots. Place one dot at a time on the four sides (this method explains the structure in b) and then pair up dots until all valence electrons are used. The group number of the main main-group elements (Groups $1A(1)$-$8A(18)$) gives the number of valence electrons. Sr is Group $2A(2)$, P is Group $5A(15)$, and S is Group $6A(16)$.
Solution:

a) $\cdot$ Sr $\cdot$ b) $\cdot\overset{\bullet\bullet}{\underset{\bullet}{P}}\cdot$ c) $\overset{\bullet\bullet}{\underset{\bullet}{:S}}\cdot$

9.14   Plan: Assuming X is an A-group element, the number of dots (valence electrons) equals the group number. Once the group number is known, the general electron configuration of the element can be written.
Solution:
a) Since there are 6 dots in the Lewis electron-dot symbol, element X has 6 valence electrons and is a Group $6A(16)$ element. Its general electron configuration is **[noble gas]**ns^2np^4, where n is the energy level.
b) Since there are 3 dots in the Lewis electron-dot symbol, element X has 3 valence electrons and is a Group $3A(13)$ element with general electron configuration **[noble gas]**ns^2np^1.

9.17   a) Because the lattice energy is the result of electrostatic attractions among the oppositely charged ions, its magnitude depends on several factors, including ionic size, ionic charge, and the arrangement of ions in the solid. For a particular arrangement of ions, the lattice energy increases as the charges on the ions increase and as their radii decrease.
b) Increasing lattice energy: **A < B < C**

9.20   Plan: Write condensed electron configurations and draw the Lewis electron-dot symbols for the atoms. The group number of the main-group elements (Groups $1A(1)$-$8A(18)$) gives the number of valence electrons. Remove electrons from the metal and add electrons to the nonmetal to attain filled outer levels. The number of electrons lost by the metal must equal the number of electrons gained by the nonmetal.
Solution:
a) Barium is a metal and loses 2 electrons to achieve a noble gas configuration:

$$Ba\ ([Xe]6s^2) \rightarrow Ba^{2+}\ ([Xe]) + 2e^-$$

$\cdot$ Ba $\cdot$ $\longrightarrow$ $\left[Ba \right]^{2+}$ + 2 e   **2e⁻**

Chlorine is a nonmetal and gains 1 electron to achieve a noble gas configuration:

$$Cl\ ([Ne]3s^23p^5) + 1e^- \rightarrow Cl^-\ ([Ne]3s^23p^6)$$

$\overset{\bullet\bullet}{\underset{\bullet\bullet}{:Cl}}\cdot$ + 1 e⁻ $\longrightarrow$ $\left[\overset{\bullet\bullet}{\underset{\bullet\bullet}{:Cl:}} \right]^-$

Two Cl atoms gain the 2 electrons lost by Ba. The ionic compound formed is **BaCl₂**.

b) Strontium is a metal and loses 2 electrons to achieve a noble gas configuration:

$$Sr\ ([Kr]5s^2) \rightarrow Sr^{2+}\ ([Kr]) + 2e^-$$

Oxygen is a nonmetal and gains 2 electrons to achieve a noble gas configuration:

$$O\ ([He]2s^22p^4) + 2e^- \rightarrow O^{2-}\ ([He]2s^22p^6)$$

One O atom gains the two electrons lost by one Sr atom. The ionic compound formed is **SrO**.

c) Aluminum is a metal and loses 3 electrons to achieve a noble gas configuration:

$$Al\ ([Ne]3s^23p^1) \rightarrow Al^{3+}\ ([Ne]) + 3e^-$$

Fluorine is a nonmetal and gains 1 electron to achieve a noble gas configuration:

$$F\ ([He]2s^22p^5) + 1e^- \rightarrow F^-\ ([He]2s^22p^6)$$

Three F atoms gains the three electrons lost by one Al atom. The ionic compound formed is **AlF₃**.

d) Rubidium is a metal and loses 1 electron to achieve a noble gas configuration:

$$Rb\ ([Kr]5s^1) \rightarrow Rb^+\ ([Kr]) + 1e^-$$

Oxygen is a nonmetal and gains 2 electrons to achieve a noble gas configuration:

$$O\ ([He]2s^22p^4) + 2e^- \rightarrow O^{2-}\ ([He]2s^22p^6)$$

One O atom gains the two electrons lost by two Rb atoms. The ionic compound formed is **Rb₂O**.

9.22   Plan: Find the charge of the known atom and use that charge to find the ionic charge of element X.
For A-group cations, ion charge = the group number; for anions, ion charge = the group number – 8.
Once the ion charge of X is known, the group number can be determined.
Solution:
a) X in XF_2 is a cation with +2 charge since the anion is F⁻ and there are two fluoride ions in the compound.
Group 2A(2) metals form +2 ions.
b) X in MgX is an anion with – 2 charge since Mg^{2+} is the cation. Elements in **Group 6A(16)** form –2 ions
(6 – 8 = –2).
c) X in X_2SO_4 must be a cation with +1 charge since the polyatomic sulfate ion has a charge of –2. X comes from
Group 1A(1).

9.24   Plan: Find the charge of the known atom and use that charge to find the ionic charge of element X.
For A-group cations, ion charge = the group number; for anions, ion charge = the group
number – 8. Once the ion charge of X is known, the group number can be determined.
Solution:
a) X in X_2O_3 is a cation with +3 charge. The oxygen in this compound has a –2 charge. To produce an electrically
neutral compound, 2 cations with +3 charge bond with 3 anions with –2 charge: 2(+3) + 3(–2) = 0. Elements in
Group 3A(13) form +3 ions.
b) The carbonate ion, CO_3^{2-}, has a –2 charge, so X has a +2 charge. **Group 2A(2)** elements form +2 ions.
c) X in Na_2X has a –2 charge, balanced with the +2 overall charge from the two Na⁺ ions. Group **6A(16)** elements
gain 2 electrons to form –2 ions with a noble gas configuration.

9.26   Plan: The magnitude of the lattice energy depends on ionic size and ionic charge. For a particular arrangement of
ions, the lattice energy increases as the charges on the ions increase and as their radii decrease.

Solution:
a) **BaS** would have the higher lattice energy since the charge on each ion (+2 for Ba and –2 for S) is twice the charge on the ions in CsCl (+1 for Cs and –1 for Cl) and lattice energy is greater when ionic charges are larger.
b) **LiCl** would have the higher lattice energy since the ionic radius of Li^+ is smaller than that of Cs^+ and lattice energy is greater when the distance between ions is smaller.

9.28   Plan: The magnitude of the lattice energy depends on ionic size and ionic charge. For a particular arrangement of ions, the lattice energy increases as the charges on the ions increase and as their radii decrease.
Solution:
a) **BaS** has the lower lattice energy because the ionic radius of Ba^{2+} is larger than Ca^{2+}. A larger ionic radius results in a greater distance between ions. The lattice energy decreases with increasing distance between ions.
b) **NaF** has the lower lattice energy since the charge on each ion (+1, –1) is half the charge on the Mg^{2+} and O^{2-} ions. Lattice energy increases with increasing ion charge.

9.30   Plan: The lattice energy of NaCl is represented by the equation $NaCl(s) \rightarrow Na^+(g) + Cl^-(g)$. Use Hess's law and arrange the given equations so that they sum up to give the equation for the lattice energy. You will need to reverse the last equation (and change the sign of $\Delta H°$); you will also need to multiply the second equation ($\Delta H°$) by ½
Solution:

	$\Delta H°$
$Na(s) \rightarrow Na(g)$	109 kJ
$1/2Cl_2(g) \rightarrow Cl(g)$	243/2 = 121.5 kJ
$Na(g) \rightarrow Na^+(g) + e^-$	496 kJ
$Cl(g) + e^- \rightarrow Cl^-(g)$	–349 kJ
$NaCl(s) \rightarrow Na(s) + 1/2Cl_2(g)$	411 kJ (Reaction is reversed; sign of $\Delta H°$ changed.)
$NaCl(s) \rightarrow Na^+(g) + Cl^-(g)$	788.5 = **788 kJ**

The lattice energy for NaCl is less than that of LiF, which is expected since lithium and fluoride ions are smaller than sodium and chloride ions, resulting in a larger lattice energy for LiF.

9.33   Plan: The electron affinity of fluorine is represented by the equation $F(g) + e^- \rightarrow F^-(g)$. Use Hess's law and arrange the given equations so that they sum up to give the equation for the lattice energy, $KF(s) \rightarrow K^+(g) + F^-(g)$. You will need to reverse the last two given equations (and change the sign of ΔH); you will also need to multiply the third equation ($\Delta H°$) by ½. Solve for EA.
Solution:
An analogous Born-Haber cycle has been described in Figure 9.7 for LiF. Use Hess's law and solve for the EA of fluorine: $\Delta H°$

$K(s) \rightarrow K(g)$	90 kJ
$K(g) \rightarrow K^+(g) + e^-$	419 kJ
$1/2F_2(g) \rightarrow F(g)$	1/2(159) = 79.5 kJ
$F(g) + e^- \rightarrow F^-(g)$	? = EA
$KF(s) \rightarrow K(s) + 1/2F_2(g)$	569 kJ (Reverse the reaction and change the sign of $\Delta H°$.)
$KF(s) \rightarrow K^+(g) + F^-(g)$	821 kJ (Reverse the reaction and change the sign of $\Delta H°$.)

821 kJ = 90 kJ + 419 kJ + 79.5 kJ + EA + 569 kJ
EA = 821 kJ – (90 + 419 + 79.5 + 569)kJ
EA = –336.5 = **–336 kJ**

9.34   When two chlorine atoms are far apart, there is no interaction between them. Once the two atoms move closer together, the nucleus of each atom attracts the electrons on the other atom. As the atoms move closer this attraction increases, but the repulsion of the two nuclei also increases. When the atoms are very close together the repulsion between nuclei is much stronger than the attraction between nuclei and electrons. The final internuclear distance for the chlorine molecule is the distance at which maximum attraction is achieved in spite of the repulsion. At this distance, the energy of the molecule is at its lowest value.

9.35   The bond energy is the energy required to overcome the attraction between H atoms and Cl atoms in one mole of HCl molecules in the gaseous state. Energy input is needed to break bonds, so bond energy is always absorbed (endothermic) and $\Delta H^\circ_{\text{bond breaking}}$ is positive. The same amount of energy needed to break the bond is released upon its formation, so $\Delta H^\circ_{\text{bond forming}}$ has the same magnitude as $\Delta H^\circ_{\text{bond breaking}}$, but opposite in sign (always exothermic and negative).

9.39   Plan: Bond strength increases as the atomic radii of atoms in the bond decrease; bond strength also increases as bond order increases.
Solution:
a) **I–I < Br–Br < Cl–Cl**. Atomic radii decrease up a group in the periodic table, so I is the largest and Cl is the smallest of the three.
b) **S–Br < S–Cl < S–H**. H has the smallest radius and Br has the largest, so the bond strength for S–H is the greatest and that for S–Br is the weakest.
c) **C–N < C=N < C≡N**. Bond strength increases as the number of electrons in the bond increases. The triple bond is the strongest and the single bond is the weakest.

9.41   Plan: Bond strength increases as the atomic radii of atoms in the bond decrease; bond strength also increases as bond order increases.
Solution:
a) The C=O bond (bond order = 2) is stronger than the C–O bond (bond order = 1).
b) O is smaller than C so the O–H bond is shorter and stronger than the C–H bond.

9.43   Reaction between molecules requires the breaking of existing bonds and the formation of new bonds. Substances with weak bonds are more reactive than are those with strong bonds because less energy is required to break weak bonds.

9.45   Plan: Write the combustion reactions of methane and of formaldehyde. The reactants requiring the smaller amount of energy to break bonds will have the greater heat of reaction. Examine the bonds in the reactant molecules that will be broken. In general, more energy is required to break double bonds than to break single bonds.
Solution:
For methane: $CH_4(g) + 2O_2(g) \rightarrow CO_2(g) + 2H_2O(l)$ which requires that 4 C–H bonds and 2 O=O bonds be broken and 2 C=O bonds and 4 O–H bonds be formed.

For formaldehyde: $CH_2O(g) + O_2(g) \rightarrow CO_2(g) + H_2O(l)$ which requires that 2 C–H bonds, 1 C=O bond, and 1 O=O bond be broken and 2 C=O bonds and 2 O–H bonds be formed.

Methane contains more C–H bonds and fewer C=O bonds than formaldehyde. Since C–H bonds take less energy to break than C=O bonds, more energy is released in the combustion of methane than of formaldehyde.

9.47   Plan: To find the heat of reaction, add the energy required to break all the bonds in the reactants to the energy released to form all bonds in the product. Remember to use a negative sign for the energy of the bonds formed since bond formation is exothermic. The bond energy values are found in Table 9.2.
Solution:
Reactant bonds broken:
1 x C=C = (1 mol)(614 kJ/mol) =   614 kJ
4 x C–H = (4 mol)(413 kJ/mol) = 1652 kJ
1 x Cl–Cl = (1 mol)(243 kJ/mol) =   243 kJ
                $\Sigma\Delta H^\circ_{\text{bonds broken}}$ = 2509 kJ

Product bonds formed:

1 x C–C = (1 mol)(–347 kJ/mol) = –347 kJ

4 x C–H = (4 mol)(–413 kJ/mol) = –1652 kJ

2 x C–Cl = (2 mol)(–339 kJ/mol = –678 kJ

$$\Sigma \Delta H^{\circ}_{\text{bonds formed}} = -2677 \text{ kJ}$$

$$\Delta H^{\circ}_{\text{rxn}} = \Sigma \Delta H^{\circ}_{\text{bonds broken}} + \Sigma \Delta H^{\circ}_{\text{bonds formed}} = 2509 \text{ kJ} + (-2677 \text{ kJ}) = \textbf{–168 kJ}$$

9.49   <u>Plan:</u> To find the heat of reaction, add the energy required to break all the bonds in the reactants to the energy released to form all bonds in the product. Remember to use a negative sign for the energy of the bonds formed since bond formation is exothermic. The bond energy values are found in Table 9.2.
<u>Solution:</u>
The reaction:

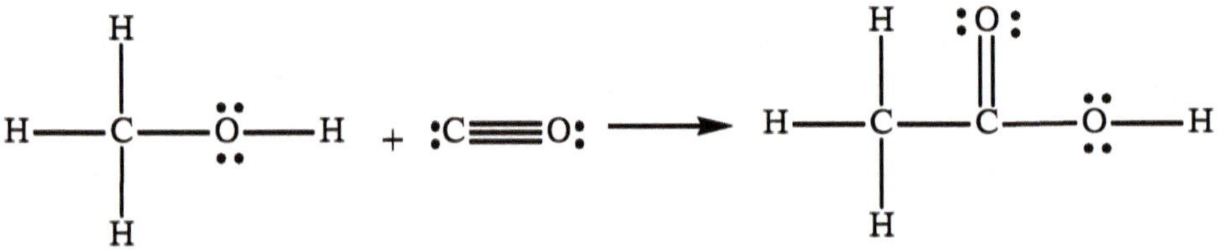

Reactant bonds broken:

1 x C–O = (1 mol)(358 kJ/mol) = 358 kJ

3 x C–H = (3 mol)(413 kJ/mol) = 1239 kJ

1 x O–H = (1 mol)(467 kJ/mol) = 467 kJ

1 x C≡O = (1 mol)(1070 kJ/mol) = 1070 kJ

$$\Sigma \Delta H^{\circ}_{\text{bonds broken}} = 3134 \text{ kJ}$$

Product bonds formed:

3 x C–H = (3 mol)(–413 kJ/mol) = –1239 kJ

1 x C–C = (1 mol)(–347 kJ/mol) = –347 kJ

1 x C=O = (1 mol)(–745 kJ/mol) = –745 kJ

1 x C–O = (1 mol)(–358 kJ/mol) = –358 kJ

1 x O–H = (1 mol)(–467 kJ/mol) = –467 kJ

$$\Sigma \Delta H^{\circ}_{\text{bonds formed}} = -3156 \text{ kJ}$$

$$\Delta H^{\circ}_{\text{rxn}} = \Sigma \Delta H^{\circ}_{\text{bonds broken}} + \Sigma \Delta H^{\circ}_{\text{bonds formed}} = 3134 \text{ kJ} + (-3156 \text{ kJ}) = \textbf{–22 kJ}$$

9.50   <u>Plan:</u> To find the heat of reaction, add the energy required to break all the bonds in the reactants to the energy released to form all bonds in the product. Remember to use a negative sign for the energy of the bonds formed since bond formation is exothermic. The bond energy values are found in Table 9.2.
<u>Solution:</u>

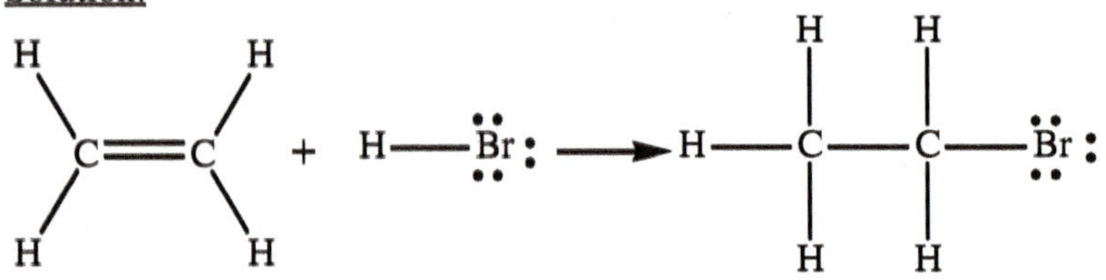

Reactant bonds broken:

1 x C=C = (1 mol)(614 kJ/mol) = 614 kJ

4 x C–H = (4 mol)(413 kJ/mol) = 1652 kJ

1 x H–Br = (1 mol)(363 kJ/mol) = 363 kJ

$$\Sigma \Delta H^{\circ}_{bonds\ broken} = 2629\ kJ$$

Product bonds formed:

5 x C–H = (5 mol)(–413 kJ/mol) = –2065 kJ

1 x C–C = (1 mol)(–347 kJ/mol) = –347 kJ

1 x C–Br = (1 mol)(–276 kJ/mol) = –276 kJ

$$\Sigma \Delta H^{\circ}_{bonds\ formed} = -2688\ kJ$$

$\Delta H^{\circ}_{rxn} = \Sigma \Delta H^{\circ}_{bonds\ broken} + \Sigma \Delta H^{\circ}_{bonds\ formed} = 2629\ kJ + (-2688\ kJ) = $ **–59 kJ**

9.51   Electronegativity increases from left to right across a period (except for the noble gases) and increases from bottom to top within a group. Fluorine (F) and oxygen (O) are the two most electronegative elements. Cesium (Cs) and francium (Fr) are the two least electronegative elements.

9.53   Ionic bonds occur between two elements of very different electronegativity, generally a metal with low electronegativity and a nonmetal with high electronegativity. Although electron sharing occurs to a very small extent in some ionic bonds, the primary force in ionic bonds is attraction of opposite charges resulting from electron transfer between the atoms. A nonpolar covalent bond occurs between two atoms with identical electronegativity values where the sharing of bonding electrons is equal. A polar covalent bond is between two atoms (generally nonmetals) of different electronegativities so that the bonding electrons are unequally shared. The H–O bond in water is **polar covalent**. The bond is between two nonmetals so it is covalent and not ionic, but atoms with different electronegativity values are involved.

9.56   Plan: Electronegativity increases from left to right across a period (except for the noble gases) and increases from bottom to top within a group.
Solution:
a) **Si < S < O**, sulfur is more electronegative than silicon since it is located further to the right in the table. Oxygen is more electronegative than sulfur since it is located nearer the top of the table.
b) **Mg < As < P**, magnesium is the least electronegative because it lies on the left side of the periodic table and phosphorus and arsenic on the right side. Phosphorus is more electronegative than arsenic because it is higher in the table.

9.58   Plan: Electronegativity increases from left to right across a period (except for the noble gases) and increases from bottom to top within a group.
Solution:
a) **N > P > Si**, nitrogen is above P in Group 5(A)15 and P is to the right of Si in Period 3.
b) **As > Ga > Ca**, all three elements are in Period 4, with As the rightmost element.

9.60   Plan: The polar arrow points toward the more electronegative atom. Electronegativity increases from left to right across a period (except for the noble gases) and increases from bottom to top within a group.
Solution:

a)   N——B   b)   N——O   c)   C——S *(none)*

d)   S——O   e)   N——H   f)   Cl——O

9.62   Plan: The more polar bond will have a greater difference in electronegativity, ΔEN.
        Solution:
        a) N: EN = 3.0; B: EN = 2.0; ΔEN_a = 3.0 – 2.0 = 1.0

        b) N: EN = 3.0; O: EN = 3.5; ΔEN_b = 3.5 – 3.0 = 0.5

        c) C: EN = 2.5; S: EN = 2.5; ΔEN_c = 2.5 – 2.5 = 0

        d) S: EN = 2.5; O: EN = 3.5; ΔEN_d = 3.5 – 2.5 = 1.0

        e) N: EN = 3.0; H: EN = 2.1; ΔEN_e = 3.0 – 2.1 = 0.9

        f) Cl: EN = 3.0; O: EN = 3.5; ΔEN_f = 3.5 – 3.0 = 0.5

        (a), (d), and (e) have greater bond polarity.

9.64   Plan: Ionic bonds occur between two elements of very different electronegativity, generally a metal with low
        electronegativity and a nonmetal with high electronegativity. Although electron sharing occurs to a very small extent
        in some ionic bonds, the primary force in ionic bonds is attraction of opposite charges resulting from electron
        transfer between the atoms. A nonpolar covalent bond occurs between two atoms with identical electronegativity
        values where the sharing of bonding electrons is equal. A polar covalent bond is between two atoms (generally
        nonmetals) of different electronegativities so that the bonding electrons are unequally shared. For polar covalent
        bonds, the larger the ΔEN, the more polar the bond.
        Solution:
        a) Bonds in S_8 are **nonpolar covalent**. All the atoms are nonmetals so the substance is covalent and bonds are
        nonpolar because all the atoms are of the same element and thus have the same electronegativity value. $\Delta EN = 0$.
        b) Bonds in RbCl are **ionic** because Rb is a metal and Cl is a nonmetal. ΔEN is large.
        c) Bonds in PF_3 are **polar covalent**. All the atoms are nonmetals so the substance is covalent. The bonds between
        P and F are polar because their electronegativity differs (by 1.9 units for P–F).
        d) Bonds in SCl_2 are **polar covalent**. S and Cl are nonmetals and differ in electronegativity (by 0.5 unit for S–Cl).
        e) Bonds in F_2 are **nonpolar covalent**. F is a nonmetal. Bonds between two atoms of the same element are
        nonpolar since $\Delta EN = 0$.
        f) Bonds in SF_2 are **polar covalent**. S and F are nonmetals that differ in electronegativity (by 1.5 units for S–F).
        Increasing bond polarity: $SCl_2 < SF_2 < PF_3$

9.66   Plan: Increasing ionic character occurs with increasing ΔEN. Electronegativity increases from left to right across
        a period (except for the noble gases) and increases from bottom to top within a group. The polar arrow points
        toward the more electronegative atom.
        Solution:
        a) H: EN = 2.1; Cl: EN = 3.0; Br: EN = 2.8; I: EN = 2.5

        ΔEN_{HBr} = 2.8 – 2.1 = 0.7; ΔEN_{HCl} = 3.0 – 2.1 = 0.9; ΔEN_{HI} = 2.5 – 2.1 = 0.4

        b) H: EN = 2.1; O: EN = 3.5; C: EN = 2.5; F: EN = 4.0

        ΔEN_{HO} = 3.5 – 2.1 = 1.4; ΔEN_{CH} = 2.5 – 2.1 = 0.4; ΔEN_{HF} = 4.0 – 2.1 = 1.9

        c) Cl: EN = 3.0; S: EN = 2.5; P: EN = 2.1; Si: EN = 1.8

        ΔEN_{SCl} = 3.0 – 2.5 = 0.5; ΔEN_{PCl} = 3.0 – 2.1 = 0.9; ΔEN_{SiCl} = 3.0 – 1.8 = 1.2

        a)   H——I   <   H——Br   <   H——Cl

        b)   H——C   <   H——O   <   H——F

        c)   S——Cl   <   P——Cl   <   Si——Cl

9.69   a) A solid metal is a shiny solid that conducts heat, is malleable, and melts at high temperatures. (Other answers include relatively high boiling point and good conductor of electricity.)
b) Metals lose electrons to form positive ions and metals form basic oxides.

9.73   Plan: Write a balanced chemical reaction. The given heat of reaction is the sum of the energy required to break all the bonds in the reactants and the energy released to form all bonds in the product. Remember to use a negative sign for the energy of the bonds formed since bond formation is exothermic. The bond energy values are found in Table 9.2. Use the ratios from the balanced reaction between the heat of reaction and acetylene and between acetylene and CO_2 and O_2 to find the amounts needed. The ideal gas law is used to convert from moles of oxygen to volume of oxygen.
Solution:

a) $C_2H_2 + 5/2 O_2 \rightarrow 2CO_2 + H_2O$ $\Delta H^{\circ}_{rxn} = -1259$ kJ/mol

H–C≡C–H + 5/2 O=O → 2 O=C=O + H–O–H

$\Delta H^{\circ}_{rxn} = \Sigma \Delta H^{\circ}_{bonds\ broken} + \Sigma \Delta H^{\circ}_{bonds\ formed}$

$\Delta H^{\circ}_{rxn} = [2\ BE_{C-H} + BE_{C\equiv C} + 5/2\ BE_{O=O}] + [4\ (-BE_{C=O}) + 2\ (-BE_{O-H})]$

-1259 kJ $= [2(413) + BE_{C\equiv C} + 5/2(498)] + [4(-799) + 2(-467)]$

-1259 kJ $= [826 + BE_{C\equiv C} + 1245] + [-4130)]$ kJ

-1259 kJ $= -2059 + BE_{C\equiv C}$ kJ

$BE_{C\equiv C} = \textbf{800. kJ/mol}$ Table 9.2 lists the value as 839 kJ/mol.

b) Heat (kJ) $= \left(500.0\ g\ C_2H_2\right)\left(\dfrac{1\ mol\ C_2H_2}{26.04\ g\ C_2H_2}\right)\left(\dfrac{-1259\ kJ}{1\ mol\ C_2H_2}\right)$

$= -2.4174347 \times 10^4 = \textbf{–2.417} \times \textbf{10}^4\ \textbf{kJ}$

c) Mass (g) of $CO_2 = \left(500.0\ g\ C_2H_2\right)\left(\dfrac{1\ mol\ C_2H_2}{26.04\ g\ C_2H_2}\right)\left(\dfrac{2\ mol\ CO_2}{1\ mol\ C_2H_2}\right)\left(\dfrac{44.01\ g\ CO_2}{1\ mol\ CO_2}\right)$

$= 1690.092 = \textbf{1690. g CO}_2$

d) Amount (mol) of $O_2 = \left(500.0\ g\ C_2H_2\right)\left(\dfrac{1\ mol\ C_2H_2}{26.04\ g\ C_2H_2}\right)\left(\dfrac{(5/2)\ mol\ O_2}{1\ mol\ C_2H_2}\right)$

$= 48.0030722\ mol\ O_2$

$PV = nRT$

Volume (L) of $O_2 = \dfrac{nRT}{P} = \dfrac{\left(48.0030722\ mol\ O_2\right)\left(0.08206\ \dfrac{L \cdot atm}{mol \cdot K}\right)(298\ K)}{18.0\ atm}$

$= 65.2145 = \textbf{65.2 L O}_2$

9.75   Plan: The heat of formation of MgCl is represented by the equation $Mg(s) + 1/2Cl_2(g) \rightarrow MgCl(s)$. Use Hess's law and arrange the given equations so that they sum up to give the equation for the heat of formation of MgCl. You will need to multiply the second equation by ½; you will need to reverse the equation for the lattice energy [$MgCl(s) \rightarrow Mg^+(g) + Cl^-(g)$] and change the sign of the given lattice energy value. Negative heats of formation are energetically favored.
Solution:

a)   1) $Mg(s) \rightarrow$ ~~Mg(g)~~ $\Delta H^{\circ}_1 = 148$ kJ

2) $1/2Cl_2(g) \rightarrow$ ~~Cl(g)~~ $\Delta H^{\circ}_2 = 1/2(243$ kJ$) = 121.5$ kJ

3) ~~Mg(g)~~ $\rightarrow$ ~~Mg$^+$(g)~~ $+ e^-$ $\Delta H^{\circ}_3 = 738$ kJ

9-13

4) $\cancel{Cl(g)} + e^- \rightarrow \cancel{Cl^-(g)}$ $\qquad\qquad$ $\Delta H_4^\circ = -349$ kJ

5) $\cancel{Mg^+(g)} + \cancel{Cl^-(g)} \rightarrow MgCl(s)$ $\qquad$ $\Delta H_5^\circ = -783.5$ kJ $(= -\Delta H_{lattice}^\circ (MgCl))$

$\overline{\qquad Mg(s) + 1/2Cl_2(g) \rightarrow MgCl(s) \quad \Delta H_f^\circ (MgCl) = ? \qquad}$

$\Delta H_f^\circ (MgCl) = \Delta H_1^\circ + \Delta H_2^\circ + \Delta H_3^\circ + \Delta H_4^\circ + \Delta H_5^\circ$

$\qquad = 148$ kJ $+ 121.5$ kJ $+ 738$ kJ $+ (-349$ kJ$) + (-783.5$ kJ$) = $ **–125 kJ**

b) **Yes**, since ΔH_f° for MgCl is negative, MgCl(s) is stable relative to its elements.

c) $2MgCl(s) \rightarrow MgCl_2(s) + Mg(s)$

$\Delta H_{rxn}^\circ = \Delta H_{rxn}^\circ = \Sigma m \Delta H_{f(products)}^\circ - \Sigma n \Delta H_{f(reactants)}^\circ$

$\Delta H_{rxn}^\circ = \{1 \Delta H_f^\circ [MgCl_2(s)] + 1 \Delta H_f^\circ [Mg(s)]\} - \{2 \Delta H_f^\circ [MgCl(s)]\}$

$\Delta H_{rxn}^\circ = [1 \text{ mol}(-641.6 \text{ kJ/mol}) + 1 \text{ mol}(0)] - [2 \text{ mol}(-125 \text{ kJ/mol})]$

$\Delta H_{rxn}^\circ = -391.6 = $ **–392 kJ**

d) **No**, ΔH_f° for MgCl$_2$ is much more negative than that for MgCl. This makes the ΔH_{rxn}° value for the above reaction very negative, and the formation of MgCl$_2$ would be favored.

9.77 $\quad$ <u>Plan:</u> Find the bond energy for an H–I bond from Table 9.2. For part a), calculate the wavelength with this energy using the relationship from Chapter 7: $E = hc/\lambda$. For part b), calculate the energy for a wavelength of 254 nm and then subtract the energy from part a) to get the excess energy.
For part c), speed can be calculated from the excess energy since $E_k = 1/2mu^2$.
<u>Solution:</u>
a) Bond energy for H–I is 295 kJ/mol (Table 9.2).

Bond energy (J/photon) $= \left(\dfrac{295 \text{ kJ}}{\text{mol}}\right)\left(\dfrac{10^3 \text{ J}}{1 \text{ kJ}}\right)\left(\dfrac{1 \text{ mol}}{6.022\text{x}10^{23} \text{ photons}}\right) = 4.898705\text{x}10^{-19}$ J/photon

$E = hc/\lambda$

$\lambda \text{ (m)} = hc/E = \dfrac{\left(6.626\text{x}10^{-34} \text{ J}\cdot\text{s}\right)\left(3.00\text{x}10^8 \text{ m/s}\right)}{\left(4.898705\text{x}10^{-19} \text{ J}\right)} = 4.057807\text{x}10^{-7}$ m

$\lambda \text{ (nm)} = \left(4.057807\text{x}10^{-7} \text{ m}\right)\left(\dfrac{1 \text{ nm}}{10^{-9} \text{ m}}\right) = 405.7807 = $ **406 nm**

b) E (HI) $= 4.898705\text{x}10^{-19}$ J

$E \text{ (254 nm)} = hc/\lambda = \dfrac{\left(6.626\text{x}10^{-34} \text{ J}\cdot\text{s}\right)\left(3.00\text{x}10^8 \text{ m/s}\right)}{254 \text{ nm}}\left(\dfrac{1 \text{ nm}}{10^{-9} \text{ m}}\right) = 7.82598\text{x}10^{-19}$ J

Excess energy $= 7.82598\text{x}10^{-19}$ J $- 4.898705\text{x}10^{-19}$ J $= 2.92728\text{x}10^{-19} = $ **2.93x10^{-19} J**

c) Mass (kg) of H $= \left(\dfrac{1.008 \text{ g H}}{\text{mol}}\right)\left(\dfrac{\text{mol}}{6.022\text{x}10^{23}}\right)\left(\dfrac{1 \text{ kg}}{10^3 \text{ g}}\right) = 1.67386\text{x}10^{-27}$ kg

$E_k = 1/2mu^2$ thus, $u = \sqrt{\dfrac{2E}{m}}$

$u = \sqrt{\dfrac{2(2.92728\text{x}10^{-19} \text{ J})}{1.67386\text{x}10^{-27} \text{ kg}}\left(\dfrac{\text{kg}\cdot\text{m}^2/\text{s}^2}{\text{J}}\right)} = 1.8701965\text{x}10^4 = $ **1.87x10^4 m/s**

9.80 $\quad$ <u>Plan:</u> Find the appropriate bond energies in Table 9.2. Calculate the wavelengths using $E = hc/\lambda$.
<u>Solution:</u>
C–Cl bond energy = 339 kJ/mol

$$\text{Bond energy (J/photon)} = \left(\frac{339 \text{ kJ}}{\text{mol}}\right)\left(\frac{10^3 \text{ J}}{1 \text{ kJ}}\right)\left(\frac{1 \text{ mol}}{6.022 \times 10^{23} \text{ photons}}\right) = 5.62936 \times 10^{-19} \text{ J/photon}$$

$$E = hc/\lambda$$

$$\lambda \text{ (m)} = hc/E = \frac{\left(6.626 \times 10^{-34} \text{ J} \cdot \text{s}\right)\left(3.00 \times 10^8 \text{ m/s}\right)}{\left(5.62936 \times 10^{-19} \text{ J}\right)} = 3.5311296 \times 10^{-7} = \mathbf{3.53 \times 10^{-7} \text{ m}}$$

O_2 bond energy = 498 kJ/mol

$$\text{Bond energy (J/photon)} = \left(\frac{498 \text{ kJ}}{\text{mol}}\right)\left(\frac{10^3 \text{ J}}{1 \text{ kJ}}\right)\left(\frac{1 \text{ mol}}{6.022 \times 10^{23} \text{ photons}}\right) = 8.269678 \times 10^{-19} \text{ J/photon}$$

$$E = hc/\lambda$$

$$\lambda \text{ (m)} = hc/E = \frac{\left(6.626 \times 10^{-34} \text{ J} \cdot \text{s}\right)\left(3.00 \times 10^8 \text{ m/s}\right)}{\left(8.269678 \times 10^{-19} \text{ J}\right)} = 2.40372 \times 10^{-7} = \mathbf{2.40 \times 10^{-7} \text{ m}}$$

9.81   Plan: Write balanced chemical equations for the formation of each of the compounds. Obtain the bond energy of fluorine from Table 9.2 (159 kJ/mol). Determine the average bond energy from ΔH = bonds broken + bonds formed. Remember that the bonds formed (Xe–F) have negative values since bond formation is exothermic.
Solution:

$$\Delta H^\circ_{rxn} = \Sigma \Delta H^\circ_{\text{bonds broken}} + \Sigma \Delta H^\circ_{\text{bonds formed}}$$

XeF_2 $Xe(g) + F_2(g) \rightarrow XeF_2(g)$

$\quad\quad\quad\quad\quad\quad \Delta H^\circ_{rxn} = -105$ kJ/mol = [1 mol F_2 (159 kJ/mol)] + [2 (–Xe–F)]

$\quad\quad\quad\quad\quad\quad -264$ kJ/mol = 2 (–Xe–F)

$\quad\quad\quad\quad\quad\quad$ Xe–F = **132 kJ/mol**

XeF_4 $Xe(g) + 2F_2(g) \rightarrow XeF_4(g)$

$\quad\quad\quad\quad\quad\quad \Delta H^\circ_{rxn} = -284$ kJ/mol = [2 mol F_2 (159 kJ/mol)] + [4 (–Xe–F)]

$\quad\quad\quad\quad\quad\quad -602$ kJ/mol = 4 (–Xe–F)

$\quad\quad\quad\quad\quad\quad$ Xe–F = 150.5 = **150. kJ/mol**

XeF_6 $Xe(g) + 3F_2(g) \rightarrow XeF_6(g)$

$\quad\quad\quad\quad\quad\quad \Delta H^\circ_{rxn} = -402$ kJ/mol = [3 mol F_2 (159 kJ/mol)] + [6 (–Xe–F)]

$\quad\quad\quad\quad\quad\quad -879$ kJ/mol = 6 (–Xe–F)

$\quad\quad\quad\quad\quad\quad$ Xe–F = 146.5 = **146 kJ/mol**

9.83   a)The presence of the very electronegative fluorine atoms bonded to one of the carbon atoms in H_3C—CF_3 makes the C–C bond polar. This polar bond will tend to undergo heterolytic rather than homolytic cleavage. More energy is required to force heterolytic cleavage.
b) Since one atom gets both of the bonding electrons in heterolytic bond breakage, this results in the formation of ions. In heterolytic cleavage a cation is formed, involving ionization energy; an anion is also formed, involving electron affinity. The bond energy of the O_2 bond is 498 kJ/mol.
ΔH = (homolytic cleavage + electron affinity + first ionization energy)
ΔH = (498/2 kJ/mol + (–141 kJ/mol) + 1314 kJ/mol) = 1422 = **1420 kJ/mol**
It would require 1420 kJ to heterolytically cleave 1 mol of O_2.

9.86   Plan: The heat of formation of SiO_2 is represented by the equation $Si(s) + O_2(g) \rightarrow SiO_2(s)$. Use Hess's law and arrange the given equations so that they sum up to give the equation for the heat of formation. The lattice energy of SiO_2 is represented by the equation
$SiO_2(s) \rightarrow Si^{4+}(g) + 2O^{2-}(g)$. You will need to reverse the lattice energy equation (and change the sign of ΔH°); you will also need to multiply the fourth given equation by 2.

Solution:

Use Hess' law. ΔH_f° of SiO_2 is found in Appendix B.

1) $Si(s) \rightarrow \cancel{Si(g)}$ $\Delta H_1^\circ = 454$ kJ

2) $\cancel{Si(g)} \rightarrow \cancel{Si^{4+}(g)} + \cancel{4e^-}$ $\Delta H_2^\circ = 9949$ kJ

3) $O_2(g) \rightarrow \cancel{2O(g)}$ $\Delta H_3^\circ = 498$ kJ

4) $\cancel{2O(g)} + \cancel{4e^-} \rightarrow \cancel{2O^{2-}(g)}$ $\Delta H_4^\circ = 2(737)$ kJ

5) $\cancel{Si^{4+}(g)} + \cancel{2O^{2-}(g)} \rightarrow SiO_2(s)$ $\Delta H_5^\circ = -\Delta H_{lattice}^\circ$ $(SiO_2) = $?

$\overline{\qquad\qquad Si(s) + O_2(g) \rightarrow SiO_2(s) \qquad\qquad \Delta H_f^\circ\ (SiO_2) = -910.9\text{ kJ}}$

$\Delta H_f^\circ = [\Delta H_1^\circ + \Delta H_2^\circ + \Delta H_3^\circ + \Delta H_4^\circ + (-\Delta H_{lattice}^\circ)]$

$-910.9 \text{ kJ} = [454 \text{ kJ} + 9949 \text{ kJ} + 498 \text{ kJ} + 2(737) \text{ kJ} + (-\Delta H_{lattice}^\circ)]$

$-\Delta H_{lattice}^\circ = -13,285.9$ kJ

$\Delta H_{lattice}^\circ = \textbf{13,286 kJ}$

9.88   Plan: Convert the bond energy in kJ/mol to units of J/photon. Use the equations $E = h\nu$, and $E = hc/\lambda$ to find the frequency and wavelength of light associated with this energy.
Solution:

Bond energy (J/photon) $= \left(\dfrac{347 \text{ kJ}}{\text{mol}}\right)\left(\dfrac{10^3 \text{ J}}{1 \text{ kJ}}\right)\left(\dfrac{1 \text{ mol}}{6.022 \times 10^{23} \text{ photons}}\right) = 5.762205 \times 10^{-19}$ J/photon

$E = h\nu$ or $\nu = \dfrac{E}{h}$

$\nu = \dfrac{E}{h} = \dfrac{5.762205 \times 10^{-19} \text{ J}}{6.626 \times 10^{-34} \text{ J s}} = 8.6963553 \times 10^{14} = \textbf{8.70} \boldsymbol{\times} \textbf{10}^{\textbf{14}} \textbf{ s}^{\textbf{-1}}$

$E = hc/\lambda$ or $\lambda = hc/E$

$\lambda \text{ (m)} = hc/E = \dfrac{\left(6.626 \times 10^{-34} \text{ J} \cdot \text{s}\right)\left(3.00 \times 10^8 \text{ m/s}\right)}{5.762205 \times 10^{-19} \text{ J}} = 3.44972 \times 10^{-7} = \textbf{3.45} \boldsymbol{\times} \textbf{10}^{\textbf{-7}} \textbf{ m}$

This is in the **ultraviolet** region of the electromagnetic spectrum.

9.90   Plan: Write the balanced equations for the reactions. Determine the heat of reaction from ΔH = bonds broken + bonds formed. Remember that the bonds formed have negative values since bond formation is exothermic.
Solution:

a) $2CH_4(g) + O_2(g) \rightarrow CH_3OCH_3(g) + H_2O(g)$

$\Delta H_{rxn}^\circ = \Sigma \Delta H_{bonds\ broken}^\circ + \Sigma \Delta H_{bonds\ formed}^\circ$

$\Delta H_{rxn}^\circ = [8 \text{ x } (BE_{C-H}) + 1 \text{ x } (BE_{O=O})] + [6 \text{ x } (BE_{C-H}) + 2 \text{ x } (BE_{C-O}) + 2 \text{ x } (BE_{O-H})]$

$\Delta H_{rxn}^\circ = [8 \text{ mol}(413 \text{ kJ/mol}) + 1 \text{ mol}(498 \text{ kJ/mol})]$

$\qquad\qquad\qquad\qquad + [6 \text{ mol}(-413 \text{ kJ/mol}) + 2 \text{ mol}(-358 \text{ kJ/mol}) + 2 \text{ mol}(-467 \text{ kJ/mol})]$

$\Delta H_{rxn}^\circ = \textbf{-326 kJ}$

$2CH_4(g) + O_2(g) \rightarrow CH_3CH_2OH(g) + H_2O(g)$

$\Delta H_{rxn}^\circ = \Sigma \Delta H_{bonds\ broken}^\circ + \Sigma \Delta H_{bonds\ formed}^\circ$

$\Delta H_{rxn}^\circ = [8 \text{ x } (BE_{C-H}) + 1 \text{ x } (BE_{O=O})] + [5 \text{ x } (BE_{C-H}) + 1 \text{ x } (BE_{C-C}) + 1 \text{ x } (BE_{C-O}) + 3 \text{ x } (BE_{O-H})]$

$\Delta H_{rxn}^\circ = [8 \text{ mol}(413 \text{ kJ/mol}) + 1 \text{ mol}(498 \text{ kJ/mol})]$

$\qquad\quad + [5 \text{ mol}(-413 \text{ kJ/mol}) + 1 \text{ mol}(-347 \text{ kJ/mol}) + 1 \text{ mol}(-358 \text{ kJ/mol}) + 3 \text{ mol}(-467 \text{ kJ/mol})]$

$\Delta H_{rxn}^\circ = \textbf{-369 kJ}$

b) The formation of gaseous **ethanol** is more exothermic.

c) The conversion reaction is $CH_3CH_2OH(g) \rightarrow CH_3OCH_3(g)$.

Use Hess's law:

$CH_3CH_2OH(g) + \cancel{H_2O(g)} \rightarrow \cancel{2CH_4(g)} + \cancel{O_2(g)}$ $\qquad\qquad$ $\Delta H^{\circ}_{rxn} = -(-369 \text{ kJ}) = 369 \text{ kJ}$

$\cancel{2CH_4(g)} + \cancel{O_2(g)} \rightarrow CH_3OCH_3(g) + \cancel{H_2O(g)}$ $\qquad\qquad$ $\Delta H^{\circ}_{rxn} = -326 \text{ kJ}$

$\overline{CH_3CH_2OH(g) \rightarrow CH_3OCH_3(g) \qquad\qquad\qquad\qquad \Delta H^{\circ}_{rxn} = -326 \text{ kJ} + 369 \text{ kJ} = \mathbf{43 \text{ kJ}}}$

CHAPTER 10 THE SHAPES OF MOLECULES

FOLLOW–UP PROBLEMS

10.1A   Plan: Count the valence electrons and follow the steps outlined in the sample problem to draw the Lewis structures.
Solution:
a) The sulfur is the central atom, as the hydrogen is never central. Each of the hydrogen atoms is placed around the sulfur. The actual positions of the hydrogen atoms are not important. The total number of valence electrons available is $[2 \times H(1e^-)] + [1 \times S(6e^-)] = 8e^-$. Connect each hydrogen atom to the sulfur with a single bond. These bonds use 4 of the electrons leaving 4 electrons. The last $4e^-$ go to the sulfur because the hydrogen atoms can take no more electrons.

$$H\text{——}\overset{\displaystyle ..}{S} :$$
$$|$$
$$H$$

Solution:
b) The aluminum has the lower group number so it is the central atom. Each of the chlorine atoms will be attached to the central aluminum. The total number of valence electrons available is $[4 \times F(7e^-)] + [1 \times Al(3e^-)] + 1e^-$ (for the negative charge) $= 32e^-$. Connecting the four chlorine atoms to the aluminum with single bonds uses $4 \times 2 = 8e^-$, leaving $32 - 8 = 24e^-$. The more electronegative chlorine atoms each need 6 electrons to complete their octets. This requires $4 \times 6 = 24e^-$. There are no more remaining electrons at this step; however, the aluminum has 8 electrons around it.

$$\left[\begin{array}{c} :\overset{..}{\underset{..}{Cl}}: \\ | \\ :\overset{..}{\underset{..}{Cl}}\text{——}Al\text{——}\overset{..}{\underset{..}{Cl}}: \\ | \\ :\overset{..}{\underset{..}{Cl}}: \end{array}\right]^{-}$$

Check: Count the electrons. Each of the five atoms has an octet.
Solution:
c) Both S and O have a lower group number than Cl, thus, one of these two elements must be central. Between S and O, S has the higher period number so it is the central atom. The total number of valence electrons available is $[2 \times Cl(7e^-)] + [1 \times S(6e^-)] + [1 \times O(6e^-)] = 26e^-$. Begin by distributing the two chlorine atoms and the oxygen atom around the central sulfur atom. Connect each of the three outlying atoms to the central sulfur with single bonds. This uses $3 \times 2 = 6e^-$, leaving $26 - 6 = 20e^-$. Each of the outlying atoms still needs 6 electrons to complete their octets. Completing these octets uses $3 \times 6 = 18$ electrons. The remaining 2 electrons are all the sulfur needs to complete its octet.

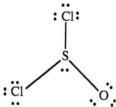

Check: Count the electrons. Each of the four atoms has an octet.

10.1B   Plan: Count the valence electrons and follow the steps outlined in the sample problem to draw the Lewis structures.

<u>Solution:</u>
a) The oxygen has the lower group number so it is the central atom. Each of the fluorine atoms will be attached to the central oxygen. The total number of valence electrons available is [2 x F(7e⁻)] + [1 x O(6e⁻)] = 20e⁻. Connecting the two fluorine atoms to the oxygen with single bonds uses 2 x 2 = 4e⁻, leaving 20 – 4 = 16e⁻. The more electronegative fluorine atoms each need 6 electrons to complete their octets. This requires 2 x 6 = 12e⁻. The 4 remaining electrons go to the oxygen.

<u>Check:</u> Count the electrons. Each of the three atoms has an octet.

<u>Solution:</u>
b) The carbon has the lower group number so it is the central atom (technically, hydrogen is in group 1A, but it can only form one bond and, thus, cannot be the central atom). Each of the hydrogen and bromine atoms will be attached to the central carbon. The total number of valence electrons available is [1 x C(4e⁻)] + [2 x H(1e⁻)] + [2 x Br(7e⁻)] = 20e⁻. Connecting the two hydrogen atoms and the two bromine atoms to the carbon with single bonds uses 4 x 2 = 8e⁻, leaving 20 – 8 = 12e⁻. The "octets" of the two hydrogen atoms (2 electrons) are filled through their bonds to the carbon. The more electronegative bromine atoms each need 6 electrons to complete their octets. This requires 2 x 6 = 12e⁻.

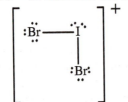

<u>Check:</u> Count the electrons. Each of the five atoms has an octet.

<u>Solution:</u>
c) Both I and Br have the same group number. Between the two atoms, I has the higher period number so it is the central atom. The total number of valence electrons available is [1 x I(7e⁻)] + [2 x Br(7e⁻)] – 1e⁻ (for the positive charge) = 20e⁻. Connecting the two bromine atoms to the iodine with single bonds uses 2 x 2 = 4e⁻, leaving 20 – 4 = 16e⁻. Each of the outlying atoms still needs 6 electrons to complete their octets. Completing these octets uses 2 x 6 = 12 electrons. The remaining 4 electrons are all the iodine needs to complete its octet.

$$
\left[\begin{array}{c} \ddot{\text{:}}\ddot{\text{Br}} \text{——} \ddot{\text{I}} \\[2mm] | \\[1mm] \text{:}\ddot{\text{Br:}} \end{array} \right]^{+}
$$

<u>Check:</u> Count the electrons. Each of the three atoms has an octet.

10.2A   <u>Plan:</u> Count the valence electrons and follow the steps outlined in the sample problem to draw the Lewis structures.
<u>Solution:</u>
a) As in CH₄O, the N and O both serve as "central" atoms. The N is placed next to the O and the H atoms are distributed around them. The N needs more electrons so it gets two of the three hydrogen atoms. You can try placing the N in the center with all the other atoms around it, but you will quickly see that you will have trouble with the oxygen. The number of valence electrons is [3 x H(1e⁻)] + [1 x N(5e⁻)] + [1 x O(6e⁻)] = 14e⁻. Four single bonds are needed (4 x 2 = 8e⁻). This leaves 6 electrons. The oxygen needs 4 electrons, and the nitrogen needs 2. These last 6 electrons serve as three lone pairs.

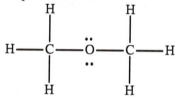

correct   incorrect

Solution:

b) The hydrogen atoms cannot be the central atoms. The problem states that there are no O–H bonds, so the oxygen must be connected to the carbon atoms. Place the O atom between the two C atoms, and then distribute the H atoms equally around each of the C atoms. The total number of valence electrons is [6 x H(1e⁻)] + [2 x C(4e⁻)] + [1 x O(6e⁻)] = 20e⁻ Draw single bonds between each of the atoms. This creates six C–H bonds, and two C–O bonds, and uses 8 x 2 = 16 electrons. The four remaining electrons will become two lone pairs on the O atom to complete its octet.

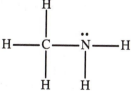

Check: Count the electrons. Both the C's and the O have octets. Each H has its pair.

10.2B   Plan: Count the valence electrons and follow the steps outlined in the sample problem to draw the Lewis structures.

Solution:

a) The hydrogen atoms cannot be the central atoms, so the two N atoms both serve as "central" atoms, bonded to each other, and the H atoms are distributed around them (2 hydrogens per N atom). The number of valence electrons is [4 x H(1e⁻)] + [2 x N(5e⁻)]= 14e⁻. Connect the two nitrogens to each other via single bonds. Connect two hydrogens to each of the two nitrogens via single bonds. These five single bonds require 5 x 2 = 10e⁻. This leaves 4 electrons. Each of the nitrogen atoms needs 2 electrons to complete its octet. These last 4 electrons serve as two lone pairs.

$$\text{H}-\overset{..}{\underset{\underset{\text{H}}{|}}{\text{N}}}-\overset{..}{\underset{\underset{\text{H}}{|}}{\text{N}}}-\text{H}$$

Solution:

b) The hydrogen atoms cannot be the central atoms. The C and the N atoms both serve as "central" atoms, bonded to each other. According to the formula, three hydrogen atoms are bonded to the C and two hydrogen atoms are bonded to the N. The number of valence electrons is [5 x H(1e⁻)] + [1 x N(5e⁻)] + [1 x C(4e⁻)] = 14e⁻. Connect the C and N atoms via a single bond. Connect the appropriate number of hydrogen atoms to the C and N. These six bonds require 6 x 2 = 12e⁻. This leaves 2 electrons. The nitrogen atom needs 2 electrons to complete its octet. These last 2 electrons serve as a lone pair on the nitrogen.

$$\text{H}-\overset{\overset{\text{H}}{|}}{\underset{\underset{\text{H}}{|}}{\text{C}}}-\overset{..}{\underset{\underset{\text{H}}{|}}{\text{N}}}-\text{H}$$

Check: Count the electrons. Both the C and the N have octets. Each H has its pair.

10.3A   Plan: Count the valence electrons and follow the steps outlined in the sample problem to draw the Lewis structures.

Solution:
a) In CO there are a total of [1 x C(4e$^-$)] + [1 x O(6e$^-$)] = 10e$^-$. The hint states that carbon has three bonds. Since oxygen is the only other atom present, these bonds must be between the C and the O. This uses 6 of the 10 electrons. The remaining 4 electrons become two lone pairs, one pair for each of the atoms.

$$: C \equiv O :$$

Check: Count the electrons. Both the C and the O have octets.

Solution:
b) In HCN there are a total of [1 x H(1e$^-$)] + [1 x C(4e$^-$)] + [1 x N(5e$^-$)] = 10 electrons. Carbon has a lower group number, so it is the central atom. Place the C between the other two atoms and connect each of the atoms to the central C with a single bond. This uses 4 of the 10 electrons, leaving 6 electrons. Distribute these 6 to nitrogen to complete its octet. However, the carbon atom is 4 electrons short of an octet. Change two lone pairs on the nitrogen atom to bonding pairs to form two more bonds between carbon and nitrogen for a total of three bonds.

$$H - C \equiv N :$$

Check: Count the electrons. Both the C and the N have octets. The H has its pair.

Solution:
c) In CO$_2$ there are a total of [1 x C(4e$^-$)] + [2 x O(6e$^-$)] = 16 electrons. Carbon has a lower group number, so it is the central atom. Placing the C between the two O atoms and adding single bonds uses 4 electrons, leaving 16 – 4 = 12e$^-$. Distributing these 12 electrons to the oxygen atoms completes those octets, but the carbon atom does not have an octet. Change one lone pair on each oxygen atom to a bonding pair to form two double bonds to the carbon atom, completing its octet.

$$\overset{..}{\underset{..}{O}} = C = \overset{..}{\underset{..}{O}}$$

<u>Check:</u> Count the electrons. Both the C and the O atoms have octets.

10.3B   <u>Plan:</u> Count the valence electrons and follow the steps outlined in the sample problem to draw the Lewis structures.

<u>Solution:</u>
a) In NO$^+$ there are a total of [1 x N(5e$^-$)] + [1 x O(6e$^-$)] – 1 e$^-$ (for the positive charge on the ion) = 10e$^-$. Connecting the N and O atoms via a single bond uses 1 x 2 = 2e$^-$, leaving 10 – 2 = 8e$^-$. The more electronegative O atom needs 6 more electrons to complete its octet. The remaining 2 electrons become a lone pair on the N atom. However, this only gives the N atom 4 electrons. Change two lone pairs on the oxygen atom to bonding pairs to form a triple bond between the N and O atoms. In this way, the atoms' octets are complete.

$$\left[: N \equiv O : \right]^+$$

Check: Count the electrons. Both the N and the O have octets.

Solution:
b) In H$_2$CO there are a total of [1 x C(4e$^-$)] + [1 x O(6e$^-$)] + [2 x H(1e$^-$)] = 12 electrons. Carbon has a lower group number, so it is the central atom (H has a lower group number, but it cannot be a central atom). Place the C between the other three atoms and connect each of the atoms to the central C with a single bond. This uses 6 of the 12 electrons, leaving 6 electrons. Distribute these 6 to oxygen to complete its octet. However, the carbon atom is 2 electrons short of an octet. Change one lone pair on the oxygen atom to a bonding pair between the oxygen and carbon.

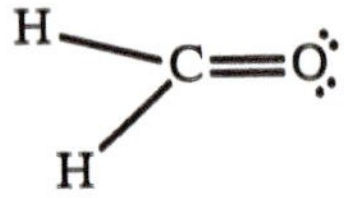

Check: Count the electrons. Both the C and the O have octets. The H atoms have 2 electrons each.

Solution:
c) In N$_2$H$_2$ there are a total of [2 x N(5e$^-$)] + [2 x H(1e$^-$)] = 12 electrons. Hydrogen cannot be a central atom, so the nitrogen atoms are the central atoms (there is more than one central atom). Connecting the nitrogen atoms via a single bond and attaching one hydrogen to each nitrogen atom uses 6 electrons, leaving six electrons. Each nitrogen needs four more electrons to complete its octet; however, there are not enough electrons to complete both octets. Four electrons can be added to one of the nitrogen atoms, but this only leaves two electrons to add to the

other nitrogen atom. The octet of the first nitrogen atom is complete, but the octet of the second nitrogen atom is not complete. Change one lone pair on the first nitrogen to a bonding pair to complete both atoms' octets.

$$H\!-\!\ddot{N}\!=\!\ddot{N}\!-\!H$$

<u>Check:</u> Count the electrons. Both N atoms have octets, and each H atom has 2 electrons.

10.4A **Plan:** Count the valence electrons and follow the steps to draw the Lewis structures. Each new resonance structure is obtained by shifting the position of a multiple bond and the electron pairs.
Solution:
H_3CNO_2 has a total of $[1 \times C(4e^-)] + [2 \times O(6e^-)] + [3 \times H(1e^-)] + [1 \times N(5e^-)] = 24$ electrons. According to the problem, the H atoms are bonded to C, and the C atom is bonded to N, which is bonded to both O atoms. Connect all of these atoms via single bonds. This uses $6 \times 2 = 12e^-$. There are 12 electrons remaining. Each of the oxygen atoms needs 6 more electrons to complete its octet. Placing 6 electrons on each of the oxygen atoms leaves the nitrogen two electrons short of an octet. Change one of the lone pairs on one of the oxygen atoms to a bonding pair to complete the octets. The other resonance structure is obtained by taking a lone pair from the *other* oxygen to create the double bond.

Formal charge of atom = no. of valence e⁻ - (no. of unshared valence e⁻ + $\frac{1}{2}$ no. of shared valence e⁻)

$FC_N = 5 - [0 + 1/2(8)] = +1$ FC_O (single-bonded) $= 6 - [6 + 1/2(2)] = -1$
$FC_C = 4 - [0 + 1/2(8)] = 0$ FC_O (double-bonded) $= 6 - [4 + 1/2(4)] = 0$
$FC_H = 1 - [0 + 1/2(2)] = 0$

10.4B **Plan:** Count the valence electrons and follow the steps to draw the Lewis structures. Each new resonance structure is obtained by shifting the position of a multiple bond and the electron pairs.
Solution:
SCN^- has a total of $[1 \times C(4e^-)] + [1 \times S(6e^-)] + [1 \times N(5e^-)] + 1 e^-$ (for the negative charge) $= 16$ electrons. According to the problem, the C is the central atom. Connect it via single bonds to both the S and the N atoms. This uses $2 \times 2 = 4e^-$. There are 12 electrons remaining. The outlying sulfur and nitrogen each require 6 electrons to complete their octets. Placing 6 electrons each on the N and on the S leaves the C atom 4 electrons short of an octet. Convert two lone pairs on the outer atoms to bonding pairs to complete the octet. The lone pairs can come from the N, from the S, or from both the S and the N. Each of the generated structures is a resonance structure.

$$\left[:\ddot{S}\!=\!C\!=\!\ddot{N}\right]^- \longleftrightarrow \left[:\ddot{S}\!-\!C\!\equiv\!N:\right]^- \longleftrightarrow \left[:S\!\equiv\!C\!-\!\ddot{N}:\right]^-$$

Structure I   Structure II   Structure III

Formal charge of atom = no. of valence e⁻ - (no. of unshared valence e⁻ + $\frac{1}{2}$ no. of shared valence e⁻)

Structure I	Structure II	Structure III
$FC_S = 6 - [4 + 1/2(4)] = 0$	$FC_S = 6 - [6 + 1/2(2)] = -1$	$FC_S = 6 - [2 + 1/2(6)] = +1$
$FC_C = 4 - [0 + 1/2(8)] = 0$	$FC_C = 4 - [0 + 1/2(8)] = 0$	$FC_C = 4 - [0 + 1/2(8)] = 0$
$FC_N = 5 - [4 + 1/2(4)] = -1$	$FC_N = 5 - [2 + 1/2(6)] = 0$	$FC_N = 5 - [6 + 1/2(2)] = -2$

Structure I is the most important structure; the formal charges are lower than those in Structure III, and the negative formal charge is on the more electronegative N atom, while the negative formal charge in Structure II is on the less electronegative S atom.

10.5A **Plan:** The presence of available *d* orbitals makes checking formal charges more important. Use the equation for formal charge: FC = # of valence e⁻ – [# unshared valence e⁻ + ½(# shared valence e⁻)]

Solution:

a) In $POCl_3$, the P is the most likely central atom because all the other elements have higher group numbers. The molecule contains: $[1 \times P(5e^-)] + [1 \times O(6e^-)] + [3 \times Cl(7e^-)] = 32$ electrons. Placing the P in the center with single bonds to all the surrounding atoms uses 8 electrons and gives P an octet. The remaining 24 can be split into 12 pairs with each of the surrounding atoms receiving three pairs. At this point, in structure **I** below, all the atoms have an octet. The central atom is P (smallest group number, highest period number) and can have more than an octet. To see how reasonable this structure is, calculate the formal (FC) for each atom. The +1 and –1 formal charges are not too unreasonable, but 0 charges are better. If one of the lone pairs is moved from the O (the atom with the negative FC) to form a double bond to the P (the atom with the positive FC), structure **II** results. The calculated formal changes in structure II are all 0 so this is a better structure even though P has 10 electrons.

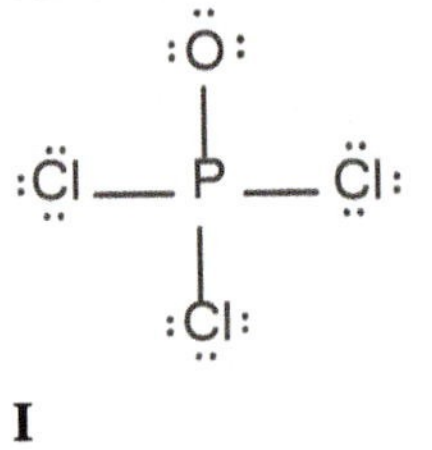

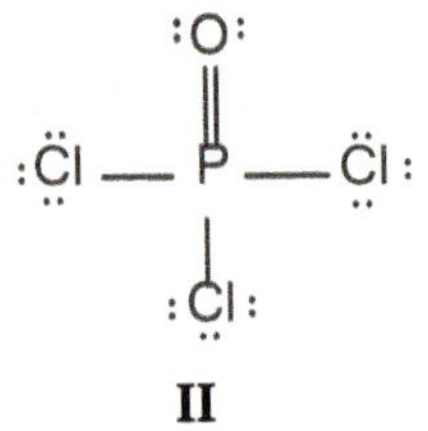

I

$FC_P = 5 - [0 + 1/2(8)] = +1$
$FC_O = 6 - [6 + 1/2(2)] = -1$
$FC_{Cl} = 7 - [6 + 1/2(2)] = 0$

II

$FC_P = 5 - [0 + 1/2(10)] = 0$
$FC_O = 6 - [4 + 1/2(4)] = 0$
$FC_{Cl} = 7 - [6 + 1/2(2)] = 0$

Solution:

b) In ClO_2, the Cl is probably the central atom because the O atoms have a lower period. The molecule contains $[1 \times Cl(7e^-)] + [2 \times O(6e^-)] = 19$ electrons. The presence of an odd number of electrons means that there will be an exception to the octet rule. Placing the O atoms around the Cl and using 4 electrons to form single bonds leaves 15 electrons, 14 of which may be separated into 7 pairs. If 3 of these pairs are given to each O, and the remaining pair plus the lone electron are given to the Cl, we have the following structure:

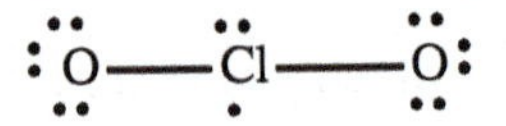

Calculating formal charges: $FC_{Cl} = 7 - [3 + 1/2(4)] = +2$ $\qquad FC_O = 6 - [6 + 1/2(2)] = -1$
The +2 charge on the Cl is a little high, so other structures should be tried. Moving a lone pair from one of the O atoms (negative FC) to form a double bond between the Cl and one of the oxygen atoms gives either structure **I** or **II** below. If both O atoms donate a pair of electrons to form a double bond, then structure **III** results. The next step is to calculate the formal charges.

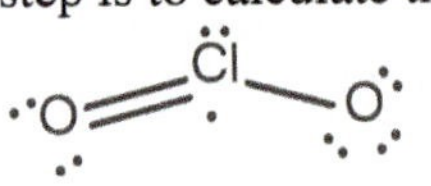

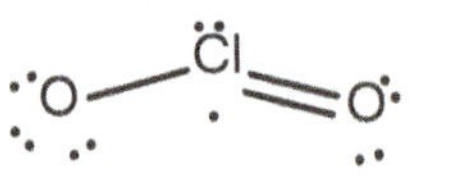

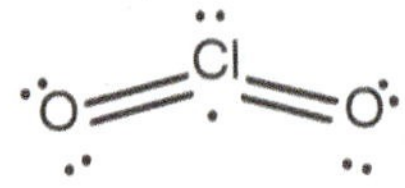

I	**II**	**III**

$FC_{Cl} = 7 - [3 + 1/2(6)] = +1$ $\qquad FC_{Cl} = 7 - [3 + 1/2(6)] = +1$ $\qquad FC_{Cl} = 7 - [3 + 1/2(8)] = 0$
The oxygen atom on the left:
$FC_O = 6 - [4 + 1/2(4)] = 0$ $\qquad FC_O = 6 - [6 + 1/2(4)] = -1$ $\qquad FC_O = 6 - [4 + 1/2(8)] = 0$
The oxygen atom on the right:
$FC_O = 6 - [6 + 1/2(2)] = -1$ $\qquad FC_O = 6 - [4 + 1/2(2)] = 0$ $\qquad FC_O = 6 - [4 + 1/2(8)] = 0$
Pick the structure with the best distribution of formal charges (**structure III**).

Solution:

c) In IBr_4^-, I is most likely to be the central atom because it has a higher period than Br. The molecule contains $[1 \times I(7e^-)] + [4 \times Br(7e^-)] + [1 \times Cl(7e^-)] + 1 e^-$ (for the negative charge) $= 36$ electrons. Drawing a single bond between I and each of the Br atoms uses $4 \times 2 = 8e^-$, leaving 28 electrons. Placing 6 electrons around each of the four outer Br atoms to complete their octets uses 24 electrons. The remaining 4 electrons are placed on the I atom, expanding its octet. This gives the structure:

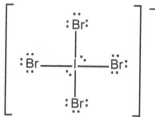

Calculating formal charges:
$FC_I = 7 - [4 + 1/2(8)] = -1$ $FC_{Br} = 7 - [6 + 1/2(2)] = 0$
The formal charges are minimized for this structure.

10.5B   Plan: The presence of available d orbitals makes checking formal charges more important. Use the equation for formal charge: FC = # of valence e^- – [# unshared valence e^- + ½(# shared valence e^-)]
Solution:
a) In BeH_2, the Be is the central atom (H cannot be a central atom). The molecule contains: $[1 \times Be(2\,e^-)] + [2 \times H(1\,e^-)] = 4$ electrons. Placing the Be in the center with single bonds to all the surrounding atoms uses all 4 electrons, leaving Be 4 atoms short of an octet. There are not any more electrons to add to Be. Nor are there any lone pairs to change to bonding pairs. Thus, BeH_2 is an exception to the octet rule in that Be does not have a complete octet.
H——Be——H

Determine the formal charges for each atom:
$FC_{Be} = 2 - [0 + 1/2(4)] = 0$ $FC_H = 1 - [0 + 1/2(2)] = 0$
Solution:

b) In I_3^-, one of the I atoms is central, and the other two I atoms are bonded to it. The molecule contains $[3 \times I(7e^-)] + 1e^-$ (for the negative charge) = 22 electrons. Placing one I atom in the center and connecting each of the other two I atoms to it by single bonds uses $2 \times 2 = 4e^-$. This leaves 18 electrons. Placing 6 electrons on each of the outer I atoms leaves $18 - 12 = 6e^-$. These remaining 6 electrons are placed, as three lone pairs on the central I atom.

$$\left[\; :\ddot{I}\!-\!\dot{\ddot{I}}\cdot\!-\!\ddot{I}: \; \right]^{-}$$

Calculating formal charges: $FC_{I\,(central)} = 7 - [6 + 1/2(4)] = -1$ $FC_{I\,(outer)} = 7 - [6 + 1/2(2)] = 0$
These are low, reasonable formal charges, and we do not need to adjust the structure.
Solution:

c) XeO_3 is a noble gas compound, thus, there will be an exception to the octet rule. The molecule contains $[1 \times Xe(8e^-)] + [3 \times O(6e^-)] = 26$ electrons. The Xe is in a higher period than O so Xe is the central atom. If it is placed in the center with a single bond to each of the three oxygen atoms, 6 electrons are used, and 20 electrons remain. The remaining electrons can be divided into 10 pairs with 3 pairs given to each O and the last pair being given to the Xe. This gives the structure:

$$\begin{array}{c} :\ddot{O}: \\ | \\ :\ddot{O}\!-\!Xe\!-\!\ddot{O}: \end{array}$$

Determine the formal charges for each atom:
$FC_{Xe} = 8 - [2 + 1/2(6)] = +3$ $FC_O = 6 - [6 + 1/2(2)] = -1$
The +3 charge on the Xe is a little high, so other structures should be tried. Moving a lone pair from one of the O atoms (negative FC) to form a double bond between the Xe and one of the oxygen atoms gives structure **I** (or one of its resonance structures). Moving two lone pair from two of the O atoms (negative FC) gives structure **II** below (or one of its resonance structures). If all three O atoms donate a pair of electrons to form a double bond, then structure **III** results. The next step is to calculate the formal charges.

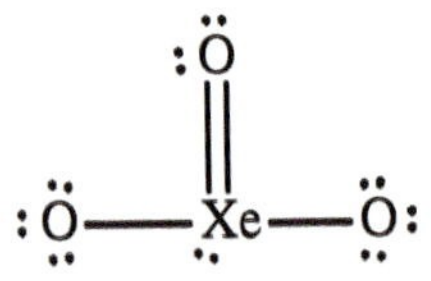

I	**II**	**III**

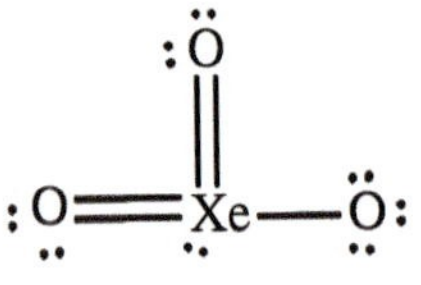

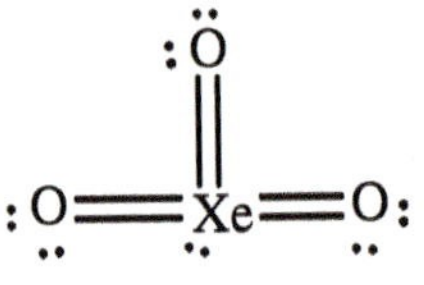

$FC_{Xe} = 8 - [2 + 1/2(8)] = +2$ $FC_{Xe} = 8 - [2 + 1/2(10)] = +1$ $FC_{Xe} = 8 - [2 + 1/2(12)] = 0$

The oxygen atom on the left:
$FC_O = 6 - [6 + 1/2(2)] = -1$ $FC_O = 6 - [4 + 1/2(4)] = 0$ $FC_O = 6 - [4 + 1/2(4)] = 0$

The oxygen atom on the right:
$FC_O = 6 - [6 + 1/2(2)] = -1$ $FC_O = 6 - [6 + 1/2(2)] = -1$ $FC_O = 6 - [4 + 1/2(4)] = 0$

The oxygen atom on the top:
$FC_O = 6 - [4 + 1/2(4)] = 0$ $FC_O = 6 - [4 + 1/2(4)] = 0$ $FC_O = 6 - [4 + 1/2(4)] = 0$

Pick the structure with the best distribution of formal charges (**structure III**).

10.6A   Plan: Draw a Lewis structure. Determine the electron arrangement by counting the electron pairs around the central atom.

Solution:

a) The Lewis structure for CS_2 is shown below. The central atom, C, has two pairs (double bonds only count once). The two pair arrangement is **linear** with the designation, AX_2. The absence of lone pairs on the C means there is no deviation in the bond angle (**180°**).

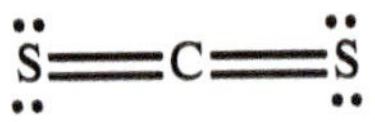

Solution:

b) Even though this is a combination of a metal with a nonmetal, it may be treated as a molecular species. The Lewis structure for $PbCl_2$ is shown below. The molecule is of the AX_2E type; the central atom has three pairs of electrons (1 lone pair and two bonding pairs). This means the *electron-group arrangement* is trigonal planar (**120°**), with a lone pair giving a **bent or V-shaped** molecule. The lone pair causes the ideal bond angle to decrease to **< 120°**.

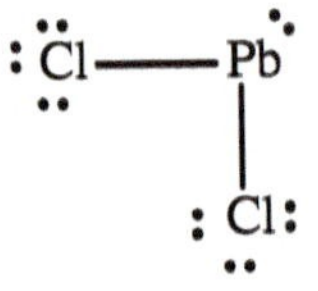

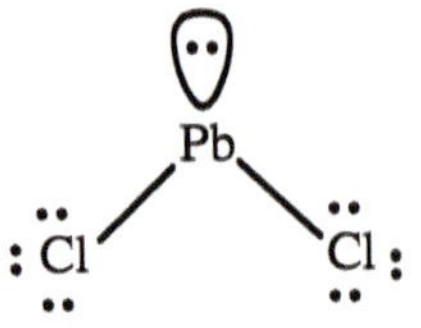

Solution:

c) The Lewis structure for the CBr_4 molecule is shown below. It has the AX_4 type formula which is a perfect **tetrahedron** (with **109.5°** bond angles) because all bonds are identical, and there are no lone pairs.

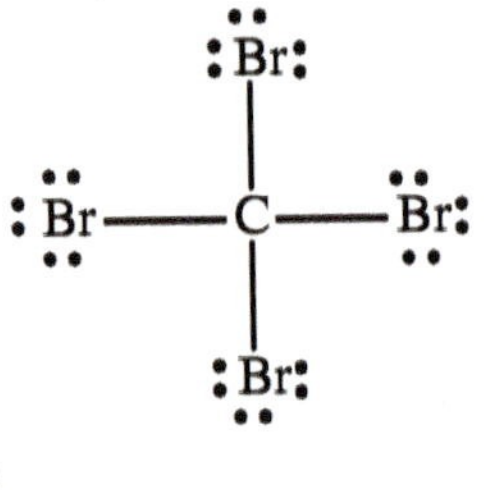

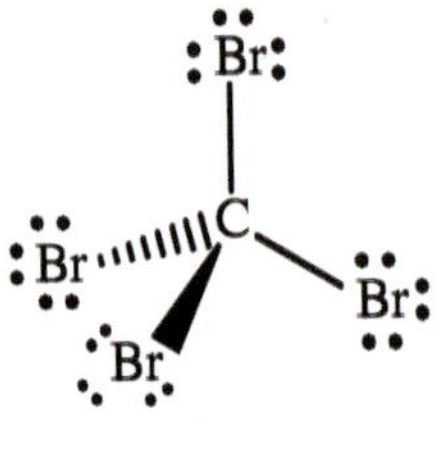

Solution:

d) The SF_2 molecule has the Lewis structure shown below. This is a AX_2E_2 molecule; the central atom is surrounded by four electron pairs, two of which are lone pairs and two of which are bonding pairs. The *electron group arrangement* is tetrahedral. The two lone pairs give a **V-shaped or bent** arrangement. The ideal tetrahedral bond angle is **decreased from the ideal 109.5°** value.

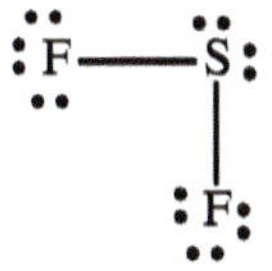

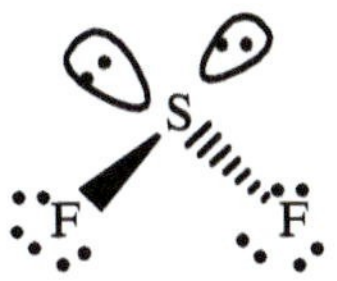

10.6B   Plan: Draw a Lewis structure. Determine the electron arrangement by counting the electron pairs around the central atom.
Solution:
a) The Lewis structure for BrO_2^- is shown below. The molecule is of the AX_2E_2 type; the central atom has four pairs of electrons (2 lone pairs and two bonding pairs; each double bond counts as one electron domain).This means the *electron-group arrangement* is tetrahedral (**109.5°**), with two lone pairs giving a **bent or V-shaped** molecule. The lone pairs cause the bond angle to decrease to **<109.5°**.

$$\left[\ddot{\underset{..}{O}}=\overset{..}{Br}=\overset{..}{O:} \right]^-$$

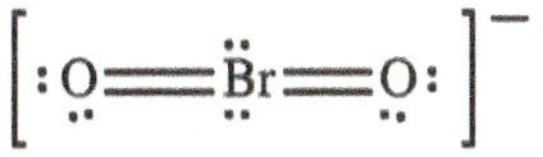

The molecular shape of BrO_2^-:

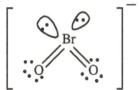

Solution:
b) The Lewis structure for AsH_3 is shown below. The molecule is of the AX_3E type; the central atom has four pairs of electrons (1 lone pair and three bonding pairs). This means the *electron-group arrangement* is tetrahedral (**109.5°**), with a lone pair giving a **trigonal pyramidal** molecule. The lone pair causes the bond angle to decrease to **<109.5°**.

$$H-\overset{..}{As}-H$$
$$|$$
$$H$$

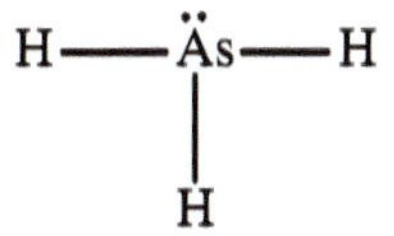

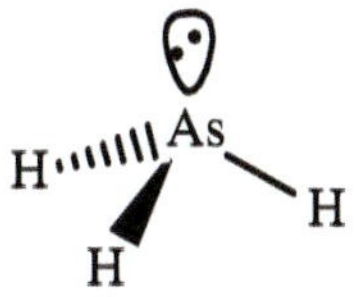

Solution:
c) The Lewis structure for N_3^- is shown below. The two pair electron arrangement is **linear** with the designation, AX_2. The molecular shape is also **linear**. The absence of lone pairs on the central N means there is no deviation in the bond angle (**180°**).

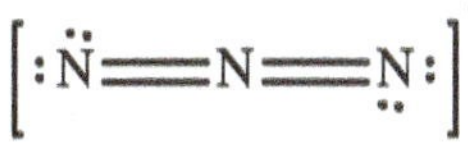

$$\left[:\overset{..}{N}=N=\overset{..}{N}: \right]^-$$

Solution:
d) The SeO_3 molecule has the Lewis structure shown below. This is an AX_3 molecule; the central atom is surrounded by three electron pairs (each double bond counts as one electron domain). The *electron group arrangement* and molecular shape is **trigonal planar**. The absence of lone pairs on the central atom means there is no deviation in the bond angle (**120°**).

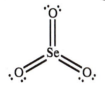

10.7A   Plan: Draw a Lewis structure. Determine the electron arrangement by counting the electron pairs around the central atom.
Solution:
a) The Lewis structure for the ICl_2^- is shown below. This is an AX_2E_3 type structure. The five pairs give a trigonal bipyramidal arrangement of electron groups. The presence of 3 lone pairs leads to a **linear** shape (**180°**). The usual distortions caused by lone pairs cancel in this case. In the trigonal bipyramidal geometry, lone pairs always occupy equatorial positions.

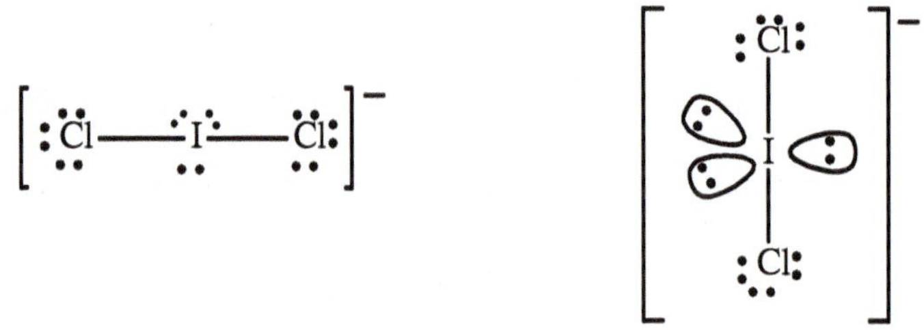

Solution:
b) The Lewis structure for the ClF_3 molecule is given below. Like ICl_2^- there are 5 pairs around the central atom; however, there are only 2 lone pairs. This gives a molecule that is **T-shaped**. The presence of the lone pairs decreases the ideal bond angles to **less than 90°**.

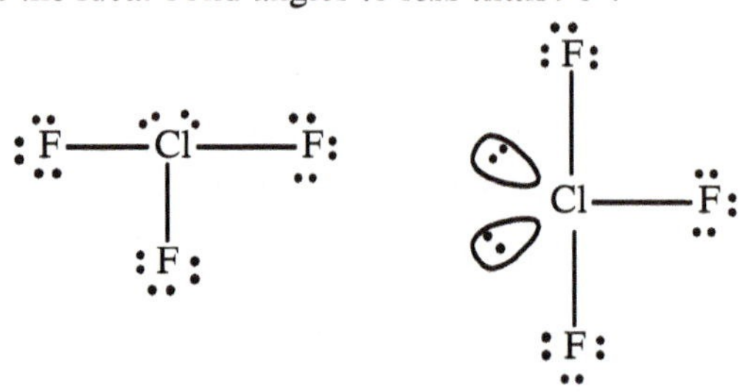

Solution:
c) The SOF_4 molecule has several possible Lewis structures, two of which are shown below. In both cases, the central atom has 5 atoms attached with no lone pairs. The formal charges work out the same in both structures. The structure on the right has an equatorial double bond. Double bonds require more room than single bonds, and equatorial positions have this extra room.

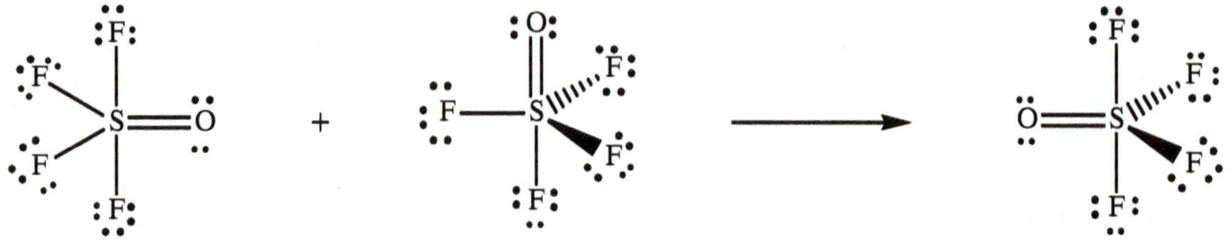

The molecule is **trigonal bipyramidal**, and the double bond causes deviation from ideal bond angles. All of the F atoms move away from the O. Thus, all angles involving the O are increased, and all other angles are decreased.

10.7B   Plan: Draw a Lewis structure. Determine the electron arrangement by counting the electron pairs around the central atom.
Solution:
a) The Lewis structure for the BrF_4^- ion is shown below. This is an AX_4E_2 type structure. The six pairs give an octahedral arrangement of electron groups. The presence of 2 lone pairs on the central atom leads to a **square planar** shape (**90°**). The usual distortions caused by lone pairs cancel in this case. In the square planar, lone pairs always occupy axial positions.

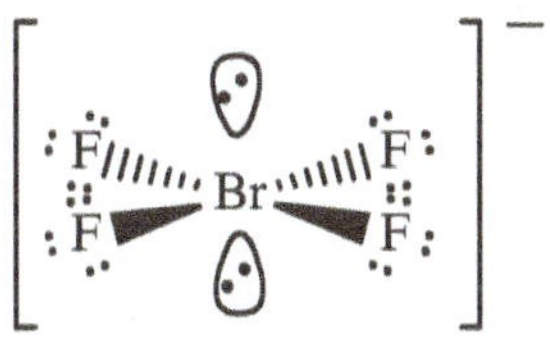

Solution:

b) The Lewis structure for the ClF_4^+ ion is given below. This is an AX_4E type structure. There are 5 pairs around the central atom; however, there is 1 lone pair. This gives a molecule that has a **seesaw** shape. The presence of the lone pair decreases the ideal bond angles to **less than 90°** (F_{ax}–Cl–F_{eq}) and **less than 120°** (F_{eq}–Cl–F_{eq}).

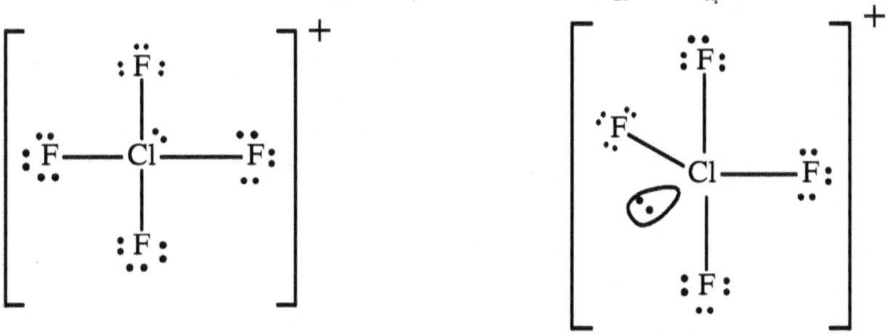

Solution:

c) The Lewis structure for the PCl_6^- is shown below. This is an AX_6 type structure. The six pairs give an octahedral arrangement of electron groups and an **octahedral** shape, with **90°** bond angles. The absence of lone pair electrons on the central atom means there is no deviation from the ideal bond angles.

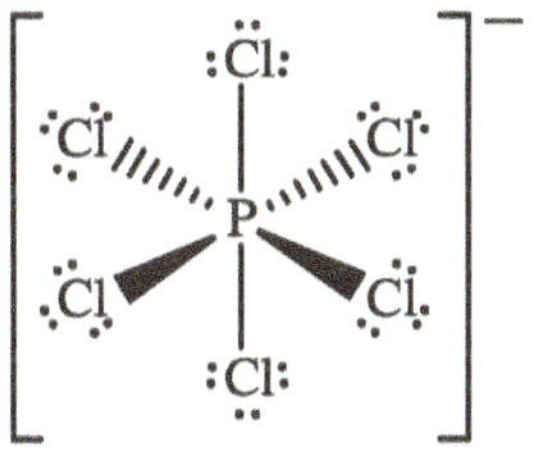

10.8A   **Plan:** Draw the Lewis structure for each of the substances, and determine the molecular geometry of each.
Solution:

a) The Lewis structure for H_2SO_4 is shown below. The double bonds ease the problem of a high formal charge on the sulfur. Sulfur is allowed to exceed an octet. The S has 4 groups around it, making it **tetrahedral**. The ideal angles around the S are **109.5°**. The double bonds move away from each other, and force the single bonds away. This opens the angle between the double bonded oxygen atoms, and results in an angle between the single bonded oxygen atoms that is less than ideal. Each single bonded oxygen atom has 4 groups around it; since two of the 4 groups are lone pairs, the shape around each of these oxygen atoms is **bent**. The lone pairs compress the H–O–S bond angle to **<109.5°**.

Solution:

b) The hydrogen atoms cannot be central so the carbons must be attached to each other. The problem states that there is a carbon-carbon triple bond. This leaves only a single bond to connect the third carbon to a triple-bonded carbon, and give that carbon an octet. The other triple-bonded carbon needs one hydrogen to complete its octet.

The remaining three hydrogen atoms are attached to the single-bonded carbon, which allows it to complete its octet. This structure is shown below. The single-bonded carbon has four groups **tetrahedrally** around it leading to bond angles ~**109.5°** (little or no deviation). The triple-bonded carbons each have two groups (the triple bond counts as one electron group) so they should be **linear (180°)**.

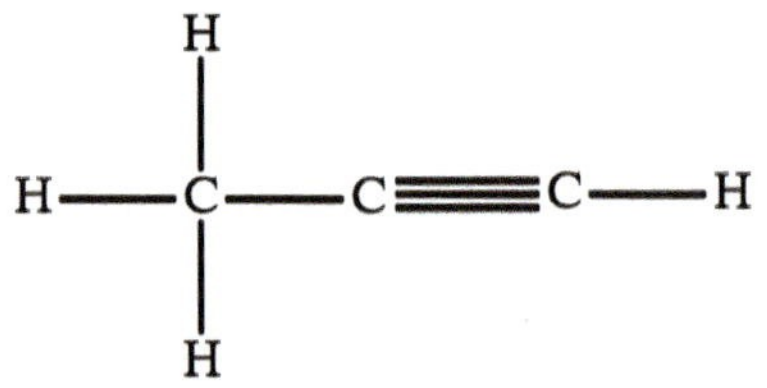

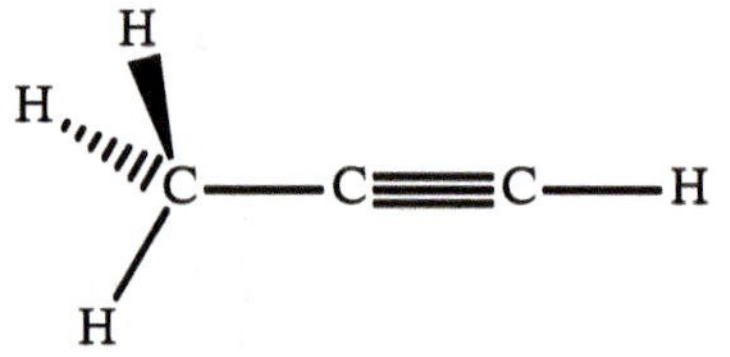

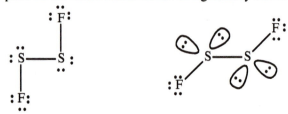

Solution:
c) Fluorine, like hydrogen, is never a central atom. Thus, the sulfuratoms must be bonded to each other. Each F has 3 lone pairs, and the sulfur atoms have 2 lone pairs. All 4 atoms now have an octet. This structure is shown below. Each sulfur has four groups around it, so the electron arrangement is **tetrahedral**. The presence of the lone pairs on the sulfur atoms results in a geometry that is **V-shaped or bent** and in a bond angle that is **< 109.5°**.

10.8B   Plan: Draw the Lewis structure for each of the substances, and determine the molecular geometry of each.
Solution:
a) The Lewis structure for CH_3NH_2 is shown below. The C has 4 groups around it (AX_4), making it **tetrahedral**. The ideal angles around the C are **109.5°**. The nitrogen also has 4 groups around it, but one of those groups is a lone pair. The lone pair makes the N an AX_3E central atom with **trigonal pyramidal** geometry. The presence of the lone pair compresses the H–N–H angle to **<109.5°**.

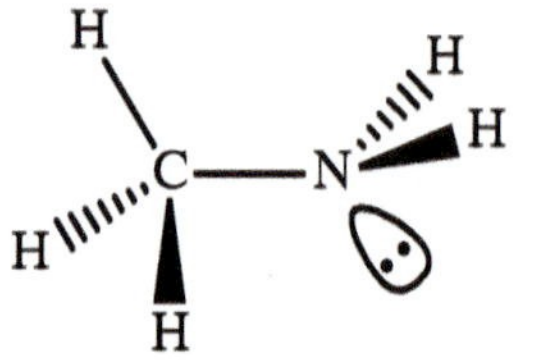

Solution:
b) The Lewis structure for C_2Cl_4 is shown below. The carbons are central and connected to each other. There are a total of 36 valence electrons. In order for all of the atoms to have a full octet, there must be a double bond between the two carbon atoms. Each of the carbon atoms has three electron groups around it, yielding a **trigonal planar** geometry. The presence of the double bond compresses each Cl–C–Cl angle to**<120°**.

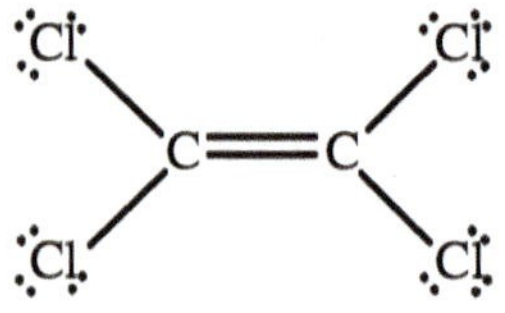

Solution:
c) The Lewis structure for Cl_2O_7 is shown below. The problem states that three oxygen atoms are bonded to the first Cl atom, which is then bonded to a fourth oxygen atom. That fourth oxygen atom is bonded, in turn, to a second Cl atom, which is bonded to 3 additional oxygen atoms. Each Cl has 4 electron groups (3 of which are double bonds to oxygen and one of which is a single bond to oxygen). This gives the Cl a **tetrahedral** geometry with bond angles very close to **109.5°**. The central oxygen atom, on the other hand, also has four electron pairs around it. In this case,

two of the electron pairs are bonding pairs and two of the pairs are lone pairs. The presence of the lone pairs give the central oxygen a **bent or V-shaped geometry** and compresses the Cl–O–Cl bond angle to **<109.5°**.

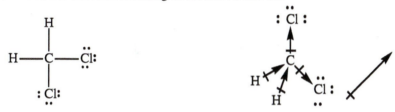

10.9A   Plan: Draw the Lewis structures, predict the shapes, and then examine the positions of the bond dipoles.
Solution:
a) Dichloromethane, CH_2Cl_2, has the Lewis structure shown below. It is tetrahedral, and if the outlying atoms were identical, it would be nonpolar. However, the chlorine atoms are more electronegative than hydrogen so there is a general shift in their direction resulting in the arrows shown.

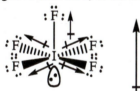

Solution:
b) Iodine oxide pentafluoride, IOF_5, has the Lewis structure shown below. The overall geometry is octahedral. All six bonds are polar, with the more electronegative O and F atoms shifting electron density away from the I. The 4 equatorial fluorines counterbalance each other. The axial F is not equivalent to the axial O. The more electronegative F results in an overall polarity in the direction of the axial F.

The lone electron pairs are left out for simplicity.
Solution:
c) Iodine pentafluoride, IF_5, has the square pyramidal Lewis structure shown below. The fluorine is more electronegative than the I, so the shift in electron density is towards the F, resulting in the dipole indicated below.

10.9B   Plan: Draw the Lewis structures, predict the shapes, and then examine the positions of the bond dipoles.
Solution:
a) Xenon tetrafluoride, XeF_4, has the Lewis structure shown below. It is square planar, and, because the outlying atoms are identical, it is nonpolar. However, the fluorine atoms are more electronegative than xenon, so the individual bonds are polar. Because of the square planar geometry, those polar bonds cancel each other out.

Solution:
b) Chlorine trifluoride, ClF₃, has the Lewis structure shown below. The overall geometry is T-shaped. All three bonds are polar, with the more electronegative F atoms shifting electron density away from the Cl. The more electronegative F results in an overall dipole indicated below.

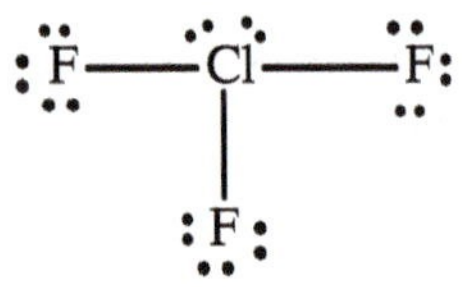

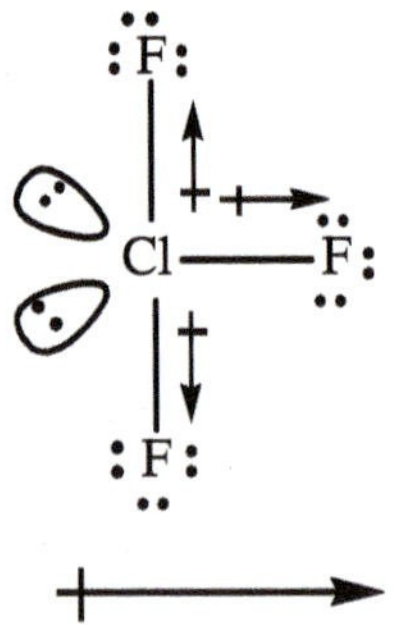

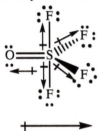

Solution:
c) Sulfur monoxide tetrafluoride, SOF₄, has the trigonal bipyramidal Lewis structure shown below. Both the F atoms and the O atom are more electronegative than the S, and F is more electronegative than O, so the shift in electron density is towards the F atoms, resulting in the dipole indicated below.

CHEMICAL CONNECTIONS BOXED READING PROBLEMS

B10.1   Plan: Examine the Lewis structure, noting the number of regions of electron density around the carbon and nitrogen atoms in the two resonance structures. The molecular shape is determined by the number of electron regions. An electron region is any type of bond (single, double, or triple) and an unshared pair of electrons.
Solution:
Resonance structure on the left:
Carbon has three electron regions (two single bonds and one double bond); three electron regions are arranged in a trigonal planar arrangement. The molecular shape around the C atom is **trigonal planar**. Nitrogen has four electron regions (three single bonds and an unshared pair of electrons); the four electron regions are arranged tetrahedrally; since one corner of the tetrahedron is occupied by an unshared electron pair, the shape around N is **trigonal pyramidal**.
Resonance structure on the right:
This C atom also has three electrons regions (two single bond and one double bond) so the molecular shape is again **trigonal planar**. The N atom also has three electron regions (two single bonds and one double bond); the molecular shape is **trigonal planar**.

B10.2   The top portion of both molecules is similar so the top portions will interact with biomolecules in a similar manner. The mescaline molecule may fit into the same nerve receptors as dopamine due to the similar molecular shape.

10.1 Plan: To be the central atom in a compound, an atom must be able to simultaneously bond to at least two other atoms.
Solution:
He, F, and H cannot serve as central atoms in a Lewis structure. Helium ($1s^2$) is a noble gas, and as such, it does not need to bond to any other atoms. Hydrogen ($1s^1$) and fluorine ($1s^2 2s^2 2p^5$) only need one electron to complete their valence shells. Thus, they can only bond to one other atom, and they do not have d orbitals available to expand their valence shells.

10.3 Plan: For an element to obey the octet rule it must be surrounded by eight electrons. To determine the number of electrons present, (1) count the individual electrons actually shown adjacent to a particular atom (lone pairs), and (2) add two times the number of bonds to that atom: number of electrons = individual electrons + 2(number of bonds).
Solution:
(a) $0 + 2(4) = 8$; (b) $2 + 2(3) = 8$; (c) $0 + 2(5) = 10$; (d) $2 + 2(3) = 8$; (e) $0 + 2(4) = 8$; (f) $2 + 2(3) = 8$;
(g) $0 + 2(3) = 6$; (h) $8 + 2(0) = 8$. **All the structures obey the octet rule except: c and g.**

10.5 Plan: Count the valence electrons and draw Lewis structures.
Solution:
Total valence electrons: SiF_4: [1 x Si(4e$^-$)] + [4 x F(7e$^-$)] = 32; $SeCl_2$: [1 x Se(6e$^-$)] + [2 x Cl(7e$^-$)] = 20;
COF_2: [1 x C(4e$^-$)] + [1 x O(6e$^-$)] + [2 x F(7e$^-$)] = 24. The Si, Se, and the C are the central atoms, because these are the elements in their respective compounds with the lower group number (in addition, we are told C is central). Place the other atoms around the central atoms and connect each to the central atom with a single bond.
SiF_4: At this point, eight electrons (2e$^-$ in four Si–F bonds) have been used with 32 – 8 = 24 remaining; the remaining electrons are placed around the fluorine atoms (three pairs each). All atoms have an octet.
$SeCl_2$: The two bonds use 4e$^-$ (2e$^-$ in two Se–Cl bonds) leaving 20 – 4 = 16e$^-$. These 16e$^-$ are used to complete the octets on Se and the Cl atoms.
COF_2: The three bonds to the C use 6e$^-$ (2e$^-$ in three bonds) leaving 24 – 6 = 18 e$^-$. These 18e$^-$ are distributed to the surrounding atoms first to complete their octets. After the 18e$^-$ are used, the central C is two electrons short of an octet. Forming a double bond to the O (change a lone pair on O to a bonding pair on C) completes the C octet.
(a) SiF_4 (b) $SeCl_2$

(c) COF_2

10.7 Plan: Count the valence electrons and draw Lewis structures.
Solution:
a) PF_3: [1 x P(5 e$^-$)] + [3 x F(7e$^-$)] = 26 valence electrons. P is the central atom. Draw single bonds from P to the three F atoms, using 2e$^-$ x 3 bonds = 6 e$^-$. Remaining e$^-$: 26 – 6 = 20 e$^-$. Distribute the 20 e$^-$ around the P and F atoms to complete their octets.

b) H_2CO_3: [2 x H(1e⁻)] + [1 x C(4e⁻)] + 3 x O(6e⁻)] = 24 valence electrons. C is the central atom with the H atoms attached to the O atoms. Place appropriate single bonds between all atoms using 2e⁻ x 5 bonds = 10e⁻ so that 24 – 10 = 14e⁻ remain. Use these 14e⁻ to complete the octets of the O atoms (the H atoms already have their two electrons). After the 14e⁻ are used, the central C is two electrons short of an octet. Forming a double bond to the O that does not have an H bonded to it (change a lone pair on O to a bonding pair on C) completes the C octet.

c) CS_2: [1 x C(4e⁻)] + [2 x S(6e⁻)] = 16 valence electrons. C is the central atom. Draw single bonds from C to the two S atoms, using 2e⁻ x 2 bonds = 4e⁻. Remaining e⁻: 16 – 4 = 12e⁻. Use these 12e⁻ to complete the octets of the surrounding S atoms; this leaves C four electrons short of an octet. Form a double bond from each S to the C by changing a lone pair on each S to a bonding pair on C.

a) PF_3 (26 valence e⁻)   b) H_2CO_3 (24 valence e⁻)

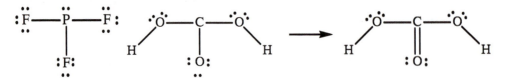

c) CS_2 (16 valence e⁻)

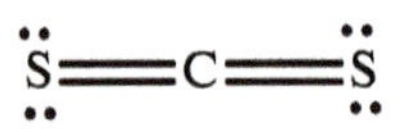

10.9   <u>Plan:</u> The problem asks for resonance structures, so there must be more than one answer for each part.
<u>Solution:</u>
a) NO_2^+ has [1 x N(5e⁻)] + [2 x O(6e⁻)] – 1e⁻ (+ charge) = 16 valence electrons. Draw a single bond from N to each O, using 2e⁻ x 2 bonds = 4e⁻; 16 – 4 = 12e⁻ remain. Distribute these 12e⁻ to the O atoms to complete their octets. This leaves N 4e⁻ short of an octet. Form a double bond from each O to the N by changing a lone pair on each O to a bonding pair on N. No resonance is required as all atoms can achieve an octet with double bonds.

$$\left[:\ddot{O}\!\!-\!\!N\!\!-\!\!\ddot{O}: \right]^+ \longrightarrow \left[\ddot{O}\!\!=\!\!N\!\!=\!\!\ddot{O} \right]^+$$

b) NO_2F has [1 x N(5e⁻)] + [2 x O(6e⁻)] + [1 x F(7e⁻)] = 24 valence electrons. Draw a single bond from N to each surrounding atom, using 2e⁻ x 3 bonds = 6e⁻; 24 – 6 = 18e⁻ remain. Distribute these 18e⁻ to the O and F atoms to complete their octets. This leaves N 2e⁻ short of an octet. Form a double bond from either O to the N by changing a lone pair on O to a bonding pair on N. There are two resonance structures since a lone pair from either of the two O atoms can be moved to a bonding pair with N:

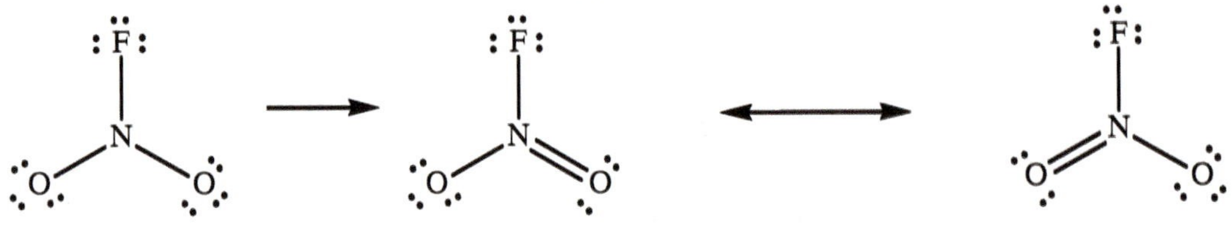

10.11   <u>Plan:</u> Count the valence electrons and draw Lewis structures. Additional structures are needed to show resonance.
<u>Solution:</u>
a) N_3^- has [3 x N(5e⁻)] + [1 e⁻(from charge)] = 16 valence electrons. Place a single bond between the nitrogen atoms. This uses 2e⁻ x 2 bonds = 4 electrons, leaving 16 – 4 = 12 electrons (6 pairs). Giving three pairs on each end nitrogen gives them an octet, but leaves the central N with only four electrons as shown below:

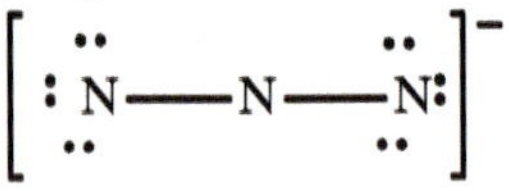

$$\left[:\ddot{N}\!\!-\!\!N\!\!-\!\!\ddot{N}: \right]^-$$

10-16

The central N needs four electrons. There are three options to do this: (1) each of the end N atoms could form a double bond to the central N by sharing one of its pairs; (2) one of the end N atoms could form a triple bond by sharing two of its lone pairs; (3) the other end N atom could form the triple bond instead.

$$\left[\ddot{N}\!=\!\!N\!=\!\ddot{N}\right]^{-} \longleftrightarrow \left[:N\!\equiv\!N\!-\!\ddot{N}:\right]^{-} \longleftrightarrow \left[:\ddot{N}\!-\!N\!\equiv\!N:\right]^{-}$$

b) NO_2^- has [1 x N(5e$^-$)] + [2 x O(6e$^-$)] + [1 e$^-$ (from charge)] = 18 valence electrons. The nitrogen should be the central atom with each of the oxygen atoms attached to it by a single bond (2e$^-$ x 2 bonds = 4 electrons). This leaves 18 – 4 = 14 electrons (seven pairs). If three pairs are given to each O and one pair is given to the N, then both O atoms have an octet, but the N atom only has six. To complete an octet the N atom needs to gain a pair of electrons from one O atom or the other (form a double bond). The resonance structures are:

$$\left[:\ddot{O}\!-\!\ddot{N}\!-\!\ddot{O}:\right]^{-} \longrightarrow \left[:\ddot{O}\!-\!\ddot{N}\!=\!\ddot{O}\right]^{-} \longleftrightarrow \left[\ddot{O}\!=\!\ddot{N}\!-\!\ddot{O}:\right]^{-}$$

10.13   Plan: Initially, the method used in the preceding problems may be used to establish a Lewis structure. The total of the formal charges must equal the charge on an ion or be equal to 0 for a compound. The formal charge only needs to be calculated once for a set of identical atoms. Formal charge (FC) = no. of valence electrons – [no. of unshared valence electrons + ½ no. of shared valence electrons].
Solution:
a) IF_5 has [1 x I(7e$^-$)] + [5 x F(7e$^-$)] = 42 valence electrons. The presence of five F atoms around the central I means that the I atom will have a minimum of ten electrons; thus, this is an exception to the octet rule. The five I–F bonds use 2e$^-$ x 5 bonds = 10 electrons leaving 42 – 10 = 32 electrons (16 pairs). Each F needs three pairs to complete an octet. The five F atoms use fifteen of the sixteen pairs, so there is one pair left for the central I. This gives:

Calculating formal charges:
FC = no. of valence electrons – [no. of unshared valence electrons + ½ no. of shared valence electrons].
For iodine: $FC_I = 7 - [2 + ½(10)] = 0$ For each fluorine: $FC_F = 7 - [6 + ½(2)] = 0$
Total formal charge = 0 = charge on the compound.
b) AlH_4^- has [1 x Al(3e$^-$)] + [4 x H(1e$^-$)] + [1e$^-$ (from charge)] = 8 valence electrons.
The four Al–H bonds use all the electrons and Al has an octet.

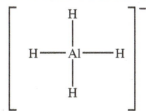

FC = no. of valence electrons – [no. of unshared valence electrons + ½ no. of shared valence electrons].
For aluminum: $FC_{Al} = 3 - [0 + ½(8)] = -1$
For each hydrogen: $FC_H = 1 - [0 + ½(2)] = 0$

10.15   Plan: Initially, the method used in the preceding problems may be used to establish a Lewis structure. The total of the formal charges must equal the charge on an ion or be equal to 0 for a compound. The formal charge only needs to be calculated once for a set of identical atoms. Formal charge (FC) = no. of valence electrons – [no. of unshared valence electrons + ½ no. of shared valence electrons].

a) CN^-: [1 x $C(4e^-)$] + [1 x $N(5e^-)$] + [1 e^- from charge] = 10 valence electrons. Place a single bond between the carbon and nitrogen atoms. This uses $2e^-$ x 1 bond = 2 electrons, leaving 10 – 2 = 8 electrons (four pairs). Giving three pairs of electrons to the nitrogen atom completes its octet but that leaves only one pair of electrons for the carbon atom which will not have an octet. The nitrogen could form a triple bond by sharing two of its lone pairs with the carbon atom. A triple bond between the two atoms plus a lone pair on each atom satisfies the octet rule and uses all ten electrons.

$$\left[\; :C\!\!\equiv\!\!N: \;\right]^-$$

FC = no. of valence electrons – [no. of unshared valence electrons + ½ no. of shared valence electrons].
$FC_C = 4 - [2 + ½(6)] = -1$; $FC_N = 5 - [2 + ½(6)] = 0$
Check: The total formal charge equals the charge on the ion (–1).
b) ClO^-: [1 x $Cl(7e^-)$] + [1 x $O(6e^-)$] + [1e^- from charge] = 14 valence electrons. Place a single bond between the chlorine and oxygen atoms. This uses $2e^-$ x 1 bond = 2 electrons, leaving 14 – 2 = 12 electrons (six pairs). Giving three pairs of electrons each to the carbon and oxygen atoms completes their octets.

$$\left[\; :\overset{..}{\underset{..}{Cl}}\!\!-\!\!\overset{..}{\underset{..}{O}}: \;\right]^-$$

FC = no. of valence electrons – [no. of unshared valence electrons + ½ no. of shared valence electrons].
$FC_{Cl} = 7 - [6 + 1/2(2)] = 0$ $FC_O = 6 - [6 + 1/2(2)] = -1$
Check: The total formal charge equals the charge on the ion (–1).

10.17   Plan: The general procedure is similar to the preceding problems, plus the oxidation number determination.
Solution:
a) BrO_3^- has [1 x $Br(7e^-)$] + 3 x $O(6e^-)$] + [1e^- (from charge)] = 26 valence electrons.
Placing the O atoms around the central Br and forming three Br–O bonds uses $2e^-$ x 3 bonds = 6 electrons and leaves 26 – 6 = 20 electrons (ten pairs). Placing three pairs on each O (3 x 3 = 9 total pairs) leaves one pair for the Br and yields structure I below. In structure I, all the atoms have a complete octet. Calculating formal charges:
$FC_{Br} = 7 - [2 + 1/2(6)] = +2$ $FC_O = 6 - [6 + 1/2(2)] = -1$
The FC_O is acceptable, but FC_{Br} is larger than is usually acceptable. Forming a double bond between any one of the O atoms gives structure II. Calculating formal charges:
$FC_{Br} = 7 - [2 + ½(8)] = +1$ $FC_O = 6 - [6 + ½(2)] = -1$ $FC_O = 6 - [4 + ½(4)] = 0$
(Double bonded O)
The FC_{Br} can be improved further by forming a second double bond to one of the other O atoms (structure III).
$FC_{Br} = 7 - [2 + ½(10)] = 0$ $FC_O = 6 - [6 + ½(2)] = -1$ $FC_O = 6 - [4 + ½(4)] = 0$
(Double bonded O atoms)

Structure III has the most reasonable distribution of formal charges.

$$\left[\; :\overset{..}{\underset{..}{O}}\!\!-\!\!\underset{\underset{:\overset{..}{\underset{..}{O}}:}{|}}{Br}\!\!-\!\!\overset{..}{\underset{..}{O}}: \;\right]^- \longrightarrow \left[\; :\overset{..}{\underset{..}{O}}\!\!-\!\!\underset{\underset{:\overset{}{\underset{..}{O}}:}{|}}{Br}\!\!=\!\!\overset{..}{\underset{..}{O}} \;\right]^- \longrightarrow \left[\; \overset{..}{\underset{..}{O}}\!\!=\!\!\underset{\underset{:\overset{}{\underset{..}{O}}:}{|}}{Br}\!\!=\!\!\overset{..}{\underset{..}{O}} \;\right]^-$$

I   II   III

–6

The oxidation numbers (O.N.) are: $O.N._{Br} = +5$ and $O.N._O = -2$.   +5 –2
Check: The total formal charge equals the charge on the ion (–1). BrO_3^-

b) SO_3^{2-} has [1 x S(6e⁻)] + [3 x O(6e⁻)] + [2e⁻ (from charge)] = 26 valence electrons.
Placing the O atoms around the central S and forming three S–O bonds uses 2e⁻ x 3 bonds = 6 electrons and leaves 26 – 6 = 20 electrons (ten pairs). Placing three pairs on each O (3 x 3 = 9 total pairs) leaves one pair for the S and yields structure I below. In structure I all the atoms have a complete octet. Calculating formal charges:
FC$_S$ = 6 – [2 + ½(6)] = +1;   FC$_O$ = 6 – [6 + ½(2)] = –1
The FC$_O$ is acceptable, but FC$_S$ is larger than is usually acceptable. Forming a double bond between any one of the O atoms (structure II) gives:
FC$_S$ = 6 – [2 + ½(8)] = 0   FC$_O$ = 6 – [6 + ½(2)] = –1   FC$_O$ = 6 – [4 + ½(4)] = 0
(Double bonded O)

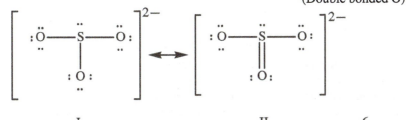

I   II   –6
Structure II has the more reasonable distribution of formal charges.   +4 –2
The oxidation numbers (O.N.) are: O.N.$_S$ = +4 and O.N.$_O$ = –2. SO_3^{2-}
Check: The total formal charge equals the charge on the ion (–2).

10.19   Plan: The octet rule states that when atoms bond, they share electrons to attain a filled outer shell of eight electrons. If an atom has fewer than eight electrons, it is electron deficient; if an atom has more than eight electrons around it, the atom has an expanded octet.
Solution:
a) BH_3 has [1 x B(3e⁻)] + [3 x H(1e⁻)] = 6 valence electrons. These are used in three B–H bonds. The B has six electrons instead of an octet; this molecule is **electron deficient**.
b) AsF_4^- has [1 x As(5e⁻)] + [4 x F(7e⁻)] + [1e⁻ (from charge)] = 34 valence electrons. Four As–F bonds use eight electrons leaving 34 – 8 = 26 electrons (13 pairs). Each F needs three pairs to complete its octet and the remaining pair goes to the As. The As has an **expanded octet** with ten electrons. The F cannot expand its octet.
c) $SeCl_4$ has [1 x Se(6e⁻)] + 4 x Cl(7e⁻)] = 34 valence electrons. The $SeCl_4$ is isoelectronic (has the same electron structure) as AsF_4^-, and so its Lewis structure looks the same. Se has an **expanded octet** of ten electrons.

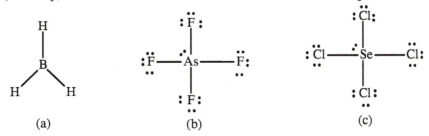

(a)   (b)   (c)

10.21   Plan: The octet rule states that when atoms bond, they share electrons to attain a filled outer shell of eight electrons. If an atom has fewer than eight electrons, it is electron deficient; if an atom has more than eight electrons around it, the atom has an expanded octet.
Solution:
a) BrF_3 has [1 x Br(7e⁻)] + [3 x F(7e⁻)] = 28 valence electrons. Placing a single bond between Br and each F uses 2e⁻ x 3 bonds = 6e⁻, leaving 28 – 6 = 22 electrons (eleven pairs). After the F atoms complete their octets with three pairs each, the Br gets the last two lone pairs. The Br has an **expanded octet** of ten electrons.
b) ICl_2^- has [1 x I(7e⁻)] + [2 x Cl(7e⁻)] + [1e⁻ (from charge)] = 22 valence electrons. Placing a single bond between I and each Cl uses 2e⁻ x 2 bond = 4e⁻, leaving 22 – 4 = 18 electrons (nine pairs). After the Cl atoms complete their octets with three pairs each, the iodine finishes with the last three lone pairs. The iodine has an **expanded octet** of ten electrons.

c) BeF_2 has $[1 \times Be(2e^-)] + [2 \times F(7e^-)] = 16$ valence electrons. Placing a single bond between Be and each of the F atoms uses $2e^- \times 2$ bonds $= 4e^-$, leaving $16 - 4 = 12$ electrons (six pairs). The F atoms complete their octets with three pairs each, and there are no electrons left for the Be. Formal charges work against the formation of double bonds. Be, with only four electrons, is **electron deficient**.

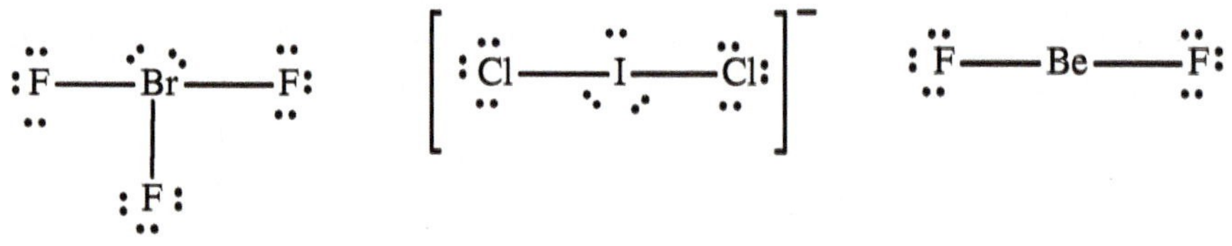

a)   b)   c)

10.23   <u>Plan:</u> Draw Lewis structures for the reactants and products.
         <u>Solution:</u>
         Beryllium chloride has the formula $BeCl_2$. $BeCl_2$ has $[1 \times Be(2e^-)] + [2 \times Cl(7e^-)] = 16$ valence electrons. Four of these electrons are used to place a single bond between Be and each of the Cl atoms, leaving $16 - 4 = 12$ electrons (six pairs). These six pairs are used to complete the octets of the Cl atoms, but Be does not have an octet – it is electron deficient.
         Chloride ion has the formula Cl^- with an octet of electrons.
         $BeCl_4^{2-}$ has $[1 \times Be(2e^-)] + [4 \times Cl(7e^-)] + [2e^-$ (from charge)$] = 32$ valence electrons. Eight of these electrons are used to place a single bond between Be and each Cl atom, leaving $32 - 8 = 24$ electrons (twelve pairs). These twelve pairs complete the octet of the Cl atoms (Be already has an octet).

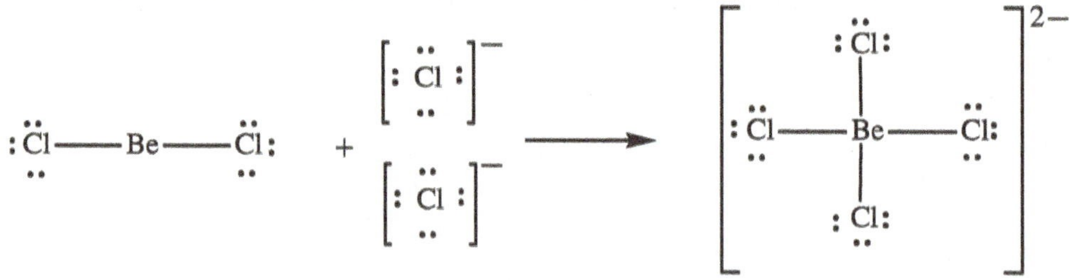

10.26   <u>Plan:</u> Use the structures in the text to determine the formal charges.
         Formal charge (FC) = no. of valence electrons – [no. of unshared valence electrons + ½ no. of shared valence electrons].
         <u>Solution:</u>
         Structure **A**: $FC_C = 4 - [0 + \frac{1}{2}(8)] = 0$; $FC_O = 6 - [4 + \frac{1}{2}(4)] = 0$; $FC_{Cl} = 7 - [6 + \frac{1}{2}(2)] = 0$
         Total FC $= 0$
         Structure **B**: $FC_C = 4 - [0 + \frac{1}{2}(8)] = 0$; $FC_O = 6 - [6 + \frac{1}{2}(2)] = -1$;
         $FC_{Cl(double\ bonded)} = 7 - [4 + \frac{1}{2}(4)] = +1$; $FC_{Cl(single\ bonded)} = 7 - [6 + \frac{1}{2}(2)] = 0$
         Total FC $= 0$
         Structure **C**: $FC_C = 4 - [0 + \frac{1}{2}(8)] = 0$; $FC_O = 6 - [6 + \frac{1}{2}(2)] = -1$;
         $FC_{Cl(double\ bonded)} = 7 - [4 + \frac{1}{2}(4)] = +1$; $FC_{Cl(single\ bonded)} = 7 - [6 + \frac{1}{2}(2)] = 0$
         Total FC $= 0$
         Structure **A** has the most reasonable set of formal charges.

10.28   The molecular shape and the electron-group arrangement are the same when there are no lone pairs on the central atom.

10.30   <u>Plan:</u> Examine a list of all possible structures, and choose the ones with four electron groups since the tetrahedral electron-group arrangement has four electron groups.
         <u>Solution:</u>

Tetrahedral	AX_4
Trigonal pyramidal	AX_3E
Bent or V shaped	AX_2E_2

10.32   <u>Plan:</u> Begin with the basic structures and redraw them.
<u>Solution:</u>
a) A molecule that is V shaped has two bonds and generally has either one (AX_2E) or two (AX_2E_2) lone electron pairs.
b) A trigonal planar molecule follows the formula AX_3 with three bonds and no lone electron pairs.
c) A trigonal bipyramidal molecule contains five bonding pairs (single bonds) and no lone pairs (AX_5).
d) A T-shaped molecule has three bonding groups and two lone pairs (AX_3E_2).
e) A trigonal pyramidal molecule follows the formula AX_3E with three bonding pairs and one lone pair.
f) A square pyramidal molecule shape follows the formula AX_5E with five bonding pairs and one lone pair.

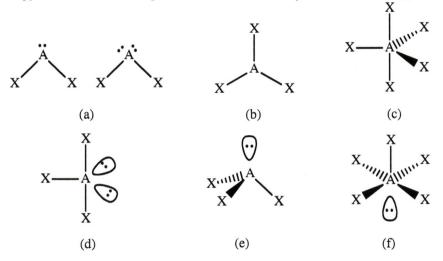

(a)   (b)   (c)

(d)   (e)   (f)

10.34   <u>Plan:</u> First, draw a Lewis structure, and then apply VSEPR.
<u>Solution:</u>
a) O_3: The molecule has $[3 \times O(6e^-)] = 18$ valence electrons. Four electrons are used to place single bonds between the oxygen atoms, leaving $18 - 4 = 14e^-$ (seven pairs). Six pairs are required to give the end oxygen atoms an octet; the last pair is distributed to the central oxygen, leaving this atom two electrons short of an octet. Form a double bond from one of the end O atoms to the central O by changing a lone pair on the end O to a bonding pair on the central O. This gives the following Lewis structure:

There are three electron groups around the central O, one of which is a lone pair. This gives a **trigonal planar** electron-group arrangement (AX_2E), a **bent** molecular shape, and an ideal bond angle of **120°**.
b) H_3O^+: This ion has $[3 \times H(1e^-)] + [1 \times O(6e^-)] - [1e^-$ (due to + charge] = eight valence electrons. Six electrons are used to place a single bond between O and each H, leaving $8 - 6 = 2e^-$ (one pair). Distribute this pair to the O atom, giving it an octet (the H atoms only get two electrons). This gives the following Lewis structure:

$$\left[\text{H} - \overset{..}{\text{O}} - \text{H} \atop \overset{|}{\text{H}} \right]^+ \qquad \left[\underset{\text{H}}{\overset{\overset{..}{\text{O}}{\text{\tiny{||||}} \text{H}}}{}} \overset{}{\underset{\text{H}}{}} \right]^+$$

There are four electron groups around the O, one of which is a lone pair. This gives a **tetrahedral** electron-group arrangement (AX_3E), a **trigonal pyramidal** molecular shape, and an ideal bond angle of **109.5°**.

c) NF$_3$: The molecule has [1 x N(5e$^-$)] + [3 x F(7e$^-$)] = 26 valence electrons. Six electrons are used to place a single bond between N and each F, leaving 26 – 6 = 20 e$^-$ (ten pairs). These ten pairs are distributed to all of the F atoms and the N atoms to give each atom an octet. This gives the following Lewis structure:

There are four electron groups around the N, one of which is a lone pair. This gives a **tetrahedral** electron-group arrangement (AX$_3$E), a **trigonal pyramidal** molecular shape, and an ideal bond angle of **109.5°**.

10.36   Plan: First, draw a Lewis structure, and then apply VSEPR.
Solution:
(a) CO$_3^{2-}$: This ion has [1 x C(4e$^-$)] + [3 x O(6e$^-$)] + [2e$^-$ (from charge)] = 24 valence electrons. Six electrons are used to place single bonds between C and each O atom, leaving 24 – 6 = 18 e$^-$ (nine pairs). These nine pairs are used to complete the octets of the three O atoms, leaving C two electrons short of an octet. Form a double bond from one of the O atoms to C by changing a lone pair on an O to a bonding pair on C. This gives the following Lewis structure:

$$\left[:\ddot{O}\!\!-\!\!C\!\!-\!\!\ddot{O}: \atop \overset{\|}{:\ddot{O}:} \right]^{2-} \qquad \left[\ddot{O} \overset{}{\diagdown}\ \overset{}{\diagup} \ddot{O} \atop \overset{C}{\overset{\|}{:\ddot{O}:}} \right]^{2-}$$

There are two additional resonance forms. There are three groups of electrons around the C, none of which are lone pairs. This gives a **trigonal planar** electron-group arrangement (AX$_3$), a **trigonal planar** molecular shape, and an ideal bond angle of **120°**.
(b) SO$_2$: This molecule has [1 x S(6e$^-$)] + [2 x S(6e$^-$)] = 18 valence electrons. Four electrons are used to place a single bond between S and each O atom, leaving 18 – 4 = 14e$^-$ (seven pairs). Six pairs are needed to complete the octets of the O atoms, leaving a pair of electrons for S. S needs one more pair to complete its octet. Form a double bond from one of the end O atoms to the S by changing a lone pair on the O to a bonding pair on the S. This gives the following Lewis structure:

$$\ddot{\ddot{O}}\!\!=\!\!\ddot{S}\!\!-\!\!\ddot{O}: \qquad \overset{\ddot{S}}{\underset{\ddot{O}\quad\ddot{O}}{\diagup\diagdown}}$$

There are three groups of electrons around the C, one of which is a lone pair.
This gives a **trigonal planar** electron-group arrangement (AX$_2$E), a **bent (V-shaped)** molecular shape, and an ideal bond angle of **120°**.
(c) CF$_4$: This molecule has [1 x C(4e$^-$)] + [4 x F(7e$^-$)] = 32 valence electrons. Eight electrons are used to place a single bond between C and each F, leaving 32 – 8 = 24 e$^-$ (twelve pairs). Use these twelve pairs to complete the octets of the F atoms (C already has an octet). This gives the following Lewis structure:

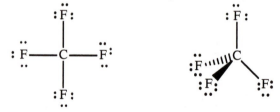

There are four groups of electrons around the C, none of which is a lone pair.
This gives a **tetrahedral** electron-group arrangement (AX$_4$), a **tetrahedral** molecular shape, and an ideal bond angle of **109.5°**.

10.38   Plan: Examine the structure shown, and then apply VSEPR.
         Solution:
         a) This structure shows three electron groups with three bonds around the central atom.
         There appears to be no distortion of the bond angles so the shape is **trigonal planar**, the classification is **AX₃**, with an ideal bond angle of **120°**.
         b) This structure shows three electron groups with three bonds around the central atom.
         The bonds are distorted down indicating the presence of a lone pair. The shape of the molecule is **trigonal pyramidal** and the classification is **AX₃E**, with an ideal bond angle of **109.5°**.
         c) This structure shows five electron groups with five bonds around the central atom.
         There appears to be no distortion of the bond angles so the shape is **trigonal bipyramidal** and the classification is **AX₅**, with ideal bond angles of **90° and 120°**.

10.40   Plan: The Lewis structures must be drawn, and VSEPR applied to the structures. Lone pairs on the central atom generally result in a deviation of the ideal bond angle.
         Solution:
         a) The ClO_2^- ion has [1 x Cl(7e⁻)] + [2 x O(6e⁻)] + [1e⁻ (from charge)] = 20 valence electrons. Four electrons are used to place a single bond between the Cl and each O, leaving 20 – 4 = 16 electrons (eight pairs). All eight pairs are used to complete the octets of the Cl and O atoms. There are two bonds (to the O atoms) and two lone pairs on the Cl for a total of four electron groups (AX₂E₂). The structure is based on a tetrahedral electron-group arrangement with an ideal bond angle of **109.5°**. The shape is **bent** (or V shaped). The presence of the lone pairs will cause the remaining angles to be **less than 109.5°**.
         b) The PF_5 molecule has [1 x P(5 e⁻)] + [5 x F(7 e⁻)] = 40 valence electrons. Ten electrons are used to place single bonds between P and each F atom, leaving 40 – 10 = 30 e⁻ (fifteen pairs). The fifteen pairs are used to complete the octets of the F atoms. There are five bonds to the P and no lone pairs (AX₅). The electron-group arrangement and the shape is **trigonal bipyramidal**. The ideal bond angles are **90° and 120°**. The absence of lone pairs means the **angles are ideal**.
         c) The SeF_4 molecule has [1 x Se(6e⁻)] + [4 x F(7e⁻)] = 34 valence electrons. Eight electrons are used to place single bonds between Se and each F atom, leaving 34 – 8 = 26e⁻ (thirteen pairs). Twelve pairs are used to complete the octets of the F atoms which leaves one pair of electrons. This pair is placed on the central Se atom. There are four bonds to the Se which also has a lone pair (AX₄E). The structure is based on a trigonal bipyramidal structure with ideal angles of **90° and 120°**. The shape is **seesaw**. The presence of the lone pairs means the angles are **less than ideal**.
         d) The KrF_2 molecule has [1 x Kr(8e⁻)] + [2 x F(7e⁻)] = 22 valence electrons. Four electrons are used to place a single bond between the Kr atom and each F atom, leaving 22 – 4 = 18 e⁻ (nine pairs). Six pairs are used to complete the octets of the F atoms. The remaining three pairs of electrons are placed on the Kr atom. The Kr is the central atom. There are two bonds to the Kr and three lone pairs (AX₂E₃). The structure is based on a trigonal bipyramidal structure with ideal angles of 90° and 120°. The shape is **linear**. The placement of the F atoms makes their ideal bond angle to be 2 x 90° = **180°**. The placement of the lone pairs is such that they cancel each other's repulsion, thus the actual **bond angle is ideal**.

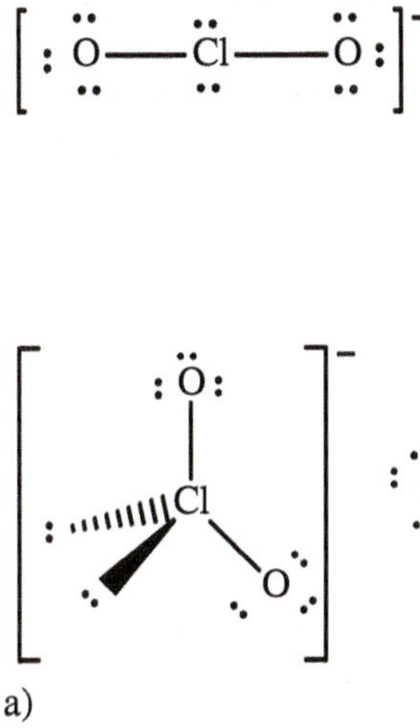

a) 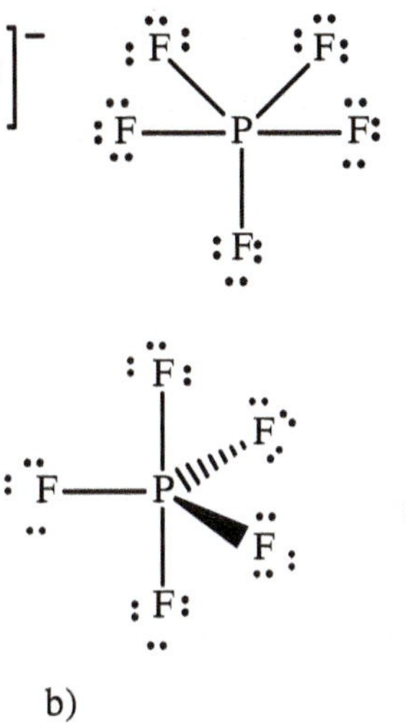   b) 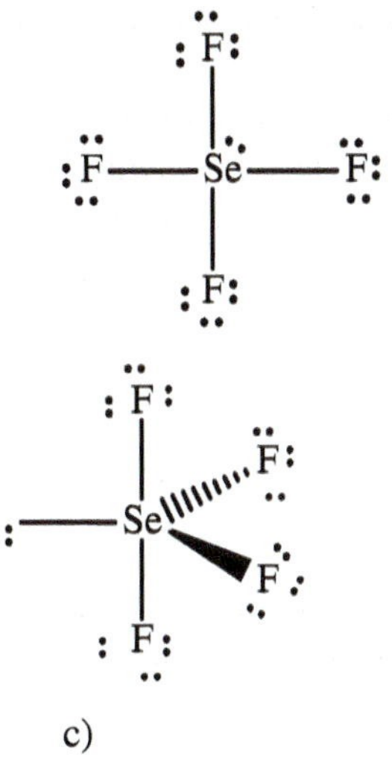   c) 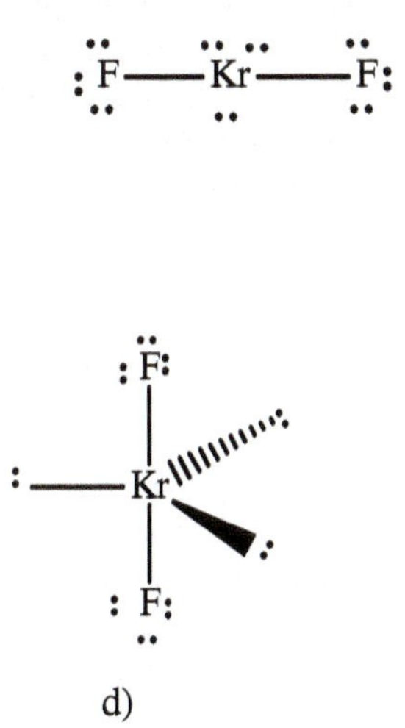   d)

10.42 Plan: The Lewis structures must be drawn, and VSEPR applied to the structures.
Solution:
a) CH_3OH: This molecule has $[1 \times C(4e^-)] + [4 \times H(1e^-)] + [1 \times O(6e^-)]$ = fourteen valence electrons. In the CH_3OH molecule, both carbon and oxygen serve as central atoms. (H can never be central.) Use eight electrons to place a single bond between the C and the O atom and three of the H atoms and another two electrons to place a single bond between the O and the last H atom. This leaves $14 - 10 = 4$ e$^-$ (two pairs). Use these two pairs to complete the octet of the O atom. C already has an octet and each H only gets two electrons. The carbon has four bonds and no lone pairs (AX_4), so it is **tetrahedral** with **no deviation** (no lone pairs) from the ideal angle of 109.5°. The oxygen has two bonds and two lone pairs (AX_2E_2), so it is **V shaped** or **bent** with the angles **less than the ideal** angle of 109.5°.

b) N_2O_4: This molecule has $[2 \times N(5e^-)] + [4 \times O(6e^-)]$ = 34 valence electrons. Use ten electrons to place a single bond between the two N atoms and between each N and two of the O atoms. This leaves $34 - 10 = 24e^-$ (twelve pairs). Use the twelve pairs to complete the octets of the oxygen atoms. Neither N atom has an octet, however. Form a double bond from one O atom to one N atom by changing a lone pair on the O to a bonding pair on the N. Do this for the other N atom as well. In the N_2O_4 molecule, both nitrogen atoms serve as central atoms. This is the arrangement given in the problem. Both nitrogen atoms are equivalent with three groups and no lone pairs (AX_3), so the arrangement is **trigonal planar** with **no deviation** (no lone pairs) from the ideal angle of 120°. The same results arise from the other resonance structures.

10.44 Plan: The Lewis structures must be drawn, and VSEPR applied to the structures.
Solution:
a) CH_3COOH has $[2 \times C(4e^-)] + [4 \times H(1e^-)] + [2 \times O(6e^-)]$ = twenty-four valence electrons. Use fourteen electrons to place a single bond between all of the atoms. This leaves $24 - 14 = 10$ e$^-$ (five pairs). Use these five pairs to complete the octets of the O atoms; the C atom bonded to the H atoms has an octet but the other C atom does not have a complete octet. Form a double bond from the O atom (not bonded to H) to the C by changing a lone pair on the O to a bonding pair on the C. In the CH_3COOH molecule, the carbons and the O with H attached serve as central atoms. The carbon bonded to the H atoms has four groups and no lone pairs (AX_4), so it is **tetrahedral** with **no deviation** from the ideal angle of 109.5°. The carbon bonded to the O atoms has three groups and no lone pairs (AX_3), so it is **trigonal planar** with **no deviation** from the ideal angle of 120°. The H bearing O has two bonds and two lone pairs (AX_2E_2), so the arrangement is **V shaped** or **bent** with an angle **less than the ideal** value of 109.5°.

b) H_2O_2 has $[2 \times H(1e^-)] + [2 \times O(6e^-)]$ = fourteen valence electrons. Use six electrons to place single bonds between the O atoms and between each O atom and an H atom. This leaves $14 - 6 = 8$ e$^-$ (four pairs). Use these four pairs to complete the octets of the O atoms. In the H_2O_2 molecule, both oxygen atoms serve as central atoms.

Both O atoms have two bonds and two lone pairs (AX_2E_2), so they are **V shaped** or **bent** with angles **less than the ideal** value of 109.5°.

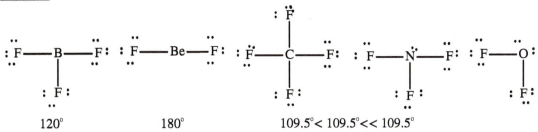

10.46   Plan: First, draw a Lewis structure, and then apply VSEPR. The presence of lone pairs on the central atom generally results in a smaller than ideal bond angle.
Solution:

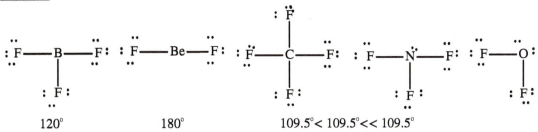

120°   180°   109.5°< 109.5°<< 109.5°

Bond angles: $OF_2 < NF_3 < CF_4 < BF_3 < BeF_2$

BeF_2 is an AX_2 type molecule, so the angle is the ideal 180°. BF_3 is an AX_3 molecule, so the angle is the ideal 120°. CF_4, NF_3, and OF_2 all have tetrahedral electron-group arrangements of the following types: AX_4, AX_3E, and AX_2E_2, respectively. The ideal tetrahedral bond angle is 109.5°, which is present in CF_4. The one lone pair in NF_3 decreases the angle a little. The two lone pairs in OF_2 decrease the angle even more.

10.48   Plan: The ideal bond angles depend on the electron-group arrangement. Deviations depend on lone pairs.
Solution:
a) The C and N have three groups, so they are **ideally 120°**, and the O has four groups, so **ideally the angle is 109.5°**. The N and O have lone pairs, so the **angles are less than ideal**.
b) All central atoms have four pairs, so ideally all the angles are **109.5°**. The lone pairs on the O **reduce** this value.
c) The B has three groups (no lone pairs) leading to an **ideal bond angle of 120°**. All the O atoms have four pairs (**ideally 109.5°**), two of which are lone, and **reduce the angle**.

10.51   Plan: The Lewis structures are needed to predict the ideal bond angles.
Solution:

The P atoms have no lone pairs in any case so the angles are ideal.

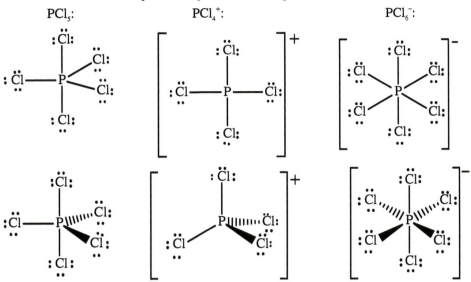

The original PCl_5 is AX_5, so the shape is trigonal bipyramidal, and the angles are 120° and 90°.
The PCl_4^+ is AX_4, so the shape is tetrahedral, and the angles are 109.5°.
The PCl_6^- is AX_6, so the shape is octahedral, and the angles are 90°.
Half the PCl_5 (trigonal bipyramidal, 120° and 90°) become tetrahedral PCl_4^+ (tetrahedral, 109.5°), and the other half become octahedral PCl_6^- (octahedral, 90°).

10.52   Molecules are polar if they have polar bonds that are not arranged to cancel each other. A polar bond is present any time there is a bond between elements with differing electronegativities.

10.55   Plan: To determine if a bond is polar, determine the electronegativity difference of the atoms participating in the bond. The greater the electronegativity difference, the more polar the bond. To determine if a molecule is polar (has a dipole moment), it must have polar bonds, and a certain shape determined by VSEPR.
Solution:

a)
Molecule	Bond	Electronegativities	Electronegativity difference
SCl_2	S–Cl	S = 2.5 Cl = 3.0	3.0 – 2.5 = 0.5
F_2	F–F	F = 4.0 F = 4.0	4.0 – 4.0 = 0.0
CS_2	C–S	C = 2.5 S = 2.5	2.5 – 2.5 = 0.0
CF_4	C–F	C = 2.5 F = 4.0	4.0 – 2.5 = 1.5
BrCl	Br–Cl	Br = 2.8 Cl = 3.0	3.0 – 2.8 = 0.2

The polarities of the bonds increase in the order: F–F = C–S < Br–Cl < S–Cl < C–F. Thus, CF_4 has the most polar bonds.

b) The F_2 and CS_2 cannot be polar since they do not have polar bonds. CF_4 is an AX_4 molecule, so it is tetrahedral with the four polar C–F bonds arranged to cancel each other giving an overall nonpolar molecule. **BrCl has a dipole moment** since there are no other bonds to cancel the polar Br–Cl bond. **SCl_2 has a dipole moment** (is polar) because it is a bent molecule, AX_2E_2, and the electron density in both S–Cl bonds is pulled towards the more electronegative chlorine atoms.

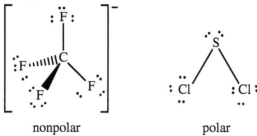

nonpolar   polar

10.57   Plan: If only two atoms are involved, only an electronegativity difference is needed. The greater the difference in electronegativity, the more polar the bond. If there are more than two atoms, the molecular geometry must be determined.
Solution:
a) All the bonds are polar covalent. The SO_3 molecule is trigonal planar, AX_3, so the bond dipoles cancel leading to a nonpolar molecule (no dipole moment). The SO_2 molecule is bent, AX_2E, so the polar bonds result in electron density being pulled towards one side of the molecule. **SO_2 has a greater dipole moment** because it is the only one of the pair that is polar.

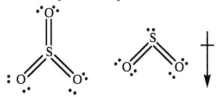

b) ICl and IF are polar, as are all diatomic molecules composed of atoms with differing electronegativities. The electronegativity difference for ICl (3.0 – 2.5 = 0.5) is less than that for IF (4.0 – 2.5 = 1.5). The greater difference means that **IF has a greater dipole moment**.

c) All the bonds are polar covalent. The SiF_4 molecule is nonpolar (has no dipole moment) because the bonds are arranged tetrahedrally, AX_4. SF_4 is AX_4E, so it has a see-saw shape, where the bond dipoles do not cancel. **SF_4 has the greater dipole moment.**

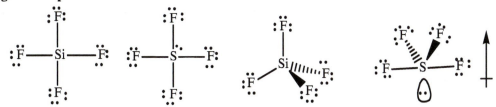

d) H_2O and H_2S have the same basic structure. They are both bent molecules, AX_2E_2, and as such, they are polar. The electronegativity difference in H_2O (3.5 – 2.1 = 1.4) is greater than the electronegativity difference in H_2S (2.5 – 2.1 = 0.4) so **H_2O has a greater dipole moment**.

10.59　Plan: Draw Lewis structures, and then apply VSEPR. A molecule has a dipole moment if polar bonds do not cancel.
Solution:
$C_2H_2Cl_2$ has [2 x C(4e⁻)] + [2 x H(1e⁻)] + [2 x Cl(7e⁻)] = 24 valence electrons. The two carbon atoms are bonded to each other. The H atoms and Cl atoms are bonded to the C atoms. Use ten electrons to place a single bond between all of the atoms. This leaves 24 – 10 = 14e⁻ (seven pairs). Use these seven pairs to complete the octets of the Cl atoms and one of the C atoms; the other C atom does not have a complete octet. Form a double bond between the carbon atoms by changing the lone pair on one C atom to a bonding pair. There are three possible structures for the compound $C_2H_2Cl_2$:

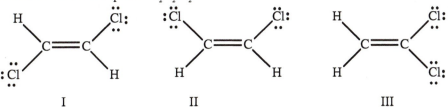

The presence of the double bond prevents rotation about the C=C bond, so the structures are "fixed." The C–Cl bonds are more polar than the C–H bonds, so the key to predicting the polarity is the positioning of the C–Cl bonds. Structure I has the C–Cl bonds arranged so that they cancel leaving I as a nonpolar molecule. Both II and III have C–Cl bonds on the same side so the bonds work together making both molecules polar. Both I and II will react with H_2 to give a compound with a Cl attached to each C (same product). Structure III will react with H_2 to give a compound with two Cl atoms on one C and none on the other (different product). **Structure I must be X** as it is the only one that is nonpolar (has no dipole moment). **Structure II must be Z** because it is polar and gives the same product as compound X. This means that **Structure III must be the remaining compound, Y. Compound Y (III) has a dipole moment.**

10.61　Plan: The Lewis structures are needed to do this problem. A single bond (bond order = 1) is weaker and longer than a double bond (bond order = 2) which is weaker and longer than a triple bond (bond order = 3). To find the heat of reaction, add the energy required to break all the bonds in the reactants to the energy released to form all bonds in the product. Remember to use a negative sign for the energy of the bonds formed since bond formation is exothermic. The bond energy values are found in Table 9.2.
Solution:
a) The H atoms cannot be central, and they are evenly distributed on the N atoms.
N_2H_4 has [2 x N(5e⁻)] + [4 x H(1e⁻)] = fourteen valence electrons, ten of which are used in the bonds between the atoms. The remaining two pairs are used to complete the octets of the N atoms.
N_2H_2 has [2 x N(5e⁻)] + (2 x H(1e⁻)] = twelve valence electrons, six of which are used in the bonds between the atoms. The remaining three pairs of electrons are not enough to complete the octets of both N atoms, so one lone pair is moved to a bonding pair between the N atoms.

N_2 has $[2 \times N(5 \, e^-)]$ = ten valence electrons, two of which are used to place a single bond between the two N atoms. Since only four pairs of electrons remain and six pairs are required to complete the octets, two lone pairs become bonding pairs to form a triple bond.

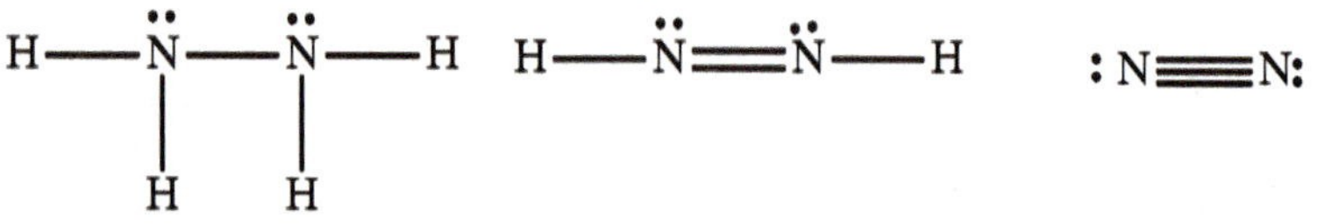

Hydrazine   Diazene   Nitrogen

The **single (bond order = 1)N–N bond is weaker and longer** than any of the others are. The **triple bond (bond order = 3) is stronger and shorter** than any of the others. The **double bond (bond order = 2) has an intermediate strength and length.**

b) N_4H_4 has $[4 \times N(5e^-)] + [4 \times H(1e^-)]$ = twenty-four valence electrons, fourteen of which are used for single bonds between the atoms. When the remaining five pairs are distributed to complete the octets, one N atom lacks two electrons. A lone pair is moved to a bonding pair for a double bond.

Reactant bonds broken:

4 N–H = 4 mol (391 kJ/mol) = 1564 kJ

2 N–N = 2 mol (160 kJ/mol) = 320 kJ

1 N=N = 1 mol (418 kJ/mol) = 418 kJ

$\Sigma \Delta H^{\circ}_{\text{bonds broken}}$ = 2302 kJ

Product bonds formed:

4 N–H = 4 mol (–391 kJ/mol) = –1564 kJ

1 N–N = 1 mol (–160 kJ/mol) = –160 kJ

1 N≡N = 1 mol (–945 kJ/mol) = –945 kJ

$\Sigma \Delta H^{\circ}_{\text{bonds formed}}$ = –2669 kJ

$\Delta H^{\circ}_{\text{rxn}} = \Sigma \Delta H^{\circ}_{\text{bonds broken}} + \Sigma \Delta H^{\circ}_{\text{bonds formed}}$ = 2302 kJ + (–2669 kJ) = **–367 kJ**

10.65 <u>Plan:</u> Use the Lewis structures shown in the text. The equation for formal charge (FC) is FC = no. of valence electrons – [no. of unshared valence electrons + ½ no. of shared valence electrons].
<u>Solution:</u>
a) Formal charges for Al_2Cl_6:

$FC_{Al} = 3 – [0 + ½(8)] = $ **–1**

$FC_{Cl, \, ends} = 7 – [6 + ½(2)] = 0$

$FC_{Cl, \, bridging} = 7 – [4 + ½(4)] = $ **+1**

(<u>Check:</u> Formal charges add to zero, the charge on the compound.)

Formal charges for I_2Cl_6:

$FC_I = 7 – [4 + ½(8)] = $ **–1**

$FC_{Cl, \, ends} = 7 – [6 + ½(2)] = 0$

$FC_{Cl, \, bridging} = 7 – [4 + ½(4)] = $ **+1**

(<u>Check:</u> Formal charges add to zero, the charge on the compound.)

b) The aluminum atoms have no lone pairs and are AX_4, so they are tetrahedral. The two tetrahedral Al atoms cannot give a planar structure. The iodine atoms in I_2Cl_6 have two lone pairs each and are AX_4E_2 so they are square planar. Placing the square planar I atoms adjacent can give a planar molecule.

10.70 <u>Plan:</u> Draw the Lewis structures, and then use VSEPR to describe propylene oxide.
<u>Solution:</u>
a)

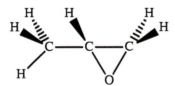

In propylene oxide, the C atoms are all AX$_4$. The C atoms do not have any unshared (lone) pairs. All of the ideal bond angles for the C atoms in propylene oxide are **109.5°** and the molecular shape around each carbon atom is **tetrahedral**.

b) In propylene oxide, the C that is not part of the three-membered ring should have an ideal angle. The atoms in the ring form an equilateral triangle. The angles in an equilateral triangle are 60°. The angles around the two carbons in the rings are reduced from the ideal 109.5° to 60°.

10.75   Plan: Ethanol burns (combusts) with O$_2$ to produce CO$_2$ and H$_2$O. To find the heat of reaction in part a), add the energy required to break all the bonds in the reactants to the energy released to form all bonds in the product. Remember to use a negative sign for the energy of the bonds formed since bond formation is exothermic. The bond energy values are found in Table 9.2. The heat of vaporization of ethanol must be included for part b). The enthalpy change in part c) is the sum of the heats of formation of the products minus the sum of the heats of formation of the reactants. The calculation for part d) is the same as in part a).

Solution:

a) $CH_3CH_2OH(g) + 3O_2(g) \rightarrow 2CO_2(g) + 3H_2O(g)$

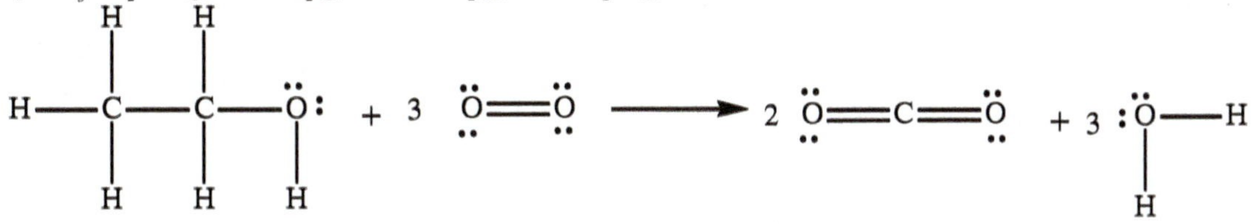

Reactant bonds broken:
1 x C–C = (1 mol)(347 kJ/mol) = 347 kJ
5 x C–H = (5 mol)(413 kJ/mol) = 2065 kJ
1 x C–O = (1 mol)(358 kJ/mol) = 358 kJ
1 x O–H = (1 mol)(467 kJ/mol) = 467 kJ
3 x O=O = (3 mol)(498 kJ/mol) = 1494 kJ
              $\Sigma\Delta H°_{\text{bonds broken}}$ = 4731 kJ

Product bonds formed:
4 x C=O = (4 mol)(–799 kJ/mol) = –3196 kJ
6 x O–H = (6 mol)(–467 kJ/mol) = –2802 kJ
              $\Sigma\Delta H°_{\text{bonds formed}}$ = –5998 kJ

$\Delta H°_{\text{rxn}} = \Sigma\Delta H°_{\text{bonds broken}} + \Sigma\Delta H°_{\text{bonds formed}}$ = 4731 kJ + (–5998 kJ) = **–1267 kJ** for each mole of ethanol burned

b) If it takes 40.5 kJ/mol to vaporize the ethanol, part of the heat of combustion must be used to convert liquid ethanol to gaseous ethanol. The new value becomes:

$\Sigma\Delta H°_{\text{combustion (liquid)}} = -1267 \text{ kJ} + (1 \text{ mol})\left[\dfrac{40.5 \text{ kJ}}{1 \text{ mol}}\right] = -1226.5 = $ **–1226 kJ per mole** of liquid ethanol burned

c) $\Delta H°_{\text{rxn}} = \Sigma m \Delta H°_{\text{f (products)}} - \Sigma n \Delta H°_{\text{f (reactants)}}$

$\Delta H°_{\text{rxn}} = \{2 \Delta H°_f [CO_2(g)] + 3 \Delta H°_f [H_2O(g)]\} - \{1 \Delta H°_f [C_2H_5OH(l)] + 3 \Delta H°_f [O_2(g)]\}$

= [(2 mol)(–393.5 kJ/mol) + (3 mol)(–241.826 kJ/mol)] – [(1 mol)(–277.63 kJ/mol) + 3 mol(0 kJ/mol)]

= –1234.848 = **–1234.8 kJ**

The two answers differ by less than 10 kJ. This is a very good agreement since average bond energies were used to calculate the answers in a) and b).

d) $C_2H_4(g) + H_2O(g) \rightarrow CH_3CH_2OH(g)$

The Lewis structures for the reaction are:

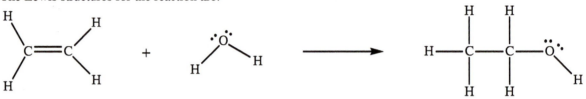

Reactant bonds broken:
1 x C=C = (1 mol)(614 kJ/mol) =   614 kJ
4 x C–H = (4 mol)(413 kJ/mol) = 1652 kJ
2 x O–H = (2 mol)(467 kJ/mol) =   934 kJ
$\Sigma \Delta H^{\circ}_{\text{bonds broken}}$ = 3200 kJ

Product bonds formed:
1 x C–C = (1 mol)(–347 kJ/mol) =   –347 kJ
5 x C–H = (5 mol)(–413 kJ/mol) = –2065 kJ
1 x C–O = (1 mol)(–358 kJ/mol) =   –358 kJ
1 x O–H = (1 mol)(–467 kJ/mol) =   –467 kJ
$\Sigma \Delta H^{\circ}_{\text{bonds formed}}$ = –3237 kJ

$\Delta H^{\circ}_{\text{rxn}} = \Sigma \Delta H^{\circ}_{\text{bonds broken}} + \Sigma \Delta H^{\circ}_{\text{bonds formed}}$ = 3200 kJ + (–3237 kJ) = **–37 kJ**

10.77   Plan: Determine the empirical formula from the percent composition (assuming 100 g of compound). Use the titration data to determine the mole ratio of acid to the NaOH. This ratio gives the number of acidic H atoms in the formula of the acid. Finally, combine this information to construct the Lewis structure.
Solution:

$\text{Moles of H} = (2.24 \text{ g H})\left(\dfrac{1 \text{ mol}}{1.008 \text{ g H}}\right) = 2.222 \text{ mol H}$

$\text{Moles of C} = (26.7 \text{ g C})\left(\dfrac{1 \text{ mol}}{12.01 \text{ g C}}\right) = 2.223 \text{ mol C}$

$\text{Moles of O} = (71.1 \text{ g O})\left(\dfrac{1 \text{ mol}}{16.00 \text{ g O}}\right) = 4.444 \text{ mol O}$

The preliminary formula is $H_{2.222}C_{2.223}O_{4.444}$.

Dividing all subscripts by the smallest subscript to obtain integer subscripts:

$H_{\frac{2.222}{2.222}}C_{\frac{2.223}{2.222}}O_{\frac{4.444}{2.222}} = HCO_2$

The empirical formula is HCO_2.

Calculate the amount of NaOH required for the titration:

$\text{mmoles of NaOH} = (50.0 \text{ mL})\left(\dfrac{1 \text{ L}}{1000 \text{ mL}}\right)\left(\dfrac{0.040 \text{ mol NaOH}}{\text{L}}\right)\left(\dfrac{1 \text{ mmol}}{0.001 \text{ mol}}\right) = 2.0 \text{ mmol NaOH}$

Thus, the ratio is 2.0 mmole base/1.0 mmole acid, or each acid molecule has two hydrogen atoms to react (diprotic). The empirical formula indicates a monoprotic acid, so the formula must be doubled to: $H_2C_2O_4$.
$H_2C_2O_4$ has [2 x H(1e⁻)] + [2 x C(4e⁻)] + [4 x O(6e⁻)] = 34 valence electrons to be used in the Lewis structure. Fourteen of these electrons are used to bond the atoms with single bonds, leaving 34 – 14 = 20 electrons or ten pairs of electrons. When these ten pairs of electrons are distributed to the atoms to complete octets, neither C atom has an octet; a lone pair from the oxygen without hydrogen is changed to a bonding pair on C.

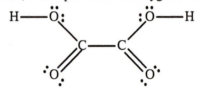

10.80 <u>Plan:</u> Write the balanced chemical equations for the reactions and draw the Lewis structures. To find the heat of reaction, add the energy required to break all the bonds in the reactants to the energy released to form all bonds in the product. Remember to use a negative sign for the energy of the bonds formed since bond formation is exothermic. The bond energy values are found in Table 9.2. Divide the heat of reaction by the number of moles of oxygen gas appearing in each reaction to get the heat of reaction per mole of oxygen.
<u>Solution:</u>

$$CH_4(g) + 2O_2(g) \rightarrow CO_2(g) + 2H_2O(g)$$

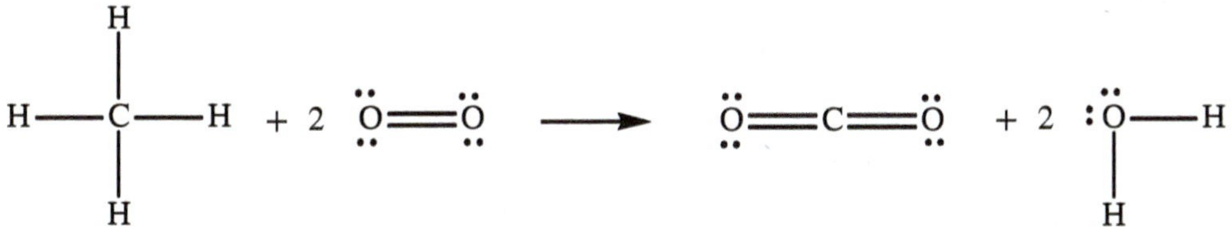

Reactant bonds broken:

4 x C–H = (4 mol)(413 kJ/mol) = 1652 kJ

2 x O=O = (2 mol)(498 kJ/mol) = 996 kJ

$\Sigma \Delta H^{\circ}_{\text{bonds broken}}$ = 2648 kJ

Product bonds formed:

2 x C=O = (2 mol)(–799 kJ/mol) = –1598 kJ

4 x O–H = (4 mol)(–467 kJ/mol) = –1868 kJ

$\Sigma \Delta H^{\circ}_{\text{bonds formed}}$ = –3466 kJ

$\Delta H^{\circ}_{\text{rxn}} = \Sigma \Delta H^{\circ}_{\text{bonds broken}} + \Sigma \Delta H^{\circ}_{\text{bonds formed}}$ = 2648 kJ + (–3466 kJ) = –818 kJ for 2 mol O_2

Per mole of O_2 = –818/2 = **–409 kJ/mol O_2**

$$2H_2S(g) + 3O_2(g) \rightarrow 2SO_2(g) + 2H_2O(g)$$

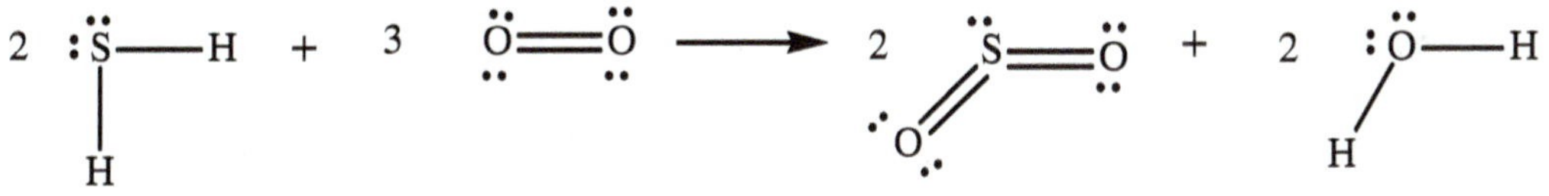

Reactant bonds broken:

4 x S–H = (4 mol)(347 kJ/mol) = 1388 kJ

3 x O=O = (3 mol)(498 kJ/mol) = 1494 kJ

$\Sigma \Delta H^{\circ}_{\text{bonds broken}}$ = 2882 kJ

Product bonds formed:

4 x S=O = (4 mol)(–552 kJ/mol) = –2208 kJ

4 x O–H = (4 mol)(–467 kJ/mol) = –1868 kJ

$\Sigma \Delta H^{\circ}_{\text{bonds formed}}$ = –4076 kJ

$\Delta H^{\circ}_{\text{rxn}} = \Sigma \Delta H^{\circ}_{\text{bonds broken}} + \Sigma \Delta H^{\circ}_{\text{bonds formed}}$ = 2882 kJ + (–4076 kJ) = –1194 kJ for 3 mol O_2

Per mole of O_2 = –1194/3 = **–398 kJ/mol O_2**

10.82 <u>Plan:</u> Draw the Lewis structure of the OH species. The standard enthalpy of formation is the sum of the energy required to break all the bonds in the reactants and the energy released to form all bonds in the product. Remember to use a negative sign for the energy of the bonds formed since bond formation is exothermic. The bond energy values are found in Table 9.2.

Solution:

a) The OH molecule has $[1 \times O(6e^-)] + [1 \times H(1e^-)] = 7$ valence electrons to be used in the Lewis structure. Two of these electrons are used to bond the atoms with a single bond, leaving $7 - 2 = 5$ electrons. Those five electrons are given to oxygen. But no atom can have an octet, and one electron is left unpaired. The Lewis structure is:

$$\cdot \ddot{\underset{\cdot\cdot}{O}}\text{——}H$$

b) The formation reaction is: $1/2 O_2(g) + 1/2 H_2(g) \rightarrow OH(g)$. The heat of reaction is:

$\Delta H^{\circ}_{rxn} = \Sigma \Delta H^{\circ}_{bonds\ broken} + \Sigma \Delta H^{\circ}_{bonds\ formed} = 39.0$ kJ

$[\frac{1}{2}\ (BE_{O=O}) + \frac{1}{2}\ (BE_{H-H})] + [BE_{O-H}] = 39.0$ kJ

$[(\frac{1}{2}\ mol)(498\ kJ/mol) + (\frac{1}{2}\ mol)(432\ kJ/mol)] + [BE_{O-H}] = 39.0$ kJ

465 kJ $+ [BE_{O-H}] = 39.0$ kJ

$BE_{O-H} = $ **–426 kJ or 426 kJ**

c) The average bond energy (from the bond energy table) is 467 kJ/mol. There are two O–H bonds in water for a total of 2 x 467 kJ/mol = 934 kJ. The answer to part b) accounts for 426 kJ of this, leaving:
934 kJ – 426 kJ = **508 kJ**

10.84   Plan: The basic Lewis structure will be the same for all species. The Cl atoms are larger than the F atoms. All of the molecules are of the type AX_5 and have trigonal bipyramidal molecular shape. The equatorial positions are in the plane of the triangle and the axial positions above and below the plane of the triangle. In this molecular shape, there is more room in the equatorial positions.
Solution:

a) The F atoms will occupy the smaller axial positions first so that the larger Cl atoms can occupy the equatorial positions which are less crowded.
b) The molecule containing only F atoms is nonpolar (has no dipole moment), as all the polar bonds would cancel. The molecules with one F or one Cl would be polar since the P–F and P–Cl bonds are not equal in polarity and thus do not cancel each other. The presence of two axial F atoms means that their polarities will cancel (as would the three Cl atoms) giving a nonpolar molecule. The molecule with three F atoms is also polar.

Polar	Nonpolar	Polar	Polar	Nonpolar
	No dipole moment			No dipole moment

10.86   Plan: Count the valence electrons and draw Lewis structures for the resonance forms.
Solution:
The $H_2C_2O_4$ molecule has $[2 \times H(1e^-)] + [2 \times C(4e^-)] + [4 \times O(6e^-)] = 34$ valence electrons to be used in the Lewis structure. Fourteen of these electrons are used to bond the atoms with a single bond, leaving $34 - 14 = 20$ electrons. If these twenty electrons are given to the oxygen atoms to complete their octet, the carbon atoms do not have octets. A lone pair from each of the oxygen atoms without hydrogen is changed to a bonding pair on C. The $HC_2O_4^-$ ion has $[1 \times H(1e^-)] + [2 \times C(4e^-)] + [4 \times O(6e^-)] + [1e^-$ (from the charge)$] = 34$ valence electrons to be used in the Lewis structure. Twelve of these electrons are used to bond the atoms with a single bond, leaving $34 - 12 = 22$ electrons. If these twenty-two electrons are given to the oxygen atoms to complete their octet, the carbon atoms do not have octets. A lone pair from two of the oxygen atoms without hydrogen is changed to a bonding pair on C. There are two resonance structures.

The $C_2O_4^{2-}$ ion has $[2 \times C(4e^-)] + [4 \times O(6e^-)] + [2e^-$ (from the charge)$] = 34$ valence electrons to be used in the Lewis structure. Ten of these electrons are used to bond the atoms with a single bond, leaving $34 - 10 = 24$ electrons. If these twenty-four electrons are given to the oxygen atoms to complete their octets, the carbon atoms do not have octets. A lone pair from two oxygen atoms is changed to a bonding pair on C. There are four resonance structures.

$H_2C_2O_4$: $HC_2O_4^-$:

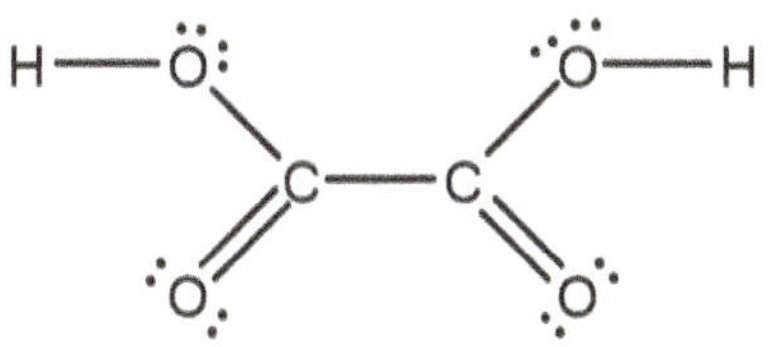

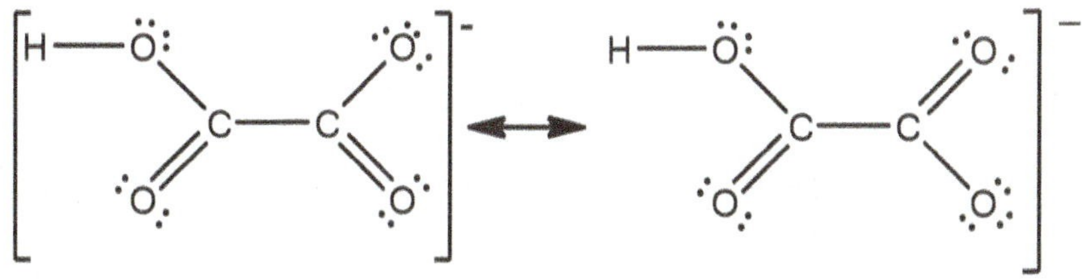

$C_2O_4^{2-}$:

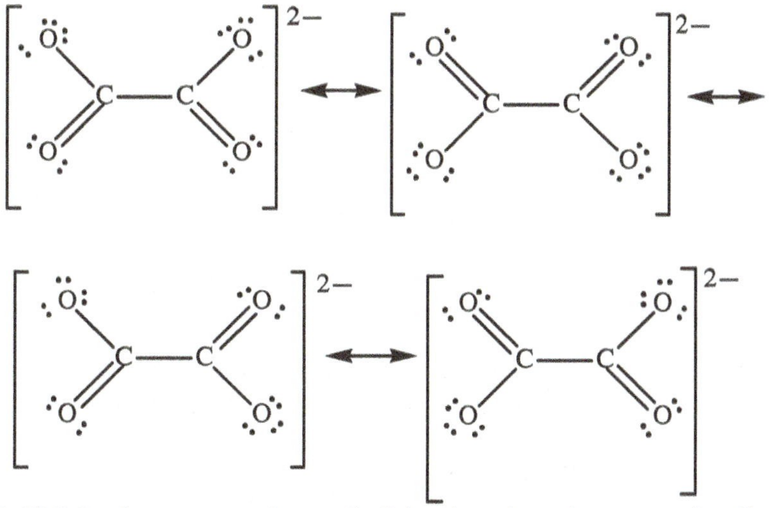

In $H_2C_2O_4$, there are two shorter C=O bonds and two longer, weaker C—O bonds.

In $HC_2O_4^-$, the C—O bonds on the side retaining the H remain as one long C—O bond and one shorter, stronger C=O bond. The C—O bonds on the other side of the molecule have resonance forms with an average bond order of 1.5, so they are intermediate in length and strength.

In $C_2O_4^{2-}$, all the carbon to oxygen bonds are resonating and have an average bond order of 1.5.

10.90   Plan: Draw the Lewis structures. Calculate the heat of reaction using the bond energies in Table 9.2.
Solution:
$$SO_3(g) + H_2SO_4(l) \rightarrow H_2S_2O_7(l)$$

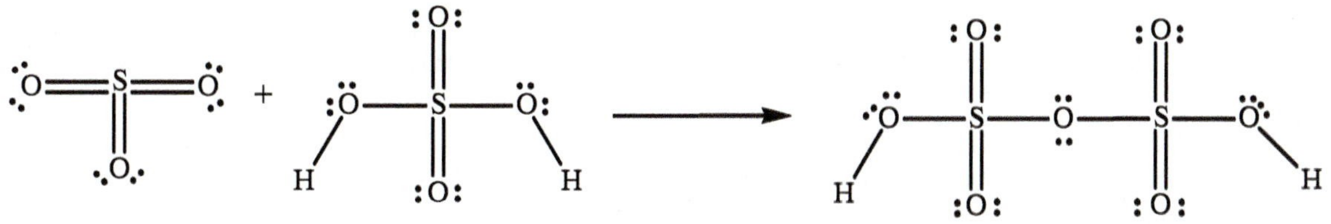

Reactant bonds broken:

$5 \times$ S=O $= (5 \text{ mol})(552 \text{ kJ/mol}) = 2760 \text{ kJ}$

$2 \times$ S–O $= (2 \text{ mol})(265 \text{ kJ/mol}) = 530 \text{ kJ}$

$\underline{2 \times \text{O–H} = (2 \text{ mol})(467 \text{ kJ/mol}) = 934 \text{ kJ}}$

$\Sigma\Delta H^\circ_{\text{bonds broken}} = 4224 \text{ kJ}$

Product bonds formed:

$4 \times S=O = (4 \text{ mol})(-552 \text{ kJ/mol}) = -2208 \text{ kJ}$

$4 \times S-O = (4 \text{ mol})(-265 \text{ kJ/mol}) = -1060 \text{ kJ}$

$\underline{2 \times O-H = (2 \text{ mol})(-467 \text{ kJ/mol}) = -934 \text{ kJ}}$

$\Sigma\Delta H^\circ_{\text{bonds formed}} = -4202 \text{ kJ}$

$\Delta H^\circ_{\text{rxn}} = \Sigma\Delta H^\circ_{\text{bonds broken}} + \Sigma\Delta H^\circ_{\text{bonds formed}} = 4224 \text{ kJ} + (-4202 \text{ kJ}) = \mathbf{22 \text{ kJ}}$

10.91   Plan: Pick the VSEPR structures for AY_3 substances. Then determine which are polar.
Solution:
The molecular shapes that have a central atom bonded to three other atoms are trigonal planar, trigonal pyramidal, and T shaped:

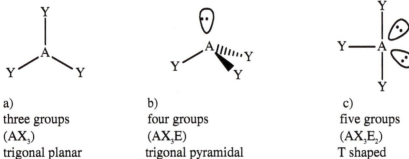

a)   b)   c)
three groups   four groups   five groups
(AX_3) (AX_3E) (AX_3E_2)
trigonal planar   trigonal pyramidal   T shaped

Trigonal planar molecules, such as a), are nonpolar, so it cannot be AY_3. Trigonal pyramidal molecules b) and T-shaped molecules c) are polar, so either could represent AY_3.

10.95   Plan: Draw the Lewis structure of each compound. Atoms 180° apart are separated by the sum of the bond's length. Atoms not at 180° apart must have their distances determined by geometrical relationships.
Solution:

(a) H——C≡C——H   (b) [structure of SF_6]   (c) [structure of PF_5]

a) C_2H_2 has $[2 \times C(4e^-)] + [2 \times H(1e^-)] = 10$ valence electrons to be used in the Lewis structure. Six of these electrons are used to bond the atoms with a single bond, leaving $10 - 6 = 4$ electrons. Giving one carbon atom the four electrons to complete its octet results in the other carbon atom not having an octet. The two lone pairs from the carbon with an octet are changed to two bonding pairs for a triple bond between the two carbon atoms. The molecular shape is linear. The H atoms are separated by two carbon-hydrogen bonds (109 pm) and a carbon-carbon triple bond (121 pm).
Total separation = 2(109 pm) + 121 pm = **339 pm**

b) SF_6 has $[1 \times S(6e^-)] + [6 \times F(7e^-)] = 48$ valence electrons to be used in the Lewis structure. Twelve of these electrons are used to bond the atoms with a single bond, leaving $48 - 12 = 36$ electrons. These thirty-six electrons are given to the fluorine atoms to complete their octets. The molecular shape is octahedral. The fluorine atoms on opposite sides of the S are separated by twice the sulfur-fluorine bond length (158 pm).
Total separation = 2(158 pm) = **316 pm**
Adjacent fluorines are at two corners of a right triangle, with the sulfur at the 90° angle. Two sides of the triangle are equal to the sulfur-fluorine bond length (158 pm). The separation of the fluorine atoms is at a distance equal to the hypotenuse of this triangle. This length of the hypotenuse may be found using the Pythagorean Theorem $(a^2 + b^2 = c^2)$. In this case a = b = 158 pm. Thus, $c^2 = (158 \text{ pm})^2 + (158 \text{ pm})^2$, and so c = 223.4457 = **223 pm.**

c) PF_5 has $[1 \times P(5e^-)] + [5 \times F(7e^-)] = 40$ valence electrons to be used in the Lewis structure. Ten of these electrons are used to bond the atoms with a single bond, leaving $40 - 10 = 30$ electrons. These thirty electrons are given to the fluorine atoms to complete their octets. The molecular shape is trigonal bipyramidal. Adjacent equatorial fluorine atoms are at two corners of a triangle with an F-P-F bond angle of $120°$. The length of the P-F bond is 156 pm. If the $120°$ bond angle is A, then the F-F bond distance is a and the P-F bond distances are b and c. The F-F bond distance can be found using the Law of Cosines: $a^2 = b^2 + c^2 - 2bc (\cos A)$.
$a^2 = (156)^2 + (156)^2 - 2(156)(156)\cos 120°$. $a = 270.1999 = \mathbf{270 \ pm}$.

CHAPTER 11 THEORIES OF COVALENT BONDING

FOLLOW–UP PROBLEMS

11.1A <u>Plan:</u> Draw a Lewis structure. Determine the number and arrangement of the electron pairs about the central atom. From this, determine the type of hybrid orbitals involved. Write the partial orbital diagram of the central atoms before and after the orbitals are hybridized.
<u>Solution:</u>
a) Be is surrounded by two electron groups (two single bonds) so hybridization around Be is *sp*.

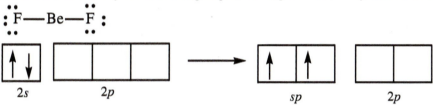

b) In SiCl$_4$, Si is surrounded by four electron groups (four single bonds) so its hybridization is *sp^3*.

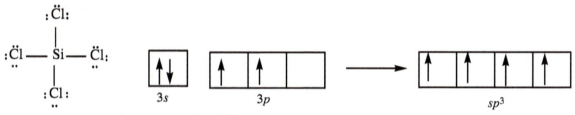

c) In XeF$_4$, xenon is surrounded by 6 electron groups (4 bonds and 2 lone pairs) so the hybridization is *sp^{3}d^2*.

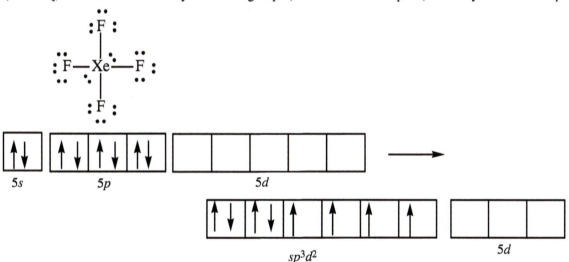

11.1B   <u>Plan:</u> Draw a Lewis structure. Determine the number and arrangement of the electron pairs about the central atom. From this, determine the type of hybrid orbitals involved. Write the partial orbital diagram of the central atoms before and after the orbitals are hybridized.

<u>Solution:</u>

a) NO_2 has 17 valence electrons and the resonance structures shown below:

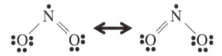

The electron group arrangement is trigonal planar, so the central N atom is sp^2 hybridized, which means one $2s$ and two $2p$ orbital are mixed. One hybrid orbital is filled with a lone pair, and two are half-filled. One electron remains in the unhybridized p orbital to form the π bond between the central N and one of the neighboring O atoms.

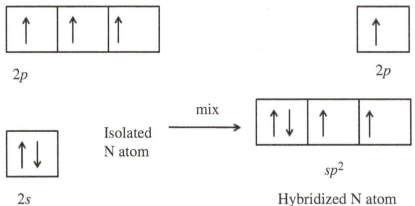

b) PCl_3 has 26 valence electrons and the Lewis structure shown below:

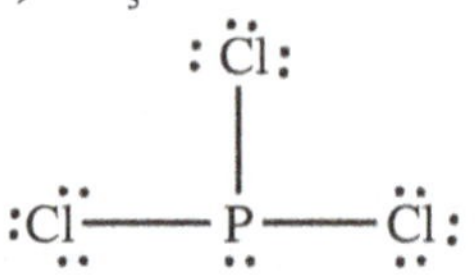

The electron-group arrangement is tetrahedral, so the central P atom is sp^3 hybridized, which means one $3s$ and three $3p$ orbital are mixed. One hybrid orbital is filled with a lone pair, and three are half-filled.

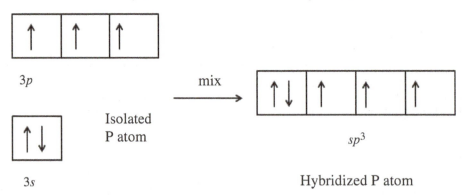

c) BrF_5 has 42 valence electrons and the Lewis structure shown below:

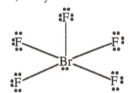

The electron-group arrangement is octahedral, so the central Br atom is sp^3d^2 hybridized, which means one $4s$, three $4p$, and two $4d$ orbital are mixed. One hybrid orbital is filled with a lone pair, and five are half-filled. Three unhybridized $4d$ orbitals remain empty.

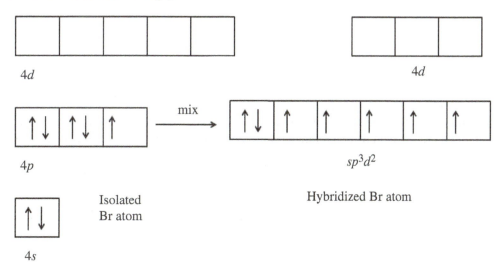

11.2A   Plan: First, determine the Lewis structure for the molecule. Then count the number of electron groups around each atom. Hybridization is sp if there are two groups, sp^2 if there are three groups, and sp^3 if there are four groups. No hybridization occurs with only one group of electrons. The bonds are then designated as sigma or pi. A single bond is a sigma bond. A double bond consists of one sigma and one pi bond. A triple bond includes one sigma bond and two pi bonds. Hybridized orbitals overlap head on to form sigma bonds whereas pi bonds form through the sideways overlap of p or d orbitals.
Solution:
a) Hydrogen cyanide has H–C≡N: as its Lewis structure. The single bond between carbon and hydrogen is a sigma bond formed by the overlap of a hybridized sp orbital on carbon with the $1s$ orbital from hydrogen. Between carbon and nitrogen are three bonds. One is a sigma bond formed by the overlap of a hybridized sp orbital on carbon with a hybridized sp orbital on nitrogen. The other two bonds between carbon and nitrogen are pi bonds formed by the overlap of p orbitals from carbon and nitrogen. One sp orbital on nitrogen is filled with a lone pair of electrons.
b) The Lewis structure of carbon dioxide, CO_2, is

$$\overset{\cdot\cdot}{\underset{\cdot\cdot}{O}}=C=\overset{\cdot\cdot}{\underset{\cdot\cdot}{O}}$$

Both oxygen atoms are sp^2 hybridized (the O atoms have three electron groups – one double bond and two lone pairs) and form a sigma bond and a pi bond with carbon. The sigma bonds are formed by the overlap of a hybridized sp orbital on carbon with a hybridized sp^2 orbital on oxygen. The pi bonds are formed by the overlap of an oxygen p orbital with a carbon p orbital. Two sp^2 orbitals on each oxygen are filled with a lone pair of electrons.

11.2B   Plan: First, determine the Lewis structure for the molecule. Then count the number of electron groups around each atom. Hybridization is sp if there are two groups, sp^2 if there are three groups, and sp^3 if there are four groups. No hybridization occurs with only one group of electrons. The bonds are then designated as sigma or pi. A single bond is a sigma bond. A double bond consists of one sigma and one pi bond. A triple bond includes one sigma bond and two pi bonds. Hybridized orbitals overlap head on to form sigma bonds whereas pi bonds form through the sideways overlap of p or d orbitals.
Solution:
a) Carbon monoxide has :C≡O: as its Lewis structure. Both C and O are sp hybridized. Between carbon and oxygen are three bonds. One is a sigma bond formed by the overlap of a hybridized sp orbital on carbon with a hybridized sp orbital on oxygen. The other two bonds between carbon and oxygen are pi bonds formed by the

overlap of unhybridized *p* orbitals from carbon and oxygen. One *sp* orbital on nitrogen is filled with a lone pair of electrons, as is one *sp* orbital on oxygen.

b) The Lewis structure of urea, H_2NCONH_2, is

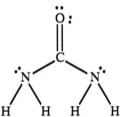

Both nitrogen atoms are *sp³*hybridized (the N atoms have four electron groups – three single bonds and one lone pair). The carbon and the oxygen are *sp²* hybridized (the C atom has three electron groups – two single bonds to nitrogen and one double bond to oxygen; the O atom also has three electron groups – two lone pairs and one double bond to carbon).One *sp³* hybrid orbital of each nitrogen forms a sigma bond with a *sp²* hybrid orbital of carbon. Two *sp³* hybrid orbitals of each nitrogen form two sigma bonds with two *s* orbitals from the hydrogen atoms (one *s* orbital from each hydrogen atom). The fourth *sp³* hybrid orbital on each of the nitrogens is filled with a lone pair of electrons. The remaining *sp²* hybrid orbital on carbon forms a sigma bond with an *sp²* hybrid orbital on the oxygen. The remaining *sp²* hybrid orbitals on oxygen are filled with lone pair electrons. Unhybridized *p* orbitals on the C and O overlap to form a pi bond.

11.3A   <u>Plan:</u> Draw the molecular orbital diagram. Determine the bond order from calculation:
BO = 1/2(# e⁻ in bonding orbitals – #e⁻ in antibonding orbitals). A bond order of zero indicates the molecule will not exist. A bond order greater than zero indicates that the molecule is at least somewhat stable and is likely to exist. H_2^{2-}has 4 electrons (1 from each hydrogen and two from the –2 charge).
<u>Solution:</u>

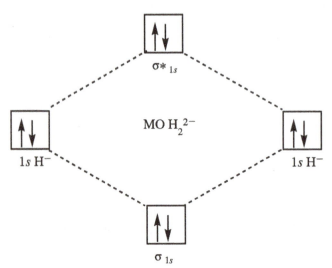

Configuration for H_2^{2-} is $(\sigma_{1s})^2(\sigma^*_{1s})^2$. Bond order of H_2^{2-} is 1/2(2 – 2) = 0. Thus, it is not likely that two H⁻ ions would combine to form the ion H_2^{2-}.

11.3B   <u>Plan:</u> Draw the molecular orbital diagram. Determine the bond order from calculation:
BO = 1/2(# e⁻ in bonding orbitals – #e⁻ in antibonding orbitals). A bond order of zero indicates the molecule will not exist. A bond order greater than zero indicates that the molecule is at least somewhat stable and is likely to exist. He_2^{2+}has 2 electrons (2 from each helium less two from the +2 charge).

Solution:

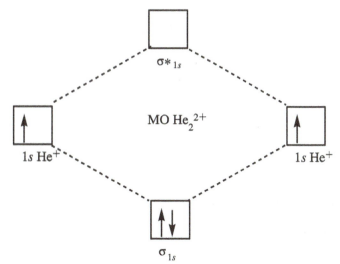

Configuration for He_2^{2+} is $(\sigma_{1s})^2$. Bond order of He_2^{2+} is $1/2(2 - 0) = 1$. Thus, we predict that He_2^{2+} does exist.

11.4A **Plan:** To find bond order it is necessary to determine the molecular orbital electron configuration from the total number of electrons. Bond order is calculated from the configuration as $1/2$(# e⁻ in bonding orbitals – #e⁻ in antibonding orbitals).
Solution:
N_2^{2+}: total electrons $= 7 + 7 - 2 = 12$
      Configuration: $(\sigma_{1s})^2(\sigma^*_{1s})^2(\sigma_{2s})^2(\sigma^*_{2s})^2(\pi_{2p})^4$
      Bond order $= 1/2(8 - 4) = 2$
N_2^-: total electrons $= 7 + 7 + 1 = 15$
      Configuration: $(\sigma_{1s})^2(\sigma^*_{1s})^2(\sigma_{2s})^2(\sigma^*_{2s})^2(\pi_{2p})^4(\sigma_{2p})^2(\pi^*_{2p})^1$
      Bond order $= 1/2(10 - 5) = 2.5$
N_2^{2-}: total electrons $= 7 + 7 + 2 = 16$
      Configuration: $(\sigma_{1s})^2(\sigma^*_{1s})^2(\sigma_{2s})^2(\sigma^*_{2s})^2(\pi_{2p})^4(\sigma_{2p})^2(\pi^*_{2p})^2$
      Bond order $= 1/2(10 - 6) = 2$
Bond energy decreases as bond order decreases: $N_2^- > N_2^{2+} = N_2^{2-}$

Bond length increases as bond energy decreases, so the order of decreasing bond length will be opposite that of decreasing bond energy.
Decreasing bond length: $N_2^{2+} = N_2^{2-} > N_2^-$

11.4B **Plan:** To find bond order it is necessary to determine the molecular orbital electron configuration from the total number of electrons. Bond order is calculated from the configuration as $1/2$(# e⁻ in bonding orbitals – #e⁻ in antibonding orbitals).
Solution:
F_2^{2-}: total electrons $= 9 + 9 + 2 = 20$
      Configuration: $(\sigma_{1s})^2(\sigma^*_{1s})^2(\sigma_{2s})^2(\sigma^*_{2s})^2(\sigma_{2p})^2(\pi_{2p})^4(\pi^*_{2p})^4(\sigma^*_{2p})^2$
      Bond order $= 1/2(10 - 10) = 0$
F_2^-: total electrons $= 9 + 9 + 1 = 19$
      Configuration: $(\sigma_{1s})^2(\sigma^*_{1s})^2(\sigma_{2s})^2(\sigma^*_{2s})^2(\sigma_{2p})^2(\pi_{2p})^4(\pi^*_{2p})^4(\sigma^*_{2p})^1$
      Bond order $= 1/2(10 - 9) = 0.5$
F_2: total electrons $= 9 + 9 = 18$
      Configuration: $(\sigma_{1s})^2(\sigma^*_{1s})^2(\sigma_{2s})^2(\sigma^*_{2s})^2(\sigma_{2p})^2(\pi_{2p})^4(\pi^*_{2p})^4$
      Bond order $= 1/2(10 - 8) = 1$

F_2^+: total electrons = 9 + 9 − 1 = 17

        Configuration: $(\sigma_{1s})^2(\sigma^*_{1s})^2(\sigma_{2s})^2(\sigma^*_{2s})^2(\sigma_{2p})^2(\pi_{2p})^4(\pi^*_{2p})^3$

        Bond order = 1/2(10 − 7) = 1.5

F_2^{2+}: total electrons = 9 + 9 − 2 = 16

        Configuration: $(\sigma_{1s})^2(\sigma^*_{1s})^2(\sigma_{2s})^2(\sigma^*_{2s})^2(\sigma_{2p})^2(\pi_{2p})^4(\pi^*_{2p})^2$

        Bond order = 1/2(10 − 6) = 2

F_2^{2-} does not exist, so it has no bond energy and is not included in the list. For the other molecule and ions, bond energy increases as bond order increases: $F_2^- < F_2 < F_2^+ < F_2^{2+}$

Bond length decreases as bond energy increases, so the order of increasing bond length will be opposite that of increasing bond energy.

Increasing bond length: $F_2^{2+} < F_2^+ < F_2 < F_2^-$

F_2^{2-} will not form a bond, so it has no bond length and is not included in the list.

END–OF–CHAPTER PROBLEMS

11.1   Plan: Table 11.1 describes the types of hybrid orbitals that correspond to the various electron-group arrangements. The number of hybrid orbitals formed by a central atom is equal to the number of electron groups arranged around that central atom.
Solution:
a)   trigonal planar: three electron groups - three hybrid orbitals: sp^2
b)   octahedral: six electron groups - six hybrid orbitals: sp^3d^2
c)   linear: two electron groups - two hybrid orbitals: sp
d)   tetrahedral: four electron groups - four hybrid orbitals: sp^3
e)   trigonal bipyramidal: five electron groups - five hybrid orbitals: sp^3d

11.3   Carbon and silicon have the same number of valence electrons, but the outer level of electrons is $n = 2$ for carbon and $n = 3$ for silicon. Thus, silicon has $3d$ orbitals in addition to $3s$ and $3p$ orbitals available for bonding in its outer level, to form up to six hybrid orbitals, whereas carbon has only $2s$ and $2p$ orbitals available in its outer level to form up to four hybrid orbitals.

11.5   Plan: The *number* of hybrid orbitals is the same as the number of atomic orbitals before hybridization. The *type* depends on the orbitals mixed. The name of the type of hybrid orbital comes from the number and type of atomic orbitals mixed. The number of each type of atomic orbital appears as a superscript in the name of the hybrid orbital.
Solution:
a) There are six unhybridized orbitals, and therefore **six** hybrid orbitals result. The type is sp^3d^2 since one s, three p, and two d atomic orbitals were mixed.
b) **Four sp^3** hybrid orbitals form from three p and one s atomic orbitals.

11.7   Plan: To determine hybridization, draw the Lewis structure and count the number of electron groups around the central nitrogen atom. Hybridize that number of orbitals. Single, double, and triple bonds all count as one electron group. An unshared pair (lone pair) of electrons or one unshared electron also counts as one electron group.
Solution:
a) The three electron groups (one double bond, one lone pair, and one unpaired electron) around nitrogen require three hybrid orbitals. The hybridization is sp^2.

b) The nitrogen has three electron groups (one single bond, one double bond, and one unpaired electron), requiring three hybrid orbitals so the hybridization is sp^2.

c) The nitrogen has three electron groups (one single bond, one double bond, and one lone pair) so the hybridization is sp^2.

11.9   Plan: To determine hybridization, draw the Lewis structure and count the number of electron groups around the central chlorine atom. Hybridize that number of orbitals. Single, double, and triple bonds all count as one electron group. An unshared pair (lone pair) of electrons or one unshared electron also counts as one electron group.
Solution:
a) The Cl has four electron groups (one lone pair, one lone electron, and two double bonds) and therefore four hybrid orbitals are required; the hybridization is sp^3. Note that in ClO_2, the π bond is formed by the overlap of d orbitals from chlorine with p orbitals from oxygen.

b) The Cl has four electron groups (one lone pair and three bonds) and therefore four hybrid orbitals are required; the hybridization is sp^3.

c) The Cl has four electron groups (four bonds) and therefore four hybrid orbitals are required; the hybridization is sp^3.

11.11   Plan: Draw the Lewis structure and count the number of electron groups around the central atom. Hybridize that number of orbitals. Single, double, and triple bonds all count as one electron group. An unshared pair (lone pair) of electrons or one unshared electron also counts as one electron group. Once the type of hybridization is known, the types of atomic orbitals that will mix to form those hybrid orbitals are also known.
Solution:
a) Silicon has four electron groups (four bonds) requiring four hybrid orbitals; four sp^3 hybrid orbitals are made from **one s and three p atomic orbitals**.

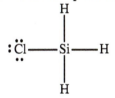

b) Carbon has two electron groups (two double bonds) requiring two hybrid orbitals; two sp hybrid orbitals are made from **one s and one p orbital**.

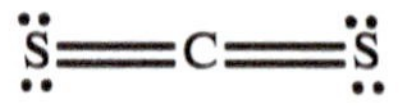

$$\ddot{S}=C=\ddot{S}$$

c) Sulfur is surrounded by five electron groups (four bonding pairs and one lone pair), requiring five hybrid orbitals; five sp^3d hybrid orbitals are formed from **one s orbital, three p orbitals, and one d orbital**.

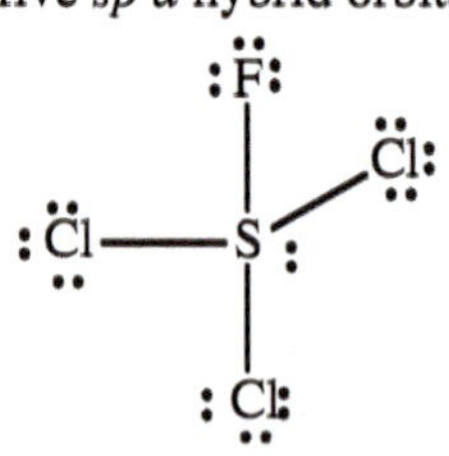

d) Nitrogen is surrounded by four electron groups (three bonding pairs and one lone pair) requiring four hybrid orbitals; four sp^3 hybrid orbitals are formed from **one s orbital and three p orbitals**.

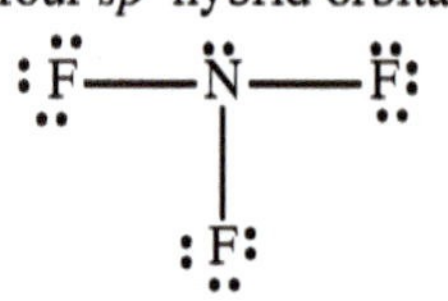

11.13   Plan: To determine hybridization, draw the Lewis structure of the reactants and products and count the number of electron groups around the central atom. Hybridize that number of orbitals. Single, double, and triple bonds all count as one electron group. An unshared pair (lone pair) of electrons or one unshared electron also counts as one electron group. Recall that sp hybrid orbitals are oriented in a linear geometry, sp^2 in a trigonal planar geometry, sp^3 in a tetrahedral geometry, sp^3d in a trigonal bipyramidal geometry, and sp^3d^2 in an octahedral geometry.
Solution:
a) The P in PH_3 has four electron groups (one lone pair and three bonds) and therefore four hybrid orbitals are required; the hybridization is sp^3. The P in the product also has four electron groups (four bonds) and again four hybrid orbitals are required. The hybridization of P remains sp^3. There is no change in hybridization. Illustration **B** best shows the hybridization of P during the reaction as $sp^3 \rightarrow sp^3$.
b) The B in BH_3 has three electron groups (three bonds) and therefore three hybrid orbitals are required; the hybridization is sp^2. The B in the product has four electron groups (four bonds) and four hybrid orbitals are required. The hybridization of B is now sp^3. The hybridization of B changes from **sp^2 to sp^3**; this is best shown by illustration **A**.

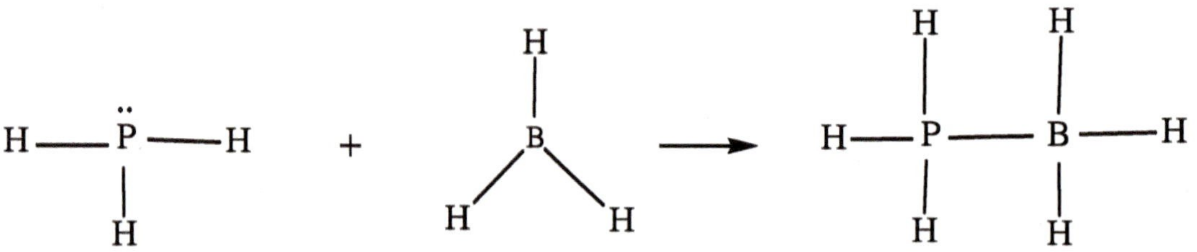

11.15   Plan: To determine hybridization, draw the Lewis structure and count the number of electron groups around the central atom. Hybridize that number of orbitals. Single, double, and triple bonds all count as one electron group. An unshared pair (lone pair) of electrons or one unshared electron also counts as one electron group. Write the electron configuration of the central atom and mix the appropriate atomic orbitals to form the hybrid orbitals.
Solution:
a) Germanium is the central atom in $GeCl_4$. Its electron configuration is $[Ar]4s^2 3d^{10}4p^2$. Ge has four electron groups (four bonds), requiring four hybrid orbitals. Hybridization is sp^3 around Ge.
One of the $4s$ electrons is moved to a $4p$ orbital and the four orbitals are hybridized.

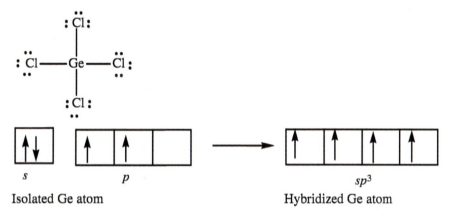

Isolated Ge atom   Hybridized Ge atom

b) Boron is the central atom in BCl₃. Its electron configuration is [He]2s²2p¹. B has three electron groups (three bonds), requiring three hybrid orbitals. Hybridization is *sp*² around B.
One of the 2*s* electrons is moved to an empty 2*p* orbital and the three atomic orbitals are hybridized. One of the 2*p* atomic orbitals is not involved in the hybridization.

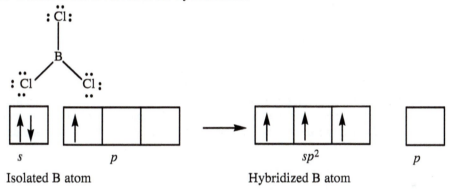

Isolated B atom   Hybridized B atom

c) Carbon is the central atom in CH₃⁺. Its electron configuration is [He]2s²2p². C has three electron groups (three bonds), requiring three hybrid orbitals. Hybridization is *sp*² around C.
One of the 2*s* electrons is moved to an empty 2*p* orbital; three orbitals are hybridized and one electron is removed to form the +1 ion.

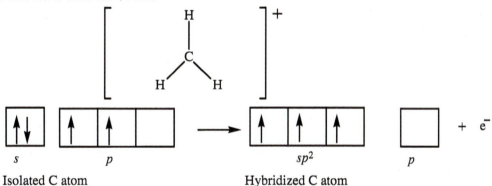

Isolated C atom   Hybridized C atom

11.17   <u>Plan:</u> To determine hybridization, draw the Lewis structure and count the number of electron groups around the central atom. Hybridize that number of orbitals. Single, double, and triple bonds all count as one electron group. An unshared pair (lone pair) of electrons or one unshared electron also counts as one electron group. Write the electron configuration of the central atom and mix the appropriate atomic orbitals to form the hybrid orbitals.
<u>Solution:</u>
a) In SeCl₂, Se is the central atom and has four electron groups (two single bonds and two lone pairs), requiring four hybrid orbitals so Se is *sp*³ hybridized. The electron configuration of Se is [Ar]4s²3d¹⁰4p⁴. The 4*s* and 4*p*

atomic orbitals are hybridized. Two sp^3 hybrid orbitals are filled with lone electron pairs and two sp^3 orbitals bond with the chlorine atoms.

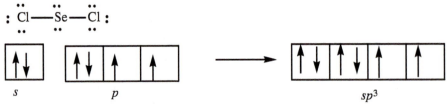

b) In H_3O^+, O is the central atom and has four electron groups (three single bonds and one lone pair), requiring four hybrid orbitals. O is sp^3 hybridized. The electron configuration of O is $[He]2s^2 2p^4$. The $2s$ and $2p$ orbitals are hybridized. One sp^3 hybrid orbital is filled with a lone electron pair and three sp^3 orbitals bond with the hydrogen atoms.

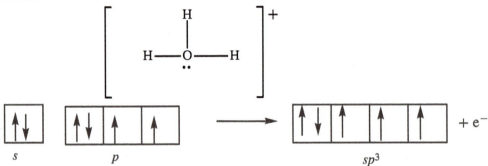

c) I is the central atom in IF_4^- with six electron groups (four single bonds and two lone pairs) surrounding it. Six hybrid orbitals are required and I has $sp^3 d^2$ hybrid orbitals. The $sp^3 d^2$ hybrid orbitals are composed of one s orbital, three p orbitals, and two d orbitals. Two $sp^3 d^2$ orbitals are filled with a lone pair and four $sp^3 d^2$ orbitals bond with the fluorine atoms.

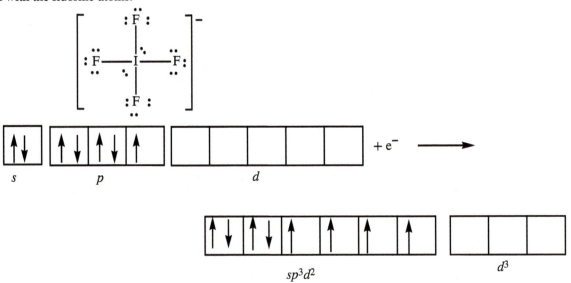

11.20   Plan: A single bond is a sigma bond which is the result of two orbitals overlapping end to end; a double bond consists of one sigma bond and one pi bond; and a triple bond consists of one sigma bond and two pi bonds. A pi bond is the result of orbitals overlapping side to side.
Solution:
a) **False**, a double bond is one sigma (σ) and one pi (π) bond.
b) **False**, a triple bond consists of one sigma (σ) and two pi (π) bonds.
c) **True**
d) **True**

e) **False**, a π bond consists of one pair of electrons; it occurs after a σ bond has been previously formed.

f) **False**, end-to-end overlap results in a bond with electron density along the bond axis.

11.21 <u>Plan:</u> To determine hybridization, draw the Lewis structure and count the number of electron groups around the central atom. Hybridize that number of orbitals. Single, double, and triple bonds all count as one electron group. An unshared pair (lone pair) of electrons or one unshared electron also counts as one electron group. A single bond is a sigma bond which is the result of two orbitals overlapping end to end; a double bond consists of one sigma bond and one pi bond; and a triple bond consists of one sigma bond and two pi bonds.
<u>Solution:</u>

a) Nitrogen is the central atom in NO_3^-. Nitrogen has three surrounding electron groups (two single bonds and one double bond), so it is sp^2 hybridized. Nitrogen forms **three σ bonds** (one each for the N–O bonds) and **one π bond** (part of the N=O double bond).

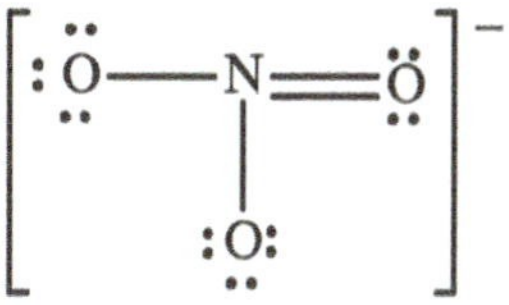

b) Carbon is the central atom in CS_2. Carbon has two surrounding electron groups (two double bonds), so it is sp hybridized. Carbon forms **two σ bonds** (one each for the C–S bonds) and **two π bonds** (part of the two C=S double bonds).

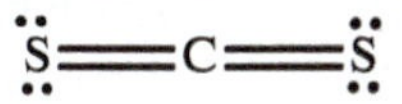

c) Carbon is the central atom in CH_2O. Carbon has three surrounding electron groups (two single bonds and one double bond), so it is sp^2 hybridized. Carbon forms **three σ bonds** (one each for the two C–H bonds and one C–O bond) and **one π bond** (part of the C=O double bond).

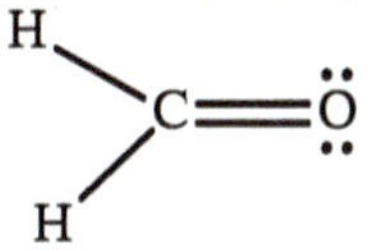

11.23 <u>Plan:</u> To determine hybridization, draw the Lewis structure and count the number of electron groups around the central nitrogen atom. Hybridize that number of orbitals. Single, double, and triple bonds all count as one electron group. An unshared pair (lone pair) of electrons or one unshared electron also counts as one electron group. A single bond is a sigma bond which is the result of two orbitals overlapping end to end; a double bond consists of one sigma bond and one pi bond; and a triple bond consists of one sigma bond and two pi bonds.
<u>Solution:</u>

a) In FNO, three electron groups (one lone pair, one single bond, and one double bond) surround the central N atom. Hybridization is sp^2 around nitrogen. One sigma bond exists between F and N, and one sigma and one pi bond exist between N and O. Nitrogen participates in a total of **2 σ and 1 π bonds**.

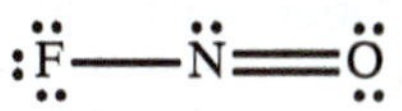

b) In C_2F_4, each carbon has three electron groups (two single bonds and one double bond) with sp^2 hybridization. The bonds between C and F are sigma bonds. The C–C bond consists of one sigma and one pi bond. Each carbon participates in a total of **three σ and one π bonds**.

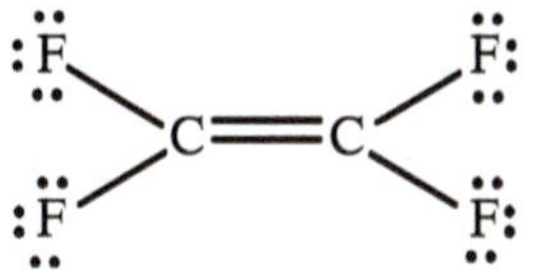

c) In $(CN)_2$, each carbon has two electron groups (one single bond and one triple bond) and is sp hybridized with a sigma bond between the two carbons and a sigma and two pi bonds comprising each C–N triple bond. Each carbon participates in a total of **two σ and two π bonds**.

11-11

11.25 <u>Plan:</u>. A single bond is a sigma bond which is the result of two orbitals overlapping end to end; a double bond consists of one sigma bond and one pi bond; and a triple bond consists of one sigma bond and two pi bonds.
<u>Solution:</u>
The double bond in 2-butene restricts rotation of the molecule, so that *cis* and *trans* structures result. The two structures are shown below:

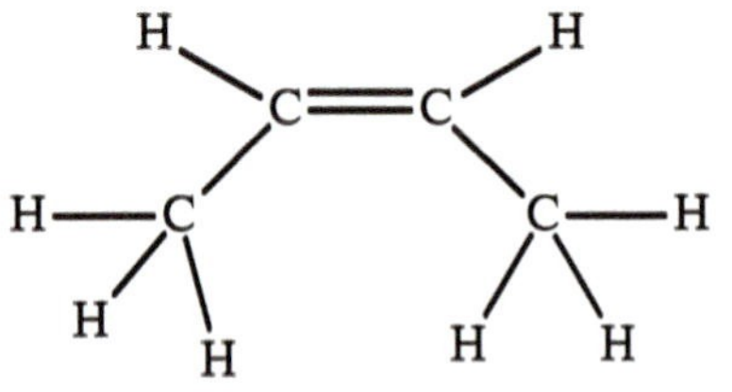

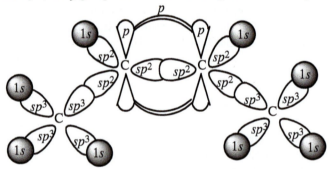

cis 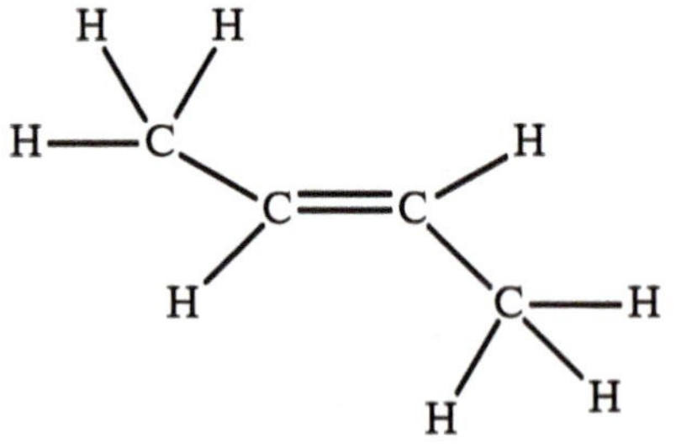   *trans*

The carbons participating in the double bond each have three surrounding groups, so they are sp^2 hybridized. The =C–H σ bonds result from the head-on overlap of a C sp^2 orbital and an H s orbital. The C–CH$_3$ bonds are also σ bonds, resulting from the head-on overlap of an sp^2 orbital and an sp^3 orbital. The C=C bond contains 1 σ bond (head on overlap of two sp^2 orbitals) and 1 π bond (sideways overlap of unhybridized p orbitals). Finally, C–H bonds in the methyl (–CH$_3$) groups are σ bonds resulting from the overlap of the sp^3 orbital of C with the s orbital of H.

11.26 Four molecular orbitals form from the four p atomic orbitals. In forming molecular orbitals, the total number of molecular orbitals must equal the number of atomic orbitals. Two of the four molecular orbitals formed are bonding orbitals and two are antibonding.

11.28 a) Bonding MOs have lower energy than antibonding MOs. The bonding MO's lower energy, even lower than its constituent atomic orbitals, accounts for the stability of a molecule in relation to its individual atoms. However, the sum of energy of the MOs must equal the sum of energy of the AOs.
b) The node is the region of an orbital where the probability of finding the electron is zero, so the nodal plane is the plane that bisects the node perpendicular to the bond axis. There is no node along the bond axis (probability is positive between the two nuclei) for the bonding MO. The antibonding MO does have a nodal plane.
c) The bonding MO has higher electron density between nuclei than the antibonding MO.

11.30 <u>Plan:</u> Like atomic orbitals, any one MO holds a maximum of two electrons. Two atomic orbitals combine to form two molecular orbitals, a bonding and an antibonding MO.
<u>Solution:</u>
a) **Two** electrons are required to fill a σ-bonding molecular orbital. Each molecular orbital requires two electrons.
b) **Two** electrons are required to fill a π-antibonding molecular orbital. There are two π-antibonding orbitals, each holding a maximum of two electrons.
c) **Four** electrons are required to fill the two σ molecular orbitals (two electrons to fill the σ-bonding and two to fill the σ-antibonding) formed from two 1s atomic orbitals.

11-12

11.32   Plan: Recall that a bonding MO has a region of high electron density between the nuclei while an antibonding MO has a node, or region of zero electron density between the nuclei. MOs formed from s orbitals, or from p orbitals overlapping end-to-end, are called σ and MOs formed by the side-to-side overlap of p orbitals are called π. A superscript star (*) is used to designate an antibonding MO. To write the electron configuration of F_2^+, determine the number of valence electrons and write the sequence of MO energy levels, following the sequence order given in the text.
   Solution:
   a) A is the π^*_{2p} molecular orbital (two p orbitals overlapping side to side with a node between them); B is the σ_{2p} molecular orbital (two p orbitals overlapping end to end with no node); C is the π_{2p} molecular orbital (two p orbitals overlapping side to side with no node); D is the σ^*_{2p} molecular orbital (two p orbitals overlapping end to end with a node).
   b) F_2^+ has thirteen valence electrons: [2 x F(7e$^-$) – 1 (from + charge)]. The MO electron configuration is $(\sigma_{2s})^2(\sigma^*_{2s})^2(\sigma_{2p})^2(\pi_{2p})^2(\pi_{2p})^2(\pi^*_{2p})^2(\pi^*_{2p})^1$. The π^*_{2p} molecular orbital, A, σ_{2p} molecular orbital, B, and π_{2p} molecular orbital, C, are all occupied by at least one electron. The σ^*_{2p} molecular orbital is unoccupied.
   c) A π^*_{2p} molecular orbital, A, has only one electron.

11.34   Plan: To write the electron configuration of Be_2^+, determine the number of electrons and write the sequence of MO energy levels, following the sequence order given in the text.
   Bond order = ½[(no. of electrons in bonding MO) – (no. of electrons in antibonding MO)]. Recall that a diamagnetic substance has no unpaired electrons.
   Solution:
   a) Be_2^+ has a total of seven electrons [2 x Be(4e$^-$) – 1 (from + charge)]. The molecular orbital configuration is $(\sigma_{1s})^2(\sigma^*_{1s})^2(\sigma_{2s})^2(\sigma^*_{2s})^1$ and bond order = ½(4 – 3) = 1/2. With a bond order of 1/2 the Be_2^+ ion will be **stable**.
   b) No, the ion has one unpaired electron in the σ^*_{2s} MO, so it is **paramagnetic**, not diamagnetic.
   c) Valence electrons would be those in the molecular orbitals at the $n = 2$ level, so the valence electron configuration is $(\sigma_{2s})^2(\sigma^*_{2s})^1$.

11.36   Plan: Write the electron configuration of each species by determining the number of electrons and writing the sequence of MO energy levels, following the sequence order given in the text. Calculate the bond order: bond order = ½[(no. of electrons in bonding MO) – (no. of electrons in antibonding MO)]. Bond energy increases as bond order increases; bond length decreases as bond order increases.
   Solution:
   C_2^- Total electrons = 6 + 6 + 1 = 13
           MO configuration: $(\sigma_{1s})^2(\sigma^*_{1s})^2(\sigma_{2s})^2(\sigma^*_{2s})^2(\pi_{2p})^4(\sigma_{2p})^1$
           Bond order = 1/2(9 – 4) = 2.5
   C_2 Total electrons = 6 + 6 = 12
           MO configuration: $(\sigma_{1s})^2(\sigma^*_{1s})^2(\sigma_{2s})^2(\sigma^*_{2s})^2(\pi_{2p})^4$
           Bond order = 1/2(8 – 4) = 2
   C_2^+ Total electrons = 6 + 6 – 1 = 11
           MO configuration: $(\sigma_{1s})^2(\sigma^*_{1s})^2(\sigma_{2s})^2(\sigma^*_{2s})^2(\pi_{2p})^3$
           Bond order = 1/2(7 – 4) = 1.5

   a) Bond energy increases as bond order increases: $C_2^+ < C_2 < C_2^-$

   b) Bond length decreases as bond energy increases, so the order of increasing bond length will be opposite that of increasing bond energy. Increasing bond length: $C_2^- < C_2 < C_2^+$

11.40   Plan: To determine hybridization, count the number of electron groups around each of the C, O, and N atoms. Hybridize that number of orbitals. Single, double, and triple bonds all count as one electron group. An unshared pair (lone pair) of electrons or one unshared electron also counts as one electron group. A single bond is a sigma bond which is the result of two orbitals overlapping end to end; a double bond consists of one sigma bond and one pi bond; and a triple bond consists of one sigma bond and two pi bonds.

Solution:
a) Each of the six C atoms in the ring has three electron groups (two single bonds and a double bond) and has sp^2 hybridization; all of the other C atoms have four electron groups (four single bonds) and have sp^3 hybridization; all of the O atoms have four electron groups (two single bonds and two lone pairs) and have sp^3 hybridization; the N atom has four electron groups (three single bonds and a lone pair) and has sp^3 hybridization.
b) Each of the single bonds is a sigma bond; each of the double bonds has one sigma bond for a total of **26 sigma bonds**.
c) The ring has three double bonds each of which is composed of one sigma bond and one pi bond; so there are three pi bonds each with two electrons for a total of **six pi electrons.**

11.42 Plan: To determine hybridization, count the number of electron groups around each C and N atom. Hybridize that number of orbitals. Single, double, and triple bonds all count as one electron group. An unshared pair (lone pair) of electrons or one unshared electron also counts as one electron group. A single bond is a sigma bond which is the result of two orbitals overlapping end to end; a double bond consists of one sigma bond and one pi bond; and a triple bond consists of one sigma bond and two pi bonds.
Solution:
a) Every single bond is a sigma bond. There is one sigma bond in each double bond as well. There are **17 σ bonds** in isoniazid. Every atom-to-atom connection contains a σ bond.
b) All carbons have three surrounding electron groups (two single and one double bond), so their hybridization is sp^2. The ring N also has three surrounding electron groups (one single bond, one double bond, and one lone pair), so its hybridization is also sp^2. The other two N atoms have four surrounding electron groups (three single bonds and one lone pair) and are sp^3 hybridized.

11.44 Plan: To determine the hybridization in each species, count the number of electron groups around the underlined atom. Hybridize that number of orbitals. Single, double, and triple bonds all count as one electron group. An unshared pair (lone pair) of electrons or one unshared electron also counts as one electron group.
Solution:
a) B changes from $sp^2 \rightarrow sp^3$. Boron in BF_3 has three electron groups with sp^2 hybridization. In BF_4^-, four electron groups surround B with sp^3 hybridization.

b) P changes from $sp^3 \rightarrow sp^3d$. Phosphorus in PCl_3 is surrounded by four electron groups (three bonds to Cl and one lone pair) for sp^3 hybridization. In PCl_5, phosphorus is surrounded by five electron groups for sp^3d hybridization.

c) C changes from $sp \rightarrow sp^2$. Two electron groups surround C in C_2H_2 and three electron groups surround C in C_2H_4.

d) Si changes from $sp^3 \rightarrow sp^3d^2$. Four electron groups surround Si in SiF_4 and six electron groups surround Si in SiF_6^{2-}.

e) **No change**, S in SO_2 is surrounded by three electron groups (one single bond, one double bond, and one lone pair) and in SO_3 is surrounded by three electron groups (two single bonds and one double bond); both have sp^2 hybridization.

11.46 <u>Plan:</u> To determine the molecular shape and hybridization, count the number of electron groups around the P, N, and C atoms. Hybridize that number of orbitals. Single, double, and triple bonds all count as one electron group. An unshared pair (lone pair) of electrons or one unshared electron also counts as one electron group.
<u>Solution:</u>

P (3 single bonds and 1 double bond)	AX_4	tetrahedral	sp^3
N (3 single bonds and 1 lone pair)	AX_3E	trigonal pyramidal	sp^3
C_1 and C_2 (4 single bonds)	AX_4	tetrahedral	sp^3
C_3 (2 single bonds and 1 double bond)	AX_3	trigonal planar	sp^2

11.51 <u>Plan:</u> To determine the hybridization, count the number of electron groups around the atoms. Hybridize that number of orbitals. Single, double, and triple bonds all count as one electron group. An unshared pair (lone pair) of electrons or one unshared electron also counts as one electron group.
<u>Solution:</u>
a) **B and D** show hybrid orbitals that are present in the molecule. B shows sp^3 hybrid orbitals, used by atoms that have four groups of electrons. In the molecule, the C atom in the CH_3 group, the S atom, and the O atom all have four groups of electrons and would have sp^3 hybrid orbitals. D shows sp^2 hybrid orbitals, used by atoms that have three groups of electrons. In the molecule, the C bonded to the nitrogen atom, the C atoms involved in the C=C bond, and the nitrogen atom all have three groups of electrons and would have sp^2 hybrid orbitals.
b) The C atoms in the C≡C bond have only two electron groups and would have **sp hybrid orbitals**. These orbitals are not shown in the picture.
c) There are **two sets of sp** hybrid orbitals, **four sets of sp^2** hybrid orbitals, and **three sets of sp^3** hybrid orbitals in the molecule.

11.52 <u>Plan:</u> Draw a resonance structure that places the double bond between the C and N atoms.
<u>Solution:</u>
The resonance gives the C–N bond some double bond character, which hinders rotation about the C–N bond. The C–N single bond is a σ bond; the resonance interaction exchanges a C–O π bond for a C–N π bond.

11.56   <u>Plan:</u> To determine hybridization, count the number of electron groups around each C and O atom. Hybridize that number of orbitals. Single, double, and triple bonds all count as one electron group. An unshared pair (lone pair) of electrons or one unshared electron also counts as one electron group. A single bond is a sigma bond which is the result of two orbitals overlapping end to end; a double bond consists of one sigma bond and one pi bond; and a triple bond consists of one sigma bond and two pi bonds.
<u>Solution:</u>
a) The six carbons in the ring each have three surrounding electron groups (two single bonds and one double bond) with sp^2 hybrid orbitals. The two carbons participating in the C=O bond are also sp^2 hybridized. The single carbon in the –CH$_3$ group has four electron groups (four single bonds) and is sp^3 hybridized. The two central oxygen atoms, one in a C–O–H configuration and the other in a C–O–C configuration, each have four surrounding electron groups (two single bonds and two lone pairs) and are sp^3 hybridized. The O atoms in the two C=O bonds have three electron groups (one double bond and two lone pairs) and are sp^2 hybridized.
Summary: C in –CH$_3$: **sp^3**, all other C atoms (8 total): **sp^2**, O in C=O (2 total): **sp^2**, O in the C–O bonds (2 total): **sp^3**.
b) The **two** C=O bonds are localized; the double bonds on the ring are delocalized as in benzene.
c) Each carbon with three surrounding groups has sp^2 hybridization and trigonal planar shape; therefore, **eight** carbons have this shape. Only **one** carbon in the CH$_3$ group has four surrounding groups with sp^3 hybridization and tetrahedral shape.

11.57   <u>Plan:</u> In the *cis* arrangement, the two H atoms are on the same side of the double bond; in the *trans* arrangement, the two H atoms are on different sides of the double bond.
<u>Solution:</u>
a) **Four** different isomeric fatty acids: *trans-cis, cis-cis, cis-trans, trans-trans.*
b) With three double bonds, there are $2^n = 2^3 = $ **8 isomers** possible.

cis-cis-cis	*trans-trans-trans*
cis-trans-cis	*trans-cis-trans*
cis-cis-trans	*trans-cis-cis*
cis-trans-trans	*trans-trans-cis*

CHAPTER 12 INTERMOLECULAR FORCES: LIQUIDS, SOLIDS, AND PHASE CHANGES

FOLLOW–UP PROBLEMS

12.1A   Plan: This is a three step process: warming the ice to 0.0°C, melting the ice, and warming the liquid water to 16.0°C. Use the molar heat capacities of ice and water and the relationship $q = n \times C \times \Delta T$ to calculate the heat use for the warming of the ice and of the water; use the heat of fusion and the relationship $q = n\Delta H_{fus}$ to calculate the heat required to melt the ice.

Solution:

The total heat required is the sum of three processes:

1) Warming the ice from −7.00°C to 0.00°C

$$q_1 = n \times C_{ice} \times \Delta T = (2.25 \text{ mol})(37.6 \text{ J/mol} \bullet °C) [0.0 - (-7.00)]°C = 592.2 \text{ J} = 0.592 \text{ kJ}$$

2) Phase change of ice at 0.00°C to water at 0.00°C

$$q_2 = n\Delta H_{fus} = (2.25 \text{ mol})\left(\frac{6.02 \text{ kJ}}{\text{mol}}\right) = 13.545 \text{ kJ} = 13.5 \text{ kJ}$$

3) Warming the liquid from 0.00°C to 16.0°C

$$q_3 = n \times C_{water} \times \Delta T = (2.25 \text{ mol})(75.4 \text{ J/mol} \bullet °C) [16.0 - (0.0)]°C = 2714.4 \text{ J} = 2.71 \text{ kJ}$$

The three heats are positive because each process takes heat from the surroundings (endothermic). The phase change requires much more energy than the two temperature change processes. The total heat is $q_1 + q_2 + q_3 = (0.5922 \text{ kJ} + 13.545 \text{ kJ} + 2.7144 \text{ kJ}) = 16.8516 = \mathbf{16.9 \text{ kJ}}$.

12.1B   Plan: This is a three step process: cooling the bromine vapor to 59.5°C, condensing the bromine vapor, and cooling the liquid bromine to 23.8°C. Use the molar heat capacities of bromine vapor and liquid bromine and the relationship $q = n \times C \times \Delta T$ to calculate the heat released by the cooling of the bromine vapor and of the liquid bromine; use the heat of vaporization and the relationship $q = n(-\Delta H_{vap})$ to calculate the heat released when bromine vapor condenses.

Solution:

$$\text{Amount (mol) of bromine} = (47.94 \text{ g } Br_2)\left(\frac{1 \text{ mol } Br_2}{159.8 \text{ g } Br_2}\right) = 0.3000 \text{ mol } Br_2$$

The total heat released is the sum of three processes:

1) Cooling the bromine vapor from 73.5°C to 59.5°C

$$q_1 = n \times C_{gas} \times \Delta T = (0.3000 \text{ mol})(36.0 \text{ J/mol°C}) [59.5 - 73.5]°C = -151.2 \text{ J} = -0.151 \text{ kJ}$$

2) Phase change of bromine vapor at 59.5°C to liquid bromine at 59.5°C

$$q_2 = n(-\Delta H_{vap}) = (0.3000 \text{ mol})\left(\frac{-29.6 \text{ kJ}}{\text{mol}}\right) = -8.88 \text{ kJ}$$

3) Cooling the liquid from 59.5°C to 23.8°C

$$q_3 = n \times C_{liquid} \times \Delta T = (0.3000 \text{ mol})(75.7 \text{ J/mol°C}) [23.8 - 59.5]°C = -810.7 \text{ J} = -0.811 \text{ kJ}$$

The three heats are negative because each process releases heat to the surroundings (exothermic). The phase change releases much more energy than the two temperature change processes. The total heat is $q_1 + q_2 + q_3 = (-0.1512 \text{ kJ} + -8.88 \text{ kJ} + -0.8107 \text{ kJ}) = -9.8419 = \mathbf{-9.84 \text{ kJ}}$.

12.2A   Plan: The variables ΔH_{vap}, P_1, T_1, and T_2 are given, so substitute them into the Clausius-Clapeyron equation and solve for P_2. Convert both temperatures from °C to K. Convert ΔH_{vap} to J so that the units cancel with R.

Solution:

$T_1 = 34.1 + 273.15 = 307.2$ K $\qquad\qquad T_2 = 85.5 + 273.15 = 358.6$ K $\qquad\qquad P_1 = 40.1$ torr

$$\ln\frac{P_2}{P_1} = \frac{-\Delta H_{vap}}{R}\left(\frac{1}{T_2} - \frac{1}{T_1}\right)$$

$$\ln\frac{P_2}{40.1\ \text{torr}} = \frac{-40.7\dfrac{\text{kJ}}{\text{mol}}}{8.314\ \text{J/mol}\bullet\text{K}}\left(\frac{1}{358.6\ \text{K}} - \frac{1}{307.2\ \text{K}}\right)\left(\frac{10^3\ \text{J}}{1\ \text{kJ}}\right) = 2.284105$$

$$\frac{P_2}{40.1\ \text{torr}} = 9.81689$$

$P_2 = (9.81689)(40.1\ \text{torr}) = 393.657 = \textbf{3.9 x 10}^{\textbf{2}}\ \textbf{torr}$

In a log term, only the numbers after the decimal point are significant. The value 2.284105 contains 3 significant figures after simplifying the expression to the right of the equal sign; this value would round to 2.28 if it was not part of an intermediate step. However, as we next perform the logarithmic function, only the 2 significant figures after the decimal are carried over through the next calculation steps.

The temperature increased, so the vapor pressure should be higher.

12.2B Plan: The variables P_1, P_2, T_1, and T_2 are given, so substitute them into the Clausius-Clapeyron equation and solve for ΔH_{vap}. Convert T_1 and T_2 from °C to K. Convert ΔH_{vap} to kJ/mol

Solution:

$T_1 = 20.2 + 273.15 = 293.35$ K $\qquad\qquad P_1 = 24.5$ kPa

$T_2 = 0.95 + 273.15 = 274.10$ K $\qquad\qquad P_2 = 10.0$ kPa

$$\ln\frac{P_2}{P_1} = \frac{-\Delta H_{vap}}{R}\left(\frac{1}{T_2} - \frac{1}{T_1}\right)$$

$$\ln\left(\frac{24.50\ \text{kPa}}{10.00\ \text{kPa}}\right) = \frac{-\Delta H_{vap}}{8.314\ \text{J/mol}\bullet\text{K}}\left(\frac{1}{293.35\text{K}} - \frac{1}{274.10\text{K}}\right)$$

$$0.896088 = \frac{-\Delta H_{vap}}{8.314\ \text{J/mol}\bullet\text{K}}(-0.000239406\ \text{K}^{-1})$$

$$\Delta H_{vap} = 31119\ \text{J/mol}\left(\frac{1\ \text{kJ}}{10^3\ \text{J}}\right) = 31.119 = \textbf{31.1 kJ/mol}$$

12.3A Plan: Refer to the phase diagram to describe the specified changes. In this problem, carbon is heated at a constant pressure.

Solution:

At the starting conditions (3 x 10^2 bar and 2000K), the carbon sample is in the form of graphite. As the sample is heated at constant pressure, it melts at about 4500 K, and then vaporizes at around 5000 K.

12.3B Plan: Refer to the phase diagram to describe the specified changes. In this problem, carbon is compressed at a constant temperature.

Solution:

At the starting conditions (4600 K and 10^4 bar), the carbon sample is in the form of graphite. As the sample is compressed at constant temperature, it melts at around 4x10^4 bar and then solidifies to diamond at around 2x10^5 bar.

12.4A Plan: Hydrogen bonding occurs in compounds in which H is directly bonded to O, N, or F. In hydrogen bonding, there is a $-$A: $^{\cdots}$H$-$B$-$ sequence in which A and B are O, N, or F.

a) The –A: H–B– sequence is present in both of the following structures:

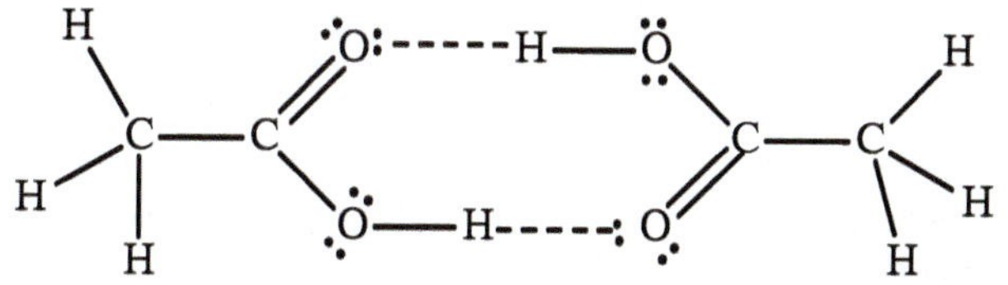

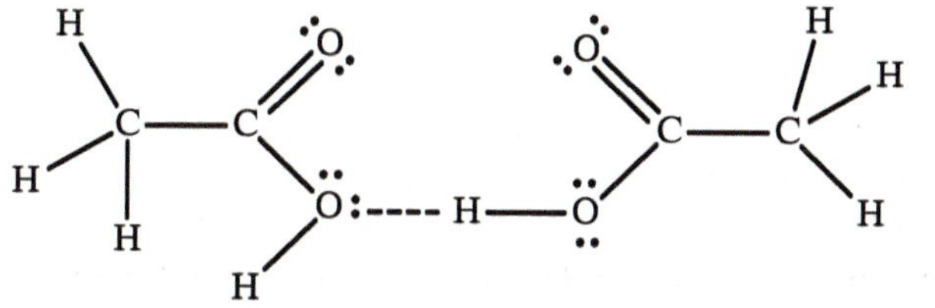

b) The –A: H–B– sequence can only be achieved in one arrangement.

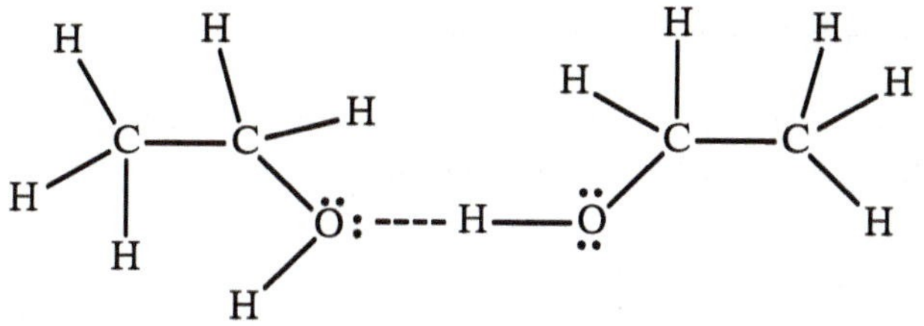

The hydrogens attached to the carbons cannot form H bonds.

c) Hydrogen bonding is not possible because there are no O–H, N–H, or F–H bonds.

12.4B   Plan: Hydrogen bonding occurs in compounds in which H is directly bonded to O, N, or F. In hydrogen bonding, there is a –A: H–B– sequence in which A and B are O, N, or F.
Solution:
a) Hydrogen bonding is not possible because there are no O–H, N–H, or F–H bonds.
b) The –A: H–B– sequence is present in the following structure (one of the possible hydrogen bonding structures):

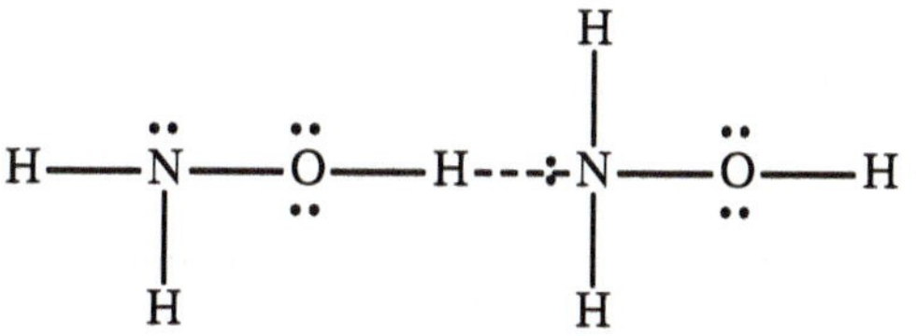

c) The –A: H–B– sequence is present in the following structure (one of the possible hydrogen bonding structures):

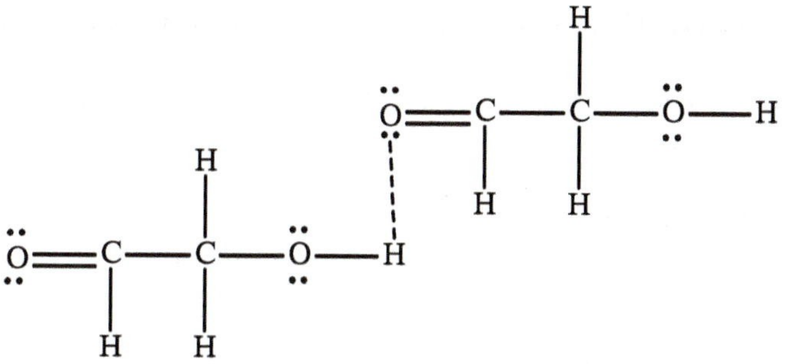

The hydrogens attached to the carbons cannot form H bonds.

12.5A   Plan: Hydrogen bonding occurs in compounds in which hydrogen is directly bonded to O, N, or F. Dipole-dipole forces occur in compounds that are polar. Dispersion forces are the dominant forces in nonpolar compounds. Boiling point increases with increasing strength of intermolecular forces; hydrogen bonds are stronger than dipole-dipole forces, which are stronger than dispersion forces. Forces generally increase in strength as molar mass increases.
Solution:
a) Both CH_3Br and CH_3F are polar molecules that experience dipole-dipole and dispersion intermolecular interactions. Because CH_3Br ($\mathcal{M}$ = 94.9 g/mol) is ~3 times larger than CH_3F ($\mathcal{M}$ = 34.0 g/mol), dispersion forces result in a higher boiling point for **CH_3Br**.
b) $CH_3CH_2CH_2OH$, n–propanol, forms hydrogen bonds and has dispersion forces whereas $CH_3CH_2OCH_3$, ethyl methyl ether, has only dipole-dipole and dispersion attractions. **$CH_3CH_2CH_2OH$** has the higher boiling point.
c) Both C_2H_6 and C_3H_8 are nonpolar and experience dispersion forces only. **C_3H_8**, with the higher molar mass, experiences greater dispersion forces and has the higher boiling point.

12.5B   Plan: Hydrogen bonding occurs in compounds in which hydrogen is directly bonded to O, N, or F. Dipole-dipole forces occur in compounds that are polar. Dispersion forces are the dominant forces in nonpolar compounds. Boiling point increases with increasing strength of intermolecular forces; hydrogen bonds are stronger than dipole-dipole forces, which are stronger than dispersion forces. Forces generally increase in strength as molar mass increases.
Solution:
a) Both CH_3CHO and CH_3CH_2OH are polar molecules that experience dipole-dipole and dispersion intermolecular interactions. Because of its O–H bond, CH_3CH_2OH also experiences hydrogen bonding. Because **CH_3CHO** does not experience the stronger hydrogen bonds, it has a lower boiling point than CH_3CH_2OH.
(b) SO_2 molecules have a bent molecular geometry; both S—O bonds are polar and the molecule is polar; SO_2 has both dispersion and dipole-dipole forces. CO_2 is a linear molecule and is nonpolar since the two C=O bonds balance each other so CO_2 has only dispersion forces. Since **CO_2** has the weaker forces, it has the lower boiling point.
c) Both $H_2NCOCH_2CH_3$ and $(CH_3)_2NCHO$ are polar molecules that experience dipole-dipole and dispersion intermolecular interactions. Because of its N–H bonds, $H_2NCOCH_2CH_3$ also experiences hydrogen bonding. Because **$(CH_3)_2NCHO$** does not experience the stronger hydrogen bonds, it has a lower boiling point than $H_2NCOCH_2CH_3$.

12.6A   Plan: To determine the number of atoms or ions in each unit cell, count the number of atoms/ions at the corners, faces, and center of the unit cell. Atoms at the eight corners are shared by eight cells for a total of 8 atoms x 1/8 atom per cell = 1 atom; atoms in the body of a cell are in that cell only; atoms at the faces are shared by two cells: 6 atoms x 1/2 atom per cell = 3 atoms. Count the number of nearest neighboring particles to obtain the coordination number.
Solution:
a) Looking at the sulfide ions, there is one ion at each corner and one ion on each face. The total number of sulfide ions is 1/8 (8 corner ions) + 1/2 (6 face ions) = **4 S^{2-} ions**. There are also **4 Pb^{2+} ions** due to the 1:1 ratio of S^{2-} ions to Pb^{2+} ions. Each ion has 6 nearest neighbor particles, so the coordination number for both ions is **6**.
b) There is one atom at each corner and one atom in the center of the unit cell. The total number of atoms is 1/8 (8 corner atoms) + 1 (1 body atom) = **2 W atoms**. Each atom has 8 nearest neighbor particles, so the coordination number for W is **8**.
c) There is one atom at each corner and one atom on each face of the unit cell. The total number of atoms is 1/8 (8 corner atoms) + 1/2 (6 face atoms) = **4 Al atoms**. Each atom has 12 nearest neighbor particles, so the coordination number for Al is **12**.

12.6B   Plan: To determine the number of atoms or ions in each unit cell, count the number of atoms/ions at the corners, faces, and center of the unit cell. Atoms at the eight corners are shared by eight cells for a total of 8 atoms x 1/8 atom per cell = 1 atom; atoms in the body of a cell are in that cell only; atoms at the faces are shared by two cells: 6 atoms x 1/2 atom per cell = 3 atoms. Count the number of nearest neighboring particles to obtain the coordination number.

Solution:
a) There is one atom at each corner and one atom in the center of the unit cell. The total number of atoms is 1/8 (8 corner atoms) + 1 (1 body atom) = **2 K atoms**. Each atom has 8 nearest neighbor particles, so the coordination number for K is **8**.
b) There is one atom at each corner and one atom on each face of the unit cell. The total number of atoms is 1/8 (8 corner atoms) + 1/2 (6 face atoms) = **4 Pt atoms**. Each atom has 12 nearest neighbor particles, so the coordination number for Pt is **12**.
c) Looking at the chloride ion, there is one ion in the center of the unit cell, so there is **1 Cl⁻ ion** in the unit cell. There is one Cs^+ ion at each corner, so there is 1/8 (8 corner ions) = **1 Cs⁺ ion**. Each ion has 8 nearest neighbor particles, so the coordination number for both ions is **8**.

12.7A   Plan: Use the radius of the Co atom to find the volume of one Co atom. Use Avogadro's number to find the volume of one mole of Co atoms, and convert from pm^3 to cm^3. Use the volume of one mole of Co *atoms* and the packing efficiency of the solid (0.74 for face centered cubic) to find the volume of one mole of Co *metal*. Calculate the density of Co metal by dividing the molar mass of Co by the volume of one mole of Co metal.
Solution:

$$V \text{ of one Co atom } = \frac{4}{3}\pi r^3 = \left(\frac{4}{3}\right)(\pi)(125 \text{ pm})^3 = 8.18123 \times 10^6 \text{ pm}^3$$

V of one mole of Co atoms =

$$\left(\frac{8.18123 \times 10^6 \text{ pm}^3}{1 \text{ Co atom}}\right)\left(\frac{6.022 \times 10^{23} \text{ Co atoms}}{1 \text{ mol Co atoms}}\right)\left(\frac{1 \text{ cm}^3}{10^{30} \text{ pm}^3}\right) = 4.9267 \text{ cm}^3/\text{mole Co atoms}$$

$$V \text{ of one mole of Co metal} = \frac{4.9267 \text{ cm}^3/\text{mol Co atoms}}{0.74} = 6.6578 \text{ cm}^3/\text{mol Co metal}$$

$$\text{Density of Co metal} = \left(\frac{58.93 \text{ g Co}}{1 \text{ mol Co}}\right)\left(\frac{1 \text{ mol Co}}{6.6578 \text{ cm}^3}\right) = 8.8513 = \textbf{8.85 g/cm}^3$$

12.7B   Plan: Use the volume of the Fe atom to find the radius of the atom; use the radius to find the edge length of the body-centered cubic cell. Find the volume of the cell by cubing the value of the edge length. Use the fact that there are two Fe atoms in the cell. The molar mass and density of iron are then used to find the number of Fe atoms in a mole (Avogadro's number).
Solution:

$$V = \frac{4}{3}\pi r^3$$

$$r = \sqrt[3]{\frac{3V}{4\pi}} = \sqrt[3]{\frac{3(8.38 \times 10^{-24} \text{ cm}^3)}{4\pi}} = 1.260042 \times 10^{-8} \text{ cm}$$

For a body-centered cubic unit cell, the edge length = $\dfrac{4r}{\sqrt{3}} = \dfrac{4(1.260042 \times 10^{-8} \text{ cm})}{\sqrt{3}}$
$$= 2.90994 \times 10^{-8} \text{ cm}$$

Volume (cm^3) of the cell = $(\text{edge length})^3 = (2.909942 \times 10^{-8} \text{ cm})^3 = 2.46407 \times 10^{-23} \text{ cm}^3$

$$\text{Avogadro's Number} = \left(\frac{2 \text{ Fe atoms}}{2.46407 \times 10^{-23} \text{ cm}^3}\right)\left(\frac{\text{cm}^3}{7.874 \text{ g}}\right)\left(\frac{55.85 \text{ g}}{\text{mol}}\right)$$

$$= 5.75711 \times 10^{23} = \textbf{5.8} \times \textbf{10}^{23} \textbf{ atoms/mol}$$

12.8A Plan: Use the relationship between radius and edge length for a body-centered cubic structure shown in Figure 12.29.
Solution:

$$\text{Edge length} = \frac{4r}{\sqrt{3}} = \frac{4(126\ \text{pm})}{\sqrt{3}} = (290.98454\ \text{pm})\left(\frac{1\ \text{nm}}{1000\ \text{pm}}\right) = \textbf{0.291 nm}$$

12.8B Plan: Use the relationship between radius and edge length for a cubic closest packed structure shown in Figure 12.29.
Solution:
$A = \sqrt{8}r$, so $r = A/\sqrt{8}$

$$r = 0.405\ \text{nm}/\sqrt{8} = (0.143\ \text{nm})\left(\frac{1000\ \text{pm}}{1\ \text{nm}}\right) = \textbf{143 pm}$$

TOOLS OF THE LABORATORY BOXED READING PROBLEMS

B12.1 Plan: The Bragg equation gives the relationship between the angle of incoming light, θ, the wavelength of the light, λ, and the distance between layers in a crystal, d.
Solution:

$n\lambda = 2d\sin\theta$

$n = 1$; $\lambda = 0.709\text{x}10^{-10}$ m; $\theta = 11.6°$

$1(0.709\text{x}10^{-10}\ \text{m}) = 2d\sin 11.6°$

$0.709\text{x}10^{-10}\ \text{m} = 0.4021558423\ d$

$d = 1.762998\text{x}10^{-10} = \textbf{1.76x10}^{\textbf{-10}}\ \textbf{m}$

B12.2 Plan: The Bragg equation gives the relationship between the angle of incoming light, θ, the wavelength of the light, λ, and the distance between layers in a crystal, d.
Solution:
a) d, the distance between the layers in the NaCl crystal, must be found. The edge length of the NaCl unit cell is equal to the sum of the diameters of the two ions:
Edge length (pm) = 204 pm (Na^+) + 362 pm (Cl^-) = 566 pm. The spacing between the layers is half the distance of the edge length: 566/2 = 283 pm.

$n = 1$; $\lambda = ?$; $\theta = 15.9°$; $d = 566$ pm

$n\lambda = 2d\sin\theta$

$$\lambda = \frac{2d\sin\theta}{n} = \frac{2(283\ \text{pm})\sin(15.9°)}{1} = 155.0609178 = \textbf{155 pm}$$

b) $n = 2$; $\lambda = 155$ pm; $\theta = ?$; $d = 566$ pm

$n\lambda = 2d\sin\theta$

$$\sin\theta = \frac{n\lambda}{2d} = \frac{2(155\ \text{pm})}{2(283\ \text{pm})}$$

$\sin\theta = 0.5477031802$

$\theta = 33.20958 = \textbf{33.2}°$

END–OF–CHAPTER PROBLEMS

12.1 The energy of attraction is a *potential* energy and denoted E_p. The energy of motion is *kinetic* energy and denoted E_k. The relative strength of E_p vs. E_k determines the phase of the substance. In the gas phase, $E_p << E_k$ because the gas particles experience little attraction for one another and the particles are moving very fast. In the solid phase, $E_p >> E_k$ because the particles are very close together and are only vibrating in place.

Two properties that differ between a gas and a solid are the volume and density. The volume of a gas expands to fill the container it is in while the volume of a solid is constant no matter what container holds the solid. Density of a gas is much less than the density of a solid. The density of a gas also varies significantly with temperature and pressure changes. The density of a solid is only slightly altered by changes in temperature and pressure. Compressibility and ability to flow are other properties that differ between gases and solids.

12.4 a) Heat of fusion refers to the change between the solid and the liquid states and heat of vaporization refers to the change between liquid and gas states. In the change from solid to liquid, the kinetic energy of the molecules must increase only enough to partially offset the intermolecular attractions between molecules. In the change from liquid to gas, the kinetic energy of the molecules must increase enough to overcome the intermolecular forces. The energy to overcome the intermolecular forces for the molecules to move freely in the gaseous state is much greater than the amount of energy needed to allow the molecules to move more easily past each other but still stay very close together.

b) The net force holding molecules together in the solid state is greater than that in the liquid state. Thus, to change solid molecules to gaseous molecules in sublimation requires more energy than to change liquid molecules to gaseous molecules in vaporization.

c) At a given temperature and pressure, the magnitude of ΔH_{vap} is the same as the magnitude of ΔH_{cond}. The only difference is in the sign: $\Delta H_{vap} = -\Delta H_{cond}$.

12.5 Plan: Intermolecular forces (nonbonding forces) are the forces that exist between molecules that attract the molecules to each other; these forces influence the physical properties of substances. Intramolecular forces (bonding forces) exist within a molecule and are the forces holding the atoms together in the molecule; these forces influence the chemical properties of substances.

Solution:

a) **Intermolecular** — Oil evaporates when individual oil molecules can escape the attraction of other oil molecules in the liquid phase.

b) **Intermolecular** — The process of butter (fat) melting involves a breakdown in the rigid, solid structure of fat molecules to an amorphous, less ordered system. The attractions between the fat molecules are weakened, but the bonds within the fat molecules are not broken.

c) **Intramolecular** — A process called oxidation tarnishes pure silver. Oxidation is a chemical change and involves the breaking of bonds and formation of new bonds.

d) **Intramolecular** — The decomposition of O_2 molecules into O atoms requires the breaking of chemical bonds, i.e., the force that holds the two O atoms together in an O_2 molecule.

Both a) and b) are physical changes, whereas c) and d) are chemical changes. In other words, intermolecular forces are involved in physical changes while intramolecular forces are involved in chemical changes.

12.7 a) **Condensation** — The water vapor in the air condenses to liquid when the temperature drops during the night.

b) **Fusion** (melting) — Solid ice melts to liquid water.

c) **Evaporation** — Liquid water on clothes evaporates to water vapor.

12.9 The propane gas molecules slow down as the gas is compressed. Therefore, much of the **kinetic energy** lost by the propane molecules is released to the surroundings upon liquefaction.

12.13 In closed containers, two processes, evaporation and condensation, occur simultaneously. Initially there are few molecules in the vapor phase, so more liquid molecules evaporate than gas molecules condense. Thus, the number of molecules in the gas phase increases, causing the vapor pressure of hexane to increase. Eventually, the number of molecules in the gas phase reaches a maximum where the number of liquid molecules evaporating equals the number of gas molecules condensing. In other words, the evaporation rate equals the condensation rate. At this point, there is no further change in the vapor pressure.

12.14   a) At the critical temperature, the molecules are moving so fast that they can no longer be condensed. This temperature decreases with weaker intermolecular forces because the forces are not strong enough to overcome molecular motion. Alternatively, as intermolecular forces increase, the **critical temperature increases**.
b) As intermolecular forces increase, the **boiling point increases** because it becomes more difficult and takes more energy to separate molecules from the liquid phase.
c) As intermolecular forces increase, the **vapor pressure decreases** for the same reason given in b). At any given temperature, strong intermolecular forces prevent molecules from easily going into the vapor phase and thus vapor pressure is decreased.
d) As intermolecular forces increase, the **heat of vaporization increases** because more energy is needed to separate molecules from the liquid phase.

12.18   When water at 100°C touches skin, the heat released is from the lowering of the temperature of the water. The specific heat of water is approximately 75 J/mol•K. When steam at 100°C touches skin, the heat released is from the condensation of the gas with a heat of condensation of approximately 41 kJ/mol. Thus, the amount of heat released from gaseous water condensing will be greater than the heat from hot liquid water cooling and the burn from the steam will be worse than that from hot water.

12.19   <u>Plan:</u> The total heat required is the sum of three processes: warming the ice to 0.00°C, the melting point; melting the ice to liquid water; warming the water to 0.500°C. The equation $q = c$ x mass x ΔT is used to calculate the heat involved in changing the temperature of the ice and of the water; the heat of fusion is used to calculate the heat involved in the phase change of ice to water.
<u>Solution:</u>
1) Warming the ice from –6.00°C to 0.00°C:

$$q_1 = c \text{ x mass x } \Delta T = (2.09 \text{ J/g°C})(22.00 \text{ g})[0.0 - (-6.00)]°C = 275.88 \text{ J}$$

2) Phase change of ice at 0.00°C to water at 0.00°C:

$$q_2 = n\left(\Delta H°_{fus}\right) = (22.0 \text{ g})\left(\frac{1 \text{ mol}}{18.02 \text{ g}}\right)\left(\frac{6.02 \text{ kJ}}{\text{mol}}\right)\left(\frac{10^3 \text{ J}}{1 \text{ kJ}}\right) = 7349.6115 \text{ J}$$

3) Warming the liquid from 0.00°C to 0.500°C:

$$q_3 = c \text{ x mass x } \Delta T = (4.184 \text{ J/g°C})(22.00 \text{ g})[0.500 - (0.0)]°C = 46.024 \text{ J}$$

The three heats are positive because each process takes heat from the surroundings (endothermic). The phase change requires much more energy than the two temperature change processes. The total heat is
$q_1 + q_2 + q_3 = (275.88 \text{ J} + 7349.6115 \text{ J} + 46.024 \text{ J}) = 7671.5155 = \textbf{7.67x10}^3 \textbf{ J.}$

12.21   <u>Plan:</u> The Clausius-Clapeyron equation gives the relationship between vapor pressure and temperature. We are given $\Delta H°_{vap}$, P_1, T_1, and T_2; these values are substituted into the equation to find the P_2, the vapor pressure.
<u>Solution:</u>

$$\ln\frac{P_2}{P_1} = \frac{-\Delta H°_{vap}}{R}\left(\frac{1}{T_2} - \frac{1}{T_1}\right) \qquad P_1 = 1.00 \text{ atm} \qquad T_1 = 122°C + 273 = 395 \text{ K}$$

$$\Delta H°_{vap} = 35.5 \text{ kJ/mol} \qquad\qquad P_2 = ? \qquad\qquad T_2 = 113°C + 273 = 386 \text{ K}$$

$$\ln\frac{P_2}{1.00 \text{ atm}} = \frac{-35.5\frac{\text{kJ}}{\text{mol}}}{8.314 \text{ J/mol}\bullet\text{K}}\left(\frac{1}{386 \text{ K}} - \frac{1}{395 \text{ K}}\right)\left(\frac{10^3 \text{ J}}{1 \text{kJ}}\right) = -0.2520440$$

$$\frac{P_2}{1.00 \text{ atm}} = 0.7772105$$

$P_2 = (0.7772105)(1.00 \text{ atm}) = 0.7772105 = \textbf{0.777 atm}$

12.23   <u>Plan:</u> The Clausius-Clapeyron equation gives the relationship between vapor pressure and temperature. We are given P_1, P_2, T_1, and T_2; these values are substituted into the equation to find $\Delta H°_{vap}$. The pressure in torr must be converted to atm.

Solution:

$$P_1 = (621 \text{ torr})\left(\frac{1 \text{ atm}}{760 \text{ torr}}\right) = 0.817105 \text{ atm} \qquad T_1 = 85.2°C + 273.2 = 358.4 \text{ K}$$

$$P_2 = 1 \text{ atm} \qquad\qquad\qquad\qquad T_2 = 95.6°C + 273.2 = 368.8 \text{ K} \qquad \Delta H^\circ_{vap} = ?$$

$$\ln\frac{P_2}{P_1} = \frac{-\Delta H^\circ_{vap}}{R}\left(\frac{1}{T_2} - \frac{1}{T_1}\right)$$

$$\ln\frac{1 \text{ atm}}{0.817105 \text{ atm}} = \frac{-\Delta H^\circ_{vap}}{8.314 \text{ J/mol} \bullet \text{K}}\left(\frac{1}{368.8 \text{ K}} - \frac{1}{358.4 \text{ K}}\right)$$

$$0.2019877 = -\Delta H_{vap}(-9.463775 \times 10^{-6})\text{J/mol}$$

$$\Delta H^\circ_{vap} = 21{,}343.248 = \mathbf{2 \times 10^4 \text{ J/mol}}$$

(The significant figures in the answer are limited by the 1 atm in the problem.)

12.25

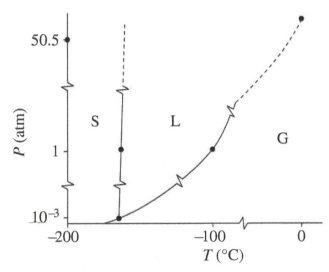

The pressure scale is distorted to represent the large range in pressures given in the problem, so the liquid-solid curve looks different from the one shown in the text. The important features of the graph include the distinction between the gas, liquid, and solid states, and the melting point T, which is located directly above the critical T. Solid ethylene is denser than liquid ethylene since the solid-liquid line slopes to the right with increasing pressure.

12.27   Plan: Refer to the phase diagram to describe the specified changes. In part a), a substance sublimes if it converts directly from a solid to a gas without passing through a liquid phase. In part b), refer to the diagram to explain the changes that occur when sulfur is heated at a constant pressure of 1 atm.
Solution:
a) Rhombic sulfur will sublime when it is heated at a pressure less than 1×10^{-4} atm.
b) At 90 °C and 1 atm, sulfur is in the solid (rhombic form). As it is heated at constant pressure, it passes through the solid (monoclinic) phase, starting at 114 °C. At about 120 °C, the solid melts to form a liquid. At about 445 °C, the liquid evaporates and changes to the gas state.

12.30   Plan: The Clausius-Clapeyron equation gives the relationship between vapor pressure and temperature. We are given ΔH°_{vap}, P_1, T_1, and T_2; these values are substituted into the equation to find P_2. Convert the temperatures from °C to K and ΔH°_{vap} from kJ/mol to J/mol to allow cancellation with the units in R.
Solution:
$P_1 = 2.3 \text{ atm} \qquad T_1 = 25.0°C + 273 = 298 \text{ K}$
$P_2 = ? \qquad\qquad T_2 = 135°C + 273 = 408 \text{ K} \qquad\qquad \Delta H^\circ_{vap} = 24.3 \text{ kJ/mol}$

$$\ln\frac{P_2}{P_1} = \frac{-\Delta H^{\circ}_{vap}}{R}\left(\frac{1}{T_2} - \frac{1}{T_1}\right)$$

$$\ln\frac{P_2}{2.3\ \text{atm}} = \frac{-24.3\dfrac{\text{kJ}}{\text{mol}}}{8.314\ \text{J/mol}\bullet\text{K}}\left(\frac{1}{408\ \text{K}} - \frac{1}{298\ \text{K}}\right)\left(\frac{10^3\ \text{J}}{1\ \text{kJ}}\right)$$

$$\ln\frac{P_2}{2.3\ \text{atm}} = 2.644311$$

$$\frac{P_2}{2.3\ \text{atm}} = 14.07374$$

$P_2 = (14.07374)(2.3\ \text{atm}) = 32.3696 = \textbf{32 atm}$

12.34 To form hydrogen bonds, the atom bonded to hydrogen must have two characteristics: small size and high electronegativity (so that the atom has a very high electron density). With this high electron density, the attraction for a hydrogen on another molecule is very strong. Selenium is much larger than oxygen (atomic radius of 119 pm vs. 73 pm) and less electronegative than oxygen (2.4 for Se and 3.5 for O) resulting in an electron density on Se in H_2Se that is too small to form hydrogen bonds.

12.36 All particles (atoms and molecules) exhibit dispersion forces, but these are the weakest of intermolecular forces. The dipole-dipole forces in polar molecules dominate the dispersion forces.

12.39 Plan: Dispersion forces are the only forces between nonpolar substances; dipole-dipole forces exist between polar substances. Hydrogen bonds only occur in substances in which hydrogen is directly bonded to either oxygen, nitrogen, or fluorine.
Solution:
a) **Hydrogen bonding** will be the strongest force between methanol molecules since they contain O–H bonds. Dipole-dipole and dispersion forces also exist.
b) **Dispersion forces** are the only forces between nonpolar carbon tetrachloride molecules and, thus, are the strongest forces.
c) **Dispersion forces** are the only forces between nonpolar chlorine molecules and, thus, are the strongest forces.

12.41 Plan: Dispersion forces are the only forces between nonpolar substances; dipole-dipole forces exist between polar substances. Hydrogen bonds only occur in substances in which hydrogen is directly bonded to either oxygen, nitrogen, or fluorine.
Solution:
a) **Dipole-dipole** interactions will be the strongest forces between methyl chloride molecules because the C–Cl bond has a dipole moment.
b) **Dispersion** forces dominate because CH_3CH_3 (ethane) is a symmetrical nonpolar molecule.
c) **Hydrogen bonding** dominates because hydrogen is bonded to nitrogen, which is one of the three atoms (N, O, or F) that participate in hydrogen bonding.

12.43 Plan: Hydrogen bonds are formed when a hydrogen atom is bonded to N, O, or F.
Solution:
a) The presence of an OH group leads to the formation of hydrogen bonds in **$CH_3CH(OH)CH_3$**. There are no hydrogen bonds in CH_3SCH_3.

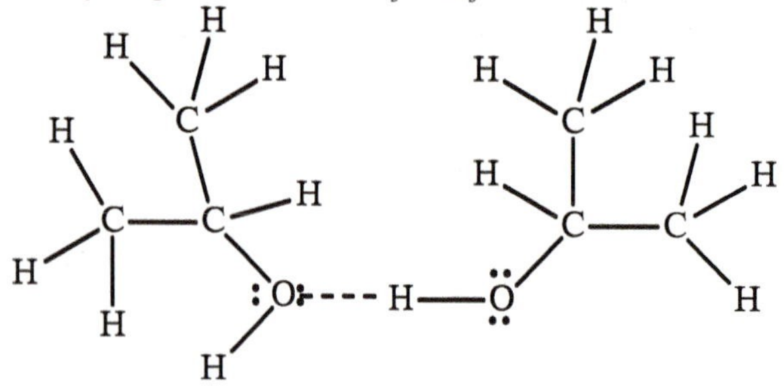

b) The presence of H attached to F in **HF** leads to the formation of hydrogen bonds. There are no hydrogen bonds in HBr.

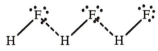

12.45   Plan: In the vaporization process, intermolecular forces between particles in the liquid phase must be broken as the particles enter the vapor phase. In other words, the question is asking for the strongest interparticle force that must be broken to vaporize the liquid. Dispersion forces are the only forces between nonpolar substances; dipole-dipole forces exist between polar substances. Hydrogen bonds only occur in substances in which hydrogen is directly bonded to either oxygen, nitrogen, or fluorine.
Solution:
a) **Dispersion**, because hexane, C₆H₁₄, is a nonpolar molecule.
b) **Hydrogen bonding**; hydrogen is bonded to oxygen in water. A single water molecule can engage in as many as four hydrogen bonds.
c) **Dispersion**, although the individual Si–Cl bonds are polar, the molecule has a symmetrical, tetrahedral shape and is therefore nonpolar.

12.47   Plan: Polarizability increases down a group and decreases from left to right because as atomic size increases, polarizability increases.
Solution:
a) **Iodide ion** has greater polarizability than the bromide ion because the iodide ion is larger. The electrons can be polarized over a larger volume in a larger atom or ion.
b) **Ethene (CH₂=CH₂)** has greater polarizability than ethane (CH₃CH₃) because the electrons involved in π bonds are more easily polarized than electrons involved in σ bonds.
c) **H₂Se** has greater polarizability than water because the selenium atom is larger than the oxygen atom.

12.49   Plan: Weaker attractive forces result in a higher vapor pressure because the molecules have a smaller energy barrier in order to escape the liquid and go into the gas phase. Decide which of the two substances in each pair has the weaker interparticle force. Dispersion forces are weaker than dipole-dipole forces, which are weaker than hydrogen bonds.
Solution:
a) **C₂H₆**   C₂H₆ is a smaller molecule exhibiting weaker dispersion forces than C₄H₁₀.
b) **CH₃CH₂F**   CH₃CH₂F has no H–F bonds (F is bonded to C, not to H), so it only exhibits dipole-dipole forces, which are weaker than the hydrogen bonding in CH₃CH₂OH.
c) **PH₃**   PH₃ has weaker intermolecular forces (dipole-dipole) than NH₃ (hydrogen bonding).

12.51   Plan: The weaker the interparticle forces, the lower the boiling point. Decide which of the two substances in each pair has the weaker interparticle force. Dispersion forces are weaker than dipole-dipole forces, which are weaker than hydrogen bonds, which are weaker than ionic forces.
Solution:
a) **HCl** would have a lower boiling point than LiCl because the dipole-dipole intermolecular forces between hydrogen chloride molecules in the liquid phase are weaker than the significantly stronger ionic forces holding the ions in lithium chloride together.
b) **PH₃** would have a lower boiling point than NH₃ because the intermolecular forces in PH₃ are weaker than those in NH₃. Hydrogen bonding exists between NH₃ molecules but weaker dipole-dipole forces hold PH₃ molecules together.
c) **Xe** would have a lower boiling point than iodine. Both are nonpolar with dispersion forces, but the forces between xenon atoms would be weaker than those between iodine molecules since the iodine molecules are more polarizable because of their larger size.

12.53   Plan: The weaker the intermolecular forces, the lower the boiling point. Decide which of the two substances in each pair has the weaker intermolecular force. Dispersion forces are weaker than dipole-dipole forces, which are weaker than hydrogen bonds, which are weaker than ionic forces.

12-11

Solution:
a) **C₄H₈**, the cyclic molecule, cyclobutane, has less surface area exposed, so its dispersion forces are weaker than the straight chain molecule, C₄H₁₀.
b) **PBr₃**, the dipole-dipole forces of phosphorous tribromide are weaker than the ionic forces of sodium bromide.
c) **HBr**, the dipole-dipole forces of hydrogen bromide are weaker than the hydrogen bonding forces of water.

12.55   The trend in both atomic size and electronegativity predicts that the trend in increasing strength of hydrogen bonds is N–H < O–H < F–H. As the atomic size decreases and electronegativity increases, the electron density of the atom increases. High electron density strengthens the attraction to a hydrogen atom on another molecule. Fluorine is the smallest of the three and the most electronegative, so its hydrogen bonds would be the strongest. Oxygen is smaller and more electronegative than nitrogen, so hydrogen bonds for water would be stronger than hydrogen bonds for ammonia.

12.59   The shape of the drop depends upon the competing cohesive forces (attraction of molecules within the drop itself) and adhesive forces (attraction between molecules in the drop and the molecules of the waxed floor). If the cohesive forces are strong and outweigh the adhesive forces, the drop will be as spherical as gravity will allow. If, on the other hand, the adhesive forces are significant, the drop will spread out. Both water (hydrogen bonding) and mercury (metallic bonds) have strong cohesive forces, whereas cohesive forces in oil (dispersion) are relatively weak. Neither water nor mercury will have significant adhesive forces to the nonpolar wax molecules, so these drops will remain nearly spherical. The adhesive forces between the oil and wax can compete with the weak, cohesive forces of the oil (dispersion) and so the oil drop spreads out.

12.61   Surface tension is defined as the energy needed to increase the surface area by a given amount, so units of energy (J) per surface area (m²) describe this property.

12.63   Plan: The stronger the intermolecular force, the greater the surface tension. Decide which of the substances has the weakest intermolecular force and which has the strongest. Dispersion forces are weaker than dipole-dipole forces, which are weaker than hydrogen bonds, which are weaker than ionic forces.
Solution:
All three molecules exhibit hydrogen bonding (H is bonded to O), but the extent of hydrogen bonding increases with the number of O–H bonds present in each molecule. HOCH₂CH(OH)CH₂OH with three O–H groups can form more hydrogen bonds than HOCH₂CH₂OH with two O–H groups, which in turn can form more hydrogen bonds than CH₃CH₂CH₂OH with only one O–H group. The greater the number of hydrogen bonds, the stronger the intermolecular forces, and the higher the surface tension.
CH₃CH₂CH₂OH < HOCH₂CH₂OH < HOCH₂CH(OH)CH₂OH

12.65   Plan: Viscosity is a measure of the resistance of a liquid to flow, and is greater for molecules with stronger intermolecular forces. The stronger the force attracting the molecules to each other, the harder it is for one molecule to move past another. Thus, the substance will not flow easily if the intermolecular force is strong. Decide which of the substances has the weakest intermolecular force and which has the strongest. Dispersion forces are weaker than dipole-dipole forces, which are weaker than hydrogen bonds, which are weaker than ionic forces.
Solution:
The ranking of decreasing viscosity is the opposite of that for increasing surface tension (Problem 12.61).
HOCH₂CH(OH)CH₂OH > HOCH₂CH₂OH > CH₃CH₂CH₂OH
The greater the number of hydrogen bonds, the stronger the intermolecular forces, and the higher the viscosity.

12.70   Water is a good solvent for polar and ionic substances and a poor solvent for nonpolar substances. Water is a polar molecule and dissolves polar substances because their intermolecular forces are of similar strength. Water is also able to dissolve ionic compounds and keep ions separated in solution through ion-dipole interactions. Nonpolar substances will not be very soluble in water since their dispersion forces are much weaker than the hydrogen bonds in water. A solute whose intermolecular attraction to a solvent molecule is less than the attraction between two solvent molecules will not dissolve because its attraction cannot replace the attraction between solvent molecules.

12.71   A single water molecule can form four hydrogen bonds. The two hydrogen atoms form a hydrogen bond each to oxygen atoms on neighboring water molecules. The two lone pairs on the oxygen atom form hydrogen bonds with hydrogen atoms on neighboring molecules.

12.74   Water exhibits strong capillary action, which allows it to be easily absorbed by the plant's roots and transported to the leaves.

12.80   The unit cell is a simple cubic cell. According to the bottom row in Figure 12.27, two atomic radii (or one atomic diameter) equal the width of the cell.

12.83   The energy gap is the energy difference between the highest filled energy level (valence band) and the lowest unfilled energy level (conduction band). In conductors and superconductors, the energy gap is zero because the valence band overlaps the conduction band. In semiconductors, the energy gap is small but greater than zero. In insulators, the energy gap is large and thus insulators do not conduct electricity.

12.85   The density of a solid depends on the atomic mass of the element (greater mass = greater density), the atomic radius (how many atoms can fit in a given volume), and the type of unit cell, which determines the packing efficiency (how much of the volume is occupied by empty space).

12.86   Plan: The simple cubic structure unit cell contains one atom since the atoms at the eight corners are shared by eight cells for a total of 8 atoms x 1/8 atom per cell = 1 atom; the body-centered cell also has an atom in the center, for a total of two atoms; the face-centered cell has six atoms in the faces which are shared by two cells: 6 atoms x ½ atom per cell = 3 atoms plus another atom from the eight corners for a total of four atoms.
Solution:
a) Ni is **face-centered cubic** since there are four atoms/unit cell.
b) Cr is **body-centered cubic** since there are two atoms/unit cell.
c) Ca is **face-centered cubic** since there are four atoms/unit cell.

12.88   Plan: Use the radius of the Ca atom to find the volume of one Ca atom. Use Avogadro's number to find the volume of one mole of Ca atoms, and convert from pm^3 to cm^3. Use the volume of one mole of Ca *atoms* and the packing efficiency of the solid (0.74 for cubic closest packing) to find the volume of one mole of Ca *metal*. Calculate the density of Ca metal by dividing the molar mass of Ca by the volume of one mole of Ca metal.
Solution:

$$V \text{ of one Ca atom} = \frac{4}{3}\pi r^3 = \left(\frac{4}{3}\right)(\pi)(197 \text{ pm})^3 = 3.2025 \times 10^7 \text{ pm}^3$$

$$V \text{ of one mole of Ca atoms} = \left(\frac{3.2025 \times 10^7 \text{ pm}^3}{1 \text{ Ca atom}}\right)\left(\frac{6.022 \times 10^{23} \text{ Ca atoms}}{1 \text{ mol Ca atoms}}\right)\left(\frac{1 \text{ cm}^3}{10^{30} \text{ pm}^3}\right) = 19.285 \text{ cm}^3$$

$$V \text{ of one mole of Ca metal} = \frac{19.285 \text{ cm}^3}{0.74} = 26.061 \text{ cm}^3/\text{mol Ca metal}$$

$$\text{Density of Ca metal} = \left(\frac{40.08 \text{ g Ca}}{1 \text{ mol Ca}}\right)\left(\frac{1 \text{ mol Ca}}{26.061 \text{ cm}^3}\right) = 1.5379 = \mathbf{1.54 \text{ g/cm}^3}$$

12.90   a) There is a change in unit cell from CdO in a sodium chloride structure to CdSe in a zinc blende structure.
b) **Yes**, the coordination number of Cd does change from six in the CdO unit cell to four in the CdSe unit cell.

12.92   Plan: Use the relationship between the edge length of the unit cell (A) and the atomic radius (r) for a body-centered cubic unit cell to solve for the edge length of the potassium unit cell.
Solution:

For a body-centered cubic unit cell, $A = \dfrac{4}{\sqrt{3}} r$

$$A = \frac{4}{\sqrt{3}} \, (227 \text{ pm}) = 524.23 = \textbf{524 pm}$$

12.94   Plan: Substances composed of individual atoms are atomic solids; molecular substances composed of covalent molecules form molecular solids; ionic compounds form ionic solids; metal elements form metallic solids; certain substances that form covalent bonds between atoms or molecules form network covalent solids.
Solution:
a) Nickel forms a **metallic solid** since nickel is a metal whose atoms are held together by metallic bonds.
b) Fluorine forms a **molecular solid** since the F_2 molecules have covalent bonds and the molecules are held to each other by dispersion forces.
c) Methanol forms a **molecular solid** since the covalently bonded CH_3OH molecules are held to each other by hydrogen bonds.
d) Tin forms a **metallic solid** since tin is a metal whose atoms are held together by metallic bonds.
e) Silicon is in the same group as carbon, so it exhibits similar bonding properties. Since diamond and graphite are both **network covalent** solids, it makes sense that Si forms the same type of bonds.
f) Xe is an **atomic** solid since individual atoms are held together by dispersion forces.

12.96   Figure P12.90 shows the face-centered cubic array of zinc blende, ZnS. Both ZnS and ZnO have a 1:1 ion ratio, so the ZnO unit cell will also contain **four** Zn^{2+} ions.

12.98   Plan: To determine the number of Zn^{2+} ions and Se^{2-} ions in each unit cell count the number of ions at the corners, faces, and center of the unit cell. Atoms at the eight corners are shared by eight cells for a total of 8 atoms x 1/8 atom per cell = 1 atom; atoms in the body of a cell are in that cell only; atoms at the faces are shared by two cells: 6 atoms x 1/2 atom per cell = 3 atoms. Add the masses of the total number of atoms in the cell to find the mass of the cell. Given the mass of one unit cell and the ratio of mass to volume (density) divide the mass, converted to grams (conversion factor is 1 amu = 1.66054×10^{-24} g), by the density to find the volume of the unit cell. Since the volume of a cube is length x width x height, the edge length is found by taking the cube root of the cell volume.
Solution:
a) Looking at selenide ions, there is one ion at each corner and one ion on each face. The total number of selenide ions is 1/8 (8 corner ions) + 1/2 (6 face ions) = **4 Se^{2-} ions**. There are also **4 Zn^{2+} ions** due to the 1:1 ratio of Se ions to Zn ions.
b) Mass of unit cell = (4 x mass of Zn atom) + (4 x mass of Se atom)
$$= (4 \times 65.41 \text{ amu}) + (4 \times 78.96 \text{ amu}) = \textbf{577.48 amu}$$

c) Volume $(\text{cm}^3) = \left(577.48 \text{ amu}\right)\left(\dfrac{1.66054 \times 10^{-24} \text{ g}}{1 \text{ amu}}\right)\left(\dfrac{\text{cm}^3}{5.42 \text{ g}}\right) = 1.76924 \times 10^{-22} = \textbf{1.77} \boldsymbol{\times} \textbf{10}^{\textbf{-22}} \textbf{ cm}^3$

d) The volume of a cube equals (length of edge)^3.

Edge length $(\text{cm}) = \sqrt[3]{1.76924 \times 10^{-22} \text{ cm}^3} = 5.6139 \times 10^{-8} = \textbf{5.61} \boldsymbol{\times} \textbf{10}^{\textbf{-8}} \textbf{ cm}$

12.100   Plan: To classify a substance according to its electrical conductivity, first locate it on the periodic table as a metal, metalloid, or nonmetal. In general, metals are conductors, metalloids are semiconductors, and nonmetals are insulators.
Solution:
a) Phosphorous is a nonmetal and an **insulator**.
b) Mercury is a metal and a **conductor**.
c) Germanium is a metalloid in Group 4A(14) and is beneath carbon and silicon in the periodic table. Pure germanium crystals are **semiconductors** and are used to detect gamma rays emitted by radioactive materials. Germanium can also be doped with phosphorous (similar to the doping of silicon) to form an n-type semiconductor or be doped with lithium to form a p-type semiconductor.

12.102   Plan: First, classify the substance as an insulator, conductor, or semiconductor. The electrical conductivity of conductors decreases with increasing temperature, whereas that of semiconductors increases with temperature. Temperature increases have little impact on the electrical conductivity of insulators.
Solution:
a) Antimony, Sb, is a metalloid, so it is a semiconductor. Its electrical conductivity **increases** as the temperature increases.
b) Tellurium, Te, is a metalloid, so it is a semiconductor. Its electrical conductivity **increases** as temperature increases.
c) Bismuth, Bi, is a metal, so it is a conductor. Its electrical conductivity **decreases** as temperature increases.

12.104   Plan: Use the molar mass and the density of Po to find the volume of one mole of Po. Divide by Avogadro's number to obtain the volume of one Po atom (and the volume of the unit cell). Since Po has a simple cubic unit cell, there is one Po atom in the cell (atoms at the eight corners are shared by eight cells for a total of 8 atoms x 1/8 atom = 1 atom per cell). Find the edge length of the cell by taking the cube root of the volume of the unit cell. The edge length of a simple cubic unit cell is twice the radius of the atom.
Solution:

$$\text{Volume (cm}^3\text{) of the unit cell} = \left(\frac{209 \text{ g Po}}{1 \text{ mol Po}}\right)\left(\frac{\text{cm}^3}{9.142 \text{ g}}\right)\left(\frac{1 \text{ mol Po}}{6.022 \times 10^{23} \text{ Po atoms}}\right)\left(\frac{1 \text{ Po atom}}{1 \text{ unit cell}}\right)$$

$$= 3.7963332 \times 10^{-23} \text{ cm}^3$$

Edge length (cm) of the unit cell $= \sqrt[3]{(3.7963332 \times 10^{-23} \text{ cm}^3)} = 3.3608937 \times 10^{-8} \text{ cm}$

$2r =$ edge length

$2r = 3.3608937 \times 10^{-8} \text{ cm}$

$r = 1.680447 \times 10^{-8} = \textbf{1.68} \times \textbf{10}^{-8} \textbf{ cm}$

12.111   A substance whose physical properties are the same in all directions is called isotropic; an anisotropic substance has properties that depend on direction. Liquid crystals flow like liquids but have a degree of order that gives them the anisotropic properties of a crystal.

12. 117   Plan: Germanium and silicon are elements in Group 4A with four valence electrons. If germanium or silicon is doped with an atom with more than four valence electrons, an n-type semiconductor is produced. If it is doped with an atom with fewer than four valence electrons, a p-type semiconductor is produced.
Solution:
a) Phosphorus has five valence electrons so an **n-type semiconductor** will form by doping Ge with P.
b) Indium has three valence electrons so a **p-type semiconductor** will form by doping Si with In.

12.119   Plan: The degree of polymerization is the number of repeat units in a polymer chain and can be found by using the equation $\mathcal{M}_{polymer} = \mathcal{M}_{repeat} \times n$, where n is the degree of polymerization.
Solution:
The molar mass of one monomer unit ($\mathcal{M}_{repeat}$), $C_6H_5CHCH_2$, is 104.14 g/mol and the molar mass of the polymer ($\mathcal{M}_{polymer}$) is 3.5×10^5 g/mol.
$\mathcal{M}_{polymer} = \mathcal{M}_{repeat} \times n$
$n = \dfrac{3.5 \times 10^5 \text{ g/mol}}{104.14 \text{ g/mol}} = 3361$ monomer units. Reporting in correct significant figures, $\boldsymbol{n = 3.4 \times 10^3}$.

12.121   Plan: Use the equation for the radius of gyration. The length of the repeat unit, l_0, is given (0.252 pm). Calculate the degree of polymerization, n, by dividing the molar mass of the polymer chain ($\mathcal{M} = 2.8 \times 10^5$ g/mol) by the molar mass of one monomer unit (CH_3CHCH_2, $\mathcal{M} = 42.08$ g/mol). Substitute the values l_0 and n into the equation.
Solution:

$\mathcal{M}_{polymer} = \mathcal{M}_{repeat} \times n$

$$n = \frac{2.8 \times 10^5 \text{ g/mol}}{42.08 \text{ g/mol}} = 6653.99$$

$$\text{Radius of gyration} = R_g = \sqrt{\frac{nI_0^2}{6}} = \sqrt{\frac{(6653.99)(0.252 \text{ pm})^2}{6}} = 8.39201 = \mathbf{8.4 \text{ pm}}$$

12.124 **Plan:** The vapor pressure of water is temperature dependent. Table 5.3 gives the vapor pressure of water at various temperatures. Use $PV = nRT$ to find the moles and then mass of water in 5.0 L of nitrogen at 22°C; then find the mass of water in the 2.5 L volume of nitrogen and subtract the two masses to calculate the mass of water that condenses.
Solution:
At 22°C the vapor pressure of water is 19.8 torr (from Table 5.3).
a) Once compressed, the N_2 gas would still be saturated with water. The vapor pressure depends on the temperature, which has not changed. Therefore, the partial pressure of water in the compressed gas remains the same at **19.8 torr**.
b) $PV = nRT$

$$\text{Pressure (atm)} = (19.8 \text{ torr})\left(\frac{1 \text{ atm}}{760 \text{ torr}}\right) = 0.026053 \text{ atm}$$

Moles of H_2O at 22°C and 5.00 L volume:

$$n = \frac{PV}{RT} = \frac{(0.026053 \text{ atm})(5.00 \text{ L})}{\left(0.0821 \frac{\text{L} \cdot \text{atm}}{\text{mol} \cdot \text{K}}\right)((273 + 22)\text{K})} = 0.0053785 \text{ mol } H_2O$$

The gas is compressed to half the volume: 5.00 L/2 = 2.50 L

$$\text{Mass (g) of } H_2O \text{ at 22°C and 5.00 L volume} = (0.0053785 \text{ mol } H_2O)\left(\frac{18.02 \text{ g } H_2O}{1 \text{ mol } H_2O}\right) = 0.096921 \text{ g}$$

Moles of H_2O at 22°C and 2.50 L volume:

$$n = \frac{PV}{RT} = \frac{(0.026053 \text{ atm})(2.50 \text{ L})}{\left(0.0821 \frac{\text{L} \cdot \text{atm}}{\text{mol} \cdot \text{K}}\right)((273 + 22)\text{K})} = 0.0026892 \text{ mol } H_2O$$

$$\text{Mass (g) of } H_2O \text{ at 22°C and 2.50 L volume} = (0.00268922 \text{ mol } H_2O)\left(\frac{18.02 \text{ g } H_2O}{1 \text{ mol } H_2O}\right) = 0.048460 \text{ g}$$

$$\text{Mass (g) of } H_2O \text{ condensed} = (0.096921 \text{ g } H_2O) - (0.048460 \text{ g } H_2O) = 0.048461 = \mathbf{0.0485 \text{ g } H_2O}$$

12.130 **Plan:** The Clausius-Clapeyron equation gives the relationship between vapor pressure and temperature. We are given P_1, P_2, T_1, and $\Delta H°_{vap}$; these values are substituted into the equation to find T_2.
Solution:

$$\ln \frac{P_2}{P_1} = -\frac{\Delta H°_{vap}}{R}\left(\frac{1}{T_2} - \frac{1}{T_1}\right) \qquad P_1 = 1.20 \times 10^{-3} \text{ torr} \qquad T_1 = 20.0°C + 273 = 293 \text{ K}$$

$$\Delta H°_{vap} = 59.1 \text{ kJ/mol} \qquad\qquad P_2 = 5.0 \times 10^{-5} \text{ torr} \qquad T_2 = ?$$

$$\ln \frac{5.0 \times 10^{-5} \text{ torr}}{1.20 \times 10^{-3} \text{ torr}} = \frac{-59.1 \text{ kJ/mol}}{8.314 \text{ J/mol} \cdot \text{K}}\left(\frac{1}{T_2} - \frac{1}{293 \text{ K}}\right)\left(\frac{10^3 \text{ J}}{1 \text{ kJ}}\right)$$

$$-3.178054 = -7108.492\left(\frac{1}{T_2} - \frac{1}{293 \text{ K}}\right)$$

$$(-3.17805)/(-7108.49) = 4.47078 \times 10^{-4} = \left(\frac{1}{T_2} - \frac{1}{293 \text{ K}}\right)$$

$$4.47078 \times 10^{-4} + 1/293 = 1/T_2$$

$$T_2 = 259.064 = \mathbf{259 \text{ K}}$$

12.132 Plan: Use the volume of the liquid water and the density of water to find the moles of water present in the volume of the greenhouse. Find the pressure of this number of moles of water vapor using the ideal gas equation. Knowing that 4.20 L results in the calculated vapor pressure, the volume of water that would give a vapor pressure corresponding to 100% relative humidity can be calculated.
Solution:

a) Moles of water $= (4.20 \text{ L}) \left(\dfrac{1 \text{ mL}}{10^{-3} \text{ L}} \right) \left(\dfrac{1.00 \text{ g}}{\text{mL}} \right) \left(\dfrac{1 \text{ mol H}_2\text{O}}{18.02 \text{ g H}_2\text{O}} \right) = 233.0744 \text{ mol}$

$PV = nRT$

$P = \dfrac{nRT}{V} = \dfrac{(233.0744 \text{ mol}) \left(0.0821 \dfrac{\text{L} \bullet \text{atm}}{\text{mol} \bullet \text{K}} \right) \left((273 + 26) \text{K} \right)}{(256 \text{ m}^3)} \left(\dfrac{10^{-3} \text{ m}^3}{1 \text{ L}} \right)$

$= 2.234956 \times 10^{-2} = \mathbf{2.23 \times 10^{-2} \text{ atm}}$

b) 25.2 torr is needed to saturate the air.

$P = (25.2 \text{ torr}) \left(\dfrac{1.00 \text{ atm}}{760 \text{ torr}} \right) = 0.0331579 \text{ atm}$

$\dfrac{4.20 \text{ L}}{2.234956 \times 10^{-2} \text{ atm}} = \dfrac{\text{x}}{0.0331579 \text{ atm}}$

Volume (L) of water needed for 100% relative humidity $= \dfrac{(0.0331579 \text{ atm})(4.20 \text{ L})}{2.234956 \times 10^{-2} \text{ atm}} = 6.23114 = \mathbf{6.23 \text{ L H}_2\text{O}}$

12.135 Plan: Hydrogen bonds only occur in substances in which hydrogen is directly bonded to either oxygen, nitrogen, or fluorine.
Solution:
a) Both furfuryl alcohol and 2-furoic acid can form hydrogen bonds since these two molecules have hydrogen directly bonded to oxygen.

2-furoic acid

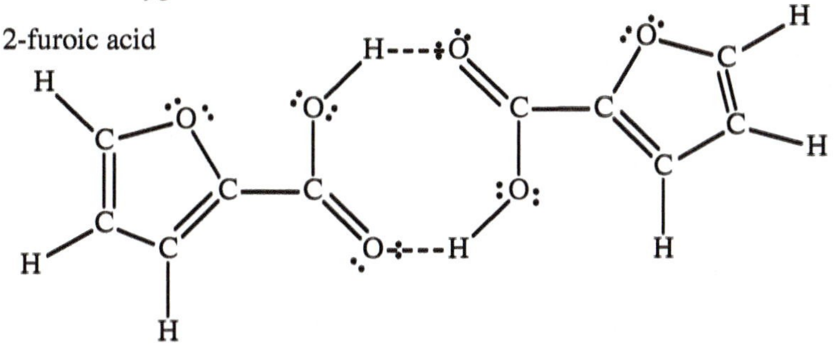

furfuryl alcohol

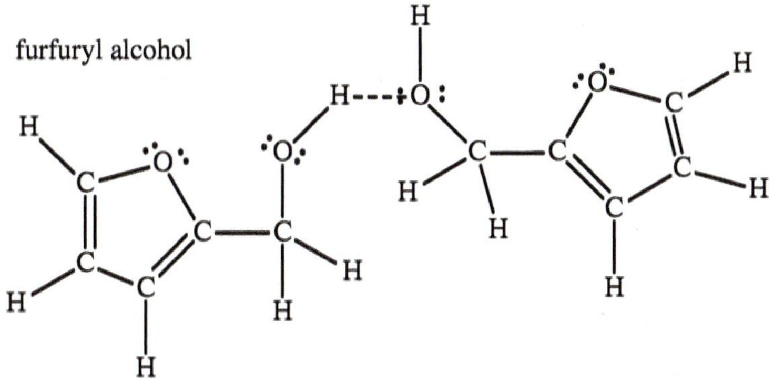

12-17

b) Both furfuryl alcohol and 2-furoic acid can form internal hydrogen bonds by forming a hydrogen bond between the O–H and the O in the ring.

2-furoic acid

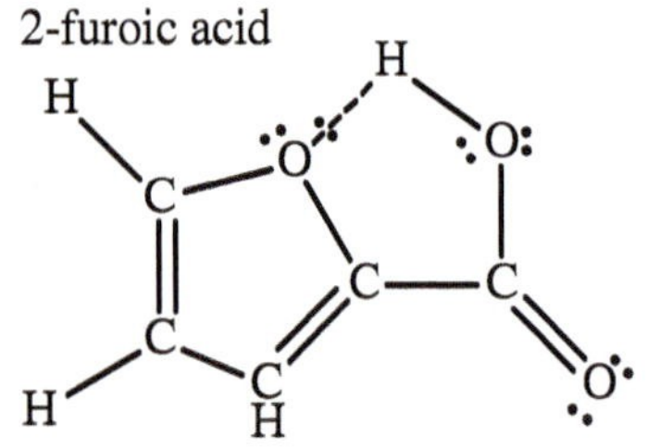

furfuryl alcohol

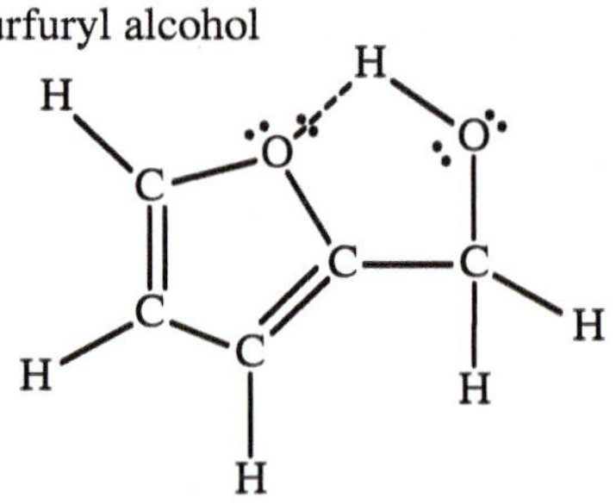

12.136 Plan: Determine the total mass of water, with a partial pressure of 31.0 torr, contained in the air. This is done by using $PV = nRT$ to find the moles of water at 31.0 torr and 22.0°C. Then calculate the total mass of water, with a partial pressure of 10.0 torr, contained in the air. The difference between these two values is the amount of water removed. The mass of water and the heat of condensation are necessary to find the amount of energy removed from the water.

Solution:

a) $PV = nRT$

$$n = \frac{PV}{RT}$$

Convert pressure to units of atm: $P \text{ (atm)} = (31.0 \text{ torr})\left(\frac{1 \text{ atm}}{760 \text{ torr}}\right) = 0.0407895 \text{ atm}$

Convert volume to units of L: $V \text{ (L)} = \left(2.4\times10^6 \text{ m}^3\right)\left(\frac{1 \text{ L}}{10^{-3} \text{ m}^3}\right) = 2.4\times10^9 \text{ L}$

Moles of H_2O at 31.0 torr = $\dfrac{\left(0.0407895 \text{ atm}\right)\left(2.4\times10^9 \text{ L}\right)}{\left(0.0821\dfrac{L\bullet atm}{mol\bullet K}\right)\left((273.2 + 22.0)K\right)} = 4{,}039{,}244.229 \text{ mol}$

Mass (metric tons) of H_2O at 31.0 torr = $\left(4{,}039{,}244.229 \text{ mol } H_2O\right)\left(\dfrac{18.02 \text{ g}}{1 \text{ mol } H_2O}\right)\left(\dfrac{1 \text{ kg}}{10^3 \text{ g}}\right)\left(\dfrac{1 \text{ ton}}{10^3 \text{ kg}}\right)$

$= 72.7872 \text{ tons } H_2O$

Convert pressure to units of atm: $P \text{ (atm)} = (10.0 \text{ torr})\left(\frac{1 \text{ atm}}{760 \text{ torr}}\right) = 0.01315789 \text{ atm}$

Moles of H_2O at 10.0 torr = $\dfrac{\left(0.01315789 \text{ atm}\right)\left(2.4\times10^9 \text{ L}\right)}{\left(0.0821\dfrac{L\bullet atm}{mol\bullet K}\right)\left((273.2 + 22.0)K\right)} = 1{,}302{,}980.7 \text{ mol}$

Mass (metric tons) of H_2O at 10.0 torr = $\left(1{,}302{,}980.7 \text{ mol } H_2O\right)\left(\dfrac{18.02 \text{ g}}{1 \text{ mol } H_2O}\right)\left(\dfrac{1 \text{ kg}}{10^3 \text{ g}}\right)\left(\dfrac{1 \text{ ton}}{10^3 \text{ kg}}\right)$

$= 23.4797 \text{ tons } H_2O$

Mass of water removed = (72.7872 metric tons H_2O) – (23.4797 metric tons H_2O) = 49.3075 = **49.3 tons H_2O**

b) The heat of condensation for water is –40.7 kJ/mol.

Heat (kJ) = $\left(49.3075 \text{ tons } H_2O\right)\left(\dfrac{10^3 \text{ kg}}{1 \text{ ton}}\right)\left(\dfrac{10^3 \text{ g}}{1 \text{ kg}}\right)\left(\dfrac{1 \text{ mol } H_2O}{18.02 \text{ g } H_2O}\right)\left(\dfrac{-40.7 \text{ kJ}}{1 \text{ mol } H_2O}\right)$

$= -1.113660\times10^8 = \mathbf{-1.11\times10^8 \text{ kJ}}$

12.137 Plan: At the boiling point, the vapor pressure equals the atmospheric pressure, so the two boiling points can be used to find the heat of vaporization for amphetamine using the Clausius-Clapeyron equation. Then use the Clausius-Clapeyron equation to find the vapor pressure of amphetamine at a different temperature, 20.°C. Then use $PV = nRT$ to find the concentration of amphetamine at this calculated pressure and 20.°C.
Solution:

$$\ln\frac{P_2}{P_1} = -\frac{\Delta H^\circ_{vap}}{R}\left(\frac{1}{T_2} - \frac{1}{T_1}\right) \qquad P_1 = 760 \text{ torr} \qquad T_1 = 201°C + 273 = 474 \text{ K}$$

$$\Delta H^\circ_{vap} = ? \qquad\qquad\qquad\qquad P_2 = 13 \text{ torr} \qquad T_2 = 83°C + 273 = 356 \text{ K}$$

$$\ln\frac{13 \text{ torr}}{760 \text{ torr}} = \frac{-\Delta H^\circ_{vap}}{8.314 \text{ J/mol}\cdot\text{K}}\left(\frac{1}{356 \text{ K}} - \frac{1}{474 \text{ K}}\right)$$

$$-4.068369076 = -0.000084109\,\Delta H^\circ_{vap}$$

$$\Delta H^\circ_{vap} = (-4.068369076)/(-0.000084109) = 48,370.199 \text{ J/mol}$$

$$\ln\frac{P_2}{P_1} = -\frac{\Delta H^\circ_{vap}}{R}\left(\frac{1}{T_2} - \frac{1}{T_1}\right) \qquad P_1 = 13 \text{ torr} \qquad T_1 = 83°C + 273 = 356 \text{ K}$$

$$\qquad\qquad\qquad\qquad\qquad\qquad P_2 = ? \qquad\qquad T_2 = 20°C + 273 = 293 \text{ K}$$

$$\qquad\qquad\qquad\qquad\qquad \Delta H^\circ_{vap} = 48,370.199 \text{ J/mol}$$

$$\ln\frac{P_2}{13 \text{ torr}} = \frac{-48,370.199 \text{ J/mol}}{8.314 \text{ J/mol}\cdot\text{K}}\left(\frac{1}{293 \text{ K}} - \frac{1}{356 \text{ K}}\right) = -3.5139112$$

$$\frac{P_2}{13 \text{ torr}} = 0.0297802$$

$$P_2 = (0.0297802)(13 \text{ torr}) = 0.3871426 \text{ torr}$$

$$P_2 \text{ (atm)} = (0.3871426 \text{ torr})\left(\frac{1 \text{ atm}}{760 \text{ torr}}\right) = 5.09398\text{x}10^{-4} \text{ atm}$$

At this pressure and a temperature of 20.°C, use the ideal gas equation to calculate the concentration of amphetamine in the air. Use a volume of 1 m³. Moles from the ideal gas equation times the molar mass gives the mass in a cubic meter.

$$\text{Moles} = n = \frac{PV}{RT} = \frac{(5.09398\text{x}10^{-4}\text{ atm})(1 \text{ m}^3)}{\left(0.0821\dfrac{\text{L}\bullet\text{atm}}{\text{mol}\bullet\text{K}}\right)(293 \text{ K})}\left(\frac{1 \text{ L}}{10^{-3} \text{ m}^3}\right) = 0.02117612 \text{ mol amphetamine}$$

$$\text{Mass (g)} = (0.02117612 \text{ mol})\left(\frac{135.20 \text{ g amphetamine}}{1 \text{ mol}}\right) = 2.8630114 = \mathbf{2.9 \text{ g/m}^3}$$

12.142 Plan: The equation $q = c$ x mass x ΔT is used to calculate the heat involved in changing the temperature of solid A, the temperature of liquid A after it melts, and the temperature of vapor A after the liquid boils; the heat of fusion is used to calculate the heat involved in the phase change of solid A to liquid A and the heat of vaporization is used to calculate the heat involved in the phase change of liquid A to A as a vapor.
Solution:

a) To heat the substance to the melting point the substance must be warmed from –40.°C to –20.°C:
$q = c$ x mass x $\Delta T = (25 \text{ g})(1.0 \text{ J/g}°\text{C})[-20. - (-40.)]°\text{C} = 500 \text{ J}$

$$\text{Time (min) required} = (500 \text{ J})\left(\frac{1 \text{ min}}{450 \text{ J}}\right) = 1.1111 = \mathbf{1.1 \text{ min}}$$

b) Phase change of the substance from solid at –20.°C to liquid at –20.°C:

$$q = n\left(\Delta H^\circ_{fus}\right) = (25 \text{ g})\left(\frac{180. \text{ J}}{1 \text{ g}}\right) = 4500 \text{ J}$$

Time (min) required $= (4500 \text{ J})\left(\dfrac{1 \text{ min}}{450 \text{ J}}\right) = \mathbf{10.\ min}$

c) To heat the substance to the melting point the substance must be warmed from $-20.°C$ to $85°C$:
$q = c \text{ x mass x } \Delta T = (25 \text{ g})(2.5 \text{ J/g°C})[85 - (-20.)]°C = 6562.5 \text{ J}$

Time (min) required $= (6562.5 \text{ J})\left(\dfrac{1 \text{ min}}{450 \text{ J}}\right) = 14.583 = \mathbf{15\ min}$

To boil the sample at $85°C$:
$q = n\left(\Delta H^{\circ}_{\text{vap}}\right) = (25 \text{ g})\left(\dfrac{500.\ \text{J}}{1 \text{ g}}\right) = 12{,}500 \text{ J}$

Time (min) required $= (12{,}500 \text{ J})\left(\dfrac{1 \text{ min}}{450 \text{ J}}\right) = 27.78 = \mathbf{28\ min}$

Warming the vapor from $85°C$ to $100.°C$:
$q = c \text{ x mass x } \Delta T = (25 \text{ g})(0.5 \text{ J/g°C})[100 - 85]°C = 187.5 \text{ J}$

Time (min) required $= (187.5 \text{ J})\left(\dfrac{1 \text{ min}}{450 \text{ J}}\right) = 0.417 = \mathbf{0.4\ min}$

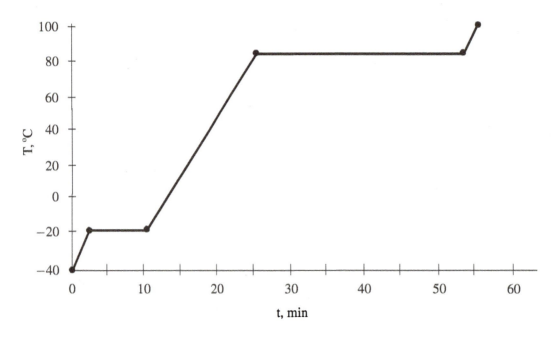

12.143   <u>Plan:</u> Balanced chemical equations are necessary. See the section on ceramic materials for the reactions.
These equations may be combined to produce an overall equation with an overall yield. Find the limiting reactant
which is then used to calculate the amount of boron nitride produced. The ideal gas law is used to calculate the
moles of NH_3.
<u>Solution:</u>

Step 1 $B(OH)_3(s) + 3NH_3(g) \rightarrow$ ~~B(NH$_2$)$_3$(s)~~ $+ 3H_2O(g)$

Step 2 ~~B(NH$_2$)$_3$(s)~~ $\rightarrow BN(s) + 2NH_3(g)$

Overall reaction: $B(OH)_3(s) + NH_3(g) \rightarrow BN(s) + 3 \ H_2O(g)$
Yields are 85.5% for step 1 and 86.8% for step 2.
Overall fractional yield $= (85.5\%/100\%)(86.8\%/100\%) = 0.74214$
Find the limiting reactant. Since the overall reaction has a 1:1 mole ratio, the reactant with the fewer moles will
be limiting.

$$\text{Moles of B(OH)}_3 = \left(1.00 \text{ t B(OH)}_3\right)\left(\frac{10^3 \text{ kg}}{1 \text{ t}}\right)\left(\frac{10^3 \text{ g}}{1 \text{ kg}}\right)\left(\frac{1 \text{ mol B(OH)}_3}{61.83 \text{ g B(OH)}_3}\right)$$

$$= 1.6173379 \times 10^4 \text{ mol B(OH)}_3$$

$$P \text{ (atm)} = \left(3.07 \times 10^3 \text{ kPa}\right)\left(\frac{1 \text{ atm}}{101.325 \text{ kPa}}\right) = 30.2985 \text{ atm}$$

$$V \text{ (L)} = \left(12.5 \text{ m}^3\right)\left(\frac{1 \text{ L}}{10^{-3} \text{ m}^3}\right) = 12,500 \text{ L}$$

$$\text{Moles of NH}_3 = PV/RT = \frac{\left(30.2985 \text{ atm}\right)\left(12,500 \text{ L}\right)}{\left(0.0821 \dfrac{\text{L} \bullet \text{atm}}{\text{mol} \bullet \text{K}}\right)\left(275 \text{ K}\right)} = 1.6774720 \times 10^4 \text{ mol NH}_3$$

B(OH)$_3$ is the limiting reactant.

$$\text{Mass (g) of BN} = \left(1.6173379 \times 10^4 \text{ mol B(OH)}_3\right)\left(\frac{1 \text{ mol BN}}{1 \text{ mol B(OH)}_3}\right)\left(\frac{24.82 \text{ g BN}}{1 \text{ mol BN}}\right)\left(\frac{74.214\%}{100\%}\right)$$

$$= 2.97912 \times 10^5 = \mathbf{2.98 \times 10^5 \text{ g BN}}$$

12.146   Plan: A body-centered cubic unit cell has eight corner atoms; 8 atoms x 1/8 atom per cell = 1 atom. In addition, the body-centered cell has an atom in the center, for a total of two atoms.
Solution:

$$\text{Mass (amu) of two Na atoms} = \left(2 \text{ Na atoms}\right)\left(\frac{22.99 \text{ amu}}{1 \text{ Na atom}}\right) = \mathbf{45.98 \text{ amu}}.$$

CHAPTER 13 THE PROPERTIES OF MIXTURES: SOLUTIONS AND COLLOIDS

FOLLOW–UP PROBLEMS

13.1A   Plan: Compare the intermolecular forces in the solutes with the intermolecular forces in the solvent. The intermolecular forces for the more soluble solute will be more similar to the intermolecular forces in the solvent than the forces in the less soluble solute.
Solution:
a) **1,4–Butanediol** is more soluble in water than butanol. Intermolecular forces in water are primarily hydrogen bonding. The intermolecular forces in both solutes also involve hydrogen bonding with the hydroxyl groups. Compared to butanol, each 1,4–butanediol molecule will form more hydrogen bonds with water because 1,4–butanediol contains two hydroxyl groups in each molecule, whereas butanol contains only one –OH group. Since 1,4–butanediol has more hydrogen bonds to water than butanol, it will be more soluble than butanol.
b) **Chloroform** is more soluble in water than carbon tetrachloride because chloroform is a polar molecule and carbon tetrachloride is nonpolar. Polar molecules are more soluble in water, a polar solvent.

13.1B   Plan: Compare the intermolecular forces in the solutes with the intermolecular forces in the solvent. The intermolecular forces for the solvent that dissolves more solute will be more similar to the intermolecular forces in the solute than the forces in the solvent that dissolves less solute.
Solution:
a) Both chloroform and chloromethane are polar molecules that experience dipole-dipole and dispersion intermolecular forces. Methanol, on the other hand, has an O–H bond and, thus, can participate in hydrogen bonding in addition to dipole-dipole and dispersion forces. **Chloroform** dissolves more chloromethane than methanol because of similar dipole-dipole forces.
b) Pentanol has a polar O–H group that can participate in hydrogen bonding, dipole-dipole and dispersion forces. However, it has a much larger non-polar section ($CH_3CH_2CH_2CH_2$-) that experiences only dispersion forces. Because the nonpolar portion of the pentanol is much larger than the polar portion, the nonpolar portion determines the overall solubility of the molecule. Thus, pentanol will be more soluble in nonpolar solvents like hexane than polar solvents like water. **Hexane** dissolves more pentanol due to dispersion forces.

13.2A   Plan: Use the relationship $\Delta H_{solution} = \Delta H_{lattice} + \Delta H_{hydration\ of\ the\ ions}$. Given $\Delta H_{solution}$ and $\Delta H_{lattice}$, $\Delta H_{hydration\ of\ the\ ions}$ can be calculated.
Solution:
The two ions in potassium nitrate are K^+ and NO_3^-. $\Delta H_{hydration\ of\ the\ ions} = \Delta H_{hydration}(K^+) + \Delta H_{hydration}(NO_3^-)$
$\Delta H_{solution} = \Delta H_{lattice} + \Delta H_{hydration\ of\ the\ ions}$
$\Delta H_{hydration\ of\ the\ ions} = \Delta H_{solution} - \Delta H_{lattice} = 34.89\ kJ/mol - 685\ kJ/mol = -650.11 = $ **–650. kJ/mol**

13.2B   Plan: Use the relationship $\Delta H_{solution} = \Delta H_{lattice} + \Delta H_{hydration\ of\ the\ ions}$. Given $\Delta H_{solution}$ and $\Delta H_{lattice}$, $\Delta H_{hydration\ of\ the\ ions}$ can be calculated. $\Delta H_{hydration\ of\ the\ ions}$ and $\Delta H_{hydration}(Na^+)$ can then be used to solve for $\Delta H_{hydration}(CN^-)$.
Solution:
The two ions in sodium cyanide are Na^+ and CN^-.
$\Delta H_{solution} = \Delta H_{lattice} + \Delta H_{hydration\ of\ the\ ions}$
$\Delta H_{hydration\ of\ the\ ions} = \Delta H_{solution} - \Delta H_{lattice} = 1.21\ kJ/mol - 766\ kJ/mol = -764.79 = -765\ kJ/mol$
$\Delta H_{hydration\ of\ the\ ions} = \Delta H_{hydration}(Na^+) + \Delta H_{hydration}(CN^-)$
$\Delta H_{hydration}(CN^-) = \Delta H_{hydration\ of\ the\ ions} - \Delta H_{hydration}(Na^+) = -765\ kJ/mol - (-410.\ kJ/mol) = $ **–355 kJ/mol**

13.3A   Plan: Solubility of a gas can be found from Henry's law: $S_{gas} = k_H \times P_{gas}$. The problem gives k_H for N_2 but not its partial pressure. To calculate the partial pressure, use the relationship from Chapter 5: $P_{gas} = X_{gas} \times P_{total}$ where X represents the mole fraction of the gas.

Solution:

To find partial pressure use the 78% N_2 given for the mole fraction:

$P_{gas} = X_{gas} \times P_{total}$

$P_{N_2} = 0.78 \times 1$ atm $= 0.78$ atm

Use Henry's law to find solubility at this partial pressure:

$S_{gas} = k_H \times P_{gas}$

$S_{N_2} = (7 \times 10^{-4}$ mol/L•atm$)(0.78$ atm$) = 5.46 \times 10^{-4}$ mol/L $= \textbf{5} \times \textbf{10}^{-4}$ **mol/L**

13.3B <u>Plan:</u> The Henry's law constant of a gas can be found from Henry's law: $S_{gas} = k_H \times P_{gas}$. The problem gives the solubility (S_{gas}) for N_2O but not its partial pressure. To calculate the partial pressure, use Dalton's law from Chapter 5: $P_{gas} = X_{gas} \times P_{total}$ where X represents the mole fraction of the gas.

<u>Solution:</u>

To find partial pressure use the 40.% N_2O given for the mole fraction:

$P_{gas} = X_{gas} \times P_{total}$

$P_{N_2O} = 0.40 \times 1.2$ atm $= 0.48$ atm

Use Henry's law to find the Henry's law constant, k_H, at 25°C:

$$k_H = \frac{S_{gas}}{P_{gas}} = \frac{1.2 \times 10^{-2} \text{ mol/L}}{0.48 \text{ atm}} = \textbf{2.5} \times \textbf{10}^{-2} \textbf{ mol/L•atm}$$

13.4A <u>Plan:</u> Molality (m) is defined as amount (mol) of solute per kg of solvent. Use the molality and the mass of solvent given to calculate the amount of glucose in moles. Then convert amount (mol) of glucose to mass (g) of glucose by multiplying by its molar mass.

<u>Solution:</u>

Mass of solvent must be converted to kg.

$$\text{Mass of solvent (kg)} = (563 \text{ g})\left(\frac{1 \text{ kg}}{10^3 \text{ g}}\right) = 0.563 \text{ kg}$$

$$\text{Mass (g) of glucose} = \left(0.563 \text{ kg solvent}\right)\left(\frac{2.40 \times 10^{-2} \text{ mol C}_6\text{H}_{12}\text{O}_6}{1 \text{ kg solvent}}\right)\left(\frac{180.16 \text{ g C}_6\text{H}_{12}\text{O}_6}{1 \text{ mol C}_6\text{H}_{12}\text{O}_6}\right)$$

$$= 2.4343 = \textbf{2.43 g C}_6\textbf{H}_{12}\textbf{O}_6$$

13.4B <u>Plan:</u> Molality (m) is defined as amount (mol) of solute per kg of solvent. Calculate the moles of solute, I_2, from the mass. Find the mass (kg) of the solvent, diethyl ether, from the given number of moles. Then divide the moles of solute by the kg of solvent to find the molality.

<u>Solution:</u>

$$\text{Amount (mol) of solute} = (15.20 \text{ g I}_2)\left(\frac{1 \text{ mol I}_2}{253.8 \text{ g I}_2}\right) = 0.0598897 \text{ mol I}_2$$

$$\text{Mass of solvent (kg)} = (1.33 \text{ mol (CH}_3\text{CH}_2)_2\text{O})\left(\frac{74.12 \text{ g (CH}_3\text{CH}_2)_2\text{O}}{1 \text{ mol (CH}_3\text{CH}_2)_2\text{O}}\right)\left(\frac{1 \text{ kg}}{1000 \text{ g}}\right) = 0.0985796 \text{ kg (CH}_3\text{CH}_2)_2\text{O}$$

$$\text{Molality } (m) = \frac{\text{mol solute}}{\text{kg solvent}} = \frac{0.0598897 \text{ mol}}{0.0985796 \text{ kg}} = .6075263 = \textbf{0.608 } \textbf{\textit{m}}$$

13.5A <u>Plan:</u> Mass percent is the mass (g) of solute per 100 g of solution. For each alcohol, divide the mass of the alcohol by the total mass. Multiply this number by 100 to obtain mass percent. To find mole fraction, first find the amount (mol) of each alcohol, then divide by the total moles.

<u>Solution:</u>
Mass percent:

$$\text{Mass \% propanol} = \frac{\text{mass of propanol}}{\text{mass of propanol} + \text{mass of ethanol}} \times 100\% = \frac{35.0 \text{ g}}{(35.0 + 150.)\text{g}} \times 100\%$$

$$= 18.9189 = \textbf{18.9\% propanol}$$

$$\text{Mass \% of ethanol} = \frac{\text{mass of ethanol}}{\text{mass of propanol} + \text{mass of ethanol}} \times 100\% = \frac{150. \text{ g}}{(35.0 + 150.)\text{g}} \times 100\%$$

$$= 81.08108 = \textbf{81.1\% ethanol}$$

Mole fraction:

$$\text{Moles of propanol} = \left(35.0 \text{ g C}_3\text{H}_7\text{OH}\right)\left(\frac{1 \text{ mol C}_3\text{H}_7\text{OH}}{60.09 \text{ g C}_3\text{H}_7\text{OH}}\right) = 0.5824596 \text{ mol propanol}$$

$$\text{Moles of ethanol} = \left(150. \text{ g C}_2\text{H}_5\text{OH}\right)\left(\frac{1 \text{ mol C}_2\text{H}_5\text{OH}}{46.07 \text{ g C}_2\text{H}_5\text{OH}}\right) = 3.2559149 \text{ mol ethanol}$$

$$X_{\text{propanol}} = \frac{0.5824596 \text{ mol propanol}}{0.5824596 \text{ mol propanol} + 3.2559149 \text{ mol ethanol}} = 0.151746 = \textbf{0.152}$$

$$X_{\text{ethanol}} = \frac{3.2559149 \text{ mol ethanol}}{0.5824596 \text{ mol propanol} + 3.2559149 \text{ mol ethanol}} = 0.84825 = \textbf{0.848}$$

13.5B   <u>Plan:</u> Mass percent is the mass (g) of solute per 100 g of solution. For each component of the mixture, divide the mass of the component by the total mass. Multiply this number by 100 to obtain mass percent. To find mole percent, first find the amount (mol) of each component, then divide by the total moles and multiply by 100%.
<u>Solution:</u>
Mass percent:

$$\text{Mass \% ethanol} = \frac{\text{mass of ethanol}}{\text{mass of ethanol} + \text{mass of iso-octane} + \text{mass of heptane}} \times 100\% =$$

$$\frac{1.87 \text{ g}}{1.87 \text{ g} + 27.4 \text{ g} + 4.10 \text{ g}} \times 100\% = 5.6038 = \textbf{5.60\% ethanol}$$

$$\text{Mass \% iso-octane} = \frac{\text{mass of iso-octane}}{\text{mass of ethanol} + \text{mass of iso-octane} + \text{mass of heptane}} \times 100\% =$$

$$\frac{27.4 \text{ g}}{1.87 \text{ g} + 27.4 \text{ g} + 4.10 \text{ g}} \times 100\% = 82.1097 = \textbf{82.1\% iso-octane}$$

$$\text{Mass \% heptane} = \frac{\text{mass of heptane}}{\text{mass of ethanol} + \text{mass of iso-octane} + \text{mass of heptane}} \times 100\% =$$

$$\frac{4.10 \text{ g}}{1.87 \text{ g} + 27.4 \text{ g} + 4.10 \text{ g}} \times 100\% = 12.2865 = \textbf{12.3\% heptane}$$

Mole percent:

$$\text{Moles of ethanol} = (1.87 \text{ g ethanol})\left(\frac{1 \text{ mole ethanol}}{46.07 \text{ g ethanol}}\right) = 0.0406 \text{ mol ethanol}$$

$$\text{Moles of iso-octane} = (27.4 \text{ g iso-octane})\left(\frac{1 \text{ mole iso-octane}}{114.22 \text{ g iso-octane}}\right) = 0.240 \text{ mol iso-octane}$$

$$\text{Moles of heptane} = (4.10 \text{ g heptane})\left(\frac{1 \text{ mole heptane}}{100.20 \text{ g heptane}}\right) = 0.0409 \text{ mol heptane}$$

$$\text{Mole percent ethanol} = \frac{\text{moles of ethanol}}{\text{moles of ethanol} + \text{moles of iso-octane} + \text{moles of heptane}} \text{ x } 100\% =$$

$$\frac{0.0406 \text{ mol}}{0.0406 \text{ mol} + 0.240 \text{ mol} + 0.0409 \text{ mol}} \text{ x } 100\% = 12.6283 = \textbf{12.6\% ethanol}$$

$$\text{Mole percent iso-octane} = \frac{\text{moles of iso-octane}}{\text{moles of ethanol} + \text{moles of iso-octane} + \text{moles of heptane}} \text{ x } 100\% =$$

$$\frac{0.240 \text{ mol}}{0.0406 \text{ mol} + 0.240 \text{ mol} + 0.0409 \text{ mol}} \text{ x } 100\% = 74.6501 = \textbf{74.6\% iso-octane}$$

$$\text{Mole percent heptane} = \frac{\text{moles of heptane}}{\text{moles of ethanol} + \text{moles of iso-octane} + \text{moles of heptane}} \text{ x } 100\% =$$

$$\frac{0.0409 \text{ mol}}{0.0406 \text{ mol} + 0.240 \text{ mol} + 0.0409 \text{ mol}} \text{ x } 100\% = 12.7216 = \textbf{12.7\% heptane}$$

(Slight differences from 100% are due to rounding.)

13.6A   Plan: To find the mass percent, molality and mole fraction of HCl, the following is needed:
  1) Moles of HCl in 1 L solution (from molarity)
  2) Mass of HCl in 1L solution (from molarity times molar mass of HCl)
  3) Mass of 1L solution (from volume times density)
  4) Mass of solvent in 1L solution (by subtracting mass of solute from mass of solution)
  5) Moles of solvent (by dividing the mass of solvent by molar mass of water)
Mass percent is calculated by dividing mass of HCl by mass of solution and multiplying by 100.
Molality is calculated by dividing moles of HCl by mass of solvent in kg.
Mole fraction is calculated by dividing mol of HCl by the sum of mol of HCl plus mol of solvent.
Solution:
Assume the volume is exactly 1 L.

1) Mole of HCl in 1 L solution $= (1.0 \text{ L})\left(\frac{11.8 \text{ mol HCl}}{1.0 \text{ L}}\right) = 11.8 \text{ mol HCl}$

2) Mass (g) of HCl in 1 L solution $= (11.8 \text{ mol HCl})\left(\frac{36.46 \text{ g HCl}}{1 \text{ mol HCl}}\right) = 430.228 \text{ g HCl}$

3) Mass (g) of 1 L solution $= (1.0 \text{ L})\left(\frac{1 \text{ mL}}{10^{-3} \text{ L}}\right)\left(\frac{1.190 \text{ g}}{1 \text{ mL}}\right) = 1190. \text{ g solution}$

4) Mass (g) of solvent in 1 L solution $= 1190. \text{ g} - 430.228 \text{ g} = 759.772 \text{ g solvent (H}_2\text{O)}$

Mass (kg) of solvent in 1 L solution $= (759.772 \text{ g})\left(\frac{1 \text{ kg}}{10^3 \text{ g}}\right) = 0.759772 \text{ kg solvent}$

5) Moles of solvent in 1 L solution $= (759.772 \text{ g H}_2\text{O})\left(\frac{1 \text{ mol H}_2\text{O}}{18.02 \text{ g H}_2\text{O}}\right) = 42.1627 \text{ mol solvent}$

Mass percent of HCl $= \dfrac{\text{mass HCl}}{\text{mass solution}}(100) = \dfrac{430.228 \text{ g HCl}}{1190 \text{ g solution}}(100) = 36.1536 = \textbf{36.2\%}$

Molality of HCl $= \dfrac{\text{mole HCl}}{\text{kg solvent}} = \dfrac{11.8 \text{ mol HCl}}{0.759772 \text{ kg}} = 15.530975 = \textbf{15.5 } \boldsymbol{m}$

Mole fraction of HCl $= \dfrac{\text{mol HCl}}{\text{mol HCl} + \text{mol H}_2\text{O}} = \dfrac{11.8 \text{ mol}}{11.8 \text{ mol} + 42.1627 \text{ mol}} = 0.21866956 = \textbf{0.219}$

13.6B Plan: To find the mass percent, molarity and mole percent of $CaBr_2$, the following is needed:
1) Mass of $CaBr_2$ and mass of H_2O in the solution
2) Total solution volume
3) Moles of H_2O in the solution

Mass percent is calculated by dividing the mass of $CaBr_2$ by the total mass of solution and multiplying by 100.
Molarity is calculated by dividing the moles of $CaBr_2$ by the volume of the solution (L).
Mole percent is calculated by dividing moles of $CaBr_2$ by the total moles in the solution and multiplying by 100.
Solution:

1) Mass of $CaBr_2$ = (5.44 mol $CaBr_2$)$\left(\dfrac{199.88 \text{ g } CaBr_2}{1 \text{ mol } CaBr_2}\right)$ = 1087.347 g $CaBr_2$

Mass of H_2O = 1 kg$\left(\dfrac{1000 \text{ g}}{1 \text{ kg}}\right)$ = 1000 g H_2O

Mass % $CaBr_2$ = $\dfrac{\text{mass of solute}}{\text{total mass of solution}}$ = $\dfrac{1087.347 \text{ g } CaBr_2}{1087.347 \text{ g } CaBr_2 + 1000 \text{ g } H_2O}$ x 100% = 52.09230

$$= \textbf{52.1 mass \% } CaBr_2$$

2) Volume(L) of solution = $(1087.347 \text{ g } CaBr_2 + 1000 \text{ g } H_2O)\left(\dfrac{1 \text{ mL solution}}{1.70 \text{ g solution}}\right)\left(\dfrac{1 \text{ L}}{1000 \text{ mL}}\right)$

$$= 1.227851 \text{ L solution}$$

Molarity = $\dfrac{\text{moles of solute}}{\text{L of solution}}$ = $\dfrac{5.44 \text{ mol } CaBr_2}{1.227851 \text{ L solution}}$ = 4.43050 = **4.43 *M***

3) Amount (mol) of H_2O = 1000 g $H_2O$$\left(\dfrac{1 \text{ mol } H_2O}{18.02 \text{ g } H_2O}\right)$ = 55.493896 mol H_2O

Mole percent = $\dfrac{\text{moles solute}}{\text{moles solute} + \text{moles solvent}}$ x 100% = $\left(\dfrac{5.44 \text{ mol}}{5.44 \text{ mol} + 55.493896 \text{ mol}}\right)$ x 100%

$$= 8.927707 = \textbf{8.93\%}$$

13.7A Plan: Raoult's law states that the vapor pressure of a solvent is proportional to the mole fraction of the solvent: $P_{solvent} = X_{solvent} \times P^{\circ}_{solvent}$. To calculate the drop in vapor pressure, a similar relationship is used with the mole fraction of the solute substituted for that of the solvent.
Solution:
Mole fraction of aspirin in methanol:

$$\dfrac{(2.00 \text{ g})\left(\dfrac{1 \text{ mol aspirin}}{180.15 \text{ g aspirin}}\right)}{(2.00 \text{ g})\left(\dfrac{1 \text{ mol aspirin}}{180.15 \text{ g aspirin}}\right) + (50.0 \text{ g})\left(\dfrac{1 \text{ mol methanol}}{32.04 \text{ g methanol}}\right)}$$

$X_{aspirin} = 7.06381 \times 10^{-3}$
$\Delta P = X_{aspirin} P^{\circ}_{methanol} = (7.06381 \times 10^{-3})(101 \text{ torr}) = 0.71344 = \textbf{0.713 torr}$

13.7B Plan: The drop in vapor pressure of a solvent is calculated by the following equation derived from Raoult's law:
$\Delta P = X_{solvent} \times P^{\circ}_{solvent}$.
Solution:

Amount (mol) menthol = (6.49 g menthol)$\left(\dfrac{1 \text{ mol menthol}}{156.26 \text{ g menthol}}\right)$ = 0.0415333 mol menthol

Amount (mol) ethanol = (25.0 g ethanol)$\left(\dfrac{1 \text{ mol ethanol}}{46.07 \text{ g ethanol}}\right)$ = 0.542652 mol ethanol

$$\Delta P = \left(\frac{0.542652 \text{ mol}}{0.0415333 \text{ mol} + 0.542652 \text{ mol}}\right)(6.87 \text{ kPa})\left(\frac{1 \text{ atm}}{101.325 \text{ kPa}}\right)\left(\frac{760 \text{ torr}}{1 \text{ atm}}\right) = 47.8658 = \mathbf{47.9 \text{ torr}}$$

13.8A Plan: Find the molality of the solution by dividing the moles of P_4 by the mass of the CS_2 in kg. The change in freezing point is calculated from $\Delta T_f = iK_f m$, where K_f is $3.83°C/m$ when CS_2 is the solvent, i is the van't Hoff factor, and m is the molality of particles in solution. Since P_4 is a covalent compound and does not ionize in water, $i = 1$. Once ΔT_f is calculated, the freezing point is determined by subtracting it from the freezing point of pure CS_2 ($-111.5°C$). The boiling point of a solution is increased relative to the pure solvent by the relationship $\Delta T_b = iK_b m$ where K_b is $2.34°C/m$ when CS_2 is the solvent, i is the van't Hoff factor, and m is the molality of particles in solution. P_4 is a nonelectrolyte (it is a molecular compound) so $i = 1$. Once ΔT_b is calculated, the boiling point is determined by adding it to the boiling point of pure CS_2 ($46.2°C$).

Solution:

Molality of P_4 solution:

$$\text{Moles of } P_4 = (8.44 \text{ g } P_4)\left(\frac{1 \text{ mol } P_4}{123.88 \text{ g } P_4}\right) = 0.0681 \text{ mol}$$

$$\text{Molality of } P_4 = \frac{\text{moles solute}}{\text{kg solvent}} = \left(\frac{0.0681 \text{ mol } P_4}{60.0 \text{ g } CS_2}\right)\left(\frac{1000 \text{ g}}{1 \text{ kg}}\right) = 1.14 \ m \ P_4$$

Freezing point:
$\Delta T_f = iK_f m = (1)(3.83°C/m)(1.14 \ m) = 4.37°C$
The freezing point is: $-111.5°C - 4.37°C = \mathbf{-115.9°C}$.
Boiling point:
$\Delta T_b = iK_b m = (1)(2.34°C/m)(1.14 \ m) = 2.67°C$
The boiling point is $46.2°C + 2.67°C = \mathbf{48.9°C}$.

13.8B Plan: The question asks for the concentration of ethylene glycol that would prevent freezing at $0.00°F$. First, calculate the change in freezing point of water, which will be $0.00°F$ subtracted from the freezing point of pure water, $32°F$. Then convert this temperature to Celsius. Use this value for ΔT in $\Delta T_f = 1.86°C/m \times$ molality of solution and solve for molality.

Solution:

Temperature conversion:

$$T(°C) = \left(T(°F) - 32°F\right)\left(\frac{5°C}{9°F}\right) = \left(32°F - 0.00°F\right)\left(\frac{5°C}{9°F}\right) = 17.7778°C$$

$$\text{Molality of } C_2H_6O_2 = \frac{\Delta T_f}{K_f} = \frac{17.7778°C}{1.86°C/m} = 9.557956 = \mathbf{9.56 \ m}$$

The minimum concentration of ethylene glycol would have to be $9.56 \ m$ in order to prevent the water from freezing at $0.00°F$.

$$\text{Mass of } C_2H_6O_2 = 4.00 \text{ kg } H_2O\left(\frac{9.557956 \text{ mol } C_2H_6O_2}{1 \text{ kg } H_2O}\right)\left(\frac{62.07 \text{ g } C_2H_6O_2}{1 \text{ mol } C_2H_6O_2}\right) = 2373.047 = \mathbf{2370 \text{ g } C_2H_6O_2}$$

13.9A Plan: Use the osmotic pressure, the temperature, and the gas constant to calculate the molarity of the solution with the equation $\Pi = MRT$. Multiply the molarity of the solution by the volume of the sample to calculate the number of moles in the sample. The molar mass is calculated by dividing the mass of the sample by the number of moles in the sample.

Solution:

$\Pi = MRT$, so $M = \dfrac{\Pi}{RT}$

$$M = \frac{8.98 \text{ torr x } \dfrac{1 \text{ atm}}{760 \text{ torr}}}{\left(0.0821 \dfrac{\text{atm} \cdot \text{L}}{\text{mol} \cdot \text{K}}\right)\left((27°C + 273.15)\text{K}\right)} = 4.79492 \times 10^{-4} \ M$$

Amount (mol) protein = $(4.79492 \times 10^{-4}$ mol/L$)$ $(12.0$ mL$)$ $(1$ L$/1000$ mL$) = 5.75390 \times 10^{-6}$ mol

$$\mathcal{M} = \frac{0.200 \text{ g}}{5.75390 \times 10^{-6} \text{ mol}} = 3.47590 \times 10^{4} \text{ g/mol} = \mathbf{3.48 \times 10^{4} \text{ g/mol}}$$

13.9B   Plan: Look up the normal freezing point for benzene in the table within this section of the textbook and calculate the freezing point depression. Find the molality using $\Delta T_{f} = 1.86°C/m \times$ molality of solution. Use the given mass of the benzene to calculate the moles of naphthalene. Divide the given mass of the naphthalene by the moles of naphthalene to calculate the molar mass.
Solution:

Molality of naphthalene $= \dfrac{\Delta T_{f}}{K_{f}} = \dfrac{5.56°C \text{ - } 4.20°C}{4.90 \text{ °C}/m} = 0.277551 \ m$ naphthalene

Amount (mol) naphthalene $= 200. \text{ g benzene} \left(\dfrac{1 \text{ kg}}{1000 \text{ g}} \right) \left(\dfrac{0.277551 \text{ mol naphthalene}}{1 \text{ kg benzene}} \right) = .0555102$ mol

Molar mass naphthalene $= \dfrac{7.01 \text{ g}}{0.0555102 \text{ mol}} = 126.283 = \mathbf{126 \text{ g/mol}}$

13.10A   Plan: Write the formula of potassium phosphate, determine if the compound is soluble and, if soluble, how many cations and anions result when one unit of the compound is placed in water. Calculate the molality of the solution. The boiling point of a solution is increased relative to the pure solvent by the relationship $\Delta T_{b} = iK_{b}m$ where K_{b} is $0.512°C/m$ for water, i is the van't Hoff factor (equal to the number of particles or ions produced when one unit of the compound is dissolved in the solvent), and m is the molality of particles in solution. Once ΔT_{b} is calculated, the boiling point is determined by adding it to the boiling point of pure H_2O (100.00°C).
Solution:
a) The formula for potassium phosphate is K_3PO_4. When it is placed in water, 3 K^+ ions and 1 PO_4^{3-} ion are formed. Scene B is the only scene that shows separate ions in a $3K^+/1PO_4^{3-}$ ratio.
b) Molality of K_3PO_4 solution:

Moles of $K_3PO_4 = (31.2 \text{ g } K_3PO_4) \left(\dfrac{1 \text{ mol } K_3PO_4}{212.27 \text{ g } K_3PO_4} \right) = 0.146983 \text{ mol } K_3PO_4$

Molality of $K_3PO_4 = \dfrac{\text{moles solute}}{\text{kg solvent}} = \left(\dfrac{0.146983 \text{ mol } K_3PO_4}{85.0 \text{ g } H_2O} \right) \left(\dfrac{1000 \text{ g}}{1 \text{ kg}} \right) = 1.72921 \ m \ K_3PO_4$

Because 4 ions total are formed when each unit of K_3PO_4 is placed in water, $i = 4$.
$\Delta T_{b} = iK_{b}m = (4)(0.512 \text{ °C}/m)(1.72921 \ m) = 3.5414°C$
The boiling point is $100.00°C + 3.5414°C = 103.5414 = \mathbf{103.54°C}$.

13.10B   Plan: Calculate the molarity of the solution, then calculate the osmotic pressure of the magnesium chloride solution. Do not forget that $MgCl_2$ is a strong electrolyte, and ionizes to yield three ions per formula unit. This will result in a pressure three times as great as a nonelectrolyte (glucose) solution of equal concentration.
Solution:

a) Volume (L) of solution $= (100. + 0.952 \text{ g solution}) \left(\dfrac{1 \text{ mL}}{1.006 \text{ g}} \right) \left(\dfrac{10^{-3} \text{ L}}{1 \text{ mL}} \right) = 0.1003499$ L

Moles of $MgCl_2 = (0.952 \text{ g } MgCl_2) \left(\dfrac{1 \text{ mol } MgCl_2}{95.21 \text{ g } MgCl_2} \right) = 9.9989 \times 10^{-3}$ mol $MgCl_2$

Molarity of $MgCl_2 = \dfrac{\text{mol } MgCl_2}{\text{volume solution}} = \dfrac{0.0099989 \text{ mol}}{0.1003499 \text{ L}} = 0.099640 \ M \ MgCl_2$

$\Pi = iMRT = 3 \left(\dfrac{0.099640 \text{ mol}}{\text{L}} \right) \left(0.0821 \dfrac{\text{L} \bullet \text{atm}}{\text{mol} \bullet \text{K}} \right) ((273.2 + 20.0) \text{ K}) = 7.1955 = \mathbf{7.20 \text{ atm}}$

b) The magnesium chloride solution has a higher osmotic pressure, thus, solvent will diffuse from the glucose solution into the magnesium chloride solution. The glucose side will lower and the magnesium chloride side will rise. **Scene C** represents this situation.

CHEMICAL CONNECTIONS BOXED READING PROBLEMS

B13.1 a) The colloidal particles in water generally have negatively charged surfaces and so repel each other, slowing the settling process. Cake alum, $Al_2(SO_4)_3$, is added to coagulate the colloids. The Al^{3+} ions neutralize the negative surface charges and allow the particles to aggregate and settle.
b) Water that contains large amounts of divalent cations (such as Ca^{2+} and Mg^{2+}) is called hard water. During cleaning, these ions combine with the fatty-acid anions in soaps to produce insoluble deposits.
c) In reverse osmosis, a pressure greater than the osmotic pressure is applied to the solution, forcing the water back through the membrane and leaving the ions behind.
d) Chlorine may give the water an unpleasant odor, and can form carcinogenic chlorinated compounds.
e) The high concentration of NaCl displaces the divalent and polyvalent ions from the ion-exchange resin.

B13.2 Plan: Osmotic pressure is calculated from the molarity of particles, the gas constant, and temperature. Convert the mass of sucrose to moles using the molar mass, and then to molarity. Sucrose is a nonelectrolyte so $i = 1$.
Solution:
$T = 273 + 20.°C = 293$ K

$$\text{Moles of sucrose} = (3.55 \text{ g sucrose})\left(\frac{1 \text{ mol sucrose}}{342.30 \text{ g sucrose}}\right) = 0.01037102 \text{ mol sucrose}$$

$$\text{Molarity} = \frac{\text{moles of sucrose}}{\text{volume of solution}} = \frac{0.01037102 \text{ mol}}{1.0 \text{ L}} = 1.037102 \times 10^{-2} \text{ } M \text{ sucrose}$$

$\Pi = iMRT = (1)(1.037102 \times 10^{-2} \text{ mol/L})(0.0821 \text{ L•atm/mol•K})(293 \text{ K}) = 0.2494780 = \textbf{0.249 atm}$
A pressure greater than 0.249 atm must be applied to obtain pure water from a 3.55 g/L solution.

END–OF–CHAPTER PROBLEMS

13.2 When a salt such as NaCl dissolves, ion-dipole forces cause the ions to separate, and many water molecules cluster around each of them in hydration shells. Ion-dipole forces hold the first shell. Additional shells are held by hydrogen bonding to inner shells.

13.4 **Sodium stearate** would be a more effective soap because the hydrocarbon chain in the stearate ion is longer than the chain in the acetate ion. A soap forms suspended particles called micelles with the polar end of the soap interacting with the water solvent molecules and the nonpolar ends forming a nonpolar environment inside the micelle. Oils dissolve in the nonpolar portion of the micelle. Thus, a better solvent for the oils in dirt is a more nonpolar substance. The long hydrocarbon chain in the stearate ion is better at dissolving oils in the micelle than the shorter hydrocarbon chain in the acetate ion.

13.7 Plan: A more concentrated solution will have more solute dissolved in the solvent. Determine the types of intermolecular forces in the solute and solvents. A solute tends to be more soluble in a solvent whose intermolecular forces are similar to its own.
Solution:
Potassium nitrate, KNO_3, is an ionic compound and can form ion-dipole forces with a polar solvent like water, thus dissolving in the water. Potassium nitrate is not soluble in the nonpolar solvent CCl_4. Because potassium nitrate dissolves to a greater extent in water, a) **KNO_3 in H_2O** will result in the more concentrated solution.

13.9 Plan: To identify the strongest type of intermolecular force, check the formula of the solute and identify the forces that could occur. Then look at the formula for the solvent and determine if the forces identified for the solute would occur with the solvent. Ionic forces are present in ionic compounds; dipole-dipole forces are present in polar substances, while nonpolar substances exhibit only dispersion forces. The strongest force is ion-dipole

followed by dipole-dipole (including H bonds). Next in strength is ion–induced dipole force and then dipole–induced dipole force. The weakest intermolecular interactions are dispersion forces.
Solution:
a) **Ion-dipole forces** are the strongest intermolecular forces in the solution of the ionic substance cesium chloride in polar water.
b) **Hydrogen bonding** (type of dipole-dipole force) is the strongest intermolecular force in the solution of polar propanone (or acetone) in polar water.
c) **Dipole–induced dipole forces** are the strongest forces between the polar methanol and nonpolar carbon tetrachloride.

13.11   Plan: To identify the strongest type of intermolecular force, check the formula of the solute and identify the forces that could occur. Then look at the formula for the solvent and determine if the forces identified for the solute would occur with the solvent. Ionic forces are present in ionic compounds; dipole-dipole forces are present in polar substances, while nonpolar substances exhibit only dispersion forces. The strongest force is ion-dipole followed by dipole-dipole (including H bonds). Next in strength is ion–induced dipole force and then dipole–induced dipole force. The weakest intermolecular interactions are dispersion forces.
Solution:
a) **Hydrogen bonding** occurs between the H atom on water and the lone electron pair on the O atom in dimethyl ether (CH_3OCH_3). However, none of the hydrogen atoms on dimethyl ether participates in hydrogen bonding because the C−H bond does not have sufficient polarity.
b) The dipole in water induces a dipole on the Ne(g) atom, so **dipole–induced dipole** interactions are the strongest intermolecular forces in this solution.
c) Nitrogen gas and butane are both nonpolar substances, so **dispersion forces** are the principal attractive forces.

13.13   Plan: $CH_3CH_2OCH_2CH_3$ is polar with dipole-dipole interactions as the dominant intermolecular forces. Examine the solutes to determine which has intermolecular forces more similar to those in diethyl ether. This solute is the one that would be more soluble.
Solution:
a) **HCl** would be more soluble since it is a covalent compound with dipole-dipole forces, whereas NaCl is an ionic solid. Dipole-dipole forces between HCl and diethyl ether are more similar to the dipole forces in diethyl ether than the ion-dipole forces between NaCl and diethyl ether.
b) **CH_3CHO** (acetaldehyde) would be more soluble. The dominant interactions in H_2O are hydrogen bonding, a stronger type of dipole-dipole force. The dominant interactions in CH_3CHO are dipole-dipole. The solute-solvent interactions between CH_3CHO and diethyl ether are more similar to the solvent intermolecular forces than the forces between H_2O and diethyl ether.
c) **CH_3CH_2MgBr** would be more soluble. CH_3CH_2MgBr has a polar end (–MgBr) and a nonpolar end (CH_3CH_2–), whereas $MgBr_2$ is an ionic compound. The nonpolar end of CH_3CH_2MgBr and diethyl ether would interact with dispersion forces, while the polar end of CH_3CH_2MgBr and the dipole in diethyl ether would interact with dipole-dipole forces.

13.16   Plan: Determine the types of intermolecular forces present in the two compounds and in water and hexane. Substances with similar types of forces tend to be soluble while substances with different type of forces tend to be insoluble.
Solution:
Gluconic acid is a very polar molecule because it has –OH groups attached to every carbon. The abundance of –OH bonds allows gluconic acid to participate in extensive H bonding with water, hence its great solubility in water. On the other hand, caproic acid has a five carbon, nonpolar, hydrophobic ("water hating") tail that does not easily dissolve in water. The dispersion forces in the nonpolar tail are more similar to the dispersion forces in hexane, hence its greater solubility in hexane.

13.18   The nitrogen bases hydrogen bond to their complimentary bases. The flat, N-containing bases stack above each other, which allow extensive interaction through dispersion forces. The exterior negatively charged sugar-phosphate chains form ion-dipole and hydrogen bonds to the aqueous surroundings, but this is of minor importance to the structure.

13.21 Dispersion forces are present between the nonpolar portions of the molecules within the bilayer. Polar groups are present to hydrogen bond or to form ion-dipole interactions with the aqueous surroundings.

13.25 For a general solvent, the energy changes needed to separate solvent into particles ($\Delta H_{solvent}$), and that needed to mix the solvent and solute particles (ΔH_{mix}) would be combined to obtain $\Delta H_{solution}$.

13.29 This compound would be very **soluble** in water. A large exothermic value in $\Delta H_{solution}$ (enthalpy of solution) means that the solution has a much lower energy state than the isolated solute and solvent particles, so the system tends to the formation of the solution. Entropy that accompanies dissolution always favors solution formation. Entropy becomes important when explaining why solids with endothermic $\Delta H_{solution}$ values (and higher energy states) are still soluble in water.

13.30 Plan: $\Delta H_{solution} = \Delta H_{lattice} + \Delta H_{hydration}$. Lattice energy values are always positive as energy is required to separate the ions from each other. Hydration energy values are always negative as energy is released when intermolecular forces between ions and water form. Since the heat of solution for KCl is endothermic, the lattice energy must be greater than the hydration energy for an overall input of energy.
Solution:

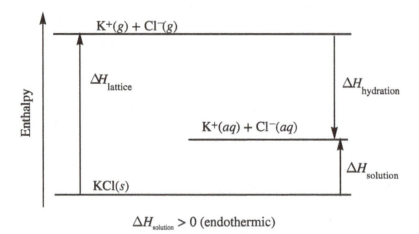

$\Delta H_{solution} > 0$ (endothermic)

13.32 Plan: Charge density is the ratio of an ion's charge (regardless of sign) to its volume. An ion's volume is related to its radius. For ions whose charges have the same sign (+ or –), ion size decreases as a group in the periodic table is ascended and as you proceed from left to right in the periodic table. Charge density increases with increasing charge and increases with decreasing size.
Solution:
a) Both ions have a +1 charge, but the volume of **Na^+** is smaller, so it has the greater charge density.
b) **Sr^{2+}** has a greater ionic charge and a smaller size (because it has a greater Z_{eff}), so it has the greater charge density.
c) **Na^+** has a smaller ion volume than Cl^-, so it has the greater charge density.
d) **O^{2-}** has a greater ionic charge and similar ion volume, so it has the greater charge density.
e) **OH^-** has a smaller ion volume than SH^- (O is smaller than S), so it has the greater charge density.
f) **Mg^{2+}** has the higher charge density because it has a smaller ion volume.
g) **Mg^{2+}** has the higher charge density because it has both a smaller ion volume and greater charge.
h) **CO_3^{2-}** has the higher charge density because it has both a smaller ion volume and greater charge.

13.34 Plan: The ion with the greater charge density will have the larger $\Delta H_{hydration}$.
Solution:
a) **Na^+** would have a larger $\Delta H_{hydration}$ than Cs^+ since its charge density is greater than that of Cs^+.
b) **Sr^{2+}** would have a larger $\Delta H_{hydration}$ than Rb^+.
c) **Na^+** would have a larger $\Delta H_{hydration}$ than Cl^-.
d) **O^{2-}** would have a larger $\Delta H_{hydration}$ than F^-.

e) **OH⁻** would have a larger $\Delta H_{\text{hydration}}$ than SH⁻.
f) **Mg²⁺** would have a larger $\Delta H_{\text{hydration}}$ than Ba²⁺.
g) **Mg²⁺** would have a larger $\Delta H_{\text{hydration}}$ than Na⁺.
h) **CO₃²⁻** would have a larger $\Delta H_{\text{hydration}}$ than NO₃⁻.

13.36 Plan: Use the relationship $\Delta H_{\text{solution}} = \Delta H_{\text{lattice}} + \Delta H_{\text{hydration}}$. Given $\Delta H_{\text{solution}}$ and $\Delta H_{\text{lattice}}$, $\Delta H_{\text{hydration}}$ can be calculated. $\Delta H_{\text{hydration}}$ increases with increasing charge density, and charge density increases with increasing charge and decreasing size.
Solution:
a) The two ions in potassium bromate are K^+ and BrO_3^-.
$\Delta H_{\text{solution}} = \Delta H_{\text{lattice}} + \Delta H_{\text{hydration}}$
$\Delta H_{\text{hydration}} = \Delta H_{\text{solution}} - \Delta H_{\text{lattice}} = 41.1 \text{ kJ/mol} - 745 \text{ kJ/mol} = -703.9 = \textbf{–704 kJ/mol}$
b) **K⁺** ion contributes more to the heat of hydration because it has a smaller size and, therefore, a greater charge density.

13.38 Plan: Entropy increases as the possible states for a system increase, which is related to the freedom of motion of its particles and the number of ways they can be arranged.
Solution:
a) Entropy **increases** as the gasoline is burned. Gaseous products at a higher temperature form.
b) Entropy **decreases** as the gold is separated from the ore. Pure gold has only the arrangement of gold atoms next to gold atoms, while the ore mixture has a greater number of possible arrangements among the components of the mixture.
c) Entropy **increases** as a solute dissolves in the solvent.

13.41 Add a pinch of the solid solute to each solution. A saturated solution contains the maximum amount of dissolved solute at a particular temperature. When additional solute is added to this solution, it will remain undissolved. An unsaturated solution contains less than the maximum amount of dissolved solute and so will dissolve added solute. A supersaturated solution is unstable and addition of a "seed" crystal of solute causes the excess solute to crystallize immediately, leaving behind a saturated solution.

13.44 Plan: The solubility of a gas in water decreases with increasing temperature and increases with increasing pressure.
Solution:
a) Increasing pressure for a gas **increases** the solubility of the gas according to Henry's law.
b) Increasing the volume of a gas causes a decrease in its pressure (Boyle's law), which **decreases** the solubility of the gas.

13.46 Plan: Solubility for a gas is calculated from Henry's law: $S_{\text{gas}} = k_H \times P_{\text{gas}}$. We know k_H and P_{gas}, so S_{gas} can be calculated with units of mol/L. To calculate the mass of oxygen gas, convert moles of O_2 to mass of O_2 using the molar mass.
Solution:
a) $S_{\text{gas}} = k_H \times P_{\text{gas}}$
$S_{\text{gas}} = \left(1.28\text{x}10^{-3}\dfrac{\text{mol}}{\text{L} \bullet \text{atm}}\right)(1.00 \text{ atm}) = 1.28\text{x}10^{-3} \text{ mol/L}$

Mass (g) of $O_2 = \left(\dfrac{1.28\text{x}10^{-3} \text{ mol } O_2}{\text{L}}\right)\left(\dfrac{32.00 \text{ g } O_2}{1 \text{ mol } O_2}\right)(2.50 \text{ L}) = 0.1024 = \textbf{0.102 g } O_2$

b) The amount of gas that will dissolve in a given volume decreases proportionately with the partial pressure of the gas.
$S_{\text{gas}} = \left(1.28\text{x}10^{-3}\dfrac{\text{mol}}{\text{L} \bullet \text{atm}}\right)(0.209 \text{ atm}) = 2.6752\text{x}10^{-4} \text{ mol/L}$

Mass (g) of $O_2 = \left(\dfrac{2.6752\text{x}10^{-4} \text{ mol } O_2}{\text{L}}\right)\left(\dfrac{32.00 \text{ g } O_2}{1 \text{ mol } O_2}\right)(2.50 \text{ L}) = 0.0214016 = \textbf{0.0214 g } O_2$

13.49   <u>Plan:</u> Solubility for a gas is calculated from Henry's law: $S_{gas} = k_H \times P_{gas}$. We know k_H and P_{gas}, so S_{gas} can be calculated with units of mol/L.
<u>Solution:</u>
$S_{gas} = k_H \times P_{gas} = (3.7 \times 10^{-2}\ \text{mol/L}\cdot\text{atm})(5.5\ \text{atm}) = 0.2035 = \textbf{0.20 mol/L}$

13.52   <u>Plan:</u> Refer to the table of concentration definitions for the different methods of expressing concentration.
<u>Solution:</u>
a) **Molarity** and **parts-by-volume** (% w/v or % v/v) include the volume of the solution.
b) **Parts-by-mass** (% w/w) include the mass of solution directly. (Others may involve the mass indirectly.)
c) **Molality** includes the mass of the solvent.

13.54   Converting between molarity and molality involves conversion between volume of solution and mass of solution. Both of these quantities are given so interconversion is possible. To convert to mole fraction requires that the mass of solvent be converted to moles of solvent. Since the identity of the solvent is not given, conversion to mole fraction is not possible if the molar mass is not known.

13.56   <u>Plan:</u> The molarity is the number of moles of solute in each liter of solution: $M = \dfrac{\text{mol of solute}}{V(\text{L of solution})}$. Convert the masses to moles and the volumes to liters and divide moles by volume.
<u>Solution:</u>

a) Molarity $= \left(\dfrac{32.3\ \text{g C}_{12}\text{H}_{22}\text{O}_{11}}{100.\ \text{mL}}\right)\left(\dfrac{1\ \text{mL}}{10^{-3}\ \text{L}}\right)\left(\dfrac{1\ \text{mol C}_{12}\text{H}_{22}\text{O}_{11}}{342.30\ \text{g C}_{12}\text{H}_{22}\text{O}_{11}}\right) = 0.943617 = \textbf{0.944 \textit{M} C}_{\textbf{12}}\textbf{H}_{\textbf{22}}\textbf{O}_{\textbf{11}}$

b) Molarity $= \left(\dfrac{5.80\ \text{g LiNO}_3}{505\ \text{mL}}\right)\left(\dfrac{1\ \text{mL}}{10^{-3}\ \text{L}}\right)\left(\dfrac{1\ \text{mol LiNO}_3}{68.95\ \text{g LiNO}_3}\right) = 0.166572 = \textbf{0.167 \textit{M} LiNO}_{\textbf{3}}$

13.58   <u>Plan:</u> Dilution calculations can be done using $M_{conc}V_{conc} = M_{dil}V_{dil}$.
<u>Solution:</u>
a) $M_{conc} = 0.240\ M$ NaOH $V_{conc} = 78.0\ \text{mL}$ $M_{dil} = ?$ $V_{dil} = 0.250\ \text{L}$

$M_{dil} = \dfrac{(M_{conc})(V_{conc})}{(V_{dil})} = \dfrac{(0.240\ M)(78.0\ \text{mL})}{(0.250\ \text{L})}\left(\dfrac{10^{-3}\ \text{L}}{1\ \text{mL}}\right) = 0.07488 = \textbf{0.0749 \textit{M}}$

b) $M_{conc} = 1.2\ M$ HNO$_3$ $V_{conc} = 38.5\ \text{mL}$ $M_{dil} = ?$ $V_{dil} = 0.130\ \text{L}$

$M_{dil} = \dfrac{(M_{conc})(V_{conc})}{(V_{dil})} = \dfrac{(1.2\ M)(38.5\ \text{mL})}{(0.130\ \text{L})}\left(\dfrac{10^{-3}\ \text{L}}{1\ \text{mL}}\right) = 0.355385 = \textbf{0.36 \textit{M}}$

13.60   <u>Plan:</u> For part a), find the number of moles of KH$_2$PO$_4$ needed to make 365 mL of a solution of this molarity. Convert moles to mass using the molar mass of KH$_2$PO$_4$. For part b), use the relationship $M_{conc}V_{conc} = M_{dil}V_{dil}$ to find the volume of 1.25 M NaOH needed.
<u>Solution:</u>

a) Moles of KH$_2$PO$_4$ $= (365\ \text{mL})\left(\dfrac{10^{-3}\ \text{L}}{1\ \text{mL}}\right)\left(\dfrac{8.55 \times 10^{-2}\ \text{mol KH}_2\text{PO}_4}{\text{L}}\right) = 0.0312075\ \text{mol}$

Mass (g) of KH$_2$PO$_4$ $= (0.0312075\ \text{mol KH}_2\text{PO}_4)\left(\dfrac{136.09\ \text{g KH}_2\text{PO}_4}{1\ \text{mol KH}_2\text{PO}_4}\right) = 4.24703 = \textbf{4.25 g KH}_{\textbf{2}}\textbf{PO}_{\textbf{4}}$

Add **4.25 g KH$_2$PO$_4$** to enough water to make 365 mL of aqueous solution.

b) $M_{conc} = 1.25\ M$ NaOH $V_{conc} = ?$ $M_{dil} = 0.335\ M$ NaOH $V_{dil} = 465$ mL

$$V_{conc} = \frac{(M_{dil})(V_{dil})}{(V_{conc})} = \frac{(0.335\ M)(465\ \text{mL})}{(1.25\ M)} = 124.62 = \textbf{125 mL}$$

Add **125 mL** of 1.25 M NaOH to enough water to make 465 mL of solution.

13.62   Plan: To find the mass of KBr needed in part a), find the moles of KBr in 1.40 L of a 0.288 M solution and convert to grams using the molar mass of KBr. To find the volume of the concentrated solution that will be diluted to 255 mL in part b), use $M_{conc}V_{conc} = M_{dil}V_{dil}$ and solve for V_{conc}.
Solution:

a) Moles of KBr = $(1.40\ \text{L})\left(\dfrac{0.288\ \text{mol KBr}}{\text{L}}\right) = 0.4032$ mol

Mass (g) of KBr = $(0.4032\ \text{mol})\left(\dfrac{119.00\ \text{g KBr}}{1\ \text{mol KBr}}\right) = 47.9808 = \textbf{48.0 g KBr}$

To make the solution, weigh **48.0 g KBr** and then dilute to 1.40 L with distilled water.

b) $M_{conc} = 0.264\ M$ LiNO$_3$ $V_{conc} = ?$ $M_{dil} = 0.0856\ M$ LiNO$_3$ $V_{dil} = 255$ mL

$$V_{conc} = \frac{(M_{dil})(V_{dil})}{(V_{conc})} = \frac{(0.0856\ M)(255\ \text{mL})}{(0.264\ M)} = 82.68182 = 82.7\ \text{mL}$$

To make the 0.0856 M solution, measure **82.7 mL** of the 0.264 M solution and add distilled water to make a total of 255 mL.

13.64   Plan: Molality, m, = $\dfrac{\text{moles of solute}}{\text{kg of solvent}}$. Convert the mass of solute to moles and divide by the mass of solvent in units of kg.
Solution:

a) Moles of glycine = 85.4 g glycine $\left(\dfrac{1\ \text{mol glycine}}{75.07\ \text{g glycine}}\right) = 1.137605$ mol

m glycine = $\dfrac{1.137605\ \text{mol glycine}}{1.270\ \text{kg}} = 0.895752 = \textbf{0.896 } \textbf{\textit{m}}\textbf{ glycine}$

b) Moles of glycerol = 8.59 g glycerol $\left(\dfrac{1\ \text{mol glycerol}}{92.09\ \text{g glycerol}}\right) = 0.093278$ mol

Volume (kg) of solvent = 77.0 g $\left(\dfrac{1\ \text{kg}}{10^3\ \text{g}}\right) = 0.0770$ kg

m glycerol = $\dfrac{0.093278\ \text{mol glycerol}}{0.0770\ \text{kg}} = 1.2114 = \textbf{1.21 } \textbf{\textit{m}}\textbf{ glycerol}$

13.66   Plan: Molality, m, = $\dfrac{\text{moles of solute}}{\text{kg of solvent}}$. Use the density of benzene to find the mass and then the moles of benzene; use the density of hexane to find the mass of hexane and convert to units of kg. Divide the moles of benzene by the mass of hexane.
Solution:

Mass (g) of benzene = $(44.0\ \text{mL C}_6\text{H}_6)\left(\dfrac{0.877\ \text{g}}{1\ \text{mL}}\right) = 38.588$ g benzene

$$\text{Moles of benzene} = (38.588 \text{ g C}_6\text{H}_6)\left(\frac{1 \text{ mol C}_6\text{H}_6}{78.11 \text{ g C}_6\text{H}_6}\right) = 0.49402 \text{ mol benzene}$$

$$\text{Mass (kg) of hexane} = (167 \text{ mL C}_6\text{H}_{14})\left(\frac{0.660 \text{ g}}{\text{mL}}\right)\left(\frac{1 \text{ kg}}{10^3 \text{ g}}\right) = 0.11022 \text{ kg hexane}$$

$$m = \frac{\text{moles of solute}}{\text{kg of solvent}} = \frac{(0.49402 \text{ mol C}_6\text{H}_6)}{(0.11022 \text{ kg C}_6\text{H}_{14})} = 4.48213 = \mathbf{4.48 \; m \; C_6H_6}$$

13.68   Plan: In part a), the total mass of the <u>solution</u> is 3.10×10^2 g, so mass$_{\text{solute}}$ + mass$_{\text{solvent}}$ = 3.10×10^2 g. Assume that you have 1000 g of the solvent water and find the mass of $C_2H_6O_2$ needed to make a 0.125 m solution. Then a ratio can be used to find the mass of $C_2H_6O_2$ needed to make 3.10×10^2 g of a 0.125 m solution. Part b) is a dilution problem. First, determine the amount of solute in your target solution and then determine the amount of the concentrated acid solution needed to get that amount of solute.
Solution:

a) Mass (g) of $C_2H_6O_2$ in 1000 g (1 kg) of H_2O = $(1 \text{ kg H}_2\text{O})\left(\dfrac{0.125 \text{ mol C}_2\text{H}_6\text{O}_2}{1 \text{ kg H}_2\text{O}}\right)\left(\dfrac{62.07 \text{ g C}_2\text{H}_6\text{O}_2}{1 \text{ mol C}_2\text{H}_6\text{O}_2}\right)$

$$= 7.75875 \text{ g C}_2\text{H}_6\text{O}_2 \text{ in 1000 g H}_2\text{O}$$

Mass (g) of this solution = 1000 g H_2O + 7.75875 g $C_2H_6O_2$ = 1007.75875 g

Mass (g) of $C_2H_6O_2$ for 3.10×10^2 g of solution = $\left(\dfrac{7.75875 \text{ g C}_2\text{H}_6\text{O}_2}{1007.75875 \text{ g solution}}\right)(3.10\times10^2 \text{ g solution})$

$$= 2.386695 \text{ g C}_2\text{H}_6\text{O}_2$$

Mass$_{\text{solvent}}$ = 3.10×10^2 g – mass$_{\text{solute}}$ = 3.10×10^2 g – 2.386695 g $C_2H_6O_2$ = 307.613305 = 308 g H_2O

Therefore, **add 2.39 g $C_2H_6O_2$ to 308 g of H_2O to make a 0.125 m solution.**

b) Mass (kg) of HNO_3 in the 2.20% solution = $(1.20 \text{ kg})\left(\dfrac{2.20\%}{100\%}\right)$ = 0.0264 kg HNO_3 (solute)

$$\text{Mass \%} = \frac{\text{mass of solute}}{\text{mass of solution}}(100)$$

Mass of 52.0% solution containing 0.0264 kg HNO_3 = $\dfrac{\text{mass of solute}(100)}{\text{mass \%}} = \dfrac{0.0264 \text{ kg}(100)}{52.0\%}$

$$= 0.050769 = 0.0508 \text{ kg}$$

Mass of water added = mass of 2.2% solution – mass of 52.0% solution

$$= 1.20 \text{ kg} – 0.050769 \text{ kg} = 1.149231 = 1.15 \text{ kg}$$

Add 0.0508 kg of the 52.0% (w/w) HNO_3 to 1.15 kg H_2O to make 1.20 kg of 2.20% (w/w) HNO_3.

13.70   Plan: You know the moles of solute (C_3H_7OH) and the moles of solvent (H_2O). Divide moles of C_3H_7OH by the total moles of C_3H_7OH and H_2O to obtain mole fraction. To calculate mass percent, convert moles of solute and solvent to mass and divide the mass of solute by the total mass of solution (solute + solvent). For molality, divide the moles of C_3H_7OH by the mass of water expressed in units of kg.
Solution:
a) Mole fraction is moles of isopropanol per total moles.

$$X_{\text{isopropanol}} = \frac{\text{moles of isopropanol}}{\text{moles of isopropanol} + \text{moles of water}} = \frac{0.35 \text{ mol isopropanol}}{(0.35 + 0.85) \text{ mol}} = 0.2916667 = \mathbf{0.29}$$

(Notice that mole fractions have no units.)

b) Mass percent = $\dfrac{\text{mass of solute}}{\text{mass of solution}}(100)$. From the mole amounts, find the masses of isopropanol and water:

$$\text{Mass (g) of isopropanol} = (0.35 \text{ mol C}_3\text{H}_7\text{OH})\left(\frac{60.09 \text{ g C}_3\text{H}_7\text{OH}}{1 \text{ mol C}_3\text{H}_7\text{OH}}\right) = 21.0315 \text{ g isopropanol}$$

$$\text{Mass (g) of water} = (0.85 \text{ mol H}_2\text{O})\left(\frac{18.02 \text{ g H}_2\text{O}}{1 \text{ mol H}_2\text{O}}\right) = 15.317 \text{ g water}$$

$$\text{Mass percent} = \frac{\text{mass of solute}}{\text{mass of solution}}(100) = \frac{21.0315 \text{ g isopropanol}}{(21.0315 + 15.317) \text{ g}}(100) = 57.860710 = \mathbf{58\%}$$

c) Molality of isopropanol is moles of isopropanol per kg of water.

$$m = \frac{\text{moles of solute}}{\text{kg of solvent}} = \frac{0.35 \text{ mol isopropanol}}{15.317 \text{ g water}}\left(\frac{10^3 \text{ g}}{1 \text{ kg}}\right) = 22.85043 = \mathbf{23}\ \boldsymbol{m} \text{ isopropanol}$$

13.72   Plan: Molality $= \dfrac{\text{moles of solute}}{\text{kg of solvent}}$. Use the density of water to convert the volume of water to mass. Multiply the

mass of water in kg by the molality to find moles of cesium bromide; convert moles to mass. To find the mole fraction, convert the masses of water and cesium bromide to moles and divide moles of cesium bromide by the total moles of cesium bromide and water. To calculate mass percent, divide the mass of cesium bromide by the total mass of solution and multiply by 100.
Solution:
The density of water is 1.00 g/mL. The mass of water is:

$$\text{Mass (g) of water} = (0.500 \text{ L})\left(\frac{1 \text{ mL}}{10^{-3} \text{ L}}\right)\left(\frac{1.00 \text{ g}}{1 \text{ mL}}\right)\left(\frac{1 \text{ kg}}{10^3 \text{ g}}\right) = 0.500 \text{ kg}$$

$$m = \frac{\text{moles of solute}}{\text{kg of solvent}}$$

$$0.400\ m \text{ CsBr} = \frac{\text{moles CsBr}}{0.500 \text{ kg H}_2\text{O}}$$

Moles of CsBr $= (0.400\ m)(0.500 \text{ kg H}_2\text{O}) = 0.200 \text{ mol}$

$$\text{Mass (g) of CsBr} = (0.200 \text{ mol CsBr})\left(\frac{212.8 \text{ g CsBr}}{1 \text{ mol CsBr}}\right) = 42.56 = \mathbf{42.6 \text{ g CsBr}}$$

$$\text{Mass (g) of water} = (0.500 \text{ kg})\left(\frac{10^3 \text{ g}}{1 \text{ kg}}\right) = 500.\text{ g water}$$

Mass of solution = mass (g) H_2O + mass of CsBr = 500. g H_2O + 42.56 g CsBr = 542.56 g

$$\text{Moles of H}_2\text{O} = (500.\text{ g H}_2\text{O})\left(\frac{1 \text{ mol H}_2\text{O}}{18.02 \text{ g H}_2\text{O}}\right) = 27.74695 \text{ mol H}_2\text{O}$$

$$X_{\text{CsBr}} = \frac{\text{mol CsBr}}{\text{mol CsBr} + \text{mol H}_2\text{O}} = \frac{0.2000 \text{ mol CsBr}}{(0.2000 + 27.74695) \text{ mol}} = 7.156\times10^{-3} = \mathbf{7.16\times10^{-3}}$$

$$\text{Mass percent CsBr} = \frac{\text{mass of CsBr}}{\text{mass of solution}}(100) = \frac{(42.56 \text{ g CsBr})}{(42.56 + 500.) \text{ g}} \times 100\% = 7.84429 = \mathbf{7.84\% \text{ CsBr}}$$

13.74   Plan: You are given the mass percent of the solution. Assuming 100. g of solution allows us to
express the mass % as the mass of solute, NH_3. To find the mass of solvent, subtract the mass of NH_3 from the
mass of solution and convert to units of kg. To find molality, convert mass of NH_3 to moles and divide by the mass
of solvent in kg. To find molarity, you will need the volume of solution. Use the density of the solution to
convert the 100. g of solution to volume in liters; divide moles of NH_3 by volume of solution. To find the mole
fraction, convert mass of solvent to moles and divide moles of NH_3 by the total moles.

Solution:
Determine some fundamental quantities:

$$\text{Mass (g) of NH}_3 = (100 \text{ g solution})\left(\frac{8.00\% \text{ NH}_3}{100\% \text{ solution}}\right) = 8.00 \text{ g NH}_3$$

Mass (g) H_2O = mass of solution − mass NH_3 = (100.00 − 8.00) g = 92.00 g H_2O

$$\text{Mass (kg) of H}_2\text{O} = (92.00 \text{ g H}_2\text{O})\left(\frac{1 \text{ kg}}{10^3 \text{ g}}\right) = 0.09200 \text{ kg H}_2\text{O}$$

$$\text{Moles of NH}_3 = (8.00 \text{ g NH}_3)\left(\frac{1 \text{ mol NH}_3}{17.03 \text{ g NH}_3}\right) = 0.469759 \text{ mol NH}_3$$

$$\text{Moles of H}_2\text{O} = (92.00 \text{ g H}_2\text{O})\left(\frac{1 \text{ mol H}_2\text{O}}{18.02 \text{ g H}_2\text{O}}\right) = 5.1054 \text{ mol H}_2\text{O}$$

$$\text{Volume (L) of solution} = (100.00 \text{ g solution})\left(\frac{1 \text{ mL solution}}{0.9651 \text{ g solution}}\right)\left(\frac{10^{-3} \text{ L}}{1 \text{ mL}}\right) = 0.103616 \text{ L}$$

Using the above fundamental quantities and the definitions of the various units:

$$\text{Molality} = m = \frac{\text{moles of solute}}{\text{kg of solvent}} = \left(\frac{0.469759 \text{ mol NH}_3}{0.09200 \text{ kg H}_2\text{O}}\right) = 5.106076 = \textbf{5.11 } \boldsymbol{m} \textbf{ NH}_3$$

$$\text{Molarity} = M = \frac{\text{moles of solute}}{\text{L of solution}} = \left(\frac{0.469759 \text{ mol NH}_3}{0.103616 \text{ L}}\right) = 4.53365 = \textbf{4.53 } \boldsymbol{M} \textbf{ NH}_3$$

$$\text{Mole fraction} = X = \frac{\text{moles of NH}_3}{\text{total moles}} = \frac{0.469759 \text{ mol NH}_3}{(0.469759 + 5.1054)\text{mol}} = 0.084259 = \textbf{0.0843}$$

13.76   <u>Plan:</u> Use the equation for parts per million, ppm. Use the given density of solution to find the mass of solution; divide the mass of each ion by the mass of solution and multiply by 1×10^6.
         Solution:

$$\text{Mass (g) of solution is } (100.0 \text{ L solution})\left(\frac{1 \text{ mL}}{10^{-3} \text{ L}}\right)\left(\frac{1.001 \text{ g}}{1 \text{ mL}}\right) = 1.001\times10^5 \text{ g}$$

$$\text{ppm} = \left(\frac{\text{mass solute}}{\text{mass solution}}\right) \times 10^6$$

$$\text{ppm Ca}^{2+} = \left(\frac{0.25 \text{ g Ca}^{2+}}{1.001\times10^5 \text{ g solution}}\right) \times 10^6 = 2.49750 = \textbf{2.5 ppm Ca}^{2+}$$

$$\text{ppm Mg}^{2+} = \left(\frac{0.056 \text{ g Mg}^{2+}}{1.001\times10^5 \text{ g solution}}\right) \times 10^6 = 0.5594406 = \textbf{0.56 ppm Mg}^{2+}$$

13.80   The "strong" in "strong electrolyte" refers to the ability of an electrolyte solution to conduct a large current. This conductivity occurs because solutes that are strong electrolytes dissociate completely into ions when dissolved in water.

13.82   The boiling point temperature is higher and the freezing point temperature is lower for the solution compared to the solvent because the addition of a solute lowers the freezing point and raises the boiling point of a liquid.

13.85   A dilute solution of an electrolyte behaves more ideally than a concentrated one. With increasing concentration, the effective concentration deviates from the molar concentration because of ionic attractions. Thus, the more dilute **0.050 m NaF** solution has a boiling point closer to its predicted value.

13.88   Plan: Strong electrolytes are substances that produce a large number of ions when dissolved in water; strong acids and bases and soluble salts are strong electrolytes. Weak electrolytes produce few ions when dissolved in water; weak acids and bases are weak electrolytes. Nonelectrolytes produce no ions when dissolved in water. Molecular compounds other than acids and bases are nonelectrolytes.
Solution:
a) **Strong electrolyte** When hydrogen chloride is bubbled through water, it dissolves and dissociates completely into H^+ (or H_3O^+) ions and Cl^- ions. HCl is a strong acid.
b) **Strong electrolyte** Potassium nitrate is a soluble salt and dissociates into K^+ and NO_3^- ions in water.
c) **Nonelectrolyte** Glucose solid dissolves in water to form individual $C_6H_{12}O_6$ molecules, but these units are not ionic and therefore do not conduct electricity. Glucose is a molecular compound.
d) **Weak electrolyte** Ammonia gas dissolves in water, but is a weak base that forms few NH_4^+ and OH^- ions.

13.90   Plan: To count solute particles in a solution of an ionic compound, count the number of ions per mole and multiply by the number of moles in solution. For a covalent compound, the number of particles equals the number of molecules.
Solution:

a) $\left(\dfrac{0.3 \text{ mol KBr}}{L}\right)\left(\dfrac{2 \text{ mol particles}}{1 \text{ mol KBr}}\right)(1 \text{ L}) = \mathbf{0.6 \text{ mol of particles}}$

Each KBr forms one K^+ ion and one Br^- ion, two particles for each KBr.

b) $\left(\dfrac{0.065 \text{ mol HNO}_3}{L}\right)\left(\dfrac{2 \text{ mol particles}}{1 \text{ mol HNO}_3}\right)(1 \text{ L}) = \mathbf{0.13 \text{ mol of particles}}$

HNO_3 is a strong acid that forms $H^+(H_3O^+)$ ions and NO_3^- ions in aqueous solution.

c) $\left(\dfrac{10^{-4} \text{ mol KHSO}_4}{L}\right)\left(\dfrac{2 \text{ mol particles}}{1 \text{ mol KHSO}_4}\right)(1 \text{ L}) = \mathbf{2 \times 10^{-4} \text{ mol of particles}}$

Each $KHSO_4$ forms one K^+ ion and one HSO_4^- ion in aqueous solution, two particles for each $KHSO_4$.

d) $\left(\dfrac{0.06 \text{ mol C}_2\text{H}_5\text{OH}}{L}\right)\left(\dfrac{1 \text{ mol particles}}{1 \text{ mol C}_2\text{H}_5\text{OH}}\right)(1 \text{ L}) = \mathbf{0.06 \text{ mol of particles}}$

Ethanol is not an ionic compound so each molecule dissolves as one particle. The number of moles of particles is the same as the number of moles of molecules, **0.06 mol** in 1 L.

13.92   Plan: The magnitude of freezing point depression is directly proportional to molality. Calculate the molality of solution by dividing the moles of solute by the mass of solvent in kg. The solution with the larger molality will have the lower freezing point.
Solution:

a) Molality of $CH_3OH = \dfrac{(11.0 \text{ g CH}_3\text{OH})}{(100. \text{ g H}_2\text{O})}\left(\dfrac{1 \text{ mol CH}_3\text{OH}}{32.04 \text{ g CH}_3\text{OH}}\right)\left(\dfrac{10^3 \text{ g}}{1 \text{ kg}}\right) = 3.4332085 = 3.43 \; m \; CH_3OH$

Molality of $CH_3CH_2OH = \dfrac{(22.0 \text{ g CH}_3\text{CH}_2\text{OH})}{(200. \text{ g H}_2\text{O})}\left(\dfrac{1 \text{ mol CH}_3\text{CH}_2\text{OH}}{46.07 \text{ g CH}_3\text{CH}_2\text{OH}}\right)\left(\dfrac{10^3 \text{ g}}{1 \text{ kg}}\right)$

$= 2.387671 = 2.39 \; m \; CH_3CH_2OH$

The molality of methanol, CH_3OH, in water is 3.43 m whereas the molality of ethanol, CH_3CH_2OH, in water is 2.39 m. Thus, **CH_3OH/H_2O solution** has the lower freezing point.

b) Molality of $H_2O = \dfrac{(20.0 \text{ g H}_2\text{O})}{(1.00 \text{ kg CH}_3\text{OH})}\left(\dfrac{1 \text{ mol H}_2\text{O}}{18.02 \text{ g H}_2\text{O}}\right) = 1.10988 = 1.11 \; m \; H_2O$

Molality of $CH_3CH_2OH = \dfrac{(20.0 \text{ g CH}_3\text{CH}_2\text{OH})}{(1.00 \text{ kg CH}_3\text{OH})}\left(\dfrac{1 \text{ mol CH}_3\text{CH}_2\text{OH}}{46.07 \text{ g CH}_3\text{CH}_2\text{OH}}\right) = 0.434122 = 0.434 \; m \; CH_3CH_2OH$

The molality of H_2O in CH_3OH is 1.11 m, whereas CH_3CH_2OH in CH_3OH is 0.434 m. Therefore, **H_2O/CH_3OH solution** has the lower freezing point.

13.94   Plan: To rank the solutions in order of increasing osmotic pressure, boiling point, freezing point, and vapor pressure, convert the molality of each solute to molality of particles in the solution. The higher the molality of particles, the higher the osmotic pressure, the higher the boiling point, the lower the freezing point, and the lower the vapor pressure at a given temperature.
Solution:

$$(I)\ (0.100\ m\ NaNO_3)\left(\frac{2\ mol\ particles}{1\ mol\ NaNO_3}\right) = 0.200\ m\ ions$$

$NaNO_3$ consists of Na^+ ions and NO_3^- ions, two particles for each $NaNO_3$.

$$(II)\ (0.100\ m\ glucose)\left(\frac{1\ mol\ particles}{1\ mol\ glucose}\right) = 0.100\ m\ molecules$$

Glucose is not an ionic compound so each molecule dissolves as one particle. The number of moles of particles is the same as the number of moles of molecules.

$$(III)\ (0.100\ m\ CaCl_2)\left(\frac{3\ mol\ particles}{1\ mol\ CaCl_2}\right) = 0.300\ m\ ions$$

$CaCl_2$ consists of Ca^{+2} ions and Cl^- ions, three particles for each $CaCl_2$.
a) Osmotic pressure: $\Pi_{II} < \Pi_I < \Pi_{III}$
b) Boiling point: $bp_{II} < bp_I < bp_{III}$
c) Freezing point: $fp_{III} < fp_I < fp_{II}$
d) Vapor pressure at 50°C: $vp_{III} < vp_I < vp_{II}$

13.96   Plan: The mole fraction of solvent affects the vapor pressure according to the equation $P_{solvent} = X_{solvent}P°_{solvent}$. Convert the masses of glycerol and water to moles and find the mole fraction of water by dividing moles of water by the total number of moles. Multiply the mole fraction of water by the vapor pressure of water to find the vapor pressure of the solution.
Solution:

$$Moles\ of\ C_3H_8O_3 = (34.0\ g\ C_3H_8O_3)\left(\frac{1\ mol\ C_3H_8O_3}{92.09\ g\ C_3H_8O_3}\right) = 0.369204\ mol\ C_3H_8O_3$$

$$Moles\ of\ H_2O = (500.0\ g\ H_2O)\left(\frac{1\ mol\ H_2O}{18.02\ g\ H_2O}\right) = 27.7469\ mol\ H_2O$$

$$X_{solvent} = \frac{mol\ H_2O}{mol\ H_2O + mol\ glycerol} = \frac{27.7469\ mol\ H_2O}{27.7469\ mol\ H_2O + 0.369204\ mol\ glycerol} = 0.9868686$$

$$P_{solvent} = X_{solvent}P°_{solvent} = (0.9868686)(23.76\ torr) = 23.447998 = \textbf{23.4 torr}$$

13.98   Plan: The change in freezing point is calculated from $\Delta T_f = iK_f m$, where K_f is 1.86°C/m for aqueous solutions, i is the van't Hoff factor, and m is the molality of particles in solution. Since urea is a covalent compound and does not ionize in water, $i = 1$. Once ΔT_f is calculated, the freezing point is determined by subtracting it from the freezing point of pure water (0.00°C).
Solution:
$\Delta T_f = iK_f m = (1)(1.86°C/m)(0.251\ m) = 0.46686°C$
The freezing point is 0.00°C – 0.46686°C = –0.46686 = **–0.467°C**.

13.100   Plan: The boiling point of a solution is increased relative to the pure solvent by the relationship $\Delta T_b = iK_b m$. Vanillin is a nonelectrolyte (it is a molecular compound) so $i = 1$. To find the molality, convert mass of vanillin to moles and divide by the mass of solvent expressed in units of kg. K_b is given (1.22°C/m).

Solution:

$$\text{Moles of vanillin} = \left(6.4 \text{ g vanillin}\right)\left(\frac{1 \text{ mol vanillin}}{152.14 \text{ g vanillin}}\right) = 0.0420665 \text{ mol}$$

$$\text{Molality of vanillin} = \frac{\text{moles of vanillin}}{\text{kg of solvent (ethanol)}} = \frac{0.042065 \text{ mol vanillin}}{50.0 \text{ g ethanol}}\left(\frac{10^3 \text{ g}}{1 \text{ kg}}\right)$$

$$= 0.8413 \, m \text{ vanillin}$$

$\Delta T_b = iK_bm = (1)(1.22°C/m)(0.8413 \, m) = 1.026386°C$

The boiling point is $78.5°C + 1.026386°C = 79.5264 = \textbf{79.5°C}$.

13.102 **Plan:** The molality of the solution can be determined from the relationship $\Delta T_f = iK_fm$ with the value $1.86°C/m$ inserted for K_f and $i = 1$ for the nonelectrolyte ethylene glycol (ethylene glycol is a covalent compound that will form one particle per molecule when dissolved). Convert the freezing point of the solution to °C and find ΔT_f by subtracting the freezing point of the solvent from the freezing point of the solution. Once the molality of the solution is known, the mass of ethylene glycol needed for a solution of that molality can be found.

Solution:

$°C = (5/9)(°F - 32.0) = (5/9)(-12.0°F - 32.0) = -24.44444°C$

$\Delta T_f = T_{f(\text{solution})} - T_{f(\text{solvent})} = (0.00 - (-24.44444))°C = 24.44444°C$

$\Delta T_f = iK_fm$

$$m = \frac{\Delta T_f}{K_f} = \frac{24.44444°C}{1.86°C/m} = 13.14217 \, m$$

Ethylene glycol will be abbreviated as EG.

$$\text{Molality of EG} = \frac{\text{moles of EG}}{\text{kg of solvent (water)}}$$

$$\text{Moles of EG} = \text{molality x kg of solvent} = \left(\frac{13.14217 \text{ mol EG}}{1 \text{ kg H}_2\text{O}}\right)\left(14.5 \text{ kg H}_2\text{O}\right) = 190.561465 \text{ mol EG}$$

$$\text{Mass (g) of ethylene glycol} = \left(190.561465 \text{ mol EG}\right)\left(\frac{62.07 \text{ g EG}}{1 \text{ mol EG}}\right)$$

$$= 1.18282 \times 10^4 = \textbf{1.18} \times \textbf{10}^\textbf{4} \textbf{ g ethylene glycol}$$

To prevent the solution from freezing, dissolve a minimum of 1.18×10^4 g ethylene glycol in 14.5 kg water.

13.104 **Plan:** Calculate the molarity of the protein (divide the mass of the protein in grams by the molar mass of the protein to calculate the moles of protein; then divide the moles of protein by the volume of the solution in L). Convert temperature from °C to K. Use these values and the ideal gas constant to calculate osmotic pressure.

Solution:

$\Pi = MRT$

$$= \left(\frac{\left(37.5 \text{ mg protein}\right)\left(\frac{1 \text{ g}}{10^3 \text{ mg}}\right)\left(\frac{1 \text{ mol protein}}{1.50 \times 10^4 \text{ g protein}}\right)}{\left(25.0 \text{ mL}\right)\left(\frac{1 \text{ L}}{10^3 \text{ mL}}\right)}\right)\left(0.0821 \frac{\text{L} \cdot \text{atm}}{\text{mol} \cdot \text{K}}\right)\left((24.0 °C + 273.15)\text{K}\right)$$

$$= 2.43960 \times 10^{-3} = \textbf{2.44} \times \textbf{10}^{-3} \textbf{ atm}$$

13.106 **Plan:** Calculate the change in the freezing point then use the relationship $T_f = iK_fm$ (assume $i = 1$) to calculate the molality. Use the molality and the mass of the water, converted to kilograms, to calculate the moles of the unknown compound. Divide the given mass of the compound by this number of moles to calculate the molar mass.

Solution:

$\Delta T_f = 0.00°C - (-1.15°C) = 1.15°C$

$\text{Molality} = \dfrac{\Delta T_f}{iK_f} = \dfrac{1.15°C}{(1)1.86°C/m} = 0.618280 \ m$

Amount (mol) of solute =

$150. \text{ g water}\left(\dfrac{1 \text{ kg}}{10^3 \text{ g}}\right)\left(\dfrac{0.618280 \text{ mol unknown compound}}{1 \text{ kg water}}\right) = 0.0927419 \text{ mol unknown compound}$

$\text{molar mass} = \dfrac{31.7 \text{ g}}{0.0927419 \text{ mol}} = 341.8087 = \textbf{342 g/mol}$

13.108   Plan: Calculate the molality, m, of the solution by dividing the moles of solute by the kg of solvent. The change in freezing point is calculated from $\Delta T_f = iK_f m$, where K_f is $1.86°C/m$ for aqueous solutions, i is the van't Hoff factor, and m is the molality of particles in solution. Ammonium phosphate, $(NH_4)_3PO_4$, is an ionic compound that dissociates in water to give 4 ions (3 ammonium ions and one phosphate ion), so $i = 4$. Once ΔT_f is calculated, the freezing point is determined by subtracting it from the freezing point of the solvent, pure water ($0.00°C$).
Solution:

$$m = \dfrac{(13.2 \text{ g } ((NH_4)_3PO_4)\left(\dfrac{1 \text{ mol } (NH_4)_3PO_4}{149.10 \text{ g } (NH_4)_3PO_4}\right)}{(45.0 \text{ g})\left(\dfrac{1 \text{ kg}}{10^3 \text{ g}}\right)} = 1.96736 \ m$$

$\Delta T_f = iK_f m = (4)(1.86°C/m)(1.96736 \ m) = 14.637 \ °C$
The freezing point is $0.00°C - 14.637°C = -14.637 = \textbf{-14.6 °C}$.

13.110   Plan: Assume 100. g of solution so that the mass of solute = mass percent. Convert the mass of solute to moles. Subtract the mass of the solute from 100. g to obtain the mass of solution. Divide moles of solute by the mass of solvent in kg to obtain molality. Use $\Delta T = iK_f m$ to find the van't Hoff factor. K_f for water = $1.86°C/m$.
Solution:
a) Assume exactly 100 g of solution.

$\text{Mass (g) of NaCl} = (100.00 \text{ g solution})\left(\dfrac{1.00\% \text{ NaCl}}{100\% \text{ solution}}\right) = 1.00 \text{ g NaCl}$

$\text{Moles of NaCl} = (1.00 \text{ g NaCl})\left(\dfrac{1 \text{ mol NaCl}}{58.44 \text{ g NaCl}}\right) = 0.0171116 \text{ mol NaCl}$

$\text{Mass of H}_2\text{O} = 100.00 \text{ g solution} - 1.00 \text{ g NaCl} = 99.00 \text{ g H}_2\text{O}$

$\text{Molality of NaCl} = \dfrac{\text{moles of NaCl}}{\text{kg of H}_2\text{O}} = \dfrac{0.0171116 \text{ mol NaCl}}{99.00 \text{ g H}_2\text{O}}\left(\dfrac{10^3 \text{ g}}{1 \text{ kg}}\right) = 0.172844 = \textbf{0.173 } \boldsymbol{m} \textbf{ NaCl}$

$\Delta T_f = T_{f(solution)} - T_{f(solvent)} = 0.000°C - (-0.593)°C = 0.593°C$
$\Delta T_f = iK_f m$

$i = \dfrac{\Delta T_f}{K_f m} = \dfrac{0.593°C}{(1.86°C/m)(0.172844 \ m)} = 1.844537 = \textbf{1.84}$

The value of i should be close to two because NaCl dissociates into two particles when dissolving in water.
b) For acetic acid, CH_3COOH:
Assume exactly 100 g of solution.

$\text{Mass (g) of CH}_3\text{COOH} = (100.00 \text{ g solution})\left(\dfrac{0.500\% \text{ CH}_3\text{COOH}}{100\% \text{ solution}}\right) = 0.500 \text{ g CH}_3\text{COOH}$

$\text{Moles of CH}_3\text{COOH} = (0.500 \text{ g CH}_3\text{COOH})\left(\dfrac{1 \text{ mol CH}_3\text{COOH}}{60.05 \text{ g CH}_3\text{COOH}}\right) = 0.0083264 \text{ mol CH}_3\text{COOH}$

$\text{Mass (g) of H}_2\text{O} = 100.00 \text{ g solution} - 0.500 \text{ g CH}_3\text{COOH} = 99.500 \text{ g H}_2\text{O}$

$$\text{Molality of CH}_3\text{COOH} = \frac{\text{moles of CH}_3\text{COOH}}{\text{kg of H}_2\text{O}} = \frac{0.0083264 \text{ mol CH}_3\text{COOH}}{99.500 \text{ g H}_2\text{O}} \left(\frac{10^3 \text{ g}}{1 \text{ kg}}\right)$$

$$= 0.083682 = \textbf{0.0837 } \textbf{\textit{m}} \textbf{ CH}_3\textbf{COOH}$$

$$\Delta T_f = T_{f(\text{solution})} - T_{f(\text{solvent})} = 0.000°C - (-0.159)°C = 0.159°C$$
$$\Delta T_f = iK_f m$$
$$i = \frac{\Delta T_f}{K_f m} = \frac{0.159°C}{(1.86°C/m)(0.083682 \text{ } m)} = 1.02153 = \textbf{1.02}$$

Acetic acid is a weak acid and dissociates to a small extent in solution, hence a van't Hoff factor that is close to 1.

13.113   Plan: The mole fraction of solvent affects the vapor pressure according to the equation $P_{\text{solvent}} = X_{\text{solvent}} P°_{\text{solvent}}$. Find the mole fraction of each substance by dividing moles of substance by the total number of moles. Multiply the mole fraction of each compound by its vapor pressure to find the vapor pressure of the compounds above the solution.
Solution:

$$X_{CH_2Cl_2} = \frac{\text{moles CH}_2\text{Cl}_2}{\text{moles CH}_2\text{Cl}_2 + \text{mol CCl}_4} = \frac{1.60 \text{ mol}}{1.60 + 1.10 \text{ mol}} = 0.592593$$

$$X_{CCl_4} = \frac{\text{moles CCl}_4}{\text{moles CH}_2\text{Cl}_2 + \text{mol CCl}_4} = \frac{1.10 \text{ mol}}{1.60 + 1.10 \text{ mol}} = 0.407407$$

$$P_A = X_A P°_A$$
$$= (0.592593)(352 \text{ torr}) = 208.593 = \textbf{209 torr CH}_2\textbf{Cl}_2$$
$$= (0.407407)(118 \text{ torr}) = 48.0740 = \textbf{48.1 torr CCl}_4$$

13.114   The fluid inside a bacterial cell is **both a solution and a colloid**. It is a solution of ions and small molecules, and a colloid of large molecules, proteins, and nucleic acids.

13.118   Soap micelles have nonpolar "tails" pointed inward and anionic "heads" pointed outward. The charges on the "heads" on one micelle repel the "heads" on a neighboring micelle because the charges are the same. This repulsion between soap micelles keeps them from coagulating.
Soap is more effective in **freshwater** than in seawater because the divalent cations in seawater combine with the anionic "head" to form an insoluble precipitate.

13.122   Plan: To find the volume of seawater needed, substitute the given information into the equation that describes the ppb concentration, account for extraction efficiency, and convert mass to volume using the density of seawater.
Solution:
1 troy ounce = 31.1 g gold

$$1.1\text{x}10^{-2} \text{ ppb} = \frac{\text{mass of gold}}{\text{mass of seawater}} \text{ x } 10^9$$

$$1.1\text{x}10^{-2} \text{ ppb} = \frac{31.1 \text{ g Au}}{\text{mass seawater}} \text{ x } 10^9$$

$$\text{Mass (g) of seawater} = \left[\frac{31.1 \text{ g}}{1.1\text{x}10^{-2}} \text{ x } 10^9\right] = 2.827273\text{x}10^{12} \text{ g (with 100\% efficiency)}$$

$$\text{Mass (g) of seawater} = \left(2.827273\text{x}10^{12} \text{ g}\right)\left(\frac{100\%}{81.5\%}\right) = 3.46905\text{x}10^{12} \text{ g seawater (81.5\% efficiency)}$$

$$\text{Volume (L) of seawater} = \left(3.46905\text{x}10^{12} \text{ g}\right)\left(\frac{1 \text{ mL}}{1.025 \text{ g}}\right)\left(\frac{10^{-3} \text{ L}}{1 \text{ mL}}\right) = 3.384439\text{x}10^9 = \textbf{3.4x}10^9 \textbf{ L}$$

13.126 <u>Plan:</u> Convert the mass of O_2 dissolved to moles of O_2. Use the density to convert the 1 kg mass of solution to volume in L. Divide moles of O_2 by volume of solution in L to obtain molarity.

<u>Solution:</u>

0.0°C:

$$\text{Moles of } O_2 = \left(\frac{14.5 \text{ mg } O_2}{1 \text{ kg } H_2O}\right)\left(\frac{10^{-3} \text{ g}}{1 \text{ mg}}\right)\left(\frac{1 \text{ mol } O_2}{32.00 \text{ g } O_2}\right) = 4.53125 \times 10^{-4} \text{ mol } O_2$$

$$\text{Volume (L) of solution} = (1 \text{ kg})\left(\frac{10^3 \text{ g}}{1 \text{ kg}}\right)\left(\frac{1 \text{ mL}}{0.99987 \text{ g}}\right)\left(\frac{10^{-3} \text{ L}}{1 \text{ mL}}\right) = 1.000130017 \text{ L}$$

$$M = \frac{\text{moles of } O_2}{\text{L of solution}} = \frac{4.53125 \times 10^{-4} \text{ mol}}{1.000130017 \text{ L}} = 4.53066 \times 10^{-4} = \mathbf{4.53 \times 10^{-4} \; M \; O_2}$$

20.0°C:

$$\text{Moles of } O_2 = \left(\frac{9.07 \text{ mg } O_2}{1 \text{ kg } H_2O}\right)\left(\frac{10^{-3} \text{ g}}{1 \text{ mg}}\right)\left(\frac{1 \text{ mol } O_2}{32.00 \text{ g } O_2}\right) = 2.834375 \times 10^{-4} \text{ mol } O_2$$

$$\text{Volume (L) of solution} = (1 \text{ kg})\left(\frac{10^3 \text{ g}}{1 \text{ kg}}\right)\left(\frac{1 \text{ mL}}{0.99823 \text{ g}}\right)\left(\frac{10^{-3} \text{ L}}{1 \text{ mL}}\right) = 1.001773138 \text{ L}$$

$$M = \frac{\text{moles of } O_2}{\text{L of solution}} = \frac{2.834375 \times 10^{-4} \text{ mol}}{1.001773138 \text{ L}} = 2.829358 \times 10^{-4} = \mathbf{2.83 \times 10^{-4} \; M \; O_2}$$

40.0°C:

$$\text{Moles of } O_2 = \left(\frac{6.44 \text{ mg } O_2}{1 \text{ kg } H_2O}\right)\left(\frac{10^{-3} \text{ g}}{1 \text{ mg}}\right)\left(\frac{1 \text{ mol } O_2}{32.00 \text{ g } O_2}\right) = 2.0125 \times 10^{-4} \text{ mol } O_2$$

$$\text{Volume (L) of solution} = (1 \text{ kg})\left(\frac{10^3 \text{ g}}{1 \text{ kg}}\right)\left(\frac{1 \text{ mL}}{0.99224 \text{ g}}\right)\left(\frac{10^{-3} \text{ L}}{1 \text{ mL}}\right) = 1.007820689 \text{ L}$$

$$M = \frac{\text{moles of } O_2}{\text{L of solution}} = \frac{2.0125 \times 10^{-4} \text{ mol}}{1.007820689 \text{ L}} = 1.996883 \times 10^{-4} = \mathbf{2.00 \times 10^{-4} \; M \; O_2}$$

13.128 <u>Plan:</u> First, find the molality from the freezing point depression using the relationship $\Delta T = iK_f m$ and then use the molality, given mass of solute and volume of water, to calculate the molar mass of the solute compound. Assume the solute is a nonelectrolyte ($i = 1$). Use the mass percent data to find the empirical formula of the compound; the molar mass is used to convert the empirical formula to the molecular formula. A Lewis structure that forms hydrogen bonds must have H atoms bonded to O atoms.

<u>Solution:</u>

a) $\Delta T_f = iK_f m = 0.000°C - (-0.201°C) = 0.201°C$

$$m = \frac{\Delta T_f}{K_f i} = \frac{0.201°C}{(1.86°C/m)(1)} = 0.1080645 \; m$$

$$\text{Mass (kg) of solvent} = (25.0 \text{ mL})\left(\frac{1.00 \text{ g}}{1 \text{ mL}}\right)\left(\frac{1 \text{ kg}}{10^3 \text{ g}}\right) = 0.0250 \text{ kg water}$$

$$m = \frac{\text{moles of solute}}{\text{kg of solvent}}$$

$$\text{Moles of solute} = (m)(\text{kg solvent}) = (0.1080656 \; m)(0.0250 \text{ kg}) = 0.0027016 \text{ mol}$$

$$\text{Molar mass} = \frac{0.243 \text{ g}}{0.0027016 \text{ mol}} = 89.946698 = \mathbf{89.9 \text{ g/mol}}$$

b) Assume that 100.00 g of the compound gives 53.31 g carbon, 11.18 g hydrogen, and $100.00 - 53.31 - 11.18 = 35.51$ g oxygen.

$$\text{Moles C} = (53.31 \text{ g C})\left(\frac{1 \text{ mol C}}{12.01 \text{ g C}}\right) = 4.43880 \text{ mol C}; \qquad \frac{4.43880}{2.219375} = 2$$

$$\text{Moles H} = (11.18 \text{ g H})\left(\frac{1 \text{ mol H}}{1.008 \text{ g H}}\right) = 11.09127 \text{ mol H}; \qquad \frac{11.09127}{2.219375} = 5$$

$$\text{Moles O} = (35.51 \text{ g O})\left(\frac{1 \text{ mol O}}{16.00 \text{ g O}}\right) = 2.219375 \text{ mol O}; \qquad \frac{2.219375}{2.219375} = 1$$

Dividing the values by the lowest amount of moles (2.219375) gives an **empirical formula of C_2H_5O** with molar mass 45.06 g/mol.

Since the molar mass of the compound, 89.9 g/mol from part a), is twice the molar mass of the empirical formula, the **molecular formula is $2(C_2H_5O)$ or $C_4H_{10}O_2$**.

c) There is more than one example in each case. Possible Lewis structures:

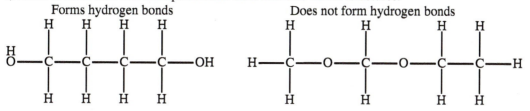

13.132 <u>Plan:</u> Find the moles of NaF required to form a 4.50×10^{-5} M solution of F^- with a volume of 5000. L. Then convert moles of NaF to mass. Then use the molarity to find the mass of F^- in 2.0 L of solution.
<u>Solution:</u>

$$\text{a) Moles of NaF} = (5000. \text{L})\left(\frac{4.50 \times 10^{-5} \text{ mol F}^-}{\text{L}}\right)\left(\frac{1 \text{ mol NaF}}{1 \text{ mol F}^-}\right) = 0.225 \text{ mol NaF}$$

$$\text{Mass (g) of NaF} = (0.225 \text{ mol})\left(\frac{41.99 \text{ g NaF}}{1 \text{ mol NaF}}\right) = 9.44775 = \textbf{9.45 g NaF}$$

$$\text{b) Mass (g) of F}^- = (2.0 \text{L})\left(\frac{4.50 \times 10^{-5} \text{ mol F}^-}{\text{L}}\right)\left(\frac{19.00 \text{ g F}^-}{1 \text{ mol F}}\right) = 0.00171 = \textbf{0.0017 g F}^-$$

13.135 <u>Plan:</u> Use the boiling point elevation of 0.45°C to calculate the molality of the solution using the relationship $\Delta T_b = iK_b m$ and then use the molality, given mass of solute and volume of water, to calculate the molar mass of the solute compound. If the solute is a nonelectrolyte, $i = 1$. If the formula is AB_2 or A_2B, then $i = 3$. For part d), use the molar mass of CaN_2O_6 to calculate the molality of the compound. Then calculate i in the boiling point elevation formula.
<u>Solution:</u>
a) $\Delta T_b = iK_b m$ $i = 1$ (nonelectrolyte)
ΔT = boiling point of solution – boiling point of solvent = $(100.45 - 100.00)$°C = 0.45°C

$$m = \frac{\Delta T_b}{K_b i} = \frac{0.45°C}{(0.512°C/m)(1)} = 0.878906 \ m = 0.878906 \text{ mol/kg}$$

$$\text{Mass (kg) of water} = (25.0 \text{ mL})\left(\frac{0.997 \text{ g}}{1 \text{ mL}}\right)\left(\frac{1 \text{ kg}}{10^3 \text{ g}}\right) = 0.0249250 \text{ kg water}$$

$$m = \frac{\text{moles of solute}}{\text{kg of solvent}}$$

Moles solute = (m)(kg solvent) = $(0.878906\ m)(0.0249250\ \text{kg})$ = 0.0219067 mol

Molar mass = $\dfrac{1.50\ \text{g}}{0.0219067\ \text{mol}}$ = 68.4722 = **68 g/mol**

b) $\Delta T_b = iK_b m$ $i = 3$ (AB$_2$ or A$_2$B)

$m = \dfrac{\Delta T_b}{K_b i} = \dfrac{0.45°\text{C}}{(0.512°\ \text{C}/m)(3)}$ = 0.29296875 m = 0.29296875 mol/kg

$m = \dfrac{\text{moles of solute}}{\text{kg of solvent}}$

Moles solute = (m)(kg solvent) = $(0.29296875\ m)(0.0249250\ \text{kg})$ = 0.00730225 mol

Molar mass = $\dfrac{1.50\ \text{g}}{0.00730225\ \text{mol}}$ = 205.416 = **2.1x10^2 g/mol**

c) The molar mass of CaN$_2$O$_6$ is 164.10 g/mol. This molar mass is less than the 2.1x10^2 g/mol calculated when the compound is assumed to be a strong electrolyte and is greater than the 68 g/mol calculated when the compound is assumed to be a nonelectrolyte. Thus, the compound is an electrolyte, since it dissociates into ions in solution. However, the ions do not dissociate completely in solution.

d) Moles of CaN$_2$O$_6$ = $\left(1.50\ \text{g CaN}_2\text{O}_6\right)\left(\dfrac{1\ \text{mol}}{164.10\ \text{g CaN}_2\text{O}_6}\right)$ = 0.0091408 mol

$m = \dfrac{\text{moles of solute}}{\text{kg of solvent}} = \dfrac{0.00914078\ \text{mol}}{0.0249250\ \text{kg}}$ = 0.3667314 m

$\Delta T_b = iK_b m$

$i = \dfrac{\Delta T_b}{K_b m} = \dfrac{(0.45°\text{C})}{(0.512°/m)(0.3667314\ m)}$ = 2.39659 = **2.4**

13.139   <u>Plan:</u> From the osmotic pressure, the molarity of the solution can be found using the relationship $\Pi = MRT$. Convert the osmotic pressure from units of torr to atm, and the temperature from °C to K. Use the molarity of the solution to find moles of solute; divide the given mass of solute in grams by the moles of solute to obtain molar mass. To find the freezing point depression, the molarity of the solution must be converted to molality by using the density of the solution to convert volume of solution to mass of solution. Then use $\Delta T_f = iK_f m$. ($i = 1$).
<u>Solution:</u>
a) $\Pi = MRT$

Osmotic pressure (atm), $\Pi = (0.340\ \text{torr})\left(\dfrac{1\ \text{atm}}{760\ \text{torr}}\right)$ = 4.47368x10^{-4} atm

$T = 25°\text{C} + 273 = 298$ K

$M = \dfrac{\Pi}{RT} = \dfrac{\left(4.47368\text{x}10^{-4}\ \text{atm}\right)}{\left(0.0821\dfrac{\text{L}\bullet\text{atm}}{\text{mol}\bullet\text{K}}\right)(298\ \text{K})}$ = 1.828544x10^{-5} M

$(M)(V)$ = moles

Moles = $\left(1.828544\text{x}10^{-5}\ M\right)(30.0\ \text{mL})\left(\dfrac{10^{-3}\ \text{L}}{1\ \text{mL}}\right)$ = 5.48563x10^{-7} mol

Molar mass = $\dfrac{(10.0\ \text{mg})\left(\dfrac{10^{-3}\ \text{g}}{1\ \text{mg}}\right)}{5.48563\text{x}10^{-7}\ \text{mol}}$ = 1.82294x10^4 = **1.82x10^4 g/mol**

b) Mass (g) of solution = $(30.0\ \text{mL})\left(\dfrac{0.997\ \text{g}}{1\ \text{mL}}\right)$ = 29.91 g

$$\text{Mass (g) of solute} = (10.0 \text{ mg})\left(\frac{10^{-3} \text{ g}}{1 \text{ mg}}\right) = 0.0100 \text{ g}$$

$$\text{Mass (kg) of solvent} = \text{mass of solution} - \text{mass of solute} = 29.91 \text{ g} - 0.0100 \text{ g}\left(\frac{1 \text{ kg}}{10^3 \text{ g}}\right) = 0.0299 \text{ kg}$$

Moles of solute = 5.48563×10^{-7} mol (from part a)

$$\text{Molality} = \frac{\text{moles of solute}}{\text{kg of solvent}} = \frac{5.48563 \times 10^{-7} \text{ mol}}{0.0299 \text{ kg}} = 1.83466 \times 10^{-5} \, m$$

$$\Delta T_f = iK_f m = (1)(1.86°C/m)(1.83466 \times 10^{-5} \, m) = 3.412 \times 10^{-5} = \mathbf{3.41 \times 10^{-5}°C}$$

(So the solution would freeze at $0 - (3.41 \times 10^{-5}°C) = -3.41 \times 10^{-5}°C$.

13.140 Plan: Henry's law expresses the relationship between gas pressure and the gas solubility (S_{gas}) in a given solvent. Use Henry's law to solve for pressure of dichloroethylene (assume that the constant (k_H) is given at 21°C), use the ideal gas law to find moles per unit volume, and convert moles/L to ng/L.
Solution:

$$S_{gas} = k_H P_{gas}$$

$$S_{gas} \text{ (mol/L)} = \left(\frac{0.65 \text{ mg C}_2\text{H}_2\text{Cl}_2}{L}\right)\left(\frac{10^{-3} \text{ g}}{1 \text{ mg}}\right)\left(\frac{1 \text{ mol C}_2\text{H}_2\text{Cl}_2}{96.94 \text{ g C}_2\text{H}_2\text{Cl}_2}\right) = 6.705178 \times 10^{-6} \text{ mol/L}$$

$$P_{gas} = \frac{S_{gas}}{k_H} = \left(\frac{6.705178 \times 10^{-6} \text{ mol/L}}{0.033 \text{ mol/L} \bullet \text{atm}}\right) = 2.031872 \times 10^{-4} \text{ atm}$$

$$PV = nRT$$

$$\frac{n}{V} = \frac{P}{RT} = \frac{\left(2.031872 \times 10^{-4} \text{ atm}\right)}{\left(0.0821 \dfrac{L \bullet \text{atm}}{\text{mol} \bullet K}\right)((273 + 21) K)} = 8.41794 \times 10^{-6} \text{ mol/L}$$

$$\text{Concentration (ng/L)} = \left(\frac{8.41794 \times 10^{-6} \text{ mol C}_2\text{H}_2\text{Cl}_2}{L}\right)\left(\frac{96.94 \text{ g C}_2\text{H}_2\text{Cl}_2}{1 \text{ mol C}_2\text{H}_2\text{Cl}_2}\right)\left(\frac{1 \text{ ng}}{10^{-9} \text{ g}}\right)$$

$$= 8.16035 \times 10^5 = \mathbf{8.2 \times 10^5 \text{ ng/L}}$$

13.144 Plan: Assume a concentration of 1 mol/m³ for both ethanol and 2-butoxyethanol in the detergent solution. Then, from Henry's law, the partial pressures of the two substances can be calculated.
Solution:

a) $S_{gas} = k_H \times P_{gas}$

$$P_{gas} = \frac{S_{gas}}{k_H}$$

$$P_{ethanol} = \left(\frac{1 \text{ mol}}{m^3}\right)\left(\frac{5 \times 10^{-6} \text{ atm} \bullet m^3}{\text{mol}}\right) = 5 \times 10^{-6} \text{ atm}$$

$$P_{2\text{-butoxyethanol}} = \left(\frac{1 \text{ mol}}{m^3}\right)\left(\frac{1.6 \times 10^{-6} \text{ atm} \bullet m^3}{\text{mol}}\right) = 1.6 \times 10^{-6} \text{ atm}$$

$$\%_{2\text{-butoxyethanol}} = (5\%)\left(\frac{1.6 \times 10^{-6} \text{ atm 2-butoxyethanol}}{5 \times 10^{-6} \text{ atm ethanol}}\right) = 1.6\%$$

"Down-the-drain" factor is $0.016 = \mathbf{0.02}$

b) $k_{H(ethanol)} = \left(\dfrac{5 \times 10^{-6} \text{ atm} \bullet m^3}{\text{mol}}\right)\left(\dfrac{1 \text{ L}}{10^{-3} \text{ m}^3}\right) = \mathbf{5 \times 10^{-3} \text{ atm} \bullet \text{L/mol}}$

c) $k_{H(ethanol)} = \left(\dfrac{0.64 \text{ Pa} \bullet m^3}{\text{mol}}\right)\left(\dfrac{1 \text{ atm}}{1.01325 \times 10^5 \text{ Pa}}\right) = 6.3163 \times 10^{-6} = \mathbf{6.3 \times 10^{-6} \text{ atm} \bullet \text{m}^3\text{/mol}}$

Considering the single significant figure in the measured value of 5×10^{-6}, the agreement is good.

Solution:

a) Mass (g) of ethanol dissolved in the blood $= (28 \text{ mL}) \left(\dfrac{40\%}{100\%} \right) \left(\dfrac{22\%}{100\%} \right) \left(\dfrac{0.789 \text{ g}}{\text{mL}} \right) = 1.944096 \text{ g}$

Concentration $= \left(\dfrac{1.944096 \text{ g ethanol}}{7.0 \text{ L}} \right) \left(\dfrac{10^{-3} \text{ L}}{1 \text{ mL}} \right) = 2.77728 \times 10^{-4} = \mathbf{2.8 \times 10^{-4}} \text{ g/mL}$

b) $\left(\dfrac{28 \text{ mL whiskey}}{2.77728 \times 10^{-4} \text{ g/mL}} \right) \left(8.0 \times 10^{-4} \text{ g/mL} \right) = 80.6545 = \mathbf{81 \text{ mL}}$

CHAPTER 14 PERIODIC PATTERNS IN THE MAIN-GROUP ELEMENTS

END–OF–CHAPTER PROBLEMS

14.1   Ionization energy is defined as the energy required to remove the outermost electron from an atom. The further the outermost electron is from the nucleus, the less energy is required to remove it from the attractive force of the nucleus. In hydrogen, the outermost electron is in the $n = 1$ level and in lithium the outermost electron is in the $n = 2$ level. Therefore, the outermost electron in lithium requires less energy to remove, resulting in a lower ionization energy.

14.3   <u>Plan:</u> Recall that to form hydrogen bonds a compound must have H directly bonded to either N, O, or F.
<u>Solution:</u>
a) **NH_3** will hydrogen bond because H is bonded to N.

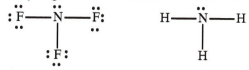

b) **CH_3CH_2OH** will hydrogen bond since H is bonded to O. CH_3OCH_3 has no O−H bonds, only C−H bonds.

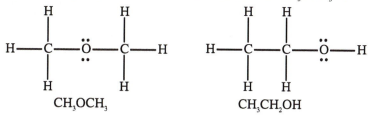

14.5   <u>Plan:</u> Active metals displace hydrogen from HCl by reducing the H^+ to H_2. In water, H^- (here in LiH) reacts as a strong base to form H_2 and OH^-.
<u>Solution:</u>
a) $2Al(s) + 6HCl(aq) \rightarrow 2AlCl_3(aq) + 3H_2(g)$
b) $LiH(s) + H_2O(l) \rightarrow LiOH(aq) + H_2(g)$

14.7   <u>Plan:</u> In metal hydrides, the oxidation state of hydrogen is –1.
<u>Solution:</u>
a)   Na = +1   B = +3   H = –1 in $NaBH_4$
         Al = +3   B = +3   H = –1 in $Al(BH_4)_3$
         Li = +1   Al = +3   H = –1 in $LiAlH_4$
b) The polyatomic ion in $NaBH_4$ is $[BH_4]^-$. There are [1 x B(3e$^-$)] + [4 x H(1e$^-$)] + [1e$^-$ from charge] = 8 valence electrons. All eight electrons are required to form the four bonds from the four hydrogen atoms to the boron atom. Boron is the central atom and has four surrounding electron groups; therefore, its shape is **tetrahedral**.

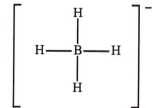

14.10 For Period 2 elements in the first four groups, the number of covalent bonds equals the number of electrons in the outer level, so it increases from one covalent bond for lithium in Group 1A(1) to four covalent bonds for carbon in Group 4A(14). For the rest of Period 2 elements, the number of covalent bonds equals the difference between 8 and the number of electrons in the outer level. So for nitrogen, $8 - 5 = 3$ covalent bonds; for oxygen, $8 - 6 = 2$ covalent bonds; for fluorine, $8 - 7 = 1$ covalent bond; and for neon, $8 - 8 = 0$, no bonds.
For elements in higher periods, the same pattern exists but with exceptions for Groups 3A(13) to 7A(17) when an expanded octet allows for more covalent bonds.

14.13 a) E must have an oxidation of $+3$ to form an oxide E_2O_3 or fluoride EF_3. E is in Group **3A(13) or 3B(3)**.
b) If E were in Group 3B(3), the oxide and fluoride would have more ionic character because 3B elements have lower electronegativity than 3A elements. The Group 3B(3) oxides would be more basic.

14.16 a) Alkali metals generally lose electrons (act as **reducing agents**) in their reactions.
b) Alkali metals have relatively low ionization energies, meaning they easily lose the outermost electron. The electron configurations of alkali metals have one more electron than a noble gas configuration, so losing an electron gives a stable electron configuration.
c) $2Na(s) + 2H_2O(l) \rightarrow 2Na^+(aq) + 2OH^-(aq) + H_2(g)$

$2Na(s) + Cl_2(g) \rightarrow 2NaCl(s)$

14.18 a) Density increases down a group. The increasing atomic size (volume) is not offset by the increasing size of the nucleus (mass), so m/V increases.
b) Ionic size increases down a group. Electron shells are added down a group, so both atomic and ionic size increase.
c) E−E bond energy decreases down a group. Shielding of the outer electron increases as the atom gets larger, so the attraction responsible for the E−E bond decreases.
d) IE_1 decreases down a group. Increased shielding of the outer electron is the cause of the decreasing IE_1.
e) ΔH_{hydr} decreases down a group. ΔH_{hydr} is the heat released when the metal salt dissolves in, or is hydrated by, water. Hydration energy decreases as ionic size increases.
Increasing down: a and b; **Decreasing down: c, d, and e**

14.20 <u>Plan:</u> Peroxides are oxides in which oxygen has a -1 oxidation state. Sodium peroxide has the formula Na_2O_2 and is formed from the elements Na and O_2.
<u>Solution:</u>
$2Na(s) + O_2(g) \rightarrow Na_2O_2(s)$

14.22 <u>Plan:</u> The problem specifies that an alkali halide is the desired product. The alkali metal is K (comes from potassium carbonate, $K_2CO_3(s)$) and the halide is I (comes from hydroiodic acid, $HI(aq)$). Treat the reaction as a double displacement reaction.
<u>Solution:</u>
$K_2CO_3(s) + 2HI(aq) \rightarrow 2KI(aq) + H_2CO_3(aq)$
However, $H_2CO_3(aq)$ is unstable and decomposes to $H_2O(l)$ and $CO_2(g)$, so the final reaction is:
$K_2CO_3(s) + 2HI(aq) \rightarrow 2KI(aq) + H_2O(l) + CO_2(g)$

14.26 Metal atoms are held together by metallic bonding, a sharing of valence electrons. Alkaline earth metal atoms have one more valence electron than alkali metal atoms, so the number of electrons shared is greater. Thus, metallic bonds in alkaline earth metals are stronger than in alkali metals. Melting requires overcoming the metallic bonds. To overcome the stronger alkaline earth metal bonds requires more energy (higher temperature) than to overcome the alkali earth metal bonds.
First ionization energy, density, and boiling points will be larger for alkaline earth metals than for alkali metals.

14.27 <u>Plan:</u> A base forms when a basic oxide, such as CaO (lime), is added to water. Alkaline earth metals reduce O_2 to form the oxide.
<u>Solution:</u>
a) $CaO(s) + H_2O(l) \rightarrow Ca(OH)_2(s)$
b) $2Ca(s) + O_2(g) \rightarrow 2CaO(s)$

14.30   Plan: The oxides of alkaline earth metals are strongly basic, but BeO is amphoteric. BeO will react with both acids and bases to form salts, but an amphoteric substance does not react with water. In part b), each chloride ion donates a lone pair of electrons to form a covalent bond with the Be in $BeCl_2$. Metal ions form similar covalent bonds with ions or molecules containing a lone pair of electrons. The difference in beryllium is that the orbital involved in the bonding is a p orbital, whereas in metal ions it is usually the d orbitals that are involved.
Solution:
a) Here, Be does not behave like other alkaline earth metals: $BeO(s) + H_2O(l) \rightarrow NR$.
b) Here, Be does behave like other alkaline earth metals:
$$BeCl_2(l) + 2\ Cl^-(solvated) \rightarrow BeCl_4^{2-}(solvated)$$

14.33   The electron removed in Group 2A(2) atoms is from the outer level s orbital, whereas in Group 3A(13) atoms the electron is from the outer level p orbital. For example, the electron configuration for Be is $1s^2 2s^2$ and for B is $1s^2 2s^2 2p^1$. It is easier to remove the p electron of B than the s electron of Be, because the energy of a p orbital is slightly higher than that of the s orbital from the same level. Even though the atomic size decreases from increasing Z_{eff}, the IE decreases from Group 2A(2) to 3A(13).

14.34   a) Compounds of Group 3A(13) elements, like boron, have only six electrons in their valence shell when combined with halogens to form three bonds. Having six electrons, rather than an octet, results in an "electron deficiency."
b) As an electron deficient central atom, B is trigonal planar. Upon accepting an electron pair to form a bond, the shape changes to tetrahedral.
$BF_3(g) + NH_3(g) \rightarrow F_3B-NH_3(g)$
$B(OH)_3(aq) + OH^-(aq) \rightarrow B(OH)_4^-(aq)$

14.36   Plan: Oxide acidity increases up a group; the less metallic an element, the more acidic is its oxide.
Solution:
$In_2O_3 < Ga_2O_3 < Al_2O_3$

14.38   Halogens typically have a −1 oxidation state in metal-halide combinations, so the apparent oxidation state of Tl = +3. However, the anion I_3^- combines with Tl in the +1 oxidation state. The anion I_3^- has [3 x (I)7e$^-$] + [1e$^-$ from the charge] = 22 valence electrons; four of these electrons are used to form the two single bonds between iodine atoms and sixteen electrons are used to give every atom an octet. The remaining two electrons belong to the central I atom; therefore the central iodine has five electron groups (two single bonds and three lone pairs) and has a general formula of AX_2E_3. The electrons are arranged in a trigonal bipyramidal with the three lone pairs in the trigonal plane. It is a linear ion with bond angles = 180°. (Tl^{3+}) (I$^-$)$_3$ does not exist because of the low strength of the Tl–I bond.
O.N. = +3 (apparent); = +1 (actual)

$$\left[:\ddot{\ddot{I}}\!\!-\!\!\dot{\ddot{I}}\!\!-\!\!\ddot{\ddot{I}}: \right]^-$$

14.43   Plan: To calculate the enthalpy of reaction, use the relationship $\Delta H^{\circ}_{rxn} = \sum m\,\Delta H^{\circ}_{products} - \sum n\,\Delta H^{\circ}_{reactants}$.
Convert the given amount of 1.0 kg of BN to moles, find the moles of B in that amount of BN, and then find the moles and then mass of borax that provides that number of moles of B.
Solution:
a) $B_2O_3(s) + 2NH_3(g) \rightarrow 2BN(s) + 3H_2O(g)$

b) $\Delta H^{\circ}_{rxn} = \sum m\,\Delta H^{\circ}_{products} - \sum n\,\Delta H^{\circ}_{reactants}$
$= \{2\,\Delta H^{\circ}_f\,[BN(s)] + 3\,\Delta H^{\circ}_f\,[H_2O(g)]\} - \{1\,\Delta H^{\circ}_f\,[B_2O_3(s)] + 2\,\Delta H^{\circ}_f\,[NH_3(g)]\}$
$= [(2\ mol)(-254\ kJ/mol) + (3\ mol)(-241.826\ mol\ kJ/mol)]$
$\qquad\qquad - [(1\ mol)(-1272\ kJ/mol) + (2\ mol)(-45.9\ kJ/mol)]$
$= 130.322 = \mathbf{1.30 \times 10^2\ kJ}$

Solution:

a)

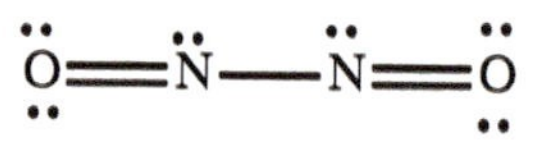

$$\ddot{\text{O}}=\ddot{\text{N}}-\ddot{\text{N}}=\ddot{\text{O}}$$

b)

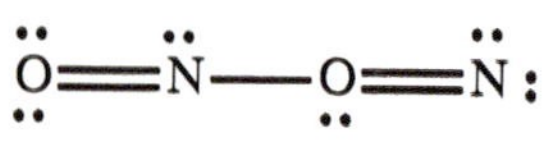

$$\ddot{\text{O}}=\ddot{\text{N}}-\ddot{\text{O}}=\ddot{\text{N}}:$$

c)

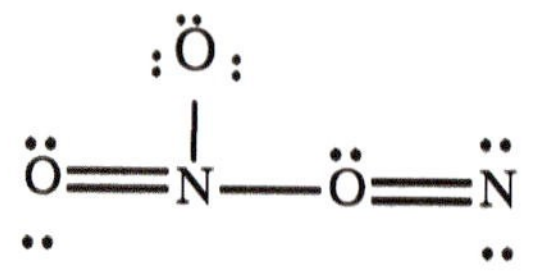

$$\ddot{\text{O}}=\ddot{\text{N}}-\ddot{\text{O}}=\ddot{\text{N}}$$
with $:\ddot{\text{O}}:$ above N

d)

$$\left[\,:\text{N}\equiv\text{O}:\,\right]^{+} \qquad \left[\ddot{\text{O}}=\text{N}\,\substack{\cdot\cdot\cdot\cdot\\ \ddot{\text{O}}:\\ :\ddot{\text{O}}:}\right]^{-}$$

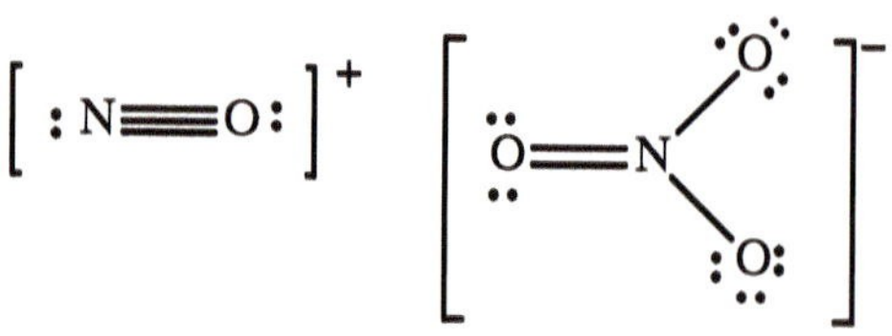

14.73 a) Thermal decomposition of KNO_3 at low temperatures:

$$2KNO_3(s) \xrightarrow{\Delta} 2KNO_2(s) + O_2(g)$$

b) Thermal decomposition of KNO_3 at high temperatures:

$$4KNO_3(s) \xrightarrow{\Delta} 2K_2O(s) + 2N_2(g) + 5O_2(g)$$

14.75 a) Both groups have elements that range from gas to metalloid to metal. Thus, their boiling points and conductivity vary in similar ways down a group.
b) The degree of metallic character and methods of bonding vary in similar ways down a group.
c) Both P and S have allotropes and both bond covalently with almost every other nonmetal.
d) Both N and O are diatomic gases at normal temperatures and pressures. Both N and O have very low melting and boiling points.
e) Oxygen, O_2, is a reactive gas whereas nitrogen, N_2, is not. Nitrogen can exist in multiple oxidation states, whereas oxygen has two oxidation states.

14.77 a) To decide what type of reaction will occur, examine the reactants. Notice that sodium hydroxide is a strong base. Is the other reactant an acid? If we separate the salt, sodium hydrogen sulfate, into the two ions, Na^+ and HSO_4^-, then it is easier to see the hydrogen sulfate ion as the acid. The sodium ions could be left out for the net ionic reaction.

$$NaHSO_4(aq) + NaOH(aq) \rightarrow Na_2SO_4(aq) + H_2O(l)$$

b) As mentioned in the book, hexafluorides are known to exist for sulfur. These will form when excess fluorine is present.

$$S_8(s) + 24F_2(g) \rightarrow 8SF_6(g)$$

c) Group 6A(16) elements, except oxygen, form hydrides in the following reaction.

$$FeS(s) + 2HCl(aq) \rightarrow H_2S(g) + FeCl_2(aq)$$

d) Tetraiodides, but not hexaiodides, of tellurium are known.

$$Te(s) + 2I_2(s) \rightarrow TeI_4(s)$$

14.79 Plan: The oxides of nonmetal elements are acidic, while the oxides metal elements are basic.
Solution:
a) Se is a nonmetal; its oxide is **acidic**.
b) N is a nonmetal; its oxide is **acidic**.
c) K is a metal; its oxide is **basic**.
d) Be is an alkaline earth metal, but all of its bonds are covalent; its oxide is **amphoteric**.
e) Ba is a metal; its oxide is **basic**.

14.81 Plan: Acid strength of binary acids increases down a group since bond energy decreases down the group.
Solution:
$$H_2O < H_2S < H_2Te$$

14.84 a) O_3, **ozone**
b) SO_3, **sulfur trioxide** (+6 oxidation state)
c) SO_2, **sulfur dioxide**
d) H_2SO_4, **sulfuric acid**
e) $Na_2S_2O_3 \cdot 5H_2O$, **sodium thiosulfate pentahydrate**

14.86　$S_2F_{10}(g) \rightarrow SF_4(g) + SF_6(g)$
O.N. of S in S_2F_{10}: $-(10 \times -1 \text{ for } F)/2 = \mathbf{+5}$
O.N. of S in SF_4: $-(4 \times -1 \text{ for } F) = \mathbf{+4}$
O.N. of S in SF_6: $-(6 \times -1 \text{ for } F) = \mathbf{+6}$

14.88　a) Bonding with very electronegative elements: **+1, +3, +5, +7**. Bonding with other elements: **–1**
b) The electron configuration for Cl is $[Ne]3s^2 3p^5$. By adding one electron to form Cl^-, Cl achieves an octet similar to the noble gas Ar. By forming covalent bonds, Cl completes or expands its octet by maintaining its electrons paired in bonds or lone pairs.
c) Fluorine only forms the –1 oxidation state because its small size and no access to d orbitals prevent it from forming multiple covalent bonds. Fluorine's high electronegativity also prevents it from sharing its electrons.

14.90　a) The Cl–Cl bond is stronger than the Br–Br bond since the chlorine atoms are smaller than the bromine atoms, so the shared electrons are held more tightly by the two nuclei.
b) The Br–Br bond is stronger than the I–I bond since the bromine atoms are smaller than the iodine atoms.
c) The Cl–Cl bond is stronger than the F–F bond. The fluorine atoms are smaller than the chlorine but they are so small that electron-electron repulsion of the lone pairs decreases the strength of the bond.

14.92　a) A substance that disproportionates serves as both an oxidizing and reducing agent. Assume that OH^- serves as the base. Write the reactants and products of the reaction, and balance like a redox reaction.
$$3\,Br_2(l) + 6OH^-(aq) \rightarrow 5Br^-(aq) + BrO_3^-(aq) + 3H_2O(l)$$
b) In the presence of base, instead of water, only the oxy*anion* (not oxo*acid*) and fluoride (not *hydro*fluoride) form. No oxidation or reduction takes place, because Cl maintains its +5 oxidation state and F maintains its –1 oxidation state.
$$ClF_5(l) + 6OH^-(aq) \rightarrow 5F^-(aq) + ClO_3^-(aq) + 3H_2O(l)$$

14.93　a) $2Rb(s) + Br_2(l) \rightarrow 2RbBr(s)$
b) $I_2(s) + H_2O(l) \rightarrow HI(aq) + HIO(aq)$
c) $Br_2(l) + 2I^-(aq) \rightarrow I_2(s) + 2Br^-(aq)$
d) $CaF_2(s) + H_2SO_4(l) \rightarrow CaSO_4(s) + 2HF(g)$

14.95　Plan: Acid strength increases with increasing electronegativity of the central atom and increasing number of oxygen atoms.
Solution:
Iodine is less electronegative than bromine, which is less electronegative than chlorine.
$HIO < HBrO < HClO < HClO_2$

14.99　$I_2 < Br_2 < Cl_2$, since Cl_2 is able to oxidize Re to the +6 oxidation state, Br_2 only to +5, and I_2 only to +4.

14.100　**Helium** is the second most abundant element in the universe. **Argon** is the most abundant noble gas in Earth's atmosphere, the third most abundant constituent after N_2 and O_2.

14.102　Whether a boiling point is high or low is a result of the strength of the forces between particles. Dispersion forces, the weakest of all the intermolecular forces, hold atoms of noble gases together. Only a relatively low temperature is required for the atoms to have enough kinetic energy to break away from the attractive force of other atoms and go into the gas phase. The boiling points are so low that all the noble gases are gases at room temperature.

14.106　Plan: To obtain the overall reaction, reverse the first reaction and the third reaction and add these two reactions to the second reaction, canceling substances that appear on both sides of the arrow. When a reaction is reversed, the sign of its enthalpy change is reversed. Add the three enthalpy values to obtain the overall enthalpy value.
Solution:

$H_3O^+(g) \rightarrow \cancel{H^+(g)} + \cancel{H_2O(g)}$　　　$\Delta H = +720$ kJ

$\cancel{H^+(g)} + \cancel{H_2O(l)} \rightarrow H_3O^+(aq)$　　　$\Delta H = -1090$ kJ

$\cancel{H_2O(g)} \rightarrow \cancel{H_2O(l)}$　　　$\Delta H = -40.7$ kJ

Overall:　　$H_3O^+(g) \rightarrow H_3O^+(aq)$　　　$\Delta H = -410.7 = \mathbf{-411}$ **kJ**

14.110 **Plan:** Examine the outer electron configuration of the alkali metals. To calculate the ΔH°_{rxn} in part b), use Hess's law.

Solution:

a) Alkali metals have an outer electron configuration of ns^1. The first electron lost by the metal is the ns electron, giving the metal a noble gas configuration. Second ionization energies for alkali metals are high because the electron being removed is from the next lower energy level and electrons in a lower level are more tightly held by the nucleus. The metal would also lose its noble gas configuration.

b) The reaction is $2CsF_2(s) \rightarrow 2CsF(s) + F_2(g)$.

You know the ΔH°_f for the formation of CsF:

$$Cs(s) + 1/2F_2(g) \rightarrow CsF(s) \qquad\qquad\qquad \Delta H^{\circ}_f = -530 \text{ kJ/mol}$$

You also know the ΔH°_f for the formation of CsF_2:

$$Cs(s) + F_2(g) \rightarrow CsF_2(s) \qquad\qquad\qquad \Delta H^{\circ}_f = -125 \text{ kJ/mol}$$

To obtain the heat of reaction for the breakdown of CsF_2 to CsF, combine the formation reaction of CsF with the reverse of the formation reaction of CsF_2, both multiplied by 2:

$2\overline{Cs(s)} + \overline{F_2(g)} \rightarrow 2CsF(s)$	$\Delta H^{\circ}_f = 2 \times (-530 \text{ kJ}) = -1060 \text{ kJ}$
$2CsF_2(s) \rightarrow 2\overline{Cs(s)} + 2F_2(g)$	$\Delta H^{\circ}_f = 2 \times (+125 \text{ kJ}) = 250 \text{ kJ}$ (Note sign change)
$2 CsF_2(s) \rightarrow 2CsF(s) + F_2(g)$	$\Delta H^{\circ}_{rxn} = -810 \text{ kJ}$

810 kJ of energy are released when two moles of CsF_2 convert to two moles of CsF, so heat of reaction for one mole of CsF is $-810/2$ or **−405 kJ/mol**.

14.112 **Plan:** To find the molecular formula, divide the molar mass of each compound by the molar mass of the empirical formula, HNO. The result of this gives the factor by which the empirical formula is multiplied to obtain the molecular formula.

Solution:

a) Empirical formula HNO has a molar mass of 31.02. Hyponitrous acid has a molar mass of 62.04 g/mol, twice the mass of the empirical formula; its molecular formula is twice as large as the empirical formula, $2(HNO) = H_2N_2O_2$. The molecular formula of nitroxyl would be the same as the empirical formula, HNO, since the molar mass of nitroxyl is the same as the molar mass of the empirical formula.

b) $H_2N_2O_2$ has $[2 \times H(1e^-)] + [2 \times N(5e^-)] + [2 \times O(6e^-)] = 24$ valence e^-. Ten electrons are used for single bonds between the atoms, leaving $24 - 10 = 14$ e^-. Sixteen electrons are needed to give every atom an octet; since only fourteen electrons are available, one double bond (between the N atoms) is needed.

HNO has $[1 \times H(1e^-)] + [1 \times N(5e^-)] + [1 \times O(6e^-)] = 12$ valence e^-. Four electrons are used for single bonds between the atoms, leaving $12 - 4 = 8e^-$. Ten electrons are needed to give every atom an octet; since only eight electrons are available, one double bond is needed between the N and O atoms.

c) In both hyponitrous acid and nitroxyl, the nitrogens are surrounded by three electron groups (one single bond, one double bond, and one unshared pair), so the electron arrangement is trigonal planar and the molecular shape is **bent**.

d)

14-8

14.116 Plan: Carbon monoxide and carbon dioxide would be formed from the reaction of coke (carbon) with the oxygen in the air. The nitrogen in the producer gas would come from the nitrogen already in the air. So, the calculation of mass of product is based on the mass of CO and CO_2 that can be produced from 1.75 metric tons of coke.
Solution:
Using 100 g of sample, the percentages simply become grams.
Since 5.0 g of CO_2 is produced for each 25 g of CO, we can calculate a mass ratio of carbon that produces each:

$$\frac{(25 \text{ g CO})\left(\dfrac{12.01 \text{ g C}}{28.01 \text{ g CO}}\right)}{(5.0 \text{ g CO}_2)\left(\dfrac{12.01 \text{ g C}}{44.01 \text{ g CO}_2}\right)} = 7.85612/1$$

Using the ratio of carbon that reacts as 7.85612:1, the total C reacting is $7.85612 + 1 = 8.85612$. The mass fraction of the total carbon that produces CO is 7.85612/8.85612 and the mass fraction of the total carbon reacting that produces CO_2 is 1.00/8.85612. To find the mass of CO produced from 1.75 metric tons of carbon with an 87% yield:

$$\text{Mass (g) of CO} = \left(\frac{7.85612}{8.85612}\right)(1.75 \text{ t})\left(\frac{28.01 \text{ t CO}}{12.01 \text{ t C}}\right)\left(\frac{87\%}{100\%}\right) = 3.1498656 \text{ t CO}$$

$$\text{Mass (g) of CO}_2 = \left(\frac{1}{8.85612}\right)(1.75 \text{ t})\left(\frac{44.01 \text{ t CO}_2}{12.01 \text{ t C}}\right)\left(\frac{87\%}{100\%}\right) = 0.6299733 \text{ t CO}_2$$

The mass of CO and CO_2 represent a total of 30% (100% – 70.% N_2) of the mass of the producer gas, so the total mass would be $(3.1498656 + 0.6299733)(100\%/30\%) = 12.59946 = \textbf{13 metric tons}$.

14.118 In a disproportionation reaction, a substance acts as both a reducing agent and oxidizing agent because an atom within the substance reacts to form atoms with higher and lower oxidation states.
     0   –1   –1/3
a) $I_2(s) + KI(aq) \rightarrow KI_3(aq)$
I in I_2 reduces to I in KI_3. I in KI oxidizes to I in KI_3. This is not a disproportionation reaction since different substances have atoms that reduce or oxidize. The *reverse* direction would be a disproportionation reaction because a single substance (I in KI) both oxidizes and reduces.
     +4   +5   +3
b) $2ClO_2(g) + H_2O(l) \rightarrow HClO_3(aq) + HClO_2(aq)$
Yes, ClO_2 disproportionates, as the chlorine reduces from +4 to +3 and oxidizes from +4 to +5.
     0   –1   +1
c) $Cl_2(g) + 2NaOH(aq) \rightarrow NaCl(aq) + NaClO(aq) + H_2O(l)$
Yes, Cl_2 disproportionates, as the chlorine reduces from 0 to –1 and oxidizes from 0 to +1.
  –3 +3   0
d) $NH_4NO_2(s) \rightarrow N_2(g) + 2H_2O(g)$
Yes, NH_4NO_2 disproportionates; the ammonium (NH_4^+) nitrogen oxidizes from –3 to 0, and the nitrite (NO_2^-) nitrogen reduces from +3 to 0.
    +6   +7   +4
e) $3MnO_4^{2-}(aq) + 2H_2O(l) \rightarrow 2MnO_4^-(aq) + MnO_2(s) + 4OH^-(aq)$
Yes, MnO_4^{2-} disproportionates; the manganese oxidizes from +6 to +7 and reduces from +6 to +4.
    +1   +3   0
f) $3 AuCl(s) \rightarrow AuCl_3(s) + 2 Au(s)$
Yes, AuCl disproportionates; the gold oxidizes from +1 to +3 and reduces from +1 to 0.

14.120 a) Group **5A(15)** elements have five valence electrons and typically form three bonds with a lone pair to complete the octet. An example is NH_3.
b) Group **7A(17)** elements readily gain an electron causing the other reactant to be oxidized. They form monatomic ions of formula X^- and oxoanions. Examples would be Cl^- and ClO^-.
c) Group **6A(16)** elements have six valence electrons and gain a complete octet by forming two covalent bonds. An example is H_2O.

d) Group **1A(1)** elements are the strongest reducing agents because they most easily lose an electron. As the least electronegative and most metallic of the elements, they are not likely to form covalent bonds. Group **2A(2)** elements have similar characteristics. Thus, either Na or Ca could be an example.

e) Group **3A(13)** elements have only three valence electrons to share in covalent bonds, but with an empty orbital they can accept an electron pair from another atom. Boron would be an example of an element of this type.

f) Group **8A(18)**, the noble gases, are the least reactive of all the elements. Xenon is an example that forms compounds, while helium does not form compounds.

14.122   <u>Plan:</u> Find ΔH°_{rxn} for the reaction $2BrF(g) \rightarrow Br_2(g) + F_2(g)$ by applying Hess's law to the equations given. Recall that when an equation is reversed, the sign of its ΔH°_{rxn} is changed.

<u>Solution:</u>

1)	$3BrF(g) \rightarrow Br_2(g) + BrF_3(l)$	$\Delta H_{rxn} = -125.3 \text{ kJ}$
2)	$5BrF(g) \rightarrow 2Br_2(g) + BrF_5(l)$	$\Delta H_{rxn} = -166.1 \text{ kJ}$
3)	$BrF_3(l) + F_2(g) \rightarrow BrF_5(l)$	$\Delta H_{rxn} = -158.0 \text{ kJ}$

Reverse equations 1 and 3, and add to equation 2:

1)	$\cancel{Br_2(g)} + \cancel{BrF_3(l)} \rightarrow 3BrF(g)$	$\Delta H_{rxn} = +125.3 \text{ kJ}$ (note sign change)
2)	$5BrF(g) \rightarrow \cancel{2Br_2(g)} + \cancel{BrF_5(l)}$	$\Delta H_{rxn} = -166.1 \text{ kJ}$
3)	$\cancel{BrF_5(l)} \rightarrow \cancel{BrF_3(l)} + F_2(g)$	$\Delta H_{rxn} = +158.0 \text{ kJ}$ (note sign change)

Total: $2BrF(g) \rightarrow Br_2(g) + F_2(g)$ **$\Delta H_{rxn} = +117.2 \text{ kJ}$**

14.127   <u>Plan:</u> Nitrite ion, NO_2^-, has $[1 \times N(5e^-)] + [2 \times O(6e^-)] + [1e^- \text{ from charge}] = 18$ valence electrons. Four electrons are used in the single bonds between the atoms, leaving $18 - 4 = 14$ electrons. Since sixteen electrons are required to complete the octets of the atoms, one double bond is needed. There are two resonance structures. Nitrogen dioxide, NO_2, has $[1 \times N(5e^-)] + [2 \times O(6e^-)] = 17$ valence electrons. Four electrons are used in the single bonds between the atoms, leaving $17 - 4 = 13$ electrons. Since sixteen electrons are required to complete the octets of the atoms, one double bond is needed and one atom must have an unpaired electron. There are two resonance structures. The nitronium ion, NO_2^+, has $[1 \times N(5e^-)] + [2 \times O(6e^-)] - [1e^- \text{ due to } + \text{ charge}] = 16$ valence electrons. Four electrons are used in the single bonds between the atoms, leaving $16 - 4 = 12$ electrons. Since sixteen electrons are required to complete the octets of the atoms, two double bonds are needed.

<u>Solution:</u>

The Lewis structures are

The nitronium ion (NO_2^+) has a linear shape because the central N atom has two surrounding electron groups, which achieve maximum repulsion at 180°. Both the nitrite ion (NO_2^-) and nitrogen dioxide (NO_2) have a central N surrounded by three electron groups. The electron-group arrangement would be trigonal planar with an ideal bond angle of 120°. The bond angle in NO_2^- is more compressed than that in NO_2 since the lone pair of electrons in NO_2^- takes up more space than the lone electron in NO_2. Therefore the bond angle in NO_2^- is smaller (115°) than that of NO_2 (134°)

14.129   <u>Plan:</u> To find the limiting reactant, find the moles of UF_6 that can be produced from the given amount of uranium and then from the given amount of ClF_3, use the mole ratios in the balanced equation. The density of ClF_3 is used to find the mass of ClF_3. The limiting reactant determines the amount of UF_6 that can be produced.

Solution:

$U(s) + 3ClF_3(l) \rightarrow UF_6(l) + 3ClF(g)$

(1 metric ton = 1 t = 1000 kg)

Moles of UF_6 from U = $(1.00 \text{ t ore})\left(\dfrac{10^3 \text{ kg}}{1 \text{ t}}\right)\left(\dfrac{10^3 \text{ g}}{1 \text{ kg}}\right)\left(\dfrac{1.55\%}{100\%}\right)\left(\dfrac{1 \text{ mol U}}{238.0 \text{ g U}}\right)\left(\dfrac{1 \text{ mol UF}_6}{1 \text{ mol U}}\right)$

= 65.12605 mol UF_6

Moles of UF_6 from ClF_3 = $(12.75 \text{ L})\left(\dfrac{1 \text{ mL}}{10^{-3} \text{ L}}\right)\left(\dfrac{1.88 \text{ g ClF}_3}{1 \text{ mL}}\right)\left(\dfrac{1 \text{ mol ClF}_3}{92.45 \text{ g ClF}_3}\right)\left(\dfrac{1 \text{ mol UF}_6}{3 \text{ mol ClF}_3}\right)$

= 86.42509 mol UF_6

Since the amount of uranium will produce less uranium hexafluoride, it is the limiting reactant.

Mass (g) of UF_6 = $(65.12605 \text{ mol UF}_6)\left(\dfrac{352.0 \text{ UF}_6}{1 \text{ mol UF}_6}\right) = 2.2924 \times 10^4 = \mathbf{2.29 \times 10^4 \text{ g UF}_6}$

14.133 **Plan:** Determine the electron configuration of each species. Partially filled orbitals lead to paramagnetism (unpaired electrons).

Solution:

O^+	$1s^2 2s^2 2p^3$	**paramagnetic**	odd number of electrons
O^-	$1s^2 2s^2 2p^5$	**paramagnetic**	odd number of electrons
O^{2-}	$1s^2 2s^2 2p^6$	diamagnetic	all orbitals filled (all electrons paired)
O^{2+}	$1s^2 2s^2 2p^2$	**paramagnetic**	Two of the $2p$ orbitals have one electron each. These electrons have parallel spins (Hund's rule).

14.134 **Plan:** To determine mass percent, divide the mass of As in 1 mole of compound by the molar mass of the compound and multiply by 100. Find the volume of the room (length x width x height) and use the toxic concentration to find the mass of As required. The mass percent of As in $CuHAsO_3$ is used to convert that mass of As to mass of compound.

Solution:

a) Mass percent = $\dfrac{\text{mass of As}}{\text{mass of compound}}(100)$

% As in $CuHAsO_3$ = $\dfrac{74.92 \text{ g As}}{187.48 \text{ g CuHAsO}_3}(100) = 39.96160 = \mathbf{39.96\% \text{ As}}$

% As in $(CH_3)_3As$ = $\dfrac{74.92 \text{ g As}}{120.02 \text{ g (CH}_3)_3\text{As}}(100) = 62.4229 = \mathbf{62.42\% \text{ As}}$

b) Volume (m^3) of room = $(12.35 \text{ m})(7.52 \text{ m})(2.98 \text{ m}) = 276.75856 \text{ m}^3$

Mass (g) of As = $(276.75856 \text{ m}^3)\left(\dfrac{0.50 \text{ mg As}}{\text{m}^3}\right)\left(\dfrac{10^{-3} \text{ g}}{1 \text{ mg}}\right) = 0.13838 \text{ g As}$

Mass (g) of $CuHAsO_3$ = $(0.13838 \text{ g As})\left(\dfrac{100 \text{ g CuHAsO}_3}{39.96160 \text{ g As}}\right) = 0.346282 = \mathbf{0.35 \text{ g CuHAsO}_3}$

CHAPTER 15 ORGANIC COMPOUNDS AND THE ATOMIC PROPERTIES OF CARBON

FOLLOW–UP PROBLEMS

15.1A a) <u>Plan:</u> For compounds with seven carbons, first start with 7 C atoms in a straight chain. Then make all arrangements with 6 carbons in a straight chain and 1 carbon branched off the chain. Examine the structures to make sure they are all different. Continue in the same manner with 5 C atoms in chain and 2 branched off the chain, then 4 C atoms in chain and 3 branched off, and 3 C atoms in chain and 4 branched off. Examine all structures to guarantee there are no duplicates.

<u>Solution:</u>

Seven carbons in chain:

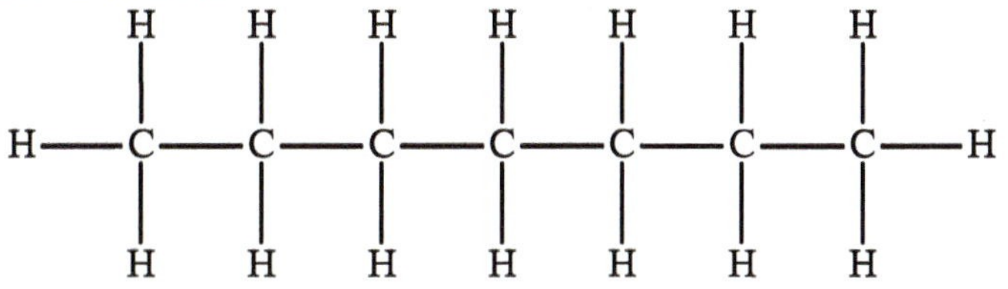

(1)

Six carbons in chain:

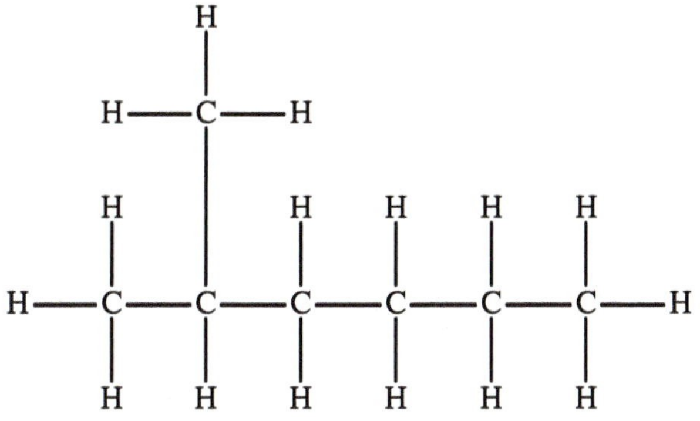

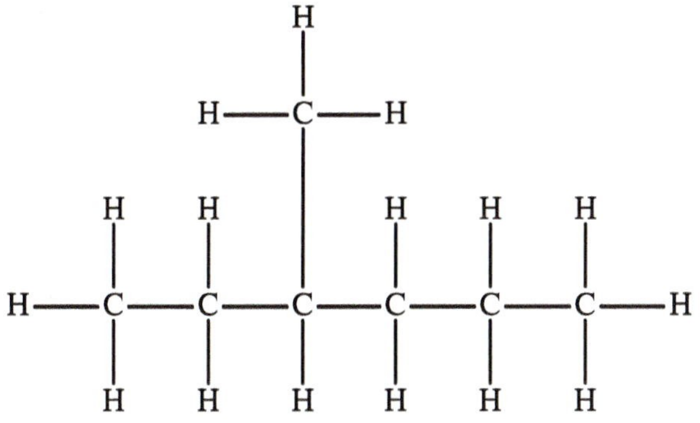

(2)   (3)

If the methyl group (—CH₃) is moved to the fourth carbon from the left, the structure is the same as (3).
If the methyl group is moved to the fifth carbon from the left, the structure is the same as (2). In addition,
if the methyl group is moved to the sixth carbon from the left, the resulting structure has 7 carbons in a
chain and is the same as (1).

Five carbons in chain:

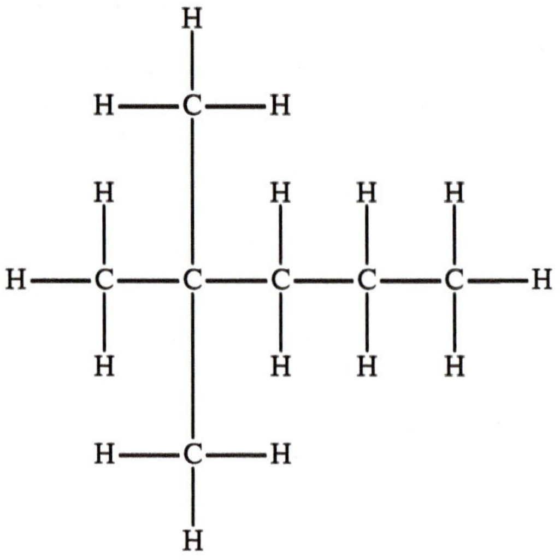

```
                H
                |
          H ——— C ——— H
                |
      H    H    H    H    H
      |    |    |    |    |
  H —— C —— C —— C —— C —— C —— H
      |    |    |    |    |
      H    H    H    H    H
                |
          H ——— C ——— H
                |
                H

                (4)
```

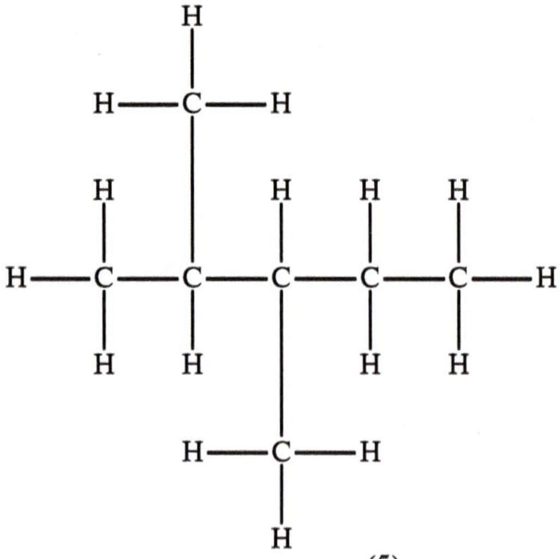

```
                H
                |
          H ——— C ——— H
                |
      H         H    H    H
      |         |    |    |
  H —— C —— C —— C —— C —— C —— H
      |    |    |    |    |
      H    H    H    H    H
                |
          H ——— C ——— H
                |
                H
                                (5)
```

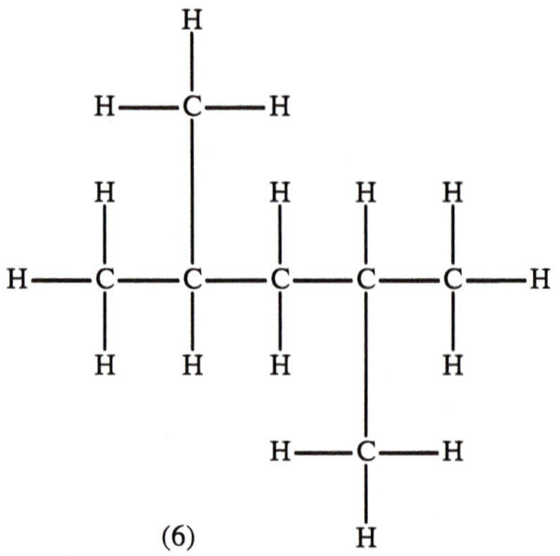

```
                H
                |
          H ——— C ——— H
                |
      H    H    H    H
      |    |    |    |
  H —— C —— C —— C —— C —— C —— H
      |    |    |    |    |
      H    H    H    H    H
                     |
                H ——— C ——— H
                     |
        (6)          H
```

```
                H
                |
          H ——— C ——— H
                |
      H    H         H    H
      |    |         |    |
  H —— C —— C —— C —— C —— C —— H
      |    |    |    |    |
      H    H    H    H    H
                |
          H ——— C ——— H
                |
                H         (7)
```

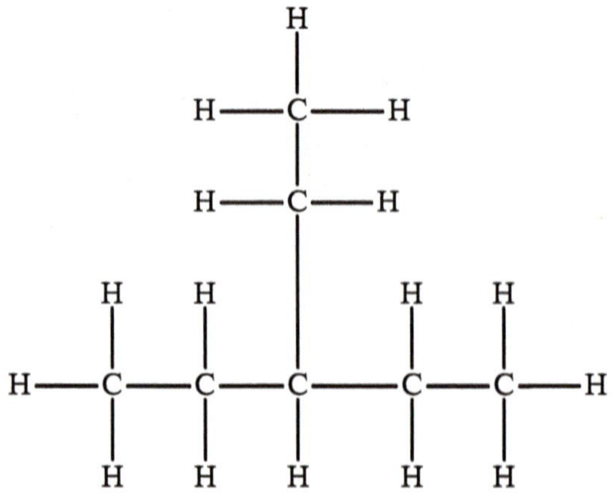

```
                H
                |
          H ——— C ——— H
                |
          H ——— C ——— H
                |
      H    H    H    H    H
      |    |    |    |    |
  H —— C —— C —— C —— C —— C —— H
      |    |    |    |    |
      H    H    H    H    H
                (8)
```

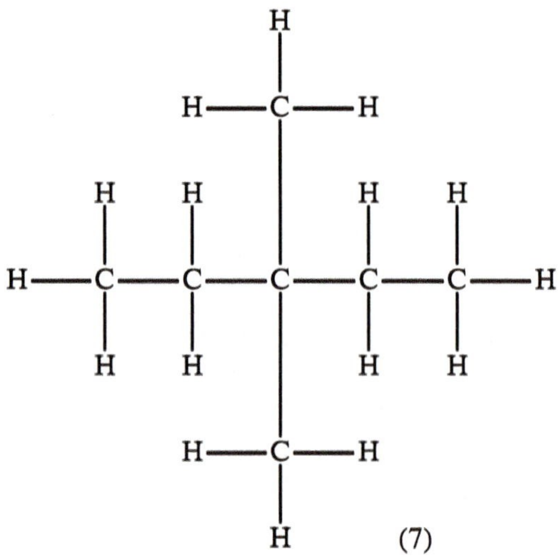

15-2

Starting with a methyl group on the second carbon from the left, add another methyl group in a systematic pattern. Structure (4) has the second methyl on the second carbon, (5) on the third carbon, (6) on the 4th carbon. These are all the unique possibilities with the first methyl group on the second carbon. Any other variation will produce a structure identical to a structure already drawn. Next, move the first methyl group to the third carbon where the only unique placement for the second methyl group is on the third carbon as shown in structure (7). Which structure above would be the same as placing the first methyl group on the third carbon and the second methyl group on the fourth carbon? Are there any more arrangements with a 5-carbon chain? Think of attaching the two extra carbons as an ethyl group (—CH₂CH₃). If the ethyl group is attached to the second carbon from the left, the longest chain becomes 6 carbons instead of 5 and the structure is the same as (3). If the ethyl group is instead attached to the third carbon, the longest chain is still 5 carbons and the ethyl group is a side chain to give structure (8).

4 carbons in chain:

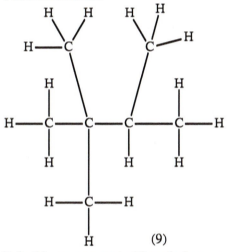

(9)

Only this arrangement of 3 methyl groups branching off the two inner carbons in a 4 C chain produces a unique structure.

Attaching 4 methyl groups off a 3 C chain is impossible without making the chain longer. So, the above nine structures are all the arrangements.

b) Plan: For a five-carbon compound start with 5 C atoms in a chain and place the triple bond in as many unique places as possible.

Solution:

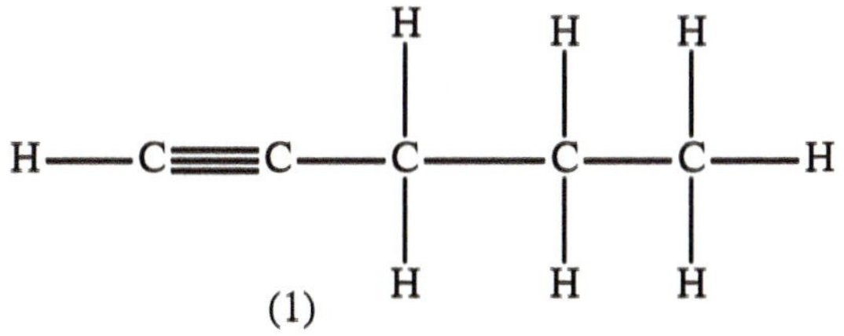

$$H-C\equiv C-\underset{\underset{H}{|}}{\overset{\overset{H}{|}}{C}}-\underset{\underset{H}{|}}{\overset{\overset{H}{|}}{C}}-\underset{\underset{H}{|}}{\overset{\overset{H}{|}}{C}}-H \qquad H-\underset{\underset{H}{|}}{\overset{\overset{H}{|}}{C}}-C\equiv C-\underset{\underset{H}{|}}{\overset{\overset{H}{|}}{C}}-\underset{\underset{H}{|}}{\overset{\overset{H}{|}}{C}}-H$$

(1) 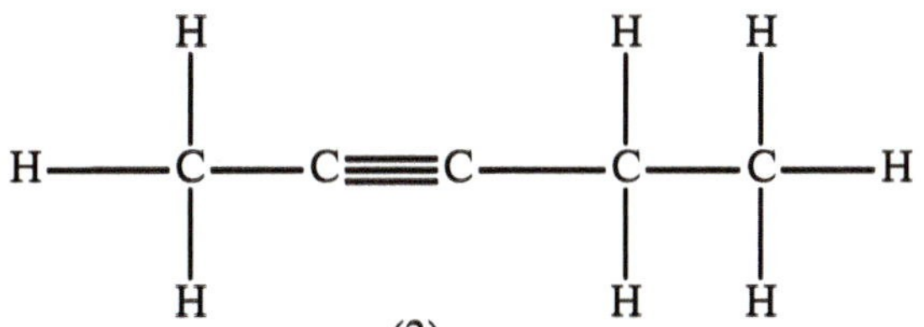   (2)

4 C atoms in chain: There are two unique placements for the triple bond: one between the first and second carbons and one between the second and third carbons in the chain.

$$C\equiv C-C-C \qquad\qquad C-C\equiv C-C$$

The fifth carbon is added as a methyl group branched off the chain. With the triple bond between the first and second carbons, the methyl group is attached to the third carbon to give structure (3).

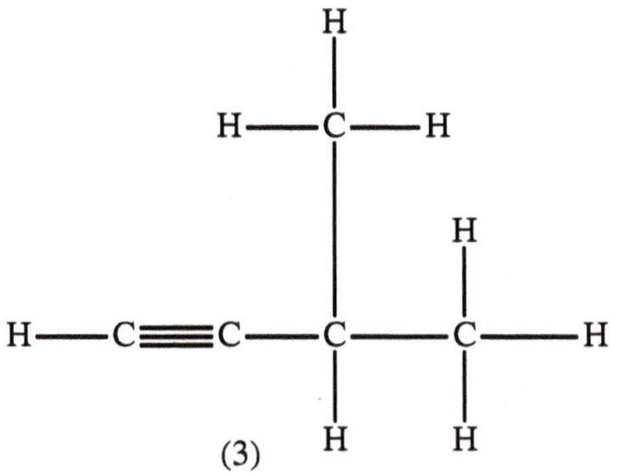

(3)

With the triple bond between the second and third carbons, the methyl group cannot be attached to either the second or the third carbons because that would create 5 bonds to a carbon and carbon has only 4 bonds. Thus, a unique structure cannot be formed with a 4 C chain where the triple bond is between the second and third carbons. There are only 3 structures with 5 carbon atoms, one triple bond, and no rings.

15.1B   a) Plan: Start with a ring consisting of 4 atoms. Make sure to include the double bond in the ring. Then work with a ring consisting of 3 atoms and 1 carbon branched off the ring. Move the position of the double bond relative to the branch to draw the 3 remaining structures. The double bond can occur within the ring or between a ring carbon and the branch carbon.
Solution:
Four carbons in the ring:

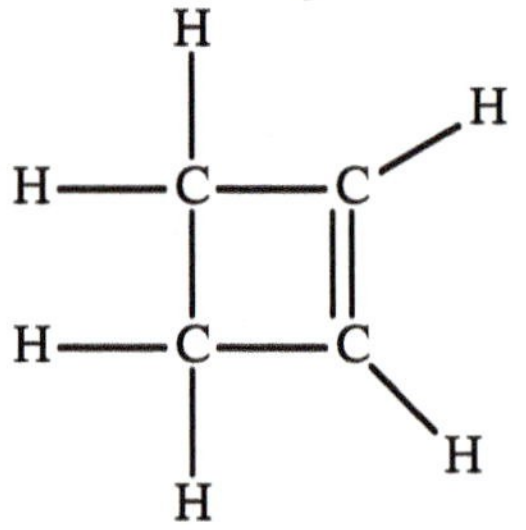

Three carbons in the ring:

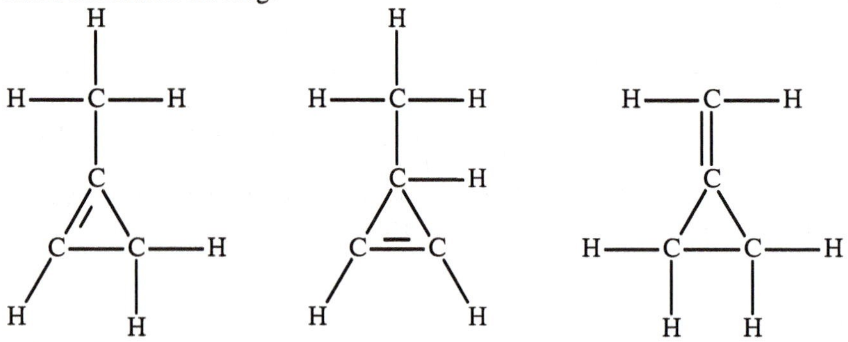

b) Plan: For a four-carbon compound start with 4 C atoms in a chain and place the two double bonds in as many unique combinations as possible (there are 2). It will not be possible to draw a molecule with 3 C atoms in a chain, 1 branch, and 2 double bonds without a C atom having more than 4 bonds.
Solution:

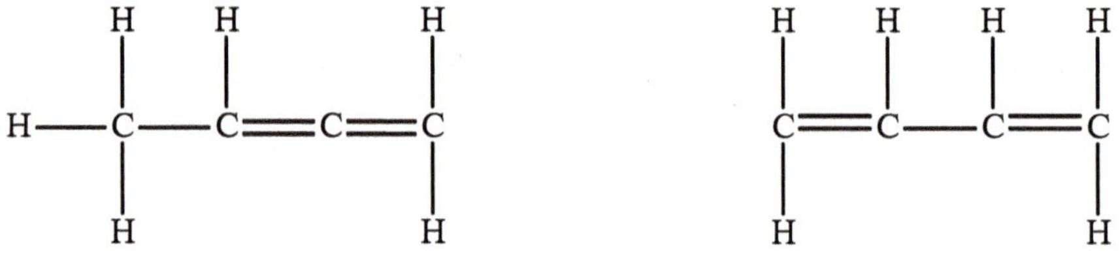

15-4

15.2A   <u>Plan:</u> Examine the structure for chain length and side groups; then use the structure to write the name. In part d), examine the structure for carbons that are bonded to four different groups. Those are the chiral carbons.
<u>Solution:</u>
a) **3,3-diethylpentane.** There are 5 single-bonded carbons in the main chain and two ethyl groups attached to carbon #3 of that chain. The end of the name, pentane, indicates a 5 C chain (pent- represents 5 C) with only single bonds between the carbons (-ane represents alkanes, only single bonds). 3,3-diethyl- means there are two ethyl groups and that each ethyl group is attached to carbon #3.

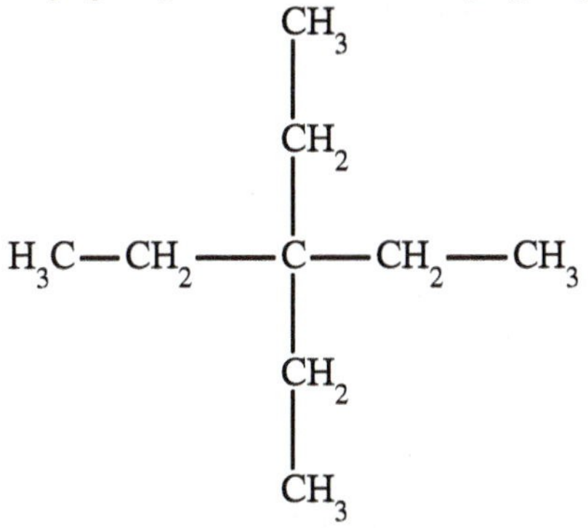

$$
\begin{array}{c}
CH_3 \\
| \\
CH_2 \\
| \\
H_3C-CH_2-\!\!\!-C-\!\!\!-CH_2-\!\!\!-CH_3 \\
| \\
CH_2 \\
| \\
CH_3
\end{array}
$$

b) **1-ethyl-2-methylcyclobutane.** The main "chain" is a four-membered ring with an ethyl group and a methyl group attached as branches. The name cyclobutane describes the 4 C ring (cyclobut-) with only single bonds (-ane). The 1-ethyl- indicates that an ethyl group ($-CH_2CH_3$) is attached to carbon #1. The 2-methyl-indicates that a methyl group ($-CH_3$) is attached to carbon #2. The lower number carbon is assigned to the ethyl group because it precedes methyl alphabetically.

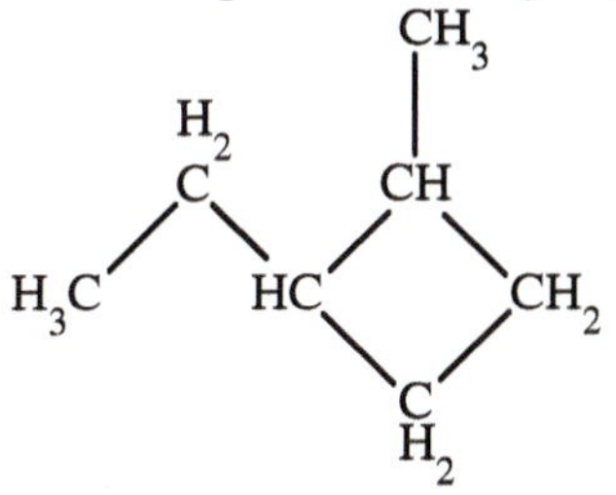

c) *trans*-**3-methyl-3-hexene.** There are 6 carbons in the main chain, with a double bond between carbons #3 and #4 (counting from the right hand side). There is also a methyl group attached to carbon #3. The name 3-hexene describes the main chain (hex- indicates that there are 6 carbons in the chain, -ene indicates that there is a double bond in the chain, 3- indicates that the double bond starts at carbon #3). The name 3-methyl- indicates that there is a methyl group on carbon #3. Finally, because there is a double bond, we have to determine whether the groups bonded to the double bond are in a *cis*- (same side of the double bond) or *trans*- (opposite sides of the double bond) configuration. In this structure, the longer carbon chains attached to the double bond are on opposite sides of the double bond, so we add the prefix *trans*- to the name of the compound.

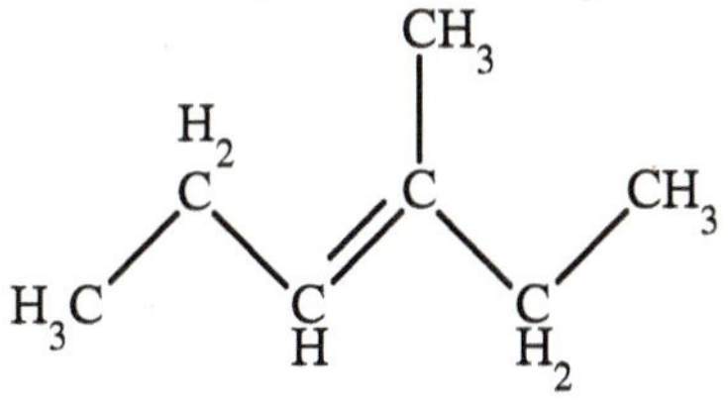

d) **1-methyl-2-propylcyclopentane.** The main "chain" in the molecule is a 5 C ring with only single bonds. A methyl group and a propyl group are attached to the ring. The name cyclopentane describes a 5-membered ring (cyclopent-) with all single bonds (-ane). The name 1-methyl- indicates that a methyl group is attached to the ring at the #1 position. The name 2-propyl- indicates that a propyl group is attached to the ring at the #2 position. The lower number carbon is assigned to the methyl group because it precedes propyl alphabetically. The chiral carbons (those bonded to four different groups) are marked with asterisks.

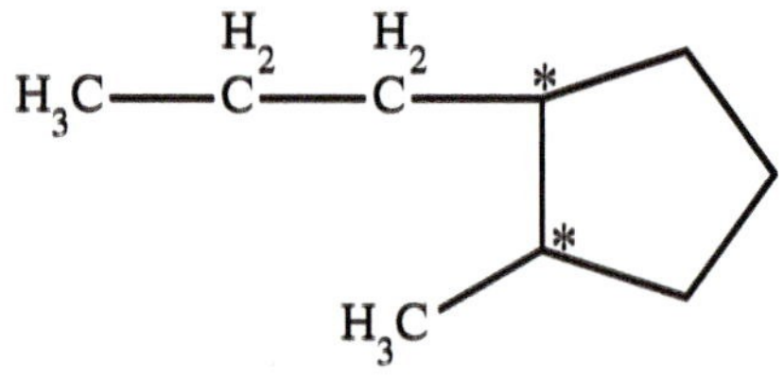

15.2B   Plan: Analyze the name for chain length and side groups; then draw the structure.

Solution:

a) 3-ethyl-3-methyloctane. The end of the name, octane, indicates an 8 C chain (oct- represents 8 C) with only single bonds between the carbons (-ane represents alkanes, only single bonds). 3-ethyl means an ethyl group ($-CH_2CH_3$) attached to carbon #3 and 3-methyl means a methyl group ($-CH_3$) attached to carbon #3.

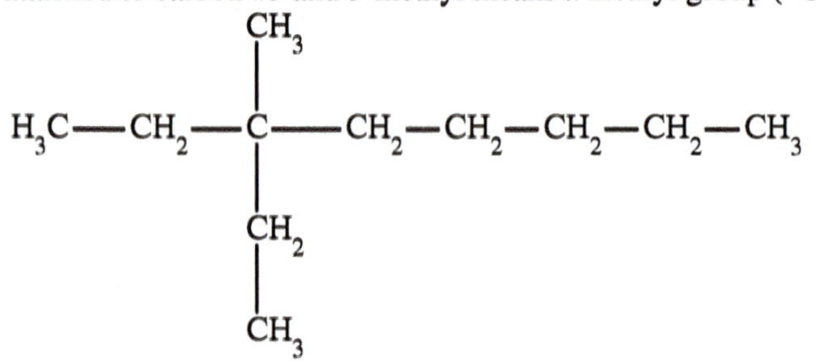

b) 1-ethyl-3-propylcyclohexane. The hexane indicates 6 C chain (hex-) and only single bonds (-ane). The cyclo-indicates that the 6 carbons are in a ring. The 1-ethyl indicates that an ethyl group ($-CH_2CH_3$) is attached to carbon #1. Select any carbon atom in the ring as carbon #1 since all the carbon atoms in the ring are equivalent. The 3-propyl indicates that a propyl group ($-CH_2CH_2CH_3$) is attached to carbon #3.

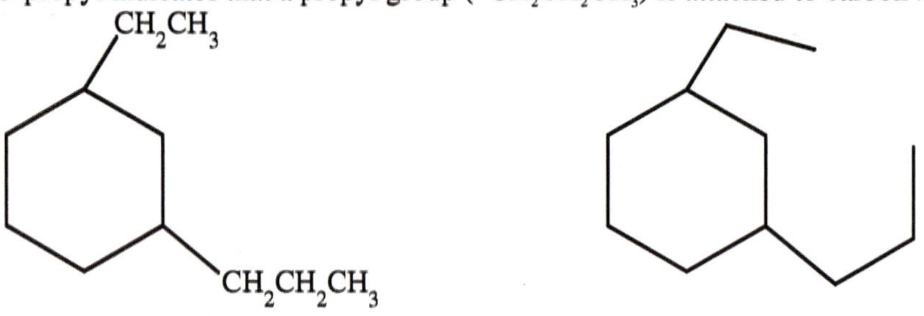

c) 3,3-diethyl-1-hexyne. The 1-hexyne indicates a 6 C chain with a triple bond (-yne) between C #1 and C #2. The 3,3-diethyl means two (di-) ethyl groups ($-CH_2CH_3$) both attached to C #3.

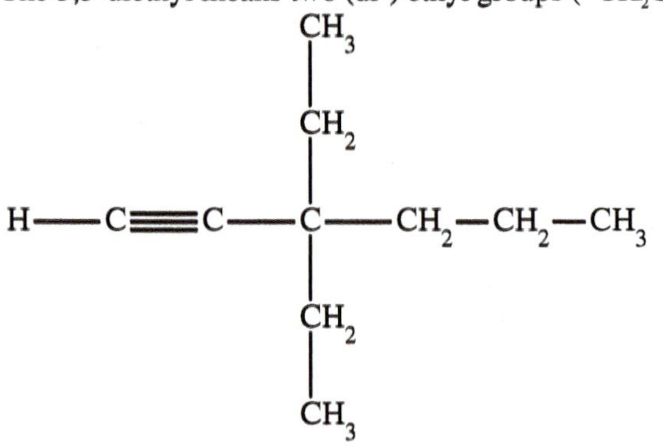

d) *trans*-3-methyl-3-heptene. The 3-heptene indicates a 7 C chain (hept-) with one double bond (-ene) between the 3rd and 4th carbons. The 3-methyl indicates a methyl group ($-CH_3$) attached to carbon #3. The *trans* indicates that the arrangement around the two carbons in the double bond gives the two smaller groups on opposite sides of the double bond. Therefore, the smaller group on the third carbon (which would be the methyl group) is above the double bond while the smaller group, H, on the fourth carbon is below the double bond.

15-6

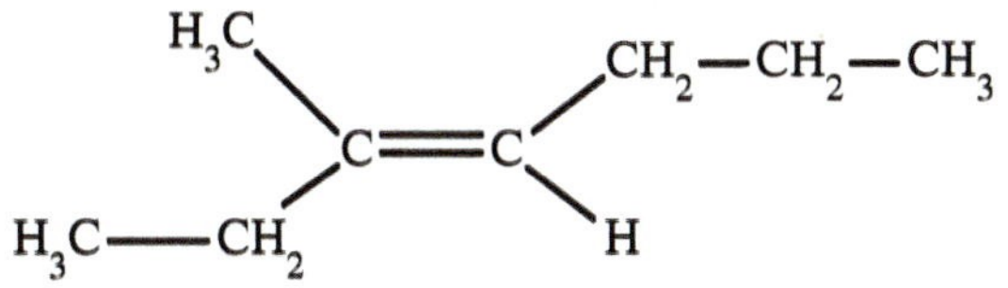

H₃C and CH₂—CH₂—CH₃ / C=C / H₃C—CH₂ and H (structure at top)

15.3A **Plan:** In an addition reaction, atoms are added to the carbon(s) in a double bond. Atoms are removed in an elimination reaction, resulting in a product with a double bond. In a substitution reaction, an atom or group of atoms substitutes for another one in the reactant.

Solution:

a) In this reaction, the Br on the carbon chain is replaced with a hydroxyl group while the carbons maintain the same number of bonds. This is a **substitution reaction**.

b) The methyl group on the left-hand side of the structure is replaced by a hydrogen atom while the carbons maintain the same number of bonds. This is a **substitution reaction**.

c) Water(in the form of a –H group and an –OH group) is added to the double bond, resulting in the product having three more atoms. Additionally, there is a second O bonded to the C in the product compared to the reactant. This is an **addition reaction**.

15.3B **Plan:** a) An addition reaction involves breaking a multiple bond, in this case the double bond in 2-butene, and adding the other reactant to the carbons in the double bond. The reactant Cl₂ will add –Cl to one of the carbons and –Cl to the other carbon.

b) A substitution reaction involves removing one atom or group from a carbon chain and replacing it with another atom or group. For 1-bromopropane the bromine will be replaced by hydroxide.

c) An elimination reaction involves removing two atoms or groups, one from each of two adjacent carbon atoms, and forming a double bond between the two carbon atoms. For 2methyl2propanol, the –OH group from the center carbon and a hydrogen from one of the terminal carbons will be removed and a double bond formed between the two carbon atoms.

Solution:

(a)

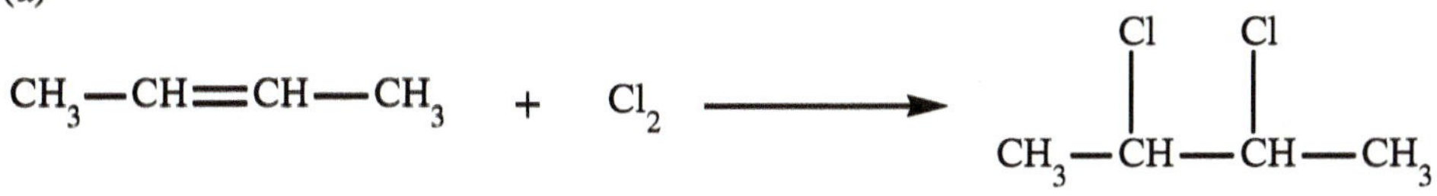

CH_3—CH=CH—CH_3 + Cl_2 ⟶ CH_3—CH—CH—CH_3 (with Cl, Cl on the two central carbons)

(b)

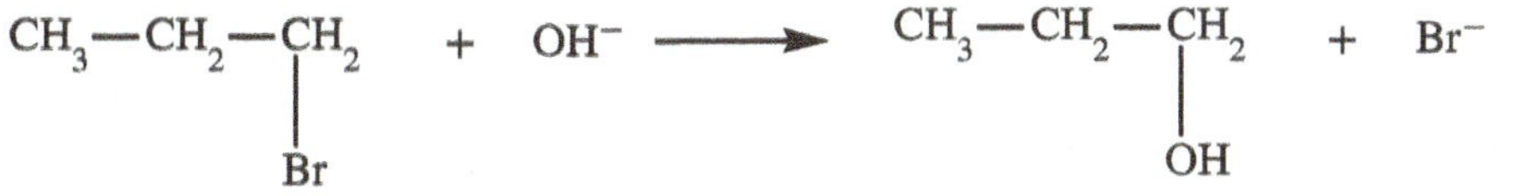

CH_3—CH_2—CH_2 (with Br) + OH^- ⟶ CH_3—CH_2—CH_2 (with OH) + Br^-

(c)

CH_3—C—CH_3 (with CH₃ above and O—H below) ⟶ CH_3—C=CH_2 (with CH₃ above) + H_2O

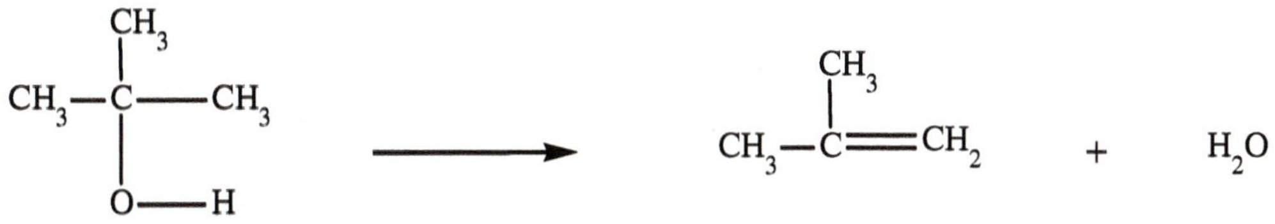

15.4A **Plan:** Determine the functional group(s) of the organic reactant(s) and then examine any inorganic reactant(s) to decide on the reaction type.

Solution:

a) This reaction is an oxidation (**elimination**) reaction because $Cr_2O_7^{2-}$ and H_2SO_4 are oxidizing agents. An alcohol group (–OH) is oxidized to a ketone group and a single bond between C and O is converted to a double bond.

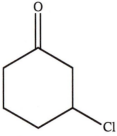

b) This is a **substitution** reaction because the bromine atoms on the organic reactant (an alkyl halide) can be replaced with the cyanide groups from the inorganic reactant.

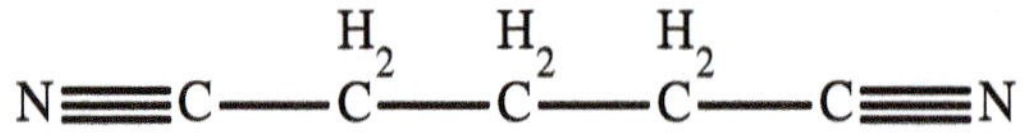

$$N\equiv C-\overset{H_2}{C}-\overset{H_2}{C}-\overset{H_2}{C}-C\equiv N$$

15.4B   Plan: Examine any inorganic compounds and the organic product to determine the organic reactant. Look for differences between the reactants and products, if possible.
Solution:
a) The organic reactant must contain a single bond between the second and third carbons and the second and third carbons each have an additional group: an H atom on one carbon and a Cl atom on the other carbon.

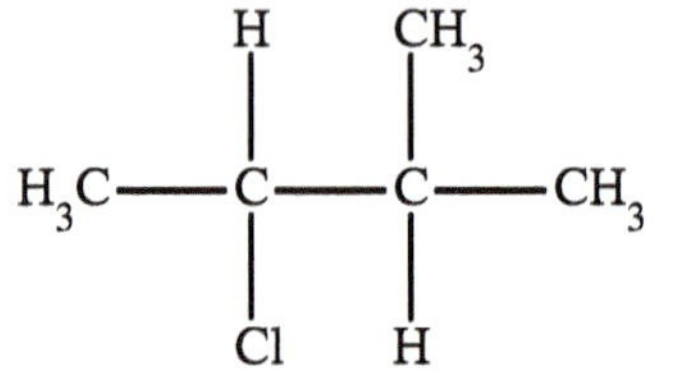

H₃C—C(H)(Cl)—C(CH₃)(H)—CH₃   or   H₃C—CH₂—C(CH₃)(Cl)—CH₃

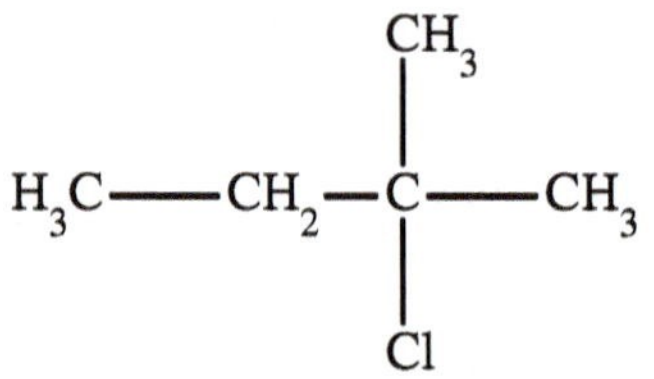

b) The third carbon from the left in the product contains the acid group. Therefore, in the reactant this carbon should have an alcohol group.

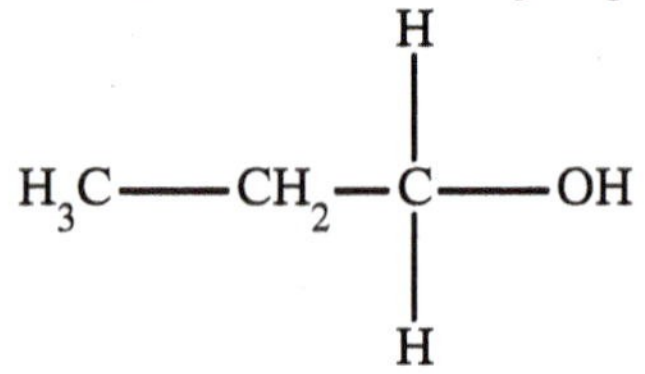

H₃C—CH₂—C(H)(H)—OH

15.5A   Plan: In part a), LiAlH₄ and H₂O act as reducing agents to convert a ketone or aldehyde to an alcohol. To form the product, change the carbonyl group, =O, to an –OH group. In part b) ethyllithium, CH₃CH₂–Li, and H₂O react to convert a ketone or aldehyde to form the alcohol *and* to add an ethyl group to the carbonyl carbon. To form the product, find the carbonyl carbon, add an ethyl group (from CH₃CH₂–Li) to the carbon and change the carbonyl group, =O, to a –OH bond.
Solution:
a)

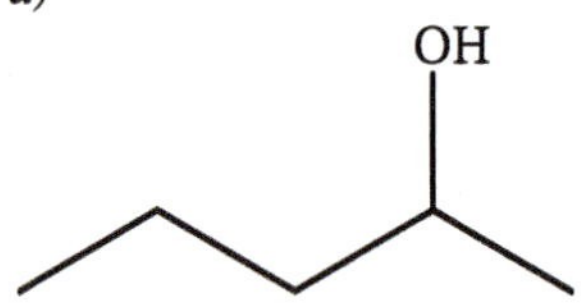

b)

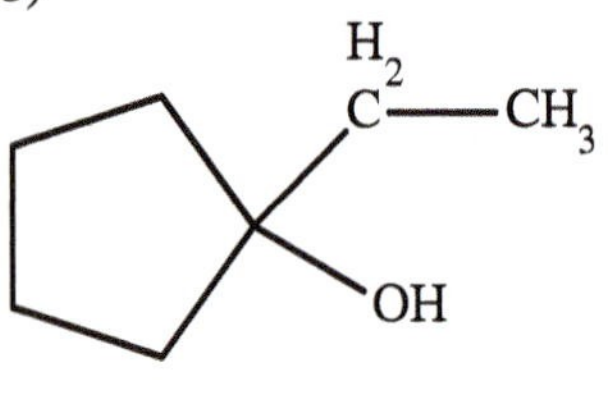

15.5B   Plan: The oxidizing agents in reaction a) indicate that the ketone group (=O) has been oxidized from an alcohol. To form the reactant, replace the C=O with an alcohol group, (C–OH). In reaction b) the reactants CH₃CH₂–Li and H₂O indicate that a ketone or aldehyde reacts to form the alcohol. To form the reactant find the carbon with the alcohol group, remove the ethyl group that came from CH₃CH₂–Li and change the –OH bond to a carbonyl group, =O.
Solution:
a) In the reactant, a hydrogen atom is lost from the oxygen and another hydrogen is lost from the adjacent carbon. A double bond forms between the carbon and the oxygen.

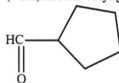

b) In the reactant, the first carbon off the ring will have a carbonyl group (–C=O) in place of the alcohol group (–OH) and the ethyl group (–CH₂CH₃).

Check: When the reactants are combined, the products are identical to those given.

15.6A   Plan: Determine the functional group(s) of the organic reactant(s) and then examine any inorganic reactant(s) to identify the reaction type. Then draw the structure of the product(s) based on the reaction type.
Solution:
a) The reactant is an ester because it contains the unit:

Reacting an ester with base results in a hydrolysis reaction in which an alcohol and a sodium carboxylate are formed.

H₃C——(H₂C)₁₄——C——Ō Na⁺   +   OH——CH₂——(CH₂)₁₄——CH₃

b) The reactant is an amide because it contains the unit:

Reacting an amide with a reducing agent like LiAlH₄ converts an amide to an amine.

H₃C——⟨benzene ring⟩——C H₂——N H——⟨benzene ring⟩——CH₃

Solution:

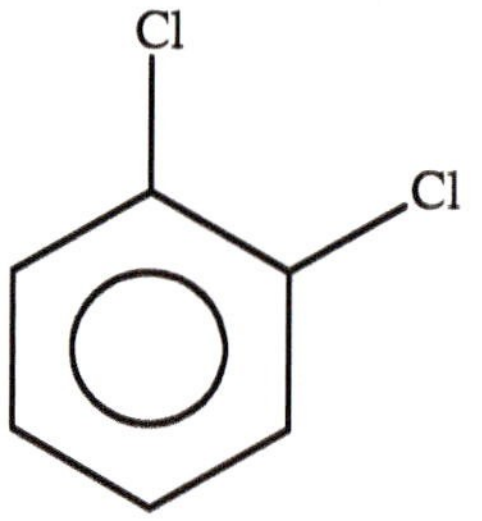

1,2-dichlorobenzene	1,3-dichlorobenzene	1,4-dichlorobenzene
(*o*-dichlorobenzene)	(*m*-dichlorobenzene)	(*p*-dichlorobenzene)

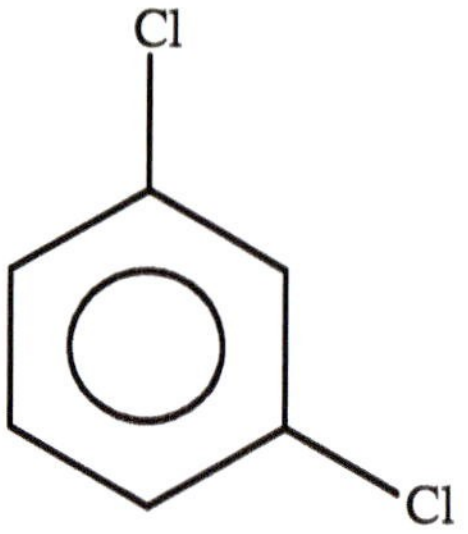

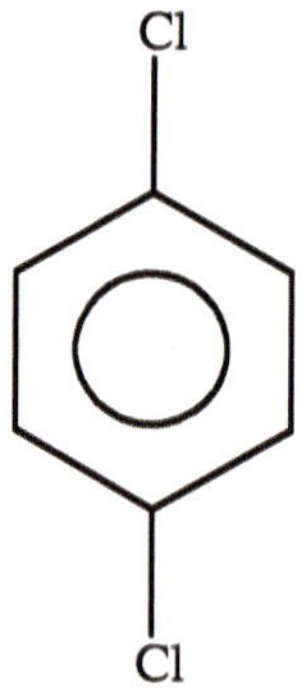

15.33 Plan: Analyzing the name gives benzene as the base structure with the following groups bonded to it: 1) on carbon #1 a hydroxy group, –OH; 2) on carbons #2 and #6 a *tert*-butyl group, –C(CH$_3$)$_3$; and 3) on carbon #4 a methyl group, –CH$_3$.

Solution:

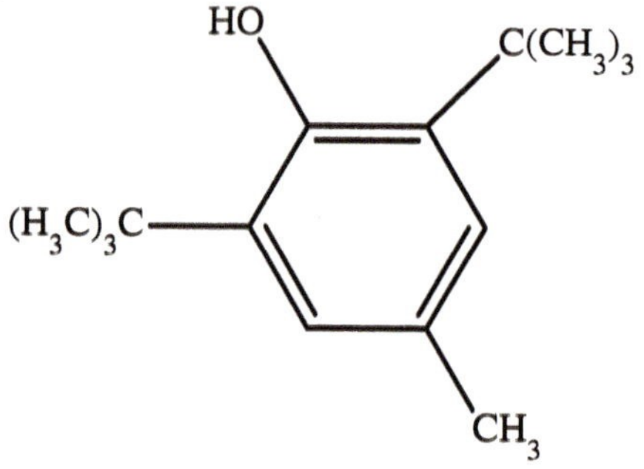

15.34 The compound 2-methyl-3-hexene has *cis-trans* isomers.

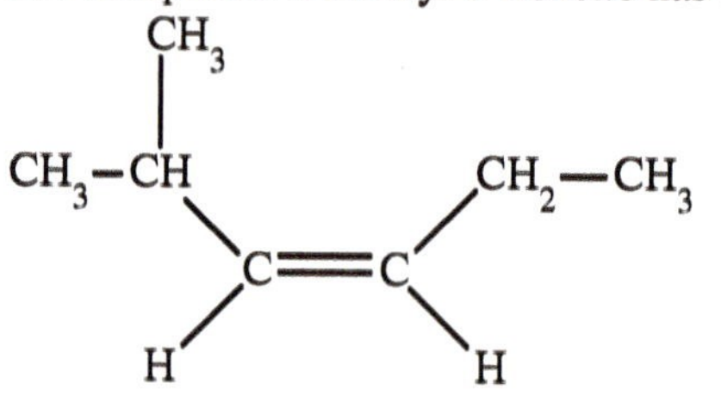

cis-2-methyl-3-hexene 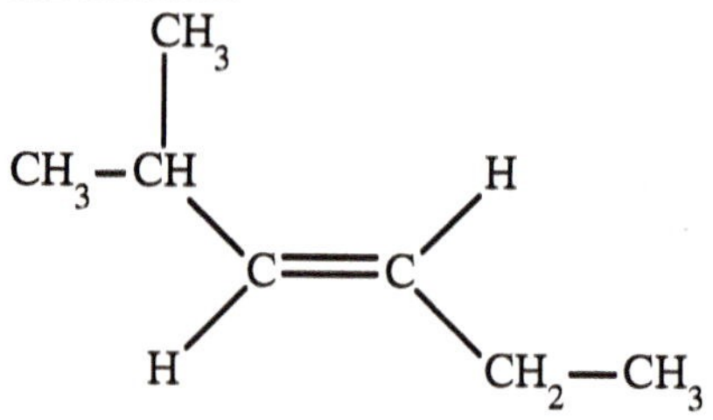   trans-2-methyl-3-hexene

The compound 2-methyl-2-hexene does not have *cis-trans* isomers because the #2 carbon atom is attached to two identical methyl (–CH$_3$) groups:

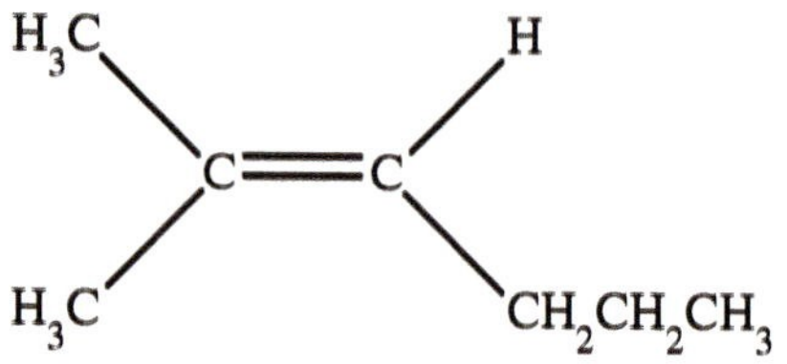

2-methyl-2-hexene

15.38   In an addition reaction, a double bond is broken to leave a single bond, and in an elimination reaction, a double bond is formed from a single bond. A double bond consists of a σ bond and a π bond. It is the **π bond** that breaks in the addition reaction and that forms in the elimination reaction.

15.40   Plan: Look for a change in the number of atoms bonded to carbon. In an addition reaction, more atoms become bonded to carbon; in an elimination reaction, fewer atoms are bonded to carbon, while in a substitution reaction, the product has the same number of atoms bonded to carbon.
Solution:
a) HBr is removed from the reactant so that there are fewer bonds to carbon in the product. This is an **elimination reaction**, and an unsaturated product is formed.
b) Hydrogen is added to the double bond, resulting in the product having two more atoms bonded to carbons. This is an **addition reaction**, resulting in a saturated product.

15.42   Plan: In an addition reaction, atoms are added to the carbons in a double bond. Atoms are removed in an elimination reaction, resulting in a product with a double bond. In a substitution reaction, an atom or group of atoms substitutes for another one in the reactant.
Solution:
a) Water (H_2O or H and OH) is added to the double bond:

$$CH_3CH_2CH=CHCH_2CH_3 + H_2O \xrightarrow{H^+} CH_3CH_2CH_2CHCH_2CH_3$$
$$|$$
$$OH$$

b) H and Br are eliminated from the molecule, resulting in a double bond:
$$CH_3CHBrCH_3 + CH_3CH_2OK \rightarrow CH_3CH=CH_2 + CH_3CH_2OH + KBr$$
c) Two chlorine atoms are substituted for two hydrogen atoms in ethane:
$$CH_3CH_3 + 2Cl_2 \xrightarrow{h\upsilon} CHCl_2CH_3 + 2HCl$$

15.44   Plan: To decide whether an organic compound is oxidized or reduced in a reaction, rely on the rules in the chapter:
A C atom is oxidized when it forms more bonds to O or fewer bonds to H because of the reaction.
A C atom is reduced when it forms fewer bonds to O or more bonds to H because of the reaction.
Solution:
a) The C atom is **oxidized** because it forms more bonds to O.
b) The C atom is **reduced** because it forms more bonds to H.
c) The C atom is **reduced** because it forms more bonds to H.

15.46   Plan: A C atom is oxidized when it forms more bonds to O or fewer bonds to H because of the reaction.
A C atom is reduced when it forms fewer bonds to O or more bonds to H because of the reaction.
Solution:
a) The reaction $CH_3CH=CHCH_2CH_2CH_3 \rightarrow CH_2CH(OH)—CH(OH)CH_2CH_2CH_3$ shows the second and third carbons in the chain gaining a bond to oxygen: C–O–H. Therefore, the 2-hexene compound has been **oxidized**.
b) The reaction shows that each carbon atom in the cyclohexane loses a bond to hydrogen to form benzene. Fewer bonds to hydrogen in the product indicates **oxidation**.

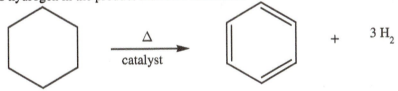

15.49   Plan: Physical properties are determined by intermolecular forces. Decide whether the compound is polar or nonpolar and if hydrogen bonding (H bonded to O, N, or F) is present. Hydrogen bonding and polarity (dipole-dipole forces) lead to higher melting and boiling points and water solubility than found in nonpolar compounds which have weaker dispersion forces.

<u>Solution:</u>

a) The structures for chloroethane and methylethylamine are given below. The compound **methylethylamine** is more soluble due to its ability to form hydrogen bonds with water. Recall that N–H, O–H, or F–H bonds are required for H bonding.

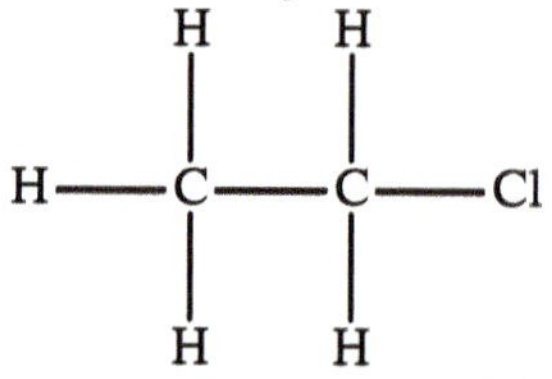

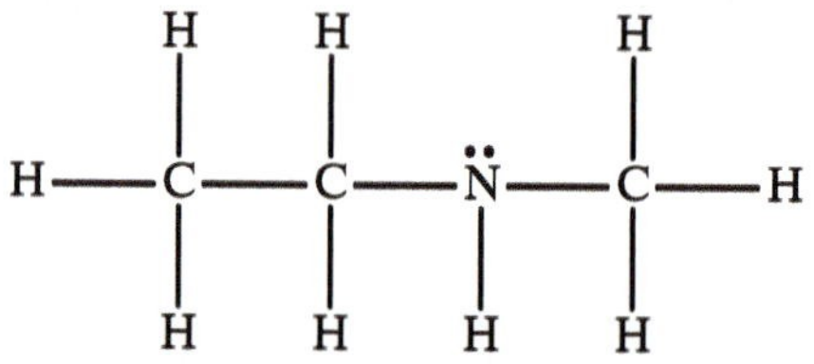

b) The compound 1–butanol is able to hydrogen bond with itself (shown below) because it contains covalent O–H bonds in the molecule. Diethyl ether molecules contain no O–H covalent bonds and experience dipole-dipole interactions instead of H bonding as intermolecular forces. Therefore, **1-butanol** has a higher melting point because H bonds are stronger intermolecular forces than dipole-dipole attractions.

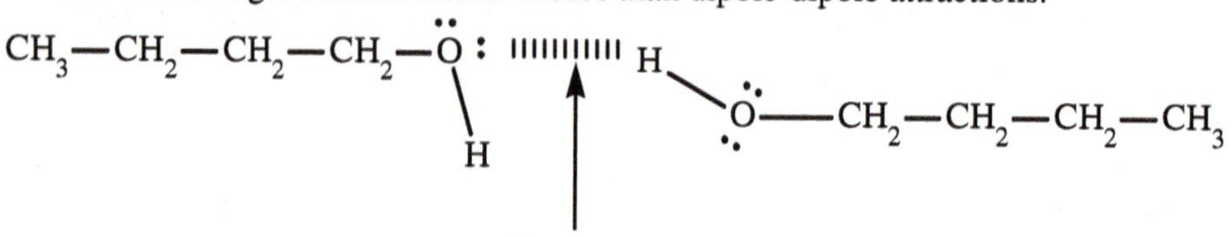

Hydrogen Bond

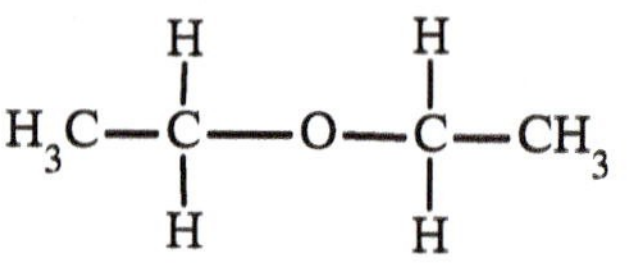

diethyl ether

c) **Propylamine** has a higher boiling point because it contains N–H bonds necessary for hydrogen bonding. Trimethylamine is a tertiary amine with no N–H bonds, and so its intermolecular forces are weaker.

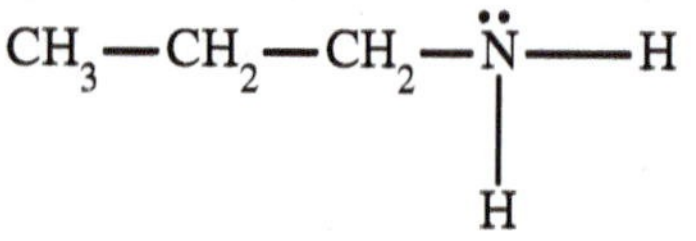

Propylamine 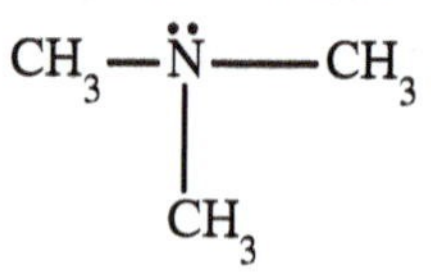   Trimethylamine

15.52   The C=C bond is nonpolar while the C=O bond is polar, since oxygen is more electronegative than carbon. Both bonds react by addition. In the case of addition to a C=O bond, an electron-rich group will bond to the carbon and an electron-poor group will bond to the oxygen, resulting in one product. In the case of addition to an alkene, the carbons are identical, or nearly so, so there will be no preference for which carbon bonds to the electron-poor group and which bonds to the electron-rich group. This may lead to two isomeric products, depending on the structure of the alkene.

When water is added to a double bond, the hydrogen is the electron-poor group and hydroxyl is the electron-rich group. For a compound with a carbonyl group, only one product results as H bonds to the O atom in the double bond and –OH bonds to the carbon atom in the double bond:

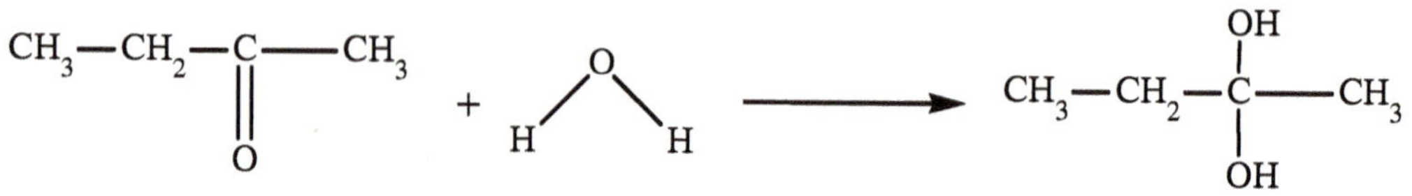

However, when water adds to a C=C, two products result since the OH can bond to either carbon in the double bond:

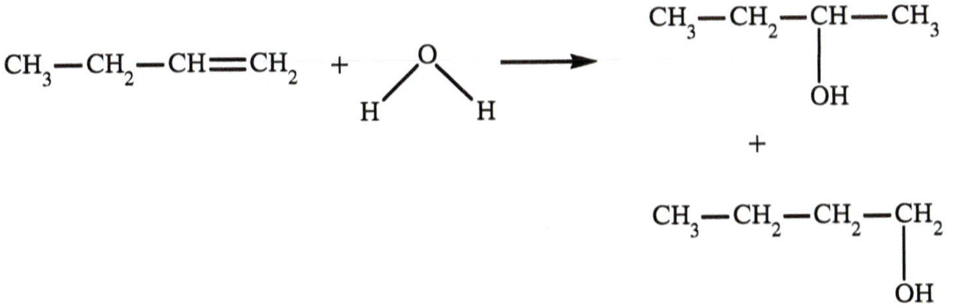

$$CH_3-CH_2-CH=CH_2 \ + \ H{-}O{-}H \longrightarrow$$

$$CH_3-CH_2-CH-CH_3$$
$$\qquad\qquad\qquad | $$
$$\qquad\qquad\quad OH$$

$$+$$

$$CH_3-CH_2-CH_2-CH_2$$
$$\qquad\qquad\qquad\qquad | $$
$$\qquad\qquad\qquad\quad OH$$

In this reaction, very little of the second product forms.

15.55 Esters and acid anhydrides form through **dehydration-condensation** reactions. Dehydration indicates that the other product is **water**. In the case of ester formation, condensation refers to the combination of the carboxylic acid and the alcohol. The ester forms and water is the other product.

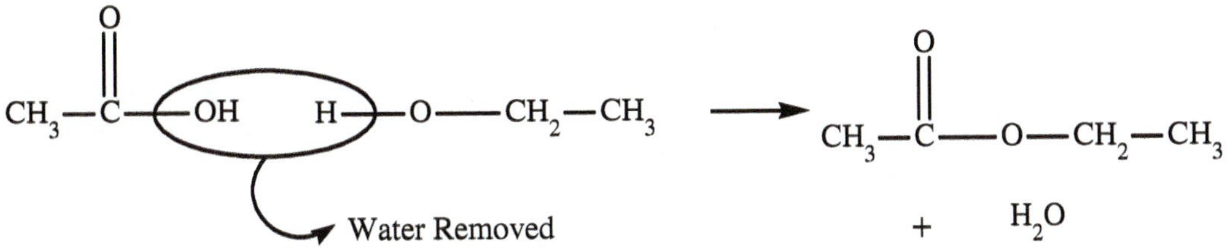

15.57 Plan: Refer to the Table of Functional Groups in the chapter.
Solution:
a) Halogens, except iodine, differ from carbon in electronegativity and form a single bond with carbon. The organic compound is an **alkyl halide**.
b) Carbon forms triple bonds with itself and nitrogen. For the bond to be polar, it must be between carbon and nitrogen. The compound is a **nitrile**.
c) **Carboxylic acids** contain a double bond to oxygen and a single bond to oxygen. Carboxylic acids dissolve in water to give acidic solutions.
d) Oxygen is commonly double bonded to carbon. A carbonyl group (C=O) that is at the end of a chain is found in an **aldehyde**.

15.59

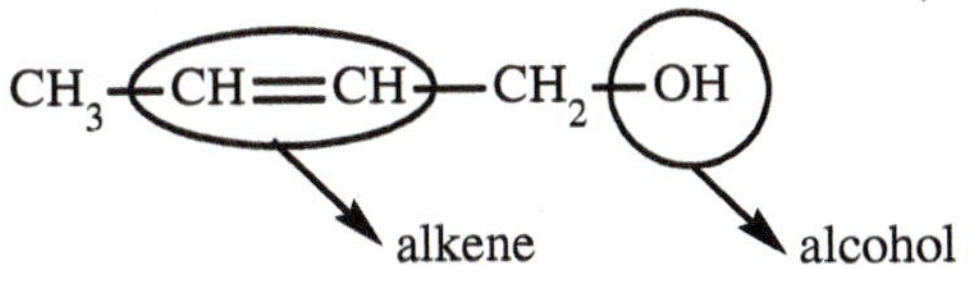

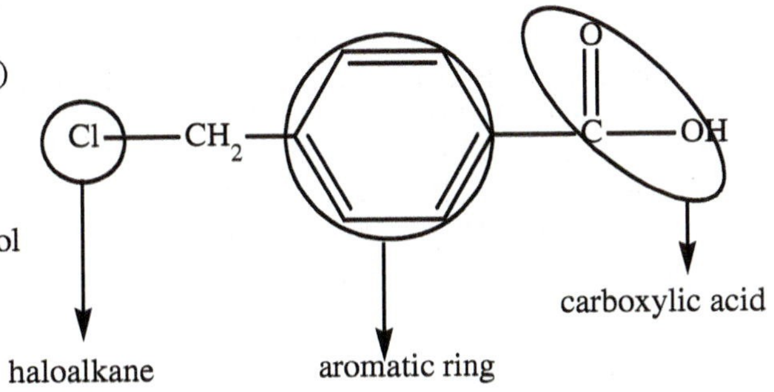

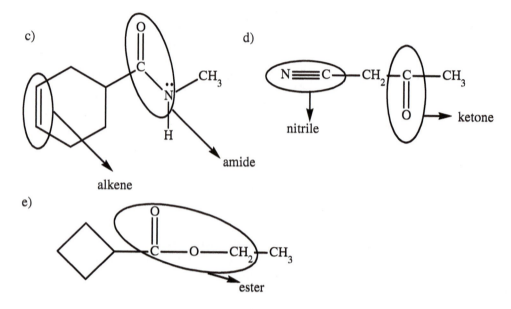

c)

alkene

amide

d)

nitrile

ketone

e)

ester

15.61   Plan: Draw the longest carbon chain first and place the –OH group at different points along the chain. Then, work down to shorter chains, with the –OH group and branches at different points along the chains.

Solution:

For $C_5H_{12}O$, the longest chain is five carbons:

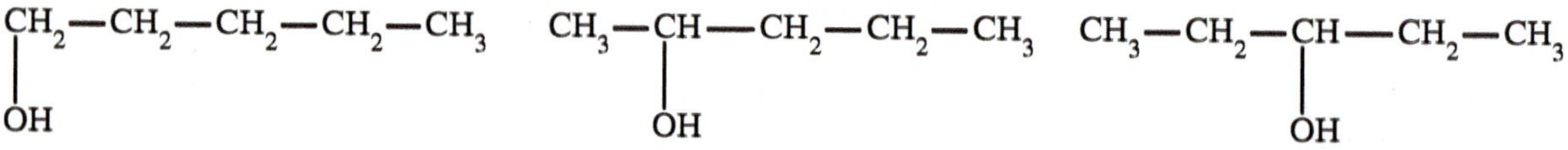

$$CH_2-CH_2-CH_2-CH_2-CH_3 \qquad CH_3-CH-CH_2-CH_2-CH_3 \qquad CH_3-CH_2-CH-CH_2-CH_3$$

with OH groups respectively.

The three structures represent all the unique positions for the alcohol group on a five-carbon chain.

Next, use a four-carbon chain and attach to side groups, –OH and –CH$_3$.

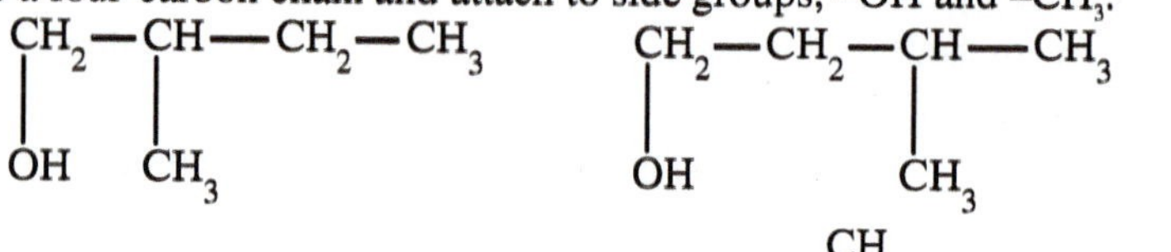

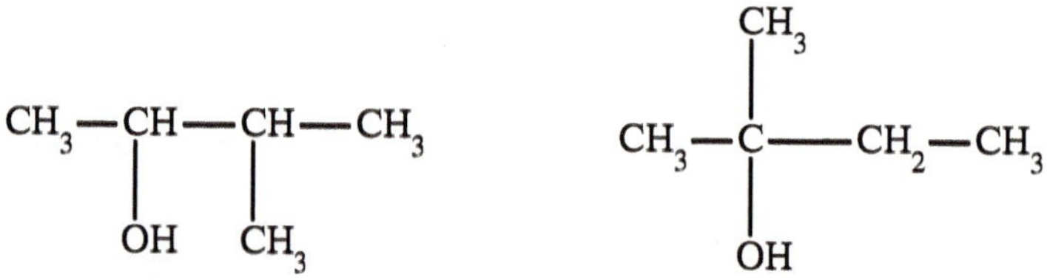

$$CH_2-CH-CH_2-CH_3 \qquad CH_2-CH_2-CH-CH_3$$

$$CH_3-CH-CH-CH_3 \qquad CH_3-C-CH_2-CH_3$$

And use a three-carbon chain with three side groups:

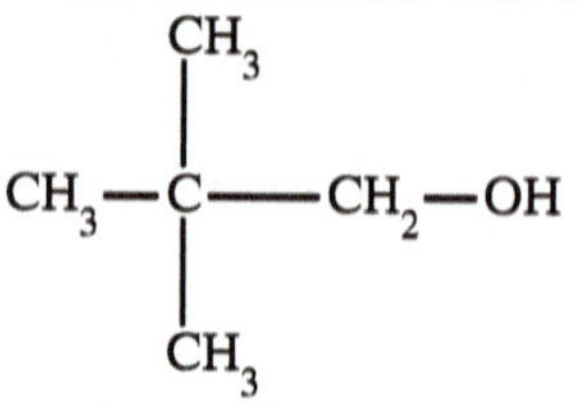

$$CH_3-C-CH_2-OH$$

The total number of different structures is eight.

15-26

15.63   Plan: First, draw all primary amines (formula R–NH₂). Next, draw all secondary amines (formula R–NH–R').
There is only one possible tertiary amine structure (formula R₃–N). Eight amines with the formula $C_4H_{11}N$ exist.
Solution:

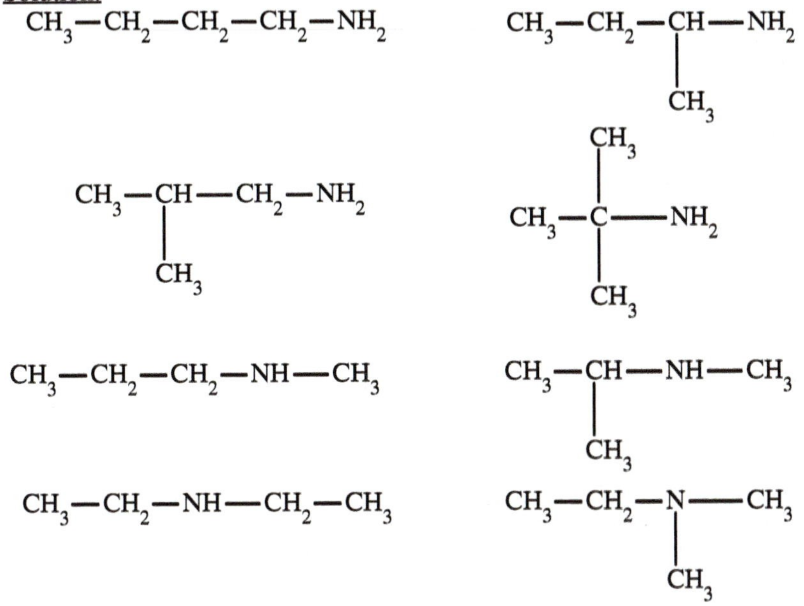

$CH_3-CH_2-CH_2-CH_2-NH_2$

$CH_3-CH_2-\underset{\underset{CH_3}{|}}{CH}-NH_2$

$CH_3-\underset{\underset{CH_3}{|}}{CH}-CH_2-NH_2$

$CH_3-\overset{\overset{CH_3}{|}}{\underset{\underset{CH_3}{|}}{C}}-NH_2$

$CH_3-CH_2-CH_2-NH-CH_3$

$CH_3-\underset{\underset{CH_3}{|}}{CH}-NH-CH_3$

$CH_3-CH_2-NH-CH_2-CH_3$

$CH_3-CH_2-\underset{\underset{CH_3}{|}}{N}-CH_3$

15.65   Plan: With mild oxidation, an alcohol group is oxidized to a carbonyl group and an aldehyde is oxidized to a
carboxylic acid.
Solution:
a) The product is 2-butanone.

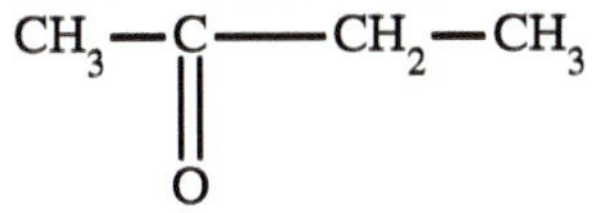

$CH_3-\overset{\overset{}{\underset{\underset{O}{||}}{C}}}{}-CH_2-CH_3$

b) The product is 2-methylpropanoic acid.

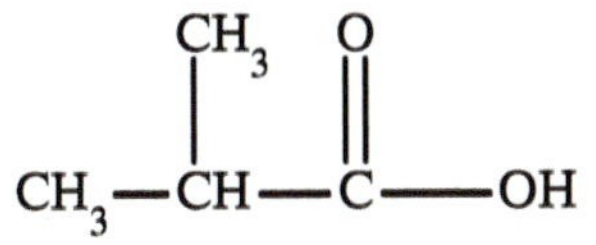

$CH_3-\underset{\underset{CH_3}{|}}{CH}-\overset{\overset{O}{||}}{C}-OH$

c) Mild oxidation of an alcohol produces a carbonyl group. The product is cyclopentanone.

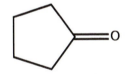

15.67   Plan: These reactions are dehydration-condensation reactions, in which H and OH groups on the two reactant
molecules react to form water and a new bond is formed between the two reactants.
Solution:
a) This reaction is a dehydration-condensation reaction to form an amide.

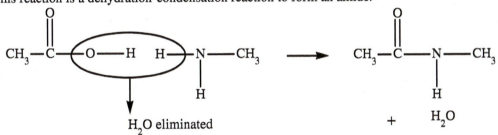

15-27

b) An alcohol and a carboxylic acid undergo dehydration-condensation to form an ester.

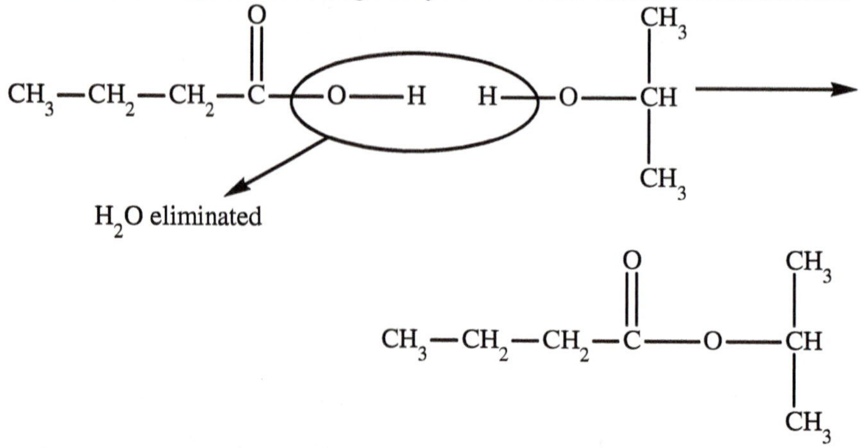

H₂O eliminated

$$CH_3-CH_2-CH_2-C(=O)-O-CH(CH_3)-CH_3$$

c) This reaction is ester formation through dehydration-condensation.

H₂O eliminated

$$H-C(=O)-O-CH_2-CH(CH_3)-CH_3$$

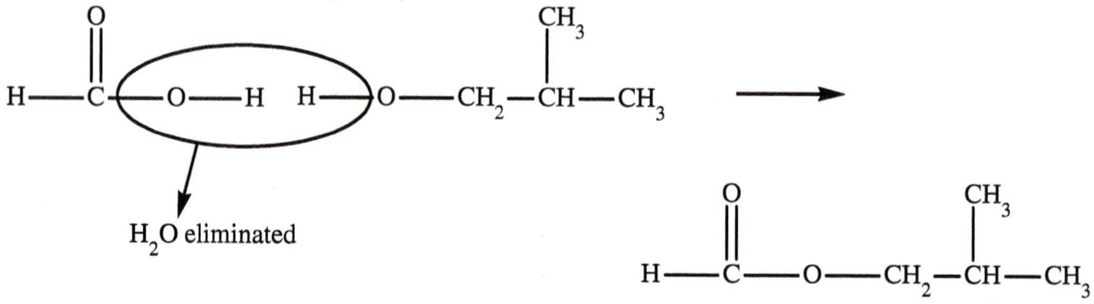

15.69   Plan: To break an ester apart, break the –C–O– single bond and add water (–H and –OH) as shown.

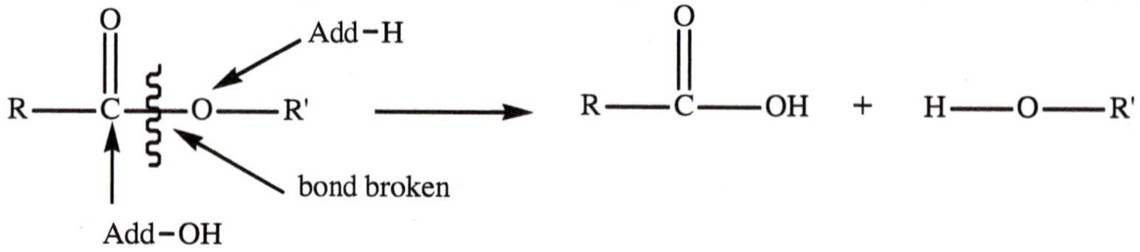

Add –H

bond broken

Add –OH

$$R-C(=O)-OH \;+\; H-O-R'$$

Solution:
(a)

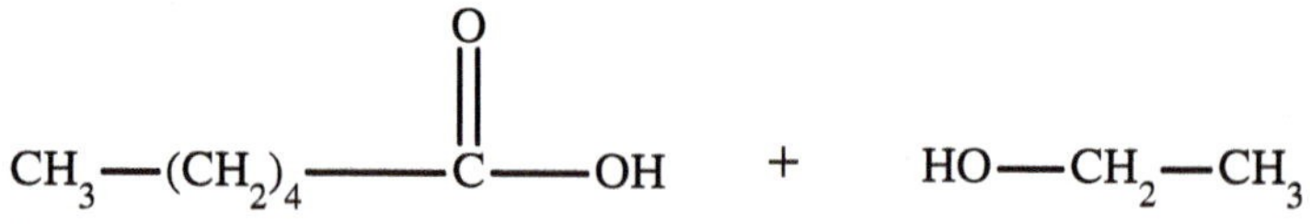

$$CH_3-(CH_2)_4-C(=O)-OH \;+\; HO-CH_2-CH_3$$

(b)

$$C(=O)-OH \;+\; HO-CH_2-CH_2-CH_3$$

(c)

$$CH_3-CH_2-OH \;+\; HO-C(=O)-CH_2-CH_2-$$

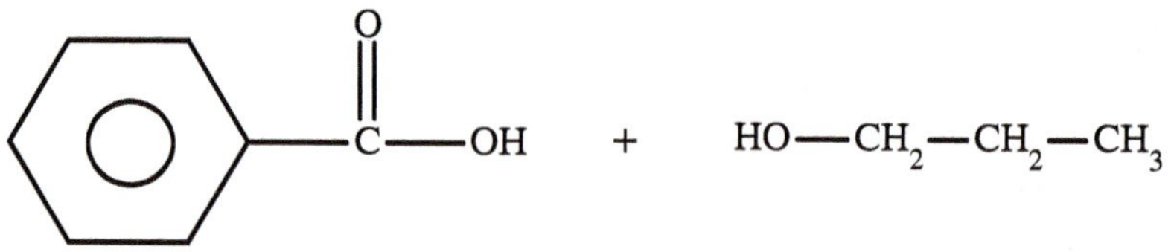

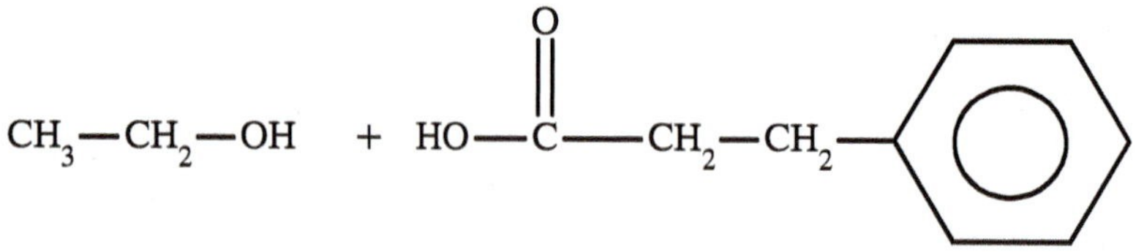

15.71 a) Substitution of Br⁻ occurs by the stronger base, OH⁻. Then a substitution reaction between the alcohol and carboxylic acid produces an ester:

$$CH_3-CH_2-Br \xrightarrow{OH^-} CH_3-CH_2-OH \xrightarrow[H^+]{CH_3-CH_2-\overset{\overset{\displaystyle O}{\|}}{C}-OH}$$

$$CH_3-CH_2-\overset{\overset{\displaystyle O}{\|}}{C}-O-CH_2-CH_3$$

b) The strong base, CN⁻, substitutes for Br. The nitrile is then hydrolyzed to a carboxylic acid.

$$CH_3-CH_2-\overset{\overset{\displaystyle Br}{|}}{CH}-CH_3 \xrightarrow{CN^-} CH_3-CH_2-\overset{\overset{\displaystyle C\equiv N}{|}}{CH}-CH_3$$

$$\xrightarrow{H_3O^+,\ H_2O} \overset{\overset{\textstyle OH}{|}}{\underset{CH_3-CH_2-CH-CH_3}{C}}=O$$

$$CH_3-CH_2-CH=CH_2 \xrightarrow{H^+,\ H_2O} CH_3-CH_2-\underset{\underset{\textstyle OH}{|}}{CH}-CH_3$$

$$\xrightarrow{Cr_2O_7^{2-},\ H^+} CH_3-CH_2-\overset{\overset{\displaystyle}{}}{\underset{\underset{\displaystyle O}{\|}}{C}}-CH_3$$

15.73 a) The product is an ester and the given reactant is an alcohol. An alcohol reacts with a carboxylic acid to make an ester. To identify the acid, break the single bond between the carbon and oxygen in the ester group.

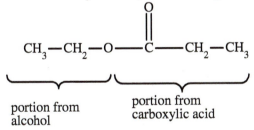

portion from alcohol   portion from carboxylic acid

The missing reactant is **propanoic acid**.

b) To form an amide, an amine must react with an ester to replace the –O–R group. To identify the amine that must be added, break the C–N bond in the amide. The amine is **ethylamine**.

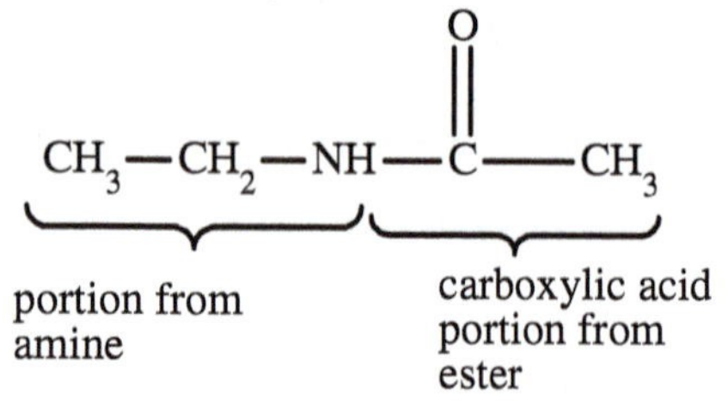

$$CH_3-CH_2-\underbrace{NH}-\overset{\overset{\textstyle O}{\|}}{C}-CH_3$$

portion from
amine

carboxylic acid
portion from
ester

15.77 **Addition reactions** and **condensation reactions** are the two reactions that lead to the two types of synthetic polymers that are named for the reactions that form them.

15.80 Polyethylene comes in a range of strengths and flexibilities. The intermolecular dispersion forces (also called London forces) that attract the long, unbranched chains of high-density polyethylene (HDPE) are strong due to the large size of the polyethylene chains. Low-density polyethylene (LDPE) has increased branching that prevents packing and weakens intermolecular dispersion forces.

15.82 Nylon is formed by the condensation reaction between an **amine and a carboxylic acid** resulting in an amide bond. Polyester is formed by the condensation reaction between a **carboxylic acid and an alcohol** to form an ester bond.

15.83 Plan: Both PVC and polypropylene are addition polymers. To draw the repeat unit, replace the double bond with a single bond, draw an additional single bond to each carbon atom, draw brackets around the molecule, and use a subscript n to denote that the monomer is repeated n times to form the polymer.
Solution:
a)

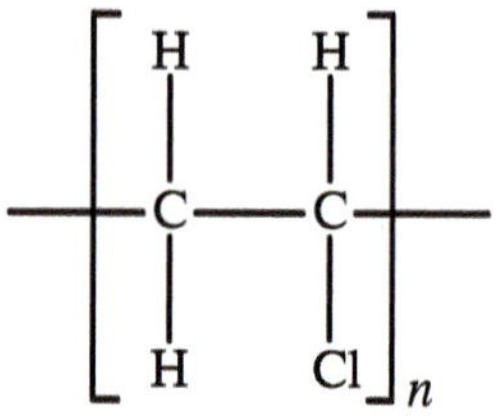

$$\begin{bmatrix} & H & & H & \\ & | & & | & \\ -\!\!&C&-&C&-\!\! \\ & | & & | & \\ & H & & Cl & \end{bmatrix}_n$$

b)

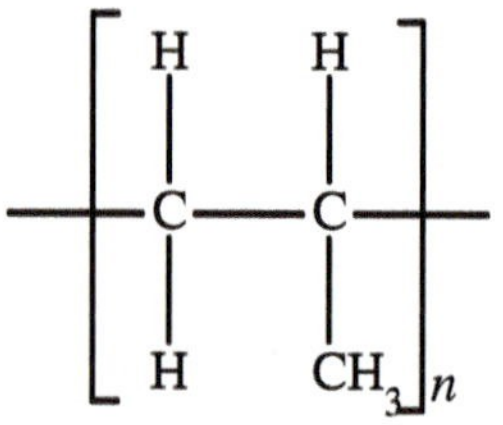

$$\begin{bmatrix} & H & & H & \\ & | & & | & \\ -\!\!&C&-&C&-\!\! \\ & | & & | & \\ & H & & CH_3 & \end{bmatrix}_n$$

15.87 a) Amino acids form **condensation** polymers, called proteins.
b) Alkenes form **addition** polymers, the simplest of which is polyethylene.
c) Simple sugars form **condensation** polymers, called polysaccharides.
d) Mononucleotides form **condensation** polymers, called nucleic acids.

15.89 The amino acid sequence in a protein determines its shape and structure, which determine its function.

15.92 The DNA base sequence contains an information template that is carried by the RNA base sequence (messenger and transfer) to create the protein amino acid sequence. In other words, the DNA sequence determines the RNA sequence, which determines the protein amino acid sequence.

15.93   Plan: Locate the specific amino acids in the Table of Amino Acids in the chapter.
         Solution:
         a) alanine   b) histidine   c) methionine

15.95   Plan: A tripeptide contains three amino acids and two peptide (amide) bonds. Find the structures of the amino
         acids in the Table of Amino Acids in the chapter. Join the three acids to give the tripeptide; water is produced.
         Solution:

a)

b) Repeat the preceding procedure, with charges on the terminal groups as are found in cell fluid.

glycine   cysteine   tyrosine

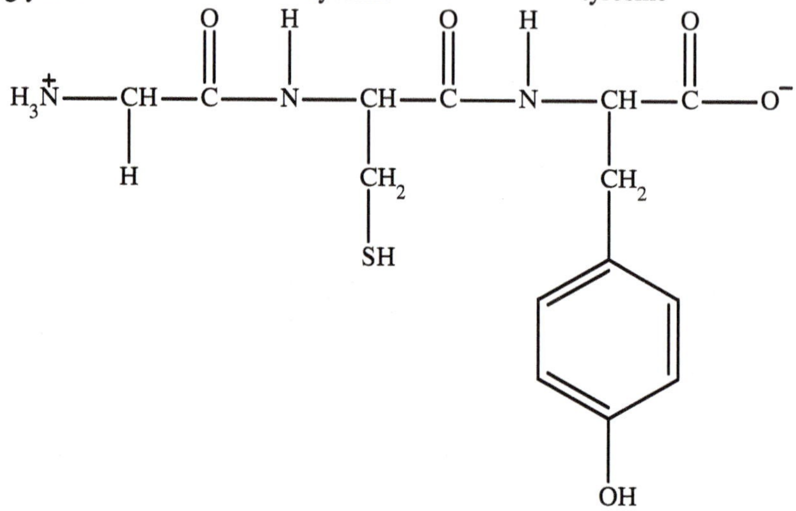

15.97   Plan: Base A always pairs with Base T; Base C always pairs with Base G.
Solution:
a) Complementary DNA strand is **AATCGG**.
b) Complementary DNA strand is **TCTGTA**.

15.99   Plan: Uracil (U) substitutes for thymine (T) in RNA. Therefore, A pairs with U. The G-C pair and A-T pair remain unchanged. A three-base sequence constitutes a word, and each word translates into an amino acid.
Solution:
The RNA sequence is derived from the DNA template sequence **ACAATGCCT**.
There are three sets of three-base sequences, so there are three words in the sequence, and **three** amino acids are coded in the sequence.

15.101   Plan: Refer to the Table of Amino Acids in the chapter. The types of forces operating in proteins were discussed in Chapter 13. Disulfide bonds form between sulfur atoms, salt links form between $-COO^-$ and $-NH_3^+$ groups, and hydrogen bonding occurs between NH– and –OH groups. Nonpolar chains interact through dispersion forces.
Solution:
a) Both side chains are part of the amino acid **cysteine**. Two cysteine R groups can form a **disulfide bond** (covalent bond).
b) The R group ($-(CH_2)_4-NH_3^+$) is found in the amino acid **lysine** and the R group ($-CH_2COO^-$) is found in the amino acid **aspartic acid**. The positive charge on the amine group in lysine is attracted to the negative charge on the acid group in aspartic acid to form a **salt link**.
c) The R group in the amino acid **asparagine** and the R group in the amino acid **serine**. The –NH– and –OH groups will **hydrogen bond**.
d) Both the R group $-CH(CH_3)-CH_3$ from **valine** and the R group $C_6H_5-CH_2-$ from **phenylalanine** are nonpolar, so their interaction is through **dispersion forces**.

15.106   a) Perform an acid-catalyzed dehydration of the alcohol (elimination), followed by bromination of the double bond (addition of Br_2):

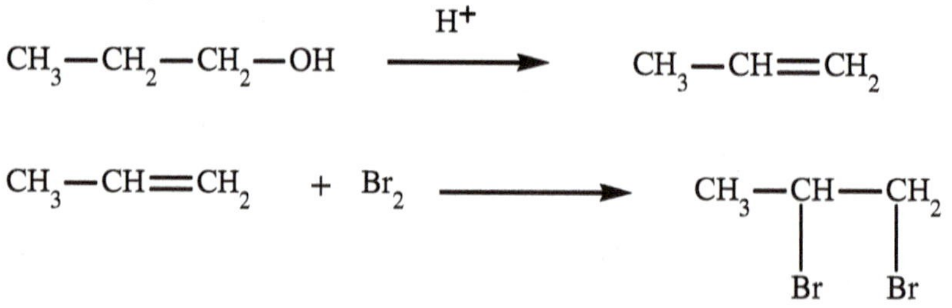

b) The product is an ester, so a carboxylic acid is needed to prepare the ester. First, oxidize one mole of ethanol to acetic acid:

$$CH_3-CH_2-OH \xrightarrow[H^+]{Cr_2O_7{}^{2-}} CH_3-\overset{\displaystyle O}{\overset{\|}{C}}-OH$$

Then, react one mole of acetic acid with a second mole of ethanol to form the ester:

$$CH_3-CH_2\underbrace{-OH\ HO}-\overset{\displaystyle O}{\overset{\|}{C}}-CH_3 \xrightarrow{H^+} CH_3-CH_2-O-\overset{\displaystyle O}{\overset{\|}{C}}-CH_3$$

H_2O eliminated

15.108 The two structures are:

$$H_2N-CH_2-CH_2-CH_2-CH_2-CH_2-NH_2 \qquad H_2N-CH_2-CH_2-CH_2-CH_2-NH_2$$
cadaverine   putrescine

The addition of cyanide, $[C\equiv N]^-$, to form a nitrile is a convenient way to increase the length of a carbon chain:

$$H-\overset{\overset{\displaystyle H}{|}}{\underset{\underset{\displaystyle Br}{|}}{C}}-\overset{\overset{\displaystyle H}{|}}{\underset{\underset{\displaystyle Br}{|}}{C}}-H\ +\ 2CN^- \longrightarrow H-\overset{\overset{\displaystyle H}{|}}{\underset{\underset{\displaystyle N\equiv C}{|}}{C}}-\overset{\overset{\displaystyle H}{|}}{\underset{\underset{\displaystyle C\equiv N}{|}}{C}}-H\ +\ 2Br^-$$

The nitrile can then be reduced to an amine through the addition of hydrogen ($NaBH_4$ is a hydrogen-rich reducing agent):

$$H-\overset{\overset{\displaystyle H}{|}}{\underset{\underset{\displaystyle N\equiv C}{|}}{C}}-\overset{\overset{\displaystyle H}{|}}{\underset{\underset{\displaystyle C\equiv N}{|}}{C}}-H \xrightarrow{\text{reduction}} H_2N-CH_2-CH_2-CH_2-CH_2-NH_2$$

15.109 <u>Plan:</u> Refer to the Table of Functional Groups in the chapter. Carbon atoms surrounded by four electron regions (four single bonds) are sp^3 hybridized. Carbon atoms surrounded by three electron regions (two single bonds and one double bond) are sp^2 hybridized. A carbon atom is chiral if it is attached to four different groups.
<u>Solution:</u>
a) Functional groups in jasmolinII:

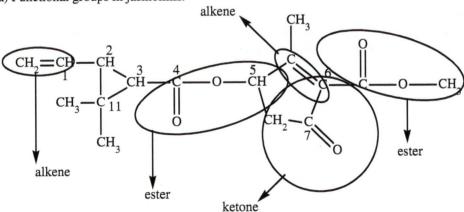

b) Carbon 1 is sp^2 hybridized.   Carbon 2 is sp^3 hybridized.
   Carbon 3 is sp^3 hybridized.   Carbon 4 is sp^2 hybridized.
   Carbon 5 is sp^3 hybridized.   Carbon 6 and 7 are sp^2 hybridized.
c) Carbons 2, 3, and 5 are chiral centers as they are each bonded to four different groups.

15.113   The resonance structures show that the bond between carbon and nitrogen will have some double bond character that restricts rotation around the bond.

15.116   Plan: Convert each mass to moles, and divide the moles by the smallest number to determine molar ratio, and thus relative numbers of the amino acids. To find the minimum molar mass, add the products of the moles of each amino acid and its molar mass to find the total mass of the amino acids and subtract the total mass of water that is eliminated during the formation of the peptide bonds.
Solution:
a) The hydrolysis process requires the addition of water to break the peptide bonds.
b)   (3.00 g gly)/(75.07 g/mol) = 0.0399627 mol glycine
            (0.90 g ala)/(89.10 g/mol) = 0.010101010 mol alanine
            (3.70 g val)/(117.15 g/mol) = 0.0315834 mol valine
            (6.90 g pro)/(115.13 g/mol) = 0.0599323 mol proline
            (7.30 g ser)/(105.10 g/mol) = 0.0694577 mol serine
            (86.00 g arg)/(174.21 g/mol) = 0.493657 mol arginine
Divide by the smallest value (0.010101010 mol alanine), and round to a whole number.
            (0.0399627 mol glycine)/(0.010101010 mol) = **4**
            (0.010101010 mol alanine)/(0.010101010 mol) = **1**
            (0.0315834 mol valine)/(0.010101010 mol) = **3**
            (0.0599323 mol proline)/(0.010101010 mol) = **6**
            (0.0694577 mol serine)/(0.010101010 mol) = **7**
            (0.493657 mol arginine)/(0.010101010 mol) = **49**
c) Mass of 70 moles of amino acids = (4 x 75.07 g/mol) + (1 x 89.09 g/mol) + (3 x 117.15 g/mol)
            + (6 x 115.13 g/mol) + (7 x 105.09 g/mol) + (49 x 174.20 g/mol) = 10,703.59 g
To link 70 moles of amino acids, 69 peptide bonds are formed by the elimination of 69 moles of H_2O.

$$\text{Mass of } H_2O \text{ eliminated} = (69 \text{ mol } H_2O)\left(\frac{18.02 \text{ g } H_2O}{1 \text{ mol } H_2O}\right) = 1243.38 \text{ g}$$

Minimum $\mathscr{M}$ of peptide = 10,703.59 g – 1243.38 g = **9,460 g/mol**

15.117   When 2-butanone is reduced, equal amounts of both isomers are produced because the reaction does not favor the production of one over the other. In a 1:1 mixture of the two stereoisomers, the light rotated to the right by the other isomer cancels the light rotated to the left by one isomer. The mixture is not optically active since there is no net rotation of light. The two stereoisomers of 2-butanol are shown below.

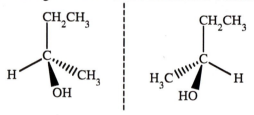

CHAPTER 16 KINETICS: RATES AND MECHANISMS OF CHEMICAL REACTIONS

FOLLOW–UP PROBLEMS

16.1A Plan: Balance the equation. The rate in terms of the change in concentration with time for each substance is expressed as $\dfrac{1}{a}\dfrac{\Delta[A]}{\Delta t}$, where a is the coefficient of the reactant or product A. The rate of the reactants is given a negative sign.

Solution:
a) The balanced equation is $4NO(g) + O_2(g) \rightarrow 2N_2O_3$.
Choose O_2 as the reference because its coefficient is 1. Four molecules of NO (nitrogen monoxide) are consumed for every one O_2 molecule, so the rate of O_2 disappearance is 1/4 the rate of NO decrease. By similar reasoning, the rate of O_2 disappearance is 1/2 the rate of N_2O_3 (dinitrogen trioxide) increase.

$$Rate = -\frac{1}{4}\frac{\Delta[NO]}{\Delta t} = -\frac{\Delta[O_2]}{\Delta t} = +\frac{1}{2}\frac{\Delta[N_2O_3]}{\Delta t}$$

b) Plan: Because NO is decreasing; its rate of concentration change is negative. Substitute the negative value into the expression and solve for $\Delta[O_2]/\Delta t$.

Solution:

$$Rate = -\frac{1}{4}\left(-1.60 \times 10^{-4}\,mol/L\bullet s\right) = -\frac{\Delta[O_2]}{\Delta t} = \mathbf{4.00 \times 10^{-5}\ mol/L\cdot s}$$

16.1B Plan: Examine the equation that expresses the rate in terms of the change in concentration with time for each substance. The number in the denominator of each fraction is the coefficient for the corresponding substance in the balanced equation. The terms that are negative in the rate equation represent reactants in the balanced equation, while the terms that are positive represent products in the balanced equation. For example, the following term, $\dfrac{1}{a}\dfrac{\Delta[A]}{\Delta t}$, describes a product (the term is positive) that has a coefficient of a in the balanced equation. In part b), use the rate equation to compare the rate of appearance of H_2O with the rate of disappearance of O_2.

Solution:
a) The balanced equation is $4NH_3 + 5O_2 \rightarrow 4NO + 6H_2O$.
b) Plan: Because H_2O is increasing; its rate of concentration change is positive. Substitute its rate into the expression and solve for $\Delta[O_2]/\Delta t$.

Solution:

$$Rate = \frac{1}{6}\left(2.52 \times 10^{-2}\ mol/L\bullet s\right) = -\frac{1}{5}\frac{\Delta[O_2]}{\Delta t}$$

$$-\frac{5}{6}\left(2.52 \times 10^{-2}\ mol/L\bullet s\right) = \frac{\Delta[O_2]}{\Delta t} = -2.10 \times 10^{-2}\ mol/L\cdot s$$

The negative value indicates that $[O_2]$ is decreasing as the reaction progresses. The rate of reaction is always expressed as a positive number, so $[O_2]$ is decreasing at a rate of $\mathbf{2.10 \times 10^{-2}\ mol/L\cdot s}$.

16.2A Plan: The reaction orders of the reactants are the exponents in the rate law. Add the individual reaction orders to obtain the overall reaction order. Use the rate law to determine how the changes listed in the problem will affect the rate.

a) The exponent of [I⁻] is 1, so the reaction is **first order with respect to I⁻**. Similarly, the reaction is **first order with respect to BrO₃⁻**, and **second order with respect to H⁺**. The overall reaction order is $(1 + 1 + 2) = 4$, or **fourth order overall**.

b) Rate $= k[I^-][BrO_3^-][H^+]^2$. If $[BrO_3^-]$ and $[I^-]$ are tripled and $[H^+]$ is doubled, rate $= k[3 \times I^-][3 \times BrO_3^-][2 \times H^+]^2$, then rate increases to $3 \times 3 \times 2^2$ or 36 times its original value. The rate **increases by a factor of 36**.

16.2B   **Plan:** The reaction orders of the reactants are the exponents in the rate law. Add the individual reaction orders to obtain the overall reaction order. Use the rate law to determine how the changes listed in the problem will affect the rate.

Solution:

a) The exponent of $[ClO_2]$ is 2, so the reaction is **second order with respect to ClO_2**. Similarly, the reaction is **first order with respect to OH⁻**. The overall reaction order is $(1 + 2) = 3$, or **third order overall**.

b) Rate $= k[ClO_2]^2[OH^-]$. If $[ClO_2]$ is halved and $[OH^-]$ is doubled, rate $= k[1/2 \times ClO_2]^2[2 \times OH^-]$, then rate increases to $(1/2)^2 \times 2$ or 1/2 its original value. The rate **decreases by a factor of 1/2**.

16.3A   **Plan:** Assume that the rate law takes the general form rate $= k[H_2]^m[I_2]^n$. To find how the rate varies with respect to $[H_2]$, find two experiments in which $[H_2]$ changes but $[I_2]$ remains constant. Take the ratio of rate laws for those two experiments to find m. To find how the rate varies with respect to $[I_2]$, find two experiments in which $[I_2]$ changes but $[H_2]$ remains constant. Take the ratio of rate laws for those two experiments to find n. Add m and n to obtain the overall reaction order. Use the rate law to solve for the value of k.

Solution:

For the reaction order with respect to $[H_2]$, compare Experiments 1 and 3:

$$\frac{\text{rate } 3}{\text{rate } 1} = \frac{[H_2]_3^m}{[H_2]_1^m}$$

$$\frac{9.3 \times 10^{-23}}{1.9 \times 10^{-23}} = \frac{[0.0550]_3^m}{[0.0113]_1^m}$$

$$4.8947 = (4.8672566)^m$$

Therefore, $m = 1$

If the reaction order was more complex, an alternate method of solving for m is:

$\log(4.8947) = m \log(4.8672566)$; $m = \log(4.8947)/\log(4.8672566) = 1$

For the reaction order with respect to $[I_2]$, compare Experiments 2 and 4:

$$\frac{\text{rate } 4}{\text{rate } 2} = \frac{[I_2]_4^n}{[I_2]_2^n}$$

$$\frac{1.9 \times 10^{-22}}{1.1 \times 10^{-22}} = \frac{[0.0056]_4^n}{[0.0033]_2^n}$$

$$1.72727 = (1.69697)^n$$

Therefore, $n = 1$

The rate law is **rate $= k[H_2][I_2]$** and is **second order overall**.

Calculation of k:

$k = \text{Rate}/([H_2][I_2])$

$k_1 = (1.9 \times 10^{-23}$ mol/L•s$)/[(0.0113$ mol/L$)(0.0011$ mol/L$)] = 1.5 \times 10^{-18}$ L/mol•s

$k_2 = (1.1 \times 10^{-22}$ mol/L•s$)/[(0.0220$ mol/L$)(0.0033$ mol/L$)] = 1.5 \times 10^{-18}$ L/mol•s

$k_3 = (9.3 \times 10^{-23}$ mol/L•s$)/[(0.0550$ mol/L$)(0.0011$ mol/L$)] = 1.5 \times 10^{-18}$ L/mol•s

$k_4 = (1.9 \times 10^{-22}$ mol/L•s$)/[(0.0220$ mol/L$)(0.0056$ mol/L$)] = 1.5 \times 10^{-18}$ L/mol•s

Average $k = $ **1.5×10^{-18} L/mol•s**

16.3B Plan: Assume that the rate law takes the general form rate = $k[H_2SeO_3]^m[I^-]^n[H^+]^p$. To find how the rate varies with respect to $[H_2SeO_3]$, find two experiments in which $[H_2SeO_3]$ changes but $[I^-]$ and $[H^+]$ remain constant. Take the ratio of rate laws for those two experiments to find m. To find how the rate varies with respect to $[I^-]$, find two experiments in which $[I^-]$ changes but $[H_2SeO_3]$ and $[H^+]$ remain constant. Take the ratio of rate laws for those two experiments to find n. To find how the rate varies with respect to $[H^+]$, find two experiments in which $[H^+]$ changes but $[H_2SeO_3]$ and $[I^-]$ remain constant. Take the ratio of rate laws for those two experiments to find p. Add m, n, and p to obtain the overall reaction order. Use the rate law to solve for the value of k.

Solution:

For the reaction order with respect to $[H_2SeO_3]$, compare Experiments 1 and 3:

$$\frac{\text{rate 3}}{\text{rate 1}} = \frac{[H_2SeO_3]_3^m}{[H_2SeO_3]_1^m}$$

$$\frac{3.94\times10^{-6}\ \text{mol/L}\bullet s}{9.85\times10^{-7}\ \text{mol/L}\bullet s} = \frac{[1.0\times10^{-2}\text{mol/L}]_3^m}{[2.5\times10^{-3}\ \text{mol/L}]_1^m}$$

$4 = (4)^m$

Therefore, **$m = 1$**

For the reaction order with respect to $[I^-]$, compare Experiments 1 and 2:

$$\frac{\text{rate 2}}{\text{rate 1}} = \frac{[I^-]_2^n}{[I^-]_1^n}$$

$$\frac{7.88\times10^{-6}\ \text{mol/L}\bullet s}{9.85\times10^{-7}\ \text{mol/L}\bullet s} = \frac{[3.0\times10^{-2}\text{mol/L}]_2^n}{[1.5\times10^{-2}\ \text{mol/L}]_1^n}$$

$8 = (2)^n$

Therefore, **$n = 3$**

For the reaction order with respect to $[H^+]$, compare Experiments 2 and 4:

$$\frac{\text{rate 4}}{\text{rate 2}} = \frac{[H+]_4^p}{[H+]_2^p}$$

$$\frac{3.15\times 10^{-5}\ \text{mol/L}\bullet s}{7.88\times10^{-6}\ \text{mol/L}\bullet s} = \frac{[3.0\times10^{-2}\text{mol/L}]_4^p}{[1.5\times10^{-2}\ \text{mol/L}]_2^p}$$

$4 = (2)^p$

Therefore, **$p = 2$**

The rate law is **rate = $k[H_2SeO_3][I^-]^3[H^+]^2$** and is sixth order overall.

b) Calculation of k:

$k = \text{Rate}/([H_2SeO_3][I^-]^3[H^+]^2)$

$k_1 = (9.85\times10^{-7}\ \text{mol/L}\bullet s)/[(2.5\times10^{-3}\ \text{mol/L})(1.5\times10^{-2}\ \text{mol/L})^3(1.5\times10^{-2}\ \text{mol/L})^2] = $ **$5.2\times10^5\ L^5/mol^5\bullet s$**

16.4A Plan: The reaction is second order in X and zero order in Y. For part a), compare the two amounts of reactant X. For part b), compare the two rate values.

Solution:

a) Since the rate law is rate = $k[X]^2$, the reaction is zero order in Y. In Experiment 2, the amount of reactant X has not changed from Experiment 1 and the amount of Y has doubled. The rate is not affected by the doubling of Y since Y is zero order. Since the amount of X is the same, the rate has not changed. The initial rate of Experiment 2 is also **0.25×10^{-5} mol/L·s**.

b) The rate of Experiment 3 is four times the rate in Experiment 1. Since the reaction is second order in X, the concentration of X must have doubled to cause a four-fold increase in rate. There should be 6 black spheres and 3 green spheres in Experiment 3.

$$\frac{1.0 \times 10^{-5}}{0.25 \times 10^{-5}} = \frac{[x]^2}{[3]^2}$$

$$4 = \frac{[x]^2}{9}$$

x = 6 black spheres

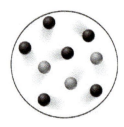

16.4B   Plan: Examine how a change in the concentration of the different reactants affects the rate in order to determine the rate law. Then use the rate law to determine the number of particles in the scene for Experiment 4.
Solution:
a) Comparing Experiments 1 and 2, we can see that the number of blue A spheres does not change while the number of yellow B spheres changes from 4 to 2. Although the number of yellow spheres changes, the rate does not change. Therefore, B has no effect on the rate, which suggests that the reaction is zero order with respect to B. Now that we know that the yellow B spheres do not affect the rate, we can look at the influence of the blue A spheres. Comparing Experiments 2 and 3, we can see that the number of blue spheres changes from 4 to 2 (experiment 3 concentration of A is half of that of experiment 2) while the rate changes from 1.6×10^{-3} mol/L•s to 8.0×10^{-4} mol/L•s (the rate of experiment 3 is half of the rate of experiment 2). The fact that the concentration of blue spheres changes in the same way the rate changes suggests that the reaction is first order with respect to the blue A spheres. Therefore, the rate law is: **rate = k[A]**.
b) The rate of Experiment 4 is twice the rate in Experiment 1. Since the reaction is first order in A, the concentration of A must have doubled to cause a two-fold increase in rate. There should be 2 x 4 particles = **8 particles of A** in the scene for Experiment 4.

16.5A   Plan: The rate expression indicates that the reaction order is two (exponent of [HI] = 2), so use the integrated second-order law. Substitute the given concentrations and the rate constant into the expression and solve for time.
Solution:

$$\frac{1}{[A]_t} - \frac{1}{[A]_0} = kt$$

$$\frac{1}{\left[0.00900 \text{ mol/L}\right]_t} - \frac{1}{\left[0.0100 \text{ mol/L}\right]_0} = (2.4 \times 10^{-21} \text{ L/mol•s})(t)$$

$$t = \frac{\dfrac{1}{0.00900 \text{ mol/L}} - \dfrac{1}{0.0100 \text{ mol/L}}}{2.4 \times 10^{-21} \text{ L/mol•s}}$$

$$t = 4.6296 \times 10^{21} \text{ s} = \textbf{4.6} \times \textbf{10}^{\textbf{21}} \textbf{ s}$$

16.5B   Plan: The problem states that the decomposition of hydrogen peroxide is a first-order, reaction, so use the first-order integrated rate law. Substitute the given concentrations and the time into the expression and solve for the rate constant.
Solution:

a) $$\ln \frac{[A]_0}{[A]_t} = kt$$

$$\ln \frac{1.28\,M}{0.85\,M} = (k)(10.0 \text{ min})$$

$$k = \frac{0.4094}{10.0 \text{ min}} = 0.04094 = \textbf{0.041 min}^{\textbf{-1}}$$

b) If you start with an initial concentration of hydrogen peroxide of 1.0 M ($[A]_0 = 1.0\ M$) and 25% of the sample decomposes, 75% of the sample remains ($[A]_t = 0.75\ M$).

$$\ln \frac{1.00\ \text{M}}{0.75\text{M}} = (0.04094\ \text{min}^{-1})(t)$$

$$t = \frac{0.2877}{0.04094\ \text{min}^{-1}} = 7.02736 = \textbf{7.0 min}$$

16.6A Plan: The initial scene contains 12 particles of Substance X. In the second scene, which occurs after 2.5 minutes have elapsed, half of the particles of Substance X (6 particles) remain. Therefore, 2.5 min is the half-life. The half-life is used to find the number of particles present at 5.0 min and 10.0 min. To find the molarity of X, moles of X is divided by the given volume.
Solution:
a) Since 2.5 minutes is the half-life, 5.0 minutes represents two half-lives:
12 particles of X $\xrightarrow{\ 2.5\ \text{min}\ }$ 6 particles of X $\xrightarrow{\ 2.5\ \text{min}\ }$ 3 particles of X
After 5.0 minutes, 3 particles of X remain; 9 particles of X have reacted to produce 9 particles of Y. Draw a scene in which there are 3 black X particles and 9 red Y particles.

b) 10.0 minutes represents 4 half-lives:
12 particles of X $\xrightarrow{\ 2.5\ \text{min}\ }$ 6 particles of X $\xrightarrow{\ 2.5\ \text{min}\ }$ 3 particles of X $\xrightarrow{\ 2.5\ \text{min}\ }$
1.5 particles of X $\xrightarrow{\ 2.5\ \text{min}\ }$ 0.75 particle of X

$$\text{Moles of X after 10.0 min} = (0.75\ \text{particle})\left(\frac{0.20\ \text{mol}}{1\ \text{particle}}\right) = 0.15\ \text{mol}$$

$$\text{Molarity} = \frac{\text{mol X}}{\text{volume}} = \frac{0.15\ \text{mol}}{0.50\ \text{L}} = \textbf{0.30}\ \boldsymbol{M}$$

16.6B Plan: The initial scene contains 16 particles of Substance A. In the second scene, which occurs after 24 minutes have elapsed, half of the particles of Substance A (8 particles) remain. Therefore, 24 min is the half-life. The half-life is used to find the amount of time that has passed when only one particle of Substance A remains and to find the number of particles of A present at 72 minutes. To find the molarity of A, moles of A is divided by the given volume.
Solution:
a) The half-life is 24 minutes. We can use that information to determine the amount of time that has passed when one particle of Substance A remains:
16 A particles $\xrightarrow{\ 24\ \text{min}\ }$ 8 A particles $\xrightarrow{\ 24\ \text{min}\ }$ → 4 A particles $\xrightarrow{\ 24\ \text{min}\ }$ 2 A particles $\xrightarrow{\ 24\ \text{min}\ }$ 1 A particle
After 4 half-lives, only one particle of Substance A remains.
4 half-lives = 4 x (24 min/half-life) = **96 min**
b) Number of half-lives at 72 min = $\dfrac{72\ \text{min}}{24\ \text{min/half-life}}$ = 3 half-lives
According to the scheme above in part a), there are 2 A particles left after three half-lives.

$$\text{Amount (mol) of A after 72 minutes} = (2\ \text{particles A})\left(\frac{0.10\ \text{mol A}}{1\ \text{particle A}}\right) = 0.20\ \text{mol A}$$

$$M = \frac{0.20\ \text{mol A}}{0.25\ \text{L}} = \textbf{0.80}\ \boldsymbol{M}$$

16.7A Plan: Rearrange the first-order half-life equation to solve for k.
Solution:

$$k = \frac{\ln 2}{t_{1/2}} = \frac{\ln 2}{13.1\ \text{h}} = 0.0529119985 = \mathbf{0.0529\ h^{-1}}$$

16.7B Plan: Rearrange the first-order half-life equation to solve for half-life. Determine the number of half-lives that pass in the 40 day period described in the problem and use this number to determine the amount of pesticide remaining.
Solution:

a) $t_{1/2} = \dfrac{\ln 2}{k} = \dfrac{\ln 2}{9 \times 10^{-2}\ \text{day}^{-1}} = 7.7016 = \mathbf{8\ days}$

b) Number of half-lives at 40 days $= \dfrac{40\ \text{days}}{8\ \text{days/half-life}} = 5$ half-lives

After each half-life, ½ of the sample remains.

After five half-lives, ½ x ½ x ½ x ½ x ½ $= (½)^5 = \dfrac{1}{32}$ **of the sample remains.**

16.8A Plan: The activation energy, rate constant at T_1, and a second temperature, T_2, are given. Substitute these values into the Arrhenius equation and solve for k_2, the rate constant at T_2.
Solution:

$$\ln \frac{k_2}{k_1} = \frac{E_a}{R}\left(\frac{1}{T_1} - \frac{1}{T_2}\right)$$

$k_1 = 0.286\ \text{L/mol} \bullet \text{s}$ $\quad T_1 = 500.\ \text{K}$ $\quad E_a = 1.00 \times 10^2\ \text{kJ/mol}$
$k_2 = ?\ \text{L/mol} \bullet \text{s}$ $\quad T_2 = 490.\ \text{K}$

$$\ln \frac{k_2}{0.286\ \text{L/mol}\cdot\text{s}} = \frac{1.00 \times 10^2\ \text{kJ/mol}}{8.314\ \text{J/mol} \bullet \text{K}}\left(\frac{1}{500.\ \text{K}} - \frac{1}{490.\ \text{K}}\right)\left(\frac{10^3\ \text{J}}{1\ \text{kJ}}\right)$$

$$\ln \frac{k_2}{0.286\ \text{L/mol}\cdot\text{s}} = -0.49093$$

$$\frac{k_2}{0.286\ \text{L/mol}\cdot\text{s}} = 0.612057$$

$k_2 = (0.612057)(0.286\ \text{L/mol} \bullet \text{s}) = 0.175048 = \mathbf{0.175\ L/mol{\bullet}s}$

16.8B Plan: The activation energy and rate constant at T_1 are given. We are asked to find the temperature at which the rate will be twice as fast (i.e., the temperature at which $k_2 = 2 \times k_1$). Substitute the given values into the Arrhenius equation and solve for T_2.
Solution:

$$\ln \frac{k_2}{k_1} = \frac{E_a}{R}\left(\frac{1}{T_1} - \frac{1}{T_2}\right)$$

$k_1 = 7.02 \times 10^{-3}\ \text{L/mol} \bullet \text{s}$ $\quad T_1 = 500.\ \text{K}$ $\quad E_a = 1.14 \times 10^5\ \text{J/mol}$
$k_2 = 2(7.02 \times 10^{-3})\ \text{L/mol} \bullet \text{s}$ $\quad T_2 = \text{unknown}$

$$\ln \frac{2(7.02 \times 10^{-3}\text{L/mol} \bullet \text{s})}{(7.02 \times 10^{-3}\text{L/mol} \bullet \text{s})} = \frac{1.14 \times 10^5\ \text{J/mol}}{8.314\ \text{J/mol} \bullet \text{K}}\left(\frac{1}{500.\ \text{K}} - \frac{1}{T_2}\right)$$

$$5.055110 \times 10^{-5}\ \text{K}^{-1} = \left(\frac{1}{500.\ \text{K}} - \frac{1}{T_2}\right)$$

$$\frac{1}{T_2} = 1.9494489 \times 10^{-3}\ \text{K}^{-1}$$

$$T_2 = \mathbf{513\ K}$$

16.9A Plan: Begin by using Sample Problem 16.9 as a guide for labeling the diagram.
Solution:
The reaction energy diagram indicates that $O(g) + H_2O(g) \rightarrow 2OH(g)$ is an endothermic process, because the energy of the product is higher than the energy of the reactants. The highest point on the curve indicates the transition state. In the transition state, an oxygen atom forms a bond with one of the hydrogen atoms on the H_2O molecule (hashed line) and the O-H bond (dashed line) in H_2O weakens. $E_{a(fwd)}$ is the sum of ΔH_{rxn} and $E_{a(rev)}$.
Transition state:

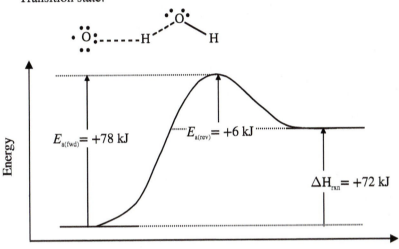

16.9B Plan: $E_{a(fwd)}$ is the sum of ΔH_{rxn} and $E_{a(rev)}$, so ΔH_{rxn} can be calculated by subtracting $E_{a(fwd)} - E_{a(rev)}$. Use Sample Problem 16.9 as a guide for drawing the diagram.
Solution:
$\Delta H_{rxn} = E_{a(fwd)} - E_{a(rev)}$
$\Delta H_{rxn} = 7\ kJ - 72\ kJ = -65\ kJ$
The negative sign of the enthalpy indicates that the reaction is exothermic. This means that the energy of the products is lower than the energy of the reactants. The highest point on the curve indicates the transition state. In the transition state, a bond is forming between the chlorine and the hydrogen (hashed line) while the bond between the hydrogen and the bromine weakens (another hashed line):
Transition state:

$$Cl----H----Br$$

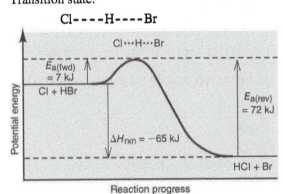

16.10A Plan: The overall reaction can be obtained by adding the three steps together. The molecularity of each step is the total number of reactant particles; the molecularities are used as the orders in the rate law for each step.
Solution:
a)
(1) $H_2O_2(aq) \rightarrow$ ~~$2OH(aq)$~~
(2) $H_2O_2(aq) +$ ~~$OH(aq)$~~ $\rightarrow H_2O(l) +$ ~~$HO_2(aq)$~~
(3) ~~$HO_2(aq)$~~ $+$ ~~$OH(aq)$~~ $\rightarrow H_2O(l) + O_2(g)$
Total: $2H_2O_2(aq) +$ ~~$2OH(aq)$~~ $+$ ~~$HO_2(aq)$~~ $\rightarrow$ ~~$2OH(aq)$~~ $+ 2H_2O(l) +$ ~~$HO_2(aq)$~~ $+ O_2(g)$
(overall) $2H_2O_2(aq) \rightarrow 2H_2O(l) + O_2(g)$

b)   (1) Unimolecular
            (2) Bimolecular
            (3) Bimolecular
c)   (1) $Rate_1 = k_1[H_2O_2]$
            (2) $Rate_2 = k_2[H_2O_2][OH]$
            (3) $Rate_3 = k_3[HO_2][OH]$

16.10B Plan: The overall reaction can be obtained by adding the three steps together. The molecularity of each step is the total number of reactant particles; the molecularities are used as the orders in the rate law for each step.
Solution:

a)   (1) $2NO(g) \leftrightarrows$ ~~$N_2O_2(g)$~~
                         (2) ~~$N_2O_2(g)$~~ $+ H_2(g) \rightarrow$ ~~$N_2O(g)$~~ $+ H_2O(g)$
                         (3) ~~$N_2O(g)$~~ $+ H_2(g) \rightarrow N_2(g) + H_2O(g)$
        Total: $2NO(g) +$ ~~$N_2O_2(g)$~~ $+ 2 H_2(g) +$ ~~$N_2O(g)$~~ $\rightarrow$ ~~$N_2O_2(g)$~~ $+$ ~~$N_2O(g)$~~ $+ 2H_2O(g) + N_2(g)$
        (overall) $2NO(g) + 2H_2(g) \rightarrow 2H_2O(g) + N_2(g)$
b)   (1) Bimolecular
                         (2) Bimolecular
                         (3) Bimolecular
c)   (1) $Rate_1 = k_1[NO]^2$
                         (2) $Rate_2 = k_2[N_2O_2][H_2]$
                         (3) $Rate_3 = k_3[N_2O][H_2]$

16.11A Plan: The overall reaction can be obtained by adding the three steps together. An intermediate is a substance that is formed in one step and consumed in a subsequent step. The overall rate law for the mechanism is determined from the slowest step (the rate-determining step) and can be compared to the experimental rate law.
Solution:

a)   (1) $H_2O_2(aq) \rightarrow$ ~~$2OH(aq)$~~
                         (2) $H_2O_2(aq) +$ ~~$OH(aq)$~~ $\rightarrow H_2O(l) +$ ~~$HO_2(aq)$~~
                         (3) ~~$HO_2(aq)$~~ $+$ ~~$OH(aq)$~~ $\rightarrow H_2O(l) + O_2(g)$
        Total: $2H_2O_2(aq) +$ ~~$2OH(aq)$~~ $+$ ~~$HO_2(aq)$~~ $\rightarrow$ ~~$2OH(aq)$~~ $+ 2H_2O(l) +$ ~~$HO_2(aq)$~~ $+ O_2(g)$
        (overall) $2H_2O_2(aq) \rightarrow 2H_2O(l) + O_2(g)$
$2OH(aq)$ and $HO_2(aq)$ are intermediates in the given mechanism. $2OH(aq)$ are produced in the first step and consumed in the second and third steps; $HO_2(aq)$ is produced in the second step and consumed in the third step. Notice that the intermediates were not included in the overall reaction.
b) The observed rate law is: rate $= k[H_2O_2]$. In order for the mechanism to be consistent with the rate law, the **first step** must be the slow step. The rate law for step one is the same as the observed rate law.

16.11B Plan: The overall reaction can be obtained by adding the three steps together. An intermediate is a substance that is formed in one step and consumed in a subsequent step. The overall rate law for the mechanism is determined from the slowest step (the rate-determining step) and can be compared to the experimental rate law.
Solution:

a)   (1) $2NO(g) \leftrightarrows$ ~~$N_2O_2(g)$~~
                         (2) ~~$N_2O_2(g)$~~ $+ H_2(g) \rightarrow$ ~~$N_2O(g)$~~ $+ H_2O(g)$
                         (3) ~~$N_2O(g)$~~ $+ H_2(g) \rightarrow N_2(g) + H_2O(g)$
        Total: $2NO(g) +$ ~~$N_2O_2(g)$~~ $+ 2 H_2(g) +$ ~~$N_2O(g)$~~ $\rightarrow$ ~~$N_2O_2(g)$~~ $+$ ~~$N_2O(g)$~~ $+ 2H_2O(g) + N_2(g)$
        (overall) $2NO(g) + 2 H_2(g) \rightarrow 2H_2O(g) + N_2(g)$
$N_2O_2(g)$ and $N_2O(g)$ are intermediates in the given mechanism. $N_2O_2(g)$ is produced in the first step and consumed in the second step; $N_2O(g)$ is produced in the second step and consumed in the third step. Notice that the intermediates were not included in the overall reaction.
b) The observed rate law is: rate $= k[NO]^2[H_2]$, and the second step is the slow, or rate-determining, step. The rate law for step two is: rate $= k_2[N_2O_2][H_2]$. This rate law is NOT the same as the observed rate law. However, since N_2O_2 is an intermediate, it must be replaced by using the first step. For an equilibrium, $rate_{forward\ rxn} = rate_{reverse\ rxn}$. For

step 1 then, $k_1[NO]^2 = k_{-1}[N_2O_2]$. Rearranging to solve for $[N_2O_2]$ gives $[N_2O_2] = (k_1/k_{-1})[NO]^2$. Substituting this value for $[N_2O_2]$ into the rate law for the second step gives the overall rate law as rate $= (k_2k_1/k_{-1})[NO]^2[H_2]$ or rate $= k[NO]^2[H_2]$, which is consistent with the observed rate law.

CHEMICAL CONNECTIONS BOXED READING PROBLEMS

B16.1   Plan: Add the two equations, canceling substances that appear on both sides of the arrow. The rate law for each step follows the format of rate $= k$[reactants]. An initial reactant that appears as a product in a subsequent step is a catalyst; a product that appears as a reactant in a subsequent step is an intermediate (produced in one step and consumed in a subsequent step).
Solution:
a) Add the two equations:

(1)   X̶(̶g̶)̶ + O$_3$(g) → X̶O̶(̶g̶)̶ + O$_2$(g)
(2)   X̶O̶(̶g̶)̶ + O(g) → X̶(̶g̶)̶ + O$_2$(g)
———————————————————
Overall O$_3$(g) + O(g) → 2O$_2$(g)

Rate laws:
Step 1   Rate$_1 = k_1[X][O_3]$
Step 2   Rate$_2 = k_2[XO][O]$
b) X acts as a **catalyst**; it is a reactant in step 1 and a product in step 2. XO acts as an **intermediate**; it was produced in step 1 and consumed in step 2.

B16.2   Plan: Replace X in the mechanism in B16.1 with NO, the catalyst. To find the rate of ozone depletion at the given concentrations, use step 1) since it is the rate-determining (slow) step.
Solution:
a)   (1) N̶O̶(̶g̶)̶ + O$_3$(g) → N̶O̶$_2$(̶g̶)̶ + O$_2$(g)
     (2) N̶O̶$_2$(̶g̶)̶ + O(g) → N̶O̶(̶g̶)̶ + O$_2$(g)
———————————————————
Overall O$_3$(g) + O(g) → 2O$_2$(g)
b) Rate$_1 = k_1[NO][O_3]$
    $= (6\times10^{-15}$ cm^3/molecule • s$)[1.0\times10^9$ molecule/cm$^3][5\times10^{12}$ molecule/cm$^3]$
    $= \mathbf{3\times10^7}$ **molecule/s**

B16.3   Plan: The *p* factor is the orientation probability factor and is related to the structural complexity of the colliding particles. The more complex the particles, the smaller the probability that collisions will occur with the correct orientation and the smaller the *p* factor.
Solution:
a) *Step 1* **with Cl** will have the higher value for the *p* factor. Since Cl is a single atom, no matter how it collides with the ozone molecule, the two particles should react, if the collision has enough energy. NO is a molecule. If the O$_3$ molecule collides with the N atom in the NO molecule, reaction can occur as a bond can form between N and O; if the O$_3$ molecule collides with the O atom in the NO molecule, reaction will not occur as the bond between N and O cannot form. The probability of a successful collision is smaller with NO.
b) The transition state would have weak bonds between the chlorine atom and an oxygen atom in ozone, and between that oxygen atom and a second oxygen atom in ozone.

END–OF–CHAPTER PROBLEMS

16.2   Rate is proportional to concentration. An increase in pressure will increase the number of gas molecules per unit volume. In other words, the gas concentration increases due to increased pressure, so the **reaction rate increases**. Increased pressure also causes more collisions between gas molecules.

16.3   The addition of more water will dilute the concentrations of all solutes dissolved in the reaction vessel. If any of these solutes are reactants, the **rate of the reaction will decrease**.

16.5 An increase in temperature affects the rate of a reaction by increasing the number of collisions, but more importantly, the energy of collisions increases. As the energy of collisions increases, more collisions result in reaction (i.e., reactants → products), so the **rate of reaction increases**.

16.8 a) For most reactions, the rate of the reaction changes as a reaction progresses. The instantaneous rate is the rate at one point, or instant, during the reaction. The average rate is the average of the instantaneous rates over a period of time. On a graph of reactant concentration vs. time of reaction, the instantaneous rate is the slope of the tangent to the curve at any one point. The average rate is the slope of the line connecting two points on the curve. The closer together the two points (shorter the time interval), the more closely the average rate agrees with the instantaneous rate.
b) The initial rate is the instantaneous rate at the point on the graph where time = 0, that is when reactants are mixed.

16.10 At time $t = 0$, no product has formed, so the B(g) curve must start at the origin. Reactant concentration (A(g)) decreases with time; product concentration (B(g)) increases with time. Many correct graphs can be drawn. Two examples are shown below. The graph on the left shows a reaction that proceeds nearly to completion, i.e., [products] >> [reactants] at the end of the reaction. The graph on the right shows a reaction that does not proceed to completion, i.e., [reactants] > [products] at reaction end.

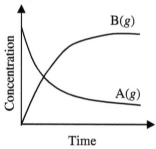 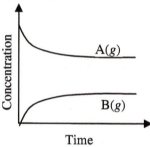

16.12 Plan: The average rate is the total change in concentration divided by the total change in time.
Solution:
a) The average rate from $t = 0$ to $t = 20.0$ s is proportional to the slope of the line connecting these two points:

$$\text{Rate} = -\frac{1}{2}\frac{\Delta[AX_2]}{\Delta t} = -\frac{1}{2}\frac{(0.0088 \text{ mol/L} - 0.0500 \text{ mol/L})}{(20.0 \text{ s} - 0 \text{ s})} = 0.00103 = \textbf{0.0010 mol/L} \bullet \textbf{s}$$

The negative of the slope is used because rate is defined as the change in product concentration with time. If a reactant is used, the rate is the negative of the change in reactant concentration. The 1/2 factor is included to account for the stoichiometric coefficient of 2 for AX_2 in the reaction.
b)

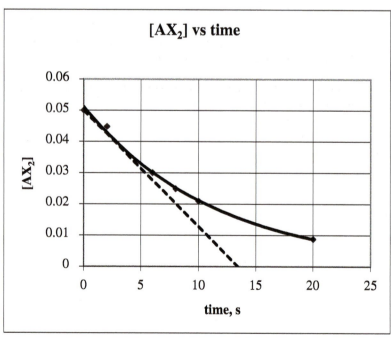

The slope of the tangent to the curve (dashed line) at $t = 0$ is approximately –0.004 mol/L•s. This initial rate is greater than the average rate as calculated in part a). The **initial rate is greater than the average rate** because rate decreases as reactant concentration decreases.

16.14 Plan: Use Equation 16.2 to describe the rate of this reaction in terms of reactant disappearance and product appearance. A negative sign is used for the rate in terms of reactant A since A is reacting and [A] is decreasing over time. Positive signs are used for the rate in terms of products B and C since B and C are being formed and [B] and [C] increase over time. Reactant A decreases twice as fast as product C increases because two molecules of A disappear for every molecule of C that appears.
Solution:
Expressing the rate in terms of each component:
$$\text{Rate} = -\frac{1}{2}\frac{\Delta[A]}{\Delta t} = \frac{\Delta[B]}{\Delta t} = \frac{\Delta[C]}{\Delta t}$$
Calculating the rate of change of [A]:
$$-\frac{1}{2}\frac{\Delta[A]}{\Delta t} = \frac{\Delta[C]}{\Delta t}$$
$$(2 \text{ mol C/L} \cdot \text{s})\left(\frac{2 \text{ mol A/L} \cdot \text{s}}{1 \text{ mol C/L} \cdot \text{s}}\right) = -4 \text{ mol/L} \cdot \text{s}$$
The negative value indicates that [A] is decreasing as the reaction progresses. The rate of reaction is always expressed as a positive number, so [A] is decreasing at a rate of **4 mol/L • s.**

16.16 Plan: Use Equation 16.2 to describe the rate of this reaction in terms of reactant disappearance and product appearance. A negative sign is used for the rate in terms of reactants A and B since A and B are reacting and [A] and [B] are decreasing over time. A positive sign is used for the rate in terms of product C since C is being formed and [C] increases over time. The 1/2 factor is included for reactant B to account for the stoichiometric coefficient of 2 for B in the reaction. Reactant A decreases half as fast as reactant B decreases because one molecule of A disappears for every two molecules of B that disappear.
Solution:
Expressing the rate in terms of each component:
$$\text{Rate} = -\frac{\Delta[A]}{\Delta t} = -\frac{1}{2}\frac{\Delta[B]}{\Delta t} = \frac{\Delta[C]}{\Delta t}$$
Calculating the rate of change of [A]:
$$(0.5 \text{ mol B/L} \cdot \text{s})\left(\frac{1 \text{ mol A/L} \cdot \text{s}}{2 \text{ mol B/L} \cdot \text{s}}\right) = -0.25 \text{ mol/L} \cdot \text{s} = -0.2 \text{ mol/L} \cdot \text{s}$$
The negative value indicates that [A] is decreasing as the reaction progresses. The rate of reaction is always expressed as a positive number, so [A] is decreasing at a rate of **0.2 mol/L • s.**

16.18 Plan: A term with a negative sign is a reactant; a term with a positive sign is a product. The inverse of the fraction becomes the coefficient of the molecule.
Solution:
N_2O_5 is the reactant; NO_2 and O_2 are products.
$$2N_2O_5(g) \rightarrow 4NO_2(g) + O_2(g)$$

16.21 Plan: Use Equation 16.2 to describe the rate of this reaction in terms of reactant disappearance and product appearance. A negative sign is used for the rate in terms of reactants N_2 and H_2 since these substances are reacting and [N_2] and [H_2] are decreasing over time. A positive sign is used for the rate in terms of the product NH_3 since it is being formed and [NH_3] increases over time.
Solution:
$$\text{Rate} = -\frac{\Delta[N_2]}{\Delta t} = -\frac{1}{3}\frac{\Delta[H_2]}{\Delta t} = \frac{1}{2}\frac{\Delta[NH_3]}{\Delta t}$$

16.22   Plan: Use Equation 16.2 to describe the rate of this reaction in terms of reactant disappearance and product appearance. A negative sign is used for the rate in terms of the reactant O_2 since it is reacting and $[O_2]$ is decreasing over time. A positive sign is used for the rate in terms of the product O_3 since it is being formed and $[O_3]$ increases over time. O_3 increases 2/3 as fast as O_2 decreases because two molecules of O_3 are formed for every three molecules of O_2 that disappear.
Solution:

a) Rate $= -\dfrac{1}{3}\dfrac{\Delta[O_2]}{\Delta t} = \dfrac{1}{2}\dfrac{\Delta[O_3]}{\Delta t}$

b) Use the mole ratio in the balanced equation:

$\left(\dfrac{2.17\times10^{-5}\ \text{mol } O_2\,/\text{L}\cdot\text{s}}{}\right)\left(\dfrac{2\ \text{mol } O_3\,/\text{L}\cdot\text{s}}{3\ \text{mol } O_2/\text{L}\cdot\text{s}}\right) = 1.446667\times10^{-5} = \mathbf{1.45\times10^{-5}\ mol/L\cdot s}$

16.23   a) k is the rate constant, the proportionality constant in the rate law. k represents the fraction of successful collisions which includes the fraction of collisions with sufficient energy and the fraction of collisions with correct orientation. k is a constant that varies with temperature.
b) m represents the order of the reaction with respect to $[A]$ and n represents the order of the reaction with respect to $[B]$. The order is the exponent in the relationship between rate and reactant concentration and defines how reactant concentration influences rate.
The order of a reactant does not necessarily equal its stoichiometric coefficient in the balanced equation. If a reaction is an elementary reaction, meaning the reaction occurs in only one step, then the orders and stoichiometric coefficients are equal. However, if a reaction occurs in a series of elementary reactions, called a mechanism, then the rate law is based on the slowest elementary reaction in the mechanism. The orders of the reactants will equal the stoichiometric coefficients of the reactants in the slowest elementary reaction but may not equal the stoichiometric coefficients in the overall reaction.
c) For the rate law rate $= k[A]\,[B]^2$ substitute in the units:

Rate (mol/L·min) $= k[A]^1[B]^2$

$k = \dfrac{\text{rate}}{[A]^1[B]^2} = \dfrac{\text{mol/L}\cdot\text{min}}{\left[\dfrac{\text{mol}}{L}\right]^1\left[\dfrac{\text{mol}}{L}\right]^2} = \dfrac{\text{mol/L}\cdot\text{min}}{\dfrac{\text{mol}^3}{L^3}}$

$k = \dfrac{\text{mol}}{L\cdot\text{min}}\left(\dfrac{L^3}{\text{mol}^3}\right)$

$\mathbf{k = L^2/mol^2\cdot min}$

16.25   a) The **rate doubles**. If rate $= k[A]^1$ and $[A]$ is doubled, then the rate law becomes rate $= k[2\times A]^1$. The rate increases by 2^1 or 2.
b) The **rate decreases by a factor of four**. If rate $= k[B]^2$ and $[B]$ is halved, then the rate law becomes rate $= k[1/2 \times B]^2$. The rate decreases to $(1/2)^2$ or 1/4 of its original value.
c) The **rate increases by a factor of nine**. If rate $= k[C]^2$ and $[C]$ is tripled, then the rate law becomes rate $= k[3 \times C]^2$. The rate increases to 3^2 or 9 times its original value.

16.26   Plan: The order for each reactant is the exponent on the reactant concentration in the rate law. The individual orders are added to find the overall reaction order.
Solution:
The orders with respect to $[BrO_3^-]$ and to $[Br^-]$ are both 1 since both have an exponent of 1. The order with respect to $[H^+]$ is 2 (its exponent in the rate law is 2). The overall reaction order is $1 + 1 + 2 = 4$.
first order with respect to BrO_3^-, first order with respect to Br^-, second order with respect to H^+, fourth order overall

16.28   a) The rate is first order with respect to $[BrO_3^-]$. If $[BrO_3^-]$ is doubled, rate $= k[2 \times BrO_3^-]$, then rate increases to 2^1 or 2 times its original value. The rate **doubles**.

b) The rate is first order with respect to [Br⁻]. If [Br⁻] is halved, rate = $k[1/2 \times Br^-]$, then rate decreases by a factor of $(1/2)^1$ or 1/2 times its original value. The rate is **halved**.

c) The rate is second order with respect to [H⁺]. If [H⁺] is quadrupled, rate = $k[4 \times H^+]^2$, then rate increases to 4^2 or **16 times** its original value.

16.30   Plan: The order for each reactant is the exponent on the reactant concentration in the rate law. The individual orders are added to find the overall reaction order.
Solution:
The order with respect to **[NO₂] is 2**, and the order with respect to **[Cl₂] is 1**. The overall order is: 2 + 1 = **3 for the overall order**.

16.32   a) The rate is second order with respect to [NO₂]. If [NO₂] is tripled, rate = $k[3 \times NO_2]^2$, then rate increases to 3^2 or 9 times its original value. The rate **increases by a factor of 9.**
b) The rate is second order with respect to [NO₂] and first order with respect to [Cl₂]. If [NO₂] and [Cl₂] are doubled, rate = $k[2 \times NO_2]^2[2 \times Cl_2]^1$, then the rate **increases by a factor of $2^2 \times 2^1 = 8$.**
c) The rate is first order with respect to [Cl₂]. If Cl₂ is halved, rate = $k[1/2 \times Cl_2]^1$, then rate decreases to 1/2 times its original value. The rate is **halved.**

16.34   Plan: The rate law is rate = $[A]^m[B]^n$ where m and n are the orders of the reactants. To find the order of each reactant, take the ratio of the rate laws for two experiments in which only the reactant in question changes. Once the rate law is known, any experiment can be used to find the rate constant k.
Solution:
a) To find the order for reactant A, first identify the reaction experiments in which [A] changes but [B] is constant. Use experiments 1 and 2 (or 3 and 4 would work) to find the order with respect to [A].
Set up a ratio of the rate laws for experiments 1 and 2 and fill in the values given for rates and concentrations and solve for m, the order with respect to [A].

$$\frac{rate_{exp\,2}}{rate_{exp\,1}} = \left(\frac{[A]_{exp\,2}}{[A]_{exp\,1}}\right)^m$$

$$\frac{45.0 \text{ mol/L} \cdot \text{min}}{5.00 \text{ mol/L} \cdot \text{min}} = \left(\frac{0.300 \text{ mol/L}}{0.100 \text{ mol/L}}\right)^m$$

$$9.00 = (3.00)^m$$
$$\log (9.00) = m \log (3.00)$$
$$m = 2$$

Using experiments 3 and 4 also gives **second order with respect to [A].**
To find the order for reactant B, first identify the reaction experiments in which [B] changes but [A] is constant. Use experiments 1 and 3 (or 2 and 4 would work) to find the order with respect to [B].
Set up a ratio of the rate laws for experiments 1 and 3 and fill in the values given for rates and concentrations and solve for n, the order with respect to [B].

$$\frac{rate_{exp\,3}}{rate_{exp\,1}} = \left(\frac{[B]_{exp\,3}}{[B]_{exp\,1}}\right)^n$$

$$\frac{10.0 \text{ mol/L} \cdot \text{min}}{5.00 \text{ mol/L} \cdot \text{min}} = \left(\frac{0.200 \text{ mol/L}}{0.100 \text{ mol/L}}\right)^n$$

$$2.00 = (2.00)^n$$
$$\log (2.00) = n \log (2.00)$$
$$n = 1$$

The reaction is **first order with respect to [B].**
b) The rate law, without a value for k, is **rate = $k[A]^2[B]$**.
c) Using experiment 1 to calculate k (the data from any of the experiments can be used):

$$Rate = k[A]^2[B]$$

$$k = \frac{rate}{[A]^2[B]} = \frac{5.00 \text{ mol/L} \cdot \text{min}}{[0.100 \text{ mol/L}]^2[0.100 \text{ mol/L}]} = \mathbf{5.00 \times 10^3 \text{ L}^2/\text{mol}^2 \cdot \text{min}}$$

16.80   a) **No,** by definition, a catalyst is a substance that increases reaction rate without being consumed. The spark provides energy that is absorbed by the H_2 and O_2 molecules to achieve the threshold energy needed for reaction.
b) **Yes,** the powdered metal acts like a heterogeneous catalyst, providing a surface upon which the reaction between O_2 and H_2 becomes more favorable because the activation energy is lowered.

16.85   Plan: An intermediate is a substance that is formed in one step and consumed in a subsequent step. The coefficients of the reactants in each elementary step are used as the orders in the rate law for the step. The overall rate law for the mechanism is determined from the slowest step (the rate-determining step) and can be compared to the actual rate law. Reactants that appear in the mechanism after the slow step do not determine the rate and therefore do not appear in the rate law.
Solution:
a) Water does not appear as a reactant in the rate-determining step.
Note that as a solvent in the reaction, the concentration of the water is assumed not to change even though some water is used up as a reactant. This assumption is valid as long as the solute concentrations are low ($\sim$1 M or less). So, even if water did appear as a reactant in the rate-determining step, it would not appear in the rate law. See rate law for step 2 below.
b)   Rate law for step (1): $rate_1 = k_1[(CH_3)_3CBr]$
        Rate law for step (2): $rate_2 = k_2[(CH_3)_3C^+]$
        Rate law for step (3): $rate_3 = k_3[(CH_3)_3COH_2^+]$
c) The intermediates are $(CH_3)_3C^+$ and $(CH_3)_3COH_2^+$. $(CH_3)_3C^+$ is formed in step 1 and consumed in step 2; $(CH_3)_3COH_2^+$ is formed in step 2 and consumed in step 3.
d) The rate-determining step is the slow step, (1). The rate law for this step is $rate = k_1[(CH_3)_3CBr]$ since the coefficient of the reactant in this slow step is 1. The rate law for this step agrees with the actual rate law with $k = k_1$.

16.87   Plan: The activation energy can be calculated using the Arrhenius equation. Although the rate constants, k_1 and k_2, are not expressly stated, the relative times give an idea of the rate. The reaction rate is proportional to the rate constant. At $T_1 = 20°C$, the rate of reaction is 1 apple/4 days while at $T_2 = 0°C$, the rate is 1 apple/16 days. Therefore, $rate_1 = 1$ apple/4 days and $rate_2 = 1$ apple/16 days are substituted for k_1 and k_2, respectively.
Solution:

$k_1 = 1/4$ $\qquad\qquad\qquad\qquad$ $T_1 = 20°C + 273 = 293$ K
$k_2 = 1/16$ $\qquad\qquad\qquad\qquad$ $T_2 = 0°C + 273 = 273$ K
$E_a = ?$

$$\ln\frac{k_2}{k_1} = \frac{E_a}{R}\left(\frac{1}{T_1} - \frac{1}{T_2}\right)$$

$$E_a = \frac{R\left(\ln\frac{k_2}{k_1}\right)}{\left(\frac{1}{T_1} - \frac{1}{T_2}\right)} = \frac{\left(8.314\,\frac{J}{mol \cdot K}\right)\left(\ln\frac{1/16}{1/4}\right)}{\left(\frac{1}{293\ K} - \frac{1}{273\ K}\right)}$$

$E_a = 4.6096266 \times 10^4$ J/mol = **4.61x10⁴ J/mol** The significant figures are based on the Kelvin temperatures.

16.91   Plan: Use the given rate law, $rate = k[H^+][sucrose]$, and enter the given values. The glucose and fructose are not in the rate law, so they may be ignored.
Solution:
a) The rate is first order with respect to [sucrose]. The [sucrose] is changed from 1.0 M to 2.5 M, or is increased by a factor of 2.5/1.0 or 2.5. Then the $rate = k[H^+][2.5 \times sucrose]$; the rate **increases by a factor of 2.5.**
b) The [sucrose] is changed from 1.0 M to 0.5 M, or is decreased by a factor of 0.5/1.0 or 0.5. Then the $rate = k[H^+][0.5 \times sucrose]$; the rate decreases by a factor of ½ or **half the original rate.**

c) The rate is first order with respect to [H$^+$]. The [H$^+$] is changed from 0.01 M to 0.0001 M, or is decreased by a factor of 0.0001/0.01 or 0.01. Then the rate = k[0.01 x H$^+$][sucrose]; the rate **decreases by a factor of 0.01**. Thus, the reaction will decrease to **1/100 the original**.

d) The [sucrose] decreases from 1.0 M to 0.1 M, or by a factor of (0.1 M/1.0 M) = 0.1. [H$^+$] increases from 0.01 M to 0.1 M, or by a factor of (0.1 M/0.01 M) = 10. Then the rate will increase by k[10 x H$^+$][0.1 x sucrose]= 1.0 times as fast. Thus, there will be **no change**.

16.92   Plan: The overall order is equal to the sum of the individual orders. Since the reaction is eleventh order overall, the sum of the exponents equals eleven. Add up the known orders and subtract that sum from eleven to find the unknown order.
Solution:
Sum of known orders = 1 + 4 + 2 + 2 = 9
Overall order – sum of known orders = 11 – 9 = 2.
The reaction is **second order** with respect to NAD.

16.95   Plan: First, find the rate constant, k, for the reaction by solving the first-order half-life equation for k. Then use the first-order integrated rate law expression to find t, the time for decay.
Solution:
Rearrange $t_{1/2} = \dfrac{\ln 2}{k}$ to $k = \dfrac{\ln 2}{t_{1/2}}$

$$k = \frac{\ln 2}{12 \text{ yr}} = 5.7762 \times 10^{-2} \text{ yr}^{-1}$$

Use the first-order integrated rate law: $\ln \dfrac{[\text{DDT}]_t}{[\text{DDT}]_0} = -kt$

$$\ln \frac{[10. \text{ ppbm}]_t}{[275 \text{ ppbm}]_0} = -(5.7762 \times 10^{-2} \text{ yr}^{-1})\, t$$

$$t = 57.3765798 = \textbf{57 yr}$$

16.98   Plan: The rate constant can be determined from the slope of the integrated rate law plot. To find the correct order, the data should be plotted as 1) [sucrose] vs. time – linear for zero order, 2) ln [sucrose] vs. time – linear for first order, and 3) 1/[sucrose] vs. time – linear for second order. Once the order is established, use the appropriate integrated rate law to find the time necessary for 75.0% of the sucrose to react. If 75.0% of the sucrose has reacted, 100–75 = 25% remains at time t. Let [sucrose]$_0$ = 100% and then [sucrose]$_t$ = 25%.
Solution:
a) All three graphs are linear, so picking the correct order is difficult. One way to select the order is to compare correlation coefficients (R^2) — you may or may not have experience with this. The best correlation coefficient is the one closest to a value of 1.00. Based on this selection criterion, the plot of ln [sucrose] vs. time for the first-order reaction is the best.
Another method when linearity is not obvious from the graphs is to examine the reaction and decide which order fits the reaction. For the reaction of one molecule of sucrose with one molecule of liquid water, the rate law would most likely include sucrose with an order of one and would not include water.
The plot for a first-order reaction is described by the equation ln [A]$_t$ = –kt + ln [A]$_0$. The slope of the plot of ln [sucrose] vs. t equals –k. The equation for the straight line in the first-order plot is $y = -0.21x - 0.6936$.
So, $k = -(-0.21 \text{ h}^{-1}) = \textbf{0.21 h}^{-1}$.
Solve the first-order half-life equation to find $t_{1/2}$:

$$t_{1/2} = \frac{\ln 2}{0.21 \text{ hr}^{-1}} = 3.3007 = \textbf{3.3 h}$$

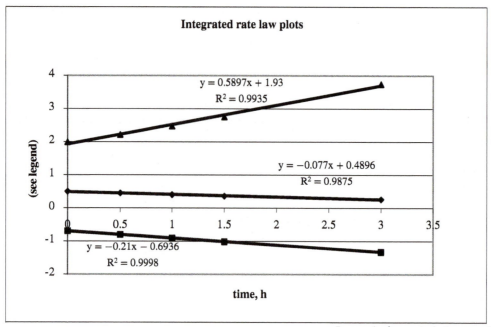

Integrated rate law plots

Legend: ◆ *y*-axis is [sucrose]
■ *y*-axis is ln [sucrose]
▲ *y*-axis is 1/[sucrose]

b) If 75% of the sucrose has been reacted, 25% of the sucrose remains. Let [sucrose]$_0$ = 100% and [sucrose]$_t$ = 25% in the first-order integrated rate law equation:

$$\ln \frac{[\text{sucrose}]_t}{[\text{sucrose}]_0} = -kt$$

$$\ln \frac{[25\%]_t}{[100\%]_0} = -(0.21\ \text{h}^{-1})\,t$$

$$t = 6.6014 = \textbf{6.6 h}$$

c) The reaction might be second order overall with first order in sucrose and first order in water. If the concentration of sucrose is relatively low, the concentration of water remains constant even with small changes in the amount of water. This gives an apparent zero-order reaction with respect to water. Thus, the reaction appears to be first order overall because the rate does not change with changes in the amount of water.

16.101 <u>Plan:</u> To find concentration of reactant at a later time, given the initial concentration and the rate constant k, use an integrated rate law expression. Since the units on k are s^{-1}, this is a first-order reaction. Use the first-order integrated rate law. Since the time unit in k is seconds, time t must also be expressed in units of seconds. To find the fraction of reactant that has decomposed, divide that amount of reactant that has decomposed ($[N_2O_5]_0 - [N_2O_5]_t$) by the initial concentration.
<u>Solution:</u>

a) Converting t in min to s: (5.00 min)$\left(\dfrac{60\ \text{s}}{1\ \text{min}}\right)$ = 300. s

$$\ln \frac{[N_2O_5]_t}{[N_2O_5]_0} = -kt \quad \text{or}$$

ln $[N_2O_5]_t$ = ln $[N_2O_5]_0 - kt$
ln $[N_2O_5]_t$ = ln [1.58 mol/L] $- (2.8 \times 10^{-3}\ \text{s}^{-1})(300.\ \text{s})$
ln $[N_2O_5]_t$ = -0.382575
$[N_2O_5]_t$ = 0.68210 = **0.68 mol/L**

b) Fraction decomposed = $\dfrac{[N_2O_5]_0 - [N_2O_5]_t}{[N_2O_5]_0} = \dfrac{1.58 - 0.68210\ \text{mol/L}}{1.58\ \text{mol/L}}$ = 0.56829 = **0.57**

16-22

16.104 Plan: To solve this problem, a clear picture of what is happening is useful. Initially only N_2O_5 is present at a pressure of 125 kPa. Then a reaction takes place that consumes the gas N_2O_5 and produces the gases NO_2 and O_2. The balanced equation gives the change in the number of moles of gas as N_2O_5 decomposes. Since the number of moles of gas is proportional to the pressure, this change mirrors the change in pressure. The total pressure at the end, 178 kPa, equals the sum of the partial pressures of the three gases.

Solution:

Balanced equation: $N_2O_5(g) \rightarrow 2NO_2(g) + 1/2O_2(g)$

Therefore, for each mole of dinitrogen pentaoxide that is consumed, 2.5 moles of gas are produced.

	$N_2O_5(g) \rightarrow$	$2NO_2(g) +$	$1/2O_2(g)$	
Initial P (kPa)	125	0	0	total $P_{initial}$ = 125 kPa
Final P (kPa)	125 – x	2x	1/2x	total P_{final} = 178 kPa

Solve for x:

$$P_{N_2O_5} + P_{NO_2} + P_{O_2} = (125 - x) + 2x + 1/2x = 178$$

x = 35.3333 kPa (unrounded)

Partial pressure of NO_2 equals 2x = 2(35.3333) = 70.667 = **71 kPa**.

Check: Substitute values for all partial pressures to find total final pressure:

(125 – 35.3333) + (2 x 35.3333) + ((1/2) x 35.3333) = 178 kPa

The result agrees with the given total final pressure.

16.108 Plan: The activation energy can be calculated using the Arrhenius equation. Although the rate constants, k_1 and k_2, are not expressly stated, the relative times give an idea of the rate. The reaction rate is proportional to the rate constant. At T_1 = 90.0°C, the rate of reaction is 1 egg/4.8 min while at T_2 = 100.0°C, the rate is 1 egg/4.5 min. Therefore, $rate_1$ = 1 egg/4.8 min and $rate_2$ = 1 egg/4.5 min are substituted for k_1 and k_2, respectively.

Solution:

k_1 = 1 egg/4.8 min T_1 = 90.0°C + 273.2 = 363.2 K

k_2 = 1 egg/4.5 min T_2 = 100.0°C + 273.2 = 373.2 K

E_a = ?

$$\ln\frac{k_2}{k_1} = \frac{E_a}{R}\left(\frac{1}{T_1} - \frac{1}{T_2}\right)$$

The number of eggs (1) is exact, and has no bearing on the significant figures.

$$E_a = \frac{R\left(\ln\frac{k_2}{k_1}\right)}{\left(\frac{1}{T_1} - \frac{1}{T_2}\right)} = \frac{\left(8.314\frac{J}{mol \cdot K}\right)\left(\ln\frac{\left(1\,egg/4.5\,min\right)}{\left(1\,egg/4.8\,min\right)}\right)}{\left(\frac{1}{363.2\,K} - \frac{1}{273.2\,K}\right)}$$

E_a = 7.2730x10³ J/mol = **7.3x10³ J/mol**

16.110 Plan: Starting with the fact that rate of formation of O (rate of step 1) equals the rate of consumption of O (rate of step 2), set up an equation to solve for [O] using the given values of k_1, k_2, [NO_2], and [O_2].

Solution:

a) $Rate_1 = k_1[NO_2]$ $Rate_2 = k_2[O][O_2]$

$Rate_1 = rate_2$

$k_1[NO_2] = k_2[O][O_2]$

$$[O] = \frac{k_1[NO_2]}{k_2[O_2]} = \frac{\left(6.0x10^{-3}\,s^{-1}\right)\left[4.0x10^{-9}\,M\right]}{\left(1.0x10^6\,L/mol \cdot s\right)\left[1.0x10^{-2}\,M\right]} = \textbf{2.4x10}^{-15}\,\textbf{M}$$

b) Since the rate of the two steps is equal, either can be used to determine rate of formation of ozone.

$Rate_2 = k_2[O][O_2] = (1.0x10^6\,L/mol \cdot s)(2.4x10^{-15}\,M)(1.0x10^{-2}\,M) = \textbf{2.4x10}^{-11}\,\textbf{mol/L} \cdot \textbf{s}$

16.113   <u>Plan:</u> This problem involves the first-order integrated rate law ($\ln [A]/[A]_0 = -kt$). The temperature must be part of the calculation of the rate constant. The concentration of the ammonium ion is directly related to the ammonia concentration. Use the given values of $[NH_3]_0$ and $[NH_3]_t$ and the calculated values of k to find time, t.
<u>Solution:</u>
a) $[NH_3]_0 = 3.0$ mol/m^3 $[NH_3]_t = 0.35$ mol/m^3 $T = 20°C$
$k_1 = 0.47e^{0.095(T-15°C)} = 0.47e^{0.095(20-15°C)} = 0.75576667$ d^{-1}

$$\ln\frac{[NH_3]_t}{[NH_3]_0} = -kt$$

$$t = -\frac{\ln\dfrac{[NH_3]_t}{[NH_3]_0}}{k}$$

$$t = -\frac{\ln\dfrac{[0.35 \text{ mol/m}^3]_t}{[3.0 \text{ mol/m}^3]_0}}{0.75576667 \text{ d}^{-1}} = 2.84272 = \mathbf{2.8\ d}$$

b) Repeating the calculation at the different temperature:
$[NH_3]_0 = 3.0$ mol/m^3 $[NH_3]_t = 0.35$ mol/m^3 $T = 10°C$
$k_1 = 0.47e^{0.095(T-15°C)} = 0.47e^{0.095(10-15°C)} = 0.292285976$ d^{-1}

$$t = -\frac{\ln\dfrac{[NH_3]_t}{[NH_3]_0}}{k}$$

$$t = -\frac{\ln\dfrac{[0.35 \text{ mol/m}^3]_t}{[3.0 \text{ mol/m}^3]_0}}{0.292285976 \text{ d}^{-1}}$$

$$t = 7.35045 = \mathbf{7.4\ d}$$

c) For NH_4^+ the rate $= k_1[NH_4^+]$
From the balanced chemical equation:

$$-\frac{\Delta[NH_4^+]}{\Delta t} = -\frac{1}{2}\frac{\Delta[O_2]}{\Delta t}$$

Thus, for O_2: Rate $= 2\,k_1[NH_4^+]$

$$\text{Rate} = (2)(0.75576667)\left[3.0 \text{ mol/m}^3\right] = 4.5346 = \mathbf{4.5\ mol/m^3}$$

16.116   <u>Plan:</u> Rate is the change in concentration divided by change in time. To find the average rate for each trial in part a), the change in concentration of $S_2O_3^{2-}$ is divided by the time required to produce the color. The rate law is rate $= k[I^-]^m[S_2O_8^{2-}]^n$ where m and n are the orders of the reactants. To find the order of each reactant, take the ratio of the rate laws for two experiments in which only the reactant in question changes. Once the rate law is known, any experiment can be used to find the rate constant k. Since several solutions are mixed, final concentrations of each solution must be found with dilution calculations using $M_iV_i = M_fV_f$ in the form:
$M_f = M_iV_i/V_f$.
<u>Solution:</u>
a) $M_f\ S_2O_3^{2-} = [(10.0 \text{ mL})(0.0050\ M)]/50.0 \text{ mL} = 0.0010\ M\ S_2O_3^{2-}$

$$-\frac{\Delta[I_2]}{\Delta t} = -\frac{1}{2}\frac{\Delta[S_2O_3^{2-}]}{\Delta t} = [1/2(0.0010\ M)]/\text{time} = [0.00050\ M]/\text{time} = \text{rate}$$

Average rates:
Rate$_1$ = $(0.00050\ M)/29.0$ s = 1.724×10^{-5} = $\mathbf{1.7\times10^{-5}\ M\ s^{-1}}$
Rate$_2$ = $(0.00050\ M)/14.5$ s = 3.448×10^{-5} = $\mathbf{3.4\times10^{-5}\ M\ s^{-1}}$
Rate$_3$ = $(0.00050\ M)/14.5$ s = 3.448×10^{-5} = $\mathbf{3.4\times10^{-5}\ M\ s^{-1}}$

b) M_r KI = M_r I$^-$ = [(10.0 mL)(0.200 M)]/50.0 mL = 0.0400 M I$^-$ (Experiment 1)

$\quad M_r$ I$^-$ = [(20.0 mL)(0.200 M)]/50.0 mL = 0.0800 M I$^-$ (Experiments 2 and 3)

$\quad M_r$ Na$_2$S$_2$O$_8$ = M_r S$_2$O$_8^{2-}$ = [(20.0 mL)(0.100 M)]/50.0 mL = 0.0400 M S$_2$O$_8^{2-}$ (Experiments 1 and 2)

$\quad M_r$ S$_2$O$_8^{2-}$ = [(10.0 mL)(0.100 M)]/50.0 mL = 0.0200 M S$_2$O$_8^{2-}$ (Experiment 3)

$\quad$ Generic rate law equation: Rate = k [I$^-$]^m[S$_2$O$_8^{2-}$]^n

To find the order for I$^-$, use experiments 1 and 2 in which [S$_2$O$_8^{2-}$] is constant while [I$^-$] changes. Set up a ratio of the rate laws for experiments 1 and 2 and fill in the values given for rates and concentrations and solve for m, the order with respect to [I$^-$].

$$\frac{\text{Rate}_2}{\text{Rate}_1} = \frac{k_2 \left[I^-\right]^m \left[S_2O_8^{2-}\right]^n}{k_1 \left[I^-\right]^m \left[S_2O_8^{2-}\right]^n}$$
$\qquad\qquad\qquad\qquad\qquad$ Molarity of S$_2$O$_8^{2-}$ is constant.

$$\frac{3.4\times10^{-5}\ Ms^{-1}}{1.7\times10^{-5}\ Ms^{-1}} = \frac{[0.0800]^m}{[0.0400]^m}$$

$$2.0 = (2.00)^m$$

$$m = 1$$

$\qquad$ The reaction is **first order with respect to I$^-$**.

To find the order for S$_2$O$_8^{2-}$, use experiments 2 and 3 in which [I$^-$] is constant while [S$_2$O$_8^{2-}$] changes. Set up a ratio of the rate laws for experiments 2 and 3 and fill in the values given for rates and concentrations and solve for n, the order with respect to [S$_2$O$_8^{2-}$].

$$\frac{\text{Rate}_3}{\text{Rate}_2} = \frac{k_3 \left[I^-\right]^m \left[S_2O_8^{2-}\right]^n}{k_2 \left[I^-\right]^m \left[S_2O_8^{2-}\right]^n}$$
$\qquad\qquad\qquad\qquad\qquad$ Molarity of I$^-$ is constant.

$$\frac{3.4\times10^{-5}\ Ms^{-1}}{3.4\times10^{-5}\ Ms^{-1}} = \frac{[0.0200]^n}{[0.0400]^n}$$

$$1.0 = (0.500)^n$$

$$n = 0$$

$\qquad$ The reaction is **zero order with respect to S$_2$O$_8^{2-}$**.

c) Rate = k[I$^-$]

$\qquad k$ = rate/[I$^-$]

$\qquad\qquad$ Using experiment 2 (unrounded rate value)

$\qquad k$ = (3.448x10^{-5} M s^{-1})/(0.0800 M) = 4.31x10^{-4} = **4.3x10^{-4} s^{-1}**

d) **Rate = (4.3x10^{-4} s^{-1})[I$^-$]**

16.118 <u>Plan:</u> This is a first-order process so use the first-order integrated rate law. The increasing cell density changes the integrated rate law from $-kt$ to $+kt$. In part a), we know t(2 h), k, and [A]$_0$ so [A]$_t$ can be found. Treat the concentration of the cells as you would molarity. Since the rate constant is expressed in units of min^{-1}, the time interval of 2 h must be converted to a time in minutes. In part b), [A]$_0$, [A]$_t$, and k are known and time t is calculated.
<u>Solution:</u>

a) Converting time in h to min: $(2\ h)\left(\dfrac{60\ \text{min}}{1\ h}\right) = 120$ min

$\ln$ [A]$_t$ = $\ln$ [A]$_0$ + kt

$\ln$ [A]$_t$ = $\ln$ [1.0x10^3] + (0.035 min^{-1})(120 min)

$\ln$ [A]$_t$ = 11.107755

[A]$_t$ = 6.6686x10^4 = **7x10^4 cells/L**

b) $\ln$ [A]$_t$ = $\ln$ [A]$_0$ + kt

$\dfrac{\ln\ [A]_t - \ln\ [A]_0}{k} = t$

$\dfrac{\ln\ [2\text{x}10^3\ \text{cells/L}] - \ln\ [1\text{x}10^3\ \text{cells/L}]}{0.035\ \text{min}^{-1}} = t$

$t = 19.804 =$ **2.0x10^1 min**

16.121 **Plan:** For part a), use the Monod equation to calculate μ for values of S between 0.0 and 1.0 kg/m³ and then graph. For parts b) and c), use the Monod equation to calculate μ at the given conditions. The value of μ is the rate constant k in the first-order integrated rate law. The increasing population density changes the integrated rate law from $-kt$ to $+kt$. We know t(1 h), k, and $[A]_0$ so $[A]_t$ can be found. Treat the density of the cells as you would molarity.

Solution:

$$\mu = \frac{\mu_{max}S}{K_s + S} = \frac{(1.5 \times 10^{-4}\ s^{-1})(0.25\ kg/m^3)}{(0.03\ kg/m^3) + (0.25\ kg/m^3)} = 1.34 \times 10^{-4}\ s^{-1}$$

$$\mu = \frac{\mu_{max}S}{K_s + S} = \frac{(1.5 \times 10^{-4}\ s^{-1})(0.50\ kg/m^3)}{(0.03\ kg/m^3) + (0.50\ kg/m^3)} = 1.42 \times 10^{-4}\ s^{-1}$$

$$\mu = \frac{\mu_{max}S}{K_s + S} = \frac{(1.5 \times 10^{-4}\ s^{-1})(0.75\ kg/m^3)}{(0.03\ kg/m^3) + (0.75\ kg/m^3)} = 1.44 \times 10^{-4}\ s^{-1}$$

$$\mu = \frac{\mu_{max}S}{K_s + S} = \frac{(1.5 \times 10^{-4}\ s^{-1})(1.0\ kg/m^3)}{(0.03\ kg/m^3) + (1.0\ kg/m^3)} = 1.46 \times 10^{-4}\ s^{-1}$$

b) $$\mu = \frac{\mu_{max}S}{K_s + S} = \frac{(1.5 \times 10^{-4}\ s^{-1})(0.30\ kg/m^3)}{(0.03\ kg/m^3) + (0.30\ kg/m^3)} = 1.3636 \times 10^{-4}\ s^{-1}$$

Converting time in h to seconds: $(1\ h)\left(\dfrac{3600\ s}{1\ h}\right) = 3600\ s$

$\ln [A]_t = \ln [A]_0 + kt$
$\ln [A]_t = \ln [5.0 \times 10^3] + (1.3636 \times 10^{-4}\ s^{-1})(3600\ s)$
$\ln [A]_t = 9.00808919$
$[A]_t = 8.1689 \times 10^3 =$ **8.2x10³ cells/m³**

c) $$\mu = \frac{\mu_{max}S}{K_s + S} = \frac{(1.5 \times 10^{-4}\ s^{-1})(0.70\ kg/m^3)}{(0.03\ kg/m^3) + (0.70\ kg/m^3)} = 1.438356 \times 10^{-4}\ s^{-1}$$

$\ln [A]_t = \ln [A]_0 + kt$
$\ln [A]_t = \ln [5.0 \times 10^3] + (1.438356 \times 10^{-4}\ s^{-1})(3600\ s)$
$\ln [A]_t = 9.03500135$
$[A]_t = 8.39172 \times 10^3 =$ **8.4x10³ cells/m³**

16.126 <u>Plan:</u> This is a first-order reaction so use the first-order integrated rate law. In part a), use the first-order half-life equation and the given value of k to find the half-life. In parts b) and c), solve the first-order integrated rate law to find the time necessary for 40.% and 90.% of the acetone to decompose. If 40.% of the acetone has decomposed, $100 - 40. = 60.\%$ remains at time t. If 90.% of the acetone has decomposed, $100 - 90. = 10.\%$ remains at time t.
<u>Solution:</u>

a) $t_{1/2} = \dfrac{\ln 2}{k} = \dfrac{\ln 2}{8.7 \times 10^{-3} \text{ s}^{-1}} = 79.672 = \mathbf{8.0 \times 10^1 \text{ s}}$

b) $\ln [\text{acetone}]_t = \ln [\text{acetone}]_0 - kt$

$\dfrac{\ln [\text{acetone}]_t - \ln [\text{acetone}]_0}{-k} = t$
$\qquad\qquad$ 40.% of acetone has decomposed; $[\text{acetone}]_t = 100 - 40. = 60.\%$

$\dfrac{\ln [60.] - \ln [100]}{-8.7 \times 10^{-3} \text{ s}^{-1}} = t = 58.71558894 = \mathbf{59 \text{ s}}$

c) $\ln [\text{acetone}]_t = \ln [\text{acetone}]_0 - kt$

$\dfrac{\ln [\text{acetone}]_t - \ln [\text{acetone}]_0}{-k} = t$
$\qquad\qquad$ 90.% of acetone has decomposed; $[\text{acetone}]_t = 100 - 90. = 10.\%$

$\dfrac{\ln [10.] - \ln [100]}{-8.7 \times 10^{-3} \text{ s}^{-1}} = t = 264.66 = \mathbf{2.6 \times 10^2 \text{ s}}$

16.129 <u>Plan:</u> Since the reaction can be treated as pseudo-first-order in NH_3, the given concentration of HOCl can be combined with the given rate constant to calculate a new rate constant in the first-order rate expression and integrated rate law. If 30% of the NH_3 has reacted, $100–30 = 70\%$ remains at time t. Let $\left[NH_3\right]_0 = 1\ M$ and then $\left[NH_3\right]_t = 70\%$ of $1\ M = 0.70\ M$.
<u>Solution:</u>

Rate = $k[NH_3][HOCl]$

Rate = 5.1×10^6 L/mol·s$[2 \times 10^{-3}$ mol/L$][NH_3] = 1.02 \times 10^4$/s$[NH_3]$

The integrated rate law for a first-order reaction is $\ln [A]_t = \ln [A]_0 - kt$

$\dfrac{\ln\left[NH_3\right]_t - \ln\left[NH_3\right]_0}{-k} = t$

$\dfrac{\ln[.70] - \ln[1]}{-1.02 \times 10^4 \text{ s}^{-1}} = t$

$t = 3.4968 \times 10^{-5} \text{ s} = \mathbf{3.5 \times 10^{-5} \text{ s}}$

16.131 <u>Plan:</u> Write a rate a law for the production of Cl_2 in Step 2 of the mechanism; eliminate the intermediate from that rate law by assuming steady-state approximation.
<u>Solution:</u>

a) Rate = $k_2[NO_2Cl][Cl]$; eliminate the intermediate Cl from the rate law:

Net rate of Cl formation = $\underset{\substack{\text{production in step} \\ \text{1 (forward reaction)}}}{k_1\left[NO_2Cl\right]} - \underset{\substack{\text{consumption in step} \\ \text{1 (reverse reaction)}}}{k_{-1}\left[NO_2\right][Cl]} - \underset{\substack{\text{consumption in step 2}}}{k_2\left[NO_2Cl\right][Cl]}$

The steady-state approximation assumes that [Cl] remains constant during the reaction so its net rate of formation is 0:

Net rate of Cl formation = $0 = k_1[NO_2Cl] - k_{-1}[NO_2][Cl] - k_2[NO_2Cl][Cl]$

$k_1[NO_2Cl] = k_{-1}[NO_2][Cl] + k_2[NO_2Cl][Cl]$

$k_1[NO_2Cl] = [Cl](k_{-1}[NO_2] + k_2[NO_2Cl])$

$[Cl] = \dfrac{k_1\left[NO_2Cl\right]}{k_{-1}\left[NO_2\right] + k_2\left[NO_2Cl\right]}$

$$\text{Rate} = k_2 [NO_2Cl][Cl] = \frac{k_1 k_2 [NO_2Cl]^2}{k_{-1}[NO_2] + k_2[NO_2Cl]}$$

b) If step 1 is the rate-determining step, then $k_2[NO_2Cl] >> k_{-1}[NO_2]$ and the rate law simplifies to

$$\text{Rate} = \frac{k_1 k_2 [NO_2Cl]^2}{k_2[NO_2Cl]} = k\frac{[NO_2Cl]^2}{[NO_2Cl]} = k[NO_2Cl]$$

c) If step 2 is the rate-determining step, then $k_2[NO_2Cl] << k_{-1}[NO_2]$ and the rate law simplifies to

$$\text{Rate} = \frac{k_1 k_2 [NO_2Cl]^2}{k_{-1}[NO_2]} = k\frac{[NO_2Cl]^2}{[NO_2]}$$

16.132   Plan: For part a), the maximum rate occurs when [S] is high so that $[S] >> K_M$; the Michaelis-Menton equation can then be simplified to rate $= k_2[E]_0$. In part b), the Michaelis-Menton equation is used to calculate the substrate concentration.
Solution:

a) $\text{Rate} = \dfrac{k_2 [E]_0 [S]}{K_M + [S]} = k_2 [E]_0$

$\text{Rate} = 4\times10^4 \text{ s}^{-1}[3\times10^{-6} M] = 0.12 = \textbf{0.1 mol/ L·s}$

b) The new rate is 25% of the rate calculated in a): 0.25(0.12 mol/L s = 0.030 mol/ L·s

$\text{Rate} = \dfrac{k_2 [E]_0 [S]}{K_M + [S]}$

$0.030 \text{ mol/L·s} = \dfrac{4\times10^4 \text{ s}^{-1}\left(3\times10^{-6} M\right)[S]}{3\times10^{-2}M + [S]}$

0.12 mol/ L·s [S] = 0.0009 mol^2/ L^2·s + 0.030 mol/L [S]

0.09 mol/ L·s [S] = 0.0009 mol^2/ L^2·s

[S] = **0.01 M**

CHAPTER 17 EQUILIBRIUM: THE EXTENT OF CHEMICAL REACTIONS

FOLLOW–UP PROBLEMS

17.1A **Plan:** First, balance the equations and then write the reaction quotient. Products appear in the numerator of the reaction quotient and reactants appear in the denominator; coefficients in the balanced reaction become exponents.
Solution:
a) Balanced equation: $4NH_3(g) + 5O_2(g) \Delta \leftrightarrows 4NO(g) + 6H_2O(g)$

Reaction quotient: $Q_c = \dfrac{[NO]^4 [H_2O]^6}{[NH_3]^4 [O_2]^5}$

b) Balanced equation: $2NO_2(g) + 7H_2(g) \leftrightarrows 2NH_3(g) + 4H_2O(l)$

Reaction quotient: $Q_c = \dfrac{[NH_3]^2}{[NO_2]^2 [H_2]^7}$

c) Balanced equation: $2KClO_3(s) \leftrightarrows 2KCl(s) + 3O_2(g)$
Reaction quotient: $Q_c = [O_2]^3$

17.1B **Plan:** First, balance the equations and then write the reaction quotient. Products appear in the numerator of the reaction quotient and reactants appear in the denominator; coefficients in the balanced reaction become exponents.
Solution:
a) Balanced equation: $CH_4(g) + CO_2(g) \leftrightarrows 2CO(g) + 2H_2 (g)$

Reaction quotient: $Q_c = \dfrac{[CO]^2 [H_2]^2}{[CH_4][CO_2]}$

b) Balanced equation: $2H_2S(g) + SO_2(g) \leftrightarrows 2S(s) + 2H_2O(g)$

Reaction quotient: $Q_c = \dfrac{[H_2O]^2}{[H_2S]^2 [SO_2]}$

c) Balanced equation: $HCN(aq) + NaOH(aq) (s) \leftrightarrows NaCN(aq) + H_2O(l)$

Reaction quotient: $Q_c = \dfrac{[NaCN]}{[HCN][NaOH]}$

17.2A **Plan:** By multiplying by a factor or reversing, determine a way for Reactions 1 and 2 to sum to make the given reaction. When a reaction is multiplied by a factor, the equilibrium constant is raised to a power equal to the factor. When a reaction is reversed, the reciprocal of the equilibrium constant is used as the new equilibrium constant.
Solution:

$$(1)\ C(s) + CO_2(g) \leftrightarrows 2CO(g) \qquad K_{c1} = 1.4 \times 10^{12}$$

Multiply by 2 $\quad \underline{(2)\ 2CO(g) + 2Cl_2(g) \leftrightarrows 2COCl_2(g) \quad K_{c2} = (0.55)^2 = 0.30}$

$$C(s) + CO_2(g) + 2Cl_2(g) \leftrightarrows 2COCl_2(g)$$

$K_{c(overall)} = K_{c1} \times K_{c2} = (1.4 \times 10^{12})(0.30) = \mathbf{4.2 \times 10^{11}}$

17.2B **Plan:** When a reaction is multiplied by a factor, the equilibrium constant is raised to a power equal to the factor. When a reaction is reversed, the reciprocal of the equilibrium constant is used as the new equilibrium constant.

Solution:

a) All coefficients have been multiplied by the factor 2. Additionally, the reaction has been reversed. Therefore, the reciprocal of the equilibrium constant should be raised to the 2 power.

For reaction a) $\quad K_c = \left(\dfrac{1}{1.3 \times 10^{-2}}\right)^2 = 5917.1598 = \mathbf{5.9 \times 10^3}$

b) All coefficients have been multiplied by the factor 1/4 so the equilibrium constant should be raised to the 1/4 power.

For reaction b) $\quad K_c = (1.3 \times 10^{-2})^{1/4} = 0.33766 = \mathbf{0.34}$

17.3A   Plan: K_p and K_c for a reaction are related through the ideal gas equation as shown in $K_p = K_c(RT)^{\Delta n}$. Find Δn_{gas}, the change in the number of moles of gas between reactants and products (calculated as products minus reactants). Then, use the given K_c to solve for K_p.
Solution:
The total number of product moles of gas is 1 and the total number of reactant moles of gas is 2.

$$\Delta n = 1 - 2 = -1$$
$$K_p = K_c(RT)^{\Delta n}$$
$$K_p = 1.67[(0.0821 \text{ atm} \cdot \text{L/mol} \cdot \text{K})(500. \text{ K})]^{-1}$$
$$K_p = 0.040682095 = \mathbf{4.07 \times 10^{-2}}$$

17.3B   Plan: K_p and K_c for a reaction are related through the ideal gas equation as shown in $K_p = K_c(RT)^{\Delta n}$. Find Δn_{gas}, the change in the number of moles of gas between reactants and products (calculated as products minus reactants). Then, use the given K_p to solve for K_c.
Solution:
The total number of product moles of gas is 3 and the total number of reactant moles of gas is 5.

$$\Delta n = 3 - 5 = -2$$
$$K_p = K_c(RT)^{\Delta n}$$
$$K_p = K_c(RT)^{-2}$$
$$K_p(RT)^2 = K_c$$
$$K_c = (3.0 \times 10^{-5}) [(0.0821 \text{ atm} \cdot \text{L/mol} \cdot \text{K})(1173 \text{ K})]^2 = 0.27822976 = \mathbf{0.28}$$

17.4A   Plan: Write the reaction quotient for the reaction and calculate Q_c for each circle. Compare Q_c to K_c to determine the direction needed to reach equilibrium. If $Q_c > K_c$, reactants are forming. If $Q_c < K_c$, products are forming.
Solution:
The reaction quotient is $\dfrac{[Y]}{[X]}$.

Circle 1: $Q_c = \dfrac{[Y]}{[X]} = \dfrac{[3]}{[9]} = 0.33$

Since $Q_c < K_c$ (0.33 < 1.4), the reaction will shift to the **right** to reach equilibrium

Circle 2: $Q_c = \dfrac{[Y]}{[X]} = \dfrac{[7]}{[5]} = 1.4$

Since $Q_c = K_c$ (1.4 = 1.4), there is **no change** in the reaction direction. The reaction is at equilibrium now.

Circle 3: $Q_c = \dfrac{[Y]}{[X]} = \dfrac{[8]}{[4]} = 2.0$

Since $Q_c > K_c$ (2.0 > 1.4), the reaction will shift to the **left** to reach equilibrium

17.4B   Plan: Write the reaction quotient for the reaction and calculate Q_c for each circle. Compare Q_c to K_c to determine the direction needed to reach equilibrium. If $Q_c > K_c$, reactants are forming. If $Q_c < K_c$, products are forming.

Solution:

The reaction quotient is $\dfrac{[D]}{[C]^2}$.

Circle 1: $Q_c = \dfrac{[D]}{[C]^2} = \dfrac{[5]}{[3]^2} = 0.56$

According to the problem, circle 1 is at equilibrium. Therefore, $K_c = 0.56$.

Circle 2: $Q_c = \dfrac{[D]}{[C]^2} = \dfrac{[6]}{[3]^2} = 0.67$

Since $Q_c > K_c$ (0.67 > 0.56), the reaction will shift to the **left** to reach equilibrium.

Circle 3: $Q_c = \dfrac{[D]}{[C]^2} = \dfrac{[7]}{[4]^2} = 0.44$

Since $Q_c < K_c$ (0.44 < 0.56), the reaction will shift to the **right** to reach equilibrium

17.5A **Plan:** To decide whether CH_3Cl or CH_4 are forming while the reaction system moves toward equilibrium, calculate Q_p and compare it to K_p. If $Q_p > K_p$, reactants are forming. If $Q_p < K_p$, products are forming.
Solution:

$$Q_p = \dfrac{P_{CH_3Cl}\,P_{HCl}}{P_{CH_4}\,P_{Cl_2}} = \dfrac{(0.24\,\text{atm})(0.47\,\text{atm})}{(0.13\,\text{atm})(0.035\,\text{atm})} = 24.7912 = 25$$

K_p for this reaction is given as 1.6×10^4. Q_p is smaller than K_p ($Q_p < K_p$) so more products will form. **CH_3Cl** is one of the products forming.

17.5B **Plan:** To determine whether the reaction is at equilibrium or, if it is not at equilibrium, which direction to will proceed, calculate Q_c and compare it to K_c. If $Q_c > K_c$, the reaction will proceed to the left. If $Q_c < K_c$, the reaction will proceed to the right.
Solution:

$$Q_c = \dfrac{[SO_3]^2}{[SO_2]^2[O_2]} = \dfrac{\left[\dfrac{1.2\,\text{mol}}{2.0\,\text{L}}\right]^2}{\left[\dfrac{3.4\,\text{mol}}{2.0\,\text{L}}\right]^2 \left[\dfrac{1.5\,\text{mol}}{2.0\,\text{L}}\right]} = 0.166089965 = 0.17$$

The system is **not at equilibrium**. K_c for this reaction is given as 4.2×10^{-2}. Q_c is larger than K_c ($Q_c > K_c$) so the reaction will proceed to the **left**.

17.6A **Plan:** The information given includes the balanced equation, initial pressures of both reactants, and the equilibrium pressure for one reactant. First, set up a reaction table showing initial partial pressures for reactants and 0 for product. The change to get to equilibrium is to react some of reactants to form some product. Use the equilibrium quantity for O_2 and the expression for O_2 at equilibrium to solve for the change. From the change find the equilibrium partial pressure for NO and NO_2. Calculate K_p using the equilibrium values.
Solution:

Pressures (atm)	2NO(g)	+	O_2(g)	$\Delta \leftrightarrows$	2NO$_2$(g)
Initial	1.000		1.000		0
Change	−2x		−x		+2x
Equilibrium	1.000 − 2x		1.000 − x		2x

At equilibrium $P_{O_2} = 0.506$ atm $= 1.000 - x$; so $x = 1.000 - 0.506 = 0.494$ atm

$P_{NO} = 1.000 - 2x = 1.000 - 2(0.494) = 0.012$ atm

$P_{NO_2} = 2x = 2(0.494) = 0.988$ atm

Use the equilibrium pressures to calculate K_p.

$$K_p = \frac{P_{NO_2}^2}{P_{NO}^2 P_{O_2}} = \frac{(0.988)^2}{(0.012)^2 (0.506)} = 1.339679 \times 10^4 = \mathbf{1.3 \times 10^4}$$

17.6B   Plan: The information given includes the balanced equation, initial concentrations of both reactants, and the equilibrium concentration for one product. First, set up a reaction table showing initial concentrations for reactants and 0 for products. The change to get to equilibrium is to react some of reactants to form some of the products. Use the equilibrium quantity for N_2O_4 and the expression for N_2O_4 at equilibrium to solve for the change. From the change find the equilibrium concentrations for NH_3, O_2, and H_2O. Calculate K_c using the equilibrium values.
Solution:

Pressures (atm)	$4NH_3(g) +$	$7O_2(g)$	$\leftrightarrows$	$2N_2O_4(g) +$	$6H_2O(g)$
Initial	2.40	2.40		0	0
Change	$-4x$	$-7x$		$+2x$	$+6x$
Equilibrium	$2.40 - 4x$	$2.40 - 7x$		$2x$	$6x$

At equilibrium $[N_2O_4] = 0.134\,M = 2x$; so $x = 0.0670\,M$

$[NH_3] = 2.40\,M - 4(0.0670\,M) = 2.13\,M$

$[O_2] = 2.40\,M - 7(0.0670\,M) = 1.93\,M$

$[H_2O] = 6(0.0670\,M) = 0.402\,M$

Use the equilibrium pressures to calculate K_c.

$$K_c = \frac{[N_2O_4]^2 [H_2O]^6}{[NH_3]^4 [O_2]^7} = \frac{[0.134]^2 [0.402]^6}{[2.13]^4 [1.93]^7} = 3.6910 \times 10^{-8} = \mathbf{3.69 \times 10^{-8}}$$

17.7A   Plan: Convert K_c to K_p for the reaction. Write the equilibrium expression for K_p and insert the atmospheric pressures for P_{N_2} and P_{O_2} as their equilibrium values. Solve for P_{NO}.
Solution:
The conversion of K_c to K_p: $K_p = K_c(RT)^{\Delta n}$. For this reaction $\Delta n = 0$, so $K_p = K_c$.

$$K_p = \frac{P_{N_2} P_{O_2}}{P_{NO}^2} = K_c = 2.3 \times 10^{30} = \frac{(0.781)(0.209)}{x^2}$$

$$x = 2.6640 \times 10^{-16} = 2.7 \times 10^{-16}\,atm$$

The equilibrium partial pressure of NO in the atmosphere is $\mathbf{2.7 \times 10^{-16}\,atm}$.

17.7B   Plan: Write the equilibrium expression for K_p and insert the partial pressures for PH_3 and P_2 as their equilibrium values. Solve for the partial pressure of H_2.
Solution:

$$K_p = \frac{\left(P_{P_2}\right)\left(P_{H_2}\right)^3}{\left(P_{PH_3}\right)^2}$$

$$P_{H_2} = \sqrt[3]{\frac{(K_p)\left(P_{PH_3}\right)^2}{\left(P_{P_2}\right)}} = \sqrt[3]{\frac{(19.6)(0.112)^2}{(0.215)}} = 1.0457 = \mathbf{1.05\,atm}$$

17.8A   Plan: Find the initial molarity of HI by dividing moles of HI by the volume. Set up a reaction table and use the variables to find equilibrium concentrations in the equilibrium expression.
Solution:

$$M_{HI} = \frac{moles\ HI}{volume} = \frac{2.50\ mol}{10.32\ L} = 0.242248\,M$$

Concentration (M)	2HI(g)	$\Delta\rightleftharpoons$	H$_2$(g)	+	I$_2$(g)
Initial	0.242248		0		0
Change	−2x		+x		+x
Equilibrium	0.242248 − 2x		x		x

Set up equilibrium expression:

$$K_c = 1.26 \times 10^{-3} = \frac{[H_2][I_2]}{[HI]^2} = \frac{[x][x]}{[0.242248 - 2x]^2}$$ Take the square root of each side.

$$3.54965 \times 10^{-2} = \frac{[x]}{[0.242248 - 2x]}$$

$$x = 8.59895 \times 10^{-3} - 7.0993 \times 10^{-2} \, x$$

$$x = 8.02895 \times 10^{-3} = 8.03 \times 10^{-3}$$

$$[H_2] = [I_2] = \mathbf{8.03 \times 10^{-3} \, M}$$

17.8B   Plan: Find the initial molarities of Cl$_2$O and H$_2$O by dividing moles by the volume of the flask. Set up a reaction table and use the variables to find equilibrium concentrations in the equilibrium expression.
Solution:

$$\text{Molarity of Cl}_2\text{O} = \frac{\text{moles Cl}_2\text{O}}{\text{volume}} = \frac{6.15 \text{ mol}}{5.00 \text{ L}} = 1.23 \, M$$

$$\text{Molarity of H}_2\text{O} = \frac{\text{moles H}_2\text{O}}{\text{volume}} = \frac{6.15 \text{ mol}}{5.00 \text{ L}} = 1.23 \, M$$

Concentration (M)	Cl$_2$O(g) +	H$_2$O(g)	$\rightleftharpoons\Delta$	2HOCl (g)
Initial	1.23	1.23		0
Change	−x	−x		+2x
Equilibrium	1.23 − x	1.23 − x		2x

Set up equilibrium expression:

$$K_c = 0.18 = \frac{[HOCl]^2}{[Cl_2O][H_2O]} = \frac{[2x]^2}{[1.23 - x][1.23 - x]}$$

Take the square root of each side.

$$0.424264068 = \frac{[2x]}{[1.23 - x]}$$

$$0.521844804 - 0.424264068x = 2x$$

$$0.521844804 = 2.424264068x$$

$$x = 0.215259000 \, M$$

$$[Cl_2O] = [H_2O] = 1.23 \, M - 0.215259 \, M = 1.014741 = \mathbf{1.01 \, M}$$

$$[HOCl] = 2(0.215259 \, M) = 0.430518 = \mathbf{0.43 \, M}$$

17.9A   Plan: Find the molarity of I$_2$ by dividing moles of I$_2$ by the volume. First set up the reaction table, then set up the equilibrium expression. To solve for the variable, x, first assume that x is negligible with respect to initial concentration of I$_2$. Check the assumption by calculating the % error. If the error is greater than 5%, calculate x using the quadratic equation. The next step is to use x to determine the equilibrium concentrations of I$_2$ and I.
Solution:

$$[I_2]_{\text{init}} = \frac{0.50 \text{ mol}}{2.5 \text{ L}} = 0.20 \, M$$

a) Equilibrium at 600 K

Concentration (M)	$I_2(g)$	$\rightleftarrows$	$2I(g)$
Initial	0.20		0
Change	−x		+2x
Equilibrium	0.20 − x		2x

Equilibrium expression: $K_c = \dfrac{[I]^2}{[I_2]} = 2.94 \times 10^{-10}$

$\dfrac{[2x]^2}{[0.20 - x]} = 2.94 \times 10^{-10}$ Assume x is negligible so $0.20 - x \approx 0.20$

$\dfrac{[2x]^2}{[0.20]} = 2.94 \times 10^{-10}$

$4x^2 = (2.94 \times 10^{-10})(0.20)$; $x = 3.834 \times 10^{-6} = 3.8 \times 10^{-6}$

Check the assumption by calculating the % error:

$\dfrac{3.8 \times 10^{-6}}{0.20}(100) = 0.0019\%$ which is smaller than 5%, so the assumption is valid.

At equilibrium $[I]_{eq} = 2x = 2(3.834 \times 10^{-6}) = 7.668 \times 10^{-6} = \textbf{7.7} \times \textbf{10}^{-6} \textbf{\textit{M}}$ and

$[I_2]_{eq} = 0.20 - x = 0.20 - 3.834 \times 10^{-6} = 0.199996 = \textbf{0.20 \textit{M}}$

b) Equilibrium at 2000 K

Equilibrium expression: $K_c = \dfrac{[I]^2}{[I_2]} = 0.209$

$\dfrac{[2x]^2}{[0.20 - x]} = 0.209$ Assume x is negligible so $0.20 - x$ is approximately 0.20

$\dfrac{[2x]^2}{[0.20]} = 0.209$

$4x^2 = (0.209)(0.20)$

$x = 0.102225 = 0.102$

Check the assumption by calculating the % error:

$\dfrac{0.102}{0.20}(100) = 51\%$ which is larger than 5% so the assumption is not valid. Solve using quadratic

equation.

$\dfrac{[2x]^2}{[0.20 - x]} = 0.209$

$4x^2 + 0.209x - 0.0418 = 0$

$x = \dfrac{-0.209 \pm \sqrt{(0.209)^2 - 4(4)(-0.0418)}}{2(4)} = 0.0793857 \text{ or } -0.1316$

Choose the positive value, x = 0.0793857

At equilibrium $[I]_{eq} = 2x = 2(0.0793857) = 0.1587714 = \textbf{0.16 \textit{M}}$ and

$[I_2]_{eq} = 0.20 - x = 0.20 - 0.0793857 = 0.1206143 = \textbf{0.12 \textit{M}}$

17.9B   Plan: First set up the reaction table, then set up the equilibrium expression. To solve for the variable, x, first assume that x is negligible with respect to initial partial pressure of PCl_5. Check the assumption by calculating the % error. If the error is greater than 5%, calculate x using the quadratic equation. The next step is to use x to determine the equilibrium partial pressure of PCl_5.

Solution:

a) Equilibrium at a PCl_5 partial pressure of 0.18 atm:

Partial Pressure (atm)	$PCl_5(g)$	$\overset{\Delta}{\rightleftarrows}$	$PCl_3(g)$ +	$Cl_2(g)$
Initial	0.18		0	0
Change	−x		+x	+x
Equilibrium	0.18 − x		x	x

Equilibrium expression: $K_p = \dfrac{\left(P_{PCl_3}\right)\left(P_{Cl_2}\right)}{\left(P_{PCl_5}\right)} = 3.4 \times 10^{-4}$

$\dfrac{(x)(x)}{(0.18 - x)} = 3.4 \times 10^{-4}$ Assume x is negligible so $0.18 - x \approx 0.18$

$\dfrac{(x)(x)}{(0.18)} = 3.4 \times 10^{-4}$

$x^2 = (3.4 \times 10^{-4})(0.18)$ so $x = 0.0078230428 = 7.8 \times 10^{-3}$

Check the assumption by calculating the % error:

$\dfrac{(7.8 \times 10^{-3})(100\%)}{(0.18)} = 4.3\%$ which is smaller than 5%, so the assumption is valid.

At equilibrium $[PCl_5]_{eq} = 0.18\,M - 7.8 \times 10^{-3}\,M = \mathbf{0.17\,M}$

b) Equilibrium at a PCl_5 partial pressure of 0.18 atm:

Partial Pressure (atm)	$PCl_5(g)$	$\overset{\Delta\Delta}{\rightleftarrows}$	$PCl_3(g)$ +	$Cl_2(g)$
Initial	0.025		0	0
Change	−x		+x	+x
Equilibrium	0.025 − x		x	x

Equilibrium expression: $K_p = \dfrac{\left(P_{PCl_3}\right)\left(P_{Cl_2}\right)}{\left(P_{PCl_5}\right)} = 3.4 \times 10^{-4}$

$\dfrac{(x)(x)}{(0.025 - x)} = 3.4 \times 10^{-4}$ Assume x is negligible so $0.025 - x \approx 0.025$

$\dfrac{(x)(x)}{(0.025)} = 3.4 \times 10^{-4}$

$x^2 = (3.4 \times 10^{-4})(0.025)$ so $x = 0.002915476 = 2.9 \times 10^{-3}$

Check the assumption by calculating the % error:

$\dfrac{(2.9 \times 10^{-3})(100\%)}{(0.025)} = 12\%$ which is larger than 5%, so the assumption is NOT valid. Solve using quadratic equation.

$\dfrac{(x)(x)}{(0.025 - x)} = 3.4 \times 10^{-4}$

$x^2 + 3.4 \times 10^{-4}x - 8.5 \times 10^{-6} = 0$

$x = \dfrac{+8.5 \times 10^{-6} \pm \sqrt{3.4 \times 10^{-4} - 4(1)(-8.5 \times 10^{-6})}}{2(1)} = 0.002750428051$ or -0.003090428051

Choose the positive value, $x = 0.0028\,M$; At equilibrium $[PCl_5]_{eq} = 0.025\,M - 0.0028\,M = \mathbf{0.022\,M}$

17.10A Plan: Calculate the initial concentrations (molarity) of each substance. For part (a), calculate Q_c and compare to given K_c. If $Q_c > K_c$ then the reaction proceeds to the left to make reactants from products. If $Q_c < K_c$ then the reaction proceeds to right to make products from reactants. For part (b), use the result of part (a) and the given equilibrium concentration of PCl_5 to find the equilibrium concentrations of PCl_3 and Cl_2.
Solution:

Initial concentrations: $[PCl_5] = \dfrac{0.1050 \text{ mol}}{0.5000 \text{ L}} = 0.2100 \text{ M}$

$$[PCl_3] = [Cl_2] = \dfrac{0.0450 \text{ mol}}{0.5000 \text{ L}} = 0.0900 \text{ M}$$

a) $Q_c = \dfrac{[PCl_3][Cl_2]}{[PCl_5]} = \dfrac{[0.0900][0.0900]}{[0.2100]} = 0.038571 = 0.0386$

Q_c, 0.0386, is less than K_c, 0.042, so the reaction will proceed to the **right** to make more products.

b) To reach equilibrium, concentrations will increase for the products, PCl_3 and Cl_2, and decrease for the reactant, PCl_5.

Concentration (M)	$PCl_5(g)$	$\rightleftharpoons$ $PCl_3(g)$ +	$Cl_2(g)$
Initial	0.2100	0.0900	0.0900
Change	−x	+x	+x
Equilibrium	0.2100 − x	0.0900 + x	0.0900 + x

$[PCl_5] = 0.2065 = 0.2100 - x$; $x = 0.0035 \text{ M}$

$[PCl_3] = [Cl_2] = 0.0900 + x = 0.0900 + 0.0035 = \mathbf{0.0935 \text{ M}}$

17.10B Plan: For part (a), calculate Q_p and compare to given K_p. If $Q_p > K_p$ then the reaction proceeds to the left to make reactants from products. If $Q_p < K_p$ then the reaction proceeds to right to make products from reactants. For part (b), set up a reaction table and use the variables to find equilibrium concentrations in the equilibrium expression.
Solution:

a) $Q_p = \dfrac{(P_{NO})^2}{(P_{N_2})(P_{O_2})} = \dfrac{(0.750)^2}{(0.500)(0.500)} = 2.25$

Q_p, 2.25, is greater than K_p, 8.44×10^3, so the reaction will proceed to the **left** to make more reactants.

b) To reach equilibrium, concentrations will increase for the reactants, N_2 and O_2, and decrease for the product, NO.

Pressure (atm)	$N_2(g)$ +	$O_2(g)$	$\rightleftharpoons$ 2NO (g)
Initial	0.500	0.500	0.750
Change	+x	+x	−2x
Equilibrium	0.500 + x	0.500 + x	0.750 − 2 x

$K_p = \dfrac{(P_{NO})^2}{(P_{N_2})(P_{O_2})} = \dfrac{(0.750 - 2x)^2}{(0.500 + x)(0.500 + x)} = 8.44 \times 10^3$

Take the square root of each side.

$\dfrac{(0.750 - 2x)}{(0.500 + x)} = 0.0919$

$0.750 - 2x = 0.04595 + 0.0919x$

$0.70405 = 2.0919x$

$x = 0.33656 \text{ M}$

$[N_2] = [O_2] = 0.500 \text{ M} + 0.33656 \text{ M} = 0.83656 = \mathbf{0.837 \text{ M}}$

$[NO] = 0.750 \text{ M} - 2(0.33656 \text{ M}) = 0.07687987 = \mathbf{0.077 \text{ M}}$

17.11A Plan: Examine each change for its impact on Q_c. Then decide how the system would respond to re-establish equilibrium.
Solution:

$$Q_c = \frac{[SiF_4][H_2O]^2}{[HF]^4}$$

a) Decreasing $[H_2O]$ leads to $Q_c < K_c$, so the reaction would shift to make more products from reactants. Therefore, the SiF_4 concentration, as a product, would **increase**.
b) Adding liquid water to this system at a temperature above the boiling point of water would result in an increase in the concentration of water vapor. The increase in $[H_2O]$ increases Q_c to make it greater than K_c. To re-establish equilibrium products will be converted to reactants and the $[SiF_4]$ will **decrease**.
c) Removing the reactant HF increases Q_c, which causes the products to react to form more reactants. Thus, $[SiF_4]$ **decreases**.
d) Removal of a solid product has no impact on the equilibrium; $[SiF_4]$ **does not change**.
Check: Look at each change and decide which direction the equilibrium would shift using Le Châtelier's principle to check the changes predicted above.
   a) Remove product, equilibrium shifts to right.
   b) Add product, equilibrium shifts to left.
   c) Remove reactant, equilibrium shifts to left.
   d) Remove solid reactant, equilibrium does not shift.

17.11B Plan: Examine each change for its impact on Q_c. Then decide how the system would respond to re-establish equilibrium.
Solution:

$$Q_c = \frac{[CO][H_2]}{[H_2O]}$$

a) Adding carbon, a solid reactant, has no impact on the equilibrium. [CO] **does not change**.
b) Removing water vapor, a reactant, increases Q_c, which causes the products to react to form more reactants. Thus, [CO] **decreases**.
c) Removing the product H_2 decreases Q_c, which causes the reactants to react to form more products. Thus, [CO] **increases**.
d) Adding water vapor, a reactant, decreases Q_c, which causes the reactants to react to form more products. Thus, [CO] **increases**.
Check: Look at each change and decide which direction the equilibrium would shift using Le Châtelier's principle to check the changes predicted above.
   a) Add solid reactant, equilibrium does not shift.
   b) Remove reactant, equilibrium shifts to the left.
   c) Remove product, equilibrium shifts to the right.
   d) Add reactant, equilibrium shifts to the right.

17.12A Plan: Changes in pressure (and volume) affect the concentration of gaseous reactants and products. A decrease in pressure, i.e., increase in volume, favors the production of more gas molecules whereas an increase in pressure favors the production of fewer gas molecules. Examine each reaction to decide whether more or fewer gas molecules will result from producing more products. If more gas molecules result, then the pressure should be increased (volume decreased) to reduce product formation. If fewer gas molecules result, then pressure should be decreased to produce more reactants.
Solution:
a) In $2SO_2(g) + O_2(g) \overset{\Delta}{\rightleftharpoons} 2SO_3(g)$ three molecules of gas form two molecules of gas, so there are fewer gas molecules in the product. **Decreasing pressure** (increasing volume) will decrease the product yield.
b) In $4NH_3(g) + 5O_2(g) \overset{\Delta}{\rightleftharpoons} 4NO(g) + 6H_2O(g)$ 9 molecules of reactant gas convert to 10 molecules of product gas. **Increasing pressure** (decreasing volume) will favor the reaction direction that produces fewer moles of gas: towards the reactants and away from products.

c) In $CaC_2O_4(s) \leftrightarrows\Delta CaCO_3(s) + CO(g)$ there are no reactant gas molecules and one product gas molecule. The yield of the products will decrease when volume decreases, which corresponds to a **pressure increase**.

17.12B **Plan:** Changes in pressure (and volume) affect the concentration of gaseous reactants and products. A decrease in pressure, i.e., increase in volume, favors the production of more gas molecules whereas an increase in pressure favors the production of fewer gas molecules. Examine each reaction to determine if a decrease in pressure will shift the reaction toward the products (resulting in an increase in the yield of products) or toward the reactants (resulting in a decrease in the yield of products).
Solution:
a) In $CH_4(g) + CO_2(g) \leftrightarrows\Delta 2CO(g) + 2H_2(g)$ two molecules of gas form four molecules of gas, so there are more gas molecules in the product. Decreasing pressure (increasing volume) will shift the reaction to the right, **increasing** the product yield.
b) In $NO(g) + CO_2(g) \leftrightarrows\Delta NO_2(g) + CO(g)$ 2 molecules of reactant gas convert to 2 molecules of product gas. Decreasing pressure (increasing volume) will have **no effect** on this reaction or on the amount of product produced because the number of moles of gas does not change.
c) In $2H_2S(g) + SO_2(g) \leftrightarrows\Delta 3S(s + 2H_2O(g)$ three molecules of reactant gas convert to 2 molecules of product gas. Decreasing pressure (increasing volume) will shift the reaction toward the reactants, **decreasing** the product yield.

17.13A **Plan:** A decrease in temperature favors the exothermic direction of an equilibrium reaction. First, identify whether the forward or reverse reaction is exothermic from the given enthalpy change. $\Delta H < 0$ means the forward reaction is exothermic, and $\Delta H > 0$ means the reverse reaction is exothermic. If the forward reaction is exothermic then a decrease in temperature will shift the equilibrium to make more products from reactants and increase K_p. If the reverse reaction is exothermic then a decrease in temperature will shift the equilibrium to make more reactants from products and decrease K_p.
Solution:
a) $\Delta H < 0$ so the forward reaction is exothermic. A decrease in temperature increases the partial pressure of products and decreases the partial pressures of reactants, so P_{H_2} **decreases**. With increases in product pressures and decreases in reactant pressures, K_p **increases**.
b) $\Delta H > 0$ so the reverse reaction is exothermic. A decrease in temperature decreases the partial pressure of products and increases the partial pressures of reactants, so P_{N_2} **increases**. K_p **decreases** with decrease in product pressures and increase in reactant pressures.
c) $\Delta H < 0$ so the forward reaction is exothermic. Decreasing temperature **increases** P_{PCl_5} and **increases** K_p.

17.13B **Plan:** A decrease in temperature favors the exothermic direction of an equilibrium reaction. First, identify whether the forward or reverse reaction is exothermic from the given enthalpy change. $\Delta H < 0$ means the forward reaction is exothermic, and $\Delta H > 0$ means the reverse reaction is exothermic. If the forward reaction is exothermic then a decrease in temperature will shift the equilibrium to make more products from reactants and increase K_p. If the reverse reaction is exothermic then an increase in temperature will shift the equilibrium to make more products from reactants and increase K_p.
Solution:
a) $\Delta H > 0$ so the reverse reaction is exothermic. An **increase in temperature** will increase the partial pressure of products and decrease the partial pressures of reactants. K_p **increases** with an increase in product pressures and a decrease in reactant pressures.
b) $\Delta H < 0$ so the forward reaction is exothermic. A **decrease in temperature** increases the partial pressure of products and decreases the partial pressures of reactants. With increases in product pressures and decreases in reactant pressures, K_p **increases**.
c) $\Delta H > 0$ so the reverse reaction is exothermic. An **increase in temperature** increases the partial pressure of products and decreases the partial pressures of reactants. K_p **increases** with an increase in product pressures and a decrease in reactant pressures.

17.14A **Plan:** Given the balanced equilibrium equation, it is possible to set up the appropriate equilibrium expression (Q_c). For the equation given $\Delta n = 0$ meaning that $K_p = K_c$. The value of K may be found for scene 1, and values for Q may be determined for the other two scenes. The reaction will shift towards the reactant side if $Q > K$, and the reaction will shift towards the product side if $Q < K$. The reaction is exothermic ($\Delta H < 0$), thus, heat may be considered a product. Increasing the temperature adds a product and decreasing the temperature removes a product.

Solution:
a) K_p requires the equilibrium value of P for each gas. The pressure may be found from
$P = nRT/V$

$$K_p = \frac{P_{CD}^2}{P_{C_2} P_{D_2}} = \frac{\left(\dfrac{n_{CD}RT}{V}\right)^2}{\left(\dfrac{n_{C_2}RT}{V}\right)\left(\dfrac{n_{D_2}RT}{V}\right)}$$

This equation may be simplified because for the sample R, T, and V are constant. Using scene 1:

$$K_p = \frac{n_{CD}^2}{n_{C_2} n_{D_2}} = \frac{(4)^2}{(2)(2)} = \mathbf{4}$$

b) Scene 2: $Q_p = \dfrac{n_{CD}^2}{n_{C_2} n_{D_2}} = \dfrac{(6)^2}{(1)(1)} = 36$

$Q > K$ so the reaction will shift to the **left** (towards the reactants).

Scene 3: $Q_p = \dfrac{n_{CD}^2}{n_{C_2} n_{D_2}} = \dfrac{(2)^2}{(3)(3)} = 0.44$

$Q < K$ so the reaction will shift to the **right** (towards the products).
c) Increasing the temperature is equivalent to adding a product (heat) to the equilibrium. The reaction will shift to consume the added heat. The reaction will shift to the left (towards the reactants). However, since there are 2 moles of gas on each side of the equation, the shift has **no effect** on total moles of gas.

17.14B Plan: Write the equilibrium expression for the reaction. Count the number of each type of particle in the first scene and use this information to calculate the value of K at T_1. Follow a similar procedure to calculate the value of K at T_2. Determine if K at T_1 is larger or smaller than K at T_2. Use this information to determine the sign of ΔH for the reaction.
Solution:

a) $K = \dfrac{[AB]}{[A][B]}$

Calculating K at T_1:

$K = \dfrac{[3]}{[2][2]} = \mathbf{0.75}$

b) Going from the scene at T_1 to the scene at T_2, the number of product molecules decreases. This decreases the value of K. The problem states that $T_2 < T_1$, so as the temperature decreases, K also decreases. The fact that both the temperature and the value of K decreased suggests that this is an endothermic reaction, with **$\Delta H > 0$**.
c) Calculating K at T_2:

$K = \dfrac{[2]}{[3][3]} = \mathbf{0.22}$

CHEMICAL CONNECTIONS BOXED READING PROBLEM

B17.1 Plan: To control the pathways, the first enzyme specific for a branch is inhibited by the end product of that branch.
Solution:
a) The enzyme that is inhibited by F is the first enzyme in that branch, which is **enzyme 3**.
b) Enzyme **6** is inhibited by I.
c) If F inhibited enzyme 1, then neither branch of the reaction would take place once enough F was produced.
d) If F inhibited enzyme 6, then the second branch would not take place when enough F was made.

END-OF-CHAPTER PROBLEMS

17.1 If the rate of the forward reaction exceeds the rate of reverse reaction, products are formed faster than they are consumed. The change in reaction conditions results in more products and less reactants. A change in reaction conditions can result from a change in concentration or a change in temperature. If concentration changes, product concentration increases while reactant concentration decreases, but the K_c remains unchanged because the *ratio* of products and reactants remains the same. If the increase in the forward rate is due to a change in temperature, the rate of the reverse reaction also increases. The equilibrium ratio of product concentration to reactant concentration is no longer the same. Since the rate of the forward reaction increases more than the rate of the reverse reaction, K_c increases (numerator, [products], is larger and denominator, [reactants], is smaller).

$$K_c = \frac{[\text{products}]}{[\text{reactants}]}$$

17.7 The equilibrium constant expression is $K = [O_2]$ (we do not include solid substances in the equilibrium expression). If the temperature remains constant, K remains constant. If the initial amount of Li_2O_2 present was sufficient to reach equilibrium, the amount of O_2 obtained will be constant, regardless of how much $Li_2O_2(s)$ is present.

17.8 a) On the graph, the concentration of HI increases at twice the rate that H_2 decreases because the stoichiometric ratio in the balanced equation is $1H_2$: 2HI. Q for a reaction is the ratio of concentrations of products to concentrations of reactants. As the reaction progresses the concentration of reactants H_2 and I_2 decrease and the concentration of product HI increases, which means that Q increases as a function of time.

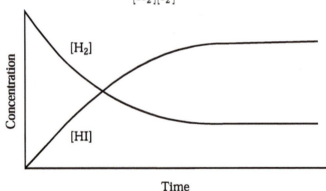

$$H_2(g) + I_2(g) \rightleftharpoons 2HI(g) \qquad Q = \frac{[HI]^2}{[H_2][I_2]}$$

The value of Q increases as a function of time until it reaches the value of K.

b) No, Q would still increase with time because the $[I_2]$ would decrease in exactly the same way as $[H_2]$ decreases.

17.11 <u>Plan:</u> Write the reaction and then the expression for Q. Remember that $Q = \dfrac{[C]^c [D]^d}{[A]^a [B]^b}$ where A and B are reactants, C and D are products, and a, b, c, and d are the stoichiometric coefficients in the balanced equation.
<u>Solution:</u>
The balanced equation for the first reaction is
$3/2H_2(g) + 1/2N_2(g) \rightleftharpoons NH_3(g)$ (1)
The coefficient in front of NH_3 is fixed at 1 mole according to the description. The reaction quotient for this reaction is $Q_1 = \dfrac{[NH_3]}{[H_2]^{3/2} [N_2]^{1/2}}$.

In the second reaction, the coefficient in front of N_2 is fixed at 1 mole.

$$3H_2(g) + N_2(g) \rightleftharpoons 2NH_3(g) \qquad (2)$$

The reaction quotient for this reaction is $Q_2 = \dfrac{[NH_3]^2}{[H_2]^3[N_2]}$

Q_2 is equal to Q_1^2.

17.12 <u>Plan:</u> Remember that $Q_c = \dfrac{[C]^c[D]^d}{[A]^a[B]^b}$ where A and B are reactants, C and D are products, and a, b, c, and d are the stoichiometric coefficients in the balanced equation.
<u>Solution:</u>

a) $4NO(g) + O_2(g) \rightleftharpoons\Delta 2N_2O_3(g)$

$$Q_c = \dfrac{[N_2O_3]^2}{[NO]^4[O_2]}$$

b) $SF_6(g) + 2SO_3(g) \rightleftharpoons\Delta 3SO_2F_2(g)$

$$Q_c = \dfrac{[SO_2F_2]^3}{[SF_6][SO_3]^2}$$

c) $2SClF_5(g) + H_2(g) \rightleftharpoons\Delta S_2F_{10}(g) + 2HCl(g)$

$$Q_c = \dfrac{[S_2F_{10}][HCl]^2}{[SClF_5]^2[H_2]}$$

17.14 <u>Plan:</u> Remember that $Q_c = \dfrac{[C]^c[D]^d}{[A]^a[B]^b}$ where A and B are reactants, C and D are products, and a, b, c, and d are the stoichiometric coefficients in the balanced equation.
<u>Solution:</u>

a) $2NO_2Cl(g) \rightleftharpoons\Delta 2NO_2(g) + Cl_2(g)$

$$Q_c = \dfrac{[NO_2]^2[Cl_2]}{[NO_2Cl]^2}$$

b) $2POCl_3(g) \rightleftharpoons\Delta 2PCl_3(g) + O_2(g)$

$$Q_c = \dfrac{[PCl_3]^2[O_2]}{[POCl_3]^2}$$

c) $4NH_3(g) + 3O_2(g) \rightleftharpoons\Delta 2N_2(g) + 6H_2O(g)$

$$Q_c = \dfrac{[N_2]^2[H_2O]^6}{[NH_3]^4[O_2]^3}$$

17.16 <u>Plan:</u> Compare each equation with the reference equation to see how the direction and coefficients have changed. If a reaction has been reversed, the K value is the reciprocal of the K value for the reference reaction. If the coefficients have been changed by a factor n, the K value is equal to the original K value raised to the nth power.
<u>Solution:</u>

a) The K for the original reaction is $K_c = \dfrac{[H_2]^2[S_2]}{[H_2S]^2}$

The given reaction $1/2S_2(g) + H_2(g) \rightleftharpoons H_2S(g)$ is the reverse reaction of the original reaction and the coefficients of the original reaction have been multiplied by a factor of 1/2. The equilibrium constant for the reverse reaction is the reciprocal $(1/K)$ of the original constant. The K value of the original reaction is raised to the 1/2 power.

$$K_{c\,(a)} = (1/K_c)^{1/2} = \frac{[H_2S]}{[S_2]^{1/2}[H_2]}$$

$$K_{c\,(a)} = (1/1.6 \times 10^{-2})^{1/2} = 7.90569 = \mathbf{7.9}$$

b) The given reaction $5H_2S(g) \rightleftharpoons 5H_2(g) + 5/2S_2(g)$ is the original reaction multiplied by 5/2. Take the original K to the 5/2 power to find K of given reaction.

$$K_{c\,(b)} = (K_c)^{5/2} = \frac{[H_2]^5[S_2]^{5/2}}{[H_2S]^5}$$

$$K_{c\,(b)} = (1.6 \times 10^{-2})^{5/2} = 3.23817 \times 10^{-5} = \mathbf{3.2 \times 10^{-5}}$$

17.18   Plan: The concentration of solids and pure liquids do not change, so their concentration terms are not written in the reaction quotient expression. Remember that stoichiometric coefficients are used as exponents in the expression for the reaction quotient.
Solution:
a) $2Na_2O_2(s) + 2CO_2(g) \rightleftharpoons 2Na_2CO_3(s) + O_2(g)$

$$Q_c = \frac{[O_2]}{[CO_2]^2}$$

b) $H_2O(l) \rightleftharpoons H_2O(g)$

$Q_c = [H_2O(g)]$ Only the gaseous water is used. The "(g)" is for emphasis.

c) $NH_4Cl(s) \rightleftharpoons NH_3(g) + HCl(g)$

$Q_c = [NH_3][HCl]$

17.20   Plan: The concentration of solids and pure liquids do not change, so their concentration terms are not written in the reaction quotient expression. Remember that stoichiometric coefficients are used as exponents in the expression for the reaction quotient.
Solution:
a) $2NaHCO_3(s) \rightleftharpoons Na_2CO_3(s) + CO_2(g) + H_2O(g)$

$Q_c = [CO_2][H_2O]$

b) $SnO_2(s) + 2H_2(g) \rightleftharpoons Sn(s) + 2H_2O(g)$

$$Q_c = \frac{[H_2O]^2}{[H_2]^2}$$

c) $H_2SO_4(l) + SO_3(g) \rightleftharpoons H_2S_2O_7(l)$

$$Q_c = \frac{1}{[SO_3]}$$

17.23   Plan: Add the two equations, canceling substances that appear on both sides of the equation. Write the Q_c expression for each of the steps and for the overall equation. Since the individual steps are added, their Q_c's are multiplied and common terms are canceled to obtain the overall Q_c.
Solution:
a) The balanced equations and corresponding reaction quotients are given below. Note the second equation must be multiplied by 2 to get the appropriate overall equation.

$$(1)\ Cl_2(g) + F_2(g) \rightleftharpoons \Delta\ \cancel{2ClF(g)} \qquad\qquad Q_1 = \frac{[ClF]^2}{[Cl_2][F_2]}$$

$$(2)\ \cancel{2ClF(g)} + 2F_2(g) \rightleftharpoons \Delta\ 2ClF_3(g) \qquad\qquad Q_2 = \frac{[ClF_3]^2}{[ClF]^2[F_2]^2}$$

$$\text{Overall: } Cl_2(g) + 3F_2(g) \rightleftharpoons \Delta\ 2ClF_3(g) \qquad\qquad Q_{overall} = \frac{[ClF_3]^2}{[Cl_2][F_2]^3}$$

b) The reaction quotient for the overall reaction, $Q_{overall}$, determined from the reaction is:

$$Q_{overall} = \frac{[ClF_3]^2}{[Cl_2][F_2]^3}$$

$$Q_{overall} = Q_1 Q_2^2 = \frac{[ClF]^2}{[Cl_2][F_2]} \quad \text{x} \quad \frac{[ClF_3]^2}{[ClF]^2[F_2]^2} = \frac{[ClF_3]^2}{[Cl_2][F_2]^3}$$

17.26   a) $K_p = K_c(RT)^{\Delta n}$. Since Δn = number of moles gaseous products – number of moles gaseous reactants, Δn is a positive integer for this reaction. If Δn is a positive integer, then $(RT)^{\Delta n}$ is greater than 1. Thus, K_c is multiplied by a number that is greater than 1 to give K_p. **K_c is smaller than K_p.**
b) Assuming that $RT > 1$ (which occurs when $T > 12.2\ K$, because 0.0821 (R) x 12.2 = 1), $K_p > K_c$ if the number of moles of gaseous products exceeds the number of moles of gaseous reactants. $K_p < K_c$ when the number of moles of gaseous reactants exceeds the number of moles of gaseous product.

17.27   <u>Plan:</u> Δn_{gas} = moles gaseous products – moles gaseous reactants.
<u>Solution:</u>
a) Number of moles of <u>gaseous</u> reactants = 0; number of moles of <u>gaseous</u> products = 3; $\Delta n_{gas} = 3 - 0 = \mathbf{3}$
b) Number of moles of <u>gaseous</u> reactants = 1; number of moles of <u>gaseous</u> products = 0; $\Delta n_{gas} = 0 - 1 = \mathbf{-1}$
c) Number of moles of <u>gaseous</u> reactants = 0; number of moles of <u>gaseous</u> products = 3; $\Delta n_{gas} = 3 - 0 = \mathbf{3}$

17.28   a) $\Delta n_{gas} = \mathbf{1}$ b) $\Delta n_{gas} = \mathbf{-3}$ c) $\Delta n_{gas} = \mathbf{1}$

17.30   First, determine Δn for the reaction and then calculate K_c using $K_p = K_c(RT)^{\Delta n}$.
a) Δn = moles gaseous products – moles gaseous reactants = 2 – 2 = 0

$$K_c = \frac{K_p}{(RT)^{\Delta n}} = \frac{49}{[(0.0821)(730.)]^0} = \mathbf{49}$$

b) Δn = moles gaseous products – moles gaseous reactants = 2 – 3 = –1

$$K_c = \frac{K_p}{(RT)^{\Delta n}} = \frac{2.5 \times 10^{10}}{[(0.0821)(500.)]^{-1}} = 1.02625 \times 10^{12} = \mathbf{1.0 \times 10^{12}}$$

17.32   First, determine Δn for the reaction and then calculate K_p using $K_p = K_c(RT)^{\Delta n}$.
a) Δn = moles gaseous products – moles gaseous reactants = 2 – 2 = 0
    $K_p = K_c(RT)^{\Delta n} = (0.77)[(0.0821)(1020.)]^0 = \mathbf{0.77}$
b) Δn = moles gaseous products – moles gaseous reactants = 2 – 3 = –1
    $K_p = K_c(RT)^{\Delta n} = (1.8 \times 10^{-56})[(0.0821)(570.)]^{-1} = 3.8464 \times 10^{-58} = \mathbf{3.8 \times 10^{-58}}$

17.34   a) The reaction is $2D \rightleftharpoons E$ and $K_c = \dfrac{[E]}{[D]^2}$.

$$\text{Concentration of D = Concentration of E} = (3\ \text{spheres})\left(\frac{0.0100\ \text{mol}}{1\ \text{sphere}}\right)\left(\frac{1}{1.00\ \text{L}}\right) = 0.0300\ M$$

$$K_c = \frac{[E]}{[D]^2} = \frac{[0.0300]}{[0.0300]^2} = 33.3333 = 33.3$$

b) In Scene B the concentrations of D and E are both 0.0300 mol/0.500 L = 0.0600 M

$$Q_c = \frac{[E]}{[D]^2} = \frac{[0.0600]}{[0.0600]^2} = 16.66666 = 16.7$$

B is not at equilibrium. Since $Q_c < K_c$, the reaction will proceed to the right.

In Scene C, the concentration of D is still 0.0600 M and the concentration of E is 0.0600 mol/0.500 L = 0.120 M

$$Q_c = \frac{[E]}{[D]^2} = \frac{[0.120]}{[0.0600]^2} = 33.3333 = 33.3$$

Since $Q_c = K_c$ in Scene C, the reaction is at equilibrium.

17.36 $\quad Q_p = \frac{P_{NO}^2 P_{Br_2}}{P_{NOBr}^2} = \frac{(0.10)^2(0.10)}{(0.10)^2} = 0.10 < K_p = 60.6$

$Q_p < K_p$ Thus, the reaction is **not** at equilibrium and will proceed to the **right** (towards the products).

17.39 $\quad$ When x mol of CH_4 reacts, 2x mol of H_2O also reacts to form x mol of CO_2 and 4x mol of H_2. This is based on the 1:2:1:4 mole ratio in the reaction. The final (equilibrium) concentration of each reactant is the initial concentration minus the amount that reacts. The final (equilibrium) concentration of each product is the initial concentration plus the amount that forms.

17.41 $\quad$ Plan: Since all equilibrium concentrations are given in molarities and the reaction is balanced, construct an equilibrium expression and substitute the equilibrium concentrations to find K_c.
Solution:

$$K_c = \frac{[HI]^2}{[H_2][I_2]} = \frac{\left[1.87\text{x}10^{-3}\right]^2}{\left[6.50\text{x}10^{-5}\right]\left[1.06\text{x}10^{-3}\right]} = 50.753 = \textbf{50.8}$$

17.42 $\quad K_c = \frac{[N_2][H_2]^3}{[NH_3]^2} = \frac{[0.114][0.342]^3}{[0.0225]^2} = 9.0077875 = \textbf{9.01}$

17.44 $\quad$ The reaction table requires that the initial $[H_2]$ and $[F_2]$ be calculated: $[H_2]$ = 0.10 mol/0.50 L = 0.20 M; $[F_2]$ = 0.050 mol/0.50 L = 0.10 M.

$x = [H_2] = [F_2]$ reacting (–x); 2x = [HF] forming (+2x)

Concentration (M)	$H_2(g)$	+	$F_2(g)$	$\rightleftharpoons$	2HF(g)
Initial	0.20		0.10		0
Change	–x		–x		+2x
Equilibrium	0.20 – x		0.10 – x		2x

17.46 $\quad C(s) + 2H_2(g) \rightleftharpoons_\Delta CH_4(g)$

$$K_p = \frac{P_{CH_4}}{P_{H_2}^2} = 0.262$$

$$P_{CH_4} = K_p P_{H_2}^2 = (0.262)(1.22)^2 = 0.38996 = \textbf{0.390 atm}$$

17.48 $2H_2S(g) \leftrightarrows\!\Delta\ 2H_2(g) + S_2(g)$

$[H_2S] = 0.45\ mol/3.0\ L = 0.15\ M$

Concentration (M)	$2H_2S(g)$	$\leftrightarrows$	$2H_2(g)$	+	$S_2(g)$
Initial	0.15		0		0
Change	−2x		+2x		+x
Equilibrium	0.15 − 2x		2x		x

$$K_c = 9.30\text{x}10^{-8} = \frac{[H_2]^2\,[S_2]}{[H_2S]^2} = \frac{[2x]^2\,[x]}{[0.15 - 2x]^2}$$

Assuming $0.15\ M - 2x \approx 0.15\ M$

$$9.30\text{x}10^{-8} = \frac{[2x]^2\,[x]}{[0.15]^2} = \frac{4x^3}{0.15^2}$$

$$x = 8.0575\text{x}10^{-4}\ M$$

$[H_2] = 2x = 2\,(8.0575\text{x}10^{-4}\ M) = 1.6115\text{x}10^{-3} = \mathbf{1.6x10^{-3}\ M}$

(Since $(1.6\text{x}10^{-3})/(0.15) < 0.05$, the assumption is OK.)

17.50 $2NO_2(g) \leftrightarrows\!\Delta\ 2NO(g) + O_2(g)$

Pressure (atm)	$2NO_2(g)$	$\leftrightarrows$	$2NO(g)$	+	$O_2(g)$
Initial	0.75		0		0
Change	− 2x		+2x		+x
Equilibrium	0.75 − 2x		2x		x

$$K_p = 4.48\text{x}10^{-13} = \frac{P_{NO}^2 P_{O_2}}{P_{NO_2}^2} = \frac{(2x)^2\,(x)}{\left(0.75 - 2x\right)^2}$$

Assume $0.75\ atm - 2x \approx 0.75\ atm$

$$4.48\text{x}10^{-13} = \frac{\left(4x^2\right)(x)}{(0.75)^2} = \frac{\left(4x^3\right)}{(0.75)^2}$$

$$x = 3.979\text{x}10^{-5}\ atm = \mathbf{4.0x10^{-5}\ atm\ O_2}$$

$$P_{NO} = 2x = 2(3.979\text{x}10^{-5}\ atm) = 7.958\text{x}10^{-5} = \mathbf{8.0x10^{-5}\ atm\ NO}$$

17.52   Initial concentrations:

$[A] = (1.75\text{x}10^{-3}\ mol)/(1.00\ L) = 1.75\text{x}10^{-3}\ M$

$[B] = (1.25\text{x}10^{-3}\ mol)/(1.00\ L) = 1.25\text{x}10^{-3}\ M$

$[C] = (6.50\text{x}10^{-4}\ mol)/(1.00\ L) = 6.50\text{x}10^{-4}\ M$

Concentration (M)	$A(g)$	$\leftrightarrows\!\Delta$	$2B(g)$	+	$C(g)$
Initial	$1.75\text{x}10^{-3}$		$1.25\text{x}10^{-3}$		$6.50\text{x}10^{-4}$
Change	− x		+ 2x		+x
Equilibrium	$1.75\text{x}10^{-3} - x$		$1.25\text{x}10^{-3} + 2x$		$6.50\text{x}10^{-4} + x$

$[A]_{eq} = 2.15\text{x}10^{-3} = 1.75\text{x}10^{-3} - x$

$\qquad x = -0.00040$

$[B]_{eq} = 1.25\text{x}10^{-3} + 2x = 1.25\text{x}10^{-3} + 2(-0.00040) = = \mathbf{4.5x10^{-4}\ M}$

$[C]_{eq} = 6.50\text{x}10^{-4} + x = 6.50\text{x}10^{-4} + (-0.00040) = \mathbf{2.5x10^{-4}\ M}$

17.54
Concentration (M)	$SCl_2(g)$	+	$2C_2H_4(g)$	$\leftrightarrows \Delta$	$S(CH_2CH_2Cl)_2(g)$
Initial	0.675		0.973		0
Change	$-x$		$-2x$		$+x$
Equilibrium	$0.675 - x$		$0.973 - 2x$		x

$[S(CH_2CH_2Cl)_2]_{eq} = x = 0.350\ M$

$[SCl_2]_{eq} = 0.675 - x = 0.675 - 0.350 = = 0.325\ M$

$[C_2H_4]_{eq} = 0.973 - 2x = 0.973 - 2(0.350) = 0.273\ M$

$$K_c = \frac{[S(CH_2CH_2Cl)_2]}{[SCl_2][C_2H_4]^2} = \frac{[0.350]}{[0.325][0.273]^2} = 14.4497$$

$K_p = K_c(RT)^{\Delta n}$ 　　　$\Delta n = 1\ mol - 3\ mol = -2$

$K_p = (14.4497)\left[(0.0821)(273.2 + 20.0)\right]^{-2} = 0.0249370 = \mathbf{0.0249}$

17.56
Pressure (atm)	$FeO(s)$	+	$CO(g)$	$\leftrightarrows \Delta$	$Fe(s)$	+	$CO_2(g)$
Initial	—		1.00		—		0
Change			$-x$				$+x$
Equilibrium			$1.00 - x$				x

$$K_p = \frac{P_{CO_2}}{P_{CO}} = 0.403 = \frac{x}{1.00 - x}$$

$x = 0.28724 = \mathbf{0.287\ atm\ CO_2}$

$1.00 - x = 1.00 - 0.28724 = 0.71276 = \mathbf{0.71\ atm\ CO}$

17.59 A positive ΔH_{rxn} indicates that the reaction is endothermic, and that heat is consumed in the reaction:
$$NH_4Cl(s) + heat \leftrightarrows \Delta NH_3(g) + HCl(g)$$
a) The addition of heat (high temperature) causes the reaction to proceed to the right to counterbalance the effect of the added heat. Therefore, more products form at a higher temperature and container (**B**) with the largest number of product molecules best represents the mixture.
b) When heat is removed (low temperature), the reaction shifts to the left to produce heat to offset that disturbance. Therefore, NH_3 and HCl molecules combine to form more reactant and container (**A**) with the smallest number of product gas molecules best represents the mixture.

17.60 Equilibrium component concentration values may change but the mass action expression of these concentrations is a constant as long as temperature remains constant. Changes in component amounts, pressures (volumes), or addition of a catalyst will not change the value of the equilibrium constant.

17.64 a) **no change**　　　　　　　b) **no change**
c) **shifts towards the products**　　d) **shifts towards the reactants**

17.66 a) **less $CH_3OH(l)$; more $CH_3OH(g)$**
b) **less CH_4 and NH_3; more HCN and H_2**

17.68 a) **more CO_2 and H_2O; less C_3H_8 and O_2**
b) **more NH_3 and O_2; less N_2 and H_2O**

17.70 a) **increase volume**　　　b) **decrease volume**

17.72 a) **decrease**　　b) **decrease**　　c) **decrease**　　d) **increase**

17.74 The van't Hoff equation shows how the equilibrium constant is affected by a change in temperature. Substitute the given variables into the equation and solve for K_2.

$$K_{298} = K_1 = 2.25 \times 10^4 \qquad T_1 = 298 \text{ K} \qquad \Delta H^{\circ}_{rxn} = -128 \text{ kJ/mol}$$

$$K_0 = K_2 = ? \qquad T_2 = (273 + 0.) = 273 \text{ K} \qquad R = 8.314 \text{ J/mol} \cdot \text{K}$$

$$\Delta H^{\circ}_{rxn} = (-128 \text{ kJ/mol})(10^3 \text{ J/1 kJ}) = -1.28 \times 10^5 \text{ J}$$

$$\ln \frac{K_2}{K_1} = -\frac{\Delta H^{\circ}_{rxn}}{R}\left(\frac{1}{T_2} - \frac{1}{T_1}\right)$$

$$\ln \frac{K_2}{2.25 \times 10^4} = -\frac{-1.28 \times 10^5 \text{ J}}{8.314 \text{ J/mol K}}\left(\frac{1}{273 \text{ K}} - \frac{1}{298 \text{ K}}\right)$$

$$\ln \frac{K_2}{2.25 \times 10^4} = 4.731088$$

$$\frac{K_2}{2.25 \times 10^4} = 1.134189 \times 10^2$$

$$K_2 = (2.25 \times 10^4)(1.134189 \times 10^2) = 2.551925 \times 10^6 = \mathbf{2.55 \times 10^6}$$

17.77 $3H_2(g) + N_2(g) \leftrightarrows^{\Delta} 2NH_3(g)$

$$P_{NH_3} = (41.49\%/100\%)(110. \text{ atm}) = 45.639 \text{ atm}$$

$$100.00\% - 41.49\% = 58.51\% \text{ N}_2 + \text{H}_2$$

$$P_{H_2} + P_{N_2} = (58.51\%/100\%)(110. \text{ atm}) = 64.361 \text{ atm}$$

$$P_{H_2} = (3/4)(64.361 \text{ atm}) = 48.27075 \text{ atm}$$

$$P_{N_2} = (1/4)(64.361 \text{ atm}) = 16.09025 \text{ atm}$$

$$K_p = \frac{\left(P_{NH_3}\right)^2}{\left(P_{H_2}\right)^3 \left(P_{N_2}\right)} = \frac{(45.639)^2}{(48.27075)^3 (16.09025)} = 1.15095 \times 10^{-3} = \mathbf{1.15 \times 10^{-3}}$$

17.79 a) **More CaCO$_3$.** Because the forward reaction is exothermic, decreasing the temperature will cause an increase in the amount of CaCO$_3$ formed as the reaction shifts to the right to produce more heat.

b) **Less CaCO$_3$.** The only gas in the equation is a reactant. Increasing the volume (decreasing the pressure) will cause the equilibrium to shift toward the reactant side and the amount of CaCO$_3$ formed decreases.

c) **More CaCO$_3$.** Increasing the partial pressure of CO$_2$ will cause more CaCO$_3$ to be formed as the reaction shifts to the right to consume the added CO$_2$.

d) **No change.** Removing half of the initial CaCO$_3$ will have no effect on the amount of CaCO$_3$ formed, because CaCO$_3$ is a solid.

17.82 a)

(1) $\quad 2Ni_3S_2(s) + 7O_2(g) \leftrightarrows^{\Delta} \cancel{6NiO(s)} + 4SO_2(g)$

(2) $\quad \cancel{6NiO(s)} + 6H_2(g) \leftrightarrows^{\Delta} \cancel{6Ni(s)} + 6H_2O(g)$

(3) $\quad \cancel{6Ni(s)} + 24CO(g) \leftrightarrows^{\Delta} 6Ni(CO)_4(g)$

Overall: $2Ni_3S_2(s) + 7O_2(g) + 6H_2(g) + 24CO(g) \leftrightarrows^{\Delta} 4SO_2(g) + 6H_2O(g) + 6Ni(CO)_4(g)$

b) As always, the solid is not included in the Q expression.

$$Q_{c(overall)} = \frac{[SO_2]^4 [H_2O]^6 [Ni(CO)_4]^6}{[O_2]^7 [H_2]^6 [CO]^{24}}$$

$$Q_1 \times Q_2 \times Q_3 = \frac{[SO_2]^4}{[O_2]^7} \times \frac{[H_2O]^6}{[H_2]^6} \times \frac{[Ni(CO)_4]^6}{[CO]^{24}} = \frac{[SO_2]^4 [H_2O]^6 [Ni(CO)_4]^6}{[O_2]^7 [H_2]^6 [CO]^{24}}$$

17.85 $n/V = M = P/RT = \dfrac{(2.0 \text{ atm})}{\left(0.0821 \dfrac{\text{L atm}}{\text{mol K}}\right)((273.2 + 25.0)\text{K})} = 0.0816919\ M$ each gas

	$H_2(g)$	$+$	$CO_2(g)$	$\rightleftharpoons$	$H_2O(g)$	$+$	$CO(g)$
Initial	0.0816919		0.0816919		0		0
Change	$-x$		$-x$		$+x$		$+x$
Equi:	$0.0816919 - x$		$0.0816919 - x$		x		x

$$K_c = \frac{[H_2O][CO]}{[H_2][CO_2]} = 0.534 = \frac{[x][x]}{[0.0816919 - x][0.0816919 - x]} = \frac{[x]^2}{[0.0816919 - x]^2}$$

$$(0.534)^{1/2} = 0.730753 = \frac{[x]}{[0.0816919 - x]}$$

$$x = 0.03449\ M$$

M of H_2 at equilibrium $= 0.0816919 - x = 0.0816919 - 0.03449 = 0.0472019$ mol/L

Mass (g) of $H_2 = (1.00\ \text{L})\left(\dfrac{0.0472019\ \text{mol}}{\text{L}}\right)\left(\dfrac{2.016\ \text{g}}{1\ \text{mol}}\right) = 0.095159 = \textbf{0.095 g } \mathbf{H_2}$

17.87 Plan: Write the equilibrium expression. You are given a value of K_c but the amounts of reactants and product are given in units of pressure. Convert K_c to K_p and use the equilibrium pressures of SO_3 and O_2 to obtain the equilibrium pressure of SO_2. For part b), set up a reaction table and solve for x. The equilibrium concentrations can then be used to find the K_c value at the higher temperature. The concentration of SO_2 is converted to pressure using the ideal gas law, $PV = nRT$.
Solution:
a) $K_p = K_c(RT)^{\Delta n}$

Δn = moles gaseous products – moles gaseous reactants $= 2 - 3 = -1$ (two mol of product, SO_3, and three mol of reactants, $2\ SO_2 + O_2$)

$$K_p = K_c(RT)^{\Delta n} = K_c(RT)^{-1} = (1.7 \times 10^8)[(0.0821\ \text{L} \cdot \text{atm/mol} \cdot \text{K})(600.\ \text{K})]^{-1} = 3.451 \times 10^6$$

$$K_p = \frac{P_{SO_3}^2}{P_{SO_2}^2 P_{O_2}} = \frac{(300.)^2}{P_{SO_2}^2 (100.)} = 3.451 \times 10^6$$

$$P_{SO_2} = 0.016149 = \textbf{0.016 atm}$$

b) Create a reaction table that describes the reaction conditions. Since the volume is 1.0 L, the moles equals the molarity. Note the 2:1:2 mole ratio between SO_2:O_2:SO_3.

Concentration (M)	$2SO_2(g)$	$+$	$O_2(g)$	$\rightleftharpoons$	$2SO_3(g)$	
Initial	0.0040		0.0028		0	
Change	$-2x$		$-x$		$+2x$	(2:1:2 mole ratio)
Equilibrium	$0.0040 - 2x$		$0.0028 - x$		$2x = 0.0020$ (given)	

$x = 0.0010$, therefore:

$[SO_2] = 0.0040 - 2x = 0.0040 - 2(0.0010) = 0.0020\ M$

$[O_2] = 0.0028 - x = 0.0028 - 0.0010 = 0.0018\ M$

$[SO_3] = 2(0.0010) = 0.0020\ M$

Substitute equilibrium concentrations into the equilibrium expression and solve for K_c.

$$K_c = \frac{[SO_3]^2}{[SO_2]^2 [O_2]} = \frac{[0.0020]^2}{[0.0020]^2 [0.0018]} = 555.5556 = \textbf{5.6} \times \textbf{10}^2$$

The pressure of SO_2 is estimated using the concentration of SO_2 and the ideal gas law (although the ideal gas law is not well behaved at high pressures and temperatures).

$$PV = nRT$$

$$P_{SO_2} = \frac{nRT}{V} = \frac{(0.0020 \text{ mol})\left(0.0821\dfrac{\text{L atm}}{\text{mol K}}\right)(1000. \text{ K})}{(1.0 \text{ L})} = 0.1642 = \textbf{0.16 atm}$$

17.89 **Plan:** Set up a reaction table to find the equilibrium amount of $CaCO_3$ after the first equilibrium is established and then the equilibrium amount after the second equilibrium is established.
Solution:
The equilibrium pressure of $CO_2 = P_{CO_2} = 0.220$ atm.

	$CaCO_3(s)$	$\rightleftharpoons\Delta$	$CaO(s)$	+	$CO_2(g)$
Initial	0.100 mol		0.100 mol		0
Change	– x		– x		+x
Equilibrium	0.100 – x		0.100 – x		x = 0.220 atm (given)

The amount of calcium carbonate solid in the container at the first equilibrium equals the original amount, 0.100 mol, minus the amount reacted to form 0.220 atm of carbon dioxide. The moles of $CaCO_3$ reacted is equal to the number of moles of carbon dioxide produced. Use the pressure of CO_2 and the ideal gas equation to calculate the moles of CO_2 produced:

$$PV = nRT$$

$$\text{Moles of } CO_2 = n = \frac{PV}{RT}$$

$$n = \frac{(0.220 \text{ atm})(10.0 \text{ L})}{\left(0.0821\dfrac{\text{L atm}}{\text{mol K}}\right)(385 \text{ K})} = 0.0696015 \text{ mol } CO_2$$

Moles of $CaCO_3$ reacted = moles of CO_2 produced = 0.0696015 mol
Moles of $CaCO_3$ remaining = initial moles – moles reacted = 0.100 mol $CaCO_3$ – 0.0696015 mol $CaCO_3$
$$= 0.0304 \text{ mol } CaCO_3 \text{ at first equilibrium}$$
As more carbon dioxide gas is added, the system returns to equilibrium by shifting to the left to convert the added carbon dioxide to calcium carbonate to maintain the partial pressure of carbon dioxide at 0.220 atm (K_p). Convert the added 0.300 atm of CO_2 to moles using the ideal gas equation. The moles of CO_2 reacted equals the moles of $CaCO_3$ formed.

$$\text{Moles of } CO_2 = n = \frac{PV}{RT}$$

$$n = \frac{(0.300 \text{ atm})(10.0 \text{ L})}{\left(0.0821\dfrac{\text{L atm}}{\text{mol K}}\right)(385 \text{ K})} = 0.09491 \text{ mol } CO_2$$

Moles of $CaCO_3$ produced = moles of CO_2 reacted = 0.09491 mol $CaCO_3$
Add the moles of $CaCO_3$ formed in the second equilibrium to the moles of $CaCO_3$ at the first equilibrium position.
Moles of $CaCO_3$ = moles at first equilibrium + moles formed in second equilibrium
$$= 0.0304 \text{ mol} + 0.09491 = 0.12531 \text{mol } CaCO_3$$

$$\text{Mass (g) of } CaCO_3 = (0.12531 \text{ mol } CaCO_3)\left(\frac{100.09 \text{ g } CaCO_3}{1 \text{ mol } CaCO_3}\right) = 12.542 = \textbf{12.5 g } CaCO_3$$

17.93 **Plan:** Use the balanced reaction to write the equilibrium expression. The equilibrium concentration of S_2F_{10} is used to write an expression for the equilibrium concentrations of SF_4 and SF_6.

Solution:

$S_2F_{10}(g) \leftrightarrows \Delta \; SF_4(g) + SF_6(g)$

The reaction is described by the following equilibrium expression:

$$K_c = \frac{[SF_4][SF_6]}{[S_2F_{10}]}$$

At the first equilibrium, $[S_2F_{10}] = 0.50 \; M$ and $[SF_4] = [SF_6] = x$ ($[SF_4]:[SF_6] = 1:1$).

$$K_c = \frac{[SF_4][SF_6]}{[S_2F_{10}]} = \frac{[x][x]}{[0.50]}$$

$$x^2 = 0.50 K_c$$

$$[SF_4] = [SF_6] = x = \sqrt{0.50 K_c}$$

At the second equilibrium, $[S_2F_{10}] = 2.5 \; M$ and $[SF_4] = [SF_6] = x$.

$$K_c = \frac{[SF_4][SF_6]}{[S_2F_{10}]} = \frac{[x][x]}{[2.5]}$$

$$x^2 = 2.5 K_c$$

$$[SF_4] = [SF_6] = x = \sqrt{2.5 K_c}$$

Thus, the concentrations of SF_4 and SF_6 increase by a factor of:

$$\frac{\sqrt{2.5 K_c}}{\sqrt{0.50 K_c}} = \frac{\sqrt{2.5}}{\sqrt{0.50}} = 2.236 = \mathbf{2.2}$$

17.95 Plan: Use the volume fraction of O_2 and CO_2 to find the partial pressure of each gas and substitute these pressures into the equilibrium expression to find the partial pressure of CO. Use $PV = nRT$ to convert the partial pressure of CO to moles per liter and then convert to pg/L.

Solution:

a) Calculate the partial pressures of oxygen and carbon dioxide because volumes are proportional to moles of gas, so volume fraction equals mole fraction. Assume that the amount of carbon monoxide gas is small relative to the other gases, so the total volume of gases equals $V_{CO_2} + V_{O_2} + V_{N_2} = 10.0 + 1.00 + 50.0 = 61.0$.

$$P_{CO_2} = \left(\frac{10.0 \; \text{mol CO}_2}{61.0 \; \text{mol gas}}\right)(4.0 \; \text{atm}) = 0.6557377 \; \text{atm}$$

$$P_{O_2} = \left(\frac{1.00 \; \text{mol O}_2}{61.0 \; \text{mol gas}}\right)(4.0 \; \text{atm}) = 0.06557377 \; \text{atm}$$

Use the partial pressures and given K_p to find P_{CO}.

$$2CO_2(g) \leftrightarrows 2CO(g) + O_2(g)$$

$$K_p = \frac{P_{CO}^2 P_{O_2}}{P_{CO_2}^2} = \frac{P_{CO}^2 (0.06557377)}{(0.6557377)^2} = 1.4 \times 10^{-28}$$

$$P_{CO} = 3.0299 \times 10^{-14} = \mathbf{3.0 \times 10^{-14} \; atm}$$

b) $PV = nRT$

$$\frac{n_{CO}}{V} = \frac{P}{RT} = \frac{\left(3.0299 \times 10^{-14} \; \text{atm}\right)}{\left(0.0821 \dfrac{\text{L atm}}{\text{mol K}}\right)(800 \; \text{K})} = 4.61312 \times 10^{-16} \; \text{mol/L}$$

Concentration (pg/L) of CO = $\left(\dfrac{4.61312 \times 10^{-16} \; \text{mol CO}}{\text{L}}\right)\left(\dfrac{28.01 \; \text{g CO}}{1 \; \text{mol CO}}\right)\left(\dfrac{1 \; \text{pg}}{10^{-12} \; \text{g}}\right) = 0.01292 = \mathbf{0.013 \; pg \; CO/L}$

17.97 **Plan:** Write a reaction table given that P_{CH_4} (init) $= P_{CO_2}$ (init) $= 10.0$ atm, substitute equilibrium values into the equilibrium expression, and solve for P_{H_2}.

Solution:

a)

Pressure (atm)	$CH_4(g)$	$+$	$CO_2(g)$	$\rightleftharpoons$	$2CO(g)$	$+$	$2H_2(g)$
Initial	10.0		10.0		0		0
Change	$-x$		$-x$		$+2x$		$+2x$
Equilibrium	$10.0 - x$		$10.0 - x$		$2x$		$2x$

$$K_p = \frac{P_{CO}^2 P_{H_2}^2}{P_{CH_4} P_{CO_2}} = \frac{(2x)^2 (2x)^2}{(10.0-x)(10.0-x)} = \frac{(2x)^4}{(10.0-x)^2} = 3.548\times10^6 \quad \text{(take square root of each side)}$$

$$\frac{(2x)^2}{(10.0-x)} = 1.8836135\times10^3$$

A quadratic is necessary:

$$4x^2 + (1.8836135\times10^3\, x) - 1.8836135\times10^4 = 0$$

$$a = 4 \quad b = 1.8836135\times10^3 \quad c = -1.8836135\times10^4$$

$$x = \frac{-b \pm \sqrt{b^2 - 4ac}}{2a}$$

$$x = \frac{-1.8836135\times10^3 \pm \sqrt{\left(1.8836135\times10^3\right)^2 - 4(4)\left(-1.8836135\times10^4\right)}}{2(4)}$$

$$x = 9.796209$$

$$P_{H_2} = 2x = 2(9.796209) = 19.592419 \text{ atm}$$

If the reaction proceeded entirely to completion, the partial pressure of H_2 would be 20.0 atm (pressure is proportional to moles, and twice as many moles of H_2 form for each mole of CH_4 or CO_2 that reacts).

The percent yield is $\dfrac{19.592418 \text{ atm}}{20.0 \text{ atm}}(100\%) = 97.96209 = \mathbf{98.0\%}$.

b) Repeat the calculations for part a) with the new K_p value. The reaction table is the same.

$$K_p = \frac{P_{CO}^2 P_{H_2}^2}{P_{CH_4} P_{CO_2}} = \frac{(2x)^2 (2x)^2}{(10.0-x)(10.0-x)} = \frac{(2x)^4}{(10.0-x)^2} = 2.626\times10^7$$

$$\frac{(2x)^2}{(10.0-x)} = 5.124451\times10^3$$

A quadratic is needed:

$$4x^2 + (5.124451\times10^3\, x) - 5.124451\times10^4 = 0$$

$$a = 4 \qquad b = 5.124451\times10^3 \qquad c = -5.124451\times10^4$$

$$x = \frac{-5.124451\times10^3 \pm \sqrt{\left(5.124451\times10^3\right)^2 - 4(4)\left(-5.124451\times10^4\right)}}{2(4)}$$

$$x = 9.923144$$

$$P_{H_2} = 2x = 2(9.923144) = 19.84629 \text{ atm}$$

If the reaction proceeded entirely to completion, the partial pressure of H_2 would be 20.0 atm (pressure is proportional to moles, and twice as many moles of H_2 form for each mole of CH_4 or CO_2 that reacts).

The percent yield is $\dfrac{19.84629 \text{ atm}}{20.0 \text{ atm}}(100\%) = 99.23145 = \mathbf{99.0\%}$.

c) van't Hoff equation:

$$K_1 = 3.548 \times 10^6 \qquad T_1 = 1200.\ K \qquad \Delta H^\circ_{rxn} = ?$$

$$K_2 = 2.626 \times 10^7 \qquad T_2 = 1300.\ K \qquad R = 8.314\ J/mol \bullet K$$

$$\ln \frac{K_2}{K_1} = -\frac{\Delta H^\circ_{rxn}}{R}\left(\frac{1}{T_2} - \frac{1}{T_1}\right)$$

$$\ln \frac{2.626 \times 10^7}{3.548 \times 10^6} = -\frac{\Delta H^\circ_{rxn}}{\left(8.314 \dfrac{J}{mol\ K}\right)}\left(\frac{1}{1200.\ K} - \frac{1}{1300.\ K}\right)$$

$$2.0016628 = \Delta H^\circ_{rxn}\ (7.710195 \times 10^{-6})$$

$$\Delta H^\circ_{rxn} = 2.0016628/7.710195 \times 10^{-6} = 2.5961247 \times 10^5 = \mathbf{2.60 \times 10^5\ J/mol}$$

(The subtraction of the $1/T$ terms limits the answer to three significant figures.)

17.99 <u>Plan:</u> Add the two reactions to obtain the overall reaction. Multiply the second equation by 2 to cancel the moles of CO produced in the first reaction. K_p for the second reaction is then $(K_p)^2$. K_p for the overall reaction is equal to the product of the K_p values for the two individual reactions. Calculate K_c using $K_p = K_c(RT)^{\Delta n}$.
<u>Solution:</u>

a)

$$2CH_4(g) + O_2(g) \rightleftharpoons \Delta\ 2CO(g) + 4H_2(g) \qquad K_p = 9.34 \times 10^{28}$$

$$\underline{2CO(g) + 2H_2O(g) \ \Delta\rightleftharpoons\ 2CO_2(g) + 2H_2(g) \qquad K_p = (1.374)^2 = 1.888}$$

$$2CH_4(g) + O_2(g) + 2H_2O(g) \ \Delta\rightleftharpoons\ 2CO_2(g) + 6H_2(g)$$

b) $K_p = (9.34 \times 10^{28})(1.888) = 1.76339 \times 10^{29} = \mathbf{1.76 \times 10^{29}}$

c) Δn = moles gaseous products – moles gaseous reactants = $8 - 5 = 3$
(8 moles of product gas – 5 moles of reactant gas)

$$K_p = K_c(RT)^{\Delta n}$$

$$K_c = \frac{K_p}{(RT)^{\Delta n}} = \frac{1.76339 \times 10^{29}}{[(0.0821\ atm \bullet L/mol \bullet K)(1000)]^3} = 3.18654 \times 10^{23} = \mathbf{3.19 \times 10^{23}}$$

d) The initial total pressure is given as 30. atm. To find the final pressure use the relationship between pressure and number of moles of gas: $n_{initial}/P_{initial} = n_{final}/P_{final}$
Total mol of gas initial = 2.0 mol CH_4 + 1.0 mol O_2 + 2.0 mol H_2O = 5.0 mol
Total mol of gas final = 2.0 mol CO_2 + 6.0 mol H_2 = 8.0 mol   (from mole ratios)

$$P_{final} = (30.\ atm\ reactants)\left(\frac{8\ mol\ products}{5\ mol\ reactants}\right) = \mathbf{48\ atm}$$

17.100 <u>Plan:</u> Write an equilibrium expression. Use the balanced equation to define x and set up a reaction table, substitute into the equilibrium expression, and solve for x, from which the pressure of N or H is calculated. Convert log K_p to K_p. Convert pressures to moles using the ideal gas law, $PV = nRT$. Convert moles to atoms using Avogadro's number.
<u>Solution:</u>

a) The initial pressure of N_2 is 200. atm.   Log K_p = –43.10; $K_p = 10^{-43.10} = 7.94328 \times 10^{-44}$

Pressure (atm)	$N_2(g)$	$\rightleftharpoons$	$2N(g)$
Initial	200.		0
Change	–x		+2x
Equilibrium	200 – x		2x

$$K_p = \frac{(P_N)^2}{(P_{N_2})} = 7.94328 \times 10^{-44}$$

$$\frac{(2x)^2}{(200. - x)} = 7.94328 \times 10^{-44} \qquad \text{Assume } 200. - x \cong 200.$$

$$\frac{(2x)^2}{(200)} = 7.94328 \times 10^{-44}$$

$$4x^2 = 1.588656 \times 10^{-41}$$

$$x = 1.992897 \times 10^{-21}$$

$$P_N = 2x = 2(1.992897 \times 10^{-21}) = 3.985795 \times 10^{-21} = \mathbf{4.0 \times 10^{-21} \ atm}$$

b) Log K_p = −17.30; $K_p = 10^{-17.30} = 5.01187 \times 10^{-18}$

Pressure (atm)	$H_2(g)$	$\leftrightarrows$	$2H(g)$
Initial	600.		0
Change	−x		+2x
Equilibrium	600 − x		2x

$$K_p = \frac{(P_H)^2}{(P_{H_2})} = 5.01187 \times 10^{-18}$$

$$\frac{(2x)^2}{(600. - x)} = 5.01187 \times 10^{-18} \qquad \text{Assume } 600. - x \cong 600.$$

$$\frac{(2x)^2}{(600)} = 5.01187 \times 10^{-18}$$

$$4x^2 = 3.007122 \times 10^{-15}$$

$$x = 2.741862 \times 10^{-8}$$

$$P_H = 2x = 2(2.741862 \times 10^{-8}) = 5.48372 \times 10^{-8} = \mathbf{5.5 \times 10^{-8}}$$

c) $PV = nRT$

$$\text{Moles of N atoms} = \frac{PV}{RT} = \frac{(3.985795 \times 10^{-21} \, \text{atm})(1.00 \ L)}{\left(0.0821 \dfrac{L \bullet atm}{mol \bullet K}\right)(1000. \, K)} = 4.85481 \times 10^{-15} \ mol$$

$$\text{Number of N atoms} = (4.85481 \times 10^{-23} \ mol \ N \ atoms)\left(\frac{6.022 \times 10^{23} \ N \ atoms}{1 \ mol \ N \ atoms}\right) = 29.2356 = \mathbf{29 \ N \ atoms/L}$$

$$\text{Moles of H atoms} = \frac{PV}{RT} = \frac{(5.48372 \times 10^{-8} \, \text{atm})(1.00 \ L)}{\left(0.0821 \dfrac{L \bullet atm}{mol \bullet K}\right)(1000. \, K)} = 6.67932 \times 10^{-10} \ mol$$

$$\text{Number of H atoms} = (6.67932 \times 10^{-10} \ mol \ H \ atoms)\left(\frac{6.022 \times 10^{23} \ H \ atoms}{1 \ mol \ H \ atoms}\right)$$

$$= 4.022 \times 10^{14} = \mathbf{4.0 \times 10^{14} \ H \ atoms/L}$$

d) The more reasonable step is $\mathbf{N_2(g) + H(g) \rightarrow NH(g) + N(g)}$. With only twenty-nine N atoms in 1.0 L, the first reaction would produce virtually no NH(g) molecules. There are orders of magnitude more N_2 molecules than N atoms, so the second reaction is the more reasonable step.

17.103   <u>Plan:</u> Write an equilibrium expression. Use the balanced equation to define x and set up a reaction table, substitute into the equilibrium expression, and solve for x, from which the equilibrium pressures of the gases are calculated. Add the equilibrium pressures of the three gases to obtain the total pressure. Use the relationship $K_p = K_c(RT)^{\Delta n}$ to find K_c.
<u>Solution:</u>

a)

Pressure (atm)	$N_2(g)$	+	$O_2(g)$	$\rightleftharpoons$	$2NO(g)$
Initial	0.780		0.210		0
Change	–x		–x		+2x
Equilibrium	0.780 – x		0.210 – x		2x

$$K_p = \frac{(P_{NO})^2}{(P_{N_2})(P_{O_2})} = 4.35\text{x}10^{-31}$$

$$\frac{(2x)^2}{(0.780 - x)(0.210 - x)} = 4.35\text{x}10^{-31}\ \text{Assume x is small because } K \text{ is small.}$$

$$\frac{(2x)^2}{(0.780)(0.210)} = 4.35\text{x}10^{-31}$$

$$x = 1.33466\text{x}10^{-16}$$

Based on the small amount of nitrogen monoxide formed, the assumption that the partial pressures of nitrogen and oxygen change to an insignificant degree holds.

P_{nitrogen} (equilibrium) = $(0.780 - 1.33466\text{x}10^{-16})$ atm = **0.780 atm N_2**

P_{oxygen} (equilibrium) = $(0.210 - 1.33466\text{x}10^{-16})$ atm = **0.210 atm O_2**

P_{NO} (equilibrium) = $2(1.33466\text{x}10^{-16})$ atm = $2.66933\text{x}10^{-16}$ = **2.67x10^{-16} atm NO**

b) The total pressure is the sum of the three partial pressures:

$$0.780\ \text{atm} + 0.210\ \text{atm} + 2.67\text{x}10^{-16}\ \text{atm} = \textbf{0.990 atm}$$

c) $K_p = K_c(RT)^{\Delta n}$

Δn = moles gaseous products – moles gaseous reactants = 2 – 2 = 0

(two moles of product NO and two moles of reactants N_2 and O_2)

$K_p = K_c(RT)$

$K_c = K_p = \textbf{4.35x10}^{\textbf{-31}}$ because there is no net increase or decrease in the number of moles of gas in the course of the reaction.

17.105   <u>Plan:</u> Use the equation $K_p = K_c(RT)^{\Delta n}$ to find K_p. The value of K_c for the formation of HI is the reciprocal of the K_c value for the decomposition of HI. Use the equation $\Delta H^{\circ}_{rxn} = \sum[\Delta H^{\circ}_{f(products)}] - \sum[\Delta H^{\circ}_{f(reactants)}]$ to find the value of ΔH°_{rxn}. Use the van't Hoff equation as a second method of calculating ΔH°_{rxn}.
<u>Solution:</u>

a) $K_p = K_c(RT)^{\Delta n}$

Δn = moles gaseous products – moles gaseous reactants = 2 – 2 = 0

(2 mol product ($1H_2 + 1I_2$) – 2 mol reactant (HI) = 0)

$K_p = K_c(RT)^0 = 1.26\text{x}10^{-3}(RT)^0 = \textbf{1.26x10}^{\textbf{-3}}$

b) The equilibrium constant for the reverse reaction is the reciprocal of the equilibrium constant for the forward reaction:

$$K_{\text{formation}} = \frac{1}{K_{\text{decomposition}}} = \frac{1}{1.26\text{x}10^{-3}} = 793.65 = \textbf{794}$$

17-26

c) $\Delta H^{\circ}_{rxn} = \sum[\Delta H^{\circ}_{f(products)}] - \sum[\Delta H^{\circ}_{f(reactants)}]$

$\Delta H^{\circ}_{rxn} = \{1\ \Delta H^{\circ}_{f}\ [H_2(g)] + 1\ \Delta H^{\circ}_{f}\ [I_2(g)]\} - \{2\ \Delta H^{\circ}_{f}\ [HI(g)]\}$

$\Delta H^{\circ}_{rxn} = [(1\ mol)(0\ kJ/mol) + (1\ mol)(0\ kJ/mol)] - [(2\ mol)(25.9\ kJ/mol)]$

$\Delta H^{\circ}_{rxn} = \mathbf{-51.8\ kJ}$

d) $\ln \dfrac{K_2}{K_1} = -\dfrac{\Delta H^{\circ}_{rxn}}{R}\left(\dfrac{1}{T_2} - \dfrac{1}{T_1}\right)$

$K_1 = 1.26\times10^{-3}$; $\quad K_2 = 2.0\times10^{-2}$, $\quad T_1 = 298\ K$; $\quad T_2 = 729\ K$

$\ln \dfrac{2.0\times10^{-2}}{1.26\times10^{-3}} = -\dfrac{\Delta H^{\circ}_{rxn}}{8.314\ J/mol\ K}\left(\dfrac{1}{729\ K} - \dfrac{1}{298\ K}\right)$

$2.764621 = 2.38629\times10^{-4}\ \Delta H^{\circ}_{rxn}$

$\Delta H^{\circ}_{rxn} = 1.1585\times10^4 = \mathbf{1.2\times10^4\ J/mol}$

17.109 Plan: Use the balanced equation to write an equilibrium expression. Find the initial concentration of each reactant from the given amounts and container volume, use the balanced equation to define x, and set up a reaction table. The equilibrium concentration of CO is known, so x can be calculated and used to find the other equilibrium concentrations. Substitute the equilibrium concentrations into the equilibrium expression to find K_c. Add the molarities of all of the gases at equilibrium, use $(M)(V)$ to find the total number of moles, and then use $PV = nRT$ to find the total pressure. To find $[CO]_{eq}$ after the pressure is doubled, set up another reaction table in which the initial concentrations are equal to the final concentrations from part a) and add in the additional CO.
Solution:
The reaction is: $CO(g) + H_2O(g) \rightleftharpoons CO_2(g) + H_2(g)$
a) Initial [CO] and initial $[H_2O] = 0.100\ mol/20.00\ L = 0.00500\ M$.

	CO	H$_2$O $\rightleftharpoons$	CO$_2$	H$_2$
Initial	0.00500 M	0.00500 M	0	0
Change	−x	−x	+x	+x
Equilibrium	0.00500 − x	0.00500 − x	x	x

$[CO]_{equilibrium} = 0.00500 - x = 2.24\times10^{-3}\ M = [H_2O]$ (given in problem)

$\qquad x = 0.00276\ M = [CO_2] = [H_2]$

$\qquad K_c = \dfrac{[CO_2][H_2]}{[CO][H_2O]} = \dfrac{[0.00276\ ][0.00276\ ]}{[0.00224][0.00224]} = 1.518176 = \mathbf{1.52}$

b) $M_{total} = [CO] + [H_2O] + [CO_2] + [H_2] = (0.00224\ M) + (0.00224\ M) + (0.00276\ M) + (0.00276\ M)$

$\qquad\qquad = 0.01000\ M$

$n_{total} = (M_{total})(V) = (0.01000\ mol/L)(20.00\ L) = 0.2000\ mol\ total$

$PV = nRT$

$P_{total} = n_{total}RT/V = \dfrac{(0.2000\ mol)\left(0.08206\dfrac{L\bullet atm}{mol\bullet K}\right)((273+900.)K)}{(20.00\ L)} = 0.9625638 = \mathbf{0.9626\ atm}$

c) Initially, an equal number of moles must be added = **0.2000 mol CO**

d) Set up a table with the initial concentrations equal to the final concentrations from part a), and then add 0.2000 mol CO/20.00 L = 0.01000 M to compensate for the added CO.

	CO	H_2O $\leftrightharpoons$	CO_2	H_2
Initial	0.00224 M	0.00224 M	0.00276 M	0.00276 M
Added CO	0.01000 M			
Change	$-x$	$-x$	$+x$	$+x$
Equilibrium	0.01224 $-x$	0.00224 $-x$	0.00276 $+x$	0.00276 $+x$

$$K_c = \frac{[CO_2][H_2]}{[CO][H_2O]} = \frac{[0.00276 + x][0.00276 + x]}{[0.01224 - x][0.00224 - x]} = 1.518176$$

$$\frac{\left[7.6176\text{x}10^{-6} + 5.52\text{x}10^{-3}x + x^2\right]}{\left[2.74176\text{x}10^{-5} - 1.448\text{x}10^{-2}x + x^2\right]} = 1.518176$$

$7.6176\text{x}10^{-6} + 5.52\text{x}10^{-3}x + x^2 = (1.518176)(2.74176\text{x}10^{-5} - 1.448\text{x}10^{-2}x + x^2)$

$7.6176\text{x}10^{-6} + 5.52\text{x}10^{-3}x + x^2 = 4.162474\text{x}10^{-5} - 0.021983x + 1.518176x^2$

$0.518176x^2 - 0.027503x + 3.400714\text{x}10^{-5} = 0$

$a = 0.518176 \qquad b = -0.027503 \qquad c = 3.400714\text{x}10^{-5}$

$$x = \frac{-b \pm \sqrt{b^2 - 4ac}}{2a}$$

$$x = \frac{-(-0.027503) \pm \sqrt{(-0.027503)^2 - 4(0.518176)(3.400714\text{x}10^{-5})}}{2(0.518176)}$$

$x = 1.31277\text{x}10^{-3}$

$[CO] = 0.01224 - x = 0.01224 - (1.31277\text{x}10^{-3}) = 0.01092723 = \textbf{0.01093 } \boldsymbol{M}$

CHAPTER 18 ACID-BASE EQUILIBRIA

FOLLOW–UP PROBLEMS

18.1A   Plan: Examine the formulas and classify each as an acid or base. Strong acids are the hydrohalic acids HCl, HBr, and HI, and oxoacids in which the number of O atoms exceeds the number of ionizable protons by at least two. Other acids are weak acids. Strong bases are soluble oxides or hydroxides of the Group 1A(1) metals and Ca, Sr, and Ba in Group 2A(2). Other bases are weak bases.
Solution:
a) Chloric acid, **HClO₃**, is the stronger acid because acid strength increases as the number of O atoms in the acid increases.
b) Hydrochloric acid, **HCl**, is one of the strong hydrohalic acids whereas acetic acid, CH_3COOH, is a weak carboxylic acid.
c) Sodium hydroxide, **NaOH**, is a strong base because Na is a Group 1A(1) metal. Methylamine, CH_3NH_2, is an organic amine and, therefore, a weak base.

18.1B   Plan: Examine the formulas and classify each as a strong acid, weak acid, strong base, or weak base. Strong acids are the hydrohalic acids HCl, HBr, and HI, and oxoacids in which the number of O atoms exceeds the number of ionizable protons by at least two. Other acids are weak acids. Strong bases are soluble oxides or hydroxides of the Group 1A(1) metals and Ca, Sr, and Ba in Group 2A(2). Other bases are weak bases.
Solution:
a) $(CH_3)_3N$ is a **weak base**. It contains a nitrogen atom with a lone pair of electrons, which classifies it as a base; however, it is not one of the strong bases.
b) Hydroiodic acid, HI, is a **strong acid** (one of the strong acids listed above).
c) HBrO is a **weak acid.** It has an ionizable hydrogen, which makes it an acid. Specifically, it is an oxoacid, in which a polyatomic ion is the anion. In the case of this oxoacid, there is only one O atom for each ionizable hydrogen, so this is a weak acid. The reaction for the dissociation of this weak acid is:
$HBrO(aq) + H_2O(l) \rightleftharpoons BrO^-(aq) + H_3O^+(aq)$. The corresponding equilibrium expression is:

$$K_a = \frac{[BrO^-][H_3O^+]}{[HBrO]}$$

d) $Ca(OH)_2$ is a **strong base** (one of the strong bases listed above).

18.2A   Plan: The product of $[H_3O^+]$ and $[OH^-]$ remains constant at 25°C because the value of K_w is constant at a given temperature. Use $K_w = [H_3O^+][OH^-] = 1.0 \times 10^{-14}$ to solve for $[H_3O^+]$.
Solution:
Calculating $[H_3O^+]$:

$$[H_3O^+] = \frac{K_w}{[OH^-]} = \frac{1.0 \times 10^{-14}}{6.7 \times 10^{-2}} = 1.4925 \times 10^{-13} = \mathbf{1.5 \times 10^{-13}\ M}$$

Since $[OH^-] > [H_3O^+]$, the solution is **basic**.

18.2B   Plan: The product of $[H_3O^+]$ and $[OH^-]$ remains constant at 25°C because the value of K_w is constant at a given temperature. Use $K_w = [H_3O^+][OH^-] = 1.0 \times 10^{-14}$ to solve for $[H_3O^+]$.
Solution:
Calculating $[OH^-]$:

$$[OH^-] = \frac{K_w}{[H_3O^+]} = \frac{1.0 \times 10^{-14}}{1.8 \times 10^{-10}} = 5.55555 \times 10^{-5} = \mathbf{5.6 \times 10^{-5}\ M}$$

Since $[OH^-] > [H_3O^+]$, the solution is **basic**.

18.3A   Plan: NaOH is a strong base that dissociates completely in water. Subtract pOH from 14.00 to find the pH, and calculate inverse logs of pOH and pH to find $[OH^-]$ and $[H_3O^+]$, and respectively. Alternatively, use the ion product constant of water (K_w) at 25°C to calculate $[H_3O^+]$ from $[OH^-]$.

Solution:

pH + pOH = 14.00

pH = 14.00 – pOH = 14.00 – 4.48 = **9.52**

pOH = –log [OH⁻]

$[OH^-] = 10^{-pOH} = 10^{-4.48} = 3.3113 \times 10^{-5} = \mathbf{3.3 \times 10^{-5}}$ *M*

pH = –log [H₃O⁺]

$[H_3O^+] = 10^{-pH} = 10^{-9.52} = 3.01995 \times 10^{-10} = \mathbf{3.0 \times 10^{-10}}$ *M*

or $[H_3O^+] = \dfrac{K_w}{[OH^-]} = \dfrac{1.0 \times 10^{-14}}{3.3113 \times 10^{-5}} = 3.01996 \times 10^{-10} = \mathbf{3.0 \times 10^{-10}}$ *M*

18.3B   Plan: HCl is a strong acid that dissociates completely in water. Subtract pH from 14.00 to find the pOH, and calculate inverse logs of pH and pOH to find [H₃O⁺] and [OH⁻], respectively.

Solution:

pH + pOH = 14.00

pOH = 14.00 – 2.28 = **11.72**

pH = –log [H₃O⁺]

$[H_3O^+] = 10^{-pH} = 10^{-2.28} = 5.2481 \times 10^{-3} = \mathbf{5.2 \times 10^{-3}}$ *M*

pOH = –log [OH⁻]

$[OH^-] = 10^{-pOH} = 10^{-11.72} = 1.9055 \times 10^{-12} = \mathbf{1.9 \times 10^{-12}}$ *M*

18.4A   Plan: Identify the conjugate pairs by first identifying the species that donates H⁺ (the acid) in either reaction direction. The other reactant accepts the H⁺ and is the base. The acid has one more H and +1 greater charge than its conjugate base.

Solution:

a) CH₃COOH has one more H⁺ than CH₃COO⁻. H₃O⁺ has one more H⁺ than H₂O. Therefore, CH₃COOH and H₃O⁺ are the acids, and CH₃COO⁻ and H₂O are the bases. The conjugate acid/base pairs are **CH₃COOH/CH₃COO⁻** and **H₃O⁺/H₂O**.

b) H₂O donates a H⁺ and acts as the acid. F⁻ accepts the H⁺ and acts as the base. In the reverse direction, HF acts as the acid and OH⁻ acts as the base. The conjugate acid/base pairs are **H₂O/OH⁻** and **HF/F⁻**.

18.4B   Plan: To derive the formula of a conjugate base, remove one H from the acid and decrease the charge by 1 (acids donate H⁺). To derive the formula of a conjugate acid, add an H and increase the charge by 1 (bases accept H⁺).

Solution:

a) Adding a H⁺ to HSO₃⁻ gives the formula of the conjugate acid: **H₂SO₃**.

b) Removing a H⁺ from C₅H₅NH⁺ gives the formula of the conjugate base: **C₅H₅N**

c) Adding a H⁺ to CO₃²⁻ gives the formula of the conjugate acid: **HCO₃⁻**.

d) Removing a H⁺ from HCN gives the formula of the conjugate base: **CN⁻**.

18.5A   Plan: The two possible reactions involve reacting the acid from one conjugate pair with the base from the other conjugate pair. The reaction that favors the products ($K_c > 1$) is the one in which the stronger acid produces the weaker acid. The reaction that favors reactants ($K_c < 1$) is the reaction is which the weaker acid produces the stronger acid.

Solution:

a) The conjugate pairs are H₂SO₃ (acid)/ HSO₃⁻ (base) and HCO₃⁻ (acid)/ CO₃²⁻ (base). Two reactions are possible:

　　　(1) **H₂SO₃ + CO₃²⁻ ⇌ HSO₃⁻ + HCO₃⁻**　　and　　(2) HSO₃⁻ + HCO₃⁻ ⇌ H₂SO₃ + CO₃²⁻

The first reaction is the reverse of the second. Both acids are weak. Of the two, H₂SO₃ is the stronger acid. Reaction (1) with the stronger acid producing the weaker acid favors products and $K_c > 1$. Reaction (2) with the weaker acid forming the stronger acid favors the reactants and $K_c < 1$. Therefore, reaction 1 is the reaction in which $K_c > 1$.

b) The conjugate pairs are HF (acid)/F⁻ (base) and HCN (acid)/CN⁻ (base). Two reactions are possible:

　　　(1) HF + CN⁻ ⇌ F⁻ + HCN　　and　　(2) F⁻ + HCN ⇌ HF + CN⁻

The first reaction is the reverse of the second. Both acids are weak. Of the two, HF is the stronger acid. Reaction (1) with the stronger acid producing the weaker acid favors products and $K_c > 1$. Reaction (2) with the weaker acid forming the stronger acid favors the reactants and $K_c < 1$. Therefore, reaction 2 is the reaction in which $K_c < 1$.

18.5B   Plan: For a), write the reaction that shows the reaction of ammonia with water; for b), write a reaction between ammonia and HCl; for c), write the reaction between the ammonium ion and NaOH to produce ammonia.
Solution:
a) The following equation describes the dissolution of ammonia in water:

$$NH_3(g) \quad + \quad H_2O(l) \quad \leftrightarrows \quad NH_4^+(aq) \quad + \quad OH^-(aq)$$
weak base   weak acid $\leftarrow$ stronger acid   strong base

Ammonia is a known weak base, so it makes sense that it accepts a H^+ from H_2O. The reaction arrow indicates that the equilibrium lies to the left because the question states, "you smell ammonia" (NH_4^+ and OH^- are odorless). NH_4^+ and OH^- are the stronger acid and base, so the reaction proceeds to the formation of the weaker acid and base.
b) The addition of excess HCl results in the following equation:

$$NH_3(g) \quad + \quad H_3O^+(aq; \text{ from HCl}) \quad \leftrightarrows \quad NH_4^+(aq) \quad + \quad H_2O(l)$$
stronger base   strong acid $\rightarrow$ weak acid   weak base

HCl is a strong acid and is much stronger than NH_4^+. Similarly, NH_3 is a stronger base than H_2O. The reaction proceeds to produce the weak acid and base, and thus the odor from NH_3 disappears.
c) The solution in (b) is mostly NH_4^+ and H_2O. The addition of excess NaOH results in the following equation:

$$NH_4^+(aq) \quad + \quad OH^-(aq; \text{ from NaOH}) \quad \leftrightarrows \quad NH_3(g) \quad + \quad H_2O(l)$$
stronger acid   strong base $\rightarrow$ weak base   weak acid

NH_4^+ and OH^- are the stronger acid and base, respectively, and drive the reaction towards the formation of the weaker base and acid, $NH_3(g)$ and H_2O, respectively. The reaction direction explains the return of the ammonia odor.

18.6A   Plan: If HA is a stronger acid than HB, $K_c > 1$ and more HA molecules will produce HB molecules. If HB is a stronger acid than HA, $K_c < 1$ and more HB molecules will produce HA molecules.
Solution:
There are more HB molecules than there are HA molecules, so the equilibrium lies to the right and **$K_c > 1$. HA is the stronger acid.**

18.6B   Plan: Because HD is a stronger acid than HC, the reaction of HD and C^- will have $K_c > 1$, and there should be more HC molecules than HD molecules at equilibrium.
Solution:
There are more green/white acid molecules in the solution than black/white acid molecules. Therefore, the green/white acid molecules represent HC, and the black/white acid molecules represent HD. **The green spheres represent C^-, and the black spheres represent D^-.** Because the reaction of the stronger acid HD with C^- will have $K_c > 1$, the reverse reaction (HC + D^-) will have **$K_c < 1$.**

18.7A   Plan: Write a balanced equation for the dissociation of NH_4^+ in water. Using the given information, construct a reaction table that describes the initial and equilibrium concentrations. Construct an equilibrium expression and make assumptions where possible to simplify the calculations. Since the pH is known, $[H_3O^+]$ can be found; that value can be substituted into the equilibrium expression.
Solution:

	$NH_4^+(aq)$	+	$H_2O(l)$	$\leftrightarrows$	$H_3O^+(aq)$	+	$NH_3(g)$
Initial	0.2 M		———		0		0
Change	−x		———		+x		+x
Equilibrium	0.2 − x		———		x		x

The initial concentration of $NH_4^+ = 0.2\ M$ because each mole of NH_4Cl completely dissociates to form one mole of NH_4^+.

$$x = [H_3O^+] = [NH_3] = 10^{-pH} = 10^{-5.0} = 1.0\times10^{-5}\ M$$

$$K_a = \frac{[NH_3][H_3O^+]}{[NH_4^+]} = \frac{x^2}{(0.2 - x)} = \frac{(1.0\times10^{-5})(1.0\times10^{-5})}{(0.2 - 1.0\times10^{-5})} = 5.00250\times10^{-10} = \mathbf{5\times10^{-10}}$$

18.7B Plan: Write a balanced equation for the dissociation of acrylic acid in water. Using the given information, construct a reaction table that describes the initial and equilibrium concentrations. Construct an equilibrium expression. Since the pH is known, $[H_3O^+]$ can be found; that value can be used to find the equilibrium concentrations of all substances, which can then be substituted into the equilibrium expression to solve for the value of K_a.
Solution:

	$H_2C=CHCOOH(aq)$	$+$	$H_2O(l)$	$\leftrightarrows$	$H_3O^+(aq)$	$+$	$H_2C=CHCOO^-(aq)$
Initial	0.3 M		———		0		0
Change	$-x$		———		$+x$		$+x$
Equilibrium	$0.3 - x$		———		x		x

According to the information given in the problem, pH at equilibrium = 2.43.
$[H_3O^+]_{eq} = 10^{-pH} = 10^{-2.43} = 3.7154\times10^{-3} = 3.7\times10^{-3}\ M = x$
Thus, $[H_3O^+] = [H_2C=CHCOO^-] = 3.7\times10^{-3}\ M$
$[H_2C=CHCOOH] = (0.30 - x) = (0.30 - 3.7\times10^{-3})\ M = 0.2963\ M$

$$K_a = \frac{[H_2CHCOO^-][H_3O^+]}{[H_2CHCOOH]}$$

$$K_a = \frac{(3.7\times10^{-3})(3.7\times10^{-3})}{(0.2963)} = 4.6203\times10^{-5} = \mathbf{4.6\times10^{-5}}$$

18.8A Plan: Write a balanced equation for the dissociation of HOCN in water. Using the given information, construct a table that describes the initial and equilibrium concentrations. Construct an equilibrium expression and solve the quadratic expression for x, the concentration of H_3O^+. Use the concentration of the hydronium ion to solve for pH.
Solution:

	$HOCN(aq)$	$+$	$H_2O(l)$	$\leftrightarrows$	$H_3O^+(aq)$	$+$	$OCN^-(aq)$
Initial	0.10 M		———		0		0
Change	$-x$		———		$+x$		$+x$
Equilibrium	$0.10 - x$		———		x		x

$$K_a = 3.5\times10^{-4} = \frac{[OCN^-][H_3O^+]}{[HOCN]} = \frac{x^2}{(0.10 - x)}$$

In this example, the dissociation of HOCN is not negligible in comparison to the initial concentration. Therefore, the equilibrium expression is solved using the quadratic formula.

$$x^2 = 3.5\times10^{-4}\,(0.10 - x)$$
$$x^2 = 3.5\times10^{-5} - 3.5\times10^{-4}\,x$$
$$x^2 + 3.5\times10^{-4}\,x - 3.5\times10^{-5} = 0 \qquad (ax^2 + bx + c = 0)$$
$$a = 1 \quad b = 3.5\times10^{-4} \quad c = -3.5\times10^{-5}$$

$$x = \frac{-b \pm \sqrt{b^2 - 4ac}}{2a}$$

$$x = \frac{-3.5\times10^{-4} \pm \sqrt{(3.5\times10^{-4})^2 - 4(1)(-3.5\times10^{-5})}}{2(1)}$$

$x = 5.7436675\times10^{-3} = \mathbf{5.7\times10^{-3}\ M\ H_3O^+}$
$pH = -\log\,[H_3O^+] = -\log\,[5.7436675\times10^{-3}] = 2.2408 = \mathbf{2.24}$

18.8B Plan: Write a balanced equation for the dissociation of C_6H_5COOH in water. Using the given information, construct a table that describes the initial and equilibrium concentrations. Use pK_a to solve for the value of K_a. Construct an equilibrium expression, use simplifying assumptions when possible to solve for x, the concentration of H_3O^+. Use the concentration of the hydronium ion to solve for pH.

Solution:

	$C_6H_5COOH(aq)$ +	$H_2O(l)$ $\leftrightarrows$	$H_3O^+(aq)$ +	$C_6H_5COO^-(aq)$
Initial	0.25 M	———	0	0
Change	−x	———	+x	+x
Equilibrium	0.25 − x	———	x	x

$K_a = 10^{-pK_a} = 10^{-4.20} = 6.3096 \times 10^{-5} = 6.3 \times 10^{-5}$

$K_a = 6.3 \times 10^{-5} = \dfrac{[C_6H_5HCOO^-][H_3O^+]}{[C_6H_5COOH]} = \dfrac{(x)(x)}{(0.25 - x)}$ \qquad Assume x is negligible so $0.25 - x \approx 0.25$

$\dfrac{(x)(x)}{(0.25)} = 6.3 \times 10^{-5}$

$x^2 = (6.3 \times 10^{-5})(0.25); \ x = 3.9686 \times 10^{-3} = 4.0 \times 10^{-3}$

Check the assumption by calculating the % error:

$\dfrac{4.0 \times 10^{-3}}{0.25} (100) = 1.6\%$ which is smaller than 5%, so the assumption is valid.

At equilibrium $[H_3O^+]_{eq} = \mathbf{4.0 \times 10^{-3}}$ **M**

pH = −log $[H_3O^+]$ = −log $[3.9686 \times 10^{-3}]$ = 2.40136 = **2.40**

18.9A Plan: Write the acid-dissociation reaction and the expression for K_a. Set up a reaction table in which x = the concentration of the dissociated acid and also $[H_3O^+]$. Use the expression for K_a to solve for x, the concentration of cyanide ion at equilibrium. Then use the initial concentration of HCN and the equilibrium concentration of CN^- to find % dissociation.

Solution:

Concentration	HCN(aq) +	$H_2O(l)$ $\leftrightarrows$	$H_3O^+(aq)$ +	$CN^-(aq)$
Initial	0.75	—	0	0
Change	−x		+x	+x
Equilibrium	0.75 − x		x	x

$K_a = 6.2 \times 10^{-10} = \dfrac{[CN^-][H_3O^+]}{[HCN]}$

$K_a = 6.2 \times 10^{-10} = \dfrac{[x][x]}{[0.75 - x]}$ \qquad Assume x is small compared to 0.75.

$K_a = 6.2 \times 10^{-10} = \dfrac{[x][x]}{[0.75]}$

$x = 2.1564 \times 10^{-5} = 2.2 \times 10^{-5} M$

Check the assumption by calculating the % error:

$\dfrac{2.2 \times 10^{-5}}{0.75} (100) = 0.0029\%$ which is smaller than 5%, so the assumption is valid.

Percent HCN dissociated = $\dfrac{[HCN]_{dissoc}}{[HCN]_{init}} (100)$

Percent HCN dissociated = $\dfrac{(2.1564 \times 10^{-5})}{0.75} (100) = .0028752 = \mathbf{0.0029\%}$

18.9B   Plan: Write the acid-dissociation reaction and the expression for K_a. Percent dissociation refers to the amount of the initial concentration of the acid that dissociates into ions. Use the percent dissociation to find the concentration of acid dissociated, which also equals $[H_3O^+]$. HA will be used as the formula of the acid. Set up a reaction table in which x = the concentration of the dissociated acid and $[H_3O^+]$. Substitute [HA], $[A^-]$, and $[H_3O^+]$ into the expression for K_a to find the value of K_a.
Solution:

$HA(aq) + H_2O(l) \rightleftarrows H_3O^+(aq) + A^-(aq)$

$\text{Percent HA} = \dfrac{\text{dissociated acid}}{\text{initial acid}}(100)$

$3.16\% = \dfrac{x}{1.5\,M}(100)$

[Dissociated acid] = x = 0.0474 M

Concentration	$HA(aq)$	$+ H_2O(l)$	$\rightleftarrows$	$H_3O^+(aq)$	$+$	$A^-(aq)$
Initial:	1.5			0		0
Change:	−x			+x		+x
Equilibrium:	1.5 − x			x		x

[Dissociated acid] = x = $[A^-] = [H_3O^+]$ = 0.0474 M

[HA] = 1.5 M − 0.0474 M = 1.4526 M

Solving for K_a.

In the equilibrium expression, substitute the concentrations above and calculate K_a.

$$K_a = \dfrac{[H_3O^+][A^-]}{[HA]} = \dfrac{(0.0474)(0.0474)}{(1.4526)} = 1.5467\times10^{-3} = \mathbf{1.5\times10^{-3}}$$

18.10A Plan: Write the balanced equation and corresponding equilibrium expression for each dissociation reaction. Calculate the equilibrium concentrations of all species and convert $[H_3O^+]$ to pH. Find the equilibrium constant values from Appendix C, $K_{a1} = 5.6\times10^{-2}$ and $K_{a2} = 5.4\times10^{-5}$.
Solution:

$HOOC-COOH(aq) + H_2O(l) \rightleftarrows HOOC-COO^-(aq) + H_3O^+(aq)$

$$K_{a1} = \dfrac{[HC_2O_4^-][H_3O^+]}{[H_2C_2O_4]} = 5.6\times10^{-2}$$

$HOOC-COO^-(aq) + H_2O(l) \rightleftarrows {}^-OOC-COO^-(aq) + H_3O^+(aq)$

$$K_{a2} = \dfrac{[C_2O_4^{2-}][H_3O^+]}{[HC_2O_4^-]} = 5.4\times10^{-5}$$

Assumptions:
1) Since $K_{a1} \gg K_{a2}$, the first dissociation produces almost all of the H_3O^+, so $[H_3O^+]_{eq} = [H_3O^+]$ from $C_2H_2O_4$.
2) Since K_{a1} (5.6×10^{-2}) is fairly large, solve the first equilibrium expression using the quadratic equation.

	$HOOC-COOH(aq)$	$+$	$H_2O(l)$	$\rightleftarrows$	$H_3O^+(aq)$	$+$	$HOOC-COO^-(aq)$
Initial	0.150 M		———		0		0
Change	−x		———		+x		+x
Equilibrium	0.150 − x		———		x		x

$$K_{a1} = \dfrac{[HC_2O_4^-][H_3O^+]}{[H_2C_2O_4]} = \dfrac{x^2}{(0.150 - x)} = 5.6\times10^{-2}$$

$$x^2 + 5.6\times10^{-2}\,x - 8.4\times10^{-3} = 0 \qquad (ax^2 + bx + c = 0)$$

$$x = \dfrac{-5.6\times10^{-2} \pm \sqrt{(5.6\times10^{-2})^2 - 4(1)(-8.4\times10^{-3})}}{2(1)}$$

$$x = 0.067833\ M\ H_3O^+$$

Therefore, **[H₃O⁺] = [HC₂O₄⁻] = 0.068 M** and **pH** = −log (0.067833) = 1.16856 = **1.17**. The oxalic acid concentration at equilibrium is [H₂C₂O₄]$_{init}$ − [H₂C₂O₄]$_{dissoc}$ = 0.150 − 0.067833 = 0.82167 = **0.082 M**.

Solve for [C₂O₄²⁻] by rearranging the K_{a2} expression:

$$K_{a2} = \frac{[C_2O_4^{2-}][H_3O^+]}{[HC_2O_4^-]} = 5.4\times10^{-5}$$

$$[C_2O_4^{2-}] = \frac{K_{a2}[HC_2O_4^-]}{[H_3O^+]} = \frac{(5.4\times10^{-5})(0.067833)}{(0.067833)} = \mathbf{5.4\times10^{-5}\ M}$$

18.10B Plan: Write the balanced equation and corresponding equilibrium expression for each dissociation reaction. Calculate the equilibrium concentrations of all species and convert [H₃O⁺] to pH. Find the equilibrium constant values from Appendix C, $K_{a1} = 4.5\times10^{-7}$ and $K_{a2} = 4.7\times10^{-11}$.

Solution:

$$H_2CO_3(aq) + H_2O(l) \rightleftharpoons HCO_3^-(aq) + H_3O^+(aq)$$

$$K_{a1} = \frac{[H_3O^+][HCO_3^-]}{[H_2CO_3]} = 4.5\times10^{-7}$$

$$HCO_3^-(aq) + H_2O(l) \rightleftharpoons CO_3^{2-}(aq) + H_3O^+(aq)$$

$$K_{a2} = \frac{[H_3O^+][CO_3^{2-}]}{[HCO_3^-]} = 4.7\times10^{-11}$$

Assumption:
1) Since $K_{a1} >> K_{a2}$, the first dissociation produces almost all of the H₃O⁺, so [H₃O⁺]$_{eq}$ = [H₃O⁺] from H₂CO₃.
2) Because K_{a1} (4.7×10⁻⁷) is fairly small, [H₂CO₃]$_{init}$ − x ≈ [H₂CO₃]$_{init}$. Thus,

   [H₂CO₃] = 0.075 M − x ≈ 0.075 M

Solve the first equilibrium expression making the assumption that x is small.

	H₂CO₃(aq)	+	H₂O(l)	$\rightleftharpoons$	H₃O⁺(aq)	+	HCO₃⁻(aq)
Initial	0.075 M		———		0		0
Change	−x		———		+x		+x
Equilibrium	0.075 − x		———		x		x

$$K_{a1} = \frac{[H_3O^+][HCO_3^-]}{[H_2CO_3]} = \frac{(x)(x)}{(0.075-x)} \approx \frac{(x)(x)}{(0.075)} = 4.5\times10^{-7}$$

$$x^2 = (0.075)(4.5\times10^{-7});\ x = 1.8371\times10^{-4} = 1.8\times10^{-4}\ M$$

Check the assumption by calculating the % error:

$$\frac{1.8\times10^{-4}}{0.075}(100) = 0.24\%\ \text{which is smaller than 5\%, so the assumption is valid.}$$

Therefore, **[H₃O⁺] = [HCO₃⁻] = 1.8×10⁻⁴ M** and **pH** = −log (1.8371×10⁻⁴) = 3.73587 = **3.74**. The carbonic acid concentration at equilibrium is [H₂CO₃]$_{init}$ − [H₂CO₃]$_{dissoc}$ = 0.075 − 1.8371×10⁻⁴ = 0.07482 = **0.075 M = [H₂CO₃]**.

Solve for [CO₃²⁻] by rearranging the K_{a2} expression:

$$K_{a2} = \frac{[H_3O^+][CO_3^{2-}]}{[HCO_3^-]} = 4.7\times10^{-11}$$

$$[CO_3^{2-}] = \frac{K_{a2}[HCO_3^-]}{[H_3O^+]} = \frac{(4.7\times10^{-11})(1.8371\times10^{-4})}{(1.8371\times10^{-4})} = \mathbf{4.7\times10^{-11}\ M}$$

18.11A Plan: Pyridine contains a nitrogen atom that accepts H⁺ from water to form OH⁻ ions in aqueous solution. Write a balanced equation and equilibrium expression for the reaction, convert pK_b to K_b, make simplifying assumptions (if valid), and solve for [OH⁻]. Calculate [H₃O⁺] using [H₃O⁺][OH⁻] = 1.0×10⁻¹⁴ and convert to pH.

Solution:

$K_b = 10^{-pK_b} = 10^{-8.77} = 1.69824 \times 10^{-9}$

	$C_5H_5N(aq)$	$+$	$H_2O(l)$	$\leftrightarrows$	$C_5H_5NH^+(aq)$	$+$	$OH^-(aq)$
Initial	0.10 M		———		0		0
Change	$-x$		———		$+x$		$+x$
Equilibrium	$0.10 - x$		———		x		x

$$K_b = \frac{[C_5N_5NH^+][OH^-]}{[C_5N_5N]} = 1.69824 \times 10^{-9}$$

Assume that $0.10 - x \approx 0.10$.

$$K_b = \frac{[C_5N_5NH^+][OH^-]}{[C_5N_5N]} = \frac{x^2}{(0.10)} = 1.69824 \times 10^{-9}$$

$$x = 1.303165 \times 10^{-5} = 1.3 \times 10^{-5} \ M = [OH^-] = [C_5H_5NH^+]$$

Since $\dfrac{[OH^-]}{[C_5H_5N_5]}(100) = \dfrac{1.303265 \times 10^{-5}}{0.10}(100) = 0.01313$ which $< 5\%$, the assumption that the dissociation of

$C_5H_5N_5$ is small is valid.

$$[H_3O^+] = \frac{K_w}{[OH^-]} = \frac{1.0 \times 10^{-14}}{1.303165 \times 10^{-5}} = 7.67362 \times 10^{-10} \ M$$

$pH = -\log (7.67362 \times 10^{-10}) = 9.1149995 = \mathbf{9.11}$ (Since pyridine is a weak base, a pH > 7 is expected.)

18.11B **Plan:** Amphetamine contains a nitrogen atom that accepts H^+ from water to form OH^- ions in aqueous solution. Write a balanced equation and equilibrium expression for the reaction, make simplifying assumptions (if valid), and solve for $[OH^-]$. Calculate $[H_3O^+]$ using $[H_3O^+][OH^-] = 1.0 \times 10^{-14}$ and convert to pH. In the information below, the symbol B will be used to represent the formula of amphetamine.

Solution:

	$B(aq)$	$+$	$H_2O(l)$	$\leftrightarrows$	$BH^+(aq)$	$+$	$OH^-(aq)$
Initial	0.075 M		———		0		0
Change	$-x$		———		$+x$		$+x$
Equilibrium	$0.075 - x$		———		x		x

$$K_b = \frac{[BH][OH^-]}{[B]} = 6.3 \times 10^{-5}$$

Assume that $0.075 - x \approx 0.075$.

$$K_b = \frac{[BH][OH^-]}{[B]} = \frac{x^2}{(0.075)} = 6.3 \times 10^{-5}$$

$$x = 0.0021737 = 2.2 \times 10^{-3} \ M = [OH^-] = [BH^+]$$

Check the assumption by calculating the % error:

$$\frac{2.2 \times 10^{-3}}{0.075}(100) = 2.9\% \text{ which is smaller than } 5\%, \text{ so the assumption is valid.}$$

$$[H_3O^+] = \frac{K_w}{[OH^-]} = \frac{1.0 \times 10^{-14}}{2.1737 \times 10^{-3}} = 4.60045 \times 10^{-12} \ M$$

$pH = -\log (4.60045 \times 10^{-12}) = 11.3372 = \mathbf{11.34}$

Since amphetamine is a weak base, a pH > 7 is expected.

18.12A **Plan:** The hypochlorite ion, ClO^-, acts as a weak base in water. Write a balanced equation and equilibrium expression for this reaction. The K_b of ClO^- is calculated from the K_a of its conjugate acid, hypochlorous acid, $HClO$ (from Appendix C, $K_a = 2.9 \times 10^{-8}$). Make simplifying assumptions (if valid), solve for $[OH^-]$, convert to $[H_3O^+]$ and calculate pH.

Solution:

	$ClO^-(aq)$	+	$H_2O(l)$	$\rightleftarrows$	$HClO(aq)$	+	$OH^-(aq)$
Initial	0.20 M		———		0		0
Change	−x		———		+x		+x
Equilibrium	0.20 − x		———		x		x

$$K_b = \frac{[HClO][OH^-]}{[ClO^-]}$$

$$K_b = \frac{K_w}{K_a} = \frac{1.0\times10^{-14}}{2.9\times10^{-8}} = 3.448276\times10^{-7}$$

Since K_b is very small, assume $[ClO^-]_{eq} = 0.20 - x \approx 0.2$.

$$K_b = \frac{[HClO][OH^-]}{[ClO^-]} = \frac{x^2}{(0.20)} = 3.448276\times10^{-7}$$

$$x = 2.6261\times10^{-4}$$

Therefore, $[HClO] = [OH^-] = 2.6\times10^{-4}\ M$.

Since $\frac{[OH^-]}{[ClO^-]}(100) = \frac{2.6261\times10^{-4}}{0.20}(100) = 0.13\%$ which $< 5\%$, the assumption that the dissociation of ClO^- is small is valid.

$$[H_3O^+] = \frac{1.0\times10^{-14}}{2.6261\times10^{-4}} = 3.8079\times10^{-11}\ M$$

$pH = -\log(3.8079\times10^{-11}) = 10.4193 = \mathbf{10.42}$ (Since hypochlorite ion is a weak base, a pH > 7 is expected.)

18.12B **Plan:** The nitrite ion, NO_2^-, acts as a weak base in water. Write a balanced equation and equilibrium expression for this reaction. The K_b of NO_2^- is calculated from the K_a of its conjugate acid, nitrous acid, HNO_2 (from Appendix C, $K_a = 7.1\times10^{-4}$). Make simplifying assumptions (if valid), solve for $[OH^-]$, convert to $[H_3O^+]$ and calculate pH.
Solution:

	$NO_2^-(aq)$	+	$H_2O(l)$	$\rightleftarrows$	$HNO_2(aq)$	+	$OH^-(aq)$
Initial	0.80 M		———		0		0
Change	−x		———		+x		+x
Equilibrium	0.80 − x		———		x		x

$$K_b = \frac{[HNO_2][OH^-]}{[NO_2^-]}$$

$$K_b = \frac{K_w}{K_a} = \frac{1.0\times10^{-14}}{7.1\times10^{-4}} = 1.40845\times10^{-11}$$

Since K_b is very small, assume $[NO_2^-]_{eq} = 0.80 - x \approx 0.8$.

$$K_b = \frac{[HNO_2][OH^-]}{[NO_2^-]} = \frac{x^2}{(0.80)} = 1.40845\times10^{-11}$$

$$x = 3.356725\times10^{-6}\ M$$

Check the assumption by calculating the % error:

$\frac{3.3\times10^{-6}}{0.80}(100) = 0.00041\%$ which is smaller than 5%, so the assumption is valid.

Therefore, $[HNO_2] = [OH^-] = 3.356725\times10^{-6}\ M$.

$$[H_3O^+] = \frac{1.0\times10^{-14}}{3.356725\times10^{-6}} = 2.97909\times10^{-9}\ M$$

$pH = -\log(2.97909\times10^{-9}) = 8.5259 = \mathbf{8.53}$
Since nitrite ion is a weak base, a pH > 7 is expected.

18.13A Plan: Examine the cations and anions in each compound. If the cation is the cation of a strong base, the cation gives a neutral solution; the cation of a weak base gives an acidic solution. An anion of a strong acid gives a neutral solution while an anion of a weak acid is basic in solution.
Solution:
a) The ions are K^+ and ClO_2^-; the K^+ is from the strong base KOH, and does not react with water. The ClO_2^- is from the weak acid $HClO_2$, so it reacts with water to produce OH^- ions. Since the base is strong and the acid is weak, the salt derived from this combination will produce a **basic** solution.
K^+ does not react with water.
$$ClO_2^-(aq) + H_2O(l) \leftrightarrows HClO_2(aq) + OH^-(aq)$$
b) The ions are $CH_3NH_3^+$ and NO_3^-; $CH_3NH_3^+$ is derived from the weak base methylamine, CH_3NH_2. Nitrate ion, NO_3^-, is derived from the strong acid HNO_3 (nitric acid). A salt derived from a weak base and strong acid produces an **acidic** solution.
NO_3^- does not react with water.
$$CH_3NH_3^+(aq) + H_2O(l) \leftrightarrows CH_3NH_2(aq) + H_3O^+(aq)$$
c) The ions are Rb^+ and Br^-. Rubidium ion is derived from rubidium hydroxide, RbOH, which is a strong base because Rb is a Group 1A(1) metal. Bromide ion is derived from hydrobromic acid, HBr, a strong hydrohalic acid. Since both the base and acid are strong, the salt derived from this combination will produce a **neutral** solution.
Neither Rb^+ nor Br^- react with water.

18.13B Plan: Examine the cations and anions in each compound. If the cation is the cation of a strong base, the cation gives a neutral solution; the cation of a weak base gives an acidic solution. An anion of a strong acid gives a neutral solution while an anion of a weak acid is basic in solution.
Solution:
a) The ions are Fe^{3+} and Br^-. The Br^- is the anion of the strong acid HBr, so it does not react with water. The Fe^{3+} ion is small and highly charged, so the hydrated ion, $Fe(H_2O)_6^{3+}$, reacts with water to produce H_3O^+. Since the base is weak and the acid is strong, the salt derived from this combination will produce an **acidic** solution.
Br^- does not react with water.
$$Fe(H_2O)_6^{3+}(aq) + H_2O(l) \leftrightarrows Fe(H_2O)_5OH^{2+}(aq) + H_3O^+(aq)$$
b) The ions are Ca^{2+} and NO_2^-; the Ca^{2+} is from the strong base $Ca(OH)_2$, and does not react with water. The NO_2^- is from the weak acid HNO_2, so it reacts with water to produce OH^- ions. Since the base is strong and the acid is weak, the salt derived from this combination will produce a **basic** solution.
Ca^{2+} does not react with water.
$$NO_2^-(aq) + H_2O(l) \leftrightarrows HNO_2(aq) + OH^-(aq)$$
c) The ions are $C_6H_5NH_3^+$ and I^-; $C_6H_5NH_3^+$ is derived from the weak base aniline, $C_6H_5NH_2$. Iodide ion, I^-, is derived from the strong acid HI (hydroiodic acid). A salt derived from a weak base and strong acid produces an **acidic** solution.
I^- does not react with water.
$$C_6H_5NH_3^+(aq) + H_2O(l) \leftrightarrows C_6H_5NH_2(aq) + H_3O^+(aq)$$

18.14A Plan: Examine the cations and anions in each compound. If the cation is the cation of a strong base, the cation gives a neutral solution; the cation of a weak base gives an acidic solution. An anion of a strong acid gives a neutral solution while an anion of a weak acid is basic in solution.
Solution:
a) The two ions that comprise this salt are cupric ion, Cu^{2+}, and acetate ion, CH_3COO^-. Metal ions are acidic in water. Assume that the hydrated cation is $Cu(H_2O)_6^{2+}$. The K_a is found in Appendix C.
$$Cu(H_2O)_6^{2+}(aq) + H_2O(l) \leftrightarrows Cu(H_2O)_5OH^+(aq) + H_3O^+(aq) \qquad K_a = 3\times10^{-8}$$
Acetate ion acts likes a base in water. The K_b is calculated from the K_a of acetic acid (1.8×10^{-5}):
$$K_b = \frac{K_w}{K_a} = \frac{1.0\times10^{-14}}{1.8\times10^{-5}} = 5.6\times10^{-10}$$
$$CH_3COO^-(aq) + H_2O(l) \leftrightarrows CH_3COOH(aq) + OH^-(aq) \qquad K_b = 5.6\times10^{-10}$$
$Cu(H_2O)_6^{2+}$ is a better proton donor than CH_3COO^- is a proton acceptor (i.e., $K_a > K_b$), so a solution of $Cu(CH_3COO)_2$ is **acidic**.

b) The two ions that comprise this salt are ammonium ion, NH_4^+, and fluoride ion, F^-. Ammonium ion is the acid

of NH_3 with $K_a = \dfrac{K_w}{K_b} = \dfrac{1.0 \times 10^{-14}}{1.76 \times 10^{-5}} = 5.7 \times 10^{-10}$.

$$NH_4^+(aq) + H_2O(l) \leftrightarrows NH_3(aq) + H_3O^+(aq) \qquad\qquad K_a = 5.7 \times 10^{-10}$$

Fluoride ion is the base with $K_b = \dfrac{K_w}{K_a} = \dfrac{1.0 \times 10^{-14}}{6.8 \times 10^{-4}} = 1.5 \times 10^{-11}$.

$$F^-(aq) + H_2O(l) \leftrightarrows HF(aq) + OH^-(aq) \qquad\qquad K_b = 1.5 \times 10^{-11}$$

Since $K_a > K_b$, a solution of NH_4F is **acidic**.

c) The ions are K^+ and $HC_6H_6O_6^-$; the K^+ is from the strong base KOH, and does not react with water. The $HC_6H_6O_6^-$ can react as an acid:

$$HC_6H_6O_6^-(aq) + H_2O(l) \leftrightarrows C_6H_6O_6^{2-}(aq) + H_3O^+(aq) \qquad K_a = 5 \times 10^{-12} \text{ (from Appendix C)}$$

$HC_6H_6O_6^-$ can also react as a base. Its K_b value can be found by using the K_a of its conjugate acid, $H_2C_6H_6O_6$.

$$HC_6H_6O_6^-(aq) + H_2O(l) \leftrightarrows H_2C_6H_6O_6(aq) + OH^-(aq) \qquad K_b = \dfrac{K_w}{K_a \text{ of } H_2C_6H_6O_6} = \dfrac{1.0 \times 10^{-14}}{1.0 \times 10^{-5}} = 1.0 \times 10^{-9}$$

Since $K_b > K_a$, a solution of $KHC_6H_6O_6$ is **basic**.

18.14B Plan: Examine the cations and anions in each compound. If the cation is the cation of a strong base, the cation gives a neutral solution; the cation of a weak base gives an acidic solution. An anion of a strong acid gives a neutral solution while an anion of a weak acid is basic in solution.
Solution:
a) The two ions that comprise this salt are sodium ion, Na^+, and bicarbonate ion, HCO_3^-. The Na^+ is from the strong base NaOH, and does not react with water.
The HCO_3^- can react as an acid:

$$HCO_3^-(aq) + H_2O(l) \leftrightarrows CO_3^{2-}(aq) + H_3O^+(aq) \qquad\qquad K_a = 4.7 \times 10^{-11}$$

HCO_3^- can also react as a base. Its K_b value can be found by using the K_a of its conjugate acid, H_2CO_3.

$$HCO_3^-(aq) + H_2O(l) \leftrightarrows H_2CO_3(aq) + OH^-(aq) \qquad\qquad K_b = \dfrac{K_w}{K_a} = \dfrac{1.0 \times 10^{-14}}{4.5 \times 10^{-7}} = 2.2 \times 10^{-8}$$

Since $K_b > K_a$, a solution of $NaHCO_3$ is **basic**.
b) The two ions that comprise this salt are anilinium ion, $C_6H_5NH_3^+$, and nitrite ion, NO_2^-.

Anilinium ion is the acid of $C_6H_5NH_2$ with $K_a = \dfrac{K_w}{K_b} = \dfrac{1.0 \times 10^{-14}}{4.0 \times 10^{-10}} = 2.5 \times 10^{-5}$.

$$C_6H_5NH_3^+(aq) + H_2O(l) \leftrightarrows C_6H_5NH_2(aq) + H_3O^+(aq) \qquad\qquad K_a = 2.5 \times 10^{-5}$$

Nitrite ion is the base with $K_b = \dfrac{K_w}{K_a} = \dfrac{1.0 \times 10^{-14}}{7.1 \times 10^{-4}} = 1.4 \times 10^{-11}$.

$$NO_2^-(aq) + H_2O(l) \leftrightarrows HNO_2(aq) + OH^-(aq) \qquad\qquad K_b = 1.4 \times 10^{-11}$$

Since $K_a > K_b$, a solution of $C_6H_5NH_3NO_2$ is **acidic**.
c) The ions are Na^+ and $H_2PO_4^-$; the Na^+ is from the strong base NaOH, and does not react with water. The $H_2PO_4^-$ can react as an acid:

$$H_2PO_4^-(aq) + H_2O(l) \leftrightarrows HPO_4^{2-}(aq) + H_3O^+(aq) \qquad K_a = 6.3 \times 10^{-8}$$

$H_2PO_4^-$ can also react as a base. Its K_b value can be found by using the K_a of its conjugate acid, H_3PO_4.

$$H_2PO_4^-(aq) + H_2O(l) \leftrightarrows H_3PO_4(aq) + OH^-(aq) \qquad\qquad K_b = \dfrac{K_w}{K_a} = \dfrac{1.0 \times 10^{-14}}{7.2 \times 10^{-3}} = 1.4 \times 10^{-12}$$

Since $K_a > K_b$, a solution of NaH_2PO_4 is **acidic**.

18.15A Plan: A Lewis acid is an electron-pair acceptor while a Lewis base is an electron-pair donor.
Solution:
a)

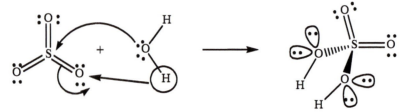

trigonal planar   tetrahedral

Hydroxide ion, OH^-, donates an electron pair and is the Lewis base; $Al(OH)_3$ accepts the electron pair and is the Lewis acid. Note the change in geometry caused by the formation of the adduct.
b)

Sulfur trioxide accepts the electron pair and is the Lewis acid. Water donates an electron pair and is the Lewis base.
c)

Co^{3+} accepts six electron pairs and is the Lewis acid. Ammonia donates an electron pair and is the Lewis base.

18.15B Plan: A Lewis acid is an electron-pair acceptor while a Lewis base is an electron-pair donor.
Solution:
a) $B(OH)_3$ is the Lewis acid because it is accepting electron pairs from water, the Lewis base.
b) Cd^{2+} accepts four electron pairs and is the Lewis acid. Each iodide ion donates an electron pair and is the Lewis base.
c) Each fluoride ion donates an electron pair to form a bond with boron in SiF_6^{2-}. The fluoride ion is the Lewis base and the boron tetrafluoride is the Lewis acid.

END–OF–CHAPTER PROBLEMS

18.2   All Arrhenius acids contain hydrogen and produce hydronium ion (H_3O^+) in aqueous solution. All Arrhenius bases contain an OH group and produce hydroxide ion (OH^-) in aqueous solution. Neutralization occurs when each H_3O^+ molecule combines with an OH^- molecule to form two molecules of H_2O. Chemists found that the ΔH_{rxn} was independent of the combination of strong acid with strong base. In other words, the reaction of any strong base with any strong acid always produced 56 kJ/mol ($\Delta H = -56$ kJ/mol). This was consistent with Arrhenius's hypothesis describing neutralization, because all other counter ions (those present from the dissociation of the strong acid and base) were spectators and did not participate in the overall reaction.

18.4   Strong acids and bases dissociate completely into their ions when dissolved in water. Weak acids only partially dissociate. The characteristic property of all weak acids is that a significant number of the acid molecules are not dissociated. For a strong acid, the concentration of hydronium ions produced by dissolving the acid is equal to the initial concentration of the undissociated acid. For a weak acid, the concentration of hydronium ions produced when the acid dissolves is less than the initial concentration of the acid.

18.5   **Plan:** Recall that an Arrhenius acid contains hydrogen and produces hydronium ion (H_3O^+) in aqueous solution.
Solution:
a) Water, H_2O, is an **Arrhenius acid** because it produces H_3O^+ ion in aqueous solution. Water is also an Arrhenius base because it produces the OH⁻ ion as well.
b) Calcium hydroxide, $Ca(OH)_2$ is a base, not an acid.
c) Phosphorous acid, H_3PO_3, is a weak **Arrhenius acid**. It is weak because the number of O atoms equals the number of ionizable H atoms.
d) Hydroiodic acid, HI, is a strong **Arrhenius acid**.

18.7   **Plan:** All Arrhenius bases contain an OH group and produce hydroxide ion (OH⁻) in aqueous solution.
Solution:
Barium hydroxide, $Ba(OH)_2$, and potassium hydroxide, KOH, (**b and d**) are Arrhenius bases because they contain hydroxide ions and form OH⁻ when dissolved in water. H_3AsO_4 and HClO, (a) and (c), are Arrhenius acids, not bases.

18.9   **Plan:** K_a is the equilibrium constant for an acid dissociation which has the generic equation
$HA(aq) + H_2O(l) \leftrightharpoons H_3O^+(aq) + A^-(aq)$. The K_a expression is $\dfrac{[H_3O^+][A^-]}{[HA]}$. [$H_2O$] is treated as a constant and omitted from the expression. Write the acid-dissociation reaction for each acid, following the generic equation, and then write the K_a expression.
Solution:
a) $HCN(aq) + H_2O(l) \leftrightharpoons H_3O^+(aq) + CN^- (aq)$
$$K_a = \frac{[H_3O^+][CN^-]}{[HCN]}$$
b) $HCO_3^- (aq) + H_2O(l) \leftrightharpoons H_3O^+(aq) + CO_3^{2-} (aq)$
$$K_a = \frac{[H_3O^+][CO_3^{2-}]}{[HCO_3^-]}$$
c) $HCOOH(aq) + H_2O(l) \leftrightharpoons H_3O^+(aq) + HCOO^- (aq)$
$$K_a = \frac{[H_3O^+][HCOO^-]}{[HCOOH]}$$

18.11   **Plan:** K_a is the equilibrium constant for an acid dissociation which has the generic equation
$HA(aq) + H_2O(l) \leftrightharpoons H_3O^+(aq) + A^-(aq)$. The K_a expression is $\dfrac{[H_3O^+][A^-]}{[HA]}$. [$H_2O$] is treated as a constant and omitted from the expression. Write the acid-dissociation reaction for each acid, following the generic equation, and then write the K_a expression.
Solution:
a) $HNO_2(aq) + H_2O(l) \leftrightharpoons H_3O^+(aq) + NO_2^-(aq)$
$$K_a = \frac{[H_3O^+][NO_2^-]}{[HNO_2]}$$
b) $CH_3COOH(aq) + H_2O(l) \leftrightharpoons H_3O^+(aq) + CH_3COO^-(aq)$
$$K_a = \frac{[H_3O^+][CH_3COO^-]}{[CH_3COOH]}$$

c) $HBrO_2(aq) + H_2O(l) \rightleftharpoons H_3O^+(aq) + BrO_2^-(aq)$

$$K_a = \frac{[H_3O^+][BrO_2^-]}{[HBrO_2]}$$

18.13 Plan: K_a values are listed in the Appendix. The larger the K_a value, the stronger the acid. The K_a value for hydroiodic acid, HI, is not shown because K_a approaches infinity for strong acids and is not meaningful.
Solution:
HI is the strongest acid (it is one of the six strong acids), and acetic acid, CH_3COOH, is the weakest:
$CH_3COOH < HF < HIO_3 < HI$

18.15 Plan: Strong acids are the hydrohalic acids HCl, HBr, HI, and oxoacids in which the number of O atoms exceeds the number of ionizable protons by two or more; these include HNO_3, H_2SO_4, and $HClO_4$. All other acids are weak acids. Strong bases are metal hydroxides (or oxides) in which the metal is a Group 1A(1) metal or Ca, Sr, or Ba in Group 2A(2). Weak bases are NH_3 and amines.
Solution:
a) Arsenic acid, H_3AsO_4, is a **weak acid**. The number of O atoms is four, which exceeds the number of ionizable H atoms, three, by one. This identifies H_3AsO_4 as a weak acid.
b) Strontium hydroxide, $Sr(OH)_2$, is a **strong base**. Soluble compounds containing OH^- ions are strong bases. Sr is a Group 2 metal.
c) HIO is a **weak acid**. The number of O atoms is one, which is equal to the number of ionizable H atoms identifying HIO as a weak acid.
d) Perchloric acid, $HClO_4$, is a **strong acid**. $HClO_4$ is one example of the type of strong acid in which the number of O atoms exceeds the number of ionizable H atoms by more than two.

18.17 Plan: Strong acids are the hydrohalic acids HCl, HBr, HI, and oxoacids in which the number of O atoms exceeds the number of ionizable protons by two or more; these include HNO_3, H_2SO_4, and $HClO_4$. All other acids are weak acids. Strong bases are metal hydroxides (or oxides) in which the metal is a Group 1A(1) metal or Ca, Sr, or Ba in Group 2A(2). Weak bases are NH_3 and amines.
Solution:
a) Rubidium hydroxide, RbOH, is a **strong base** because Rb is a Group 1A(1) metal.
b) Hydrobromic acid, HBr, is a **strong acid**, because it is one of the listed hydrohalic acids.
c) Hydrogen telluride, H_2Te, is a **weak acid**, because H is not bonded to an oxygen or halide.
d) Hypochlorous acid, HClO, is a **weak acid**. The number of O atoms is one, which is equal to the number of ionizable H atoms identifying HClO as a weak acid.

18.22 Plan: The lower the concentration of hydronium (H_3O^+) ions, the higher the pH. pH increases as K_a or the molarity of acid decreases. Recall that $pK_a = -\log K_a$.
Solution:
a) At equal concentrations, the acid with the larger K_a will ionize to produce more hydronium ions than the acid with the smaller K_a. The solution of an **acid with the smaller $K_a = 4 \times 10^{-5}$** has a lower $[H_3O^+]$ and higher pH.
b) pK_a is equal to $-\log K_a$. The smaller the K_a, the larger the pK_a is. So the **acid with the larger pK_a**, 3.5, has a lower $[H_3O^+]$ and higher pH.
c) **Lower concentration** of the same acid means lower concentration of hydronium ions produced. The 0.01 M solution has a lower $[H_3O^+]$ and higher pH.
d) At the same concentration, strong acids dissociate to produce more hydronium ions than weak acids. The 0.1 M solution of a **weak acid** has a lower $[H_3O^+]$ and higher pH.
e) Bases produce OH^- ions in solution, so the concentration of hydronium ion for a solution of a base solution is lower than that for a solution of an acid. The 0.01 M **base solution** has the higher pH.
f) pOH equals $-\log [OH^-]$. At 25°C, the equilibrium constant for water ionization, K_w, equals 1×10^{-14} so $14 = pH + pOH$. As pOH decreases, pH increases. The solution of **pOH = 6.0** has the higher pH.

18.23　Plan: Part a) can be approached two ways. Because NaOH is a strong base, the $[OH^-]_{eq} = [NaOH]_{init}$. One method involves calculating $[H_3O^+]$ using $K_w = [H_3O^+][OH^-]$, then calculating pH from the relationship $pH = -\log [H_3O^+]$. The other method involves calculating pOH and then using $pH + pOH = 14.00$ to calculate pH. Part b) also has two acceptable methods analogous to those in part a); only one method will be shown.

Solution:

a) First method:

$K_w = [H_3O^+][OH^-]$

$[H_3O^+] = \dfrac{K_w}{[OH^-]} = \dfrac{1.0 \times 10^{-14}}{0.0111} = 9.0090 \times 10^{-13}\ M$

$pH = -\log [H_3O^+] = -\log (9.0090 \times 10^{-13}) = 12.04532 = \mathbf{12.05}$

Second method:

$pOH = -\log [OH^-] = -\log (0.0111) = 1.954677$

$pH = 14.00 - pOH = 14.00 - 1.954677 = 12.04532 = \mathbf{12.05}$

With a pH > 7, the solution is **basic**.

b) For a strong acid such as HCl:

$[H_3O^+] = [HCl] = 1.35 \times 10^{-3}\ M$

$pH = -\log (1.35 \times 10^{-3}) = 2.869666$

$pOH = 14.00 - 2.869666 = 11.130334 = \mathbf{11.13}$

With a pH < 7, the solution is **acidic**.

18.25　Plan: HI is a strong acid, so $[H_3O^+] = [HI]$ and the pH can be calculated from the relationship $pH = -\log [H_3O^+]$. $Ba(OH)_2$ is a strong base, so $[OH^-] = 2 \times [Ba(OH)_2]$ and $pOH = -\log [OH^-]$.

Solution:

a) $[H_3O^+] = [HI] = 6.14 \times 10^{-3}\ M$.

　　$pH = -\log (6.14 \times 10^{-3}) = 2.211832 = \mathbf{2.212}$. Solution is **acidic**.

b) $[OH^-] = 2 \times [Ba(OH)_2] = 2(2.55\ M) = 5.10\ M$

　　$pOH = -\log (5.10) = -0.70757 = \mathbf{-0.708}$. Solution is **basic**.

18.27　Plan: The relationships are: $pH = -\log [H_3O^+]$ and $[H_3O^+] = 10^{-pH}$; $pOH = -\log [OH^-]$ and $[OH^-] = 10^{-pOH}$; and $14 = pH + pOH$.

Solution:

a) $[H_3O^+] = 10^{-pH} = 10^{-9.85} = 1.4125375 \times 10^{-10} = \mathbf{1.4 \times 10^{-10}}\ \mathbf{M\ H_3O^+}$

　　$pOH = 14.00 - pH = 14.00 - 9.85 = \mathbf{4.15}$

　　$[OH^-] = 10^{-pOH} = 10^{-4.15} = 7.0794578 \times 10^{-5} = \mathbf{7.1 \times 10^{-5}}\ \mathbf{M\ OH^-}$

b) $pH = 14.00 - pOH = 14.00 - 9.43 = \mathbf{4.57}$

　　$[H_3O^+] = 10^{-pH} = 10^{-4.57} = 2.691535 \times 10^{-5} = \mathbf{2.7 \times 10^{-5}}\ \mathbf{M\ H_3O^+}$

　　$[OH^-] = 10^{-pOH} = 10^{-9.43} = 3.7153523 \times 10^{-10} = \mathbf{3.7 \times 10^{-10}}\ \mathbf{M\ OH^-}$

18.29　Plan: The relationships are: $pH = -\log [H_3O^+]$ and $[H_3O^+] = 10^{-pH}$; $pOH = -\log [OH^-]$ and $[OH^-] = 10^{-pOH}$; and $14 = pH + pOH$.

Solution:

a) $[H_3O^+] = 10^{-pH} = 10^{-4.77} = 1.69824 \times 10^{-5} = \mathbf{1.7 \times 10^{-5}}\ \mathbf{M\ H_3O^+}$

　　$pOH = 14.00 - pH = 14.00 - 4.77 = \mathbf{9.23}$

　　$[OH^-] = 10^{-pOH} = 10^{-9.23} = 5.8884 \times 10^{-10} = \mathbf{5.9 \times 10^{-10}}\ \mathbf{M\ OH^-}$

b) $pH = 14.00 - pOH = 14.00 - 5.65 = \mathbf{8.35}$

　　$[H_3O^+] = 10^{-pH} = 10^{-8.35} = 4.46684 \times 10^{-9} = \mathbf{4.5 \times 10^{-9}}\ \mathbf{M\ H_3O^+}$

　　$[OH^-] = 10^{-pOH} = 10^{-5.65} = 2.23872 \times 10^{-6} = \mathbf{2.2 \times 10^{-6}}\ \mathbf{M\ OH^-}$

18.31　Plan: The pH is increasing, so the solution is becoming more basic. Therefore, OH^- ion is added to increase the pH. Since 1 mole of H_3O^+ will react with 1 mole of OH^-, the difference in $[H_3O^+]$ would be equal to the $[OH^-]$ added. Use the relationship $[H_3O^+] = 10^{-pH}$ to find $[H_3O^+]$ at each pH.

Solution:
$$[H_3O^+] = 10^{-pH} = 10^{-3.15} = 7.07946 \times 10^{-4} \ M \ H_3O^+$$
$$[H_3O^+] = 10^{-pH} = 10^{-3.65} = 2.23872 \times 10^{-4} \ M \ H_3O^+$$
Add $(7.07946 \times 10^{-4} \ M - 2.23872 \times 10^{-4} \ M) = 4.84074 \times 10^{-4} = \mathbf{4.8 \times 10^{-4} \ mol \ of \ OH^- \ per \ liter}$.

18.33 <u>Plan:</u> The pH is increasing, so the solution is becoming more basic. Therefore, OH^- ion is added to increase the pH. Since 1 mole of H_3O^+ will react with 1 mole of OH^-, the difference in $[H_3O^+]$ would be equal to the $[OH^-]$ added. Use the relationship $[H_3O^+] = 10^{-pH}$ to find $[H_3O^+]$ at each pH.
<u>Solution:</u>
$$[H_3O^+] = 10^{-pH} = 10^{-4.52} = 3.01995 \times 10^{-5} \ M \ H_3O^+$$
$$[H_3O^+] = 10^{-pH} = 10^{-5.25} = 5.623413 \times 10^{-6} \ M \ H_3O^+$$
$$3.01995 \times 10^{-5} \ M - 5.623413 \times 10^{-6} \ M = 2.4576 \times 10^{-5} \ M \ OH^- \ \text{must be added.}$$

$$\frac{2.4576 \times 10^{-5} \ mol}{L} (5.6 \ L) = 1.3763 \times 10^{-4} = \mathbf{1.4 \times 10^{-4} \ mol \ of \ OH^-}$$

18.36 <u>Plan:</u> Apply Le Chatelier's principle in part a). In part b), given that the pH is 6.80, $[H_3O^+]$ can be calculated by using the relationship $[H_3O^+] = 10^{-pH}$. The problem specifies that the solution is neutral (pure water), meaning $[H_3O^+] = [OH^-]$. A new K_w can then be calculated.
<u>Solution:</u>
a) Heat is absorbed in an endothermic process: $2H_2O(l) + \text{heat} \rightarrow H_3O^+(aq) + OH^-(aq)$. As the temperature increases, the reaction shifts to the formation of products. Since the products are in the numerator of the K_w expression, rising temperature **increases** the value of K_w.
b) $[H_3O^+] = 10^{-pH} = 10^{-6.80} = 1.58489 \times 10^{-7} \ M \ H_3O^+ = \mathbf{1.6 \times 10^{-7} \ M} \ [H_3O^+] = [OH^-]$
$K_w = [H_3O^+][OH^-] = (1.58489 \times 10^{-7})(1.58489 \times 10^{-7}) = 2.511876 \times 10^{-14} = \mathbf{2.5 \times 10^{-14}}$
For a neutral solution: $\mathbf{pH = pOH = 6.80}$

18.37 The Brønsted-Lowry theory defines acids as proton donors and bases as proton acceptors, while the Arrhenius definition looks at acids as containing ionizable H atoms and at bases as containing hydroxide ions. In both definitions, an acid produces hydronium ions and a base produces hydroxide ions when added to water. Ammonia, NH_3, and carbonate ion, CO_3^{2-}, are two Brønsted-Lowry bases that are not Arrhenius bases because they do not contain hydroxide ions. Brønsted-Lowry acids must contain an ionizable H atom in order to be proton donors, so a Brønsted-Lowry acid that is not an Arrhenius acid cannot be identified. (Other examples are also acceptable.)

18.40 An amphoteric substance can act as either an acid or a base. In the presence of a strong base (OH^-), the dihydrogen phosphate ion acts like an acid by donating hydrogen:
$H_2PO_4^-(aq) + OH^-(aq) \rightarrow H_2O(aq) + HPO_4^{2-}(aq)$
In the presence of a strong acid (HCl), the dihydrogen phosphate ion acts like a base by accepting hydrogen:
$H_2PO_4^-(aq) + HCl(aq) \rightarrow H_3PO_4(aq) + Cl^-(aq)$

18.41 <u>Plan:</u> K_a is the equilibrium constant for an acid dissociation which has the generic equation

$HA(aq) + H_2O(l) \rightleftharpoons H_3O^+(aq) + A^-(aq)$. The K_a expression is $\dfrac{[H_3O^+][A^-]}{[HA]}$. $[H_2O]$ is treated as a constant and omitted from the expression. Write the acid-dissociation reaction for each acid, following the generic equation, and then write the K_a expression.
<u>Solution:</u>
a) When phosphoric acid is dissolved in water, a proton is donated to the water and dihydrogen phosphate ions are generated.
$$H_3PO_4(aq) + H_2O(l) \rightleftharpoons H_2PO_4^-(aq) + H_3O^+(aq)$$
$$K_a = \frac{[H_3O^+][H_2PO_4^-]}{[H_3PO_4]}$$

18-16

b) Benzoic acid is an organic acid and has only one proton to donate from the carboxylic acid group. The H atoms bonded to the benzene ring are not acidic hydrogens.

$$C_6H_5COOH(aq) + H_2O(l) \leftrightarrows C_6H_5COO^-(aq) + H_3O^+(aq)$$

$$K_a = \frac{[H_3O^+][C_6H_5COO^-]}{[C_6H_5COOH]}$$

c) Hydrogen sulfate ion donates a proton to water and forms the sulfate ion.

$$HSO_4^-(aq) + H_2O(l) \leftrightarrows SO_4^{2-}(aq) + H_3O^+(aq)$$

$$K_a = \frac{[H_3O^+][SO_4^{2-}]}{[HSO_4^-]}$$

18.43   Plan: To derive the conjugate base, remove one H from the acid and decrease the charge by 1 (acids donate H^+). Since each formula in this problem is neutral, the conjugate base will have a charge of –1.
Solution:
a) **Cl^-**   b) **HCO_3^-**   c) **OH^-**

18.45   Plan: To derive the conjugate acid, add an H and increase the charge by 1 (bases accept H^+).
Solution:
a) **NH_4^+**   b) **NH_3**   c) **$C_{10}H_{14}N_2H^+$**

18.47   Plan: The acid donates the proton to form its conjugate base; the base accepts a proton to form its conjugate acid.
Solution:
a) HCl   + H_2O $\leftrightarrows$ Cl^- + H_3O^+
   acid   base   conjugate base   conjugate acid
        Conjugate acid-base pairs: HCl/Cl^- and H_3O^+/H_2O
b) $HClO_4$ + H_2SO_4 $\leftrightarrows$ ClO_4^- + $H_3SO_4^+$
   acid   base   conjugate base   conjugate acid
        Conjugate acid-base pairs: $HClO_4/ClO_4^-$ and $H_3SO_4^+/H_2SO_4$
Note: Perchloric acid is able to protonate another strong acid, H_2SO_4, because perchloric acid is a stronger acid. ($HClO_4$'s oxygen atoms exceed its hydrogen atoms by one more than H_2SO_4.)
c) HPO_4^{2-} + H_2SO_4 $\leftrightarrows$ $H_2PO_4^-$ + HSO_4^-
   base   acid   conjugate acid   conjugate base
        Conjugate acid-base pairs: H_2SO_4/HSO_4^- and $H_2PO_4^-/HPO_4^{2-}$

18.49   Plan: The acid donates the proton to form its conjugate base; the base accepts a proton to form its conjugate acid.
Solution:
a) NH_3 + H_3PO_4 $\leftrightarrows$ NH_4^+ + $H_2PO_4^-$
   base   acid   conjugate acid   conjugate base
        Conjugate acid-base pairs: $H_3PO_4/H_2PO_4^-$; NH_4^+/NH_3
b) CH_3O^- + NH_3 $\leftrightarrows$ CH_3OH + NH_2^-
   base   base   conjugate acid   conjugate base
        Conjugate acid-base pairs: NH_3/NH_2^-; CH_3OH/CH_3O^-
c) HPO_4^{2-} + HSO_4^- $\leftrightarrows$ $H_2PO_4^-$ + SO_4^{2-}
   base   acid   conjugate acid   conjugate base
        Conjugate acid-base pairs: HSO_4^-/SO_4^{2-}; $H_2PO_4^-/HPO_4^{2-}$

18.51   Plan: Write total ionic equations (show all soluble ionic substances as dissociated into ions) and then remove the spectator ions to write the net ionic equations. The (aq) subscript denotes that each species is soluble and dissociates in water. The acid donates the proton to form its conjugate base; the base accepts a proton to form its conjugate acid.

Solution:

a) $\overline{Na^+(aq)} + OH^-(aq) + \overline{Na^+(aq)} + H_2PO_4^-(aq) \rightleftharpoons H_2O(l) + \overline{2Na^+(aq)} + HPO_4^{2-}(aq)$

   Net: $OH^-(aq) + H_2PO_4^-(aq) \rightleftharpoons H_2O(l) + HPO_4^{2-}(aq)$

    base   acid   conjugate acid conjugate base

   Conjugate acid-base pairs: $H_2PO_4^-/HPO_4^{2-}$ and H_2O/OH^-

b) $\overline{K^+(aq)} + HSO_4^-(aq) + \overline{2K^+(aq)} + CO_3^{2-}(aq) \rightleftharpoons \overline{2K^+(aq)} + SO_4^{2-}(aq) + \overline{K^+(aq)} + HCO_3^-(aq)$

   Net: $HSO_4^-(aq) + CO_3^{2-}(aq) \rightleftharpoons SO_4^{2-}(aq) + HCO_3^-(aq)$

    acid   base   conjugate base conjugate acid

   Conjugate acid-base pairs: HSO_4^-/SO_4^{2-} and HCO_3^-/CO_3^{2-}

18.53 Plan: The two possible reactions involve reacting the acid from one conjugate pair with the base from the other conjugate pair. The reaction that favors the products ($K_c > 1$) is the one in which the stronger acid produces the weaker acid. The reaction that favors reactants ($K_c < 1$) is the reaction in which the weaker acid produces the stronger acid.

   Solution:

   The conjugate pairs are H_2S (acid)/HS^- (base) and HCl (acid)/Cl^- (base). Two reactions are possible:

    (1) $HS^- + HCl \rightleftharpoons H_2S + Cl^-$ and (2) $H_2S + Cl^- \rightleftharpoons HS^- + HCl$

   The first reaction is the reverse of the second. HCl is a strong acid and H_2S a weak acid. Reaction (1) with the stronger acid producing the weaker acid favors products and $K_c > 1$. Reaction (2) with the weaker acid forming the stronger acid favors the reactants and $K_c < 1$.

18.55 Plan: An acid-base reaction that favors the products ($K_c > 1$) is one in which the stronger acid produces the weaker acid. Use the figure to decide which of the two acids is the stronger acid.

   Solution:

   a) $HCl + NH_3 \rightleftharpoons NH_4^+ + Cl^-$

   strong acid stronger base weak acid weaker base

     HCl is ranked above NH_4^+ in the list of conjugate acid-base pair strength and is the stronger acid. NH_3 is ranked above Cl^- and is the stronger base. NH_3 is shown as a "stronger" base because it is stronger than Cl^-, but is not considered a "strong" base. The reaction proceeds toward the production of the weaker acid and base, i.e., the reaction as written proceeds to the right and $K_c > 1$. The stronger acid is more likely to donate a proton than the weaker acid.

   b) $H_2SO_3 + NH_3 \rightleftharpoons HSO_3^- + NH_4^+$

   stronger acid stronger base weaker base weaker acid

     H_2SO_3 is ranked above NH_4^+ and is the stronger acid. NH_3 is a stronger base than HSO_3^-. The reaction proceeds toward the production of the weaker acid and base, i.e., the reaction as written proceeds to the right and $K_c > 1$.

18.57 Plan: An acid-base reaction that favors the reactants ($K_c < 1$) is one in which the weaker acid produces the stronger acid. Use the figure to decide which of the two acids is the weaker acid.

   Solution:

   a) $NH_4^+ + HPO_4^{2-} \rightleftharpoons NH_3 + H_2PO_4^-$

   weaker acid weaker base stronger base stronger acid

     $K_c < 1$ The reaction proceeds toward the production of the weaker acid and base, i.e., the reaction as written proceeds to the left.

   b) $HSO_3^- + HS^- \rightleftharpoons H_2SO_3 + S^{2-}$

   weaker base weaker acid stronger acid stronger base

     $K_c < 1$ The reaction proceeds toward the production of the weaker acid and base, i.e., the reaction as written proceeds to the left.

18.59 a) The concentration of a strong acid is **very different** before and after dissociation since a strong acid exhibits 100% dissociation. After dissociation, the concentration of the strong acid approaches 0, or $[HA] \approx 0$.

   b) A weak acid dissociates to a very small extent (<<100%), so the acid concentration after dissociation is **nearly the same** as before dissociation.

   c) Same as b), but the percent, or extent, of dissociation is greater than in b).

   d) Same as a)

18.60   **No**, HCl and CH$_3$COOH are never of equal strength because HCl is a strong acid with $K_a > 1$ and CH$_3$COOH is a weak acid with $K_a < 1$. The K_a of the acid, not the concentration of H$_3$O$^+$ in a solution of the acid, determines the strength of the acid.

18.61   Plan: We are given the percent dissociation of the original HA solution (33%), and we know that the percent dissociation increases as the acid is diluted. Thus, we calculate the percent dissocation of each diluted sample and see which is greater than 33%. To determine percent dissociation, we use the following formula:
Percent HA dissociated = ([HA]$_{dissoc}$/[HA]$_{init}$) X 100, with [HA]$_{dissoc}$ equal to the number of H$_3$O$^+$ (or A$^-$) ions and [HA]$_{init}$ equal to the number of HA *plus* the number of H$_3$O$^+$ (or A$^-$)
Solution:
Calculating the percent dissociation of each diluted solution:
Solution 1. Percent dissociation = [4/(5+4)] X 100 = 44%
Solution 2. Percent dissociation = [2/(7 + 2)] X 100 = 22%
Solution 3. Percent dissociation = [3/(6 + 3)] X 100 = 33%
Therefore, scene 1 represents the diluted solution.

18.64   Plan: Write the acid-dissociation reaction and the expression for K_a. Set up a reaction table and substitute the given value of [H$_3$O$^+$] for x; solve for K_a.
Solution:
Butanoic acid dissociates according to the following equation:
$$CH_3CH_2CH_2COOH(aq) \ + \ H_2O(l) \ \leftrightarrows \ H_3O^+(aq) \ + \ CH_3CH_2CH_2COO^-(aq)$$

Initial:	0.15 M	0	0
Change:	−x	+x	+x
Equilibrium:	0.15 − x	x	x

According to the information given in the problem, [H$_3$O$^+$]$_{eq}$ = 1.51x10^{-3} M = x
Thus, [H$_3$O$^+$] = [CH$_3$CH$_2$CH$_2$COO$^-$] = 1.51x10^{-3} M
[CH$_3$CH$_2$CH$_2$COOH] = (0.15 − x) = (0.15 − 1.51x10^{-3}) M = 0.14849 M

$$K_a = \frac{[H_3O^+][CH_3CH_2CH_2COO^-]}{[CH_3CH_2CH_2COOH]}$$

$$K_a = \frac{(1.51x10^{-3})(1.51x10^{-3})}{(0.14849)} = 1.53552x10^{-5} = \mathbf{1.5x10^{-5}}$$

18.66   Plan: Write the acid-dissociation reaction and the expression for K_a. Set up a reaction table in which x = the concentration of the dissociated HNO$_2$ and also [H$_3$O$^+$]. Use the expression for K_a to solve for x ([H$_3$O$^+$]).
Solution:
For a solution of a weak acid, the acid-dissociation equilibrium determines the concentrations of the weak acid, its conjugate base and H$_3$O$^+$. The acid-dissociation reaction for HNO$_2$ is:

Concentration	HNO$_2$(aq)	+	H$_2$O(l)	⇄	H$_3$O$^+$(aq)	+	NO$_2^-$(aq)
Initial	0.60		—		0		0
Change	−x				+x		+x
Equilibrium	0.60 − x				x		x

(The H$_3$O$^+$ contribution from water has been neglected.)

$$K_a = 7.1x10^{-4} = \frac{[H_3O^+][NO_2^-]}{[HNO_2]}$$

$$K_a = 7.1x10^{-4} = \frac{(x)(x)}{(0.60 - x)} \qquad \text{Assume x is small compared to 0.60: } 0.60 - x = 0.60$$

$$K_a = 7.1x10^{-4} = \frac{(x)(x)}{(0.60)}$$

x = 0.020639767
Check assumption that x is small compared to 0.60:

$$\frac{0.020639767}{0.60}(100) = 3.4\% \text{ error, so the assumption is valid.}$$

$[H_3O^+] = [NO_2^-] = \textbf{2.1x10}^{-2}\textbf{\textit{ M}}$

The concentration of hydroxide ion is related to concentration of hydronium ion through the equilibrium for water: $2H_2O(l) \leftrightarrows H_3O^+(aq) + OH^-(aq)$ with $K_w = 1.0x10^{-14}$

$K_w = 1.0x10^{-14} = [H_3O^+][OH^-]$

$[OH^-] = 1.0x10^{-14}/0.020639767 = 4.84502x10^{-13} = \textbf{4.8x10}^{-13}\textbf{\textit{ M}}\textbf{ OH}^-$

18.68 **Plan:** Write the acid-dissociation reaction and the expression for K_a. Set up a reaction table in which x = the concentration of the dissociated acid and also $[H_3O^+]$. Use the expression for K_a to solve for x ($[H_3O^+]$). K_a is found from the pK_a by using the relationship $K_a = 10^{-pK_a}$.

Solution:

$K_a = 10^{-pK_a} = 10^{-2.87} = 1.34896x10^{-3}$

Concentration	$ClCH_2COOH(aq)$	$+ H_2O(l) \leftrightarrows$	$H_3O^+(aq)$	$+ ClCH_2COO^-(aq)$
Initial	1.25		0	0
Change	$-x$		$+x$	$+x$
Equilibrium	$1.25 - x$		x	x

$$K_a = 1.34896x10^{-3} = \frac{[H_3O^+][ClCH_2COO^-]}{[ClCH_2COOH]}$$

$$K_a = 1.34896x10^{-3} = \frac{(x)(x)}{\left(1.25 - x\right)} \qquad \text{Assume x is small compared to 1.25.}$$

$$K_a = 1.34896x10^{-3} = \frac{(x)(x)}{(1.25)}$$

$x = 0.04106337$

Check assumption that x is small compared to 1.25:

$$\frac{0.04106337}{1.25}(100) = 3.3\%. \text{ The assumption is good.}$$

$[H_3O^+] = [ClCH_2COO^-] = \textbf{0.041 \textit{M}}$

$[ClCH_2COOH] = 1.25 - 0.04106337 = 1.20894 = \textbf{1.21 \textit{M}}$

$pH = -\log [H_3O^+] = -\log (0.04106337) = 1.3865 = \textbf{1.39}$

18.70 **Plan:** Write the acid-dissociation reaction and the expression for K_a. Percent dissociation refers to the amount of the initial concentration of the acid that dissociates into ions. Use the percent dissociation to find the concentration of acid dissociated, which also equals $[H_3O^+]$. HA will be used as the formula of the acid. Set up a reaction table in which x = the concentration of the dissociated acid and $[H_3O^+]$. pH and $[OH^-]$ are determined from $[H_3O^+]$. Substitute [HA], $[A^-]$, and $[H_3O^+]$ into the expression for K_a to find the value of K_a.

Solution:

a) $HA(aq) + H_2O(l) \leftrightarrows H_3O^+(aq) + A^-(aq)$

$$\text{Percent HA} = \frac{\text{dissociated acid}}{\text{initial acid}}(100)$$

$$3.0\% = \frac{x}{0.20}(100)$$

[Dissociated acid] = x = $6.0x10^{-3}$ M

Concentration	$HA(aq)$	$+ H_2O(l) \leftrightarrows$	$H_3O^+(aq)$	$+ A^-(aq)$
Initial:	0.20		0	0
Change:	$-x$		$+x$	$+x$
Equilibrium:	$0.20 - x$		x	x

[Dissociated acid] = x = $[A^-]$ = $[H_3O^+]$ = $\textbf{6.0x10}^{-3}\textbf{\textit{ M}}$

$pH = -\log [H_3O^+] = -\log (6.0x10^{-3}) = 2.22185 = \textbf{2.22}$

$K_w = 1.0x10^{-14} = [H_3O^+][OH^-]$

$$[OH^-] = \frac{K_w}{[H_3O^+]} = \frac{1.0 \times 10^{-14}}{6.0 \times 10^{-3}} = 1.6666667 \times 10^{-12} = \mathbf{1.7 \times 10^{-12}}\ M$$

$$pOH = -\log [OH^-] = -\log (1.6666667 \times 10^{-12}) = 11.7782 = \mathbf{11.78}$$

b) In the equilibrium expression, substitute the concentrations above and calculate K_a.

$$K_a = \frac{[H_3O^+][A^-]}{[HA]} = \frac{(6.0 \times 10^{-3})(6.0 \times 10^{-3})}{(0.20 - 6.0 \times 10^{-3})} = 1.85567 \times 10^{-4} = \mathbf{1.9 \times 10^{-4}}$$

18.72 **Plan:** Write the acid-dissociation reaction and the expression for K_a. Calculate the molarity of HX by dividing moles by volume. Convert pH to $[H_3O^+]$, set up a reaction table in which x = the concentration of the dissociated acid and also $[H_3O^+]$, and substitute into the equilibrium expression to find K_a.

$$\text{Concentration } (M) \text{ of HX} = \left(\frac{0.250 \text{ mol}}{655 \text{ mL}} \right)\left(\frac{1 \text{ mL}}{10^{-3} \text{ L}} \right) = 0.381679\ M$$

Concentration	HX(aq) + H$_2$O(l)	$\rightleftarrows$	H$_3$O$^+$(aq)	+	X$^-$(aq)
Initial:	0.381679		0		0
Change:	−x		+x		+x
Equilibrium:	0.381679 − x		x		x

$[H_3O^+] = 10^{-pH} = 10^{-3.54} = 2.88403 \times 10^{-4}\ M = x$

Thus, $[H_3O^+] = [X^-] = 2.88403 \times 10^{-4}\ M$, and $[HX] = (0.381679 - 2.88403 \times 10^{-4})\ M$

$$K_a = \frac{[H_3O^+][X^-]}{[HX]} = \frac{(2.88403 \times 10^{-4})(2.88403 \times 10^{-4})}{(0.381679 - 2.88403 \times 10^{-4})} = 2.18087 \times 10^{-7} = \mathbf{2.2 \times 10^{-7}}$$

18.74 **Plan:** Write the acid-dissociation reaction and the expression for K_a. Set up a reaction table in which x = the concentration of the dissociated acid and also $[H_3O^+]$. Use the expression for K_a to solve for x ($[H_3O^+]$). OH$^-$ and then pOH can be found from $[H_3O^+]$.
Solution:
a)

Concentration	HZ(aq)	+	H$_2$O(l)	$\rightleftarrows$	H$_3$O$^+$(aq)	+	Z$^-$(aq)
Initial	0.075		—		0		0
Change	−x				+x		+x
Equilibrium	0.075 − x				x		x

(The H$_3$O$^+$ contribution from water has been neglected.)

$$K_a = 2.55 \times 10^{-4} = \frac{[H_3O^+][Z^-]}{[HZ]}$$

$$K_a = 2.55 \times 10^{-4} = \frac{(x)(x)}{(0.075 - x)} \qquad \text{Assume x is small compared to 0.075.}$$

$$K_a = 2.55 \times 10^{-4} = \frac{(x)(x)}{(0.075)}$$

$[H_3O^+] = x = 4.3732 \times 10^{-3}$

Check assumption that x is small compared to 0.075:

$$\frac{4.3732 \times 10^{-3}}{0.075}(100) = 6\% \text{ error, so the assumption is not valid.}$$

Since the error is greater than 5%, it is not acceptable to assume x is small compared to 0.075, and it is necessary to use the quadratic equation.

$$K_a = 2.55 \times 10^{-4} = \frac{(x)(x)}{(0.075 - x)}$$

$$x^2 + 2.55 \times 10^{-4}\ x - 1.9125 \times 10^{-5} = 0$$

$$a = 1 \qquad \qquad b = 2.55 \times 10^{-4} \qquad \qquad c = -1.9125 \times 10^{-5}$$

$$x = \frac{-b \pm \sqrt{b^2 - 4ac}}{2a}$$

$$x = \frac{-(2.55 \times 10^{-4}) \pm \sqrt{(2.55 \times 10^{-4})^2 - 4(1)(-1.9125 \times 10^{-5})}}{2(1)}$$

x = 0.00425 or –0.004503

(The –0.004503 value is not possible.)

pH = –log [H_3O^+] = –log (0.00425) = 2.3716 = **2.37**

b)

Concentration	HZ(aq)	+	$H_2O(l)$	⇌	H_3O^+(aq)	+	Z^-(aq)
Initial	0.045		—		0		0
Change	–x				+x		+x
Equilibrium	0.045 – x				x		x

(The H_3O^+ contribution from water has been neglected.)

$$K_a = 2.55 \times 10^{-4} = \frac{[H_3O^+][Z^-]}{[HZ]}$$

$$K_a = 2.55 \times 10^{-4} = \frac{(x)(x)}{(0.045 - x)} \qquad \text{Assume x is small compared to 0.045.}$$

$$K_a = 2.55 \times 10^{-4} = \frac{(x)(x)}{(0.045)}$$

[H_3O^+] = x = 3.3875 × 10⁻³

Check assumption that x is small compared to 0.045:

$$\frac{3.3875 \times 10^{-3}}{0.045}(100) = 7.5\% \text{ error, so the assumption is not valid.}$$

Since the error is greater than 5%, it is not acceptable to assume x is small compared to 0.045, and it is necessary to use the quadratic equation.

$$K_a = 2.55 \times 10^{-4} = \frac{(x)(x)}{(0.045 - x)}$$

$$x^2 = (2.55 \times 10^{-4})(0.045 - x) = 1.1475 \times 10^{-5} - 2.55 \times 10^{-4} x$$

$$x^2 + 2.55 \times 10^{-4} x - 1.1475 \times 10^{-5} = 0$$

$$a = 1 \qquad b = 2.55 \times 10^{-4} \qquad c = -1.1475 \times 10^{-5}$$

$$x = \frac{-b \pm \sqrt{b^2 - 4ac}}{2a}$$

$$x = \frac{-2.55 \times 10^{-4} \pm \sqrt{(2.55 \times 10^{-4})^2 - 4(1)(-1.1475 \times 10^{-5})}}{2(1)}$$

$$x = 3.26238 \times 10^{-3} \ M \ H_3O^+$$

$$[OH^-] = \frac{K_w}{[H_3O^+]} = \frac{1.0 \times 10^{-14}}{3.26238 \times 10^{-3}} = 3.0652468 \times 10^{-12} \ M$$

$$pOH = -\log [OH^-] = -\log (3.0652468 \times 10^{-12}) = 11.51353 = \mathbf{11.51}$$

18.76   <u>Plan:</u> Write the acid-dissociation reaction and the expression for K_a. Set up a reaction table in which x = the concentration of the dissociated acid and also [H_3O^+]. Use the expression for K_a to solve for x ([H_3O^+]). OH⁻ and then pOH can be found from [H_3O^+].

<u>Solution:</u>

a)

Concentration	HY(aq)	+	$H_2O(l)$	⇌	H_3O^+(aq)	+	Y^-(aq)
Initial	0.175		—		0		0
Change	–x				+x		+x
Equilibrium	0.175 – x				x		x

(The H_3O^+ contribution from water has been neglected.)

$$K_a = 1.50 \times 10^{-4} = \frac{[H_3O^+][Y^-]}{[HY]}$$

$$K_a = 1.50 \times 10^{-4} = \frac{(x)(x)}{\left(0.175 - x\right)} \qquad \text{Assume x is small compared to 0.175.}$$

$$K_a = 1.50 \times 10^{-4} = \frac{(x)(x)}{(0.175)}$$

$$[H_3O^+] = x = 5.1235 \times 10^{-3} \ M$$

Check assumption that x is small compared to 0.175:

$$\frac{5.1235 \times 10^{-3}}{0.175}(100) = 3\% \text{ error, so the assumption is valid.}$$

$$pH = -\log [H_3O^+] = -\log (5.1235 \times 10^{-3}) = 2.29043 = \mathbf{2.290}$$

b) Concentration

	HX(aq)	+	H$_2$O(l)	$\rightleftharpoons$	H$_3$O$^+$$(aq)$	+	X$^-$$(aq)$
Initial	0.175		—		0		0
Change	−x				+x		+x
Equilibrium	0.175 − x				x		x

(The H_3O^+ contribution from water has been neglected.)

$$K_a = 2.00 \times 10^{-2} = \frac{[H_3O^+][X^-]}{[HX]}$$

$$K_a = 2.00 \times 10^{-2} = \frac{(x)(x)}{\left(0.175 - x\right)} \qquad \text{Assume x is small compared to 0.175.}$$

$$K_a = 2.00 \times 10^{-2} = \frac{(x)(x)}{(0.175)}$$

$$[H_3O^+] = x = 5.9161 \times 10^{-2} \ M$$

Check assumption that x is small compared to 0.175:

$$\frac{5.9161 \times 10^{-2}}{0.175}(100) = 34\% \text{ error, so the assumption is not valid.}$$

Since the error is greater than 5%, it is not acceptable to assume x is small compared to 0.175, and it is necessary to use the quadratic equation.

$$K_a = 2.00 \times 10^{-2} = \frac{(x)(x)}{\left(0.175 - x\right)}$$

$$x^2 = (2.00 \times 10^{-2})(0.175 - x) = 0.0035 - 2.00 \times 10^{-2}x$$

$$x^2 + 2.00 \times 10^{-2}x - 0.0035 = 0$$

$$a = 1 \qquad b = 2.00 \times 10^{-2} \qquad c = -0.0035$$

$$x = \frac{-b \pm \sqrt{b^2 - 4ac}}{2a}$$

$$x = \frac{-2.00 \times 10^{-2} \pm \sqrt{(2.00 \times 10^{-2})^2 - 4(1)(-0.0035)}}{2(1)}$$

$$x = 5.00 \times 10^{-2} \ M \ H_3O^+$$

$$[OH^-] = \frac{K_w}{[H_3O^+]} = \frac{1.0 \times 10^{-14}}{5.00 \times 10^{-2}} = 2.00 \times 10^{-13} \ M$$

$$pOH = -\log [OH^-] = -\log (2.00 \times 10^{-13}) = 12.69897 = \mathbf{12.699}$$

18.78 <u>Plan:</u> Write the acid-dissociation reaction and the expression for K_a. Set up a reaction table in which x = the concentration of the dissociated acid and also [H$_3$O$^+$]. Use the expression for K_a to solve for x, the concentration of benzoate ion at equilibrium. Then use the initial concentration of benzoic acid and the equilibrium concentration of benzoate to find % dissociation.
<u>Solution:</u>

Concentration	C$_6$H$_5$COOH(*aq*)	+	H$_2$O(*l*)	⇆	H$_3$O$^+$(*aq*)	+	C$_6$H$_5$COO$^-$(*aq*)
Initial	0.55		—		0		0
Change	−x				+x		+x
Equilibrium	0.55 − x				x		x

$$K_a = 6.3 \times 10^{-5} = \frac{[H_3O^+][C_6H_5COO^-]}{[C_6H_5COOH]}$$

$$K_a = 6.3 \times 10^{-5} = \frac{[x][x]}{[0.55 - x]} \qquad \text{Assume x is small compared to 0.55.}$$

$$K_a = 6.3 \times 10^{-5} = \frac{[x][x]}{[0.55]}$$

$$x = 5.8864 \times 10^{-3} \ M$$

$$\text{Percent C}_6\text{H}_5\text{COOH dissociated} = \frac{[C_6H_5COOH]_{\text{dissociated}}}{[C_6H_5COOH]_{\text{initial}}} (100)$$

$$\text{Percent C}_6\text{H}_5\text{COOH dissociated} = \frac{5.8864 \times 10^{-3} \ M}{0.55 \ M}(100) = 1.07025 = \textbf{1.1\%}$$

18.80 <u>Plan:</u> Write balanced chemical equations and corresponding equilibrium expressions for dissociation of hydrosulfuric acid, H$_2$S, and HS$^-$. Since $K_{a1} \gg K_{a2}$, assume that almost all of the H$_3$O$^+$ comes from the first dissociation. Set up reaction tables in which x = the concentration of dissociated acid and [H$_3$O$^+$].
<u>Solution:</u>

$$H_2S(aq) + H_2O(l) \leftrightarrows H_3O^+(aq) + HS^-(aq) \qquad \qquad HS^-(aq) + H_2O(l) \leftrightarrows H_3O^+(aq) + S^{2-}(aq)$$

$$K_{a1} = 9 \times 10^{-8} = \frac{[H_3O^+][HS^-]}{[H_2S]} \qquad \qquad \qquad K_{a2} = 1 \times 10^{-17} = \frac{[H_3O^+][S^{2-}]}{[HS^-]}$$

Concentration	H$_2$S(*aq*)	+	H$_2$O(*l*)	⇆	H$_3$O$^+$(*aq*)	+	HS$^-$(*aq*)
Initial	0.10		—		0		0
Change	−x				+x		+x
Equilibrium	0.10 − x				x		x

$$K_{a1} = 9 \times 10^{-8} = \frac{[H_3O^+][HS^-]}{[H_2S]}$$

$$K_{a1} = 9 \times 10^{-8} = \frac{[x][x]}{[0.10 - x]} \qquad \text{Assume x is small compared to 0.10.}$$

$$K_{a1} = 9 \times 10^{-8} = \frac{[x][x]}{[0.10]}$$

$$x = 9.48683 \times 10^{-5}$$

$$[H_3O^+] = [HS^-] = x = \textbf{9} \times \textbf{10}^{-5} \ \textbf{M}$$

$$pH = -\log [H_3O^+] = -\log (9.48683 \times 10^{-5}) = 4.022878 = \textbf{4.0}$$

$$[OH^-] = \frac{K_w}{[H_3O^+]} = \frac{1.0 \times 10^{-14}}{9.48683 \times 10^{-5}} = 1.05409 \times 10^{-10} = \textbf{1} \times \textbf{10}^{-10} \ \textbf{M}$$

$$pOH = -\log [OH^-] = -\log (1.05409 \times 10^{-10}) = 9.9771 = \textbf{10.0}$$

$$[H_2S] = 0.10 - x = 0.10 - 9.48683 \times 10^{-5} = 0.099905 = \textbf{0.10} \ \textbf{M}$$

Concentration is limited to one significant figure because K_a is given to only one significant figure. The pH is given to what appears to be two significant figures because the number before the decimal point (4) represents the exponent and the number after the decimal point represents the significant figures in the concentration. Calculate $[S^{2-}]$ by using the K_{a2} expression and assuming that $[HS^-]$ and $[H_3O^+]$ come mostly from the first dissociation. This new calculation will have a new x value.

Concentration	$HS^-(aq)$	$+$	$H_2O(l)$	$\leftrightarrows$	$H_3O^+(aq)$	$+$	$S^{2-}(aq)$
Initial	9.48683×10^{-5}		—		9.48683×10^{-5}		0
Change	$-x$				$+x$		$+x$
Equilibrium	$9.48683 \times 10^{-5} - x$				$9.48683 \times 10^{-5} + x$		x

$$K_{a2} = 1 \times 10^{-17} = \frac{[H_3O^+][S^{2-}]}{[HS^-]}$$

$$K_{a2} = 1 \times 10^{-17} = \frac{(9.48683 \times 10^{-5} + x)(x)}{(9.48683 \times 10^{-5} - x)} \qquad \text{Assume x is small compared to } 9.48683 \times 10^{-5}.$$

$$K_{a2} = 1 \times 10^{-17} = \frac{(9.48683 \times 10^{-5})(x)}{(9.48683 \times 10^{-5})}$$

$$x = [S^{2-}] = \mathbf{1 \times 10^{-17} \ M}$$

The small value of x means that it is not necessary to recalculate the $[H_3O^+]$ and $[HS^-]$ values.

18.83 Plan: Write the acid-dissociation reaction and the expression for K_a. Set up a reaction table in which x = the concentration of the dissociated acid and also $[H_3O^+]$. Use the expression for K_a to solve for x, the concentration of formate ion at equilibrium. Then use the initial concentration of formic acid and the equilibrium concentration of formate to find % dissociation.

Solution:

Concentration	$HCOOH(aq)$	$+$ $H_2O(l)$	$\leftrightarrows$ $H_3O^+(aq)$	$+$ $HCOO^-(aq)$
Initial	0.75		0	0
Change	$-x$		$+x$	$+x$
Equilibrium	$0.75 - x$		x	x

$$K_a = 1.8 \times 10^{-4} = \frac{[H_3O^+][HCOO^-]}{[HCOOH]}$$

$$K_a = 1.8 \times 10^{-4} = \frac{(x)(x)}{(0.75 - x)} \qquad \text{Assume x is small compared to 0.75.}$$

$$K_a = 1.8 \times 10^{-4} = \frac{(x)(x)}{(0.75)}$$

$$x = 1.161895 \times 10^{-2}$$

$$\text{Percent HCOOH dissociated} = \frac{[HCOOH]_{dissociated}}{[HCOOH]_{initial}}(100)$$

$$\text{Percent HCOOH dissociated} = \frac{1.161895 \times 10^{-2} \ M}{0.75 \ M}(100) = 1.54919 = \mathbf{1.5\%}$$

18.84 Electronegativity increases left to right across a period. As the nonmetal becomes more electronegative, the acidity of the binary hydride increases. The electronegative nonmetal attracts the electrons more strongly in the polar bond, shifting the electron density away from H^+ and making the H^+ more easily transferred to a surrounding water molecule to make H_3O^+.

18.87   The two factors that explain the greater acid strength of $HClO_4$ are:
1) Chlorine is more electronegative than iodine, so chlorine more strongly attracts the electrons in the bond with oxygen. This makes the H in $HClO_4$ less tightly held by the oxygen than the H in HIO.
2) Perchloric acid has more oxygen atoms than HIO, which leads to a greater shift in electron density from the hydrogen atom to the oxygen atoms making the H in $HClO_4$ more susceptible to transfer to a base.

18.88   Plan: For oxyacids, acid strength increases with increasing number of oxygen atoms and increasing electronegativity of the nonmetal in the acid. For binary acids, acid strength increases with increasing electronegativity across a row and increases with increasing size of the nonmetal down a column.
Solution:
a) Selenic acid, H_2SeO_4, is the stronger acid because it contains more oxygen atoms.
b) Phosphoric acid, H_3PO_4, is the stronger acid because P is more electronegative than As.
c) Hydrotelluric acid, H_2Te, is the stronger acid because Te is larger than S and so the Te–H bond is weaker.

18.90   Plan: For oxyacids, acid strength increases with increasing number of oxygen atoms and increasing electronegativity of the nonmetal in the acid. For binary acids, acid strength increases with increasing electronegativity across a row and increases with increasing size of the nonmetal in a column.
Solution:
a) H_2Se, hydrogen selenide, is a stronger acid than H_3As, arsenic hydride, because Se is more electronegative than As.
b) $B(OH)_3$, boric acid also written as H_3BO_3, is a stronger acid than $Al(OH)_3$, aluminum hydroxide, because boron is more electronegative than aluminum.
c) $HBrO_2$, bromous acid, is a stronger acid than HBrO, hypobromous acid, because there are more oxygen atoms in $HBrO_2$ than in HBrO.

18.92   Plan: Acidity increases as the value of K_a increases. Determine the ion formed from each salt and compare the corresponding K_a values from Appendix C.
Solution:
a) Copper(II) bromide, $CuBr_2$, contains Cu^{2+} ion with $K_a = 3 \times 10^{-8}$. Aluminum bromide, $AlBr_3$, contains Al^{3+} ion with $K_a = 1 \times 10^{-5}$. The concentrations of Cu^{2+} and Al^{3+} are equal, but the K_a of $AlBr_3$ is almost three orders of magnitude greater. Therefore, **0.5 *M* $AlBr_3$** is the stronger acid and would have the lower pH.
b) Zinc chloride, $ZnCl_2$, contains the Zn^{2+} ion with $K_a = 1 \times 10^{-9}$. Tin(II) chloride, $SnCl_2$, contains the Sn^{2+} ion with $K_a = 4 \times 10^{-4}$. Since both solutions have the same concentration, and K_a $(Sn^{2+}) > K_a$ (Zn^{2+}), **0.3 *M* $SnCl_2$** is the stronger acid and would have the lower pH.

18.94   Plan: A higher pH (more basic solution) results when an acid has a smaller K_a (from the Appendix). Determine the ion formed from each salt and compare the corresponding K_a values from Appendix C.
Solution:
a) The **$Ni(NO_3)_2$** solution has a higher pH than the $Co(NO_3)_2$ solution because K_a of Ni^{2+} (1×10^{-10}) is smaller than the K_a of Co^{2+} (2×10^{-10}). Note that nitrate ion is the conjugate base of a strong acid and therefore does not influence the pH of the solution.
b) The **$Al(NO_3)_3$** solution has a higher pH than the $Cr(NO_3)_2$ solution because K_a of Al^{3+} (1×10^{-5}) is smaller than the K_a of Cr^{3+} (1×10^{-4}).

18.96   All Brønsted-Lowry bases contain at least one lone pair of electrons. This lone pair binds with an H^+ and allows the base to act as a proton-acceptor.

18.99   Plan: K_b is the equilibrium constant for a base dissociation which has the generic equation
$B(aq) + H_2O(l) \rightleftharpoons BH^+(aq) + OH^-(aq)$. The K_b expression is $\dfrac{[BH^+][OH^-]}{[B]}$. $[H_2O]$ is treated as a constant

and omitted from the expression. Write the base-dissociation reaction for each base, showing the base accepting a proton from water, and then write the K_b expression.

Solution:
a) $C_5H_5N(aq) + H_2O(l) \leftrightarrows C_5H_5NH^+(aq) + OH^-(aq)$

$$K_b = \frac{[C_5H_5NH^+][OH^-]}{[C_5H_5N]}$$

b) $CO_3^{2-}(aq) + H_2O(l) \leftrightarrows HCO_3^-(aq) + OH^-(aq)$

$$K_b = \frac{[HCO_3^-][OH^-]}{[CO_3^{2-}]}$$

The bicarbonate can then also dissociate as a base, but this occurs to an insignificant amount in a solution of carbonate ions.

18.101 Plan: K_b is the equilibrium constant for a base dissociation which has the generic equation
$B(aq) + H_2O(l) \leftrightarrows BH^+(aq) + OH^-(aq)$. The K_b expression is $\frac{[BH^+][OH^-]}{[B]}$. [H$_2$O] is treated as a constant and omitted from the expression. Write the base-dissociation reaction for each base, showing the base accepting a proton from water, and then write the K_b expression.
Solution:
a) $HONH_2(aq) + H_2O(l) \leftrightarrows OH^-(aq) + HONH_3^+(aq)$

$$K_b = \frac{[HONH_3^+][OH^-]}{[HONH_2]}$$

b) $HPO_4^{2-}(aq) + H_2O(l) \leftrightarrows H_2PO_4^-(aq) + OH^-(aq)$

$$K_b = \frac{[H_2PO_4^-][OH^-]}{[HPO_4^{2-}]}$$

18.103 Plan: Write the balanced equation for the base reaction and the expression for K_b. Set up a reaction table in which x = the concentration of reacted base and also [OH$^-$]. Use the expression for K_b to solve for x, [OH$^-$], and then calculate [H$_3$O$^+$] and pH.
Solution:
The formula of dimethylamine has two methyl (CH$_3$–) groups attached to a nitrogen:

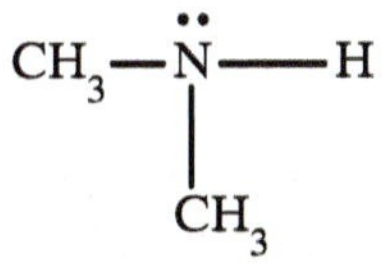

$$CH_3 - \overset{..}{N} - H$$
$$|$$
$$CH_3$$

The nitrogen has a lone pair of electrons that will accept the proton from water in the base-dissociation reaction:
The value for the dissociation constant is from Appendix C.

Concentration	$(CH_3)_2NH(aq) + H_2O(l) \leftrightarrows$	$OH^-(aq) +$	$(CH_3)_2NH_2^+(aq)$
Initial	0.070	0	0
Change	–x	+x	+x
Equilibrium	0.070 – x	x	x

$$K_b = 5.9 \times 10^{-4} = \frac{[(CH_3)_2NH_2^+][OH^-]}{[(CH_3)_2NH]}$$

$$K_b = 5.9 \times 10^{-4} = \frac{[x][x]}{[0.070 - x]} \qquad \text{Assume } 0.070 - x = 0.070$$

$$5.9 \times 10^{-4} = \frac{[x][x]}{[0.070]}$$

$$x = 6.4265 \times 10^{-3} \ M$$

18-27

Check assumption that x is small compared to 0.070:

$$\frac{6.4265 \times 10^{-3}}{0.070}(100) = 9\% \text{ error, so the assumption is not valid.}$$

The problem will need to be solved as a quadratic.

$$5.9 \times 10^{-4} = \frac{[x][x]}{\left[0.070 - x\right]}$$

$$x^2 = (5.9 \times 10^{-4})(0.070 - x) = 4.13 \times 10^{-5} - 5.9 \times 10^{-4}\,x$$
$$x^2 + 5.9 \times 10^{-4}\,x - 4.13 \times 10^{-5} = 0$$
$$a = 1 \qquad b = 5.9 \times 10^{-4} \qquad c = -4.13 \times 10^{-5}$$

$$x = \frac{-b \pm \sqrt{b^2 - 4ac}}{2a}$$

$$x = \frac{-5.9 \times 10^{-4} \pm \sqrt{\left(5.9 \times 10^{-4}\right)^2 - 4(1)\left(-4.13 \times 10^{-5}\right)}}{2(1)} = 6.13827 \times 10^{-3} \; M \; OH^-$$

$$[H_3O]^+ = \frac{K_w}{[OH^-]} = \frac{1.0 \times 10^{-14}}{6.13827 \times 10^{-3}} = 1.629124 \times 10^{-12} \; M \; H_3O^+$$

$$pH = -\log[H_3O^+] = -\log(1.629124 \times 10^{-12}) = 11.7880 = \textbf{11.79}$$

18.105 Plan: Write the balanced equation for the base reaction and the expression for K_b. Set up a reaction table in which x = the concentration of reacted base and also [OH⁻]. Use the expression for K_b to solve for x, [OH⁻], and then calculate [H₃O⁺] and pH.
Solution:

Concentration	$HOCH_2CH_2NH_2(aq)$ + $H_2O(l)$ $\leftrightarrows$	$OH^-(aq)$ +	$HOCH_2CH_2NH_3^+(aq)$
Initial	0.25	0	0
Change	–x	+x	+x
Equilibrium	0.25 – x	x	x

$$K_b = 3.2 \times 10^{-5} = \frac{\left[HOCH_2CH_2NH_3^+\right]\left[OH^-\right]}{\left[HOCH_2CH_2NH_2\right]}$$

$$K_b = 3.2 \times 10^{-5} = \frac{[x][x]}{\left[0.25 - x\right]} \qquad \text{Assume x is small compared to 0.25.}$$

$$K_b = 3.2 \times 10^{-5} = \frac{(x)(x)}{(0.25)}$$

$$x = 2.8284 \times 10^{-3} \; M \; OH^-$$

Check assumption that x is small compared to 0.25:

$$\frac{2.8284 \times 10^{-3}}{0.25}(100) = 1\% \text{ error, so the assumption is valid.}$$

$$[H_3O]^+ = \frac{K_w}{[OH^-]} = \frac{1.0 \times 10^{-14}}{2.8284 \times 10^{-3}} = 3.535568 \times 10^{-12} \; M \; H_3O^+$$

$$pH = -\log[H_3O^+] = -\log(3.535568 \times 10^{-12}) = 11.4515 = \textbf{11.45}$$

18.107 Plan: The K_b of a conjugate base is related to the K_a of the conjugate acid through the equation $K_w = K_a \times K_b$.
Solution:
a) Acetate ion, CH_3COO^-, is the conjugate base of acetic acid, CH_3COOH.
$$K_w = K_a \times K_b$$

$$K_b \text{ of } CH_3COO^- = \frac{K_w}{K_a} = \frac{1.0 \times 10^{-14}}{1.8 \times 10^{-5}} = 5.55556 \times 10^{-10} = \textbf{5.6} \times \textbf{10}^{-10}$$

b) Anilinium ion is the conjugate acid of the weak base aniline, $C_6H_5NH_2$.

$$K_a \text{ of } C_6H_5NH_3^+ = \frac{K_w}{K_b} = \frac{1.0 \times 10^{-14}}{4.0 \times 10^{-10}} = \mathbf{2.5 \times 10^{-5}}$$

18.109 Plan: The K_b of a conjugate base is related to the K_a of the conjugate acid through the equation $K_w = K_a \times K_b$.
Solution:
a) $HClO_2$ is the conjugate acid of chlorite ion, ClO_2^-.

$$K_b \text{ of } ClO_2^- = \frac{K_w}{K_a} = \frac{1.0 \times 10^{-14}}{1.1 \times 10^{-2}} = 9.0909 \times 10^{-13}$$

$pK_b = -\log (9.0909 \times 10^{-13}) = 12.04139 = \mathbf{12.04}$
b) $(CH_3)_2NH$ is the conjugate base of $(CH_3)_2NH_2^+$.

$$K_a \text{ of } (CH_3)_2NH_2^+ = \frac{K_w}{K_b} = \frac{1.0 \times 10^{-14}}{5.9 \times 10^{-4}} = 1.694915 \times 10^{-11}$$

$pK_a = -\log (1.694915 \times 10^{-11}) = 10.77085 = \mathbf{10.77}$

18.111 Plan: In part a), potassium cyanide, when placed in water, dissociates into potassium ions, K^+, and cyanide ions, CN^-. Potassium ion is the conjugate acid of a strong base, KOH, so K^+ does not react with water. Cyanide ion is the conjugate base of a weak acid, HCN, so it does react with a base-dissociation reaction. To find the pH first set up a reaction table and use K_b for CN^- to calculate $[OH^-]$. Find the K_b for CN^- from the equation $K_w = K_a \times K_b$.
In part b), the salt triethylammonium chloride in water dissociates into two ions: $(CH_3CH_2)_3NH^+$ and Cl^-. Chloride ion is the conjugate base of a strong acid so it will not influence the pH of the solution. Triethylammonium ion is the conjugate acid of a weak base, so an acid-dissociation reaction determines the pH of the solution. To find the pH first set up a reaction table and use K_a for $(CH_3CH_2)_3NH^+$ to calculate $[H_3O^+]$. Find the K_a for $(CH_3CH_2)_3NH^+$ from the equation $K_w = K_a \times K_b$.
Solution:
a) $CN^-(aq) + H_2O(l) \rightleftarrows HCN(aq) + OH^-(aq)$

Concentration (M)	$CN^-(aq)$	$+ H_2O(l)$	$\rightleftarrows$	$HCN(aq)$	$+ OH^-(aq)$
Initial	0.150	—		0	0
Change	−x			+x	+x
Equilibrium	0.150 − x			x	x

$$K_b \text{ of } CN^- = \frac{K_w}{K_a} = \frac{1.0 \times 10^{-14}}{6.2 \times 10^{-10}} = 1.612903 \times 10^{-5}$$

$$K_b = 1.612903 \times 10^{-5} = \frac{[HCN][OH^-]}{[CN^-]}$$

$$K_b = 1.612903 \times 10^{-5} = \frac{[x][x]}{[0.150 - x]} \qquad \text{Assume x is small compared to 0.150.}$$

$$K_b = 1.612903 \times 10^{-5} = \frac{(x)(x)}{(0.150)}$$

$x = 1.555 \times 10^{-3} M\ OH^-$
Check assumption that x is small compared to 0.150:

$$\frac{1.555 \times 10^{-3}}{0.150}(100) = 1\% \text{ error, so the assumption is valid.}$$

$$[H_3O]^+ = \frac{K_w}{[OH^-]} = \frac{1.0 \times 10^{-14}}{1.555 \times 10^{-3}} = 6.430868 \times 10^{-12} M\ H_3O^+$$

$pH = -\log [H_3O^+] = -\log (6.430868 \times 10^{-12}) = 11.19173 = \mathbf{11.19}$

b) $(CH_3CH_2)_3NH^+(aq) + H_2O(l) \leftrightarrows (CH_3CH_2)_3N(aq) + H_3O^+(aq)$

Concentration (M)	$(CH_3CH_2)_3NH^+(aq) + H_2O(l)$	$\leftrightarrows$	$(CH_3CH_2)_3N(aq) + H_3O^+(aq)$	
Initial	0.40	—	0	0
Change	−x		+x	+x
Equilibrium	0.40 − x		x	x

$$K_a \text{ of } (CH_3CH_2)_3NH^+ = \frac{K_w}{K_b} = \frac{1.0\text{x}10^{-14}}{5.2\text{x}10^{-4}} = 1.9230769\text{x}10^{-11}$$

$$K_a = 1.9230769\text{x}10^{-11} = \frac{[H_3O^+][(CH_3CH_2)_3N]}{[(CH_3CH_2)_3NH^+]}$$

$$K_a = 1.9230769\text{x}10^{-11} = \frac{(x)(x)}{(0.40 - x)} \qquad \text{Assume x is small compared to 0.40.}$$

$$K_a = 1.9230769\text{x}10^{-11} = \frac{(x)(x)}{(0.40)}$$

$[H_3O^+] = x = 2.7735\text{x}10^{-6} M$

Check assumption that x is small compared to 0.40:

$\dfrac{2.7735\text{x}10^{-6}}{0.40}(100) = 0.0007\%$ error, so the assumption is valid.

$pH = -\log [H_3O^+] = -\log (2.7735\text{x}10^{-6}) = 5.55697 = \mathbf{5.56}$

18.113 Plan: In part a), potassium formate, when placed in water, dissociates into potassium ions, K^+, and formate ions, $HCOO^-$. Potassium ion is the conjugate acid of a strong base, KOH, so K^+ does not react with water. Formate ion is the conjugate base of a weak acid, HCOOH, so it does react with a base-dissociation reaction. To find the pH first set up a reaction table and use K_b for $HCOO^-$ to calculate $[OH^-]$. Find the K_b for $HCOO^-$ from the equation $K_w = K_a \text{ x } K_b$. In part b), the salt ammonium bromide in water dissociates into two ions: NH_4^+ and Br^-. Bromide ion is the conjugate base of a strong acid so it will not influence the pH of the solution. Ammonium ion is the conjugate acid of the weak base NH_3, so an acid-dissociation reaction determines the pH of the solution. To find the pH first set up a reaction table and use K_a for NH_4^+ to calculate $[H_3O^+]$. Find the K_a for NH_4^+ from the equation $K_w = K_a \text{ x } K_b$.

Solution:

a) $HCOO^-(aq) + H_2O(l) \leftrightarrows HCOOH(aq) + OH^-(aq)$

Concentration (M)	$HCOO^-(aq)$ +	$H_2O(l)$	$\leftrightarrows$	$HCOOH(aq)$ +	$OH^-(aq)$
Initial	0.65	—		0	0
Change	−x			+x	+x
Equilibrium	0.65 − x			x	x

$$K_b \text{ of } HCOO^- = \frac{K_w}{K_a} = \frac{1.0\text{x}10^{-14}}{1.8\text{x}10^{-4}} = 5.55556\text{x}10^{-11}$$

$$K_b = 5.55556\text{x}10^{-11} = \frac{[HCOOH][OH^-]}{[HCOO^-]}$$

$$K_b = 5.55556\text{x}10^{-11} = \frac{[x][x]}{\left[0.65 - x\right]} \qquad \text{Assume x is small compared to 0.65.}$$

$$K_b = 5.55556\text{x}10^{-11} = \frac{(x)(x)}{(0.65)}$$

$x = 6.00925\text{x}10^{-6} M \; OH^-$

Check assumption that x is small compared to 0.65:

$\dfrac{6.00925\text{x}10^{-6}}{0.65}(100) = 0.0009\%$ error, so the assumption is valid.

$$[H_3O]^+ = \frac{K_w}{[OH^-]} = \frac{1.0 \times 10^{-14}}{6.00925 \times 10^{-6}} = 1.66410 \times 10^{-9} \, M \, H_3O^+$$

$$pH = -\log [H_3O^+] = -\log (1.66410 \times 10^{-9}) = 8.7788 = \mathbf{8.78}$$

b) $NH_4^+(aq) + H_2O(l) \leftrightarrows H_3O^+(aq) + NH_3(aq)$

Concentration (M)	$NH_4^+(aq)$	+	$H_2O(l)$	$\leftrightarrows$	$NH_3(aq)$	+	$H_3O^+(aq)$
Initial	0.85		—		0		0
Change	−x				+x		+x
Equilibrium	0.85 − x				x		x

$$K_a \text{ of } NH_4^+ = \frac{K_w}{K_b} = \frac{1.0 \times 10^{-14}}{1.76 \times 10^{-5}} = 5.681818 \times 10^{-10}$$

$$K_a = 5.681818 \times 10^{-10} = \frac{[H_3O^+][NH_3]}{[NH_4^+]}$$

$$K_a = 5.681818 \times 10^{-10} = \frac{[x][x]}{[0.85 - x]} \qquad \text{Assume x is small compared to 0.85.}$$

$$K_a = 5.681818 \times 10^{-10} = \frac{[x][x]}{[0.85]}$$

$[H_3O^+] = x = 2.1976 \times 10^{-5} \, M$

Check assumption that x is small compared to 0.85:

$$\frac{2.1976 \times 10^{-5}}{0.85} (100) = 0.003\% \text{ error, so the assumption is valid.}$$

$$pH = -\log [H_3O^+] = -\log (2.1976 \times 10^{-5}) = 4.65805 = \mathbf{4.66}$$

18.115 Plan: First, calculate the initial molarity of ClO⁻ from the mass percent. Then, set up reaction table with base dissociation of ClO⁻. Find the K_b for ClO⁻ from the equation $K_w = K_a \times K_b$, using the K_a for HClO from Appendix C.
Solution:

$$\text{Molarity of ClO}^- = \left(\frac{1 \text{ mL solution}}{10^{-3} \text{ L solution}} \right) \left(\frac{1.0 \text{ g solution}}{1 \text{ mL solution}} \right) \left(\frac{6.5\% \text{ NaClO}}{100\% \text{ Solution}} \right) \left(\frac{1 \text{ mol NaClO}}{74.44 \text{ g NaClO}} \right) \left(\frac{1 \text{ mol ClO}^-}{1 \text{ mol NaClO}} \right)$$

$$= 0.873186 \, M \, ClO^-$$

The sodium ion is from a strong base; therefore, it will not affect the pH, and can be ignored.

Concentration (M)	$ClO^-(aq)$	+	$H_2O(l)$	$\leftrightarrows$	$HClO(aq)$	+	$OH^-(aq)$
Initial	0.873186		—		0		0
Change	−x				+x		+x
Equilibrium	0.873186 − x				x		x

$$K_b \text{ of } ClO^- = \frac{K_w}{K_a} = \frac{1.0 \times 10^{-14}}{2.9 \times 10^{-8}} = 3.448275862 \times 10^{-7}$$

$$K_b = 3.448275862 \times 10^{-7} = \frac{[HClO][OH^-]}{[ClO^-]}$$

$$K_b = 3.448275862 \times 10^{-7} = \frac{[x][x]}{[0.873186 - x]} \qquad \text{Assume x is small compared to 0.873186.}$$

$$K_b = 3.448275862 \times 10^{-7} = \frac{(x)(x)}{(0.873186)}$$

$x = 5.4872 \times 10^{-4} = \mathbf{5.5 \times 10^{-4} \, M \, OH^-}$

Check assumption that x is small compared to 0.873186:

$$\frac{5.4872 \times 10^{-4}}{0.873186} (100) = 0.006\% \text{ error, so the assumption is valid.}$$

$$[H_3O]^+ = \frac{K_w}{[OH^-]} = \frac{1.0 \times 10^{-14}}{5.4872 \times 10^{-4}} = 1.82242 \times 10^{-11} \ M \ H_3O^+$$

$$pH = -\log[H_3O^+] = -\log(1.82242 \times 10^{-11}) = 10.73935 = \mathbf{10.74}$$

18.118 Sodium fluoride, NaF, contains the cation of a strong base, NaOH, and anion of a weak acid, HF. This combination yields a salt that is basic in aqueous solution as the F^- ion acts as a base:

$$F^-(aq) + H_2O(l) \ \leftrightarrows \ HF(aq) + OH^-(aq)$$

Sodium chloride, NaCl, is the salt of a strong base, NaOH, and strong acid, HCl. This combination yields a salt that is neutral in aqueous solution as neither Na^+ or Cl^- react in water to change the $[H_3O^+]$.

18.120 Plan: For each salt, first break into the ions present in solution and then determine if either ion acts as a weak acid or weak base to change the pH of the solution. Cations are neutral if they are from a strong base; other cations will be weakly acidic. Anions are neutral if they are from a strong acid; other anions are weakly basic.
Solution:

a) $KBr(s) \xrightarrow{\ H_2O\ } K^+(aq) + Br^-(aq)$

$\quad$ K^+ is the conjugate acid of a strong base, so it does not influence pH.
$\quad$ Br^- is the conjugate base of a strong acid, so it does not influence pH.
$\quad$ Since neither ion influences the pH of the solution, it will remain at the **neutral** pH of pure water.

b) $NH_4I(s) \xrightarrow{\ H_2O\ } NH_4^+(aq) + I^-(aq)$

$\quad$ NH_4^+ is the conjugate acid of a weak base, so it will act as a weak acid in solution and produce H_3O^+ as represented by the acid-dissociation reaction:

$$NH_4^+(aq) + H_2O(l) \leftrightarrows NH_3(aq) + H_3O^+(aq)$$

$\quad$ I^- is the conjugate base of a strong acid, so it will not influence the pH.
$\quad$ The production of H_3O^+ from the ammonium ion makes the solution of NH_4I **acidic**.

c) $KCN(s) \xrightarrow{\ H_2O\ } K^+(aq) + CN^-(aq)$

$\quad$ K^+ is the conjugate acid of a strong base, so it does not influence pH.
$\quad$ CN^- is the conjugate base of a weak acid, so it will act as a weak base in solution and impact pH by the base-dissociation reaction:

$$CN^-(aq) + H_2O(l) \leftrightarrows HCN(aq) + OH^-(aq)$$

$\quad$ Hydroxide ions are produced in this equilibrium so solution will be **basic**.

18.122 Plan: For each salt, first break into the ions present in solution and then determine if either ion acts as a weak acid or weak base to change the pH of the solution. Cations are neutral if they are from a strong base; other cations will be weakly acidic. Anions are neutral if they are from a strong acid; other anions are weakly basic.
Solution:

a) The two ions that comprise sodium carbonate, Na_2CO_3, are sodium ion, Na^+, and carbonate ion, CO_3^{2-}.

$\quad$ $Na_2CO_3(s) \xrightarrow{\ H_2O\ } 2Na^+(aq) + CO_3^{2-}(aq)$

$\quad$ Sodium ion is derived from the strong base NaOH. Carbonate ion is derived from the weak acid HCO_3^-. A salt derived from a strong base and a weak acid produces a **basic** solution.
$\quad$ Na^+ does not react with water.

$$CO_3^{2-}(aq) + H_2O(l) \leftrightarrows HCO_3^-(aq) + OH^-(aq)$$

b) The two ions that comprise calcium chloride, $CaCl_2$, are calcium ion, Ca^{2+}, and chloride ion, Cl^-.

$\quad$ $CaCl_2(s) \xrightarrow{\ H_2O\ } Ca^{2+}(aq) + 2Cl^-(aq)$

$\quad$ Calcium ion is derived from the strong base $Ca(OH)_2$. Chloride ion is derived from the strong acid HCl. A salt derived from a strong base and strong acid produces a **neutral** solution.
$\quad$ Neither Ca^{2+} nor Cl^- reacts with water.

c) The two ions that comprise cupric nitrate, $Cu(NO_3)_2$, are the cupric ion, Cu^{2+}, and the nitrate ion, NO_3^-.

$\quad$ $Cu(NO_3)_2(s) \xrightarrow{\ H_2O\ } Cu^{2+}(aq) + 2NO_3^-(aq)$

$\quad$ Small metal ions are acidic in water (assume the hydration of Cu^{2+} is 6):
$\quad$ $Cu(H_2O)_6^{2+}(aq) + H_2O(l) \leftrightarrows Cu(H_2O)_5OH^+(aq) + H_3O^+(aq)$
$\quad$ Nitrate ion is derived from the strong acid HNO_3. Therefore, NO_3^- does not react with water. A solution of cupric nitrate is **acidic**.

18.124   <u>Plan:</u> For each salt, first break into the ions present in solution and then determine if either ion acts as a weak acid or weak base to change the pH of the solution. Cations are neutral if they are from a strong base; other cations will be weakly acidic. Anions are neutral if they are from a strong acid; other anions are weakly basic.
<u>Solution:</u>
a) A solution of strontium bromide is **neutral** because Sr^{2+} is the conjugate acid of a strong base, $Sr(OH)_2$, and Br^- is the conjugate base of a strong acid, HBr, so neither change the pH of the solution.
b) A solution of barium acetate is **basic** because CH_3COO^- is the conjugate base of a weak acid and therefore forms OH^- in solution whereas Ba^{2+} is the conjugate acid of a strong base, $Ba(OH)_2$, and does not influence solution pH. The base-dissociation reaction of acetate ion is
$$CH_3COO^-(aq) + H_2O(l) \leftrightharpoons CH_3COOH(aq) + OH^-(aq).$$
c) A solution of dimethylammonium bromide is **acidic** because $(CH_3)_2NH_2^+$ is the conjugate acid of a weak base and therefore forms H_3O^+ in solution whereas Br^- is the conjugate base of a strong acid and does not influence the pH of the solution. The acid-dissociation reaction for methylammonium ion is
$$(CH_3)_2NH_2^+(aq) + H_2O(l) \leftrightharpoons (CH_3)_2NH(aq) + H_3O^+(aq).$$

18.126   <u>Plan:</u> For each salt, first break into the ions present in solution and then determine if either ion acts as a weak acid or weak base to change the pH of the solution. Cations are neutral if they are from a strong base; other cations will be weakly acidic. Anions are neutral if they are from a strong acid; other anions are weakly basic.
<u>Solution:</u>
a) The two ions that comprise ammonium phosphate, $(NH_4)_3PO_4$, are the ammonium ion, NH_4^+, and the phosphate ion, PO_4^{3-}.

$NH_4^+(aq) + H_2O(l) \leftrightharpoons NH_3(aq) + H_3O^+(aq)$ $K_a = K_w/K_b (NH_3) = 5.7 \times 10^{-10}$
$PO_4^{3-}(aq) + H_2O(l) \leftrightharpoons HPO_4^{2-}(aq) + OH^-(aq)$ $K_b = K_w/K_{a3} (H_3PO_4) = 2.4 \times 10^{-2}$

A comparison of K_a and K_b is necessary since both ions are derived from a weak base and weak acid. The K_a of NH_4^+ is determined by using the K_b of its conjugate base, NH_3 (Appendix). The K_b of PO_4^{3-} is determined by using the K_a of its conjugate acid, HPO_4^{2-}. The K_a of HPO_4^{2-} comes from K_{a3} of H_3PO_4 (Appendix). Since $K_b > K_a$, a solution of $(NH_4)_3PO_4$ is **basic**.
b) The two ions that comprise sodium sulfate, Na_2SO_4, are sodium ion, Na^+, and sulfate ion, SO_4^{2-}. The sodium ion is derived from the strong base NaOH. The sulfate ion is derived from the weak acid, HSO_4^-.
$$SO_4^{2-}(aq) + H_2O(l) \leftrightharpoons HSO_4^-(aq) + OH^-(aq)$$
A solution of sodium sulfate is **basic**.
c) The two ions that comprise lithium hypochlorite, LiClO, are lithium ion, Li^+, and hypochlorite ion, ClO^-. Lithium ion is derived from the strong base LiOH. Hypochlorite ion is derived from the weak acid, HClO (hypochlorous acid).
$$ClO^-(aq) + H_2O(l) \leftrightharpoons HClO(aq) + OH^-(aq)$$
A solution of lithium hypochlorite is **basic**.

18.128   <u>Plan:</u> For each salt, first break into the ions present in solution and then determine if either ion acts as a weak acid or weak base to change the pH of the solution. Cations are neutral if they are from a strong base; other cations will be weakly acidic. Anions are neutral if they are from a strong acid; other anions are weakly basic. Use K_a and K_b values to rank the pH; the larger the K_a value, the lower the pH and the larger the K_b value, the higher the pH.
<u>Solution:</u>
a) Order of increasing pH: **$Fe(NO_3)_2 < KNO_3 < K_2SO_3 < K_2S$** (assuming concentrations equivalent)
Iron(II) nitrate, $Fe(NO_3)_2$, is an acidic solution because the iron ion is a small, highly charged metal ion that acts as a weak acid and nitrate ion is the conjugate base of a strong acid, so it does not influence pH.
Potassium nitrate, KNO_3, is a neutral solution because potassium ion is the conjugate acid of a strong base and nitrate ion is the conjugate base of a strong acid, so neither influences solution pH.
Potassium sulfite, K_2SO_3, and potassium sulfide, K_2S, are similar in that the potassium ion does not influence solution pH, but the anions do because they are conjugate bases of weak acids. K_a for HSO_3^- is 6.5×10^{-8}, so K_b for SO_3^- is 1.5×10^{-7}, which indicates that sulfite ion is a weak base. K_a for HS^- is 1×10^{-17} (see the table of K_a values for polyprotic acids), so sulfide ion has a K_b equal to 1×10^3. Sulfide ion is thus a strong base. The solution of a strong base will have a greater concentration of hydroxide ions (and higher pH) than a solution of a weak base of equivalent concentrations.

b) In order of increasing pH: **NaHSO$_4$ < NH$_4$NO$_3$ < NaHCO$_3$ < Na$_2$CO$_3$**

In solutions of ammonium nitrate, only the ammonium will influence pH by dissociating as a weak acid:

$$NH_4^+(aq) + H_2O(l) \leftrightarrows NH_3(aq) + H_3O^+(aq)$$
$$\text{with } K_a = 1.0 \times 10^{-14}/1.8 \times 10^{-5} = 5.6 \times 10^{-10}$$

Therefore, the solution of ammonium nitrate is acidic.

In solutions of sodium hydrogen sulfate, only HSO_4^- will influence pH. The hydrogen sulfate ion is amphoteric so both the acid and base dissociations must be evaluated for influence on pH. As a base, HSO_4^- is the conjugate base of a strong acid, so it will not influence pH. As an acid, HSO_4^- is the conjugate acid of a weak base, so the acid dissociation applies:

$$HSO_4^-(aq) + H_2O(l) \leftrightarrows SO_4^{2-}(aq) + H_3O^+(aq) \quad K_{a2} = 1.2 \times 10^{-2}$$

In solutions of sodium hydrogen carbonate, only the HCO_3^- will influence pH and it, like HSO_4^-, is amphoteric:

As an acid: $HCO_3^-(aq) + H_2O(l) \leftrightarrows CO_3^{2-}(aq) + H_3O^+(aq)$
$K_a = 4.7 \times 10^{-11}$, the second K_a for carbonic acid

As a base: $HCO_3^-(aq) + H_2O(l) \leftrightarrows H_2CO_3(aq) + OH^-(aq)$
$K_b = 1.0 \times 10^{-14}/4.5 \times 10^{-7} = 2.2 \times 10^{-8}$

Since $K_b > K_a$, a solution of sodium hydrogen carbonate is basic.

In a solution of sodium carbonate, only CO_3^{2-} will influence pH by acting as a weak base:

$$CO_3^{2-}(aq) + H_2O(l) \leftrightarrows HCO_3^-(aq) + OH^-(aq)$$
$$K_b = 1.0 \times 10^{-14}/4.7 \times 10^{-11} = 2.1 \times 10^{-4}$$

Therefore, the solution of sodium carbonate is basic.

Two of the solutions are acidic. Since the K_a of HSO_4^- is greater than that of NH_4^+, the solution of sodium hydrogen sulfate has a lower pH than the solution of ammonium nitrate, assuming the concentrations are relatively close.

Two of the solutions are basic. Since the K_b of CO_3^{2-} is greater than that of HCO_3^-, the solution of sodium carbonate has a higher pH than the solution of sodium hydrogen carbonate, assuming concentrations are not extremely different.

18.130   Both methoxide ion and amide ion produce OH^- in aqueous solution. In water, the strongest base possible is OH^-. Since both bases produce OH^- in water, both bases appear equally strong.
$$CH_3O^-(aq) + H_2O(l) \rightarrow OH^-(aq) + CH_3OH(aq)$$
$$NH_2^-(aq) + H_2O(l) \rightarrow OH^-(aq) + NH_3(aq)$$

18.132   Ammonia, NH_3, is a more basic solvent than H_2O. In a more basic solvent, weak acids like HF act like strong acids and are 100% dissociated.

18.134   A Lewis acid is defined as an electron-pair acceptor, while a Brønsted-Lowry acid is a proton donor. If only the proton in a Brønsted-Lowry acid is considered, then every Brønsted-Lowry acid fits the definition of a Lewis acid since the proton is accepting an electron pair when it bonds with a base. There are Lewis acids that do not include a proton, so all Lewis acids are not Brønsted-Lowry acids.

A Lewis base is defined as an electron-pair donor and a Brønsted-Lowry base is a proton acceptor. In this case, the two definitions are essentially the same.

18.135   a) **No**, a weak Brønsted-Lowry base is not necessarily a weak Lewis base. For example, the following equation shows that the weak Bronsted-Lowry base NH_3 is a good Lewis base.
$$Ni(H_2O)_6^{2+}(aq) + 6\,NH_3(aq) \leftrightarrows Ni(NH_3)_6^{2+}(aq) + 6H_2O(l)$$
b) The **cyanide ion** has a lone pair to donate from either the C or the N, and donates an electron pair to the $Cu(H_2O)_6^{2+}$ complex. It is the Lewis base for the forward direction of this reaction. In the reverse direction, **water** donates one of the electron pairs on the oxygen to the $Cu(CN)_4^{2-}$ and is the Lewis base.
c) Because $K_c > 1$, the reaction proceeds in the direction written (left to right) and is driven by the stronger Lewis base, the **cyanide ion**.

18.138   Plan: A Lewis acid is an electron-pair acceptor and therefore must be able to accept an electron pair. A Lewis base is an electron-pair donor and therefore must have an electron pair to donate.

Solution:

a) Cu^{2+} is a **Lewis acid** because it accepts electron pairs from molecules such as water.

b) Cl^- is a **Lewis base** because it has lone pairs of electrons it can donate to a Lewis acid.

c) Tin(II) chloride, $SnCl_2$, is a compound with a structure similar to carbon dioxide, so it will act as a **Lewis acid** to form an additional bond to the tin.

d) Oxygen difluoride, OF_2, is a **Lewis base** with a structure similar to water, where the oxygen has lone pairs of electrons that it can donate to a Lewis acid.

18.140 Plan: A Lewis acid is an electron-pair acceptor and therefore must be able to accept an electron pair. A Lewis base is an electron-pair donor and therefore must have an electron pair to donate.

Solution:

a) The boron atom in boron trifluoride, BF_3, is electron deficient (has six electrons instead of eight) and can accept an electron pair; it is a **Lewis acid**.

b) The sulfide ion, S^{2-}, can donate any of four electron pairs and is a **Lewis base**.

c) The Lewis dot structure for the sulfite ion, SO_3^{2-}, shows lone pairs on the sulfur and on the oxygen atoms. The sulfur atom has a lone electron pair that it can donate more easily than the electronegative oxygen in the formation of an adduct. The sulfite ion is a **Lewis base**.

d) Sulfur trioxide, SO_3, acts as a **Lewis acid**.

18.142 Plan: A Lewis acid is an electron-pair acceptor while a Lewis base is an electron-pair donor.

Solution:

a) Sodium ion is the Lewis acid because it is accepting electron pairs from water, the Lewis base.

$$Na^+ \quad + \quad 6H_2O \quad \leftrightarrows \quad Na(H_2O)_6^+$$
Lewis acid　　　　Lewis base　　　　adduct

b) The oxygen from water donates a lone pair to the carbon in carbon dioxide. Water is the Lewis base and carbon dioxide the Lewis acid.

$$CO_2 \quad + \quad H_2O \quad \leftrightarrows \quad H_2CO_3$$
Lewis acid　　　　Lewis base　　　　adduct

c) Fluoride ion donates an electron pair to form a bond with boron in BF_4^-. The fluoride ion is the Lewis base and the boron trifluoride is the Lewis acid.

$$F^- \quad + \quad BF_3 \quad \leftrightarrows \quad BF_4^-$$
Lewis base　　　　Lewis acid　　　　adduct

18.144 Plan: In an Arrhenius acid-base reaction, H^+ ions react with OH^- ions to produce H_2O. In a Brønsted-Lowry acid-base reaction, an acid donates H^+ to a base. In a Lewis acid-base reaction, an electron pair is donated by the base and accepted by the acid.

Solution:

a) Since neither H^+ nor OH^- is involved, this is not an Arrhenius acid-base reaction. Since there is no exchange of protons, this is not a Brønsted-Lowry reaction. This reaction is only classified as **Lewis acid-base reaction**, where Ag^+ is the acid and NH_3 is the base.

b) Again, no OH^- is involved, so this is not an Arrhenius acid-base reaction. This is an exchange of a proton, from H_2SO_4 to NH_3, so it is a **Brønsted-Lowry acid-base reaction**. Since the Lewis definition is most inclusive, anything that is classified as a Brønsted-Lowry (or Arrhenius) reaction is automatically classified as a **Lewis acid-base reaction**.

c) This is not an acid-base reaction.

d) For the same reasons listed in a), this reaction is only classified as **Lewis acid-base reaction**, where $AlCl_3$ is the acid and Cl^- is the base.

18.147 Plan: Calculate the $[H_3O^+]$ using the pH values given. Determine the value of K_w from the pK_w given. The $[H_3O^+]$ is combined with the K_w value at 37°C to find $[OH^-]$ using $K_w = [H_3O^+][OH^-]$.

Solution:

$$K_w = 10^{-pK_w} = 10^{-13.63} = 2.34423 \times 10^{-14}$$
$$K_w = [H_3O^+][OH^-] = 2.34423 \times 10^{-14} \text{ at } 37°C$$

[H_3O^+] range

High value (low pH) = 10^{-pH} = $10^{-7.35}$ = 4.46684×10^{-8} = 4.5×10^{-8} M H_3O^+

Low value (high pH) = 10^{-pH} = $10^{-7.45}$ = 3.54813×10^{-8} = 3.5×10^{-8} M H_3O^+

Range: **3.5×10^{-8} to 4.5×10^{-8} M H_3O^+**

[OH^-] range

K_w = [H_3O^+][OH^-] = 2.34423×10^{-14} at 37°C

[OH^-] = $\dfrac{K_w}{[H_3O]^+}$

High value (high pH) = $\dfrac{2.34423 \times 10^{-14}}{3.54813 \times 10^{-8}}$ = 6.60695×10^{-7} = 6.6×10^{-7} M OH^-

Low value (low pH) = $\dfrac{2.34423 \times 10^{-14}}{4.46684 \times 10^{-8}}$ = 5.24807×10^{-7} = 5.2×10^{-7} M OH^-

Range: **5.2×10^{-7} to 6.6×10^{-7} M OH^-**

18.148 a) Acids will vary in the amount they dissociate (acid strength) depending on the acid-base character of the solvent. Water and methanol have different acid-base characters.

b) The K_a is the measure of an acid's strength. A stronger acid has a smaller pK_a. Therefore, phenol is a stronger acid in water than it is in methanol. In other words, water more readily accepts a proton from phenol than does methanol, i.e., methanol is a weaker base than water.

c) C_6H_5OH(*solvated*) + CH_3OH(*l*) ⇌ $CH_3OH_2^+$(*solvated*) + $C_6H_5O^-$(*solvated*)

The term "*solvated*" is analogous to "*aqueous.*" "*Aqueous*" would be incorrect in this case because the reaction does not take place in water.

d) In the autoionization process, one methanol molecule is the proton donor while another methanol molecule is the proton acceptor.

CH_3OH(*l*) + CH_3OH(*l*) ⇌ CH_3O^-(*solvated*) + $CH_3OH_2^+$(*solvated*)

In this equation "(*solvated*)" indicates that the molecules are solvated by methanol.

The equilibrium constant for this reaction is the autoionization constant of methanol:

K = [CH_3O^-][$CH_3OH_2^+$]

18.151 Plan: A Lewis acid is an electron-pair acceptor and a Lewis base is an electron-pair donor. Recall that n is the main energy level and l is the orbital type.

Solution:

a) $SnCl_4$ is the Lewis acid accepting an electron pair from $(CH_3)_3N$, the Lewis base.

b) Tin is the element in the Lewis acid accepting the electron pair. The electron configuration of tin is $[Kr]5s^2 4d^{10} 5p^2$. The four bonds to tin are formed by sp^3 hybrid orbitals, which completely fill the $5s$ and $5p$ orbitals. The **$5d$** orbitals are empty and available for the bond with trimethylamine.

18.152 Plan: A 10-fold dilution means that the chemist takes 1 mL of the 1.0×10^{-5} M solution and dilutes it to 10 mL (or dilute 10 mL to 100 mL). The chemist then dilutes the diluted solution in a 1:10 ratio, and repeats this process for the next two successive dilutions. $M_1V_1 = M_2V_2$ can be used to find the molarity after each dilution. After each dilution, find [H_3O^+] and calculate the pH.

Solution:

Hydrochloric acid is a strong acid that completely dissociates in water. Therefore, the concentration of H_3O^+ is the same as the starting acid concentration: [H_3O^+] = [HCl]. The original solution pH:

pH = $-\log$ (1.0×10^{-5}) = **5.00 = pH**

Dilution 1: $M_1V_1 = M_2V_2$

(1.0×10^{-5} M)(1.0 mL) = (x)(10. mL)

[H_3O^+]$_{HCl}$ = 1.0×10^{-6} M H_3O^+

pH = $-\log$ (1.0×10^{-6}) = **6.00**

Dilution 2:

$$(1.0\text{x}10^{-6}\ M)(1.0\ \text{mL}) = (x)(10.\ \text{mL})$$

$$[H_3O^+]_{HCl} = 1.0\text{x}10^{-7}\ M\ H_3O^+$$

Once the concentration of strong acid is close to the concentration of H_3O^+ from water autoionization, the $[H_3O^+]$ in the solution does not equal the initial concentration of the strong acid. The calculation of $[H_3O^+]$ must be based on the water ionization equilibrium:

$$H_2O(l) + H_2O(l) \rightleftarrows H_3O^+(aq) + OH^-(aq)\ \text{with}\ K_w = 1.0\text{x}10^{-14}\ \text{at}\ 25°C.$$

The dilution gives an initial $[H_3O^+]$ of $1.0\text{x}10^{-7}\ M$. Assuming that the initial concentration of hydroxide ions is zero, a reaction table is set up.

Concentration (M)	$2H_2O(l)$	$\rightleftarrows$	$H_3O^+(aq)$	$+$	$OH^-(aq)$
Initial	—		$1\text{x}10^{-7}$		0
Change	—		$+x$		$+x$
Equilibrium	—		$1\text{x}10^{-7} + x$		x

$$K_w = [H_3O^+][OH^-] = (1\text{x}10^{-7} + x)(x) = 1.0\text{x}10^{-14}$$

Set up as a quadratic equation: $x^2 + 1.0\text{x}10^{-7}\ x - 1.0\text{x}10^{-14} = 0$

$$a = 1 \quad b = 1.0\text{x}10^{-7} \quad c = -1.0\text{x}10^{-14}$$

$$x = \frac{-1.0\text{x}10^{-7} \pm \sqrt{\left(1.0\text{x}10^{-7}\right)^2 - 4(1)\left(-1.0\text{x}10^{-14}\right)}}{2(1)}$$

$$x = 6.18034\text{x}10^{-8}$$

$$[H_3O^+] = (1.0\text{x}10^{-7} + x)\ M = (1.0\text{x}10^{-7} + 6.18034\text{x}10^{-8})\ M = 1.618034\text{x}10^{-7}\ M\ H_3O^+$$

$$pH = -\log [H_3O^+] = -\log (1.618034\text{x}10^{-7}) = 6.79101 = \mathbf{6.79}$$

Dilution 3:

$$(1.0\text{x}10^{-7}\ M)(1.0\ \text{mL}) = (x)(10.\ \text{mL})$$

$$[H_3O^+]_{HCl} = 1.0\text{x}10^{-8}\ M\ H_3O^+$$

The dilution gives an initial $[H_3O^+]$ of $1.0\text{x}10^{-8}\ M$. Assuming that the initial concentration of hydroxide ions is zero, a reaction table is set up.

Concentration (M)	$2H_2O(l)$	$\rightleftarrows$	$H_3O^+(aq)$	$+$	$OH^-(aq)$
Initial	—		$1\text{x}10^{-8}$		0
Change	—		$+x$		$+x$
Equilibrium	—		$1\text{x}10^{-8} + x$		x

$$K_w = [H_3O^+][OH^-] = (1\text{x}10^{-8} + x)(x) = 1.0\text{x}10^{-14}$$

Set up as a quadratic equation: $x^2 + 1.0\text{x}10^{-8}\ x - 1.0\text{x}10^{-14} = 0$

$$a = 1 \quad b = 1.0\text{x}10^{-8} \quad c = -1.0\text{x}10^{-14}$$

$$x = \frac{-1.0\text{x}10^{-8} \pm \sqrt{\left(1.0\text{x}10^{-8}\right)^2 - 4(1)\left(-1.0\text{x}10^{-14}\right)}}{2(1)}$$

$$x = 9.51249\text{x}10^{-8}$$

$$[H_3O^+] = (1.0\text{x}10^{-8} + x)\ M = (1.0\text{x}10^{-8} + 9.51249\text{x}10^{-8})\ M = 1.051249\text{x}10^{-7}\ M\ H_3O^+$$

$$pH = -\log [H_3O^+] = -\log (1.051249\text{x}10^{-7}) = 6.97829 = \mathbf{6.98}$$

Dilution 4:

$$(1.0\text{x}10^{-8}\ M)(1.0\ \text{mL}) = (x)(10.\ \text{mL})$$

$$[H_3O^+]_{HCl} = 1.0\text{x}10^{-9}\ M\ H_3O^+$$

The dilution gives an initial $[H_3O^+]$ of $1.0\text{x}10^{-9}\ M$. Assuming that the initial concentration of hydroxide ions is zero, a reaction table is set up.

Concentration (M)	$2H_2O(l)$	$\rightleftharpoons$	$H_3O^+(aq)$	+	$OH^-(aq)$
Initial	—		1×10^{-9}		0
Change	—		$+x$		$+x$
Equilibrium	—		$1\times10^{-9}+x$		x

$K_w = [H_3O^+][OH^-] = (1\times10^{-9}+x)(x) = 1.0\times10^{-14}$

Set up as a quadratic equation: $x^2 + 1.0\times10^{-9}\,x - 1.0\times10^{-14} = 0$

$$a = 1 \quad b = 1.0\times10^{-9} \quad c = -1.0\times10^{-14}$$

$$x = \frac{-1.0\times10^{-9} \pm \sqrt{\left(1.0\times10^{-9}\right)^2 - 4(1)\left(-1.0\times10^{-14}\right)}}{2(1)}$$

$x = 9.95012\times10^{-8}$

$[H_3O^+] = (1.0\times10^{-9}+x)\,M = (1.0\times10^{-9} + 9.95012\times10^{-8})\,M = 1.005012\times10^{-7}\,M\ H_3O^+$

$pH = -\log[H_3O^+] = -\log(1.005012\times10^{-7}) = 6.9978 = \textbf{7.00}$

As the HCl solution is diluted, the pH of the solution becomes closer to 7.0. Continued dilutions will not significantly change the pH from 7.0. Thus, a solution with a basic pH cannot be made by adding acid to water.

18.158 Plan: Determine the hydrogen ion concentration from the pH. The molarity and the volume will give the number of moles, and with the aid of Avogadro's number, the number of ions may be found.
Solution:

$M\ H_3O^+ = 10^{-pH} = 10^{-6.2} = 6.30957\times10^{-7}\,M$

$$\left(\frac{6.30957\times10^{-7}\ \text{mol } H_3O^+}{L}\right)\left(\frac{10^{-3}\ L}{1\ mL}\right)\left(\frac{1250.\ mL}{d}\right)\left(\frac{7\ d}{1\ wk}\right)\left(\frac{6.022\times10^{23}\ H_3O^+}{1\ \text{mol } H_3O^+}\right) = 3.32467\times10^{18} = \textbf{3}\times\textbf{10}^{\textbf{18}}\ \textbf{H}_3\textbf{O}^+$$

The pH has only one significant figure, and limits the significant figures in the final answer.

18.161 Plan: Determine K_b using the relationship $K_b = 10^{-pK_b}$. Write the base-dissociation equation and set up a reaction table in which x = the amount of OH^- produced. Use the K_b expression to find x. From $[OH^-]$, $[H_3O^+]$ and then pH can be calculated.
Solution:

$K_b = 10^{-pK} = 10^{-5.91} = 1.23027\times10^{-6}$

	$TRIS(aq)$	+	$H_2O(l)$	$\rightleftharpoons$	$OH^-(aq)$	+	$HTRIS^+(aq)$
Initial	0.075		—		0		0
Change	$-x$				$+x$		$+x$
Equilibrium	$0.075 - x$				x		x

$$K_b = 1.23027\times10^{-6} = \frac{[HTRIS^+][OH^-]}{[TRIS]}$$

$$K_b = 1.23027\times10^{-6} = \frac{[x][x]}{\left[0.075 - x\right]} \qquad \text{Assume } x \text{ is small compared to } 0.075.$$

$$K_b = 1.23027\times10^{-6} = \frac{[x][x]}{[0.075]}$$

$x = [OH^-] = 3.03760\times10^{-4}\,M\ OH^-$

Check assumption that x is small compared to 0.075:

$$\frac{3.03760\times10^{-4}}{0.075}(100) = 0.40\%\ \text{error, so the assumption is valid.}$$

$$[H_3O]^+ = \frac{K_w}{[OH^-]} = \frac{1.0\times10^{-14}}{3.03760\times10^{-4}} = 3.292073\times10^{-11}\,M$$

$pH = -\log[H_3O^+] = -\log(3.292073\times10^{-11}) = 10.4825 = \textbf{10.48}$

18.163 The pH is dependent on the *molar* concentration of H_3O^+. Convert % w/v to molarity, and use the K_a of acetic acid to determine $[H_3O^+]$ from the equilibrium expression.

Convert % w/v to molarity using the molecular weight of acetic acid (CH_3COOH):

$$Molarity = \left(\frac{5.0 \text{ g } CH_3COOH}{100 \text{ mL solution}}\right)\left(\frac{1 \text{ mol } CH_3COOH}{60.05 \text{ g } CH_3COOH}\right)\left(\frac{1 \text{ mL}}{10^{-3} \text{ L}}\right) = 0.832639 \text{ M } CH_3COOH$$

Acetic acid dissociates in water according to the following equation and equilibrium expression:

$$CH_3COOH(aq) + H_2O(l) \leftrightarrows CH_3COO^-(aq) + H_3O^+(aq)$$

Initial	0.832639	—	0	0
Change	–x		+x	+x
Equilibrium	0.832639 – x		x	x

$$K_a = 1.8 \times 10^{-5} = \frac{[H_3O^+][CH_3COO^-]}{[CH_3COOH]}$$

$$K_a = 1.8 \times 10^{-5} = \frac{[x][x]}{[0.832639 - x]} \qquad \text{Assume x is small compared to 0.832639.}$$

$$K_a = 1.8 \times 10^{-5} = \frac{[x][x]}{[0.832639]}$$

$$x = 3.8714 \times 10^{-3} \text{ M} = [H_3O^+]$$

Check assumption: $[3.871 \times 10^{-3}/0.832639] \times 100\% = 0.46\%$, therefore the assumption is good.

$pH = -\log [H_3O^+] = -\log (3.8714 \times 10^{-3}) = 2.412132 = \textbf{2.41}$

18.166 Plan: Assuming that the pH in the specific cellular environment is equal to the optimum pH for the enzyme, the hydronium ion concentrations are $[H_3O^+] = 10^{-pH}$.

Solution:

Salivary amylase, mouth: $[H_3O^+] = 10^{-6.8} = 1.58489 \times 10^{-7} = \textbf{2} \times \textbf{10}^{-7} \textbf{M}$

Pepsin, stomach: $[H_3O^+] = 10^{-2.0} = \textbf{1} \times \textbf{10}^{-2} \textbf{M}$

Trypsin, pancreas: $[H_3O^+] = 10^{-9.5} = 3.1623 \times 10^{-10} = \textbf{3} \times \textbf{10}^{-10} \textbf{M}$

18.170 The freezing point depression equation is required to determine the molality of the solution.

$$\Delta T = [0.00 - (-1.93°C)] = 1.93°C = iK_f m$$

Temporarily assume $i = 1$.

$$m = \frac{\Delta T}{iK_f} = \frac{1.93°C}{(1)(1.86°C/m)} = 1.037634 \text{ m} = 1.037634 \text{ M}$$

This molality is the total molality of all species in the solution, and is equal to their molarity.

From the equilibrium:

$$ClCH_2COOH(aq) + H_2O(l) \leftrightarrows H_3O^+(aq) + ClCH_2COO^-(aq)$$

Initial	1.000 M	x	x
Change	–x	+x	+x
Equilibrium	1.000 – x	x	x

The total concentration of all species is:

$[ClCH_2COOH] + [H_3O^+] + [ClCH_2COO^-] = 1.037634 \text{ M}$

$[1.000 - x] + [x] + [x] = 1.000 + x = 1.037634 \text{ M}$

$x = 0.037634 \text{ M}$

$$K_a = \frac{[H_3O^+][CH_3COO^-]}{[CH_3COOH]}$$

$$K_a = \frac{(0.037634)(0.037634)}{(1.000 - 0.037634)} = 0.0014717 = \textbf{0.00147}$$

Check the assumption by calculating the percent error in $[HBrO]_{eq}$.

$$\text{Percent error} = \frac{2.144761 \times 10^{-5}}{0.2000}(100) = 0.01\%; \text{ this is well below the 5\% maximum.}$$

b) When $[HBrO] = [BrO^-]$, the solution contains significant concentrations of both the weak acid and its conjugate base. Use the equilibrium expression for reaction 2a to find pH.

Since $[HBrO] = [BrO^-]$, their ratio equals 1.

$$K_a = \frac{[H_3O^+][BrO^-]}{[HBrO]}$$

$$[H_3O^+] = K_a \frac{[HBrO]}{[BrO^-]} = (2.3 \times 10^{-9})[1] = 2.3 \times 10^{-9}\ M$$

$$pH = -\log [H_3O^+] = -\log (2.3 \times 10^{-9}) = 8.63827 = \mathbf{8.64}$$

Note that when $[HBrO] = [BrO^-]$, the titration is at the midpoint (half the volume to the equivalence point) and $pH = pK_a$.

c) At the equivalence point, the total number of moles of HBrO present initially in solution equals the number of moles of base added. Therefore, reaction 1 goes to completion to produce that number of moles of BrO^-. The solution consists of BrO^- and water. Calculate the concentration of BrO^-, and then find the pH using the base dissociation equilibrium, reaction 2b.

First, find equivalence point volume of NaOH.

$$\text{Volume (mL) of NaOH} = \left(\frac{0.2000\ \text{mol HBrO}}{L}\right)\left(\frac{10^{-3}\ L}{1\ \text{mL}}\right)(20.00\ \text{mL})\left(\frac{1\ \text{mol NaOH}}{1\ \text{mol HBrO}}\right)\left(\frac{1\ L}{0.1000\ \text{mol NaOH}}\right)\left(\frac{1\ \text{mL}}{10^{-3}\ L}\right)$$

$$= 40.00\ \text{mL NaOH added}$$

All of the HBrO present at the beginning of the titration is neutralized and converted to BrO^- at the equivalence point. Calculate the concentration of BrO^-.

Initial moles of HBrO: $(0.2000\ M)(0.02000\ L) = 0.004000\ \text{mol}$

Moles of added NaOH: $(0.1000\ M)(0.04000\ L) = 0.004000\ \text{mol}$

Amount (mol)	HBrO(aq)	+	OH⁻(aq)	→	H₂O(l)	+	BrO⁻(aq)
Before addition	0.004000 mol		—		—		0
Addition	—		0.004000 mol		—		—
Change	– 0.004000 mol		– 0.004000 mol		—		+0.004000 mol
After addition	0		0		—		0.004000 mol

At the equivalence point, 40.00 mL of NaOH solution has been added (see calculation above) to make the total volume of the solution (20.00 + 40.00) mL = 60.00 mL.

$$[BrO^-] = \left(\frac{0.004000\ \text{mol BrO}^-}{60.00\ \text{mL}}\right)\left(\frac{1\ \text{mL}}{10^{-3}\ L}\right) = 0.06666667\ M$$

Set up reaction table with reaction 2b, since only BrO^- and water are present initially:

Concentration (M)	BrO⁻(aq)	+	H₂O(l)	⇌	HBrO(aq)	+	OH⁻(aq)
Initial	0.06666667		—		0		0
Change	– x		—		+x		+x
Equilibrium	0.06666667 – x		—		x		x

$$K_b = K_w/K_a = (1.0 \times 10^{-14}/2.3 \times 10^{-9}) = 4.347826 \times 10^{-6}$$

$$K_b = \frac{[OH^-][HBrO]}{[BrO^-]} = \frac{[x][x]}{[0.06666667 - x]} = 4.347826 \times 10^{-6}$$

Assume that x is negligible, since $[BrO^-] \gg K_b$.

$$4.347826 \times 10^{-6} = \frac{[x][x]}{[0.06666667]}$$

$$x = [OH^-] = 5.3838191 \times 10^{-4} = 5.4 \times 10^{-4}\ M$$

Check the assumption by calculating the percent error in $[BrO^-]_{eq}$.

Percent error = $\dfrac{5.3838191 \times 10^{-4}}{0.06666667}$ (100) = 0.8%, which is well below the 5% maximum.

pOH = –log (5.3838191×10^{-4}) = 3.26891

pH = 14 – pOH = 14 – 3.26891 = 10.73109 = **10.73**

d) After the equivalence point, the concentration of excess strong base determines the pH. Find the concentration of excess base and use it to calculate the pH.

Initial moles of HBrO: (0.2000 M)(0.02000 L) = 0.004000 mol

Moles of added NaOH: 2 x 0.004000 mol = 0.008000 mol NaOH

Amount (mol)	HBrO(aq)	+	OH$^-$(aq)	→	H$_2$O(l)	+	BrO$^-$(aq)
Before addition	0.004000 mol		—		—		0
Addition	—		0.008000 mol		—		—
Change	– 0.004000 mol		– 0.004000 mol		—		+0.004000 mol
After addition	0		0.004000 mol		—		0.004000 mol

Excess NaOH: 0.004000 mol

Volume (mL) of added NaOH = (0.0080000 mol NaOH)$\left(\dfrac{1 \text{ L}}{0.1000 \text{ mol NaOH}}\right)\left(\dfrac{1 \text{ mL}}{10^{-3} \text{ L}}\right)$

= 80.00 mL

Total volume: 20.00 mL + 80.00 mL = 100.0 mL

[NaOH] = 0.004000 mol NaOH/0.1000 L = 0.0400 M

pOH = –log (0.0400) = 1.3979

pH = 14 – pOH = 14 – 1.3979 = **12.60**

e) Plot the pH values calculated in the preceding parts of this problem as a function of the volume of titrant.

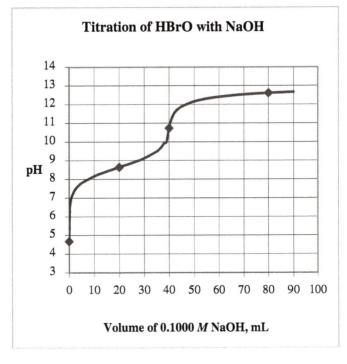

Titration of HBrO with NaOH

pH vs. Volume of 0.1000 M NaOH, mL

The plot and pH values follow the pattern for a weak acid vs. strong base titration. The pH at the midpoint of the titration does equal pK_a. The equivalence point should be, and is, greater than 7.

Moles of added NaOH: (0.1500 M)(0.02200 L) = 0.003300 mol

Amount (mol)	$C_6H_5COOH(aq)$	+	$OH^-(aq)$	$\rightarrow$	$H_2O(l)$	+	$C_6H_5COO^-(aq)$
Before addition	0.003000 mol		—		—		0
Addition	—		0.003300 mol		—		—
Change	– 0.003000 mol		– 0.003000 mol		—		+0.003000 mol
After addition	0		0.000300 mol		—		0.003000 mol

Excess NaOH: 0.000300 mol
Total volume: 22.00 mL + 30.00 mL = 52.00 mL (0.05200 L)
[NaOH] = 0.000300 mol NaOH/0.05200 L = 0.00576923 M
pOH = –log (0.0056923) = 2.2388
pH = 14 – pOH = 14 – 2.2388 = **11.76** (pH is shown to two decimal places)

19.5A   Plan: Write the formula of the salt and the reaction showing the equilibrium of a saturated solution. The ion-product expression can be written from the stoichiometry of the solution reaction as the coefficients in the reaction become exponents in the ion-product expression.
Solution:
a) The formula of calcium sulfate is $CaSO_4$. The equilibrium reaction is:
$$CaSO_4(s) \leftrightarrows Ca^{2+}(aq) + SO_4^{2-}(aq)$$
Ion–product expression: $K_{sp} = [Ca^{2+}][SO_4^{2-}]$
b) Chromium(III) carbonate is $Cr_2(CO_3)_3$.
$$Cr_2(CO_3)_3(s) \leftrightarrows 2Cr^{3+}(aq) + 3CO_3^{2-}(aq)$$
Ion–product expression: $K_{sp} = [Cr^{3+}]^2[CO_3^{2-}]^3$
c) Magnesium hydroxide is $Mg(OH)_2$.
$$Mg(OH)_2(s) \leftrightarrows Mg^{2+}(aq) + 2OH^-(aq)$$
Ion–product expression: $K_{sp} = [Mg^{2+}][OH^-]^2$
d) Aluminum hydroxide is $Al(OH)_3$.
$$Al(OH)_3(s) \leftrightarrows Al^{3+}(aq) + 3OH^-(aq)$$
Ion-product expression: $K_{sp} = [Al^{3+}][OH^-]^3$

19.5B   Plan: Examine the ion-product expressions. The exponents in the ion-product expression are the subscripts of those ions in the chemical formula.
Solution:
a) The compound is **lead(II) chromate**. Its formula is **$PbCrO_4$**.
b) The compound is **iron(II) sulfide**. Its formula is **FeS**.
c) The compound is **strontium fluoride**. Its formula is **SrF_2**.
d) The compound is **copper(II) phosphate**. Its formula is **$Cu_3(PO_4)_2$**.

19.6A   Plan: Calculate the solubility of CaF_2 as molarity and use molar ratios to find the molarity of Ca^{2+} and F^- dissolved in solution. Calculate K_{sp} from [Ca^{2+}] and [F^-] using the ion-product expression.
Solution: Convert the solubility to molar solubility:

$$Molarity = \left(\frac{1.5 \times 10^{-4} \text{ g } CaF_2}{10.0 \text{ mL}}\right)\left(\frac{1 \text{ mL}}{10^{-3} \text{ L}}\right)\left(\frac{1 \text{ mol } CaF_2}{78.08 \text{ g } CaF_2}\right) = 1.9211 \times 10^{-4} \text{ } M \text{ } CaF_2$$

[Ca^{2+}] = [CaF_2] = 1.9211×10^{-4} M because there is 1 mol of calcium ions in each mol of CaF_2.
[F^-] = 2[CaF_2] = 3.8422×10^{-4} M because there are 2 mol of fluoride ions in each mol of CaF_2.
The solubility equilibrium is:
$$CaF_2(s) \leftrightarrows Ca^{2+}(aq) + 2F^-(aq) \quad K_{sp} = [Ca^{2+}][F^-]^2$$
Calculate K_{sp} using the solubility product expression from above and the saturated concentrations of calcium and fluoride ions.
$K_{sp} = [Ca^{2+}][F^-]^2 = (1.9211 \times 10^{-4})(3.8422 \times 10^{-4})^2 = 2.836024 \times 10^{-11} = $ **2.8×10^{-11}**
The K_{sp} for CaF_2 is 2.8×10^{-11} at 18°C.

19.6B Plan: Calculate the solubility of Ag_3PO_4 as molarity and use molar ratios to find the molarity of Ag^+ and PO_4^{3-} dissolved in solution. Calculate K_{sp} from $[Ag^+]$ and $[PO_4^{3-}]$ using the ion-product expression.

Solution: Convert the solubility to molar solubility:

$$\text{Molarity} = \left(\frac{3.2 \times 10^{-4} \text{ g } Ag_3PO_4}{50. \text{ mL}}\right)\left(\frac{1000 \text{ mL}}{1 \text{ L}}\right)\left(\frac{1 \text{ mol } Ag_3PO_4}{418.7 \text{ g } Ag_3PO_4}\right) = 1.5 \times 10^{-5} M Ag_3PO_4$$

$[Ag^+] = 3[Ag_3PO_4] = 4.5 \times 10^{-5} M$ because there are 3 mol of silver ions in each mol of Ag_3PO_4.
$[PO_4^{3-}] = [Ag_3PO_4] = 1.5 \times 10^{-5} M$ because there is 1 mol of phosphate ions in each mol of Ag_3PO_4.
The solubility equilibrium is:

$$Ag_3PO_4 (s) \leftrightarrows 3 Ag^+(aq) + PO_4^{3-}(aq) \qquad K_{sp} = [Ag^+]^3[PO_4^{3-}]$$

Calculate K_{sp} using the solubility product expression from above and the saturated concentrations of silver and phosphate ions.

$$K_{sp} = [Ag^+]^3[PO_4^{3-}] = (4.5 \times 10^{-5})^3(1.5 \times 10^{-5}) = 1.3669 \times 10^{-18} = \mathbf{1.4 \times 10^{-18}}$$

The K_{sp} for Ag_3PO_4 is 1.4×10^{-18} at 20°C.

19.7A Plan: Write the solubility reaction for $Mg(OH)_2$ and set up a reaction table, where S is the unknown molar solubility of the Mg^{2+} ion. Use the ion-product expression to solve for the concentration of $Mg(OH)_2$ in a saturated solution (also called the solubility of $Mg(OH)_2$).

Solution:

Concentration (M)	$Mg(OH)_2(s)$	$\leftrightarrows$	$Mg^{2+}(aq)$	+	$2OH^-(aq)$
Initial	—		0		0
Change	—		$+S$		$+2S$
Equilibrium	—		S		$2S$

$$K_{sp} = [Mg^{2+}][OH^-]^2 = (S)(2S)^2 = 4S^3 = 6.3 \times 10^{-10}$$
$$S = 5.4004114 \times 10^{-4} = 5.4 \times 10^{-4} M Mg(OH)_2$$

The solubility of $Mg(OH)_2$ is equal to S, the concentration of magnesium ions at equilibrium, so the molar solubility of magnesium hydroxide in pure water is $\mathbf{5.4 \times 10^{-4} M}$.

19.7B Plan: Write the solubility reaction for $Ca_3(PO_4)_2$ and set up a reaction table, where S is the unknown molar solubility of $Ca_3(PO_4)_2$. Use the ion-product expression to solve for the concentration of $Ca_3(PO_4)_2$ in a saturated solution (also called the solubility of $Ca_3(PO_4)_2$).

Solution:

Concentration (M)	$Ca_3(PO_4)_2(s)$	$\leftrightarrows$	$3Ca^{2+}(aq)$	+	$2PO_4^{3-}(aq)$
Initial	—		0		0
Change	—		$+3S$		$+2S$
Equilibrium	—		$3S$		$2S$

$$K_{sp} = [Ca^{2+}]^3[PO_4^{3-}]^2 = (3S)^3(2S)^2 = 108S^5 = 1.2 \times 10^{-29}$$
$$S = 6.4439 \times 10^{-7} = 6.4 \times 10^{-7} M Ca_3(PO_4)_2$$

The solubility of $Ca_3(PO_4)_2$ in pure water, S, is $\mathbf{6.4 \times 10^{-7} M}$.

19.8A Plan: Write the solubility reaction of $BaSO_4$. For part (a) set up a reaction table in which $[Ba^{2+}] = [SO_4^{2-}] = S$, which also equals the solubility of $BaSO_4$. Then, solve for S using the ion-product expression. For part (b), there is an initial concentration of sulfate, so set up the reaction table including this initial $[SO_4^{2-}]$. Solve for the solubility, S, which equals $[Ba^{2+}]$ at equilibrium.

Solution:
a) Set up reaction table.

Concentration (M)	$BaSO_4(s)$	$\leftrightarrows$	$Ba^{2+}(aq)$	+	$SO_4^{2-}(aq)$
Initial	—		0		0
Change	—		$+S$		$+S$
Equilibrium	—		S		S

$K_{sp} = [Ba^{2+}][SO_4^{2-}] = S^2 = 1.1 \times 10^{-10}$

$S = 1.0488 \times 10^{-5} = \textbf{1.0} \times \textbf{10}^{\textbf{-5}} \textbf{\textit{M}}$

The molar solubility of $BaSO_4$ in pure water is 1.0×10^{-5} M.

b) Set up another reaction table with initial $[SO_4^{2-}] = 0.10$ M (from the Na_2SO_4).

Concentration (M)	$BaSO_4(s)$	$\leftrightarrows$	$Ba^{2+}(aq)$	+	$SO_4^{2-}(aq)$
Initial	—		0		0.10
Change	—		+S		+S
Equilibrium	—		S		0.10 + S

$K_{sp} = [Ba^{2+}][SO_4^{2-}] = S(0.10 + S) = 1.1 \times 10^{-10}$

Assume that $0.10 + S$ is approximately equal to 0.10, which appears to be a good assumption based on the fact that $0.10 > 1 \times 10^{-10}$, K_{sp}

$K_{sp} = S(0.10) = 1.1 \times 10^{-10}$

$S = 1.1 \times 10^{-9}$ M

Molar solubility of $BaSO_4$ in 0.10 M Na_2SO_4 is $\textbf{1.1} \times \textbf{10}^{\textbf{-9}}$ $\textbf{\textit{M}}$.

The solubility of $BaSO_4$ decreases when sulfate ions are already present in the solution. The calculated decrease is from 10^{-5} M to 10^{-9} M, for a 10,000-fold decrease. This decrease is expected to be large because of the high concentration of sulfate ions.

19.8B   <u>Plan:</u> Write the solubility reaction of CaF_2. For part (a) set up a reaction table in which $[Ca^{2+}] = S$, which also equals the solubility of CaF_2. Then, solve for S using the ion-product expression. For part (b), there is an initial concentration of calcium, so set up the reaction table including this initial $[Ca^{2+}]$. Solve for the solubility, S, which equals $[Ca^{2+}]$ at equilibrium. For part (c), there is an initial concentration of fluoride, so set up the reaction table including this initial $[F^-]$. Solve for the solubility, S, which equals $[Ca^{2+}]$ at equilibrium.
<u>Solution:</u>

a) Set up reaction table.

Concentration (M)	$CaF_2(s)$	$\leftrightarrows$	$Ca^{2+}(aq)$	+	$2F^-(aq)$
Initial	—		0		0
Change	—		+S		+2S
Equilibrium	—		S		2S

$K_{sp} = [Ca^{2+}][F^-]^2 = (S)(2S)^2 = 4S^3 = 3.2 \times 10^{-11}$

$S = \textbf{2.0} \times \textbf{10}^{\textbf{-4}} \textbf{\textit{M}}$

The molar solubility of CaF_2 in pure water is $\textbf{2.0} \times \textbf{10}^{\textbf{-4}}$ $\textbf{\textit{M}}$.

b) Set up another reaction table with initial $[Ca^{2+}] = 0.20$ M (from the $CaCl_2$).

Concentration (M)	$CaF_2(s)$	$\leftrightarrows$	$Ca^{2+}(aq)$	+	$2F^-(aq)$
Initial	—		0.20		0
Change	—		+S		+2S
Equilibrium	—		0.20 + S		2S

$K_{sp} = [Ca^{2+}][F^-]^2 = (0.20 + S)(2S)^2 = 1.1 \times 10^{-10}$

Assume that $0.20 + S$ is approximately equal to 0.20, which appears to be a good assumption based on the fact that $0.20 > 3.2 \times 10^{-11}$, K_{sp}.

$K_{sp} = (0.20)(2S)^2 = 3.2 \times 10^{-11}$

$S = 6.3 \times 10^{-6}$ M

Molar solubility of CaF_2 in 0.20 M $CaCl_2$ is $\textbf{6.3} \times \textbf{10}^{\textbf{-6}}$ $\textbf{\textit{M}}$.

c) Set up another reaction table with initial $[F^-] = 0.40$ M (from the NiF_2; there are two fluoride ions per NiF_2 unit, so an NiF_2 concentration of 0.20 M gives a fluoride ion concentration of 0.40 M.).

Concentration (M)	$CaF_2(s)$	$\leftrightarrows$	$Ca^{2+}(aq)$	+	$2F^-(aq)$
Initial	—		0		0.40
Change	—		+S		+2S
Equilibrium	—		S		0.40 + 2S

$K_{sp} = [Ca^{2+}][F^-]^2 = (S)(0.40 + 2S)^2 = 1.1 \times 10^{-10}$

Assume that $0.40 + 2S$ is approximately equal to 0.40, which appears to be a good assumption based on the fact that $0.40 > 3.2 \times 10^{-11}$, K_{sp}.

$K_{sp} = (S)(0.40)^2 = 3.2 \times 10^{-11}$

$S = 2.0 \times 10^{-10} M$

Molar solubility of CaF_2 in 0.20 M NiF_2 is **$2.0 \times 10^{-10} M$.**

The solubility of CaF_2 decreases when either calcium or fluoride ions are already present in the solution.

19.9A   Plan: First, write the solubility reaction for the salt. Then, check the ions produced when the salt dissolves to see if they will react with acid. Three cases are possible:

1) If OH^- is produced, then addition of acid will neutralize the hydroxide ions and shift the solubility equilibrium toward the products. This causes more salt to dissolve. Write the solubility and neutralization reactions.

2) If the anion from the salt is the conjugate base of a weak acid, it will react with the added acid in a neutralization reaction. Solubility of the salt increases as the anion is neutralized. Write the solubility and neutralization reactions.

3) If the anion from the salt is the conjugate base of a strong acid, it does not react with a strong acid. The solubility of the salt is unchanged by the addition of acid. Write the solubility reaction.

Solution:

a) Calcium fluoride, CaF_2

Solubility reaction: $CaF_2(s) \leftrightarrows Ca^{2+}(aq) + 2F^-(aq)$

Fluoride ion is the conjugate base of HF, a weak acid. Thus, it will react with H_3O^+ from the strong acid, HNO_3.

Neutralization reaction: $F^-(aq) + H_3O^+(aq) \rightarrow HF(aq) + H_2O(l)$

The neutralization reaction decreases the concentration of fluoride ions, which causes the solubility equilibrium to shift to the right and more CaF_2 dissolves. The **solubility of CaF_2 increases** with the addition of HNO_3.

b) Iron(III) hydroxide, $Fe(OH)_3$

Solubility reaction: $Fe(OH)_3(s) + H_2O(l) \leftrightarrows Fe^{3+}(aq) + 3OH^-(aq)$

The hydroxide ion reacts with the added acid:

Neutralization reaction: $OH^-(aq) + H_3O^+(aq) \rightarrow 2H_2O(l)$

The neutralization reactions decrease the concentration of hydroxide in the solubility equilibrium, which causes a shift to the right, and more $Fe(OH)_3$ dissolves. The addition of HNO_3 will **increase the solubility** of $Fe(OH)_3$.

c) Silver iodide, AgI

Solubility reaction: $AgI(s) \leftrightarrows Ag^+(aq) + I^-(aq)$

The iodide ion is the conjugate base of a strong acid, HI. So, I^- will not react with added acid. The **solubility of AgI will not change** with added HNO_3.

19.9B   Plan: First, write the solubility reaction for the salt. Then, check the ions produced when the salt dissolves to see if they will react with acid. Three cases are possible:

1) If OH^- is produced, then addition of acid will neutralize the hydroxide ions and shift the solubility equilibrium toward the products. This causes more salt to dissolve. Write the solubility and neutralization reactions.

2) If the anion from the salt is the conjugate base of a weak acid, it will react with the added acid in a neutralization reaction. Solubility of the salt increases as the anion is neutralized. Write the solubility and neutralization reactions.

3) If the anion from the salt is the conjugate base of a strong acid, it does not react with a strong acid. The solubility of the salt is unchanged by the addition of acid. Write the solubility reaction.

Solution:

a) Silver cyanide, AgCN

Solubility reaction: $AgCN(s) \leftrightarrows Ag^+(aq) + CN^-(aq)$

Cyanide ion is the conjugate base of HCN, a weak acid. Thus, it will react with H_3O^+ from the strong acid, HBr.

Neutralization reaction: $CN^-(aq) + H_3O^+(aq) \rightarrow HCN(aq) + H_2O(l)$

The neutralization reaction decreases the concentration of cyanide ions, which causes the solubility equilibrium to shift to the right and more AgCN dissolves. The **solubility of AgCN increases** with the addition of HBr.

b) copper(I) chloride, CuCl

Solubility reaction: $CuCl (s) \leftrightarrows Cu^+(aq) + Cl^-(aq)$

The chloride ion is the conjugate base of a strong acid, HCl. So, Cl^- will not react with added acid. The **solubility of CuCl will not change** with added HBr.

c) Magnesium phosphate, $Mg_3(PO_4)_2$

Solubility reaction: $Mg_3(PO_4)_2(s) \leftrightarrows Mg^{2+}(aq) + 2PO_4^{3-}(aq)$

Phosphate ion is the conjugate base of HPO_4^{2-}, a weak acid. Thus, it will react with H_3O^+ from the strong acid, HBr.

Neutralization reaction: $PO_4^{3-}(aq) + H_3O^+(aq) \rightarrow HPO_4^{2-}(aq) + H_2O(l)$

HPO_4^{2-}, in turn, is the conjugate base of $H_2PO_4^-$, a weak acid. Thus, it will react with H_3O^+ from the strong acid, HBr.

Neutralization reaction: $HPO_4^{2-}(aq) + H_3O^+(aq) \rightarrow H_2PO_4^-(aq) + H_2O(l)$

$H_2PO_4^-$, in turn, is the conjugate base of H_3PO_4, a weak acid. Thus, it will react with H_3O^+ from the strong acid, HBr.

Neutralization reaction: $H_2PO_4^-(aq) + H_3O^+(aq) \rightarrow H_3PO_4(aq) + H_2O(l)$

Each of these neutralization reactions ultimately decreases the concentration of phosphate ions, which causes the solubility equilibrium to shift to the right and more $Mg_3(PO_4)_2$ dissolves. The **solubility of $Mg_3(PO_4)_2$ increases** with the addition of HBr.

19.10A Plan: First, write the solubility equilibrium equation and ion-product expression. Use the given concentrations of calcium and phosphate ions to calculate Q_{sp}. Compare Q_{sp} to K_{sp}.

If $K_{sp} < Q_{sp}$, precipitation occurs. If $K_{sp} \geq Q_{sp}$ then, precipitation will not occur.

Solution:

Write the solubility equation:

$Ca_3(PO_4)_2(s) \leftrightarrows 3Ca^{2+}(aq) + 2PO_4^{3-}(aq)$

and ion-product expression: $Q_{sp} = [Ca^{2+}]^3[PO_4^{3-}]^2$

$\qquad [Ca^{2+}] = [PO_4^{3-}] = 1.0 \times 10^{-9}\ M$

$Q_{sp} = [Ca^{2+}]^3[PO_4^{3-}]^2 = (1.0 \times 10^{-9})^3(1.0 \times 10^{-9})^2 = 1.0 \times 10^{-45}$

Compare K_{sp} and Q_{sp}. $K_{sp} = 1.2 \times 10^{-29} > 1.0 \times 10^{-45} = Q_{sp}$

Precipitation will not occur because concentrations are below the level of a saturated solution as shown by the value of Q_{sp}.

19.10B Plan: First, write the solubility equilibrium equation and ion-product expression. Find the concentrations of the lead and sulfide ions in the final mixture. Use these concentrations to calculate Q_{sp}. Compare Q_{sp} to K_{sp}.

If $K_{sp} < Q_{sp}$, precipitation occurs. If $K_{sp} \geq Q_{sp}$ then, precipitation will not occur.

Solution:

Write the solubility equation:

$PbS(s) \leftrightarrows Pb^{2+}(aq) + S^{2-}(aq)$

and ion-product expression: $Q_{sp} = [Pb^{2+}][S^{2-}]$

25 L of a solution containing Pb^{2+} are mixed with 0.500 L of a solution containing S^{2-}. The final volume is 25.5 L.

Amount (mol) of $Pb^{2+} = (25\ L)\left(\dfrac{0.015\ g\ Pb^{2+}}{1\ L}\right)\left(\dfrac{1\ mol\ Pb^{2+}}{207.2\ g\ Pb^{2+}}\right) = 0.00180985\ mol\ Pb^{2+}$

Molarity of $Pb^{2+} = \dfrac{0.00180985\ mol\ Pb^{2+}}{25.5\ L} = 7.09743 \times 10^{-5}\ M$

Amount (mol) of S^{2-}: $(0.500\ L)\left(\dfrac{0.10\ mol\ S^{2-}}{1\ L}\right) = 0.050\ mol\ S^{2-}$

Molarity of $S^{2-} = \dfrac{0.050\ mol\ S^{2-}}{25.5\ L} = 2.0 \times 10^{-3}\ M$

$Q_{sp} = [Pb^{2+}][S^{2-}] = (7.09743 \times 10^{-5})(2.0 \times 10^{-3}) = 1.41949 \times 10^{-7} = 1.4 \times 10^{-7}$

Compare K_{sp} and Q_{sp}. $K_{sp} = 3 \times 10^{-25} < 1.4 \times 10^{-7} = Q_{sp}$

Precipitation will occur because concentrations are above the level of a saturated solution as shown by the value of Q_{sp}.

19.11A **Plan:** First, write the solubility equilibrium equation and ion-product expression. For b) use the given amounts of nickel (II) and hydroxide ions to calculate Q_{sp}. Compare Q_{sp} to K_{sp}. For c) check the ions produced when the salt dissolves to see if they will react with acid or if the hydroxide ion of a strong base is part of the ion-product expression, in which case, it will influence the solubility of the solid through the common ion effect.
Solution:
a) Write the solubility equation:
$Ni(OH)_2(s) \leftrightarrows Ni^{2+}(aq) + 2OH^-(aq)$
Scene 3 has the same relative number of ions as in the formula of $Ni(OH)_2$. The Ni^{2+} and OH^- ions are in a 1:2 ratio in Scene 3.

b) Write the ion-product expression: $Q_{sp} = [Ni^{2+}][OH^-]^2$
Calculate K_{sp} using Scene 3: $K_{sp} = [Ni^{2+}][OH^-]^2 = [2][4]^2 = 32$
Calculate Q_{sp} using Scene 1: $Q_{sp} = [Ni^{2+}][OH^-]^2 = [3][4]^2 = 48$
Calculate Q_{sp} using Scene 2: $Q_{sp} = [Ni^{2+}][OH^-]^2 = [4][2]^2 = 16$
Q_{sp} exceeds K_{sp} in Scene 1 (48 > 32) so additional solid will form in **Scene 1.**
c) Hydroxide ion is one of the products of the solubility equilibrium reaction. The hydroxide ion reacts with added acid:
Neutralization reaction: $OH^-(aq) + H_3O^+(aq) \rightarrow 2H_2O(l)$
The neutralization reaction decreases the concentration of $OH^-(aq)$ in the solubility equilibrium, which causes a shift to the right, and more $Ni(OH)_2(s)$ dissolves. Addition of base (OH^-) shifts the equilibrium to the left due to the common-ion effect and the mass of $Ni(OH)_2$ increases.

19.11B **Plan:** First, write the solubility equilibrium equation and ion-product expression. For b) use the given amounts of lead(II) and chloride ions to calculate Q_{sp}. Compare Q_{sp} to K_{sp}. For c) check the ions produced when the salt dissolves to see if they will react with acid or if the anion of the acid is part of the ion-product expression, in which case, it will influence the solubility of the solid through the common ion effect.
Solution:
a) Write the solubility equation:
$PbCl_2(s) \leftrightarrows Pb^{2+}(aq) + 2Cl^-(aq)$
Scene 1 has the same relative number of ions as in the formula of $PbCl_2$. The Pb^{2+} and Cl^- ions are in a 1:2 ratio in Scene 1.
b) Write the ion-product expression: $Q_{sp} = [Pb^{2+}][Cl^-]^2$
Calculate Q_{sp} using Scene 1: $Q_{sp} = [Pb^{2+}][Cl^-]^2 = [3][6]^2 = 108$
Calculate Q_{sp} using Scene 2: $Q_{sp} = [Pb^{2+}][Cl^-]^2 = [4][5]^2 = 100$
Calculate K_{sp} using Scene 3: $K_{sp} = [Pb^{2+}][Cl^-]^2 = [5][5]^2 = 125$
Q_{sp} exceeds K_{sp} in Scene 3 (125 > 108) so additional solid will form in **Scene 3.**
c) Chloride ion is one of the products of the solubility equilibrium reaction. Added chloride ion (from HCl) will shift the reaction to the left, due to the common ion effect. As a result of the reaction shifting to the left, the **mass of solid $PbCl_2$ will increase**.

19.12A **Plan:** Compare the K_{sp} values for the two salts. Since $CaSO_4$ is more soluble, calculate the concentration of sulfate ions in equilibrium with the Ca^{2+} concentration.
Solution:
The solubility equilibrium for $CaSO_4$ is
$CaSO_4(s) \leftrightarrows Ca^{2+}(aq) + SO_4^{2-}(aq)$
and $K_{sp} = [Ca^{2+}][SO_4^{2-}] = 2.4 \times 10^{-5}$
$$[SO_4^{2-}] = \frac{K_{sp}}{[Ca^{2+}]} = \left(\frac{2.4 \times 10^{-5}}{0.025\ M}\right) = 9.6 \times 10^{-4}\ M$$

19-15

19.12B <u>Plan:</u> Compare the K_{sp} values for the two salts. Since BaF_2 is more soluble, calculate the concentration of fluoride ions in equilibrium with the Ba^{2+} concentration.
<u>Solution:</u>
The solubility equilibrium for BaF_2 is
$BaF_2(s) \leftrightarrows Ba^{2+}(aq) + 2F^-(aq)$
and $K_{sp} = [Ba^{2+}][F^-]^2 = 1.5 \times 10^{-6}$

$$[F^-] = \sqrt{\frac{K_{sp}}{[Ba^{2+}]}} = \sqrt{\frac{1.5 \times 10^{-6}}{(0.020)}} = 8.66025 \times 10^{-3} = \mathbf{8.7 \times 10^{-3}\ \textit{M}}$$

19.13A <u>Plan:</u> Write the complex-ion formation equilibrium reaction. Calculate the initial concentrations of $Fe(H_2O)_6^{3+}$ and CN^-. The approach to complex ion equilibria problems is slightly different than for solubility equilibria because formation constants are generally large while solubility product constants are generally very small. The best mathematical approach is to assume that the equilibrium reaction goes to completion and then calculate back to find the equilibrium concentrations of reactants. So, assume that all of the $Fe(H_2O)_6^{3+}$ reacts to form $Fe(CN)_6^{3-}$ and calculate the concentration of $Fe(CN)_6^{3-}$ formed from the given concentrations of $Fe(H_2O)_6^{3+}$ and CN^- and the concentration of the excess reactant. Then use the complex ion formation equilibrium to find the equilibrium concentration of $Fe(H_2O)_6^{3+}$.
<u>Solution:</u>
Equilibrium reaction: $Fe(H_2O)_6^{3+}(aq) + 6CN^-(aq) \leftrightarrows Fe(CN)_6^{3-}(aq) + 6H_2O(l)$
Initial concentrations from a simple dilution calculation:

$$M_f = M_i V_i / V_f$$

$$[Fe(H_2O)_6^{3+}] = \frac{(3.1 \times 10^{-2}\ M)(25.5\ mL)}{25.5 + 35.0\ mL} = 0.0130661157\ M$$

$$[CN^-] = \frac{(1.5\ M)(35.0\ mL)}{25.5 + 35.0\ mL} = 0.867768595\ M$$

Set up a reaction table:

Concentration (M)	$Fe(H_2O)_6^{3+}(aq)$	+	$6CN^-(aq)$	$\leftrightarrows$	$Fe(CN)_6^{3-}(aq)$	+	$6H_2O(l)$
Initial	0.0130661157		0.867768595		0		—
Change	$-0.0130661157 + x$		$-6(0.013066115)$		$+0.013066115$		—
Equilibrium	x		0.789371905		0.013066115		—

$$K_f = 4.0 \times 10^{43} = \frac{\left[Fe(CN)_6^{3-}\right]}{\left[Fe(H_2O)_6^{3+}\right]\left[CN^-\right]^6} = \frac{[0.0130661]}{[x][0.789371905]^6}$$

$$x = 1.35019 \times 10^{-45} = \mathbf{1.4 \times 10^{-45}\ \textit{M}}$$

The concentration of $Fe(H_2O)_6^{3+}$ at equilibrium is $1.4 \times 10^{-45}\ M$. This concentration is so low that it is impossible to calculate it using the initial concentrations minus a variable x. The variable would have to be so close to the initial concentration that the initial concentration of x cannot be calculated to enough significant figures (43 in this case) to get a difference in concentrations of $1 \times 10^{-45}\ M$. Thus, the approach above is the best to calculate the very low equilibrium concentration of $Fe(H_2O)_6^{3+}$.

19.13B <u>Plan:</u> Write the complex-ion formation equilibrium reaction. Calculate the initial concentration of $Al(H_2O)_6^{3+}$. The approach to complex ion equilibria problems is slightly different than for solubility equilibria because formation constants are generally large while solubility product constants are generally very small. The best mathematical approach is to assume that the equilibrium reaction goes to completion and then calculate back to find the equilibrium concentrations of reactants. So, assume that all of the $Al(H_2O)_6^{3+}$ reacts to form AlF_6^{3-} and calculate the concentration of AlF_6^{3-} formed from the given concentrations of $Al(H_2O)_6^{3+}$ and F^-. Then use the complex ion formation equilibrium to find the equilibrium concentration of $Al(H_2O)_6^{3+}$.
<u>Solution:</u>
Equilibrium reaction: $Al(H_2O)_6^{3+}(aq) + 6F^-(aq) \leftrightarrows AlF_6^{3-}(aq) + 6H_2O(l)$

Initial concentrations:

$$[Al(H_2O)_6^{3+}] = \frac{2.4 \text{ g AlCl}_3 \times \dfrac{1 \text{ mol AlCl}_3}{133.33 \text{ g AlCl}_3}}{0.250 \text{ L}} = 0.0720018 \, M$$

$[F^-] = 0.560 \, M$

Set up a reaction table:

Concentration (M)	$Al(H_2O)_6^{3+}(aq)$	+	$6F^-(aq)$	$\leftrightarrows$	$AlF_6^{3-}(aq)$	+	$6H_2O(l)$
Initial	0.0720018		0.560		0		—
Change	$-0.0720018 + x$		$-6(0.0720018)$		$+0.0720018$		—
Equilibrium	x		0.1279892		0.0720018		—

$$K_f = 4 \times 10^{19} = \frac{\left[AlF_6^{3-}\right]}{\left[Al(H_2O)_6^{3+}\right]\left[F^-\right]^6} = \frac{(0.0720018)}{(x)(0.1279892)^6} \qquad x = 4.0949 \times 10^{-16} = 4 \times 10^{-16} \, M$$

The concentration of $Al(H_2O)_6^{3+}$ at equilibrium is **$4 \times 10^{-16} \, M$**

19.14A **Plan:** Write equations for the solubility equilibrium and formation of the silver-ammonia complex ion. Add the two equations to get the overall reaction. Set up a reaction table for the overall reaction with the given value for initial $[NH_3]$. Write equilibrium expressions from the overall balanced reaction and calculate $K_{overall}$ from K_f and K_{sp} values. Insert the equilibrium concentration values from the reaction table into the equilibrium expression and calculate solubility.

Solution:

Equilibria:

$$
\begin{array}{ll}
AgBr(s) \leftrightarrows \cancel{Ag^+(aq)} + Br^-(aq) & K_{sp} = 5.0 \times 10^{-13} \\
\cancel{Ag^+(aq)} + 2NH_3(aq) \leftrightarrows Ag(NH_3)_2^+(aq) & K_f = 1.7 \times 10^7 \\
\hline
AgBr(s) + 2NH_3(aq) \leftrightarrows Ag(NH_3)_2^+(aq) + Br^-(aq) & K_{overall} = K_{sp}K_f
\end{array}
$$

Set up reaction table:

Concentration (M)	$AgBr(s)$	+	$2NH_3(aq)$	$\leftrightarrows$	$Ag(NH_3)_2^+(aq)$	+	$Br^-(aq)$
Initial	—		1.0		0		0
Change	—		$-2S$		$+S$		$+S$
Equilibrium	—		$1.0 - 2S$		S		S

$$K_{overall} = \frac{\left[Ag(NH_3)_2^+\right]\left[Br^-\right]}{\left[NH_3\right]^2} = K_{sp}K_f = (5.0 \times 10^{-13})(1.7 \times 10^7) = 8.5 \times 10^{-6}$$

Calculate the solubility of AgBr:

$$K_{overall} = \frac{[S][S]}{\left[1.0 - 2S\right]^2} = 8.5 \times 10^{-6}$$

Assume that $1.0 - 2S$ is approximately equal to 1.0, which appears to be a good assumption based on the fact that $1.0 >> 8.5 \times 10^{-6}$, $K_{overall}$.

$$\frac{[S][S]}{[1.0]^2} = 8.5 \times 10^{-6}$$

$S = 2.9154759 \times 10^{-3} = 2.9 \times 10^{-3} \, M$

The solubility of AgBr in ammonia is **less** than its solubility in hypo (sodium thiosulfate).

Since the formation constant for $Ag(NH_3)_2^+$ is less than the formation constant of $Ag(S_2O_3)_2^{3-}$, the addition of ammonia will increase the solubility of AgBr less than the addition of thiosulfate ion increases its solubility.

19.14B **Plan:** Write equations for the solubility equilibrium and formation of the lead(II)-hydroxide complex ion. Add the two equations to get the overall reaction. Set up a reaction table for the overall reaction with the given value for initial $[OH^-]$. Write equilibrium expressions from the overall balanced reaction and calculate $K_{overall}$ from K_f and K_{sp} values. Insert the equilibrium concentration values from the reaction table into the equilibrium expression and calculate solubility.

Solution:
Equilibria:

$$PbCl_2(s) \rightleftharpoons \cancel{Pb^{2+}(aq)} + 2Cl^-(aq) \qquad K_{sp} = 1.7\text{x}10^{-5}$$

$$\cancel{Pb^{2+}(aq)} + 3OH^-(aq) \rightleftharpoons Pb(OH)_3^-(aq) \qquad K_f = 8\text{x}10^{13}$$

$$PbCl_2(s) + 3OH^-(aq) \rightleftharpoons Pb(OH)_3^-(aq) + 2Cl^-(aq) \qquad K_{overall} = K_{sp}K_f$$

Set up reaction table:

Concentration (M)	$PbCl_2$ (s)	+	$3OH^-$ (aq)	$\rightleftharpoons$	$Pb(OH)_3^-$ (aq)	+	$2Cl^-$
Initial	—		0.75		0		0
Change	—		$-3S$		$+S$		$+2S$
Equilibrium	—		$0.75 - 3S$		S		$2S$

$$K_{overall} = \frac{[Pb(OH)_3^-][Cl^-]^2}{[OH^-]^3} = K_{sp}K_f = (1.7\text{x}10^{-5})(8\text{x}10^{13}) = 1.36\text{x}10^9$$

Calculate the solubility of $PbCl_2$:

$$K_{overall} = \frac{[S][2S]^2}{[0.75 - 3S]^3} = 1.36\text{x}10^9$$

$$\frac{4S^3}{[0.75 - 3S]^3} = 1.36\text{x}10^9 \quad \text{Take the cube root of both sides of the equation}$$

$$\frac{1.5874S}{0.75 - 3S} = 1.1079\text{x}10^3$$

$$S = 523.4503 - 2093.801S$$

$$S = 0.24988 = \mathbf{0.25\ M}$$

CHEMICAL CONNECTIONS BOXED READING PROBLEMS

B19.1 Plan: Consult Figure 19.5 for the colors and pH ranges of the indicators.
Solution:
Litmus paper indicates the pH is below 7. The result from thymol blue, which turns yellow at a pH above 2.5, indicates that the pH is above 2.5. Bromphenol blue is the best indicator as it is green in a fairly narrow range of $3.5 < pH < 4$. Methyl red turns red below a pH of 4.3. Therefore, a reasonable estimate for the rainwater pH is **3.5 to 4.**

B19.2 Plan: Find the volume of the rain received by multiplying the surface area of the lake by the depth of rain. Find the volume of the lake before the rain. Express both volumes in liters. The pH of the rain is used to find the molarity of H_3O^+; this molarity multiplied by the volume of rain gives the moles of H_3O^+. The moles of H_3O^+ divided by the volume of the lake plus rain gives the molarity of H_3O^+ and the pH of the lake.
Solution:
a) To find the volume of rain, multiply the surface area in square inches by the depth of rain. Convert the volume in in^3 to cm^3 and then to L.

$$\text{Volume (L) of rain} = (10.0\ \text{acres})\left(\frac{4.840\text{x}10^3\ yd^2}{1\ \text{acre}}\right)\left(\frac{36\ \text{in}}{1\ \text{yd}}\right)^2(1.00\ \text{in})\left(\frac{2.54\ \text{cm}}{1\ \text{in}}\right)^3\left(\frac{1\ \text{mL}}{1\ cm^3}\right)\left(\frac{10^{-3}\ \text{L}}{1\ \text{mL}}\right)$$

$$= 1.027902\text{x}10^6\ \text{L}$$

At $pH = 4.20$, $[H_3O^+] = 10^{-4.20} = 6.3095734\text{x}10^{-5}\ M$

$$\text{Moles of } H_3O^+ = (1.027902\text{x}10^6\ \text{L})\left(\frac{6.3095734\text{x}10^{-5}\ \text{mol}}{\text{L}}\right) = 64.8562 = \mathbf{65\ mol}$$

b) Volume (L) of the lake $= (10.0\ \text{acres})\left(\frac{4.840\text{x}10^3\ yd^2}{1\ \text{acre}}\right)\left(\frac{36\ \text{in}}{1\ \text{yd}}\right)^2(10.0\ \text{ft})\left(\frac{12\ \text{in}}{1\ \text{ft}}\right)\left(\frac{2.54\ \text{cm}}{1\ \text{in}}\right)^3\left(\frac{1\ \text{mL}}{1\ cm^3}\right)\left(\frac{10^{-3}\ \text{L}}{1\ \text{mL}}\right)$

$$= 1.23348\text{x}10^8\ \text{L}$$

Total volume of lake after rain $= 1.23348\text{x}10^8\ \text{L} + 1.027902\text{x}10^6\ \text{L} = 1.243759\text{x}10^8\ \text{L}$

$$[H_3O^+] = \frac{\text{mol } H_3O^+}{L} = \frac{64.8562 \text{ mol}}{1.243759 \times 10^8 \text{ L}} = 5.214531 \times 10^{-7} M$$

$$pH = -\log (5.214531 \times 10^{-7}) = 6.2827847 = \mathbf{6.28}$$

c) Each mol of H_3O^+ requires one mole of HCO_3^- for neutralization.

$$\text{Mass (g)} = (64.8562 \text{ mol } H_3O^+) \left(\frac{1 \text{ mol } HCO_3^-}{1 \text{ mol } H_3O^+} \right) \left(\frac{61.02 \text{ g } HCO_3^-}{1 \text{ mol } HCO_3^-} \right)$$

$$= 3.97575 \times 10^3 = \mathbf{4.0 \times 10^3 \text{ g } HCO_3^-}$$

END–OF–CHAPTER PROBLEMS

19.2 The weak-acid component neutralizes added base and the weak-base component neutralizes added acid so that the pH of the buffer solution remains relatively constant. The components of a buffer do not neutralize one another when they are a conjugate acid-base pair.

19.6 The buffer-component ratio refers to the ratio of concentrations of the acid and base that make up the buffer. When this ratio is equal to 1, the buffer resists changes in pH with added acid to the same extent that it resists changes in pH with added base. The buffer range extends equally in both the acidic and basic direction. When the ratio shifts with higher [base] than [acid], the buffer is more effective at neutralizing added acid than base so the range extends further in the acidic than basic direction. The opposite is true for a buffer where [acid] > [base]. Buffers with a ratio equal to 1 have the greatest buffer range. The more the buffer-component ratio deviates from 1, the smaller the buffer range.

19.8 Plan: Remember that the weak-acid buffer component neutralizes added base and the weak-base buffer component neutralizes added acid.
Solution:
a) The buffer-component ratio and pH **increase** with added base. The OH^- reacts with HA to decrease its concentration and increase [NaA]. The ratio [NaA]/[HA] thus increases. The pH of the buffer will be more basic because the concentration of base, A^-, has increased and the concentration of acid, HA, decreased.
b) Buffer-component ratio and pH **decrease** with added acid. The H_3O^+ reacts with A^- to decrease its concentration and increase [HA]. The ratio [NaA]/[HA] thus decreases. The pH of the buffer will be more acidic because the concentration of base, A^-, has decreased and the concentration of acid, HA, increased.
c) Buffer-component ratio and pH **increase** with the added sodium salt. The additional NaA increases the concentration of both NaA and HA, but the relative increase in [NaA] is greater. Thus, the ratio increases and the solution becomes more basic. Whenever base is added to a buffer, the pH always increases, but only slightly if the amount of base is not too large.
d) Buffer-component ratio and pH **decrease**. The concentration of HA increases more than the concentration of NaA, so the ratio is less and the solution is more acidic.

19.10 a) **Buffer 3** has equal, high concentrations of both HA and A^-. It has the highest buffering capacity.
b) **All** of the buffers have the same pH range. The practical buffer range is pH = p$K_a \pm 1$, and is independent of concentration.
c) **Buffer 2** has the greatest amount of weak base and can therefore neutralize the greatest amount of added acid.

19.12 Plan: The buffer components are propanoic acid and propanoate ion, the concentrations of which are known. The sodium ion is a spectator ion and is ignored because it is not involved in the buffer. Write the propanoic acid-dissociation reaction and its K_a expression. Set up a reaction table in which x equals the amount of acid that dissociates; solving for x will result in $[H_3O^+]$, from which the pH can be calculated. Alternatively, the pH can be calculated from the Henderson-Hasselbalch equation.
Solution:

Concentration (M)	$CH_3CH_2COOH(aq) + H_2O(l)$		$\rightleftharpoons$ $CH_3CH_2COO^-(aq)$	$+ H_3O^+(aq)$
Initial	0.15	—	0.35	0
Change	– x	—	+x	+x
Equilibrium	0.15 – x	—	0.35 + x	x

Assume that x is negligible with respect to both 0.15 and 0.35 since both concentrations are much larger than K_a.

$$K_a = 1.3 \times 10^{-5} = \frac{[H_3O^+][CH_3CH_2COO^-]}{[CH_3CH_2COOH]} = \frac{[x][0.35+x]}{[0.15-x]} = \frac{[x][0.35]}{[0.15]}$$

$$x = [H_3O^+] = K_a \frac{[CH_3CH_2COOH]}{[CH_3CH_2COO^-]} = (1.3 \times 10^{-5})\left(\frac{0.15}{0.35}\right) = 5.57143 \times 10^{-6} = \mathbf{5.6 \times 10^{-6}\ M}$$

Check assumption: Percent error = $(5.6 \times 10^{-6}/0.15)100\%$ = 0.0037%. The assumption is valid.

$$pH = -\log [H_3O^+] = -\log (5.57143 \times 10^{-6}) = 5.2540 = \mathbf{5.25}$$

Another solution path to find pH is using the Henderson-Hasselbalch equation:

$$pH = pK_a + \log\left(\frac{[base]}{[acid]}\right) \qquad\qquad pK_a = -\log (1.3 \times 10^{-5}) = 4.886$$

$$pH = 4.886 + \log\left(\frac{[CH_3CH_2COO^-]}{[CH_3CH_2COOH]}\right) = 4.886 + \log\left(\frac{[0.35]}{[0.15]}\right)$$

$$pH = 5.25398 = \mathbf{5.25}$$

19.14 **Plan:** The buffer components are HNO_2 and NO_2^-, the concentrations of which are known. The potassium ion is a spectator ion and is ignored because it is not involved in the buffer. Write the HNO_2 acid-dissociation reaction and its K_a expression. Set up a reaction table in which x equals the amount of acid that dissociates; solving for x will result in $[H_3O^+]$, from which the pH can be calculated. Alternatively, the pH can be calculated from the Henderson-Hasselbalch equation.

Solution:

Concentration (M)	$HNO_2(aq)$	+ $H_2O(l)$	$\leftrightarrows$	$NO_2^-(aq)$	+ $H_3O^+(aq)$
Initial	0.55	—		0.75	0
Change	−x	—		+x	+x
Equilibrium	0.55 − x	—		0.75 + x	x

Assume that x is negligible with respect to both 0.55 and 0.75 since both concentrations are much larger than K_a.

$$K_a = 7.1 \times 10^{-4} = \frac{[H_3O^+][NO_2^-]}{[HNO_2]} = \frac{[x][0.75+x]}{[0.55-x]} = \frac{[x][0.75]}{[0.55]}$$

$$x = [H_3O^+] = K_a \frac{[HNO_2]}{[NO_2^-]} = (7.1 \times 10^{-4})\frac{[0.55]}{[0.75]} = 5.2066667 \times 10^{-4} = \mathbf{5.2 \times 10^{-4}\ M}$$

Check assumption: Percent error = $(5.2066667 \times 10^{-4}/0.55)100\%$ = 0.095%. The assumption is valid.

$$pH = -\log [H_3O^+] = -\log (5.2066667 \times 10^{-4}) = 3.28344 = \mathbf{3.28}$$

Verify the pH using the Henderson-Hasselbalch equation.

$$pH = pK_a + \log\left(\frac{[base]}{[acid]}\right) \qquad\qquad pK_a = -\log(7.1 \times 10^{-4}) = 3.149$$

$$pH = 3.149 + \log\left(\frac{[NO_2^-]}{[HNO_2]}\right) = 3.149 + \log\left(\frac{[0.75]}{[0.55]}\right)$$

$$pH = 3.2837 = \mathbf{3.28}$$

19.16 **Plan:** The buffer components are formic acid, HCOOH, and formate ion, $HCOO^-$, the concentrations of which are known. The sodium ion is a spectator ion and is ignored because it is not involved in the buffer. Write the HCOOH acid-dissociation reaction and its K_a expression. Set up a reaction table in which x equals the amount of acid that dissociates; solving for x will result in $[H_3O^+]$, from which the pH can be calculated. Alternatively, the pH can be calculated from the Henderson-Hasselbalch equation.

Solution:

$K_a = 10^{-pK_a} = 10^{-3.74} = 1.8197 \times 10^{-4}$

Concentration (M)	HCOOH(aq)	+ H$_2$O(l)	⇆	HCOO$^-$(aq)	+ H$_3$O$^+$(aq)
Initial	0.45	—		0.63	0
Change	−x	—		+x	+x
Equilibrium	0.45 − x	—		0.63 + x	x

Assume that x is negligible because both concentrations are much larger than K_a.

$$K_a = 1.8197 \times 10^{-4} = \frac{[H_3O^+][HCOO^-]}{[HCOOH]} = \frac{[x][0.63+x]}{[0.45-x]} = \frac{[x][0.63]}{[0.45]}$$

$$x = [H_3O^+] = K_a \frac{[HCOOH]}{[HCOO^-]} = (1.8197 \times 10^{-4})\frac{[0.45]}{[0.63]} = 1.29979 \times 10^{-4} = 1.3 \times 10^{-4}\ M$$

Check assumption: Percent error = $(1.29979 \times 10^{-4}/0.45)100\% = 0.029\%$. The assumption is valid.

$$pH = -\log[H_3O^+] = -\log(1.29979 \times 10^{-4}) = 3.886127 = \textbf{3.89}$$

Verify the pH using the Henderson-Hasselbalch equation.

$$pH = pK_a + \log\left(\frac{[\text{base}]}{[\text{acid}]}\right)$$

$$pH = 3.74 + \log\left(\frac{[HCOO^-]}{[HCOOH]}\right) = 3.74 + \log\left(\frac{[0.63]}{[0.45]}\right)$$

$$pH = 3.8861 = 3.89$$

19.18   Plan: The buffer components phenol, C$_6$H$_5$OH, and phenolate ion, C$_6$H$_5$O$^-$, the concentrations of which are known. The sodium ion is a spectator ion and is ignored because it is not involved in the buffer. Write the C$_6$H$_5$OH acid-dissociation reaction and its K_a expression. Set up a reaction table in which x equals the amount of acid that dissociates; solving for x will result in [H$_3$O$^+$], from which the pH can be calculated. Alternatively, the pH can be calculated from the Henderson-Hasselbalch equation.

Solution:

$K_a = 10^{-pK_a} = 10^{-10.00} = 1.0 \times 10^{-10}$

Concentration (M)	C$_6$H$_5$OH(aq)	+ H$_2$O(l)	⇆	C$_6$H$_5$O$^-$(aq) +	H$_3$O$^+$(aq)
Initial	1.2	—		1.3	0
Change	−x	—		+x	+x
Equilibrium	1.2 − x	—		1.3 + x	x

Assume that x is negligible with respect to both 1.0 and 1.2 because both concentrations are much larger than K_a.

$$K_a = 1.0 \times 10^{-10} = \frac{[H_3O^+][C_6H_5O^-]}{[C_6H_5OH]} = \frac{[x][1.3+x]}{[1.2-x]} = \frac{[x][1.3]}{[1.2]}$$

$$x = [H_3O^+] = K_a \frac{[C_6H_5OH]}{[C_6H_5O^-]} = (1.0 \times 10^{-10})\left(\frac{1.2}{1.3}\right) = 9.23077 \times 10^{-11}\ M$$

Check assumption: Percent error = $(9.23077 \times 10^{-11}/1.2)100\% = 7.7 \times 10^{-9}\%$. The assumption is valid.

$$pH = -\log(9.23077 \times 10^{-11}) = 10.03476 = \textbf{10.03}$$

Verify the pH using the Henderson-Hasselbalch equation:

$$pH = pK_a + \log\left(\frac{[\text{base}]}{[\text{acid}]}\right)$$

$$pH = 10.00 + \log\left(\frac{[C_6H_5O^-]}{[C_6H_5OH]}\right) = 10.00 + \log\left(\frac{[1.3]}{[1.2]}\right)$$

$$\textbf{pH = 10.03}$$

19.20   Plan: The buffer components ammonia, NH$_3$, and ammonium ion, NH$_4^+$, the concentrations of which are known. The chloride ion is a spectator ion and is ignored because it is not involved in the buffer. Write the NH$_4^+$ acid-dissociation reaction and its K_a expression. Set up a reaction table in which x equals the amount of acid that

dissociates; solving for x will result in $[H_3O^+]$, from which the pH can be calculated. Alternatively, the pH can be calculated from the Henderson-Hasselbalch equation. The K_a of NH_4^+ will have to be calculated from the pK_b.
Solution:
$14 = pK_a + pK_b$
$pK_a = 14 - pK_b = 14 - 4.75 = 9.25$
$K_a = 10^{-pK_a} = 10^{-9.25} = 5.62341325 \times 10^{-10}$

Concentration (M)	$NH_4^+(aq)$ +	$H_2O(l)$	$\rightleftharpoons$	$NH_3(aq)$ +	$H_3O^+(aq)$
Initial	0.15	—		0.25	0
Change	−x	—		+x	+x
Equilibrium	0.15 − x	—		0.25 + x	x

Assume that x is negligible with respect to both 0.25 and 0.15 because both concentrations are much larger than K_a.

$$K_a = .62341325 \times 10^{-10} = \frac{[NH_3][H_3O^+]}{[NH_4^+]} = \frac{[0.25+x][H_3O^+]}{[0.15-x]} = \frac{[0.25][H_3O^+]}{[0.15]}$$

$$X = [H_3O^+] = K_a \frac{[NH_4^+]}{[NH_3]} = (5.62341325 \times 10^{-10})\left(\frac{0.15}{0.25}\right) = 3.374048 \times 10^{-10}\ M$$

Check assumption: Percent error = $(3.374048 \times 10^{-10}/0.15)100\% = 2 \times 10^{-7}\%$. The assumption is valid.

$$pH = -\log [H_3O^+] = -\log [3.374048 \times 10^{-10}] = 9.4718 = \mathbf{9.47}$$

Verify the pH using the Henderson-Hasselbalch equation.

$$pH = pK_a + \log\left(\frac{[base]}{[acid]}\right)$$

$$pH = 9.25 + \log\left(\frac{[NH_3]}{[NH_4^+]}\right) = 9.25 + \log\left(\frac{[0.25]}{[0.15]}\right)$$

$$\mathbf{pH = 9.47}$$

19.22   Plan: The buffer components are HCO_3^- from the salt $KHCO_3$ and CO_3^{2-} from the salt K_2CO_3. Choose the K_a value that corresponds to the equilibrium with these two components. The potassium ion is a spectator ion and is ignored because it is not involved in the buffer. Write the acid-dissociation reaction and its K_a expression. Set up a reaction table in which x equals the amount of acid that dissociates; solving for x will result in $[H_3O^+]$, from which the pH can be calculated. Alternatively, the pH can be calculated from the Henderson-Hasselbalch equation.
Solution:
a) K_{a1} refers to carbonic acid, H_2CO_3, losing one proton to produce HCO_3^-. This is not the correct K_a because H_2CO_3 is not involved in the buffer. K_{a2} is the correct K_a to choose because it is the equilibrium constant for the loss of the second proton to produce CO_3^{2-} from HCO_3^-.
b) Set up the reaction table and use K_{a2} to calculate pH.

Concentration (M)	$HCO_3^-(aq)$ +	$H_2O(l)$	$\rightleftharpoons$	$CO_3^{2-}(aq)$ +	$H_3O^+(aq)$
Initial	0.22	—		0.37	0
Change	−x	—		+x	+x
Equilibrium	0.22 − x	—		0.37 + x	x

Assume that x is negligible with respect to both 0.22 and 0.37 because both concentrations are much larger than K_a.

$$K_a = 4.7 \times 10^{-11} = \frac{[H_3O^+][CO_3^{2-}]}{[HCO_3^-]} = \frac{[x][0.37+x]}{[0.22-x]} = \frac{[x][0.37]}{[0.22]}$$

$$[H_3O^+] = K_a\frac{[HCO_3^-]}{[CO_3^{2-}]} = (4.7 \times 10^{-11})\left(\frac{0.22}{0.37}\right) = 2.79459 \times 10^{-11}\ M$$

Check assumption: Percent error = $(2.79459 \times 10^{-11}/0.22)100\% = 1.3 \times 10^{-8}\%$. The assumption is valid.

$$pH = -\log [H_3O^+] = -\log (2.79459 \times 10^{-11}) = 10.5537 = \mathbf{10.55}$$

Verify the pH using the Henderson-Hasselbalch equation.

$$pH = pK_a + \log\left(\frac{[\text{base}]}{[\text{acid}]}\right) \qquad\qquad pK_a = -\log\,(4.7\text{x}10^{-11}) = 10.328$$

$$pH = 10.328 + \log\left(\frac{[CO_3^{2-}]}{[HCO_3^{-}]}\right) = 10.328 + \log\left(\frac{[0.37]}{[0.22]}\right)$$

$$\mathbf{pH = 10.55}$$

19.24   <u>Plan:</u> Given the pH and K_a of an acid, the buffer-component ratio can be calculated from the Henderson-Hasselbalch equation. Convert K_a to pK_a.
<u>Solution:</u>
$$pK_a = -\log K_a = -\log\,(1.3\text{x}10^{-5}) = 4.8860566$$

$$pH = pK_a + \log\left(\frac{[\text{base}]}{[\text{acid}]}\right)$$

$$5.44 = 4.8860566 + \log\left(\frac{[Pr^-]}{[HPr]}\right)$$

$$0.5539467 = \log\left(\frac{[Pr^-]}{[HPr]}\right) \qquad \text{Raise each side to } 10^x.$$

$$\frac{[Pr^-]}{[HPr]} = 3.5805 = \mathbf{3.6}$$

19.26   <u>Plan:</u> Given the pH and K_a of an acid, the buffer-component ratio can be calculated from the Henderson-Hasselbalch equation. Convert K_a to pK_a.
<u>Solution:</u>
$$pK_a = -\log K_a = -\log\,(2.3\text{x}10^{-9}) = 8.63827$$

$$pH = pK_a + \log\left(\frac{[\text{base}]}{[\text{acid}]}\right)$$

$$7.95 = 8.63827 + \log\left(\frac{[BrO^-]}{[HBrO]}\right)$$

$$-0.68827 = \log\left(\frac{[BrO^-]}{[HBrO]}\right) \qquad \text{Raise each side to } 10^x.$$

$$\frac{[BrO^-]}{[HBrO]} = 0.204989 = \mathbf{0.20}$$

19.28   <u>Plan:</u> Determine the pK_a of the acid from the concentrations of the conjugate acid and base, and the pH of the solution. This requires the Henderson-Hasselbalch equation. Set up a reaction table that shows the stoichiometry of adding the strong base NaOH to the weak acid in the buffer. Calculate the new concentrations of the buffer components and use the Henderson-Hasselbalch equation to find the new pH.
<u>Solution:</u>
$$pH = pK_a + \log\left(\frac{[\text{base}]}{[\text{acid}]}\right)$$

$$3.35 = pK_a + \log\left(\frac{[A^-]}{[HA]}\right) = pK_a + \log\left(\frac{[0.1500]}{[0.2000]}\right)$$

$$3.35 = pK_a - 0.1249387$$
$$pK_a = 3.474939 = 3.47$$

Determine the moles of conjugate acid (HA) and conjugate base (A$^-$) using $(M)(V)$ = moles.

$$\text{Moles of HA} = (0.5000\ \text{L})\left(\frac{0.2000\ \text{mol HA}}{1\ \text{L}}\right) = 0.1000\ \text{mol HA}$$

$$\text{Moles of A}^- = (0.5000 \text{ L})\left(\frac{0.1500 \text{ mol A}^-}{1 \text{ L}}\right) = 0.07500 \text{ mol A}^-$$

The reaction is:

	HA(aq)	+	NaOH(aq)	$\rightarrow$	Na$^+$(aq)	+	A$^-$(aq)	+	H$_2$O(l)
Initial	0.1000 mol		0.0015 mol				0.07500 mol		
Change	–0.0015 mol		–0.0015 mol				+ 0.0015 mol		
Final	0.0985 mol		0 mol				0.0765 mol		

NaOH is the limiting reagent. The addition of 0.0015 mol NaOH produces an additional 0.0015 mol A$^-$ and consumes 0.0015 mol of HA.
Then:

$$[\text{A}^-] = \frac{0.0765 \text{ mol A}^-}{0.5000 \text{ L}} = 0.153 \text{ } M \text{ A}^-$$

$$[\text{HA}] = \frac{0.0985 \text{ mol HA}}{0.5000 \text{ L}} = 0.197 \text{ } M \text{ HA}$$

$$\text{pH} = \text{p}K_a + \log\left(\frac{[\text{base}]}{[\text{acid}]}\right)$$

$$\text{pH} = 3.474939 + \log\left(\frac{[0.153]}{[0.197]}\right) = 3.365164 = \textbf{3.37}$$

19.30 <u>Plan:</u> Determine the pK_a of the acid from the concentrations of the conjugate acid and base, and the pH of the solution. This requires the Henderson-Hasselbalch equation. Set up a reaction table that shows the stoichiometry of adding the strong base Ba(OH)$_2$ to the weak acid in the buffer. Calculate the new concentrations of the buffer components and use the Henderson-Hasselbalch equation to find the new pH.
<u>Solution:</u>

$$\text{pH} = \text{p}K_a + \log\left(\frac{[\text{base}]}{[\text{acid}]}\right)$$

$$8.77 = \text{p}K_a + \log\left(\frac{[\text{Y}^-]}{[\text{HY}]}\right) = \text{p}K_a + \log\left(\frac{[0.220]}{[0.110]}\right)$$

$$8.77 = \text{p}K_a + 0.3010299957$$
$$\text{p}K_a = 8.46897 = 8.47$$

Determine the moles of conjugate acid (HY) and conjugate base (Y$^-$) using $(M)(V)$ = moles.

$$\text{Moles of HY} = (0.350 \text{ L})\left(\frac{0.110 \text{ mol HY}}{1 \text{ L}}\right) = 0.0385 \text{ mol HY}$$

$$\text{Moles of Y}^- = (0.350 \text{ L})\left(\frac{0.220 \text{ mol Y}^-}{1 \text{ L}}\right) = 0.077 \text{ mol Y}^-$$

The reaction is:

	2HY(aq)	+	Ba(OH)$_2$(aq)	$\rightarrow$	Ba^{2+}(aq)	+	2Y$^-$(aq)	+	2H$_2$O(l)
Initial	0.0385 mol		0.0015 mol				0.077 mol		
Change	–0.0030 mol		–0.0015 mol				+0.0030 mol		
Final	0.0355 mol		0 mol				0.0800 mol		

Ba(OH)$_2$ is the limiting reagent. The addition of 0.0015 mol Ba(OH)$_2$ will produce 2 x 0.0015 mol Y$^-$ and consume 2 x 0.0015 mol of HY.
Then:

$$[\text{Y}^-] = \frac{0.0800 \text{ mol Y}^-}{0.350 \text{ L}} = 0.228571 \text{ } M \text{ Y}^-$$

$$[\text{HY}] = \frac{0.0355 \text{ mol HY}}{0.350 \text{ L}} = 0.101429 \text{ } M \text{ HY}$$

$$\text{pH} = \text{p}K_a + \log\left(\frac{[\text{Y}^-]}{[\text{HY}]}\right)$$

$$pH = 8.46897 + \log\left(\frac{[0.228571]}{[0.101429]}\right) = 8.82183 = \textbf{8.82}$$

19.32   Plan: The hydrochloric acid will react with the sodium acetate, $NaC_2H_3O_2$, to form acetic acid, $HC_2H_3O_2$. Calculate the number of moles of HCl and $NaC_2H_3O_2$. Set up a reaction table that shows the stoichiometry of the reaction of HCl and $NaC_2H_3O_2$. All of the HCl will be consumed to form $HC_2H_3O_2$, and the number of moles of $C_2H_3O_2^-$ will decrease. Find the new concentrations of $NaC_2H_3O_2$ and $HC_2H_3O_2$ and use the Henderson-Hasselbalch equation to find the pH of the buffer. Add 0.15 to find the pH of the buffer after the addition of the KOH. Use the Henderson-Hasselbalch equation to find the [base]/[acid] ratio needed to achieve that pH.
Solution:

a)   Initial moles of HCl $= \left(\frac{0.452\,\text{mol HCl}}{L}\right)\left(\frac{10^{-3}\,L}{1\,\text{mL}}\right)(204\,\text{mL}) = 0.092208\,\text{mol HCl}$

Initial moles of $NaC_2H_3O_2 = \left(\frac{0.400\ \text{mol } NaC_2H_3O_2}{L}\right)(0.500\,L) = 0.200\,\text{mol } NaC_2H_3O_2$

	HCl	+	$NaC_2H_3O_2$	$\rightarrow$	$HC_2H_3O_2$	+	NaCl
Initial	0.092208 mol		0.200 mol		0 mol		
Change	−0.092208 mol		−0.092208 mol		+0.092208 mol		
Final	0 mol		0.107792 mol		0.092208 mol		

Total volume $= 0.500\,L + (204\,\text{mL})(10^{-3}\,L/1\,\text{mL}) = 0.704\,L$

$$[HC_2H_3O_2] = \frac{0.092208\ \text{mol}}{0.704\ L} = 0.1309773\ M$$

$$[C_2H_3O_2^-] = \frac{0.107792\ \text{mol}}{0.704\ L} = 0.1531136\ M$$

$$pK_a = -\log K_a = -\log(1.8\times10^{-5}) = 4.744727495$$

$$pH = pK_a + \log\left(\frac{[C_2H_3O_2^-]}{[HC_2H_3O_2]}\right)$$

$$pH = 4.744727495 + \log\left(\frac{[0.1531136]}{[0.1309773]}\right) = 4.812545 = \textbf{4.81}$$

b) The addition of base would increase the pH, so the new pH is $(4.81 + 0.15) = 4.96$.
The new $[C_2H_3O_2^-]/[HC_2H_3O_2]$ ratio is calculated using the Henderson-Hasselbalch equation.

$$pH = pK_a + \log\left(\frac{[C_2H_3O_2^-]}{[HC_2H_3O_2]}\right)$$

$$4.96 = 4.744727495 + \log\left(\frac{[C_2H_3O_2^-]}{[HC_2H_3O_2]}\right)$$

$$0.215272505 = \log\left(\frac{[C_2H_3O_2^-]}{[HC_2H_3O_2]}\right)$$

$$\frac{[C_2H_3O_2^-]}{[HC_2H_3O_2]} = 1.64162$$

From part a), we know that $[HC_2H_3O_2] + [C_2H_3O_2^-] = (0.1309773\ M + 0.1531136\ M) = 0.2840909\ M$. Although the *ratio* of $[C_2H_3O_2^-]$ to $[HC_2H_3O_2]$ can change when acid or base is added, the *absolute amount* does not change unless acetic acid or an acetate salt is added.
Given that $[C_2H_3O_2^-]/[HC_2H_3O_2] = 1.64162$ and $[HC_2H_3O_2] + [C_2H_3O_2^-] = 0.2840909\ M$, solve for $[C_2H_3O_2^-]$ and substitute into the second equation.
$[C_2H_3O_2^-] = 1.64162[HC_2H_3O_2]$ and $[HC_2H_3O_2] + 1.64162[HC_2H_3O_2] = 0.2840909\ M$
$[HC_2H_3O_2] = 0.1075441\ M$ and $[C_2H_3O_2^-] = 0.176547\ M$
Moles of $C_2H_3O_2^-$ needed $= (0.176547\ \text{mol } C_2H_3O_2^-/L)(0.500\ L) = 0.0882735\ \text{mol}$
Moles of $C_2H_3O_2^-$ initially $= (0.1531136\ \text{mol } C_2H_3O_2^-/L)(0.500\ L) = 0.0765568\ \text{mol}$

This would require the addition of (0.0882735 mol – 0.0765568 mol) = 0.0117167 mol $C_2H_3O_2^-$
The KOH added reacts with $HC_2H_3O_2$ to produce additional $C_2H_3O_2^-$:

$$HC_2H_3O_2 + KOH \rightarrow C_2H_3O_2^- + K^+ + H_2O(l)$$

To produce 0.0117167 mol $C_2H_3O_2^-$ would require the addition of 0.0117167 mol KOH.

Mass (g) of KOH = $(0.0117167 \text{ mol KOH})\left(\dfrac{56.11 \text{ g KOH}}{1 \text{ mol KOH}}\right)$ = 0.657424 = **0.66 g KOH**

19.34   Plan: Select conjugate pairs with K_a values close to the desired $[H_3O^+]$.
Solution:
a) For pH ≈ 4.5, $[H_3O^+] = 10^{-4.5} = 3.2\times10^{-5}$ M. Some good selections are the $HOOC(CH_2)_4COOH/$
$HOOC(CH_2)_4COOH^-$ conjugate pair with K_a equal to 3.8×10^{-5} or $C_6H_5CH_2COOH/C_6H_5CH_2COO^-$ conjugate pair with K_a equal to 4.9×10^{-5}. From the base list, the $C_6H_5NH_2/C_6H_5NH_3^+$ conjugate pair comes close with $K_a = K_w/K_b = 1.0\times10^{-14}/4.0\times10^{-10} = 2.5\times10^{-5}$.
b) For pH ≈ 7.0, $[H_3O^+] = 10^{-7.0} = 1.0\times10^{-7}$ M. Two choices are the $H_2PO_4^-/HPO_4^{2-}$ conjugate pair with K_a of 6.3×10^{-8} and the $H_2AsO_4^-/HAsO_4^{2-}$ conjugate pair with K_a of 1.1×10^{-7}.

19.36   Plan: Select conjugate pairs with pK_a values close to the desired pH. Convert pH to $[H_3O^+]$ for easy comparison to K_a values in the Appendix. Determine an appropriate base by $[OH^-] = K_w/[H_3O^+]$.
Solution:
a) For pH ≈ 3.5 ($[H_3O^+] = 10^{-pH} = 10^{-3.5} = 3.2\times10^{-4}$), the best selection is the $HOCH_2CH(OH)COOH/$
$HOCH_2CH(OH)COOH^-$ conjugate pair with a $K_a = 2.9\times10^{-4}$. The $CH_3COOC_6H_4COOH/CH_3COOC_6H_4COO^-$ pair, with $K_a = 3.6\times10^{-4}$, is also a good choice. The $[OH^-] = K_w/[H_3O^+] = 1.0\times10^{-14}/3.2\times10^{-4} = 3.1\times10^{-11}$, results in no reasonable K_b values from the Appendix.
b) For pH ≈ 5.5 ($[H_3O^+] = 10^{-pH} = 3\times10^{-6}$), no K_{a1} gives an acceptable pair; the K_{a2} values for adipic acid, malonic acid, and succinic acid are reasonable. The $[OH^-] = K_w/[H_3O^+] = 1.0\times10^{-14}/3\times10^{-6} = 3\times10^{-9}$; the K_b selection is $C_5H_5N/C_5H_5NH^+$.

19.39   Plan: Given the pH and K_a of an acid, the buffer-component ratio can be calculated from the Henderson-Hasselbalch equation. Convert K_a to pK_a.
Solution:
The value of the K_a from the Appendix: $K_a = 6.3\times10^{-8}$ (We are using K_{a2} since we are dealing with the equilibrium in which the second hydrogen ion is being lost.)
$pK_a = -\log K_a = -\log (6.3\times10^{-8}) = 7.200659451$
Use the Henderson-Hasselbalch equation:

$$pH = pK_a + \log\left(\frac{[HPO_4^{2-}]}{[H_2PO_4^-]}\right)$$

$$7.40 = 7.200659451 + \log\left(\frac{[HPO_4^{2-}]}{[H_2PO_4^-]}\right)$$

$$0.19934055 = \log\left(\frac{[HPO_4^{2-}]}{[H_2PO_4^-]}\right)$$

$$\frac{[HPO_4^{2-}]}{[H_2PO_4^-]} = 1.582486 = \mathbf{1.6}$$

19.41   a) The initial pH is lowest for the flask solution of the strong acid, followed by the weak acid and then the weak base. In other words, *strong acid*–strong base < *weak acid*–strong base < strong acid–*weak base* in terms of initial pH.
b) At the equivalence point, the moles of H_3O^+ equal the moles of OH^-, regardless of the type of titration. However, the strong acid–strong base equivalence point occurs at pH = 7.00 because the resulting cation-anion combination does not react with water. An example is the reaction NaOH + HCl → H_2O + NaCl. Neither Na^+ nor Cl^- ions dissociate in water.

The weak acid–strong base equivalence point occurs at pH > 7, because the anion of the weak acid is weakly basic, whereas the cation of the strong base does not react with water. An example is the reaction $HCOOH + NaOH \rightarrow HCOO^- + H_2O + Na^+$. The conjugate base, $HCOO^-$, reacts with water according to this reaction: $HCOO^- + H_2O \rightarrow HCOOH + OH^-$.

The strong acid–weak base equivalence point occurs at pH < 7, because the anion of the strong acid does not react with water, whereas the cation of the weak base is weakly acidic. An example is the reaction $HCl + NH_3 \rightarrow NH_4^+ + Cl^-$. The conjugate acid, NH_4^+, dissociates slightly in water: $NH_4^+ + H_2O \rightarrow NH_3 + H_3O^+$.

In rank order of pH at the equivalence point, strong acid–*weak base* < *strong acid*–strong base < *weak acid*–strong base.

19.43   At the very center of the buffer region of a weak acid–strong base titration, the concentration of the weak acid and its conjugate base are equal. If equal values for concentration are put into the Henderson-Hasselbalch equation, the [base]/[acid] ratio is 1, the log of 1 is 0, and the pH of the solution equals the pK_a of the weak acid.

$$pH = pK_a + \log\left(\frac{[base]}{[acid]}\right)$$

$$pH = pK_a + \log 1$$

19.45   To see a distinct color in a mixture of two colors, you need one color to be about 10 times the intensity of the other. For this to take place, the concentration ratio $[HIn]/[In^-]$ needs to be greater than 10:1 or less than 1:10. This will occur when $pH = pK_a - 1$ or $pH = pK_a + 1$, respectively, giving a transition range of about two units.

19.47   The equivalence point in a titration is the point at which the number of moles of OH^- equals the number of moles of H_3O^+ (be sure to account for stoichiometric ratios, e.g., one mol of $Ca(OH)_2$ produces two moles of OH^-). The end point is the point at which the added indicator changes color. If an appropriate indicator is selected, the end point is close to the equivalence point, but not normally the same. Using an indicator that changes color at a pH after the equivalence point means the equivalence point is reached first. However, if an indicator is selected that changes color at a pH before the equivalence point, then the end point is reached first.

19.49   Plan: The reaction occurring in the titration is the neutralization of H_3O^+ (from HCl) by OH^- (from NaOH):
$$HCl(aq) + NaOH(aq) \rightarrow H_2O(l) + NaCl(aq) \quad \text{or, omitting spectator ions:}$$
$$H_3O^+(aq) + OH^-(aq) \rightarrow 2H_2O(l)$$
For the titration of a strong acid with a strong base, the pH before the equivalence point depends on the excess concentration of acid and the pH after the equivalence point depends on the excess concentration of base. At the equivalence point, there is not an excess of either acid or base so the pH is 7.0. The equivalence point occurs when 40.00 mL of base has been added. Use $(M)(V)$ to determine the number of moles of acid and base. Note that the NaCl product is a neutral salt that does not affect the pH.

Solution:
The initial number of moles of HCl = (0.1000 mol HCl/L)(10^{-3} L/1 mL)(40.00 mL) = 4.000×10^{-3} mol HCl

a) At 0 mL of base added, the concentration of hydronium ion equals the original concentration of HCl.
$$pH = -\log (0.1000\ M) = \textbf{1.0000}$$

b) Determine the moles of NaOH added:
Moles of added NaOH = (0.1000 mol NaOH/L)(10^{-3} L/1 mL)(25.00 mL) = 2.500×10^{-3} mol NaOH

	$HCl(aq)$	+	$NaOH(aq)$	$\rightarrow$	$H_2O(l)$	+	$NaCl(aq)$
Initial	4.000×10^{-3} mol		2.500×10^{-3} mol		–		0
Change	-2.500×10^{-3} mol		-2.500×10^{-3} mol		–		$+2.500 \times 10^{-3}$ mol
Final	1.500×10^{-3} mol		0				2.500×10^{-3} mol

The volume of the solution at this point is [(40.00 + 25.00) mL](10^{-3} L/1 mL) = 0.06500 L
The molarity of the excess HCl is (1.500×10^{-3} mol HCl)/(0.06500 L) = 0.023077 M
$$pH = -\log (0.023077) = \textbf{1.6368}$$

c) Determine the moles of NaOH added:
Moles of added NaOH = (0.1000 mol NaOH/L)(10^{-3} L/1 mL)(39.00 mL) = 3.900×10^{-3} mol NaOH

	$HCl(aq)$	+	$NaOH(aq)$	$\rightarrow$	$H_2O(l)$	+	$NaCl(aq)$
Initial	4.000×10^{-3} mol		3.900×10^{-3} mol		–		0
Change	-3.900×10^{-3} mol		-3.900×10^{-3} mol		–		$+3.900 \times 10^{-3}$ mol
Final	1.000×10^{-4} mol		0				3.900×10^{-3} mol

The volume of the solution at this point is [(40.00 + 39.00) mL](10^{-3} L/1 mL) = 0.07900 L
The molarity of the excess HCl is (1.00x 10^{-4}mol HCl)/(0.07900 L) = 0.0012658 M
$$pH = -\log (0.0012658) = \textbf{2.898}$$
d) Determine the moles of NaOH added:
Moles of added NaOH = (0.1000 mol NaOH/L)(10^{-3} L/1 mL)(39.90 mL) = 3.990x10^{-3} mol NaOH

	HCl(aq)	+	NaOH(aq)	→	H$_2$O(l)	+	NaCl(aq)
Initial	4.000x10^{-3} mol		3.990x10^{-3} mol		–		0
Change	–3.990x10^{-3} mol		–3.990x10^{-3} mol		–		+3.990x10^{-3} mol
Final	1.000x10^{-5} mol		0				3.990x10^{-3} mol

The volume of the solution at this point is [(40.00 + 39.90) mL](10^{-3} L/1 mL) = 0.07990 L
The molarity of the excess HCl is (1.0x10^{-5} mol HCl)/(0.07990 L) = 0.000125156 M
$$pH = -\log (0.000125156) = \textbf{3.903}$$
e) Determine the moles of NaOH added:
Moles of added NaOH = (0.1000 mol NaOH/L)(10^{-3} L/1 mL)(40.00 mL) = 4.000x10^{-3} mol NaOH

	HCl(aq)	+	NaOH(aq)	→	H$_2$O(l)	+	NaCl(aq)
Initial	4.000x10^{-3} mol		4.000x10^{-3} mol		–		0
Change	–4.000x10^{-3} mol		–4.000x10^{-3} mol		–		+4.000x10^{-3} mol
Final	0		0				4.000x10^{-3} mol

The NaOH will react with an equal amount of the acid and 0.0 mol HCl will remain. This is the equivalence point of a strong acid–strong base titration, thus, the pH is **7.00**. Only the neutral salt NaCl is in solution at the equivalence point.
f) The NaOH is now in excess. It will be necessary to calculate the excess base after reacting with the HCl. The excess strong base will give the pOH, which can be converted to the pH.
Determine the moles of NaOH added:
Moles of added NaOH = (0.1000 mol NaOH/L)(10^{-3} L/1 mL)(40.10 mL) = 4.010x10^{-3} mol NaOH
The HCl will react with an equal amount of the base, and 1.0x10^{-5} mol NaOH will remain.

	HCl(aq)	+	NaOH(aq)	→	H$_2$O(l)	+	NaCl(aq)
Initial	4.000x10^{-3} mol		4.010x10^{-3} mol		–		0
Change	–4.000x10^{-3} mol		–4.000x10^{-3} mol		–		+4.000x10^{-3} mol
Final	0		1.000x10^{-5} mol				4.000x10^{-3} mol

The volume of the solution at this point is [(40.00 + 40.10) mL](10^{-3} L/1 mL) = 0.08010 L
The molarity of the excess NaOH is (1.0x10^{-5} mol NaOH)/(0.08010 L) = 0.00012484 M
$$pOH = -\log (0.00012484) = 3.9036$$
$$pH = 14.00 - pOH = 14.00 - 3.9036 = 10.09637 = \textbf{10.10}$$
g) Determine the moles of NaOH added:
Moles of NaOH = (0.1000 mol NaOH/L)(10^{-3} L/1 mL)(50.00 mL) = 5.000x10^{-3} mol NaOH
The HCl will react with an equal amount of the base, and 1.000x10^{-3} mol NaOH will remain.

	HCl(aq)	+	NaOH(aq)	→	H$_2$O(l)	+	NaCl(aq)
Initial	4.000x10^{-3} mol		5.000x10^{-3} mol		–		0
Change	–4.000x10^{-3} mol		–4.000x10^{-3} mol		–		+4.000x10^{-3} mol
Final	0		1.000x10^{-3} mol				4.000x10^{-3} mol

The volume of the solution at this point is [(40.00 + 50.00) mL](10^{-3} L/ 1 mL) = 0.09000 L
The molarity of the excess NaOH is (1.000x10^{-3} mol NaOH)/(0.09000 L) = 0.011111 M
$$pOH = -\log (0.011111) = 1.95424$$
$$pH = 14.00 - pOH = 14.00 - 1.95424 = 12.04576 = \textbf{12.05}$$

19.51   Plan: This is a titration between a weak acid and a strong base. The pH before addition of the base is dependent on the K_a of the acid (labeled HBut). Prior to reaching the equivalence point, the added base reacts with the acid to form butanoate ion (labeled But⁻). The equivalence point occurs when 20.00 mL of base is added to the acid because at this point, moles acid = moles base. Addition of base beyond the equivalence point is simply the addition of excess OH⁻.

Solution:

a) At 0 mL of base added, the concentration of $[H_3O^+]$ is dependent on the dissociation of butanoic acid:

	HBut	+	H_2O	$\leftrightarrows$	H_3O^+	+	But^-
Initial	0.100 M				0		0
Change	−x				+x		+x
Equilibrium	0.100 − x				x		x

$$K_a = 1.54 \times 10^{-5} = \frac{[H_3O^+][But^-]}{[HBut]} = \frac{x^2}{0.1000 - x} = \frac{x^2}{0.1000}$$

$$x = [H_3O^+] = 1.2409674 \times 10^{-3} \ M$$

$$pH = -\log [H_3O^+] = -\log (1.2409674 \times 10^{-3}) = 2.9062 = \mathbf{2.91}$$

b) The initial number of moles of HBut = $(M)(V)$ = (0.1000 mol HBut/L)(10^{-3} L/1 mL)(20.00 mL)
$$= 2.000 \times 10^{-3} \ \text{mol HBut}$$

Determine the moles of NaOH added:

Moles of added NaOH = (0.1000 mol NaOH/L)(10^{-3} L/1 mL)(10.00 mL) = 1.000×10^{-3} mol NaOH
The NaOH will react with an equal amount of the acid, and 1.000×10^{-3} mol HBut will remain. An equal number of moles of But^- will form.

	HBut(aq)	+	NaOH(aq)	$\rightarrow$	$H_2O(l)$	+	But^-(aq)	+	Na^+(aq)
Initial	2.000×10^{-3} mol		1.000×10^{-3} mol		–		0		–
Change	-1.000×10^{-3} mol		-1.000×10^{-3} mol		–		$+1.000 \times 10^{-3}$ mol		–
Final	1.000×10^{-3} mol		0		–		1.000×10^{-3} mol		

The volume of the solution at this point is [(20.00 + 10.00) mL](10^{-3} L/1 mL) = 0.03000 L
The molarity of the excess HBut is (1.000×10^{-3} mol HBut)/(0.03000 L) = 0.03333 M
The molarity of the But^- formed is (1.000×10^{-3} mol But^- /(0.03000 L) = 0.03333 M
Using a reaction table for the equilibrium reaction of HBut:

	HBut	+	H_2O	$\leftrightarrows$	H_3O^+	+	But^-
Initial	0.03333 M		–		0		0.03333 M
Change	−x				+x		+x
Equilibrium	0.03333 − x				x		0.03333 + x

$$K_a = 1.54 \times 10^{-5} = \frac{[H_3O^+][But^-]}{[HBut]} = \frac{x(0.0333 + x)}{0.03333 - x} = \frac{x(0.03333)}{0.03333}$$

$$x = [H_3O^+] = 1.54 \times 10^{-5} \ M$$

$$pH = -\log [H_3O^+] = -\log (1.54 \times 10^{-5}) = 4.812479 = \mathbf{4.81}$$

c) Determine the moles of NaOH added:

Moles of added NaOH = (0.1000 mol NaOH/L)(10^{-3} L/1 mL)(15.00 mL) = 1.500×10^{-3} mol NaOH
The NaOH will react with an equal amount of the acid, and 5.00×10^{-4} mol HBut will remain, and 1.500×10^{-3} moles of But^- will form.

	HBut(aq)	+	NaOH(aq)	$\rightarrow$	$H_2O(l)$	+	But^-(aq)	+	Na^+(aq)
Initial	2.000×10^{-3} mol		1.500×10^{-3} mol		–		0		–
Change	-1.500×10^{-3} mol		-1.500×10^{-3} mol		–		$+1.500 \times 10^{-3}$ mol		–
Final	5.000×10^{-4} mol		0		–		1.500×10^{-3} mol		

The volume of the solution at this point is [(20.00 + 15.00) mL](10^{-3} L/1 mL) = 0.03500 L
The molarity of the excess HBut is (5.00×10^{-4} mol HBut)/(0.03500 L) = 0.0142857 M
The molarity of the But^- formed is (1.500×10^{-3} mol But^-)/(0.03500 L) = 0.042857 M
Using a reaction table for the equilibrium reaction of HBut:

	HBut	+	H_2O	$\leftrightarrows$	H_3O^+	+	But^-
Initial	0.0142857 M		–		0		0.042857 M
Change	−x				+x		+x
Equilibrium	0.0142857 − x				x		0.042857 + x

$$K_a = 1.54 \times 10^{-5} = \frac{[H_3O^+][But^-]}{[HBut]} = \frac{x(0.042857 + x)}{0.0142857 - x} = \frac{x(0.042857)}{0.0142857}$$

$$x = [H_3O^+] = 5.1333 \times 10^{-6}\ M$$
$$pH = -\log\ [H_3O^+] = -\log\ (5.1333 \times 10^{-6}) = 5.2896 = \mathbf{5.29}$$

d) Determine the moles of NaOH added:

Moles of added NaOH = (0.1000 mol NaOH/L)(10^{-3} L/1 mL)(19.00 mL) = 1.900×10^{-3} mol NaOH
The NaOH will react with an equal amount of the acid, and 1.00×10^{-4} mol HBut will remain, and 1.900×10^{-3} moles of But$^-$ will form.

	HBut(aq)	+	NaOH(aq)	→	H$_2$O(l)	+	But$^-$(aq)	+	Na$^+$(aq)
Initial	2.000×10^{-3} mol		1.900×10^{-3} mol		–		0		–
Change	-1.900×10^{-3} mol		-1.900×10^{-3} mol		–		$+1.900 \times 10^{-3}$ mol		–
Final	1.000×10^{-4} mol		0				1.900×10^{-3} mol		

The volume of the solution at this point is [(20.00 + 19.00) mL](10^{-3} L/1 mL) = 0.03900 L
The molarity of the excess HBut is (1.00×10^{-4} mol HBut)/(0.03900 L) = 0.0025641 M
The molarity of the But$^-$ formed is (1.900×10^{-3} mol But$^-$)/(0.03900 L) = 0.0487179 M
Using a reaction table for the equilibrium reaction of HBut:

	HBut	+	H$_2$O	⇆	H$_3$O$^+$	+	But$^-$
Initial	0.0025641 M		–		0		0.0487179 M
Change	−x		–		+x		+x
Equilibrium	0.0025641 − x				+x		0.0487179 + x

$$K_a = 1.54 \times 10^{-5} = \frac{[H_3O^+][But^-]}{[HBut]} = \frac{x(0.0487179 + x)}{0.0025641 - x} = \frac{x(0.0487179)}{0.0025641}$$

$$x = [H_3O^+] = 8.1052632 \times 10^{-7}\ M$$
$$pH = -\log\ [H_3O^+] = -\log\ (8.1052632 \times 10^{-7}) = 6.09123 = \mathbf{6.09}$$

e) Determine the moles of NaOH added:

Moles of addedNaOH = (0.1000 mol NaOH/L)(10^{-3} L/1 mL)(19.95 mL) = 1.995×10^{-3} mol NaOH
The NaOH will react with an equal amount of the acid, and 5×10^{-6} mol HBut will remain, and 1.995×10^{-3} moles of But$^-$ will form.

	HBut(aq)	+	NaOH(aq)	→	H$_2$O(l)	+	But$^-$(aq)	+	Na$^+$(aq)
Initial	2.000×10^{-3} mol		1.995×10^{-3} mol		–		0		–
Change	-1.995×10^{-3} mol		-1.995×10^{-3} mol		–		$+1.995 \times 10^{-3}$ mol		–
Final	5.000×10^{-6} mol		0				1.995×10^{-3} mol		

The volume of the solution at this point is [(20.00 + 19.95) mL](10^{-3} L/1 mL) = 0.03995 L
The molarity of the excess HBut is (5×10^{-6} mol HBut)/(0.03995 L) = 0.000125156 M
The molarity of the But$^-$ formed is (1.995×10^{-3} mol But$^-$)/(0.03995 L) = 0.0499374 M
Using a reaction table for the equilibrium reaction of HBut:

	HBut	+	H$_2$O	⇆	H$_3$O$^+$	+	But$^-$
Initial	0.000125156 M		–		0		0.0499374 M
Change	−x		–		+x		+x
Equilibrium	0.000125156 − x				x		0.0499374 + x

$$K_a = 1.54 \times 10^{-5} = \frac{[H_3O^+][But^-]}{[HBut]} = \frac{x(0.0499374 + x)}{0.000125156 - x} = \frac{x(0.0499374)}{0.000125156}$$

$$x = [H_3O^+] = 3.859637 \times 10^{-8}\ M$$
$$pH = -\log\ [H_3O^+] = -\log\ (3.859637 \times 10^{-8}) = 7.41345 = \mathbf{7.41}$$

f) Determine the moles of NaOH added:

Moles of added NaOH = (0.1000 mol NaOH/L)(10^{-3} L/1 mL)(20.00 mL) = 2.000×10^{-3} mol NaOH
The NaOH will react with an equal amount of the acid, and 0 mol HBut will remain, and 2.000×10^{-3} moles of But$^-$ will form. This is the equivalence point.

	HBut(aq)	+	NaOH(aq)	→	H$_2$O(l)	+	But$^-$(aq)	+	Na$^+$(aq)
Initial	2.000×10^{-3} mol		2.000×10^{-3} mol		–		0		–
Change	-2.000×10^{-3} mol		-2.000×10^{-3} mol		–		$+2.000 \times 10^{-3}$ mol		–
Final	0		0				2.000×10^{-3} mol		

The K_b of But$^-$ is now important.

The volume of the solution at this point is $[(20.00 + 20.00) \text{ mL}](10^{-3} \text{ L/1 mL}) = 0.04000 \text{ L}$

The molarity of the But⁻ formed is $(2.000 \times 10^{-3} \text{ mol But}^-)/(0.04000 \text{ L}) = 0.05000 \ M$

$$K_b = K_w/K_a = (1.0 \times 10^{-14})/(1.54 \times 10^{-5}) = 6.49351 \times 10^{-10}$$

Using a reaction table for the equilibrium reaction of But⁻:

	But⁻	+	H₂O	⇌	HBut	+	OH⁻
Initial	0.05000 M		–		0		0
Change	–x		–		+x		+x
Equilibrium	0.05000 – x				x		x

$$K_b = 6.49351 \times 10^{-10} = \frac{[\text{HBut}][\text{OH}^-]}{[\text{But}^-]} = \frac{[x][x]}{[0.05000 - x]} = \frac{[x][x]}{[0.05000]}$$

$[\text{OH}^-] = x = 5.6980304 \times 10^{-6} \ M$

$\text{pOH} = -\log (5.6980304 \times 10^{-6}) = 5.244275238$

$\text{pH} = 14.00 - \text{pOH} = 14.00 - 5.244275238 = 8.7557248 = \textbf{8.76}$

g) After the equivalence point, the excess strong base is the primary factor influencing the pH.

Determine the moles of NaOH added:

Moles of added NaOH = $(0.1000 \text{ mol NaOH/L})(10^{-3} \text{ L/1 mL})(20.05 \text{ mL}) = 2.005 \times 10^{-3} \text{ mol NaOH}$

The NaOH will react with an equal amount of the acid, 0 mol HBut will remain, and 5×10^{-6} moles of NaOH will be in excess. There will be 2.000×10^{-3} mol of But⁻ produced, but this weak base will not affect the pH compared to the excess strong base, NaOH.

	HBut(aq)	+	NaOH(aq)	→	H₂O(l)	+	But⁻(aq)	+	Na⁺(aq)
Initial	2.000×10⁻³ mol		2.005×10⁻³ mol		–		0		–
Change	–2.000×10⁻³ mol		–2.000×10⁻³ mol		–		+2.000×10⁻³ mol		–
Final	0		5.000×10⁻⁶ mol				2.000×10⁻³ mol		

The volume of the solution at this point is $[(20.00 + 20.05) \text{ mL}](10^{-3} \text{ L/1 mL}) = 0.04005 \text{ L}$

The molarity of the excess OH⁻ is $(5 \times 10^{-6} \text{ mol OH}^-)/(0.04005 \text{ L}) = 1.2484 \times 10^{-4} \ M$

$\text{pOH} = -\log (1.2484 \times 10^{-4}) = 3.9036$

$\text{pH} = 14.00 - \text{pOH} = 14.00 - 3.9036 = 10.0964 = \textbf{10.10}$

h) Determine the moles of NaOH added:

Moles of added NaOH = $(0.1000 \text{ mol NaOH/L})(10^{-3} \text{ L/1 mL})(25.00 \text{ mL}) = 2.500 \times 10^{-3} \text{ mol NaOH}$

The NaOH will react with an equal amount of the acid, 0 mol HBut will remain, and 5.00×10^{-4} moles of NaOH will be in excess.

	HBut(aq)	+	NaOH(aq)	→	H₂O(l)	+	But⁻(aq)	+	Na⁺(aq)
Initial	2.000×10⁻³ mol		2.500×10⁻³ mol		–		0		–
Change	–2.000×10⁻³ mol		–2.000×10⁻³ mol		–		+2.000×10⁻³ mol		–
Final	0		5.000×10⁻⁴ mol				2.000×10⁻³ mol		

The volume of the solution at this point is $[(20.00 + 25.00) \text{ mL}](10^{-3} \text{ L/1 mL}) = 0.04500 \text{ L}$

The molarity of the excess OH⁻ is $(5.00 \times 10^{-4} \text{ mol OH}^-/(0.04500 \text{ L}) = 1.1111 \times 10^{-2} \ M$

$\text{pOH} = -\log (1.1111 \times 10^{-2}) = 1.9542$

$\text{pH} = 14.00 - \text{pOH} = 14.00 - 1.9542 = 12.0458 = \textbf{12.05}$

19.53   Plan: Use $(M)(V)$ to find the initial moles of acid and then use the mole ratio in the balanced equation to find moles of base; dividing moles of base by the molarity of the base gives the volume. At the equivalence point, the conjugate base of the weak acid is present; set up a reaction table for the base dissociation in which x = the amount of dissociated base. Use the K_b expression to solve for x from which pOH and then pH is obtained.
Solution:
a) The balanced chemical equation is:

$$\text{NaOH}(aq) + \text{CH}_3\text{COOH}(aq) \rightarrow \text{Na}^+(aq) + \text{CH}_3\text{COO}^-(aq) + \text{H}_2\text{O}(l)$$

The sodium ions on the product side are written as separate species because they have no effect on the pH of the solution. Calculate the volume of NaOH needed:

Volume (mL) of NaOH =

$$\left(\frac{0.0520 \text{ mol CH}_3\text{COOH}}{L}\right)\left(\frac{10^{-3} \text{ L}}{1 \text{ mL}}\right)(42.2 \text{ mL})\left(\frac{1 \text{ mol NaOH}}{1 \text{ mol CH}_3\text{COOH}}\right)\left(\frac{L}{0.0372 \text{ mol NaOH}}\right)\left(\frac{1 \text{ mL}}{10^{-3} \text{ L}}\right)$$

$$= 58.989247 = \textbf{59.0 mL NaOH}$$

Determine the moles of initially CH_3COOH present:

$$\text{Moles of CH}_3\text{COOH} = \left(\frac{0.0520 \text{ mol CH}_3\text{COOH}}{L}\right)\left(\frac{10^{-3} \text{ L}}{1 \text{ mL}}\right)(42.2 \text{ mL}) = 0.0021944 \text{ mol CH}_3\text{COOH}$$

At the equivalence point, 0.0021944 mol NaOH will be added so the moles acid = moles base.
The NaOH will react with an equal amount of the acid, 0 mol CH_3COOH will remain, and 0.0021944 moles of CH_3COO^- will be formed.

	$CH_3COOH(aq)$	+	$NaOH(aq)$	$\rightarrow$	$H_2O(l)$	+	$CH_3COO^-(aq)$	+	$Na^+(aq)$
Initial	0.0021944 mol		0.0021944 mol		–		0		–
Change	−0.0021944 mol		−0.0021944 mol		–		+0.0021944 mol		–
Final	0		0				0.0021944 mol		

Determine the liters of solution present at the equivalence point:

$$\text{Volume} = [(42.2 + 58.989247) \text{ mL}](10^{-3} \text{ L}/1 \text{ mL}) = 0.101189247 \text{ L}$$

Concentration of CH_3COO^- at equivalence point:

$$\text{Molarity} = (0.0021944 \text{ mol CH}_3\text{COO}^-)/(0.101189247 \text{ L}) = 0.0216861 \text{ M}$$

Calculate K_b for CH_3COO^-: $K_a \text{ CH}_3\text{COOH} = 1.8 \times 10^{-5}$

$$K_b = K_w/K_a = (1.0 \times 10^{-14})/(1.8 \times 10^{-5}) = 5.556 \times 10^{-10}$$

Using a reaction table for the equilibrium reaction of CH_3COO^-:

	CH_3COO^-	+	H_2O	$\leftrightarrows$	CH_3COOH	+	OH^-
Initial	0.0216861 M		–		0		0
Change	−x				+x		+x
Equilibrium	0.0216861 − x				x		x

Determine the hydroxide ion concentration from the K_b, and then determine the pH from the pOH.

$$K_b = 5.556 \times 10^{-10} = \frac{[CH_3COOH][OH^-]}{[CH_3COO^-]} = \frac{[x][x]}{[0.0216861 - x]} = \frac{[x][x]}{[0.0216861]}$$

$[OH^-] = x = 3.471138 \times 10^{-6} \text{ M}$

$\text{pOH} = -\log (3.471138 \times 10^{-6}) = 5.459528$

$\text{pH} = 14.00 - \text{pOH} = 14.00 - 5.459528 = 8.54047 = \textbf{8.540}$

b) In the titration of a diprotic acid such as H_2SO_3, two OH^- ions are required to react with the two H^+ ions of each acid molecule. Because of the large difference in K_a values, each mole of H^+ is titrated separately, so H_2SO_3 molecules lose one H^+ before any HSO_3^- ions do.
The balanced chemical equation for the neutralization of H_2SO_3 at the first equivalence point is:

$$NaOH(aq) + H_2SO_3(aq) \rightarrow Na^+(aq) + HSO_3^-(aq) + H_2O(l)$$

The sodium ions on the product side are written as separate species because they have no effect on the pH of the solution.
Calculate the volume of NaOH needed to reach the first equivalence point:

Volume (mL) of NaOH =

$$\left(\frac{0.0850 \text{ mol H}_2\text{SO}_3}{L}\right)\left(\frac{10^{-3} \text{ L}}{1 \text{ mL}}\right)(28.9 \text{ mL})\left(\frac{1 \text{ mol NaOH}}{1 \text{ mol H}_2\text{SO}_3}\right)\left(\frac{L}{0.0372 \text{ mol NaOH}}\right)\left(\frac{1 \text{ mL}}{10^{-3} \text{ L}}\right)$$

$$= 66.034946 = \textbf{66.0 mL NaOH}$$

Determine the moles of HSO_3^- produced at the first equivalence point:

$$\text{Moles of HSO}_3^- = \left(\frac{0.0850 \text{ mol H}_2\text{SO}_3}{L}\right)\left(\frac{10^{-3} \text{ L}}{1 \text{ mL}}\right)(28.9 \text{ mL})\left(\frac{1 \text{ mol HSO}_3^{2-}}{1 \text{ mol H}_2\text{SO}_3}\right) = 0.0024565 \text{ mol HSO}_3^-$$

Determine the liters of solution present at the first equivalence point:

$$\text{Volume} = [(28.9 + 66.034946) \text{ mL}](10^{-3} \text{ L}/1 \text{ mL}) = 0.094934946 \text{ L}$$

Determine the concentration of HSO_3^- at the first equivalence point:

Molarity = (0.0024565 moles HSO_3^-)/(0.094934946 L) = 0.0258756 M

HSO_3^- is an amphoteric substance. To calculate the pH at the first equivalence point, determine whether HSO_3^- is a stronger acid or a stronger base.

The K_a for HSO_3^- is 6.5 x 10^{-8}.

Calculate K_b for HSO_3^-: K_{a1} for H_2SO_3 = 1.4x10^{-2}

$K_b = K_w/K_{a1}$ = (1.0x10^{-14})/(1.4x10^{-2}) = 7.142857x10^{-13}

Because HSO_3^- is a stronger acid than it is a base (K_a is larger than K_b for HSO_3^-), at the first equilvance point, it will behave as an acid and donate a hydrogen ion.

For the first equivalence point:

Using a reaction table for the equilibrium reaction of HSO_3^-:

	HSO_3^-	+	H_2O	⇌	H_3O^+	+	SO_3^{2-}
Initial	0.0258756 M				0		0
Change	−x				+x		+x
Equilibrium	0.0258756 − x				x		x

Determine the hydronium ion concentration from the K_a, and then determine the pH from the hydronium ion concentration.

$$K_a = 6.5\text{x}10^{-8} = \frac{[H_3O^+][SO_3^{2-}]}{[HSO_3^-]} = \frac{[x][x]}{[0.0258756 - x]} = \frac{[x][x]}{[0.0258756]}$$

(assuming that x is small compared with 0.0258756 M)

$[H_3O^+]$ = x = 4.101114x10^{-5} M

pH at first equivalence point = −log (4.101114x10^{-5}) = 4.3870982 = **4.387**

The balanced chemical equation for the neutralization of HSO_3^- at the second equivalence point is:

$NaOH(aq)$ + $HSO_3^-(aq)$ → $Na^+(aq)$ + $SO_3^{2-}(aq)$ + $H_2O(l)$

A total of 66.034946 mL were required to reach the first equivalence point. It will require an equal volume of NaOH to reach the second equivalence point from the first equivalence point. (2 x 66.0 mL = **132 mL**)

Determine the total volume of solution present at the second equivalence point.

Volume = [(28.9 + 66.034946 + 66.034946) mL](10^{-3} L/1 mL) = 0.160969892 L

0.0024565 mol HSO_3^- were present at the first equivalence point. An equal number of moles of SO_3^{2-} will be present at the second equivalence point.

Determine the concentration of SO_3^{2-} at the second equivalence point.

Molarity = (0.0024565 moles SO_3^{2-})/(0.160969892 L) = 0.0152606 M

SO_3^{2-} does not have a hydrogen ion to donate, so it acts as a base and accepts a hydrogen. Because it acts as a base, we must calculate its K_b.

Calculate K_b for SO_3^{2-}: K_a HSO_3^- (K_{a2})= 6.5x10^{-8}

$K_b = K_w/K_a$ = (1.0x10^{-14})/(6.5x10^{-8}) = 1.53846x10^{-7}

For the second equivalence point:

Using a reaction table for the equilibrium reaction of SO_3^{2-}:

	SO_3^{2-}	+	H_2O	⇌	HSO_3^-	+	OH^-
Initial	0.0152606 M				0		0
Change	−x				+x		+x
Equilibrium	0.0152606 − x				x		x

Determine the hydroxide ion concentration from the K_b, and then determine the pH from the pOH.

$$K_b = 1.53846\text{x}10^{-7} = \frac{[HSO_3^-][OH^-]}{[SO_3^{2-}]} = \frac{[x][x]}{[0.0152606 - x]} = \frac{[x][x]}{[0.0152606]}$$

(assuming x is small compared with 0.0152606 M)

$[OH^-]$ = x = 4.84539x10^{-5} M

pOH = −log (4.84539x10^{-5}) = 4.31467

pH = 14.00 − pOH = 14.00 − 4.31467 = 9.68533 = **9.685**

19.55   <u>Plan:</u> Use $(M)(V)$ to find the initial moles of base and then use the mole ratio in the balanced equation to find moles of acid; dividing moles of acid by the molarity of the acid gives the volume. At the equivalence point, the conjugate acid of the weak base is present; set up a reaction table for the acid dissociation in which x = the amount of dissociated acid. Use the K_a expression to solve for x from which pH is obtained.

<u>Solution:</u>

a) The balanced chemical equation is:

$$HCl(aq) + NH_3(aq) \rightarrow NH_4^+(aq) + Cl^-(aq)$$

The chloride ions on the product side are written as separate species because they have no effect on the pH of the solution. Calculate the volume of HCl needed:

Volume (mL) of HCl =

$$\left(\frac{0.234 \text{ mol NH}_3}{L}\right)\left(\frac{10^{-3} \text{ L}}{1 \text{ mL}}\right)(65.5 \text{ mL})\left(\frac{1 \text{ mol HCl}}{1 \text{ mol NH}_3}\right)\left(\frac{L}{0.125 \text{ mol HCl}}\right)\left(\frac{1 \text{ mL}}{10^{-3} \text{ L}}\right) = 122.616 = \textbf{123 mL HCl}$$

Determine the moles of NH_3 present:

$$\text{Moles} = \left(\frac{0.234 \text{ mol NH}_3}{L}\right)\left(\frac{10^{-3} \text{ L}}{1 \text{ mL}}\right)(65.5 \text{ mL}) = 0.015327 \text{ mol NH}_3$$

At the equivalence point, 0.015327 mol HCl will be added so the moles acid = moles base.

The HCl will react with an equal amount of the base, 0 mol NH_3 will remain, and 0.015327 moles of NH_4^+ will be formed.

	$HCl(aq)$	+	$NH_3(aq)$	$\rightarrow$	$NH_4^+(aq)$	+	$Cl^-(aq)$
Initial	0.015327 mol		0.015327 mol		0		–
Change	–0.015327 mol		–0.015327 mol		+0.015327 mol		–
Final	0		0		0.015327 mol		

Determine the liters of solution present at the equivalence point:

Volume = [(65.5 + 122.616) mL](10^{-3} L/1 mL) = 0.188116 L

Concentration of NH_4^+ at equivalence point:

Molarity = (0.015327 mol NH_4^+)/(0.188116 L) = 0.081476 M

Calculate K_a for NH_4^+: K_b NH_3 = 1.76×10^{-5}

$$K_a = K_w/K_b = (1.0 \times 10^{-14})/(1.76 \times 10^{-5}) = 5.6818 \times 10^{-10}$$

Using a reaction table for the equilibrium reaction of NH_4^+:

	NH_4^+	+	H_2O	$\leftrightarrows$	NH_3	+	H_3O^+
Initial	0.081476 M		–		0		0
Change	–x				+x		+x
Equilibrium	0.081476 – x				x		x

Determine the hydrogen ion concentration from the K_a, and then determine the pH.

$$K_a = 5.6818 \times 10^{-10} = \frac{[H_3O^+][NH_3]}{[NH_4^+]} = \frac{[x][x]}{[0.081476 - x]} = \frac{[x][x]}{[0.081476]}$$

$$x = [H_3O^+] = 6.803898 \times 10^{-6} \, M$$

$$pH = -\log [H_3O^+] = -\log (6.803898 \times 10^{-6}) = 5.1672 = \textbf{5.17}$$

b) The balanced chemical equation is:

$$HCl(aq) + CH_3NH_2(aq) \rightarrow CH_3NH_3^+(aq) + Cl^-(aq)$$

The chloride ions on the product side are written as separate species because they have no effect on the pH of the solution. Calculate the volume of HCl needed:

Volume (mL) of HCl =

$$\left(\frac{1.11 \text{ mol CH}_3\text{NH}_2}{L}\right)\left(\frac{10^{-3} \text{ L}}{1 \text{ mL}}\right)(21.8 \text{ mL})\left(\frac{1 \text{ mol HCl}}{1 \text{ mol CH}_3\text{NH}_2}\right)\left(\frac{L}{0.125 \text{ mol HCl}}\right)\left(\frac{1 \text{ mL}}{10^{-3} \text{ L}}\right)$$

$$= 193.584 = \textbf{194 mL HCl}$$

Determine the moles of CH_3NH_2 present:

$$\text{Moles} = \left(\frac{1.11 \text{ mol } CH_3NH_2}{L}\right)\left(\frac{10^{-3} \text{ L}}{1 \text{ mL}}\right)\left(21.8 \text{ mL}\right) = 0.024198 \text{ mol } CH_3NH_2$$

At the equivalence point, 0.024198 mol HCl will be added so the moles acid = moles base.
The HCl will react with an equal amount of the base, 0 mol CH_3NH_2 will remain, and 0.024198 moles of $CH_3NH_3^+$ will be formed.

	$HCl(aq)$	+	$CH_3NH_2(aq)$	$\rightarrow$	$CH_3NH_3^+(aq)$	+	$Cl^-(aq)$
Initial	0.024198 mol		0.024198 mol		0		–
Change	–0.024198 mol		–0.024198 mol		+0.024198 mol		–
Final	0		0		0.024198 mol		

Determine the liters of solution present at the equivalence point:

Volume = $[(21.8 + 193.584) \text{ mL}](10^{-3} \text{ L}/1 \text{ mL}) = 0.215384 \text{ L}$

Concentration of $CH_3NH_3^+$ at equivalence point:

Molarity = $(0.024198 \text{ mol } CH_3NH_3^+)/(0.215384 \text{ L}) = 0.1123482 \, M$

Calculate K_a for $CH_3NH_3^+$: $K_b \, CH_3NH_2 = 4.4 \times 10^{-4}$

$K_a = K_w/K_b = (1.0 \times 10^{-14})/(4.4 \times 10^{-4}) = 2.2727 \times 10^{-11}$

Using a reaction table for the equilibrium reaction of $CH_3NH_3^+$:

	$CH_3NH_3^+$	+	H_2O	$\rightleftharpoons$	CH_3NH_2	+	H_3O^+
Initial	0.1123482 M		–		0		0
Change	–x				+x		+x
Equilibrium	0.1123482 – x				x		x

Determine the hydrogen ion concentration from the K_a, and then determine the pH.

$$K_a = 2.2727 \times 10^{-11} = \frac{[H_3O^+][CH_3NH_2]}{[CH_3NH_3^+]} = \frac{[x][x]}{[0.1123482 - x]} = \frac{[x][x]}{[0.1123482]}$$

$x = [H_3O^+] = 1.5979 \times 10^{-6} \, M$

$pH = -\log [H_3O^+] = -\log (1.5979 \times 10^{-6}) = 5.7964 = \mathbf{5.80}$

19.57   Plan: Indicators have a pH range that is approximated by $pK_a \pm 1$. Find the pK_a of the indicator by using the relationship $pK_a = -\log K_a$.
Solution:
The pK_a of cresol red is $-\log (3.5 \times 10^{-9}) = 8.5$, so the indicator changes color over an approximate range of 8.5 ± 1 or **7.5 to 9.5**.

19.59   Plan: Choose an indicator that changes color at a pH close to the pH of the equivalence point.
Solution:
a) The equivalence point for a strong acid–strong base titration occurs at pH = 7.0. **Bromthymol blue** is an indicator that changes color around pH 7.
b) The equivalence point for a weak acid–strong base is above pH 7. Estimate the pH at equivalence point from equilibrium calculations.
At the equivalence point, all of the HCOOH and NaOH have been consumed; the solution is 0.050 M HCOO⁻. (The volume doubles because equal volumes of base and acid are required to reach the equivalence point. When the volume doubles, the concentration is halved.) The weak base HCOO⁻ undergoes a base reaction:

Concentration, M	$COOH^-(aq)$	+	$H_2O(l)$	$\rightleftharpoons$	$HCOOH(aq)$	+	$OH^-(aq)$
Initial	0.050 M		—		0		0
Change	–x				+x		+x
Equilibrium	0.050 – x				x		x

The K_a for HCOOH is 1.8×10^{-4}, so $K_b = 1.0 \times 10^{-14}/1.8 \times 10^{-4} = 5.5556 \times 10^{-11}$

$$K_b = 5.5556 \times 10^{-11} = \frac{[HCOOH][OH^-]}{[HCOO^-]} = \frac{[x][x]}{[0.050 - x]} = \frac{[x][x]}{[0.050]}$$

$[OH^-] = x = 1.666673 \times 10^{-6} M$

$pOH = -\log(1.666673 \times 10^{-6}) = 5.7781496$

$pH = 14.00 - pOH = 14.00 - 5.7781496 = 8.2218504 = 8.22$

Choose **thymol blue** or **phenolphthalein**.

19.61 Plan: Choose an indicator that changes color at a pH close to the pH of the equivalence point.
Solution:
a) The equivalence point for a weak base–strong acid is below pH 7. Estimate the pH at equivalence point from equilibrium calculations.
At the equivalence point, the solution is 0.25 M $(CH_3)_2NH_2^+$. (The volume doubles because equal volumes of base and acid are required to reach the equivalence point. When the volume doubles, the concentration is halved.)
$K_a = K_w/K_b = (1.0 \times 10^{-14})/(5.9 \times 10^{-4}) = 1.69491525 \times 10^{-11}$

Concentration, M	$(CH_3)_2NH_2^+(aq)$ +	$H_2O(l)$	$\rightleftharpoons$	$(CH_3)_2NH(aq)$ +	$H_3O^+(aq)$
Initial	0.25 M	—		0	0
Change	−x			+x	+x
Equilibrium	0.25 − x			x	x

$$K_a = 1.69491525 \times 10^{-11} = \frac{[H_3O^+][(CH_3)_2NH]}{[(CH_3)_2NH_2^+]} = \frac{x^2}{0.25 - x} = \frac{x^2}{0.25}$$

$x = [H_3O^+] = 2.0584674 \times 10^{-6} M$

$pH = -\log[H_3O^+] = -\log(2.0584674 \times 10^{-6}) = 5.686456 = 5.69$

Methyl red is an indicator that changes color around pH 5.7.
b) This is a strong acid–strong base titration; thus, the equivalence point is at pH = 7.00. **Bromthymol blue** is an indicator that changes color around pH 7.

19.64 Fluoride ion in BaF_2 is the conjugate base of the weak acid HF. The base hydrolysis reaction of fluoride ion
$$F^-(aq) + H_2O(l) \rightleftharpoons HF(aq) + OH^-(aq)$$
therefore is influenced by the pH of the solution. As the pH increases, $[OH^-]$ increases and the equilibrium shifts to the left to decrease $[OH^-]$ and increase $[F^-]$. As the pH decreases, $[OH^-]$ decreases and the equilibrium shifts to the right to increase $[OH^-]$ and decrease $[F^-]$. The changes in $[F^-]$ influence the solubility of BaF_2.
Chloride ion is the conjugate base of a strong acid so it does not react with water. Thus, its concentration is not influenced by pH, and solubility of $BaCl_2$ does not change with pH.

19.66 Consider the reaction $AB(s) \rightleftharpoons A^+(aq) + B^-(aq)$, where $Q_{sp} = [A^+][B^-]$. If $Q_{sp} > K_{sp}$, then there are more ions dissolved than expected at equilibrium, and the equilibrium shifts to the left and the compound AB precipitates. The excess ions precipitate as solid from the solution.

19.67 Plan: Write an equation that describes the solid compound dissolving to produce its ions. The ion-product expression follows the equation $K_{sp} = [M^{n+}]^p[X^{z-}]^q$ where p and q are the subscripts of the ions in the compound's formula.
Solution:
a) $Ag_2CO_3(s) \rightleftharpoons 2Ag^+(aq) + CO_3^{2-}(aq)$
    Ion-product expression: $K_{sp} = [Ag^+]^2[CO_3^{2-}]$
b) $BaF_2(s) \rightleftharpoons Ba^{2+}(aq) + 2F^-(aq)$
    Ion-product expression: $K_{sp} = [Ba^{2+}][F^-]^2$
c) $CuS(s) + H_2O(l) \rightleftharpoons Cu^{2+}(aq) + S^{2-}(aq)$
    Ion-product expression: $K_{sp} = [Cu^{2+}][S^{2-}]$

19.69 Plan: Write an equation that describes the solid compound dissolving to produce its ions. The ion-product expression follows the equation $K_{sp} = [M^{n+}]^p[X^{z-}]^q$ where p and q are the subscripts of the ions in the compound's formula.
Solution:
a) $CaCrO_4(s) \rightleftharpoons Ca^{2+}(aq) + CrO_4^{2-}(aq)$
    Ion-product expression: $K_{sp} = [Ca^{2+}][CrO_4^{2-}]$

b) $AgCN(s) \leftrightarrows Ag^+(aq) + CN^-(aq)$

Ion-product expression: $K_{sp} = \mathbf{[Ag^+][CN^-]}$

c) $Ag_3PO_4(s) + H_2O(l) \leftrightarrows 3Ag^+(aq) + PO_4^{3-}(aq)$

Ion-product expression: $K_{sp} = \mathbf{[Ag^+]^3[PO_4^{3-}]}$

19.71 Plan: Write an equation that describes the solid compound dissolving in water and then write the ion-product expression. Write a reaction table, where S is the molar solubility of Ag_2CO_3. Substitute the given solubility, S, into the ion-expression and solve for K_{sp}.

Solution:

Concentration (M)	$Ag_2CO_3(s)$	$\leftrightarrows$	$2Ag^+(aq)$	$+$	$CO_3^{2-}(aq)$
Initial	—		0		0
Change	—		$+2S$		$+S$
Equilibrium	—		$2S$		S

$S = [Ag_2CO_3] = 0.032\ M$ so $[Ag^+] = 2S = 0.064\ M$ and $[CO_3^{2-}] = S = 0.032\ M$

$K_{sp} = [Ag^+]^2[CO_3^{2-}] = (0.064)^2(0.032) = 1.31072\times10^{-4} = \mathbf{1.3\times10^{-4}}$

19.73 Plan: Write an equation that describes the solid compound dissolving in water and then write the ion-product expression. Write a reaction table, where S is the molar solubility of $Ag_2Cr_2O_7$. Substitute the given solubility, S, converted from mass/volume to molarity, into the ion-expression and solve for K_{sp}.

Solution:

The solubility of $Ag_2Cr_2O_7$, converted from g/100 mL to M is:

$$\text{Molar solubility} = S = \left(\frac{8.3\times10^{-3}\ \text{g}\ Ag_2Cr_2O_7}{100\ \text{mL}}\right)\left(\frac{1\ \text{mL}}{10^{-3}\ \text{L}}\right)\left(\frac{1\ \text{mol}\ Ag_2Cr_2O_7}{431.8\ \text{g}\ Ag_2Cr_2O_7}\right) = 0.00019221862\ M$$

The equation for silver dichromate, $Ag_2Cr_2O_7$, is:

Concentration (M)	$Ag_2Cr_2O_7(s)$	$\leftrightarrows$	$2Ag^+(aq)$	$+$	$Cr_2O_7^{2-}(aq)$
Initial	—		0		0
Change	—		$+2S$		$+S$
Equilibrium	—		$2S$		S

$2S = [Ag^+] = 2(0.00019221862\ M) = 0.00038443724\ M$

$S = [Cr_2O_7^{2-}] = 0.00019221862\ M$

$K_{sp} = [Ag^+]^2[Cr_2O_7^{2-}] = (2S)^2(S) = (0.00038443724)^2(0.00019221862) = 2.8408\times10^{-11} = \mathbf{2.8\times10^{-11}}$

19.75 Plan: Write the equation that describes the solid compound dissolving in water and then write the ion-product expression. Set up a reaction table that expresses $[Sr^{2+}]$ and $[CO_3^{2-}]$ in terms of S, substitute into the ion-product expression, and solve for S. In part b), the $[Sr^{2+}]$ that comes from the dissolved $Sr(NO_3)_2$ must be included in the reaction table.

Solution:

a) The equation and ion-product expression for $SrCO_3$ is:

$SrCO_3(s) \leftrightarrows Sr^{2+}(aq) + CO_3^{2-}(aq)$ $K_{sp} = [Sr^{2+}][CO_3^{2-}]$

The solubility, S, in pure water equals $[Sr^{2+}]$ and $[CO_3^{2-}]$

Write a reaction table, where S is the molar solubility of $SrCO_3$:

Concentration (M)	$SrCO_3(s)$	$\leftrightarrows$	$Sr^{2+}(aq)$	$+$	$CO_3^{2-}(aq)$
Initial	—		0		0
Change	—		$+S$		$+S$
Equilibrium	—		S		S

$K_{sp} = 5.4\times10^{-10} = [Sr^{2+}][CO_3^{2-}] = [S][S] = S^2$

$S = 2.32379\times10^{-5} = \mathbf{2.3\times10^{-5}\ M}$

b) In $0.13\ M\ Sr(NO_3)_2$, the initial concentration of Sr^{2+} is $0.13\ M$.

Equilibrium $[Sr^{2+}] = 0.13 + S$ and equilibrium $[CO_3^{2-}] = S$ where S is the solubility of $SrCO_3$.

Concentration (M)	$SrCO_3(s)$	$\leftrightarrows$	$Sr^{2+}(aq)$	$+$	$CO_3^{2-}(aq)$

	Initial	—	0.13	0
	Change	—	$+S$	$+S$
	Equilibrium	—	$0.13 + S$	S

$$K_{sp} = 5.4\text{x}10^{-10} = [Sr^{2+}][CO_3^{2-}] = (0.13 + S)S$$

This calculation may be simplified by assuming S is small and setting $0.13 + S = 0.13$.

$$K_{sp} = 5.4\text{x}10^{-10} = (0.13)S$$
$$S = 4.1538\text{x}10^{-9} = \mathbf{4.2\text{x}10^{-9}\ M}$$

19.77 Plan: Write the equation that describes the solid compound dissolving in water and then write the ion-product expression. Set up a reaction table that expresses $[Ca^{2+}]$ and $[IO_3^-]$ in terms of S, substitute into the ion-product expression, and solve for S. The $[Ca^{2+}]$ that comes from the dissolved $Ca(NO_3)_2$ and the $[IO_3^-]$ that comes from $NaIO_3$ must be included in the reaction table.
Solution:
a) The equilibrium is: $Ca(IO_3)_2(s) \leftrightarrows Ca^{2+}(aq) + 2IO_3^-(aq)$. From the Appendix, $K_{sp}(Ca(IO_3)_2) = 7.1\text{x}10^{-7}$.
Write a reaction table that reflects an initial concentration of $Ca^{2+} = 0.060\ M$. In this case, Ca^{2+} is the common ion.

Concentration (M)	$Ca(IO_3)_2(s)$	$\leftrightarrows$	$Ca^{2+}(aq)$	$+$	$2IO_3^-(aq)$
Initial	—		0.060		0
Change	—		$+S$		$+2S$
Equilibrium	—		$0.060 + S$		$2S$

Assume that $0.060 + S \approx 0.060$ because the amount of compound that dissolves will be negligible in comparison to $0.060\ M$.

$$K_{sp} = [Ca^{2+}][IO_3^-]^2 = (0.060)(2S)^2 = 7.1\text{x}10^{-7}$$
$$S = 1.71998\text{x}10^{-3} = 1.7\text{x}10^{-3}\ M$$

Check assumption: $(1.71998\text{x}10^{-3}\ M)/(0.060\ M) \times 100\% = 2.9\% < 5\%$, so the assumption is good.
S represents both the molar solubility of Ca^{2+} and $Ca(IO_3)_2$, so the molar solubility of $Ca(IO_3)_2$ is $\mathbf{1.7\text{x}10^{-3}\ M}$.
b) Write a reaction table that reflects an initial concentration of $IO_3^- = 0.060\ M$. IO_3^- is the common ion.

Concentration (M)	$Ca(IO_3)_2(s)$	$\leftrightarrows$	$Ca^{2+}(aq)$	$+$	$2IO_3^-(aq)$
Initial	—		0		0.060
Change	—		$+S$		$+2S$
Equilibrium	—		S		$0.060 + 2S$

The equilibrium concentration of Ca^{2+} is S, and the IO_3^- concentration is $0.060 + 2S$.

Assume that $0.060 + 2S \approx 0.060$

$$K_{sp} = [Ca^{2+}][IO_3^-]^2 = (S)(0.060)^2 = 7.1\text{x}10^{-7}$$
$$S = 1.97222\text{x}10^{-4} = 2.0\text{x}10^{-4}\ M$$

Check assumption: $(1.97222\text{x}10^{-4}\ M)/(0.060\ M) \times 100\% = 0.3\% < 5\%$, so the assumption is good.
S represents both the molar solubility of Ca^{2+} and $Ca(IO_3)_2$, so the molar solubility of $Ca(IO_3)_2$ is $\mathbf{2.0\text{x}10^{-4}\ M}$.

19.79 Plan: The larger the K_{sp}, the larger the molar solubility if the number of ions are equal.
Solution:
a) $\mathbf{Mg(OH)_2}$ with $K_{sp} = 6.3\text{x}10^{-10}$ has higher molar solubility than $Ni(OH)_2$ with $K_{sp} = 6\text{x}10^{-16}$.
b) $\mathbf{PbS}$ with $K_{sp} = 3\text{x}10^{-25}$ has higher molar solubility than CuS with $K_{sp} = 8\text{x}10^{-34}$.
c) $\mathbf{Ag_2SO_4}$ with $K_{sp} = 1.5\text{x}10^{-5}$ has higher molar solubility than MgF_2 with $K_{sp} = 7.4\text{x}10^{-9}$.

19.81 Plan: The larger the K_{sp}, the more water soluble the compound if the number of ions are equal.
Solution:
a) $\mathbf{CaSO_4}$ with $K_{sp} = 2.4\text{x}10^{-5}$ is more water soluble than $BaSO_4$ with $K_{sp} = 1.1\text{x}10^{-10}$.
b) $\mathbf{Mg_3(PO_4)_2}$ with $K_{sp} = 5.2\text{x}10^{-24}$ is more water soluble than $Ca_3(PO_4)_2$ with $K_{sp} = 1.2\text{x}10^{-29}$.
c) $\mathbf{PbSO_4}$ with $K_{sp} = 1.6\text{x}10^{-8}$ is more water soluble than $AgCl$ with $K_{sp} = 1.8\text{x}10^{-10}$.

19.83 Plan: If a compound contains an anion that is the weak conjugate base of a weak acid, the concentration of that anion, and thus the solubility of the compound, is influenced by pH.

Solution:
a) $AgCl(s) \leftrightarrows Ag^+(aq) + Cl^-(aq)$
The chloride ion is the anion of a strong acid, so it does not react with H_3O^+. The solubility is not affected by pH.
b) $SrCO_3(s) \leftrightarrows Sr^{2+}(aq) + CO_3^{2-}(aq)$
The strontium ion is the cation of a strong base, so pH will not affect its solubility.
The carbonate ion is the conjugate base of a weak acid and will act as a base:
$\qquad CO_3^{2-}(aq) + H_2O(l) \leftrightarrows HCO_3^-(aq) + OH^-(aq)$
$\qquad$ and $HCO_3^-(aq) + H_2O(l) \leftrightarrows H_2CO_3(aq) + OH^-(aq)$
The H_2CO_3 will decompose to $CO_2(g)$ and $H_2O(l)$. The gas will escape and further shift the equilibrium. Changes in pH will change the $[CO_3^{2-}]$, so the solubility of $SrCO_3$ is affected. **Solubility increases with addition of H_3O^+ (decreasing pH).** A decrease in pH will decrease $[OH^-]$, causing the base equilibrium to shift to the right which decreases $[CO_3^{2-}]$, causing the solubility equilibrium to shift to the right, dissolving more solid.

19.85 Plan: If a compound contains an anion that is the weak conjugate base of a weak acid, the concentration of that anion, and thus the solubility of the compound, is influenced by pH.
Solution:
a) $Fe(OH)_2(s) \leftrightarrows Fe^{2+}(aq) + 2OH^-(aq)$
The hydroxide ion reacts with added H_3O^+:
$\qquad OH^-(aq) + H_3O^+(aq) \rightarrow 2H_2O(l)$
The added H_3O^+ consumes the OH^-, driving the equilibrium toward the right to dissolve more $Fe(OH)_2$. **Solubility increases with addition of H_3O^+ (decreasing pH).**
b) $CuS(s) + H_2O(l) \leftrightarrows Cu^{2+}(aq) + S^{2-}(aq)$
S^{2-} is the anion of a weak acid, so it reacts with added H_3O^+. **Solubility increases with addition of H_3O^+ (decreasing pH).**

19.87 Plan: Find the initial molar concentrations of Cu^{2+} and OH^-. The molarity of the KOH is calculated by converting mass to moles and dividing by the volume. Put these concentrations in the ion-product expression, solve for Q_{sp}, and compare Q_{sp} with K_{sp}. If $Q_{sp} > K_{sp}$, precipitate forms.
Solution:
The equilibrium is: $Cu(OH)_2(s) \leftrightarrows Cu^{2+}(aq) + 2OH^-(aq)$. The ion-product expression is $K_{sp} = [Cu^{2+}][OH^-]^2$ and, from the Appendix, K_{sp} equals 2.2×10^{-20}.

$[Cu^{2+}] = \left(\dfrac{1.0 \times 10^{-3} \text{ mol } Cu(NO_3)_2}{L} \right) \left(\dfrac{1 \text{ mol } Cu^{2+}}{1 \text{ mol } Cu(NO_3)_2} \right) = 1.0 \times 10^{-3} \ M \ Cu^{2+}$

$[OH^-] = \left(\dfrac{0.075 \text{ g KOH}}{1.0 \text{ L}} \right) \left(\dfrac{1 \text{ mol KOH}}{56.11 \text{ g KOH}} \right) \left(\dfrac{1 \text{ mol } OH^-}{1 \text{ mol KOH}} \right) = 1.33666 \times 10^{-3} \ M \ OH^-$

$Q_{sp} = [Cu^{2+}][OH^-]^2 = (1.0 \times 10^{-3})(1.33666 \times 10^{-3})^2 = 1.786660 \times 10^{-9}$
Q_{sp} is greater than K_{sp} ($1.8 \times 10^{-9} > 2.2 \times 10^{-20}$), so **Cu(OH)$_2$ will precipitate.**

19.89 Plan: Find the initial molar concentrations of Ba^{2+} and IO_3^-. The molarity of the $BaCl_2$ is calculated by converting mass to moles and dividing by the volume. Put these concentrations in the ion-product expression, solve for Q_{sp}, and compare Q_{sp} with K_{sp}. If $Q_{sp} > K_{sp}$, precipitate forms.
Solution:
The equilibrium is: $Ba(IO_3)_2(s) \leftrightarrows Ba^{2+}(aq) + 2IO_3^-(aq)$. The ion-product expression is $K_{sp} = [Ba^{2+}][IO_3^-]^2$ and, from the Appendix, K_{sp} equals 1.5×10^{-9}.

$[Ba^{2+}] = \left(\dfrac{7.5 \text{ mg } BaCl_2}{500. \text{ mL}} \right) \left(\dfrac{10^{-3} \text{ g}}{1 \text{ mg}} \right) \left(\dfrac{1 \text{ mL}}{10^{-3} \text{ L}} \right) \left(\dfrac{1 \text{ mol } BaCl_2}{208.2 \text{ g } BaCl_2} \right) \left(\dfrac{1 \text{ mol } Ba^{2+}}{1 \text{ mol } BaCl_2} \right) = 7.204611 \times 10^{-5} \ M \ Ba^{2+}$

$[IO_3^-] = \left(\dfrac{0.023 \text{ mol } NaIO_3}{L} \right) \left(\dfrac{1 \text{ mol } IO_3^-}{1 \text{ mol } NaIO_3} \right) = 0.023 \ M \ IO_3^-$

$Q_{sp} = [Ba^{2+}][IO_3^-]^2 = (7.204611 \times 10^{-5})(0.023)^2 = 3.81124 \times 10^{-8}$
Since $Q_{sp} > K_{sp}$ ($3.8 \times 10^{-8} > 1.5 \times 10^{-9}$), **Ba(IO$_3$)$_2$ will precipitate.**

19.92 Plan: When $Fe(NO_3)_3$ and $Cd(NO_3)_2$ mix with NaOH, the insoluble compounds $Fe(OH)_3$ and $Cd(OH)_2$ form. The compound with the smaller value of K_{sp} precipitates first. Calculate the initial concentrations of Fe^{3+} and Cd^{2+} from the dilution formula $M_{conc}V_{conc} = M_{dil}V_{dil}$. Use the ion-product expressions to find the minimum OH^- concentration required to cause precipitation of each compound.
Solution:
a) **$Fe(OH)_3$** will precipitate first because its K_{sp} (1.6×10^{-39}) is smaller than the K_{sp} for $Cd(OH)_2$ at 7.2×10^{-15}.
The precipitation reactions are:

$$Fe^{3+}(aq) + 3OH^-(aq) \rightarrow Fe(OH)_3(s) \qquad\qquad K_{sp} = [Fe^{3+}][\,OH^-]^3$$
$$Cd^{2+}(aq) + 2OH^-(aq) \rightarrow Cd(OH)_2(s) \qquad\qquad K_{sp} = [Cd^{2+}][\,OH^-]^2$$

The concentrations of Fe^{3+} and Cd^{2+} in the mixed solution are found from $M_{conc}V_{conc} = M_{dil}V_{dil}$
$[Fe^{3+}] = [(0.50\ M)(50.0\ mL)]/[(50.0 + 125)\ mL] = 0.142857\ M\ Fe^{3+}$
$[Cd^{2+}] = [(0.25\ M)(125\ mL)]/[(50.0 + 125)\ mL] = 0.178571\ M\ Cd^{2+}$
The hydroxide ion concentration required to precipitate the metal ions comes from the metal ion concentrations and the K_{sp}.

$$[OH^-]_{Fe} = \sqrt[3]{\frac{K_{sp}}{[Fe^{3+}]}} = \sqrt[3]{\frac{1.6 \times 10^{-39}}{[0.142857]}} = 2.237 \times 10^{-13} = 2.2 \times 10^{-13}\ M$$

$$[OH^-]_{Cd} = \sqrt{\frac{K_{sp}}{[Cd^{2+}]}} = \sqrt{\frac{7.2 \times 10^{-15}}{[0.178571]}} = 2.0079864 \times 10^{-7} = 2.0 \times 10^{-7}\ M$$

A lower hydroxide ion concentration is required to precipitate the Fe^{3+}.
b) The two ions are separated by adding just enough NaOH to precipitate the iron(III) hydroxide, but precipitating no more than 0.01% of the cadmium. The Fe^{3+} is found in the solid precipitate while the Cd^{2+} remains in solution.
c) A hydroxide concentration between the values calculated in part a) will work. The best separation would be when $Q_{sp} = K_{sp}$ for $Cd(OH)_2$. This occurs when $[OH^-] = $ **$2.0 \times 10^{-7}\ M$**.

19.95 In the context of this equilibrium only, the increased solubility with added OH^- appears to be a violation of Le Châtelier's principle. Adding OH^- should cause the equilibrium to shift towards the left, decreasing the solubility of $Zn(OH)_2$. Before accepting this conclusion, other possible equilibria must be considered. Zinc is a metal ion and hydroxide ion is a ligand, so it is possible that a complex ion forms between the zinc ion and hydroxide ion:
$$Zn^{2+}(aq) + nOH^-(aq) \leftrightharpoons Zn(OH)_n^{2-n}(aq)$$
This decreases the concentration of Zn^{2+}, shifting the solubility equilibrium to the right to dissolve more $Zn(OH)_2$.

19.96 Plan: In many cases, a hydrated metal complex (e.g., $Hg(H_2O)_4^{2+}$) will exchange ligands when placed in a solution of another ligand (e.g., CN^-).
Solution:
$$Hg(H_2O)_4^{2+}(aq) + 4CN^-(aq) \leftrightharpoons Hg(CN)_4^{2-}(aq) + 4H_2O(l)$$
Note that both sides of the equation have the same "overall" charge of -2. The mercury complex changes from $+2$ to -2 because water is a neutral *molecular* ligand, whereas cyanide is an *ionic* ligand.

19.98 Plan: In many cases, a hydrated metal complex (e.g., $Ag(H_2O)_4^+$) will exchange ligands when placed in a solution of another ligand (e.g., $S_2O_3^{2-}$).
Solution:
The two water ligands are replaced by two thiosulfate ion ligands. The $+1$ charge from the silver ion plus the -4 charge from the two thiosulfate ions gives a net charge on the complex ion of -3.
$$Ag(H_2O)_2^+(aq) + 2S_2O_3^{2-}(aq) \leftrightharpoons Ag(S_2O_3)_2^{3-}(aq) + 2H_2O(l)$$

19.100 Plan: Write the formation reaction and the K_f expression. The initial concentrations of Ag^+ and $S_2O_3^{2-}$ may be determined from $M_{conc}V_{conc} = M_{dil}V_{dil}$. Set up a reaction table and use the limiting reactant to find the amounts of species in the mixture, assuming a complete reaction. A second reaction table is then written, with x representing the amount of complex ion that dissociates. Use the K_f expression to solve for x.

Solution:

$Ag^+(aq) + 2S_2O_3^{2-}(aq) \leftrightarrows Ag(S_2O_3)_2^{3-}(aq)$

$[Ag^+] = (0.044\ M)(25.0\ mL)/((25.0 + 25.0)\ mL) = 0.022\ M\ Ag^+$

$[S_2O_3^{2-}] = (0.57\ M)(25.0\ mL)/((25.0 + 25.0)\ mL) = 0.285\ M\ S_2O_3^{2-}$

The reaction gives:

Concentration (M)	$Ag^+(aq)$	+	$2S_2O_3^{2-}(aq)$	$\rightarrow$	$Ag(S_2O_3)_2^{3-}(aq)$	
Initial	0.022		0.285		0	
Change	−0.022		−2(0.022)		+0.022	1:2:1 mole ratio
Equilibrium	0		0.241		0.022	

To reach equilibrium:

Concentration (M)	$Ag^+(aq)$	+	$2S_2O_3^{2-}(aq)$	$\leftrightarrows$	$Ag(S_2O_3)_2^{3-}(aq)$
Initial	0		0.241		0.022
Change	+x		+2x		−x
Equilibrium	+ x		0.241 + 2x		0.022 − x

K_f is large, so $[Ag(S_2O_3)_2^{3-}] \approx 0.022\ M$ and $[S_2O_3^{2-}]_{equil} \approx 0.241\ M$

$$K_f = 4.7 \times 10^{13} = \frac{\left[Ag(S_2O_3)_2^{3-}\right]}{\left[Ag^+\right]\left[S_2O_3^{2-}\right]^2} = \frac{[0.022]}{[x][0.241]^2}$$

$x = [Ag^+] = 8.0591778 \times 10^{-15} = \mathbf{8.1 \times 10^{-15}\ M}$

19.102 **Plan:** Write the ion-product equilibrium reaction and the complex-ion equilibrium reaction. Add the two reactions to yield an overall reaction; multiply the two constants to obtain $K_{overall}$. Write a reaction table where $S = [Cr(OH)_3]_{dissolved} = [Cr(OH)_4^-]$.
Solution:

Solubility-product:	$Cr(OH)_3(s) \leftrightarrows Cr^{3+}(aq) + 3OH^-(aq)$	$K_{sp} = 6.3 \times 10^{-31}$
Complex-ion	$Cr^{3+}(aq) + 4OH^-(aq) \leftrightarrows Cr(OH)_4^-(aq)$	$K_f = 8.0 \times 10^{29}$
Overall:	$Cr(OH)_3(s) + OH^-(aq) \leftrightarrows Cr(OH)_4^-(aq)$	$K = K_{sp}K_f = 0.504$

At pH 13.0, the pOH is 1.0 and $[OH^-] = 10^{-1.0} = 0.1\ M$.
Reaction table:

Concentration (M)	$Cr(OH)_3(s)$ +	$OH^-(aq)$	$\leftrightarrows$	$Cr(OH)_4^-(aq)$
Initial	——	0.1		0
Change	——	− S		+S
Equilibrium	——	0.1 − S		S

Assume that $0.1 - S \approx 0.1$.

$$K_{overall} = 0.504 = \frac{\left[Cr(OH)_4^-\right]}{\left[OH^-\right]} = \frac{[S]}{[0.1]}$$

$S = [Cr(OH)_4^-] = 0.0504 = \mathbf{0.05\ M}$

19.104 **Plan:** First, calculate the initial moles of Zn^{2+} and CN^-, then set up reaction table assuming that the reaction first goes to completion, and then calculate back to find the reactant concentrations.
Solution:
The complex formation equilibrium is:

$$Zn^{2+}(aq) + 4CN^-(aq) \leftrightarrows Zn(CN)_4^{2-}(aq) \qquad K_f = 4.2 \times 10^{19}$$

$$\text{Moles of } Zn^{2+} = \left(0.84\ g\ ZnCl_2\right)\left(\frac{1\ mol\ ZnCl_2}{136.31\ g\ ZnCl_2}\right)\left(\frac{1\ mol\ Zn^{2+}}{1\ mol\ ZnCl_2}\right) = 0.0061624\ mol\ Zn^{2+}$$

$$\text{Moles of } CN^- = \left(\frac{0.150\ mol\ NaCN}{L}\right)\left(\frac{10^{-3}\ L}{1\ mL}\right)\left(245\ mL\right)\left(\frac{1\ mol\ CN^-}{1\ mol\ NaCN}\right) = 0.03675\ mol\ CN^-$$

The Zn^{2+} is limiting because there are significantly fewer moles of this ion, thus, $[Zn^{2+}] = 0$.

Moles of CN⁻ reacting = $(0.0061624 \text{ mol Zn}^{2+})\left(\dfrac{4 \text{ mol CN}^-}{1 \text{ mol Zn}^{2+}}\right) = 0.0246496 \text{ mol CN}^-$

Moles of CN⁻ remaining are: $0.03675 - 0.0246496 = 0.0121004 \text{ mol CN}^-$

$[\text{CN}^-] = \dfrac{(0.0121004 \text{ mol CN}^-)}{(245 \text{ mL})}\left(\dfrac{1 \text{ mL}}{10^{-3} \text{ L}}\right) = 0.0493894 \; M \; \text{CN}^-$

The Zn^{2+} will produce an equal number of moles of the complex with the concentration:

$[\text{Zn(CN)}_4^{2-}] = \left(\dfrac{0.0061624 \text{ mol Zn}^{2+}}{245 \text{ mL}}\right)\left(\dfrac{1 \text{ mL}}{10^{-3} \text{ L}}\right)\left(\dfrac{1 \text{ mol Zn(CN)}_4^{2-}}{1 \text{ mol Zn}^{2+}}\right) = 0.025153 \; M \; \text{Zn(CN)}_4^{2-}$

Concentration (M)	$\text{Zn}^{2+}(aq)$	+	$4\text{CN}^-(aq)$	⇆	$\text{Zn(CN)}_4^{2-}(aq)$
Initial	0		0.0493894		0.025153
Change	+x		+4x		−x
Equilibrium	x		0.0493894 + 4x		0.025153 − x

Assume the −x and the +4x do not significantly change the associated concentrations.

$K_f = 4.2\text{x}10^{19} = \dfrac{[\text{Zn(CN)}_4^{2-}]}{[\text{Zn}^{2+}][\text{CN}^-]^4} = \dfrac{[0.025153 - x]}{[x][0.0493894 + 4x]^4} = \dfrac{[0.025153]}{[x][0.0493894]^4}$

$x = 1.006481\text{x}10^{-16} = 1.0\text{x}10^{-16}$
$[\text{Zn}^{2+}] = \mathbf{1.0\text{x}10^{-16} \; M \; Zn^{2+}}$
$[\text{Zn(CN)}_4^{2-}] = 0.025153 - x = 0.025153 - 1.0\text{x}10^{-16} = 0.025153 = \mathbf{0.025 \; M \; Zn(CN)_4^{2-}}$
$[\text{CN}^-] = 0.0493894 + 4x = 0.0493894 + 4(1.0\text{x}10^{-16}) = 0.0493894 = \mathbf{0.049 \; M \; CN^-}$

19.106 Plan: The NaOH will react with the benzoic acid, C_6H_5COOH, to form the conjugate base benzoate ion, $C_6H_5COO^-$. Calculate the number of moles of NaOH and C_6H_5COOH. Set up a reaction table that shows the stoichiometry of the reaction of NaOH and C_6H_5COOH. All of the NaOH will be consumed to form $C_6H_5COO^-$, and the number of moles of C_6H_5COOH will decrease. Find the new concentrations of C_6H_5COOH and $C_6H_5COO^-$ and use the Henderson-Hasselbalch equation to find the pH of this buffer. Once the pH of the benzoic acid/benzoate buffer is known, the Henderson-Hasselbalch equation can be used to find the ratio of formate ion and formic acid that will produce a buffer of that same pH. From the ratio, the volumes of HCOOH and NaOH are calculated.
Solution:
The K_a for benzoic acid is $6.3\text{x}10^{-5}$ (from the Appendix). The pK_a is $-\log(6.3\text{x}10^{-5}) = 4.201$. The reaction of benzoic acid with sodium hydroxide is:
$C_6H_5COOH(aq) + NaOH(aq) \rightarrow Na^+(aq) + C_6H_5COO^-(aq) + H_2O(l)$

Moles of $C_6H_5COOH = \left(\dfrac{0.200 \text{ mol } C_6H_5COOH}{L}\right)\left(\dfrac{10^{-3} \text{ L}}{1 \text{ mL}}\right)(475 \text{ mL}) = 0.0950 \text{ mol } C_6H_5COOH$

Moles of NaOH = $\left(\dfrac{2.00 \text{ mol NaOH}}{L}\right)\left(\dfrac{10^{-3} \text{ L}}{1 \text{ mL}}\right)(25 \text{ mL}) = 0.050 \text{ mol NaOH}$

NaOH is the limiting reagent:
The reaction table gives:

	$C_6H_5COOH(aq)$	+ NaOH(aq)	→ $Na^+(aq)$	+ $C_6H_5COO^-(aq)$	+ $H_2O(l)$
Initial	0.0950 mol	0.050 mol	—	0	—
Reacting	−0.050 mol	−0.050 mol		+ 0.050 mol	
Final	0.045 mol	0 mol		0.050 mol	

The concentrations after the reactions are:

$$[C_6H_5COOH] = \left(\frac{0.045 \text{ mol } C_6H_5COOH}{(475 + 25)\text{mL}} \right)\left(\frac{1 \text{ mL}}{10^{-3}\text{ L}} \right) = 0.090 \text{ } M \text{ } C_6H_5COOH$$

$$[C_6H_5COO^-] = \left(\frac{0.050 \text{ mol } C_6H_5COO^-}{(475 + 25)\text{mL}} \right)\left(\frac{1 \text{ mL}}{10^{-3}\text{ L}} \right) = 0.10 \text{ } M \text{ } C_6H_5COO^-$$

Calculating the pH from the Henderson-Hasselbalch equation:

$$pH = pK_a + \log\left(\frac{[C_6H_5CO^-]}{[C_6H_5COOH]} \right) = 4.201 + \log\left(\frac{[0.10]}{[0.090]} \right) = 4.24676 = 4.2$$

Calculations on formic acid (HCOOH) also use the Henderson-Hasselbalch equation. The K_a for formic acid is 1.8×10^{-4} and the $pK_a = -\log(1.8 \times 10^{-4}) = 3.7447$.
The formate to formic acid ratio may now be determined:

$$pH = pK_a + \log\left(\frac{[HCOO^-]}{[HCOOH]} \right)$$

$$4.24676 = 3.7447 + \log\left(\frac{[HCOO^-]}{[HCOOH]} \right)$$

$$0.50206 = \log\left(\frac{[HCOO^-]}{[HCOOH]} \right)$$

$$\left(\frac{[HCOO^-]}{[HCOOH]} \right) = 3.177313$$

$$[HCOO^-] = 3.177313 \text{ } [HCOOH]$$

Since the conjugate acid and the conjugate base are in the same volume, the mole ratio and the molarity ratios are identical.

Moles $HCOO^- = 3.177313$ mol HCOOH

The total volume of the solution is $(500. \text{ mL})(10^{-3} \text{ L}/1 \text{ mL}) = 0.500$ L

Let V_a = volume of acid solution added, and V_b = volume of base added. Thus:

$V_a + V_b = 0.500$ L

The reaction between the formic acid and the sodium hydroxide is:

$HCOOH(aq) + NaOH(aq) \rightarrow HCOONa(aq) + H_2O(l)$

The moles of NaOH added equal the moles of HCOOH reacted and the moles of HCOONa formed.

Moles NaOH = $(2.00 \text{ mol NaOH/L})(V_b) = 2.00V_b$ mol

Total moles HCOOH = $(0.200 \text{ mol HCOOH/L})(V_a) = 0.200V_a$ mol

The stoichiometric ratios in this reaction are all 1:1.

Moles HCOOH remaining after the reaction = $(0.200V_a - 2.00V_b)$ mol

Moles $HCOO^-$ = moles HCOONa = moles NaOH = $2.00 \text{ } V_b$

Using these moles and the mole ratio determined for the buffer gives:

Moles $HCOO^- = 3.177313$ mol HCOOH

$2.00V_b$ mol = $3.177313(0.200V_a - 2.00V_b)$ mol

$2.00V_b = 0.6354626V_a - 6.354626V_b$

$8.354626 \text{ } V_b = 0.6354626 \text{ } V_a$

The volume relationship given above gives $V_a = (0.500 - V_b)$ L.

$8.354626 \text{ } V_b = 0.6354626 (0.500 - V_b)$

$8.354626 \text{ } V_b = 0.3177313 - 0.6354626 \text{ } V_b$

$8.9900886 \text{ } V_b = 0.3177313$

$V_b = 0.0353424 =$ **0.035 L NaOH**

$V_a = 0.500 - 0.0353424 = 0.4646576 =$ **0.465 L HCOOH**

Limitations due to the significant figures lead to a solution with only an approximately correct pH.

19.108 **Plan:** A formate buffer contains formate (HCOO⁻) as the base and formic acid (HCOOH) as the acid. The Henderson-Hasselbalch equation gives the component ratio, [HCOO⁻]/[HCOOH]. The ratio is used to find the volumes of acid and base required to prepare the buffer.

Solution:

From the Appendix, the K_a for formic acid is 1.8×10^{-4} and the $pK_a = -\log(1.8 \times 10^{-4}) = 3.7447$.

a) $pH = pK_a + \log\left(\dfrac{[HCOO^-]}{[HCOOH]}\right)$

$3.74 = 3.7447 + \log\left(\dfrac{[HCOO^-]}{[HCOOH]}\right)$

$-0.0047 = \log\left(\dfrac{[HCOO^-]}{[HCOOH]}\right)$

$\left(\dfrac{[HCOO^-]}{[HCOOH]}\right) = 0.989236 = \mathbf{0.99}$

b) To prepare solutions, set up equations for concentrations of formate and formic acid with x equal to the volume, in L, of 1.0 M HCOOH added. The equations are based on the neutralization reaction between HCOOH and NaOH that produces HCOO⁻.

$$HCOOH(aq) + NaOH(aq) \rightarrow HCOO^-(aq) + Na^+(aq) + H_2O(l)$$

$$[HCOO^-] = (1.0\,M\ NaOH)\left(\frac{(0.700-x)L\ NaOH}{0.700\ L\ solution}\right)\left(\frac{1\ mol\ HCOO^-}{1\ mol\ NaOH}\right)$$

$$[HCOOH] = (1.0\ M\ HCOOH)\left(\frac{x\ L\ HCOOH}{0.700\ L\ solution}\right) - (1.0\ M\ NaOH)\left(\frac{(0.700-x)L\ NaOH}{0.700\ L\ solution}\right)\left(\frac{1\ mol\ HCOO^-}{1\ mol\ NaOH}\right)$$

The component ratio equals 0.99 (from part a)). Simplify the above equations and plug into ratio:

$$\frac{[HCOO^-]}{[HCOOH]} = \frac{\left[\left(\dfrac{0.700-x}{0.700}\right)M\ HCOO^-\right]}{\left[\left(\dfrac{x-(0.700-x)}{0.700}\right)M\ HCOOH\right]} = \frac{0.700-x}{2x-0.700} = 0.989236$$

Solving for x:

$x = 0.46751 = 0.468$ L

Mixing **0.468 L of 1.0 M HCOOH** and $0.700 - 0.468 = \mathbf{0.232\ L\ of\ 1.0\ M\ NaOH}$ gives a buffer of pH 3.74.

c) The final concentration of HCOOH from the equation in part b):

$$[HCOOH] = (1.0\ M\ HCOOH)\left(\frac{0.468\ L\ HCOOH}{0.700\ L\ solution}\right) - (1.0\ M\ NaOH)\left(\frac{0.232\ L\ NaOH}{0.700\ L\ solution}\right)\left(\frac{1\ mole\ HCOO^-}{1\ mole\ NaOH}\right)$$

$= 0.33714 = \mathbf{0.34\ M\ HCOOH}$

19.111 **Plan:** The minimum urate ion concentration necessary to cause a deposit of sodium urate is determined by the K_{sp} for the salt. Convert solubility in g/100. mL to molar solubility and calculate K_{sp}. Substituting [Na⁺] and K_{sp} into the ion-product expression allows one to find [Ur⁻].

Solution:

Molar solubility of NaUr:

$$[NaUr] = \left(\frac{0.085\ g\ NaUr}{100.\ mL}\right)\left(\frac{1\ mL}{10^{-3}\ L}\right)\left(\frac{1\ mol\ NaUr}{190.10\ mol\ NaUr}\right) = 4.4713309 \times 10^{-3}\ M\ NaUr$$

$4.4713309 \times 10^{-3}\ M\ NaUr = [Na^+] = [Ur^-]$

$K_{sp} = [\text{Na}^+][\text{Ur}^-] = (4.4713309\text{x}10^{-3})(4.4713309\text{x}10^{-3}) = 1.99927998\text{x}10^{-5} \, M$
When $[\text{Na}^+] = 0.15 \, M$:
$K_{sp} = 1.99927998\text{x}10^{-5} \, M = [0.15][\text{Ur}^-]$
$[\text{Ur}^-] = 1.33285\text{x}10^{-4}$
The minimum urate ion concentration that will cause precipitation of sodium urate is **$1.3\text{x}10^{-4} \, M$**.

19.114 <u>Plan:</u> Substitute the given molar solubility of KCl into the ion-product expression to find the K_{sp} of KCl. Determine the total concentration of chloride ion in each beaker after the HCl has been added. This requires the moles originally present and the moles added. Determine a Q_{sp} value to see if K_{sp} is exceeded. If $Q_{sp} < K_{sp}$, nothing will precipitate.
<u>Solution:</u>
a) The solubility equilibrium for KCl is: $\text{KCl}(s) \leftrightarrows \text{K}^+(aq) + \text{Cl}^-(aq)$
The solubility of KCl is 3.7 M.
$K_{sp} = [\text{K}^+][\text{Cl}^-] = (3.7)(3.7) = 13.69 = \mathbf{14}$

b) Find the moles of Cl^-:
       Original moles from the KCl:

$$\text{Moles of K}^+ = \text{moles of Cl}^- = \left(\frac{3.7 \text{ mol KCl}}{1 \text{ L}}\right)\left(\frac{10^{-3} \text{ L}}{1 \text{ mL}}\right)(100. \text{ mL})\left(\frac{1 \text{ mol Cl}^- \text{ ion}}{1 \text{ mol KCl}}\right) = 0.37 \text{ mol Cl}^-$$

       Original moles from the 6.0 M HCl in the first beaker:

$$\text{Moles of Cl}^- = \left(\frac{6.0 \text{ mol HCl}}{1 \text{ L}}\right)\left(\frac{10^{-3} \text{ L}}{1 \text{ mL}}\right)(100. \text{ mL})\left(\frac{1 \text{ mol Cl}^-}{1 \text{ mol HCl}}\right) = 0.60 \text{ mol Cl}^-$$

       This results in $(0.37 + 0.60)$ mol $= 0.97$ mol Cl^-.
       Original moles from the 12 M HCl in the second beaker:

$$\text{Moles of Cl}^- = \left(\frac{12 \text{ mol HCl}}{1 \text{ L}}\right)\left(\frac{10^{-3} \text{ L}}{1 \text{ mL}}\right)(100. \text{ mL})\left(\frac{1 \text{ mol Cl}^-}{1 \text{ mol HCl}}\right) = 1.2 \text{ mol Cl}^-$$

       This results in $(0.37 + 1.2)$ mol $= 1.57$ mol Cl^-
Volume of mixed solutions $= (100. \text{ mL} + 100. \text{ mL})(10^{-3} \text{ L}/1 \text{ mL}) = 0.200$ L
       After the mixing:
           $[\text{K}^+] = (0.37 \text{ mol K}^+)/(0.200 \text{ L}) = 1.85 \, M \text{ K}^+$
           From 6.0 M HCl in the first beaker:
               $[\text{Cl}^-] = (0.97 \text{ mol Cl}^-)/(0.200 \text{ L}) = 4.85 \, M \text{ Cl}^-$
       From 12 M HCl in the second beaker:
               $[\text{Cl}^-] = (1.57 \text{ mol Cl}^-)/(0.200 \text{ L}) = 7.85 \, M \text{ Cl}^-$
Determine a Q_{sp} value to see if K_{sp} is exceeded. If $Q_{sp} < K_{sp}$, nothing will precipitate.
       From 6.0 M HCl in the first beaker:
           $Q_{sp} = [\text{K}^+][\text{Cl}^-] = (1.85)(4.85) = 8.9725 = 9.0 < 14$, so no KCl will precipitate.
       From 12 M HCl in the second beaker:
           $Q_{sp} = [\text{K}^+][\text{Cl}^-] = (1.85)(7.85) = 14.5225 = 15 > 14$, so KCl will precipitate.
The mass of KCl that will precipitate when 12 M HCl is added:
       Equal amounts of K and Cl will precipitate. Let x be the molarity change.
           $K_{sp} = [\text{K}^+][\text{Cl}^-] = (1.85 - \text{x})(7.85 - \text{x}) = 13.69$
      x $= 0.08659785 = 0.09$ This is the change in the molarity of each of the ions.

$$\text{Mass (g) of KCl} = \left(\frac{0.08659785 \text{ mol K}^+}{\text{L}}\right)(0.200 \text{ L})\left(\frac{1 \text{ mol KCl}}{1 \text{ mol K}^+}\right)\left(\frac{74.55 \text{ g KCl}}{1 \text{ mol KCl}}\right) = 1.291174 = \mathbf{1 \text{ g KCl}}$$

19.117 <u>Plan:</u> Use the Henderson-Hasselbalch equation to find the ratio of $[\text{HCO}_3^-]/[\text{H}_2\text{CO}_3]$ that will produce a buffer with a pH of 7.40 and a buffer of 7.20.

<u>Solution:</u>

a) $K_{a1} = 4.5 \times 10^{-7}$

$pK_a = -\log K_a = -\log (4.5 \times 10^{-7}) = 6.34679$

$$pH = pK_a + \log\left(\frac{[HCO_3^-]}{[H_2CO_3]}\right)$$

$$7.40 = 6.34679 + \log\left(\frac{[HCO_3^-]}{[H_2CO_3]}\right)$$

$$1.05321 = \log\left(\frac{[HCO_3^-]}{[H_2CO_3]}\right)$$

$$\frac{[HCO_3^-]}{[H_2CO_3]} = 11.30342352$$

$$\frac{[H_2CO_3]}{[HCO_3^-]} = 0.0884688 = \mathbf{0.088}$$

b) $\qquad pH = pK_a + \log\left(\frac{[HCO_3^-]}{[H_2CO_3]}\right)$

$$7.20 = 6.34679 + \log\left(\frac{[HCO_3^-]}{[H_2CO_3]}\right)$$

$$0.85321 = \log\left(\frac{[HCO_3^-]}{[H_2CO_3]}\right)$$

$$\frac{[HCO_3^-]}{[H_2CO_3]} = 7.131978$$

$$\frac{[H_2CO_3]}{[HCO_3^-]} = 0.14021 = \mathbf{0.14}$$

19.118 <u>Plan:</u> The buffer components will be TRIS, $(HOCH_2)_3CNH_2$, and its conjugate acid TRISH$^+$, $(HOCH_2)_3CNH_3^+$. The conjugate acid is formed from the reaction between TRIS and HCl. Since HCl is the limiting reactant in this problem, the concentration of conjugate acid will equal the starting concentration of HCl, 0.095 M. The concentration of TRIS is the initial concentration minus the amount reacted. Once the concentrations of the TRIS-TRISH$^+$ acid-base pair are known, the Henderson-Hasselbalch equation can be used to find the pH.

<u>Solution:</u>

$$\text{Moles of TRIS} = \left(43.0 \text{ g TRIS}\right)\left(\frac{1 \text{ mol TRIS}}{121.14 \text{ g TRIS}}\right) = 0.354961 \text{ mol}$$

$$\text{Moles of HCl added} = \left(\frac{0.095 \text{ mol HCl}}{L}\right)\left(1.00 \text{ L}\right) = 0.095 \text{ mol HCl} = \text{mol TRISH}^+$$

$$(HOCH_2)_3CNH_2(aq) + HCl(aq) \leftrightarrows (HOCH_2)_3CNH_3^+(aq) + Cl^-(aq)$$

	$(HOCH_2)_3CNH_2$	HCl	$(HOCH_2)_3CNH_3^+$	Cl^-
Initial	0.354961 mol	0.095 mol	0	0
Reacting	−0.095 mol	−0.095 mol	+0.095 mol	—
Final	0.259961 mol	0 mol	0.095 mol	

Since there is 1.00 L of solution, the moles of TRIS and TRISH$^+$ equal their molarities.

pK_a of TRISH$^+$ = $14 - pK_b = 14 - 5.91 = 8.09$

$$pH = pK_a + \log\left(\frac{[TRIS]}{[TRISH^+]}\right) = 8.09 + \log\left(\frac{[0.259961]}{[0.095]}\right) = 8.527185 = \mathbf{8.53}$$

19.123 Plan: An indicator changes color when the buffer-component ratio of the two forms of the indicator changes from a value greater than 1 to a value less than 1. The pH at which the ratio equals 1 is equal to pK_a. The midpoint in the pH range of the indicator is a good estimate of the pK_a of the indicator.
Solution:

$$pK_a = (3.4 + 4.8)/2 = 4.1 \qquad\qquad K_a = 10^{-4.1} = 7.943 \times 10^{-5} = \mathbf{8 \times 10^{-5}}$$

19.125 Plan: A spreadsheet will help you to quickly calculate $\Delta pH / \Delta V$ and average volume for each data point. At the equivalence point, the pH changes drastically when only a small amount of base is added, therefore, $\Delta pH / \Delta V$ is at a maximum at the equivalence point.
Solution:
a) Example calculation: For the first two lines of data: $\Delta pH = 1.22 - 1.00 = 0.22$; $\Delta V = 10.00 - 0.00 = 10.00$

$$\frac{\Delta pH}{\Delta V} = \frac{0.22}{10.00} = 0.022 \qquad\qquad V_{average}(mL) = (0.00 + 10.00)/2 = 5.00$$

V(mL)	pH	$\dfrac{\Delta pH}{\Delta V}$	$V_{average}$(mL)
0.00	1.00		
10.00	1.22	0.022	5.00
20.00	1.48	0.026	15.00
30.00	1.85	0.037	25.00
35.00	2.18	0.066	32.50
39.00	2.89	0.18	37.00
39.50	3.20	0.62	39.25
39.75	3.50	1.2	39.63
39.90	3.90	2.67	39.83
39.95	4.20	6	39.93
39.99	4.90	18	39.97
40.00	7.00	200	40.00
40.01	9.40	200	40.01
40.05	9.80	10	40.03
40.10	10.40	10	40.08
40.25	10.50	0.67	40.18
40.50	10.79	1.2	40.38
41.00	11.09	0.60	40.75
45.00	11.76	0.17	43.00
50.00	12.05	0.058	47.50
60.00	12.30	0.025	55.00
70.00	12.43	0.013	65.00
80.00	12.52	0.009	75.00

b)

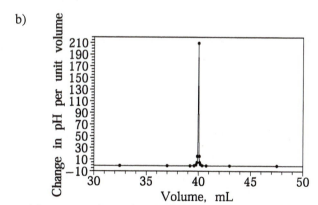

Maximum slope (equivalence point) is at V = 40.00 mL

19.127 Use HLac to indicate lactic acid and Lac⁻ to indicate the lactate ion. The Henderson-Hasselbalch equation gives the pH of the buffer. Determine the final concentrations of the buffer components from $M_{conc}V_{conc} = M_{dil}V_{dil}$. Determine the pK_a of the acid from the K_a.

$$pK_a = -\log K_a = -\log (1.38\times10^{-4}) = 3.86012$$

Determine the molarity of the diluted buffer component as $M_{dil} = M_{conc}V_{conc}/V_{dil}$.

[HLac] = [(0.85 M) (225 mL)]/[(225 + 435) mL] = 0.28977 M HLac

[Lac⁻] = [(0.68 M) (435 mL)]/[(225 + 435) mL] = 0.44818 M Lac⁻

$$pH = pK_a + \log\left(\frac{[\text{Lac}^-]}{[\text{HLac}]}\right)$$

$$pH = 3.86012 + \log\left(\frac{[0.44818]}{[0.28977]}\right) = 4.049519 = \mathbf{4.05}$$

19.133 **Plan:** To determine which species are present from a buffer system of a polyprotic acid, check the pK_a values for the one that is closest to the pH of the buffer. The two components involved in the equilibrium associated with this K_a are the principle species in the buffer. Use the Henderson-Hasselbalch equation to find the ratio of the phosphate species that will produce a buffer with a pH of 7.4.
Solution:
For carbonic acid, pK_{a1} [$-\log (8\times10^{-7}) = 6.1$] is closer to the pH of 7.4, so H_2CO_3 and HCO_3^- are the species present. For phosphoric acid, pK_{a2} [$-\log (2.3\times10^{-7}) = 6.6$] is closest to the pH, so $H_2PO_4^-$ and HPO_4^{2-} are the principle species present.

$$H_2PO_4^-(aq) + H_2O(l) \rightleftharpoons HPO_4^{2-}(aq) + H_3O^+(aq)$$

$$pH = pK_a + \log\left(\frac{[HPO_4^{2-}]}{[H_2PO_4^-]}\right)$$

$$7.4 = 6.6383 + \log\left(\frac{[HPO_4^{2-}]}{[H_2PO_4^-]}\right)$$

$$\frac{[HPO_4^{2-}]}{[H_2PO_4^-]} = 5.77697 = \mathbf{5.8}$$

19.135 **Plan:** Find the moles of quinidine initially present in the sample by dividing its mass in grams by the molar mass. Use the molar ratio between quinidine and HCl to find the moles of HCl that would react with the moles of quinidine and subtract the reacted HCl from the initial moles of HCl to find the excess. Use the molar ratio between HCl and NaOH to find the volume of NaOH required to react with the excess HCl. Then use the molar ratio between NaOH and quinidine to find the volume of NaOH required to react with the quinidine.
Solution:
a) To find the concentration of HCl after neutralizing the quinidine, calculate the concentration of quinidine and the amount of HCl required to neutralize it, remembering that the mole ratio for the neutralization is 2 mol HCl/1 mol quinidine.

$$\text{Moles of quinidine} = \left(33.85 \text{ mg quinidine}\right)\left(\frac{10^{-3}\text{ g}}{1\text{ mg}}\right)\left(\frac{1\text{ mol quinidine}}{324.41\text{ g quinidine}}\right) = 1.0434327\times10^{-4}\text{ mol quinidine}$$

$$\text{Moles of HCl excess} = \left(6.55 \text{ mL}\right)\left(\frac{10^{-3}\text{ L}}{1\text{ mL}}\right)\left(\frac{0.150\text{ mol HCl}}{L}\right)$$

$$- \left(1.0434327\times10^{-4}\text{ mol quinidine}\right)\left(\frac{2\text{ mol HCl}}{1\text{ mol quinidine}}\right) = 7.7381346\times10^{-4}\text{ mol HCl}$$

$$\text{Volume (mL) of NaOH needed} = \left(7.7381346\times10^{-4}\text{ mol HCl}\right)\left(\frac{1\text{ mol NaOH}}{1\text{ mol HCl}}\right)\left(\frac{1\text{ L}}{0.0133\text{ mol NaOH}}\right)\left(\frac{1\text{ mL}}{10^{-3}\text{ L}}\right)$$

$$= 58.18146 = \mathbf{58.2\text{ mL NaOH solution}}$$

b) Use the moles of quinidine and the concentration of the NaOH to determine the milliliters.

$$\text{Volume} = (1.0434327 \times 10^{-4} \text{ mol quinidine}) \left(\frac{1 \text{ mol NaOH}}{1 \text{ mol quinidine}} \right) \left(\frac{1 \text{ L}}{0.0133 \text{ mol NaOH}} \right) \left(\frac{1 \text{ mL}}{10^{-3} \text{ L}} \right)$$

$$= 7.84536 = \textbf{7.85 mL NaOH solution}$$

c) When quinidine (QNN) is first acidified, it has the general form QNH^+NH^+. At the first equivalence point, one of the acidified nitrogen atoms has completely reacted, leaving a singly protonated form, $QNNH^+$. This form of quinidine can react with water as either an acid or a base, so both must be considered. If the concentration of quinidine at the first equivalence point is greater than K_{b1}, then the $[OH^-]$ at the first equivalence point can be estimated as:

$$[OH^-] = \sqrt{K_{b1}K_{b2}} = \sqrt{(4.0 \times 10^{-6})(1.0 \times 10^{-10})} = 2.0 \times 10^{-8} \text{ M}$$
$$[H_3O^+] = K_w/[OH^-] = (1.0 \times 10^{-14})/(2.0 \times 10^{-8}) = 5.0 \times 10^{-7} \text{ M}$$
$$pH = -\log [H_3O^+] = -\log (5.0 \times 10^{-7} \text{ M}) = 6.3010 = \textbf{6.30}$$

19.137   Plan: The Henderson-Hasselbalch equation demonstrates that the pH changes when the ratio of acid to base in the buffer changes (pK_a is constant at a given temperature).
Solution:

$$pH = pK_a + \log \left(\frac{[A^-]}{[HA]} \right)$$

The pH of the A^-/HA buffer cannot be calculated because the identity of "A" and, thus, the value of pK_a are unknown. However, the change in pH can be described:

$$\Delta pH = \log \left(\frac{[A^-]}{[HA]} \right)_{final} - \log \left(\frac{[A^-]}{[HA]} \right)_{initial}$$

Since both $[HA]$ and $[A^-] = 0.10$ M, $\log \left(\frac{[A^-]}{[HA]} \right)_{initial} = 0$ because $[HA] = [A^-]$, and $\log (1) = 0$.

So the change in pH is equal to the concentration ratio of base to acid after the addition of H_3O^+.
Consider the buffer prior to addition to the medium.

$H_3O^+ (aq)$	+	$A^- (aq)$	$\rightarrow$	$HA (aq)$
0.0010 mol		0.10 mol		0.10 mol
−0.0010 mol		−0.0010 mol		+0.0010 mol
0		0.099 mol		0.101 mol

When 0.0010 mol H_3O^+ is added to 1 L of the undiluted buffer, the $[A^-]/[HA]$ ratio changes from 0.10/0.10 to (0.099)/(0.101). The change in pH is:

$$\Delta pH = \log (0.099/0.101) = -0.008686$$

If the undiluted buffer changes 0.009 pH units with addition of 0.0010 mol H_3O^+, how much can the buffer be diluted and still not change by 0.05 pH units ($\Delta pH < 0.05$)?
Let x = fraction by which the buffer can be diluted. Assume 0.0010 mol H_3O^+ is added to 1 L.

$$\log \frac{[\text{base}]}{[\text{acid}]} = \log \left(\frac{(0.10x - 0.0010)}{(0.10x + 0.0010)} \right) = -0.05$$

$$\left(\frac{(0.10x - 0.0010)}{(0.10x + 0.0010)} \right) = 10^{-0.05} = 0.89125$$

$$0.10x - 0.0010 = 0.89125 (0.10x + 0.0010)$$
$$x = 0.173908 = 0.17$$

The buffer concentration can be decreased by a factor of 0.17, or **170 mL** of buffer can be diluted to 1 L of medium. At least this amount should be used to adequately buffer the pH change.

19.141   Plan: Use the ideal gas law to calculate the moles of CO_2 dissolved in water. Use the K_a expression for H_2CO_3 to find the $[H_3O^+]$ associated with that CO_2 concentration.

<u>Solution:</u>

Carbon dioxide dissolves in water to produce H_3O^+ ions:

$$CO_2(g) \leftrightarrows CO_2(aq)$$
$$CO_2(aq) + H_2O(l) \leftrightarrows H_2CO_3(aq)$$
$$H_2CO_3(aq) \leftrightarrows H_3O^+(aq) + HCO_3^-(aq) \qquad K_{a1} = 4.5 \times 10^{-7}$$

The molar concentration of CO_2, $[CO_2]$, depends on how much $CO_2(g)$ from the atmosphere can dissolve in pure water.

Since air is not pure CO_2, account for the volume fraction of air (0.040 L/100 L) when determining the moles.

$$\text{Volume (L) of } CO_2 = (88 \text{ mL}) \left(\frac{10^{-3} \text{ L}}{1 \text{ mL}}\right) \left(\frac{0.040\%}{100\%}\right) = 3.520 \times 10^{-5} \text{ L } CO_2$$

$$\text{Moles of dissolved } CO_2 = \frac{PV}{RT} = \frac{(1 \text{ atm})(3.520 \times 10^{-5} \text{ L})}{\left(0.0821 \dfrac{\text{L} \cdot \text{atm}}{\text{mol} \cdot \text{K}}\right)((273+25)\text{K})} = 1.438743 \times 10^{-6} \text{ mol } CO_2$$

$$[CO_2] = (1.438743 \times 10^{-6} \text{ mol } CO_2)/[(100 \text{ mL})(10^{-3} \text{ L}/1 \text{ mL})] = 1.438743 \times 10^{-5} M \text{ } CO_2$$

$$K_{a1} = 4.5 \times 10^{-7} = \frac{[H_3O^+][HCO_3^-]}{[H_2CO_3]} = \frac{[H_3O^+][HCO_3^-]}{[CO_2]} \qquad \text{Let } x = [H_3O^+] = [HCO_3^-]$$

$$4.5 \times 10^{-7} = \frac{[x][x]}{[1.438743 \times 10^{-5} - x]} \qquad \text{Assume that x is small compared to } 1.438743 \times 10^{-5}$$

$$4.5 \times 10^{-7} = \frac{[x][x]}{[1.438743 \times 10^{-5}]}$$

$$x = 2.544473 \times 10^{-6}$$

Check assumption that x is small compared to 1.438743×10^{-5}:

$$\frac{2.544473 \times 10^{-6}}{1.438743 \times 10^{-5}} (100) = 18\% \text{ error, so the assumption is not valid.}$$

Since the error is greater than 5%, it is not acceptable to assume x is small compared to 1.438743×10^{-5}, and it is necessary to use the quadratic equation.

$$4.5 \times 10^{-7} = \frac{[x][x]}{[1.438743 \times 10^{-5} - x]}$$

$$x^2 + 4.5 \times 10^{-7}x - 6.474344 \times 10^{-12} = 0$$

$$a = 1 \qquad b = 4.5 \times 10^{-7} \qquad c = -6.474344 \times 10^{-12}$$

$$x = \frac{-b \pm \sqrt{b^2 - 4ac}}{2a}$$

$$x = \frac{-4.5 \times 10^{-7} \pm \sqrt{(4.5 \times 10^{-7})^2 - 4(1)(-6.474344 \times 10^{-12})}}{2(1)}$$

$$x = 2.329402 \times 10^{-6} M = [H_3O^+]$$

$$pH = -\log(2.329402 \times 10^{-6}) = 5.632756 = \mathbf{5.63}$$

19.143 Initial concentrations of Pb^{2+} and $Ca(EDTA)^{2-}$ before reaction based on mixing 100. mL of 0.10 M $Na_2Ca(EDTA)$ with 1.5 L blood:

$$[Pb^{2+}] = \left(\frac{120 \text{ } \mu g \text{ } Pb^{2+}}{100 \text{ mL}}\right)\left(\frac{1 \text{ mL}}{10^{-3} \text{ L}}\right)\left(\frac{1.5 \text{ L blood}}{1.6 \text{ L mixture}}\right)\left(\frac{10^{-6} \text{ g}}{1 \text{ } \mu g}\right)\left(\frac{1 \text{ mol } Pb^{2+}}{207.2 \text{ g } Pb^{2+}}\right) = 5.42953668 \times 10^{-6} M \text{ } Pb^{2+}$$

$$M_{conc}V_{conc} = M_{dil}V_{dil}$$

$[Ca(EDTA)^{2-}] = M_{conc}V_{conc}/V_{dil} = [(0.10\ M\ (100\ mL)(10^{-3}\ L/1\ mL)]/(1.6\ L)$
$\qquad = 6.25\times10^{-3}\ M$

Set up a reaction table assuming the reaction goes to completion:

Concentration (M)	$[Ca(EDTA)]^{2-}(aq)$	+	$Pb^{2+}(aq)$	⇌	$[Pb(EDTA)]^{2-}(aq)$	+	$Ca^{2+}(aq)$
Initial	6.25×10^{-3}		5.42953668×10^{-6}		0		0
React	-5.42953668×10^{-6}		-5.42953668×10^{-6}		$+5.42953668\times10^{-6}$		$+5.42953668\times10^{-6}$
	6.24457×10^{-3}		0		5.42953668×10^{-6}		5.42953668×10^{-6}

Now set up a reaction table for the equilibrium process:

Concentration (M)	$[Ca(EDTA)]^{2-}(aq)$	+	$Pb^{2+}(aq)$	⇌	$[Pb(EDTA)]^{2-}(aq)$	+	$Ca^{2+}(aq)$
Initial	6.24457×10^{-3}		0		5.42953668×10^{-6}		5.42953668×10^{-6}
Change	$+x$		$+x$		$-x$		$-x$
Equilibrium	$6.24457\times10^{-3} + x$		x		$5.4295366\times10^{-6} - x$		$5.4295366\times10^{-6} - x$

$$K_c = 2.5\times10^{7} = \frac{\left[Pb(EDTA)^{2-}\right]\left[Ca^{2+}\right]}{\left[Ca(EDTA)^{2-}\right]\left[Pb^{2+}\right]} = \frac{\left[5.42953668\times10^{-6}\right]\left[5.42953668\times10^{-6}\right]}{\left[6.24457\times10^{-3}\right][x]}$$

$x = [Pb^{2+}] = 1.8883522\times10^{-16}\ M$

$$\text{Mass (μg) of } Pb^{2+}\text{ in 100 mL} = \left(\frac{1.8883522\times10^{-16}\text{ mol }Pb^{2+}}{L}\right)\left(\frac{10^{-3}\ L}{1\ mL}\right)(100\ mL)\left(\frac{207.2\text{ g }Pb^{2+}}{1\text{ mol }Pb^{2+}}\right)\left(\frac{1\ μg}{10^{-6}\ g}\right)$$

$$= 3.9126658\times10^{-9}\ μg\ Pb^{2+}$$

The final concentration is **3.9×10^{-9} μg/100 mL**.

19.145 Plan: Convert the solubility of NaCl from g/L to mol/l (molarity). Use the solubility to find the K_{sp} value for NaCl. Find the moles of Na^+ and Cl^- in the original solution; find the moles of added Cl^- (from the added HCl). The molarity of the Na^+ and Cl^- ions are then found by dividing moles of each by the total volume after mixing. Using the molarities of the two ions, determine a Q value and compare this value to the K_{sp} to determine if precipitation will occur.
Solution:

$$\text{Concentration (M) of NaCl} = \left(\frac{317\text{ g NaCl}}{L}\right)\left(\frac{1\text{ mol NaCl}}{58.44\text{ g NaCl}}\right) = 5.42436687\ M\text{ NaCl}$$

Determine the K_{sp} from the molarity just calculated.
$NaCl(s) ⇌ Na^+(aq) + Cl^-(aq)$
$K_{sp} = [Na^+][Cl^-] = S^2 = (5.42436687)^2 = 29.42375594 = 29.4$

$$\text{Moles of } Cl^-\text{ initially} = \left(\frac{5.42436687\text{ mol NaCl}}{L}\right)(0.100\ L)\left(\frac{1\text{ mol }Cl^-}{1\text{ mol NaCl}}\right) = 0.542436687\text{ mol }Cl^-$$

This is the same as the moles of Na^+ in the solution.

$$\text{Moles of } Cl^-\text{ added} = \left(\frac{8.65\text{ mol HCl}}{L}\right)\left(\frac{10^{-3}\ L}{1\ mL}\right)(28.5\ mL)\left(\frac{1\text{ mol }Cl^-}{1\text{ mol HCl}}\right) = 0.246525\text{ mol }Cl^-$$

0.100 L of saturated solution contains 0.542 mol each Na^+ and Cl^-, to which you are adding 0.246525 mol of additional Cl^- from HCl.
Volume of mixed solutions = 0.100 L + (28.5 mL)(10^{-3} L/1 mL) = 0.1285 L
Molarity of Cl^- in mixture = [(0.542436687 + 0.246525) mol Cl^-]/(0.1285 L) = 6.13978 $M\ Cl^-$
Molarity of Na^+ in mixture = (0.542436687 mol Na^+)/(0.1285 L) = 4.22130 $M\ Na^+$
Determine a Q value and compare this value to the K_{sp} to determine if precipitation will occur.
$\qquad Q_{sp} = [Na^+][Cl^-] = (4.22130)(6.13978) = 25.9179 = 25.9$
Since $Q_{sp} < K_{sp}$, no NaCl will precipitate.

19.146　**Plan:** A buffer contains a weak acid conjugate base pair. A K_a expression is used to calculate the pH of a weak acid while a K_b expression is used to calculate the pH of a weak base. The Henderson-Hasselbalch equation is used to calculate the pH when both the weak acid and conjugate base are present (a buffer).
Solution:
a) For the solution to be a buffer, both HA and A^- must be present in the solution. This situation occurs in **A** and **D**.
b) Scene A:

The amounts of HA and A^- are equal.

$$pH = pK_a + \log\left(\frac{[A^-]}{[HA]}\right) \qquad \left(\frac{[A^-]}{[HA]}\right) = 1 \text{ when the amounts of HA and } A^- \text{ are equal}$$

$$pH = pK_a + \log 1$$
$$pH = pK_a = -\log(4.5\times10^{-5}) = 4.346787 = \textbf{4.35}$$

Scene B:

Only A^- is present at a concentration of $0.10\ M$.
The K_b for A^- is needed.
$$K_b = K_w/K_a = 1.0\times10^{-14}/4.5\times10^{-5} = 2.222\times10^{-10}$$
$$A^-(aq) + H_2O(l) \rightleftharpoons OH^-(aq) + HA(aq)$$

Initial:	0.10 M	0	0
Change:	−x	−x	−x
Equilibrium:	0.10 − x	x	x

$$K_b = 2.222\times10^{-10} = \frac{[HA][OH^-]}{[A^-]}$$

$$K_b = 2.222\times10^{-10} = \frac{[x][x]}{[0.10-x]} \qquad \text{Assume that x is small compared to 0.10}$$

$$K_b = 2.222\times10^{-10} = \frac{(x)(x)}{(0.10)}$$

$$x = 4.7138095\times10^{-6}\ M\ OH^-$$

Check assumption: $(4.7138095\times10^{-6}/0.10) \times 100\% = 0.005\%$ error, so the assumption is valid.
$$[H_3O]^+ = K_w/[OH^-] = (1.0\times10^{-14})/(4.7138095\times10^{-6}) = 2.1214264\times10^{-9}\ M\ H_3O^+$$
$$pH = -\log[H_3O^+] = -\log(2.1214264\times10^{-9}) = 8.67337 = \textbf{8.67}$$

Scene C:

This is a $0.10\ M$ HA solution. The hydrogen ion, and hence the pH, can be determined from the K_a.

Concentration	HA(aq) + H₂O(l)	⇌	H₃O⁺(aq) +	A⁻(aq)
Initial	0.10 M	—	0	0
Change	−x		+x	+x
Equilibrium	0.10 − x		x	x

(The H_3O^+ contribution from water has been neglected.)

$$K_a = 4.5\times10^{-5} = \frac{[H_3O^+][A^-]}{[HA]}$$

$$K_a = 4.5\times10^{-5} = \frac{(x)(x)}{(0.10-x)} \qquad \text{Assume that x is small compared to 0.10.}$$

$$K_a = 4.5\times10^{-5} = \frac{(x)(x)}{(0.10)}$$

$$[H_3O^+] = x = 2.12132\times10^{-3}$$

Check assumption: $(2.12132\times10^{-3}/0.10) \times 100\% = 2\%$ error, so the assumption is valid.
$$pH = -\log[H_3O^+] = -\log(2.12132\times10^{-3}) = 2.67339 = \textbf{2.67}$$

Scene D:

This is a buffer with a ratio of $[A^-]/[HA] = 5/3$.

$$pH = pK_a + \log\left(\frac{[A^-]}{[HA]}\right)$$

$$pH = -\log(4.5 \times 10^{-5}) + \log\left[\frac{5}{3}\right] = 4.568636 = \textbf{4.57}$$

c) The initial stage in the titration would only have HA present. The amount of HA will decrease, and the amount of A^- will increase until only A^- remains. The sequence will be: **C, A, D, and B**.

d) At the equivalence point, all the HA will have reacted with the added base. This occurs in scene **B**.

CHAPTER 20 THERMODYNAMICS: ENTROPY, FREE ENERGY, AND THE DIRECTION OF CHEMICAL REACTIONS

FOLLOW–UP PROBLEMS

20.1A **Plan:** Particles with more freedom of motion have higher entropy. In general the entropy of gases is greater than that of liquids, and the entropy of liquids is greater than that of solids. Entropy increases with temperature. For substances in the same phase, entropy increases with atomic size and molecular complexity. If the entropy of the products is greater than that of the reactants, ΔS is positive.
Solution:
a) $PCl_5(g)$. For substances with the same type of atoms and in the same physical state, entropy increases with increasing number of atoms per molecule because more types of molecular motion are available.
b) $BaCl_2(s)$. Entropy increases with increasing atomic size. The Ba^{2+} ion and Cl^- ion are larger than the Ca^{2+} ion and F^- ion, respectively.
c) $Br_2(g)$. Entropy increases from solid $\rightarrow$ liquid $\rightarrow$ gas.

20.1B **Plan:** Particles with more freedom of motion have higher entropy. In general the entropy of gases is greater than that of liquids, and the entropy of liquids is greater than that of solids. Entropy increases with temperature. For substances in the same phase, entropy increases with atomic size and molecular complexity. If the entropy of the products is greater than that of the reactants, ΔS is positive.
Solution:
a) $LiBr(aq)$. For substances with the same number of atoms and in the same physical state, entropy increases with increasing atomic size. LiBr has the lower molar mass of the two substances and, therefore, the lower entropy.
b) **Quartz.** Quartz has a crystalline structure and the particles have less freedom (and lower entropy) in that structure than in glass, an amorphous solid.
c) **Cyclohexane.** Ethylcyclobutane has a side chain, which has more freedom of motion than the atoms in the ring. In cyclohexane, there are no side chains. Freedom of motion is restricted by the ring structure.

20.2A **Plan:** Predict the sign of ΔS°_{rxn} by comparing the randomness of the products with the randomness of the reactants. Calculate ΔS°_{rxn} using Appendix B values and the relationship $\Delta S^\circ_{rxn} = \sum m\, S^\circ_{products} - \sum n\, S^\circ_{reactants}$.
Solution:
a) $4NO\,(g) \rightarrow N_2O(g) + N_2O_3(g)$
The ΔS°_{rxn} is predicted to decrease ($\Delta S^\circ_{rxn} < 0$) because four moles of random, gaseous product are transformed into two moles of random, gaseous product (the change in gas moles is –2).

$\Delta S^\circ_{rxn} = \sum m\, S^\circ_{products} - \sum n\, S^\circ_{reactants}$

$\Delta S^\circ_{rxn} = [(1\ mol\ N_2O_3)(S^\circ\ of\ N_2O_3) + (1\ mol\ N_2O)(S^\circ\ of\ N_2O)]$
$\qquad\qquad - [(4\ mol\ NO)(S^\circ\ of\ NO)]$
$\Delta S^\circ_{rxn} = [(1\ mol\ N_2O_3)(314.7\ J/mol\bullet K) + (1\ mol\ N_2O)(219.7\ J/mol\bullet K)]$
$\qquad\qquad - [(4\ mol\ NO)(210.65\ J/mol\bullet K)]$
$\qquad\quad = -308.2\ \textbf{J/K}$
$\Delta S^\circ_{rxn} < 0$ as predicted.
b) $CH_3OH(g) \rightarrow CO(g) + 2H_2(g)$
The change in gaseous moles is +2, so the sign of ΔS°_{rxn} is predicted to be greater than zero.
$\Delta S^\circ_{rxn} = [(1\ mol\ CO)(S^\circ\ of\ CO) + (2\ mol\ H_2)(S^\circ\ of\ H_2)]$
$\qquad\qquad - [(1\ mol\ CH_3OH)(S^\circ\ of\ CH_3OH)]$

$\Delta S^\circ_{rxn} = [(1 \text{ mol CO})(197.5 \text{ J/mol}\cdot\text{K}) + (2 \text{ mol H}_2)(130.6 \text{ J/mol}\cdot\text{K})]$
$$- [(1 \text{ mol CH}_3\text{OH})(238 \text{ J/mol}\cdot\text{K})]$$
$$= 220.7 = \textbf{221 J/K}$$
$\Delta S^\circ_{rxn} > 0$ as predicted.

20.2B   Plan: Predict the sign of ΔS°_{rxn} by comparing the randomness of the products with the randomness of the reactants. Calculate ΔS°_{rxn} using Appendix B values and the relationship $\Delta S^\circ_{rxn} = \sum m S^\circ_{products} - \sum n S^\circ_{reactants}$.
Solution:

a) $2\text{NaOH}(s) + \text{CO}_2(g) \rightarrow \text{Na}_2\text{CO}_3(s) + \text{H}_2\text{O}(l)$

The ΔS°_{rxn} is predicted to decrease ($\Delta S^\circ_{rxn} < 0$) because the more random, gaseous reactant is transformed into a more ordered, liquid product.

$\Delta S^\circ_{rxn} = \sum m S^\circ_{products} - \sum n S^\circ_{reactants}$

$\Delta S^\circ_{rxn} = [(1 \text{ mol Na}_2\text{CO}_3)(S^\circ \text{ of Na}_2\text{CO}_3) + (1 \text{ mol H}_2\text{O})(S^\circ \text{ of H}_2\text{O})]$
$$- [(2 \text{ mol NaOH})(S^\circ \text{ of NaOH}) + (1 \text{ mol CO}_2)(S^\circ \text{ of CO}_2)]$$

$\Delta S^\circ_{rxn} = [(1 \text{ mol Na}_2\text{CO}_3)(139 \text{ J/mol}\cdot\text{K}) + (1 \text{ mol H}_2\text{O})(69.940 \text{ J/mol}\cdot\text{K})]$
$$- [(2 \text{ mol NaOH})(64.454 \text{ J/mol}\cdot\text{K}) + (1 \text{ mol CO}_2)(213.7 \text{ J/mol}\cdot\text{K})]$$
$$= -133.668 = \textbf{- 134 J/K}$$

$\Delta S^\circ_{rxn} < 0$ as predicted.

b) $2\text{Fe}(s) + 3\text{H}_2\text{O}(g) \rightarrow \text{Fe}_2\text{O}_3(s) + 3\text{H}_2(g)$

The change in gaseous moles is zero, so the sign of ΔS°_{rxn} is difficult to predict. Iron(III) oxide has greater entropy than Fe because it is more complex, but this is offset by the greater molecular complexity of H_2O versus H_2.

$\Delta S^\circ_{rxn} = [(1 \text{ mol Fe}_2\text{O}_3)(S^\circ \text{ of Fe}_2\text{O}_3) + (3 \text{ mol H}_2)(S^\circ \text{ of H}_2)]$
$$- [(2 \text{ mol Fe})(S^\circ \text{ of Fe}) + (3 \text{ mol H}_2\text{O})(S^\circ \text{ of H}_2\text{O})]$$

$\Delta S^\circ_{rxn} = [(1 \text{ mol Fe}_2\text{O}_3)(87.400 \text{ J/mol}\cdot\text{K}) + (3 \text{ mol H}_2)(130.6 \text{ J/mol}\cdot\text{K})]$
$$- [(2 \text{ mol Fe})(27.3 \text{ J/mol}\cdot\text{K}) + (3 \text{ mol H}_2\text{O})(188.72 \text{ J/mol}\cdot\text{K})]$$
$$= -141.56 = \textbf{-141.6 J/K}$$

The negative ΔS°_{rxn} shows that the greater entropy of H_2O versus H_2 does outweigh the greater entropy of Fe_2O_3 versus Fe.

20.3A   Plan: Write the balanced equation for the reaction and calculate the ΔS°_{rxn} using Appendix B. Determine the ΔS_{surr} by first finding ΔH°_{rxn}. Add ΔS_{surr} to ΔS°_{rxn} to verify that ΔS_{univ} is positive.
Solution:

$\text{P}_4(s) + 6\text{Cl}_2(g) \rightarrow 4\text{PCl}_3(g)$

$\Delta S^\circ_{rxn} = [(4 \text{ mol PCl}_3)(S^\circ \text{ of PCl}_3)] - [(1 \text{ mol P}_4)(S^\circ \text{ of P}_4) + (6 \text{ mol Cl}_2)(S^\circ \text{ of Cl}_2)]$

$\Delta S^\circ_{rxn} = [(4 \text{ mol PCl}_3)(312 \text{ J/mol}\cdot\text{K})]$
$$- [(1 \text{ mol P}_4)(41.1 \text{ J/mol}\cdot\text{K}) + (6 \text{ mol Cl}_2)(223.0 \text{J/mol}\cdot\text{K})]$$

$\Delta S^\circ_{rxn} = \textbf{-131 J/K}$ (The entropy change is expected to be negative because the change in gas moles is negative.)

$\Delta H^\circ_{rxn} = \sum m \Delta H^\circ_{f(products)} - \sum n \Delta H^\circ_{f(reactants)}$

$\Delta H^\circ_{rxn} = [(4 \text{ mol PCl}_3)(\Delta H^\circ_f \text{ of PCl}_3)]$
$$- [(1 \text{ mol P}_4)(\Delta H^\circ_f \text{ of P}_4) + (6 \text{ mol Cl}_2)(\Delta H^\circ_f \text{ of Cl}_2)]$$

$\Delta H^\circ_{rxn} = [(4 \text{ mol PCl}_3)(-287 \text{ kJ/mol})]$
$$- [(1 \text{ mol P}_4)(0 \text{ kJ/mol}) + (6 \text{ mol Cl}_2)(0 \text{ kJ/mol})]$$

$\Delta H^\circ_{rxn} = \textbf{-1148 kJ}$

$\Delta S_{surr} = -\dfrac{\Delta H^\circ_{rxn}}{T} = -\dfrac{-1148 \text{ kJ}}{298 \text{ K}} = 3.8523 \text{ kJ/K}(10^3 \text{ J/1 kJ}) = \textbf{3850 J/K}$

$\Delta S_{univ} = \Delta S^\circ_{rxn} + \Delta S_{surr} = (-131 \text{ J/K}) + (3850 \text{ J/K}) = 3719 \text{ J/K}$

Because ΔS_{univ} is positive, the reaction **is spontaneous** at 298 K.

20.3B Plan: Write the balanced equation for the reaction and calculate the $\Delta S^{\circ}_{\mathrm{rxn}}$ using Appendix B. Determine the ΔS_{surr} by first finding $\Delta H^{\circ}_{\mathrm{rxn}}$. Add ΔS_{surr} to $\Delta S^{\circ}_{\mathrm{rxn}}$ to verify that ΔS_{univ} is positive.

Solution:

$2\mathrm{FeO}(s) + 1/2\mathrm{O_2}(g) \rightarrow \mathrm{Fe_2O_3}(s)$

$\Delta S^{\circ}_{\mathrm{rxn}} = [(1 \text{ mol } \mathrm{Fe_2O_3})(S^{\circ} \text{ of } \mathrm{Fe_2O_3})] - [(2 \text{ mol FeO})(S^{\circ} \text{ of FeO}) + (1/2 \text{ mol } \mathrm{O_2})(S^{\circ} \text{ of } \mathrm{O_2})]$

$\Delta S^{\circ}_{\mathrm{rxn}} = [(1 \text{ mol } \mathrm{Fe_2O_3})(87.400 \text{ J/mol•K})]$

$\qquad\qquad\qquad - [(2 \text{ mol FeO})(60.75 \text{ J/mol•K}) + (1/2 \text{ mol } \mathrm{O_2}) (205.0 \text{ J/mol•K})]$

$\Delta S^{\circ}_{\mathrm{rxn}} = -136.6 \text{ J/K}$ (entropy change is expected to be negative because gaseous reactant is converted to solid product).

$\Delta H^{\circ}_{\mathrm{rxn}} = \sum m \, \Delta H^{\circ}_{\mathrm{f(products)}} - \sum n \, \Delta H^{\circ}_{\mathrm{f(reactants)}}$

$\Delta H^{\circ}_{\mathrm{rxn}} = [(1 \text{ mol } \mathrm{Fe_2O_3})(\Delta H^{\circ}_{\mathrm{f}} \text{ of } \mathrm{Fe_2O_3})]$

$\qquad\qquad\qquad - [(2 \text{ mol FeO})(\Delta H^{\circ}_{\mathrm{f}} \text{ of FeO}) + (1/2 \text{ mol } \mathrm{O_2})(\Delta H^{\circ}_{\mathrm{f}} \text{ of } \mathrm{O_2})]$

$\Delta H^{\circ}_{\mathrm{rxn}} = [(1 \text{ mol } \mathrm{Fe_2O_3})(-825.5 \text{ kJ/mol})]$

$\qquad\qquad\qquad - [(2 \text{ mol FeO})(-272.0 \text{ kJ/mol}) + (1/2 \text{ mol } \mathrm{O_2})(0 \text{ kJ/mol})]$

$\Delta H^{\circ}_{\mathrm{rxn}} = -281.5 \text{ kJ}$

$\Delta S_{\mathrm{surr}} = -\dfrac{\Delta H^{\circ}_{\mathrm{rxn}}}{T} = -\dfrac{-281.5 \text{ kJ}}{298 \text{ K}} = 0.94463 \text{ kJ/K}(10^3 \text{ J/1 kJ}) = 944.63 \text{ J/K}$

$\Delta S_{\mathrm{univ}} = \Delta S^{\circ}_{\mathrm{rxn}} + \Delta S_{\mathrm{surr}} = (-136.6 \text{ J/K}) + (944.63 \text{ J/K}) = 808.03 = 808 \text{ J/K}$

Because ΔS_{univ} is positive, the reaction is **spontaneous** at 298 K.

This process is also known as rusting. Common sense tells us that rusting occurs spontaneously. Although the entropy change of the system is negative, the increase in entropy of the surroundings is large enough to offset $\Delta S^{\circ}_{\mathrm{rxn}}$.

20.4A Plan: Calculate the $\Delta H^{\circ}_{\mathrm{rxn}}$ using $\Delta H^{\circ}_{\mathrm{f}}$ values from Appendix B. Calculate $\Delta S^{\circ}_{\mathrm{rxn}}$ from tabulated S° values and then use the relationship $\Delta G^{\circ}_{\mathrm{rxn}} = \Delta H^{\circ}_{\mathrm{rxn}} - T\Delta S^{\circ}_{\mathrm{rxn}}$.

Solution:

$\Delta H^{\circ}_{\mathrm{rxn}} = \sum m \, \Delta H^{\circ}_{\mathrm{f(products)}} - \sum n \, \Delta H^{\circ}_{\mathrm{f(reactants)}}$

$\Delta H^{\circ}_{\mathrm{rxn}} = [(2 \text{ mol NOCl})(\Delta H^{\circ}_{\mathrm{f}} \text{ of NOCl})] - [(2 \text{ mol NO})(\Delta H^{\circ}_{\mathrm{f}} \text{ of NO}) + (1 \text{ mol } \mathrm{Cl_2})(\Delta H^{\circ}_{\mathrm{f}} \text{ of } \mathrm{Cl_2})]$

$\Delta H^{\circ}_{\mathrm{rxn}} = [(2 \text{ mol NOCl})(51.71 \text{ kJ/mol})] - [(2 \text{ mol NO})(90.29 \text{ kJ/mol}) + (1 \text{ mol } \mathrm{Cl_2})(0 \text{ kJ/mol})]$

$\Delta H^{\circ}_{\mathrm{rxn}} = -77.16 \text{ kJ}$

$\Delta S^{\circ}_{\mathrm{rxn}} = \sum m \, S^{\circ}_{\mathrm{products}} - \sum n \, S^{\circ}_{\mathrm{reactants}}$

$\Delta S^{\circ}_{\mathrm{rxn}} = [(2 \text{ mol NOCl})(S^{\circ} \text{ of NOCl})] - [(2 \text{ mol NO})(S^{\circ} \text{ of NO}) + (1 \text{ mol } \mathrm{Cl_2})(S^{\circ} \text{ of } \mathrm{Cl_2})]$

$\Delta S^{\circ}_{\mathrm{rxn}} = [(2 \text{ mol NOCl})(261.6 \text{ J/mol•K})] - [(2 \text{ mol NO})(210.65 \text{ J/mol•K}) + (1 \text{ mol } \mathrm{Cl_2})(223.0 \text{ J/mol•K})]$

$\Delta S^{\circ}_{\mathrm{rxn}} = -121.1 \text{ J/K}$

$\Delta G^{\circ}_{\mathrm{rxn}} = \Delta H^{\circ}_{\mathrm{rxn}} - T\Delta S^{\circ}_{\mathrm{rxn}} = -77.16 \text{ kJ} - [(298 \text{ K})(-121.1 \text{ J/K})(1 \text{ kJ}/10^3 \text{ J})] = -41.0722 = \mathbf{-41.1 \text{ kJ}}$

20.4B Plan: Calculate the $\Delta H^{\circ}_{\mathrm{rxn}}$ using $\Delta H^{\circ}_{\mathrm{f}}$ values from Appendix B. Calculate $\Delta S^{\circ}_{\mathrm{rxn}}$ from tabulated S° values and then use the relationship $\Delta G^{\circ}_{\mathrm{rxn}} = \Delta H^{\circ}_{\mathrm{rxn}} - T\Delta S^{\circ}_{\mathrm{rxn}}$.

Solution:

$\Delta H^{\circ}_{\mathrm{rxn}} = \sum m \, \Delta H^{\circ}_{\mathrm{f(products)}} - \sum n \, \Delta H^{\circ}_{\mathrm{f(reactants)}}$

$\Delta H^{\circ}_{\mathrm{rxn}} = [(4 \text{ mol NO})(\Delta H^{\circ}_{\mathrm{f}} \text{ of NO}) + (6 \text{ mol } \mathrm{H_2O})(\Delta H^{\circ}_{\mathrm{f}} \text{ of } \mathrm{H_2O})]$

$\qquad\qquad\qquad - [(4 \text{ mol } \mathrm{NH_3})(\Delta H^{\circ}_{\mathrm{f}} \text{ of } \mathrm{NH_3}) + (5 \text{ mol } \mathrm{O_2})(\Delta H^{\circ}_{\mathrm{f}} \text{ of } \mathrm{O_2})]$

$\Delta H^{\circ}_{\mathrm{rxn}} = [(4 \text{ mol NO})(90.29 \text{ kJ/mol}) + (6 \text{ mol } \mathrm{H_2O})(-241.826 \text{ kJ/mol})]$

$\qquad\qquad\qquad - [(4 \text{ mol } \mathrm{NH_3})(-45.9 \text{ kJ/mol}) + (5 \text{ mol } \mathrm{O_2})(0 \text{ kJ/mol})]$

$\Delta H^{\circ}_{\mathrm{rxn}} = -906.196 \text{ kJ}$

$$\Delta S^\circ_{rxn} = \sum m\, S^\circ_{products} - \sum n\, S^\circ_{reactants}$$

$$\Delta S^\circ_{rxn} = [(4\ mol\ NO)(S^\circ\ of\ NO) + (6\ mol\ H_2O)(S^\circ\ of\ H_2O)]$$
$$- [(4\ mol\ NH_3)(S^\circ\ of\ NH_3) + (5\ mol\ O_2)(S^\circ\ of\ O_2)]$$

$$\Delta S^\circ_{rxn} = [(4\ mol\ NO)(210.65\ J/mol\bullet K) + (6\ mol\ H_2O)(188.72\ J/mol\bullet K)]$$
$$- [(4\ mol\ NH_3)(193\ J/mol\bullet K) + (5\ mol\ O_2)(205.0\ J/mol\bullet K)]$$

$$\Delta S^\circ_{rxn} = 177.92\ J/K$$

$$\Delta G^\circ_{rxn} = \Delta H^\circ_{rxn} - T\Delta S^\circ_{rxn} = -906.196\ kJ - [(298\ K)(177.92\ J/K)(1\ kJ/10^3\ J)] = -959.21616 = \textbf{--959 kJ}$$

20.5A Plan: Use ΔG°_f values from Appendix B to calculate ΔG°_{rxn} using the relationship

$$\Delta G^\circ_{rxn} = \sum m\, \Delta G^\circ_{f(products)} - \sum n\, \Delta G^\circ_{f(reactants)}.$$

Solution:

a) $\Delta G^\circ_{rxn} = [(2\ mol\ NOCl)(\Delta G^\circ_f\ of\ NOCl)] - [(2\ mol\ NO)(\Delta G^\circ_f\ of\ NO) + (1\ mol\ Cl_2)(\Delta G^\circ_f\ of\ Cl_2)]$

$\Delta G^\circ_{rxn} = [(2\ mol\ NOCl)(66.07\ kJ/mol)] - [(2\ mol\ NO)(86.60\ kJ/mol) + (1\ mol\ Cl_2)(0\ kJ/mol)]$

$\Delta G^\circ_{rxn} = \textbf{--41.06 kJ}$

b) $\Delta G^\circ_{rxn} = [(2\ mol\ Fe)(\Delta G^\circ_f\ of\ Fe) + (3\ mol\ H_2O)(\Delta G^\circ_f\ of\ H_2O)] -$
$$[(3\ mol\ H_2)(\Delta G^\circ_f\ of\ H_2) + (1\ mol\ Fe_2O_3)(\Delta G^\circ_f\ of\ Fe_2O_3)]$$

$\Delta G^\circ_{rxn} = [(2\ mol\ Fe)(0\ kJ/mol) + (3\ mol\ H_2O)(-228.60\ kJ/mol)] -$
$$[(3\ mol\ H_2)(0\ kJ/mol) + (1\ mol\ Fe_2O_3)(-743.6\ kJ/mol)]$$

$\Delta G^\circ_{rxn} = \textbf{57.8 kJ}$

20.5B Plan: Use ΔG°_f values from Appendix B to calculate ΔG°_{rxn} using the relationship

$$\Delta G^\circ_{rxn} = \sum m\, \Delta G^\circ_{f(products)} - \sum n\, \Delta G^\circ_{f(reactants)}.$$

Solution:

a) $\Delta G^\circ_{rxn} = [(4\ mol\ NO)(\Delta G^\circ_f\ of\ NO) + (6\ mol\ H_2O)(\Delta G^\circ_f\ of\ H_2O)]$
$$- [(4\ mol\ NH_3)(\Delta G^\circ_f\ of\ NH_3) + (5\ mol\ O_2)(\Delta G^\circ_f\ of\ O_2)]$$

$\Delta G^\circ_{rxn} = [(4\ mol\ NO)(86.60\ kJ/mol) + (6\ mol\ H_2O)(-228.60\ kJ/mol)]$
$$- [(4\ mol\ NH_3)(-16\ kJ/mol) + (5\ mol\ O_2)(0\ kJ/mol)]$$

$\Delta G^\circ_{rxn} = -961.2 = \textbf{--961 kJ}$

b) $\Delta G^\circ_{rxn} = [(2\ mol\ CO)(\Delta G^\circ_f\ of\ CO)] - [(2\ mol\ C)(\Delta G^\circ_f\ of\ C) + (1\ mol\ O_2)(\Delta G^\circ_f\ of\ O_2)]$

$\Delta G^\circ_{rxn} = [(2\ mol\ CO)(-137.2\ kJ/mol)] - [(2\ mol\ C)(0\ kJ/mol) + (1\ mol\ O_2)(0\ kJ/mol)]$

$\Delta G^\circ_{rxn} = \textbf{--274.4 kJ}$

20.6A Plan: Predict the sign of ΔS°_{rxn} by comparing the randomness of the products with the randomness of the reactants. Use the relationship $\Delta G^\circ_{rxn} = \Delta H^\circ_{rxn} - T\Delta S^\circ_{rxn}$ to answer b).
Solution:

a) The reaction is $X_2Y_2(g) \rightarrow X_2(g) + Y_2(g)$. Since there are more moles of gaseous product than there are of gaseous reactant, entropy increases and $\Delta S > 0$.

b) The reaction is only spontaneous above 325°C or in other words, at high temperatures. In the relationship $\Delta G^\circ_{rxn} = \Delta H^\circ_{rxn} - T\Delta S^\circ_{rxn}$, when $\Delta S > 0$ so that $-T\Delta S^\circ$ is < 0, ΔG° will only be negative at high T if $\Delta H^\circ > 0$.

20.6B Plan: Predict the sign of ΔS°_{rxn} by comparing the randomness of the products with the randomness of the reactants. Use the relationship $\Delta G^\circ_{rxn} = \Delta H^\circ_{rxn} - T\Delta S^\circ_{rxn}$ to answer b).
Solution:

a) A solid forms a gas and a liquid, so $\Delta S > 0$. A crystalline array breaks down, so $\Delta H > 0$.

b) For the reaction to occur spontaneously ($\Delta G < 0$), $-T\Delta S°$ must be greater than ΔH, which would occur only at higher T.

20.7A   Plan: Use the equation $\Delta G° = \Delta H° - T\Delta S°$ to determine if the reaction is spontaneous ($\Delta G° < 0$). Then examine the same equation to determine the effect of raising the temperature on the spontaneity of the reaction.
Solution:
a) $\Delta G°_{rxn} = \Delta H°_{rxn} - T\Delta S°_{rxn} = -192.7$ kJ $- [(298$ K$)(-308.2$ J/K$)(1$ kJ/10^3 J$)] = -100.8564 = -100.9$ kJ
Because $\Delta G < 0$, the reaction is **spontaneous** at 298 K.
b) As temperature increases, $-T\Delta S°$ becomes more positive, so the reaction becomes **less spontaneous** at higher temperatures.
c) $\Delta G°_{rxn} = \Delta H°_{rxn} - T\Delta S°_{rxn} = -192.7$ kJ $- [(773$ K$)(-308.2$ J/K$)(1$ kJ/10^3 J$)] = 45.5386 = $ **45.5 kJ**

20.7B   Plan: Examine the equation $\Delta G° = \Delta H° - T\Delta S°$ and determine which combination of enthalpy and entropy will describe the given reaction.
Solution:
Two choices can already be eliminated:
1) When $\Delta H > 0$ (endothermic reaction) and $\Delta S < 0$ (entropy decreases), the reaction is always nonspontaneous, regardless of temperature, so this combination does not describe the reaction.
2) When $\Delta H < 0$ (exothermic reaction) and $\Delta S > 0$ (entropy increases), the reaction is always spontaneous, regardless of temperature, so this combination does not describe the reaction.
Two combinations remain: 3) $\Delta H° > 0$ and $\Delta S° > 0$, or 4) $\Delta H° < 0$ and $\Delta S° < 0$. If the reaction becomes spontaneous at $-40°C$, this means that $\Delta G°$ becomes negative at lower temperatures. Case 3) becomes spontaneous at higher temperatures, when the $-T\Delta S°$ term is larger than the positive enthalpy term. By process of elimination, Case 4) describes the reaction. At a lower temperature, the negative $\Delta H°$ becomes larger than the positive $(-T\Delta S°)$ value, so $\Delta G°$ becomes negative.

20.8A   Plan: To find the temperature at which the reaction becomes spontaneous, use
$\Delta G°_{rxn} = 0 = \Delta H°_{rxn} - T\Delta S°_{rxn}$ and solve for temperature.
Solution:
The reaction will become spontaneous when ΔG changes from being positive to being negative. This point occurs when ΔG is 0.
$\Delta G°_{rxn} = \Delta H°_{rxn} - T\Delta S°_{rxn}$
$0 = -192.7 \times 10^3$ J $- (T)(-308.2$ J/K$)$
$T = 625.2$ K $- 273.15 = $ **352.0°C**

20.8B   Plan: To find the temperature at which the reaction becomes spontaneous, use
$\Delta G°_{rxn} = 0 = \Delta H°_{rxn} - T\Delta S°_{rxn}$ and solve for temperature. $\Delta H°_{rxn}$ can be calculated from the individual $\Delta H°_f$ values of the reactants and products by using the relationship
$\Delta H°_{rxn} = \sum m \Delta H°_{f\,(products)} - \sum n \Delta H°_{f\,(reactants)}$. $\Delta S°_{rxn}$ can be calculated from the individual $S°$ values of the reactants and products by using the relationship $\Delta S°_{rxn} = \sum m S°_{products} - \sum n S°_{reactants}$.
Solution:
$CaO(s) + CO_2(g) \rightarrow CaCO_3(s)$
$\Delta H°_{rxn} = \sum m \Delta H°_{f\,(products)} - \sum n \Delta H°_{f\,(reactants)}$
$\Delta H°_{rxn} = [(1$ mol $CaCO_3)(\Delta H°_f$ of $CaCO_3)]$
$\qquad\qquad - [(1$ mol $CaO)(\Delta H°_f$ of $CaO) + (1$ mol $CO_2)(\Delta H°_f$ of $CO_2)]$
$\Delta H°_{rxn} = [(1$ mol $CaCO_3)(-1206.9$ kJ/mol$)]$
$\qquad\qquad - [(1$ mol $CaO)(-635.1$ kJ/mol$) + (1$ mol $CO_2)(-393.5$ kJ/mol$)]$
$\Delta H°_{rxn} = -178.3$ kJ
$\Delta S°_{rxn} = \sum m S°_{products} - \sum n S°_{reactants}$

$\Delta S_{rxn}^{\circ} = [(1 \text{ mol } CaCO_3)(S^{\circ} \text{ of } CaCO_3)] - [(1 \text{ mol } CaO)(S^{\circ} \text{ of } CaO) + (1 \text{ mol } CO_2)(S^{\circ} \text{ of } CO_2)]$

$\Delta S_{rxn}^{\circ} = [(1 \text{mol } CaCO_3)(92.9 \text{ J/mol•K})]$

$\qquad\qquad\qquad - [(1 \text{ mol } CaO)(38.2 \text{ J/mol•K}) + (1 \text{ mol } CO_2)(213.7 \text{ J/mol•K})]$

$\Delta S_{rxn}^{\circ} = -159.0 \text{ J/K} = -0.159 \text{ kJ/K}$

$\Delta G_{rxn}^{\circ} = 0 = \Delta H_{rxn}^{\circ} - T\Delta S_{rxn}^{\circ}$

$\Delta H_{rxn}^{\circ} = T\Delta S_{rxn}^{\circ}$

$T = \dfrac{\Delta H^{\circ}}{\Delta S^{\circ}} = \dfrac{-178.3 \text{ kJ}}{-0.1590 \text{ kJ/K}} = 1121.384 = 1121 \text{ K}$

The reaction becomes spontaneous at temperatures < **1121 K**.

20.9A Plan: First find ΔG°, then calculate K from $\Delta G^{\circ} = -RT \ln K$. Calculate ΔG° using ΔG_f° values in the relationship $\Delta G_{rxn}^{\circ} = \sum m \Delta G_{f\,(products)}^{\circ} - \sum n \Delta G_{f\,(reactants)}^{\circ}$.
Solution:

$2C(graphite) + O_2(g) \leftrightarrows 2CO(g)$

$\Delta G_{rxn}^{\circ} = \sum m \Delta G_{f\,(products)}^{\circ} - \sum n \Delta G_{f\,(reactants)}^{\circ}$

$\Delta G_{rxn}^{\circ} = [(2 \text{ mol } CO)(-137.2 \text{ kJ/mol})] - [(2 \text{ mol } C)(0 \text{ kJ/mol}) + (1 \text{ mol } O_2)(0 \text{ kJ/mol})] = -274.4 \text{ kJ}$

$\ln K = -\dfrac{\Delta G^{\circ}}{RT} = -\dfrac{-274.4 \text{ kJ/mol}}{(8.314 \text{ J/mol•K})(298 \text{ K})}\left(\dfrac{1000 \text{ J}}{1 \text{ kJ}}\right) = 110.7536 = 111$

$K = e^{111} = 1.6095 \times 10^{48} = \mathbf{1.6 \times 10^{48}}$

20.9B Plan: The equilibrium constant, K, is related to ΔG° through the equation $\Delta G^{\circ} = -RT \ln K$.
Solution:

$\Delta G^{\circ} = -RT \ln K = -(8.314 \text{ J/mol•K})(298 \text{ K}) \ln (2.22 \times 10^{-15}) = 8.35964 \times 10^4 \text{ J/mol} = \mathbf{83.6 \text{ kJ/mol}}$

20.10A Plan: Write the equilibrium expression for the reaction and calculate Q_c for each scene. Remember that each particle represents 0.10 mol and that the volume is 1.0 L. A reaction that is proceeding to the right will have $\Delta G^{\circ} < 0$ and a reaction that is proceeding to the left will have $\Delta G^{\circ} > 0$. A reaction at equilibrium has $\Delta G^{\circ} = 0$.
Solution:

a) $A(g) + 3B(g) \leftrightarrows AB_3(g)$

b) $Q_c = \dfrac{[AB_3]}{[A][B]^3}$

Mixture 1: $Q_c = \dfrac{[AB_3]}{[A][B]^3} = \dfrac{[0.20]}{[0.40][0.80]^3} = 0.98$

Mixture 2: $Q_c = \dfrac{[AB_3]}{[A][B]^3} = \dfrac{[0.30]}{[0.30][0.50]^3} = 8$

Mixture 3: $Q_c = \dfrac{[AB_3]}{[A][B]^3} = \dfrac{[0.40]}{[0.20][0.20]^3} = 250$

Mixture 2 is at equilibrium since $Q_c = K_c$

c) $Q_c < K_c$ for Mixture 1 and the reaction is proceeding right to reach equilibrium; thus $\Delta G^{\circ} < 0$. Mixture 2 is at equilibrium and $\Delta G^{\circ} = 0$. Mixture 3 proceeds to the left to reach equilibrium since $Q_c > K_c$ and $\Delta G^{\circ} > 0$. The ranking for most positive to most negative is **3 > 2 > 1.**

20.10B Plan: Write the equilibrium expression for the reaction and calculate Q_c for each scene. A reaction that is proceeding to the right will have $\Delta G^{\circ} < 0$ and a reaction that is proceeding to the left will have $\Delta G^{\circ} > 0$. A reaction at equilibrium has $\Delta G^{\circ} = 0$.

Solution:

a) $X_2(g) + 2Y_2(g) \leftrightharpoons 2XY_2(g)$

$$Q_c = \frac{[XY_2]^2}{[X_2][Y_2]^2}$$

Mixture 1: $Q_c = \dfrac{[XY_2]^2}{[X_2][Y_2]^2} = \dfrac{[5]^2}{[2][1]^2} = 12.5$

Mixture 2: $Q_c = \dfrac{[XY_2]^2}{[X_2][Y_2]^2} = \dfrac{[4]^2}{[2][2]^2} = 2$

Mixture 3: $Q_c = \dfrac{[XY_2]^2}{[X_2][Y_2]^2} = \dfrac{[2]^2}{[4][2]^2} = 0.25$

Mixture 2 is at equilibrium since $Q_c = K_c$.

b) $Q_c > K_c$ for Mixture 1 and the reaction is proceeding left to reach equilibrium; thus $\Delta G° > 0$. Mixture 2 is at equilibrium and $\Delta G° = 0$. Mixture 3 proceeds to the right to reach equilibrium since $Q_c < K_c$ and $\Delta G° < 0$. The ranking for most negative to most positive is **3 < 2 < 1.**

c) Any reaction mixture moves spontaneously towards equilibrium so both changes have a negative $\Delta G°$.

20.11A Plan: The equilibrium constant, K, is related to $\Delta G°$ through the equation $\Delta G° = -RT \ln K$. The free energy of the reaction under non-standard state conditions is calculated using $\Delta G = \Delta G° + RT \ln Q$.

Solution:

a) $\ln K = -\dfrac{\Delta G°}{RT} = -\dfrac{-33.5 \text{ kJ/mol}}{(8.314 \text{ J/mol} \bullet \text{K})(298 \text{ K})}\left(\dfrac{1000 \text{ J}}{1 \text{ kJ}}\right) = 13.5213$

$K = e^{13.5213} = 7.4512 \times 10^5 = \mathbf{7.45 \times 10^5}$

b) $Q = \dfrac{[C_2H_5Cl]}{[C_2H_4][HCl]} = \dfrac{(1.5)}{(0.50)(1.0)} = 3.0$

$\Delta G = \Delta G° + RT \ln Q = (-33.5 \text{ kJ/mol}) + (8.314 \text{ J/mol}\bullet\text{K}) (1 \text{ kJ}/1000 \text{ J}) (298 \text{ K}) \ln (3.0)$

$= -30.7781 = \mathbf{-30.8 \text{ kJ/mol}}$

20.11B Plan: Write a balanced equation for the dissociation of hypobromous acid in water. The free energy of the reaction at standard state is calculated using $\Delta G° = -RT \ln K$. The free energy of the reaction under non-standard state conditions is calculated using $\Delta G = \Delta G° + RT \ln Q$.

Solution:

$HBrO(aq) + H_2O(l) \leftrightharpoons BrO^-(aq) + H_3O^+(aq)$

a) $\Delta G° = -RT \ln K = -(8.314 \text{ J/mol}\bullet\text{K})(298)\ln (2.3 \times 10^{-9}) = 4.927979 \times 10^4 = \mathbf{4.9 \times 10^4 \text{ J/mol} = 49 \text{ kJ/mol}}$

b) $Q = \dfrac{[H_3O^+][BrO^-]}{[HBrO]} = \dfrac{[6.0 \times 10^{-4}][0.10]}{[0.20]}$

$\Delta G = \Delta G° + RT \ln Q = (4.927979 \times 10^4 \text{ J/mol}) + (8.314 \text{ J/mol}\bullet\text{K}) (298 \text{ K}) \ln \dfrac{[6.0 \times 10^{-4}][0.10]}{[0.20]}$

$= 2.9182399 \times 10^4 = \mathbf{2.9 \times 10^4 \text{ J/mol} = 29 \text{ kJ/mol}}$

The value of K_a is very small, so it makes sense that $\Delta G°$ is a positive number. The natural log of a negative exponent gives a negative number ($\ln 3.0 \times 10^{-4}$), so the value of ΔG decreases with concentrations lower than the standard state 1 M values.

CHEMICAL CONNECTIONS BOXED READING PROBLEMS

B20.1   Plan: Convert mass of glucose (1 g) to moles and use the ratio between moles of glucose and moles of ATP to find the moles and then molecules of ATP formed. Do the same calculation with tristearin.
Solution:
a) Molecules of ATP/g glucose =

$$(1 \text{ g glucose})\left(\frac{1 \text{ mol glucose}}{180.16 \text{ g glucose}}\right)\left(\frac{36 \text{ mol ATP}}{1 \text{ mol glucose}}\right)\left(\frac{6.022\times10^{23} \text{ molecules ATP}}{1 \text{ mol ATP}}\right)$$

$$= 1.20333\times10^{23} = \mathbf{1.203\times10^{23} \text{ molecules ATP/g glucose}}$$

b) Molecules of ATP/g tristearin =

$$(1 \text{ g tristearin})\left(\frac{1 \text{ mol tristearin}}{897.50 \text{ g tristearin}}\right)\left(\frac{458 \text{ mol ATP}}{1 \text{ mol tristearin}}\right)\left(\frac{6.022\times10^{23} \text{ molecules ATP}}{1 \text{ mol ATP}}\right)$$

$$= 3.073065\times10^{23} = \mathbf{3.073\times10^{23} \text{ molecules ATP/g tristearin}}$$

B20.2   Plan: Add the two reactions to obtain the overall process; the values of the two reactions are then added to obtain for the overall reaction.
Solution:

$$\text{creatine phosphate} \rightarrow \text{creatine} + \text{\sout{phosphate}} \qquad \Delta G° = -43.1 \text{ kJ/mol}$$
$$\underline{\text{ADP} + \text{\sout{phosphate}} \rightarrow \text{ATP}} \qquad\qquad \Delta G° = +30.5 \text{ kJ/mol}$$
$$\text{creatine phosphate} + \text{ADP} \rightarrow \text{creatine} + \text{ATP}$$
$$\Delta G = -43.1 \text{ kJ/mol} + 30.5 \text{ kJ/mol} = \mathbf{-12.6 \text{ kJ/mol}}$$

END–OF–CHAPTER PROBLEMS

20.2   A spontaneous process occurs by itself (possibly requiring an initial input of energy), whereas a nonspontaneous process requires a continuous supply of energy to make it happen. It is possible to cause a nonspontaneous process to occur, but the process stops once the energy source is removed. A reaction that is found to be nonspontaneous under one set of conditions may be spontaneous under a different set of conditions (different temperature, different concentrations).

20.5   Vaporization is the change of a liquid substance to a gas so $\Delta S_{vaporization} = S_{gas} - S_{liquid}$. Fusion is the change of a solid substance into a liquid so $\Delta S_{fusion} = S_{liquid} - S_{solid}$. Vaporization involves a greater change in volume than fusion. Thus, the transition from liquid to gas involves a greater entropy change than the transition from solid to liquid.

20.6   In an exothermic process, the *system* releases heat to its *surroundings*. The entropy of the surroundings increases because the temperature of the surroundings increases ($\Delta S_{surr} > 0$). In an endothermic process, the system absorbs heat from the surroundings and the surroundings become cooler. Thus, the entropy of the surroundings decreases ($\Delta S_{surr} < 0$). A chemical cold pack for injuries is an example of a spontaneous, endothermic chemical reaction as is the melting of ice cream at room temperature.

20.8   Plan: A spontaneous process is one that occurs by itself without a continuous input of energy.
Solution:
a) **Spontaneous,** evaporation occurs because a few of the liquid molecules have enough energy to break away from the intermolecular forces of the other liquid molecules and move spontaneously into the gas phase.
b) **Spontaneous,** a lion spontaneously chases an antelope without added force. This assumes that the lion has not just eaten.
c) **Spontaneous,** an unstable substance decays spontaneously to a more stable substance.

20.10   Plan: A spontaneous process is one that occurs by itself without a continuous input of energy.
Solution:
a) **Spontaneous,** with a small amount of energy input, methane will continue to burn without additional energy (the reaction itself provides the necessary energy) until it is used up.

b) **Spontaneous**, the dissolved sugar molecules have more states they can occupy than the crystalline sugar, so the reaction proceeds in the direction of dissolution.

c) **Not spontaneous,** a cooked egg will not become raw again, no matter how long it sits or how many times it is mixed.

20.12 <u>Plan:</u> Particles with more freedom of motion have higher entropy. Therefore, $S_{gas} > S_{liquid} > S_{solid}$. If the products of the process have more entropy than the reactants, ΔS_{sys} is positive. If the products of the process have less entropy than the reactants, ΔS_{sys} is negative.

<u>Solution:</u>

a) ΔS_{sys} **positive,** melting is the change in state from solid to liquid. The solid state of a particular substance always has lower entropy than the same substance in the liquid state. Entropy increases during melting.

b) ΔS_{sys} **negative,** the entropy of most salt solutions is greater than the entropy of the solvent and solute separately, so entropy decreases as a salt precipitates.

c) ΔS_{sys} **negative,** dew forms by the condensation of water vapor to liquid. Entropy of a substance in the gaseous state is greater than its entropy in the liquid state. Entropy decreases during condensation.

20.14 <u>Plan:</u> Particles with more freedom of motion have higher entropy. Therefore, $S_{gas} > S_{liquid} > S_{solid}$. If the products of the process have more entropy than the reactants, ΔS_{sys} is positive. If the products of the process have less entropy than the reactants, ΔS_{sys} is negative.

<u>Solution:</u>

a) ΔS_{sys} **positive,** the process described is liquid alcohol becoming gaseous alcohol. The gas molecules have greater entropy than the liquid molecules.

b) ΔS_{sys} **positive,** the process described is a change from solid to gas, an increase in possible energy states for the system.

c) ΔS_{sys} **positive,** the perfume molecules have more possible locations in the larger volume of the room than inside the bottle. A system that has more possible arrangements has greater entropy.

20.16 <u>Plan:</u> ΔS_{sys} is the entropy of the products – the entropy of the reactants. Use the fact that $S_{gas} > S_{liquid} > S_{solid}$; also, the greater the number of particles of a particular phase of matter, the higher the entropy.

<u>Solution:</u>

a) ΔS_{sys} **negative,** reaction involves a gaseous reactant and no gaseous products, so entropy decreases. The number of particles also decreases, indicating a decrease in entropy.

b) ΔS_{sys} **negative,** gaseous reactants form solid product and number of particles decreases, so entropy decreases.

c) ΔS_{sys} **positive,** when a solid salt dissolves in water, entropy generally increases since the entropy of the aqueous mixture has higher entropy than the solid.

20.18 <u>Plan:</u> ΔS_{sys} is the entropy of the products – the entropy of the reactants. Use the fact that $S_{gas} > S_{liquid} > S_{solid}$; also, the greater the number of particles of a particular phase of matter, the higher the entropy.

<u>Solution:</u>

a) ΔS_{sys} **positive,** the reaction produces gaseous CO_2 molecules that have greater entropy than the physical states of the reactants.

b) ΔS_{sys} **negative,** the reaction produces a net decrease in the number of gaseous molecules, so the system's entropy decreases.

c) ΔS_{sys} **positive,** the reaction produces a gas from a solid.

20.20 <u>Plan:</u> Particles with more freedom of motion have higher entropy. In general the entropy of gases is greater than that of liquids, and the entropy of liquids is greater than that of solids. Entropy increases with temperature. For substances in the same phase, entropy increases with atomic size and molecular complexity. If the entropy of the products is greater than that of the reactants, ΔS is positive.

<u>Solution:</u>

a) ΔS_{sys} **positive,** decreasing the pressure increases the volume available to the gas molecules so entropy of the system increases.

b) ΔS_{sys} **negative,** gaseous nitrogen molecules have greater entropy (more possible states) than dissolved nitrogen molecules.

c) ΔS_{sys} **positive,** dissolved oxygen molecules have lower entropy than gaseous oxygen molecules.

20.22 Plan: Particles with more freedom of motion have higher entropy. In general the entropy of gases is greater than that of liquids, and the entropy of liquids is greater than that of solids. Entropy increases with temperature. For substances in the same phase, entropy increases with atomic size and molecular complexity.
Solution:
a) **Butane** has the greater molar entropy because it has two additional C—H bonds that can vibrate and has greater rotational freedom around its bond. The presence of the double bond in 2-butene restricts rotation.
b) **Xe(g)** has the greater molar entropy because entropy increases with atomic size.
c) **CH$_4$(g)** has the greater molar entropy because gases in general have greater entropy than liquids.

20.24 Plan: Particles with more freedom of motion have higher entropy. In general the entropy of gases is greater than that of liquids, and the entropy of liquids is greater than that of solids. Entropy increases with temperature. For substances in the same phase, entropy increases with atomic size and molecular complexity.
Solution:
a) Ethanol, **C$_2$H$_5$OH(l)**, is a more complex molecule than methanol, CH$_3$OH, and has the greater molar entropy.
b) When a salt dissolves, there is an increase in the number of possible states for the ions. Thus, **KClO$_3$(aq)** has the greater molar entropy.
c) **K(s)** has greater molar entropy because K(s) has greater mass than Na(s).

20.26 Plan: Particles with more freedom of motion have higher entropy. In general the entropy of gases is greater than that of liquids, and the entropy of liquids is greater than that of solids. Entropy increases with temperature. For substances in the same phase, entropy increases with atomic size and molecular complexity.
Solution:
a) **Diamond < graphite < charcoal**. Diamond has an ordered, three-dimensional crystalline shape, followed by graphite with an ordered two-dimensional structure, followed by the amorphous (disordered) structure of charcoal.
b) **Ice < liquid water < water vapor**. Entropy increases as a substance changes from solid to liquid to gas.
c) **O atoms < O$_2$ < O$_3$**. Entropy increases with molecular complexity because there are more modes of movement (e.g., bond vibration) available to the complex molecules.

20.28 Plan: Particles with more freedom of motion have higher entropy. In general the entropy of gases is greater than that of liquids, and the entropy of liquids is greater than that of solids. Entropy increases with temperature. For substances in the same phase, entropy increases with atomic size and molecular complexity.
Solution:
a) **ClO$_4^-$(aq) > ClO$_3^-$(aq) > ClO$_2^-$(aq)**. The decreasing order of molar entropy follows the order of decreasing molecular complexity.
b) **NO$_2$(g) > NO(g) > N$_2$(g)**. N$_2$ has lower molar entropy than NO because N$_2$ consists of two of the same atoms while NO consists of two different atoms. NO$_2$ has greater molar entropy than NO because NO$_2$ consists of three atoms while NO consists of only two.
c) **Fe$_3$O$_4$(s) > Fe$_2$O$_3$(s) > Al$_2$O$_3$(s)**. Fe$_3$O$_4$ has greater molar entropy than Fe$_2$O$_3$ because Fe$_3$O$_4$ is more complex and more massive. Fe$_2$O$_3$ and Al$_2$O$_3$ contain the same number of atoms but Fe$_2$O$_3$ has greater molar entropy because iron atoms are more massive than aluminum atoms.

20.31 A system at equilibrium does not spontaneously produce more products or more reactants. For either reaction direction, the entropy change of the system is exactly offset by the entropy change of the surroundings. Therefore, for system at equilibrium, $\Delta S_{univ} = \Delta S_{sys} + \Delta S_{surr} = 0$. However, for a system moving to equilibrium, $\Delta S_{univ} > 0$, because the second law states that for any spontaneous process, the entropy of the universe increases.

20.32 Plan: Since entropy is a state function, the entropy changes can be found by summing the entropies of the products and subtracting the sum of the entropies of the reactants.
Solution:

ΔS_{rxn}° = [(2 mol HClO)(S° of HClO)] – [(1 mol H$_2$O)(S° of H$_2$O) + (1 mol Cl$_2$O)(S° of Cl$_2$O)]

Rearranging this expression to solve for S° of Cl$_2$O gives: S° of Cl$_2$O = 2(S° of HClO) – S° of H$_2$O – ΔS_{rxn}°

20.33 Plan: To calculate the standard entropy change, use the relationship $\Delta S^{\circ}_{rxn} = \sum m S^{\circ}_{products} - \sum n S^{\circ}_{reactants}$.

To predict the sign of entropy recall that in general $S_{gas} > S_{liquid} > S_{solid}$, and entropy increases as the number of particles of a particular phase of matter increases, and with increasing atomic size and molecular complexity.

Solution:

a) Prediction: ΔS° **negative** because number of moles of (Δn) gas decreases.

$\Delta S^{\circ} = [(1 \text{ mol } N_2O)(S^{\circ} \text{ of } N_2O) + (1 \text{ mol } NO_2)(S^{\circ} \text{ of } NO_2)] - [(3 \text{ mol } NO)(S^{\circ} \text{ of } NO)]$

$\Delta S^{\circ} = [(1 \text{ mol})(219.7 \text{ J/mol•K}) + (1 \text{ mol})(239.9 \text{ J/mol•K})] - [(3 \text{ mol})(210.65 \text{ J/mol•K})]$

$\Delta S^{\circ} = -172.35 = \mathbf{-172.4 \text{ J/K}}$

b) Prediction: Sign difficult to predict because $\Delta n = 0$, but **possibly ΔS° positive** because water vapor has greater complexity than H_2 gas.

$\Delta S^{\circ} = [(2 \text{ mol Fe})(S^{\circ} \text{ of Fe}) + (3 \text{ mol } H_2O)(S^{\circ} \text{ of } H_2O)] - [(3 \text{ mol } H_2)(S^{\circ} \text{ of } H_2) + (1 \text{ mol } Fe_2O_3)(S^{\circ} \text{ of } Fe_2O_3)]$

$\Delta S^{\circ} = [(2 \text{ mol})(27.3 \text{ J/mol•K}) + (3 \text{ mol})(188.72 \text{ J/mol•K})] - [(3 \text{ mol})(130.6 \text{ J/mol•K}) + (1 \text{ mol})(87.400 \text{ J/mol•K})]$

$\Delta S^{\circ} = 141.56 = \mathbf{141.6 \text{ J/K}}$

c) Prediction: ΔS° **negative** because a gaseous reactant forms a solid product and also because the number of moles of gas (Δn) decreases.

$\Delta S^{\circ} = [(1 \text{ mol } P_4O_{10})(S^{\circ} \text{ of } P_4O_{10})] - [(1 \text{ mol } P_4)(S^{\circ} \text{ of } P_4) + (5 \text{ mol } O_2)(S^{\circ} \text{ of O})]$

$\Delta S^{\circ} = [(1 \text{ mol})(229 \text{ J/mol•K})] - [(1 \text{ mol})(41.1 \text{ J/mol•K}) + (5 \text{ mol})(205.0 \text{ J/mol•K})]$

$\Delta S^{\circ} = -837.1 = \mathbf{-837 \text{ J/K}}$

20.35 Plan: Write the balanced equation. To calculate the standard entropy change, use the relationship $\Delta S^{\circ}_{rxn} = \sum m S^{\circ}_{products} - \sum n S^{\circ}_{reactants}$. To predict the sign of entropy recall that in general $S_{gas} > S_{liquid} > S_{solid}$, entropy increases as the number of particles of a particular phase of matter increases, and entropy increases with increasing atomic size and molecular complexity.

Solution:

The balanced combustion reaction is:

$2C_2H_6(g) + 7O_2(g) \rightarrow 4CO_2(g) + 6H_2O(g)$

$\Delta S^{\circ} = [(4 \text{ mol } CO_2)(S^{\circ} \text{ of } CO_2) + (6 \text{ mol } H_2O)(S^{\circ} \text{ of } H_2O)] - [(2 \text{ mol } C_2H_6)(S^{\circ} \text{ of } C_2H_6) + (7 \text{ mol } O_2)(S^{\circ} \text{ of } O_2)]$

$\Delta S^{\circ} = [(4 \text{ mol})(213.7 \text{ J/mol•K}) + (6 \text{ mol})(188.72 \text{ J/mol•K})] - [(2 \text{ mol})(229.5 \text{ J/mol•K}) + (7 \text{ mol})(205.0 \text{ J/mol•K})]$

$\Delta S^{\circ} = 93.12 = \mathbf{93.1 \text{ J/K}}$

The entropy value is not per mole of C_2H_6 but per two moles. Divide the calculated value by two to obtain entropy per mole of C_2H_6.

Yes, the positive sign of ΔS° is expected because there is a net increase in the number of gas molecules from nine moles as reactants to ten moles as products.

20.37 Plan: Write the balanced equation. To calculate the standard entropy change, use the relationship $\Delta S^{\circ}_{rxn} = \sum m S^{\circ}_{products} - \sum n S^{\circ}_{reactants}$. To predict the sign of entropy recall that in general $S_{gas} > S_{liquid} > S_{solid}$, entropy increases as the number of particles of a particular phase of matter increases, and entropy increases with increasing atomic size and molecular complexity.

Solution:

The balanced chemical equation for the described reaction is:

$2NO(g) + 5H_2(g) \rightarrow 2NH_3(g) + 2H_2O(g)$

Because the number of moles of gas decreases, i.e., $\Delta n = 4 - 7 = -3$, the entropy is expected to decrease.

$\Delta S^{\circ} = [(2 \text{ mol } NH_3)(S^{\circ} \text{ of } NH_3) + (2 \text{ mol } H_2O)(S^{\circ} \text{ of } H_2O)] - [(2 \text{ mol } NO)(S^{\circ} \text{ of } NO) + (5 \text{ mol } H_2)(S^{\circ} \text{ of } H_2)]$

$\Delta S^{\circ} = [(2 \text{ mol})(193 \text{ J/mol•K}) + (2 \text{ mol})(188.72 \text{ J/mol•K})] - [(2 \text{ mol})(210.65 \text{ J/mol•K}) + (5 \text{ mol})(130.6 \text{ J/mol•K})]$

$\Delta S^{\circ} = -310.86 = \mathbf{-311 \text{ J/K}}$

Yes, the calculated entropy matches the predicted decrease.

20.39 Plan: Write the balanced equation. To calculate the standard entropy change (part a), use the relationship

$\Delta S^{\circ}_{rxn} = \sum m S^{\circ}_{products} - \sum n S^{\circ}_{reactants}$.

To calculate ΔS° of the universe in part b, first calculate ΔH° of the reaction using the relationship

$\Delta H^{\circ}_{rxn} = \sum m \Delta H^{\circ}_{f(products)} - \sum n \Delta H^{\circ}_{f(reactants)}$

Use ΔH° of the reaction to calculate ΔS of the surroundings.

$$\Delta S_{surr} = -\frac{\Delta H^{\circ}_{rxn}}{T}$$

Add ΔS of the surroundings and ΔS° of the reaction to calculate ΔS of the universe. If ΔS of the universe is greater than zero, the reaction is spontaneous at the given temperature.

$$\Delta S_{univ} = \Delta S^{\circ}_{rxn} + \Delta S_{surr}$$

Solution:

a) The reaction for forming Cu_2O from copper metal and oxygen gas is:

$$2Cu(s) + 1/2 O_2(g) \rightarrow Cu_2O(s)$$

$\Delta S^{\circ} = [(1 \text{ mol } Cu_2O)(S^{\circ} \text{ of } Cu_2O)] - [(2 \text{ mol } Cu)(S^{\circ} \text{ of } Cu) + (1/2 \text{ mol } O_2)(S^{\circ} \text{ of } O_2)]$

$\Delta S^{\circ} = [(1 \text{ mol})(93.1 \text{ J/mol} \cdot K)] - [(2 \text{ mol})(33.1 \text{ J/mol} \cdot K) + (1/2 \text{ mol})(205.0 \text{ J/mol} \cdot K)]$

$\Delta S^{\circ} = -75.6 \text{ J/K}$

b) $\quad \Delta H^{\circ}_{rxn} = \sum m \Delta H^{\circ}_{f(products)} - \sum n \Delta H^{\circ}_{f(reactants)}$

$\Delta H^{\circ}_{rxn} = [(1 \text{ mol } Cu_2O)(\Delta H^{\circ}_f \text{ of } Cu_2O)]$

$\qquad\qquad - [(2 \text{ mol } Cu)(\Delta H^{\circ}_f \text{ of } Cu) + (1/2 \text{ mol } O_2)(\Delta H^{\circ}_f \text{ of } O_2)]$

$\Delta H^{\circ}_{rxn} = [(1 \text{ mol } Cu_2O)(-168.6 \text{ kJ/mol})]$

$\qquad\qquad - [(2 \text{ mol } Cu)(0 \text{ kJ/mol}) + (1/2 \text{ mol } O_2)(0 \text{ kJ/mol})]$

$\Delta H^{\circ}_{rxn} = -168.6 \text{ kJ}$

$\Delta S_{surr} = -\dfrac{\Delta H^{\circ}_{rxn}}{T} = -\dfrac{-168.6 \text{ kJ}}{298 \text{ K}} = 0.56577 \text{ kJ/K}(10^3 \text{ J/1 kJ}) = 565.77 \text{ J/K}$

$\Delta S_{univ} = \Delta S^{\circ}_{rxn} + \Delta S_{surr} = (-75.6 \text{ J/K}) + (565.77 \text{ J/K}) = 490.17 = \textbf{490. J/K}$

Because ΔS_{univ} is positive, the reaction is spontaneous at 298 K.

20.41 Plan: Write the balanced equation. To calculate the standard entropy change (part a), use the relationship

$$\Delta S^{\circ}_{rxn} = \sum m S^{\circ}_{products} - \sum n S^{\circ}_{reactants}.$$

To calculate ΔS° of the universe in part b, first calculate ΔH° of the reaction using the relationship

$$\Delta H^{\circ}_{rxn} = \sum m \Delta H^{\circ}_{f(products)} - \sum n \Delta H^{\circ}_{f(reactants)}$$

Use ΔH° of the reaction to calculate ΔS of the surroundings.

$$\Delta S_{surr} = -\frac{\Delta H^{\circ}_{rxn}}{T}$$

Add ΔS of the surroundings and ΔS° of the reaction to calculate ΔS of the universe. If ΔS of the universe is greater than zero, the reaction is spontaneous at the given temperature.

$$\Delta S_{univ} = \Delta S^{\circ}_{rxn} + \Delta S_{surr}$$

Solution:

a) One mole of methanol is formed from its elements in their standard states according to the following equation:

$C(g) + 2H_2(g) + 1/2 O_2(g) \rightarrow CH_3OH(l)$

$\Delta S^{\circ} = [(1 \text{ mol } CH_3OH)(S^{\circ} \text{ of } CH_3OH)] - [(1 \text{ mol } C)(S^{\circ} \text{ of } C) + (2 \text{ mol } H_2)(S^{\circ} \text{ of } H_2) + (1/2 \text{ mol } O_2)(S^{\circ} \text{ of } O_2)]$

$\Delta S^{\circ} = [(1 \text{ mol})(127 \text{ J/mol} \cdot K)] - [(1 \text{ mol})(5.686 \text{ J/mol} \cdot K) + (2 \text{ mol})(130.6 \text{ J/mol} \cdot K) + (1/2 \text{ mol})(205.0 \text{ J/mol} \cdot K)]$

$\Delta S^{\circ} = -242.386 = -242 \text{ J/K}$

b) $\quad \Delta H^{\circ}_{rxn} = \sum m \Delta H^{\circ}_{f(products)} - \sum n \Delta H^{\circ}_{f(reactants)}$

$\Delta H^{\circ}_{rxn} = [(1 \text{ mol } CH_3OH)(\Delta H^{\circ}_f \text{ of } CH_3OH)]$

$\qquad\qquad - [(1 \text{ mol } C)(\Delta H^{\circ}_f \text{ of } C) + (2 \text{ mol } H_2)(\Delta H^{\circ}_f \text{ of } H_2) + (1/2 \text{ mol } O_2)(\Delta H^{\circ}_f \text{ of } O_2)]$

$\Delta H^{\circ}_{rxn} = [(1 \text{ mol } CH_3OH)(-238.6 \text{ kJ/mol})]$

$\qquad\qquad - [(1 \text{ mol } C)(0 \text{ kJ/mol}) + (2 \text{ mol } H_2)(0 \text{ kJ/mol}) + (1/2 \text{ mol } O_2)(0 \text{ kJ/mol})]$

$\Delta H^{\circ}_{rxn} = -238.6 \text{ kJ}$

$$\Delta S_{surr} = -\frac{\Delta H^{\circ}_{rxn}}{T} = -\frac{-238.6 \text{ kJ}}{298 \text{ K}} = 0.80067 \text{ kJ/K}(10^3 \text{ J/1 kJ}) = 800.67 \text{ J/K}$$

$\Delta S_{univ} = \Delta S^{\circ}_{rxn} + \Delta S_{surr} = (-242.386 \text{ J/K}) + (800.67 \text{ J/K}) = 558.284 = \textbf{558 J/K}$

Because ΔS_{univ} is positive, the reaction is spontaneous at 298 K.

20.44 Plan: Write the balanced equation. To calculate the standard entropy change, use the relationship

$\Delta S^{\circ}_{rxn} = \sum m\, S^{\circ}_{products} - \sum n\, S^{\circ}_{reactants}$.

Solution:

Complete combustion of a hydrocarbon includes oxygen as a reactant and carbon dioxide and water as the products.

$C_2H_2(g) + 5/2 O_2(g) \rightarrow 2CO_2(g) + H_2O(g)$

$\Delta S^{\circ} = [(2 \text{ mol } CO_2)(S^{\circ} \text{ of } CO_2) + (1 \text{ mol } H_2O)(S^{\circ} \text{ of } H_2O)] - [(1 \text{ mol } C_2H_2)(S^{\circ} \text{ of } C_2H_2) + (5/2 \text{ mol } O_2)(S^{\circ} \text{ of } O_2)]$

$\Delta S^{\circ} = [(2 \text{ mol})(213.7 \text{ J/mol·K}) + (1 \text{ mol})(188.72 \text{ J/mol·K})]$
$\qquad - [(1 \text{ mol})(200.85 \text{ J/mol·K}) + (5/2 \text{ mol})(205.0 \text{ J/mol·K})]$

$\Delta S^{\circ} = -97.23 = \textbf{-97.2 J/K}$

20.46 A spontaneous process has $\Delta S_{univ} > 0$. Since the Kelvin temperature is always positive, ΔG_{sys} must be negative ($\Delta G_{sys} < 0$) for a spontaneous process.

20.48 Plan: Examine the provided diagrams. Determine whether bonds are being formed or broken in order to determine the sign of ΔH. Determine the relative number of particles before and after the reaction in order to determine the sign of ΔS.

Solution:

a) **The sign of ΔH is negative.** Bonds are being formed, so energy is released and ΔH is negative.

b) **The sign of ΔS is negative.** There are fewer particles in the system after the reaction, so entropy decreases.

c) **The sign of ΔS_{surr} is positive.** $\Delta S_{surr} = -\frac{\Delta H^{\circ}_{rxn}}{T}$ Since ΔH_{rxn} is negative, ΔS_{surr} will be positive.

d) Because both ΔH and ΔS are negative, **this reaction will become more spontaneous (ΔG will become more negative) as temperature decreases.**

20.49 **ΔH° is positive and ΔS° is positive.** The reaction is endothermic ($\Delta H^{\circ} > 0$) and requires a lot of heat from its surroundings to be spontaneous. The removal of heat from the surroundings results in $\Delta S^{\circ} < 0$. The only way an endothermic reaction can proceed spontaneously is if $\Delta S^{\circ} > 0$, effectively offsetting the decrease in surroundings entropy. In summary, the values of ΔH° and ΔS° are both positive for this reaction. Melting is an example.

20.50 For a given substance, the entropy changes greatly from one phase to another, e.g., from liquid to gas. However, the entropy changes little within a phase. As long as the substance does not change phase, the value of ΔS° is relatively unaffected by temperature.

20.51 Plan: ΔG° can be calculated with the relationship $\sum m\, \Delta G^{\circ}_{f\,(products)} - \sum n\, \Delta G^{\circ}_{f\,(reactants)}$.

Solution:

a) $\Delta G^{\circ} = [(2 \text{ mol MgO})(\Delta G^{\circ}_f \text{ of MgO})] - [(2 \text{ mol Mg})(\Delta G^{\circ}_f \text{ of Mg}) + (1 \text{ mol } O_2)(\Delta G^{\circ}_f \text{ of } O_2)]$

Both $Mg(s)$ and $O_2(g)$ are the standard-state forms of their respective elements, so their ΔG°_f values are zero.

$\Delta G^{\circ} = [(2 \text{ mol})(-569.0 \text{ kJ/mol})] - [(2 \text{ mol})(0) + (1 \text{ mol})(0)] = \textbf{-1138.0 kJ}$

b) $\Delta G^{\circ} = [(2 \text{ mol } CO_2)(\Delta G^{\circ}_f \text{ of } CO_2) + (4 \text{ mol } H_2O)(\Delta G^{\circ}_f \text{ of } H_2O)]$
$\qquad - [(2 \text{ mol } CH_3OH)(\Delta G^{\circ}_f \text{ of } CH_3OH) + (3 \text{ mol } O_2)(\Delta G^{\circ}_f \text{ of } O_2)]$

$\Delta G^{\circ} = [(2 \text{ mol})(-394.4 \text{ kJ/mol}) + (4 \text{ mol})(-228.60 \text{ kJ/mol})] - [(2 \text{ mol})(-161.9 \text{ kJ/mol}) + (3 \text{ mol})(0)]$

$\Delta G^{\circ} = \textbf{-1379.4 kJ}$

c) $\Delta G° = [(1 \text{ mol } BaCO_3)(\Delta G_f° \text{ of } BaCO_3)] - [(1 \text{ mol } BaO)(\Delta G_f° \text{ of } BaO) + (1 \text{ mol } CO_2)(\Delta G_f° \text{ of } CO_2)]$

$\Delta G° = [(1 \text{ mol})(-1139 \text{ kJ/mol})] - [(1 \text{ mol})(-520.4 \text{ kJ/mol}) + (1 \text{ mol})(-394.4 \text{ kJ/mol})]$

$\Delta G° = -224.2 = \mathbf{-224 \text{ kJ}}$

20.53 Plan: $\Delta H_{rxn}°$ can be calculated from the individual $\Delta H_f°$ values of the reactants and products by using the relationship $\Delta H_{rxn}° = \sum m \Delta H_{f \text{(products)}}° - \sum n \Delta H_{f \text{(reactants)}}°$. $\Delta S_{rxn}°$ can be calculated from the individual $S°$ values of the reactants and products by using the relationship $\Delta S_{rxn}° = \sum m S_{products}° - \sum n S_{reactants}°$. Once $\Delta H_{rxn}°$ and $\Delta S_{rxn}°$ are known, $\Delta G°$ can be calculated with the relationship $\Delta G_{rxn}° = \Delta H_{rxn}° - T\Delta S_{rxn}°$. $\Delta S_{rxn}°$ values in J/K must be converted to units of kJ/K to match the units of $\Delta H_{rxn}°$.

Solution:

a) $\Delta H_{rxn}° = [(2 \text{ mol } MgO)(\Delta H_f° \text{ of } MgO)] - [(2 \text{ mol } Mg)(\Delta H_f° \text{ of } Mg) + (1 \text{ mol } O_2)(\Delta H_f° \text{ of } O_2)]$

$\Delta H_{rxn}° = [(2 \text{ mol})(-601.2 \text{ kJ/mol})] - [(2 \text{ mol})(0 \text{ kJ/mol}) + (1 \text{ mol})(0 \text{ kJ/mol})]$

$\Delta H_{rxn}° = -1202.4 \text{ kJ}$

$\Delta S_{rxn}° = [(2 \text{ mol } MgO)(S° \text{ of } MgO)] - [(2 \text{ mol } Mg)(S° \text{ of } Mg) + (1 \text{ mol } O_2)(S° \text{ of } O_2)]$

$\Delta S_{rxn}° = [(2 \text{ mol})(26.9 \text{ J/mol•K})] - [(2 \text{ mol})(32.69 \text{ J/mol•K}) + (1 \text{ mol})(205.0 \text{ J/mol•K})]$

$\Delta S_{rxn}° = -216.58 \text{ J/K}$

$\Delta G_{rxn}° = \Delta H_{rxn}° - T\Delta S_{rxn}° = -1202.4 \text{ kJ} - [(298 \text{ K})(-216.58 \text{ J/K})(1 \text{ kJ}/10^3 \text{ J})] = -1137.859 = \mathbf{-1138 \text{ kJ}}$

b) $\Delta H_{rxn}° = [(2 \text{ mol } CO_2)(\Delta H_f° \text{ of } CO_2) + (4 \text{ mol } H_2O)(\Delta H_f° \text{ of } H_2O)]$
$- [(2 \text{ mol } CH_3OH)(\Delta H_f° \text{ of } CH_3OH) + (3 \text{ mol } O_2)(\Delta H_f° \text{ of } O_2)]$

$\Delta H_{rxn}° = [(2 \text{ mol})(-393.5 \text{ kJ/mol}) + (4 \text{ mol})(-241.826 \text{ kJ/mol})] - [(2 \text{ mol})(-201.2 \text{ kJ/mol}) + (3 \text{ mol})(0 \text{ kJ/mol})]$

$\Delta H_{rxn}° = -1351.904 \text{ kJ}$

$\Delta S_{rxn}° = [(2 \text{ mol } CO_2)(S° \text{ of } CO_2) + (4 \text{ mol } H_2O)(S° \text{ of } H_2O)]$
$- [(2 \text{ mol } CH_3OH)(S° \text{ of } CH_3OH) + (3 \text{ mol } O_2)(S° \text{ of } O_2)]$

$\Delta S_{rxn}° = [(2 \text{ mol})(213.7 \text{ J/mol•K}) + (4 \text{ mol})(188.72 \text{ J/mol•K})]$
$- [(2 \text{ mol})(238 \text{ J/mol•K}) + (3 \text{ mol})(205.0 \text{ J/mol•K})] = 91.28 \text{ J/K}$

$\Delta G_{rxn}° = \Delta H_{rxn}° - T\Delta S_{rxn}° = -1351.904 \text{ kJ} - [(298 \text{ K})(91.28 \text{ J/K})(1 \text{ kJ}/10^3 \text{ J})] = -1379.105 = \mathbf{-1379 \text{ kJ}}$

c) $\Delta H_{rxn}° = [(1 \text{ mol } BaCO_3)(\Delta H_f° \text{ of } BaCO_3)] - [(1 \text{ mol } BaO)(\Delta H_f° \text{ of } BaO) + (1 \text{ mol } CO_2)(\Delta H_f° \text{ of } CO_2)]$

$\Delta H_{rxn}° = [(1 \text{ mol})(-1219 \text{ kJ/mol})] - [(1 \text{ mol})(-548.1 \text{ kJ/mol}) + (1 \text{ mol})(-393.5 \text{ kJ/mol})]$

$\Delta H_{rxn}° = -277.4 \text{ kJ}$

$\Delta S_{rxn}° = [(1 \text{ mol } BaCO_3)(S° \text{ of } BaCO_3)] - [(1 \text{ mol } BaO)(S° \text{ of } BaO) + (1 \text{ mol } CO_2)(S° \text{ of } CO_2)]$

$\Delta S_{rxn}° = [(1 \text{ mol})(112 \text{ J/mol•K})] - [(1 \text{ mol})(72.07 \text{ J/mol•K}) + (1 \text{ mol})(213.7 \text{ J/mol•K})]$

$\Delta S_{rxn}° = -173.77 \text{ J/K}$

$\Delta G_{rxn}° = \Delta H_{rxn}° - T\Delta S_{rxn}° = -277.4 \text{ kJ} - [(298 \text{ K})(-173.77 \text{ J/K})(1 \text{ kJ}/10^3 \text{ J})] = -225.6265 = \mathbf{-226 \text{ kJ}}$

20.55 Plan: $\Delta G_{rxn}°$ can be calculated with the relationship $\sum m \Delta G_{f \text{(products)}}° - \sum n \Delta G_{f \text{(reactants)}}°$. Alternatively, $\Delta G_{rxn}°$ can be calculated with the relationship $\Delta G_{rxn}° = \Delta H_{rxn}° - T\Delta S_{rxn}°$. Entropy decreases (is negative) when there are fewer moles of gaseous products than there are of gaseous reactants.

Solution:

a) Entropy decreases ($\mathbf{\Delta S°}$ **negative**) because the number of moles of gas decreases from reactants (1 1/2 mol) to products (1 mole). The oxidation (combustion) of CO requires initial energy input to start the reaction, but then releases energy (exothermic, $\mathbf{\Delta H°}$ **negative**) which is typical of all combustion reactions.

b) Method 1: Calculate $\Delta G_{rxn}°$ from $\Delta G_f°$ values of products and reactants.

$\Delta G_{rxn}° = \sum m \Delta G_{f \text{(products)}}° - \sum n \Delta G_{f \text{(reactants)}}°$

$\Delta G_{rxn}° = [(1 \text{ mol } CO_2)(\Delta G_f° \text{ of } CO_2)] - [(1 \text{ mol } CO)(\Delta G_f° \text{ of } CO) + (1/2 \text{ mol})(\Delta G_f° \text{ of } O_2)]$

$\Delta G_{rxn}° = [(1 \text{ mol})(-394.4 \text{ kJ/mol})] - [(1 \text{ mol})(-137.2 \text{ kJ/mol}) + (1/2 \text{ mol})(0 \text{ kJ/mol})] = \mathbf{-257.2 \text{ kJ}}$

Method 2: Calculate ΔG_{rxn}° from ΔH_{rxn}° and ΔS_{rxn}° at 298 K (the degree superscript indicates a reaction at standard state, given in the Appendix at 25°C).

$\Delta H_{rxn}^{\circ} = \sum m \Delta H_{f(products)}^{\circ} - \sum n \Delta H_{f(reactants)}^{\circ}$

$\Delta H_{rxn}^{\circ} = [(1 \text{ mol } CO_2)(\Delta H_f^{\circ} \text{ of } CO_2)] - [(1 \text{ mol } CO)(\Delta H_f^{\circ} \text{ of } CO) + (1/2 \text{ mol})(\Delta H_f^{\circ} \text{ of } O_2)]$

$\Delta H_{rxn}^{\circ} = [(1 \text{ mol})(-393.5 \text{ kJ/mol})] - [(1 \text{ mol})(-110.5 \text{ kJ/mol}) + (1/2 \text{ mol})(0 \text{ kJ/mol})] = -283.0 \text{ kJ}$

$\Delta S_{rxn}^{\circ} = \sum m S_{products}^{\circ} - \sum n S_{reactants}^{\circ}$

$\Delta S_{rxn}^{\circ} = [(1 \text{ mol } CO_2)(S^{\circ} \text{ of } CO_2)] - [(1 \text{ mol } CO)(S^{\circ} \text{ of } CO) + (1/2 \text{ mol})(S^{\circ} \text{ of } O_2)]$

$\Delta S_{rxn}^{\circ} = [(1 \text{mol})(213.7 \text{ J/mol•K})] - [(1 \text{mol})(197.5 \text{ J/mol•K}) + (1/2 \text{ mol})(205.0 \text{ J/mol•K})]$

$\Delta S_{rxn}^{\circ} = -86.3 \text{ J/K}$

$\Delta G_{rxn}^{\circ} = \Delta H_{rxn}^{\circ} - T\Delta S_{rxn}^{\circ} = (-283.0 \text{ kJ}) - [(298 \text{ K})(-86.3 \text{ J/K})(1 \text{ kJ}/10^3 \text{ J})] = -257.2826 = \textbf{-257.3 kJ}$

20.57 Plan: Use the relationship $\Delta G_{rxn}^{\circ} = \Delta H_{rxn}^{\circ} - T\Delta S_{rxn}^{\circ}$ to find ΔS_{rxn}°, knowing ΔH_{rxn}° and ΔG_{rxn}°. This relationship is also used to find ΔG_{rxn}° at a different temperature.
Solution:
Reaction is $Xe(g) + 3F_2(g) \rightarrow XeF_6(g)$
a) $\Delta G_{rxn}^{\circ} = \Delta H_{rxn}^{\circ} - T\Delta S_{rxn}^{\circ}$

$\Delta S^{\circ} = \dfrac{\Delta H^{\circ} - \Delta G^{\circ}}{T} = \dfrac{-402 \text{ kJ/mol} - (-280. \text{ kJ/mol})}{298 \text{ K}} = -0.40939597 = \textbf{-0.409 kJ/mol•K}$

b) $\Delta G_{rxn}^{\circ} = \Delta H_{rxn}^{\circ} - T\Delta S_{rxn}^{\circ} = (-402 \text{ kJ/mol}) - [(500. \text{ K})(-0.40939597 \text{ kJ/mol•K})] = -197.302 = \textbf{-197 kJ/mol}$

20.59 Plan: ΔH_{rxn}° can be calculated from the individual ΔH_f° values of the reactants and products by using the relationship $\Delta H_{rxn}^{\circ} = \sum m \Delta H_{f(products)}^{\circ} - \sum n \Delta H_{f(reactants)}^{\circ}$. ΔS_{rxn}° can be calculated from the individual S° values of the reactants and products by using the relationship $\Delta S_{rxn}^{\circ} = \sum m S_{products}^{\circ} - \sum n S_{reactants}^{\circ}$. Once ΔH_{rxn}° and ΔS_{rxn}° are known, ΔG° can be calculated with the relationship $\Delta G_{rxn}^{\circ} = \Delta H_{rxn}^{\circ} - T\Delta S_{rxn}^{\circ}$. ΔS_{rxn}° values in J/K must be converted to units of kJ/K to match the units of ΔH_{rxn}°. The temperature at which a reaction becomes spontaneous can be calculated by setting ΔG to zero in the free energy equation (assuming the reaction is at equilibrium) and solving for T.
Solution:
a) $\Delta H_{rxn}^{\circ} = [(1 \text{ mol } CO)(\Delta H_f^{\circ} \text{ of } CO) + (2 \text{ mol } H_2)(\Delta H_f^{\circ} \text{ of } H_2)] - [(1 \text{ mol } CH_3OH)(\Delta H_f^{\circ} \text{ of } CH_3OH)]$

$\Delta H_{rxn}^{\circ} = [(1 \text{ mol})(-110.5 \text{ kJ/mol}) + (2 \text{ mol})(0 \text{ kJ/mol})] - [(1 \text{ mol})(-201.2 \text{ kJ/mol})]$

$\Delta H_{rxn}^{\circ} = \textbf{90.7 kJ}$

$\Delta S_{rxn}^{\circ} = [(1 \text{ mol } CO)(S^{\circ} \text{ of } CO) + (2 \text{ mol } H_2)(S^{\circ} \text{ of } H_2)] - [(1 \text{ mol } CH_3OH)(S^{\circ} \text{ of } CH_3OH)]$

$\Delta S_{rxn}^{\circ} = [(1 \text{ mol})(197.5 \text{ J/mol•K}) + (2 \text{ mol})(130.6 \text{ J/mol•K})] - [(1 \text{ mol})(238 \text{ J/mol•K})]$

$\Delta S_{rxn}^{\circ} = 220.7 = \textbf{221 J/K}$

b) $\Delta G_{rxn}^{\circ} = \Delta H_{rxn}^{\circ} - T\Delta S_{rxn}^{\circ}$

$T_1 = 28 + 273 = 301 \text{ K} \quad \Delta G^{\circ} = 90.7 \text{ kJ} - [(301 \text{ K})(220.7 \text{ J/K})(1 \text{ kJ}/10^3 \text{ J})] = 24.2693 = \textbf{24.3 kJ}$
$T_2 = 128 + 273 = 401 \text{ K} \quad \Delta G^{\circ} = 90.7 \text{ kJ} - [(401 \text{ K})(220.7 \text{ J/K})(1 \text{ kJ}/10^3 \text{ J})] = 2.1993 = \textbf{2.2 kJ}$
$T_3 = 228 + 273 = 501 \text{ K} \quad \Delta G^{\circ} = 90.7 \text{ kJ} - [(501 \text{ K})(220.7 \text{ J/K})(1 \text{ kJ}/10^3 \text{ J})] = -19.8707 = \textbf{-19.9 kJ}$

c) For the substances in their standard states, the reaction is nonspontaneous at 28°C, near equilibrium at 128°C, and spontaneous at 228°C. Reactions with positive values of ΔH_{rxn}° and ΔS_{rxn}° become spontaneous at high temperatures.

d) The reaction will become spontaneous when ΔG changes from being positive to being negative. This point occurs when ΔG is 0.

$\Delta G^{\circ}_{rxn} = \Delta H^{\circ}_{rxn} - T\Delta S^{\circ}_{rxn}$

$0 = 90.7 \times 10^3 \text{ J} - (T)(220.7 \text{ J/K})$

$T = 411 \text{ K}$

At temperatures above 411 K, this reaction is spontaneous. (Because both ΔH and ΔS are positive, the reaction becomes spontaneous above this temperature.)

20.61 Plan: ΔH°_{rxn} can be calculated from the individual ΔH°_f values of the reactants and products by using the relationship $\Delta H^{\circ}_{rxn} = \sum m \Delta H^{\circ}_{f(products)} - \sum n \Delta H^{\circ}_{f(reactants)}$. ΔS°_{rxn} can be calculated from the individual S° values of the reactants and products by using the relationship $\Delta S^{\circ}_{rxn} = \sum m S^{\circ}_{products} - \sum n S^{\circ}_{reactants}$. Once ΔH°_{rxn} and ΔS°_{rxn} are known, ΔG° can be calculated with the relationship $\Delta G^{\circ}_{rxn} = \Delta H^{\circ}_{rxn} - T\Delta S^{\circ}_{rxn}$. ΔS°_{rxn} values in J/K must be converted to units of kJ/K to match the units of ΔH°_{rxn}. To find the temperature at which the reaction becomes spontaneous, use $\Delta G^{\circ}_{rxn} = 0 = \Delta H^{\circ}_{rxn} - T\Delta S^{\circ}_{rxn}$ and solve for temperature.
Solution:

a) The reaction for this process is $H_2(g) + 1/2O_2(g) \rightarrow H_2O(g)$. The coefficients are written this way (instead of $2H_2(g) + O_2(g) \rightarrow 2H_2O(g)$) because the problem specifies thermodynamic values "per (1) mol H_2," not per 2 mol H_2.

$\Delta H^{\circ}_{rxn} = \sum m \Delta H^{\circ}_{f(products)} - \sum n \Delta H^{\circ}_{f(reactants)}$

$\Delta H^{\circ}_{rxn} = [(1 \text{ mol } H_2O)(\Delta H^{\circ}_f \text{ of } H_2O)] - [(1 \text{ mol } H_2)(\Delta H^{\circ}_f \text{ of } H_2) + (1/2 \text{ mol } O_2)(\Delta H^{\circ}_f \text{ of } O_2)]$

$\Delta H^{\circ}_{rxn} = [(1 \text{ mol } H_2O)(-241.826 \text{ kJ/mol})] - [(1 \text{ mol } H_2)(0 \text{ kJ/mol}) + (1/2 \text{ mol } O_2)(0 \text{ kJ/mol})]$

$\Delta H^{\circ}_{rxn} = \textbf{-241.826 kJ}$

$\Delta S^{\circ}_{rxn} = \sum m S^{\circ}_{products} - \sum n S^{\circ}_{reactants}$

$\Delta S^{\circ}_{rxn} = [(1 \text{ mol } H_2O)(S^{\circ} \text{ of } H_2O)] - [(1 \text{ mol } H_2)(S^{\circ} \text{ of } H_2) + (1/2 \text{ mol } O_2)(S^{\circ} \text{ of } O_2)]$

$\Delta S^{\circ}_{rxn} = [(1 \text{ mol})(188.72 \text{ J/mol·K})] - [(1 \text{ mol})(130.6 \text{ J/mol·K}) + (1/2 \text{ mol})(205.0 \text{ J/mol·K})]$

$\Delta S^{\circ}_{rxn} = -44.38 = \textbf{-44.4 J/K = -0.0444 kJ/K}$

$\Delta G^{\circ}_{rxn} = \Delta H^{\circ}_{rxn} - T\Delta S^{\circ}_{rxn}$

$\Delta G^{\circ}_{rxn} = -241.826 \text{ kJ} - [(298 \text{ K})(-0.04438 \text{ kJ/K})]$

$\Delta G^{\circ}_{rxn} = -228.6008 \text{ kJ} = \textbf{-228.6 kJ}$

b) Because $\Delta H < 0$ and $\Delta S < 0$, the reaction will become nonspontaneous at higher temperatures because the positive $(-T\Delta S)$ term becomes larger than the negative ΔH term.

c) The reaction becomes spontaneous below the temperature where $\Delta G^{\circ}_{rxn} = 0$

$\Delta G^{\circ}_{rxn} = 0 = \Delta H^{\circ}_{rxn} - T\Delta S^{\circ}_{rxn}$

$\Delta H^{\circ}_{rxn} = T\Delta S^{\circ}_{rxn}$

$T = \dfrac{\Delta H^{\circ}}{\Delta S^{\circ}} = \dfrac{-241.826 \text{ kJ}}{-0.04438 \text{ kJ/K}} = 5448.986 = \textbf{5.45x10}^3 \textbf{ K}$

20.63 a) An equilibrium constant that is much less than one indicates that very little product is made to reach equilibrium. The reaction, thus, is not spontaneous in the forward direction and ΔG° is a relatively large positive value.
b) A large negative ΔG° indicates that the reaction is quite spontaneous and goes almost to completion. At equilibrium, much more product is present than reactant so $K > 1$. Q depends on initial conditions, not equilibrium conditions, so its value cannot be predicted from ΔG°.

20.66 The standard free energy change, ΔG°, occurs when all components of the system are in their standard states. Standard state is defined as 1 atm for gases, 1 M for solutes, and pure solids and liquids. Standard state does not specify a temperature because standard state can occur at any temperature. $\Delta G^{\circ} = \Delta G$ when all concentrations equal 1 M and all partial pressures equal 1 atm. This occurs because the value of $Q = 1$ and $\ln Q = 0$ in the equation $\Delta G = \Delta G^{\circ} + RT \ln Q$.

20.67   **Plan:** For each reaction, first find $\Delta G°$, then calculate K from $\Delta G° = -RT \ln K$. Calculate $\Delta G°$ using $\Delta G_f°$ values in the relationship $\Delta G_{rxn}° = \sum m \Delta G_{f\,(products)}° - \sum n \Delta G_{f\,(reactants)}°$.
Solution:
a) $MgCO_3(s) \leftrightarrows Mg^{2+}(aq) + CO_3^{2-}(aq)$

$\Delta G_{rxn}° = \sum m \Delta G_{f\,(products)}° - \sum n \Delta G_{f\,(reactants)}°$

$\Delta G_{rxn}° = [(1\ mol\ Mg^{2+})(-456.01\ kJ/mol) + (1\ mol\ CO_3^{2-})(-528.10\ kJ/mol)]$
$\qquad\qquad - [(1\ mol\ MgCO_3)(-1028\ kJ/mol)] = 43.89\ kJ$

$$\ln K = \frac{\Delta G°}{-RT} = \left(\frac{43.89\ kJ/mol}{-(8.314\ J/mol \bullet K)(298\ K)} \right) \left(\frac{10^3\ J}{1\ kJ} \right) = -17.7149$$

$K = e^{-17.7149} = 2.0254274 \times 10^{-8} = \mathbf{2.0 \times 10^{-8}}$

b) $H_2(g) + O_2(g) \leftrightarrows H_2O_2(l)$

$\Delta G_{rxn}° = [(1\ mol\ H_2O_2)(-120.4\ kJ/mol)] - [(1\ mol\ H_2)(0\ kJ/mol) + (1\ mol\ O_2)(0\ kJ/mol)] = -120.4\ kJ/mol$

$$\ln K = \frac{\Delta G°}{-RT} = \left(\frac{-120.4\ kJ/mol}{-(8.314\ J/mol \bullet K)(298\ K)} \right) \left(\frac{10^3\ J}{1\ kJ} \right) = 48.59596$$

$K = e^{48.59596} = 1.2733777 \times 10^{21} = \mathbf{1.27 \times 10^{21}}$

20.69   **Plan:** For each reaction, first find $\Delta G°$, then calculate K from $\Delta G° = -RT \ln K$. Calculate $\Delta G°$ using $\Delta G_f°$ values in the relationship $\Delta G_{rxn}° = \sum m \Delta G_{f\,(products)}° - \sum n \Delta G_{f\,(reactants)}°$.
Solution:
a) $\Delta G_{rxn}° = [(1\ mol\ NaCN)(\Delta G_f°\ of\ NaCN) + (1\ mol\ H_2O)(\Delta G_f°\ of\ H_2O)]$
$\qquad\qquad - [(1\ mol\ HCN)(\Delta G_f°\ of\ HCN) + (1\ mol\ NaOH)(\Delta G_f°\ of\ NaOH)]$

$NaCN(aq)$ and $NaOH(aq)$ are not listed in Appendix B.

Converting the equation to net ionic form will simplify the problem:

$\qquad\qquad HCN(aq) + OH^-(aq) \leftrightarrows CN^-(aq) + H_2O(l)$

$\Delta G_{rxn}° = [(1\ mol\ CN^-)(\Delta G_f°\ of\ CN^-) + (1\ mol\ H_2O)(\Delta G_f°\ of\ H_2O)]$
$\qquad\qquad - [(1\ mol\ HCN)(\Delta G_f°\ of\ HCN) + (1\ mol\ OH^-)(\Delta G_f°\ of\ OH^-)]$

$\Delta G_{rxn}° = [(1\ mol)(166\ kJ/mol) + (1\ mol)(-237.192\ kJ/mol)]$
$\qquad\qquad - [(1\ mol)(112\ kJ/mol) + (1\ mol)(-157.30\ kJ/mol)]$

$\Delta G_{rxn}° = -25.892\ kJ$

$$\ln K = \frac{\Delta G°}{-RT} = \left(\frac{-25.892\ kJ/mol}{-(8.314\ J/mol \bullet K)(298\ K)} \right) \left(\frac{10^3\ J}{1\ kJ} \right) = 10.45055$$

$K = e^{10.45055} = 3.4563 \times 10^4 = \mathbf{3.46 \times 10^4}$

b) $SrSO_4(s) \leftrightarrows Sr^{2+}(aq) + SO_4^{2-}(aq)$

$\Delta G_{rxn}° = [(1\ mol\ Sr^{2+})(-557.3\ kJ/mol) + (1\ mol\ SO_4^{2-})(-741.99\ kJ/mol)]$
$\qquad\qquad - [(1\ mol\ SrSO_4)(-1334\ kJ/mol)] = 34.71\ kJ$

$$\ln K = \frac{\Delta G°}{-RT} = \left(\frac{34.71\ kJ/mol}{-(8.314\ J/mol \bullet K)(298\ K)} \right) \left(\frac{10^3\ J}{1\ kJ} \right) = -14.00968$$

$K = e^{-14.00968} = 8.23518 \times 10^{-7} = \mathbf{8.2 \times 10^{-7}}$

20.71 Plan: At the normal boiling point, defined as the temperature at which the vapor pressure of the liquid equals 1 atm, the phase change from liquid to gas is at equilibrium. For a system at equilibrium, the change in Gibbs free energy is zero. Since the gas is at 1 atm and the liquid assumed to be pure, the system is at standard state and $\Delta G° = 0$. The temperature at which this occurs can be found from $\Delta G°_{rxn} = 0 = \Delta H°_{rxn} - T\Delta S°_{rxn}$. $\Delta H°_{rxn}$ can be calculated from the individual $\Delta H°_f$ values of the reactants and products by using the relationship $\Delta H°_{rxn} = \sum m\, \Delta H°_{f(products)} - \sum n\, \Delta H°_{f(reactants)}$. $\Delta S°_{rxn}$ can be calculated from the individual $S°$ values of the reactants and products by using the relationship $\Delta S°_{rxn} = \sum m\, S°_{products} - \sum n\, S°_{reactants}$.

Solution:

$Br_2(l) \leftrightarrows Br_2(g)$

$\Delta H°_{rxn} = \sum m\, \Delta H°_{f(products)} - \sum n\, \Delta H°_{f(reactants)}$

$\Delta H°_{rxn} = [(1\ mol\ Br_2)(\Delta H°_f\ of\ Br_2(g))] - [(1\ mol\ Br_2)(\Delta H°_f\ of\ Br_2(l))]$

$\Delta H°_{rxn} = [(1\ mol)(30.91\ kJ/mol)] - [(1\ mol)(0\ kJ/mol)] = 30.91\ kJ$

$\Delta S°_{rxn} = \sum m\, S°_{products} - \sum n\, S°_{reactants}$

$\Delta S°_{rxn} = [(1\ mol\ Br_2)(S°\ of\ Br_2(g))] - [(1\ mol\ Br_2)(S°\ of\ Br_2(l))]$

$\Delta S°_{rxn} = [(1\ mol)(245.38\ J/K{\cdot}mol)] - [(1\ mol)(152.23\ J/K{\cdot}mol)] = 93.15\ J/K = 0.09315\ kJ/K$

$\Delta G°_{rxn} = 0 = \Delta H°_{rxn} - T\Delta S°_{rxn}$

$\Delta H°_{rxn} = T\Delta S°_{rxn}$

$T = \dfrac{\Delta H°}{\Delta S°} = \dfrac{30.91\ kJ}{0.09315\ kJ/K} = 331.830 = \textbf{331.8 K}$

20.73 Plan: Write the balanced equation. First find $\Delta G°$, then calculate K from $\Delta G° = -RT \ln K$. Calculate $\Delta G°$ using $\Delta G°_f$ values in the relationship $\Delta G°_{rxn} = \sum m\, \Delta G°_{f(products)} - \sum n\, \Delta G°_{f(reactants)}$.

Solution:

The solubility reaction for Ag_2S is

$\qquad Ag_2S(s) \leftrightarrows 2Ag^+(aq) + S^{2-}(aq)$

$\Delta G°_{rxn} = \sum m\, \Delta G°_{f(products)} - \sum n\, \Delta G°_{f(reactants)}$

$\Delta G°_{rxn} = [(2\ mol\ Ag^+)(\Delta G°_f\ of\ Ag^+) + (1\ mol\ S^{2-})(\Delta G°_f\ of\ S^{2-})] - [(1\ mol\ Ag_2S)(\Delta G°_f\ of\ Ag_2S)]$

$\Delta G°_{rxn} = [(2\ mol)(77.111\ kJ/mol) + (1\ mol)(83.7\ kJ/mol)] - [(1\ mol)(-40.3\ kJ/mol)]$

$\Delta G°_{rxn} = 278.222\ kJ$

$\ln K = \dfrac{\Delta G°}{-RT} = \left(\dfrac{278.222\ kJ/mol}{-(8.314\ J/mol \bullet K)(298\ K)}\right)\left(\dfrac{10^3\ J}{1\ kJ}\right) = -112.296232$

$K = e^{-112.296232} = 1.6996759 \times 10^{-49} = \textbf{1.70} \times \textbf{10}^{-49}$

20.75 Plan: First find $\Delta G°$, then calculate K from $\Delta G° = -RT \ln K$. Calculate $\Delta G°$ using $\Delta G°_f$ values in the relationship $\Delta G°_{rxn} = \sum m\, \Delta G°_{f(products)} - \sum n\, \Delta G°_{f(reactants)}$. Recognize that $I_2(s)$, not $I_2(g)$, is the standard state for iodine.

Solution:

$\Delta G°_{rxn} = \sum m\, \Delta G°_{f(products)} - \sum n\, \Delta G°_{f(reactants)}$

$\Delta G°_{rxn} = [(2\ mol\ ICl)(\Delta G°_f\ of\ ICl)] - [(1\ mol\ I_2)(\Delta G°_f\ of\ I_2) + (1\ mol\ Cl_2)(\Delta G°_f\ of\ Cl_2)]$

$\Delta G°_{rxn} = [(2\ mol)(-6.075\ kJ/mol)] - [(1\ mol)(19.38\ kJ/mol) + (1\ mol)(0\ kJ/mol)]$

$\Delta G°_{rxn} = -31.53\ kJ$

$\ln K_p = \dfrac{\Delta G°}{-RT} = \left(\dfrac{-31.53\ kJ/mol}{-(8.314\ J/mol \bullet K)(298\ K)}\right)\left(\dfrac{10^3\ J}{1\ kJ}\right) = 12.726169$

$K_p = e^{12.726169} = 3.3643794 \times 10^5 = \textbf{3.36} \times \textbf{10}^5$

20.77 Plan: The equilibrium constant, K, is related to $\Delta G°$ through the equation $\Delta G° = -RT \ln K$.
Solution:
$\Delta G° = -RT \ln K = -(8.314 \text{ J/mol} \cdot \text{K})(298 \text{ K}) \ln (1.7 \times 10^{-5}) = 2.72094 \times 10^4 \text{ J/mol} = \mathbf{2.7 \times 10^4 \text{ J/mol}}$
The large positive $\Delta G°$ indicates that it would not be possible to prepare a solution with the concentrations of lead and chloride ions at the standard-state concentration of 1 M. A Q calculation using 1 M solutions will confirm this:

$$PbCl_2(s) \leftrightarrows Pb^{2+}(aq) + 2Cl^-(aq)$$
$$Q = [Pb^{2+}][Cl^-]^2$$
$$= (1 \, M)(1 \, M)^2$$
$$= 1$$

Since $Q > K_{sp}$, it is impossible to prepare a standard-state solution of $Pb^{2+}(aq)$ and $Cl^-(aq)$.

20.79 Plan: The equilibrium constant, K, is related to $\Delta G°$ through the equation $\Delta G° = -RT \ln K$. ΔG is found by using the relationship $\Delta G = \Delta G° + RT \ln Q$.
Solution:
a) $\Delta G° = -RT \ln K = -(8.314 \text{ J/mol} \cdot \text{K})(298 \text{ K}) \ln (9.1 \times 10^{-6}) = 2.875776 \times 10^4 = \mathbf{2.9 \times 10^4 \text{ J/mol}}$
b) Since $\Delta G°_{rxn}$ is positive, the reaction direction as written is nonspontaneous. The reverse direction, formation of reactants, is spontaneous, so the reaction proceeds to the left.
c) Calculate the value for Q and then use it to find ΔG.

$$Q = \frac{\left[Fe^{2+}\right]^2 \left[Hg^{2+}\right]^2}{\left[Fe^{3+}\right]^2 \left[Hg_2^{2+}\right]} = \frac{[0.010]^2 [0.025]^2}{[0.20]^2 [0.010]} = 1.5625 \times 10^{-4}$$

$\Delta G = \Delta G° + RT \ln Q = 2.875776 \times 10^4 \text{ J/mol} + (8.314 \text{ J/mol} \cdot \text{K})(298) \ln (1.5625 \times 10^{-4})$

$= 7.044187 \times 10^3 = \mathbf{7.0 \times 10^3 \text{ J/mol}}$

Because $\Delta G_{298} > 0$ and $Q > K$, the reaction proceeds to the left to reach equilibrium.

20.81 Plan: Start by writing the balanced equation for the reaction. Use the balanced equation to write the reaction quotient expression for the reaction. For part a, count the number of each type of particle present in the individual scenes. Using the given information that each particle represents 0.10 mol and the volume of the container is 0.10 L, calculate the molarity of each species in each individual scene. Use these concentrations and the reaction quotient expression to calculate the value of the reaction quotient for each scene to determine which scene is at equilibrium. For part b, the scene that is at equilibrium has $\Delta G = 0$. Use the values of Q and the properties of logarithms to determine the relative values of ΔG for the other scenes.
Solution:
a) The balanced equation for the reaction is:

$$2A(g) \leftrightarrows A_2(g)$$

The corresponding reaction quotient expression is:

$$Q = \frac{[A_2]}{[A]^2}$$

	Number of A Particles	Number of A$_2$ Particles
Scene 1	3	3
Scene 2	5	2
Scene 3	1	4

Scene 1:

$$\text{Molarity of A:} \quad \frac{(3 \text{ particles})\left(\dfrac{0.10 \text{ mol}}{\text{particle}}\right)}{0.10 \text{ L}} = 3.0 \text{ M}$$

$$\text{Molarity of A}_2: \quad \frac{(3 \text{ particles})\left(\dfrac{0.10 \text{ mol}}{\text{particle}}\right)}{0.10 \text{ L}} = 3.0 \text{ M}$$

$$Q = \frac{(3.0 \text{ M})}{(3.0 \text{ M})^2} = 0.33$$

Scene 2:

$$\text{Molarity of A:} \quad \frac{(5 \text{ particles})\left(\dfrac{0.10 \text{ mol}}{\text{particle}}\right)}{0.10 \text{ L}} = 5.0 \text{ M}$$

$$\text{Molarity of A}_2: \quad \frac{(2 \text{ particles})\left(\dfrac{0.10 \text{ mol}}{\text{particle}}\right)}{0.10 \text{ L}} = 2.0 \text{ M}$$

$$Q = \frac{(2.0 \text{ M})}{(5.0 \text{ M})^2} = 0.080$$

Scene 3:

$$\text{Molarity of A:} \quad \frac{(1 \text{ particle})\left(\dfrac{0.10 \text{ mol}}{\text{particle}}\right)}{0.10 \text{ L}} = 1.0 \text{ M}$$

$$\text{Molarity of A}_2: \quad \frac{(4 \text{ particles})\left(\dfrac{0.10 \text{ mol}}{\text{particle}}\right)}{0.10 \text{ L}} = 4.0 \text{ M}$$

$$Q = \frac{(4.0 \text{ M})}{(1.0 \text{ M})^2} = 4.0$$

The reaction quotient for scene 1 is equal to the equilibrium constant ($K = 0.33$), so **scene 1 is at equilibrium.**

b) $\Delta G = \Delta G° + RT \ln Q$

Scene 1 is at equilibrium, so its $\Delta G = 0$. Scene 2 has $Q = 0.08$. Since $Q < 1$, $\ln Q$ is negative, which makes ΔG more negative than it is for the equilibrium scene. Scene 3 has $Q = 4.0$. Since $Q > 1$, $\ln Q$ is positive and ΔG is positive for scene 3.

Scene 3 has the most positive ΔG, followed by Scene 1 ($\Delta G = 0$) and Scene 2 (ΔG is negative).

20.83 Plan: To decide when production of ozone is favored, both the signs of $\Delta H°_{rxn}$ and $\Delta S°_{rxn}$ for ozone are needed. The values of $\Delta H°_f$ and $S°$ can be used. Once $\Delta H°_{rxn}$ and $\Delta S°_{rxn}$ are known, $\Delta G°$ can be calculated with the relationship $\Delta G°_{rxn} = \Delta H°_{rxn} - T\Delta S°_{rxn}$. $\Delta S°_{rxn}$ values in J/K must be converted to units of kJ/K to match the units of $\Delta H°_{rxn}$. ΔG is found by using the relationship $\Delta G = \Delta G° + RT \ln Q$.

Solution:

a) Formation of O_3 from O_2: $3O_2(g) \leftrightarrows 2O_3(g)$ or per mole of ozone: $3/2 O_2(g) \leftrightarrows O_3(g)$.

$\Delta H°_{rxn} = \sum m \Delta H°_{f (products)} - \sum n \Delta H°_{f (reactants)}$

$\Delta H°_{rxn} = [(1 \text{ mol } O_3)(\Delta H°_f \text{ of } O_3)] - [(3/2 \text{ mol } O_2)(\Delta H°_f \text{ of } O_2)]$

$\Delta H°_{rxn} = [(1 \text{ mol})(143 \text{ kJ/mol})] - [(3/2 \text{ mol})(0 \text{ kJ/mol})] = 143 \text{ kJ}$

$\Delta S°_{rxn} = \sum m S°_{products} - \sum n S°_{reactants}$

$\Delta S°_{rxn} = [(1 \text{ mol } O_3)(S° \text{ of } O_3)] - [(3/2 \text{ mol } O_2)(S° \text{ of } O_2)]$

$\Delta S°_{rxn} = [(1 \text{ mol})(238.82 \text{ J/mol·K})] - [(3/2 \text{ mol})(205.0 \text{ J/mol·K})] = -68.68 \text{ J/K} = -0.06868 \text{ kJ/K}$

The positive sign for ΔH°_{rxn} and the negative sign for ΔS°_{rxn} indicates the formation of ozone is favored at **no temperature**. The reaction is nonspontaneous at all temperatures.

b) $\Delta G^\circ_{rxn} = \Delta H^\circ_{rxn} - T\Delta S^\circ_{rxn}$

$\Delta G^\circ_{rxn} = 143$ kJ $- [(298$ K$)(-0.06868$ kJ/K$)] = 163.46664 = $ **163 kJ** for the formation of one mole of O_3.

c) Calculate the value for Q and then use to find ΔG.

$$Q = \frac{[O_3]}{[O_2]^{3/2}} = \frac{\left[5 \times 10^{-7} \text{ atm}\right]}{\left[0.21 \text{ atm}\right]^{3/2}} = 5.195664 \times 10^{-6}$$

$\Delta G = \Delta G^\circ + RT \ln Q = 163.46664$ kJ/mol $+ (8.314$ J/mol•K$)(298)(1$ kJ/10^3 J$) \ln (5.195664 \times 10^{-6})$

$= 133.3203215 = $ **1×10^2 kJ/mol**

20.86

	ΔS_{rxn}	ΔH_{rxn}	ΔG_{rxn}	Comment
(a)	+	–	–	**Spontaneous**
(b)	(+)	0	–	Spontaneous
(c)	–	+	(+)	Not spontaneous
(d)	0	(–)	–	Spontaneous
(e)	(–)	0	+	**Not spontaneous**
(f)	+	+	(–)	$T\Delta S > \Delta H$

a) The reaction is always spontaneous when $\Delta G_{rxn} < 0$, so there is no need to look at the other values other than to check the answer.

b) Because $\Delta G_{rxn} = \Delta H - T\Delta S = -T\Delta S$, ΔS must be positive for ΔG_{rxn} to be negative.

c) The reaction is always nonspontaneous when $\Delta G_{rxn} > 0$, so there is no need to look at the other values other than to check the answer.

d) Because $\Delta G_{rxn} = \Delta H - T\Delta S = \Delta H$, ΔH must be negative for ΔG_{rxn} to be negative.

e) Because $\Delta G_{rxn} = \Delta H - T\Delta S = -T\Delta S$, ΔS must be negative for ΔG_{rxn} to be positive.

f) Because $T\Delta S > \Delta H$, the subtraction of a larger positive term causes ΔG_{rxn} to be negative.

20.90 Plan: Write the equilibrium expression for the reaction. The equilibrium constant, K, is related to ΔG° through the equation $\Delta G^\circ = -RT \ln K$. Once K is known, the ratio of the two species is known.
Solution:

a) For the reaction, $K = \dfrac{[Hb \bullet CO][O_2]}{[Hb \bullet O_2][CO]}$ since the problem states that $[O_2] = [CO]$; the K expression simplifies to:

$K = \dfrac{[Hb \bullet CO]}{[Hb \bullet O_2]}$

$\ln K_p = \dfrac{\Delta G^\circ}{-RT} = \left(\dfrac{-14 \text{ kJ/mol}}{-(8.314 \text{ J/mol} \bullet \text{K})((273+37)\text{K})}\right)\left(\dfrac{10^3 \text{ J}}{1 \text{ kJ}}\right) = 5.431956979$

$K = e^{5.431956979} = 228.596 = $ **2.3×10^2** $= \dfrac{[Hb \bullet CO]}{[Hb \bullet O_2]}$

b) By increasing the concentration of oxygen, the equilibrium can be shifted in the direction of $Hb \bullet O_2$. Administer oxygen-rich air to counteract the CO poisoning.

20.92 Plan: Sum the two reactions to yield an overall reaction. Use the relationship $\Delta G^\circ_{rxn} = \Delta H^\circ_{rxn} - T\Delta S^\circ_{rxn}$ to calculate ΔG°_{rxn}. ΔH°_{rxn} and ΔS°_{rxn} will have to be calculated first.

Solution:

$$UO_2(s) + 4HF(g) \rightarrow \cancel{UF_4(s)} + 2H_2O(g)$$

$$\cancel{UF_4(s)} + F_2(g) \rightarrow UF_6(s)$$

$$\overline{UO_2(s) + 4\,HF(g) + F_2(g) \rightarrow UF_6(g) + 2H_2O(g)\ \text{(overall process)}}$$

$\Delta H^\circ_{rxn} = \sum m\,\Delta H^\circ_{f\,(products)} - \sum n\,\Delta H^\circ_{f\,(reactants)}$

$\Delta H^\circ_{rxn} = [(1\ \text{mol}\ UF_6)(\Delta H^\circ_f\ \text{of}\ UF_6) + (2\ \text{mol}\ H_2O)(\Delta H^\circ_f\ \text{of}\ H_2O)]$
$\qquad - [(1\ \text{mol}\ UO_2)(\Delta H^\circ_f\ \text{of}\ UO_2) + (4\ \text{mol}\ HF)(\Delta H^\circ_f\ \text{of}\ HF) + (1\ \text{mol}\ F_2)(\Delta H^\circ_f\ \text{of}\ F_2)]$

$\Delta H^\circ_{rxn} = [(1\ \text{mol})(-2197\ \text{kJ/mol}) + (2\ \text{mol})(-241.826\ \text{kJ/mol})]$
$\qquad - [(1\ \text{mol})(-1085\ \text{kJ/mol}) + (4\ \text{mol})(-273\ \text{kJ/mol}) + (1\ \text{mol})(0\ \text{kJ/mol})]$

$\Delta H^\circ_{rxn} = -503.652\ \text{kJ}$

$\Delta S^\circ_{rxn} = \sum m\,S^\circ_{products} - \sum n\,S^\circ_{reactants}$

$\Delta S^\circ_{rxn} = [(1\ \text{mol}\ UF_6)(S^\circ\ \text{of}\ UF_6) + (2\ \text{mol}\ H_2O)(S^\circ\ \text{of}\ H_2O)]$
$\qquad - [(1\ \text{mol}\ UO_2)(S^\circ\ \text{of}\ UO_2) + (4\ \text{mol}\ HF)(S^\circ\ \text{of}\ HF) + (1\ \text{mol}\ F_2)(S^\circ\ \text{of}\ F_2)]$

$\Delta S^\circ_{rxn} = [(1\ \text{mol})(225\ \text{J/mol·K}) + (2\ \text{mol})(188.72\ \text{J/mol·K})]$
$\qquad - [(1\ \text{mol})(77.0\ \text{J/mol·K}) + (4\ \text{mol})(173.67\ \text{J/mol·K}) + (1\ \text{mol})(202.7\ \text{J/mol·K})]$

$\Delta S^\circ_{rxn} = -371.94\ \text{J/K}$

$\Delta G^\circ_{rxn} = \Delta H^\circ_{rxn} - T\Delta S^\circ_{rxn} = (-503.652\ \text{kJ}) - ((273 + 85)\text{K})(-371.94\ \text{J/K})(\text{kJ}/10^3\ \text{J}) = -370.497 = \mathbf{-370.\ kJ}$

20.94   Plan: ΔG°_{rxn} can be calculated with the relationship $\sum m\,\Delta G^\circ_{f\,(products)} - \sum n\,\Delta G^\circ_{f\,(reactants)}$. ΔG is found by using the relationship $\Delta G = \Delta G^\circ + RT \ln Q$.

Solution:

a) $2N_2O_5(g) + 6F_2(g) \rightarrow 4NF_3(g) + 5O_2(g)$

b) $\Delta G^\circ_{rxn} = \sum m\,\Delta G^\circ_{f\,(products)} - \sum n\,\Delta G^\circ_{f\,(reactants)}$

$\Delta G^\circ_{rxn} = [(4\ \text{mol}\ NF_3)(\Delta G^\circ_f\ \text{of}\ NF_3) + (5\ \text{mol}\ O_2)(\Delta G^\circ_{rxn}\ \text{of}\ O_2)]$
$\qquad - [(2\ \text{mol}\ N_2O_5)(\Delta G^\circ_{rxn}\ \text{of}\ N_2O_5)) + (6\ \text{mol}\ F_2)\,(\Delta G^\circ_{rxn}\ \text{of}\ F_2)]$

$\Delta G^\circ_{rxn} = [(4\ \text{mol})(-83.3\ \text{kJ/mol})) + (5\ \text{mol})(0\ \text{kJ/mol})] - [(2\ \text{mol})(118\ \text{kJ/mol}) + (6\ \text{mol})(0\ \text{kJ/mol})]$

$\Delta G^\circ_{rxn} = -569.2 = \mathbf{-569\ kJ}$

c) Calculate the value for Q and then use to find ΔG.

$$Q = \frac{[NF_3]^4 [O_2]^5}{[N_2O_5]^2 [F_2]^6} = \frac{[0.25\,\text{atm}]^4 [0.50\,\text{atm}]^5}{[0.20\,\text{atm}]^2 [0.20\,\text{atm}]^6} = 47.6837$$

$\Delta G = \Delta G^\circ + RT \ln Q = -569.2\ \text{kJ/mol} + (1\ \text{kJ}/10^3\ \text{J})(8.314\ \text{J/mol·K})(298) \ln(47.6837)$
$\qquad = -559.625 = \mathbf{-5.60 \times 10^2\ kJ/mol}$

20.96   Plan: ΔH°_{rxn} can be calculated from the individual ΔH°_f values of the reactants and products by using the relationship $\Delta H^\circ_{rxn} = \sum m\,\Delta H^\circ_{f\,(products)} - \sum n\,\Delta H^\circ_{f\,(reactants)}$. ΔS°_{rxn} can be calculated from the individual S° values of the reactants and products by using the relationship $\Delta S^\circ_{rxn} = \sum m\,S^\circ_{products} - \sum n\,S^\circ_{reactants}$. ΔG°_{rxn} can be calculated with the relationship $\sum m\,\Delta G^\circ_{f\,(products)} - \sum n\,\Delta G^\circ_{f\,(reactants)}$.

Solution:

The reaction for the hydrogenation of ethene is $C_2H_4(g) + H_2(g) \rightarrow C_2H_6(g)$.

$\Delta H^\circ_{rxn} = \sum m\,\Delta H^\circ_{f\,(products)} - \sum n\,\Delta H^\circ_{f\,(reactants)}$

$\Delta H^\circ_{rxn} = [(1\ \text{mol}\ C_2H_6)(\Delta H^\circ_f\ \text{of}\ C_2H_6)] - [(1\ \text{mol}\ C_2H_4)(\Delta H^\circ_f\ \text{of}\ C_2H_4) + (1\ \text{mol}\ H_2)(\Delta H^\circ_f\ \text{of}\ H_2)]$
$\qquad = [(1\ \text{mol})(-84.667\ \text{kJ/mol})] - [(1\ \text{mol})(52.47\ \text{kJ/mol}) + (1\ \text{mol})(0\ \text{kJ/mol})] = -137.137 = \mathbf{-137.14\ kJ}$

$\Delta S^{\circ}_{rxn} = \sum m S^{\circ}_{products} - \sum n S^{\circ}_{reactants}$

$\Delta S^{\circ}_{rxn} = [(1 \text{ mol } C_2H_6)(S^{\circ} \text{ of } C_2H_6)] - [(1 \text{ mol } C_2H_4)(S^{\circ} \text{ of } C_2H_4)) + (1 \text{ mol } H_2)(S^{\circ} \text{ of } H_2)]$

$\quad = [(1 \text{ mol})(229.5 \text{ J/mol·K})] - [(1 \text{ mol})(219.22 \text{ J/mol·K}) + (1 \text{ mol})(130.6 \text{ J/mol·K})] = -120.32 = \mathbf{-120.3 \text{ J/K}}$

$\Delta G^{\circ}_{rxn} = \sum m \Delta G^{\circ}_{f(products)} - \sum n \Delta G^{\circ}_{f(reactants)}$

$\Delta G^{\circ}_{rxn} = [(1 \text{ mol } C_2H_6)(\Delta G^{\circ}_f \text{ of } C_2H_6))] - [(1 \text{ mol } C_2H_4)(\Delta G^{\circ}_f \text{ of } C_2H_4) + (1 \text{ mol } H_2)(\Delta G^{\circ}_f \text{ of } H_2)]$

$\quad = (1 \text{ mol})(-32.89 \text{ kJ/mol}) - [(1 \text{ mol})(68.36 \text{ kJ/mol}) + (1 \text{ mol})(0 \text{ kJ/mol})] = \mathbf{-101.25 \text{ kJ}}$

20.104 <u>Plan:</u> Use the relationship $\Delta G^{\circ} = -RT \ln K$ to calculate ΔG°. Use the relationship $\Delta G = \Delta G^{\circ} + RT \ln Q$ to calculate ΔG.

<u>Solution:</u>

The given equilibrium is: $\text{G6P} \rightleftharpoons \text{F6P} \qquad K = 0.510 \text{ at } 298 \text{ K}$

a) $\Delta G^{\circ} = -RT \ln K$

$\quad = -(8.314 \text{ J/mol·K})(298 \text{ K}) \ln (0.510) = 1.6682596 \times 10^3 = \mathbf{1.67 \times 10^3 \text{ J/mol}}$

b) $Q = [\text{F6P}]/[\text{G6P}] = 10.0$

$\Delta G = \Delta G^{\circ} + RT \ln Q$

$\Delta G = 1.6682596 \times 10^3 \text{ J/mol} + (8.314 \text{ J/mol·K})(298 \text{ K}) \ln 10.0$

$\Delta G = 7.3730799 \times 10^3 = \mathbf{7.37 \times 10^3 \text{ J/mol}}$

c) Repeat the calculation in part b) with $Q = 0.100$

$\Delta G = 1.6682596 \times 10^3 \text{ J/mol} + (8.314 \text{ J/mol·K})(298 \text{ K}) \ln 0.100$

$\Delta G = -4.0365607 \times 10^3 = \mathbf{-4.04 \times 10^3 \text{ J/mol}}$

d) $\Delta G = \Delta G^{\circ} + RT \ln Q$

$(-2.50 \text{ kJ/mol})(10^3 \text{ J/1 kJ}) = 1.6682596 \times 10^3 \text{ J/mol} + (8.314 \text{ J/mol·K})(298 \text{ K}) \ln Q$

$(-4.1682596 \times 10^3 \text{ J/mol}) = (8.314 \text{ J/mol·K})(298 \text{ K}) \ln Q$

$\ln Q = (-4.1682596 \times 10^3 \text{ J/mol})/[(8.314 \text{ J/mol·K})(298 \text{ K})] = -1.68239696$

$Q = 0.18592778 = \mathbf{0.19}$

20.108 <u>Plan:</u> First calculate ΔH° and ΔS°. According to the relationship $\ln K = \dfrac{\Delta G^{\circ}}{-RT}$, since $\ln K_p = 0$ when $K_p = 1.00$,

$\Delta G^{\circ} = 0$. Use the relationship $\Delta G^{\circ} = 0 = \Delta H^{\circ} - T\Delta S^{\circ}$ to find the temperature. For part b), calculate ΔG° at

the higher temperature with the relationship $\Delta G^{\circ} = \Delta H^{\circ} - T\Delta S^{\circ}$ and then calculate K with $\ln K = \dfrac{\Delta G^{\circ}}{-RT}$.

<u>Solution:</u>

a) $\Delta H^{\circ} = \sum m \Delta H^{\circ}_{f(products)} - \sum n \Delta H^{\circ}_{f(reactants)}$

$\Delta H^{\circ} = [(2 \text{ mol } NH_3)(\Delta H^{\circ}_f \text{ of } NH_3)] - [(1 \text{ mol } N_2)(\Delta H^{\circ}_f \text{ of } N_2) + (3 \text{ mol } H_2)(\Delta H^{\circ}_f \text{ of } H_2)]$

$\Delta H^{\circ} = [(2 \text{ mol})(-45.9 \text{ kJ/mol})] - [(1 \text{ mol})(0 \text{ kJ/mol}) + (3 \text{ mol})(0 \text{ kJ/mol})] = -91.8 \text{ kJ}$

$\Delta S^{\circ} = \sum m S^{\circ}_{products} - \sum n S^{\circ}_{reactants}$

$\Delta S^{\circ} = [(2 \text{ mol } NH_3)(S^{\circ} \text{ of } NH_3)] - [(1 \text{ mol } N_2)(S^{\circ} \text{ of } N_2) + (3 \text{ mol } H_2)(S^{\circ} \text{ of } H_2)]$

$\Delta S^{\circ} = [(2 \text{ mol})(193 \text{ J/mol·K})] - [(1 \text{ mol})(191.50 \text{ J/mol·K}) + (3 \text{ mol})(130.6 \text{ J/mol·K})]$

$\Delta S^{\circ} = -197.3 \text{ J/K}$

$\ln K = \dfrac{\Delta G^{\circ}}{-RT}$

Since $\ln K_p = 0$ when $K_p = 1.00$, $\Delta G° = 0$.

$\Delta G° = 0 = \Delta H° - T\Delta S°$

$\Delta H° = T\Delta S°$

$T = \dfrac{\Delta H°}{\Delta S°} = \dfrac{-91.8 \text{ kJ}}{-197.3 \text{ J/K}}\left(\dfrac{10^3 \text{ J}}{1 \text{ kJ}}\right) = 465.281 = \textbf{465 K}$

b) $\Delta G° = \Delta H° - T\Delta S° = (-91.8 \text{ kJ})(10^3 \text{ J/1 kJ}) - (673 \text{ K})(-197.3 \text{ J/K}) = 4.09829 \times 10^4 \text{ J}$

$\ln K = \dfrac{\Delta G°}{-RT} = \left(\dfrac{4.09829 \times 10^4 \text{ J/mol}}{-(8.314 \text{ J/mol} \bullet \text{K})(673 \text{ K})}\right) = -7.32449$

$\qquad K = e^{-7.32449} = 6.591958 \times 10^{-4} = \textbf{6.59} \times \textbf{10}^{-4}$

c) The reaction rate is higher at the higher temperature. The time required (kinetics) overshadows the lower yield (thermodynamics).

20.112 $\quad$ Plan: We know the mass (10.0 g) and the initial (6.15 L) and final (11.5 L) volumes of the gas. The mass must be converted to amount in moles. We use Equation 20.3 to calculate the entropy change.
Solution:

Amount (mol) of $CO_2 = (10.0 \text{ g } CO_2)\left(\dfrac{1 \text{ mol } CO_2}{44.0 \text{ g } CO_2}\right) = 0.227 \text{ mol } CO_2$

$\Delta S_{sys} = nR \ln \dfrac{V_{final}}{V_{intial}} = 0.227 \text{ mol} \times 8.314 \text{ J/mol} \bullet \text{K} \times \ln \dfrac{11.5 \text{ L}}{6.15 \text{ L}} = \textbf{1.18 J/K}$

20.114 $\quad$ Plan: We know the melting point (1535°C), and the value of ΔH_{fus} ; the temperature must be converted to kelvins. Use Equation 20.4 to calculate the entropy change for the phase change.
Solution:

$T_f \text{ (K)} = 1535°C + 273.15 = 1808 \text{ K}$

$\Delta S_{fus} = \dfrac{\Delta H_{fus}}{T_f} = \dfrac{13.8 \text{ kJ}}{1808 \text{ K}} = 0.00763 \text{ kJ/K} = \textbf{7.63 J/K}$

20.116 $\quad$ Plan: We know the mass (235 g), the initial (32.0°C) and final (125°C) temperatures, and the constant pressure molar heat capacity of copper (24.5 J/mol·K). The mass must be converted to amount in moles and the temperatures must be converted to kelvins. We then use Equation 20.6 to calculate the entropy change.
Solution:

$T_1 \text{ (K)} = 32.0°C + 273.15 = 305 \text{ K}$ $\qquad\qquad$ $T_2 \text{ (K)} = 125°C + 273.15 = 398 \text{ K}$

Amount (mol) of Cu $= (235 \text{ g Cu})\left(\dfrac{1 \text{ mol Cu}}{63.55 \text{ g Cu}}\right) = 3.70 \text{ mol Cu}$

$\Delta S_{sys} = nC_p \ln \dfrac{T_2}{T_1} = 3.70 \text{ mol} \times 24.5 \text{ J/mol} \bullet \text{K} \times \ln \dfrac{398 \text{ K}}{305 \text{ K}} = \textbf{24.1 J/K}$

CHAPTER 21 ELECTROCHEMISTRY: CHEMICAL CHANGE AND ELECTRICAL WORK

FOLLOW–UP PROBLEMS

21.1A   Plan: Follow the steps for balancing a redox reaction in acidic solution:

        1. Divide into half-reactions

        2. For each half-reaction balance

            a) Atoms other than O and H,

            b) O atoms with H_2O,

            c) H atoms with H^+ and

            d) Charge with e^-.

        3. Multiply each half-reaction by an integer that will make the number of electrons lost equal to the number of electrons gained.

        4. Add the half-reactions and cancel substances appearing as both reactants and products. Then, add another step for basic solution.

        5. Add hydroxide ions to neutralize H^+. Cancel water.

   Solution:

        1. Divide into half-reactions: group the reactants and products with similar atoms.

            $MnO_4^-(aq) \rightarrow MnO_4^{2-}(aq)$

            $I^-(aq) \rightarrow IO_3^-(aq)$

        2. For each half-reaction balance

            a) Atoms other than O and H

            Mn and I are balanced so no changes needed.

            b) O atoms with H_2O

            $MnO_4^-(aq) \rightarrow MnO_4^{2-}(aq)$ O already balanced

            $I^-(aq) + 3H_2O(l) \rightarrow IO_3^-(aq)$ Add 3 H_2O to balance oxygen.

            c) H atoms with H^+

            $MnO_4^-(aq) \rightarrow MnO_4^{2-}(aq)$ H already balanced

            $I^-(aq) + 3H_2O(l) \rightarrow IO_3^-(aq) + 6H^+(aq)$ Add 6 H^+ to balance hydrogen.

            d) Charge with e^-; Total charge of reactants is –1 and of products is –2, so add 1 e^- to reactants to balance charge:

            $MnO_4^-(aq) + e^- \rightarrow MnO_4^{2-}(aq)$

            Total charge is –1 for reactants and +5 for products, so add 6 e^- as product:

            $I^-(aq) + 3H_2O(l) \rightarrow IO_3^-(aq) + 6 H^+(aq) + 6e^-$

       3. Multiply each half-reaction by an integer that will make the number of electrons lost equal to the number of electrons gained. One electron is gained and 6 are lost so reduction must be multiplied by 6 for the number of electrons to be equal.

            $6\ \{MnO_4^-(aq) + e^- \rightarrow MnO_4^{2-}(aq)\}$

            $I^-(aq) + 3H_2O(l) \rightarrow IO_3^-(aq) + 6H^+(aq) + 6 e^-$

       4. Add half-reactions and cancel substances appearing as both reactants and products.

            $6MnO_4^-(aq) + \cancel{6e^-} \rightarrow 6MnO_4^{2-}(aq)$

            $\underline{I^-(aq) + 3H_2O(l) \rightarrow IO_3^-(aq) + 6H^+(aq) + \cancel{6e^-}}$

   Overall: $6MnO_4^-(aq) + I^-(aq) + 3H_2O(l) \rightarrow 6MnO_4^{2-}(aq) + IO_3^-(aq) + 6H^+(aq)$

   5. Add hydroxide ions to neutralize H^+. Cancel water. The 6 H^+ are neutralized by adding 6 OH^-. The same number of hydroxide ions must be added to the reactants to keep the balance of O and H atoms on both sides of the reaction.

$6MnO_4^-(aq) + I^-(aq) + 3H_2O(l) + 6OH^-(aq) \rightarrow 6MnO_4^{2-}(aq) + IO_3^-(aq) + 6H^+(aq) + 6\ OH^-(aq)$
The neutralization reaction produces water: $6\ \{H^+ + OH^- \rightarrow H_2O\}$.
$6MnO_4^-(aq) + I^-(aq) + 3H_2O(l) + 6OH^-(aq) \rightarrow 6MnO_4^{2-}(aq) + IO_3^-(aq) + 6H_2O(l)$
Cancel water:
$6MnO_4^-(aq) + I^-(aq) + \cancel{3H_2O(l)} + 6OH^-(aq) \rightarrow 6MnO_4^{2-}(aq) + IO_3^-(aq) + \cancel{6}H_2O(l)$
Balanced reaction is
$6MnO_4^-(aq) + I^-(aq) + 6OH^-(aq) \rightarrow 6MnO_4^{2-}(aq) + IO_3^-(aq) + 3H_2O(l)$
Balanced reaction including spectator ions is
$6KMnO_4(aq) + KI(aq) + 6KOH(aq) \rightarrow 6K_2MnO_4(aq) + KIO_3(aq) + 3H_2O(l)$

21.1B   Plan: Follow the steps for balancing a redox reaction in acidic solution:
1. Divide into half-reactions
2. For each half-reaction balance
   a) Atoms other than O and H,
   b) O atoms with H_2O,
   c) H atoms with H^+ and
   d) Charge with e^-.
3. Multiply each half-reaction by an integer that will make the number of electrons lost equal to the number of electrons gained.
4. Add the half-reactions and cancel substances appearing as both reactants and products. Then, add another step for basic solution.
5. Add hydroxide ions to neutralize H^+. Cancel water.

Solution:
1. Divide into half-reactions: group the reactants and products with similar atoms.
   $Cr(OH)_3(aq) \rightarrow CrO_4^{2-}(aq)$
   $IO_3^-(aq) \rightarrow I^-(aq)$
2. For each half-reaction balance
   a) Atoms other than O and H
   Cr and I are balanced so no changes needed.
   b) O atoms with H_2O
   $H_2O(l) + Cr(OH)_3(aq) \rightarrow CrO_4^{2-}(aq)$ Add 1 H_2O to balance oxygen.
   $IO_3^-(aq) \rightarrow I^-(aq) + 3H_2O(l)$ Add 3 H_2O to balance oxygen.
   c) H atoms with H^+
   $H_2O(l) + Cr(OH)_3(aq) \rightarrow CrO_4^{2-}(aq) + 5H^+(aq)$ Add 5 H^+ to balance hydrogen.
   $6H^+(aq) + IO_3^-(aq) \rightarrow I^-(aq) + 3H_2O(l)$ Add 6 H^+ to balance hydrogen.
   d) Charge with e^-; total charge of reactants is 0 and of products is +3, so add 3 e^- to products to balance charge:
   $H_2O(l) + Cr(OH)_3(aq) \rightarrow CrO_4^{2-}(aq) + 5H^+(aq) + 3e^-$
   Total charge is –1 for products and +5 for reactants, so add 6 e^- as reactant:
   $6e^- + 6H^+(aq) + IO_3^-(aq) \rightarrow I^-(aq) + 3H_2O(l)$
3. Multiply each half-reaction by an integer that will make the number of electrons lost equal to the number of electrons gained. Six electrons are gained and 3 are lost so oxidation must be multiplied by 2 for the number of electrons to be equal.
   $2\ \{H_2O(l) + Cr(OH)_3(aq) \rightarrow CrO_4^{2-}(aq) + 5H^+(aq) + 3e^-\}$
   $6e^- + 6H^+(aq) + IO_3^-(aq) \rightarrow I^-(aq) + 3H_2O(l)$
4. Add half-reactions and cancel substances appearing as both reactants and products.
   $\cancel{2H_2O(l)} + 2Cr(OH)_3(aq) \rightarrow 2CrO_4^{2-}(aq) + \cancel{10}4H^+(aq) + \cancel{6e^-}$
   $\underline{\cancel{6e^-} + \cancel{6}H^+(aq) + IO_3^-(aq) \rightarrow I^-(aq) + \cancel{3}1H_2O(l)}$
Overall: $2Cr(OH)_3(aq) + IO_3^-(aq) \rightarrow 2CrO_4^{2-}(aq) + I^-(aq) + H_2O(l) + 4H^+(aq)$
5. Add hydroxide ions to neutralize H^+. Cancel water. The 4 H^+ are neutralized by adding 4 OH^-. The same number of hydroxide ions must be added to the reactants to keep the balance of O and H atoms on both sides of the reaction.
   $2Cr(OH)_3(aq) + IO_3^-(aq) + 4OH^-(aq) \rightarrow 2CrO_4^{2-}(aq) + I^-(aq) + H_2O(l) + 4H^+(aq) + 4OH^-(aq)$

The neutralization reaction produces water: $4 \{H^+ + OH^- \to H_2O\}$.

$2Cr(OH)_3(aq) + IO_3^-(aq) + 4OH^-(aq) \to 2CrO_4^{2-}(aq) + I^-(aq) + 5H_2O(l)$

Balanced reaction is

$2Cr(OH)_3(aq) + IO_3^-(aq) + 4OH^-(aq) \to 2CrO_4^{2-}(aq) + I^-(aq) + 5H_2O(l)$

Balanced reaction including spectator ions is

$2Cr(OH)_3(aq) + NaIO_3(aq) + 4NaOH(aq) \to 2Na_2CrO_4(aq) + NaI(aq) + 5H_2O(l)$

21.2A   Plan: Given the solution and electrode compositions, the two half cells involve the transfer of electrons 1) between chromium in $Cr_2O_7^{2-}$ and Cr^{3+} and 2) between Sn and Sn^{2+}. The negative electrode is the anode so the tin half-cell is where oxidation occurs. The graphite electrode with the chromium ion/chromate solution is where reduction occurs. In the cell diagram, show the electrodes and the solutes involved in the half-reactions. Include the salt bridge and wire connection between electrodes. Set up the two half-reactions and balance. (Note that the $Cr^{3+}/Cr_2O_7^{2-}$ half-cell is in acidic solution.) Write the cell notation placing the anode half-cell first, then the salt bridge, then the cathode half-cell.
Solution: Cell diagram:

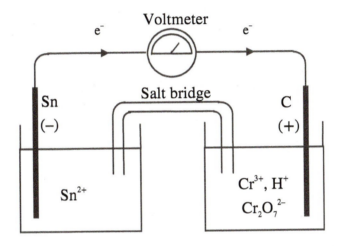

Balanced equations:

Anode is Sn/Sn^{2+} half-cell. Oxidation of Sn produces Sn^{2+}:

   $Sn(s) \to Sn^{2+}(aq)$

All that needs to be balanced is charge:

   $Sn(s) \to Sn^{2+}(aq) + 2e^-$

Cathode is the $Cr^{3+}/Cr_2O_7^{2-}$ half-cell. Check the oxidation number of chromium in each substance to determine which is reduced. Cr^{3+} oxidation number is +3 and chromium in $Cr_2O_7^{2-}$ has oxidation number +6. Going from +6 to +3 involves gain of electrons so $Cr_2O_7^{2-}$ is reduced.

   $Cr_2O_7^{2-}(aq) \to Cr^{3+}(aq)$

Balance Cr:

   $Cr_2O_7^{2-}(aq) \to 2Cr^{3+}(aq)$

Balance O:

   $Cr_2O_7^{2-}(aq) \to 2Cr^{3+}(aq) + 7H_2O(l)$

Balance H:

   $Cr_2O_7^{2-}(aq) + 14H^+(aq) \to 2Cr^{3+}(aq) + 7H_2O(l)$

Balance charge:

   $Cr_2O_7^{2-}(aq) + 14H^+(aq) + 6e^- \to 2Cr^{3+}(aq) + 7H_2O(l)$

Add two half-reactions, multiplying the tin half-reaction by 3 to equalize the number of electrons transferred.

   $3\{Sn(s) \to Sn^{2+}(aq) + 2e^-\}$

   $\underline{Cr_2O_7^{2-}(aq) + 14H^+(aq) + 6e^- \to 2\ Cr^{3+}(aq) + 7H_2O(l)}$

   $3Sn(s) + Cr_2O_7^{2-}(aq) + 14H^+(aq) \to 3Sn^{2+}(aq) + 2Cr^{3+}(aq) + 7H_2O(l)$

Cell notation:

   $Sn(s) \mid Sn^{2+}(aq) \parallel H^+(aq), Cr_2O_7^{2-}(aq), Cr^{3+}(aq) \mid C(graphite)$

21.2B Plan: Given the solution and electrode compositions, the two half cells involve the transfer of electrons 1) between Cl^- and ClO_3^- and 2) between Ni and Ni^{2+}. The negative electrode is the anode so the nickel half-cell is where oxidation occurs. The graphite electrode with the chloride ion/chlorate solution is where reduction occurs. In the cell diagram, show the electrodes and the solutes involved in the half-reactions. Include the salt bridge and wire connection between electrodes. Set up the two half-reactions and balance. (Note that the Cl^-/ClO_3^- half-cell is in acidic solution.) Write the cell notation placing the anode half-cell first, then the salt bridge, then the cathode half-cell.

Solution: Cell diagram:

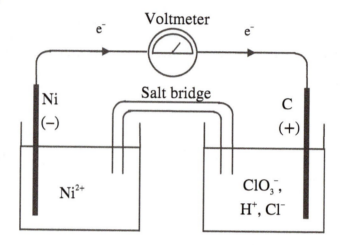

Balanced equations:

Anode is Ni/Ni^{2+} half-cell. Oxidation of Ni produces Ni^{2+}:

$$Ni(s) \rightarrow Ni^{2+}(aq)$$

All that needs to be balanced is charge:

$$Ni(s) \rightarrow Ni^{2+}(aq) + 2e^-$$

Cathode is the Cl^-/ClO_3^- half-cell. Check the oxidation number of chlorine in each substance to determine which is reduced. Cl^- oxidation number is −1 and chlorine in ClO_3^- has oxidation number +5. Going from +5 to −1 involves gain of electrons so ClO_3^- is reduced.

$$ClO_3^-(aq) \rightarrow Cl^-(aq)$$

Cl is balanced. Balance O:

$$ClO_3^-(aq) \rightarrow Cl^-(aq) + 3H_2O(l)$$

Balance H:

$$6H^+(aq) + ClO_3^-(aq) \rightarrow Cl^-(aq) + 3H_2O(l)$$

Balance charge:

$$6H^+(aq) + ClO_3^-(aq) + 6e^- \rightarrow Cl^-(aq) + 3H_2O(l)$$

Add two half-reactions, multiplying the nickel half-reaction by 3 to equalize the number of electrons transferred.

$$3\{Ni(s) \rightarrow Ni^{2+}(aq) + 2e^-\}$$
$$6H^+(aq) + ClO_3^-(aq) + 6e^- \rightarrow Cl^-(aq) + 3H_2O(l)$$

$$3Ni(s) + ClO_3^-(aq) + 6H^+(aq) \rightarrow 3Ni^{2+}(aq) + Cl^-(aq) + 3H_2O(l)$$

Cell notation:

$$Ni(s) \mid Ni^{2+}(aq) \parallel H^+(aq), ClO_3^-(aq), Cl^-(aq) \mid C(graphite)$$

21.3A Plan: Use the relationship $E^o_{cell} = E^o_{cathode} - E^o_{anode}$. E^o values are found in Appendix D. Spontaneous reactions have $E^o_{cell} > 0$.

Solution:

Oxidation:	$2\{Ag(s) \rightarrow Ag^+(aq) + e^-\}$	$E^o(anode) = 0.80$ V
Reduction:	$Cu^{2+}(aq) + 2e^- \rightarrow Cu(s)$	$E^o(cathode) = 0.34$ V

Overall reaction: $Cu^{2+}(s) + 2Ag(s) \rightarrow Cu(s) + 2Ag^+(aq)$

$E°_{cell} = 0.34\ V -\ (0.80\ V) = \mathbf{-0.46\ V}$

Reaction is **not spontaneous** under standard-state conditions because $E°_{cell}$ is negative.

21.3B Plan: Use the relationship $E°_{cell} = E°_{cathode} - E°_{anode}$. $E°$ values are found in Appendix D. Spontaneous reactions have $E°_{cell} > 0$.
Solution:

Oxidation:	$2\{Fe^{2+}(aq) \rightarrow Fe^{3+}(aq) + e^-\}$	$E°(anode) = 0.77\ V$
Reduction:	$Cl_2(g) + 2e^- \rightarrow 2Cl^-(aq)$	$E°(cathode) = 1.36\ V$
Overall reaction:	$Cl_2(g) + 2Fe^{2+}(aq) \rightarrow 2Cl^{-}(aq) + 2Fe^{3+}(aq)$	

$E°_{cell} = 1.36\ V -\ (0.77\ V) = \mathbf{+0.59\ V}$

Reaction is **spontaneous** under standard-state conditions because $E°_{cell}$ is positive.

21.4A Plan: Divide the reaction into half-reactions showing that Br_2 is reduced and V^{3+} is oxidized. Use the equation $E°_{cell} = E°_{cathode} - E°_{anode}$ to solve for $E°_{anode}$.
Solution:
Half-reactions:

Reduction (cathode): $Br_2(aq) + 2e^- \rightarrow 2\ Br^-(aq)$ $E°_{cathode} = 1.07\ V$ from Appendix D

Oxidation (anode): $2V^{3+}(aq) + 2H_2O(l) \rightarrow 2VO^{2+}(aq) + 4H^+(aq) + 2e^-$

Overall:

$Br_2(aq) + 2V^{3+}(aq) + 2H_2O(l) \rightarrow 2VO^{2+}(aq) + 4H^+(aq) + 2Br^-(aq)$ $E°_{cell} = 0.73\ V$ (given)

$E°_{cell} = E°_{cathode} - E°_{anode}$

$E°_{anode} = E°_{cathode} - E°_{cell} = 1.07\ V - 0.73\ V = \mathbf{0.34\ V}$

21.4B Plan: Divide the reaction into half-reactions showing that nitrate is reduced and V^{3+} is oxidized. Use the equation $E°_{cell} = E°_{cathode} - E°_{anode}$ to solve for $E°_{anode}$.
Solution:
Half-reactions:

Reduction (cathode): $4H^+(aq) + NO_3^-(aq) + 3e^- \rightarrow NO(g) + 2H_2O(l)$

Oxidation (anode): $3\{V^{3+}(aq) + H_2O(l) \rightarrow VO^{2+}(aq) + 2H^+(aq) + e^-\}$ $E°(anode) = 0.34\ V$
from Follow Up Problem 21.4A

Overall:

$NO_3^-(aq) + 3V^{3+}(aq) + H_2O(l) \rightarrow 3VO^{2+}(aq) + 2H^+(aq) + NO(g)$ $E°_{cell} = 0.62\ V$ (given)

$E°_{cell} = E°_{cathode} - E°_{anode}$

$E°_{cathode} = E°_{cell} + E°_{anode} = 0.62\ V + 0.34\ V = \mathbf{0.96\ V}$

21.5A Plan: To write a spontaneous reaction, examine two reduction half-reactions. The half-reaction with the smaller $E°$ is reversed (to become an oxidation). Reducing strength increases with decreasing $E°$.
Solution:
a) Combining reactions (1) and (2), reaction (1) is reversed because it has the smaller $E°$.

Oxidation:	$3\{2Ag(s) + 2OH^-(aq) \rightarrow Ag_2O(s) + H_2O(l) + 2e^-\}$	$E°(anode) = 0.34\ V$
Reduction:	$BrO_3^-(g) + 3H_2O(l) + 6e^- \rightarrow Br^-(aq) + 6OH^-(aq)$	$E°(cathode) = 0.61\ V$

Overall reaction: $6Ag(s) + BrO_3^-(g) \rightarrow 3Ag_2O(s) + Br^-(aq)$

$E°_{cell} = 0.61\ V -\ (0.34\ V) = \mathbf{+0.27\ V}$

Combining reactions (1) and (3), reaction (3) is reversed because it has the smaller $E°$.

Oxidation:	$Zn(s) + 2OH^-(aq) \rightarrow Zn(OH)_2(s) + 2e^-$	$E°(anode) = -1.25\ V$
Reduction:	$Ag_2O(s) + H_2O(l) + 2e^- \rightarrow 2Ag(s) + 2OH^-(aq)$	$E°(cathode) = 0.34\ V$
Overall reaction:	$Zn(s) + Ag_2O(s) + H_2O(l) \rightarrow Zn(OH)_2(s) + 2Ag(s)$	

$E°_{cell} = 0.34\ V -\ (-1.25\ V) = \mathbf{+1.59\ V}$

Combining reactions (2) and (3), reaction (3) is reversed because it has the smaller $E°$.

Oxidation: $3\{Zn(s) + 2OH^-(aq) \rightarrow Zn(OH)_2(s) + 2e^-\}$ $E°(\text{anode}) = -1.25$ V

Reduction: $BrO_3^-(g) + 3H_2O(l) + 6e^- \rightarrow Br^-(aq) + 6OH^-(aq)$ $E°(\text{cathode}) = 0.61$ V

Overall reaction: $3Zn(s) + BrO_3^-(g) + 3H_2O(l) \rightarrow 3Zn(OH)_2(s) + Br^-(aq)$

$$E°_{cell} = 0.61 \text{ V} - (-1.25 \text{ V}) = \mathbf{+1.86 \text{ V}}$$

b) Oxidizing agents are substances that cause another substance to be oxidized. In other words, oxidizing agents are, themselves, reduced. The more likely a substance is to be reduced (the stronger the oxidizing agent), the greater the value of its reducing potential, $E°$. Examining, the reduction half-reactions given in the problem statement, we can see that $Zn(OH)_2$ has the lowest reducing potential, Ag_2O has the next highest reducing potential, and BrO_3^- has the greatest reducing potential. Therefore, we can rank these substances by their **strengths as oxidizing agents: $BrO_3^- > Ag_2O > Zn(OH)_2$.**

Reducing agents are substances that cause another substance to be reduced. In other words, reducing agents are, themselves, oxidized. The more likely a substance is to be oxidized (the stronger the reducing agent), the greater the value of its oxidizing potential. The oxidizing potentials of substances in this problem can be obtained by reversing the provided reduction half-reactions and changing the sign of the half-reaction potential. When we do this, we find that Ag has an oxidizing potential of -0.34 V, Br^- has an oxidizing potential of -0.61 V, and Zn has an oxidizing potential of 1.25 V. The more positive the oxidizing potential, the stronger the reducing agent. Therefore, we can rank these substances by their **strengths as reducing agents: $Zn > Ag > Br^-$.**

21.5B **Plan:** To determine if the reaction is spontaneous, divide into half-reactions and calculate $E°_{cell}$. If $E°_{cell}$ is negative, the reaction is not spontaneous, so reverse the reaction to obtain the spontaneous reaction. Reducing strength increases with decreasing $E°$.

Solution: Divide into half-reactions and balance:

Reduction: $Fe^{2+}(aq) + 2e^- \rightarrow Fe(s)$ $E°_{cathode} = -0.44$ V

Oxidation: $2\{Fe^{2+}(aq) \rightarrow 2Fe^{3+}(aq) + e^-\}$ $E°_{anode} = 0.77$ V

The first half-reaction is reduction, so it is the cathode half-cell. The second half-reaction is oxidation, so it is the anode half-cell. Find the half-reactions in Appendix D.

$E°_{cell} = E°_{cathode} - E°_{anode} = -0.44 \text{ V} - 0.77 \text{ V} = -1.21$ V

The reaction is not spontaneous as written, so reverse the reaction:

$Fe(s) + 2Fe^{3+}(aq) \rightarrow 3Fe^{2+}(aq)$

$E°_{cell}$ is now **+1.21 V**, so the reversed reaction is spontaneous under standard state conditions.

When a substance acts as a reducing agent, it is oxidized. Both Fe and Fe^{2+} can be oxidized, so they can act as reducing agents. Since Fe^{3+} cannot lose more electrons, it cannot act as a reducing agent. The stronger reducing agent between Fe and Fe^{2+} is the one with the smaller standard reduction potential. $E°$ for Fe is -0.44, which is less than $E°$ for Fe^{2+}, $+0.77$. Therefore, Fe is a stronger reducing agent than Fe^{2+}. Ranking all three in order of decreasing reducing strength gives **$Fe > Fe^{2+} > Fe^{3+}$.**

21.6A **Plan:** Determine the half-reactions and calculate the $E°_{cell}$. Find the number of moles of electrons transferred in the balanced equation. Both K and $\Delta G°$ can be calculated using the relationships $\log K = \dfrac{nE°_{cell}}{0.0592 \text{ V}}$ and $\Delta G° = -nFE°_{cell}$.

Solution:

The balanced equation is:

$2MnO_4^-(aq) + 4H_2O(l) + 3Cu(s) \rightarrow 2MnO_2(s) + 8OH^-(aq) + 3Cu^{2+}(aq)$

The half reactions are:

Reduction: $2\{MnO_4^-(aq) + 2H_2O(l) + 3e^- \rightarrow MnO_2(s) + 8OH^-(aq)\}$ $E°(\text{cathode}) = 0.59$ V

Oxidation: $3\{Cu(s) \rightarrow Cu^{2+}(aq) + 2e^-\}$ $E°(\text{anode}) = 0.34$ V

$E°_{cell} = E°_{cathode} - E°_{anode} = 0.59 \text{ V} - 0.34 \text{ V} = 0.25$ V

In this reaction, 3 moles of copper atoms are oxidized to become 3 moles of copper(II) ions. This process requires the loss of 3 x 2 = 6 electrons. Therefore, 6 electrons are transferred in this reaction, and n = 6.

$$\log K = \frac{nE^\circ_{cell}}{0.0592\,V} = \frac{(6)(0.25\,V)}{0.0592\,V} = 25.3378$$

$$K = 10^{25.3378} = 2.1769\text{x}10^{25} = \mathbf{2.2x10^{25}}$$

$$\Delta G^\circ = -nFE^\circ_{cell} = -\,(6\text{ mol e}^-/\text{mol rxn})(96{,}485\text{ C/mol e}^-)(0.25\text{ J/C}) = -1.4473\text{x}10^5 = \mathbf{-1.4 \times 10^5\ J/mol\ rxn}$$
$$\mathbf{= -1.4 \times 10^2\ kJ/mol\ rxn}$$

21.6B   Plan: Reaction is $Cd(s) + Cu^{2+}(aq) \rightarrow Cd^{2+}(aq) + Cu(s)$. Given ΔG°, both K and E°_{cell} can be calculated using the relationships $\Delta G^\circ = -RT \ln K$ and $\Delta G^\circ = -nFE^\circ_{cell}$.

Solution:

$$\ln K = -\frac{\Delta G^\circ}{RT} = -\left(\frac{-143\text{ kJ}}{\left(8.314\text{ J/mol}\bullet K\right)\left((273+25)K\right)}\right)\left(\frac{10^3\text{ J}}{1\text{ kJ}}\right) = 57.71779791$$

$$K = 1.1655238\text{x}10^{25} = \mathbf{1.2x10^{25}}$$

$$\Delta G^\circ = -nFE^\circ_{cell} \qquad\qquad n = 2\text{ mol e}^-\text{ for this reaction}$$

$$E^\circ_{cell} = -\frac{\Delta G^\circ}{nF} = -\left(\frac{-143\text{ kJ}}{(2\text{ mol})(96485\text{ C/mol})}\right)\left(\frac{10^3\text{ J}}{1\text{ kJ}}\right)\left(\frac{C}{J/V}\right) = 0.7410478 = \mathbf{0.741\ V}$$

Note that an alternative way to calculate E°_{cell} is to use $E^\circ_{cell} = \dfrac{0.0592\,V}{n}\log K$

21.7A   Plan: Write a balanced equation for the spontaneous reaction between Cr and Sn (we know the reaction is spontaneous because we are told the reaction occurs in a voltaic cell). Determine the number of moles of electrons transferred, and calculate E°_{cell}. Use the Nernst equation, $E_{cell} = E^\circ_{cell} - \dfrac{0.0592}{n}\log Q$ to find E_{cell}.

Solution:
Determining the cell reaction and E°_{cell} :
The Cr/Cr^{3+} reduction half-reaction has a smaller E°. Therefore, it is reversed (and becomes the oxidation half-reaction).

Oxidation:	$2\{Cr(s) \rightarrow Cr^{3+}(aq) + 3e^-\}$	E° (anode) = –0.74 V
Reduction:	$3\{Sn^{2+}(aq) + 2e^- \rightarrow Sn(s)]\}$	E° (cathode) = –0.14 V
Overall	$2Cr(s) + 3Sn^{2+}(aq) \rightarrow 2Cr^{3+}(aq) + 3Sn(s)$	(6 moles of electrons transferred)

$$E^\circ_{cell} = E^\circ_{cathode} - E^\circ_{anode} = -0.14\text{ V} - (-0.74\text{ V}) = 0.60\text{ V}$$

For the reaction $Q = [Cr^{2+}]/[Sn^{2+}]$, so the Nernst equation is:

$$E_{cell} = E^\circ_{cell} - \frac{0.0592}{n}\log\frac{\left[Cr^{3+}\right]^2}{\left[Sn^{2+}\right]^3}.$$

Substituting in values from the problem:

$$E_{cell} = 0.60\text{ V} - \frac{0.0592}{6}\log\frac{(1.60)^2}{(0.20)^3} = 0.60\text{ V} - \frac{0.0592}{6}\log(320) = 0.60\text{ V} - 0.024717 = 0.575283 = \mathbf{0.58\ V}$$

21.7B   Plan: The problem is asking for the concentration of iron ions when $E_{cell} = E^\circ_{cell} + 0.25\text{ V}$.

Use the Nernst equation, $E_{cell} = E^\circ_{cell} - \dfrac{0.0592}{n}\log Q$ to find $[Fe^{2+}]$.

Solution:
Determining the cell reaction and E°_{cell} :

Oxidation: $\qquad$ $Fe(s) \rightarrow Fe^{2+}(aq) + 2e^-$ $\qquad\qquad\qquad$ $E^{\circ} = -0.44$ V

Reduction: $\qquad$ $Cu^{2+}(aq) + 2e^- \rightarrow Cu(s)$ $\qquad\qquad\qquad$ $E^{\circ} = 0.34$ V

Overall $\; Fe(s) + Cu^{2+}(aq) \rightarrow Fe^{2+}(aq) + Cu(s)$

$$E^{\circ}_{cell} = E^{\circ}_{cathode} - E^{\circ}_{anode} = 0.34 \text{ V} - (-0.44 \text{ V}) = 0.78 \text{ V}$$

For the reaction $Q = [Fe^{2+}]/[Cu^{2+}]$, so the Nernst equation is:

$$E_{cell} = E^{\circ}_{cell} - \frac{0.0592}{n} \log \frac{[Fe^{2+}]}{[Cu^{2+}]}$$

Substituting in values from the problem:

$$E_{cell} = E^{\circ}_{cell} + 0.25 \text{ V} = 0.78 \text{ V} + 0.25 \text{ V} = 1.03 \text{ V}$$

$$1.03 \text{ V} = 0.78 \text{ V} - \frac{0.0592}{n} \log \frac{[Fe^{2+}]}{[Cu^{2+}]}$$

$$0.25 \text{ V} = -\frac{0.0592}{2} \log \frac{[Fe^{2+}]}{[0.30]}$$

$$-8.44595 = \log \frac{[Fe^{2+}]}{[0.30]}$$

$$3.58141 \times 10^{-9} = \frac{[Fe^{2+}]}{[0.30]}$$

$$[Fe^{2+}] = 1.074423 \times 10^{-9} = \mathbf{1.1 \times 10^{-9}\ \textit{M}}$$

21.8A $\quad$ Plan: Half-cell B contains a higher concentration of nickel ions, so the ions will be reduced to decrease the concentration while in half-cell A, with a lower $[Ni^{2+}]$, nickel metal will be oxidized to increase the concentration of nickel ions. In the overall cell reaction, the lower $[Ni^{2+}]$ appears as a product and the higher $[Ni^{2+}]$ appears as a reactant. This means that $Q = [Ni^{2+}]_{lower}/[Ni^{2+}]_{higher}$. Use the Nernst equation with $E^{\circ}_{cell} = 0$ to find E_{cell}. Oxidation of nickel metal occurs at the anode, half-cell A, which is negative.
Solution:

$$E_{cell} = E^{\circ}_{cell} - \frac{0.0592}{n} \log Q$$

$$E_{cell} = E^{\circ}_{cell} - \frac{0.0592}{n} \log \frac{[Ni^{2+}]_{lower}}{[Ni^{2+}]_{higher}} \qquad n = 2 \text{ mol } e^- \text{ for the reduction of } Ni^{2+} \text{ to Ni}$$

$$E_{cell} = 0 - \frac{0.0592}{2} \log \frac{(0.015)}{(0.40)} = 0.04221 = \mathbf{0.042\ V}$$

21.8B $\quad$ Plan: Half-cell B contains a higher concentration of gold ions, so the ions will be reduced to decrease the concentration while in half-cell A, with a lower $[Au^{3+}]$, gold metal will be oxidized to increase the concentration of gold ions. In the overall cell reaction, the lower $[Au^{3+}]$ appears as a product and the higher $[Au^{3+}]$ appears as a reactant. This means that $Q = [Au^{3+}]_{lower}/[Au^{3+}]_{higher}$. Use the Nernst equation with $E^{\circ}_{cell} = 0$ to find E_{cell}.
Solution:

$$E_{cell} = E^{\circ}_{cell} - \frac{0.0592}{n} \log Q$$

$$E_{cell} = E^{\circ}_{cell} - \frac{0.0592}{n} \log \frac{[Au^{3+}]_{lower}}{[Au^{3+}]_{higher}} \qquad n = 3 \text{ mol } e^- \text{ for the reduction of } Au^{3+} \text{ to Au}$$

$$E_{cell} = 0 - \frac{0.0592}{3} \log \frac{[7.0 \times 10^{-4}]}{[2.5 \times 10^{-2}]} = 0.0306427 = \mathbf{0.031\ V}$$

Oxidation of gold metal occurs at the anode, **half-cell A**, which is negative.

21.9A   Plan: In the electrolysis, the more easily reduced metal ion will be reduced at the cathode. Compare Cs^+ and Li^+ as to their location on the periodic table to find which has the higher ionization energy (IE increases up and across the table). Higher ionization energy of the metal means more easily reduced. At the anode, the more easily oxidized nonmetal ion will be oxidized. For oxidation, compare the electronegativity of the two nonmetals. The ion from the less electronegative element will be more easily oxidized. Note that the standard reduction potentials cannot be used in the case of molten salts because the $E°$ values are based on aqueous solutions of ions, not liquid salts.
Solution:
Li is above Cs, so Li^+ is more easily reduced than Cs^+. The reaction at the cathode is
$$Li^+(l) + e^- \rightarrow Li(s)$$
F is the most electronegative element, so I^- must be more easily oxidized than F^-. The reaction at the anode is
$$2 I^- (l) \rightarrow I_2(g) + 2e^-$$
To add the two half-reaction, the number of electrons must be equal:
$$2 \{Li^+(l) + e^- \rightarrow Li(s)\}$$
$$\underline{2I^- (l) \rightarrow I_2(g) + 2e^-}$$
Overall: $2Li^+(l) + 2I^-(l) \rightarrow 2Li(s) + I_2(g)$
This is the overall reaction. **$Li(s)$ forms at the cathode, and $I_2(g)$ forms at the anode.**

21.9B   Plan: In the electrolysis, the more easily reduced metal ion will be reduced at the cathode. Compare K^+ and Al^{3+} as to their location on the periodic table to find which has the higher ionization energy (IE increases up and across the table). Higher ionization energy of the metal means more easily reduced. At the anode, the more easily oxidized nonmetal ion will be oxidized. For oxidation, compare the electronegativity of the two nonmetals. The ion from the less electronegative element will be more easily oxidized. Note that the standard reduction potentials cannot be used in the case of molten salts because the $E°$ values are based on aqueous solutions of ions, not liquid salts.
Solution:
Al is above and to the right of K, so Al^{3+} is more easily reduced than K^+. The reaction at the cathode is
$$Al^{3+}(l) + 3e^- \rightarrow Al(s)$$
F is the most electronegative element, so Br^- must be more easily oxidized than F^-. The reaction at the anode is
$$2 Br^- (l) \rightarrow Br_2(g) + 2e^-$$
To add the two half-reaction, the number of electrons must be equal:
$$2 \{Al^{3+}(l) + 3e^- \rightarrow Al(s)\}$$
$$\underline{3 \{2Br^- (l) \rightarrow Br_2(g) + 2e^-\}}$$
Overall: $2Al^{3+}(l) + 6Br^- (l) \rightarrow 2Al(s) + 3Br_2(g)$
This is the overall reaction. **$Al(s)$ forms at the cathode, and $Br_2(g)$ forms at the anode.**

21.10A   Plan: In aqueous $Pb(NO_3)_2$, the species present are $Pb^{2+}(aq)$, $NO_3^-(aq)$, H_2O, and very small amounts of H^+ and OH^-. The possible half-reactions are reduction of either Pb^{2+} or H_2O and oxidation of either NO_3^- or H_2O. Whichever reduction and oxidation half-reactions are more spontaneous will take place, with consideration of the overvoltage.
Solution: The two possible reductions are

$Pb^{2+}(aq) + 2e^- \rightarrow Pb(s)$		$E° = -0.13$ V
$2H_2O(l) + 2e^- \rightarrow H_2(g) + 2OH^-(aq)$		$E = -0.42$ V

The reduction of lead ions occurs because it has a higher reduction potential (more spontaneous) than reduction of water.
The two possible oxidations are

$NO_3^-(aq) \rightarrow$ no reaction (nitrate compounds are always soluble)
$2H_2O(l) \rightarrow O_2(g) + 4H^+(aq) + 4e^-$ $\qquad E = 0.82$ V

Water will be oxidized at the anode since no reaction is possible for the nitrate ion.
The two half-reactions that are predicted are

Cathode:	$Pb^{2+}(aq) + 2e^- \rightarrow Pb(s)$	$E° = -0.13$ V
Anode:	$2H_2O(l) \rightarrow O_2(g) + 4H^+(aq) + 4e^-$	$E = 0.82$ V

Balance elements other than O and H:

$$ClO_3^-(aq) \rightarrow Cl^-(aq) \qquad\qquad \text{chlorine is balanced}$$
$$2I^-(aq) \rightarrow I_2(s) \qquad\qquad \text{iodine now balanced}$$

Balance O by adding H_2O:

$$ClO_3^-(aq) \rightarrow Cl^-(aq) + 3H_2O(l) \qquad \text{add three waters to add three O atoms to product}$$
$$2I^-(aq) \rightarrow I_2(s) \qquad\qquad \text{no change}$$

Balance H by adding H^+:

$$ClO_3^-(aq) + 6H^+(aq) \rightarrow Cl^-(aq) + 3H_2O(l) \qquad \text{add six } H^+ \text{ to reactants}$$
$$2I^-(aq) \rightarrow I_2(s) \qquad\qquad \text{no change}$$

Balance charge by adding e^-:

$$ClO_3^-(aq) + 6H^+(aq) + 6e^- \rightarrow Cl^-(aq) + 3\,H_2O(l) \qquad \text{add } 6e^- \text{ to reactants for a } -1 \text{ charge on each side}$$
$$2I^-(aq) \rightarrow I_2(s) + 2e^- \qquad\qquad \text{add } 2e^- \text{ to products for a } -2 \text{ charge on each side}$$

Multiply each half-reaction by an integer to equalize the number of electrons:

$$ClO_3^-(aq) + 6H^+(aq) + 6e^- \rightarrow Cl^-(aq) + 3H_2O(l) \qquad \text{multiply by one to give } 6e^-$$
$$3\{2I^-(aq) \rightarrow I_2(s) + 2e^-\} \text{ or} \qquad\qquad \text{multiply by three to give } 6e^-$$
$$6I^-(aq) \rightarrow 3I_2(s) + 6e^-$$

Add half-reactions to give balanced equation in acidic solution:

$$ClO_3^-(aq) + 6H^+(aq) + 6I^-(aq) \rightarrow Cl^-(aq) + 3H_2O(l) + 3I_2(s)$$

Check balancing:

Reactants:		Products:	
1 Cl		1 Cl	
3 O		3 O	
6 H		6 H	
6 I		6 I	
−1 charge		−1 charge	

Oxidizing agent is ClO_3^- and reducing agent is I^-.

b) Divide into half-reactions:

$$MnO_4^-(aq) \rightarrow MnO_2(s)$$
$$SO_3^{2-}(aq) \rightarrow SO_4^{2-}(aq)$$

Balance elements other than O and H:

$$MnO_4^-(aq) \rightarrow MnO_2(s) \qquad\qquad \text{Mn is balanced}$$
$$SO_3^{2-}(aq) \rightarrow SO_4^{2-}(aq) \qquad\qquad \text{S is balanced}$$

Balance O by adding H_2O:

$$MnO_4^-(aq) \rightarrow MnO_2(s) + 2H_2O(l) \qquad\qquad \text{add two } H_2O \text{ to products}$$
$$SO_3^{2-}(aq) + H_2O(l) \rightarrow SO_4^{2-}(aq) \qquad\qquad \text{add one } H_2O \text{ to reactants}$$

Balance H by adding H^+:

$$MnO_4^-(aq) + 4H^+(aq) \rightarrow MnO_2(s) + 2H_2O(l) \qquad\qquad \text{add four } H^+ \text{ to reactants}$$
$$SO_3^{2-}(aq) + H_2O(l) \rightarrow SO_4^{2-}(aq) + 2H^+(aq) \qquad\qquad \text{add two } H^+ \text{ to products}$$

Balance charge by adding e^-:

$$MnO_4^-(aq) + 4H^+(aq) + 3e^- \rightarrow MnO_2(s) + 2H_2O(l) \qquad \text{add } 3e^- \text{ to reactants for a 0 charge on each side}$$
$$SO_3^{2-}(aq) + H_2O(l) \rightarrow SO_4^{2-}(aq) + 2H^+(aq) + 2e^- \qquad \text{add } 2e^- \text{ to products for a } -2 \text{ charge on each side}$$

Multiply each half-reaction by an integer to equalize the number of electrons:

$$2\{MnO_4^-(aq) + 4H^+(aq) + 3e^- \rightarrow MnO_2(s) + 2H_2O(l)\} \text{ or} \qquad \text{multiply by two to give } 6e^-$$
$$2MnO_4^-(aq) + 8H^+(aq) + 6e^- \rightarrow 2MnO_2(s) + 4H_2O(l)$$
$$3\{SO_3^{2-}(aq) + H_2O(l) \rightarrow SO_4^{2-}(aq) + 2H^+(aq) + 2e^-\} \text{ or} \qquad \text{multiply by three to give } 6e^-$$
$$3SO_3^{2-}(aq) + 3H_2O(l) \rightarrow 3SO_4^{2-}(aq) + 6H^+(aq) + 6e^-$$

Add half-reactions and cancel substances that appear as both reactants and products:

$$2\,MnO_4^-(aq) + 8H^+(aq) + 3SO_3^{2-}(aq) + \cancel{3H_2O(l)} \rightarrow 2MnO_2(s) + \cancel{4}H_2O(l) + 3SO_4^{2-}(aq) + \cancel{6H^+(aq)}$$

The balanced equation in acidic solution is:

$$2MnO_4^-(aq) + 2H^+(aq) + 3SO_3^{2-}(aq) \rightarrow 2MnO_2(s) + H_2O(l) + 3SO_4^{2-}(aq)$$

To change to basic solution, add OH^- to both sides of equation to neutralize H^+

$$2MnO_4^-(aq) + 2H^+(aq) + 2OH^-(aq) + 3SO_3^{2-}(aq) \rightarrow 2MnO_2(s) + H_2O(l) + 3SO_4^{2-}(aq) + 2OH^-(aq)$$
$$2MnO_4^-(aq) + \cancel{2}H_2O(l) + 3SO_3^{2-}(aq) \rightarrow 2MnO_2(s) + \cancel{H_2O(l)} + 3SO_4^{2-}(aq) + 2OH^-(aq)$$

Balanced equation in basic solution:

$$2MnO_4^-(aq) + H_2O(l) + 3SO_3^{2-}(aq) \rightarrow 2MnO_2(s) + 3SO_4^{2-}(aq) + 2OH^-(aq)$$

Check balancing:

Reactants:	2 Mn	Products:	2 Mn
	18 O		18 O
	2 H		2 H
	3 S		3 S
	–8 charge		–8 charge

Oxidizing agent is MnO_4^- and reducing agent is SO_3^{2-}.

c) Divide into half-reactions:

$$MnO_4^-(aq) \rightarrow Mn^{2+}(aq)$$
$$H_2O_2(aq) \rightarrow O_2(g)$$

Balance elements other than O and H:

$MnO_4^-(aq) \rightarrow Mn^{2+}(aq)$ Mn is balanced

$H_2O_2(aq) \rightarrow O_2(g)$ No other elements to balance

Balance O by adding H_2O:

$MnO_4^-(aq) \rightarrow Mn^{2+}(aq) + 4H_2O(l)$ add four H_2O to products

$H_2O_2(aq) \rightarrow O_2(g)$ O is balanced

Balance H by adding H^+:

$MnO_4^-(aq) + 8H^+(aq) \rightarrow Mn^{2+}(aq) + 4H_2O(l)$ add eight H^+ to reactants

$H_2O_2(aq) \rightarrow O_2(g) + 2H^+(aq)$ add two H^+ to products

Balance charge by adding e^-:

$MnO_4^-(aq) + 8H^+(aq) + 5e^- \rightarrow Mn^{2+}(aq) + 4H_2O(l)$ add $5e^-$ to reactants for +2 on each side

$H_2O_2(aq) \rightarrow O_2(g) + 2H^+(aq) + 2e^-$ add $2e^-$ to products for 0 charge on each side

Multiply each half-reaction by an integer to equalize the number of electrons:

$2\{MnO_4^-(aq) + 8H^+(aq) + 5e^- \rightarrow Mn^{2+}(aq) + 4H_2O(l)\}$ or   multiply by two to give $10e^-$

$2MnO_4^-(aq) + 16H^+(aq) + 10e^- \rightarrow 2Mn^{2+}(aq) + 8H_2O(l)$

$5\{H_2O_2(aq) \rightarrow O_2(g) + 2H^+(aq) + 2e^-\}$ or   multiply by five to give $10e^-$

$5H_2O_2(aq) \rightarrow 5O_2(g) + 10H^+(aq) + 10e^-$

Add half-reactions and cancel substances that appear as both reactants and products:

$$2MnO_4^-(aq) + \cancel{16}H^+(aq) + 5H_2O_2(aq) \rightarrow 2Mn^{2+}(aq) + 8H_2O(l) + 5O_2(g) + \cancel{10H^+(aq)}$$

The balanced equation in acidic solution:

$$2MnO_4^-(aq) + 6H^+(aq) + 5H_2O_2(aq) \rightarrow 2Mn^{2+}(aq) + 8H_2O(l) + 5O_2(g)$$

Check balancing:

Reactants:	2 Mn	Products:	2 Mn
	18 O		18 O
	16 H		16 H
	+4 charge		+4 charge

Oxidizing agent is MnO_4^- and reducing agent is H_2O_2.

21.14   Plan: Divide the reaction into the two half-reactions, balance elements other than oxygen and hydrogen, and then balance oxygen by adding H_2O and hydrogen by adding H^+. Balance the charge by adding electrons and multiply each half-reaction by an integer so that the number of electrons lost equals the number of electrons gained. Add the half-reactions together, canceling substances that appear on both sides. For basic solutions, add one OH^- ion to each side of the equation for every H^+ ion present to form H_2O and cancel excess H_2O molecules. The substance that gains electrons is the oxidizing agent while the substance that loses electrons is the reducing agent.

Solution:

a) Balance the reduction half-reaction:

$Cr_2O_7^{2-}(aq) \rightarrow 2Cr^{3+}(aq)$ balance Cr

$Cr_2O_7^{2-}(aq) \rightarrow 2Cr^{3+}(aq) + 7H_2O(l)$ balance O by adding H_2O

$Cr_2O_7^{2-}(aq) + 14H^+(aq) \rightarrow 2Cr^{3+}(aq) + 7H_2O(l)$ balance H by adding H^+

$Cr_2O_7^{2-}(aq) + 14H^+(aq) + 6e^- \rightarrow 2Cr^{3+}(aq) + 7H_2O(l)$ balance charge by adding $6e^-$

Balance the oxidation half-reaction:

$$Zn(s) \rightarrow Zn^{2+}(aq) + 2e^-$$ balance charge by adding 2e⁻

Add the two half-reactions multiplying the oxidation half-reaction by three to equalize the electrons:

$$Cr_2O_7^{2-}(aq) + 14H^+(aq) + 6e^- \rightarrow 2Cr^{3+}(aq) + 7H_2O(l)$$
$$3Zn(s) \rightarrow 3Zn^{2+}(aq) + 6e^-$$

Add half-reactions and cancel substances that appear as both reactants and products:

$$Cr_2O_7^{2-}(aq) + 14H^+(aq) + 3Zn(s) \rightarrow 2Cr^{3+}(aq) + 7H_2O(l) + 3Zn^{2+}(aq)$$

Oxidizing agent is $Cr_2O_7^{2-}$ and reducing agent is Zn.

b) Balance the reduction half-reaction:

$$MnO_4^-(aq) \rightarrow MnO_2(s) + 2H_2O(l)$$ balance O by adding H_2O
$$MnO_4^-(aq) + 4H^+(aq) \rightarrow MnO_2(s) + 2H_2O(l)$$ balance H by adding H^+
$$MnO_4^-(aq) + 4H^+(aq) + 3e^- \rightarrow MnO_2(s) + 2H_2O(l)$$ balance charge by adding 3 e⁻

Balance the oxidation half-reaction:

$$Fe(OH)_2(s) + H_2O(l) \rightarrow Fe(OH)_3(s)$$ balance O by adding H_2O
$$Fe(OH)_2(s) + H_2O(l) \rightarrow Fe(OH)_3(s) + H^+(aq)$$ balance H by adding H^+
$$Fe(OH)_2(s) + H_2O(l) \rightarrow Fe(OH)_3(s) + H^+(aq) + e^-$$ balance charge by adding 1e⁻

Add half-reactions after multiplying oxidation half-reaction by 3:

$$MnO_4^-(aq) + 4H^+(aq) + 3e^- \rightarrow MnO_2(s) + 2H_2O(l)$$
$$3Fe(OH)_2(s) + 3H_2O(l) \rightarrow 3Fe(OH)_3(s) + 3H^+(aq) + 3e^-$$

Add half-reactions and cancel substances that appear as both reactants and products:

$$MnO_4^-(aq) + \cancel{4}H^+(aq) + 3Fe(OH)_2(s) + \cancel{3}H_2O(l) \rightarrow MnO_2(s) + \cancel{2H_2O(l)} + 3Fe(OH)_3(s) + \cancel{3H^+(aq)}$$
$$MnO_4^-(aq) + H^+(aq) + 3Fe(OH)_2(s) + H_2O(l) \rightarrow MnO_2(s) + 3Fe(OH)_3(s)$$

Add OH⁻ to both sides to neutralize the H⁺ and convert $H^+ + OH^- \rightarrow H_2O$:

$$MnO_4^-(aq) + H^+(aq) + OH^-(aq) + 3Fe(OH)_2(s) + H_2O(l) \rightarrow MnO_2(s) + 3Fe(OH)_3(s) + OH^-(aq)$$
$$MnO_4^-(aq) + 3Fe(OH)_2(s) + 2H_2O(l) \rightarrow MnO_2(s) + 3Fe(OH)_3(s) + OH^-(aq)$$

Oxidizing agent is MnO_4^- and reducing agent is $Fe(OH)_2$.

c) Balance the reduction half-reaction:

$$2NO_3^-(aq) \rightarrow N_2(g)$$ balance N
$$2NO_3^-(aq) \rightarrow N_2(g) + 6H_2O(l)$$ balance O by adding H_2O
$$2NO_3^-(aq) + 12H^+(aq) \rightarrow N_2(g) + 6H_2O(l)$$ balance H by adding H^+
$$2NO_3^-(aq) + 12H^+(aq) + 10e^- \rightarrow N_2(g) + 6H_2O(l)$$ balance charge by adding 10e⁻

Balance the oxidation half-reaction:

$$Zn(s) \rightarrow Zn^{2+}(aq) + 2e^-$$ balance charge by adding 2e⁻

Add the half-reactions after multiplying the reduction half-reaction by one and the oxidation half-reaction by five:

$$2NO_3^-(aq) + 12H^+(aq) + 10e^- \rightarrow N_2(g) + 6H_2O(l)$$
$$5Zn(s) \rightarrow 5Zn^{2+}(aq) + 10e^-$$

Add half-reactions and cancel substances that appear as both reactants and products:

$$2NO_3^-(aq) + 12H^+(aq) + 5Zn(s) \rightarrow N_2(g) + 6H_2O(l) + 5Zn^{2+}(aq)$$

Oxidizing agent is NO_3^- and reducing agent is Zn.

21.16   Plan: Divide the reaction into the two half-reactions, balance elements other than oxygen and hydrogen, and then balance oxygen by adding H_2O and hydrogen by adding H^+. Balance the charge by adding electrons and multiply each half-reaction by an integer so that the number of electrons lost equals the number of electrons gained. Add the half-reactions together, canceling substances that appear on both sides. For basic solutions, add one OH⁻ ion to each side of the equation for every H⁺ ion present to form H_2O and cancel excess H_2O molecules. The substance that gains electrons is the oxidizing agent while the substance that loses electrons is the reducing agent.

Solution:

a) Balance the reduction half-reaction:

$$NO_3^-(aq) \rightarrow NO(g) + 2H_2O(l)$$ balance O by adding H_2O
$$NO_3^-(aq) + 4H^+(aq) \rightarrow NO(g) + 2H_2O(l)$$ balance H by adding H^+
$$NO_3^-(aq) + 4H^+(aq) + 3e^- \rightarrow NO(g) + 2H_2O(l)$$ balance charge by adding 3e⁻

Balance oxidation half-reaction:

$4Sb(s) \rightarrow Sb_4O_6(s)$ balance Sb

$4Sb(s) + 6H_2O(l) \rightarrow Sb_4O_6(s)$ balance O by adding H_2O

$4Sb(s) + 6H_2O(l) \rightarrow Sb_4O_6(s) + 12H^+(aq)$ balance H by adding H^+

$4Sb(s) + 6H_2O(l) \rightarrow Sb_4O_6(s) + 12H^+(aq) + 12e^-$ balance charge by adding $12e^-$

Multiply each half-reaction by an integer to equalize the number of electrons:

$4\{NO_3^-(aq) + 4H^+(aq) + 3e^- \rightarrow NO(g) + 2H_2O(l)\}$ multiply by four to give $12e^-$

$1\{4Sb(s) + 6H_2O(l) \rightarrow Sb_4O_6(s) + 12H^+(aq) + 12e^-\}$ multiply by one to give $12e^-$

This gives:

$4NO_3^-(aq) + 16H^+(aq) + 12e^- \rightarrow 4NO(g) + 8H_2O(l)$

$4Sb(s) + 6H_2O(l) \rightarrow Sb_4O_6(s) + 12H^+(aq) + 12e^-$

Add half-reactions. Cancel common reactants and products:

$4 NO_3^-(aq) + \cancel{16}H^+(aq) + 4Sb(s) + \cancel{6H_2O}(l) \rightarrow 4NO(g) + 8H_2O(l) + Sb_4O_6(s) + \cancel{12H^+(aq)}$

Balanced equation in acidic solution:

$4NO_3^-(aq) + 4H^+(aq) + 4Sb(s) \rightarrow 4NO(g) + 2H_2O(l) + Sb_4O_6(s)$

Oxidizing agent is NO_3^- and reducing agent is Sb.

b) Balance reduction half-reaction:

$BiO_3^-(aq) \rightarrow Bi^{3+}(aq) + 3H_2O(l)$ balance O by adding H_2O

$BiO_3^-(aq) + 6H^+(aq) \rightarrow Bi^{3+}(aq) + 3H_2O(l)$ balance H by adding H^+

$BiO_3^-(aq) + 6H^+(aq) + 2e^- \rightarrow Bi^{3+}(aq) + 3H_2O(l)$ balance charge to give +3 on each side

Balance oxidation half-reaction:

$Mn^{2+}(aq) + 4H_2O(l) \rightarrow MnO_4^-(aq)$ balance O by adding H_2O

$Mn^{2+}(aq) + 4H_2O(l) \rightarrow MnO_4^-(aq) + 8H^+(aq)$ balance H by adding H^+

$Mn^{2+}(aq) + 4H_2O(l) \rightarrow MnO_4^-(aq) + 8H^+(aq) + 5e^-$ balance charge to give +2 on each side

Multiply each half-reaction by an integer to equalize the number of electrons:

$5\{BiO_3^-(aq) + 6H^+(aq) + 2e^- \rightarrow Bi^{3+}(aq) + 3H_2O(l)\}$ multiply by five to give $10e^-$

$2\{Mn^{2+}(aq) + 4H_2O(l) \rightarrow MnO_4^-(aq) + 8H^+(aq) + 5e^-\}$ multiply by two to give $10e^-$

This gives:

$5BiO_3^-(aq) + 30H^+(aq) + 10e^- \rightarrow 5Bi^{3+}(aq) + 15H_2O(l)$

$2Mn^{2+}(aq) + 8H_2O(l) \rightarrow 2MnO_4^-(aq) + 16H^+(aq) + 10e^-$

Add half-reactions. Cancel H_2O and H^+ in reactants and products:

$5BiO_3^-(aq) + \cancel{30}H^+(aq) + 2Mn^{2+}(aq) + \cancel{8H_2O}(l) \rightarrow 5Bi^{3+}(aq) + \cancel{15}H_2O(l) + 2MnO_4^-(aq) + \cancel{16H^+(aq)}$

Balanced reaction in acidic solution:

$5BiO_3^-(aq) + 14H^+(aq) + 2Mn^{2+}(aq) \rightarrow 5Bi^{3+}(aq) + 7H_2O(l) + 2MnO_4^-(aq)$

BiO_3^- is the oxidizing agent and Mn^{2+} is the reducing agent.

c) Balance the reduction half-reaction:

$Pb(OH)_3^-(aq) \rightarrow Pb(s) + 3H_2O(l)$ balance O by adding H_2O

$Pb(OH)_3^-(aq) + 3H^+(aq) \rightarrow Pb(s) + 3H_2O(l)$ balance H by adding H^+

$Pb(OH)_3^-(aq) + 3H^+(aq) + 2e^- \rightarrow Pb(s) + 3H_2O(l)$ balance charge to give 0 on each side

Balance the oxidation half-reaction:

$Fe(OH)_2(s) + H_2O(l) \rightarrow Fe(OH)_3(s)$ balance O by adding H_2O

$Fe(OH)_2(s) + H_2O(l) \rightarrow Fe(OH)_3(s) + H^+(aq)$ balance H by adding H^+

$Fe(OH)_2(s) + H_2O(l) \rightarrow Fe(OH)_3(s) + H^+(aq) + e^-$ balance charge to give 0 on each side

Multiply each half-reaction by an integer to equalize the number of electrons:

$1\{Pb(OH)_3^-(aq) + 3H^+(aq) + 2e^- \rightarrow Pb(s) + 3H_2O(l)\}$ multiply by 1 to give $2e^-$

$2\{Fe(OH)_2(s) + H_2O(l) \rightarrow Fe(OH)_3(s) + H^+(aq) + e^-\}$ multiply by 2 to give $2e^-$

This gives:

$Pb(OH)_3^-(aq) + 3H^+(aq) + 2e^- \rightarrow Pb(s) + 3H_2O(l)$

$2Fe(OH)_2(s) + 2H_2O(l) \rightarrow 2Fe(OH)_3(s) + 2H^+(aq) + 2e^-$

Add the two half-reactions. Cancel H_2O and H^+:

$Pb(OH)_3^-(aq) + \cancel{3}H^+(aq) + 2Fe(OH)_2(s) + \cancel{2H_2O}(l) \rightarrow Pb(s) + 3H_2O(l) + 2Fe(OH)_3(s) + \cancel{2H^+(aq)}$

$Pb(OH)_3^-(aq) + H^+(aq) + 2Fe(OH)_2(s) \rightarrow Pb(s) + H_2O(l) + 2Fe(OH)_3(s)$

Add one OH⁻ to both sides to neutralize H⁺:

$$Pb(OH)_3{}^-(aq) + H^+(aq) + OH^-(aq) + 2Fe(OH)_2(s) \rightarrow Pb(s) + H_2O(l) + 2\,Fe(OH)_3(s) + OH^-(aq)$$
$$Pb(OH)_3{}^-(aq) + \cancel{H_2O(l)} + 2Fe(OH)_2(s) \rightarrow Pb(s) + \cancel{H_2O(l)} + 2Fe(OH)_3(s) + OH^-(aq)$$

Balanced reaction in basic solution:

$$Pb(OH)_3{}^-(aq) + 2Fe(OH)_2(s) \rightarrow Pb(s) + 2Fe(OH)_3(s) + OH^-(aq)$$

$Pb(OH)_3{}^-$ is the oxidizing agent and $Fe(OH)_2$ is the reducing agent.

21.18   Plan: Divide the reaction into the two half-reactions, balance elements other than oxygen and hydrogen, and then balance oxygen by adding H_2O and hydrogen by adding H^+. Balance the charge by adding electrons and multiply each half-reaction by an integer so that the number of electrons lost equals the number of electrons gained. Add the half-reactions together, canceling substances that appear on both sides. For basic solutions, add one OH⁻ ion to each side of the equation for every H^+ ion present to form H_2O and cancel excess H_2O molecules. The substance that gains electrons is the oxidizing agent while the substance that loses electrons is the reducing agent.

Solution:

a) Balance reduction half-reaction:

$MnO_4{}^-(aq) \rightarrow Mn^{2+}(aq) + 4H_2O(l)$	balance O by adding H_2O
$MnO_4{}^-(aq) + 8H^+(aq) \rightarrow Mn^{2+}(aq) + 4H_2O(l)$	balance H by adding H^+
$MnO_4{}^-(aq) + 8H^+(aq) + 5e^- \rightarrow Mn^{2+}(aq) + 4H_2O(l)$	balance charge by adding 5e⁻

Balance oxidation half-reaction:

$As_4O_6(s) \rightarrow 4AsO_4{}^{3-}(aq)$	balance As
$As_4O_6(s) + 10H_2O(l) \rightarrow 4AsO_4{}^{3-}(aq)$	balance O by adding H_2O
$As_4O_6(s) + 10H_2O(l) \rightarrow 4AsO_4{}^{3-}(aq) + 20H^+(aq)$	balance H by adding H^+
$As_4O_6(s) + 10H_2O(l) \rightarrow 4AsO_4{}^{3-}(aq) + 20H^+(aq) + 8e^-$	balance charge by adding 8e⁻

Multiply reduction half-reaction by 8 and oxidation half-reaction by 5 to transfer 40 e⁻ in overall reaction.

$$8MnO_4{}^-(aq) + 64H^+(aq) + 40e^- \rightarrow 8Mn^{2+}(aq) + 32H_2O(l)$$
$$5As_4O_6(s) + 50H_2O(l) \rightarrow 20AsO_4{}^{3-}(aq) + 100H^+(aq) + 40e^-$$

Add the half-reactions and cancel H_2O and H^+:

$$5As_4O_6(s) + 8MnO_4{}^-(aq) + \cancel{64H^+(aq)} + 50H_2O(l) \rightarrow 20AsO_4{}^{3-}(aq) + 8Mn^{2+}(aq) + \cancel{32H_2O(l)} + \cancel{100}H^+(aq)$$

Balanced reaction in acidic solution:

$$5As_4O_6(s) + 8MnO_4{}^-(aq) + 18H_2O(l) \rightarrow 20AsO_4{}^{3-}(aq) + 8Mn^{2+}(aq) + 36H^+(aq)$$

Oxidizing agent is $MnO_4{}^-$ and reducing agent is As_4O_6.

b) The reaction gives only one reactant, P_4. Since both products contain phosphorus, divide the half-reactions so each includes P_4 as the reactant.

Balance reduction half-reaction:

$P_4(s) \rightarrow 4PH_3(g)$	balance P
$P_4(s) + 12H^+(aq) \rightarrow 4PH_3(g)$	balance H by adding H^+
$P_4(s) + 12H^+(aq) + 12e^- \rightarrow 4PH_3(g)$	balance charge by adding 12 e⁻

Balance oxidation half-reaction:

$P_4(s) \rightarrow 4HPO_3{}^{2-}(aq)$	balance P
$P_4(s) + 12H_2O(l) \rightarrow 4HPO_3{}^{2-}(aq)$	balance O by adding H_2O
$P_4(s) + 12H_2O(l) \rightarrow 4HPO_3{}^{2-}(aq) + 20H^+(aq)$	balance H by adding H^+
$P_4(s) + 12H_2O(l) \rightarrow 4HPO_3{}^{2-}(aq) + 20H^+(aq) + 12e^-$	balance charge by adding 12 e⁻

Add two half-reactions and cancel H^+:

$$2P_4(s) + \cancel{12H^+(aq)} + 12H_2O(l) \rightarrow 4HPO_3{}^{2-}(aq) + 4PH_3(g) + \cancel{20}H^+(aq)$$

Balanced reaction in acidic solution:

$$2P_4(s) + 12H_2O(l) \rightarrow 4HPO_3{}^{2-}(aq) + 4PH_3(g) + 8H^+(aq) \text{ or}$$
$$P_4(s) + 6H_2O(l) \rightarrow 2HPO_3{}^{2-}(aq) + 2PH_3(g) + 4H^+(aq)$$

P_4 is both the oxidizing agent and reducing agent.

c) Balance the reduction half-reaction:

$MnO_4{}^-(aq) \rightarrow MnO_2(s) + 2H_2O(l)$	balance O by adding H_2O
$MnO_4{}^-(aq) + 4H^+(aq) \rightarrow MnO_2(s) + 2H_2O(l)$	balance H by adding H^+
$MnO_4{}^-(aq) + 4H^+(aq) + 3e^- \rightarrow MnO_2(s) + 2H_2O(l)$	balance charge by adding 3e⁻

Balance oxidation half-reaction:

$$CN^-(aq) + H_2O(l) \rightarrow CNO^-(aq)$$ balance O by adding H_2O

$$CN^-(aq) + H_2O(l) \rightarrow CNO^-(aq) + 2H^+(aq)$$ balance H by adding H^+

$$CN^-(aq) + H_2O(l) \rightarrow CNO^-(aq) + 2H^+(aq) + 2e^-$$ balance charge by adding $2e^-$

Multiply the oxidation half-reaction by three and reduction half-reaction by two to transfer $6e^-$ in overall reaction.

$$2MnO_4^-(aq) + 8H^+(aq) + 6e^- \rightarrow 2MnO_2(s) + 4H_2O(l)$$
$$3CN^-(aq) + 3H_2O(l) \rightarrow 3CNO^-(aq) + 6H^+(aq) + 6e^-$$

Add the two half-reactions. Cancel the H_2O and H^+:

$$2MnO_4^-(aq) + 3CN^-(aq) + 8H^+(aq) + 3H_2O(l) \rightarrow 2MnO_2(s) + 3CNO^-(aq) + 6H^+(aq) + 4H_2O(l)$$
$$2MnO_4^-(aq) + 3CN^-(aq) + 2H^+(aq) \rightarrow 2MnO_2(s) + 3CNO^-(aq) + H_2O(l)$$

Add 2 OH^- to both sides to neutralize H^+ and form H_2O:

$$2MnO_4^-(aq) + 3CN^-(aq) + 2H^+(aq) + 2OH^-(aq) \rightarrow 2MnO_2(s) + 3CNO^-(aq) + H_2O(l) + 2OH^-(aq)$$
$$2MnO_4^-(aq) + 3CN^-(aq) + 2H_2O(l) \rightarrow 2MnO_2(s) + 3CNO^-(aq) + H_2O(l) + 2OH^-(aq)$$

Balanced reaction in basic solution:

$$2MnO_4^-(aq) + 3CN^-(aq) + H_2O(l) \rightarrow 2MnO_2(s) + 3CNO^-(aq) + 2OH^-(aq)$$

Oxidizing agent is MnO_4^- and reducing agent is CN^-.

21.21 a) Balance reduction half-reaction:

$$NO_3^-(aq) \rightarrow NO_2(g) + H_2O(l)$$ balance O by adding H_2O

$$NO_3^-(aq) + 2H^+(aq) \rightarrow NO_2(g) + H_2O(l)$$ balance H by adding H^+

$$NO_3^-(aq) + 2H^+(aq) + e^- \rightarrow NO_2(g) + H_2O(l)$$ balance charge to give 0 on each side

Balance oxidation half-reaction:

$$Au(s) + 4Cl^-(aq) \rightarrow AuCl_4^-(aq)$$ balance Cl

$$Au(s) + 4Cl^-(aq) \rightarrow AuCl_4^-(aq) + 3e^-$$ balance charge to –4 on each side

Multiply each half-reaction by an integer to equalize the number of electrons:

$$3\{NO_3^-(aq) + 2H^+(aq) + e^- \rightarrow NO_2(g) + H_2O(l)\}$$ multiply by three to give $3e^-$

$$1\{ Au(s) + 4Cl^-(aq) \rightarrow AuCl_4^-(aq) + 3e^-\}$$ multiply by one to give $3e^-$

This gives:

$$3NO_3^-(aq) + 6H^+(aq) + 3e^- \rightarrow 3NO_2(g) + 3H_2O(l)$$
$$Au(s) + 4Cl^-(aq) \rightarrow AuCl_4^-(aq) + 3e^-$$

Add half-reactions:

$$Au(s) + 3NO_3^-(aq) + 4Cl^-(aq) + 6H^+(aq) \rightarrow AuCl_4^-(aq) + 3NO_2(g) + 3H_2O(l)$$

b) Oxidizing agent is **NO_3^-** and reducing agent is **Au**.

c) The HCl provides chloride ions that combine with the unstable gold ion to form the stable ion, $AuCl_4^-$.

21.22 Plan: The oxidation half-cell (anode) is shown on the left while the reduction half-cell (cathode) is shown on the right. Remember that oxidation is the loss of electrons and electrons leave the oxidation half-cell and move towards the positively charged cathode. If a metal is reduced, it will plate out on the cathode.
Solution:
a) **A** is the anode because by convention the anode is shown on the left.
b) **E** is the cathode because by convention the cathode is shown on the right.
c) **C** is the salt bridge providing electrical connection between the two solutions.
d) **A** is the anode, so oxidation takes place there. Oxidation is the loss of electrons, meaning that electrons are leaving the anode.
e) **E** is assigned a positive charge because it is the cathode.
f) **E** gains mass because the reduction of the metal ion produces the solid metal which plates out on E.

21.25 An active electrode is a reactant or product in the cell reaction, whereas an inactive electrode is neither a reactant nor a product. An inactive electrode is present only to conduct electricity when the half-cell reaction does not include a metal. Platinum and graphite are commonly used as inactive electrodes.

21.26   a) The metal **A** is being oxidized to form the metal cation. To form positive ions, an atom must always lose electrons, so this half-reaction is always an oxidation.
b) The metal ion **B** is gaining electrons to form the metal **B**, so it is displaced.
c) The anode is the electrode at which oxidation takes place, so metal **A** is used as the anode.
d) Acid oxidizes metal **B** and metal **B** oxidizes metal **A**, so acid will oxidize metal **A** and **bubbles will form** when metal **A** is placed in acid. The same answer results if strength of reducing agents is considered. The fact that metal **A** is a better reducing agent than metal **B** indicates that if metal **B** reduces acid, then metal **A** will also reduce acid.

21.27   Plan: The oxidation half-cell (anode) is shown on the left while the reduction half-cell (cathode) is shown on the right. Remember that oxidation is the loss of electrons and electrons leave the oxidation half-cell and move towards the positively charged cathode. Anions from the salt bridge flow into the oxidation half-cell, while cations from the salt bridge flow into the reduction half-cell.
Solution:
a) Electrons flow from the anode to the cathode, so **from the iron half-cell to the nickel half-cell**, left to right in the figure. By convention, the anode appears on the left and the cathode on the right.
b) Oxidation occurs at the anode, which is the electrode in the **iron** half-cell.
c) Electrons enter the reduction half-cell, the **nickel** half-cell in this example.
d) Electrons are consumed in the reduction half-reaction. Reduction takes place at the cathode, **nickel** electrode.
e) The anode is assigned a negative charge, so the **iron** electrode is negatively charged.
f) Metal is oxidized in the oxidation half-cell, so the **iron** electrode will decrease in mass.
g) The solution must contain nickel ions, so any nickel salt can be added. **$1\ M\ NiSO_4$** is one choice.
h) KNO_3 is commonly used in salt bridges, the ions being **K^+ and NO_3^-**. Other salts are also acceptable answers.
i) **Neither**, because an inactive electrode could not replace either electrode since both the oxidation and the reduction half-reactions include the metal as either a reactant or a product.
j) Anions will move towards the half-cell in which positive ions are being produced. The oxidation half-cell produces Fe^{2+}, so salt bridge anions move **from right** (nickel half-cell) **to left** (iron half-cell).
k)   Oxidation half-reaction: $Fe(s) \rightarrow Fe^{2+}(aq) + 2e^-$
         Reduction half-reaction: $Ni^{2+}(aq) + 2e^- \rightarrow Ni(s)$
         Overall cell reaction: $Fe(s) + Ni^{2+}(aq) \rightarrow Fe^{2+}(aq) + Ni(s)$

21.29   Plan: The anode, at which the oxidation takes place, is the negative electrode. Electrons flow from the anode to the cathode. Anions from the salt bridge flow into the oxidation half-cell, while cations from the salt bridge flow into the reduction half-cell.
Solution:
a) If the zinc electrode is negative, it is the anode and oxidation takes place at the zinc electrode:
         $Zn(s) \rightarrow Zn^{2+}(aq) + 2e^-$
Reduction half-reaction: $Sn^{2+}(aq) + 2e^- \rightarrow Sn(s)$
Overall reaction: $Zn(s) + Sn^{2+}(aq) \rightarrow Zn^{2+}(aq) + Sn(s)$
b)

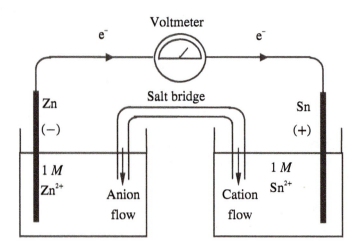

21.31   Plan: The cathode, at which the reduction takes place, is the positive electrode. Electrons flow from the anode to the cathode. Anions from the salt bridge flow into the oxidation half-cell, while cations from the salt bridge flow into the reduction half-cell.
Solution:
a) The cathode is assigned a positive charge, so the iron electrode is the cathode.
Reduction half-reaction: $Fe^{2+}(aq) + 2e^- \rightarrow Fe(s)$
Oxidation half-reaction: $Mn(s) \rightarrow Mn^{2+}(aq) + 2e^-$
Overall cell reaction: $Fe^{2+}(aq) + Mn(s) \rightarrow Fe(s) + Mn^{2+}(aq)$
b)

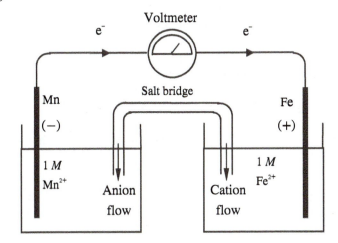

21.33   Plan: In cell notation, the oxidation components of the anode compartment are written on the left of the salt bridge and the reduction components of the cathode compartment are written to the right of the salt bridge. A double vertical line separates the anode from the cathode and represents the salt bridge. A single vertical line separates species of different phases. Anode ‖ Cathode
Solution:
a) Al is oxidized, so it is the anode and appears first in the cell notation. There is a single vertical line separating the solid metals from their solutions.
        $Al(s) \mid Al^{3+}(aq) \parallel Cr^{3+}(aq) \mid Cr(s)$
b) Cu^{2+} is reduced, so Cu is the cathode and appears last in the cell notation. The oxidation of SO_2 does not include a metal, so an inactive electrode must be present. Hydrogen ion must be included in the oxidation half-cell.
        $Pt \mid SO_2(g) \mid SO_4^{2-}(aq), H^+(aq) \parallel Cu^{2+}(aq) \mid Cu(s)$

21.36   A negative E°_{cell} indicates that the cell reaction is not spontaneous, $\Delta G^\circ > 0$. The reverse reaction is spontaneous with $E^\circ_{cell} > 0$.

21.37   Similar to other state functions, the sign of E° changes when a reaction is reversed. Unlike ΔG°, ΔH° and S°, E° is an intensive property, the ratio of energy to charge. When the coefficients in a reaction are multiplied by a factor, the values of ΔG°, ΔH° and S° are multiplied by the same factor. However, E° does not change because both the energy and charge are multiplied by the factor and their ratio remains unchanged.

21.38   Plan: Divide the balanced equation into reduction and oxidation half-reactions and add electrons. Add water and hydroxide ion to the half-reaction that includes oxygen. Use the relationship $E^\circ_{cell} = E^\circ_{cathode} - E^\circ_{anode}$ to find the unknown E° value.
Solution:
a) Oxidation: $Se^{2-}(aq) \rightarrow Se(s) + 2e^-$
   Reduction: $2SO_3^{2-}(aq) + 3H_2O(l) + 4e^- \rightarrow S_2O_3^{2-}(aq) + 6OH^-(aq)$

b) $E^{\circ}_{cell} = E^{\circ}_{cathode} - E^{\circ}_{anode}$

$E^{\circ}_{anode} = E^{\circ}_{cathode} - E^{\circ}_{cell} = -0.57 \text{ V} - 0.35 \text{ V} = \mathbf{-0.92 \text{ V}}$

21.40 Plan: The greater (more positive) the reduction potential, the greater the strength as an oxidizing agent.
Solution:
a) From Appendix D:

$Fe^{3+}(aq) + e^- \rightarrow Fe^{2+}(aq)$	$E^{\circ} = 0.77 \text{ V}$
$Br_2(l) + 2e^- \rightarrow 2Br^-(aq)$	$E^{\circ} = 1.07 \text{ V}$
$Cu^{2+}(aq) + e^- \rightarrow Cu(s)$	$E^{\circ} = 0.34 \text{ V}$

When placed in order of decreasing strength as oxidizing agents: $\mathbf{Br_2 > Fe^{3+} > Cu^{2+}}$.
b) From Appendix D:

$Ca^{2+}(aq) + 2e- \rightarrow Ca(s)$	$E^{\circ} = -2.87 \text{ V}$
$Cr_2O_7^{2-}(aq) + 14H^+(aq) \ 6e^- \rightarrow 2Cr^{3+}(aq) + 7H_2O(l)$	$E^{\circ} = 1.33 \text{ V}$
$Ag^+(aq) + e^- \rightarrow Ag(s)$	$E^{\circ} = 0.80 \text{ V}$

When placed in order of increasing strength as oxidizing agents: $\mathbf{Ca^{2+} < Ag^+ < Cr_2O_7^{2-}}$.

21.42 Plan: Use the relationship $E^{\circ}_{cell} = E^{\circ}_{cathode} - E^{\circ}_{anode}$. E° values are found in Appendix D. Spontaneous reactions have $E^{\circ}_{cell} > 0$.
Solution:
a)
Oxidation:	$Co(s) \rightarrow Co^{2+}(aq) + 2e^-$	$E^{\circ} = -0.28 \text{ V}$
Reduction:	$2H^+(aq) + 2e^- \rightarrow H_2(g)$	$E^{\circ} = 0.00 \text{ V}$
Overall reaction:	$Co(s) + 2H^+(aq) \rightarrow Co^{2+}(aq) + H_2(g)$	

$E^{\circ}_{cell} = 0.00 \text{ V} - (-0.28 \text{ V}) = \mathbf{0.28 \text{ V}}$

Reaction is **spontaneous** under standard-state conditions because E°_{cell} is positive.

b)
Oxidation:	$Hg_2^{2+}(aq) \rightarrow 2Hg^{2+}(aq) + 2e^-$	$E^{\circ} = +0.92 \text{ V}$
Reduction:	$Hg_2^{2+}(aq) + 2e^- \rightarrow 2Hg(l)$	$E^{\circ} = +0.79 \text{ V}$
Overall:	$2Hg_2^{2+}(aq) \rightarrow 2Hg^{2+}(aq) + 2Hg(l)$	
	or $Hg_2^{2+}(aq) \rightarrow Hg^{2+}(aq) + Hg(l)$	$E^{\circ}_{cell} = 0.79 \text{ V} - 0.92 \text{ V} = \mathbf{-0.13 \text{ V}}$

Negative E°_{cell} indicates reaction is **not spontaneous** under standard-state conditions.

21.44 Plan: Use the relationship $E^{\circ}_{cell} = E^{\circ}_{cathode} - E^{\circ}_{anode}$. E° values are found in Appendix D. Spontaneous reactions have $E^{\circ}_{cell} > 0$.
Solution:
a)
Oxidation:	$3\{Cd(s) \rightarrow Cd^{2+}(aq) + 2e^-\}$	$E^{\circ} = -0.40 \text{ V}$
Reduction:	$Cr_2O_7^{2-}(aq) + 14H^+(aq) + 6e^- \rightarrow 2Cr^{3+}(aq) + 7H_2O(l)$	$E^{\circ} = +1.33 \text{ V}$
Overall:	$Cr_2O_7^{2-}(aq) + 3Cd(s) + 14H^+(aq) \rightarrow 2Cr^{3+}(aq) + 3Cd^{2+}(aq) + 7H_2O(l)$	

$E^{\circ}_{cell} = +1.33 \text{ V} - (-0.40 \text{ V}) = \mathbf{+1.73 \text{ V}}$

The reaction is **spontaneous**.

b)
Oxidation:	$Pb(s) \rightarrow Pb^{2+}(aq) + 2e^-$	$E^{\circ} = -0.13 \text{ V}$
Reduction:	$Ni^{2+}(aq) + 2e^- \rightarrow Ni(s)$	$E^{\circ} = -0.25 \text{ V}$
Overall:	$Pb(s) + Ni^{2+}(aq) \rightarrow Pb^{2+}(aq) + Ni(s)$	

$E^{\circ}_{cell} = -0.25 \text{ V} - (-0.13 \text{ V}) = \mathbf{-0.12 \text{ V}}$

The reaction is **not spontaneous**.

21.46 Plan: Spontaneous reactions have $E^{\circ}_{cell} > 0$. All three reactions are written as reductions. When two half-reactions are paired, one half-reaction must be reversed and written as an oxidation. Reverse the half-reaction that will result in a positive value of E°_{cell} using the relationship $E^{\circ}_{cell} = E^{\circ}_{cathode} - E^{\circ}_{anode}$. To balance each reaction, multiply each half-reaction by an integer so that the number of electrons lost equals the number of electrons gained and then add the half-reactions. The greater (more positive) the reduction potential, the greater the strength as an oxidizing agent.

Solution:

Adding (1) and (2) to give a spontaneous reaction involves converting (1) to oxidation:

Oxidation: $2\{Al(s) \rightarrow Al^{3+}(aq) + 3e^-\}$ $E° = -1.66$ V

Reduction: $3\{N_2O_4(g) + 2e^- \rightarrow 2NO_2^-(aq)\}$ $E° = +0.867$ V

$3N_2O_4(g) + 2Al(s) \rightarrow 6NO_2^-(aq) + 2Al^{3+}(aq)$

$$E°_{cell} = 0.867 \text{ V} - (-1.66 \text{ V}) = \mathbf{2.53 \text{ V}}$$

Oxidizing agents: $N_2O_4 > Al^{3+}$; reducing agents: $Al > NO_2^-$

Adding (1) and (3) to give a spontaneous reaction involves converting (1) to oxidation:

Oxidation: $2\{Al(s) \rightarrow Al^{3+}(aq) + 3e^-\}$ $E° = -1.66$ V

Reduction: $3\{SO_4^{2-}(aq) + H_2O(l) + 2e^- \rightarrow SO_3^{2-}(aq) + 2OH^-(aq)\}$ $E° = +0.93$ V

$2Al(s) + 3SO_4^{2-}(aq) + 3H_2O(l) \rightarrow 2Al^{3+}(aq) + 3SO_3^{2-}(aq) + 6OH^-(aq)$

$$E°_{cell} = 0.93 \text{ V} - (-1.66 \text{ V}) = \mathbf{2.59 \text{ V}}$$

Oxidizing agents: $SO_4^{2-} > Al^{3+}$; reducing agents: $Al > SO_3^{2-}$

Adding (2) and (3) to give a spontaneous reaction involves converting (2) to oxidation:

Oxidation: $2NO_2^-(aq) \rightarrow N_2O_4(g) + 2e^-$ $E° = 0.867$ V

Reduction: $SO_4^{2-}(aq) + H_2O(l) + 2e^- \rightarrow SO_3^{2-}(aq) + 2OH^-(aq)$ $E° = 0.93$ V

$SO_4^{2-}(aq) + 2NO_2^-(aq) + H_2O(l) \rightarrow SO_3^{2-}(aq) + N_2O_4(g) + 2OH^-(aq)$

$$E°_{cell} = 0.93 \text{ V} - 0.867 \text{ V} = \mathbf{0.06 \text{ V}}$$

Oxidizing agents: $SO_4^{2-} > N_2O_4$; reducing agents: $NO_2^- > SO_3^{2-}$

Rank oxidizing agents (substance being reduced) in order of increasing strength:

     $\mathbf{Al^{3+} < N_2O_4 < SO_4^{2-}}$

Rank reducing agents (substance being oxidized) in order of increasing strength:

     $\mathbf{SO_3^{2-} < NO_2^- < Al}$

21.48   <u>Plan:</u> Spontaneous reactions have $E°_{cell} > 0$. All three reactions are written as reductions. When two half-reactions are paired, one half-reaction must be reversed and written as an oxidation. Reverse the half-reaction that will result in a positive value of $E°_{cell}$ using the relationship $E°_{cell} = E°_{cathode} - E°_{anode}$. To balance each reaction, multiply each half-reaction by an integer so that the number of electrons lost equals the number of electrons gained and then add the half-reactions. The greater (more positive) the reduction potential, the greater the strength as an oxidizing agent.

Solution:

Adding (1) and (2) to give a spontaneous reaction involves converting (2) to oxidation:

Oxidation: $Pt(s) \rightarrow Pt^{2+}(aq) + 2e^-$ $E° = +1.20$ V

Reduction: $2HClO(aq) + 2H^+(aq) + 2e^- \rightarrow Cl_2(g) + 2H_2O(l)$ $E° = +1.63$ V

     $2HClO(aq) + Pt(s) + 2H^+(aq) \rightarrow Cl_2(g) + Pt^{2+}(aq) + 2H_2O(l)$

     $E°_{cell} = 1.63 \text{ V} - 1.20 \text{ V} = \mathbf{0.43 \text{ V}}$

     Oxidizing agents: $HClO > Pt^{2+}$; reducing agents: $Pt > Cl_2$

Adding (1) and (3) to give a spontaneous reaction involves converting (3) to oxidation:

Oxidation: $Pb(s) + SO_4^{2-}(aq) \rightarrow PbSO_4(s) + 2e^-$ $E° = -0.31$ V

Reduction: $2HClO(aq) + 2H^+(aq) + 2e^- \rightarrow Cl_2(g) + 2H_2O(l)$ $E° = +1.63$ V

     $2HClO(aq) + Pb(s) + SO_4^{2-}(aq) + 2H^+(aq) \rightarrow Cl_2(g) + PbSO_4(s) + 2H_2O(l)$

     $E°_{cell} = 1.63 \text{ V} - (-0.31 \text{ V}) = \mathbf{1.94 \text{ V}}$

     Oxidizing agents: $HClO > PbSO_4$; reducing agents: $Pb > Cl_2$

Adding (2) and (3) to give a spontaneous reaction involves converting (3) to oxidation:

Oxidation: $Pb(s) + SO_4^{2-}(aq) \rightarrow PbSO_4(s) + 2e^-$ $E° = -0.31$ V

Reduction: $Pt^{2+}(aq) + 2e^- \rightarrow Pt(s)$ $E° = 1.20$ V

$$Pt^{2+}(aq) + Pb(s) + SO_4^{2-}(aq) \rightarrow Pt(s) + PbSO_4(s)$$
$$E_{cell}^{\circ} = 1.20 \text{ V} - (-0.31 \text{ V}) = \mathbf{1.51 \text{ V}}$$

Oxidizing agents: $Pt^{2+} > PbSO_4$; reducing agents: $Pb > Pt$

Order of increasing strength as oxidizing agent: **$PbSO_4 < Pt^{2+} < HClO$**

Order of increasing strength as reducing agent: **$Cl_2 < Pt < (Pb + SO_4^{2-})$**

21.50   Metal A + Metal B salt → solid colored product on metal A

   Conclusion: Product is solid metal B. B is undergoing reduction and plating out on A. A is a better reducing agent than B.

   Metal B + acid → gas bubbles

   Conclusion: Product is H_2 gas produced as result of reduction of H^+. B is a better reducing agent than acid.

   Metal A + Metal C salt → no reaction

   Conclusion: C is not undergoing reduction. C must be a better reducing agent than A.

   Since C is a better reducing agent than A, which is a better reducing agent than B and B reduces acid, then **C would also reduce acid to form H_2 bubbles**.

   The order of strength of reducing agents is: **C > A > B**.

21.53   At the negative (anode) electrode, oxidation occurs so the overall cell reaction is

   $A(s) + B^+(aq) \rightarrow A^+(aq) + B(s)$ with $Q = [A^+]/[B^+]$.

   a) The reaction proceeds to the right because with $E_{cell} > 0$ (voltaic cell), the spontaneous reaction occurs. As the cell operates, **[A⁺] increases and [B⁺] decreases**.

   b) E_{cell} **decreases** because the cell reaction takes place to approach equilibrium, $E_{cell} = 0$.

   c) E_{cell} and E_{cell}° are related by the Nernst equation: $E_{cell} = E_{cell}^{\circ} - \dfrac{RT}{nF} \ln \dfrac{[A^+]}{[B^+]}$.

   $E_{cell} = E_{cell}^{\circ}$ when $\dfrac{RT}{nF} \ln \dfrac{[A^+]}{[B^+]} = 0$. This occurs when $\ln \dfrac{[A^+]}{[B^+]} = 0$. Recall that $e^0 = 1$, so **[A⁺] must equal [B⁺]** for E_{cell} to equal E_{cell}°.

   d) **Yes**, it is possible for E_{cell} to be less than E_{cell}° when **[A⁺] > [B⁺]**.

21.55   In a concentration cell, the overall reaction takes place to decrease the concentration of the more concentrated electrolyte. The more concentrated electrolyte is reduced, so it is in the **cathode** compartment.

21.56   <u>Plan:</u> The equilibrium constant can be found by using $\ln K = \dfrac{nFE_{cell}^{\circ}}{RT}$ or $\log K = \dfrac{nE_{cell}^{\circ}}{0.0592}$. Use E° values from the Appendix to calculate E_{cell}° ($E_{cell}^{\circ} = E_{cathode}^{\circ} - E_{anode}^{\circ}$) and then calculate K. The substances given in the problem must be the reactants in the equation.

   <u>Solution:</u>

   a) Oxidation: $Ni(s) \rightarrow Ni^{2+}(aq) + 2e^-$ $\qquad\qquad\qquad E^{\circ} = -0.25$ V

   Reduction: $2\{Ag^+(aq) + 1e^- \rightarrow Ag(s)\}$ $\qquad\qquad E^{\circ} = +0.80$ V

   $\qquad\qquad Ni(s) + 2Ag^+(aq) \rightarrow Ni^{2+}(aq) + 2Ag(s)$

   $E_{cell}^{\circ} = E_{cathode}^{\circ} - E_{anode}^{\circ} = 0.80 \text{ V} - (-0.25 \text{ V}) = 1.05$ V; two electrons are transferred.

   $$\log K = \frac{nE_{cell}^{\circ}}{0.0592} = \frac{2(1.05 \text{ V})}{0.0592 \text{ V}} = 35.47297$$

   $K = 10^{35.47297}$

   $K = 2.97146 \times 10^{35} = \mathbf{3 \times 10^{35}}$

   b) Oxidation: $3\{Fe(s) \rightarrow Fe^{2+}(aq) + 2e^-\}$ $\qquad\qquad E^{\circ} = -0.44$ V

   Reduction: $2\{Cr^{3+}(aq) + 3e^- \rightarrow Cr(s)\}$ $\qquad\qquad E^{\circ} = -0.74$ V

$$3Fe(s) + 2Cr^{3+}(aq) \rightarrow 3Fe^{2+}(aq) + 2Cr(s)$$

$E^\circ_{cell} = E^\circ_{cathode} - E^\circ_{anode} = -0.74 \text{ V} - (-0.44 \text{ V}) = -0.30 \text{ V}$; six electrons are transferred.

$$\log K = \frac{nE^\circ_{cell}}{0.0592} = \frac{6(-0.30 \text{ V})}{0.0592 \text{ V}} = -30.4054$$

$$K = 10^{-30.4054}$$

$$K = 3.936 \times 10^{-31} = \mathbf{4 \times 10^{-31}}$$

21.58   Plan: The equilibrium constant can be found by using $\ln K = \dfrac{nFE^\circ_{cell}}{RT}$ or $\log K = \dfrac{nE^\circ_{cell}}{0.0592}$. Use E° values from the Appendix to calculate E°_{cell} ($E^\circ_{cell} = E^\circ_{cathode} - E^\circ_{anode}$) and then calculate K. The substances given in the problem must be the reactants in the equation.

Solution:

a) Oxidation: $2\{Ag(s) \rightarrow Ag^+(aq) + 1e^-\}$ $E^\circ = +0.80 \text{ V}$

  Reduction: $Mn^{2+}(aq) + 2e^- \rightarrow Mn(s)$ $E^\circ = -1.18 \text{ V}$

   $2Ag(s) + Mn^{2+}(aq) \rightarrow 2Ag^+(aq) + Mn(s)$

   $E^\circ_{cell} = E^\circ_{cathode} - E^\circ_{anode} = -1.18 \text{ V} - (0.80 \text{ V}) = -1.98 \text{V}$; two electrons are transferred.

$$\log K = \frac{nE^\circ_{cell}}{0.0592} = \frac{2(-1.98 \text{ V})}{0.0592 \text{ V}} = -66.89189$$

$$K = 10^{-66.89189}$$

$$K = 1.2826554 \times 10^{-67} = \mathbf{1 \times 10^{-67}}$$

b) Oxidation: $2Br^-(aq) \rightarrow Br_2(l) + 2e^-$ $E^\circ = 1.07 \text{ V}$

  Reduction: $Cl_2(g) + 2e^- \rightarrow 2Cl^-(aq)$ $E^\circ = 1.36 \text{ V}$

   $2Br^-(aq) + Cl_2(g) \rightarrow Br_2(l) + 2Cl^-(aq)$

   $E^\circ_{cell} = E^\circ_{cathode} - E^\circ_{anode} = 1.36 \text{ V} - 1.07 \text{ V} = 0.29 \text{ V}$; two electrons transferred.

$$\log K = \frac{nE^\circ_{cell}}{0.0592} = \frac{2(0.29 \text{ V})}{0.0592 \text{ V}} = 9.797297$$

$$K = 10^{9.797297}$$

$$K = 6.2704253 \times 10^9 = \mathbf{6 \times 10^9}$$

21.60   Plan: Use $\Delta G^\circ = -nFE^\circ_{cell}$ to calculate ΔG°. Substitute J/C for V in the unit for E°_{cell}.
Solution:
a) $\Delta G^\circ = -nFE^\circ_{cell} = -$ (2 mol e$^-$)(96,485 C/mol e$^-$)(1.05 J/C) $= -2.026185 \times 10^5 = \mathbf{-2.03 \times 10^5}$ J
b) $\Delta G^\circ = -nFE^\circ_{cell} = -$ (6 mol e$^-$)(96,485 C/mol e$^-$)(-0.30 J/C) $= 1.73673 \times 10^5 = \mathbf{1.7 \times 10^5}$ J

21.62   Plan: Use $\Delta G^\circ = -nFE^\circ_{cell}$ to calculate ΔG°. Substitute J/C for V in the unit for E°_{cell}.
Solution:
a) $\Delta G^\circ = -nFE^\circ_{cell} = -$ (2 mol e$^-$)(96,485 C/mol e$^-$)(-1.98 J/C) $= 3.820806 \times 10^5 = \mathbf{3.82 \times 10^5}$ J
b) $\Delta G^\circ = -nFE^\circ_{cell} = -$ (2 mol e$^-$)(96,485 C/mol e$^-$)(0.29 J/C) $= -5.59613 \times 10^4 = \mathbf{-5.6 \times 10^4}$ J

21.64   Plan: Use $E^\circ_{cell} = \dfrac{0.0592 \text{ V}}{n} \log K$ to find E°_{cell} and then $\Delta G^\circ = -RT \ln K$ to find ΔG°.

Solution:
$T = (273 + 25)\text{K} = 298 \text{ K}$

$$E^{\circ}_{cell} = \frac{0.0592 \text{ V}}{n} \log K = \frac{0.0592 \text{ V}}{1} \log(5.0 \times 10^4) = 0.278179 = \textbf{0.28 V}$$

$$\Delta G^{\circ} = -RT \ln K = -(8.314 \text{ J/mol} \cdot \text{K})(298 \text{ K}) \ln (5.0 \times 10^4) = -2.68067797 \times 10^4 = \textbf{-2.7x10}^4 \textbf{ J}$$

21.66 Plan: Use $E^{\circ}_{cell} = \dfrac{0.0592 \text{ V}}{n} \log K$ to find E°_{cell} and then $\Delta G^{\circ} = -RT \ln K$ to find ΔG°.

Solution:

$T = (273 + 25)\text{K} = 298 \text{ K}$

$$E^{\circ}_{cell} = \frac{0.0592 \text{ V}}{n} \log K = \frac{0.0592}{2} \log 65 = 0.0536622 = \textbf{0.054 V}$$

$$\Delta G^{\circ} = -RT \ln K = -(8.314 \text{ J/mol} \cdot \text{K})(298 \text{ K}) \ln (65) = -1.03423 \times 10^4 = \textbf{-1.0x10}^4 \textbf{ J}$$

21.68 Plan: The standard reference half-cell is the H_2/H^+ cell. Since this is a voltaic cell, a spontaneous reaction is occurring. For a spontaneous reaction between H_2/H^+ and Cu/Cu^{2+}, Cu^{2+} must be reduced and H_2 must be oxidized. Write the balanced reaction and calculate E°_{cell}. Use the Nernst equation, $E_{cell} = E^{\circ}_{cell} - \dfrac{0.0592}{n} \log Q$, to find $[Cu^{2+}]$ when $E_{cell} = 0.22$ V.

Solution:

Oxidation: $H_2(g) \rightarrow 2H^+(aq) + 2e^-$ $E^{\circ} = 0.00$ V

Reduction: $Cu^{2+}(aq) + 2e^- \rightarrow Cu(s)$ $E^{\circ} = 0.34$ V

$\qquad Cu^{2+}(aq) + H_2(g) \rightarrow Cu(s) + 2H^+(aq)$

$E^{\circ}_{cell} = E^{\circ}_{cathode} - E^{\circ}_{anode} = 0.34 \text{ V} - 0.00 \text{ V} = 0.34$ V

$$E_{cell} = E^{\circ}_{cell} - \frac{0.0592}{n} \log Q$$

$$E_{cell} = E^{\circ}_{cell} - \frac{0.0592}{n} \log \frac{[H^+]^2}{[Cu^{2+}]P_{H_2}}$$

For a standard hydrogen electrode $[H^+] = 1.0$ M and $[H_2] = 1.0$ atm.

$$0.22 \text{ V} = 0.34 \text{ V} - \frac{0.0592}{2} \log \frac{1.0}{[Cu^{2+}]1.0}$$

$$0.22 \text{ V} - 0.34 \text{ V} = -\frac{0.0592}{2} \log \frac{1.0}{[Cu^{2+}]1.0}$$

$$-0.12 \text{ V} = -\frac{0.0592}{2} \log \frac{1.0}{[Cu^{2+}]1.0}$$

$$4.054054 = \log \frac{1.0}{[Cu^{2+}]1.0} \qquad \text{Raise each side to } 10^x.$$

$$1.132541 \times 10^4 = \frac{1}{[Cu^{2+}]}$$

$$[Cu^{2+}] = 8.82970 \times 10^{-5} = \textbf{8.8 x 10}^{-5} \textbf{ M}$$

21.70 Plan: Since this is a voltaic cell, a spontaneous reaction is occurring. For a spontaneous reaction between Ni/Ni^{2+} and Co/Co^{2+}, Ni^{2+} must be reduced and Co must be oxidized. Write the balanced reaction and calculate E°_{cell}. Use the Nernst equation, $E_{cell} = E^{\circ}_{cell} - \dfrac{0.0592}{n} \log Q$, to find E_{cell} at the given ion concentrations. Then the Nernst equation can be used to calculate $[Ni^{2+}]$ at the given E_{cell}. To calculate equilibrium concentrations, recall that at equilibrium $E_{cell} = 0.00$.

Solution:

a) Oxidation: $Co(s) \rightarrow Co^{2+}(aq) + 2e^-$ $\qquad\qquad\qquad\qquad E° = -0.28$ V

Reduction: $Ni^{2+}(aq) + 2e^- \rightarrow Ni(s)$ $\qquad\qquad\qquad\qquad E° = -0.25$ V

$\qquad\qquad Ni^{2+}(aq) + Co(s) \rightarrow Ni(s) + Co^{2+}(aq)$

$\qquad\qquad E^{o}_{cell} = E^{o}_{cathode} - E^{o}_{anode} = -0.25$ V $- (-0.28$ V$) = 0.03$ V

$E_{cell} = E^{o}_{cell} - \dfrac{0.0592}{n} \log Q$

$E_{cell} = E^{o}_{cell} - \dfrac{0.0592}{n} \log \dfrac{[Co^{2+}]}{[Ni^{2+}]} \qquad\qquad n = 2e^-$

$E_{cell} = 0.03$ V $- \dfrac{0.0592}{2} \log \dfrac{[0.20]}{[0.80]}$

$E_{cell} = 0.047820976$ V $= \mathbf{0.05}$ **V**

b) From part a), notice that an increase in $[Co^{2+}]$ leads to a decrease in cell potential. Therefore, the concentration of cobalt ion must increase further to bring the potential down to 0.03 V. Thus, the new concentrations will be $[Co^{2+}] = 0.20$ M + x and $[Ni^{2+}] = 0.80$ M - x (there is a 1:1 mole ratio).

$E_{cell} = E^{o}_{cell} - \dfrac{0.0592}{n} \log \dfrac{[Co^{2+}]}{[Ni^{2+}]}$

0.03 V $= 0.03$ V $- \dfrac{0.0592}{2} \log \dfrac{[0.20 + x]}{[0.80 - x]}$

$0 = - \dfrac{0.0592}{2} \log \dfrac{[0.20 + x]}{[0.80 - x]}$

$0 = \log \dfrac{[0.20 + x]}{[0.80 - x]} \qquad\qquad$ Raise each side to 10^x.

$1 = \dfrac{[0.20 + x]}{[0.80 - x]}$

$0.20 + x = 0.80 - x$

$x = 0.30$ M

$[Ni^{2+}] = 0.80 - x = 0.80 - 0.30 = \mathbf{0.50}$ ***M***

c) At equilibrium $E_{cell} = 0.00$; to decrease the cell potential to 0.00, $[Co^{2+}]$ increases and $[Ni^{2+}]$ decreases.

0.00 V $= 0.03$ V $- \dfrac{0.0592}{2} \log \dfrac{[0.20 + x]}{[0.80 - x]}$

-0.03 V $= - \dfrac{0.0592}{2} \log \dfrac{[0.20 + x]}{[0.80 - x]}$

$1.0135135 = \log \dfrac{[0.20 + x]}{[0.80 - x]} \qquad\qquad$ Raise each side to 10^x.

$10.316051 = \dfrac{[0.20 + x]}{[0.80 - x]}$

$x = 0.71163$

$[Co^{2+}] = 0.20 + 0.71163 = 0.91163 = \mathbf{0.91}$ ***M***

$[Ni^{2+}] = 0.80 - 0.71163 = 0.08837 = \mathbf{0.09}$ ***M***

21.72   Plan: The overall cell reaction proceeds to increase the 0.10 M H^+ concentration and decrease the 2.0 M H^+ concentration. Use the Nernst equation to calculate E_{cell}. $E^o_{cell} = 0$ V for a concentration cell since the half-reactions are the same.
Solution:

Half-cell **A is the anode** because it has the lower concentration.

Oxidation: $H_2(g; 0.95$ atm$) \rightarrow 2H^+(aq; 0.10\ M) + 2e^-$ $E°$(anode) = 0.00 V
Reduction: $2H^+(aq; 2.0\ M) + 2e^- \rightarrow H_2(g; 0.60$ atm$)$ $E°$(cathode) = 0.00 V

$2H^+(aq; 2.0\ M) + H_2(g; 0.95$ atm$) \rightarrow 2H^+(aq; 0.10\ M) + H_2(g; 0.60$ atm$)$

$E^o_{cell} = 0.00$ V $n = 2e^-$

$$E_{cell} = E^o_{cell} - \frac{0.0592}{n} \log Q$$

Q for the cell equals $\dfrac{[H^+]^2_{anode}\ P_{H_2 cathode}}{[H^+]^2_{cathode}\ P_{H_2 anode}}$

$$Q = \frac{(0.10)^2\,(0.60)}{(2.0)^2\,(0.95)} = 0.00157895$$

$$E_{cell} = 0.00\text{ V} - \frac{0.0592}{2}\ \log (0.00157895) = 0.0829283 = \textbf{0.083 V}$$

21.74   Electrons flow from the anode, where oxidation occurs, to the cathode, where reduction occurs. The electrons always flow from the anode to the cathode, no matter what type of cell.

21.76   A D-sized battery is much larger than an AAA-sized battery, so the D-sized battery contains a greater amount of the cell components. The potential, however, is an intensive property and does not depend on the amount of the cell components. (Note that amount is different from concentration.) The total amount of charge a battery can produce does depend on the amount of cell components, so the D-sized battery produces more charge than the AAA-sized battery.

21.78   The Teflon spacers keep the two metals separated so the copper cannot conduct electrons that would promote the corrosion of the iron skeleton. Oxidation of the iron by oxygen causes rust to form and the metal to corrode.

21.81   Plan: Sacrificial anodes are metals with $E°$ values that are more negative than that for iron, –0.44 V, so they are more easily oxidized than iron.
Solution:
a) $E°$(aluminum) = –1.66 V. Yes, except aluminum resists corrosion because once a coating of its oxide covers it, no more aluminum corrodes. Therefore, it would not be a good choice.
b) $E°$(magnesium) = –2.37 V. Yes, magnesium is appropriate to act as a sacrificial anode.
c) $E°$(sodium) = –2.71 V. Yes, except sodium reacts with water, so it would not be a good choice.
d) $E°$(lead) = –0.13 V. No, lead is not appropriate to act as a sacrificial anode because its $E°$ value is too high.
e) $E°$(nickel) = –0.25 V. No, nickel is inappropriate as a sacrificial anode because its $E°$ value is too high.
f) $E°$(zinc) = –0.76 V. Yes, zinc is appropriate to act as a sacrificial anode.
g) $E°$(chromium) = –0.74 V. Yes, chromium is appropriate to act as a sacrificial anode.

21.83 $3Cd^{2+}(aq) + 2Cr(s) \rightarrow 3Cd(s) + 2Cr^{3+}(aq)$
         $E^o_{cell} = -0.40$ V $- (-0.74$ V$) = 0.34$ V

To reverse the reaction requires 0.34 V with the cell in its standard state. A 1.5 V supplies more than enough potential, so the cadmium metal oxidizes to Cd^{2+} and chromium plates out.

21.85 The oxidation number of nitrogen in the nitrate ion, NO_3^-, is +5 and cannot be oxidized further since nitrogen has only five electrons in its outer level. In the nitrite ion, NO_2^-, on the other hand, the oxidation number of nitrogen is +3, so it can be oxidized at the anode to the +5 state.

21.87 Plan: Oxidation occurs at the anode, while reduction occurs at the cathode.
Solution:
a) At the anode, bromide ions are oxidized to form bromine (**Br₂**). $2Br^-(l) \rightarrow Br_2(l) + 2e^-$
b) At the cathode, sodium ions are reduced to form sodium metal (**Na**). $Na^+(l) + e^- \rightarrow Na(s)$

21.89 Plan: Oxidation occurs at the anode, while reduction occurs at the cathode. Decide which anion is more likely to be oxidized and which cation is more likely to be reduced. The less electronegative anion holds its electrons less tightly and is more likely to be oxidized; the cation with the higher ionization energy has the greater attraction for electrons and is more likely to be reduced.
Solution:
Either iodide ions or fluoride ions can be oxidized at the anode. The ion that more easily loses an electron will form. Since I is less electronegative than F, I^- will more easily lose its electron and be oxidized at the anode. The product at the **anode is I₂** gas. The iodine is a gas because the temperature is high to melt the salts.
Either potassium or magnesium ions can be reduced at the cathode. Magnesium has greater ionization energy than potassium because magnesium is located up and to the right of potassium on the periodic table. The greater ionization energy means that magnesium ions will more readily add an electron (be reduced) than potassium ions. The product at the cathode is **magnesium** (liquid).

21.91 Plan: Oxidation occurs at the anode, while reduction occurs at the cathode. Decide which anion is more likely to be oxidized and which cation is more likely to be reduced. The less electronegative anion holds its electrons less tightly and is more likely to be oxidized; the cation with the higher ionization energy has the greater attraction for electrons and is more likely to be reduced.
Solution:
Bromine gas forms at the anode because the electronegativity of bromine is less than that of chlorine. **Calcium** metal forms at the cathode because its ionization energy is greater than that of sodium.

21.93 Plan: Compare the electrode potentials of the species with those of water. The reduction half-reaction with the more positive electrode potential occurs at the cathode, and the oxidation half-reaction with the more negative electrode potential occurs at the anode.
Solution:
Possible reductions:
$Cu^{2+}(aq) + 2e^- \rightarrow Cu(s)$ $\quad\quad\quad\quad\quad\quad E° = +0.34$ V
$Ba^{2+}(aq) + 2e^- \rightarrow Ba(s)$ $\quad\quad\quad\quad\quad\quad E° = -2.90$ V
$Al^{3+}(aq) + 3e^- \rightarrow Al(s)$ $\quad\quad\quad\quad\quad\quad E° = -1.66$ V
$2H_2O(l) + 2e^- \rightarrow H_2(g) + 2OH^-(aq)$ $\quad\quad E = -1$ V with overvoltage
Copper can be prepared by electrolysis of its aqueous salt since its reduction half-cell potential is more positive than the potential for the reduction of water. The reduction of copper is more spontaneous than the reduction of water. Since the reduction potentials of Ba^{2+} and Al^{3+} are more negative and therefore less spontaneous than the reduction of water, these ions cannot be reduced in the presence of water since the water is reduced instead.
Possible oxidations:
$2Br^-(aq) \rightarrow Br_2(l) + 2e^-$ $\quad\quad\quad\quad\quad\quad E° = +1.07$ V
$2H_2O(l) \rightarrow O_2(g) + 4H^+(aq) + 4e^-$ $\quad\quad E = 1.4$ V with overvoltage
Bromine can be prepared by electrolysis of its aqueous salt because its reduction half-cell potential is more negative than the potential for the oxidation of water with overvoltage. The more negative reduction potential for Br^- indicates that its oxidation is more spontaneous than the oxidation of water.

21.95 Plan: Compare the electrode potentials of the species with those of water. The reduction half-reaction with the more positive electrode potential occurs at the cathode, and the oxidation half-reaction with the more negative electrode potential occurs at the anode.

$$\text{Moles of Zn} = (0.75 \text{ g Zn})\left(\frac{80\%}{100\%}\right)\left(\frac{1 \text{ mol Zn}}{65.41 \text{ g Zn}}\right) = 0.00917291 \text{ mol Zn}$$

a) Time (days) = $(0.00917291 \text{ mol Zn})\left(\frac{2 \text{ mol } e^-}{1 \text{ mol Zn}}\right)\left(\frac{96,485\,C}{1 \text{ mol } e^-}\right)\left(\frac{A}{C_{\!/\!s}}\right)\left(\frac{1\,\mu A}{10^{-6}\,A}\right)\left(\frac{1}{0.85\,\mu A}\right)\left(\frac{1 \text{ h}}{3600 \text{ s}}\right)\left(\frac{1 \text{ day}}{24 \text{ h}}\right)$

$$= 2.410262 \times 10^4 = \textbf{2.4} \times \textbf{10}^4 \textbf{ days}$$

b) Mass (g) of Ag = $(0.00917291 \text{ mol Zn})\left(\frac{1 \text{ mol Ag}_2\text{O}}{1 \text{ mol Zn}}\right)\left(\frac{100\%}{95\%}\right)\left(\frac{2 \text{ mol Ag}}{1 \text{ mol Ag}_2\text{O}}\right)\left(\frac{107.9 \text{ g Ag}}{1 \text{ mol Ag}}\right)$

$$= 2.0836989 = \textbf{2.1 g Ag}$$

c) Cost = $(2.0836989 \text{ g Ag})\left(\frac{95\%}{100\%}\right)\left(\frac{1 \text{ troy oz}}{31.10 \text{ g Ag}}\right)\left(\frac{\$23.00}{\text{troy oz}}\right)\left(\frac{}{2.410262 \times 10^4 \text{ days}}\right)$

$$= 6.073818 \times 10^{-5} = \textbf{\$6.1} \times \textbf{10}^{-5}\textbf{/day}$$

21.121 Plan: Since the cells are voltaic cells, the reactions occurring are spontaneous and will have a positive E°_{cell}. Write the two half-reactions. When two half-reactions are paired, one half-reaction must be reversed and written as an oxidation. Reverse the half-reaction that will result in a positive value of E°_{cell} using the relationship $E^\circ_{\text{cell}} = E^\circ_{\text{cathode}} - E^\circ_{\text{anode}}$. E° values are found in Appendix D. The oxidation occurs at the negative electrode (the anode). Use the Nernst equation to find cell potential at concentrations other than 1 M.

Solution:

a) Cell with SHE and Pb/Pb^{2+}:

$\qquad$ Oxidation: $\text{Pb}(s) \rightarrow \text{Pb}^{2+}(aq) + 2e^-$ $\qquad\qquad$ $E^\circ = -0.13 \text{ V}$

$\qquad$ Reduction: $2\text{H}^+(aq) + 2e^- \rightarrow \text{H}_2(g)$ $\qquad\quad$ $E^\circ = 0.0 \text{ V}$

$E^\circ_{\text{cell}} = E^\circ_{\text{cathode}} - E^\circ_{\text{anode}} = 0.0 \text{ V} - (-0.13 \text{ V}) = \textbf{0.13 V}$

Cell with SHE and Cu/Cu^{2+}:

$\qquad$ Oxidation: $\text{H}_2(g) \rightarrow 2\text{ H}^+(aq) + 2e^-$ $\qquad\qquad$ $E^\circ = 0.0 \text{ V}$

$\qquad$ Reduction: $\text{Cu}^{2+}(aq) + 2e^- \rightarrow \text{Cu}(s)$ $\qquad\quad$ $E^\circ = 0.34 \text{ V}$

$E^\circ_{\text{cell}} = E^\circ_{\text{cathode}} - E^\circ_{\text{anode}} = 0.34 \text{ V} - 0.00 \text{ V} = \textbf{0.34 V}$

b) The anode (negative electrode) in cell with SHE and Pb/Pb^{2+} is **Pb**.

The anode in cell with SHE and Cu/Cu^{2+} is **platinum** in the SHE.

c) The precipitation of PbS decreases [Pb^{2+}]. Use Nernst equation to see how this affects potential. Cell reaction is: $\qquad$ $\text{Pb}(s) + 2\text{H}^+(aq) \rightarrow \text{Pb}^{2+}(aq) + \text{H}_2(g)$

$$E_{\text{cell}} = E^\circ_{\text{cell}} - \frac{0.0592}{n} \log Q$$

$$E_{\text{cell}} = E^\circ_{\text{cell}} - \frac{0.0592}{n} \log \frac{[\text{Pb}^{2+}]P_{\text{H}_2}}{[\text{H}^+]^2} \qquad\qquad n = 2e^-$$

$$E_{\text{cell}} = E^\circ_{\text{cell}} - \frac{0.0592}{2} \log \frac{[\text{Pb}^{2+}]P_{\text{H}_2}}{[\text{H}^+]^2}$$

Decreasing the concentration of lead ions gives a negative value for the term:

$$\frac{0.0592}{2} \log \frac{[\text{Pb}^{2+}]P_{\text{H}_2}}{[\text{H}^+]^2}$$

When this negative value is subtracted from E°_{cell}, cell potential **increases**.

d) The [H$^+$] = 1.0 M and the H$_2$ = 1 atm in the SHE.

$\qquad$ Cell reaction: $\text{Cu}^{2+}(aq) + \text{H}_2(g) \rightarrow \text{Cu}(s) + 2\text{H}^+(aq)$

$$E_{cell} = E^{\circ}_{cell} - \frac{0.0592}{n} \log Q$$

$$E_{cell} = E^{\circ}_{cell} - \frac{0.0592}{n} \log \frac{[H^+]^2}{[Cu^{2+}]P_{H_2}} \qquad n = 2e^-$$

$$E_{cell} = 0.34 \text{ V} - \frac{0.0592}{2} \log \frac{(1)}{(1 \times 10^{-16})(1 \text{ atm})}$$

$$E_{cell} = -0.1336 = \mathbf{-0.13 \text{ V}}$$

21.124 The three steps equivalent to the overall reaction $M^+(aq) + e^- \rightarrow M(s)$ are:

         1) $M^+(aq) \rightarrow M^+(g)$ Energy is $-\Delta H_{hydration}$

         2) $M^+(g) + e^- \rightarrow M(g)$ Energy is $-IE$ or $-\Delta H_{ionization}$

         3) $M(g) \rightarrow M(s)$ Energy is $-\Delta H_{deposition}$

The energy for step 3 is similar for all three elements, so the difference in the energy for the overall reaction depends on the values for $-\Delta H_{hydration}$ and $-IE$. The lithium ion has a more negative hydration energy than Na^+ and K^+ because it is a smaller ion with large charge density that holds the water molecules more tightly. The amount of energy required to remove the waters surrounding the lithium ion offsets the lower ionization energy to make the overall energy for the reduction of lithium larger than expected.

21.125 The key factor is that the table deals with electrode potentials in aqueous solution. The very high and low standard electrode potentials involve extremely reactive substances, such as F_2 (a powerful oxidant), and Li (a powerful reductant). These substances react directly with water, rather than according to the desired half-reactions. An alternative (essentially equivalent) explanation is that any aqueous cell with a voltage of more than 1.23 V has the ability to electrolyze water into hydrogen and oxygen. When two electrodes with 6 V across them are placed in water, electrolysis of water will occur.

21.127 Plan: Write the half-reaction for the reduction of Al^{3+}. Convert mass of Al to moles and use the mole ratio in the balanced reaction to find the number of moles of electrons required for every mole of Al produced. The Faraday constant is used to find the charge of the electrons in coulombs. To find the time, the charge is divided by the current. To calculate the electrical power, multiply the time by the current and voltage, remembering that 1 A = 1 C/s (thus, 100,000 A is 100,000 C/s) and 1 V = 1 J/C (thus, 5.0 V = 5.0 J/C). Change units of J to kW•h. To find the cost of the electricity, convert the mass of aluminum from lb to kg and use the kW•h per 1000 kg of aluminum calculated in part b) to find the kW•h for that mass of aluminum, keeping in mind the 90.% efficiency.

Solution:

a) Aluminum half-reaction: $Al^{3+}(aq) + 3 e^- \rightarrow Al(s)$, so $n = 3$. Remember that 1 A = 1 C/s.

$$\text{Time (s)} = \left(1000 \text{ kg Al}\right)\left(\frac{10^3 \text{ g}}{1 \text{ kg}}\right)\left(\frac{1 \text{ mol Al}}{26.98 \text{ g Al}}\right)\left(\frac{3 \text{ mol e}^-}{1 \text{ mol Al}}\right)\left(\frac{96,485 \text{ C}}{1 \text{ mol e}^-}\right)\left(\frac{A}{C/s}\right)\left(\frac{1}{100,000 \text{ A}}\right)$$

$$= 1.0728503 \times 10^5 = \mathbf{1.073 \times 10^5 \text{ s}}$$

The molar mass of aluminum limits the significant figures.

b)

$$\text{Power} = \left(1.0728503 \times 10^5 \text{ s}\right)\left(\frac{100,000 \text{ C}}{s}\right)\left(\frac{5.0 \text{ J}}{C}\right)\left(\frac{1 \text{ kJ}}{10^3 \text{ J}}\right)\left(\frac{1 \text{ kW} \cdot \text{h}}{3.6 \times 10^3 \text{ kJ}}\right) = 1.4900699 \times 10^4 = \mathbf{1.5 \times 10^4 \text{ kW} \cdot \text{h}}$$

c) Cost = $\left(1 \text{ lb Al}\right)\left(\frac{1 \text{ kg}}{2.205 \text{ lb}}\right)\left(\frac{1.4900699 \times 10^4 \text{ kW} \cdot \text{h}}{1000 \text{ kg Al}}\right)\left(\frac{0.123 \text{ cents}}{1 \text{ kW h}}\right)\left(\frac{100\%}{90.\%}\right) = 0.923551 = \mathbf{0.92 \text{ ¢/lb Al}}$

21.129 Plan: When considering two substances, the stronger reducing agent will reduce the other substance.

Solution:

Statement: Metal D + hot water $\rightarrow$ reaction   Conclusion: D reduces water to produce $H_2(g)$. D is a stronger reducing agent than H^+.

Statement: D + E salt → no reaction   Conclusion: D does not reduce E salt, so E reduces D salt. E is better reducing agent than D.

Statement: D + F salt → reaction   Conclusion: D reduces F salt. D is better reducing agent than F.
If E metal and F salt are mixed, the salt F would be reduced producing F metal because E has the greatest reducing strength of the three metals (E is stronger than D and D is stronger than F). The ranking of increasing reducing strength is **F < D < E.**

21.131 Plan: Examine the change in oxidation numbers in the equations to find n, the moles of electrons transferred. Use $\Delta G = -nFE$ to calculate ΔG. Substitute J/C for V in the unit for E. Convert ΔG to units of kJ and divide by the total mass of reactants to obtain the ratio.
Solution:
a) Cell I: Oxidation number (O.N.) of H changes from 0 to +1, so one electron is lost from each of four hydrogen atoms for a total of four electrons. O.N. of oxygen changes from 0 to –2, indicating that two electrons are gained by each of the two oxygen atoms for a total of four electrons. There is a transfer of **four electrons** in the reaction.
$$\Delta G = -nFE = -(4 \text{ mol e}^-)(96{,}485 \text{ C/mol e}^-)(1.23 \text{ J/C}) = -4.747062 \times 10^5 = \mathbf{-4.75 \times 10^5 \text{ J}}$$
Cell II: In $Pb(s) \rightarrow PbSO_4$, O.N. of Pb changes from 0 to +2 and in $PbO_2 \rightarrow PbSO_4$, O.N. of Pb changes from +4 to +2. There is a transfer of **two electrons** in the reaction.
$$\Delta G = -nFE = -(2 \text{ mol e}^-)(96{,}485 \text{ C/mol e}^-)(2.04 \text{ J/C}) = -3.936588 \times 10^5 = \mathbf{-3.94 \times 10^5 \text{ J}}$$
Cell III: O.N. of each of two Na atoms changes from 0 to +1 and O.N. of Fe changes from +2 to 0. There is a transfer of **two electrons** in the reaction.
$$\Delta G = -nFE = -(2 \text{ mol e}^-)(96{,}485 \text{ C/mol e}^-)(2.35 \text{ J/C}) = -4.534795 \times 10^5 = \mathbf{-4.53 \times 10^5 \text{ J}}$$

b) Cell I: Mass of reactants $= \left(2 \text{ mol H}_2\right)\left(\dfrac{2.016 \text{ g H}_2}{1 \text{ mol H}_2}\right) + \left(1 \text{ mol O}_2\right)\left(\dfrac{32.00 \text{ g O}_2}{1 \text{ mol O}_2}\right) = 36.032 \text{ g}$

$$\frac{w_{max}}{\text{reactant mass}} = \left(\frac{-4.747062 \times 10^5 \text{ J}}{36.032 \text{ g}}\right)\left(\frac{1 \text{ kJ}}{10^3 \text{ J}}\right) = -13.17457 = \mathbf{-13.2 \text{ kJ/g}}$$

Cell II: Mass of reactants =

$$\left(1 \text{ mol Pb}\right)\left(\frac{207.2 \text{ g Pb}}{1 \text{ mol Pb}}\right) + \left(1 \text{ mol PbO}_2\right)\left(\frac{239.2 \text{ g PbO}_2}{1 \text{ mol PbO}_2}\right) + \left(2 \text{ mol H}_2SO_4\right)\left(\frac{98.09 \text{ g H}_2SO_4}{1 \text{ mol H}_2SO_4}\right)$$
$$= 642.58 \text{ g}$$

$$\frac{w_{max}}{\text{reactant mass}} = \left(\frac{-3.936588 \times 10^5 \text{ J}}{642.58 \text{ g}}\right)\left(\frac{1 \text{ kJ}}{10^3 \text{ J}}\right) = -0.612622 = \mathbf{-0.613 \text{ kJ/g}}$$

Cell III: Mass of reactants $= \left(2 \text{ mol Na}\right)\left(\dfrac{22.99 \text{ g Na}}{1 \text{ mol Na}}\right) + \left(1 \text{ mol FeCl}_2\right)\left(\dfrac{126.75 \text{ g FeCl}_2}{1 \text{ mol FeCl}_2}\right) = 172.73 \text{ g}$

$$\frac{w_{max}}{\text{reactant mass}} = \left(\frac{-4.534795 \times 10^5 \text{ J}}{172.73 \text{ g}}\right)\left(\frac{1 \text{ kJ}}{10^3 \text{ J}}\right) = -2.625366 = \mathbf{-2.63 \text{ kJ/g}}$$

Cell I has the highest ratio (most energy released per gram) because the reactants have very low mass while Cell II has the lowest ratio because the reactants are very massive.

21.135 Plan: Write the balanced equation. Multiply the current and time to calculate total charge in coulombs. Remember that the unit 1 A is 1 C/s, so the time must be converted to seconds. From the total charge, the number of electrons transferred to form copper is calculated by dividing total charge by the Faraday constant. Each mole of copper deposited requires two moles of electrons, so divide moles of electrons by two to get moles of copper. Then convert to grams of copper. The initial concentration of Cu^{2+} is 1.00 M (standard condition) and initial volume is 345 mL. Use this to calculate the initial moles of copper ions, then subtract the number of moles of copper ions converted to copper metal and divide by the cell volume to find the remaining $[Cu^{2+}]$.

a) Since the cell is a voltaic cell, write a spontaneous reaction. The reduction of Cu^{2+} is more spontaneous than the reduction of Sn^{2+}: $Cu^{2+}(aq) + Sn(s) \rightarrow Cu(s) + Sn^{2+}(aq)$

$$\text{Mass (g) of Cu} = (0.17 \text{ A})\left(\frac{C/s}{A}\right)\left(\frac{3600 \text{ s}}{1 \text{ h}}\right)(48.0 \text{ h})\left(\frac{1 \text{ mol e}^-}{96,485 \text{ C}}\right)\left(\frac{1 \text{ mol Cu}}{2 \text{ mol e}^-}\right)\left(\frac{63.55 \text{ g Cu}}{1 \text{ mol Cu}}\right)$$

$$= 9.674275 = \textbf{9.7 g Cu}$$

b) Initial moles of $Cu^{2+} = \left(1.00\frac{\text{mol Cu}^{2+}}{\text{L}}\right)(345 \text{ mL})\left(\frac{10^{-3} \text{ L}}{1 \text{ mL}}\right) = 0.345 \text{ mol Cu}^{2+}$

Moles of Cu^{2+} reduced $= (9.674275 \text{ g Cu})\left(\frac{1 \text{ mol Cu}}{63.55 \text{ g Cu}}\right) = 0.1522309205 \text{ mol Cu}^{2+}$

Remaining moles of Cu^{2+} = initial moles − moles reduced = 0.345 mol − 0.1522309205 = 0.1927691 mol Cu^{2+}

$$M \text{ Cu}^{2+} = \frac{0.1927691 \text{ mol Cu}^{2+}}{(345 \text{ mL})\left(\frac{10^{-3} \text{ L}}{1 \text{ mL}}\right)} = 0.558751 = \textbf{0.56 } \textit{M} \textbf{ Cu}^{2+}$$

21.136 Plan: Write the balanced equation and determine the half reactions in the anode and cathode compartments. Use the Nernst equation to calculate the H^+ ion concentration within the value of Q. Calculate the pH from this concentration.

Solution:

Oxidation: $\underline{H_2(g) \rightarrow 2H^+(aq) + 2e^-}$ $\qquad\qquad\qquad$ $E°$(anode) = 0.00 V

Reduction: $\underline{2[Ag^+(aq) + e^- \rightarrow Ag(s)]}$ $\qquad\qquad\qquad$ $E°$(cathode) = 0.80 V

$\qquad\qquad$ $2Ag^+(aq) + H_2(g) \rightarrow 2Ag(s) + 2H^+(aq)$

$E°_{cell} = E°_{cathode} - E°_{anode} = 0.80 \text{ V} - 0.00 \text{ V} = 0.80 \text{ V}$ $\qquad\qquad$ $n = 2 \text{ mol } \underline{e}^-$

$E_{cell} = E°_{cell} - \dfrac{0.0592}{n} \log Q = E°_{cell} - \dfrac{0.0592}{n} \log \dfrac{[H^+]^2}{P_{H_2}[Ag^+]^2}$

$0.915 \text{ V} = 0.80 \text{ V} - \dfrac{0.0592 \text{ V}}{2} \log \dfrac{[H^+]^2}{(1.00)[0.100]^2}$

$-3.885135 = \log \dfrac{[H^+]^2}{0.0100}$

take the inverse log of both sides

$0.000130276 = \dfrac{[H^+]^2}{0.0100}$

$[H^+]^2 = 1.30276 \times 10^{-6}$

$[H^+] = 1.141385 \times 10^{-3}$

pH = $-\log[H^+] = -\log[1.141385 \times 10^{-3}] = 2.942568 = \textbf{2.94}$

21.137 Plan: Examine each reaction to determine which reactant is the oxidizing agent; the oxidizing agent is the reactant that gains electrons in the reaction, resulting in a decrease in its oxidation number.

Solution:

$\qquad$ From reaction between $U^{3+} + Cr^{3+} \rightarrow Cr^{2+} + U^{4+}$, find that Cr^{3+} oxidizes U^{3+}.

$\qquad$ From reaction between $Fe + Sn^{2+} \rightarrow Sn + Fe^{2+}$, find that Sn^{2+} oxidizes Fe.

From the fact that there is no reaction that occurs between Fe and U^{4+}, find that Fe^{2+} oxidizes U^{3+}.
From reaction between $Cr^{3+} + Fe \rightarrow Cr^{2+} + Fe^{2+}$, find that Cr^{3+} oxidizes Fe.
From reaction between $Cr^{2+} + Sn^{2+} \rightarrow Sn + Cr^{3+}$, find that Sn^{2+} oxidizes Cr^{2+}.
Notice that nothing oxidizes Sn, so Sn^{2+} must be the strongest oxidizing agent. Both Cr^{3+} and Fe^{2+} oxidize U^{3+}, so U^{4+} must be the weakest oxidizing agent. Cr^{3+} oxidizes iron so Cr^{3+} is a stronger oxidizing agent than Fe^{2+}.
The half-reactions in order from strongest to weakest oxidizing agent:

$$Sn^{2+}(aq) + 2e^- \rightarrow Sn(s)$$
$$Cr^{3+}(aq) + e^- \rightarrow Cr^{2+}(aq)$$
$$Fe^{2+}(aq) + 2e^- \rightarrow Fe(s)$$
$$U^{4+}(aq) + e^- \rightarrow U^{3+}(aq)$$

21.141 Plan: Write a balanced equation that gives a positive E°_{cell} for a spontaneous reaction. Calculate the E°_{cell} and use the Nernst equation to find the silver ion concentration that results in the given E_{cell}.

Solution:

a) The calomel half-cell is the anode and the silver half-cell is the cathode. The overall reaction is:

$$2Ag^+(aq) + 2Hg(l) + 2Cl^-(aq) \rightarrow 2Ag(s) + Hg_2Cl_2(s)$$

$E^{\circ}_{cell} = E^{\circ}_{cathode} - E^{\circ}_{anode} = 0.80\ V - 0.24\ V = 0.56\ V$ with $n = 2$.

Use the Nernst equation to find $[Ag^+]$ when $E_{cell} = 0.060\ V$.

$$E_{cell} = E^{\circ}_{cell} - \frac{0.0592}{n} \log Q$$

$$E_{cell} = E^{\circ}_{cell} - \frac{0.0592\ V}{2} \log \frac{1}{[Ag^+]^2 [Cl^-]^2}$$

$$0.060\ V = 0.56\ V - \frac{0.0592V}{2} \log \frac{1}{[Ag^+]^2 [Cl^-]^2}$$

The problem suggests assuming that $[Cl^-]$ is constant. Assume it is 1.00 M.

$$-0.50\ V = -\frac{0.0592V}{2} \log \frac{1}{[Ag^+]^2 [1.00]^2}$$

$$16.89189 = \log \frac{1}{[Ag^+]^2 [1.00]^2}$$

$$10^{16.89189} = \frac{1}{[Ag^+]^2}$$

$$7.7963232 \times 10^{16} = \frac{1}{[Ag^+]^2}$$

$$7.7963232 \times 10^{16}[Ag^+]^2 = 1$$

$$[Ag^+]^2 = 1.282656 \times 10^{-17}$$

$$[Ag^+] = 3.581419 \times 10^{-9} = \mathbf{3.6 \times 10^{-9}}\ \boldsymbol{M}$$

b) Again use the Nernst equation and assume $[Cl^-] = 1.00\ M$.

$$E_{cell} = E^{\circ}_{cell} - \frac{0.0592\ V}{2} \log \frac{1}{[Ag^+]^2 [Cl^-]^2}$$

$$0.53\ V = 0.56\ V - \frac{0.0592\ V}{2} \log \frac{1}{[Ag^+]^2 [Cl^-]^2}$$

$$-0.03\ \text{V} = -\frac{0.0592\ \text{V}}{2}\ \log\frac{1}{\left[\text{Ag}^+\right]^2\left[\text{Cl}^-\right]^2}$$

$$1.0135135 = \log\frac{1}{\left[\text{Ag}^+\right]^2\left[1.00\right]^2}$$

$$10^{1.0135135} = \frac{1}{\left[\text{Ag}^+\right]^2}$$

$$10.31605146 = \frac{1}{\left[\text{Ag}^+\right]^2}$$

$$10.31605146\left[\text{Ag}^+\right]^2 = 1$$

$$\left[\text{Ag}^+\right]^2 = 0.0969363$$

$$\left[\text{Ag}^+\right] = 0.311346 = \mathbf{0.3\ M}$$

21.143 <u>Plan:</u> Use the Nernst equation to write the relationship between E_{cell}° and the cell potential for both the waste stream and for the silver standard.

<u>Solution:</u>

a) The reaction is $\text{Ag}^+(aq)\ \rightarrow\ \text{Ag}(s)\ +\ 1e^-$

$$E_{\text{cell}} = E_{\text{cell}}^\circ - \frac{0.0592}{n}\ \log Q$$

Nonstandard cell: $\qquad E_{\text{waste}} = E_{\text{cell}}^\circ - \left(\frac{0.0592}{1\ e^-}\right)\ \log\left[\text{Ag}^+\right]_{\text{waste}}$

Standard cell: $\qquad E_{\text{standard}} = E_{\text{cell}}^\circ - \left(\frac{0.0592}{1\ e^-}\right)\ \log\left[\text{Ag}^+\right]_{\text{standard}}$

b) To find $\left[\text{Ag}^+\right]_{\text{waste}}$: $\quad E_{\text{cell}}^\circ = E_{\text{standard}} + \left(\frac{0.0592}{1\ e^-}\right)\ \log\left[\text{Ag}^+\right]_{\text{standard}} = E_{\text{waste}} + \left(\frac{0.0592}{1\ e^-}\right)\ \log\left[\text{Ag}^+\right]_{\text{waste}}$

$$E_{\text{waste}} - E_{\text{standard}} = -\left(\frac{0.0592}{1\ e^-}\right)\ \log\frac{\left[\text{Ag}^+\right]_{\text{waste}}}{\left[\text{Ag}^+\right]_{\text{standard}}}$$

$$E_{\text{standard}} - E_{\text{waste}} = (0.0592\ \text{V})(\log\left[\text{Ag}^+\right]_{\text{waste}} - \log\left[\text{Ag}^+\right]_{\text{standard}})$$

$$\frac{E_{\text{standard}} - E_{\text{waste}}}{0.0592\ \text{V}} = (\log\left[\text{Ag}^+\right]_{\text{waste}} - \log\left[\text{Ag}^+\right]_{\text{standard}})$$

$$\log\left[\text{Ag}^+\right]_{\text{waste}} = \frac{E_{\text{standard}} - E_{\text{waste}}}{0.0592\ \text{V}} + \log\left[\text{Ag}^+\right]_{\text{standard}}$$

$$\left[\text{Ag}^+\right]_{\text{waste}} = \left[\text{antilog}\left(\frac{E_{\text{standard}} - E_{\text{waste}}}{0.0592\ \text{V}}\right)\right]\left(\left[\text{Ag}^+\right]_{\text{standard}}\right)$$

c) Convert M to ng/L for both $\left[\text{Ag}^+\right]_{\text{waste}}$ and $\left[\text{Ag}^+\right]_{\text{standard}}$:

$$E_{\text{waste}} - E_{\text{standard}} = (-0.0592\ \text{V})\ \log\frac{\left[\text{Ag}^+\right]_{\text{waste}}}{\left[\text{Ag}^+\right]_{\text{standard}}}$$

If both silver ion concentrations are in the same units, in this case ng/L, the "conversions" cancel and the equation derived in part b) applies if the standard concentration is in ng/L.

$$\text{Conc.}\left(\text{Ag}^+\right)_{\text{waste}} = \left[\text{antilog}\left(\frac{E_{\text{standard}} - E_{\text{waste}}}{0.0592\ \text{V}}\right)\right]\left(\text{Conc.}\left(\text{Ag}^+\right)_{\text{standard}}\right)$$

21-37

d) Plug the values into the answer for part c).

$$[Ag^+]_{waste} = \left[anti\log\left(\frac{-0.003}{0.0592\ V}\right) \right](1000.\ ng/L) = 889.8654 = \textbf{900 ng/L}$$

e) Temperature is included in the RT/nF term, which equals 0.0592 V/n at 25°C. To account for different temperatures, insert the RT/nF term in place of 0.0592 V/n.

$$E_{standard} + \left(\frac{2.303RT}{nF}\right)\log[Ag^+]_{standard} = E_{waste} + \left(\frac{2.303RT}{nF}\right)\log[Ag^+]_{waste}$$

$$E_{standard} - E_{waste} = \left(\frac{2.303R}{nF}\right)(T_{waste}\log[Ag^+]_{waste} - T_{standard}\log[Ag^+]_{standard})$$

$$(E_{standard} - E_{waste})\left(\frac{nF}{2.303R}\right) = T_{waste}\log[Ag^+]_{waste} - T_{standard}\log[Ag^+]_{standard}$$

$$(E_{standard} - E_{waste})\left(\frac{nF}{2.303R}\right) + T_{standard}\log[Ag^+]_{standard} = T_{waste}\log[Ag^+]_{waste}$$

$$\log[Ag^+]_{waste} = \left(\frac{(E_{standard} - E_{waste})(nF/2.303\,R) + T_{standard}\log[Ag^+]_{standard}}{T_{waste}}\right)$$

$$[Ag^+]_{waste} = anti\log\left(\frac{(E_{standard} - E_{waste})(nF/2.303\,R) + T_{standard}\log[Ag^+]_{standard}}{T_{waste}}\right)$$

21.145 **Plan:** Multiply the current in amperes by the time in seconds to obtain coulombs. Convert coulombs to moles of electrons with the Faraday constant and use the mole ratio in the balanced half-reactions to convert moles of electrons to moles and then mass of reactants. Divide the total mass of reactants by the mass of the battery to find the mass percentage that consists of reactants.
Solution:
a) Determine the total charge the cell can produce.

$$\text{Capacity (C)} = (300.\ mA \bullet h)\left(\frac{10^{-3}\ A}{1\ mA}\right)\left(\frac{3600\ s}{1\ h}\right)\left(\frac{1\ C}{1\ A \bullet s}\right) = \textbf{1.08 x 10}^3\ \textbf{C}$$

b) The half-reactions are:
$$Cd^0 \rightarrow Cd^{2+} + 2e^- \quad \text{and} \quad NiO(OH) + H_2O(l) + e^- \rightarrow Ni(OH)_2 + OH^-$$
Assume 100% conversion of reactants.

$$\text{Mass (g) of Cd} = (1080\ C)\left(\frac{1\ mol\ e^-}{96,485\ C}\right)\left(\frac{1\ mol\ Cd}{2\ mol\ e^-}\right)\left(\frac{112.4\ g\ Cd}{1\ mol\ Cd}\right) = 0.62907 = \textbf{0.629 g Cd}$$

$$\text{Mass (g) of NiO(OH)} = (1080\ C)\left(\frac{1\ mol\ e^-}{96,485\ C}\right)\left(\frac{1\ mol\ NiO(OH)}{1\ mol\ e^-}\right)\left(\frac{91.70\ g\ NiO(OH)}{1\ mol\ NiO(OH)}\right)$$
$$= 1.026439 = \textbf{1.03 g NiO(OH)}$$

$$\text{Mass (g) of H}_2\text{O} = (1080\ C)\left(\frac{1\ mol\ e^-}{96,485\ C}\right)\left(\frac{1\ mol\ H_2O}{1\ mol\ e^-}\right)\left(\frac{18.02\ g\ H_2O}{1\ mol\ H_2O}\right) = 0.20170596 = \textbf{0.202 g H}_2\textbf{O}$$

Total mass of reactants = 0.62907 g Cd + 1.026439 g NiO(OH) + 0.20170596 g H₂O
$$= 1.857215 = \textbf{1.86 g reactants}$$
c) Mass % reactants = $\dfrac{1.85721\ g}{18.3\ g}(100) = 10.14872 = \textbf{10.1\%}$

21.147 **Plan:** For a list of decreasing reducing strength, place the elements in order of increasing (more positive) $E°$. Metals with potentials lower than that of water (–0.83 V) can displace hydrogen from water by reducing the hydrogen in water. Metals with potentials lower than that of hydrogen (0.00 V) can displace hydrogen from acids by reducing the H^+ in acid. Metals with potentials above that of hydrogen (0.00 V) cannot displace (reduce) hydrogen.

Solution:
Reducing agent strength: Li > Ba > Na > Al > Mn > Zn > Cr > Fe > Ni > Sn > Pb > Cu > Ag > Hg > Au
These can displace H_2 from water: Li, Ba, Na, Al, and Mn.
These can displace H_2 from acid: Li, Ba, Na, Al, Mn, Zn, Cr, Fe, Ni, Sn, and Pb.
These cannot displace H_2: Cu, Ag, Hg, and Au.

21.150 a) The reference half-reaction is: $Cu^{2+}(aq) + 2e^- \rightarrow Cu(s)$ $\quad E° = 0.34$ V

Before the addition of the ammonia, $E_{cell} = 0$. The addition of ammonia lowers the concentration of copper ions through the formation of the complex $Cu(NH_3)_4^{2+}$. The original copper ion concentration is $[Cu^{2+}]_{original,}$ and the copper ion concentration in the solution containing ammonia is $[Cu^{2+}]_{ammonia}$.

The Nernst equation is used to determine the copper ion concentration in the cell containing ammonia.

The reaction is $Cu^{2+}_{initial}(aq) + Cu(s) \rightarrow Cu(s) + Cu^{2+}_{ammonia}(aq)$.

The half-cell with the larger concentration of copper ion (no ammonia added) is the reduction and the half-cell with the lower concentration of copper ion due to the addition of ammonia and formation of the complex is the oxidation.

$$E_{cell} = E°_{cell} - \frac{0.0592}{n} \log Q$$

$$0.129 \text{ V} = 0.00 \text{ V} - \frac{0.0592 \text{ V}}{2} \log \frac{[Cu^{2+}]_{ammonia}}{[Cu^{2+}]_{original}}$$

$$0.129 \text{ V} = -\frac{0.0592 \text{ V}}{2} \log \frac{[Cu^{2+}]_{ammonia}}{[0.0100]_{original}}$$

$$-4.358108108 = \log \frac{[Cu^{2+}]_{ammonia}}{[0.0100]_{original}}$$

$$4.3842155 \times 10^{-5} = \frac{[Cu^{2+}]_{ammonia}}{[0.0100]_{original}}$$

$[Cu^{2+}]_{ammonia} = 4.3842155 \times 10^{-7}$ M

This is the concentration of the copper ion that is not in the complex. The concentration of the complex and of the uncomplexed ammonia must be determined before K_f may be calculated.

The original number of moles of copper and the original number of moles of ammonia are found from the original volumes and molarities:

$$\text{Original moles of copper} = \left(\frac{0.0100 \text{ mol Cu(NO}_3)_2}{L}\right)\left(\frac{1 \text{ mol Cu}^{2+}}{1 \text{ mol Cu(NO}_3)_2}\right)\left(\frac{10^{-3} \text{ L}}{1 \text{ mL}}\right)(90.0 \text{ mL})$$

$$= 9.00 \times 10^{-4} \text{ mol Cu}^{2+}$$

$$\text{Original moles of ammonia} = \left(\frac{0.500 \text{ mol NH}_3}{L}\right)\left(\frac{10^{-3} \text{ L}}{1 \text{ mL}}\right)(10.0 \text{ mL}) = 5.00 \times 10^{-3} \text{ mol NH}_3$$

Determine the moles of copper still remaining uncomplexed.

$$\text{Remaining moles of copper} = \left(\frac{4.3842155 \times 10^{-7} \text{ mol Cu}^{2+}}{L}\right)\left(\frac{10^{-3} \text{ L}}{1 \text{ mL}}\right)(100.0 \text{ mL}) = 4.3842155 \times 10^{-8} \text{ mol Cu}$$

The difference between the original moles of copper and the copper ion remaining in solution is the copper in the complex (= moles of complex). The molarity of the complex may now be found.

Moles copper in complex = $(9.00 \times 10^{-4} - 4.3842155 \times 10^{-8})$ mol Cu^{2+} = 8.9995616×10^{-4} mol Cu^{2+}

$$\text{Molarity of complex} = \left(\frac{8.9995616 \times 10^{-4} \text{ mol Cu}^{2+}}{100.0 \text{ mL}}\right)\left(\frac{1 \text{ mol Cu(NH}_3)_4^{2+}}{1 \text{ mol Cu}^{2+}}\right)\left(\frac{1 \text{ mL}}{10^{-3} \text{ L}}\right)$$

$$= 8.9995616 \times 10^{-3} \ M \text{ Cu(NH}_3)_4^{2+}$$

The concentration of the remaining ammonia is found as follows:

$$\text{Molarity of ammonia} = \left(\frac{\left(5.00 \times 10^{-3} \text{ mol NH}_3\right) - \left(8.9995616 \times 10^{-4} \text{ mol Cu}^{2+}\right)\left(\frac{4 \text{ mol NH}_3}{1 \text{ mol Cu}^{2+}}\right)}{100.0 \text{ mL}}\right)\left(\frac{1 \text{ mL}}{10^{-3} \text{ L}}\right)$$

$$= 0.014001754 \ M \text{ ammonia}$$

The K_f equilibrium is:

$$\text{Cu}^{2+}(aq) + 4\text{NH}_3(aq) \leftrightarrows \text{Cu(NH}_3)_4^{2+}(aq)$$

$$K_f = \frac{\left[\text{Cu(NH}_3)_4^{2+}\right]}{\left[\text{Cu}^{2+}\right]\left[\text{NH}_3\right]^4} = \frac{\left[8.9995616 \times 10^{-3}\right]}{\left[4.3842155 \times 10^{-7}\right]\left[0.014001754\right]^4} = 5.34072 \times 1011 = \mathbf{5.3 \times 10^{-11}}$$

b) The K_f will be used to determine the new concentration of free copper ions.

Moles uncomplexed ammonia before the addition of new ammonia =

$$(0.014001754 \text{ mol NH}_3/\text{L})(10^{-3} \text{ L}/1 \text{ mL})(100.0 \text{ mL}) = 0.0014001754 \text{ mol NH}_3$$

Moles ammonia added = 5.00×10^{-3} mol NH$_3$ (same as original moles of ammonia)

From the stoichiometry:

	Cu^{2+}(aq)	+	4NH$_3$(aq)	→	Cu(NH$_3$)$_4^{2+}$(aq)
Initial moles	4.3842155 x 10^{-8} mol		0.001400175 mol		8.9995616 x 10^{-4} mol
Added moles			5.00x10^{-3} mol		
Cu^{2+} is limiting	−(4.3842155 x 10^{-8} mol)		−4(4.3842155 x 10^{-8} mol)		+(4.3842155 x 10^{-8} mol)
After the reaction	0		0.006400 mol		9.00000x10^{-4} mol

Determine concentrations before equilibrium:

$[\text{Cu}^{2+}] = 0$

$[\text{NH}_3] = (0.006400 \text{ mol NH}_3/110.0 \text{ mL})(1 \text{ mL}/10^{-3} \text{ L}) = 0.0581818 \ M \text{ NH}_3$

$[\text{Cu(NH}_3)_4^{2+}] = (9.00000 \times 10^{-4} \text{ mol Cu(NH}_3)_4^{2+}/110.0 \text{ mL})(1 \text{ mL}/10^{-3} \text{ L})$

$= 0.008181818 \ M \text{ Cu(NH}_3)_4^{2+}$

Now allow the system to come to equilibrium:

	Cu^{2+}(aq)	+	4NH$_3$(aq)	⇆	Cu(NH$_3$)$_4^{2+}$(aq)
Initial molarity	0		0.0581818		0.008181818
Change	+x		+4x		−x
Equilibrium	x		0.0581818 + 4 x		0.008181818 − x

$$K_f = \frac{\left[\text{Cu(NH}_3)_4^{2+}\right]}{\left[\text{Cu}^{2+}\right]\left[\text{NH}_3\right]^4} = \frac{\left[0.008181818 - x\right]}{[x]\left[0.0581818 + 4x\right]^4} = 5.34072 \times 10^{11}$$

Assume − x and + 4x are negligible when compared to their associated numbers:

$$K_f = 5.34072 \times 10^{11} = \frac{[0.008181818]}{[x][0.0581818]^4}$$

$$x = [\text{Cu}^{2+}] = 1.3369 \times 10^{-9} \ M \text{ Cu}^{2+}$$

Use the Nernst equation to determine the new cell potential:

$$E = 0.00 \text{ V} - \frac{0.0592 \text{ V}}{2} \log \frac{\left[Cu^{2+}\right]_{ammonia}}{\left[Cu^{2+}\right]_{original}}$$

$$E = -\frac{0.0592 \text{ V}}{2} \log \frac{\left[1.3369 \text{x} 10^{-9}\right]}{[0.0100]}$$

$$E = 0.203467 = \textbf{0.20 V}$$

c) The first step will be to do a stoichiometry calculation of the reaction between copper ions and hydroxide ions.

$$\text{Moles of OH}^- = \left(\frac{0.500 \text{ mol NaOH}}{L}\right)\left(\frac{1 \text{ mol OH}^-}{1 \text{ mol NaOH}}\right)\left(\frac{10^{-3} \text{ L}}{1 \text{ mL}}\right)\left(10.0 \text{ mL}\right) = 5.00 \text{ x } 10^{-3} \text{ mol OH}^-$$

The initial moles of copper ions were determined earlier: $9.00 \text{ x } 10^{-4} \text{ mol Cu}^{2+}$
The reaction:

	$Cu^{2+}(aq)$	+	$2OH^-(aq)$	$\rightarrow$	$Cu(OH)_2(s)$
Initial moles	$9.00 \text{ x } 10^{-4}$ mol		$5.00 \text{ x } 10^{-3}$ mol		
Cu^{2+} is limiting	$-(9.00 \text{ x } 10^{-4}$ mol$)$		$-2(9.00 \text{ x } 10^{-4}$ mol$)$		
After the reaction	0		0.0032 mol		

Determine concentrations before equilibrium:

$$[Cu^{2+}] = 0$$

$$[NH_3] = (0.0032 \text{ mol OH}^-/100.0 \text{ mL})(1 \text{ mL}/10^{-3} \text{ L}) = 0.032 \ M \text{ OH}^-$$

Now allow the system to come to equilibrium:

	$Cu(OH)_2(s)$	$\leftrightarrows$	$Cu^{2+}(aq)$	+	$2OH^-(aq)$
Initial molarity			0.0		0.032
Change			+x		+2x
Equilibrium			x		0.032 + 2 x

$$K_{sp} = 2.2 \text{ x } 10^{-20} = [Cu^{2+}][OH^-]^2$$

$$K_{sp} = 2.2 \text{ x } 10^{-20} = [x][0.032 + 2x]^2$$

Assume 2x is negligible compared to 0.032 M.

$$K_{sp} = 2.2 \text{ x } 10^{-20} = [x][0.032]^2$$

$$x = [Cu^{2+}] = 2.1484375 \text{ x } 10^{-17} = 2.1 \text{ x } 10^{-17} \ M$$

Use the Nernst equation to determine the new cell potential:

$$E = 0.00 \text{ V} - \frac{0.0592 \text{ V}}{2} \log \frac{\left[Cu^{2+}\right]_{hydroxide}}{\left[Cu^{2+}\right]_{original}}$$

$$E = -\frac{0.0592 \text{ V}}{2} \log \frac{\left[2.1484375 \text{x} 10^{-17}\right]}{[0.0100]}$$

$$E = 0.434169 = \textbf{0.43 V}$$

d) Use the Nernst equation to determine the copper ion concentration in the half-cell containing the hydroxide ion.

$$E = 0.00 \text{ V} - \frac{0.0592 \text{ V}}{2} \log \frac{\left[Cu^{2+}\right]_{hydroxide}}{\left[Cu^{2+}\right]_{original}}$$

$$0.340 = -\frac{0.0592 \text{ V}}{2} \log \frac{\left[Cu^{2+}\right]_{hydroxide}}{[0.0100]}$$

$$-11.486486 = \log \frac{\left[Cu^{2+}\right]_{hydroxide}}{[0.0100]}$$

$$3.2622257 \times 10^{-12} = \frac{\left[Cu^{2+}\right]_{hydroxide}}{[0.0100]}$$

$$\left[Cu^{2+}\right]_{hydroxide} = 3.2622257 \times 10^{-14} \ M$$

Now use the K_{sp} relationship:

$$K_{sp} = [Cu^{2+}][OH^-]^2 = 2.2 \times 10^{-20}$$

$$K_{sp} = 2.2 \times 10^{-20} = [3.2622257 \times 10^{-14}][OH^-]^2$$

$$[OH^-]^2 = 6.743862 \times 10^{-7}$$

$$[OH^-] = 8.2121 \times 10^{-4} = 8.2 \times 10^{-4} \ M \ OH^- = \textbf{8.2} \times \textbf{10}^{-4} \ \textbf{\textit{M} NaOH}$$

21.153 a) The chemical equation for the combustion of octane is:

$$2C_8H_{18}(l) + 25O_2(g) \rightarrow 16CO_2(g) + 18H_2O(g)$$

The heat of reaction may be determined from heats of formation.

$$\Delta H^\circ_{rxn} = \sum m \, \Delta H^\circ_{f(products)} - \sum n \, \Delta H^\circ_{f(reactants)}$$

$$\Delta H^\circ_{rxn} = [(16 \text{ mol } CO_2)(\Delta H^\circ_f \text{ of } CO_2) + (18 \text{ mol } H_2O)(\Delta H^\circ_f \text{ of } H_2O)]$$
$$- [(2 \text{ mol } C_8H_{18})(\Delta H^\circ_f \text{ of } C_8H_{18}) + (25 \text{ mol } O_2)(\Delta H^\circ_f \text{ of } O_2)]$$

$$\Delta H^\circ_{rxn} = [(16 \text{ mol})(-393.5 \text{ kJ/mol}) + (18 \text{ mol})(-241.826 \text{ kJ/mol})]$$
$$- [(2 \text{ mol})(-250.1 \text{ kJ/mol}) + (25 \text{ mol})(0 \text{ kJ/mol})]$$

$$\Delta H^\circ_{rxn} = -10148.868 = -10148.9 \text{ kJ}$$

The energy from 1.00 gal of gasoline is:

$$\text{Energy (kJ)} = \left(1.00 \text{ gal}\right)\left(\frac{4 \text{ qt}}{1 \text{ gal}}\right)\left(\frac{1 \text{ L}}{1.057 \text{ qt}}\right)\left(\frac{1 \text{ mL}}{10^{-3} \text{ L}}\right)\left(\frac{0.7028 \text{ g}}{\text{mL}}\right)\left(\frac{1 \text{ mol } C_8H_{18}}{114.22 \text{ g } C_8H_{18}}\right)\left(\frac{-10148.9 \text{ kJ}}{2 \text{ mol } C_8H_{18}}\right)$$

$$= -1.18158 \times 10^5 = \textbf{--1.18} \times \textbf{10}^5 \ \textbf{kJ}$$

b) The energy from the combustion of hydrogen must be found using the balanced chemical equation and the heats of formation.

$$H_2(g) + 1/2 O_2(g) \rightarrow H_2O(g)$$

With the reaction written this way, the heat of reaction is simply the heat of formation of water vapor, and no additional calculations are necessary.

$$\Delta H^\circ = -241.826 \text{ kJ}$$

The moles of hydrogen needed to produce the energy from part a) are:

$$\text{Moles of } H_2 = \left(-1.18158 \times 10^5 \text{ kJ}\right)\left(\frac{1 \text{ mol } H_2}{-241.826 \text{ kJ}}\right) = 488.6075 \text{ mol}$$

Finally, use the ideal gas equation to determine the volume.

$$V = \frac{nRT}{P} = \frac{\left(488.6075 \text{ mol } H_2\right)\left(0.0821 \frac{\text{L} \cdot \text{atm}}{\text{mol} \cdot \text{K}}\right)\left((273 + 25)K\right)}{\left(1.00 \text{ atm}\right)} = 1.195417 \times 10^4 = \textbf{1.20} \times \textbf{10}^4 \ \textbf{L}$$

c) This part of the problem requires the half-reaction for the electrolysis of water to produce hydrogen gas.

$$2H_2O(l) + 2e^- \rightarrow H_2(g) + 2OH^-(aq)$$

Use 1 A = 1 C/s

$$\text{Time (s)} = \left(488.6075 \text{ mol } H_2\right)\left(\frac{2 \text{ mol } e^-}{1 \text{ mol } H_2}\right)\left(\frac{96,485 \text{ C}}{1 \text{ mol } e^-}\right)\left(\frac{s}{1.00 \times 10^3 \text{ C}}\right) = 9.42866 \times 10^4 = \textbf{9.43} \times \textbf{10}^4 \ \textbf{seconds}$$

d) Find the coulombs involved in the electrolysis of 488.6075 moles of H_2.

$$\text{Coulombs} = \left(488.6075 \text{ mol } H_2\right)\left|\frac{2 \text{ mol } e^-}{1 \text{ mol } H_2}\right|\left|\frac{96485 \text{ C}}{1 \text{ mol } e^-}\right| = 94,286,589 \text{ C}$$

Joules = C x V = 94,286,589 C x 6.00 V = 565,719,534 J

$$\text{Power (kW} \bullet \text{h)} = \left(565,719,534 \text{ J}\right)\left|\frac{1 \text{ kW} \bullet \text{h}}{3.6 \times 10^6 \text{ J}}\right| = 157.144 = \mathbf{157 \text{ kW} \bullet \text{h}}$$

e) The process is only 88.0% efficient, additional electricity is necessary to produce sufficient hydrogen. This is the purpose of the (100%/88.0%) factor.

$$\text{Cost} = \left(157.144 \text{ kW} \bullet \text{h}\right)\left(\frac{100\%}{88\%}\right)\left(\frac{0.123 \text{ cents}}{1 \text{ kW} \bullet \text{h}}\right) = 21.964445 = \mathbf{22.0 \text{ ¢}}$$

CHAPTER 22 THE ELEMENTS IN NATURE AND INDUSTRY

END–OF–CHAPTER PROBLEMS

22.2   Iron forms iron(III) oxide, Fe_2O_3 (commonly known as hematite). Calcium forms calcium carbonate, $CaCO_3$, (commonly known as limestone). Sodium is commonly found in sodium chloride (halite), $NaCl$. Zinc is commonly found in zinc sulfide (sphalerite), ZnS.

22.3   a) Differentiation refers to the processes involved in the formation of Earth into regions (core, mantle, and crust) of differing composition. Substances separated according to their densities, with the more dense material in the core and less dense in the crust.
b) The four most abundant elements are oxygen, silicon, aluminum, and iron in order of decreasing abundance.
c) Oxygen is the most abundant element in the crust and mantle but is not found in the core. Silicon, aluminum, calcium, sodium, potassium, and magnesium are also present in the crust and mantle but not in the core.

22.7   Plants produced O_2, slowly increasing the oxygen concentration in the atmosphere and creating an oxidative environment for metals. Evidence of Fe(II) deposits pre-dating Fe(III) deposits suggests this hypothesis is true. The decay of plant material and its incorporation into the crust increased the concentration of carbon in the crust and created large fossil-fuel deposits.

22.9   Fixation refers to the process of converting a substance in the atmosphere into a form more readily usable by organisms. Examples are the fixation of carbon by plants in the form of carbon dioxide and of nitrogen by bacteria in the form of nitrogen gas. Fixation of carbon dioxide gas by plants converts the CO_2 into carbohydrates during photosynthesis. Fixation of nitrogen gas by nitrogen-fixing bacteria involves the conversion of N_2 to ammonia and ammonium ions.

22.12   Atmospheric nitrogen is utilized by three fixation pathways: atmospheric, industrial, and biological. Atmospheric fixation requires high-temperature reactions (e.g., lightning) to convert N_2 into NO and other oxidized species. Industrial fixation involves mainly the formation of ammonia, NH_3, from N_2 and H_2. Biological fixation occurs in nitrogen-fixing bacteria that live in the roots of legumes. "Human activity" is referred to as "industrial fixation." Human activity is a significant factor, contributing about 17% of the nitrogen removed.

22.14   a) No gaseous phosphorus compounds are involved in the phosphorus cycle, so the atmosphere is not included in the phosphorus cycle.
b) Two roles of organisms in phosphorus cycle: 1) Plants excrete acid from their roots to convert PO_4^{3-} ions into more soluble $H_2PO_4^-$ ions, which the plant can absorb. 2) Through excretion and decay after death, organisms return soluble phosphate compounds to the cycle.

22.17   Plan: First determine the moles of F present in 100. kg of fluorapatite, using the molar mass of $Ca_5(PO_4)_3F$ (504.31 g/mol); take 15% of this amount. Convert mol F to mol SiF_4, and use the ideal gas law to determine the volume of the SiF_4 gas. For part b), convert moles of SiF_4 to moles of Na_2SiF_6 using the mole ratio in the given equation and convert moles of Na_2SiF_6 to moles and then mass of F. The definition of ppm ("parts per million") states that 1 ppm $F^- = (1$ g $F^-/(10^6$ g $H_2O))$. Find the volume of water that can be fluoridated with the calculated mass of F.
Solution:
a) Moles of F =

$$\left(100.\text{ kg }Ca_5(PO_4)_3F\right)\left(\frac{10^3\text{ g }Ca_5(PO_4)_3F}{1\text{ kg }Ca_5(PO_4)_3F}\right)\left(\frac{1\text{ mol }Ca_5(PO_4)_3F}{504.31\text{ g }Ca_5(PO_4)_3F}\right)\left(\frac{1\text{ mol F}}{1\text{ mol }Ca_5(PO_4)_3F}\right)\left(\frac{15\%}{100\%}\right)$$

$$= 29.74361\text{ mol F}$$

$$\text{Moles of SiF}_4 = (29.74361 \text{ mol F}) \left(\frac{1 \text{ mol SiF}_4}{4 \text{ mol F}} \right) = 7.43590 \text{ mol SiF}_4$$

$$PV = nRT$$

$$V\,(\text{L}) = \frac{nRT}{P} = \frac{(7.43590 \text{ mol SiF}_4)\left(0.0821\dfrac{\text{L}\bullet\text{atm}}{\text{mol}\bullet\text{K}}\right)((273+1450.)\text{K})}{(1.00 \text{ atm})} = 1051.86977 = \mathbf{1.1x10^3\ L}$$

b) Moles of $\text{Na}_2\text{SiF}_6 = (7.43590 \text{ mol SiF}_4) \left(\dfrac{1 \text{ mol Na}_2\text{SiF}_6}{2 \text{ mol SiF}_4} \right) = 3.71795 \text{ mol Na}_2\text{SiF}_6$

$$\text{Mass (g) of F}^- = (3.71795 \text{ mol Na}_2\text{SiF}_6) \left(\frac{6 \text{ mol F}^-}{1 \text{ mol Na}_2\text{SiF}_6} \right) \left(\frac{19.00 \text{ g F}^-}{1 \text{ mol F}^-} \right) = 423.8463 = \mathbf{4.2x10^2\ g\ F^-}$$

If 1 g F⁻ will fluoridate 10^6 g H_2O to a level of 1 ppm, how many grams (converted to mL using density, then converted from mL to L to m³) of water can 423.8463 g F⁻ fluoridate to the 1 ppm level? Necessary conversion factors: 1 m³ = 1000 L, density of water = 1.00 g/mL .

$$\text{Volume (m}^3) = (423.8463 \text{ g F}^-) \left(\frac{10^6 \text{ g H}_2\text{O}}{1 \text{ g F}^-} \right) \left(\frac{\text{mL H}_2\text{O}}{1.00 \text{ g H}_2\text{O}} \right) \left(\frac{1 \text{ cm}^3}{1 \text{ mL}} \right) \left(\frac{10^{-2} \text{ m}}{1 \text{ cm}} \right)^3 = 423.8463 = \mathbf{4.2x10^2\ m^3}$$

22.18 **Plan:** Since the 50. t contains 2.0% Fe_2O_3 by mass, 100 – 2.0 = 98% by mass of the 50. t contains $\text{Ca}_3(\text{PO}_4)_2$. Find the moles of $\text{Ca}_3(\text{PO}_4)_2$ in the sample and convert moles of $\text{Ca}_3(\text{PO}_4)_2$ to moles and then mass of P_4, remembering that the production of P_4 has a 90% yield.
Solution:
a) The iron ions form an insoluble salt, $\text{Fe}_3(\text{PO}_4)_2$, that decreases the yield of phosphorus. This salt is of limited value.
b) Use the conversion factor 1 metric ton (t) = $1\text{x}10^3$ kg.

$$\text{Moles of Ca}_3(\text{PO}_4)_2 = (50.\text{ t}) \left(\frac{98\%}{100\%} \right) \left(\frac{10^3 \text{ kg}}{1 \text{ t}} \right) \left(\frac{10^3 \text{ g}}{1 \text{ kg}} \right) \left(\frac{1 \text{ mol Ca}_3(\text{PO}_4)_2}{310.18 \text{ g Ca}_3(\text{PO}_4)_2} \right) = 157972.79 \text{ mol Ca}_3(\text{PO}_4)_2$$

Mass (metric tons) of P_4 =

$$(157972.79 \text{ mol Ca}_3(\text{PO}_4)_2) \left(\frac{1 \text{ mol P}_4}{2 \text{ mol Ca}_3(\text{PO}_4)_2} \right) \left(\frac{123.88 \text{ g P}_4}{1 \text{ mol P}_4} \right) \left(\frac{1 \text{ kg}}{10^3 \text{ g}} \right) \left(\frac{1 \text{ t}}{10^3 \text{ kg}} \right) \left(\frac{90.\%}{100\%} \right)$$
$$= 8.80635 = \mathbf{8.8\ t\ P_4}$$

22.20 a) Roasting involves heating the mineral in air (O_2) at high temperatures to convert the mineral to the oxide.
b) Smelting is the reduction of the metal oxide to the free metal using heat and a reducing agent such as coke.
c) Flotation is a separation process in which the ore is removed from the gangue by exploiting the different abilities of the two to interact with detergent. The gangue sinks to the bottom and the lighter ore-detergent mix is skimmed off the top.
d) Refining is the final step in the purification process to yield the pure metal with no impurities.

22.25 a) Slag is a waste product of iron metallurgy formed by the reaction:
$$\text{CaO}(s) + \text{SiO}_2(s) \rightarrow \text{CaSiO}_3(l)$$
In other words, slag is a by-product of steelmaking and contains the impurity SiO_2.
b) Pig iron, used to make cast iron products, is the impure product of iron metallurgy (containing 3–4% C) that is purified to steel.
c) Steel refers to the products of iron metallurgy, specifically alloys of iron containing small amounts of other elements including 1–1.5% carbon.
d) Basic-oxygen process refers to the process used to purify pig iron to form steel. The pig iron is melted and oxygen gas under high pressure is passed through the liquid metal. The oxygen oxidizes impurities to their oxides, which then react with calcium oxide to form a liquid that is decanted. Molten steel is left after the basic-oxygen process.

22.27   Iron and nickel are more easily oxidized than copper, so they are separated from the copper in the roasting step and conversion to slag. In the electrorefining process, all three metals are oxidized into solution, but only Cu^{2+} ions are reduced at the cathode to form $Cu(s)$.

22.30   The potassium in the molten potassium salt is reduced by molten sodium. Ordinarily, sodium would not be a good reducing agent for the more active potassium. But the reduction is carried out above the boiling point of potassium, producing gaseous potassium:
$$Na(l) + K^+(l) \leftrightarrows Na^+(l) + K(g)$$
The potassium gas is removed as it is produced. Le Châtelier's principle states that the system shifts toward formation of more K as the gaseous metal leaves the cell.

22.31   a) Aqueous salt solutions are mixtures of ions and water. When two half-reactions are possible at an electrode, the one with the more positive electrode potential occurs. In this case, the two half-reactions are:

$M^+ + e^- \rightarrow M^0$	$E^{\circ}_{red} = -3.05$ V, -2.93 V, and -2.71 V for Li^+, K^+, and Na^+, respectively
$2H_2O + 2e^- \rightarrow H_2 + 2OH^-$	$E_{red} = -0.42$ V, with an overvoltage of about -1 V

In all of these cases, it is energetically more favorable to reduce H_2O to H_2 than to reduce M^+ to M.

b) The question asks if Ca could chemically reduce RbX, i.e., convert Rb^+ to Rb^0. In order for this to occur, Ca^0 loses electrons ($Ca^0 \rightarrow Ca^{2+} + 2e^-$) and each Rb^+ gains an electron ($2Rb^+ + 2e^- \rightarrow 2Rb^0$). The reaction is written as follows:
$$2RbX + Ca \rightarrow CaX_2 + 2Rb$$
where $\Delta H = IE_1(Ca) + IE_2(Ca) - 2IE_1(Rb) = 590 + 1145 - 2(403) = +929$ kJ/mol.
Recall that Ca^0 acts as a *reducing agent* for the Rb^+ ion because it *oxidizes*. The energy required to remove an electron is the ionization energy. It requires more energy to ionize calcium's electrons, so it seems unlikely that Ca^0 could reduce Rb^+. Based on values of IE and a positive ΔH for the forward reaction, it seems more reasonable that Rb^0 would reduce Ca^{2+}.

c) If the reaction is carried out at a temperature greater than 688°C (the boiling point of rubidium), the product mixture will contain gaseous Rb. This can be removed from the reaction vessel, causing a shift in equilibrium to form more Rb product. If the reaction is carried out between 688°C and 1484°C (bp for Ca), then Ca remains in the molten phase and remains separated from gaseous Rb.

d) The reaction of calcium with molten CsX is written as follows:
$$2CsX + Ca \rightarrow CaX_2 + 2Cs$$
where $\Delta H = IE_1(Ca) + IE_2(Ca) - 2IE_1(Cs) = 590 + 1145 - 2(376) = +983$ kJ/mol. This reaction is more unfavorable than for Rb, but Cs has a lower boiling point of 671°C. If the reaction is carried out between 671°C and 1484°C, then calcium can be used to separate gaseous Cs from molten CsX.

22.32   Plan: Write the balanced equation for the process. For every two moles of Na metal produced, one mole of Cl_2 is produced. Calculate the amount of chlorine gas from the reaction stoichiometry, then use the ideal gas law to find the volume of chlorine gas. For part b), write the balanced equation for the oxidation of Cl^- to Cl_2; convert moles of Cl_2 produced to moles of electrons required and then convert to coulombs with the conversion factor 96,485 C = 1 mole of e^-. Convert coulombs to time with the conversion factor 1 C = 1 A•s.
Solution:
a) $2NaCl(l) \rightarrow 2Na(l) + Cl_2(g)$

$$\text{Moles of } Cl_2 = \left(31.0 \text{ kg Na}\right)\left(\frac{10^3 \text{ g}}{1 \text{ kg}}\right)\left(\frac{1 \text{ mol Na}}{22.99 \text{ g Na}}\right)\left(\frac{1 \text{ mol } Cl_2}{2 \text{ mol Na}}\right) = 674.2062 \text{ mol } Cl_2$$

$$PV = nRT$$

$$V \text{ (L)} = \frac{nRT}{P} = \frac{\left(674.2062 \text{ mol } Cl_2\right)\left(0.0821 \frac{L \bullet atm}{mol \bullet K}\right)\left((273 + 540.)K\right)}{\left(1.0 \text{ atm}\right)} = 4.50014 \times 10^4 = \mathbf{4.5 \times 10^4 \text{ L}}$$

b) Two moles of electrons are passed through the cell for each mole of Cl_2 produced:
$$2Cl^-(l) \rightarrow Cl_2(g) + 2e^-$$

$$\text{Coulombs} = (674.2062 \text{ mol } Cl_2)\left(\frac{2 \text{ mol } e^-}{1 \text{ mol } Cl_2}\right)\left(\frac{96,485 \text{ C}}{1 \text{ mol } e^-}\right) = 1.30101570\text{x}10^8 = \mathbf{1.30\text{x}10^8 \text{ C}}$$

c) Time (s) $= (1.30101570\text{x}10^8 \text{ C})\left(\frac{A \bullet s}{1 \text{ C}}\right)\left(\frac{1}{77.0 \text{ A}}\right) = 1.689631\text{x}10^6 = \mathbf{1.69\text{x}10^6 \text{ s}}$

22.35 a) Mg^{2+} is more difficult to reduce than H_2O, so $H_2(g)$ would be produced instead of Mg metal. $Cl_2(g)$ forms at the anode due to overvoltage.
b) The ΔH_f° for $MgCl_2(s)$ is -641.6 kJ/mol: $Mg(s) + Cl_2(g) \rightarrow MgCl_2(s) + heat$.
High temperature favors the reverse reaction (which is endothermic), so high temperatures favor the formation of magnesium metal and chlorine gas.

22.37 a) Sulfur dioxide is the reducing agent and is oxidized to the +6 state (as sulfate ion, SO_4^{2-}).
b) The sulfate ion formed reacts as a base in the presence of acid to form the hydrogen sulfate ion.
$$SO_4^{2-}(aq) + H^+(aq) \rightarrow HSO_4^-(aq)$$
c) Skeleton redox equation: $SO_2(g) + H_2SeO_3(aq) \rightarrow Se(s) + HSO_4^-(aq)$
    Reduction half-reaction: $H_2SeO_3(aq) \rightarrow Se(s)$
    balance O and H $H_2SeO_3(aq) + 4H^+(aq) \rightarrow Se(s) + 3H_2O(l)$
    balance charge $H_2SeO_3(aq) + 4H^+(aq) + 4e^- \rightarrow Se(s) + 3H_2O(l)$
    Oxidation half-reaction: $SO_2(g) \rightarrow HSO_4^-(aq)$
    balance O and H $SO_2(g) + 2H_2O(l) \rightarrow HSO_4^-(aq) + 3H^+(aq)$
    balance charge $SO_2(g) + 2H_2O(l) \rightarrow HSO_4^-(aq) + 3H^+(aq) + 2e^-$
    Multiply oxidation half-reaction by 2 $2SO_2(g) + 4H_2O(l) \rightarrow 2HSO_4^-(aq) + 6H^+(aq) + 4e^-$
    Add the two half-reactions:

$$H_2SeO_3(aq) + \cancel{4H^+}(aq) + \cancel{4e^-} \rightarrow Se(s) + \cancel{3}H_2O(l)$$
$$2SO_2(g) + \cancel{4}H_2O(l) \rightarrow 2HSO_4^-(aq) + \cancel{6}H^+(aq) + \cancel{4e^-}$$
$$\overline{H_2SeO_3(aq) + 2SO_2(g) + H_2O(l) \rightarrow Se(s) + 2HSO_4^-(aq) + 2H^+(aq)}$$

22.42 a) The oxidation state of copper in Cu_2S is +1 (sulfur is –2).
The oxidation state of copper in Cu_2O is +1 (oxygen is –2).
The oxidation state of copper in Cu is 0 (oxidation state is always 0 in the elemental form).
b) The reducing agent is the species that is *oxidized*. Sulfur changes oxidation state from –2 in Cu_2S to +4 in SO_2. Therefore, Cu_2S is the reducing agent, leaving Cu_2O as the oxidizing agent.

22.47 Plan: Free energy, ΔG°, is the measure of the ability of a reaction to proceed spontaneously. ΔG° can be calculated with the relationship $\sum m \Delta G_{f\text{(products)}}^\circ - \sum n \Delta G_{f\text{(reactants)}}^\circ$.
Solution:
The direct reduction of ZnS follows the reaction:
$2ZnS(s) + C(\text{graphite}) \rightarrow 2Zn(s) + CS_2(g)$
$\Delta G_{rxn}^\circ = [(2 \text{ mol } Zn)(\Delta G_f^\circ \text{ of } Zn) + (1 \text{ mol } CS_2)(\Delta G_f^\circ \text{ of } CS_2)]$
        $- [(2 \text{ mol } ZnS)(\Delta G_f^\circ \text{ of } ZnS) + (1 \text{ mol } C)(\Delta G_f^\circ \text{ of } C)]$
$\Delta G_{rxn}^\circ = [(2 \text{ mol})(0 \text{ kJ/mol}) + (1 \text{ mol})(66.9 \text{ kJ/mol})] - [(2 \text{ mol})(-198 \text{ kJ/mol}) + (1 \text{ mol})(0 \text{ kJ/mol})]$
$\Delta G_{rxn}^\circ = 462.9 = +463 \text{ kJ}$
Since ΔG_{rxn}° is positive, this reaction is not spontaneous at standard-state conditions. The stepwise reduction involves the conversion of ZnS to ZnO, followed by the final reduction step:
$2ZnO(s) + C(s) \rightarrow 2Zn(s) + CO_2(g)$
$\Delta G_{rxn}^\circ = [(2 \text{ mol } Zn)(\Delta G_f^\circ \text{ of } Zn) + (1 \text{ mol } CO_2)(\Delta G_f^\circ \text{ of } CO_2)]$
        $- [(2 \text{ mol } ZnO)(\Delta G_f^\circ \text{ of } ZnO) + (1 \text{ mol } C)(\Delta G_f^\circ \text{ of } C)]$
$\Delta G_{rxn}^\circ = [(2 \text{ mol})(0 \text{ kJ/mol}) + (1 \text{ mol})(-394.4 \text{ kJ/mol})] - [(2 \text{ mol})(-318.2 \text{ kJ/mol}) + (1 \text{ mol})(0 \text{ kJ/mol})]$
$\Delta G_{rxn}^\circ = +242.0 \text{ kJ}$
This reaction is also not spontaneous, but the oxide reaction is less unfavorable.

22.48 The formation of sulfur trioxide is very slow at ordinary temperatures. Increasing the temperature can speed up the reaction, but the reaction is exothermic, so increasing the temperature decreases the yield. Recall that yield, or the extent to which a reaction proceeds, is controlled by the thermodynamics of the reaction. Adding a catalyst increases the rate of the formation reaction but does not impact the thermodynamics, so a lower temperature can be used to enhance the yield. Catalysts are used in such a reaction to allow control of both rate and yield of the reaction.

22.51 a) The chlor-alkali process yields Cl_2, H_2, and NaOH.
b) The mercury-cell process yields higher purity NaOH, but produces Hg-polluted waters that are discharged into the environment.

22.52 Plan: $\Delta G°$ at 25°C can be calculated with the relationship $\sum m \Delta G°_{f(products)} - \sum n \Delta G°_{f(reactants)}$. Calculate $\Delta G°$ at 500.°C with the relationship $\Delta G°_{rxn} = \Delta H°_{rxn} - T\Delta S°_{rxn}$. $\Delta H°_{rxn}$ and $\Delta S°_{rxn}$ must be calculated first. The equilibrium constant K is calculated by using $\Delta G° = -RT \ln K$. Finally, to find the temperature at which the reaction becomes spontaneous, use $\Delta G°_{rxn} = 0 = \Delta H°_{rxn} - T\Delta S°_{rxn}$ and solve for temperature.
Solution:

a) The balanced equation for the oxidation of SO_2 to SO_3 is $2SO_2(g) + O_2(g) \rightarrow 2SO_3(g)$.

$\Delta G°_{rxn} = \sum m \Delta G°_{f(products)} - \sum n \Delta G°_{f(reactants)}$

$\Delta G°_{rxn} = [(2 \text{ mol } SO_3)(\Delta G°_f \text{ of } SO_3)] - [(2 \text{ mol } SO_2)(\Delta G°_f \text{ of } SO_2) + (1 \text{ mol } O_2)(\Delta G°_f \text{ of } O_2)]$

$\Delta G°_{rxn} = [(2 \text{ mol})(-371 \text{ kJ/mol})] - [(2 \text{ mol})(-300.2 \text{ kJ/mol}) + (1 \text{ mol } O_2)(0 \text{ kJ/mol})]$

$\Delta G°_{rxn} = -141.6 = -142 \text{ kJ}$

Since $\Delta G°_{rxn}$ is negative, **the reaction is spontaneous at 25°C.**
b) The rate of the reaction is very slow at 25°C, so the reaction does not produce significant amounts of SO_3 at room temperature.

c) $\Delta H°_{rxn} = \sum m \Delta H°_{f(products)} - \sum n \Delta H°_{f(reactants)}$

$\Delta H°_{rxn} = [(2 \text{ mol } SO_3)(\Delta H°_f \text{ of } SO_3)] - [(2 \text{ mol } SO_2)(\Delta H°_f \text{ of } SO_2) + (1 \text{ mol } O_2)(\Delta H°_f \text{ of } O_2)]$

$\Delta H°_{rxn} = [(2 \text{ mol})(-396 \text{ kJ/mol})] - [(2 \text{ mol})(-296.8 \text{ kJ/mol}) + (1 \text{ mol})(0 \text{ kJ/mol})]$

$\Delta H°_{rxn} = -198.4 = -198 \text{ kJ}$

$\Delta S°_{rxn} = \sum m S°_{products} - \sum n S°_{reactants}$

$\Delta S°_{rxn} = [(2 \text{ mol } SO_3)(S° \text{ of } SO_3)] - [(2 \text{ mol } SO_2)(S° \text{ of } SO_2) + (1 \text{ mol } O_2)(S° \text{ of } O_2)]$

$\Delta S°_{rxn} = [(2 \text{ mol})(256.66 \text{ J/mol}\cdot\text{K})] - [(2 \text{ mol})(248.1 \text{ J/mol}\cdot\text{K}) + (1 \text{ mol})(205.0 \text{ J/mol}\cdot\text{K})]$

$\Delta S°_{rxn} = -187.88 = -187.9 \text{ J/K}$

$\Delta G°_{500} = \Delta H°_{rxn} - T\Delta S°_{rxn} = -198.4 \text{ kJ} - ((273 + 500.) \text{ K})(-187.88 \text{ J/K})(1 \text{ kJ}/10^3 \text{ J}) = -53.16876 = \mathbf{-53 \text{ kJ}}$
   The reaction is spontaneous at 500°C since free energy is negative.
d) The equilibrium constant at 500°C is smaller than the equilibrium constant at 25°C because the free energy at 500°C indicates the reaction does not go as far to completion as it does at 25°C.
Equilibrium constants can be calculated from $\Delta G° = -RT \ln K$.

At 25°C: $\ln K = \dfrac{\Delta G°}{-RT} = -\dfrac{(-142 \text{ kJ})\left(\dfrac{10^3 \text{ J}}{1 \text{ kJ}}\right)}{(8.314 \text{ J/mol} \cdot \text{K})((273 + 25)\text{K})} = 57.31417694$

$K_{25} = e^{57.31417694} = 7.784501 \times 10^{24} = \mathbf{7.8 \times 10^{24}}$

At 500°C: $\ln K = \dfrac{\Delta G°}{-RT} = -\dfrac{(-53 \text{ kJ})\left(\dfrac{10^3 \text{ J}}{1 \text{ kJ}}\right)}{(8.314 \text{ J/mol} \cdot \text{K})((273 + 500)\text{K})} = 8.246817$

$K_{500} = e^{8.246817} = 3.815 \times 10^3 = \mathbf{3.8 \times 10^3}$

e) Temperature below which the reaction is spontaneous at standard state can be calculated by setting $\Delta G°$ equal to zero and using the enthalpy and entropy values at 25°C to calculate the temperature.

$$\Delta G°_{rxn} = 0 = \Delta H°_{rxn} - T\Delta S°_{rxn}$$

$$\Delta H°_{rxn} = T\Delta S°_{rxn}$$

$$T = \frac{\Delta H°}{\Delta S°} = \frac{-198 \text{ kJ}}{-187.9 \text{ J/K}\left(\dfrac{1 \text{ kJ}}{10^3 \text{ J}}\right)} = 1.05375 \times 10^3 = \mathbf{1.05 \times 10^3 \text{ K}}$$

22.53 **Plan:** This problem deals with the stoichiometry of electrolysis. The balanced oxidation half-reaction for the chlor-alkali process is given in the chapter:

$$2Cl^-(aq) \rightarrow Cl_2(g) + 2e^-$$

Use the Faraday constant, F, $(1\ F = 9.6485 \times 10^4$ C/mol e^-) and the fact that one mol of Cl_2 produces two mol e^- or $2\ F$, to convert coulombs to moles of Cl_2.
Solution:

$$\text{Mass (lb) of } Cl_2 = \left(3 \times 10^4 \text{ A}\right)\left(\frac{C/s}{A}\right)\left(\frac{3600 \text{ s}}{1 \text{ h}}\right)(8 \text{ h})\left(\frac{1 \text{ mol } e^-}{96,485 \text{ C}}\right)\left(\frac{1 \text{ mol } Cl_2}{2 \text{ mol } e^-}\right)\left(\frac{70.90 \text{ g } Cl_2}{1 \text{ mol } Cl_2}\right)\left(\frac{1 \text{ kg}}{10^3 \text{ g}}\right)\left(\frac{2.205 \text{ lb}}{1 \text{ kg}}\right)$$

$$= 699.9689 = \mathbf{7 \times 10^2 \text{ lb } Cl_2}$$

22.56 **Plan:** Write the balanced equation. Use the stoichiometry in the equation to find the moles of H_3PO_4 produced by the reaction of the given amount of P_4O_{10}. Divide moles of H_3PO_4 by the volume to obtain its molarity. Since H_3PO_4 is a weak acid, use its equilibrium expression to find the $[H_3O^+]$ and then pH.
Solution:
a) Because P_4O_{10} is a drying agent, water is incorporated in the formation of phosphoric acid, H_3PO_4.

$$P_4O_{10}(s) + 6H_2O(l) \rightarrow 4H_3PO_4(l)$$

b) Moles of $H_3PO_4 = \left(8.5 \text{ g } P_4O_{10}\right)\left(\frac{1 \text{ mol } P_4O_{10}}{283.88 \text{ g } P_4O_{10}}\right)\left(\frac{4 \text{ mol } H_3PO_4}{1 \text{ mol } P_4O_{10}}\right) = 0.11976892 \text{ mol } H_3PO_4$

$$\text{Molarity of } H_3PO_4 = \left(\frac{0.11976892 \text{ mol } H_3PO_4}{0.750 \text{ L}}\right) = 0.1596919\ M\ H_3PO_4$$

Phosphoric acid is a weak acid and only partly dissociates to form H_3O^+ in water, based on its K_a. Since $K_{a1} \gg K_{a2} \gg K_{a3}$, assume that the H_3O^+ contributed by the second and third dissociation is negligible in comparison to the H_3O^+ contributed by the first dissociation.

$$H_3PO_4(l) + H_2O(l) \rightarrow H_3O^+(aq) + H_2PO_4^-(aq)$$

$$K_a = 7.2 \times 10^{-3} = \frac{[H_3O^+][H_2PO_4^-]}{[H_3PO_4]}$$

$$K_a = 7.2 \times 10^{-3} = \frac{[x][x]}{(0.1596919 - x)} \qquad \text{Assume x is small compared to } 0.1596919.$$

$$K_a = 7.2 \times 10^{-3} = \frac{[x][x]}{(0.1596919)}$$

$x = 0.0339084\ M$
Check assumption that x is small compared to 0.1596919:

$$\frac{0.0339084}{0.1596919}(100) = 21\% \text{ error, so the assumption is not valid.}$$

The problem will need to be solved as a quadratic.
$x^2 = (7.2 \times 10^{-3})(0.1596919 - x) = 1.14978 \times 10^{-3} - 7.2 \times 10^{-3}x$
$x^2 + 7.2 \times 10^{-3}x - 1.14978 \times 10^{-3} = 0$

$$a = 1 \qquad b = 7.2 \times 10^{-3} \qquad c = -1.14978 \times 10^{-3}$$

$$x = \frac{-7.2 \times 10^{-3} \pm \sqrt{(7.2 \times 10^{-3})^2 - 4(1)(-1.14978 \times 10^{-3})}}{2(1)}$$

$x = 3.049897 \times 10^{-2}\ M\ H_3O^+$

$pH = -\log [H_3O^+] = -\log (3.049897 \times 10^{-2}) = 1.51571 = \mathbf{1.52}$

22.58 **Plan:** Write a balanced equation for the production of iron and use the reaction stoichiometry to calculate the amount of CO_2 produced when the given amount of Fe is produced. Then write a balanced equation for the combustion of gasoline (C_8H_{18}.) Find the total volume of gasoline, use the density to convert volume to mass of gasoline, and use the reaction stoichiometry to calculate the amount of CO_2 produced.
Solution:
a) The reaction taking place in the blast furnace to produce iron is:
$$Fe_2O_3(s) + 3CO(g) \rightarrow 2Fe(s) + 3CO_2(g)$$

Mass (g) of $CO_2 = (8400.\,t\ Fe)\left(\dfrac{2000\ lb}{1\ t}\right)\left(\dfrac{1\ kg}{2.205\ lb}\right)\left(\dfrac{10^3\ g}{1\ kg}\right)\left(\dfrac{1\ mol\ Fe}{55.85\ g\ Fe}\right)\left(\dfrac{3\ mol\ CO_2}{2\ mol\ Fe}\right)\left(\dfrac{44.01\ g\ CO_2}{1\ mol\ CO_2}\right)$

$= 9.005755 \times 10^9 = \mathbf{9.006 \times 10^9\ g\ CO_2}$

b) Combustion reaction for octane:
$$2C_8H_{18}(l) + 25O_2(g) \rightarrow 16CO_2(g) + 18H_2O(l)$$

Mass (g) of $CO_2 =$

$(1.0 \times 10^6\ autos)\left(\dfrac{5.0\ gal}{1\ auto}\right)\left(\dfrac{4\ qt}{1\ gal}\right)\left(\dfrac{1\ L}{1.057\ qt}\right)\left(\dfrac{1\ mL}{10^{-3}\ L}\right)\left(\dfrac{0.74\ g}{mL}\right)\left(\dfrac{1\ mol\ C_8H_{18}}{114.22\ g\ C_8H_{18}}\right)\left(\dfrac{16\ mol\ CO_2}{2\ mol\ C_8H_{18}}\right)\left(\dfrac{44.01\ g\ CO_2}{1\ mol\ CO_2}\right)$

$= 4.31604 \times 10^{10} = 4.3 \times 10^{10}\ g\ CO_2$

The CO_2 production by automobiles is much greater than that from steel production.

22.60 **Plan:** Step 2 of "Isolation of Magnesium" describes this reaction. Sufficient OH^- must be added to precipitate $Mg(OH)_2$. The necessary amount of OH^- can be determined from the K_{sp} of $Mg(OH)_2$. For part b), use the K_{sp} of $Ca(OH)_2$ to determine the amount of OH^- that a saturated solution of $Ca(OH)_2$ provides. Substitute this amount into the K_{sp} expression for $Mg(OH)_2$ to determine how much Mg^{2+} remains in solution with this amount of OH^-. Calculate the fraction of $[Mg^{2+}]$ remaining and subtract from 1 to obtain the fraction of $[Mg^{2+}]$ that precipitated.
Solution:
a) $Mg(OH)_2(s) \rightarrow Mg^{2+}(aq) + 2OH^-(aq)$
$K_{sp} = 6.3 \times 10^{-10} = [Mg^{2+}][OH^-]^2$

$[OH^-] = \sqrt{\dfrac{6.3 \times 10^{-10}}{[Mg^{2+}]}} = \sqrt{\dfrac{6.3 \times 10^{-10}}{0.052}} = 1.100699 \times 10^{-4} = 1.1 \times 10^{-4}\ M$

Thus, if $[OH^-] > 1.1 \times 10^{-4}\ M$ (i.e., if pH > 10.04), $Mg(OH)_2$ will precipitate.

b) $Ca(OH)_2(s) \rightarrow Ca^{2+}(aq) + 2OH^-(aq)$
$K_{sp} = 6.5 \times 10^{-6} = [Ca^{2+}][OH^-]^2 = (x)(2x)^2 = 4x^3$
$x = 0.011756673;\ [OH^-] = 2x = 0.023513347\ M$
$K_{sp}\ (Mg(OH)_2) = 6.3 \times 10^{-10} = [Mg^{2+}][OH^-]^2$

$[Mg^{2+}] = \dfrac{6.3 \times 10^{-10}}{(0.023513347)^2} = 1.1394930 \times 10^{-6}\ M$

This concentration is the amount of the original $0.052\ M\ Mg^{2+}$ that was not precipitated. The percent precipitated is the difference between the remaining $[Mg^{2+}]$ and the initial $[Mg^{2+}]$ divided by the initial concentration.
Fraction Mg^{2+} remaining = $(1.1394930 \times 10^{-6}\ M)/(0.052\ M) = 2.19134 \times 10^{-5}$
Mg^{2+} precipitated = $1 - 2.19134 \times 10^{-5} = 0.9999781 = 1$ (To the limit of the significant figures, all the magnesium has precipitated.)

22.61 Plan: The equation $\Delta G^{\circ}_{rxn} = \Delta H^{\circ}_{rxn} - T\Delta S^{\circ}_{rxn}$ will be used to calculate ΔG°_{rxn} at the two temperatures. ΔH°_{rxn} and ΔS°_{rxn} will have to be calculated first. Then use $\Delta G^{\circ} = -RT \ln K$ to determine K. Subscripts indicate the temperature, and (1) or (2) indicate the reaction.
Solution:

For the first reaction (1):

$\Delta H^{\circ}_{rxn} = \sum m \Delta H^{\circ}_{f(products)} - \sum n \Delta H^{\circ}_{f(reactants)}$

$\Delta H^{\circ}_{rxn} = [(4 \text{ mol NO})(\Delta H^{\circ}_f \text{ of NO}) + (6 \text{ mol } H_2O)(\Delta H^{\circ}_f \text{ of } H_2O)]$

$\qquad\qquad - [(4 \text{ mol } NH_3)(\Delta H^{\circ}_f \text{ of } NH_3) + (5 \text{ mol } O_2)(\Delta H^{\circ}_f \text{ of } O_2)]$

$\Delta H^{\circ}_{rxn} = [(4 \text{ mol})(90.29 \text{ kJ/mol}) + (6 \text{ mol})(-241.826 \text{ kJ/mol})] - [(4 \text{ mol})(-45.9 \text{ kJ/mol}) + (5 \text{ mol})(0 \text{ kJ/mol})]$

$\Delta H^{\circ}_{rxn} = -906.196 \text{ kJ}$

$\Delta S^{\circ}_{rxn} = [(4 \text{ mol NO})(S^{\circ} \text{ of NO}) + (6 \text{ mol } H_2O)(S^{\circ} \text{ of } H_2O)]$

$\qquad\qquad - [(4 \text{ mol } NH_3)(S^{\circ} \text{ of } NH_3) + (5 \text{ mol } O_2)(S^{\circ} \text{ of } O_2)]$

$\Delta S^{\circ}_{rxn} = [(4 \text{ mol})(210.65 \text{ J/K} \bullet \text{mol}) + (6 \text{ mol})(188.72 \text{ J/K} \bullet \text{mol})]$

$\qquad\qquad - [(4 \text{ mol})(193 \text{ J/K} \bullet \text{mol}) + (5 \text{ mol } O_2)(205.0 \text{ J/K} \bullet \text{mol})](1 \text{ kJ}/10^3 \text{ J})$

$\Delta S^{\circ}_{rxn} = 0.17792 \text{ kJ/K}$

$\Delta G^{\circ}_{rxn} = \Delta H^{\circ}_{rxn} - T\Delta S^{\circ}_{rxn}$

$\Delta G^{\circ}_{25} = -906.196 \text{ kJ} - [(273 + 25)\text{K}](0.17792 \text{ kJ/K}) = -959.216 \text{ kJ}$

$\Delta G^{\circ}_{900} = -906.196 \text{ kJ} - [(273 + 900.)\text{K}](0.17792 \text{ kJ/K}) = -1114.896 \text{ kJ}$

For the second reaction (2):

$\Delta H^{\circ}_{rxn} = [(2 \text{ mol } N_2)(\Delta H^{\circ}_f \text{ of } N_2) + (6 \text{ mol } H_2O)(\Delta H^{\circ}_f \text{ of } H_2O)]$

$\qquad\qquad - [(4 \text{ mol } NH_3)(\Delta H^{\circ}_f \text{ of } NH_3) + (3 \text{ mol } O_2)(\Delta H^{\circ}_f \text{ of } O_2)]$

$\Delta H^{\circ}_{rxn} = [(2 \text{ mol})(0 \text{ kJ/mol}) + (6 \text{ mol})(-241.826 \text{ kJ/mol})] - [(4 \text{ mol})(-45.9 \text{ kJ/mol}) + (3 \text{ mol})(0 \text{ kJ/mol})]$

$\Delta H^{\circ}_{rxn} = -1267.356 \text{ kJ}$

$\Delta S^{\circ}_{rxn} = [(2 \text{ mol } N_2)(S^{\circ} \text{ of } N_2) + (6 \text{ mol } H_2O)(S^{\circ} \text{ of } H_2O)]$

$\qquad\qquad - [(4 \text{ mol } NH_3)(S^{\circ} \text{ of } NH_3) + (3 \text{ mol } O_2)(S^{\circ} \text{ of } O_2)]$

$\Delta S^{\circ}_{rxn} = [(2 \text{ mol})(191.5 \text{ J/K} \bullet \text{mol}) + (6 \text{ mol})(188.72 \text{ J/K} \bullet \text{mol})]$

$\qquad\qquad - [(4 \text{ mol})(193 \text{ J/K} \bullet \text{mol}) + (3 \text{ mol})(205.0 \text{ J/K} \bullet \text{mol})](1 \text{ kJ}/10^3 \text{ J})$

$\Delta S^{\circ}_{rxn} = 0.12832 \text{ J/K}$

$\Delta G^{\circ}_{rxn} = \Delta H^{\circ}_{rxn} - T\Delta S^{\circ}_{rxn}$

$\Delta G^{\circ}_{25} = -1267.356 \text{ kJ} - [(273 + 25)\text{K}](0.12832 \text{ kJ/K}) = -1305.595 \text{ kJ}$

$\Delta G^{\circ}_{900} = -1267.356 \text{ kJ} - [(273 + 900.)\text{K}](0.12832 \text{ kJ/K}) = -1417.875 \text{ kJ}$

a) $\ln K_{25}(1) = \dfrac{\Delta G^{\circ}}{-RT} = -\dfrac{(-959.216 \text{ kJ/mol})}{(8.314 \text{ J/mol} \bullet \text{K})((273 + 25)\text{K})}\left(\dfrac{10^3 \text{ J}}{1 \text{ kJ}}\right) = 387.15977$

$\qquad\qquad K_{25}(1) = 1.38 \times 10^{168} = \mathbf{1 \times 10^{168}}$

$\ln K_{25}(2) = \dfrac{\Delta G^{\circ}}{-RT} = -\dfrac{(-1305.595 \text{ kJ/mol})}{(8.314 \text{ J/mol} \bullet \text{K})((273 + 25)\text{K})}\left(\dfrac{10^3 \text{ J}}{1 \text{ kJ}}\right) = 526.9655$

$\qquad\qquad K_{25}(2) = 7.2146 \times 10^{228} = \mathbf{7 \times 10^{228}}$

b) $\ln K_{900}(1) = \dfrac{\Delta G^{\circ}}{-RT} = -\dfrac{(-1114.896 \text{ kJ/mol})}{(8.314 \text{ J/mol} \bullet \text{K})((273+900)\text{K})}\left(\dfrac{10^3 \text{ J}}{1 \text{ kJ}}\right) = 114.3210817$

$$K_{900}(1) = 4.4567 \times 10^{49} = \mathbf{4.6 \times 10^{49}}$$

$\ln K_{900}(2) = \dfrac{\Delta G^{\circ}}{-RT} = -\dfrac{(-1417.875 \text{ kJ/mol})}{(8.314 \text{ J/mol} \bullet \text{K})((273+900)\text{K})}\left(\dfrac{10^3 \text{ J}}{1 \text{ kJ}}\right) = 145.388$

$$K_{900}(2) = 1.38422 \times 10^{63} = \mathbf{1.4 \times 10^{63}}$$

c) Cost (\$) Pt $= \left(1.01 \times 10^7 \text{ t HNO}_3\right)\left(\dfrac{175 \text{ mg Pt}}{1 \text{ t HNO}_3}\right)\left(\dfrac{10^{-3} \text{ g}}{1 \text{ mg}}\right)\left(\dfrac{1 \text{ kg}}{10^3 \text{ g}}\right)\left(\dfrac{32.15 \text{ Troy Oz}}{1 \text{ kg}}\right)\left(\dfrac{\$1557}{1 \text{ troy oz}}\right)$

$$= 8.847672 \times 10^7 = \mathbf{\$8.8 \times 10^7}$$

d) Cost (\$) Pt $= \left(1.01 \times 10^7 \text{ t HNO}_3\right)\left(\dfrac{175 \text{ mg Pt}}{1 \text{ t HNO}_3}\right)\left(\dfrac{72\%}{100\%}\right)\left(\dfrac{10^{-3} \text{ g}}{1 \text{ mg}}\right)\left(\dfrac{1 \text{ kg}}{10^3 \text{ g}}\right)\left(\dfrac{32.15 \text{ troy oz}}{1 \text{ kg}}\right)\left(\dfrac{\$1557}{1 \text{ troy oz}}\right)$

$$= 6.3703238 \times 10^7 = \mathbf{\$6.4 \times 10^7}$$

22.64 Plan: Balance each reaction. Acidity is a measure of the concentration of H_3O^+ (H^+), so any reaction that produces H^+ will increase the acidity.
Solution:
a) $2H_2O(l) + 2FeS_2(s) + 7O_2(g) \rightarrow 2Fe^{2+}(aq) + 4SO_4^{2-}(aq) + 4H^+(aq)$ Increases acidity.
b) $4H^+(aq) + 4Fe^{2+}(aq) + O_2(g) \rightarrow 4Fe^{3+}(aq) + 2H_2O(l)$
c) $Fe^{3+}(aq) + 3H_2O(l) \rightarrow Fe(OH)_3(s) + 3H^+(aq)$ Increases acidity.
d) $8H_2O(l) + FeS_2(s) + 14Fe^{3+}(aq) \rightarrow 15Fe^{2+}(aq) + 2SO_4^{2-}(aq) + 16H^+(aq)$ Increases acidity.

22.65 Plan: The carbon fits into interstitial positions. The cell size will increase slightly to accommodate the carbons atoms. The increase in size is assumed to be negligible. The mass of the carbon added depends upon the carbon in the unit cell.
Solution:
Ferrite:

$$\left(\dfrac{7.86 \text{ g}}{\text{cm}^3}\right) + \left(\dfrac{7.86 \text{ g}}{\text{cm}^3}\right)\left(\dfrac{0.0218\%}{100\%}\right) = 7.86171 = \mathbf{7.86 \text{ g/cm}^3}$$

Austenite:

$$\left(\dfrac{7.40 \text{ g}}{\text{cm}^3}\right) + \left(\dfrac{7.40 \text{ g}}{\text{cm}^3}\right)\left(\dfrac{2.08\%}{100\%}\right) = 7.55392 = \mathbf{7.55 \text{ g/cm}^3}$$

22.67 Plan: Decide what is reduced and what is oxidized of the two ions present in molten NaOH. Then, write half-reactions and balance them. Then write the overall equation for the electrolysis of NaOH and also for the reaction of Na with water. Examine the molar ratios involving water and Na in these equations.
Solution:
a) At the cathode, sodium ions are reduced: $Na^+(l) + e^- \rightarrow Na(l)$
At the anode, hydroxide ions are oxidized: $4OH^-(l) \rightarrow O_2(g) + 2H_2O(g) + 4e^-$
b) The overall cell reaction is $4Na^+(l) + 4OH^-(l) \rightarrow 4Na(l) + O_2(g) + 2H_2O(g)$ and the reaction between sodium metal and water is $2Na(l) + 2H_2O(g) \rightarrow 2NaOH(l) + H_2(g)$.
For each mole of water produced, two moles of sodium are produced, but for each mole of water that reacts only one mole of sodium reacts. So, the maximum amount of sodium that reacts with water is half of the sodium produced. In the balanced cell reaction, four moles of electrons are transferred for four moles of sodium. Since 1/2 of the Na reacts with H_2O, the maximum efficiency is 1/2 mol Na/mol e^-, or **50%**.

22.70 **Plan:** Convert mass of CO_2 to moles of O_2 using the reaction stoichiometry in the balanced equation. $PV = nRT$ is used to convert moles of O_2 to volume. Temperature must be in units of Kelvin.
Solution:
a) The reaction for the fixation of carbon dioxide includes CO_2 and H_2O as reactants and $(CH_2O)_n$ and O_2 as products. The balanced equation is:
$$nCO_2(g) + nH_2O(l) \rightarrow (CH_2O)_n(s) + nO_2(g)$$
b) The moles of carbon dioxide fixed equal the moles of O_2 produced.

$$\text{Mole } O_2/\text{day} = \left(\frac{48 \text{ g } CO_2}{\text{day}}\right)\left(\frac{1 \text{ mol } CO_2}{44.01 \text{ g } CO_2}\right)\left(\frac{1 \text{ mol } O_2}{1 \text{ mol } CO_2}\right) = 1.090661 \text{ mol } O_2/\text{day}$$

To find the volume of oxygen, use the ideal gas equation. Temperature must be converted to K.

$$T \text{ (in K)} = \frac{5}{9}\left[T(°F) - 32\right] + 273.15 = \frac{5}{9}\left[T(78.°) - 32\right] + 273.15 = 298.7056 \text{ K}$$

$$PV = nRT$$

$$V = \frac{nRT}{P} = \frac{(1.090661 \text{ mol } O_2)(0.0821 \text{ L} \bullet \text{atm/mol} \bullet \text{K})(298.7056 \text{ K})}{(1.0 \text{ atm})} = 26.74708 = \mathbf{27 \text{ L}}$$

c) The moles of air containing 1.090661 mol of CO_2 is:

$$\text{Mole } CO_2/\text{day} = \left(\frac{48 \text{ g } CO_2}{\text{day}}\right)\left(\frac{1 \text{ mol } CO_2}{44.01 \text{ g } CO_2}\right)\left(\frac{100 \text{ mol \% air}}{0.040 \text{ mol \% } CO_2}\right) = 2.7266530 \times 10^3 \text{ mol air/day}$$

$$V = \frac{nRT}{P} = \frac{(2.7266530 \times 10^3 \text{ mol air})(0.0821 \text{ L} \bullet \text{atm/mol} \bullet \text{K})(298.7056 \text{ K})}{(1.0 \text{ atm})} = 6.6867 \times 10^4 = \mathbf{6.7 \times 10^4 \text{ L}}$$

22.71 This problem requires a series of conversion steps:
Concentration =

$$\left(\frac{210. \text{ kg } (NH_4)_2SO_4}{1000. \text{ m}^3}\right)\left(\frac{10^3 \text{ g}}{1 \text{ kg}}\right)\left(\frac{1 \text{ mol } (NH_4)_2SO_4}{132.15 \text{ g } (NH_4)_2SO_4}\right)\left(\frac{2 \text{ mol } NO_3^-}{1 \text{ mol } (NH_4)_2SO_4}\right)\left(\frac{37\%}{100\%}\right)\left(\frac{62.01 \text{ g } NO_3^-}{1 \text{ mol } NO_3^-}\right)\left(\frac{1 \text{ mg}}{10^{-3} \text{ g}}\right)\left(\frac{10^{-3} \text{ m}^3}{1 \text{ L}}\right)$$

$$= 72.9198 = \mathbf{73 \text{ mg/L}}$$

This assumes the plants in the field absorb none of the fertilizer.

22.72 **Plan:** Write a balanced equation. Determine which one of the three reactants is the limiting reactant by determining the amount of cryolite produced by each reactant. The amount of $Al(OH)_3$ requires conversion of kg to g; the amount of NaOH requires conversion from amount of solution to mass in grams by using the density; the amount of HF requires conversion from volume to moles using the ideal gas law.
Solution:
The balanced chemical equation for this reaction is:
$$6HF(g) + Al(OH)_3(s) + 3NaOH(aq) \rightarrow Na_3AlF_6(aq) + 6H_2O(l)$$
Assuming $Al(OH)_3$ is the limiting reactant:

$$\text{Moles of } Na_3AlF_6 = (365 \text{ kg } Al(OH)_3)\left(\frac{10^3 \text{ g}}{1 \text{ kg}}\right)\left(\frac{1 \text{ mol } Al(OH)_3}{78.00 \text{ g } Al(OH)_3}\right)\left(\frac{1 \text{ mol } Na_3AlF_6}{1 \text{ mol } Al(OH)_3}\right)$$

$$= 4.6795 \times 10^3 \text{ mol } Na_3AlF_6$$

Assuming NaOH is the limiting reactant:
Moles of Na_3AlF_6 =

$$(1.20 \text{ m}^3)\left(\frac{1 \text{ L}}{10^{-3} \text{ m}^3}\right)\left(\frac{1 \text{ mL}}{10^{-3} \text{ L}}\right)\left(\frac{1.53 \text{ g}}{\text{mL}}\right)\left(\frac{50.0\% \text{ NaOH}}{100\%}\right)\left(\frac{1 \text{ mol NaOH}}{40.00 \text{ g NaOH}}\right)\left(\frac{1 \text{ mol } Na_3AlF_6}{3 \text{ mol NaOH}}\right)$$

$$= 7.650 \times 10^3 \text{ mol } Na_3AlF_6$$

Assuming HF is the limiting reactant:
$PV = nRT$

Moles of HF $= \dfrac{PV}{RT} = \dfrac{(305 \text{ kPa})(265 \text{ m}^3)}{(0.0821 \text{ L} \cdot \text{atm/mol} \cdot \text{K})((273.2 + 91.5)\text{K})}\left(\dfrac{10^3 \text{ Pa}}{1 \text{ kPa}}\right)\left(\dfrac{1 \text{ atm}}{1.01325 \times 10^5 \text{ Pa}}\right)\left(\dfrac{1 \text{ L}}{10^{-3} \text{ m}^3}\right)$

$= 26{,}640.979 \text{ mol HF}$

Moles of $Na_3AlF_6 = (26{,}640.979 \text{ mol HF})\left(\dfrac{1 \text{ mol Na}_3\text{AlF}_6}{6 \text{ mol HF}}\right) = 4.4402 \times 10^3 \text{ mol Na}_3\text{AlF}_6$

Because the 265 m³ of HF produces the smallest amount of Na_3AlF_6, it is the limiting reactant. Convert moles Na_3AlF_6 to g Na_3AlF_6, using the molar mass, convert to kg, and multiply by 95.6% to determine the final yield.

Mass (g) of cryolite $= \left(4.4402 \times 10^3 \text{ mol Na}_3\text{AlF}_6\right)\left(\dfrac{209.95 \text{ g Na}_3\text{AlF}_6}{1 \text{ mol Na}_3\text{AlF}_6}\right)\left(\dfrac{1 \text{ kg}}{10^3 \text{ g}}\right)\left(\dfrac{95.6\%}{100\%}\right)$

$= 891.195 = \textbf{891 kg Na}_3\textbf{AlF}_6$

22.73 Plan: The rate of effusion is inversely proportional to the square root of the molar mass of the molecule (Graham's law).
Solution:

a) $\dfrac{\text{Rate}_H}{\text{Rate}_D} = \sqrt{\dfrac{\text{molar mass}_D}{\text{molar mass}_H}} = \sqrt{\dfrac{4.00 \text{ g/mol}}{2.016 \text{ g/mol}}} = 1.40859$

The time for D_2 to effuse is 1.41 times greater than that for H_2.
$\text{Time}_{D_2} = (1.40859)(16.5 \text{ min}) = 23.2417 = \textbf{23.2 min}$
b) Set x equal to the number of effusion steps. The ratio of mol H_2 to mol D_2 is 99:1.
Set up the equation to solve for x: $99/1 = (1.40859)^x$.
When solving for an exponent, take the log of both sides.
$\log (99) = \log (1.40859)^x$
Remember that $\log (a^b) = b \log (a)$
$\log (99) = x \log (1.40859)$
$x = 13.4129$
To separate H_2 and D_2 to 99% purity requires **14 effusion steps**.

22.75 Plan: Use the stoichiometric relationships found in the balanced chemical equation to find the mass of Al_2O_3, mass of graphite, and moles of CO_2. Use the ideal gas law to convert moles of CO_2 into volume.
Solution:
a) $2Al_2O_3(\text{in Na}_3\text{AlF}_6) + 3C(\text{gr}) \rightarrow 4Al(l) + 3CO_2(g)$

Mass (t) of $Al_2O_3 = (1 \text{ t Al})\left(\dfrac{10^3 \text{ kg}}{1 \text{ t}}\right)\left(\dfrac{10^3 \text{ g}}{1 \text{ kg}}\right)\left(\dfrac{1 \text{ mol Al}}{26.98 \text{ g Al}}\right)\left(\dfrac{2 \text{ mol Al}_2\text{O}_3}{4 \text{ mol Al}}\right)\left(\dfrac{101.96 \text{ g Al}_2\text{O}_3}{1 \text{ mol Al}_2\text{O}_3}\right)\left(\dfrac{1 \text{ kg}}{10^3 \text{ g}}\right)\left(\dfrac{1 \text{ t}}{10^3 \text{ kg}}\right)$

$= 1.8895478 = \textbf{1.890 t Al}_2\textbf{O}_3$

Therefore, **1.890 t of Al_2O_3** are consumed in the production of 1 t of pure Al.

b) Mass (t) of $C = (1 \text{ t Al})\left(\dfrac{10^3 \text{ kg}}{1 \text{ t}}\right)\left(\dfrac{10^3 \text{ g}}{1 \text{ kg}}\right)\left(\dfrac{1 \text{ mol Al}}{26.98 \text{ g Al}}\right)\left(\dfrac{3 \text{ mol C}}{4 \text{ mol Al}}\right)\left(\dfrac{12.01 \text{ g C}}{1 \text{ mol C}}\right)\left(\dfrac{1 \text{ kg}}{10^3 \text{ g}}\right)\left(\dfrac{1 \text{ t}}{10^3 \text{ kg}}\right)$

$= 0.3338584 = \textbf{0.3339 t C}$

Therefore, **0.3339 t of C** is consumed in the production of 1 t of pure Al, assuming 100% efficiency.
c) The percent yield with respect to Al_2O_3 is **100%** because the actual plant requirement of 1.89 t Al_2O_3 equals the theoretical amount calculated in part a).
d) The amount of graphite used in reality to produce 1 t of Al is greater than the amount calculated in part b). In other words, a 100% efficient reaction takes only 0.3339 t of graphite to produce a ton of Al, whereas real production requires more graphite and is less than 100% efficient. Calculate the efficiency using a simple ratio:
$(0.45 \text{ t})(x) = (0.3338584 \text{ t})(100\%)$
$x = 74.19076 = \textbf{74\%}$

e) For every four moles of Al produced, three moles of CO_2 are produced.

$$\text{Moles of C} = \left(1 \text{ t Al}\right)\left(\frac{10^3 \text{ kg}}{1 \text{ t}}\right)\left(\frac{10^3 \text{ g}}{1 \text{ kg}}\right)\left(\frac{1 \text{ mol Al}}{26.98 \text{ g Al}}\right)\left(\frac{3 \text{ mol CO}_2}{4 \text{ mol Al}}\right) = 2.7798\text{x}10^4 \text{ mol CO}_2$$

The problem states that 1 atm is exact. Use the ideal gas law to calculate volume, given moles, temperature, and pressure.

$PV = nRT$

$$V = \frac{nRT}{P} = \left(\frac{\left(2.7798\text{x}10^4 \text{ mol CO}_2\right)\left(0.08206 \text{ L} \bullet \text{atm/mol} \bullet \text{K}\right)\left((273+960.)\text{K}\right)}{1 \text{ atm}}\right)\left(\frac{10^{-3} \text{ m}^3}{1 \text{ L}}\right)$$

$$= 2.812601\text{x}10^3 = \mathbf{2.813\text{x}10^3 \text{ m}^3}$$

22.80   Plan: Solubility of a salt can be calculated from its K_{sp}. K_{sp} for $Ca_3(PO_4)_2$ is $1.2\text{x}10^{-29}$. Phosphate is derived from a weak acid, so the pH of the solution impacts exactly where the acid-base equilibrium lies. Phosphate can gain one H^+ to form HPO_4^{2-}. Gaining another H^+ gives $H_2PO_4^-$ and a last H^+ added gives H_3PO_4. The K_a values for phosphoric acid are $K_{a1} = 7.2\text{x}10^{-3}$, $K_{a2} = 6.3\text{x}10^{-8}$, $K_{a3} = 4.2\text{x}10^{-13}$. To find the ratios of the various forms of phosphate, use the equilibrium expressions and the concentration of H^+.

Solution:

a) $Ca_3(PO_4)_2(s) \leftrightarrows Ca^{2+}(aq) + 2PO_4^{3-}(aq)$

Initial	—	0	0
Change	−x	+3x	+2x
Equilibrium	—	3x	2x

$K_{sp} = 1.2\text{x}10^{-29} = [Ca^{2+}]^3[PO_4^{3-}]^2 = (3x)^3(2x)^2 = 108x^5$

$x = 6.4439\text{x}10^{-7} = \mathbf{6.4\text{x}10^{-7} \textit{ M} \, Ca_3(PO_4)_2}$

b) $[H_3O^+] = 10^{-4.5} = 3.162\text{x}10^{-5} \, M$

$H_3PO_4(aq) + H_2O(l) \leftrightarrows H_2PO_4^-(aq) + H_3O^+(aq)$

$$K_{a1} = \frac{[H_2PO_4^-][H_3O^+]}{[H_3PO_4]}$$

$$\frac{[H_2PO_4^-]}{[H_3PO_4]} = \frac{K_{a1}}{[H_3O^+]} = \frac{7.2\text{x}10^{-3}}{3.162\text{x}10^{-5}} = 227.70$$

$H_2PO_4^-(aq) + H_2O(l) \leftrightarrows HPO_4^{2-}(aq) + H_3O^+(aq)$

$$K_{a2} = \frac{[HPO_4^{2-}][H_3O^+]}{[H_2PO_4^-]}$$

$$\frac{[HPO_4^{2-}]}{[H_2PO_4^-]} = \frac{K_{a2}}{[H_3O^+]} = \frac{6.3\text{x}10^{-8}}{3.162\text{x}10^{-5}} = 1.9924\text{x}10^{-3}$$

$HPO_4^{2-}(aq) + H_2O(l) \leftrightarrows PO_4^{3-}(aq) + H_3O^+(aq)$

$$K_{a3} = \frac{[PO_4^{3-}][H_3O^+]}{[HPO_4^{2-}]}$$

$$\frac{[PO_4^{3-}]}{[HPO_4^{2-}]} = \frac{K_{a3}}{[H_3O^+]} = \frac{4.2\text{x}10^{-13}}{3.162\text{x}10^{-5}} = 1.32827\text{x}10^{-8}$$

From the ratios, the concentration of $H_2PO_4^-$ is at least 100 times more than the concentration of any other species, so assume that dihydrogen phosphate is the dominant species and find the value for $[PO_4^{3-}]$.

$$\frac{[PO_4^{3-}]}{[HPO_4^{2-}]} = 1.32827\text{x}10^{-8} \quad \text{and} \quad \frac{[HPO_4^{2-}]}{[H_2PO_4^-]} = 1.9924\text{x}10^{-3}$$

$[PO_4^{3-}] = (1.32827\text{x}10^{-8})[HPO_4^{2-}]$ and $[HPO_4^{2-}] = (1.9924\text{x}10^{-3})[H_2PO_4^-]$

$[PO_4^{3-}] = (1.32827\times10^{-8})[HPO_4^{2-}] = (1.32827\times10^{-8})(1.9924\times10^{-3})[H_2PO_4^-]$

$[PO_4^{3-}] = (2.6464\times10^{-11})[H_2PO_4^-]$

Substituting this into the K_{sp} expression gives

$K_{sp} = [Ca^{2+}]^3[PO_4^{3-}]^2 = 1.2\times10^{-29}$

$K_{sp} = [Ca^{2+}]^3\{(2.6464\times10^{-11})[H_2PO_4^-]\}^2$

Rearranging gives

$K_{sp}/(2.6464\times10^{-11})^2 = [Ca^{2+}]^3[H_2PO_4^-]^2$

The concentration of calcium ions is still represented as 3x and the concentration of dihydrogen phosphate ion as 2x, since each $H_2PO_4^-$ comes from one PO_4^{3-}.

$K_{sp}/(2.6464\times10^{-11})^2 = (3x)^3(2x)^2$

$(1.2\times10^{-29})/(2.6464\times10^{-11})^2 = 108x^5$

$x = 1.0967\times10^{-2} = \mathbf{1.1\times10^{-2}}\ \boldsymbol{M}$

Acid rain increases the leaching of PO_4^{3-} into the groundwater, due to the protonation of PO_4^{3-} to form HPO_4^{2-} and $H_2PO_4^-$. As shown in calculations a) and b), solubility increases from $6.4\times10^{-7}\ M$ (in pure water) to $1.1\times10^{-2}\ M$ (in acidic rainwater).

22.81   a) The mole % of oxygen missing may be estimated from the discrepancy between the actual and ideal formula.

$$\text{Mol \% O missing} = \left|\frac{(2.000 - 1.98)\,\text{mol O}}{2.000\,\text{mol O}}\right| \times 100\% = \mathbf{1.00\%}$$

b) The molar mass is determined like a normal molar mass except that the quantity of oxygen is not an integer value.

Molar mass = 1 mol Pb (207.2 g/mol) + 1.98 mol O (16.00 g/mol) = 238.88 = **238.9 g/mol**

22.83   Plan: Convert the unit cell edge length to cm and cube that length to obtain the volume of the cell. A face-centered cubic structure contains four atoms of silver. Divide that number by Avogadro's number to find the moles of silver in the cell and multiply by the molar mass to obtain the mass of silver in the cell. Density is obtained by dividing the mass of silver by the volume of the cell. The calculation will now be repeated replacing the atomic mass of silver with the "atomic mass" of sterling silver. The "atomic mass" of sterling silver is simply a weighted average of the masses of the atoms present (Ag and Cu).
Solution:

$$\text{Edge length (cm)} = (408.6\ \text{pm})\left(\frac{10^{-12}\ \text{m}}{1\ \text{pm}}\right)\left(\frac{1\ \text{cm}}{10^{-2}\ \text{m}}\right) = 4.086\times10^{-8}\ \text{cm}$$

Volume (cm³) of cell = $(4.086\times10^{-8}\ \text{cm})^3 = 6.82174\times10^{-23}\ \text{cm}^3$

$$\text{Mass (g) of silver} = \left(\frac{4\ \text{Ag atoms}}{\text{unit cell}}\right)\left(\frac{1\ \text{mol Ag}}{6.022\times10^{23}\ \text{Ag atoms}}\right)\left(\frac{107.9\ \text{g Ag}}{1\ \text{mol Ag}}\right) = 7.1671\times10^{-22}\ \text{g}$$

$$\text{Density of silver} = \frac{7.1671\times10^{-22}\ \text{g}}{6.82174\times10^{-23}\ \text{cm}^3} = 10.506264 = \mathbf{10.51\ \text{g/cm}^3}$$

$$\text{Atomic mass of sterling silver} = \left(\frac{107.9\ \text{g}}{\text{mol}}\right)\left(\frac{(100.0 - 7.5)\%}{100\%}\right) + \left(\frac{63.55\ \text{g}}{\text{mol}}\right)\left(\frac{7.5\%}{100\%}\right) = 104.57375\ \text{g/mol}$$

$$\text{Mass (g) of sterling silver} = \left(\frac{4\ \text{Ag atoms}}{\text{unit cell}}\right)\left(\frac{1\ \text{mol Ag}}{6.022\times10^{23}\ \text{Ag atoms}}\right)\left(\frac{104.57375\ \text{g Ag}}{1\ \text{mol Ag}}\right) = 6.946\times10^{-22}\ \text{g}$$

$$\text{Density of sterling silver} = \frac{6.9461\times10^{-22}\ \text{g}}{6.82174\times10^{-23}\ \text{cm}^3} = 10.1823 = \mathbf{10.2\ \text{g/cm}^3}$$

CHAPTER 23 TRANSITION ELEMENTS AND THEIR COORDINATION COMPOUNDS

FOLLOW–UP PROBLEMS

23.1A **Plan:** Locate the element on the periodic table and use its position and atomic number to write a partial electron configuration. Add or subtract electrons to obtain the configuration for the ion. Partial electron configurations do not include the noble gas configuration and any filled f sublevels, but do include outer level electrons (n-level) and the $n-1$ level d orbitals. Remember that ns electrons are removed before $n-1$ electrons.
Solution:
a) Ag is in the fifth row of the periodic table, so $n = 5$, and it is in group 1B(11). The partial electron configuration of the element is $5s^1 4d^{10}$. For the ion, Ag^+, remove one electron from the $5s$ orbital. The partial electron configuration of Ag^+ is **$4d^{10}$**.
b) Cd is in the fifth row and group 2B(12), so the partial electron configuration of the element is $5s^2 4d^{10}$. Remove two electrons from the $5s$ orbital for the ion. The partial electron configuration of Cd^{2+} is **$4d^{10}$**.
c) Ir is in the sixth row and group 8B(9). The partial electron configuration of the element is $6s^2 5d^7$. Remove two electrons from the $6s$ orbital and one from $5d$ for ion Ir^{3+}. The partial electron configuration of Ir^{3+} is **$5d^6$**.

23.1B **Plan:** Locate the element on the periodic table and use its position and atomic number to write a condensed electron configuration for its atom. Compare the partial electron configurations of the atom and the ion (provided in the problem statement) to determine how many electrons have been removed from the atom to produce the ion. The number of electrons removed is the charge on the ion. Because electrons are being removed, the charge will be positive.
Solution:
a) Ta has $Z = 73$. The condensed electron configuration of its atom is: $[Xe] 6s^2 4f^{14} 5d^3$. Three electrons were removed to produce its ion, which has a condensed electron configuration of $[Xe] 4f^{14} 5d^2$. Therefore, the charge on the Ta ion is $+3$, **Ta^{3+}**.
b) Mn has $Z = 25$. The condensed electron configuration of its atom is: $[Ar] 4s^2 3d^5$. Four electrons were removed to produce its ion, which has a condensed electron configuration of $[Ar] 3d^3$. Therefore, the charge on the Mn ion is $+4$, **Mn^{4+}**.
c) Os has $Z = 76$. The condensed electron configuration of its atom is: $[Xe] 6s^2 4f^{14} 5d^6$. Three electrons were removed to produce its ion, which has a condensed electron configuration of $[Xe] 4f^{14} 5d^5$. Therefore, the charge on the Os ion is $+3$, **Os^{3+}**.

23.2A **Plan:** Find Er on the periodic table and write its electron configuration. Remove 3 electrons for the electron configuration of Er^{3+}. Place the outer level electrons in an electron diagram, making sure to follow the Aufbau principle and Hund's rule. Count the number of unpaired electrons.
Solution:
Er is a lanthanide element in row 6 and with atomic number 68. The electron configuration of the element is $[Xe] 6s^2 5d^1 4f^{11}$. Remove three electrons: 2 from the $6s$ and 1 from $5d$ for the Er^{3+} ion, $[\mathbf{Xe}]\mathbf{4}f^{11}$.

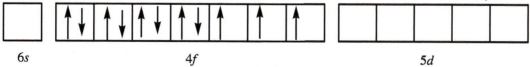

      6s   4f   5d

The number of unpaired electrons is **three**, all in the $4f$ sublevel.

23.2B **Plan:** Find Dy on the periodic table and write its electron configuration. Remove 3 electrons for the electron configuration of Dy^{3+}. Place the outer level electrons in an electron diagram, making sure to follow the Aufbau principle and Hund's rule. Count the number of unpaired electrons.

Solution:

Dy is a lanthanide element in row 6 and with atomic number 66. The electron configuration of the element is $[Xe]6s^2 5d^1 4f^9$. Remove three electrons: 2 from the $6s$ and 1 from $5d$ for the Dy^{3+} ion, $\mathbf{[Xe]4f^9}$.

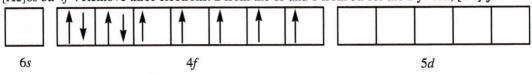

The number of unpaired electrons is **five**, all in the $4f$ sublevel.

23.3A   Plan: Use the charges on the counter ions and the ligands to calculate the charge on the central metal ion, remembering that all the charges in a neutral compound must add to zero and that the charges for an ion must add to the overall charge on that ion. The coordination number, or number of ligand atoms bonded to the metal ion, is found by examining the bonded entities inside the square brackets to determine if they are monodentate, bidentate, or polydentate.
Solution:
a) The counter ion is NO_3^-, so the complex ion is $[Co(en)_2Br_2]^+$. Each bromo ligand has a –1 charge and ethylenediamine is neutral, so cobalt has a **+3** charge: $+3 + 2(0) + 2(-1) = +1$. Each ethylenediamine ligand is bidentate and each bromo ligand is monodentate, so the coordination number is **6**.
b) The counter ion is Mg^{2+}, so the complex ion is $[CrCl_5Br]^{3-}$. Each bromo ligand and each chloro ligand has a –1 charge, so Cr has a **+3** charge: $+3 + 5(-1) + 1(-1) = -3$. Each chloro ligand and each bromo ligand is monodentate, so the coordination number is **6**.

23.3B   Plan: Examine the ligands. Determine their charges and whether they are mono-, bi-, or poly-dentate. Determine the number of ligands that are necessary for the given coordination number. Write the formula of the complex ion and determine its charge. Balance the charge on the complex ion with the counter ion.
Solution:
a) The ligand in this case is the cyano ligand. It is monodentate and has a charge of –1. Because cyano is monodentate, 6 cyano ligands are needed to fulfill the coordination number of 6. Therefore, the formula of the complex ion is $[Mn(CN)_6]^{4-}$. The charge on the complex ion is –4: $(1\ Mg)(+2) + (6\ CN)(-1) = -4$. Four potassium ions, K^+, are needed to balance the charge: $\mathbf{K_4[Mn(CN)_6]}$.
b) The ligand in this case is the ethylenediamine ligand. It is bidentate and is neutral. Because ethylenediamine is bidentate, 2 ethylenediamine ligands are needed to fulfill the coordination number of 4. Therefore, the formula of the complex ion is $[Cu(en)_2]^{2+}$. The charge on the complex ion is +2: $(1\ Cu)(+2) + (2\ en)(0) = +2$. One sulfate ion, SO_4^{2-}, is needed to balance the charge: $\mathbf{[Cu(en)_2]SO_4}$.

23.4A   Plan: Name the compound by following rules outlined in the chapter. Use Table 23.7 for the names of ligands. Write the formula from the name by breaking down the name into pieces that follow the naming rules.
Solution:
a) The compound consists of the cation $[Cr(H_2O)_5Br]^{2-}$ and anion Cl^-. The metal ion in the cation is chromium and its charge is found from the charges on the ligands and the total charge on the ion.

   total charge = (charge on Cr) + 5(charge on H_2O) + (charge on Br^-)
          -2 = (charge on Cr) + 5(0) + (–1)
   charge on Cr = +3

The name of the cation will end with chromium(III) to indicate the oxidation state of the chromium ion since there is more than one possible oxidation state for chromium. The water ligand is named aqua. There are five water ligands, so pentaaqua describes the $(H_2O)_5$ ligands. The bromide ligand is named bromo. The ligands are named in alphabetical order, so aqua comes before bromo. The name of the cation is pentaaquabromochromium(III). Add the chloride for the anion to complete the name: **pentaaquabromochromium(III) chloride**.
b) The compound consists of a cation, barium ion (Ba^{2+}), and the anion hexacyanocobaltate(III). The formula of the anion consists of six (from hexa) cyanide (from cyano) ligands and cobalt (from cobaltate) in the +3 oxidation state (from (III)). Putting the formula together gives $[Co(CN)_6]^{n+}$. To find the charge on the complex ion, calculate from charges on the ligands and the metal ion:

total charge = 6(charge on CN^-) + (charge on cobalt ion)
$$= 6(-1) + (+3)$$
$$= -3$$

The formula of complex ion is $[Co(CN)_6]^{3-}$. Combining this with the cation, Ba^{2+}, gives the formula for the compound: **$Ba_3[Co(CN)_6]_2$**. The three barium ions give a +6 charge and two anions give a –6 charge, so the net result is a neutral salt.

23.4B   Plan: Name the compound by following rules outlined in the chapter. Use Table 23.7 for the names of ligands. Write the formula from the name by breaking down the name into pieces that follow the naming rules.
Solution:
a) The compound consists of an anion, sulfate ion (SO_4^{2-}), and the cation diaquabis(ethylenediamine)cobalt(III). The formula of the cation consists of two (from di) water (from aqua) ligands, two (from bis) ethylenediamine ligands, and cobalt in the +3 oxidation state (from (III)). Putting the formula together gives $[Co(H_2O)_2(en)_2]^{n+}$. To find the charge on the complex ion, calculate from charges on the ligands and the metal ion:
total charge = 2(charge on H_2O) + 2(charge on en) + (charge on cobalt ion)
$$= 2(0) + 2(0) + (+3)$$
$$= +3$$

The formula of complex ion is $[Co(H_2O)_2(en)_2]^{3+}$. Combining this with the anion, SO_4^{2-}, gives the formula for the compound: **$[Co(H_2O)_2(en)_2]_2(SO_4)_3$**. The three sulfate ions give a –6 charge and two cations give a +6 charge, so the net result is a neutral salt.
b) The compound consists of the cation Mg^{2+} and anion $[Cr(NH_3)_2Cl_4]^-$. The metal ion in the complex ion is chromium and its charge is found from the charges on the ligands and the total charge on the ion.
total charge = (charge on Cr) + 2(charge on NH_3) + 4(charge on Cl^-)
$$-1 = \text{(charge on Cr)} + 2(0) + 4(-1)$$
$$\text{charge on Cr} = +3$$

The name of the anion will end with chromate(III) to indicate the oxidation state of the chromium ion since there is more than one possible oxidation state for chromium. The –ate ending indicates that the complex ion is an anion. The ammonia ligand is named ammine. There are two ammonia ligands, so diammine describes the $(NH_3)_2$ ligands. The chloride ligand is named chloro. There are four chlorine ligands, so tetrachloro describes the Cl_4 ligands. The ligands are named in alphabetical order, so ammine comes before chloro. The name of the anion is diamminetetrachlorochromate(III). Add the magnesium for the cation to complete the name. The cation is named before the anion: **magnesium diamminetetrachlorochromate(III)**.

23.5A   Plan: The given complex ion has a coordination number of 6 (*en* is bidentate), so it will have an octahedral arrangement of ligands. Stereoisomers can be either geometric or optical. For geometric isomers, check if the ligands can be arranged either next to (*cis*) or opposite (*trans*) each other. Then, check if the mirror image of any of the geometric isomers is not superimposable. If mirror image is not superimposable, the structure is an optical isomer.
Solution:
The complex ion contains two NH_3 ligands and two chloride ligands, both of which can be arranged in either the *cis* or *trans* geometry. The possible combinations are 1) *cis*-NH_3 and *cis*-Cl, 2) *trans*-NH_3 and *cis*-Cl, and 3) *cis*-NH_3 and *trans*-Cl. Both NH_3 and Cl *trans* is not possible because the two bonds to ethylenediamine can be arranged only in this *cis*-position, which leaves only one set of *trans* positions.

cis-NH₃ cis-Cl   trans-NH₃ cis-Cl   cis-NH₃ trans-Cl

The mirror images of the second two structures are superimposable since two of the ligands, either ammonia or chloride ion, are arranged in the *trans* position. When both types of ligands are in the *cis* arrangement, the mirror image is not a superimposable optical isomer:

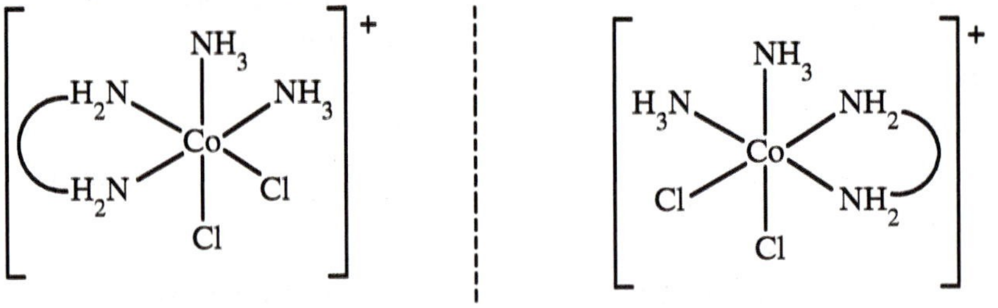

cis-NH₃ *cis*-Cl   *cis*-NH₃ *cis*-Cl

There are four stereoisomers of $[Co(NH_3)_2(en)Cl_2]^+$.

23.5B   Plan: The given complex ion has a coordination number of 6, so it will have an octahedral arrangement of ligands. Stereoisomers can be either geometric or optical. For geometric isomers, check if the ligands can be arranged either next to (*cis*) or opposite (*trans*) each other. Then, check if the mirror image of any of the geometric isomers is not superimposable. If mirror image is not superimposable, the structure is an optical isomer.
Solution:
The complex ion contains two H_2O ligands, two NH_3 ligands and two chloride ligands, all of which can be arranged in either the *cis* or *trans* geometry. The possible combinations are drawn below:

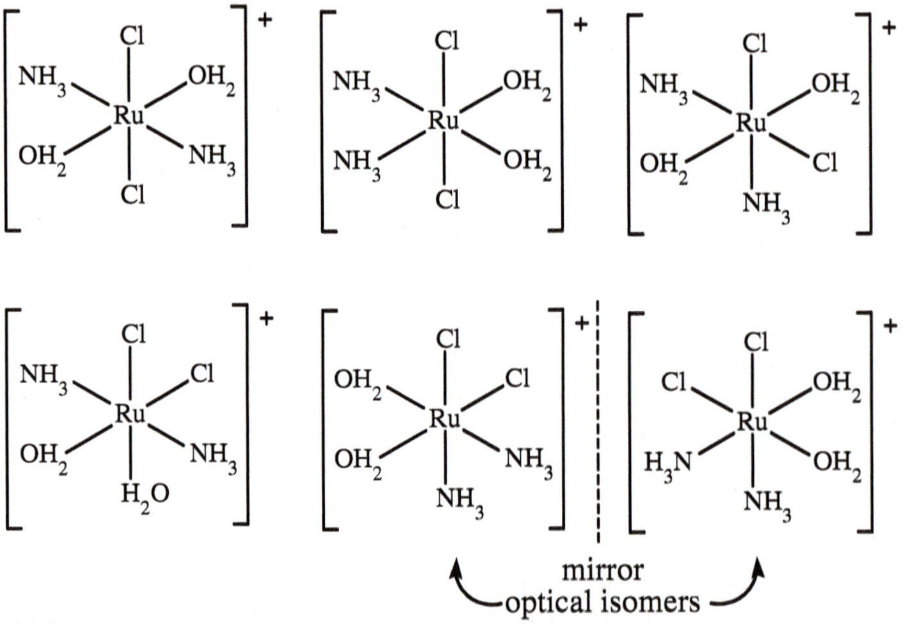

mirror
optical isomers

23.6A   Plan: Compare the two ions for oxidation state of the metal ion and the relative ability of ligands to split the *d*-orbital energies. The splitting strength of the ligands is obtained from the spectrochemical series. The greater the splitting strength of the ligands, the greater the energy of absorbed light.
Solution:
The oxidation number of vanadium in both ions is the same, so compare the two ligands. Ammonia is a stronger field ligand than water, so the complex ion $[V(NH_3)_6]^{3+}$ absorbs visible light of higher energy than $[V(H_2O)_6]^{3+}$ absorbs.

23.6B   Plan: Each of these ions includes cobalt in the +3 oxidation state. In order to distinguish between them, compare the relative ability of ligands to split the *d*-orbital energies. The splitting strength of the ligands is obtained from the spectrochemical series. The greater the splitting strength of the ligands, the greater the energy of absorbed light. Place the ligands in order from lowest energy of absorbed light (absorbs closer to red light) to highest energy of absorbed light (absorbs closer to violet light). Then use an artist's wheel to determine the color of observed light, which is complementary to the color of the absorbed light.

Solution:

Four different ligands are present in the complex ions: en, NH_3, Cl^-, and H_2O. Putting these ions in order from weaker field ligands to stronger field ligands gives: $Cl^- < H_2O < NH_3 <$ en. Putting the complex ions in order from those with weaker field ligands (which absorb lower energy light) to those with stronger field ligands (which absorb higher energy light) gives: (absorbs closer to red light) $[Co(NH_3)_4Cl_2]^+ < [Co(NH_3)_5Cl]^{2+} < [Co(NH_3)_5(H_2O)]^{3+} < [Co(en)_3]^{3+}$ (absorbs closer to violet light).

The possible *observed* colors of the complexes are: red, green, purple, and yellow. Correlate the observed colors with absorbed colors (see Figure 23.15):

Observed Color	Absorbed Color
Red	Green
Green	Red
Purple	Yellow
Yellow	Purple

Putting the possible absorbed colors in order from lowest energy (highest wavelength) to highest energy (lowest wavelength) gives: red < yellow < green < purple. Correlate this order of absorbed colors with the order of complex ions listed above.

Absorbed Color	Complex Ion	Observed Color
Red	$[Co(NH_3)_4Cl_2]^+$	Green
Yellow	$[Co(NH_3)_5Cl]^{2+}$	Purple
Green	$[Co(NH_3)_5(H_2O)]^{3+}$	Red
Purple	$[Co(en)_3]^{3+}$	Yellow

23.7A **Plan:** Determine the charge on the manganese ion in the complex ion and the number of electrons in its *d* orbitals. Since it is an octahedral ligand, the *d* orbitals split into three lower energy orbitals and two higher energy orbitals. Check the spectrochemical series to see if CN^- is a strong or weak field ligand. If it is a strong field ligand, fill the three lower energy orbitals before placing any electrons in the higher energy *d*-orbitals. If it is a weak field ligand, place one electron in each of the five *d* orbitals, low energy before high energy, before pairing any electrons. After filling orbitals, count the number of unpaired electrons. The complex ion is low-spin if the ligand is a strong field ligand and high-spin if the ligand is weak field.

Solution:

Find the charge on manganese:

total charge = (charge on Mn) + 6(charge on CN^-)

charge on Mn = $-3 - [6(-1)] = +3$

The electron configuration of Mn is $[Ar]4s^23d^5$; the electron configuration of Mn^{3+} is $[Ar]3d^4$.

The ligand is CN^-, which is a strong field ligand, so the four *d* electrons will fill the lower energy *d* orbitals before any are placed in the higher energy *d*-orbitals:

Higher energy *d*-orbitals

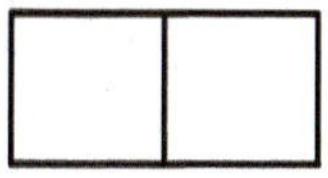

Lower energy *d*-orbitals

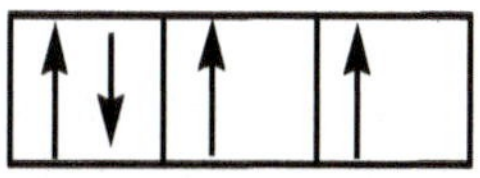

Two electrons are unpaired in $[Mn(CN)_6]^{3-}$. The complex is **low spin** since the cyanide ligand is a strong field ligand. The splitting energy is greater than the electron pairing energy.

23.7B   <u>Plan:</u> Determine the charge on the cobalt ion in the complex ion and the number of electrons in its d orbitals. Since it is an octahedral ligand, the d orbitals split into three lower energy orbitals and two higher energy orbitals. Check the spectrochemical series to see if F^- is a strong or weak field ligand. If it is a strong field ligand, fill the three lower energy orbitals before placing any electrons in the higher energy d-orbitals. If it is a weak field ligand, place one electron in each of the five d orbitals, low energy before high energy, before pairing any electrons. After filling orbitals, count the number of unpaired electrons. The complex ion is low-spin if the ligand is a strong field ligand and high-spin if the ligand is weak field.

<u>Solution:</u>

Find the charge on cobalt:

$$\text{total charge} = (\text{charge on Co}) + 6(\text{charge on } F^-)$$

charge on Co $= -3 - [6(-1)] = +3$

The electron configuration of Co is $[Ar]4s^2 3d^7$; the electron configuration of Co^{3+} is $[Ar]3d^6$.

The ligand is F^-, which is a weak field ligand, so the six d electrons will enter each of the 5 orbitals before pairing up.

Higher energy d-orbitals

Lower energy d-orbitals   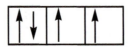

Four electrons are unpaired in $[Co(F)_6]^{3-}$. The complex is **high spin** since the fluoride ligand is a weak field ligand. The splitting energy is smaller than the electron pairing energy.

CHEMICAL CONNECTIONS BOXED READING PROBLEMS

B23.1   a) The central metal ion in chlorophyll is Mg^{2+} (oxidation state $= +2$). The central metal ion in heme is Fe^{2+} (oxidation state $= +2$). The central metal ion in vitamin B_{12} is Co^{3+} (oxidation state $= +3$).

b) All of these compounds have the central metal in the center of a large planar ring containing four N donor atoms.

B23.2   Zinc ions prefer a tetrahedral environment. The other ions listed prefer other environments (Ni^{2+} forms square planar complex ions while Fe^{2+} and Mn^{2+} form octahedral complex ions). If these ions are placed in a tetrahedral environment in place of the zinc, they cannot function as well as zinc since the enzyme catalyst would have a different shape.

END–OF–CHAPTER PROBLEMS

23.2   a) All transition elements in Period 5 will have a "base" configuration of $[Kr]5s^2$, and will differ in the number of d electrons (x) that the configuration contains. Therefore, the general electron configuration is **$1s^2 2s^2 2p^6 3s^2 3p^6 4s^2 3d^{10} 4p^6 5s^2 4d^x$**.

b) A general electron configuration for Period 6 transition elements includes f sublevel electrons, which are lower in energy than the d sublevel. The configuration is **$1s^2 2s^2 2p^6 3s^2 3p^6 4s^2 3d^{10} 4p^6 5s^2 4d^{10} 5p^6 6s^2 4f^{14} 5d^x$**.

23.4   The maximum number of unpaired d electrons is **five** since there are five d orbitals. An example of an atom with five unpaired d electrons is Mn with electron configuration $[Ar]4s^2 3d^5$. An ion with five unpaired electrons is Mn^{2+} with electron configuration $[Ar]3d^5$.

23.6   a) One would expect that the elements would increase in size as they increase in mass from Period 5 to 6. Because there are fourteen inner transition elements in Period 6, the effective nuclear charge increases significantly. As effective charge increases, the atomic size decreases or "contracts." This effect is significant enough that Zr^{4+} and Hf^{4+} are almost the same size but differ greatly in atomic mass.
b) The size increases from Period 4 to 5, but stays fairly constant from Period 5 to 6.
c) Atomic mass increases significantly from Period 5 to 6, but atomic radius (and thus volume) hardly increases, so Period 6 elements are very dense.

23.9   a) A paramagnetic substance is attracted to a magnetic field, while a diamagnetic substance is slightly repelled by one.
b) Ions of transition elements often have unfilled d orbitals whose unpaired electrons make the ions paramagnetic. Ions of main-group elements usually have a noble gas configuration with no partially filled levels. When orbitals are filled, electrons are paired and the ion is diamagnetic.
c) The d orbitals in the transition element ions are not filled, which allows an electron from a lower energy d orbital to move to a higher energy d orbital. The energy required for this transition is relatively small and falls in the visible wavelength range. All orbitals are filled in a main-group element ion, so enough energy would have to be added to move an electron to a higher energy level, not just another orbital within the same energy level. This amount of energy is relatively large and outside the visible range of wavelengths.

23.10   Plan: The transition elements in Periods 4 and 5 have a general electron configuration of [noble gas] $ns^2 (n-1)d^x$. Transition elements in Periods 6 and 7 have a general electron configuration of [noble gas] $ns^2 (n-2)f^{14} (n-1)d^x$.
Solution:
a) Vanadium is in Period 4 and Group 5B(5). Electron configuration of V is $1s^2 2s^2 2p^6 3s^2 3p^6 4s^2 3d^3$ or $[Ar]4s^2 3d^3$.
b) Yttrium is in Period 5 and Group 3B(3). Electron configuration of Y is $1s^2 2s^2 2p^6 3s^2 3p^6 4s^2 3d^{10} 4p^6 5s^2 4d^1$ or $[Kr]5s^2 4d^1$.
c) Mercury is in Period 6 and Group 2B(12). Electron configuration of Hg is $[Xe]6s^2 4f^{14} 5d^{10}$.

23.12   Plan: The transition elements in Periods 4 and 5 have a general electron configuration of [noble gas] $ns^2 (n-1)d^x$. Transition elements in Periods 6 and 7 have a general electron configuration of [noble gas] $ns^2 (n-2)f^{14} (n-1)d^x$.
Solution:
a) Osmium is in Period 6 and Group 8B(8). Electron configuration of Os is $[Xe]6s^2 4f^{14} 5d^6$.
b) Cobalt is in Period 4 and Group 8B(9). Electron configuration of Co is $[Ar]4s^2 3d^7$.
c) Silver is in Period 5 and Group 1B(11). Electron configuration of Ag is $[Kr]5s^1 4d^{10}$. Note that the filled d orbital is the preferred arrangement, so the configuration is not $5s^2 4d^9$.

23.14   Plan: Write the electron configuration of the atom and then remove the electrons as indicated by the charge of the metal ion. Transition metals lose their ns orbital electrons first in forming cations. After losing the ns electrons, additional electrons may be lost from the $(n-1)d$ orbitals.
Solution:
a) The two $4s$ electrons and one $3d$ electron are removed to form Sc^{3+}:
Sc: $[Ar]4s^2 3d^1$; Sc^{3+}: [Ar] or $1s^2 2s^2 2p^6 3s^2 3p^6$. There are **no unpaired electrons**.
b) The single $4s$ electron and one $3d$ electron are removed to form Cu^{2+}:
Cu: $[Ar]4s^1 3d^{10}$; Cu^{2+}: $[Ar]3d^9$. There is **one unpaired electron**.
c) The two $4s$ electrons and one $3d$ electron are removed to form Fe^{3+}:
Fe: $[Ar]4s^2 3d^6$; Fe^{3+}: $[Ar]3d^5$. There are **five unpaired electrons** since each of the five d electrons occupies its own orbital.
d) The two $5s$ electrons and one $4d$ electron are removed to form Nb^{3+}:
Nb: $[Kr]5s^2 4d^3$; Nb^{3+}: $[Kr]4d^2$. There are **two unpaired electrons**.

23.16   Plan: For Groups 3B(3) to 7B(7), the highest oxidation state is equal to the group number. The highest oxidation state occurs when both ns electrons and all $(n-1)d$ electrons have been removed.
Solution:
a) Tantalum, Ta, is in Group 5B(5), so the highest oxidation state is **+5**. The electron configuration of Ta is $[Xe]6s^2 4f^{14} 5d^3$ with a total of five electrons in the $6s$ and $5d$ orbitals.

b) Zirconium, Zr, is in Group 4B(4), so the highest oxidation state is **+4**. The electron configuration of Zr is $[Kr]5s^24d^2$ with a total of four electrons in the $5s$ and $4d$ orbitals.

c) Manganese, Mn, is in Group 7B(7), so the highest oxidation state is **+7**. The electron configuration of Mn is $[Ar]4s^23d^5$ with a total of seven electrons in the $4s$ and $3d$ orbitals.

23.18 Plan: For Groups 3B(3) to 7B(7), the highest oxidation state is equal to the group number. The highest oxidation state occurs when both ns electrons and all $(n–1)d$ electrons have been removed.
Solution:
The elements in Group 6B(6) exhibit an oxidation state of +6. These elements have a total of six electrons in the outermost s orbital and d orbital: $ns^1(n–1)d^5$. These elements include **Cr, Mo, and W**. Sg (Seaborgium) is also in Group 6B(6), but its lifetime is so short that chemical properties, like oxidation states within compounds, are impossible to measure.

23.20 Plan: Transition elements in their lower oxidation states act more like metals.
Solution:
The oxidation state of chromium in CrF_2 is +2 and in CrF_6 is +6 (use –1 oxidation state of fluorine to find oxidation state of Cr). **CrF_2** exhibits greater metallic behavior than CrF_6 because the chromium is in a lower oxidation state in CrF_2 than in CrF_6.

23.22 While atomic size increases slightly down a group of transition elements, the nuclear charge increases much more, so the first ionization energy generally increases. The reduction potential for Mo is lower, so **it is more difficult to oxidize Mo** than Cr. In addition, the ionization energy of Mo is higher than that of Cr, so it is more difficult to remove electrons from, i.e., oxidize, Mo.

23.24 Plan: Oxides of transition metals become less basic (or more acidic) as oxidation state increases.
Solution:
The oxidation state of chromium in CrO_3 is +6 and in CrO is +2, based on the –2 oxidation state of oxygen. The oxide of the higher oxidation state, **CrO_3**, produces a more acidic solution.

23.28 a) The f block contains **seven** orbitals, so if one electron occupied each orbital, a maximum of **seven** electrons would be unpaired.
b) The maximum number of unpaired electrons corresponds to a half-filled f subshell.

23.30 Plan: The inner transition elements have a general electron configuration of [noble gas] $ns^2 (n – 2)f^x (n – 1)d^0$. Inner transition metals lose their ns orbital electrons first in forming cations. After losing the ns electrons, additional electrons may be lost from the $(n – 1)d$ orbitals.
Solution:
a) Lanthanum is a transition element in Period 6 with atomic number 57. La: **$[Xe]6s^25d^1$**
b) Cerium is in the lanthanide series in Period 6 with atomic number 58. Ce: $[Xe]6s^24f^15d^1$, so Ce^{3+}: **$[Xe]4f^1$**. Note that cerium is one of the three lanthanide elements that have one electron in a $5d$ orbital.
c) Einsteinium is in the actinide series in Period 7 with atomic number 99. Es: **$[Rn]7s^25f^{11}$**
d) Uranium is in the actinide series in Period 7 with atomic number 92. U: $[Rn]7s^25f^36d^1$. Removing four electrons gives U^{4+} with configuration **$[Rn]5f^2$**.

23.32 Plan: Write the electron configuration of the atom and then remove electrons as indicated by the charge of the metal ion. Electrons are removed first from the $6s$ orbital and then from the $4f$ orbital.
Solution:
a) Europium is in the lanthanide series with atomic number 63. The configuration of Eu is $[Xe]6s^24f^7$. The stability of the half-filled f sublevel explains why the configuration is not $[Xe]6s^24f^65d^1$. The two $6s$ electrons are removed to form the Eu^{2+} ion, followed by electron removal in the f block to form the other two ions:

   Eu^{2+}: **$[Xe]4f^7$**
   Eu^{3+}: **$[Xe]4f^6$**
   Eu^{4+}: **$[Xe]4f^5$**

The stability of the half-filled f sublevel makes Eu^{2+} most stable.

b) Terbium is in the lanthanide series with atomic number 65. The configuration of Tb is $[Xe]6s^24f^9$. The two $6s$ electrons are removed to form the Tb^{2+} ion, followed by electron removal in the f block to form the other two ions:

Tb^{2+}: $[\textbf{Xe}]\textbf{4}f^9$

Tb^{3+}: $[\textbf{Xe}]\textbf{4}f^8$

Tb^{4+}: $[\textbf{Xe}]\textbf{4}f^7$

Tb would demonstrate a +4 oxidation state because it has the half-filled sublevel.

23.34   The lanthanide element **gadolinium, Gd**, has electron configuration $[Xe]6s^24f^75d^1$ with **eight** unpaired electrons. The ion Gd^{3+} has **seven** unpaired electrons: $[Xe]4f^7$.

23.37   The coordination number indicates the number of ligand atoms bonded to the central metal ion. The oxidation number represents the number of electrons lost to form the ion. The coordination number is unrelated to the oxidation number.

23.39   Coordination number of two indicates **linear** geometry.
Coordination number of four indicates either **tetrahedral** or **square planar** geometry.
Coordination number of six indicates **octahedral** geometry.

23.42   The -*ate* ending signifies that the complex ion has a negative charge.

23.45   <u>Plan:</u> Use the naming rules for coordination compounds given in the text.
<u>Solution:</u>
a) The oxidation state of nickel is found from the total charge on the ion (+2 because two Cl^- charges equals –2) and the charge on ligands:

charge on nickel = +2 – 6(0 charge on water) = +2

Name nickel as nickel(II) to indicate oxidation state. Ligands are six (hexa-) waters (aqua). Put together with chloride anions to give **hexaaquanickel(II) chloride**.
b) The cation is $[Cr(en)_3]^{n+}$ and the anion is ClO_4^-, the perchlorate ion. The charge on the cation is +3 to make a neutral salt in combination with the –3 charge of the three perchlorate ions. The ligand is ethylenediamine, which has 0 charge. The charge of the cation equals the charge on chromium ion, so chromium(III) is included in the name. The three ethylenediamine ligands, abbreviated en, are indicated by the prefix tris- because the name of the ligand includes a numerical indicator, di-. The complete name is **tris(ethylenediamine)chromium(III) perchlorate**.
c) The cation is K^+ and the anion is $[Mn(CN)_6]^{4-}$. The charge of 4– is deduced from the four potassium +1 ions in the formula. The oxidation state of Mn is $–4 – \{6(–1)\} = +2$. The name of CN^- ligand is cyano and six ligands are represented by the prefix hexa-. The name of manganese anion is manganate(II). The -ate suffix on the complex ion is used to indicate that it is an anion. The full name of compound is **potassium hexacyanomanganate(II)**.

23.47   <u>Plan:</u> The charge of the central metal atom was determined in Problem 23.45 because the Roman numeral indicating oxidation state is part of the name. The coordination number, or number of ligand atoms bonded to the metal ion, is found by examining the bonded entities inside the square brackets to determine if they are unidentate, bidentate, or polydentate.
<u>Solution:</u>
a) The Roman numeral "II" indicates a **+2** oxidation state. There are six water molecules bonded to Ni and each ligand is unidentate, so the coordination number is **6**.
b) The Roman numeral "III" indicates a **+3** oxidation state. There are three ethylenediamine molecules bonded to Cr, but each ethylenediamine molecule contains two donor N atoms (bidentate). Therefore, the coordination number is **6**.
c) The Roman numeral "II" indicates a **+2** oxidation state. There are six unidentate cyano molecules bonded to Mn, so the coordination number is **6**.

23.49   <u>Plan:</u> Use the naming rules for coordination compounds given in the text.
<u>Solution:</u>
a) The cation is K^+, potassium. The anion is $[Ag(CN)_2]^-$ with the name dicyanoargentate(I) ion for the two cyanide ligands and the name of silver in an anion, argentate(I). The Roman numeral (I) indicates the oxidation number on Ag. O.N. for $Ag = –1 – \{2(–1)\} = +1$ since the complex ion has a charge of –1 and the cyanide ligands are also –1. The complete name is **potassium dicyanoargentate(I)**.

b) The cation is Na$^+$, sodium. Since there are two +1 sodium ions, the anion is [CdCl$_4$]$^{2-}$ with a charge of 2–. The anion is the tetrachlorocadmate(II) ion. With four –1 chloride ligands, the oxidation state of cadmium is +2 and the name of cadmium in an anion is cadmate. The complete name is **sodium tetrachlorocadmate(II)**.
c) The cation is [Co(NH$_3$)$_4$(H$_2$O)Br]$^{2+}$. The 2+ charge is deduced from the two Br$^-$ ions. The cation has the name tetraammineaquabromocobalt(III) ion, with four ammonia ligands (tetraammine), one water ligand (aqua) and one bromide ligand (bromo). The oxidation state of cobalt is +3: 2 – {4(0) + 1(0) + 1(–1)}. The oxidation state is indicated by (III), following cobalt in the name. The anion is Br$^-$, bromide. The complete name is **tetraammineaquabromocobalt(III) bromide.**

23.51   Plan: The charge of the central metal atom was determined in Problem 23.49 because the Roman numeral indicating oxidation state is part of the name. The coordination number, or number of ligand atoms bonded to the metal ion, is found by examining the bonded entities inside the square brackets to determine if they are unidentate, bidentate, or polydentate.
Solution:
a) The counter ion is K$^+$, so the complex ion is [Ag(CN)$_2$]$^-$. Each cyano ligand has a –1 charge, so silver has a **+1** charge: +1 + 2(–1) = –1. Each cyano ligand is unidentate, so the coordination number is **2**.
b) The counter ion is Na$^+$, so the complex ion is [CdCl$_4$]$^{2-}$. Each chloride ligand has a –1 charge, so Cd has a **+2** charge: +2 + 4(–1) = –2. Each chloride ligand is unidentate, so the coordination number is **4**.
c) The counter ion is Br$^-$, so the complex ion is [Co(NH$_3$)$_4$(H$_2$O)Br]$^{2+}$. Both the ammine and aqua ligands are neutral. The bromide ligand has a –1 charge, so Co has a **+3** charge: +3 + 4(0) + 0 + (–1) = +2. Each ligand is unidentate, so the coordination number is **6**.

23.53   Plan: Use the rules given in the chapter.
Solution:
a) The cation is tetramminezinc ion. The tetraammine indicates four NH$_3$ ligands. Zinc has an oxidation state of +2, so the charge on the cation is +2. The anion is SO$_4^{2-}$. Only one sulfate is needed to make a neutral salt. The formula of the compound is **[Zn(NH$_3$)$_4$]SO$_4$.**
b) The cation is pentaamminechlorochromium(III) ion. The ligands are five NH$_3$ from pentaammine, and one chloride from chloro. The chromium ion has a charge of +3, so the complex ion has a charge equal to +3 from chromium, plus 0 from ammonia, plus –1 from chloride for a total of +2. The anion is chloride, Cl$^-$. Two chloride ions are needed to make a neutral salt. The formula of compound is **[Cr(NH$_3$)$_5$Cl]Cl$_2$.**
c) The anion is bis(thiosulfato)argentate(I). Argentate(I) indicates silver in the +1 oxidation state, and bis(thiosulfato) indicates two thiosulfate ligands, S$_2$O$_3^{2-}$. The total charge on the anion is +1 plus 2(–2) to equal –3. The cation is sodium, Na$^+$. Three sodium ions are needed to make a neutral salt. The formula of compound is **Na$_3$[Ag(S$_2$O$_3$)$_2$].**

23.55   Plan: The coordination number, or number of ligand atoms bonded to the metal ion, is found by examining the bonded entities inside the square brackets to determine if they are unidentate, bidentate, or polydentate. Coordination compounds act like electrolytes, i.e., they dissolve in water to yield charged species, the counter ion, and the complex ion. However, the complex ion itself does not dissociate. The "number of individual ions per formula unit" refers to the number of ions that would form per coordination compound upon dissolution in water.
Solution:
a) The counter ion is SO$_4^{2-}$, so the complex ion is [Zn(NH$_3$)$_4$]$^{2+}$. Each ammine ligand is unidentate, so the coordination number is **4**. Each molecule dissolves in water to form one SO$_4^{2-}$ ion and one [Zn(NH$_3$)$_4$]$^{2+}$ ion, so **two ions** form per formula unit.
b) The counter ion is Cl$^-$, so the complex ion is [Cr(NH$_3$)$_5$Cl]$^{2+}$. Each ligand is unidentate, so the coordination number is **6**. Each molecule dissolves in water to form two Cl$^-$ ions and one [Cr(NH$_3$)$_5$Cl]$^{2+}$ ion, so **three ions** form per formula unit.
c) The counter ion is Na$^+$, so the complex ion is [Ag(S$_2$O$_3$)$_2$]$^{3-}$. Assuming that the thiosulfate ligand is unidentate, the coordination number is **2**. Each molecule dissolves in water to form three Na$^+$ ions and one [Ag(S$_2$O$_3$)$_2$]$^{3-}$ ion, so **four ions** form per formula unit.

23.57   Plan: Follow the naming rules given in the chapter.
        Solution:
        a) The cation is hexaaquachromium(III) with the formula $[Cr(H_2O)_6]^{3+}$. The total charge of the ion equals the
        charge on chromium because water is a neutral ligand. Six water ligands are indicated by a *hexa* prefix to
        aqua. The anion is SO_4^{2-}. To make the neutral salt requires three sulfate ions for two cations. The compound
        formula is **$[Cr(H_2O)_6]_2(SO_4)_3$**.
        b) The anion is tetrabromoferrate(III) with the formula $[FeBr_4]^-$. The total charge on the ion equals +3 charge on
        iron plus 4 x –1 charge on each bromide ligand for –1 overall. The cation is barium, Ba^{2+}. Two anions are
        needed for each barium ion to make a neutral salt. The compound formula is **$Ba[FeBr_4]_2$**.
        c) The anion is bis(ethylenediamine)platinum(II) ion. Charge on platinum is +2 and this equals the total charge on
        the complex ion because ethylenediamine is a neutral ligand. The *bis* preceding ethylenediamine indicates two
        ethylenediamine ligands, which are represented by the abbreviation "en." The cation is $[Pt(en)_2]^{2+}$. The anion is
        CO_3^{2-}. One carbonate combines with one cation to produce a neutral salt. The compound formula is **$[Pt(en)_2]CO_3$**.

23.59   Plan: The coordination number, or number of ligand atoms bonded to the metal ion, is found by examining the
        bonded entities inside the square brackets to determine if they are unidentate, bidentate, or polydentate. Coordination
        compounds act like electrolytes, i.e., they dissolve in water to yield charged species, the counter ions, and the
        complex ion. However, the complex ion itself does not dissociate. The "number of individual ions per formula unit"
        refers to the number of ions that would form per coordination compound upon dissolution in water.
        Solution:
        a) The counter ion is SO_4^{2-}, so the complex ion is $[Cr(H_2O)_6]^{3+}$. Each aqua ligand is unidentate, so the
        coordination number is **6**. Each molecule dissolves in water to form three SO_4^{2-} ions and two $[Cr(H_2O)_6]^{3+}$ ions,
        so **five ions** form per formula unit.
        b) The counter ion is Ba^{2+}, so the complex ion is $[FeBr_4]^-$. Each bromo ligand is unidentate, so the coordination
        number is **4**. Each molecule dissolves in water to form one Ba^{2+} ion and two $[FeBr_4]^-$ ions, so **three ions** form per
        formula unit.
        c) The counter ion is CO_3^{2-}, so the complex ion is $[Pt(en)_2]^{2+}$. Each ethylenediamine ligand is bidentate, so the
        coordination number is **4**. Each molecule dissolves in water to form one CO_3^{2-} ion and one $[Pt(en)_2]^{2+}$ ion,
        so **two ions** form per formula unit.

23.61   Plan: Ligands that form linkage isomers have two different possible donor atoms.
        Solution:
        a) The nitrite ion **forms linkage isomers** because it can bind to the metal ion through the lone pair on the N
        atom or any lone pair on either O atom. Resonance Lewis structures are:

        b) Sulfur dioxide molecules **form linkage isomers** because the lone pair on the S atom or any lone pair on either
        O atom can bind the central metal ion.

        c) Nitrate ions have an N atom with no lone pair and three O atoms, all with lone pairs than can bond to the
        metal ion. But all of the O atoms are equivalent so these ions **do not form linkage isomers**.

23.63   Plan: Types of isomers for coordination compounds are coordination isomers with different arrangements of ligands and counter ions, linkage isomers with different donor atoms from the same ligand bound to the metal ion, geometric isomers with differences in ligand arrangement relative to other ligands, and optical isomers with mirror images that are not superimposable.
Solution:
a) Platinum ion, Pt^{2+}, is a d^8 ion so the ligand arrangement is square planar. A *cis* and *trans* geometric isomer exist for this complex ion:

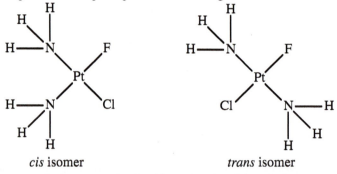

cis isomer   trans isomer

No optical isomers exist because the mirror images of both compounds are superimposable on the original molecules. In general, a square planar molecule is superimposable on its mirror image.

b) A *cis* and *trans* geometric isomer exist for this complex ion. No optical isomers exist because the mirror images of both compounds are superimposable on the original molecules.

cis isomer   trans isomer

c) Three geometric isomers exist for this molecule, although they are not named *cis* or *trans* because all the ligands are different. A different naming system is used to indicate the relation of one ligand to another.

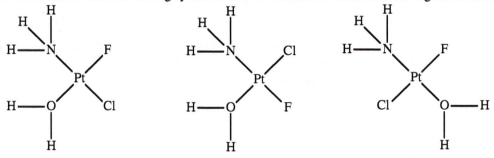

23.65   Plan: Types of isomers for coordination compounds are coordination isomers with different arrangements of ligands and counter ions, linkage isomers with different donor atoms from the same ligand bound to the metal ion, geometric isomers with differences in ligand arrangement relative to other ligands, and optical isomers with mirror images that are not superimposable.

Solution:

a) Platinum ion, Pt^{2+}, is a d^8 ion, so the ligand arrangement is square planar. The ligands are two Cl^- and two Br^-, so the arrangement can be either both ligands *trans* or both ligands *cis* to form geometric isomers.

b) The complex ion can form linkage isomers with the NO_2 ligand. Either the N or an O may be the donor.

c) In the octahedral arrangement, the two iodide ligands can be either *trans* to each other, 180° apart, or *cis* to each other, 90° apart.

cis   *trans*

23.67   The traditional formula does not correctly indicate that at least some of the chloride ions serve as counter ions. The NH_3 molecules serve as ligands, so n could equal 6 to satisfy the coordination number requirement. This complex has the formula $[Cr(NH_3)_6]Cl_3$. This is a correct formula because the name "chromium(III)" means that the chromium has a +3 charge, and this balances with the −3 charge provided by the three chloride counter ions. However, when this compound dissociates in water, it produces four ions ($[Cr(NH_3)_6]^{3+}$ and three Cl^-, whereas NaCl only produces two ions. Other possible compounds are $n = 5$, $CrCl_3 \cdot 5NH_3$, with formula $[Cr(NH_3)_5Cl]Cl_2$; $n = 4$, $CrCl_3 \cdot 4NH_3$, with formula $[Cr(NH_3)_4Cl_2]Cl$; $n = 3$, $CrCl_3 \cdot 3NH_3$, with formula $[Cr(NH_3)_3Cl_3]$. The compound **$[Cr(NH_3)_4Cl_2]Cl$** has a coordination number equal to 6 and produces two ions when dissociated in water, so it has an electrical conductivity similar to an equimolar solution of NaCl.

23.69   Plan: First find the charge on the palladium ion, then arrange ligands and counter ions to form the complex.
Solution:
a) Charge on palladium $= - [(+1 \text{ on } K^+) + (0 \text{ on } NH_3) + 3(-1 \text{ on } Cl^-)] = +2$
Palladium(II) forms four-coordinate complexes. The four ligands in the formula are one NH_3 and three Cl^- ions. The formula of the complex ion is $[Pd(NH_3)Cl_3]^-$. Combined with the potassium cation, the compound formula is **$K[Pd(NH_3)Cl_3]$**.
b) Charge on palladium $= - [2(-1 \text{ on } Cl^-) + 2(0 \text{ on } NH_3)] = +2$
Palladium(II) forms four-coordinate complexes. The four ligands are two chloride ions and two ammonia molecules. The formula is **$[PdCl_2(NH_3)_2]$**.
c) Charge on palladium $= - [2(+1 \text{ on } K^+) + 6(-1 \text{ on } Cl^-)] = +4$
Palladium(IV) forms six-coordinate complexes. The six ligands are the six chloride ions. The formula is **$K_2[PdCl_6]$**.

d) Charge on palladium $= -[4(0 \text{ on } NH_3) + 4(-1 \text{ on } Cl^-)] = +4$
Palladium(IV) forms six-coordinate complexes. The ammonia molecules have to be ligands. The other two ligand bonds are formed with two of the chloride ions. The remaining two chloride ions are the counter ions. The formula is **$[Pd(NH_3)_4Cl_2]Cl_2$**.

23.71   a) Four empty orbitals of equal energy are "created" to receive the donated electron pairs from four ligands. The four orbitals are hybridized from an s, two p, and one d orbital from the previous n level to form four **dsp^2** orbitals.
b) One s and three p orbitals become four **sp^3** hybrid orbitals.

23.74   Absorption of **orange** or **yellow** light gives a blue solution.

23.75   a) The crystal field splitting energy is the energy difference between the two sets of d orbitals that result from the bonding of ligands to a central transition metal atom.
b) In an octahedral field of ligands, the ligands approach along the x, y, and z axes. The $d_{x^2-y^2}$ and d_{z^2} orbitals are located along the x, y, and z axes, so ligand interaction is higher in energy than the other orbital-ligand interactions. The other orbital-ligand interactions are lower in energy because the d_{xy}, d_{yz}, and d_{xz} orbitals are located between the x, y, and z axes.
c) In a tetrahedral field of ligands, the ligands do not approach along the x, y, and z axes. The ligand interaction is greater for the d_{xy}, d_{yz}, and d_{xz} orbitals and lesser for the $d_{x^2-y^2}$ and d_{z^2} orbitals. The crystal field splitting is reversed, and the d_{xy}, d_{yz}, and d_{xz} orbitals are higher in energy than the $d_{x^2-y^2}$ and d_{z^2} orbitals.

23.78   If Δ is greater than $E_{pairing}$, electrons will preferentially pair spins in the lower energy d orbitals before adding as unpaired electrons to the higher energy d orbitals. If Δ is less than $E_{pairing}$, electrons will preferentially add as unpaired electrons to the higher d orbitals before pairing spins in the lower energy d orbitals. The first case gives a complex that is low-spin and less paramagnetic than the high-spin complex formed in the latter case.

23.80   Plan: To determine the number of d electrons in a central metal ion, first write the electron configuration for the metal atom. Examine the formula of the complex to determine the charge on the central metal ion, and then write the ion's configuration by removing the correct number of electrons, beginning with the ns electrons and then the $(n-1)d$ electrons.
Solution:
a) Electron configuration of Ti: $[Ar]4s^2 3d^2$
        Charge on Ti: Each chloride ligand has a -1 charge, so Ti has a $+4$ charge $\{+4 + 6(-1)\} = 2-$ ion.
Both of the $4s$ electrons and both $3d$ electrons are removed.
        Electron configuration of Ti^{4+}: $[Ar]$
        Ti^{4+} has **no d electrons**.
b) Electron configuration of Au: $[Xe]6s^1 4f^{14} 5d^{10}$
Charge on Au: The complex ion has a -1 charge $([AuCl_4]^-)$ since K has a $+1$ charge. Each chloride ligand has a -1 charge, so Au has a $+3$ charge $\{+3 + 4(-1)\} = 1-$ ion. The $6s$ electron and two d electrons are removed.
        Electron configuration of Au^{3+}: $[Xe]4f^{14} 5d^8$
        Au^{3+} has **eight d electrons**.
c) Electron configuration of Rh: $[Kr]5s^2 4d^7$
        Charge on Rh: Each chloride ligand has a -1 charge, so Rh has a $+3$ charge $\{+3 + 6(-1)\} = 3-$ ion.
The $5s$ electrons and one $4d$ electron are removed.
        Electron configuration of Rh^{3+}: $[Kr]4d^6$ Rh^{3+} has **six d electrons**.

23.82   Plan: To determine the number of d electrons in a central metal ion, first write the electron configuration for the metal atom. Examine the formula of the complex to determine the charge on the central metal ion, and then write the ion's configuration by removing the correct number of electrons, beginning with the ns electrons and then the $(n-1)d$ electrons.
Solution:
a) $[(+2 \text{ on } Ca^{2+}) + 6(-1 \text{ on } F^-) + Ir = 0]$
Charge on iridium $= -[(+2 \text{ on } Ca^{2+}) + 6(-1 \text{ on } F^-)] = +4$
        Configuration of Ir is $[Xe]6s^2 4f^{14} 5d^7$.
        Configuration of Ir^{4+} is $[Xe]4f^{14} 5d^5$. **Five d electrons in Ir^{4+}.**

b) [Hg + 4(–1 on I⁻)] = –2
  Charge on mercury = – [4(–1 on I⁻)] – 2 = +2
        Configuration of Hg is $[Xe]6s^2 4f^{14} 5d^{10}$.
        Configuration of Hg^{2+} is $[Xe]4f^{14}5d^{10}$. **Ten** d electrons in Hg^{2+}.
  c) [Co + (–4 on EDTA)] = – 2
  Charge on cobalt = – [–4 on EDTA] – 2 = +2
        Configuration of Co is $[Ar]4s^2 3d^7$.
        Configuration of Co^{2+} is $[Ar]3d^7$. **Seven** d electrons in Co^{2+}.

23.84

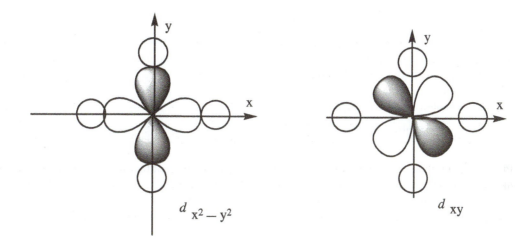

The unequal shading of the lobes of the d orbitals is a consequence of the quantum mechanical derivation of these orbitals, and does not affect the current discussion.
In an octahedral field of ligands, the ligands approach along the x, y, and z axes. The $d_{x^2-y^2}$ orbital is located *along* the x and y axes, so ligand interaction is greater. The d_{xy} orbital is offset from the x and y axes by 90°, so ligand interaction is less. The greater interaction of the $d_{x^2-y^2}$ orbital results in its higher energy.

23.86 Plan: Determine the electron configuration of the ion, which gives the number of d electrons. If there are only 1, 2, or 3 d electrons, the complex is always high-spin since there are not enough electrons to pair. If there are 8 or 9 d electrons, the complex is always high-spin since the higher energy d orbitals will always contain two (d^8) or one (d^9) unpaired electrons.
  Solution:
  a) Ti: $[Ar]4s^2 3d^2$. The electron configuration of Ti^{3+} is $[Ar]3d^1$. With only one electron in the d orbitals, the titanium(III) ion **cannot form** high- and low-spin complexes – all complexes will contain one unpaired electron and have the same spin.
  b) Co: $[Ar]4s^2 3d^7$. The electron configuration of Co^{2+} is $[Ar]3d^7$ and will form high- and low-spin complexes with seven electrons in the d orbital.
  c) Fe: $[Ar]4s^2 3d^6$. The electron configuration of Fe^{2+} is $[Ar]3d^6$ and will form high- and low-spin complexes with six electrons in the d orbital.
  d) Cu: $[Ar]4s^1 3d^{10}$. The electron configuration of Cu^{2+} is $[Ar]3d^9$, so in complexes with both strong- and weak-field ligands, one electron will be unpaired and the spin in both types of complexes is identical. Cu^{2+} **cannot form** high- and low-spin complexes.

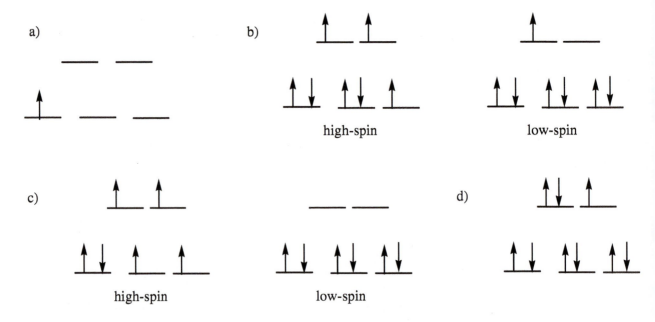

a)

b)

high-spin   low-spin

c)

high-spin   low-spin

d)

23.88   <u>Plan:</u> To draw the orbital-energy splitting diagram, first determine the number of d electrons in the transition metal ion. Examine the formula of the complex ion to determine the electron configuration of the metal ion, remembering that the ns electrons are lost first. Determine the coordination number from the number of ligands, recognizing that six ligands result in an octahedral arrangement and four ligands result in a tetrahedral or square planar arrangement. Weak-field ligands give the maximum number of unpaired electrons (high-spin) while strong-field ligands lead to electron pairing (low-spin).
<u>Solution:</u>
a) Electron configuration of Cr: $[Ar]4s^1 3d^5$
Charge on Cr: The aqua ligands are neutral, so the charge on Cr is +3.
Electron configuration of Cr^{3+}: $[Ar]3d^3$
Six ligands indicate an octahedral arrangement. Using Hund's rule, fill the lower energy t_{2g} orbitals first, filling empty orbitals before pairing electrons within an orbital.
b) Electron configuration of Cu: $[Ar]4s^1 3d^{10}$
Charge on Cu: The aqua ligands are neutral, so Cu has a +2 charge.
Electron configuration of Cu^{2+}: $[Ar]3d^9$
Four ligands and a d^9 configuration indicate a square planar geometry (only filled d sublevel ions exhibit tetrahedral geometry). Use Hund's rule to fill in the nine d electrons. Therefore, the correct orbital-energy splitting diagram shows one unpaired electron.
c) Electron configuration of Fe: $[Ar]4s^2 3d^6$
Charge on Fe: Each fluoride ligand has a –1 charge for a total charge of –6, so Fe has a +3 charge to make the overall complex charge equal to –3.
Electron configuration of Fe^{3+}: $[Ar]3d^5$
Six ligands indicate an octahedral arrangement. Use Hund's rule to fill the orbitals.
F^- is a weak-field ligand, so the splitting energy, Δ, is not large enough to overcome the resistance to electron pairing. The electrons remain unpaired, and the complex is called high-spin.

(a)   (b)   (c)

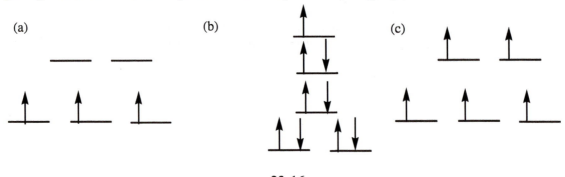

23.90   Plan: To draw the orbital-energy splitting diagram, first determine the number of d electrons in the transition metal ion. Examine the formula of the complex ion to determine the electron configuration of the metal ion, remembering that the ns electrons are lost first. Determine the coordination number from the number of ligands, recognizing that six ligands result in an octahedral arrangement and four ligands result in a tetrahedral or square planar arrangement. Weak-field ligands give the maximum number of unpaired electrons (high-spin) while strong-field ligands lead to electron pairing (low-spin).
Solution:
a) Electron configuration of Mo: $[Kr]5s^1 4d^5$
Charge on Mo: Each chloride ligand has a –1 charge for a total charge of –6, so Mo has a +3 charge to make the overall complex charge equal to –3.
Electron configuration of Mo^{3+}: $[Kr]4d^3$
Six ligands indicate an octahedral arrangement. Using Hund's rule, fill the lower energy t_{2g} orbitals first, filling empty orbitals before pairing electrons within an orbital.
b) Electron configuration of Ni: $[Ar]4s^2 3d^8$
Charge on Ni: The aqua ligands are neutral, so the charge on Ni is +2.
Electron configuration of Ni^{2+}: $[Ar]3d^8$
Six ligands indicate an octahedral arrangement. Use Hund's rule to fill the orbitals.
H_2O is a weak-field ligand, so the splitting energy, Δ, is not large enough to overcome the resistance to electron pairing. One electron occupies each of the five d orbitals before pairing in the t_{2g} orbitals, and the complex is called high-spin.
c) Electron configuration of Ni: $[Ar]4s^2 3d^8$
Charge on Ni: Each cyanide ligand has a –1 charge for a total charge of –4, so Ni has a +2 charge to make the overall complex charge equal to –2.
Electron configuration of Ni^{2+}: $[Ar]3d^8$
The coordination number is 4 and the complex is square planar.
The complex is low-spin because CN^- is a strong-field ligand. Electrons pair in one set of orbitals before occupying orbitals of higher energy.

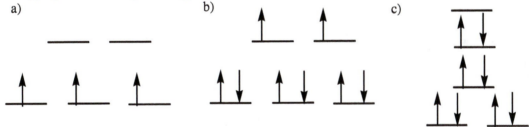

a)   b)   c)

23.92   Plan: The spectrochemical series describes the spectrum of splitting energy, Δ. The greater the crystal field strength of the ligand, the greater the crystal field splitting energy, Δ, and the greater the energy of light absorbed.
Solution:
NO_2^- is a stronger field ligand than NH_3, which is a stronger field ligand than H_2O. NO_2^- produces the largest Δ, followed by NH_3, and then H_2O with the smallest Δ value. The energy of light absorbed increases as Δ increases since more energy is required to excite an electron from a lower energy orbital to a higher energy orbital when Δ is very large.
$$[Cr(H_2O)_6]^{3+} < [Cr(NH_3)_6]^{3+} < [Cr(NO_2)_6]^{3-}$$

23.94   Plan: A weaker field ligand will result in a smaller Δ in the complex and lower energy light absorbed. When a particular color of light is absorbed, the complementary color is seen.
Solution:
A violet complex absorbs yellow-green light. The light absorbed by a complex with a weaker ligand would be at a lower energy and longer wavelength. Light of lower energy than yellow-green light is yellow, orange, or red light. The color observed would be **blue** or **green**.

23.97 <u>Plan:</u> A weaker field ligand will result in a smaller Δ in the complex and lower energy light absorbed. When a particular color of light is absorbed, the complementary color is seen.
<u>Solution:</u>
The aqua ligand is weaker than the ammine ligand. The weaker ligand results in a lower splitting energy and absorbs a lower energy of visible light. The green hexaaqua complex appears green because it absorbs red light (opposite side of the color wheel). The hexaammine complex appears violet because it absorbs yellow light, which is higher in energy (shorter λ) than red light.

23.101 The electron configuration of Hg is $[Xe]6s^2 4f^{14} 5d^{10}$ and that of Hg$^+$ is $[Xe]6s^1 4f^{14} 5d^{10}$. The electron configuration of Cu is $[Ar]4s^1 3d^{10}$ and that of Cu$^+$ is $[Ar]3d^{10}$. In the mercury(I) ion, there is one electron in the $6s$ orbital that can form a covalent bond with the electron in the $6s$ orbital of another Hg$^+$ ion. In the copper(I) ion, there are no electrons in the s orbital to bond with another copper(I) ion.

23.102 <u>Plan:</u> The coordination number, or number of ligand atoms bonded to the metal ion, is found by examining the bonded entities inside the square brackets to determine if they are unidentate, bidentate, or polydentate. The oxidation of the central metal ion is found by determining the charges of the ligands and insuring that the charges of the metal ion, ligands, and counter ion add to zero. Coordination compounds act like electrolytes, i.e., they dissolve in water to yield charged species, the counter ions and the complex ion. However, the complex ion itself does not dissociate. The "number of individual ions per formula unit" refers to the number of ions that would form per coordination compound upon dissolution in water.
<u>Solution:</u>
a) The coordination number of cobalt is **6**. The two Cl⁻ ligands are unidentate and the two ethylenediamine ligands are bidentate (each en ligand forms two bonds to the metal), so a total of six ligand atoms are connected to the central metal ion.
b) The counter ion is Cl⁻, so the complex ion is $[Co(en)_2Cl_2]^+$. Each chloride ligand has a -1 charge and each en ligand is neutral, so cobalt has a **+3** charge: $+3 + 2(0) + 2(-1) = +1$.
c) One mole of complex dissolves in water to yield one mole of $[Co(en)_2Cl_2]^+$ ions and one mole of Cl⁻ ions. Therefore, each formula unit yields **two** individual ions.
d) One mole of compound dissolves to form one mole of Cl⁻ ions, which reacts with the Ag$^+$ ion (from AgNO$_3$) to form **one mole of AgCl precipitate**.

23.109 <u>Plan:</u> Types of isomers for coordination compounds are coordination isomers with different arrangements of ligands and counter ions, linkage isomers with different donor atoms from the same ligand bound to the metal ion, geometric isomers with differences in ligand arrangement relative to other ligands, and optical isomers with mirror images that are not superimposable.
<u>Solution:</u>
This compound exhibits geometric (*cis-trans*) and linkage isomerism. The SCN⁻ ligand can bond to the metal through either the S atom or the N atom.

cis-diamminedithiocyanatoplatinum(II)

trans-diamminedithiocyanatoplatinum(II)

cis-diamminediisothiocyanatoplatinum(II)

trans-diamminediisothiocyanatoplatinum(II)

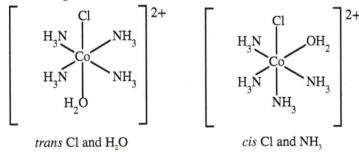

cis-diamminethiocyanatoisothiocyanatoplatinum(II)   *trans*-diamminethiocyanatoisothiocyanatoplatinum(II)

23.110   <u>Plan:</u> Types of isomers for coordination compounds are coordination isomers with different arrangements of ligands and counter ions, linkage isomers with different donor atoms from the same ligand bound to the metal ion, geometric isomers with differences in ligand arrangement relative to other ligands, and optical isomers with mirror images that are not superimposable.
<u>Solution:</u>
$[Co(NH_3)_4(H_2O)Cl]^{2+}$ tetraammineaquachlorocobalt(III) ion
2 geometric isomers

trans Cl and H$_2$O   *cis* Cl and NH$_3$

$[Cr(H_2O)_3Br_2Cl]$ triaquadibromochlorochromium(III)
3 geometric isomers

Br's *trans*   Br's *cis*   Br's *cis*
                     H$_2$O's facial   H$_2$O's meridional

(Unfortunately meridional and facial isomers are not covered in the text. Facial (fac) isomers have three adjacent corners of the octahedron occupied by similar groups. Meridional (mer) isomers have three similar groups around the outside of the complex.)
$[Cr(NH_3)_2(H_2O)_2Br_2]^+$ diamminediaquadibromochromium(III) ion
6 isomers (5 geometric)

All pairs are *trans*   Only NH$_3$'s are *trans*

23-19

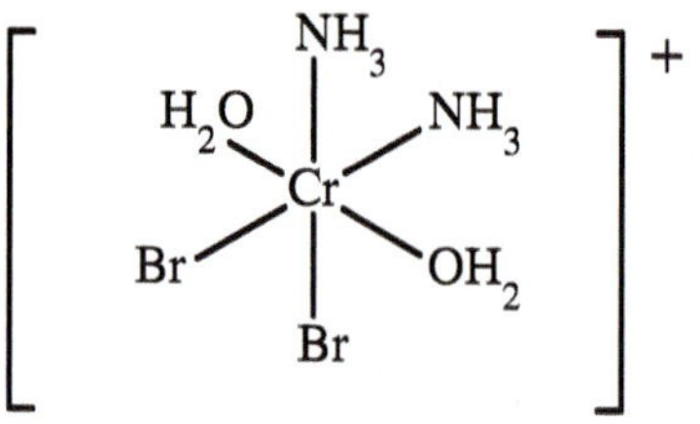

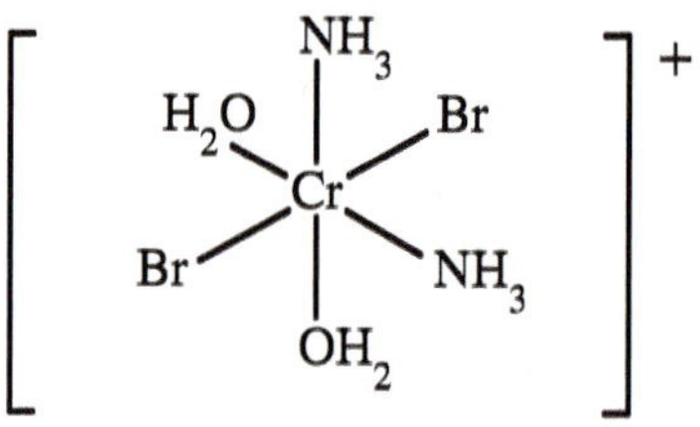

Only H$_2$O's are *trans*   Only Br's are *trans*

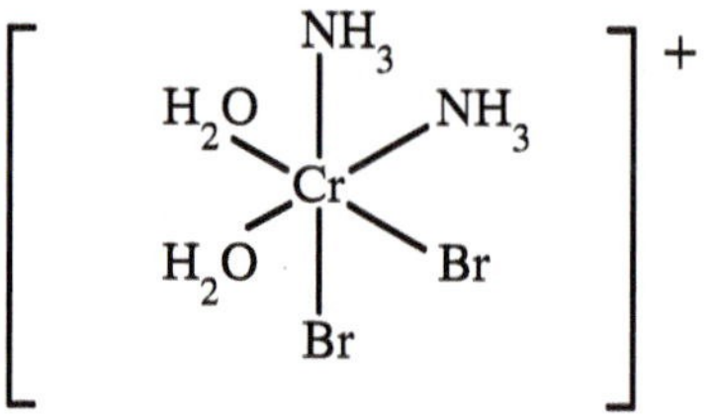

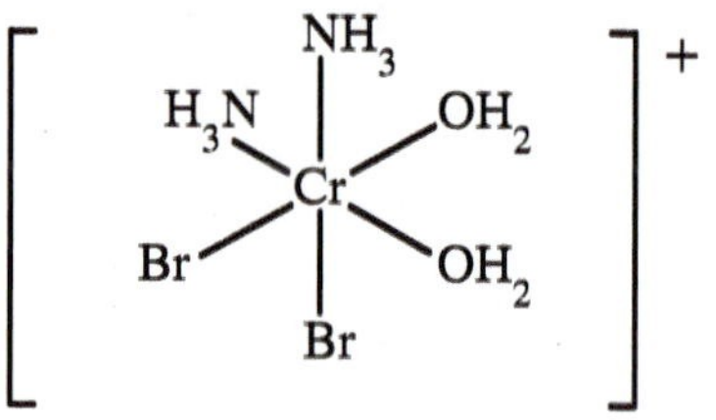

All pairs are *cis*. These are optical isomers of each other.

23.115   <u>Plan:</u> Sketch the structure of each complex ion and look for a plane of symmetry. A complex ion with a plane of symmetry does not have optical isomers.
<u>Solution:</u>
a) A plane that includes both ammonia ligands and the zinc ion is a plane of symmetry. The complex **does not have optical isomers** because it has a plane of symmetry.
b) The Pt^{2+} ion is d^8, so the complex is square planar. Any square planar complex has a plane of symmetry, so the complex **does not have optical isomers**.
c) The *trans* octahedral complex has the two chloride ions opposite each other. A plane of symmetry can be passed through the two chlorides and platinum ion, so the *trans* complex **does not have optical isomers**.
d) **No optical isomers** for same reason as in part c).
e) The *cis* isomer does not have a plane of symmetry, so it **does have optical isomers**.

23.116   <u>Plan:</u> Assume a 100-g sample (making the percentages of each element equal to the mass present in grams) and convert the mass of each element to moles using the element's molar mass. Divide each mole amount by the smallest mole amount to determine whole-number ratios of the elements to determine the empirical formula. Once the empirical formula is known, use the formula of the triethylphosphine to deduce the molecular formula.
<u>Solution:</u>

$$\text{Moles of Pt} = \left(38.8 \text{ g Pt}\right)\left(\frac{1 \text{ mol Pt}}{195.1 \text{ g Pt}}\right) = 0.198872 \text{ mol Pt}; \qquad \frac{0.198872}{0.198872} = 1$$

$$\text{Moles of Cl} = \left(14.1 \text{ g Cl}\right)\left(\frac{1 \text{ mol Cl}}{35.45 \text{ g Cl}}\right) = 0.397743 \text{ mol Cl}; \qquad \frac{0.397743}{0.198872} = 2$$

$$\text{Moles of C} = \left(28.7 \text{ g C}\right)\left(\frac{1 \text{ mol C}}{12.01 \text{ g C}}\right) = 2.389675 \text{ mol C}; \qquad \frac{2.389675}{0.198872} = 12$$

$$\text{Moles of P} = \left(12.4 \text{ g P}\right)\left(\frac{1 \text{ mol P}}{30.97 \text{ g P}}\right) = 0.400387 \text{ mol P}; \qquad \frac{0.400387}{0.198872} = 2$$

$$\text{Moles of H} = \left(6.02 \text{ g H}\right)\left(\frac{1 \text{ mol H}}{1.008 \text{ g H}}\right) = 5.972222 \text{ mol H}; \qquad \frac{5.972222}{0.198872} = 30$$

The empirical formula for the compound is $PtCl_2C_{12}P_2H_{30}$. Each triethylphosphine ligand, $P(C_2H_5)_3$ accounts for one phosphorus atom, six carbon atoms, and fifteen hydrogen atoms. According to the empirical formula, there are two triethylphosphine ligands: 2 $P(C_2H_5)_3$ "equals" $C_{12}P_2H_{30}$. The compound must have two $P(C_2H_5)_3$ ligands and two chloro ligands per Pt ion. The formula is $[Pt[P(C_2H_5)_3]_2Cl_2]$. The central Pt ion has four ligands and is square planar, existing as either a *cis* or *trans* compound.

cis-dichlorobis(triethylphosphine)platinum(II)   *trans*-dichlorobis(triethylphosphine)platinum(II)

23.118 Plan: A reaction is favored in terms of entropy if there is an increase in entropy. Entropy increases if there are more moles of product than of reactant.
Solution:
a) The first reaction shows no change in the number of particles. In the second reaction, the number of reactant particles is greater than the number of product particles. A decrease in the number of particles means a decrease in entropy, while no change in number of particles indicates little change in entropy. Based on entropy change only, the first reaction is favored.
b) The ethylenediamine complex is more stable with respect to ligand exchange with water because the entropy change is unfavorable.

CHAPTER 24 NUCLEAR REACTIONS AND THEIR APPLICATIONS

FOLLOW–UP PROBLEMS

24.1A **Plan:** Write a skeleton equation that shows fluorine-20 undergoing beta decay. Conserve mass and atomic number by ensuring the superscripts and subscripts equal one another on both sides of the equation. Determine the identity of the daughter nuclide by using the periodic table.
Solution:
Fluorine-20 has $Z = 9$. Its symbol is $^{20}_{9}F$. When it undergoes beta decay, a beta particle, $^{0}_{-1}\beta$, and a daughter nuclide, $^{A}_{Z}X$ are produced.
$$^{20}_{9}F \rightarrow {}^{A}_{Z}X + {}^{0}_{-1}\beta$$

To conserve atomic number, Z must equal 10. Element is neon.
To conserve mass number, A must equal 20.
The identity of $^{A}_{Z}X$ is $^{20}_{10}Ne$.

The balanced equation is: $^{20}_{9}F \rightarrow {}^{20}_{10}Ne + {}^{0}_{-1}\beta$

24.1B **Plan:** Write a skeleton equation that shows an unknown nuclide, $^{A}_{Z}X$, undergoing positron decay, $^{0}_{1}\beta$, to form tellurium–124, $^{124}_{52}Te$. Conserve mass and atomic number by ensuring the superscripts and subscripts equal one another on both sides of the equation. Determine the identity of X by using the periodic table to identify the element with atomic number equal to Z.
Solution:
The unknown nuclide yields tellurium-124 and a β^{+} particle:
$$^{A}_{Z}X \rightarrow {}^{124}_{52}Te + {}^{0}_{1}\beta$$
To conserve atomic number, Z must equal 53. Element is iodine.
To conserve mass number, A must equal 124.
The identity of $^{A}_{Z}X$ is $^{124}_{53}I$.

The balanced equation is: $^{124}_{53}I \rightarrow {}^{124}_{52}Te + {}^{0}_{1}\beta$
Check: $A = 124 = 124 + 0$ and $Z = 53 = 52 + 1$.

24.2A **Plan:** Look at the N/Z ratio, the ratio of the number of neutrons to the number of protons. If the N/Z ratio falls in the band of stability, the nuclide is predicted to be stable. For stable nuclides of elements with atomic number greater than 20, the ratio of number of neutrons to number of protons (N/Z) is greater than one. In addition, the ratio increases gradually as atomic number increases. Also check for exceptionally stable numbers of neutrons and/or protons – the "magic" numbers of 2, 8, 20, 28, 50, 82, and ($N = 126$). Also, even numbers of protons and or neutrons are related to stability whereas odd numbers are related to instability.
Solution:
a) $^{10}_{5}B$ appears **stable** because its N/Z ratio $(10 - 5)/5 = 1.00$ is in the band of stability.

b) $^{58}_{23}V$ appears unstable/**radioactive** because its N/Z ratio $(58 - 23)/23 = 1.52$ is too high and is above the band of stability. Additionally, this nuclide has both odd $N(35)$ and $Z(23)$.

24.2B **Plan:** Nuclear stability is found in nuclides with an N/Z ratio that falls within the band of stability. Nuclides with an even N and Z, especially those nuclides that have magic numbers, are exceptionally stable. Examine the two nuclides to see which of these criteria can explain the difference in stability.

24.6 A neutron-rich nuclide decays to convert neutrons to protons while a neutron-poor nuclide decays to convert protons to neutrons. The conversion of neutrons to protons occurs by beta decay:

$$^{1}_{0}n \rightarrow {}^{1}_{1}p + {}^{0}_{-1}\beta$$

The conversion of protons to neutrons occurs by either positron decay:

$$^{1}_{1}p \rightarrow {}^{1}_{0}n + {}^{0}_{1}\beta$$

or electron capture:

$$^{1}_{1}p + {}^{0}_{-1}e \rightarrow {}^{1}_{0}n$$

Neutron-rich nuclides, with a high N/Z, undergo β decay. Neutron-poor nuclides, with a low N/Z, undergo positron decay or electron capture.

24.8 Plan: In a balanced nuclear equation, the total of mass numbers and the total of charges on the left side and the right side must be equal.
Solution:

a) $^{234}_{92}U \rightarrow {}^{4}_{2}He + {}^{230}_{90}Th$ Mass: $234 = 4 + 230$;   Charge: $92 = 2 + 90$

b) $^{232}_{93}Np + {}^{0}_{-1}e \rightarrow {}^{232}_{92}U$ Mass: $232 + 0 = 232$;   Charge: $93 + (-1) = 92$

c) $^{12}_{7}N \rightarrow {}^{0}_{1}\beta + {}^{12}_{6}C$ Mass: $12 = 0 + 12$;   Charge: $7 = 1 + 6$

24.10 Plan: In a balanced nuclear equation, the total of mass numbers and the total of charges on the left side and the right side must be equal.
Solution:

a) The process converts a neutron to a proton, so the mass number is the same, but the atomic number increases by one.

$$^{27}_{12}Mg \rightarrow {}^{0}_{-1}\beta + {}^{27}_{13}Al \qquad \text{Mass: } 27 = 0 + 27; \qquad \text{Charge: } 12 = -1 + 13$$

b) Positron emission decreases atomic number by one, but not mass number.

$$^{23}_{12}Mg \rightarrow {}^{0}_{1}\beta + {}^{23}_{11}Na \qquad \text{Mass: } 23 = 0 + 23; \qquad \text{Charge: } 12 = 1 + 11$$

c) The electron captured by the nucleus combines with a proton to form a neutron, so mass number is constant, but atomic number decreases by one.

$$^{103}_{46}Pd + {}^{0}_{-1}e \rightarrow {}^{103}_{45}Rh \qquad \text{Mass: } 103 + 0 = 103; \qquad \text{Charge: } 46 + (-1) = 45$$

24.12 Plan: In a balanced nuclear equation, the total of mass numbers and the total of charges on the left side and the right side must be equal.
Solution:

a) In other words, an unknown nuclide decays to give Ti-48 and a positron.

$$^{48}_{23}V \rightarrow {}^{48}_{22}Ti + {}^{0}_{1}\beta \qquad \text{Mass: } 48 = 48 + 0; \qquad \text{Charge: } 23 = 22 + 1$$

b) In other words, an unknown nuclide captures an electron to form Ag-107.

$$^{107}_{48}Cd + {}^{0}_{-1}e \rightarrow {}^{107}_{47}Ag \qquad \text{Mass: } 107 + 0 = 107; \qquad \text{Charge: } 48 + (-1) = 47$$

c) In other words, an unknown nuclide decays to give Po-206 and an alpha particle.

$$^{210}_{86}Rn \rightarrow {}^{206}_{84}Po + {}^{4}_{2}He \qquad \text{Mass: } 210 = 206 + 4; \qquad \text{Charge: } 86 = 84 + 2$$

24.14 Plan: In a balanced nuclear equation, the total of mass numbers and the total of charges on the left side and the right side must be equal.
Solution:

a) In other words, an unknown nuclide captures an electron to form Ir-186.

$$^{186}_{78}Pt + {}^{0}_{-1}e \rightarrow {}^{186}_{77}Ir \qquad \text{Mass: } 186 + 0 = 186; \qquad \text{Charge: } 78 + (-1) = 77$$

b) In other words, an unknown nuclide decays to give Fr-221 and an alpha particle.

$$^{225}_{89}Ac \rightarrow {}^{221}_{87}Fr + {}^{4}_{2}He \qquad \text{Mass: } 225 = 221 + 4; \qquad \text{Charge: } 89 = 87 + 2$$

c) In other words, an unknown nuclide decays to give I-129 and a beta particle.

$$^{129}_{52}\text{Te} \rightarrow {}^{129}_{53}\text{I} + {}^{0}_{-1}\beta \qquad \text{Mass: } 129 = 129 + 0; \qquad \text{Charge: } 52 = 53 + (-1)$$

24.16   Plan: Look at the N/Z ratio, the ratio of the number of neutrons to the number of protons. If the N/Z ratio falls in the band of stability, the nuclide is predicted to be stable. For stable nuclides of elements with atomic number greater than 20, the ratio of number of neutrons to number of protons (N/Z) is greater than one. In addition, the ratio increases gradually as atomic number increases. Also check for exceptionally stable numbers of neutrons and/or protons – the "magic" numbers of 2, 8, 20, 28, 50, 82, and ($N = 126$). Also, even numbers of protons and or neutrons are related to stability whereas odd numbers are related to instability.
Solution:
a) $^{20}_{8}\text{O}$ appears stable because its Z (8) value is a magic number, but its N/Z ratio $(20 - 8)/8 = 1.50$ is too high and this nuclide is above the band of stability; $^{20}_{8}\text{O}$ is unstable.
b) $^{59}_{27}\text{Co}$ might look unstable because its Z value is an odd number, but its N/Z ratio $(59 - 27)/27 = 1.19$ is in the band of stability, so $^{59}_{27}\text{Co}$ appears stable.
c) $^{9}_{3}\text{Li}$ appears unstable because its N/Z ratio $(9 - 3)/3 = 2.00$ is too high and is above the band of stability.

24.18   Plan: Look at the N/Z ratio, the ratio of the number of neutrons to the number of protons. If the N/Z ratio falls in the band of stability, the nuclide is predicted to be stable. For stable nuclides of elements with atomic number greater than 20, the ratio of number of neutrons to number of protons (N/Z) is greater than one. In addition, the ratio increases gradually as atomic number increases. Also check for exceptionally stable numbers of neutrons and/or protons – the "magic" numbers of 2, 8, 20, 28, 50, 82, and ($N = 126$). Also, even numbers of protons and or neutrons are related to stability whereas odd numbers are related to instability.
Solution:
a) For the element iodine $Z = 53$. For iodine-127, $N = 127 - 53 = 74$. The N/Z ratio for ^{127}I is $74/53 = 1.4$. Of the examples of stable nuclides given in the book, ^{107}Ag has the closest atomic number to iodine. The N/Z ratio for ^{107}Ag is 1.3. Thus, it is likely that iodine with six additional protons is stable with an N/Z ratio of 1.4.
b) Tin is element number 50 ($Z = 50$). The N/Z ratio for ^{106}Sn is $(106 - 50)/50 = 1.1$. The nuclide ^{106}Sn is unstable with an N/Z ratio that is too low.
c) For ^{68}As, $Z = 33$ and $N = 68 - 33 = 35$ and $N/Z = 1.1$. The ratio is within the range of stability, but the nuclide is most likely unstable because there is an odd number of both protons and neutrons.

24.20   Plan: Calculate the N/Z ratio for each nuclide. A neutron-rich nuclide decays to convert neutrons to protons while a neutron-poor nuclide decays to convert protons to neutrons. Neutron-rich nuclides, with a high N/Z, undergo β decay. Neutron-poor nuclides, with a low N/Z, undergo positron decay or electron capture. For $Z < 20$, β^{+} emission is more common; for $Z > 80$, e^{-} capture is more common. Alpha decay is the most common means of decay for a heavy, unstable nucleus ($Z > 83$).
Solution:
a) $^{238}_{92}\text{U}$: Nuclides with $Z > 83$ decay through α decay.
b) The N/Z ratio for $^{48}_{24}\text{Cr}$ is $(48 - 24)/24 = 1.00$. This number is below the band of stability because N is too low and Z is too high. To become more stable, the nucleus decays by converting a proton to a neutron, which is positron decay. Alternatively, a nucleus can capture an electron and convert a proton into a neutron through electron capture.
c) The N/Z ratio for $^{50}_{25}\text{Mn}$ is $(50 - 25)/25 = 1.00$. This number is below the band of stability, so the nuclide undergoes positron decay or electron capture.

24.22   Plan: Calculate the N/Z ratio for each nuclide. A neutron-rich nuclide decays to convert neutrons to protons while a neutron-poor nuclide decays to convert protons to neutrons. Neutron-rich nuclides, with a high N/Z, undergo β decay. Neutron-poor nuclides, with a low N/Z, undergo positron decay or electron capture. For $Z < 20$, β^{+} emission is more common; for $Z > 80$, e^{-} capture is more common. Alpha decay is the most common means of decay for a heavy, unstable nucleus ($Z > 83$).

Solution:

a) For carbon-15, $N/Z = 9/6 = 1.5$, so the nuclide is neutron-rich. To decrease the number of neutrons and increase the number of protons, carbon-15 decays by beta decay.

b) The N/Z ratio for ^{120}Xe is $66/54 = 1.2$. Around atomic number 50, the ratio for stable nuclides is larger than 1.2, so ^{120}Xe is proton-rich. To decrease the number of protons and increase the number of neutrons, the xenon-120 nucleus either undergoes positron emission or electron capture.

c) Thorium-224 has an N/Z ratio of $134/90 = 1.5$. All nuclides of elements above atomic number 83 are unstable and decay to decrease the number of both protons and neutrons. Alpha decay by thorium-224 is the most likely mode of decay.

24.24   Plan: Stability results from a favorable N/Z ratio, even numbers of N and/or Z, and the occurrence of magic numbers.
Solution:

The N/Z ratio of $^{52}_{24}\text{Cr}$ is $(52 - 24)/24 = 1.17$, which is within the band of stability. The fact that Z is even does not account for the variation in stability because all isotopes of chromium have the same Z. However, $^{52}_{24}\text{Cr}$ has 28 neutrons, so N is both an even number and a magic number for this isotope only.

24.28   The equation for the nuclear reaction is $^{235}_{92}\text{U} \rightarrow {}^{207}_{82}\text{Pb} + \underline{\quad} {}^{0}_{-1}\beta + \underline{\quad} {}^{4}_{2}\text{He}$

To determine the coefficients, notice that the beta particles will not impact the mass number. Subtracting the mass number for lead from the mass number for uranium will give the total mass number for the alpha particles released, $235 - 207 = 28$. Each alpha particle is a helium nucleus with mass number 4. The number of helium atoms is determined by dividing the total mass number change by 4, $28/4 = 7$ helium atoms or seven alpha particles. The equation is now

$$^{235}_{92}\text{U} \rightarrow {}^{207}_{82}\text{Pb} + \underline{\quad} {}^{0}_{-1}\beta + 7\,{}^{4}_{2}\text{He}$$

To find the number of beta particles released, examine the difference in number of protons (atomic number) between the reactant and products. Uranium, the reactant, has 92 protons. The atomic number in the products, lead atom and 7 helium nuclei, total 96. To balance the atomic numbers, four electrons (beta particles) must be emitted to give the total atomic number for the products as $96 - 4 = 92$, the same as the reactant. In summary, seven alpha particles and four beta particles are emitted in the decay of uranium-235 to lead-207.

$$^{235}_{92}\text{U} \rightarrow {}^{207}_{82}\text{Pb} + {}^{0}_{-1}\beta + 7\,{}^{4}_{2}\text{He}$$

24.31   No, it is not valid to conclude that $t_{1/2}$ equals 1 min because the number of nuclei is so small (six nuclei). Decay rate is an average rate and is only meaningful when the sample is macroscopic and contains a large number of nuclei, as in the second case. Because the second sample contains 6×10^{12} nuclei, the conclusion that $t_{1/2} = 1$ min is valid.

24.33   Plan: Specific activity of a radioactive sample is its decay rate per gram. Calculate the specific activity by dividing the number of particles emitted per second (disintegrations per second = dps) by the mass of the sample. Convert disintegrations per second to Ci by using the conversion factor between the two units.
Solution:
$1 \text{ Ci} = 3.70 \times 10^{10} \text{ dps}$

$$\text{Specific activity (Ci/g)} = \left(\frac{1.56 \times 10^6 \text{ dps}}{1.65 \text{ mg}} \right)\left(\frac{1 \text{ mg}}{10^{-3} \text{ g}} \right)\left(\frac{1 \text{ Ci}}{3.70 \times 10^{10} \text{ dps}} \right) = 2.55528 \times 10^{-2} = \mathbf{2.56 \times 10^{-2} \text{ Ci/g}}$$

24.35   Plan: Specific activity of a radioactive sample is its decay rate per gram. Calculate the specific activity by dividing the number of particles emitted per second (disintegrations per second = dps) by the mass of the sample. Convert disintegrations per second to Bq by using the conversion factor between the two units.

Solution:
A becquerel is a disintegration per second (dps).

$$\text{Specific activity (Bq/g)} = \left|\dfrac{\left(\dfrac{7.4\times10^4\ d}{min}\right)\left(\dfrac{1\ min}{60\ s}\right)}{8.58\ \mu g\left(\dfrac{10^{-6}\ g}{1\ \mu g}\right)}\right|\left(\dfrac{1\ Bq}{1\ dps}\right) = 1.43745\times10^8 = \mathbf{1.4\times10^8\ Bq/g}$$

24.37 Plan: The decay constant is the rate constant for the first-order reaction.
Solution:

$$\text{Decay rate} = -\dfrac{\Delta N}{\Delta t} = kN$$

$$-\dfrac{-1\ atom}{day} = k(1\times10^{12}\ atom)$$

$$k = \mathbf{1\times10^{-12}\ d^{-1}}$$

24.39 Plan: The rate constant, k, relates the number of radioactive nuclei to their decay rate through the equation $A = kN$. The number of radioactive nuclei is calculated by converting moles to atoms using Avogadro's number. The decay rate is 1.39×10^5 atoms/yr or more simply, $1.39\times10^5\ \text{yr}^{-1}$ (the disintegrations are assumed).
Solution:

$$\text{Decay rate} = A = -\dfrac{\Delta N}{\Delta t} = kN$$

$$-\dfrac{-1.39\times10^5\ atoms}{1.00\ yr} = k\left(\dfrac{1.00\times10^{-12}\ mol}{}\right)\left(\dfrac{6.022\times10^{23}\ atoms}{1\ mol}\right)$$

$$1.39\times10^5\ atom/yr = k(6.022\times10^{11}\ atom)$$

$$k = (1.39\times10^5\ atom/yr)/6.022\times10^{11}\ atom$$

$$k = 2.30820\times10^{-7} = \mathbf{2.31\times10^{-7}\ yr^{-1}}$$

24.41 Plan: Radioactive decay is a first-order process, so the integrated rate law is $\ln N_t = \ln N_0 - kt$
First find the value of k from the half-life and use the integrated rate law to find N_t. The time unit in the time and the k value must agree.
Solution:

$$t_{1/2} = 1.01\ yr \qquad t = 3.75\times10^3\ h$$

$$t_{1/2} = \dfrac{\ln 2}{k} \quad \text{or } k = \dfrac{\ln 2}{t_{1/2}}$$

$$k = \dfrac{\ln 2}{1.01\ yr} = 0.686284\ yr^{-1}$$

$$\ln N_t = \ln N_0 - kt$$

$$\ln N_t = \ln [2.00\ mg] - (0.686284\ yr^{-1})(3.75\times10^3\ h)\left(\dfrac{1\ d}{24\ h}\right)\left(\dfrac{1\ yr}{365\ d}\right)$$

$$\ln N_t = 0.399361$$

$$N_t = e^{0.399361}$$

$$N_t = 1.49087 = \mathbf{1.49\ mg}$$

24.43 Plan: Lead-206 is a stable daughter of ^{238}U. Since all of the ^{206}Pb came from ^{238}U, the starting amount of ^{238}U was (270 μmol + 110 μmol) = 380 μmol = N_0. The amount of ^{238}U at time t (current) is 270 μmol = N_t. Find k from the first-order rate expression for half-life, and then substitute the values into the integrated rate law and solve for t.

Solution:

$$t_{1/2} = \frac{\ln 2}{k} \quad \text{or } k = \frac{\ln 2}{t_{1/2}}$$

$$k = \frac{\ln 2}{4.5 \times 10^9 \text{ yr}} = 1.540327 \times 10^{-10} \text{yr}^{-1}$$

$$\ln N_t = \ln N_0 - kt \qquad \text{or} \qquad \ln \frac{N_0}{N_t} = kt$$

$$\ln \frac{380 \text{ } \mu\text{mol}}{270 \text{ } \mu\text{mol}} = (1.540327 \times 10^{-10} \text{yr}^{-1})(t)$$

$$0.3417492937 = (1.540327 \times 10^{-10} \text{yr}^{-1})(t)$$

$$t = 2.21868 \times 10^9 = \mathbf{2.2 \times 10^9 \, yr}$$

24.45 **Plan:** The specific activity of the potassium-40 is the decay rate per mL of milk. Use the conversion factor $1 \text{ Ci} = 3.70 \times 10^{10}$ disintegrations per second (dps) to find the disintegrations per mL per s; convert the time unit to min and change the volume to 8 oz.

Solution:

$$\text{Activity} = \left(\frac{6 \times 10^{-11} \text{ mCi}}{\text{mL}}\right)\left(\frac{10^{-3} \text{ Ci}}{1 \text{ mCi}}\right)\left(\frac{3.70 \times 10^{10} \text{ dps}}{1 \text{ Ci}}\right)\left(\frac{60 \text{ s}}{1 \text{ min}}\right)\left(\frac{1000 \text{ mL}}{1.057 \text{ qt}}\right)\left(\frac{1 \text{ qt}}{4 \text{ cups}}\right)\left(\frac{1 \text{ cup}}{8 \text{ oz}}\right)(8 \text{ oz})$$

$$= 31.50426 = \mathbf{30 \, dpm}$$

24.47 **Plan:** Both N_t and N_0 are given: the number of nuclei present currently, N_t, is found from the moles of ^{232}Th. Each fission track represents one nucleus that disintegrated, so the number of nuclei disintegrated is added to the number of nuclei currently present to determine the initial number of nuclei, N_0. The rate constant, k, is calculated from the half-life. All values are substituted into the first-order decay equation to find t.

Solution:

$$t_{1/2} = \frac{\ln 2}{k} \quad \text{or } k = \frac{\ln 2}{t_{1/2}}$$

$$k = \frac{\ln 2}{1.4 \times 10^{10} \text{ yr}} = 4.95105129 \times 10^{-11} \text{yr}^{-1}$$

$$N_t = (3.1 \times 10^{-15} \text{ mol Th})\left(\frac{6.022 \times 10^{23} \text{ Th atoms}}{1 \text{ mol Th}}\right) = 1.86682 \times 10^9 \text{ atoms Th}$$

$$N_0 = 1.86682 \times 10^9 \text{ atoms} + 9.5 \times 10^4 \text{ atoms} = 1.866915 \times 10^9 \text{ atoms}$$

$$\ln N_t = \ln N_0 - kt \qquad \text{or} \qquad \ln \frac{N_0}{N_t} = kt$$

$$\ln \frac{1.866915 \times 10^9 \text{ atoms}}{1.86682 \times 10^9 \text{ atoms}} = (4.95105129 \times 10^{-11} \text{yr}^{-1})(t)$$

$$5.088738 \times 10^{-5} = (4.95105129 \times 10^{-11} \text{yr}^{-1})(t)$$

$$t = 1.027809 \times 10^6 = \mathbf{1.0 \times 10^6 \, yr}$$

24.50 Both gamma radiation and neutron beams have no charge, so neither is deflected by electric or magnetic fields. Neutron beams differ from gamma radiation in that a neutron has mass approximately equal to that of a proton. Researchers observed that a neutron beam could induce the emission of protons from a substance. Gamma rays do not cause such emissions.

24.52   Protons are repelled from the target nuclei due to the interaction of like (positive) charges. Higher energy is required to overcome the repulsion.

24.53   Plan: In a balanced nuclear equation, the total of mass numbers and the total of charges on the left side and the right side must be equal. In the shorthand notation, the nuclide to the left of the parentheses is the reactant while the nuclide written to the right of the parentheses is the product. The first particle inside the parentheses is the projectile particle while the second substance in the parentheses is the ejected particle.
Solution:
a) An alpha particle is a reactant with ^{10}B and a neutron is one product. The mass number for the reactants is $10 + 4 = 14$. So, the missing product must have a mass number of $14 - 1 = 13$. The total atomic number for the reactants is $5 + 2 = 7$, so the atomic number for the missing product is 7.

$$^{10}_{5}B \; + \; ^{4}_{2}He \; \rightarrow \; ^{1}_{0}n \; + \; ^{13}_{7}N$$

b) A deuteron (^{2}H) is a reactant with ^{28}Si and ^{29}P is one product. For the reactants, the mass number is $28 + 2 = 30$ and the atomic number is $14 + 1 = 15$. The given product has mass number 29 and atomic number 15, so the missing product particle has mass number 1 and atomic number 0. The particle is thus a neutron.

$$^{28}_{14}Si \; + \; ^{2}_{1}H \; \rightarrow \; ^{1}_{0}n \; + \; ^{29}_{15}P$$

c) The products are two neutrons and ^{244}Cf with a total mass number of $2 + 244 = 246$, and an atomic number of 98. The given reactant particle is an alpha particle with mass number 4 and atomic number 2. The missing reactant must have mass number of $246 - 4 = 242$ and atomic number $98 - 2 = 96$. Element 96 is Cm.

$$^{242}_{96}Cm \; + \; ^{4}_{2}He \; \rightarrow 2 \; ^{1}_{0}n \; + \; ^{244}_{98}Cf$$

24.58   Ionizing radiation is more dangerous to children because their rapidly dividing cells are more susceptible to radiation than an adult's slowly dividing cells.

24.60   Plan: The rad is the amount of radiation energy absorbed in J per body mass in kg: 1 rad = 0.01 J/kg. Change the mass unit from pounds to kilograms. The conversion factor between rad and gray is 1 rad = 0.01 Gy.
Solution:

a) Dose (rad) $= \left(\dfrac{3.3 \times 10^{-7} \, J}{135 \, lb} \right) \left(\dfrac{2.205 \, lb}{1 \, kg} \right) \left(\dfrac{1 \, rad}{1 \times 10^{-2} \, J/kg} \right) = 5.39 \times 10^{-7} = \mathbf{5.4 \times 10^{-7} \, rad}$

b) Gray (rad) $= \left(5.39 \times 10^{-7} \, rad \right) \left(\dfrac{0.01 \, gy}{1 \, rad} \right) = 5.39 \times 10^{-9} = \mathbf{5.4 \times 10^{-9} \, Gy}$

24.62   Plan: Multiply the number of particles by the energy of one particle to obtain the total energy absorbed. Convert the energy to dose in grays with the conversion factor 1 rad = 0.01 J/kg = 0.01 Gy. To find the millirems, convert grays to rads and multiply rads by RBE to find rems. Convert rems to mrems. Convert the dose to sieverts with the conversion factor 1 rem = 0.01 Sv.
Solution:

a) Energy (J) absorbed $= \left(6.0 \times 10^{5} \, \beta \right) \left(8.74 \times 10^{-14} \, J/\beta \right) = 5.244 \times 10^{-8} \, J$

Dose (Gy) $= \dfrac{5.244 \times 10^{-8} \, J}{70. \, kg} \left(\dfrac{1 \, rad}{0.01 \, J/kg} \right) \left(\dfrac{0.01 \, Gy}{1 \, rad} \right) = 7.4914 \times 10^{-10} = \mathbf{7.5 \times 10^{-10} \, Gy}$

b) rems = rads x RBE $= \left(7.4914 \times 10^{-10} \, Gy \right) \left(\dfrac{1 \, rad}{0.01 \, Gy} \right) (1.0) \left(\dfrac{1 \, mrem}{10^{-3} \, rem} \right) = 7.4914 \times 10^{-5} = \mathbf{7.5 \times 10^{-5} \, mrem}$

Sv $= \left(7.4914 \times 10^{-5} \, mrem \right) \left(\dfrac{10^{-3} \, rem}{1 \, mrem} \right) \left(\dfrac{0.01 \, Sv}{1 \, rem} \right) = 7.4914 \times 10^{-10} = \mathbf{7.5 \times 10^{-10} \, Sv}$

the time and disintegrations per second (Bq) to find the number of ^{60}Co atoms that disintegrate, which equals the number of β particles emitted. The dose in rads is calculated as energy absorbed per body mass.

$$\text{Dose} = \left(\frac{475 \text{ Bq}}{1.858 \text{ g}}\right)\left(\frac{10^3 \text{ g}}{1 \text{ kg}}\right)\left(\frac{1 \text{ dps}}{1 \text{ Bq}}\right)\left(\frac{5.05 \times 10^{-14} \text{ J}}{1 \text{ disint.}}\right)(24.0 \text{ min})\left(\frac{60 \text{ s}}{1 \text{ min}}\right)\left(\frac{1 \text{ rad}}{0.01 \text{ J}/\text{kg}}\right)$$

$$= 1.8591 \times 10^{-3} = \textbf{1.86} \times \textbf{10}^{-3} \textbf{ rad}$$

24.67   NAA does not destroy the sample while chemical analysis does. Neutrons bombard a non-radioactive sample, "activating" or energizing individual atoms within the sample to create radioisotopes. The radioisotopes decay back to their original state (thus, the sample is not destroyed) by emitting radiation that is different for each isotope.

24.73   Energy is released when a nuclide forms from nucleons. The nuclear binding energy is the amount of energy holding the nucleus together. Energy is absorbed to break the nucleus into nucleons and is released when nucleons "come together."

24.75   Plan: The conversion factors are: $1 \text{ MeV} = 10^6 \text{ eV}$ and $1 \text{ eV} = 1.602 \times 10^{-19} \text{ J}$.
Solution:

a) Energy (eV) $= (0.01861 \text{ MeV})\left(\dfrac{10^6 \text{ eV}}{1 \text{ MeV}}\right) = \textbf{1.861} \times \textbf{10}^{4} \textbf{ eV}$

b) Energy (J) $= (0.01861 \text{ MeV})\left(\dfrac{10^6 \text{ eV}}{1 \text{ MeV}}\right)\left(\dfrac{1.602 \times 10^{-19} \text{ J}}{1 \text{ eV}}\right) = 2.981322 \times 10^{-15} = \textbf{2.981} \times \textbf{10}^{-15} \textbf{ J}$

24.77   Plan: Convert moles of ^{239}Pu to atoms of ^{239}Pu using Avogadro's number. Multiply the number of atoms by the energy per atom (nucleus) and convert the MeV to J using the conversion $1 \text{ eV} = 1.602 \times 10^{-19}$ J.
Solution:

$$\text{Number of atoms} = \left(1.5 \text{ mol}^{239}\text{Pu}\right)\left(\frac{6.022 \times 10^{23} \text{ atoms}}{\text{mol}}\right) = 9.033 \times 10^{23} \text{ atoms}$$

$$\text{Energy (J)} = \left(9.033 \times 10^{23} \text{ atoms}\right)\left(\frac{5.243 \text{ MeV}}{1 \text{ atom}}\right)\left(\frac{10^6 \text{ eV}}{1 \text{ MeV}}\right)\left(\frac{1.602 \times 10^{-19} \text{ J}}{1 \text{ eV}}\right) = 7.587075 \times 10^{11} = \textbf{7.6} \times \textbf{10}^{11} \textbf{ J}$$

24.79   Plan: Oxygen-16 has eight protons and eight neutrons. First find the Δm for the nucleus by subtracting the given mass of one oxygen atom from the sum of the masses of eight ^{1}H atoms and eight neutrons. Use the conversion factor $1 \text{ amu} = 931.5 \text{ MeV}$ to convert Δm to binding energy in MeV and divide the binding energy by the total number of nucleons (protons and neutrons) in the oxygen nuclide to obtain binding energy per nucleon. Convert Δm of one oxygen atom to MeV using the conversion factor for binding energy/atom. To obtain binding energy per mole of oxygen, use the relationship $\Delta E = \Delta mc^2$. Δm must be converted to units of kg/mol.
Solution:

Mass of 8 ^{1}H atoms = 8 x 1.007825 = 8.062600 amu

Mass of 8 neutrons = 8 x 1.008665 = 8.069320 amu

Total mass = 16.131920 amu

$\Delta m = 16.131920 - 15.994915 = 0.137005$ amu/^{16}O = 0.137005 g/mol ^{16}O

a) Binding energy (MeV/nucleon) $= \left(\dfrac{0.137005 \text{ amu} \,^{16}\text{O}}{16 \text{ nucleons}}\right)\left(\dfrac{931.5 \text{ MeV}}{1 \text{ amu}}\right) = 7.976259844 = \textbf{7.976 MeV/nucleon}$

b) Binding energy (MeV/atom) $= \left(\dfrac{0.137005 \text{ amu} \,^{16}\text{O}}{1 \text{ atom}}\right)\left(\dfrac{931.5 \text{ MeV}}{1 \text{ amu}}\right) = 127.6201575 = \textbf{127.6 MeV/atom}$

c) $\Delta E = \Delta mc^2$

$$\text{Binding energy (kJ/mol)} = \left(\frac{0.137005 \text{ g}\ ^{16}\text{O}}{\text{mol}}\right)\left(\frac{1 \text{ kg}}{10^3 \text{ g}}\right)\left(2.99792\text{x}10^8 \text{ m/s}\right)^2\left(\frac{1 \text{ J}}{\text{kg}\bullet\text{m}^2\big/\text{s}^2}\right)\left(\frac{1 \text{ kJ}}{10^3 \text{ J}}\right)$$

$$= 1.23133577\text{x}10^{10} = \mathbf{1.23134\text{x}10^{10}\ kJ/mol}$$

24.81   Plan: Cobalt-59 has 27 protons and 32 neutrons. First find the Δm for the nucleus by subtracting the given mass of one cobalt atom from the sum of the masses of 27 ^{1}H atoms and 32 neutrons. Use the conversion factor 1 amu = 931.5 MeV to convert Δm to binding energy in MeV and divide the binding energy by the total number of nucleons (protons and neutrons) in the cobalt nuclide to obtain binding energy per nucleon. Convert Δm of one cobalt atom to MeV using the conversion factor for binding energy/atom. To obtain binding energy per mole of cobalt, use the relationship $\Delta E = \Delta mc^2$. Δm must be converted to units of kg/mol.
Solution:
Mass of 27 ^{1}H atoms = 27 x 1.007825 = 27.211275 amu

Mass of 32 neutrons = 32 x 1.008665 = 32.27728 amu

Total mass = 59.488555 amu

$\Delta m = 59.488555 - 58.933198 = 0.555357$ amu/^{59}Co $= 0.555357$ g/mol^{59}Co

a) Binding energy (MeV/nucleon) $= \left(\dfrac{0.555357 \text{ amu}\ ^{59}\text{Co}}{59 \text{ nucleons}}\right)\left(\dfrac{931.5 \text{ MeV}}{1 \text{ amu}}\right) = 8.768051619 = \mathbf{8.768\ MeV/nucleon}$

b) Binding energy (MeV/atom) $= \left(\dfrac{0.555357 \text{ amu}\ ^{59}\text{Co}}{1 \text{ atom}}\right)\left(\dfrac{931.5 \text{ MeV}}{1 \text{ amu}}\right) = 517.3150 = \mathbf{517.3\ MeV/atom}$

c) Use $\Delta E = \Delta mc^2$

$$\text{Binding energy (kJ/mol)} = \left(\frac{0.555357 \text{ g}\ ^{59}\text{Co}}{\text{mol}}\right)\left(\frac{1 \text{ kg}}{10^3 \text{ g}}\right)\left(2.99792\text{x}10^8 \text{ m/s}\right)^2\left(\frac{1 \text{ J}}{\text{kg}\bullet\text{m}^2\big/\text{s}^2}\right)\left(\frac{1 \text{ kJ}}{10^3 \text{ J}}\right)$$

$$= 4.9912845\text{x}10^{10} = \mathbf{4.99128\text{x}10^{10}\ kJ/mol}$$

24.85   In both radioactive decay and fission, radioactive particles are emitted, but the process leading to the emission is different. Radioactive decay is a spontaneous process in which unstable nuclei emit radioactive particles and energy. Fission occurs as the result of high-energy bombardment of nuclei with small particles that cause the nuclides to break into smaller nuclides, radioactive particles, and energy.
In a chain reaction, all fission events are not the same. The collision between the small particle emitted in the fission and the large nucleus can lead to splitting of the large nuclei in a number of ways to produce several different products.

24.88   The water serves to slow the neutrons so that they are better able to cause a fission reaction. Heavy water ($_1^2$H$_2$O or D$_2$O) is a better moderator because it does not absorb neutrons as well as light water ($_1^1$H$_2$O) does, so more neutrons are available to initiate the fission process. However, D$_2$O does not occur naturally in great abundance, so production of D$_2$O adds to the cost of a heavy water reactor. In addition, if heavy water does absorb a neutron, it becomes *tritiated*, i.e., it contains the isotope tritium, $_1^3$H, which is radioactive.

24.93   Plan: Use the masses given in the problem to calculate the mass change (reactant – products) for the reaction. The conversion factor between amu and kg is 1 amu = 1.66054x10^{-27} kg. Use the relationship $E = \Delta mc^2$ to convert the mass change to energy.

Solution:

a) $^{243}_{96}Cm \rightarrow ^{239}_{94}Pu + ^{4}_{2}He$

Δm (amu) = 243.0614 amu − (4.0026 + 239.0522) amu = 0.0066 amu

Δm (kg) = (0.0066 amu)$\left(\dfrac{1.661\times10^{-24}\text{ g}}{1\text{ amu}}\right)\left(\dfrac{1\text{ kg}}{10^3\text{ g}}\right)$ = 1.09626×10^{-29} = $\mathbf{1.1\times10^{-29}}$ **kg**

b) $E = \Delta mc^2 = \left(1.09626\times10^{-29}\text{ kg}\right)\left(2.99792\times10^8\text{ m/s}\right)^2\left(\dfrac{1\text{ J}}{kg\cdot m^2/s^2}\right)$ = 9.85266×10^{-13} = $\mathbf{9.9\times10^{-13}}$ **J**

c) E released = $\left(\dfrac{9.85266\times10^{-13}\text{ J}}{\text{reaction}}\right)\left(\dfrac{6.022\times10^{23}\text{ reactions}}{\text{mol}}\right)\left(\dfrac{1\text{ kJ}}{10^3\text{ J}}\right)$ = 5.93317×10^8 = $\mathbf{5.9\times10^8}$ **kJ/mol**

This is approximately one million times larger than a typical heat of reaction.

24.95   Plan: Determine k for ^{14}C using the half-life (5730 yr). Determine the mass of carbon in 4.58 g of $CaCO_3$.
Divide the given activity of the C in d/min by the mass of carbon to obtain the activity in d/min•g; this is A_t and is compared to the activity of a living organism (A_0 = 15.3 d/min•g) in the integrated rate law, solving for t.
Solution:

$k = \dfrac{\ln 2}{t_{1/2}} = \dfrac{\ln 2}{5730\text{ yr}} = 1.2096809\times10^{-4}\text{yr}^{-1}$

Mass (g) of C = $\left(4.58\text{ g CaCO}_3\right)\left(\dfrac{1\text{ mol CaCO}_3}{100.09\text{ g CaCO}_3}\right)\left(\dfrac{1\text{ mol C}}{1\text{ mol CaCO}_3}\right)\left(\dfrac{12.01\text{ g C}}{1\text{ mol C}}\right)$ = 0.5495634 g C

$A_t = \dfrac{3.2\text{ d/min}}{0.5495634\text{ g}}$ = 5.8228 d/min•g

Using the integrated rate law:

$\ln\left(\dfrac{A_t}{A_0}\right) = -kt$ A_0 = 15.3 d/min•g (the ratio of $^{12}C{:}^{14}C$ in living organisms)

$\ln\left(\dfrac{5.8228\text{ d/min}\cdot\text{g}}{15.3\text{ d/min}\cdot\text{g}}\right) = -(1.2096809\times10^{-4}\text{yr}^{-1})(t)$

$t = 7986.17 = \mathbf{8.0\times10^3 yr}$

24.98   Plan: Determine how many grams of AgCl are dissolved in 1 mL of solution. The activity of the radioactive Ag^+ indicates how much AgCl dissolved, given a starting sample with a specific activity (175 nCi/g). Convert g/mL to mol/L (molar solubility) using the molar mass of AgCl.
Solution:

Concentration = $\left(\dfrac{1.25\times10^{-2}\text{ Bq}}{\text{mL}}\right)\left(\dfrac{1\text{ dps}}{1\text{ Bq}}\right)\left(\dfrac{1\text{ Ci}}{3.70\times10^{10}\text{ dps}}\right)\left(\dfrac{1\text{ nCi}}{10^{-9}\text{ Ci}}\right)\left(\dfrac{1\text{ g AgCl}}{175\text{ nCi}}\right)$ = 1.93050×10^{-6} g AgCl/mL

Molarity = $\left(\dfrac{1.93050\times10^{-6}\text{g AgCl}}{\text{mL}}\right)\left(\dfrac{1\text{ mol AgCl}}{143.4\text{ g AgCl}}\right)\left(\dfrac{1\text{ mL}}{10^{-3}\text{ L}}\right)$ = 1.34623×10^{-5} = $\mathbf{1.35\times10^{-5}}M$ **AgCl**

24.100 Plan: Determine the value of k from the half-life. Then determine the fraction from the integrated rate law.
Solution:

$k = \dfrac{\ln 2}{t_{1/2}} = \dfrac{\ln 2}{7.0\times10^8\text{ yr}} = 9.90210\times10^{-10}\text{ yr}^{-1}$

$\ln\dfrac{N_0}{N_t} = kt = (9.90210\times10^{-10}\text{ yr}^{-1})(2.8\times10^9\text{ yr}) = 2.772588$

$$\frac{N_0}{N_t} = 15.99998844$$

$$\frac{N_t}{N_0} = 0.062500 = \mathbf{6.2 \times 10^{-2}}$$

24.102 **Plan:** Find the rate constant, k, using any two data pairs (the greater the time between the data points, the greater the reliability of the calculation). Calculate $t_{1/2}$ using k. Once k is known, use the integrated rate law to find the percentage lost after 2 h. The percentage of isotope *remaining* is the fraction remaining after 2.0 h (N_t, where $t = 2.0$ h) divided by the initial amount (N_0), i.e., fraction remaining is N_t/N_0. Solve the first-order rate expression for N_t/N_0, and then subtract from 100% to get fraction *lost*.
Solution:

a) $\ln \dfrac{N_t}{N_0} = -kt$

$\ln \left(\dfrac{495 \text{ photons/s}}{5000 \text{ photons/s}} \right) = -k(20 \text{ h})$

$-2.312635 = -k(20 \text{ h})$

$k = 0.11563 \text{ h}^{-1}$

$t_{1/2} = \dfrac{\ln 2}{k} = \dfrac{\ln 2}{0.11563 \text{ h}^{-1}} = 5.9945 = \mathbf{5.99 \text{ h}}$ (Assuming the times are exact, and the emissions have three significant figures.)

b) $\ln \dfrac{N_t}{N_0} = -kt$

$\ln \dfrac{N_t}{N_0} = -(0.11563 \text{ h}^{-1})(2.0 \text{ h}) = -0.23126$

$\dfrac{N_t}{N_0} = 0.793533$ $\qquad\qquad$ $\dfrac{N_t}{N_0} \times 100\% = 79.3533\%$

The fraction lost upon preparation is $100\% - 79.3533\% = 20.6467\% = \mathbf{21\%}$.

24.104 **Plan:** Use the given relationship for the fraction remaining after time t, where $t = 10.0$ yr, 10.0×10^3 yr, and 10.0×10^4 yr.
Solution:

a) Fraction remaining after 10.0 yr $= \left(\frac{1}{2}\right)^{t/t_{1/2}} = \left(\frac{1}{2}\right)^{10.0/5730} = 0.998791 = \mathbf{0.999}$

b) Fraction remaining after 10.0×10^3 yr $= \left(\frac{1}{2}\right)^{10.0 \times 10^3/5730} = 0.298292 = \mathbf{0.298}$

c) Fraction remaining after 10.0×10^4 yr $= \left(\frac{1}{2}\right)^{10.0 \times 10^4/5730} = 5.5772795 \times 10^{-6} = \mathbf{5.58 \times 10^{-6}}$

d) Radiocarbon dating is more reliable for b) because a significant quantity of ^{14}C has decayed and a significant quantity remains. Therefore, a change in the amount of ^{14}C would be noticeable. For the fraction in a), very little ^{14}C has decayed and for c) very little ^{14}C remains. In either case, it will be more difficult to measure the change so the error will be relatively large.

24.106 **Plan:** At one half-life, the fraction of sample is 0.500. Find n for which $(0.900)^n = 0.500$.
Solution:
$(0.900)^n = 0.500$

$n \ln (0.900) = \ln (0.500)$

$n = (\ln 0.500)/(\ln 0.900) = 6.578813 = \mathbf{6.58 \text{ h}}$

___: The *production rate* of radon gas (volume/hour) is also the *decay rate* of ^{226}Ra. The decay rate, or activity, is proportional to the number of radioactive nuclei decaying, or the number of atoms in 1.000 g of ^{226}Ra, using the relationship $A = kN$. Calculate the number of atoms in the sample, and find k from the half-life. Convert the activity in units of nuclei/time (also disintegrations per unit time) to volume/time using the ideal gas law.

Solution:

$$^{226}_{88}\text{Ra} \rightarrow {}^{4}_{2}\text{He} + {}^{222}_{86}\text{Rn}$$

$$k = \frac{\ln 2}{t_{1/2}} = \frac{\ln 2}{1599 \text{ yr}} = 4.33879178\text{x}10^{-4} \text{ yr}^{-1}(1 \text{ yr}/8766 \text{ h}) = 4.94510515\text{x}10^{-8} \text{ h}^{-1}$$

The mass of ^{226}Ra is 226.025402 amu/atom or 226.025402 g/mol.

$$N = \left(1.000 \text{ g Ra}\right)\left(\frac{1 \text{ mol Ra}}{226.025402 \text{ g Ra}}\right)\left(\frac{6.022\text{x}10^{23} \text{ Ra atoms}}{1 \text{ mol Ra}}\right) = 2.6643023\text{x}10^{21} \text{ Ra atoms}$$

$$A = kN = (4.94510515\text{x}10^{-8} \text{ h}^{-1})(2.6643023\text{x}10^{21} \text{ Ra atoms}) = 1.3175255\text{x}10^{14} \text{ Ra atoms/h}$$

This result means that $1.318\text{x}10^{14}$ ^{226}Ra nuclei are decaying into ^{222}Rn nuclei every hour. Convert atoms of ^{222}Rn into volume of gas using the ideal gas law.

$$\text{Moles of Rn/h} = \left(\frac{1.3175255\text{x}10^{14} \text{ Ra atoms}}{\text{h}}\right)\left(\frac{1 \text{ atom Rn}}{1 \text{ atom Ra}}\right)\left(\frac{1 \text{ mol Rn}}{6.022\text{x}10^{23} \text{ Rn atoms}}\right)$$

$$= 2.1878537\text{x}10^{-10} \text{ mol Rn/h}$$

$$V = \frac{nRT}{P} = \frac{\left(2.1878537\text{x}10^{-10} \text{ mol Rn/h}\right)\left(0.08206 \frac{\text{L} \cdot \text{atm}}{\text{mol} \cdot \text{K}}\right)\left(273.15 \text{ K}\right)}{1 \text{ atm}}$$

$$= 4.904006\text{x}10^{-9} = \mathbf{4.904\text{x}10^{-9} \text{ L/h}}$$

Therefore, radon gas is produced at a rate of $4.904\text{x}10^{-9}$ L/h. Note: Activity could have been calculated as decay in moles/time, removing Avogadro's number as a multiplication and division factor in the calculation.

24.113 Plan: Determine k from the half-life and then use the integrated rate law, solving for time.

Solution:

$$k = \frac{\ln 2}{t_{1/2}} = \frac{\ln 2}{29 \text{ yr}} = 0.0239016 \text{ yr}^{-1}$$

$$\ln \frac{N_t}{N_0} = -kt$$

$$\ln \left(\frac{1.0\text{x}10^4 \text{ particles}}{7.0\text{x}10^4 \text{ particles}}\right) = -(0.0239016 \text{ yr}^{-1})(t)$$

$$-1.945910 = -(0.0239016 \text{ yr}^{-1})(t)$$

$$t = 81.413378 = \mathbf{81 \text{ yr}}$$

24.115 Plan: Convert pCi to Bq using the conversion factors 1 Ci = $3.70\text{x}10^{10}$ Bq and 1 pCi = 10^{-12} Ci. For part b), use the first-order integrated rate law to find the activity at the later time ($t = 9.5$ days). You will first need to calculate k from the half-life expression. For part c), solve for the time at which N_t = the EPA recommended level.

Solution:

a) Activity (Bq/L) = $\left(\frac{4.0 \text{ pCi}}{\text{L}}\right)\left(\frac{10^{-12} \text{ Ci}}{1 \text{ pCi}}\right)\left(\frac{3.70\text{x}10^{10} \text{ Bq}}{1 \text{ Ci}}\right) = 0.148 = 0.15 \text{ Bq/L}$

The safe level is **0.15 Bq/L.**

b) $k = \dfrac{\ln 2}{t_{1/2}} = \dfrac{\ln 2}{3.82\ \text{d}} = 0.181452\ \text{d}^{-1}$

$\ln \dfrac{N_t}{N_0} = -kt$

$\ln \dfrac{N_t}{41.5\ \text{pCi/L}} = -(0.181452\ \text{d}^{-1})(9.5\ \text{d}) = -1.723794$

$\dfrac{N_t}{41.5\ \text{pCi/L}} = 0.17838806$

$N_t = 7.403104\ \text{pCi/L}$

Activity (Bq/L) $= \left(\dfrac{7.403104\ \text{pCi}}{\text{L}}\right)\left(\dfrac{10^{-12}\ \text{Ci}}{1\ \text{pCi}}\right)\left(\dfrac{3.70\,\text{x}10^{10}\ \text{Bq}}{1\ \text{Ci}}\right) = 0.2739148 = \mathbf{0.27\ Bq/L}$

c) The desired activity is 0.15 Bq/L, however, the room air currently contains 0.27 Bq/L.

$\ln \dfrac{N_t}{N_0} = -kt$

$\ln \dfrac{0.15\ \text{Bq/L}}{0.2739148\ \text{Bq/L}} = -(0.181452\ \text{d}^{-1})(t)$

$-0.602182 = -(0.181452\ \text{d}^{-1})(t)$

$t = 3.318685 = \mathbf{3.3\ d}$

It takes 3.3 d more to reach the recommended EPA level. A total of 12.8 (3.3 + 9.5) d is required to reach recommended levels when the room was initially measured at 41.5 pCi/L.

24.118 <u>Plan:</u> Convert mCi to Ci to disintegrations per second; multiply the dps by the energy of each disintegration in MeV and convert to energy in J. Recall that 1 rad = 0.01 J/kg. The mass of the child must be converted from lb to kg.
<u>Solution:</u>

Energy (J/s) $= (1.0\ \text{mCi})\left(\dfrac{10^{-3}\ \text{Ci}}{1\ \text{mCi}}\right)\left(\dfrac{3.70\text{x}10^{10}\ \text{dps}}{1\ \text{Ci}}\right)\left(\dfrac{5.59\ \text{MeV}}{1\ \text{disint.}}\right)\left(\dfrac{1.602\text{x}10^{-13}\ \text{J}}{1\ \text{MeV}}\right) = 3.3134166\text{x}10^{-5}\ \text{J/s}$

Convert the mass of the child to kg:

$(54\ \text{lb})\left(\dfrac{1\ \text{kg}}{2.205\ \text{lb}}\right) = 24.4897960\ \text{kg}$

Time for 1.0 mrad to be absorbed:

Time (s) $= (1.0\ \text{mrad})\left(\dfrac{10^{-3}\ \text{rad}}{1\ \text{mrad}}\right)\left(\dfrac{0.01\ \text{J/kg}}{1\ \text{rad}}\right)\left(\dfrac{24.4897960\ \text{kg}}{3.3134166\text{x}10^{-5}\ \text{J/s}}\right) = 7.3911008 = \mathbf{7.4\ s}$

24.121 Because the 1941 wine has a little over twice as much tritium in it, just over one half-life has passed between the two wines. Therefore, the older wine was produced before 1929 (1941– 12.26) but not much earlier than that. To find the number of years back in time, use the first-order rate expression, where $N_0 = 2.32\ N$, $N_t = N$ and $t =$ years transpired between the manufacture date and 1941.

$k = \dfrac{\ln 2}{t_{1/2}} = \dfrac{\ln 2}{12.26\ \text{yr}} = 0.05653729\ \text{yr}^{-1}$

$\ln \dfrac{N_t}{N_0} = -kt$

$$\frac{N}{2.32\ N} = -(0.05653729\ \text{yr}^{-1})(t)$$

$$-0.841567 = -(0.05653729\ \text{yr}^{-1})(t)$$

$$t = 14.92857 = 14.9\ \text{yr}$$

The wine was produced in (1941 − 15) = **1926**.

24.124　Plan: Determine the change in mass for the reaction by subtracting the masses of the products from the masses of the reactants. Use conversion factors to convert the mass change in amu to energy in eV and then to J.
Solution:

Δm = mass of reactants − mass of products

　　= (14.003074 + 1.008665) − (14.003241 + 1.007825) = 0.000673 amu

$$\text{Energy (eV)} = (0.000673\ \text{amu})\left(\frac{931.5\ \text{MeV}}{1\ \text{amu}}\right)\left(\frac{10^6\ \text{eV}}{1\ \text{MeV}}\right) = 6.268995 \times 10^5 = \textbf{6.27} \times \textbf{10}^{\textbf{5}}\ \textbf{eV}$$

$$\text{Energy (kJ/mol)} = \left(6.268995 \times 10^5\ \text{eV}\right)\left(\frac{1.602 \times 10^{-19}\ \text{J}}{1\ \text{eV}}\right)\left(\frac{1\ \text{kJ}}{10^3\ \text{J}}\right)\left(\frac{6.022 \times 10^{23}}{1\ \text{mol}}\right)$$

$$= 6.0478524 \times 10^7 = \textbf{6.05} \times \textbf{10}^{\textbf{7}}\ \textbf{kJ/mol}$$

24.127　a) Kinetic energy $= 1/2\ mv^2 = \dfrac{3}{2}\left(\dfrac{R}{N_A}\right)T$

$$\text{Energy} = \left(\frac{3}{2}\right)\frac{\left(8.314\ \text{J/mol}\cdot\text{K}\right)\left(1.00 \times 10^6\ \text{K}\right)}{6.022 \times 10^{23}\ \text{atom/mol}} = 2.07090668 \times 10^{-17} = \textbf{2.07} \times \textbf{10}^{\textbf{−17}}\ \textbf{J/atom}\ {}_{1}^{1}\textbf{H}$$

b) A kilogram of ^{1}H will annihilate a kilogram of anti-H; thus, two kilograms will be converted to energy:

Energy $= mc^2 = (2.00\ \text{kg})(2.99792 \times 10^8\ \text{m/s})^2(\text{J}/(\text{kg}\bullet\text{m}^2/\text{s}^2)) = 1.7975048 \times 10^{17}\ \text{J}$

$$\text{Number of atoms} = \left(\frac{1.7975048 \times 10^{17}\ \text{J}}{1\ \text{kg H}}\right)\left(\frac{1\ \text{kg}}{10^3\ \text{g}}\right)\left(\frac{1.0078\ \text{g H}}{1\ \text{mol H}}\right)\left(\frac{1\ \text{mol H}}{6.022 \times 10^{23}\ \text{atoms H}}\right)\left(\frac{1\ \text{H atom}}{2.0709066 \times 10^{-17}\ \text{J}}\right)$$

$$= 1.4525903 \times 10^7 = \textbf{1.45} \times \textbf{10}^{\textbf{7}}\ \textbf{H atoms}$$

c) $4{}_{1}^{1}\text{H} \rightarrow {}_{2}^{4}\text{He} + 2{}_{1}^{0}\beta$　　(Positrons have the same mass as electrons.)

$\Delta m = [4(1.007825\ \text{amu})] - [4.00260\ \text{amu} + 2(0.000549\ \text{amu})] = 0.027602\ \text{amu}\ /{}_{2}^{4}\text{He}$

$$\Delta m = \left(\frac{0.027602\ \text{g}}{\text{mol He}}\right)\left(\frac{1\ \text{kg}}{10^3\ \text{g}}\right)\left(\frac{1\ \text{mol He}}{4\ \text{mol H}}\right)\left(\frac{1\ \text{mol H}}{6.022 \times 10^{23}\ \text{H atoms}}\right)\left(1.4525903 \times 10^7\ \text{H atoms}\right)$$

$$= 1.6644967 \times 10^{-22}\ \text{kg}$$

Energy $= (\Delta m)c^2 = (1.6644967 \times 10^{-22}\ \text{kg})\ (2.99792 \times 10^8\ \text{m/s})^2(\text{J}/(\text{kg}\bullet\text{m}^2/\text{s}^2)$

$$= 1.4959705 \times 10^{-5} = \textbf{1.4960} \times \textbf{10}^{\textbf{−5}}\ \textbf{J}$$

d) Calculate the energy generated in part b):

$$\text{Energy} = \left(\frac{1.7975048 \times 10^{17}\ \text{J}}{1\ \text{kg H}}\right)\left(\frac{1\ \text{kg}}{10^3\ \text{g}}\right)\left(\frac{1.0078\ \text{g H}}{1\ \text{mol H}}\right)\left(\frac{1\ \text{mol H}}{6.022 \times 10^{23}\ \text{atoms H}}\right) = 3.0081789 \times 10^{-10}\ \text{J}$$

Energy increase $= (1.4959705 \times 10^{-5} - 3.0081789 \times 10^{-10})\ \text{J} = 1.4959404 \times 10^{-5} = \textbf{1.4959} \times \textbf{10}^{\textbf{−5}}\ \textbf{J}$

e) $3{}_{1}^{1}\text{H} \rightarrow {}_{2}^{3}\text{He} + 1{}_{1}^{0}\beta$

$\Delta m = [3(1.007825\ \text{amu})] - [3.01603\ \text{amu} + 0.000549\ \text{amu}] = 0.006896\ \text{amu}/{}_{2}^{3}\text{He} = 0.006896\ \text{g/mol}\ {}_{2}^{3}\text{He}$

$$\Delta m = \left(\frac{0.006896 \text{ g}}{\text{mol He}}\right)\left(\frac{1 \text{ kg}}{10^3 \text{ g}}\right)\left(\frac{1 \text{ mol He}}{3 \text{ mol H}}\right)\left(\frac{1 \text{ mol H}}{6.022\text{x}10^{23} \text{ H atoms}}\right)\left(1.4525903\text{x}10^7 \text{ H atoms}\right)$$

$$= 5.54470426\text{x}10^{-23} \text{ kg}$$

Energy $= (\Delta m)c^2 = (5.54470426\text{x}10^{-23} \text{ kg})(2.99792\text{x}10^8 \text{ m/s})^2(\text{J/(kg}\bullet\text{m}^2/\text{s}^2)) = 4.9833164\text{x}10^{-6} = \textbf{4.983x10}^{-6}\textbf{ J}$

No, the Chief Engineer should advise the Captain to keep the current technology.

24.130 Plan: The difference in the energies of the two α particles gives the energy of the γ ray released to get from excited state I to the ground state. Use $E = hc/\lambda$ to determine the wavelength. The energy of the gamma ray must be converted from MeV to J. The energy of the 2% α particle is equal to the highest energy α particle minus the energy of the two γ rays (the rays are from excited state II to excited state I, and from excited state I to the ground state).
Solution:
a) Energy of the γ ray = (4.816 – 4.773) MeV = **0.043 MeV**

$$\lambda = \frac{hc}{E} = \frac{\left(6.626\text{x}10^{-34} \text{ J}\bullet\text{s}\right)\left(2.99792\text{x}10^8 \text{ m/s}\right)}{\left(0.043 \text{ MeV}\right)}\left(\frac{1 \text{ MeV}}{1.602\text{x}10^{-13} \text{ J}}\right) = 2.8836\text{x}10^{-11} = \textbf{2.9x10}^{-11}\textbf{ m}$$

b) (4.816 – 0.043 – 0.060) MeV = **4.713 MeV**

24.134 Plan: Multiply each of the half-lives by 20 (the number of half-lives is considered to be exact).
Solution:
a) ^{242}Cm   20(163 d) = **3.26x10³ d**

b) ^{214}Po   20(1.6x10⁻⁴ s) = **3.2x10⁻³ s**

c) ^{232}Th   20(1.39x10¹⁰ yr) = **2.78x10¹¹ yr**

24.136 a) When 1.00 kg of antimatter annihilates 1.00 kg of matter, the change in mass is:

$\Delta m = 0 – 2.00 \text{ kg} = -2.00 \text{ kg}$. The energy released is calculated from $\Delta E = \Delta mc^2$.

$$\Delta E = (-2.00 \text{ kg})(2.99792\text{x}10^8 \text{ m/s})^2(\text{J/(kg} \bullet \text{ m}^2/\text{s}^2)) = -1.7975049\text{x}10^{17} = \textbf{-1.80x10}^{17}\textbf{ J}$$

The negative value indicates the energy is released.
b) Assuming that four hydrogen atoms fuse to form the two protons and two neutrons in one helium atom and release two positrons, the energy released can be calculated from the binding energy of helium-4.

$$4\,^1_1\text{H} \rightarrow\,^4_2\text{He} + 2\,^0_{+1}\beta$$

$\Delta m = [4(1.007825 \text{ amu})] – [4.00260 \text{ amu} + 2(0.000549 \text{ amu})]$

$= 0.02760 \text{ amu per }\,^4_2\text{He}$ formed

$$\text{Total } \Delta m = \left(1.00\text{x}10^5 \text{ H atoms}\right)\left(\frac{0.02760 \text{ amu}}{^4\text{He}}\right)\left(\frac{1\,^4 \text{He}}{4 \text{ H atoms}}\right) = 6.90\text{x}10^2 \text{ amu}$$

$$\text{Energy} = \left(6.90 \text{ x } 10^2 \text{ amu}\right)\left(\frac{931.5 \text{ MeV}}{1 \text{ amu}}\right)\left(\frac{1.602\text{x}10^{-16} \text{ kJ}}{1 \text{ MeV}}\right) = 1.02966\text{x}10^{-10} \text{ kJ per antiH collision}$$

$$\text{AntiH atoms} = \left(1.00 \text{ kg}\right)\left(\frac{10^3 \text{ g}}{1 \text{ kg}}\right)\left(\frac{1 \text{ mol antiH}}{1.008 \text{ g antiH}}\right)\left(\frac{6.022\text{x}10^{23} \text{ antiH}}{1 \text{ mol antiH}}\right) = 5.9742\text{x}10^{26} \text{ antiH}$$

$$\text{Energy released} = \left(\frac{1.02966\text{x}10^{-10} \text{ kJ}}{\text{antiH}}\right)\left(5.9742\text{x}10^{26} \text{ antiH}\right) = 6.15139\text{x}10^{16} = \textbf{6.15x10}^{16}\textbf{ kJ}$$

c) From the above calculations, the procedure in part b) with excess hydrogen produces more energy per kilogram of antihydrogen.

Einstein's equation is $E = mc^2$, which is modified to $E = \Delta mc^2$ to reflect a mass difference. The speed of light, c, is 2.99792×10^8 m/s. The mass of exactly one amu is 1.66054×10^{-27} kg (inside back cover of text). When the quantities are multiplied together, the unit will be kg•m^2/s^2, which is also the unit of joules. Convert J to MeV using the conversion factor 1.602×10^{-13} J = 1 MeV.

Solution:

$\Delta E = (\Delta m)c^2$

$$\Delta E = \left[\left(1\,\text{amu} \right) \left(\frac{1.66054 \times 10^{-27}\,\text{kg}}{1\,\text{amu}} \right) \right] \left(2.99792 \times 10^8\,\text{m/s} \right)^2 \left(\frac{\text{J}}{\text{kg} \cdot \text{m}^2 / \text{s}^2} \right) \left(\frac{1\,\text{MeV}}{1.602 \times 10^{-13}\,\text{J}} \right)$$

$= 9.3159448 \times 10^2 = \mathbf{9.316 \times 10^2\ MeV}$

24.141 Plan: The rate of formation of plutonium-239 depends on the rate of decay of neptunium-239 with a half-life of 2.35 d. Calculate k from the half-life equation and use the integrated rate law to find the time necessary to react 90% of the neptunium-239 (10% left).

Solution:

$$k = \frac{\ln 2}{t_{1/2}} = \frac{\ln 2}{2.35\ \text{d}} = 0.294956\ \text{d}^{-1}$$

$$\ln \frac{N_t}{N_0} = -kt$$

$$\ln \left(\frac{(1.00\ \text{kg})\left(10.\% / 100\% \right)}{1.00\ \text{kg}} \right) = -(0.294956\ \text{d}^{-1})(t)$$

$$-2.3025851 = -(0.294956\ \text{d}^{-1})(t)$$

$$t = 7.806538 = \mathbf{7.81\ d}$$